I HOPE YOU FIND IN THIS BOOK THE ASSISTANCE NEEDED TO BE A SUCCESSFUL AND COMPASSIONATE NURSE.

IN MEMORY OF CNA LUIS CID BALDERAS.
1993 - 2021

SSU STUDENT, NURSING PROGRAM FALL OF 2020.

D0821650

Anatomy & Physiology for Emergency Care

Second Edition

Frederic H. Martini, Ph.D.

Edwin F. Bartholomew, M.S.

Bryan E. Bledsoe, D.O., F.A.C.E.P.

with

William C. Ober, M.D.
Art Coordinator and Illustrator

Claire W. Garrison, R.N.
Illustrator

Kathleen Welch, M.D.
Clinical Consultant

Ralph T. Hutchings
Biomedical Photographer

PEARSON

Prentice Hall

Upper Saddle River, New Jersey 07458

Library of Congress Cataloging-in-Publication Data

Martini, Frederic
anatomy and physiology for emergency care / Frederic Martini, Edwin Bartholomew, Bryan
Bledsoe.—2nd ed.
  p. ; cm.
Rev. ed. of: Anatomy & physiology for emergency care / Frederic H. Martini, Edwin F.
Bartholomew, Bryan E. Bledsoe. c2002.
Includes bibliographical references and index.
ISBN-13: 978-0-13-234298-8
ISBN-10: 0-13-234298-7
1. Human physiology 2. Human anatomy. 3. Emergency medical services.
I. Bartholomew, Edwin F. II. Bledsoe, Bryan. III. Martini, Frederic. Anatomy &
physiology for emergency care. IV. Title.
  [DNLM: 1. Anatomy. 2. Emergency Medical Services. 3. Physiology.   QS 4 M3855a 2008]
QP36.M42 2008
612—dc22   2007014847

Publisher: Julie Levin Alexander
Publisher's Assistant: Regina Bruno
Executive Editor: Marlene McHugh Pratt
Senior Acquisitions Editor: Stephen Smith
Associate Editor: Monica Moosang
Editorial Assistant: Patricia Linard
Senior Managing Editor for Development: Lois Berlowitz
Developmental Editor: Josephine Cepeda
Director of Marketing: Karen Allman
Executive Marketing Manager: Katrin Beacom
Marketing Specialist: Michael Sirinides
Managing Editor for Production: Patrick Walsh
Production Liaison: Faye Gemmellaro
Production Editor: Emily Bush, Carlisle Publishing Services
Manufacturing Manager: Ilene Sanford
Manufacturing Buyer: Pat Brown
Senior Design Coordinator: Christopher Weigand
Cover Designer: Christopher Weigand
Interior Designer: Patrick Devine
Composition: Carlisle Publishing Services
Printing and Binding: Worldcolor Dubuque
Cover Printer: Phoenix Color

Notice: Our knowledge in clinical sciences is constantly changing. The authors and the publisher of this volume have taken care that the information contained herein is accurate and compatible with the standards generally accepted at the time of the publication. Nevertheless, it is difficult to ensure that all information given is entirely accurate for all circumstances. The authors and the publisher disclaim any liability, loss, or damage incurred as a consequence, directly or indirectly, of the use and application of any of the contents of this volume.

Many of the designations used by manufacturers and sellers to distinguish their products are claimed as trademarks. Where those designations appear in this book, and the publisher was aware of a trademark claim, the designations have been printed in initial caps or all caps.

Pearson Education Ltd.
Pearson Education Singapore, Pte. Ltd.
Pearson Education Canada, Ltd.
Pearson Education—Japan
Pearson Education Australia Pty. Ltd.

Pearson Education North Asia Ltd.
Pearson Educación de Mexico, S.A. de C.V.
Pearson Education Malaysia, Pte., Ltd.
Pearson Education, Upper Saddle River, New Jersey

10 9 8 7 6 5 4

ISBN 10:   0-13-234298-7
ISBN 13: 978-0-13-234298-8

Frederic (Ric) H. Martini, Ph.D. (Author) received his Ph.D. from Cornell University in comparative and functional anatomy for work on the pathophysiology of stress. His publications include journal articles, technical reports, magazine articles, and a book for naturalists about the biology and geology of tropical islands. He is the author or coauthor of six other undergraduate texts on anatomy and physiology or anatomy. He has been affiliated with the University of Hawaii (Hilo and Manoa) and the Shoals Marine Laboratory, a joint venture between Cornell University and the University of New Hampshire. Dr. Martini, President of the Human Anatomy and Physiology Society (2005–2006), is a member of the American Physiological Society, the American Association of Anatomists, the Society for Integrative and Comparative Biology, the Western Society of Naturalists, and the International Society of Vertebrate Morphologists.

Edwin F. Bartholomew, M.S. (Author) received his undergraduate degree from Bowling Green State University in Ohio and his M.S. from the University of Hawaii. His interests range widely, from human anatomy and physiology to paleontology, the marine environment, and "backyard" aquaculture. Mr. Bartholomew has taught human anatomy and physiology at both the secondary and undergraduate levels. In addition, he has taught a wide variety of other science courses (from botany to zoology) at Maui Community College. He is presently teaching at historic Lahainaluna High School, the oldest high school west of the Rockies. He has written journal articles, a weekly newspaper column, and many magazine articles. Working with Dr. Martini, he coauthored *Structure and Function of the Human Body* (Prentice Hall, 1999) and The *Human Body in Health and Disease* (Prentice Hall, 2000). Mr. Bartholomew is a member of the Human Anatomy and Physiology Society, the National Association of Biology Teachers, the National Science Teachers Association, the Hawaii Science Teachers Association, and the American Association for the Advancement of Science.

Dr. Bryan E. Bledsoe (Author) is an emergency physician with a special interest in pre-hospital care. He received his B.S. degree from the University of Texas at Arlington and his medical degree from the University of North Texas Health Sciences Center/Texas College of Osteopathic Medicine. He completed his internship at Texas Tech University and residency training at Scott and White Memorial Hospital/Texas A&M College of Medicine. Dr. Bledsoe is board certified in emergency medicine, and he is an Adjunct Associate Professor of Emergency Medicine at The George Washington University Medical Center in Washington, DC. Prior to attending medical school, Dr. Bledsoe worked as an EMT, a paramedic, and a paramedic instructor. He completed EMT training in 1974 and paramedic training in 1976 and worked for 6 years as a field paramedic in Fort Worth, Texas. In 1979, he joined the faculty of the University of North Texas Health Sciences Center and served as coordinator of EMT and paramedic education programs at the university. Dr. Bledsoe is active in emergency medicine and EMS research. He is a popular speaker at state, national, and international seminars and writes regularly for numerous EMS journals.

Dr. Bledsoe has authored several EMS books published by Brady including *Paramedic Care: Principles & Practice, Essentials of Paramedic Care, Intermediate Emergency Care: Principles & Practice, Anatomy & Physiology for Emergency Care, Prehospital Emergency Pharmacology,* and *Pocket Reference for ALS Providers.* He is married to Emma Bledsoe. They have two children, Bryan and Andrea, and a grandson, Andrew, and live on a ranch south of Dallas, Texas. He enjoys saltwater fishing and warm latitudes.

Kathleen Welch, M.D. (Clinical Consultant) received her M.D. from the University of Washington in Seattle and did her residency at the University of North Carolina in Chapel Hill. For two years, she served as Director of Maternal and Child Health at the LBJ Tropical Medical Center in American Samoa and subsequently was a member of the Department of Family Practice at the Kaiser Permanente Clinic in Lahaina, Hawaii. She has been in private practice since 1987. Dr. Welch is a Fellow of the American Academy of Family Practice, a member of the Hawaii Medical Association, the Maui County Medical Society, and the Human Anatomy and Physiology Society.

William C. Ober, M.D. (Art Coordinator and Illustrator) received his undergraduate degree from Washington and Lee University and his M.D. from the University of Virginia. While in medical school, he also studied in the Department of Art as Applied to Medicine at Johns Hopkins University. After graduation, Dr. Ober completed a residency in Family Practice and is currently an instructor in the Department of Sports Medicine at the University of Virginia. He is also part of the Core Faculty at Shoals

Marine Laboratory, where he teaches biological illustration in the summer program. The textbooks illustrated by his company, Medical & Scientific Illustration, have won numerous design and illustration awards.

Claire W. Garrison, R.N. (Illustrator) practiced pediatric and obstetric nursing before turning to medical illustration as a full-time career. She returned to school at Mary Baldwin College where she received her degree with distinction in studio art. Following a 5-year apprenticeship, she has worked as Dr. Ober's partner in Medical & Scientific Illustration since 1986. She is on the Core Faculty at Shoals Marine Laboratory and co-teaches the Biological Illustration course.

Ralph T. Hutchings (Biomedical Photographer) was associated with the Royal College of Surgeons for 20 years. An engineer by training, he has focused for years on photographing the structure of the human body. The result has been a series of color atlases, including the *Color Atlas of Human Anatomy, the Color Atlas of Surface Anatomy,* and *The Human Skeleton* (all published by Mosby-Yearbook Publishing). Mr. Hutchings makes his home in North London, where he tries to balance the demands of his photographic assignments with his hobbies of early motor cars and airplanes.

DEDICATION

This book is dedicated to all who take the time to

study the marvelous structure and function of the

human body. Also, it is dedicated to those who make

the ultimate gift to science by allowing their bodies to

be used for medical education and research. Their

gift, in death, ultimately helps the living.

—Bryan E. Bledsoe

Contents in Brief

v

Contents

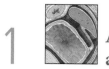

Contents

5 **The Integumentary System 122**

6 **The Skeletal System 150**

7 **The Muscular System 206**

8   **The Nervous System 258**

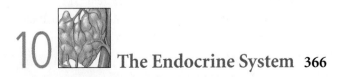

9 The General and Special Senses 326

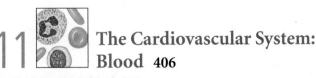

10 The Endocrine System 366

11 The Cardiovascular System: Blood 406

Contents

15 The Respiratory System 546

16 The Digestive System 584

17 Nutrition and Metabolism 628

18 The Urinary System 658

Contents

Preface

Before one can learn the various aspects of prehospital emergency care, one must first learn and understand normal human anatomy and physiology. This text, *Anatomy and Physiology for Emergency Care*, Second Edition, provides the necessary anatomy and physiology instruction to embark upon prehospital emergency care studies.

This text is based upon the popular college text *Essentials of Anatomy and Physiology* by Fredric H. Martini and Edwin F. Bartholomew. We have taken this work and added clinical correlations and applications specific to emergency care. Now, as an emergency medical services student, you can begin to learn selected disease processes and injury patterns while you learn normal human anatomy and physiology.

This text has three primary goals:

1. **Build a foundation of essential knowledge in human anatomy and physiology.** Constructing this knowledge requires answers to questions such as: What structure is that? How does it work? What happens when it does not work?

2. **Provide a framework for interpreting and applying information that can be used in problem-solving.** This framework is based on the fact that certain themes and patterns appear again and again in the study of anatomy and physiology. These themes and patterns provide the hooks on which to organize and hang the information you will find in the text. Problems that require interpreting and applying information include such general questions as: How does a change in one body system affect the others? How does aging affect body systems?

3. **Provide an introduction to common injuries and illnesses in a manner that will reinforce basic anatomy and physiology principles.** You will tend to remember material better if presented in a context that directly relates to your interest and field of study. In this text we have taken selected anatomy and physiology principles and carefully integrated clinical concepts, usually of an emergency nature, that will help you to retain the material and see how abnormal anatomy and physiology (pathophysiology) are directly related to normal anatomy and physiology.

In addition, the need for well-trained professionals in the allied health field is greater than ever. To succeed in an allied health career, you must master the same skills that you need to succeed in this course. You must do more than develop a large technical vocabulary and retain a large volume of detailed information. You must learn how to learn—how to organize new information, connect it to what you already know, and then apply it as needed.

Can a text simplify this process? No textbook can give you more time. A text can, however, help you make better use of your time. By presenting the material in a clear, logical way that stresses concept organization, this textbook provides you with a strong foundation of essential knowledge and a framework for integrating and applying that knowledge. For example, certain themes and patterns appear again and again in the study of anatomy and physiology. The conceptual material will be much easier to deal with if you learn to recognize those patterns and organize the information accordingly. With this realization in mind, we have taken extra care to highlight the important patterns, create a sensible framework, and organize new information around that framework.

The User's Guide that follows this page describes the many learning aids that are built into this text. If you use them, they will help you improve your study skills and will make you more efficient at learning and integrating new information. These features were developed through feedback from students and instructors on campuses across the United States, Canada, Australia, New Zealand, and Europe. Many students, in person or by phone or e-mail, have told us that this system really works for them when other presentation styles have not. Take the time to examine this User's Guide carefully, and ask your instructor if you have questions about any of the book's learning aids. If you invest the time now, you can learn to use the book properly from the outset. Doing so will ensure that you will get the most from the time you invest in this course. In addition, becoming a more efficient learner will set you forth on a lifetime of learning that will make you a more valuable future professional, constantly improving and growing long after you've taken your last exam.

■ New for the Second Edition

Anatomy and Physiology for Emergency Care, Second Edition, has been revised and updated. Most important, the clinical correlation material has been placed throughout the text next to the related topic being discussed. This is a vast improvement over the earlier edition where all of the clinical material was placed as an appendix to each chapter. Now, the flow of the text is better and the page numbers continuous.

Numerous new clinical discussions have been added as well as new art and photographs. The practice of prehospital care is always changing and the text has been updated to reflect these changes.

We hope that you find *Anatomy and Physiology for Emergency Care*, Second Edition, a valuable tool for your prehospital care and allied health education. Your ability to work as a member of the health care team is dependent upon you having an excellent understanding of human anatomy and physiology.

■ Acknowledgments

Reviewers for the Second Edition

Brandon R. Beck, AS, NREMT-P, CCEMT-P
EMS Instructor/Clinical Coordinator
Gadsden State Community College
Gadsden, AL

Pamela L. Bradshaw, RN, MSNCNP, EMS-I
Collins Career Center
Chesapeake, OH

Richard Campbell, FFP, LP, NREMT-P
Brookhaven College
Farmers Branch Fire Department
Farmers Branch, TX

Robert L. Darr, NREMT-P/FF
MEMS
Little Rock, AR

Mary Fuglaar, LP, EMS-I
Clinical Division
Fort Bend County Emergency Medical Service
Rosenberg, TX

Christopher D. Jones
Limited Service Faculty, Youngstown State University
Paramedic, Rural/Metro Ambulance
Youngstown, OH

Mark D. Minton, EMT-P, RN, B.S.
Gadsden State Community College
Gadsden, AL

Greg Mullen, MS, NREMT-P
National EMS Academy
Lafayette, LA

Kenneth W. Navarro
CE Coordinator
Emergency Medicine Education
University of Texas Southwestern Medical Center at Dallas
Dallas, TX

James Shiplet, AAS, LP, EMSC
EMS Education Coordinator
Collin County Community College
McKinney, TX

Frederic H. Martini, Ph.D.
Edwin F. Bartholomew, M.S.
Bryan E. Bledsoe, D.O., F.A.C.E.P.

Visual Introduction
Award-Winning Art Program

Superb Anatomical Art

Anatomical art features an artist's drawing with a photograph of the "real thing" to help students compare structures as they might appear in a laboratory or clinical setting, or includes an illustration on top of a photograph to help students identify the location of structures.

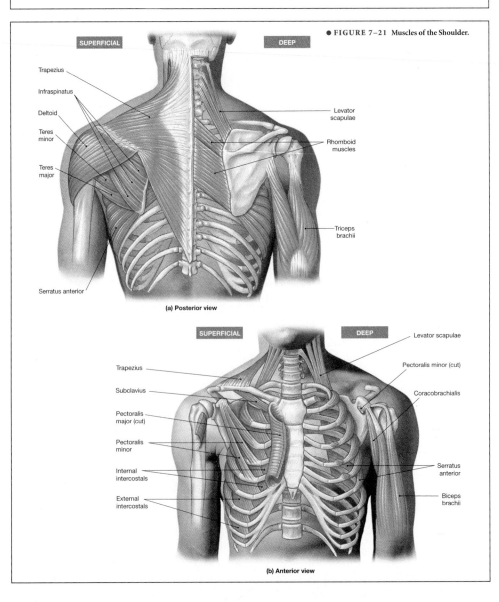

● **FIGURE 6–41 The Pelvis.** The different colors indicate the components of (**a**) the pelvis in an anterior view and (**b**) the right coxal bone in a lateral view. The photograph in (**c**) shows an anterior view of the pelvis of an adult male.

● **FIGURE 7–21 Muscles of the Shoulder.**

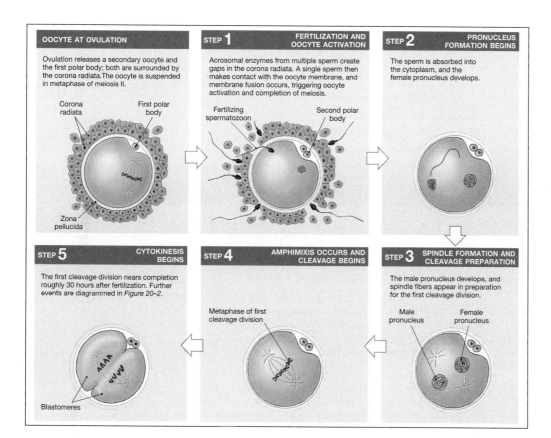

OOCYTE AT OVULATION

Ovulation releases a secondary oocyte and the first polar body; both are surrounded by the corona radiata. The oocyte is suspended in metaphase of meiosis II.

Corona radiata
First polar body
Zona pellucida

STEP 1 FERTILIZATION AND OOCYTE ACTIVATION

Acrosomal enzymes from multiple sperm create gaps in the corona radiata. A single sperm then makes contact with the oocyte membrane, and membrane fusion occurs, triggering oocyte activation and completion of meiosis.

Fertilizing spermatozoon
Second polar body

STEP 2 PRONUCLEUS FORMATION BEGINS

The sperm is absorbed into the cytoplasm, and the female pronucleus develops.

STEP 5 CYTOKINESIS BEGINS

The first cleavage division nears completion roughly 30 hours after fertilization. Further events are diagrammed in *Figure 20–2*.

Blastomeres

STEP 4 AMPHIMIXIS OCCURS AND CLEAVAGE BEGINS

Metaphase of first cleavage division

STEP 3 SPINDLE FORMATION AND CLEAVAGE PREPARATION

The male pronucleus develops, and spindle fibers appear in preparation for the first cleavage division.

Male pronucleus
Female pronucleus

Step-by-Step Diagrams

Step-by-Step Diagrams coordinate with the authors' narrative descriptions and break down complex processes into easy-to-follow steps.

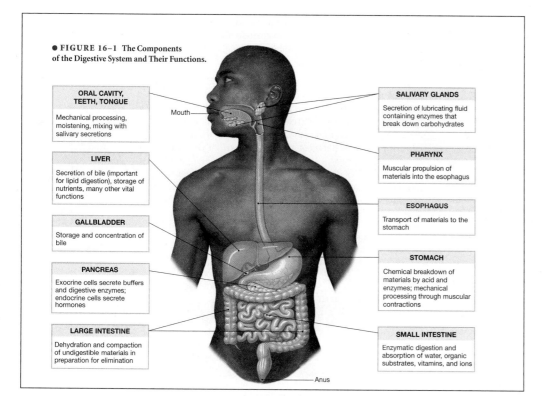

● **FIGURE 16–1** The Components of the Digestive System and Their Functions.

ORAL CAVITY, TEETH, TONGUE

Mechanical processing, moistening, mixing with salivary secretions

Mouth

LIVER

Secretion of bile (important for lipid digestion), storage of nutrients, many other vital functions

GALLBLADDER

Storage and concentration of bile

PANCREAS

Exocrine cells secrete buffers and digestive enzymes; endocrine cells secrete hormones

LARGE INTESTINE

Dehydration and compaction of undigestible materials in preparation for elimination

SALIVARY GLANDS

Secretion of lubricating fluid containing enzymes that break down carbohydrates

PHARYNX

Muscular propulsion of materials into the esophagus

ESOPHAGUS

Transport of materials to the stomach

STOMACH

Chemical breakdown of materials by acid and enzymes; mechanical processing through muscular contractions

SMALL INTESTINE

Enzymatic digestion and absorption of water, organic substrates, vitamins, and ions

Anus

Conceptual Overviews

Conceptual Overviews present "big-picture" overviews of structure and function relationships.

Practical Reference for Future Careers

Clinical Notes

Clinical Notes present relevant information on diseases and disorders to help students prepare for future workplace situations. The discussions embedded in the running narrative explain how pathologies relate to normal physiological function, and the larger "boxed" versions focus on core medical or social topics.

Clinical Note—continued
IV FLUID THERAPY

The type of IV fluid used depends upon the patient's needs. In the prehospital setting, isotonic fluids are usually used, because the patient's underlying electrolyte and hydration status is unknown (Figure 3–11●). However, once the patient arrives in the emergency department, blood electrolyte studies are used to guide fluid selection and administration. The IV fluids most frequently used in prehospital care include:

■ *Lactated Ringer's.* Lactated Ringer's solution is an isotonic electrolyte solution that contains sodium chloride, potassium chloride, calcium chloride, and sodium lactate in water.
■ *Normal saline.* Normal saline is an isotonic electrolyte solution that contains sodium chloride in water.

■ *5% dextrose in water (D_5W).* D_5W is a hypotonic glucose solution used to keep a vein patent and to dilute concentrated medications. While D_5W initially increases intravascular volume, glucose molecules rapidly diffuse across the vascular membrane and increase the amount of free water.

Both lactated Ringer's and normal saline are used for fluid replacement because of their immediate ability to expand the circulating fluid volume. However, due to the movement of electrolytes and water, two-thirds of either solution will be lost to the extravascular space within one hour (Figure 3–12●).

Dextrose-Containing Solutions

Several IV fluids contain dextrose (d-glucose) in varying concentrations. The most commonly used of these are 50% dextrose ($D_{50}W$) and 5% dextrose in water (D_5W). Other dextrose solutions include 25% dextrose ($D_{25}W$) and 10% dextrose and water ($D_{10}W$). Some dextrose-containing solutions will also contain electrolytes, usually sodium and chloride. Examples of these include:

■ 2.5% dextrose and 0.45% sodium chloride ($D_{2.5}NS$)
■ 5% dextrose and 0.20% sodium chloride (D_5NS)
■ 5% dextrose and 0.33% sodium chloride (D_5NS)
■ 5% dextrose and 0.45% sodium chloride (D_5NS)
■ 5% dextrose and 0.9% sodium chloride (D_5NS)
■ 10% dextrose and 0.9% sodium chloride ($D_{10}NS$)
■ 5% dextrose in lactated Ringer's (D_5LR)

The high-concentration solutions, such as $D_{50}W$ and $D_{25}W$, are used for glucose replacement in documented hypoglycemia. $D_{10}W$ is used in patients such as chronic alcoholics who require calorie replacement in addition to water and electrolytes. D_5W and similar solutions are usually used for diluting intravenous medications and for conditions where an IV is started at a "to keep open" (TKO) or "keep vein open" (KVO) rate.

The solubility of dextrose is 1 gram per milliliter of water. Based on this property, dextrose-containing solutions are usually measured in weight-in-volume percentages. This system of measurement indicates the number of grams of dextrose in 100 mL of solution (water). A fully

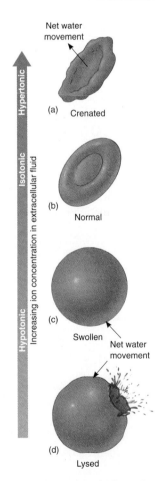

Net water movement

Hypertonic

Isotonic

Hypotonic

Increasing ion concentration in extracellular fluid

(a) Crenated

(b) Normal

(c) Swollen

Net water movement

(d) Lysed

● **FIGURE 3–11** **Effects of IV Fluids on Circulating Red Blood Cells (Erythrocytes).** Hypertonic fluids cause cell shrinkage (crenation), while hypotonic fluids cause intracellular swelling that eventually leads to cell lysis.

● **FIGURE 3–12** **Emergency Fluid Resuscitation.** In the prehospital setting, it is best to administer just enough fluids to maintain the systolic pressure between 70–80 mmHg.

Related Clinical Terms

Related Clinical Terms featured at the end of each chapter help students review vocabulary that they will likely use in their chosen careers. Each term is followed by a definition and, in many cases, a pronunciation guide.

Related Clinical Terms

botulism A disease characterized by severe, potentially fatal paralysis of skeletal muscles, that results from the consumption of a bacterial toxin.

carpal tunnel syndrome Inflammation of the sheath that surrounds the flexor tendons of the palm and leads to nerve compression and pain.

compartment syndrome Ischemia (defined shortly) that results from accumulated blood and fluid trapped within limb muscle compartments formed by partitions of dense connective tissue.

fibrosis A process in which a tissue is replaced by fibrous connective tissue. Fibrosis makes muscles weaker and less flexible.

hernia A condition involving an organ or a body part that protrudes through an abnormal opening in the wall of a body cavity.

intramuscular (IM) injection The administration of a drug by injecting it into the mass of a large skeletal muscle.

ischemia (is-KĒ-mē-uh) A deficiency of blood ("blood starvation") in a body part due to compression of regional blood vessels.

muscle cramps Prolonged, involuntary, painful muscular contractions.

muscular dystrophies (DIS-trō-fēz) A varied collection of inherited diseases that produce progressive muscle weakness and deterioration.

myalgia (mī-AL-jē-uh) Muscular pain; a common symptom of a wide variety of conditions and infections.

myasthenia gravis (mī-as-THĒ-nē-uh GRA-vis) A general muscular weakness that results from a reduction in the number of ACh receptors on the motor end plate.

myoma A benign tumor of muscle tissue.

myositis (mi-ō-SĪ-tis) Inflammation of muscle tissue.

polio A viral disease in which the destruction of motor neurons produces paralysis and atrophy of motor units.

rigor mortis A state following death during which muscles are locked in the contracted position, which makes the body extremely stiff.

sarcoma A malignant tumor of mesoderm-derived tissue (muscle, bone, or other connective tissue).

strains Tears or breaks in muscles.

tendinitis Inflammation of the connective tissue that surrounds a tendon.

tetanus A disease caused by a bacterial toxin that results in sustained, powerful contractions of skeletal muscles throughout the body.

Critical Thinking and Clinical Applications Questions

Questions located at the end of each chapter offer hypothetical and clinical situations related to the body system covered in the chapter. Students can test and refresh their knowledge by applying the information they've learned to a specific clinical situation.

Level 2: Reviewing Concepts

35. A graded potential:
    (a) decreases with distance from the point of stimulation.
    (b) spreads passively because of local currents.
    (c) may involve either depolarization or hyperpolarization.
    (d) a, b, and c are correct.

36. The loss of positive ions from the interior of a neuron produces:
    (a) depolarization.
    (b) threshold.
    (c) hyperpolarization.
    (d) an action potential.

37. What would happen if the ventral root of a spinal nerve was damaged or transected?

38. Which major part of the brain is associated with respiratory and cardiac activity?

39. Why is response time in a monosynaptic reflex much faster than response time in a polysynaptic reflex?

40. Compare the general effects of the sympathetic and parasympathetic divisions of the ANS.

Level 3: Critical Thinking and Clinical Applications

41. If neurons in the central nervous system lack centrioles and are unable to divide, how can a person develop brain cancer?

42. A police officer has just stopped Bill on suspicion of driving while intoxicated. The officer asks Bill to walk the yellow line on the road and then asks him to place the tip of his index finger on the tip of his nose. How would these activities indicate Bill's level of sobriety? Which part of the brain is being tested by these activities?

43. In some severe cases of stomach ulcers, the branches of the vagus nerve (N X) that lead to the stomach are surgically severed. How might this procedure control the ulcers?

44. Improper use of crutches can produce a condition known as crutch paralysis, which is characterized by a lack of response by the extensor muscles of the arm and a condition known as wrist drop. Which nerve is involved?

45. While playing football, Ramon is tackled hard and suffers an injury to his left leg. As he tries to get up, he finds that he cannot flex his left hip or extend the knee. Which nerve is damaged, and how would this damage affect sensory perception in his left leg?

New! Key Notes

Key Notes call out core concepts in anatomy and physiology. These factual statements are highlighted with an icon and appear at appropriate points throughout the chapters. The complete set of Key Notes also appears in Appendix V at the back of the book.

> **Key Note**
>
> The adrenal glands produce hormones that adjust metabolic activities at specific sites, which affects either the pattern of nutrient utilization, mineral ion balance, or the rate of energy consumption by active tissues.

Concept Check Questions

Concept Check Questions appear at the end of chapter sections and help students test their comprehension before moving on to the next topic.

> **CONCEPT CHECK QUESTIONS**
>
> 1. If a sample of bone marrow has fewer than normal numbers of megakaryocytes, what body process would you expect to be impaired as a result?
> 2. Two alternate pathways of interacting clotting proteins lead to coagulation, or blood clotting. How is each pathway initiated?
> 3. How do inadequate levels of vitamin K affect blood clotting (coagulation)?
>
> *Answers begin on p. 792.*

Figure Locator Dots

Figure Locator Dots appear with every figure reference in the narrative and function as placeholders to help students return to reading after viewing a figure.

Concept Links

Concept Links provide quick visual signals that new material is related to topics presented earlier and offer specific page numbers to facilitate review.

■ The Anatomy and Organization of the Heart

The heart lies near the anterior chest wall, directly behind the sternum (Figure 12–2a●). It is enclosed by the *mediastinum,* which is the connective tissue mass that divides the thoracic cavity into two pleural cavities (see Figure 1–11c, p. 21) and also contains the thymus, esophagus, and trachea.

The heart is surrounded by the **pericardial** (per-i-KAR-dē-al) **cavity.** The lining of the pericardial cavity is a serous membrane called the **pericardium.** ∞ p. 113 To visualize the relationship between the heart and the pericardial cavity, imagine pushing your fist toward the center of a large balloon (Figure 12–2b●). The balloon represents the pericardium, and your fist represents the heart. Your wrist, where the balloon folds back on itself, corresponds to the **base** of the heart (see Figure 12–2a). The air space inside the balloon corresponds to the pericardial cavity.

Also Available for Students

Student Workbook

By Greg Mullen
ISBN: 0-13-614021-1
The student workbook includes a variety of assessment questions and labeling exercises to reinforce key concepts from the text.

Companion Website

Open-access Companion Website (www.prenhall.com/bledsoe) includes chapter-correlated quizzes, labeling activities, animations, essay questions, and Web links.

Instructor Supplements

Instructor's Resource CD-ROM

ISBN: 0-13-158995-4
The Instructor's Resource CD-ROM includes the following three components:

Instructor's Manual

Includes objectives, lecture outlines, teaching strategies, teaching tips, and handouts for evaluation and reinforcement.

TestGen

Organized by chapter and question type, this test bank includes hundreds of questions for designing a variety of tests and quizzes. It includes references to text pages where answers can be found or supported.

PowerPoint® Slides

New PowerPoint® slide program provides the basis of dynamic classroom presentations.

WebCT, Course Compass, and Blackboard

As an optional supplement to the text, WebCT, Course Compass, and Blackboard courses are also available to provide ready-to-assign materials.

1 An Introduction to Anatomy and Physiology

EMERGENCY MEDICAL SERVICES (EMS) is unique among the health care professions in that Emergency Medical Technicians (EMTs) and paramedics often function in a relatively austere, yet constantly changing environment. EMS is a mixture of public safety, public health, and health care. Above all, though,

EMTs and paramedics are health care professionals. As with all health care professions, the road to becoming an EMT or paramedic begins with a thorough understanding of relevant human anatomy and physiology.

Chapter Outline

Chapter Objectives

1. Describe the basic functions of living organisms. (pp. 3–4)
2. Define anatomy and physiology, and describe the various specialties within each discipline. (pp. 4, 5)
3. Identify the major levels of organization in living organisms. (p. 5)
4. Identify the organ systems of the human body and the major components of each system. (pp. 7–13)
5. Explain the significance of homeostasis. (p. 7)
6. Describe how negative and positive feedback is involved in homeostatic regulation. (pp. 7, 14–15)
7. Use anatomical terms to describe body sections, body regions, and relative positions. (pp. 16–20)
8. Identify the major body cavities and their subdivisions. (pp. 20–23)

Vocabulary Development

bios life; *biology*
cardium heart; *pericardium*
dorsum back; *dorsal*
homeo- unchanging; *homeostasis*
-logy study of; *biology*

medianus situated in the middle; *median*
paries wall; *parietal*
pathos disease; *pathology*
peri- around; *perimeter*

pronus inclined forward; *prone*
-stasis standing; *homeostasis*
supinus lying on the back; *supine*
venter belly or abdomen; *ventral*

THE WORLD AROUND US contains an enormous diversity of living organisms that vary widely in appearance and lifestyle. One aim of **biology**—the study of life—is to discover the common patterns that underlie this diversity. Such discoveries show that all living things perform the following basic functions:

■ *Responsiveness.* Organisms respond to changes in their immediate environment; this property is also called *irritability.* You move your hand away from a hot stove, your dog barks at approaching strangers, fish are alarmed by loud noises, and tiny amoebas glide toward potential prey. Organisms also make longer-term changes as they adjust to their environments. For example, an animal may grow a heavier coat of fur as winter approaches, or it may migrate to a warmer climate. The capacity to make such adjustments is termed *adaptability.*

■ *Growth.* Over a lifetime, organisms increase in size through the growth of *cells,* the simplest units of life. Single-celled creatures grow by getting larger; whereas more complex organisms grow primarily by increasing the number of cells. Familiar organisms, such as dogs, cats, and people, are com-

posed of trillions of cells. In such multicellular organisms, individual cells become specialized to perform particular functions. This specialization is called *differentiation.*

■ *Reproduction.* Organisms reproduce, and create subsequent generations of similar organisms.

■ *Movement.* Organisms are capable of producing movement, which may be internal (transporting food, blood, or other materials within the body) or external (moving through the environment).

■ *Metabolism.* Organisms rely on complex chemical reactions to provide the energy required for responsiveness, growth, reproduction, and movement. They must also synthesize complex chemicals, such as proteins. *Metabolism* refers to all of the chemical operations that take place in the body. Normal metabolic operations require the absorption of materials from the environment. To generate energy efficiently, most cells require various nutrients obtained in food, as well as oxygen, which is a gas. *Respiration* refers to the absorption, transport, and use of oxygen by cells. Metabolic operations often generate unneeded or potentially harmful waste products that must be eliminated through the process of *excretion.*

For very small organisms, absorption, respiration, and excretion involve the movement of materials across exposed surfaces. But creatures larger than a few millimeters in width seldom absorb nutrients directly from their environment. For example, humans cannot absorb steaks, apples, or ice cream without processing them first. That processing, called *digestion*, occurs in specialized structures in which complex foods are broken down into simpler components that can be transported and absorbed easily. Respiration and excretion are also more complicated for large organisms. Humans have specialized structures responsible for gas exchange (lungs) and excretion (kidneys). Although digestion, respiration, and excretion occur in different parts of the body, the cells of the body cannot travel to one place for nutrients, another for oxygen, and a third to get rid of waste products. Instead, individual cells remain where they are but communicate with other areas of the body through an internal transport system—the circulation. For example, the blood absorbs the waste products released by each of your cells and carries those wastes to the kidneys for excretion.

Biology includes many subspecialties. This text considers two biological subjects: **anatomy** (ah-NAT-o-mē) and **physiology** (fiz-ē-OL-o-jē). Over the course of 20 chapters, you will become familiar with the basic anatomy and physiology of the human body.

■ The Sciences of Anatomy and Physiology

The word *anatomy* has Greek origins, as do many other anatomical terms and phrases. **Anatomy,** which means "a cutting open," is the study of internal and external structure and the physical relationships between body parts. **Physiology,** another word derived from Greek, is the study of how living organisms perform their vital functions. The two subjects are interrelated. Anatomical information provides clues about probable functions, and physiological mechanisms can be explained only in terms of their underlying anatomy.

The link between structure and function is always present but not always understood. For example, the anatomy of the heart was clearly described in the fifteenth century, but almost 200 years passed before anyone realized that it pumped blood. This text will familiarize you with basic anatomy and give you an appreciation of the physiological processes that make human life possible. The information will enable you to understand many kinds of disease processes and will help you make informed decisions about your own health.

Anatomy

Anatomy can be divided into gross (macroscopic) anatomy or microscopic anatomy on the basis of the degree of structural detail under consideration. Other anatomical specialties focus on specific processes, such as respiration, or medical applications, such as developing artificial limbs.

Gross Anatomy

Gross anatomy, or *macroscopic anatomy,* considers features visible with the unaided eye. There are many ways to approach gross anatomy. **Surface anatomy** refers to the study of general form and superficial markings. **Regional anatomy** considers all of the superficial and internal features in a specific region of the body, such as the head, neck, or trunk. **Systemic anatomy** considers the structure of major *organ systems,* which are groups of organs that function together in a coordinated manner. For example, the heart, blood, and blood vessels form the *cardiovascular system,* which circulates oxygen and nutrients throughout the body.

Microscopic Anatomy

Microscopic anatomy concerns structures that cannot be seen without magnification. The boundaries of microscopic anatomy are established by the limits of the equipment used. A light microscope reveals basic details about cell structure; whereas an electron microscope can visualize individual molecules only a few nanometers (nm, 1 millionth of a millimeter) across. As we proceed through the text, we will consider details at all levels, from macroscopic to microscopic. (Readers unfamiliar with the terms used to describe measurements and weights should consult the reference tables in Appendix 3.)

Microscopic anatomy can be subdivided into specialties that consider features within a characteristic range of sizes. **Cytology** (sī-TOL-o-jē) analyzes the internal structure of individual **cells.** The trillions of living cells in our bodies are composed of chemical substances in various combinations, and our lives depend on the chemical processes that occur in those cells. For this reason we will consider basic chemistry (Chapter 2) before examining cell structure (Chapter 3).

Histology (his-TOL-o-jē) takes a broader perspective and examines **tissues,** which are groups of specialized cells and cell products that work together to perform specific functions (Chapter 4). Tissues combine to form **organs,** such as the heart, kidney, liver, and brain. Many organs can be examined without a microscope, so at the organ level we cross the boundary into gross anatomy.

Physiology

Physiology is the study of the function of anatomical structures. **Human physiology** is the study of the functions of the human body. These functions are complex and much more difficult to examine than most anatomical structures. As a result, the science of physiology includes even more specialties than does the science of anatomy.

The cornerstone of human physiology is **cell physiology,** which is the study of the functions of living cells. Cell physiology includes events at the chemical or molecular levels—both within cells and between cells. **Special physiology** is the study of the physiology of specific organs. Examples include renal physiology (kidney function) and cardiac physiology (heart function). **Systemic physiology** considers all aspects of the function of specific organ systems. Respiratory physiology and reproductive physiology are examples. **Pathological physiology,** or **pathology** (pah-THOL-o-jē), is the study of the effects of diseases on organ or system functions. (The Greek word *pathos* means "disease.") Modern medicine depends on an understanding of both normal and pathological physiology, of understanding not only what has gone wrong but also how to correct it.

Special topics in physiology address specific functions of the human body as a whole. These specialties focus on physiological interactions among multiple-organ systems. Exercise physiology, for example, studies the physiological adjustments to exercise.

Key Note

All physiological functions are performed by anatomical structures. These functions follow the same physical and mechanical principles that can be seen in the world at large.

CONCEPT CHECK QUESTIONS

1. How are vital functions such as growth, responsiveness, reproduction, and movement dependent on metabolism?
2. Would a histologist more likely be considered a specialist in microscopic anatomy or in gross anatomy? Why?

Answers begin on p. 792.

■ Levels of Organization

To understand the human body, we must examine its organization at several different levels, from the submicroscopic to the macroscopic. Figure 1–1● presents the relationships among the various levels of organization, using the cardiovascular system as an example.

■ *Chemical (or Molecular) Level. Atoms,* the smallest stable units of matter, combine to form *molecules* with complex shapes. Even at this simplest level, a molecule's specialized shape determines its function. This is the chemical, or molecular, level of organization.

■ *Cellular Level.* Different molecules can interact to form larger structures, each of which has a specific function in a cell. (Cells are the smallest living units in the body and make up the cellular level of organization.) For example, different types of protein filaments interact to produce the contractions of muscle cells in the heart.

■ *Tissue Level.* A *tissue* is composed of similar cells that work together to perform a specific function. Heart muscle cells form *cardiac muscle tissue,* an example of the tissue level of organization.

■ *Organ Level.* An *organ* consists of two or more different tissues that work together to perform specific functions. An example of the organ level of organization is the *heart,* a hollow, three-dimensional organ with walls composed of layers of cardiac muscle and other tissues.

■ *Organ System Level.* Organs interact in *organ systems.* Each time the heart contracts, it pushes blood into a network of blood vessels. Together, the heart, blood, and blood vessels form the *cardiovascular system,* an example of the organ system level of organization.

■ *Organism Level.* All of the organ systems of the body work together to maintain life and health. This brings us to the highest level of organization, that of the *organism*—in this case, a human being.

The organization at each level determines both the structural characteristics and the functions of higher levels. For example, the arrangement of atoms and molecules at the chemical level creates the protein filaments that, at the cellular level, give cardiac muscle cells the ability to contract powerfully. At the tissue level, these cells are linked, which forms cardiac muscle tissue. The structure of the tissue ensures that the contractions are coordinated and produce a heartbeat. When that beat occurs, the internal anatomy of the heart, an organ, enables it to function as a pump. The heart is filled with blood and connected to the blood vessels, and the pumping

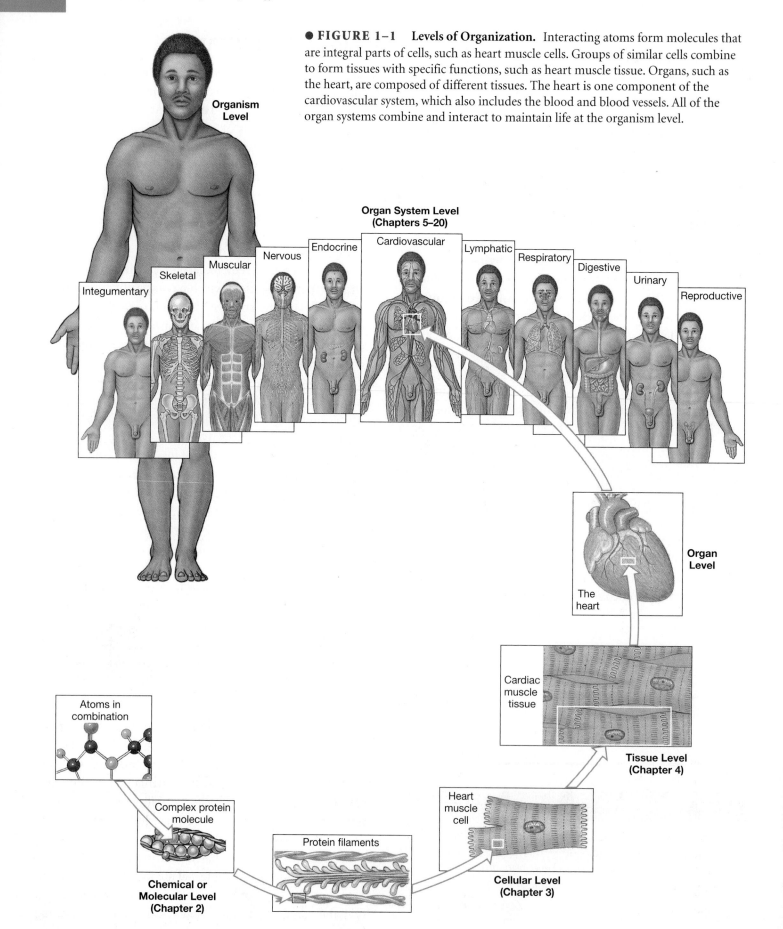

● **FIGURE 1–1** **Levels of Organization.** Interacting atoms form molecules that are integral parts of cells, such as heart muscle cells. Groups of similar cells combine to form tissues with specific functions, such as heart muscle tissue. Organs, such as the heart, are composed of different tissues. The heart is one component of the cardiovascular system, which also includes the blood and blood vessels. All of the organ systems combine and interact to maintain life at the organism level.

Organism Level

Organ System Level (Chapters 5–20)

Integumentary

Skeletal

Muscular

Nervous

Endocrine

Cardiovascular

Lymphatic

Respiratory

Digestive

Urinary

Reproductive

Organ Level

The heart

Cardiac muscle tissue

Tissue Level (Chapter 4)

Atoms in combination

Complex protein molecule

Chemical or Molecular Level (Chapter 2)

Protein filaments

Heart muscle cell

Cellular Level (Chapter 3)

action circulates blood through the vessels of the cardiovascular system. Through interactions with the respiratory, digestive, urinary, and other systems, the cardiovascular system performs a variety of functions essential to the survival of the organism.

Something that affects a system will ultimately affect each of the system's components. For example, the heart cannot pump blood effectively after massive blood loss. If the heart cannot pump and blood cannot flow, oxygen and nutrients cannot be distributed. Very soon, the cardiac muscle tissue begins to break down as individual muscle cells die from oxygen and nutrient starvation. These changes will not be restricted to the cardiovascular system; cells, tissues, and organs throughout the body will be damaged.

■ An Introduction to Organ Systems

Figure 1–2● introduces the 11 organ systems in the human body and their major functions. These organ systems are (1) the integumentary system, (2) the skeletal system, (3) the muscular system, (4) the nervous system, (5) the endocrine system, (6) the cardiovascular system, (7) the lymphatic system, (8) the respiratory system, (9) the digestive system, (10) the urinary system, and (11) the reproductive system.

Key Note

The body can be divided into 11 organ systems, which all work together, but the boundaries between them are not absolute.

■ Homeostasis and System Integration

Organ systems are interdependent, interconnected, and occupy a relatively small space. The cells, tissues, organs, and systems of the body function together in a shared environment. Just as the inhabitants of a large city breathe the same air and drink water provided by the local water company, the cells in the human body absorb oxygen and nutrients from the body fluids that surround them. All living cells are in contact with blood or some other body fluid, and any change in the composition of these fluids will affect them in some way.

For example, changes in the temperature or salt content of the blood could cause anything from a minor adjustment (heart muscle tissue contracts more often, and the heart rate goes up) to a total disaster (the heart stops beating altogether).

Homeostatic Regulation

A variety of physiological mechanisms act to prevent potentially dangerous changes in the environment inside the body. **Homeostasis** (hō-mē-ō-STĀ-sis; *homeo,* unchanging + *stasis,* standing) refers to the existence of a stable internal environment. To survive, every living organism must maintain homeostasis. The term **homeostatic regulation** refers to the adjustments in physiological systems that preserve homeostasis.

Homeostatic regulation usually involves (1) a **receptor** that is sensitive to a particular environmental change or *stimulus;* (2) a **control center,** or *integration center,* which receives and processes information from the receptor; and (3) an **effector,** which responds to the commands of the control center and whose activity opposes or reinforces the stimulus. You are probably already familiar with several examples of homeostatic regulation, although not in those terms. As an example, consider the operation of the thermostat in a house or apartment (Figure 1–3●).

The thermostat is a control center that monitors room temperature. The gauge on the thermostat establishes the set point, the "ideal" room temperature—in this example, 22°C (about 72°F). The function of the thermostat is to keep room temperature within acceptable limits, usually within a degree or so of the set point. The thermostat receives information from a receptor, which is a thermometer exposed to air in the room, and it controls one of two effectors: a heater or an air conditioner. In the summer, for example, a rise in temperature above the set point causes the thermostat to turn on the air conditioner, which then cools the room; when the temperature at the thermometer returns to the set point, the thermostat turns off the air conditioner. The essential feature of temperature control by a thermostat can be summarized very simply: a variation outside the desired range triggers an automatic response that corrects the situation. This method of homeostatic regulation is called *negative feedback,* because an effector activated by the control center opposes, or *negates,* the original stimulus.

Negative Feedback

The essential feature of **negative feedback** is this: regardless of whether the stimulus (such as temperature) rises or falls at the receptor, *a variation outside normal limits triggers an automatic response that corrects the situation.*

● **FIGURE 1–2**
The Organ Systems of the Human Body.

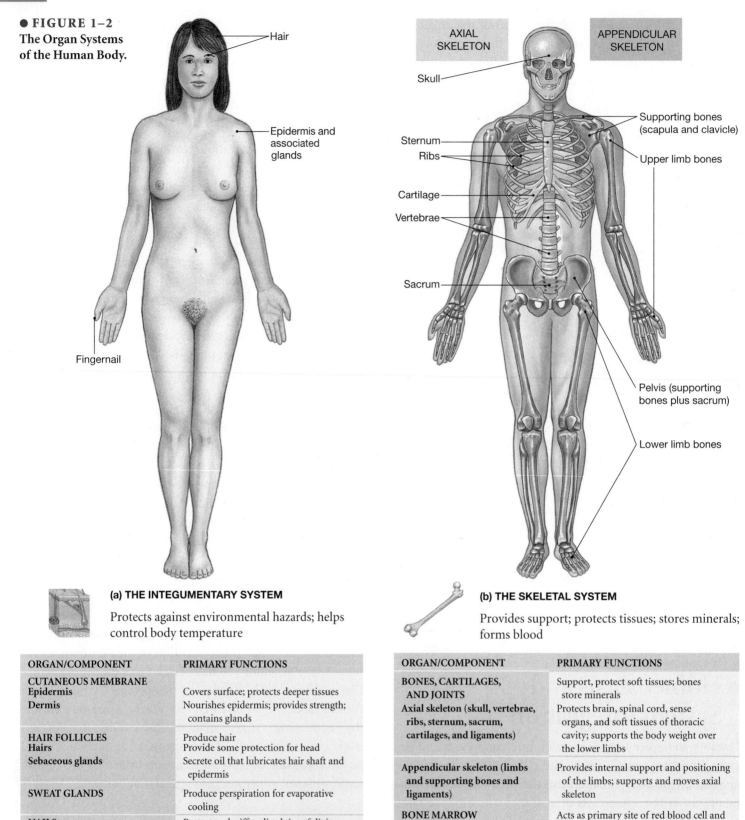

(a) THE INTEGUMENTARY SYSTEM

Protects against environmental hazards; helps control body temperature

ORGAN/COMPONENT	PRIMARY FUNCTIONS
CUTANEOUS MEMBRANE **Epidermis** **Dermis**	Covers surface; protects deeper tissues Nourishes epidermis; provides strength; contains glands
HAIR FOLLICLES **Hairs** **Sebaceous glands**	Produce hair Provide some protection for head Secrete oil that lubricates hair shaft and epidermis
SWEAT GLANDS	Produce perspiration for evaporative cooling
NAILS	Protect and stiffen distal tips of digits
SENSORY RECEPTORS	Provide sensations of touch, pressure, temperature, and pain
SUBCUTANEOUS LAYER	Stores lipids; attaches skin to deeper structures

(b) THE SKELETAL SYSTEM

Provides support; protects tissues; stores minerals; forms blood

ORGAN/COMPONENT	PRIMARY FUNCTIONS
BONES, CARTILAGES, AND JOINTS **Axial skeleton (skull, vertebrae, ribs, sternum, sacrum, cartilages, and ligaments)**	Support, protect soft tissues; bones store minerals Protects brain, spinal cord, sense organs, and soft tissues of thoracic cavity; supports the body weight over the lower limbs
Appendicular skeleton (limbs and supporting bones and ligaments)	Provides internal support and positioning of the limbs; supports and moves axial skeleton
BONE MARROW	Acts as primary site of red blood cell and white blood cell production

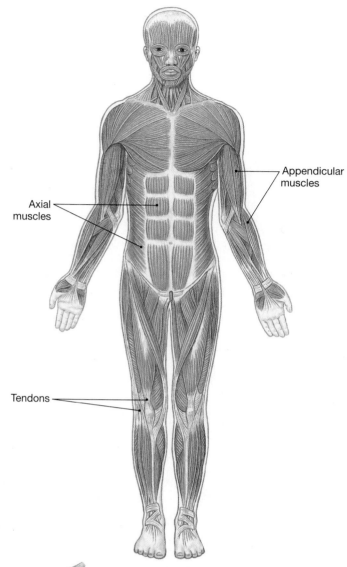

Axial muscles

Appendicular muscles

Tendons

(c) THE MUSCULAR SYSTEM

Allows for locomotion; provides support; produces heat

ORGAN/COMPONENT	PRIMARY FUNCTIONS
SKELETAL MUSCLES (700)	Provide skeletal movement; control entrances and exits of digestive tract; produce heat; support skeletal position; protect soft tissues
Axial muscles	Support and position axial skeleton
Appendicular muscles	Support, move, and brace limbs
TENDONS	Harness forces of contraction to perform specific tasks

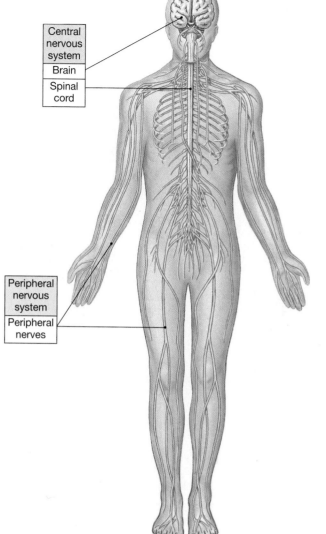

Central nervous system
Brain
Spinal cord

Peripheral nervous system
Peripheral nerves

(d) THE NERVOUS SYSTEM

Directs immediate response to stimuli, usually by coordinating the activities of other organ systems

ORGAN/COMPONENT	PRIMARY FUNCTIONS
CENTRAL NERVOUS SYSTEM (CNS)	Acts as control center for nervous system; processes information; provides short-term control over activities of other systems
Brain	Performs complex integrative functions; controls both voluntary and autonomic activities
Spinal cord	Relays information to and from brain; performs less complex integrative functions; directs many simple involuntary activities
PERIPHERAL NERVOUS SYSTEM (PNS)	Links CNS with other systems and with sense organs

● **FIGURE 1–2** **Continued.**

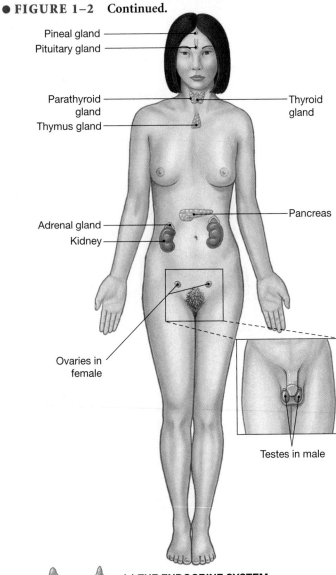

Pineal gland
Pituitary gland

Parathyroid gland
Thymus gland

Thyroid gland

Adrenal gland
Kidney

Pancreas

Ovaries in female

Testes in male

(e) THE ENDOCRINE SYSTEM
Directs long-term changes in activities of other organ systems

ORGAN/COMPONENT	PRIMARY FUNCTIONS
PINEAL GLAND	May control timing of reproduction and set day–night rhythms
PITUITARY GLAND	Controls other endocrine glands; regulates growth and fluid balance
THYROID GLAND	Controls tissue metabolic rate; regulates calcium levels
PARATHYROID GLANDS	Regulate calcium levels (with thyroid)
THYMUS	Controls maturation of lymphocytes
ADRENAL GLANDS	Adjust water balance, tissue metabolism, and cardiovascular and respiratory activity
KIDNEYS	Control red blood cell production and regulate blood pressure
PANCREAS	Regulates blood glucose levels
GONADS Testes	Support male sexual characteristics and reproductive functions (see Figure 1–2k)
Ovaries	Support female sexual characteristics and reproductive functions (see Figure 1–2l)

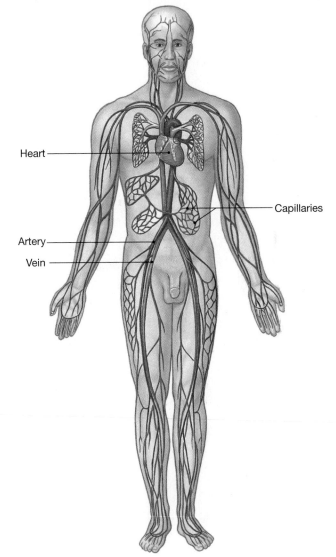

Heart

Capillaries

Artery
Vein

(f) THE CARDIOVASCULAR SYSTEM
Transports cells and dissolved materials, including nutrients, wastes, and gases

ORGAN/COMPONENT	PRIMARY FUNCTIONS
HEART	Propels blood; maintains blood pressure
BLOOD VESSELS Arteries Capillaries	Distribute blood throughout the body Carry blood from heart to capillaries Permit diffusion between blood and interstitial fluids
Veins	Return blood from capillaries to the heart
BLOOD	Transports oxygen, carbon dioxide, and blood cells; delivers nutrients and hormones; removes waste products; assists in temperature regulation and defense against disease

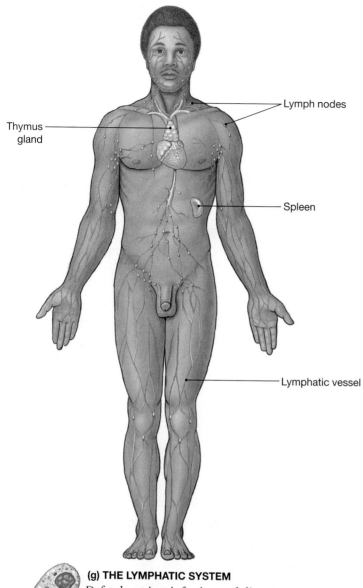

Thymus gland

Lymph nodes

Spleen

Lymphatic vessel

Nasal cavity

Sinuses

Pharynx

Trachea

Larynx

Bronchi

Lung

Diaphragm

(g) THE LYMPHATIC SYSTEM
Defends against infection and disease; returns tissue fluid to the bloodstream

(h) THE RESPIRATORY SYSTEM
Delivers air to sites where gas exchange can occur between the air and circulating blood

ORGAN/COMPONENT	PRIMARY FUNCTIONS
LYMPHATIC VESSELS	Carry lymph (water and proteins) and lymphocytes from peripheral tissues to veins of the cardiovascular system
LYMPH NODES	Monitor the composition of lymph; contain cells that engulf pathogens and stimulate immune response
SPLEEN	Monitors circulating blood; contains cells that engulf pathogens and stimulate immune response
THYMUS	Controls development and maintenance of one class of lymphocytes (T cells)

ORGAN/COMPONENT	PRIMARY FUNCTIONS
NASAL CAVITIES, PARANASAL SINUSES	Filter, warm, and humidify air; detect smells
PHARYNX	Conducts air to larynx; the pharynx is a chamber shared with the digestive tract (*see Figure 1–2i*)
LARYNX	Protects opening to trachea and contains vocal cords
TRACHEA	Filters air, traps particles in mucus; cartilages keep airway open
BRONCHI	(Same functions as trachea)
LUNGS	Responsible for air movement through volume changes that result from movements of ribs and diaphragm; include airways and alveoli
Alveoli	Act as sites of gas exchange between air and blood

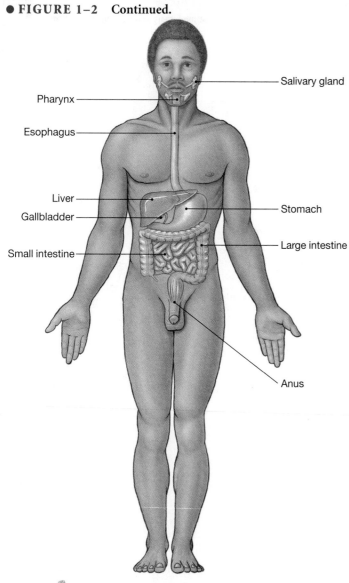

Salivary gland

Pharynx

Esophagus

Liver

Gallbladder

Small intestine

Stomach

Large intestine

Anus

(i) THE DIGESTIVE SYSTEM

Processes food and absorbs nutrients

ORGAN/COMPONENT	PRIMARY FUNCTIONS
SALIVARY GLANDS	Provide buffers and lubrication; produce enzymes that begin digestion
PHARYNX	Conducts solid food and liquids to esophagus; is a chamber shared with respiratory tract *(see Figure 1–2h)*
ESOPHAGUS	Delivers food to stomach
STOMACH	Secretes acids and enzymes
SMALL INTESTINE	Secretes digestive enzymes, buffers, and hormones; absorbs nutrients
LIVER	Secretes bile; regulates nutrient composition of blood
GALLBLADDER	Stores bile for release into small intestine
PANCREAS	Secretes digestive enzymes and buffers; secretes hormones *(see Figure 1–2e)*
LARGE INTESTINE	Removes water from fecal material; stores wastes

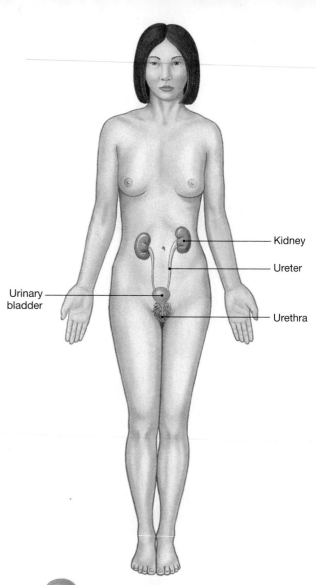

Kidney

Ureter

Urinary bladder

Urethra

(j) THE URINARY SYSTEM

Eliminates excess water, salts, and waste products

ORGAN/COMPONENT	PRIMARY FUNCTIONS
KIDNEYS	Form and concentrate urine; regulate blood pH and ion concentrations; perform endocrine functions *(see Figure 1–2e)*
URETERS	Conduct urine from kidneys to urinary bladder
URINARY BLADDER	Stores urine for eventual elimination
URETHRA	Conducts urine to exterior

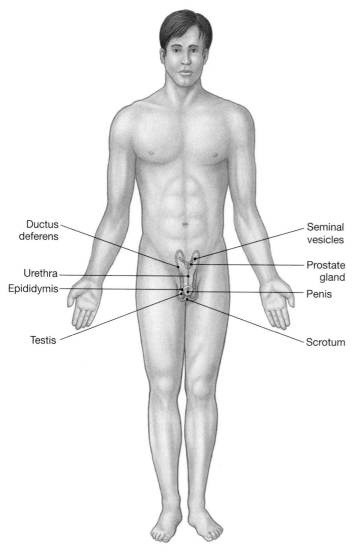

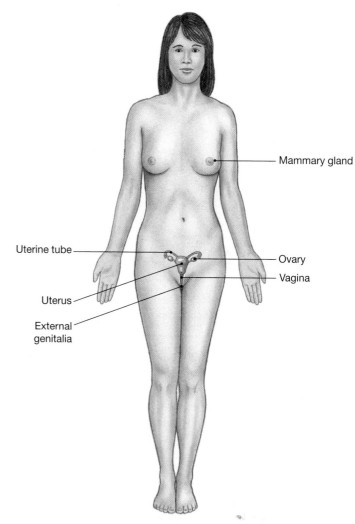

(k) THE MALE REPRODUCTIVE SYSTEM

Produces sex cells and hormones

ORGAN/COMPONENT	PRIMARY FUNCTIONS
TESTES	Produce sperm and hormones *(see Figure 1–2e)*
ACCESSORY ORGANS Epididymis	Acts as site of sperm maturation
Ductus deferens (sperm duct)	Conducts sperm between epididymis and prostate gland
Seminal vesicles	Secrete fluid that makes up much of the volume of semen
Prostate gland	Secretes fluid and enzymes
Urethra	Conducts semen to exterior
EXTERNAL GENITALIA Penis	Contains erectile tissue; deposits sperm in vagina of female; produces pleasurable sensations during sexual activities
Scrotum	Surrounds the testes and controls their temperature

(l) THE FEMALE REPRODUCTIVE SYSTEM

Produces sex cells and hormones

ORGAN/COMPONENT	PRIMARY FUNCTIONS
OVARIES	Produce oocytes and hormones *(see Figure 1–2e)*
UTERINE TUBES	Deliver oocyte or embryo to uterus; normal site of fertilization
UTERUS	Site of embryonic development and exchange between maternal and embryonic bloodstreams
VAGINA	Site of sperm deposition; acts as birth canal at delivery; provides passageway for fluids during menstruation
EXTERNAL GENITALIA Clitoris	Contains erectile tissue; produces pleasurable sensations during sexual activities
Labia	Contain glands that lubricate entrance to vagina
MAMMARY GLANDS	Produce milk that nourishes newborn infant

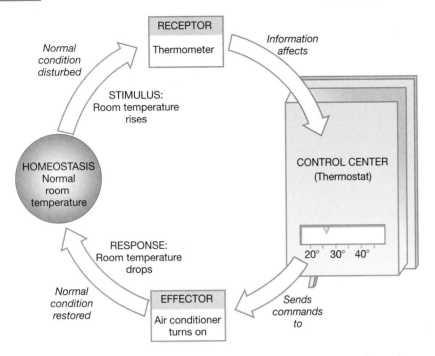

● **FIGURE 1–3** **The Control of Room Temperature.** In response to input from a receptor (a thermometer), a thermostat (the control center) triggers a response from an effector (in this case, an air conditioner) that restores normal temperature. When room temperature rises above the set point, the thermostat turns on the air conditioner, and the temperature returns to normal.

Most homeostatic mechanisms in the body involve negative feedback. For example, consider the control of body temperature, a process called *thermoregulation* (Figure 1–4●). Thermoregulation involves altering the relationship between heat loss, which occurs primarily at the body surface, and heat production, which occurs in all active tissues. In the human body, skeletal muscles are the most important generators of body heat.

The cells of the thermoregulatory control center are located in the brain. Temperature receptors are located in the skin and in cells in the control center. The thermoregulatory center has a normal set point near 37°C (98.6°F). If body temperature rises above 37.2°C, activity in the control center targets two effectors: (1) smooth muscles in the walls of blood vessels that supply the skin and (2) sweat glands. The muscle tissue relaxes and the blood vessels widen, or dilate, which increases blood flow at the body surface, and the sweat glands accelerate their secretion. The skin then acts like a radiator; it loses heat to the environment, and the evaporation of sweat speeds the process. When body temperature returns to normal, the control center becomes inactive, and superficial blood flow and sweat gland activity decrease to normal resting levels.

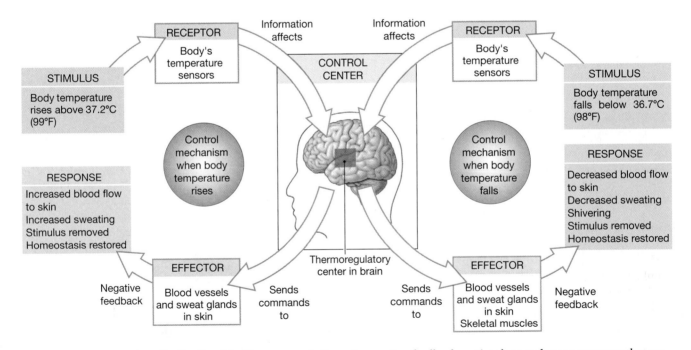

● **FIGURE 1–4** **Negative Feedback in Thermoregulation.** In negative feedback, a stimulus produces a response that opposes the original stimulus. Body temperature is regulated by a control center in the brain that functions as a thermostat with a set point of 37°C. If body temperature climbs above 37.2°C, heat loss is increased through enhanced blood flow to the skin and increased sweating. If body temperature falls below 36.7°C, heat loss is decreased through decreases in blood flow to the skin and sweating, and heat is produced by shivering.

If temperature at the control center falls below 36.7°C, the control center targets the same two effectors and skeletal muscles. This time, blood flow to the skin declines, and sweat gland activity decreases. This combination reduces the rate of heat loss to the environment. Because heat production continues, body temperature gradually rises; once the set point has been reached, the thermoregulatory center turns itself "off," and both blood flow and sweat gland activity in the skin increase to normal resting levels. Additional heat may be generated by shivering, which is caused by random contractions of skeletal muscles.

Homeostatic mechanisms using negative feedback usually ignore minor variations, and they maintain a normal range rather than a fixed value. In the previous example, body temperature oscillates around the ideal set-point temperature. Thus, for any single individual, any measured value (such as body temperature) can vary from moment to moment or day to day. The variability among individuals is even greater, because each person has slightly different homeostatic set points. It is, therefore, impractical to define "normal" homeostatic conditions very precisely. By convention, physiological values are reported either as average values obtained by sampling a large number of individuals or as a range that includes 95 percent or more of the sample population. However, 5 percent of normal adults have a body temperature outside the "normal" range (below 36.7°C or above 37.2°C). Still, these temperatures are perfectly normal for them, and the variations have no clinical significance.

Positive Feedback

In **positive feedback**, *the initial stimulus produces a response that reinforces that stimulus.* For example, suppose a thermostat was wired so that when the temperature rose, the thermostat would turn on the heater rather than the air conditioner. In that case, the initial stimulus (rising room temperature) would cause a response (heater turns on) that strengthens the stimulus. Room temperature would continue to rise until someone switched off the thermostat, unplugged the heater, or intervened in some other way before the house caught fire and burned down. This kind of escalating cycle is called a *positive feedback loop.*

In the body, positive feedback loops are involved in the regulation of a potentially dangerous or stressful process that must be completed quickly (Figure 1–5●). For example, the immediate danger from a severe cut is blood loss, which can lower blood pressure and reduce the pumping efficiency of the heart. Damage to cells in the cut blood vessel wall releases chemicals that begin the multistep process of blood clotting. As clotting gets under way, each step releases chemicals that accelerate the process. This escalating process is a positive feedback loop that ends with the formation of a blood clot, which patches the vessel wall and stops the bleeding. Blood clotting will be examined more closely in Chapter 11. Labor and delivery, another example of positive feedback in action, will be discussed in Chapter 20.

Homeostasis and Disease

The human body is amazingly effective in maintaining homeostasis. Nevertheless, an infection, an injury, or a genetic abnormality can sometimes have effects so severe that homeostatic mechanisms cannot fully compensate for them. When homeostatic regulation fails, organ systems begin to malfunction, and the individual experiences the symptoms of illness, or **disease.**

Key Note

Physiological systems work together to maintain a stable internal environment, which is the foundation of homeostasis. In doing so they monitor and adjust the volume and composition of body fluids, and keep body temperature within normal limits. If they cannot do so, internal conditions become increasingly abnormal and survival becomes uncertain.

CONCEPT CHECK QUESTIONS

1. Why is homeostatic regulation important to humans?
2. Why is positive feedback helpful in blood clotting but unsuitable for the regulation of body temperature?
3. What happens to the body when homeostasis breaks down?

Answers begin on p. 792.

● **FIGURE 1–5** **Positive Feedback.** In positive feedback, a stimulus produces a response that reinforces the original stimulus. Positive feedback is important in accelerating processes that must proceed to completion rapidly. In this example, positive feedback accelerates blood clotting until bleeding stops.

■ The Language of Anatomy

Early anatomists faced serious communication problems. For example, stating that a bump is "on the back" does not give very precise information about its location. So anatomists created maps of the human body. Prominent anatomical structures serve as landmarks, distances are measured (in centimeters or inches), and specialized directional terms are used. In effect, anatomy uses a language of its own that must be learned almost at the start of your study.

A familiarity with Latin and Greek word roots and their combinations makes anatomical terms more understandable. As new terms are introduced in the text, notes on their pronunciation and the relevant word roots will be provided. Additional information on foreign word roots, prefixes, suffixes, and combining forms can be found inside the front cover.

Latin and Greek terms are not the only foreign words imported into the anatomical vocabulary over the centuries, and the vocabulary continues to expand. Many anatomical structures and clinical conditions were initially named after either the

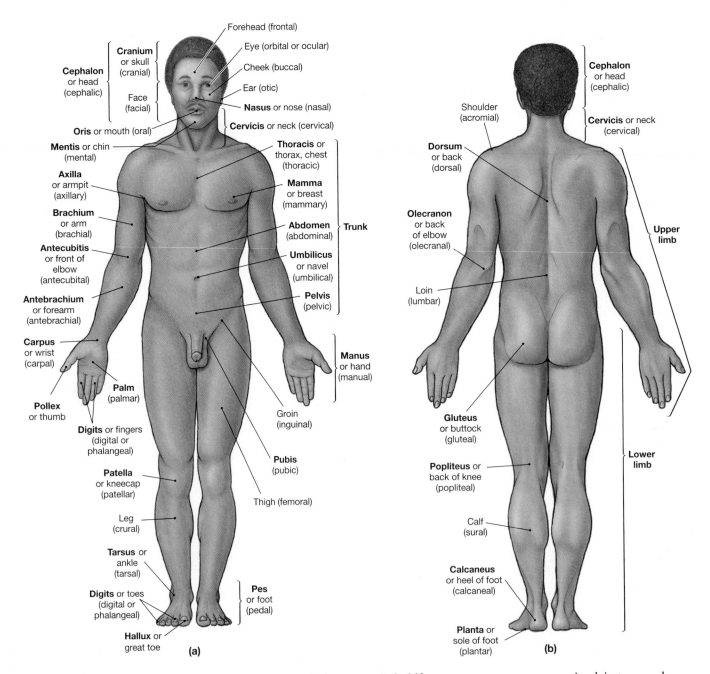

● **FIGURE 1–6 Anatomical Landmarks.** Anatomical terms are in boldface type, common names are in plain type, and anatomical adjectives are in parentheses.

discoverer or, in the case of diseases, the most famous victim. Although most such commemorative names, or *eponyms,* have been replaced by more precise terms, many are still in use.

Surface Anatomy

With the exception of the skin, none of the organ systems can be seen from the body surface. Therefore, you must create your own mental maps and extract information from the terms given in Figures 1–6● and 1–7●. Learning these terms now will make subsequent chapters more understandable.

Anatomical Landmarks

Standard anatomical illustrations show the human form in the **anatomical position,** with the hands at the sides, palms forward, and feet together (Figure 1–6). A person lying in the anatomical position is said to be **supine** (soo-PĪN) when face up and **prone** when face down.

Important anatomical landmarks are also presented in Figure 1–6. The anatomical terms are in boldface, the common names in plain type, and the anatomical adjectives in parentheses. Understanding these terms and their origins can help you remember both the location of a particular structure and its name. For example, the term *brachium* refers to the arm, and later chapters will discuss the brachial artery, brachial nerve, and so forth. You might remember this term more easily if you know that the Latin word *brachium* is also the source of Old English and French words meaning "to embrace."

Anatomical Regions

Major regions of the body are listed in Table 1–1 and shown in Figure 1–6. Anatomists and clinicians often need to use regional terms as well as specific landmarks to describe a general area of interest or injury. Two methods are used to map the surface of the abdomen and pelvis. Clinicians refer to four **abdominopelvic quadrants** formed by a pair of imaginary perpendicular lines that intersect at the *umbilicus* (navel). This simple method, shown in Figure 1–7a●, is useful for describing the location of aches, pains, and injuries, which can help a doctor determine the possible cause. For example, tenderness in the right lower quadrant (RLQ) is a symptom of appendicitis, whereas tenderness in the right upper quadrant (RUQ) may indicate gallbladder or liver problems.

Anatomists like to use more precise regional distinctions to describe the location and orientation of internal organs. They recognize nine **abdominopelvic regions** (Figure 1–7b●). Figure 1–7c● shows the relationships among quadrants, regions, and internal organs.

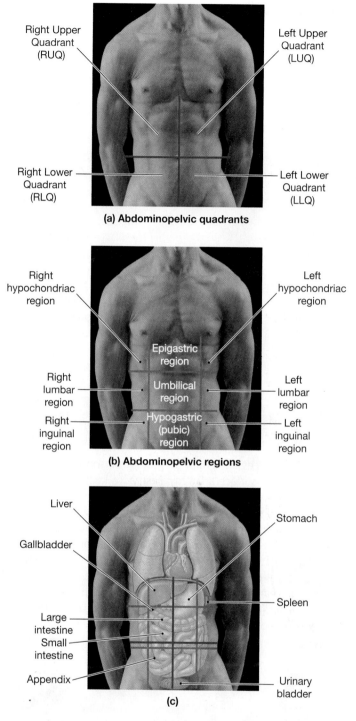

(a) Abdominopelvic quadrants

(b) Abdominopelvic regions

(c)

● **FIGURE 1–7   Abdominopelvic Quadrants and Regions.** **(a)** Two imaginary perpendicular lines divide the area into four abdominopelvic quadrants. These quadrants, or their abbreviations, are most often used in clinical discussions. **(b)** More precise regional descriptions are provided by referring to one of nine abdominopelvic regions. **(c)** The usefulness of quadrants or regions results from the known relationships between superficial anatomical landmarks and underlying organs.

TABLE 1–1 *Regions of the Human Body (see Figure 1–6)*

STRUCTURE	REGION
Cephalon (head)	Cephalic region
Cervicis (neck)	Cervical region
Thoracis (thorax or chest)	Thoracic region
Abdomen	Abdominal region
Pelvis	Pelvic region
Loin (lower back)	Lumbar region
Buttock	Gluteal region
Pubis (anterior pelvis)	Pubic region
Groin	Inguinal region
Axilla (armpit)	Axillary region
Brachium (arm)	Brachial region
Antebrachium (forearm)	Antebrachial region
Manus (hand)	Manual region
Thigh	Femoral region
Leg (anterior)	Crural region
Calf	Sural region
Pes (foot)	Pedal region
Planta (sole)	Plantar region

Clinical Note
SURFACE ANATOMY

It is essential that prehospital personnel have a good grasp of human surface anatomy. First, such knowledge will help determine what underlying structures might be injured. Second, using standard surface anatomical terms will allow the paramedic to describe an injury accurately to a remote medical direction physician (Figure 1–8●). Finally, post-call documentation requires a detailed report of the call and the care delivered. It is not uncommon for prehospital personnel to be questioned about a particular call months or years after it occurred. Only through use of proper anatomical terms and documentation can you accurately recall the circumstances of the run and the patient's injuries. ■

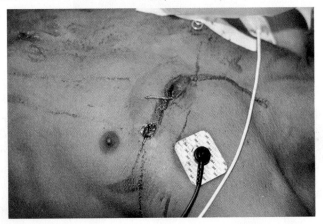

● **FIGURE 1–8** Gunshot Wound.

Anatomical Directions

Figure 1–9● and Table 1–2 present the principal directional terms and some examples of their use. There are many different terms, and some can be used interchangeably. For example, *anterior* refers to the front of the body, when viewed in the anatomical position; in humans, this term is equivalent to *ventral,* which refers to the belly. Likewise, the terms *posterior* and *dorsal* refer to the back of the human body. Terms that appear frequently in later chapters have been emphasized. Remember that *left* and *right* always refer to the left and right sides of the *subject,* not of the observer.

Key Note
Anatomical descriptions refer to an individual in the anatomical position: standing, with the hands at the sides, palms facing forward, and feet together.

Sectional Anatomy

Sometimes the only way to understand the relationships among the parts of a three-dimensional object is to slice through it and look at the internal organization. An understanding of sectional views has become increasingly important since the development of electronic imaging techniques that enable us to see inside the living body without resorting to surgery.

Planes and Sections

Any slice through a three-dimensional object can be described with reference to three primary **sectional planes,** indicated in Figure 1–10● and Table 1–3:

1. *Transverse Plane.* The **transverse plane** lies at right angles to the long (head-to-foot) axis of the body, and divides the body into superior and inferior portions. A cut in this plane is called a transverse section, or cross section.
2. *Frontal Plane.* The **frontal plane,** or *coronal plane,* runs along the long axis of the body. The frontal plane extends laterally (side to side), and divides the body into **anterior** and **posterior** portions.
3. *Sagittal Plane.* The **sagittal plane** also runs along the long axis of the body, but it extends anteriorly and posteriorly (front to back). A sagittal plane divides the body into *left* and *right* portions. A cut that passes along the body's midline and divides the body into left and right halves is a **midsagittal section.** (Note that a midsagittal section does not cut through the legs.)

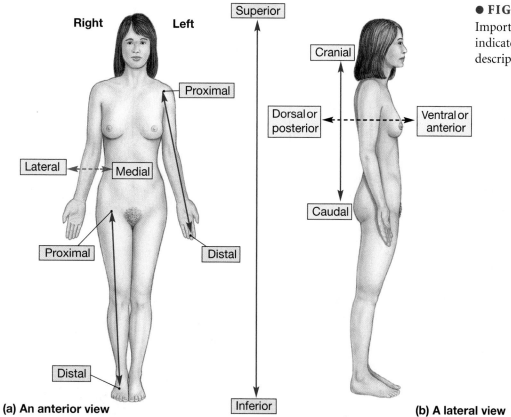

● **FIGURE 1–9**   **Directional References.**
Important directional terms used in this text are indicated by arrows; definitions and descriptions are included in Table 1–2.

(a) An anterior view

(b) A lateral view

TABLE 1–2	*Directional Terms (see Figure 1–9)*	
TERM	**REGION OR REFERENCE**	**EXAMPLE**
ANTERIOR	The front; before	The navel is on the *anterior* (*ventral*) surface of the trunk.
VENTRAL	The belly side (equivalent to anterior when referring to the human body)	
POSTERIOR	The back; behind	The shoulder blade is located *posterior* (*dorsal*) to the rib cage.
DORSAL	The back (equivalent to posterior when referring to the human body)	
CRANIAL OR CEPHALIC	The head	The *cranial*, or *cephalic*, border of the pelvis is superior to the thigh.
SUPERIOR	Above; at a higher level (in the human body, toward the head)	The nose is *superior* to the chin.
CAUDAL	The tail (coccyx in humans)	The hips are *caudal* to the waist.
INFERIOR	Below; at a lower level	The knees are *inferior* to the hips.
MEDIAL	Toward the body's longitudinal axis	The *medial* surfaces of the thighs may be in contact; moving medially from the arm across the chest surface brings you to the sternum.
LATERAL	Away from the body's longitudinal axis	The thigh articulates with the *lateral* surface of the pelvis; moving laterally from the nose brings you to the eyes.
PROXIMAL	Toward an attached base	The thigh is *proximal* to the foot; moving proximally from the wrist brings you to the elbow.
DISTAL	Away from an attached base	The fingers are *distal* to the wrist; moving distally from the elbow brings you to the wrist.
SUPERFICIAL	At, near, or relatively close to the body surface	The scalp is *superficial* to the skull.
DEEP	Farther from the body surface	The bone of the thigh is *deep* to the surrounding skeletal muscles.

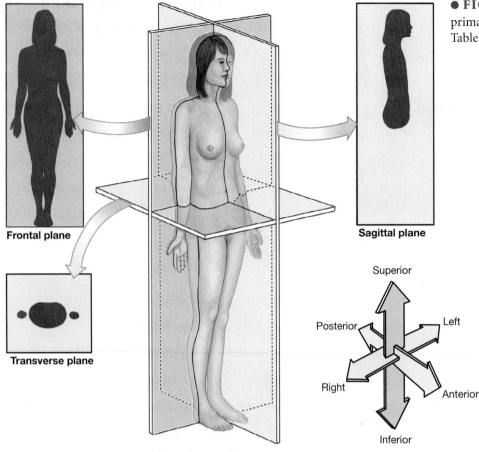

● **FIGURE 1–10** **Planes of Section.** The three primary planes of section are indicated here. Table 1–3 defines and describes them.

Frontal plane

Sagittal plane

Transverse plane

Superior

Posterior

Left

Right

Anterior

Inferior

TABLE 1–3	*Terms That Indicate Planes of Section (see Figure 1–10)*		
ORIENTATION PLANE	**ADJECTIVE**	**DIRECTIONAL REFERENCE**	**DESCRIPTION**
PARALLEL TO LONG AXIS	Sagittal	Sagittally	A *sagittal section* separates right and left portions. You examine a sagittal section, but you section sagittally.
	Midsagittal		In a *midsagittal section*, the plane passes through the midline, divides the body in half and separates right and left sides.
	Frontal or coronal	Frontally or coronally	A *frontal*, or *coronal*, *section* separates anterior and posterior portions of the body; *coronal* usually refers to sections that pass through the skull.
PERPENDICULAR TO LONG AXIS	Transverse or horizontal	Transversely or horizontally	A *transverse*, or *horizontal*, *section* separates superior and inferior portions of the body.

Body Cavities

Viewed in sections, the human body is not a solid object, like a rock, in which all of the parts are fused together. Many vital organs are suspended in internal chambers called *body cavities*. These cavities have two essential functions: (1) they protect delicate organs, such as the brain and spinal cord, from accidental shocks and cushion them from the jolting that occurs when we walk, jump, or run; and (2) they permit significant changes in the size and shape of internal organs. For example, because they are inside body cavities, the lungs, heart, stomach, intestines, urinary bladder, and many other

organs can expand and contract without distorting surrounding tissues or disrupting the activities of nearby organs.

The **ventral body cavity,** or *coelom* (SĒ-lōm; *koila,* cavity), appears early in embryonic development. It contains organs of the respiratory, cardiovascular, digestive, urinary, and reproductive systems. As these internal organs develop, their relative positions change, and the ventral body cavity is gradually subdivided. The **diaphragm** (DĪ-uh-fram), which is a flat muscular sheet, divides the ventral body cavity into a superior **thoracic cavity,** bounded by the chest wall, and an inferior **abdominopelvic cavity,** enclosed by the abdominal wall and by the bones and muscles of the pelvis. The boundaries between the divisions of the ventral body cavity are shown in Figure 1–11●.

Many of the organs in these cavities change size and shape as they perform their functions. For example, the lungs inflate and deflate as you breathe, and your stomach swells during each meal and shrinks between meals. These organs are surrounded by moist internal spaces that permit expansion and limited movement while preventing friction. The internal organs within the thoracic and abdominopelvic cavities are called **viscera** (VIS-e-ruh). A delicate layer called a *serous membrane* lines the walls of these internal cavities and covers the surfaces of the enclosed viscera. Serous membranes are moistened by a watery fluid that coats the opposing surfaces and reduces friction. The portion of a serous membrane that covers a visceral organ is called the *visceral* layer; the opposing layer that lines the inner surface of the body wall or chamber is called the *parietal* layer.

THE THORACIC CAVITY. The thoracic cavity contains three internal chambers: a single *pericardial cavity* and a pair of *pleural cavities* (see Figure 1–11a, c). Each of these cavities is lined by shiny, slippery serous membranes. The heart projects into a space known as the **pericardial cavity.** The relationship between the heart and the pericardial cavity resembles that of a fist pushing into a balloon (see Figure 1–11b). The wrist corresponds to the *base* (attached portion) of the heart, and the balloon corresponds to the serous membrane that lines the pericardial cavity. The serous membrane is called the **pericardium** (*peri-*, around + *cardium*, heart). The layer that covers the heart is the **visceral pericardium,** and the opposing surface is the **parietal pericardium.**

The pericardium lies within the **mediastinum** (mē-dē-as-TĪ-num or mē-dē-AS-ti-num) (see Figure 1–11c). The connective tissue of the mediastinum surrounds the pericardial cavity and heart, the large arteries and veins attached to the heart, and the thymus, trachea, and esophagus.

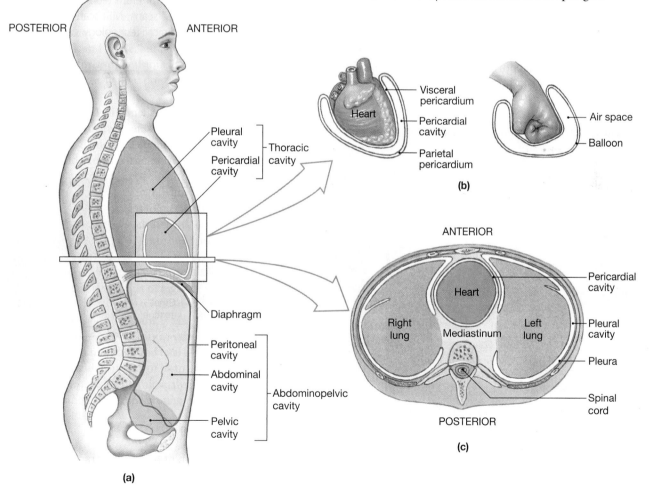

● **FIGURE 1–11   The Ventral Body Cavity and Its Subdivisions.** (a) A lateral view that shows the ventral body cavity, which is divided by the muscular diaphragm into a superior thoracic (chest) cavity and an inferior abdominopelvic cavity. (b) The heart is suspended within the pericardial cavity like a fist pushed into a balloon. The attachment site, which corresponds to the wrist of the hand, lies at the connection between the heart and major blood vessels. (c) A transverse section through the ventral body cavity, that shows the central location of the pericardial cavity within the thoracic cavity. Notice how the mediastinum divides the thoracic cavity into two pleural cavities.

Chapter Review

Access more review material online at *www.prenhall.com/bledsoe*. There you will find quiz questions, labeling activities, animations, essay questions, and web links.

Key Terms

anatomical position 17
anatomy 4
diaphragm 20
frontal plane 18

homeostasis 7
negative feedback 7
peritoneum 22
physiology 4

positive feedback 15
sagittal plane 18
transverse plane 18
viscera 21

Related Clinical Terms

abdominopelvic quadrant One of four divisions of the anterior abdominal surface.

abdominopelvic region One of nine divisions of the anterior abdominal surface.

auscultation (aws-kul-TĀ-shun) Listening to a patient's body sounds using a stethoscope.

CT, CAT (computerized [axial] tomography) An imaging technique that uses X-rays to reconstruct the body's three-dimensional structure.

disease A malfunction of organs or organ systems that results from a failure in homeostatic regulation.

echogram An image created by ultrasound.

histology (his-TOL-o-jē) The study of tissues.

MRI (magnetic resonance imaging) An imaging technique that employs a magnetic field and radio waves to make subtle structural differences visible.

PET (positron emission tomography) **scan** An imaging technique that shows the chemical functioning, as well as the structure, of an organ.

radiologist A physician who specializes in performing and analyzing radiological procedures.

radiology (rā-dē-OL-o-jē) The study of radioactive energy and radioactive substances and their use in the diagnosis and treatment of disease.

ultrasound An imaging technique that uses brief bursts of high-frequency sound waves reflected by internal structures.

X-rays High-energy radiation that can penetrate living tissues.

Summary Outline

1. **Biology** is the study of life; one of its goals is to discover the unity and patterns that underlie the diversity of living organisms.

2. All living things, from single *cells* to large multicellular *organisms,* perform the same basic functions: they respond to changes in their environment; they grow and reproduce to create future generations; they are capable of producing movement; and they absorb materials from the environment. Organisms absorb and consume oxygen during respiration, and they discharge waste products during excretion. Digestion occurs in specialized structures of the body to break down complex foods. The circulation forms an internal transportation system between areas of the body.

THE SCIENCES OF ANATOMY AND PHYSIOLOGY 4

Anatomy 4

1. **Anatomy** is the study of internal and external structure and the physical relationships among body parts. **Physiology** is the study of how living organisms perform vital functions. All specific functions are performed by specific structures.

2. **Gross (macroscopic) anatomy** considers features visible without a microscope. It includes **surface anatomy** (general form and superficial markings), **regional anatomy** (superficial and internal features in a specific area of the body), and **systemic anatomy** (structure of major organ systems).

3. The boundaries of **microscopic anatomy** are established by the equipment used. **Cytology** analyzes the internal structure of individual **cells**. **Histology** examines **tissues** (groups of cells that have specific functional roles). Tissues combine to form organs, which are anatomical units with specific functions.

Physiology 5

4. **Human physiology** is the study of the functions of the human body. It is based on **cell physiology,** which is the study of the functions of living cells. **Special physiology** studies the physiology of specific organs. **System physiology** considers all aspects of the function of specific organ systems. **Pathological physiology (pathology)** studies the effects of diseases on organ or system functions.

Key Note 5

LEVELS OF ORGANIZATION 5

1. Anatomical structures and physiological mechanisms are arranged in a series of interacting levels of organization. *(Figure 1–1)*

AN INTRODUCTION TO ORGAN SYSTEMS 7

1. The major organs of the human body are arranged into 11 organ systems. The organ systems of the human body are the *integumentary, skeletal, muscular, nervous, endocrine, cardiovascular, lymphatic, respiratory, digestive, urinary,* and *reproductive systems. (Figure 1–2)*

Key Note 7

HOMEOSTASIS AND SYSTEM INTEGRATION 7

1. **Homeostasis** is the tendency for physiological systems to stabilize internal conditions; through **homeostatic regulation** these systems adjust to preserve homeostasis.

Homeostatic Regulation 7

2. Homeostatic regulation usually involves a **receptor** sensitive to a particular stimulus and an **effector** whose activity affects the same stimulus.

3. **Negative feedback** is a corrective mechanism involving an action that directly opposes a variation from normal limits. *(Figures 1–3, 1–4)*

4. In **positive feedback** the initial stimulus produces a response that reinforces the stimulus. *(Figure 1–5)*

Homeostasis and Disease 15

5. Symptoms of **disease** appear when failure of homeostatic regulation causes organ systems to malfunction.

Key Note 15

THE LANGUAGE OF ANATOMY 16

Surface Anatomy 17

1. Standard anatomical illustrations show the body in the **anatomical position.** If the figure is shown lying down, it can be either **supine** (face up) or **prone** (face down). *(Figure 1–6; Table 1–1)*

2. **Abdominopelvic quadrants** and **abdominopelvic regions** represent two different approaches to describing anatomical regions of the body. *(Figures 1–7 and 1–8)*

3. The use of special directional terms provides clarity when describing anatomical structures. *(Figure 1–9; Table 1–2)*

Key Note 18

Sectional Anatomy 18

4. The three **sectional planes** (**frontal** or **coronal plane, sagittal plane,** and **transverse plane**) describe relationships between the parts of the three-dimensional human body. *(Figure 1–10; Table 1–3)*

5. **Body cavities** protect delicate organs and permit changes in the size and shape of visceral organs. The **ventral body cavity** surrounds developing respiratory, cardiovascular, digestive, urinary, and reproductive organs. *(Figure 1–11)*

6. The **diaphragm** divides the ventral body cavity into the superior **thoracic** and inferior **abdominopelvic cavities.** The thoracic cavity contains two **pleural cavities** (each of which contain a lung) and a **pericardial cavity** (which surrounds the heart). The abdominopelvic cavity consists of the **abdominal cavity** and the **pelvic cavity.** It contains the *peritoneal cavity,* which is an internal chamber lined by *peritoneum,* which is a *serous membrane.*

7. Important **radiological procedures** (which can provide detailed information about internal systems) include **X-rays, CT scans, MRI,** and **ultrasound.** Each technique has advantages and disadvantages. *(Figures 1–12, 1–13)*

Review Questions

Level 1: Reviewing Facts and Terms

Match each item in column A with the most closely related item in column B. Place letters for answers in the spaces provided.

COLUMN A

___ 1. cytology
___ 2. physiology
___ 3. histology
___ 4. metabolism
___ 5. homeostasis
___ 6. muscle
___ 7. heart
___ 8. endocrine
___ 9. temperature regulation
___ 10. blood clot formation
___ 11. supine
___ 12. prone
___ 13. ventral body cavity
___ 14. thoracic body cavity
___ 15. pericardium

COLUMN B

a. study of tissues
b. constant internal environment
c. face up
d. study of functions
e. positive feedback
f. system
g. study of cells
h. negative feedback
i. heart and lungs
j. all chemical activity in body
k. thoracic and abdominopelvic
l. tissue
m. serous membrane
n. organ
o. face down

16. The process by which an organism increases the size and/or number of its cells is called:
    (a) reproduction.
    (b) adaptation.
    (c) growth.
    (d) metabolism.

17. When a variation outside normal limits triggers a response that restores the normal condition, the regulatory mechanism involves:
    (a) negative feedback.
    (b) positive feedback.
    (c) compensation.
    (d) adaptation.

18. The terms that apply to the front of the body in the anatomical position are:
    (a) posterior, dorsal.
    (b) back, front.
    (c) medial, lateral.
    (d) anterior, ventral.

19. A cut through the body that passes perpendicular to the long axis of the body and divides the body into superior and inferior portions is known as a _____ section.
    (a) sagittal
    (b) transverse
    (c) coronal
    (d) frontal

20. The diaphragm, a flat muscular sheet, divides the ventral body cavity into a superior _____ cavity and an inferior _____ cavity.
    (a) pleural, pericardial
    (b) abdominal, pelvic
    (c) thoracic, abdominopelvic
    (d) cranial, thoracic

21. The mediastinum is the region between the:
    (a) lungs and heart.
    (b) two pleural cavities.
    (c) thorax and abdomen.
    (d) heart and pericardium.

Level 2: Reviewing Concepts

22. What basic functions are performed by all living things?

23. Beginning at the molecular level, list in correct sequence the levels of organization, from the simplest level to the most complex.

24. What is homeostatic regulation, and what is its physiological importance?

25. How does negative feedback differ from positive feedback?

26. Describe the position of the body when it is in the anatomical position.

27. As a surgeon, you perform an invasive procedure that necessitates cutting through the peritoneum. Are you more likely to be operating on the heart or on the stomach?

28. In which body cavity would each of the following organs or systems be found?
    (a) cardiovascular, digestive, and urinary systems
    (b) heart, lungs
    (c) stomach, intestines

Level 3: Critical Thinking and Clinical Applications

29. A hormone called *calcitonin*, which is produced by the thyroid gland, is released in response to increased levels of calcium ions in the blood. If this hormone acts through negative feedback, what effect will its release have on blood calcium levels?

30. An anatomist wishes to make detailed comparisons of medial surfaces of the left and right sides of the brain. This work requires sections that will show the entire medial surface. Which kind of sections should be ordered from the lab for this investigation?

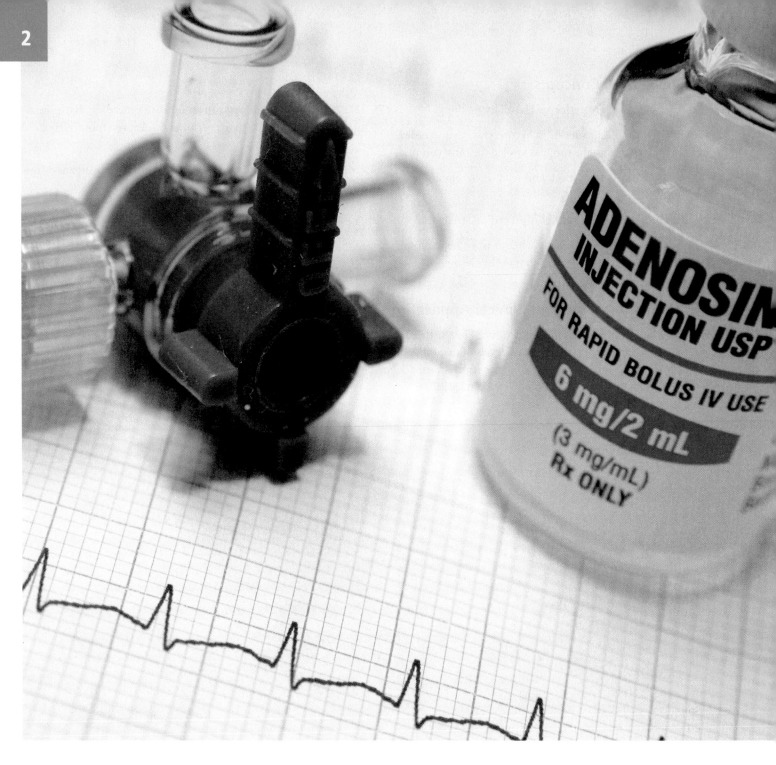

2 The Chemical Level of Organization

THE BODY IS a network of marvelously intertwined biochemical processes. In disease, these processes can go awry. Medical practice often involves correcting or changing some of these bio-chemical processes in order to restore the patient to health. Fundamentally chemicals, medications are an important tool in achieving this goal.

Chapter Outline

Chapter Objectives

1. Describe an atom and an element. (p. 29)
2. Compare the ways in which atoms combine to form molecules and compounds. (p. 32)
3. Use chemical notation to symbolize chemical reactions. (pp. 34, 35)
4. Distinguish among the three major types of chemical reactions that are important for studying physiology. (p. 36)
5. Describe the important role of enzymes in metabolism. (pp. 36–37)
6. Distinguish between organic and inorganic compounds. (p. 37)
7. Explain how the chemical properties of water make life possible. (p. 38)
8. Describe the pH scale and the role of buffers in body fluids. (pp. 39, 40–41)
9. Describe the physiological roles of inorganic compounds. (pp. 37–41)
10. Discuss the structure and functions of carbohydrates, lipids, proteins, nucleic acids, and high-energy compounds. (pp. 42–52)

Vocabulary Development

anabole a building up; *anabolism*
endo- inside; *endergonic*
exo- outside; *exergonic*
glyco- sugar; *glycogen*
hydro- water + **lysis** breakdown; *hydrolysis*

katabole a throwing down; *catabolism*
katalysis dissolution; *catalysis*
lipos fat; *lipids*
metabole change; *metabolism*

sakcharon sugar
 + **mono-** single; *monosaccharide*
 + **di-** two; *disaccharide*
 + **poly-** many; *polysaccharide*

OUR STUDY OF THE HUMAN BODY begins at the most basic level of organization, that of individual atoms and molecules. The characteristics of all living and nonliving things—people, elephants, oranges, oceans, rocks, and air—result from the types of atoms involved and the ways those atoms combine and interact. **Chemistry** is the science that investigates *matter* and its interactions. A familiarity with basic chemistry helps us understand how the properties of atoms can affect the anatomy and physiology of the cells, tissues, organs, and organ systems that make up the human body.

■ Matter: Atoms and Molecules

Matter is anything that takes up space and has mass. *Mass* is a physical property that determines the weight of an object in Earth's gravitational field. On our planet, the weight of an object is essentially the same as its mass. However, the two are not always the same. For example, in orbit you would be weightless, but your mass would remain unchanged. Matter occurs in one of three familiar states: solid (such as a rock), liquid (such as water), or gas (such as the atmosphere).

All matter is composed of substances called **elements.** Elements cannot be changed or broken down into simpler substances, whether by chemical processes, heating, or other ordinary physical means. The smallest, stable unit of matter is an **atom.** Atoms are so small that atomic measurements are most conveniently reported in billionths of a meter or *nanometers* (NA-nō-mē-terz) (nm). The very largest atoms approach half of one-billionth of a meter (0.5 nm) in diameter. One million atoms placed side-by-side would span a period on this page, but the line would be far too thin to be visible with anything but the most powerful of microscopes.

The Structure of an Atom

Atoms contain three major types of subatomic particles: protons, neutrons, and electrons. Protons and neutrons are

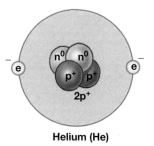

Helium (He)

● **FIGURE 2–1   A Diagram of Atomic Structure.** The atom shown here—helium—contains two of each type of subatomic particle: two protons, two neutrons, and two electrons.

similar in size and mass, but **protons** (p^+) have a positive electrical charge; whereas **neutrons** (n^0) are neutral—that is, uncharged. **Electrons** (e^-) are much lighter—only 1/1836 as massive as protons—and have a negative electrical charge. Figure 2–1● is a diagram of a simple atom: the element *helium*. This atom contains two protons, two neutrons, and two electrons.

All atoms contain protons and electrons, normally in equal numbers. The number of protons in an atom is known as its **atomic number.** A chemical element is a substance that consists entirely of atoms with the same atomic number. For example, all atoms of the element helium contain two protons.

Table 2–1 lists the 13 most abundant elements in the human body. Each element is universally known by its own abbreviation, or chemical symbol. Most of the symbols are easily connected with the English names of the elements, but a few, such as Na for sodium, are abbreviations of their original Latin names—in this example, *natrium*. (Appendix II, the *periodic table*, gives the chemical symbols and atomic numbers of each element.)

Hydrogen, the simplest element, has an atomic number of 1 because its atom contains one proton. The proton is located in the center of the atom and forms the **nucleus.** Hydrogen atoms seldom contain neutrons, but whenever neutrons are present in any type of atom, they are also located in the nucleus. In a hydrogen atom, a single electron orbits the nucleus at high speed, which forms an **electron cloud** (Figure 2–2a●). To simplify matters, this cloud is usually represented as a spherical **electron shell** (Figure 2–2b●).

Isotopes

The atoms of a given element can differ in terms of the number of neutrons in the nucleus. Such atoms of an element are called **isotopes.** The presence or absence of neutrons generally has no effect on the chemical properties of an atom of a

TABLE 2–1   *The Principal Elements in the Human Body*

ELEMENT (% OF BODY WEIGHT)	SIGNIFICANCE
OXYGEN, O (65)	A component of water and other compounds; oxygen gas is essential for respiration
CARBON, C (18.6)	Found in all organic molecules
HYDROGEN, H (9.7)	A component of water and most other compounds in the body
NITROGEN, N (3.2)	Found in proteins, nucleic acids, and other organic compounds
CALCIUM, Ca (1.8)	Found in bones and teeth; important for membrane function, nerve impulses, muscle contraction, and blood clotting
PHOSPHORUS, P (1)	Found in bones and teeth, nucleic acids, and high-energy compounds
POTASSIUM, K (0.4)	Important for proper membrane function, nerve impulses, and muscle contraction
SODIUM, Na (0.2)	Important for membrane function, nerve impulses, and muscle contraction
CHLORINE, Cl (0.2)	Important for membrane function and water absorption
MAGNESIUM, Mg (0.06)	Required for activation of several enzymes
SULFUR, S (0.04)	Found in many proteins
IRON, Fe (0.007)	Essential for oxygen transport and energy capture
IODINE, I (0.0002)	A component of hormones of the thyroid gland

particular element. As a result, isotopes can be distinguished from one another only by their **mass number**—this is the total number of protons and neutrons in the nucleus. The nuclei of some isotopes may be unstable. Unstable isotopes are radioactive; that is, they spontaneously emit subatomic particles or radiation in measurable amounts. These *radioisotopes* are sometimes used in diagnostic procedures.

Atomic Weight

Atomic mass numbers are useful because they tell us the number of protons and neutrons in the nuclei of different atoms. However, they do not tell us the *actual* mass of an atom, because they do not take into account the masses of electrons and the slight difference between the masses of a proton and

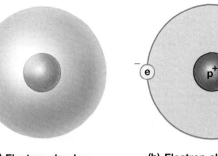

(a) Electron cloud or space-filling model   **(b) Electron-shell model**

● **FIGURE 2–2   Hydrogen Atoms.** A typical hydrogen atom consists of a nucleus that contains one proton and no neutrons, around which a single electron orbits. (**a**) This space-filling model of a hydrogen atom depicts the three-dimensional electron cloud formed by the single orbiting electron. (**b**) In a two-dimensional electron-shell model, it is easier to visualize the atom's components.

neutron. They also do not tell us the mass of a "typical" atom, since any element consists of a mixture of isotopes. It is, therefore, useful to know the *average mass* of an element's atoms, and this value is that element's **atomic weight.** Atomic weight takes into account the mass of the subatomic particles and the relative proportions of any isotopes. For example, even though the atomic number of hydrogen is 1, the atomic weight of hydrogen is 1.0079. In this case, the atomic number and atomic weight differ primarily because a few hydrogen atoms have a mass number of 2 (one proton plus one neutron), and an even smaller number have a mass number of 3 (one proton plus two neutrons). The atomic weights of the elements are included in Appendix II.

Electrons and Electron Shells

Atoms are electrically neutral; every positively charged proton is balanced by a negatively charged electron. These electrons occupy an orderly series of electron shells around the nucleus, and only the electrons in the outer shell can interact with other atoms. *The number of electrons in an atom's outer electron shell determines the chemical properties of that element.*

Atoms with an unfilled outer electron shell are unstable— that is, they will react with other atoms, usually in ways that give them full outer electron shells. An atom with a filled outer shell is stable and will not interact with other atoms. The first electron shell (the one closest to the nucleus) is filled when it contains two electrons. A hydrogen atom has one electron in this electron shell (see Figure 2–2b) and, thus, hydrogen atoms can react with many other atoms. A helium atom has two electrons in this electron shell (see Figure 2–1). Because

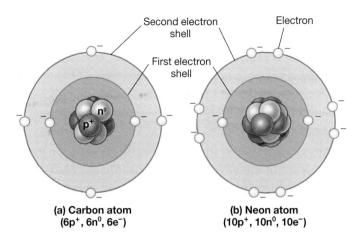

(a) Carbon atom ($6p^+$, $6n^0$, $6e^-$)   **(b) Neon atom** ($10p^+$, $10n^0$, $10e^-$)

● **FIGURE 2–3   The Electron Shells of Atoms.** The first electron shell can hold only two electrons; the second shell can hold up to eight electrons. (**a**) In a carbon atom, which has six protons and six electrons, the first shell is full, but the second shell contains only four electrons. (**b**) In a neon atom, which has 10 protons and 10 electrons, both the first and second electron shells are filled. Notice that the nuclei of carbon and neon contain neutrons as well as protons.

its outer electron shell is full, a helium atom is stable. Helium is called an inert gas because its atoms will neither react with one another nor combine with atoms of other elements.

The second electron shell can contain up to eight electrons. Carbon, with an atomic number of 6, has six electrons. In a carbon atom the first shell is filled (two electrons), and the second shell contains four electrons (Figure 2–3a●). In a neon atom (atomic number 10), the second shell is filled (Figure 2–3b●; neon is another inert gas).

Key Note

All matter is composed of atoms in various combinations. The chemical rules that govern the interactions among atoms alone and in combination establish the foundations of physiology at the cellular level.

Chemical Bonds and Chemical Compounds

An atom with a full outer electron shell is very stable and not reactive. The atoms that are most important to biological systems are *un*stable because those atoms can interact to form larger structures (see Table 2–1). Atoms with unfilled outer electron shells can achieve stability by sharing, gaining, or losing electrons through chemical reactions with other atoms. This often involves the formation of **chemical bonds,** which hold the participating atoms together once the reaction has ended.

Chemical bonding produces *molecules* and *compounds.* **Molecules** are chemical structures that contain more than one atom bonded together by shared electrons. A **compound** is any chemical substance made up of atoms of two or more elements, regardless of how the participating atoms achieve stability. A compound is a new chemical substance with properties that can be quite different from those of its component elements. For example, a mixture of hydrogen and oxygen gases is highly flammable, but chemically combining hydrogen and oxygen atoms produces a compound—water—that can put out a fire's flames.

Ionic Bonds

Atoms are electrically neutral because the number of protons (each with a +1 charge) equals the number of electrons (each with a $^-1$ charge). If an atom loses an electron, it then exhibits a charge of +1 because there is one proton without a corresponding electron; losing a second electron would leave the atom with a charge of +2. Similarly, adding one or two extra electrons to the atom gives it a charge of $^-1$ or $^-2$ respectively.

TABLE 2–2 *The Most Common Ions in Body Fluids*	
CATIONS	**ANIONS**
Na^+ (sodium)	Cl^- (chloride)
K^+ (potassium)	HCO_3^- (bicarbonate)
Ca^{2+} (calcium)	HPO_4^{2-} (biphosphate)
Mg^{2+} (magnesium)	SO_4^{2-} (sulfate)

Atoms or molecules that have an electric charge are called **ions.** Ions with a positive charge (+) are **cations** (KAT-ī-onz); those with a negative charge (−) are **anions** (AN-ī-onz). Table 2–2 lists several important ions in body fluids.

Ionic (ī-ON-ik) **bonds** are chemical bonds created by the electrical attraction between anions and cations. Ionic bonds are formed in a process called *ionic bonding,* as shown in Figure 2–4a●. In the example shown, a sodium atom donates an electron to a chlorine atom. This loss of an electron creates

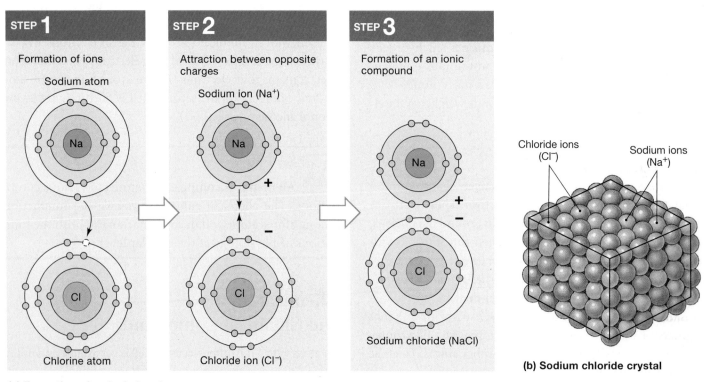

STEP 1

Formation of ions

Sodium atom

Na

Chlorine atom

STEP 2

Attraction between opposite charges

Sodium ion (Na⁺)

Na

+

−

Chloride ion (Cl⁻)

STEP 3

Formation of an ionic compound

Na

+

−

Cl

Sodium chloride (NaCl)

Chloride ions (Cl⁻)

Sodium ions (Na⁺)

(a) Formation of an ionic bond

(b) Sodium chloride crystal

● **FIGURE 2–4   Ionic Bonding. (a) Step 1:** A sodium atom loses an electron, which is accepted by a chlorine atom. **Step 2:** Because the sodium (Na⁺) and chloride (Cl⁻) ions have opposite charges, they are attracted to one another. **Step 3:** The association of sodium and chloride ions forms the ionic compound sodium chloride. **(b)** Large numbers of sodium and chloride ions form a crystal of sodium chloride (table salt).

a *sodium ion* with a +1 charge and a *chloride ion* with a −1 charge. The two ions do not move apart after the electron transfer because the positively charged sodium ion is attracted to the negatively charged chloride ion. In this case, the combination of oppositely charged ions forms the *ionic compound* **sodium chloride,** which is the chemical name for the crystals we know as table salt (Figure 2–4b●).

Covalent Bonds

Another way atoms can fill their outer electron shells is by sharing electrons with other atoms. The result is a molecule held together by **covalent** (kō-VA-lent) **bonds** (Figure 2–5●).

As an example, consider hydrogen. Individual hydrogen atoms, as diagrammed in Figure 2–2, are not found in nature. Instead, we find hydrogen molecules (Figure 2–5a●). In chemical shorthand, molecular hydrogen is indicated by H_2, where H is the chemical symbol for hydrogen, and the subscript 2 indicates the number of atoms. Molecular hydrogen is a gas present in the atmosphere in very small quantities. The two hydrogen atoms share their electrons, and each electron whirls around both nuclei. The sharing of one pair of electrons creates a **single covalent bond.**

Oxygen, with an atomic number of 8, has two electrons in its first electron shell and six in the second. Oxygen atoms (Figure 2–5b●) reach stability by sharing two pairs of electrons, which forms a **double covalent bond.** Molecular oxygen (O_2) is an atmospheric gas that is very important to living organisms; our cells would die without a constant supply of oxygen.

In our bodies, the chemical processes that consume oxygen also produce carbon dioxide (CO_2) as a waste product. The oxygen atoms in a carbon dioxide molecule form double covalent bonds with the carbon atom, as shown in Figure 2–5c●.

Covalent bonds are very strong because the shared electrons tie the atoms together. In most covalent bonds the atoms remain electrically neutral because the electrons are shared equally. Such bonds are called **nonpolar covalent bonds.** Nonpolar covalent bonds between carbon atoms create the stable framework of the large molecules that make up most of the structural components of the human body.

Elements differ in how strongly they hold or attract shared electrons. An unequal sharing between atoms of different elements creates a **polar covalent bond.** Such bonding often forms a *polar molecule* because one end,

or pole, has a slight negative charge and the other a slight positive charge. For example, in a molecule of water, an oxygen atom forms covalent bonds with two hydrogen atoms. However, the oxygen atom has a much stronger attraction for the shared electrons than do the hydrogen atoms, so those electrons spend most of their time with the oxygen atom. Because of the two extra electrons, the oxygen atom develops a slight negative charge (see Figure 2–6a●). At the same time, the hydrogen atoms develop a slight positive charge because their electrons are away part of the time.

Hydrogen Bonds

In addition to ionic and covalent bonds, weaker attractive forces act between adjacent molecules and between atoms within a large molecule. The most important of these weak attractive forces is the **hydrogen bond.** A hydrogen bond is the attraction between a slight positive charge on the hydrogen atom of one polar covalent bond and a weak negative charge on an oxygen or nitrogen atom of another polar covalent bond. The polar covalent bond that contains the oxygen or nitrogen atom can be in a different molecule from, or in the same molecule as, the hydrogen atom. For example, water molecules are attracted to each other through hydrogen bonding (Figure 2–6a).

Hydrogen bonds are too weak to create molecules, but they can alter molecular shapes or pull molecules together. For example, the attraction between water molecules at a free surface slows the rate of evaporation and creates the phenomenon

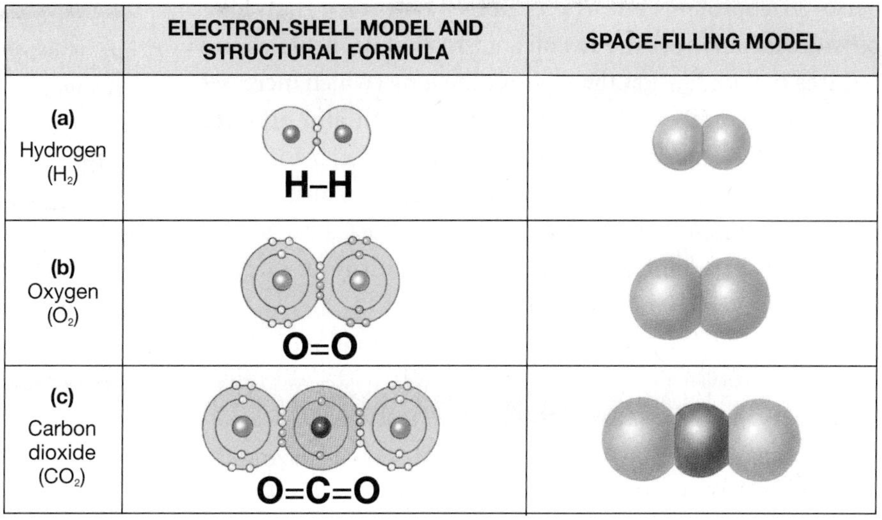

	ELECTRON-SHELL MODEL AND STRUCTURAL FORMULA	SPACE-FILLING MODEL
(a) Hydrogen (H_2)	H–H	
(b) Oxygen (O_2)	O=O	
(c) Carbon dioxide (CO_2)	O=C=O	

● **FIGURE 2–5** **Covalent Bonds.** (a) In a molecule of hydrogen, two hydrogen atoms share their electrons such that each has a filled outer electron shell. This sharing creates a single covalent bond. (b) A molecule of oxygen consists of two oxygen atoms that share two pairs of electrons. The result is a double covalent bond. (c) In a molecule of carbon dioxide, a central carbon atom forms double covalent bonds with a pair of oxygen atoms.

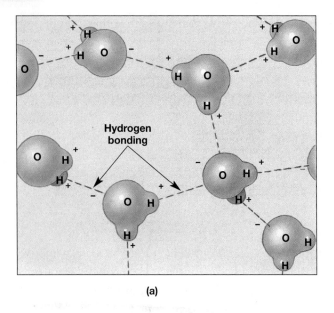

(a) (b)

● **FIGURE 2–6 Hydrogen Bonds. (a)** The unequal sharing of electrons in a water molecule causes each of its two hydrogen atoms to have a slight positive charge and its oxygen atom to have a slight negative charge. Attraction between a hydrogen atom of one water molecule and the oxygen atom of another is a hydrogen bond (indicated by dashed lines). **(b)** Hydrogen bonding between water molecules at a free surface creates surface tension and restricts evaporation.

known as **surface tension** (Figure 2–6b●). Surface tension acts as a barrier that keeps small objects from entering the water; it is the reason that insects can walk across the surface of a pond or puddle. Similarly, a layer of watery tears keeps small dust particles from touching the surface of the eye.

Surfactant

Surfactants are chemicals that serve as wetting agents. That is, they lower the surface tension of a liquid and allow easy spreading of the liquid across a surface. A substance named **surfactant** is found in living tissues. Surfactant is a complex chemical substance that contains phospholipids and a number of glycoproteins and is produced by the Type II alveolar cells. Surfactant lines the alveoli and smallest bronchioles and reduces surface tension throughout the lung (which increases lung compliance). It also stabilizes the alveoli and prevents them from collapsing. Surfactant is not produced by the fetus until approximately 28–32 weeks of gestation. Thus, premature babies often suffer from pulmonary disease because of the lack of surfactant. Surfactant has a high turnover rate and is replaced every 10 hours or so.

→ **CONCEPT CHECK QUESTIONS**

1. Oxygen and neon are both gases at room temperature. Oxygen combines readily with other elements, but neon does not. Why?
2. How is it possible for two samples of hydrogen to contain the same number of atoms but have different weights?
3. Which kind of bond holds atoms in a water molecule together? Which kind of bond attracts water molecules to each other?

Answers begin on p.792.

■ Chemical Notation

Complex chemical compounds and reactions are most easily described with a simple form of "chemical shorthand" known as **chemical notation.** The rules of chemical notation are summarized in Table 2–3.

■ Chemical Reactions

Cells remain alive by controlling chemical reactions. In a **chemical reaction,** new chemical bonds form between atoms or existing bonds between atoms are broken. These changes occur as atoms in the reacting substances, or **reactants,** are rearranged to form different substances, or **products.** (See item 4 in Table 2–3.)

In effect, each cell is a chemical factory. For example, growth, maintenance and repair, secretion, and muscle contraction all involve complex chemical reactions. Cells use chemical reactions to provide the energy required to maintain homeostasis and to perform essential functions. **Metabolism** (me-TAB-ō-lizm; *metabole*, change) refers to all of the chemical reactions in the body.

Basic Energy Concepts

Knowledge of some basic relationships between matter and energy is essential for understanding any discussion of chemical reactions. **Work** is movement or a change in the physical structure of matter. In your body, work includes movements like walking or running and also the synthesis of

TABLE 2-3 *Rules of Chemical Notation*

1. The abbreviation of an element indicates one atom of that element:

   H = an atom of hydrogen; O = an atom of oxygen

2. A number that proceed the abbreviation of an element indicates more than one atom:

   2H = two individual atoms of hydrogen
   2O = two individual atoms of oxygen

3. A subscript that follows the abbreviation of an element indicates a molecule with that number of atoms:

   H_2 = a hydrogen molecule composed of two hydrogen atoms
   O_2 = an oxygen molecule composed of two oxygen atoms

4. In a description of a chemical reaction, the interacting participants are called **reactants,** and the reaction generates one or more **products.** An arrow indicates the direction of the reaction, from reactants (usually on the left) to products (usually on the right). In the following reaction, two atoms of hydrogen combine with one atom of oxygen to produce a single molecule of water.

   $$2H + O \longrightarrow H_2O$$

5. A superscript plus or minus sign that follows the abbreviation for an element indicates an ion. A single plus sign indicates an ion with a charge of +1 (loss of one electron). A single minus sign indicates an ion with a charge of $^-1$ (gain of one electron). If more than one electron has been lost or gained, the charge on the ion is indicated by a number that proceed the plus or minus.

   Na^+ = one sodium ion (has lost 1 electron)
   Cl^- = one chloride ion (has gained 1 electron)
   Ca^{2+} = one calcium ion (has lost 2 electrons)

6. Chemical reactions neither create nor destroy atoms—they merely rearrange them into new combinations. Therefore, the numbers of atoms of each element must always be the same on both sides of the equation. When this is the case, the equation is balanced.

   Unbalanced: $H_2 + O_2 \longrightarrow H_2O$
   Balanced: $2H_2 + O_2 \longrightarrow 2H_2O$

organic molecules and the conversion of liquid water to water vapor (evaporation). **Energy** is the capacity to perform work. There are two major types of energy: *kinetic energy* and *potential energy.*

Kinetic energy is the energy of motion. When you fall off a ladder, kinetic energy does the damage. **Potential energy** is stored energy. It may result from an object's position (you standing on a ladder) or from its physical or chemical structure (a stretched spring or a charged battery). Kinetic energy must be used in climbing the ladder, in stretching the spring, or in charging the battery. The potential energy is converted back into kinetic energy when you fall, the spring recoils, or the battery discharges. The kinetic energy can then be used to perform work.

Clinical Note
KINETIC ENERGY AND INJURY

Kinetic energy is the energy contained by a body in motion. The kinetic energy of an object is directly proportional to the square of its speed. That is, for a twofold increase in speed, the kinetic energy will increase by a factor of four. For a threefold increase in speed, the kinetic energy will increase by a factor of nine. For a fourfold increase in speed, the kinetic energy will increase by a factor of sixteen. The kinetic energy is dependent upon the square of the speed of the object in motion.

A moving automobile has a tremendous amount of kinetic injury. The energy can be released slowly as occurs when the driver applies the brakes and stops in 100 meters. Or, the energy can be released rapidly when the vehicle strikes a bridge and stops in less than one meter. Thus, not only speed can cause injury. Stopping can cause injury as well. In a motor vehicle collision some of the kinetic energy is transferred to the occupants. Likewise, something that stops the car, such as a bridge or a pedestrian, absorbs massive amounts of kinetic energy.

There is a limit on the amount of kinetic energy a human body can absorb. Small amounts of kinetic energy, such as occur with walking, are readily absorbed and do not cause injury. As the energy increases, as occurs in a fall from a bicycle, the body may sustain damage from the energy. However, it may not be fatal. Absorbing large quantities of kinetic energy, as occurs in a high-speed motor-vehicle collision can cause severe injuries. At a certain energy level, the forces that result are fatal. ■

Energy cannot be destroyed; it can only be converted from one form to another. A conversion between potential energy and kinetic energy is not 100 percent efficient. Each time an energy exchange occurs, some of the energy is released as heat. *Heat* is an increase in random molecular motion. The temperature of an object is directly related to the average kinetic energy of its molecules. Heat can never be completely converted to work or to any other form of energy, and cells cannot capture it or use it to perform work.

Living cells perform work in many forms, and the cells' energy exchanges produce heat. For example, when skeletal muscle cells contract, they perform work; potential energy (the positions of protein filaments and the covalent bonds between molecules inside the cells) is converted into kinetic energy, and heat is released. The amount of heat is related to the amount of work done. As a result, when you exercise, your body temperature rises.

Key Note

When energy is exchanged, heat is produced. Heat raises local temperatures, but cells cannot capture it or use it to perform work.

Types of Reactions

Three types of chemical reactions are important to the study of physiology: *decomposition reactions, synthesis reactions,* and *exchange reactions*.

Decomposition Reactions

A **decomposition reaction** breaks a molecule into smaller fragments. Such reactions occur during digestion, when food molecules are broken into smaller pieces. You could diagram a typical decomposition reaction as:

$$AB \longrightarrow A + B$$

Decomposition reactions that involve water are important in the breakdown of complex molecules in the body. In **hydrolysis** (hī-DROL-i-sis; *hydro-*, water + *lysis*, dissolution), one of the bonds in a complex molecule is broken, and the components of a water molecule (H and OH) are added to the resulting fragments:

$$A—B—C—D—E + H_2O \longrightarrow$$
$$A—B—C—H + HO—D—E$$

Catabolism (kah-TAB-o-lizm; *katabole*, a throwing down) refers to the decomposition reactions of complex molecules within cells. When a covalent bond—a form of potential energy—is broken, it releases kinetic energy that can perform work. Cells can harness some of that energy to power essential functions such as growth, movement, and reproduction.

Synthesis Reactions

Synthesis (SIN-the-sis) is the opposite of decomposition. A synthesis reaction assembles larger molecules from smaller components. These relatively simple reactions could be diagrammed as:

$$A + B \longrightarrow AB$$

A and B could be individual atoms that combine to form a molecule, or they could be individual molecules that combine to form even larger products. Synthesis always involves the formation of new chemical bonds, whether the reactants are atoms or molecules.

Dehydration synthesis, or *condensation,* is the formation of a complex molecule by the removal of water:

$$A—B—C—H + HO—D—E \longrightarrow$$
$$A—B—C—D—E + H_2O$$

Dehydration synthesis is, therefore, the opposite of hydrolysis. We will encounter examples of both reactions in later sections.

Anabolism (a-NAB-ō-lizm; *anabole,* a building up) is the synthesis of new compounds in the body. Because it takes energy to create a chemical bond, anabolism is usually an "uphill" process. Living cells are constantly balancing their chemical activities, and catabolism provides the energy needed to support anabolism as well as other vital functions.

Exchange Reactions

In an **exchange reaction,** parts of the reacting molecules are shuffled around, as follows:

$$AB + CD \longrightarrow AD + CB$$

Although the reactants and products contain the same components (A, B, C, and D), the components are present in different combinations. In an exchange reaction, the reactant molecules AB and CD break apart (a decomposition), and then the resulting components interact to form AD and CB (a synthesis).

Reversible Reactions

Many important biological reactions are freely reversible. Such reactions can be diagrammed as:

$$A + B \longleftrightarrow AB$$

This equation indicates that two reactions occur simultaneously, one a synthesis (A + B $\longrightarrow$ AB) and the other a decomposition (AB $\longrightarrow$ A + B). At **equilibrium** (ē-kwi-LIB-rē-um), the rates of the two reactions are in balance. As fast as a molecule of AB forms, another degrades into A + B. As a result, the numbers of A, B, and AB molecules present at any given moment do not change. Altering the concentrations of one or more of these molecules will temporarily upset the equilibrium. For example, adding additional molecules of A and B will accelerate the synthesis reaction (A + B $\longrightarrow$ AB). As the concentration of AB rises, however, so does the rate of the decomposition reaction (AB $\longrightarrow$ A + B), until a new equilibrium is established.

Key Note

In biological reactions, things tend to even out, unless something prevents this from happening. Most reversible reactions quickly reach equilibrium, where opposing reaction rates are balanced. If reactants are added or removed, reaction rates change until a new equilibrium is established.

Enzymes and Chemical Reactions

Most chemical reactions do not occur spontaneously, or they occur so slowly that they would be of little value to cells. Before a reaction can proceed, enough energy must be provided to activate the reactants. The amount of energy required to start a reaction is called the **activation energy** (Figure 2–7a●). Although many reactions can be activated by changes in temperature or pH, such changes are deadly to cells. For example, to break down a complex sugar in the laboratory, you must boil it in an acid solution. Cells, however, avoid such harsh requirements by using special molecules called **enzymes** to speed up the reactions that support life. Enzymes belong to a class of substances called **catalysts** (KAT-uh-lists; *katalysis,* dissolution), which are compounds that accelerate chemical reactions without themselves being permanently changed. A cell makes an enzyme molecule to promote each specific reaction.

Enzymes promote chemical reactions by lowering the activation energy requirements (Figure 2–7b●). Lowering the activation energy affects only the rate of a reaction, not the

direction of the reaction or the products that are formed. An enzyme cannot bring about a reaction that would otherwise be impossible.

It takes activation energy to start a chemical reaction, but once it has begun, the reaction as a whole may absorb or release energy, generally in the form of heat, as it proceeds to completion. If the amount of energy released is greater than the activation energy needed to start the reaction, there will be a net release of energy. Reactions that release energy are said to be **exergonic** (*exo-*, outside). If more energy is required to begin the reaction than is released as it proceeds, the reaction as a whole will absorb energy. Such reactions are called **endergonic** (*endo-*, inside). Exergonic reactions are relatively common in the body; they are responsible for generating the heat that maintains your body temperature.

Key Note

Most of the chemical reactions that sustain life cannot occur under homeostatic conditions unless appropriate enzymes are present.

CONCEPT CHECK QUESTIONS

1. In living cells, glucose, a six-carbon molecule, is converted into two three-carbon molecules by a reaction that yields energy. How would you classify this reaction?
2. If the product of a reversible reaction is continuously removed, what will be the effect on the equilibrium?
3. Why are enzymes needed in our cells?

Answers begin on p. 792.

■ Inorganic Compounds

The rest of this chapter focuses on nutrients and metabolites. **Nutrients** are the essential elements and molecules that are obtained from the diet. **Metabolites** (me-TAB-ō-līts) include all of the molecules synthesized or broken down by chemical reactions inside our bodies. Like all chemical substances, nutrients and metabolites can be broadly categorized as inorganic or organic. Generally speaking, **inorganic compounds** are small molecules that do not contain carbon and hydrogen atoms. **Organic compounds** are primarily composed of carbon and hydrogen atoms, and they can be much larger and more complex than inorganic compounds.

The most important inorganic substances in the human body are carbon dioxide, oxygen, water, inorganic acids and bases, and salts.

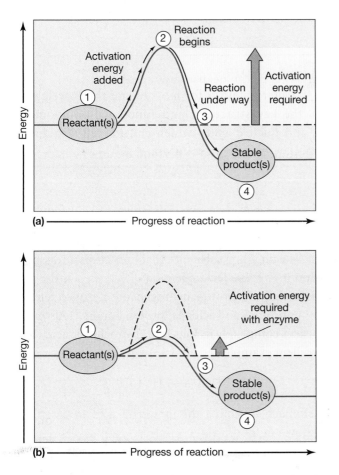

● **FIGURE 2–7   Enzymes and Activation Energy.** (a) Before a reaction can begin on its own, considerable activation energy must be provided. In this diagram, the activation energy represents the energy required for the reaction to proceed from point 1 to point 2. (**b**) The activation energy requirement of the reaction is much lower in the presence of an appropriate enzyme. The enzyme enables the reaction to take place much more rapidly without the need for extreme conditions that would harm cells.

Carbon Dioxide and Oxygen

Cells produce carbon dioxide (CO_2) through normal metabolic activity. It is transported in the blood and released into the air in the lungs. Oxygen (O_2), which is an atmospheric gas, is absorbed at the lungs, transported in the blood, and consumed by cells throughout the body. The chemical structures of these compounds were introduced earlier in the chapter. ∞ p. 33

Water and Its Properties

Water (H_2O) is the single most important constituent of the body, and accounts for almost two-thirds of total body weight. A change in body water content can have fatal consequences because virtually all physiological systems will be affected.

The hydrogen bonds that occur between adjacent water molecules give liquid water some unique properties. Three general properties of water are particularly important to our discussion of the human body:

■ *Water is an excellent solvent.* Water dissolves a remarkable variety of inorganic and organic molecules, which creates a solution. As the molecules dissolve, they break apart, and release ions or smaller molecules that become uniformly dispersed throughout the solution. The chemical reactions within living cells occur in solution, and the watery component of blood, called plasma, carries dissolved nutrients and waste products throughout the body. Most chemical reactions in the body occur in solution.

■ *Water has a very high heat capacity. Heat capacity* is the ability to absorb and retain heat. It takes a lot of energy to change the temperature of a quantity of water. Once the water has reached a particular temperature, it will change temperature only slowly. As a result, body temperature is stabilized, and the water in our cells remains a liquid over a wide range of environmental temperatures. Hydrogen bonding also explains why a large amount of energy is required to change liquid water to a gas, or ice to liquid. When water finally changes from a liquid to a gas, it carries a great deal of heat away with it. This feature accounts for the cooling effect of perspiration on the skin.

■ *Water is an essential reactant in the chemical reactions of living systems.* Chemical reactions in our bodies occur in water, and water molecules are also participants in some reactions. During the dehydration synthesis of large molecules, water molecules are released; during hydrolysis, complex molecules are broken down by the addition of water molecules. ∞ p. 36

Solutions

A **solution** consists of a uniform mixture of a fluid *solvent* and dissolved *solutes*. Within organisms the solvent is usually water, which forms an *aqueous solution,* and the solutes may be inorganic or organic. Inorganic compounds held together by ionic bonds undergo **ionization** (ī-on-i-ZĀ-shun), or *dissociation* (dis-sō-sē-Ā-shun), in solution. In this process, ionic bonds are broken apart as individual ions interact with the positive or negative ends of polar water molecules (Figure 2–8a●). As shown in Figure 2–8b●, the result is a mixture of cations and anions surrounded by so many water molecules that they are unable to reform their original bonds.

An aqueous solution that contains anions and cations can also conduct an electrical current. Electrical forces across cell membranes affect the functioning of all cells, and small electrical currents carried by ions are essential to muscle contraction and nerve function. Chapters 7 and 8 will discuss these processes in more detail.

🔒 **Key Note**

Water accounts for most of your body weight; proteins, the key structural and functional components of cells, and nucleic acids, which control cell structure and function, work only in solution.

Inorganic Acids and Bases

The body contains both inorganic and organic acids and bases. An **acid** is any substance that breaks apart (dissociates) in solution to *release* hydrogen ions. (Because a hydrogen ion consists solely of a proton, hydrogen ions are often referred to simply as protons and acids as "proton donors.") A *strong acid* dissociates completely in solution. Hydrochloric acid (HCl) is an excellent example:

$$HCl \longrightarrow H^+ + Cl^-$$

The stomach produces this powerful acid to assist in the breakdown of food.

A **base** is a substance that removes hydrogen ions from a solution. Many common bases are compounds that dissociate in solution to liberate a hydroxide ion (OH^-). Hydroxide ions have a strong affinity for hydrogen ions and quickly react with them to form water molecules. A *strong base* dissociates completely in solution. For example, sodium hydroxide (NaOH) dissociates in solution as follows:

$$NaOH \longrightarrow Na^+ + OH^-$$

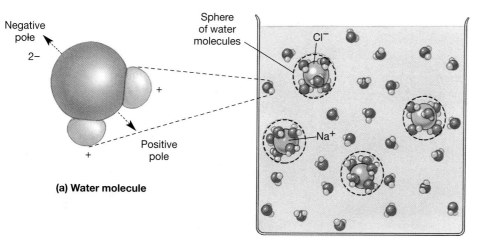

(a) Water molecule

(b) Sodium chloride in solution

● **FIGURE 2–8** **Water Molecules and Their Role in Solutions.** **(a)** In a water molecule, oxygen forms polar covalent bonds with two hydrogen atoms. Because the hydrogen atoms are located near one end of the molecule, the molecule has an uneven distribution of charges, which creates positive and negative poles. **(b)** Ionic compounds dissociate in water as the polar water molecules disrupt the ionic bonds. The ions remain in solution because the sphere of water that molecules surround them prevents the ionic bonds from re-forming.

Strong bases have a variety of industrial and household uses; drain openers and lye are two familiar examples.

Weak acids and weak bases do not dissociate completely in solution. The human body contains weak bases that are important in counteracting acids produced during cellular metabolism.

Hydrogen Ions and pH

The concentration of hydrogen ions in blood or other body fluids is important because hydrogen ions are extremely reactive. In excessive numbers, they will break chemical bonds, change the shapes of complex molecules, and disrupt cell and tissue functions. The concentration of hydrogen ions must, therefore, be precisely regulated.

The concentration of hydrogen ions is usually reported in terms of the **pH** of the solution. The pH value is a number between 0 and 14. Pure water has a pH of 7. A solution with a pH of 7 is called *neutral* because it contains equal numbers of hydrogen ions and hydroxide ions. A solution with a pH below 7 is

acidic (a-SI-dik) because there are more hydrogen ions than hydroxide ions. A pH above 7 is called *basic,* or *alkaline* (Al-kah-lin), because hydroxide ions outnumber hydrogen ions. The pH values of some common liquids are indicated in Figure 2–9●. The pH of blood and most body fluids normally ranges from 7.35 to 7.45. Variations in pH outside this range can damage cells and disrupt normal cellular functions. For example, a blood pH below 7 can produce coma, and a blood pH higher than 7.8 usually causes uncontrollable, sustained muscular contractions.

Buffers and pH

Buffers are compounds that stabilize pH by either removing or replacing hydrogen ions. Antacids such as Alka-Seltzer®, Rolaids®, and Tums® are buffers that tie up excess hydrogen ions in the stomach. A variety of buffers, including sodium bicarbonate, are responsible for keeping the pH of most body fluids between 7.35 and 7.45. We will consider the role of buffers and pH control in Chapter 18.

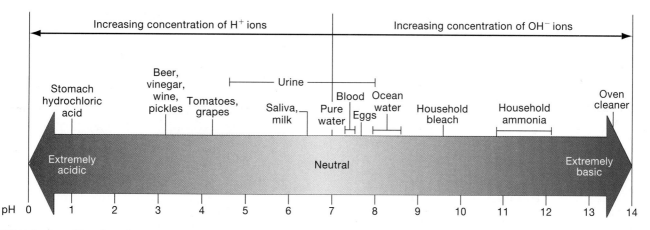

● **FIGURE 2–9** **pH and Hydrogen Ion Concentration.** An increase or decrease of one unit corresponds to a tenfold change in H^+ concentration.

Clinical Note
UNDERSTANDING pH

Acid-base balance is a dynamic relationship that reflects the relative concentration of hydrogen ions (H⁺) in the body. Hydrogen ions are acidic, and their concentration must be maintained within fairly strict limits. Any deviation in the hydrogen ion concentration adversely affects all of the biochemical events in the body. Because of this, determination of the body's hydrogen ion concentration is important in emergency care.

Water molecules have a slight tendency to undergo reversible ionization to yield a *hydrogen ion (H⁺)* and a *hydroxide ion (OH⁻)* to form the following equilibrium:

$$H_2O \leftrightarrows H^+ + OH^-$$

The ionization of water is very weak. In fact, only 1 out of every 10 million molecules of water is ionized at any given moment. Although water has only a very slight tendency to ionize, the products of ionization (H⁺ and OH⁻) have profound biological effects. The number of hydrogen ions and hydroxide ions in an aqueous (water) solution at 25°C (77°F) always equals the fixed number 1.0×10^{-14} M. When the concentrations of both hydrogen ions and hydroxide ions are exactly equal, as in pure water, the solution is said to be *neutral*. When neutral, the number of hydrogen ions equals 1.0×10^{-7} M and the number of hydroxide ions equals 1.0×10^{-7} M. Because of this, we know that when the number of hydrogen ions present is very high, then the concentration of hydroxide ions must be very low. Likewise, when the concentration of hydroxide ions is high, then the concentration of hydrogen ions must be low. Again, regardless of the balance between hydrogen ions and hydroxide ions, the total number of ions must be 1.0×10^{-14} M.

The need for a much simpler system of describing hydrogen ion concentrations quickly became evident. In 1909 the Swedish chemist H.P.L Sørenson proposed that the hydrogen ion concentration be expressed in terms of the logarithm (log) of the reciprocal of the number. This system allows simple numbers to represent extremely large numbers of hydrogen ions. The system was called the pH system, which is an abbreviation for *potential of hydrogen*. Because of the large numbers involved, the use of logarithms simplifies discussion of hydrogen ion concentrations. It is important to remember that the pH system represents the reciprocal of the hydrogen ion concentration. Thus, the greater the number of hydrogen ions, the lower the pH. Conversely, the fewer the number of hydrogen ions, the higher the pH. The pH scale can be defined as:

$$pH = \log 1/[H^+]$$

Because the logarithm of 1 is 0, then the equation can be simplified as:

$$pH = -\log [H^+]$$

In precisely neutral solution at 25°C (77°F), where the hydrogen ion concentration equals 1.0×10^{-7} M, the pH would be calculated as:

$$pH = \log 1/1.0 \times 10^{-7} = \log [1.0 \times 10^{-7}] = \log 1.0 + \log 10^{-7}$$
$$pH = 0 + 7$$
$$pH = 7$$

Remember, the value of 7.0 for the pH of a precisely neutral solution is not an arbitrary number but is derived from the absolute value of the hydrogen ion concentration at 25°C. The pH scale varies from 0 to 14. A pH of 14 indicates that only hydroxide ions are present, while a pH of 0 indicates that only hydrogen ions are present.

Any pH greater than 7.0 is considered *alkaline* while any pH less than 7.0 is considered *acidic*. It is also important to note that the pH scale is logarithmic, not arithmetic. For example, if two solutions differ in pH by 1 pH unit, then one solution has 10 times the hydrogen ion concentration of the other. Thus, small changes in pH reflect massive changes in the number of hydrogen ions present.

Occasionally, the pOH is used to denote the alkalinity of a solution. The pOH is similar to pH, except it represents the number of hydroxide ions present, not hydrogen ions (Table 2–4). It is helpful to remember that pH and pOH are related to each other in a very simple way:

$$pH + pOH = 14$$

The pH of the body is closely regulated between 7.35 and 7.45. Thus, alkalosis is a pH greater than 7.45, while acidosis is a pH of less than 7.35. Small changes in pH are corrected by the body's

---

TABLE 2–4 *The pH Scale*

ACIDIC CONCENTRATIONS		ALKALINE CONCENTRATIONS	
[H⁺], Moles	pH	[OH⁻], Moles	pOH
1.0	0	10^{-14}	14
0.1	1	10^{-13}	13
0.01	2	10^{-12}	12
0.001	3	10^{-11}	11
10^{-4}	4	10^{-10}	10
10^{-5}	5	10^{-9}	9
10^{-6}	6	10^{-8}	8
10^{-7}	7	10^{-7}	7
10^{-8}	8	10^{-6}	6
10^{-9}	9	10^{-5}	5
10^{-10}	10	10^{-4}	4
10^{-11}	11	0.001	3
10^{-12}	12	0.01	2
10^{-13}	13	0.1	1
10^{-14}	14	1.0	0

compensatory mechanisms, but significant changes are poorly tolerated. In fact, a change in pH of 0.5 units can be fatal.

In medical practice, the pH values of blood and urine are measured most frequently. Blood pH is obtained through an arterial blood gas (ABG) sample. For this, a small amount of arterial blood is removed from the radial, brachial, or femoral artery. This is quickly introduced into a blood gas machine where the pH is carefully measured. In the blood gas machine, a special glass electrode selectively measures hydrogen ion concentration. The signal from this electrode is amplified and compared with the signal generated by a control solution having a known pH. Based on this comparison, the pH can be determined. Because of this, it is important for ABG machines to be carefully maintained and calibrated with the control solutions checked on a daily basis. In addition to pH, an ABG analysis measures the partial pressure of oxygen and carbon dioxide in the blood. Also, some machines will measure the bicarbonate concentration and the amount of hemoglobin present. A more detailed discussion of acid-base balance and the body's compensatory buffering mechanisms is presented in Chapter 18.

What Is a Logarithm?

A *logarithm* in mathematics is the *exponent*, or power, to which a stated number, called the *base*, must be raised to yield a specific number. For example, in the expression $10^2 = 100$, the logarithm of 100 to the base 10 is 2. This is properly written as $\log_{10} 100 = 2$. Logarithms were initially created to help simplify arithmetical operations that involve very large numbers. *Common logarithms* use the number 10 as the base number. However, logarithms can be used with base numbers other than 10. This can be illustrated by considering a sequence of powers to the number 2: 2^1, 2^2, 2^3, 2^4, 2^5, and 2^6, which correspond to the sequence of numbers 2, 4, 8, 16, 32, and 64. The exponents 1, 2, 3, 4, 5, and 6 are the logarithms of these numbers to the base 2.

In medicine, logarithms are used to describe chemical concentrations. Unless specified otherwise, always assume that a logarithm is in base 10 ($\log_{10}$). If another logarithm is used, it will be specified in the notation (e.g., $\log_5$). The most frequent use of logarithms in medicine is in the pH system that describes the hydrogen ion concentration of a solution. In pure water, the hydrogen ion concentration is equal to 0.0000001, or 10^{-7} moles per liter. Because the pH scale describes the reciprocal of the hydrogen ion concentration, the numbers are positive even though the concentration of hydrogen ions is less than 1. This simplifies discussion and calculations. Thus, a pH of 3.0 reflects a hydogen ion concentration of 0.001, or 10^{-3} moles per liter. Likewise, a pH of 10 reflects a hydrogen ion concentration of 0.0000000001, or 10^{-10} moles per liter. A change in pH of 1 unit reflects a tenfold change in hydrogen ion concentration. Likewise, a change of 2 pH units indicates a hundredfold change in hydrogen ion concentration. Thus, the use of logarithms simplifies descriptions and calculations of these extremely large or small numbers. ■

Salts

A **salt** is an ionic compound that consists of any cation except a hydrogen ion and any anion except a hydroxide ion. Salts are held together by ionic bonds, and in water they dissociate, and release cations and anions. For example, table salt (NaCl) in solution dissociates into Na^+ and Cl^- ions; these are the most abundant ions in body fluids.

Salts are examples of **electrolytes** (e-LEK-trō-līts), which are inorganic compounds whose ions can conduct an electrical current in solution. For example, sodium ions (Na^+), potassium ions (K^+), calcium ions (Ca^{2+}), and chloride ions (Cl^-) are released by the dissociation of electrolytes in blood and other body fluids. Alterations in the concentrations of these ions in body fluids will disturb almost every vital function. For example, declining potassium levels will lead to general muscular paralysis, and rising concentrations will cause weak and irregular heartbeats.

CONCEPT CHECK QUESTIONS

1. Why does water resist changes in temperature?
2. What is the difference between an acid and a base?
3. Why is an extreme change in pH of body fluids undesirable?
4. How does an antacid decrease stomach discomfort?

Answers begin on p. 792.

■ Organic Compounds

Organic compounds always contain the elements carbon and hydrogen and generally oxygen as well. Many organic molecules are made up of long chains of carbon atoms linked by covalent bonds. These carbon atoms often form additional covalent bonds with hydrogen or oxygen atoms and less often with nitrogen, phosphorus, sulfur, iron, or other elements.

Many organic molecules are soluble in water. Although inorganic acids and bases were discussed previously, there are also important organic acids and bases. For example, active muscle tissues generate *lactic acid,* which is an organic acid that must be neutralized to prevent potentially dangerous pH changes in body fluids.

This section focuses on four major classes of large organic molecules: *carbohydrates, lipids, proteins,* and *nucleic acids.* We will also consider the high-energy compounds that are vital to the survival of our cells. In addition, the human body contains small quantities of many other organic compounds whose structures and functions will be considered in later chapters.

Carbohydrates

A **carbohydrate** (kar-bō-HĪ-drāt) is an organic molecule that contains carbon, hydrogen, and oxygen in a ratio near 1:2:1. Familiar carbohydrates include the sugars and starches that make up roughly half of the typical U.S. diet. Our tissues can break down most carbohydrates, and although they sometimes have other functions, carbohydrates are most important as sources of energy. Despite their importance as an energy source, however, carbohydrates account for less than 3 percent of total body weight. The three major types of carbohydrates are *monosaccharides, disaccharides,* and *polysaccharides.*

Monosaccharides

A **simple sugar,** or **monosaccharide** (mon-ō-SAK-uh-rīd; *mono-,* single + *sakcharon,* sugar), is a carbohydrate that contains from three to seven carbon atoms. Included within this group is **glucose** (GLOO-kōs) ($C_6H_{12}O_6$), which is the most important metabolic "fuel" in the body (Figure 2–10●). Glu-

cose and other monosaccharides dissolve readily in water and are rapidly distributed throughout the body by blood and other body fluids.

Disaccharides and Polysaccharides

Carbohydrates other than simple sugars are complex molecules composed of monosaccharide building blocks. Two monosaccharides joined together form a **disaccharide** (dī-SAK-uh-rīd; *di-,* two). Disaccharides such as *sucrose* (table sugar) have a sweet taste and, like monosaccharides, are quite soluble in water. The formation of sucrose (Figure 2–11a●) involves dehydration synthesis, a process introduced earlier in the chapter. ∞ p. 36 Dehydration synthesis, or condensation, links molecules together by removing a water molecule. The breakdown of sucrose into simple sugars is an example of hydrolysis, which is the functional opposite of dehydration synthesis (Figure 2–11b●).

Many foods contain disaccharides, but all carbohydrates except monosaccharides must be disassembled through hydrolysis before they can provide useful energy. Most sweet junk foods, such as candy and soft drinks, abound in simple sugars (commonly fructose) and disaccharides (generally sucrose). Some people cannot tolerate sugar for medical reasons; others avoid it because they do not want to gain weight (excess sugars are stored as fat). Many such people use artificial sweeteners in their foods and beverages. These compounds have a very sweet taste but either cannot be broken down in the body or are used in such small amounts that their breakdown does not contribute to the overall energy balance of the body.

Larger carbohydrate molecules are called **polysaccharides** (pol-ē-SAK-uh-rīdz; *poly-,* many). They result when repeated dehydration synthesis reactions add additional monosaccharides or disaccharides. *Starches* are glucose-based polysaccharides that are important in our diets. Most starches are manufactured by plants. Your digestive tract can break these molecules into simple sugars. Starches found in potatoes and grains are important energy sources. In contrast, *cellulose,* a component of the cell walls of plants, is a polysaccharide that our bodies cannot digest. The cellulose of foods such as celery contributes to the bulk of digestive wastes but is useless as an energy source.

Glycogen (GLĪ-kō-jen), or *animal starch,* is a polysaccharide that is composed of interconnected glucose molecules (Figure 2–11c●). Like most other large polysaccharides, glycogen will not dissolve in water or other body fluids. Liver and muscle tissues make and store glycogen. When these tissues have a high demand for energy, glycogen molecules are broken down into glucose; when demands are low, the tissues absorb glucose from the bloodstream and rebuild glycogen reserves.

Table 2–5 summarizes information about the carbohydrates in the body.

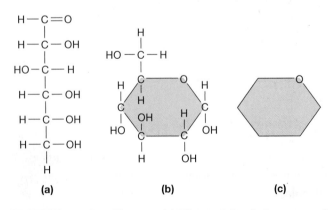

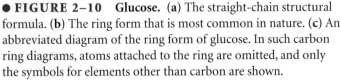

●**FIGURE 2–10**   **Glucose.** (**a**) The straight-chain structural formula. (**b**) The ring form that is most common in nature. (**c**) An abbreviated diagram of the ring form of glucose. In such carbon ring diagrams, atoms attached to the ring are omitted, and only the symbols for elements other than carbon are shown.

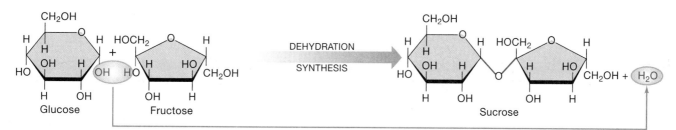

(a) During dehydration synthesis, two molecules are joined by the removal of a water molecule.

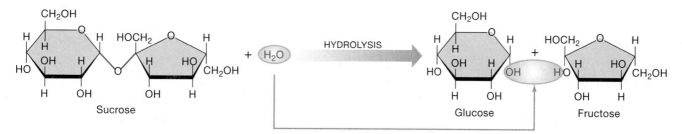

(b) Hydrolysis reverses the steps of dehydration synthesis; a complex molecule is broken down by the addition of a water molecule.

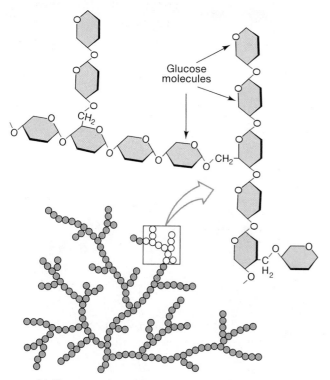

(c) Glycogen, a branching chain of glucose molecules, is stored in muscle cells and liver cells.

● **FIGURE 2–11   Complex Sugars and Glycogen. (a, b)** The formation and breakdown of complex sugars (such as disaccharides) are functionally opposite reactions. **(c)** The structure of glycogen, a polysaccharide.

Lipids

Lipids (*lipos,* fat) also contain carbon, hydrogen, and oxygen, but because they have relatively less oxygen than carbohydrates, the ratios do not approximate 1:2:1. In addition, lipids may contain small quantities of other elements, including phosphorus, nitrogen, or sulfur. Familiar lipids include *fats, oils,* and *waxes.* Most lipids are insoluble in water, but special transport mechanisms carry them in the circulating blood.

Lipids form essential structural components of all cells. In addition, lipid deposits are important as energy reserves. Based on equal weights, lipids provide roughly twice as much energy as carbohydrates when broken down in the body. When the supply of lipids exceeds the demand for energy, the excess is stored in fat deposits. For this reason there has been great interest in developing *fat substitutes,* such as Olestra®, that provide less energy but have the same taste and texture as lipids.

TABLE 2–5	*Carbohydrates in the Body*		
STRUCTURE	**EXAMPLES**	**PRIMARY FUNCTIONS**	**REMARKS**
MONOSACCHARIDES (SIMPLE SUGARS)	Glucose, fructose	Energy source	Manufactured in the body and obtained from food; found in body fluids
DISACCHARIDES	Sucrose, lactose, maltose	Energy source	Sucrose is table sugar, lactose is present in milk; all must be broken down to monosaccharides before absorption
POLYSACCHARIDES	Glycogen	Storage of glucose molecules	Glycogen is in animal cells; other starches and cellulose are in plant cells

TABLE 2-6 *Representative Lipids and Their Functions in the Body*

LIPID TYPE	EXAMPLES	PRIMARY FUNCTIONS	REMARKS
FATTY ACIDS	Lauric acid	Energy sources	Absorbed from food or synthesized in cells; transported in the blood for use in many tissues
FATS	Monoglycerides, diglycerides, triglycerides	Energy source, energy storage, insulation, and physical protection	Stored in fat deposits; must be broken down into fatty acids and glycerol before they can be used as an energy source
STEROIDS	Cholesterol	Structural component of cell membranes, hormones, digestive secretions in bile	All have the same carbon-ring framework
PHOSPHOLIPIDS	Lecithin	Structural components of cell membranes	Composed of fatty acids and nonlipid molecules

Lipids normally account for 12–18 percent of total body weight of adult men, and 18–24 percent of that of adult women. There are many kinds of lipids in the body. The major types are *fatty acids, fats, steroids,* and *phospholipids* (Table 2–6).

Fatty Acids

Fatty acids are long chains of carbon atoms with attached hydrogen atoms that end in a *carboxylic* (kar-bok-SIL-ik) *acid group* (—COOH). The word *carboxyl* should help you remember that a carbon and a hydroxyl (—OH) group are the important structural features of fatty acids. When a fatty acid is in solution, only the carboxyl end dissolves in water. The carbon chain, known as the hydrocarbon *tail* of the fatty acid, is relatively insoluble. Figure 2–12a● shows a representative fatty acid, *lauric acid.*

In a **saturated** fatty acid, such as lauric acid, the four single covalent bonds of each carbon atom permit each neighboring carbon to link to each other and to two hydrogen atoms. If any of the carbon-to-carbon bonds are double covalent bonds, then fewer hydrogen atoms are present and the fatty acid is **unsaturated.** The structures of saturated and unsaturated fatty acids are shown in Figure 2–12b●. A *monounsaturated* fatty acid has a lone double bond in the hydrocarbon tail. A *polyunsaturated* fatty acid contains multiple double bonds.

Both saturated and unsaturated fatty acids can be broken down for energy, but a diet that contains large amounts of saturated fatty acids increases the risk of heart disease and other circulatory problems. Butter, fatty meat, and ice cream are popular dietary sources of saturated fatty acids. Vegetable oils such as olive oil or corn oil contain a mixture of unsaturated fatty acids.

Fats

Unlike simple sugars, individual fatty acids cannot be strung together in a chain by dehydration synthesis. But they can be attached to another compound, **glycerol** (GLIS-er-ol), to make a **fat**

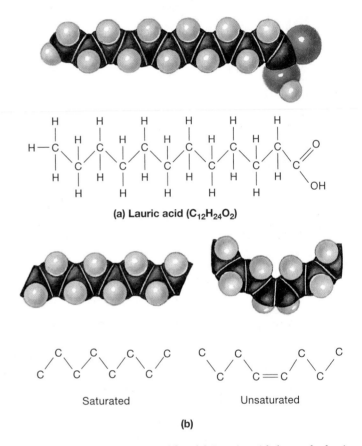

(a) Lauric acid ($C_{12}H_{24}O_2$)

Saturated   Unsaturated

(b)

● **FIGURE 2–12   Fatty Acids.** (a) Lauric acid shows the basic structure of a fatty acid: a long chain of carbon atoms and a carboxylic acid group (—COOH). (b) A fatty acid is either saturated or unsaturated. Unsaturated fatty acids have double covalent bonds that cause sharp bends in the molecule.

through a similar reaction. In a **triglyceride** (trī-GLI-se-rīd), a glycerol molecule is attached to three fatty acids (see Figure 2–13●). Triglycerides are the most common fats in the body.

In addition to serving as an random energy reserve, fat deposits under the skin serve as insulation, and a mass of fat around a delicate organ, such as a kidney, provides a protective

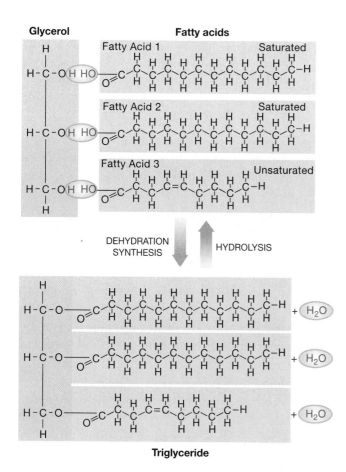

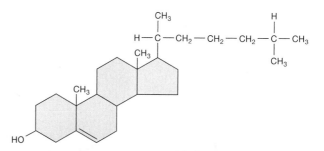
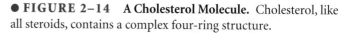

● **FIGURE 2–14** **A Cholesterol Molecule.** Cholesterol, like all steroids, contains a complex four-ring structure.

● **FIGURE 2–13** **Triglyceride Formation.** The formation of a triglyceride involves the attachment of three fatty acids to the carbons of a glycerol molecule. This example shows the attachment of one unsaturated and two saturated fatty acids to a glycerol molecule.

cushion. *Saturated fats*—triglycerides that contain saturated fatty acids—are usually solid at room temperature. *Unsaturated fats*, which are triglycerides that contain unsaturated fatty acids, are usually liquid at room temperature. Such liquid fats are oils.

Steroids

Steroids are large lipid molecules composed of four connected rings of carbon atoms. They differ in the carbon chains that are attached to this basic structure. **Cholesterol** (ko-LES-ter-ol; *chole-*, bile + *stereos,* solid) is probably the best-known steroid (Figure 2–14●). All of our cells are surrounded by *cell membranes* that contain cholesterol, and some chemical messengers, or *hormones,* are derived from cholesterol. Examples include the sex hormones testosterone and estrogen.

The cholesterol needed to maintain cell membranes and manufacture steroid hormones comes from two sources. One source is the diet; animal products such as meat, cream, and egg yolks are especially rich in cholesterol. The second is the body itself, for the liver can synthesize large amounts of cholesterol. The body's ability to synthesize this steroid can make it difficult to control blood cholesterol levels by dietary restriction alone. This difficulty can have serious repercussions because a strong link exists between high blood cholesterol concentrations and heart disease. Current nutritional advice suggests limiting cholesterol intake to under 300 mg per day; this amount represents a 40 percent reduction for the average adult in the U. S. The connection between blood cholesterol levels and heart disease will be examined more closely in Chapters 11 and 12.

Phospholipids

Phospholipids (FOS-fō-lip-idz) consist of a glycerol and two fatty acids (a diglyceride) linked to a nonlipid group by a phosphate group (PO_4^{3-}) (Figure 2–15●). The nonlipid portion of a phospholipid is soluble in water; whereas the fatty acid portion is relatively insoluble. Phospholipids are the most abundant lipid components of cell membranes.

> ### → CONCEPT CHECK QUESTIONS
>
> 1. A food contains organic molecules with the elements C, H, and O in a ratio of 1:2:1. To what class of compounds do these molecules belong, and what are their major functions in the body?
> 2. When two monosaccharides undergo a dehydration synthesis reaction, which type of molecule is formed?
> 3. Which kind of lipid would be found in a sample of fatty tissue taken from beneath the skin?
> 4. Which lipids would you find in human cell membranes?
>
> *Answers begin on p. 792.*

Proteins

Proteins are the most abundant organic components of the human body and in many ways the most important. There are at least 400,000 different kinds of proteins, and they account for about 20 percent of total body weight. All proteins contain carbon, hydrogen, oxygen, and nitrogen; smaller quantities of sulfur may also be present.

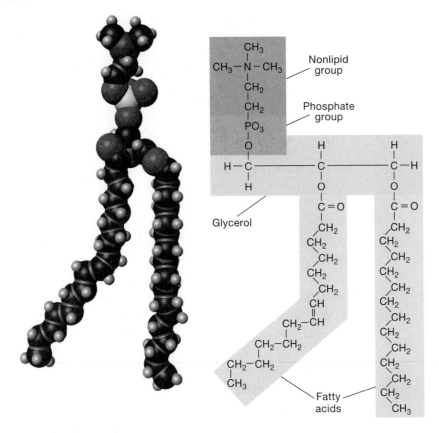

● **FIGURE 2–15   A Phospholipid Molecule.**
In a phospholipid, a phosphate group links a nonlipid molecule to a glycerol with two fatty acids (a diglyceride). This phospholipid is lecithin.

hazards. In addition, proteins known as antibodies protect us from disease, and special clotting proteins restrict bleeding after an injury to the cardiovascular system.

Key Note

Proteins are the most abundant organic components of the body, and they are the key to both anatomical structure and physiological function. Proteins determine cell shape and tissue properties, and almost all cell functions are performed by proteins and by interactions between proteins and their immediate environment.

Protein Function

Proteins perform a variety of functions, which can be grouped into seven major categories:

1. *Support.* Structural proteins create a three-dimensional framework for the body, and provide strength, organization, and support for cells, tissues, and organs.
2. *Movement.* Contractile proteins are responsible for muscular contraction; related proteins are responsible for the movement of individual cells.
3. *Transport.* Insoluble lipids, respiratory gases, minerals such as iron, and several hormones are carried in the blood attached to transport proteins. Other specialized proteins transport materials between different parts of a cell.
4. *Buffering.* Proteins provide a considerable buffering action, which helps to prevent potentially dangerous changes in pH in cells and tissues.
5. *Metabolic regulation.* Enzymes accelerate chemical reactions in living cells. The sensitivity of enzymes to environmental factors is extremely important in controlling the pace and direction of metabolic operations.
6. *Coordination, communication, and control.* Protein hormones can influence the metabolic activities of every cell in the body or affect the function of specific organs or organ systems.
7. *Defense.* The tough, waterproof proteins of the skin, hair, and nails protect the body from environmental

Protein Structure

Proteins are long chains of organic molecules called **amino acids.** The human body contains significant quantities of the 20 different amino acids that are the building blocks of proteins. Each amino acid consists of a central carbon atom bonded to a hydrogen atom, an amino group ($-NH_2$), a carboxylic acid group ($-COOH$), and a variable R group or side chain (Figure 2–16a●). The R group may be a straight chain or a ring of atoms. The name *amino acid* refers to the presence of the *amino* group and the carboxylic *acid* group, which all amino acids have in common. The different R groups distinguish one amino acid from another, which gives each its own chemical properties.

A typical protein contains 1000 amino acids, but the largest protein complexes may have 100,000 or more amino acids. The individual amino acids are strung together like beads on a string, with the carboxylic acid group of one amino acid attached to the amino group of another. This connection is called a **peptide bond** (Figure 2–16b●). **Peptides** are molecules made up of amino acids held together by peptide bonds. If a molecule consists of two amino acids, it is called a *dipeptide. Polypeptides* are long chains of amino acids. Polypeptides that contain more than 100 amino acids are usually called proteins.

At its most basic level, the *structure* of a protein is established by the sequence of its amino acids (Figure 2–17a●). The *characteristics* of a particular protein are determined in part by the R groups on its amino acids. But the properties of a protein

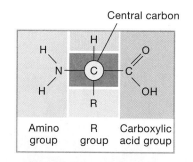

Central carbon

(a) Structure of an amino acid

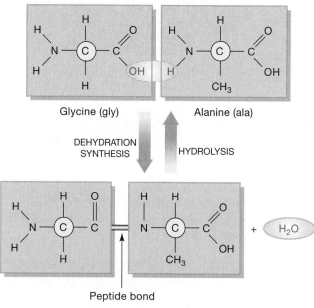

Glycine (gly)   Alanine (ala)

DEHYDRATION SYNTHESIS   HYDROLYSIS

Peptide bond

(b) Peptide bond formation

● **FIGURE 2–16** **Amino Acids and the Formation of Peptide Bonds.** (a) Each amino acid consists of a central carbon atom to which four different groups are attached: a hydrogen atom, an amino group ($-NH_2$), a carboxylic acid group ($-COOH$), and a variable group generally designated R. (b) Peptides form when a dehydration synthesis reaction creates a peptide bond between the carboxyl group of one amino acid and the amino group of another. In this example, glycine and alanine are linked to form a dipeptide.

are more than just the sum of the properties of its parts, for polypeptides can have highly complex shapes. Interactions between the R groups of the amino acids, the formation of hydrogen bonds at different parts of the chain, and interactions between the polypeptide chain and surrounding water molecules contribute to the complex three-dimensional shapes of large proteins.

In a *globular protein,* such as *myoglobin,* the peptide chain folds back on itself, which creates a rounded mass (Figure 2–17b●). Myoglobin is a protein found in muscle cells.

Complex proteins may consist of several protein subunits. An example is *hemoglobin,* a globular protein found inside red blood cells (Figure 2–17c●). Another protein with several protein sub-

units is keratin, an example of a *fibrous protein* (Figure 2–17d●). In fibrous proteins the polypeptide strands are wound together as in a rope. Fibrous proteins are flexible but very strong.

The shape of a protein determines its functional properties. The 20 common amino acids can be linked in an astonishing number of combinations, which creates proteins of enormously varied shape and function. Small differences can have large effects; changing one amino acid in a protein that contains 10,000 or more amino acids may make it incapable of performing its normal function. For example, several cancers and *sickle cell anemia,* a blood disorder, result from single changes in the amino acid sequences of complex proteins.

The shape of a protein—and, thus, its function—can also be altered by small changes in the ionic composition, temperature, or pH of its surroundings. For example, very high body temperatures (over 43°C, or 110°F) cause death because at these temperatures proteins undergo **denaturation,** a change in their three-dimensional shape. Denatured proteins are nonfunctional, and the loss of structural proteins and enzymes causes irreparable damage to organs and organ systems. You see denaturation in progress each time you fry an egg. As the temperature rises, the structure of the abundant proteins dissolved in the clear egg white changes, and eventually the egg proteins form an insoluble white mass.

Enzyme Function

Among the most important of all the body's proteins are enzymes. These molecules catalyze the reactions that sustain life: Almost everything that happens inside the human body does so because a specific enzyme makes it possible.

Figure 2–18● shows a simple model of enzyme function. The reactants in an enzymatic reaction, called **substrates,** interact to form a specific **product.** Before an enzyme can function as a catalyst—to accelerate a chemical reaction without itself being permanently changed or consumed—the substrates must bind to a special region of the enzyme called the **active site.** This binding depends on the complementary shapes of the two molecules, much as a key fits into a lock. The shape of the active site is determined by the three-dimensional shape of the enzyme molecule. Once the reaction is completed and the products are released, the enzyme is free to catalyze another reaction. Enzymes work quickly; an enzyme that provides energy during a muscular contraction performs its reaction sequence 100 times per second.

Each enzyme works best at an optimal temperature and pH. As temperatures rise or pH shifts outside normal limits, proteins change shape and enzyme function deteriorates.

Each enzyme catalyzes only one type of reaction. This *specificity* is determined by the ability of its active sites to

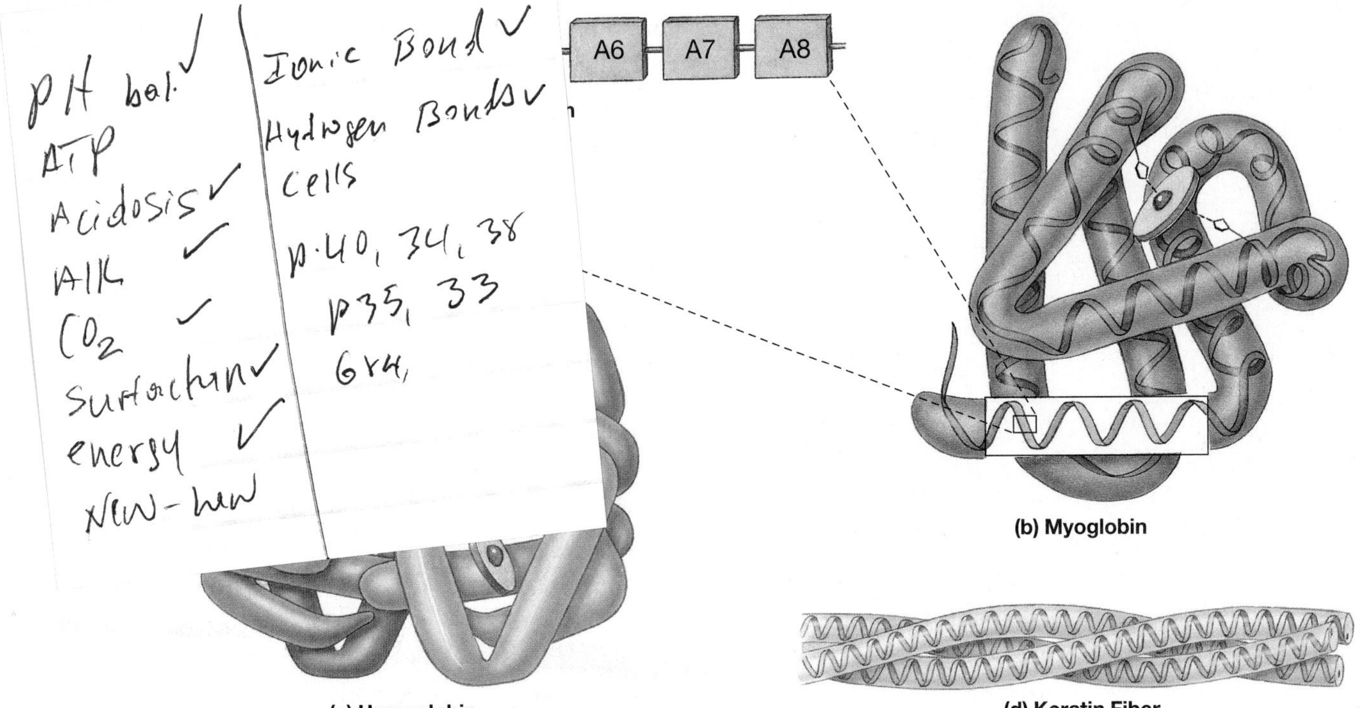

[handwritten notes]
pH bal. ✓   Ionic Bond ✓
ATP   Hydrogen Bonds ✓
Acidosis ✓   Cells
Alk ✓   p. 40, 34, 38
CO₂ ✓   p35, 33
Surfactant ✓   644,
energy ✓
New-hen

A6   A7   A8

(b) Myoglobin

(c) Hemoglobin

(d) Keratin Fiber

● **FIGURE 2–17   Protein Structure. (a)** The shape of a polypeptide is determined by its sequence of amino acids. **(b)** Attraction between R groups plays a large role in forming globular proteins. Myoglobin is a globular protein involved in the storage of oxygen in muscle tissue. **(c)** A single hemoglobin molecule contains four globular subunits, each structurally similar to myoglobin. Hemoglobin transports oxygen in the blood. **(d)** In keratin, three fibrous subunits intertwine like the strands of a rope.

bind only to substrates with particular shapes and charges. The complex reactions that support life proceed in a series of interlocking steps, each step controlled by a different enzyme. Such a reaction sequence is called a *pathway*. We will consider important pathways in later chapters.

→ **CONCEPT CHECK QUESTIONS**

1. Proteins are chains of which small organic molecules?
2. Why does boiling a protein affect its structural and functional properties?

Answers begin on p. 792.

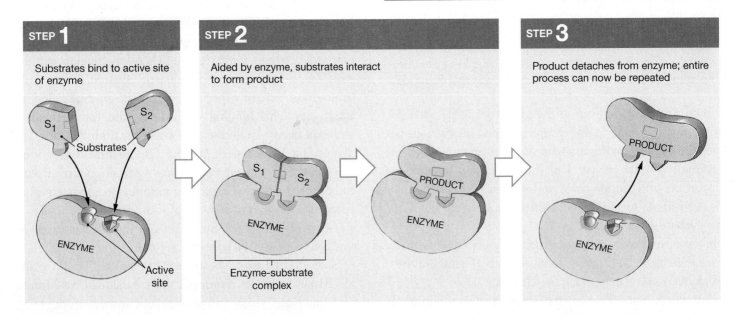

STEP 1
Substrates bind to active site of enzyme

S₁   S₂
Substrates
ENZYME
Active site

STEP 2
Aided by enzyme, substrates interact to form product

S₁   S₂
ENZYME
Enzyme-substrate complex

PRODUCT
ENZYME

STEP 3
Product detaches from enzyme; entire process can now be repeated

PRODUCT
ENZYME

● **FIGURE 2–18   A Simplified View of Enzyme Structure and Function.** Each enzyme contains a specific active site on its exposed surface. Because the structure of the enzyme has not been affected, the entire process can be repeated.

Nucleic Acids

Nucleic (noo-KLĀ-ik) **acids** are large organic molecules composed of carbon, hydrogen, oxygen, nitrogen, and phosphorus. Nucleic acids store and process information at the molecular level inside cells. There are two classes of nucleic acid molecules: **deoxyribonucleic** (dē-oks-ē-rī-bō-noo-KLĀ-ik) **acid,** or **DNA**; and **ribonucleic** (rī-bō-noo-KLĀ-ik) **acid,** or **RNA.**

The DNA in our cells determines our inherited characteristics, including eye color, hair color, and blood type. It affects all aspects of body structure and function because DNA molecules encode the information needed to build proteins. By directing the synthesis of structural proteins, DNA controls the shape and physical characteristics of our bodies. By controlling the manufacture of enzymes, DNA regulates not only protein synthesis but also all aspects of cellular metabolism, including the creation and destruction of lipids, carbohydrates, and other vital molecules.

Several forms of RNA cooperate to manufacture specific proteins using the information provided by DNA. The functional relationships between DNA and RNA will be detailed in Chapter 3.

Structure of Nucleic Acids

A nucleic acid is made up of subunits called **nucleotides.** Each single nucleotide has three basic components: a *sugar*, a *phosphate group* (PO_4^{3-}), and a *nitrogenous (nitrogen-containing) base* (Figure 2–19a●). The sugar is always a five-carbon sugar, either **ribose** (in RNA) or **deoxyribose** (in DNA). There are five nitrogenous bases: **adenine (A), guanine (G), cytosine (C), thymine (T),** and **uracil (U).** Both RNA and DNA contain adenine, guanine, and cytosine. Uracil is found only in RNA, and thymine only in DNA (Figure 2–19b●).

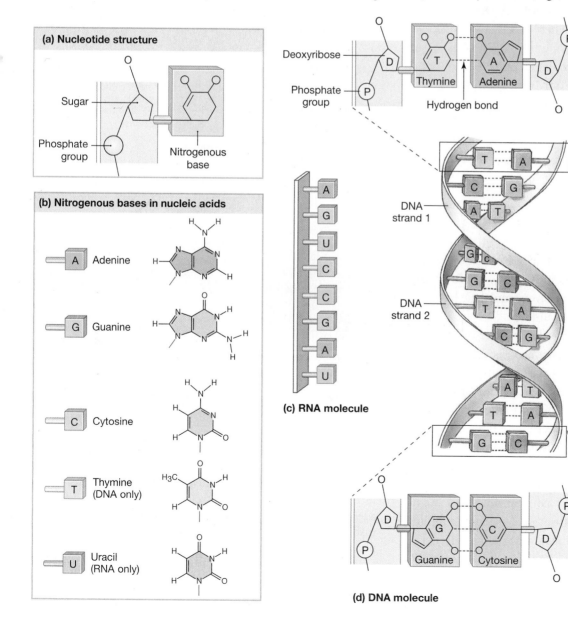

●**FIGURE 2–19** The **Structure of Nucleic Acids.** Nucleic acids are long chains of nucleotides. (**a**) A nucleotide is composed of a five-carbon sugar attached to both a phosphate group and a nitrogenous base. (**b**) Shown here are the structures of the five nitrogenous bases. (**c**) An RNA molecule consists of a single nucleotide chain. Its shape is determined by the sequence of nucleotides and the interactions between them. (**d**) A DNA molecule consists of a pair of nucleotide chains linked by hydrogen bonding between complementary base pairs.

(a) Nucleotide structure

Sugar

Phosphate group

Nitrogenous base

(b) Nitrogenous bases in nucleic acids

A Adenine

G Guanine

C Cytosine

T Thymine (DNA only)

U Uracil (RNA only)

Deoxyribose

Phosphate group

Thymine   Adenine

Hydrogen bond

DNA strand 1

DNA strand 2

(c) RNA molecule

Guanine   Cytosine

(d) DNA molecule

TABLE 2–7 *A Comparison of RNA and DNA*

CHARACTERISTIC	RNA	DNA
SUGAR	Ribose	Deoxyribose
NITROGENOUS BASES	Adenine	Adenine
	Guanine	Guanine
	Cytosine	Cytosine
	Uracil	Thymine
NUMBER OF NUCLEOTIDES IN A TYPICAL MOLECULE	Varies from fewer than 100 nucleotides to about 50,000	Always more than 45 million nucleotides
SHAPE OF MOLECULE	Single strand	Paired strands coiled in a double helix
FUNCTION	Performs protein synthesis as directed by DNA	Stores genetic information that controls protein synthesis

Important structural differences between RNA and DNA are listed in Table 2–7. A molecule of RNA consists of a single chain of nucleotides (see Figure 2–19c●). A DNA molecule consists of two nucleotide chains held together by weak hydrogen bonds between the opposing nitrogenous bases (see Figure 2–19d●). Because of their shapes, adenine can bond only with thymine, and cytosine only with guanine. As a result, adenine-thymine and cytosine-guanine are known as **complementary base pairs.**

The two strands of DNA twist around one another in a **double helix** that resembles a spiral staircase; the stair steps correspond to the nitrogenous base pairs.

High-Energy Compounds

The energy that powers a cell is obtained by the breakdown (catabolism) of organic molecules such as glucose. To be useful, that energy must be transferred from molecule to molecule or from one part of the cell to another.

The usual method of energy transfer involves the creation of **high-energy bonds** by enzymes within cells. A high-energy bond is a covalent bond that stores an unusually large amount of energy (as does a tightly wound rubber band attached to the propeller of a model airplane). When that bond is later broken, the energy is released under controlled conditions (as when the propeller is turned by the unwinding rubber band). In our cells, a high-energy bond usually connects a phosphate group (PO_4^{3-}) to an organic molecule, which results in a **high-energy compound.** Most high-energy compounds are derived from nucleotides, the building blocks of nucleic acids. The most important high-energy compound in the body is **adenosine triphosphate,** or **ATP.** ATP is composed of the nucleotide *adenosine monophosphate (AMP)* and two phosphate groups (Figure 2–20●).

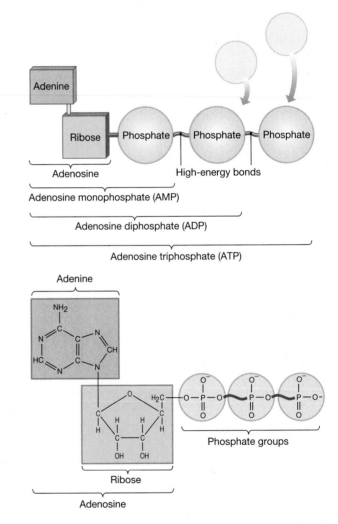

● **FIGURE 2–20** **The Structure of ATP.** An ATP molecule is formed by the linkage of two phosphate groups to the nucleotide adenosine monophosphate (AMP). (The adenosine portion is made up of adenine and the five-carbon sugar ribose.) The two phosphate groups are connected to AMP by high-energy bonds. Cells most often store energy by attaching a third phosphate group to ADP. Removing the phosphate group releases the energy for cellular work, including the synthesis of other molecules.

● **FIGURE 2–21** **Energy Flow and the Recycling of ATP within Cells.**

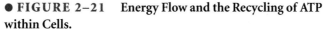

The addition of the two phosphate groups requires a significant amount of energy.

In ATP, a high-energy bond connects a phosphate group to **adenosine diphosphate (ADP).** Within our cells, the conversion of ADP to ATP represents the primary method of energy storage, and the reverse reaction provides a means for the release of energy. The arrangement can be summarized as:

$$ADP + \text{phosphate group} + \text{energy} \leftrightarrow ATP + H_2O$$

Throughout life, our cells continuously generate ATP from ADP and use the energy provided by that ATP to perform vital functions, such as the synthesis of protein molecules or the contraction of muscles. Figure 2–21● shows the relationship between energy flow and the recycling of ADP and ATP within cells.

Figure 2–22● and Table 2–8 review the major chemical compounds we have discussed in this chapter.

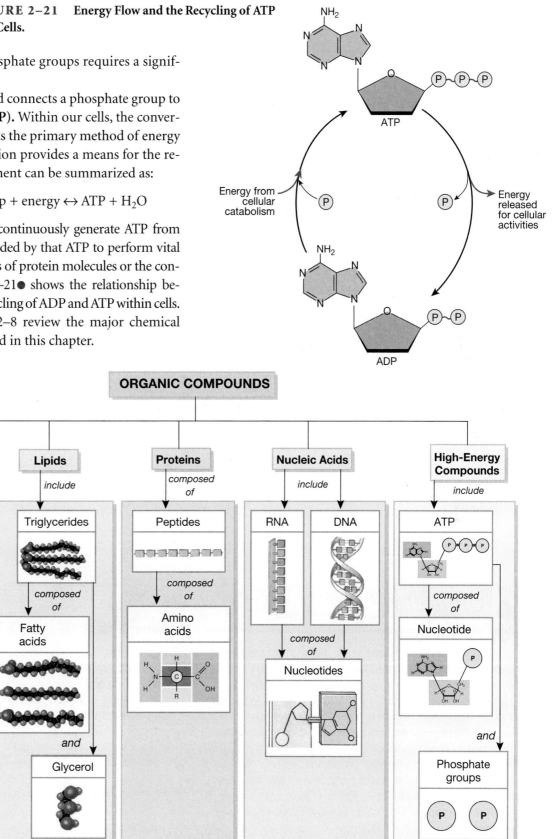

● **FIGURE 2–22** **A Structural Overview of Organic Compounds in the Body.** Each of the classes of organic compounds is composed of simple structural subunits. Specific compounds within each class are listed above the basic subunits. Fatty acids are the main subunits of all lipids except steroids, such as cholesterol; only one type of lipid, the triglyceride, is represented here.

TABLE 2–8 *The Structure and Function of Biologically Important Compounds*

CLASS	BUILDING BLOCKS	SOURCES	FUNCTIONS
INORGANIC **Water**	Hydrogen and oxygen atoms	Absorbed as liquid water or generated by metabolism	Solvent; transport medium for dissolved materials and heat; cooling through evaporation; medium for chemical reactions; reactant in hydrolysis
Acids, bases, salts	H^+, OH^-, various anions and cations	Obtained from the diet or generated by metabolism	Structural components; buffers; sources of ions
Dissolved gases	Oxygen, carbon, nitrogen, and other atoms	Atmosphere	O_2 required for normal cellular metabolism; CO_2 generated by cells as a waste product
ORGANIC **Carbohydrates**	C, H, and O; CHO in a 1:2:1 ratio	Obtained from the diet or manufactured in the body	Energy source; some structural role when attached to lipids or proteins; energy storage
Lipids	C, H, O, sometimes N or P; CHO not in 1:2:1 ratio	Obtained from the diet or manufactured in the body	Energy source; energy storage; insulation; structural components; chemical messengers; protection
Proteins	C, H, O, N, often S	20 common amino acids; roughly half can be manufactured in the body, others must be obtained from the diet	Catalysts for metabolic reactions; structural components; movement; transport; buffers; defense; control and coordination of activities
Nucleic acids	C, H, O, N, and P; nucleotides composed of phosphates, sugars, and nitrogenous bases	Obtained from the diet or manufactured in the body	Storage and processing of genetic information
High-energy compounds	Nucleotides joined to phosphates by high-energy bonds	Synthesized by all cells	Storage or transfer of energy

Clinical Note
PHYSICAL PROPERTIES OF EMERGENCY MEDICATIONS

Many medications are used in prehospital care. Unlike in the hospital setting, medications used in the field are often exposed to environmental factors. In most cases, this rarely constitutes a problem. However, some medications are affected by environmental factors. Because of this, prehospital personnel must be familiar with the physical properties of the medications they use. Also, knowledge of the physical properties of prehospital medications helps in understanding their mechanism of action.

The three states of matter are *solid, liquid, or gas*. Two major factors affect the physical state of matter. The first factor is the intensity of the intermolecular forces that hold compounds together: solids have the strongest forces while gases have the weakest. The other factor is temperature. As the temperature rises, a substance will go from a solid to a liquid and then to a gaseous state.

As the material passes through these three phases from solid to liquid to gas, it absorbs heat, and its *enthalpy,* or heat content, increases.

Thus, the enthalpy of a liquid is greater than that of its solid, and the enthalpy of a gas is greater than that of its liquid, because heat is absorbed during melting and vaporization. The temperature at which a solid becomes a liquid is its *melting point.* The temperature at which a liquid becomes a gas is its *boiling point.* However, prior to reaching the boiling point, some liquids will begin to convert to the gaseous state through *evaporation.* Chemicals vary in their tendency to change physical states. The tendency to assume the gaseous state is *volatility.* Chemicals with high volatility rapidly evaporate and assume the gaseous state, while chemicals with low volatility tend to remain in the liquid state.

Most medications used in prehospital care are in a liquid state. This allows the medication to be administered into the body without having to be absorbed by the gastrointestinal tract. The boiling point and the freezing point of most emergency medicines will not be encountered in routine usage. However, some medications pose problems. For example, the medication mannitol is occasionally used in prehospital care to treat increased intracranial pressure, acute glaucoma, and blood transfusion reactions. Mannitol is a sugar and is used as an osmotic diuretic. It is supplied in liquid form in vials that contain a 20 percent solution. At temperatures less than 10°C (50°F), mannitol will begin to crystallize. Initially, these crystals cannot be seen by the naked eye. If the temperature of the drug drops below 5°C (40°F), the entire vial will crystallize as the drug changes to the solid state. The higher the concentration of mannitol, the greater its tendency to crystallize. Because of this, mannitol should always be administered through a filter to prevent any crystals, including microscopic crystals, from entering the patient's circulation. Remelting mannitol with boiling water or in a microwave oven can potentially damage the drug and should not be attempted. Mannitol, like all EMS medications, should be maintained at the appropriate temperature.

The volatility of certain drugs is important in emergency care. For example, the general anesthetic agents are highly volatile, which allows their rapid delivery to patients. For example, the initial prototype anesthetic, ether, is highly volatile. In early anesthetic practice, it was dripped slowly into an ether mask and the patient inhaled the drug, which resulted in anesthesia (Figure 2–23●). However, because ether is extremely flammable and hazardous to surgical personnel, it is rarely used in modern medicine.

An anesthetic agent occasionally used in modern medicine is *methoxyflurane (Pentrane)*. In high doses, methoxyflurane is a general anesthetic, and in lower doses, it is an extremely effective analgesic agent. Although not approved for prehospital use in the U.S., methoxyflurane is frequently used throughout Australian EMS as an analgesic agent. At room temperature, methoxyflurane is a liquid; however, it is highly volatile and rapidly evaporates. For analgesia, 3 to 6 milliliters of a 0.5 percent solution of methoxyflurane are placed on the absorbent wick of a methoxyflurane (Penthrane) inhaler (Figure 2–24●). The inhaler is then handed to the patient, who breathes in and out through the mouthpiece. The onset of analgesia is rapid, and the effects of the drug quickly dissipate when removed.

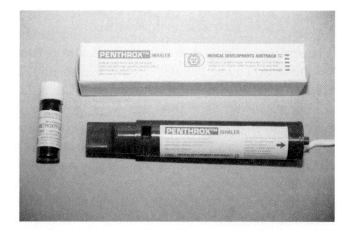

● **FIGURE 2–24   Penthrox (methoxyflurane) Inhaler Used in Australia for Analgesia.** Methoxyflurane, a volatile anesthetic and analgesic agent, is placed onto a wick in the device. The patient inhales through the mouthpiece. Supplemental oxygen can be provided.

Because of this, it can be used in trauma, obstetrics, and other types of acute pain. A port on the methoxyflurane inhaler allows supplemental oxygen if required.

Nitrous oxide is another general anesthetic agent that has analgesic properties at lower concentrations. It is frequently used in the United States, Canada, and the United Kingdom for rapid analgesia. Because nitrous oxide can be an asphyxiant when used alone, it must always be administered with oxygen. Anesthesia is obtained when the level of nitrous oxide approaches 70 percent. Analgesia is obtained at a concentration of approximately 50 percent. In the U.S., nitrous oxide and oxygen are administered from different cylinders. These gases are fed into a blender that mixes the correct 50 percent nitrous oxide/50 percent oxygen mixture. If the level of oxygen falls below 50 percent, then the unit shuts down. Unfortunately, requiring separate oxygen and nitrous oxide cylinders makes the unit much heavier, bulkier, and harder to use in emergency care. In the Commonwealth countries (former colonies of the United Kingdom) premixing nitrous oxide and oxygen into one cylinder is allowed (Entonox, Dolonox). This system is less bulky, less expensive, and easier to use in prehospital care.

One physical property of nitrous oxide affects its use in prehospital care. Nitrous oxide exists as a gas over liquid in pressurized steel containers at room temperature. However, at approximately −3.3°C (26°F), nitrous oxide returns to its liquid form. When this occurs, drug delivery ceases until the cylinder is warmed. Occasionally, inverting the cylinder two or three times may free enough gas to complete patient care. In the U.S. system (Nitronox), when nitrous oxide delivery ceases, the device allows the continued flow of pure oxygen. In the single-cylinder system, when the nitrous oxide liquefies, the mixture separates, and the gaseous oxygen occupies the top of the cylinder while the liquid nitrous oxide occupies the bottom of the cylinder. Administration of nitrous oxide through the single cylinder system in cold environments can cause decreased concentrations of the drug as it liquefies and leaves the mixture. Gentle rewarming of the cylinder at 20°C (68°F) and completely inverting the cylinder will restore the mixture. ■

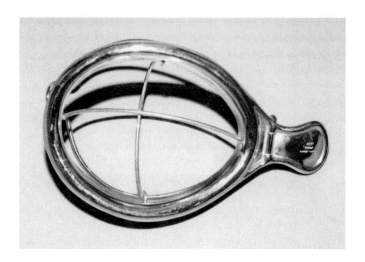

● **FIGURE 2–23   Ether Mask Used to Administer Ether as an Anesthetic.** A cloth was placed in the frame and volatile ether was dripped onto the cloth for inhalation by the patient.

■ Chemicals and Cells

The human body is more than a random collection of chemicals. The biochemical building blocks discussed in this chapter are the components of cells. ∞ p. 5 Each cell behaves like a miniature organism, and responds to internal and external stimuli. A lipid membrane separates the cell from its environment, and internal membranes create compartments with specific functions. Proteins form an internal supporting framework and act as enzymes to accelerate and control the chemical reactions that maintain homeostasis. Nucleic acids direct the synthesis of all cellular proteins, including the enzymes that enable the cell to synthesize a wide variety of other substances. Carbohydrates provide energy for vital activities and form part of specialized compounds in combination with proteins or lipids. The next chapter considers the combination of these compounds within a living, functional cell.

→ **CONCEPT CHECK QUESTIONS**

1. A large organic molecule composed of the sugar ribose, nitrogenous bases, and phosphate groups is which kind of nucleic acid?
2. How are DNA and RNA similar?
3. What are the products of the hydrolysis of ATP?

Answers begin on p. 792.

Chapter Review

Access more review material online at *www.prenhall.com/bledsoe*. There you will find quiz questions, labeling activities, animations, essay questions, and web links.

Key Terms

atom 29
buffer 39
carbohydrate 42
covalent bond 33
decomposition reaction 36
electrolytes 41

electron 30
element 29
enzyme 37
ion 32
isotope 30
lipid 43

metabolism 34
molecule 32
neutron 30
nucleic acid 49
protein 45
proton 30

Related Clinical Terms

cholesterol A steroid and an important component of cellular membranes; in high concentrations it increases the risk of heart disease.

familial hypercholesterolemia A genetic disorder that results in high cholesterol levels in blood and cholesterol buildup in body tissues, especially the walls of blood vessels.

galactosemia A metabolic disorder that results from the lack of an enzyme that converts galactose, a monosaccharide in milk, to glucose within cells. Affected individuals have elevated galactose levels in the blood and urine. High levels of galactose during childhood can cause abnormalities in ner-

vous system development and liver function, and cataracts.

nuclear imaging A procedure in which an image is created on a photographic plate or video screen by the radiation emitted by injected radioisotopes.

omega-3 fatty acids Fatty acids, abundant in fish flesh and fish oils, that have a double bond three carbon atoms away from the end of the hydrocarbon chain. Their presence in the diet has been linked to reduced risks of heart disease and other conditions.

phenylketonuria (PKU) A metabolic disorder that results from a defect in the enzyme that normally converts the amino acid

phenylalanine to tyrosine, another amino acid. If the resulting elevated phenylalanine levels are not detected in infancy, mental retardation can result from damage to the developing nervous system.

radioisotopes Isotopes with unstable nuclei, which spontaneously emit subatomic particles or radiation in measurable amounts.

radiopharmaceuticals Drugs that incorporate radioactive atoms; administered to expose specific target tissues to radiation.

tracer A radioisotope-labeled compound that can be tracked in the body by the radiation it releases.

Summary Outline

MATTER: ATOMS AND MOLECULES 29

1. **Atoms** are the smallest units of matter; they consist of **protons, neutrons,** and **electrons.** *(Figure 2–1)*

The Structure of an Atom 29

2. An **element** consists entirely of atoms with the same number of protons (**atomic number**). Within an atom, an **electron cloud** surrounds the nucleus. *(Figure 2–2); Table 2–1)*

3. The atomic mass of an atom is equal to the total number of protons and neutrons in its nucleus. **Isotopes** are atoms of the same element whose nuclei contain different numbers of neutrons. The **atomic weight** of an element takes into account the abundance of its various isotopes.

4. Electrons occupy a series of **electron shells** around the nucleus. The number of electrons in the outermost electron shell determines an atom's chemical properties. *(Figure 2–3)*

Key Note 31

Chemical Bonds and Chemical Compounds 31

5. An **ionic bond** results from the attraction between **ions**—atoms that have gained or lost electrons. **Cations** are positively charged, and **anions** are negatively charged. *(Figure 2–4; Table 2–2)*

6. Atoms can combine to form a **molecule;** combinations of atoms of different elements form a **compound.** Some atoms share electrons to form a molecule held together by **covalent bond.**

7. Sharing one pair of electrons creates a single covalent bond; sharing two pairs forms a **double covalent bond**. An unequal sharing of electrons creates a **polar covalent bond.** *(Figure 2–5)*

8. A **hydrogen bond** is the attraction between a hydrogen atom with a slight positive charge and a negatively-charged atom in another molecule or within the same molecule. Hydrogen bonds can affect the shapes and properties of molecules. *(Figure 2–6)*

CHEMICAL NOTATION 34

1. Chemical notation allows us to describe reactions between reactants that generate one or more products. *(Table 2–3)*

CHEMICAL REACTIONS 34

1. **Metabolism** refers to all the **chemical reactions** in the body. Our cells capture, store, and use energy to maintain homeostasis and support essential functions.

Basic Energy Concepts 34

2. **Work** involves movement of an object or a change in its physical structure, and **energy** is the capacity to perform work. There are two major types of energy: kinetic and potential.

3. **Kinetic energy** is the energy of motion. **Potential energy** is stored energy that results from the position or structure of an

object. Conversions from potential to kinetic energy are not 100 percent efficient; every energy exchange produces **heat.**

Key Note 35

Types of Reactions 36

4. A chemical reaction may be classified as a **decomposition, synthesis,** or **exchange reaction**.

5. Cells gain energy to power their functions through **catabolism,** the breakdown of complex molecules. Much of this energy supports **anabolism,** the synthesis of new organic molecules.

6. Reversible reactions consist of simultaneous synthesis and decomposition reactions. At **equilibrium** the rates of these two opposing reactions are in balance.

Key Note 36

Enzymes and Chemical Reactions 36

7. **Activation energy** is the amount of energy required to start a reaction. Proteins called **enzymes** control many chemical reactions within our bodies. Enzymes are **catalysts** that participate in reactions without themselves being permanently changed. *(Figure 2–7)*

8. **Exergonic** reactions release heat; **endergonic** reactions absorb heat.

Key Note 37

INORGANIC COMPOUNDS 37

1. **Nutrients** and **metabolites** can be broadly classified as **organic** or **inorganic compounds.**

Carbon Dioxide and Oxygen 38

2. Living cells in the body consume oxygen and generate carbon dioxide.

Water and Its Properties 38

3. Water is the most important inorganic component of the body.

4. Water is an excellent solvent, has a high heat capacity, and participates in the metabolic reactions of the body.

5. Many inorganic compounds undergo **ionization,** or dissociation, in water to form ions. *(Figure 2–8)*

Key Note 38

Inorganic Acids and Bases 38

6. An **acid** releases hydrogen ions into a solution, and a **base** removes hydrogen ions from a solution.

7. The **pH** of a solution indicates the concentration of hydrogen ions it contains. Solutions can be classified as neutral ($pH = 7$), acidic ($pH < 7$), or basic (alkaline) ($pH > 7$) on the basis of pH. *(Figure 2–9; Table 2–4)*

8. **Buffers** maintain pH within normal limits (7.35–7.45 in most body fluids) by releasing or absorbing hydrogen ions.

Salts 41

9. A **salt** is an ionic compound whose cation is not H^+ and whose anion is not OH^-. Salts are **electrolytes,** compounds that dissociate in water and conduct an electrical current.

ORGANIC COMPOUNDS 41

1. Organic compounds contain carbon and hydrogen and usually oxygen as well. Large and complex organic molecules include carbohydrates, lipids, proteins, and nucleic acids.

Carbohydrates 42

2. **Carbohydrates** are most important as an energy source for metabolic processes. The three major types are **monosaccharides** (simple sugars), **disaccharides,** and **polysaccharides.** *(Figures 2–10 and 2–11; Table 2–5)*

Lipids 43

3. **Lipids** are water-insoluble molecules that include fats, oils, and waxes. There are four important classes of lipids: **fatty acids, fats, steroids,** and **phospholipids.** *(Table 2–6)*

4. **Triglycerides (fats)** consist of three fatty acid molecules attached to a molecule of **glycerol.** *(Figure 2–12, 2–13)*

5. Cholesterol is a precursor of steroid hormones and is a component of cell membranes. *(Figure 2–14)*

6. Phospholipids are the most abundant components of cell membranes. *(Figure 2–15)*

Proteins 45

7. Proteins perform a great variety of functions in the body. Important types of proteins include **structural proteins, contractile proteins, transport proteins, enzymes, hormones,** and **antibodies.**

Key Note 46

8. Proteins are chains of **amino acids** linked by **peptide bonds.** The sequence of amino acids and the interactions of their R groups influence the final shape of a protein molecule. *(Figures 2–16, 2–17)*

9. The shape of a protein determines its function. Each protein works best at an optimal combination of temperature and pH.

10. The reactants in an enzymatic reaction, called **substrates,** interact to form a **product** by binding to the enzyme at the active site. *(Figure 2–18)*

Nucleic Acids 49

11. **Nucleic acids** store and process information at the molecular level. There are two kinds of nucleic acids: **deoxyribonucleic acid (DNA)** and **ribonucleic acid (RNA).** *(Figure 2–19; Table 2–7)*

12. Nucleic acids are chains of nucleotides. Each nucleotide contains a sugar, a **phosphate group,** and a **nitrogenous base.** The sugar is always **ribose** or **deoxyribose.** The nitrogenous bases found in DNA are **adenine, guanine, cytosine,** and **thymine.** In RNA, **uracil** replaces thymine.

High-Energy Compounds 50

13. Cells store energy in **high-energy compounds.** The most important high-energy compound is **ATP (adenosine triphosphate).** When energy is available, cells make ATP by adding a phosphate group to ADP. When energy is needed, ATP is broken down to ADP and phosphate. *(Figures 2–20, 2–21)*

14. Biologically important compounds are composed of simple structural subunits. *(Figures 2–22 through 2–24; Table 2–8)*

CHEMICALS AND CELLS 54

1. Biochemical building blocks form cells.

Review Questions

Level 1: Reviewing Facts and Terms

Match each item in column A with the most closely related item in column B. Place the letters for answers in the spaces provided.

COLUMN A

____ 1. atomic number

____ 2. covalent bond

____ 3. ionic bond

____ 4. catabolism

____ 5. anabolism

____ 6. exchange reaction

____ 7. reversible reaction

____ 8. acid

____ 9. enzyme

____ 10. buffer

____ 11. organic compounds

____ 12. inorganic compounds

COLUMN B

a. synthesis

b. catalyst

c. sharing of electrons

d. $A + B \leftrightarrow AB$

e. stabilize pH

f. number of protons

g. decomposition

h. carbohydrates, lipids, proteins

i. loss or gain of electrons

j. water, salts

k. donor

l. $AB + CD \rightarrow AD + CB$

13. In atoms, protons and neutrons are found:
    (a) only in the nucleus.
    (b) outside the nucleus.
    (c) inside and outside the nucleus.
    (d) in the electron cloud.

14. The number and arrangement of electrons in an atom's outer electron shell determine its:
    (a) atomic weight.
    (b) atomic number.
    (c) electrical properties.
    (d) chemical properties.

15. The bond between sodium and chlorine in the compound sodium chloride (NaCl) is:
    (a) an ionic bond.
    (b) a single covalent bond.
    (c) a nonpolar covalent bond.
    (d) a double covalent bond.

16. What is the role of enzymes in chemical reactions?

17. List the six most abundant elements in the body.

18. What four major classes of organic compounds are found in the body?

19. List seven major functions performed by proteins.

Level 2: Reviewing Concepts

20. Oxygen has eight protons, eight neutrons, and eight electrons. What is its atomic mass?
    (a) 8
    (b) 16
    (c) 24
    (d) 32

21. Of the following selections, the one that contains only inorganic compounds is:
    (a) water, electrolytes, oxygen, carbon dioxide.
    (b) oxygen, carbon dioxide, water, sugars.
    (c) water, electrolytes, salts, nucleic acids.
    (d) carbohydrates, lipids, proteins, vitamins.

22. Glucose and fructose are examples of:
    (a) monosaccharides (simple sugars).
    (b) isotopes.
    (c) lipids.
    (d) monosaccharides, isotopes, and lipids.

23. Explain the differences among (1) nonpolar covalent bonds, (2) polar covalent bonds, and (3) ionic bonds.

24. Why does pure water have a neutral pH?

25. A biologist analyzes a sample that contains an organic molecule and finds the following constituents: carbon, hydrogen, oxygen, nitrogen, and phosphorus. On the basis of this information, is the molecule a carbohydrate, a lipid, a protein, or a nucleic acid?

Level 3: Critical Thinking and Clinical Applications

26. The element sulfur has an atomic number of 16 and an atomic mass of 32. How many neutrons are in the nucleus of a sulfur atom? Assuming that sulfur forms covalent bonds with hydrogen, how many hydrogen atoms could bond to one sulfur atom?

27. An important buffer system in the human body involves carbon dioxide (CO_2) and bicarbonate ions (HCO_3^-) as shown:

$$CO_2 + H_2O \leftrightarrow H_2CO_3 \leftrightarrow H^+ + HCO_3^-$$

If a person becomes excited and exhales large amounts of CO_2, how will his or her body's pH be affected?

3 Cell Structure and Function

ALL CELLS REQUIRE a constant supply of oxygen and essential nutrients. At a cellular level, a deficiency of either oxygen or essential nutrients will quickly result in cellular dysfunction and eventually cellular death. Many of the emergencies that EMTs are called upon to treat, such as drowning, are ultimately due to lack of oxygen at the cellular level.

Chapter Outline

Chapter Objectives

1. List the main points of the cell theory. (p. 59)
2. Describe the functions of the cell membrane and the structures that enable it to perform those functions. (pp. 59–60)
3. Describe the various mechanisms that cells use to transport substances across the cell membrane. (pp. 64–75)
4. Describe the organelles of a typical cell and indicate their specific functions. (pp. 76–81)
5. Explain the functions of the cell nucleus. (pp. 81–82)
6. Summarize the process of protein synthesis. (pp. 83–86)
7. Describe the process of mitosis and explain its significance. (p. 87)
8. Define differentiation and explain its importance. (pp. 88–89)

Vocabulary Development

aero- air; *aerobic*
ana- apart; *anaphase*
chondrion granule; *mitochondrion*
chroma color; *chromosome*
cyto- cell; *cytoplasm*
endo- inside; *endocytosis*
exo- outside; *exocytosis*
hemo- blood; *hemolysis*
hyper- above; *hypertonic*
hypo- below; *hypotonic*

inter- between; *interphase*
interstitium something standing between; *interstitial fluid*
iso- equal; *isotonic*
kinesis motion; *cytokinesis*
meta- after; *metaphase*
micro- small; *microtubules*
mitos thread; *mitosis*
osmos thrust; *osmosis*
phagein to eat; *phagocyte*

pinein to drink; *pinocytosis*
podon foot; *pseudopod*
pro- before; *prophase*
pseudo- false; *pseudopod*
ptosis a falling away; *apoptosis*
reticulum network; *endoplasmic reticulum*
soma body; *lysosome*
telos end; *telophase*
tonos tension; *isotonic*

JUST AS ATOMS ARE the building blocks of molecules, **cells** are the building blocks of the human body. Over the years, biologists have developed the **cell theory**, which includes the following four basic concepts:

1. Cells are the building blocks of all plants and animals.
2. Cells are the smallest functioning units of life.
3. Cells are produced through the division of pre-existing cells.
4. Each cell maintains homeostasis.

An individual organism maintains homeostasis only through the combined and coordinated actions of many different types of cells. Figure 3–1● shows some of the variety of cell shapes and sizes in the human body.

The trillions of human body cells form and maintain anatomical structures, and they perform physiological functions as different as running and thinking. An understanding of how the human body functions, thus, requires a familiarity with the nature of cells.

■ Studying Cells

The study of the structure and function of cells is called **cytology** (sī-TOL-ō-jē; *cyto-*, cell + *-logy*, the study of). What we have learned since the 1950s has provided new insights into the physiology of cells and their means of homeostatic control. This knowledge resulted from improved equipment

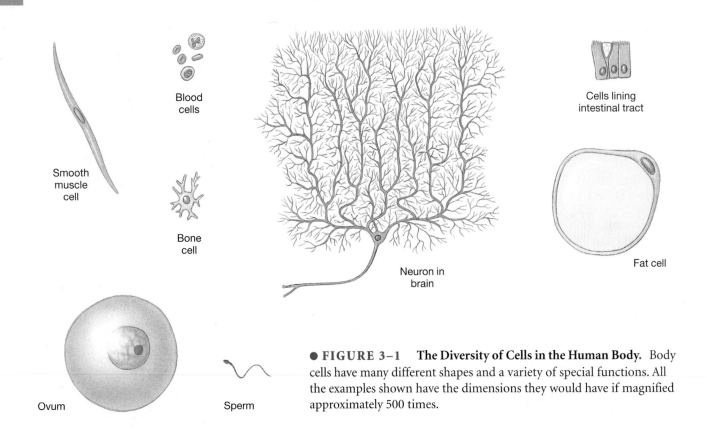

Smooth muscle cell

Blood cells

Bone cell

Neuron in brain

Cells lining intestinal tract

Fat cell

Ovum

Sperm

● **FIGURE 3–1   The Diversity of Cells in the Human Body.** Body cells have many different shapes and a variety of special functions. All the examples shown have the dimensions they would have if magnified approximately 500 times.

for viewing cells and new experimental techniques, not only from biology but also from chemistry and physics.

The two most common methods used to study cell and tissue structure are light microscopy and electron microscopy. Before the 1950s, cells were viewed through light microscopes. Using a series of glass lenses, *light microscopy* can magnify cellular structures about 1000 times. Light microscopy typically involves looking at thin sections sliced from a larger piece of tissue. A photograph taken through a light microscope is called a *light micrograph (LM)*.

Many fine details of intracellular structure are too small to be seen with a light microscope. These details remained a mystery until cell biologists began using *electron microscopy,* a technique that replaced light with a focused beam of electrons. *Transmission electron micrographs (TEMs)* are photographs of very thin sections, and they can reveal fine details of cell membranes and intracellular structures. *Scanning electron micrographs (SEMs)* provide less magnification but reveal the three-dimensional nature of cell structures. An SEM provides a surface view of a cell, a portion of a cell, or extracellular structures rather than a detailed sectional view.

You will see examples of light micrographs and both kinds of electron micrographs in figures throughout this text. The abbreviations LM, TEM, and SEM are followed by a number that indicates the total magnification of the image. For example, "LM × 160" indicates that the structures shown in a light micrograph have been magnified 160 times. (See Appendix 3 for the size ranges and scales included in the study of anatomy and physiology.)

An Overview of Cell Anatomy

The "typical" cell is like the "average" person; any such description masks enormous individual variations. Our model (or representative) cell shares features with most cells of the body without being identical to any specific one (Figure 3–2●). Table 3–1 summarizes the structures and functions of a cell's parts.

Our model body cell is surrounded by a watery medium known as the **extracellular fluid.** The extracellular fluid in most tissues is called **interstitial** (in-ter-STISH-ul) **fluid** (*interstitium,* something standing between). A **cell membrane** separates the cell contents, or **cytoplasm,** from the extracellular fluid. The cytoplasm surrounds the **nucleus**—the control center for cellular operations.

The cytoplasm can be subdivided into a liquid—the *cytosol*—and intracellular structures collectively known as *organelles.* Our discussion of the cell begins at its boundary with the external environment—the cell membrane. It then proceeds to the cytosol and the individual organelles and concludes with the nucleus.

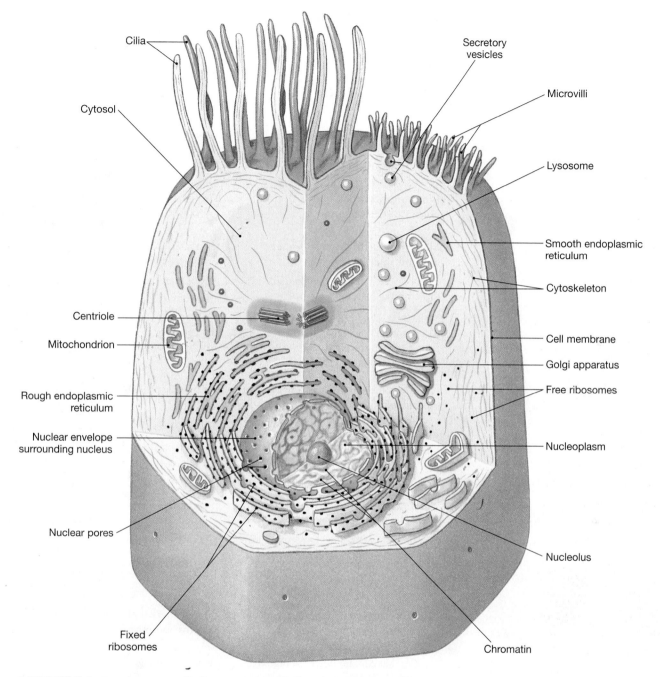

● **FIGURE 3–2**   **Anatomy of a Representative Cell.**   See Table 3–1 for an overview of the functions associated with the various structures shown.

■ The Cell Membrane

As noted, the outer boundary of the cell is formed by a cell membrane, or *plasma membrane.* Its general functions include:

■ *Physical isolation.* The cell membrane is a physical barrier that separates the inside of the cell from the surrounding extracellular fluid. Conditions inside and outside the cell are very different, and those differences must be maintained to preserve homeostasis.

■ *Regulation of exchange with the environment.* The cell membrane controls the entry of ions and nutrients, the elimination of wastes, and the release of secretions.

■ *Sensitivity.* The cell membrane is the first part of the cell affected by changes in the extracellular fluid. It also contains a variety of receptors that enable the cell to recognize and respond to specific molecules in its environment.

■ *Structural support.* Specialized connections between cell membranes, or between membranes and extracellular materials, give tissues a stable structure.

TABLE 3–1 *Components of a Representative Cell*

APPEARANCE	COMPOSITION	FUNCTION
Cell membrane	Lipid bilayer, contains phospholipids, steroids, and proteins	Provides isolation, protection, sensitivity, and support; controls entrance/exit of materials
Cytosol	Fluid component of cytoplasm	Distributes materials by diffusion
NONMEMBRANOUS ORGANELLES Cytoskeleton: Microtubule Microfilament	Proteins organized in fine filaments or slender tubes	Provides strength and support; enables movement of cellular structures and materials
Microvilli	Membrane extensions that contain microfilaments	Increase surface area to facilitate absorption of extracellular materials
Cilia	Membrane extensions that contain microtubules	Move materials over cell surface
Centrioles	Two centrioles, at right angles; each composed of microtubules	Essential for movement of chromosomes during cell division
Ribosomes	RNA + proteins; fixed ribosomes bound to endoplasmic reticulum; free ribosomes scattered in cytoplasm	Synthesize proteins
Proteasomes	Cylindrical structures that contain proteases (protein-breaking enzymes)	Break down and recycle damaged or abnormal intracellular proteins
MEMBRANOUS ORGANELLES Endoplasmic reticulum (ER):	Network of membranous channels that extend throughout the cytoplasm	Synthesizes secretory products; provides intracellular storage and transport
Rough ER	Has ribosomes attached to membranes	Packages newly-synthesized proteins
Smooth ER	Lacks attached ribosomes	Synthesizes lipids and carbohydrates
Golgi apparatus	Stacks of flattened membranes that contain chambers	Stores, alters, and packages secretory products; forms lysosomes
Lysosomes	Vesicles that contain powerful digestive enzymes	Remove damaged organelles or pathogens within cells
Peroxisomes	Vesicles that contain degradative enzymes	Catabolize fats and other organic compounds; neutralize toxic compounds generated in the process
Mitochondria	Double membrane, with inner folds (cristae) that enclose important metabolic enzymes	Produce 95% of the ATP required by the cell
Nucleus	Nucleoplasm that contains DNA, nucleotides, enzymes, and proteins; surrounded by double membrane (nuclear envelope)	Controls metabolism; stores and processes genetic information; controls protein synthesis
Nucleolus	Dense region in nucleoplasm that contains DNA and RNA	Synthesizes RNA and assembles ribosomal subunits

Membrane Structure

The cell membrane is extremely thin and delicate, and ranges from 6 nm to 10 nm in thickness. This membrane contains lipids, proteins, and carbohydrates (Figure 3–3●).

Membrane Lipids

Phospholipids are a major component of cell membranes. ∞ p. 45 In a phospholipid, a phosphate group (PO_4^{3-}) serves as a link between a diglyceride (a glycerol molecule bonded to two fatty acid "tails") and a nonlipid "head." The phospholipids in a cell membrane lie in two distinct layers, with the **hydrophilic** (hī-drō-FI-lik; *hydro-*, water + *philos*, loving; soluble in water) heads on the outside, and the **hydrophobic** (hī-drō-FŌB-ik; *hydro-*, + *phobos* fear; insoluble in water) tails on the inside. For this reason, the cell membrane is often called a **phospholipid bilayer** (see Figure 3–3). Mixed in with the fatty acid tails are cholesterol molecules and small quantities of other lipids.

The hydrophobic lipid tails will not associate with water or charged molecules, and this characteristic enables the cell membrane to act as a selective physical barrier. Lipid-soluble molecules and compounds such as oxygen and carbon diox-ide are able to cross the lipid portion of a cell membrane, but ions and water-soluble compounds cannot. Consequently, the cell membrane isolates the cytoplasm from the surrounding extracellular fluid.

Membrane Proteins

Several types of proteins are associated with the cell membrane. The most common of these membrane proteins span the width of the membrane one or more times and are known as *transmembrane proteins.* Other membrane proteins are either partially embedded in the membrane's phospholipid bilayer or loosely bound to its inner or outer surface. Membrane proteins may function as *receptors, channels, carriers, enzymes, anchors,* or *identifiers.* Table 3–2 provides a functional description and an example of each class of membrane protein.

Cell membranes are neither rigid nor are their inner and outer surfaces structurally the same. Although some embedded proteins are always confined to specific areas of the membrane, others drift from place to place along its surface like ice cubes in a punch bowl. In addition, the composition of the cell membrane can change over time, as components of the membrane are added or removed.

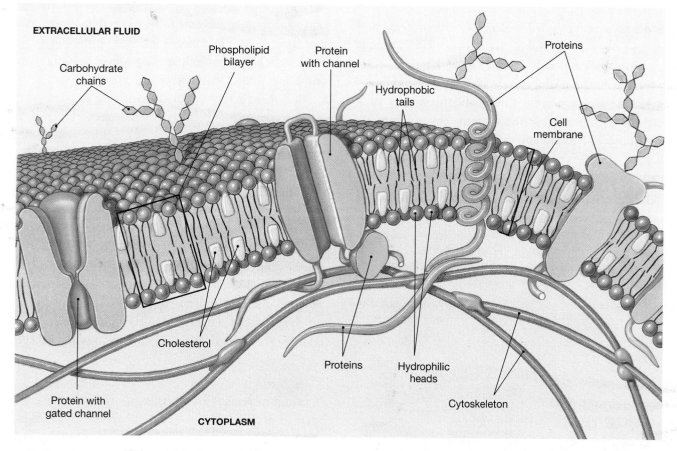

● **FIGURE 3–3**   The Cell Membrane.

TABLE 3-2 *Types of Membrane Proteins*

CLASS	FUNCTION	EXAMPLE
RECEPTOR PROTEINS	Sensitive to specific extracellular materials that bind to them and trigger a change in a cell's activity.	Binding of the hormone insulin to membrane receptors increases the rate of glucose absorption by the cell.
CHANNEL PROTEINS	Central pore, or channel, permits water, ions, and other solutes to bypass lipid portion of cell membrane.	Calcium ion movement through channels is crucial to muscle contraction and the conduction of nerve impulses.
CARRIER PROTEINS	Bind and transport solutes across the cell membrane. This process may or may not require energy.	Carrier proteins bring glucose into the cytoplasm and also transport sodium, potassium, and calcium ions into and out of the cell.
ENZYMES	Catalyze reactions in the extracellular fluid or within the cell.	Dipeptides are broken down into amino acids by enzymes on the exposed membranes of cells that line the intestinal tract.
ANCHORING PROTEINS	Attach the cell membrane to other structures and stabilize its position.	Inside the cell, anchor proteins bind to the cytoskeleton (network of supporting filaments). Outside the cell, anchor proteins attach the cell to extracellular protein fibers or to another cell.
RECOGNITION (IDENTIFIER) PROTEINS	Identify a cell as self or nonself, normal or abnormal, to the immune system.	One group of such recognition proteins is the major histocompatibility complex (MHC) discussed in Chapter 14.

Membrane Carbohydrates

Carbohydrates form complex molecules with proteins and lipids on the outer surface of the membrane. The carbohydrate portions of molecules, such as *glycoproteins* and *glycolipids*, function as cell lubricants and adhesives, act as receptors for extracellular compounds, and form part of a recognition system that keeps the immune system from attacking the body's own cells and tissues.

CONCEPT CHECK QUESTIONS

1. Which component of the cell membrane is primarily responsible for the membrane's ability to form a physical barrier between the cell's internal and external environments?
2. Which type of membrane protein allows water and small ions to pass through the cell membrane?

Answers begin on p. 792.

Membrane Transport

The **permeability** of the cell membrane is the property that determines precisely which substances can enter or leave the cytoplasm. If nothing can cross a membrane, it is described as *impermeable*. If any substance can cross without difficulty, the membrane is *freely permeable*. Cell membranes are *selectively permeable*, which permits the free passage of some materials and restricts the passage of others. Whether or not a substance can cross the cell membrane is based on the substance's size, electrical charge, molecular shape, lipid solubility, or some combination of these factors.

Movement across the membrane may be passive or active. **Passive processes** move ions or molecules across the cell membrane without any energy expenditure by the cell. **Active processes** require that the cell expend energy, usually in the form of adenosine triphosphate (ATP).

In the sections that follow, we will first consider two types of passive processes: (1) *diffusion*, which includes a special type of diffusion called *osmosis*; and (2) *filtration*. Then we will examine some types of *carrier-mediated transport*, which includes both active and passive processes. We will consider *facilitated diffusion*, which is a passive carrier-mediated process, and *active transport*, which is an active carrier-mediated process. Finally, we will examine two active processes that involve *vesicular transport*: endocytosis and exocytosis.

Diffusion

Ions and molecules are in constant motion; they collide and bounce off one another and off obstacles in their paths. **Diffusion** is the net movement of molecules from an area of relatively high concentration (of many collisions) to an area of relatively low concentration (of fewer collisions). The difference between the high and low concentrations represents a **concentration gradient,** and diffusion is often described as proceeding "down a concentration gradient" or "downhill." As a result of the process of diffusion, molecules eventually become uniformly distributed, and concentration gradients are eliminated.

All of us have experienced the effects of diffusion, which occurs in air as well as in water. The smell of fresh flowers in a vase sweetens the air in a large room, just as a drop of ink spreads to color an entire glass of water. Both examples begin with an extremely high concentration of molecules in a very localized area. Consider a colored sugar cube dropped into a beaker of water (Figure 3–4●). As the cube dissolves, its sugar and dye molecules establish a steep concentration gradient with the surrounding clear water. Eventually, dissolved molecules of both types spread through the water until they are distributed evenly.

Diffusion is important in body fluids because it tends to eliminate local concentration gradients. For example, every cell in your body generates carbon dioxide, and its intracellular (within the cell) concentration is relatively high. Carbon dioxide concentrations are lower in the surrounding extracellular fluid, and lower still in the circulating blood. Because cell membranes are freely permeable to carbon dioxide, it can diffuse down its concentration gradient—and disperse from the cell's interior into the extracellular fluid, and from the extracellular fluid into the bloodstream, for delivery to and elimination from the lungs.

DIFFUSION ACROSS CELL MEMBRANES. In extracellular fluids of the body, water and dissolved solutes diffuse freely. A cell membrane, however, acts as a barrier that selectively restricts diffusion. Some substances can pass through easily, whereas others cannot penetrate the membrane at all. An ion or molecule can independently diffuse across a cell membrane in one of two ways: (1) by moving across the lipid portion of the membrane; or (2) by passing through a channel protein in the membrane. Therefore, the primary factors that determine whether a substance can diffuse across a cell membrane are its lipid solubility and its size relative to the sizes of membrane channels (Figure 3–5●).

Alcohol, fatty acids, and steroids can enter cells easily because they can diffuse through the lipid portions of the membrane. Dissolved gases such as oxygen and carbon dioxide also enter and leave cells by diffusing through the phospholipid bilayer.

Ions and most water-soluble compounds are not lipid-soluble, so they must pass through membrane channels to enter the cytoplasm. These channels are very small, about 0.8 nm in diameter. Water molecules can enter or exit freely, as can ions such as sodium and potassium; but even a small organic molecule, such as glucose, is too big to fit through the channels.

OSMOSIS: A SPECIAL TYPE OF DIFFUSION. The diffusion of water across a membrane is called **osmosis** (oz-MŌ-sis; *osmos,* thrust). Both intracellular and extracellular fluids are solutions that contain a variety of dissolved materials, or **solutes.** Each solute tends to diffuse as if it were the only substance in solution. Thus, changes in the concentration of potassium ions, for example, have no effect on the rate or direction of sodium ion diffusion. Some ions and molecules

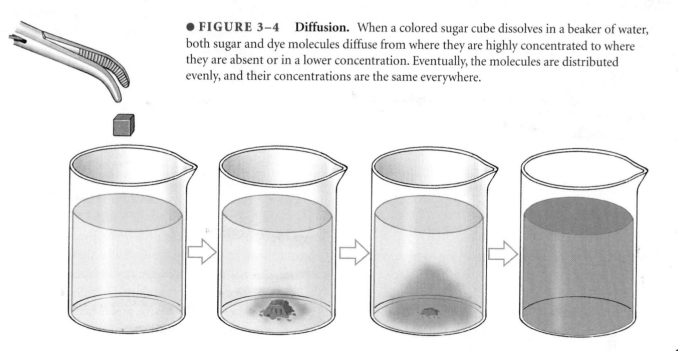

●**FIGURE 3–4** **Diffusion.** When a colored sugar cube dissolves in a beaker of water, both sugar and dye molecules diffuse from where they are highly concentrated to where they are absent or in a lower concentration. Eventually, the molecules are distributed evenly, and their concentrations are the same everywhere.

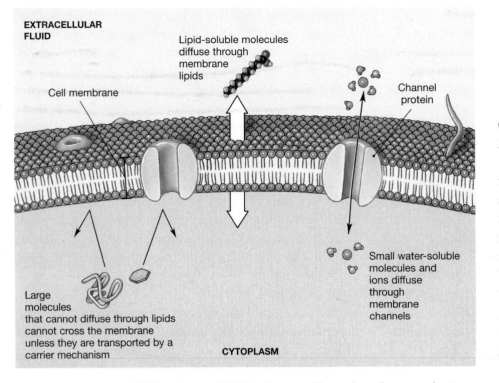

EXTRACELLULAR
FLUID

Lipid-soluble molecules
diffuse through
membrane
lipids

Cell membrane

Channel
protein

Large
molecules
that cannot diffuse through lipids
cannot cross the membrane
unless they are transported by a
carrier mechanism

Small water-soluble
molecules and
ions diffuse
through
membrane
channels

CYTOPLASM

● **FIGURE 3–5**   **Diffusion Across Cell Membranes.** The path a substance takes in crossing a cell membrane depends on the substance's size and lipid solubility.

(solutes) diffuse into the cytoplasm; others diffuse out; and a few, such as proteins, are unable to diffuse across a cell membrane. But if we ignore the individual identities and simply count ions and molecules, we find that the total concentration of ions and molecules on the inside of the cell membrane equals the total on the outside.

This state of solute equilibrium persists because *the cell membrane is freely permeable to water.* Whenever a solute concentration gradient exists across a cell membrane, a concentration gradient for water exists also. Because the dissolved solute molecules occupy space that would otherwise be taken up by water molecules, the higher the solute concentration, the lower the water concentration. As a result, *water molecules tend to flow across a membrane toward the solution that contains the higher solute concentration,* because this movement is down the concentration gradient for water molecules. Water movement continues until water concentrations—and, thus, total solute concentrations—are the same on either side of the membrane.

Three characteristics of osmosis are important to remember:

1. Osmosis is the diffusion of water molecules across a membrane.
2. Osmosis occurs across a selectively permeable membrane that is freely permeable to water but is not freely permeable to solutes.

3. In osmosis, water flows across a membrane toward the solution that has the higher concentration of solutes, because that is where the concentration of water is lower.

OSMOSIS AND OSMOTIC PRESSURE. Figure 3–6● diagrams the process of osmosis. Step 1 shows two solutions (A and B), with different solute concentrations, separated by a selectively permeable membrane. As osmosis occurs, water molecules cross the membrane until the solute concentrations in the two solutions are identical (Step 2a). Thus, the volume of solution B increases at the expense of that of solution A. The greater the initial difference in solute concentrations, the stronger the osmotic flow. The **osmotic pressure** of a solution is an indication of the force of water movement *into that solution* as a result of solute concentration. As the solute concentration of a solution increases, so does its osmotic pressure. Osmotic pressure can be measured in several ways. For example, a strong enough opposing pressure can prevent the entry of water molecules. Pushing against a fluid generates *hydrostatic pressure*. In Step 2b, hydrostatic pressure opposes the osmotic pressure of solution B, so no net osmotic flow occurs.

Solutions of various solute concentrations are described as *isotonic, hypotonic,* or *hypertonic* with regard to their effects on the shape or tension of the plasma membrane of living cells. Although the effects of various osmotic solutions are difficult to see in most tissues, they are readily observable in red blood cells.

Figure 3–7a● shows the appearance of red blood cells in an isotonic solution. An **isotonic** (*iso-,* equal + *tonos,* tension) solution is one that does not cause a net movement of water into or out of the cell. In other words, an equilibrium exists, and as one water molecule moves out of the cell, another moves in to replace it.

When a red blood cell is placed in a **hypotonic** (*hypo-,* below) solution, water will flow into the cell, which causes it to swell up like a balloon (Figure 3–7b●). Eventually the cell may burst, or *lyse.* In the case of red blood cells, this event is known as **hemolysis** (*hemo-,* blood + *lysis,* breakdown). Red blood cells in a **hypertonic** (*hyper-,* above) solution will lose water by osmosis. As they do, they shrivel and dehydrate. The shrinking of red blood cells is called **crenation** (Figure 3–7c●).

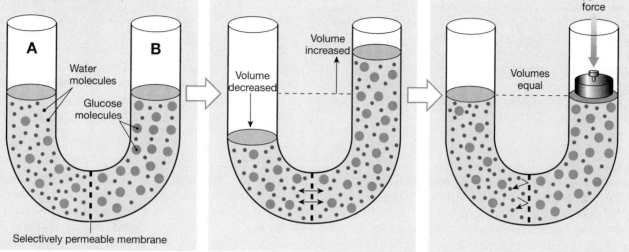

Two solutions containing different solute concentrations are separated by a selectively permeable membrane. Water molecules (small blue dots) begin to cross the membrane toward solution B, the solution with the higher concentration of solutes (larger pink circles).

At equilibrium, the solute concentrations on the two sides of the membrane are equal. The volume of solution B has increased at the expense of that of solution A.

Osmosis can be prevented by resisting the change in volume. The osmotic pressure of solution B is equal to the amount of hydrostatic pressure required to stop the osmotic flow.

A
B
Water molecules
Glucose molecules
Selectively permeable membrane

Volume decreased
Volume increased

Applied force
Volumes equal

● **FIGURE 3–6** **Osmosis.** The osmotic flow of water can create osmotic pressure across a membrane. The osmotic pressure of solution B is equal to the amount of hydrostatic pressure required to stop the osmotic flow.

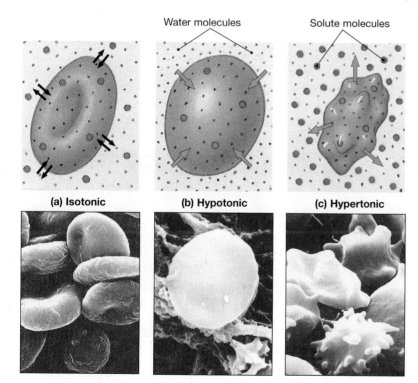

Water molecules
Solute molecules

(a) Isotonic
(b) Hypotonic
(c) Hypertonic

● **FIGURE 3–7** **Osmotic Flow Across a Cell Membrane.** Blue arrows indicate the direction of net osmotic water movement. **(a)** Because these red blood cells are immersed in an isotonic saline solution, no osmotic flow occurs and the cells have their normal appearance. **(b)** Immersion in a hypotonic saline solution results in the osmotic flow of water into the cells. The swelling may continue until the cell membrane ruptures. **(c)** Exposure to a hypertonic solution results in the movement of water out of the cells. The red blood cells shrivel and become crenated. (SEMs × 833)

It is often necessary to give patients large volumes of fluid after severe blood loss or dehydration. One commonly administered fluid is a 0.9 percent (0.9g/dL) solution of sodium chloride (NaCl). This solution, which approximates the normal *osmotic concentration* (total solute concentration) of extracellular fluids, is called *normal saline.* It is used because sodium and chloride are the most abundant ions in the body's extracellular fluid. Because little net movement of either type of ion occurs across cell membranes, normal saline is essentially isotonic with respect to body cells.

Prehospital personnel must have a good understanding of human cellular function. Many of the fluids and medications used in emergency medicine directly affect the cells. A major component of emergency medical care is monitoring and replacing the various fluids and electrolytes of the body. A decrease in the absolute volume of body fluids is referred to as *dehydration*. An increase in fluid volume is called *overhydration* and can result in edema and heart failure. In addition to fluids, the concentration of essential electrolytes can be disturbed in injury and illness. Optimal body function depends on a normal concentration of body fluids and electrolytes. Because of this, the fluid and electrolyte balance of the body must be maintained within fairly narrow limits. In critical situations, such as shock, the rapid replacement of lost fluids can be life-saving.

Water moves through the various membranes of the body by the process of osmosis (Figure 3–8●). The rate at which the fluid moves depends upon the differences in the concentration of solutes in the fluid (also called *tonicity*). The total number of particles in a solution is measured in terms of *osmoles*. In general, the osmole is too large a unit for expressing osmotic activity of solutes in the body fluids. Instead, the term *milliOsmole (mOsm)*, which equals 1/1000 osmole, is commonly used. Normal body tonicity is approximately 280–310 milliOsmoles/liter. The terms *osmolality* and *osmolarity* are often used to describe the number of solute particles in a solution. *Osmolality* is the number of osmoles per kilogram of water. Osmolarity is the number of osmoles per liter of solution. In dilute solutions, such as body fluids, these two terms can be used almost synonymously because the differences are small. In practice, osmolarity is used more frequently than *osmolality*.

Sodium, the most abundant ion in the extracellular fluid, is responsible for the osmotic balance of the extracellular fluid compartment. Water follows sodium into the extracellular fluid. Potassium is the most abundant ion in the intracellular fluid compartment. Generally, the osmolarity of intracellular fluid does not change very rapidly. However, when there is a change in the osmolarity of extracellular fluid, water will move from the intracellular to the extracellular compartment, or vice versa, until osmotic equilibrium is regained.

Within the extracellular compartment, movement of water between the plasma in the intravascular space and fluid in the interstitial space is primarily a function of forces at play in the capillary beds. In general, the movement of water and solutes across a cell membrane is governed by *osmotic pressure*. Osmotic pressure is the pressure exerted by the concentration of solutes on one side of a semipermeable membrane, such as a cell membrane or the thin wall of a capillary. Osmotic pressure can be thought of as a "pull" rather than a "push" because a hypertonic concentration of solutes tends to pull water from the other side of the membrane.

Generally, there is a "two-way street" as solutes move out of a space while water moves into the space to balance the concentration of solutes on both sides of the membrane. However, a somewhat different osmotic mechanism operates between the plasma inside a capillary and the fluid in the interstitial space outside of the capillary. Blood plasma generates *oncotic force*, which is sometimes called *colloid osmotic pressure*. Plasma proteins are colloids, which are large particles that do not readily move across the capillary membrane. The most abundant of these is the plasma protein *albumin*. Because of their size, colloids tend to remain within the capillary. At the same time, there is usually very little water in the interstitial space. The small amount of water that does get into the interstitial space is usually taken up by the lymphatic system. Therefore, because there is little water outside the capillary and because plasma proteins do not readily move outside of the capillary, the forces governing water's movement between the capillary and the interstitial space are almost all on one side, governed by the plasma on the inside of the capillary.

Another force inside the capillaries is *hydrostatic pressure*, which is the blood pressure, or force against the vessel walls, created by contractions of the heart. Hydrostatic pressure tends to force some water out of the plasma and across the capillary wall into the interstitial space, a process that is called *filtration*. Hydrostatic pressure (a force that favors filtration, and pushes water out of the capillary) and *oncotic force* (a force that opposes filtration, and pulls water into the capillary) together are responsible for *net filtration*, which is described in *Starling's hypothesis*:

Net filtration = (forces that favor filtration) – (forces that oppose filtration)

Net filtration in a capillary is normally zero. It works this way: as plasma enters the capillary at the arterial end, hydrostatic pressure forces water to cross the capillary membrane into the interstitial space. This loss of water increases the relative concentration of plasma proteins and hence the colloid osmotic pressure. By the time the plasma reaches the venous end of the capillary, the oncotic force exerted by the increased concentration of plasma proteins is great enough to pull the water from the interstitial space back into the capillary. The outcome is that water is retained in the intravascular space and does not remain in the interstitial space.

In prehospital emergency care, the exact concentration of fluid and electrolytes is not known. Because of this, intravenous (IV) replacement fluids whose fluid and electrolyte concentrations are similar to those of the body are used initially. Later, when the patient's fluid and electrolyte status is known, IV fluids that will correct any detected underlying fluid and electrolyte problems are administered. ■

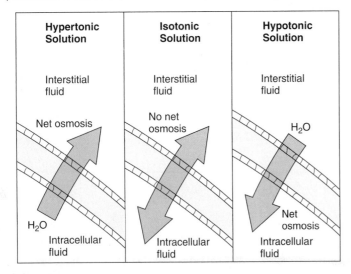

● **FIGURE 3–8** **Osmosis.** Fluid movement between body fluid compartments depends upon the difference in solute concentrations. During osmosis water moves from an area of low solute concentration to an area of high solute concentration.

Key Note

Things tend to even out, unless something—like a cell membrane—prevents this from happening. In the absence of a cell membrane, or across a freely-permeable membrane, diffusion will quickly eliminate concentration gradients. Osmosis will attempt to eliminate concentration gradients across membranes that are permeable to water, but not to the solutes involved.

Filtration

In the passive process called **filtration,** hydrostatic pressure forces water across a membrane. If solute molecules are small enough to fit through membrane pores, they will be carried along with the water. In the body, the heart pushes blood through the circulatory system and generates hydrostatic pressure, or *blood pressure.* Filtration occurs across the walls of small blood vessels, and pushes water and dissolved nutrients into the tissues of the body. Filtration across specialized blood vessels in the kidneys is an essential step in the production of urine.

Carrier-Mediated Transport

In **carrier-mediated transport,** membrane proteins bind specific ions or organic substrates and carry them across the cell membrane. These proteins share several characteristics with enzymes. They may be used repeatedly and they can only bind to specific substrates; the carrier protein that transports glucose, for example, will not carry other simple sugars.

Carrier-mediated transport can be passive (no ATP required) or active (ATP-dependent). In *passive transport,* solutes are typically carried from an area of high concentration to an area of low concentration. *Active transport* mechanisms may follow or oppose an existing concentration gradient.

Many carrier proteins transport one ion or molecule at a time, but some move two solutes simultaneously. In *cotransport,* the carrier transports the two substances in the same direction, either into or out of the cell. In *countertransport,* one substance is moved into the cell while the other is moved out. Two major examples of carrier-mediated transport—*facilitated diffusion* and *active transport*—are discussed next.

FACILITATED DIFFUSION. Many essential nutrients, including glucose and amino acids, are insoluble in lipids and too large to fit through membrane channels. However, these compounds can be passively transported across the membrane by carrier proteins in a process called **facilitated diffusion** (Figure 3–9●). First, the molecule to be transported binds to a **receptor site** on the carrier protein. Then the shape of the protein changes, and moves the molecule to the inside of the cell membrane, where it is released into the cytoplasm.

As in the case of simple diffusion, no ATP is expended in facilitated diffusion, and the molecules move from an area of higher concentration to one of lower concentration. Facilitated diffusion differs from simple diffusion, however, in that the rate of transport cannot increase indefinitely; only a limited number of carrier proteins are available in the membrane. Once all of them are operating, any further increase in the solute concentration in the extracellular fluid will have no effect on the rate of movement into the cell.

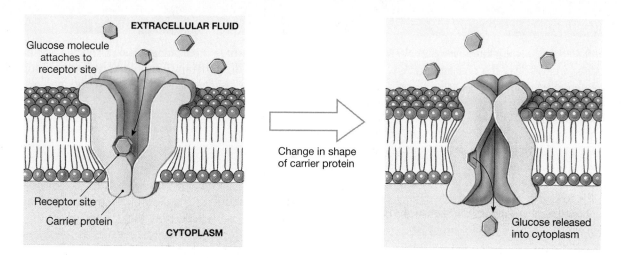

●**FIGURE 3–9   Facilitated Diffusion.** In this process, an extracellular molecule, such as glucose, binds to a receptor site on a carrier protein. The binding alters the shape of the protein, which then releases the molecule to diffuse into the cytoplasm.

ACTIVE TRANSPORT. In **active transport,** the high-energy bond in ATP provides the energy needed to move ions or molecules across the membrane. Despite the energy cost, active transport offers one great advantage: it is not dependent on a concentration gradient. This means that the cell can import or export specific materials *regardless of their intracellular or extracellular concentrations.*

All cells contain carrier proteins called **ion pumps** that actively transport the cations sodium (Na^+), potassium (K^+), calcium (Ca^{2+}), and magnesium (Mg^{2+}) across cell membranes. Specialized cells can transport other ions, including iodide (I^-), chloride (Cl^-), and iron (Fe^{2+}). Many of these carrier proteins move a specific cation or anion in one direction only, either into or out of the cell. In a few cases, one carrier protein can move more than one ion at a time. If one kind of ion moves in one direction and the other moves in the opposite direction (countertransport), the carrier protein is called an **exchange pump.**

A major function of exchange pumps is to maintain cell homeostasis. Sodium and potassium ions are the principal cations in body fluids. Sodium ion concentrations are high in the extracellular fluids but low in the cytoplasm. The distribution of potassium in the body is just the opposite—low in the extracellular fluids and high in the cytoplasm. Because of the presence of channel proteins in the membrane that are always open (so-called *leak channels*), sodium ions slowly diffuse into the cell, and potassium ions diffuse out.

Homeostasis within the cell depends on maintaining sodium and potassium ion concentration gradients with the extracellular fluid. The **sodium-potassium exchange pump**

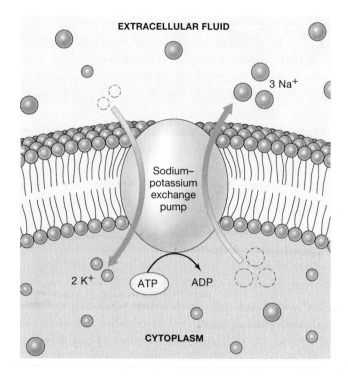

● **FIGURE 3–10**   **The Sodium-Potassium Exchange Pump.** The operation of the sodium-potassium exchange pump is an example of active transport because its operation requires the release of energy through the conversion of ATP to ADP.

maintains these gradients by ejecting sodium ions and recapturing lost potassium ions. For each ATP molecule consumed, three sodium ions are ejected and two potassium ions are reclaimed by the cell (Figure 3–10●). The energy demands are impressive: the sodium-potassium exchange pump may use up to 40 percent of the ATP produced by a resting cell.

Clinical Note
IV FLUID THERAPY

IV fluid therapy is the introduction of fluids and other substances into the venous side of the circulatory system. An important part of emergency care, it is used to replace blood lost through hemorrhage, to replace electrolytes or fluids, and to introduce medications directly into the vascular system.

Numerous types of intravenous fluids are available, each designed to treat a particular type of fluid and electrolyte imbalance. The two primary reasons for starting an IV in the emergency setting are: (1) fluid volume replacement, and (2) intravenous access for drug administration. The following is a discussion of common intravenous fluids and their physiological actions after they enter the body.

IV Fluids

IV fluids are chemically prepared, sterile solutions tailored to the body's specific needs. They replace the body's lost fluids or aid in

the delivery of IV medications. They also keep a vein patent when no fluid or drug therapy is required. IV fluids are supplied in four different forms: crystalloids, colloids, blood, and oxygen-carrying fluids.

CRYSTALLOIDS

Crystalloids, the most commonly used IV fluid type in emergency medicine, contain water and electrolytes. They are classified based upon their tonicity and their relation to the tonicity of the body. Table 3–3 illustrates the electrolyte concentration and tonicity of common IV crystalloids. Note that the tonicity of the fluids is measured in terms of osmolarity (reflected as milliOsmoles per liter). Classes of crystalloids include:

■ *Isotonic solutions.* Isotonic solutions have a tonicity equal to that of blood plasma. In a normally hydrated patient, they will not cause any significant fluid or electrolyte shift.

TABLE 3-3 *Contents and Characteristics of Common Emergency Intravenous Fluids*

	APPROXIMATE IONIC CONCENTRATIONS (mEq/L) AND CALORIES PER LITER							
	IONIC CONCENTRATIONS (mEq/L)							
	SODIUM	POTASSIUM	CALCIUM	CHLORIDE	LACTATE	CALORIES PER LITER	OSMOLARITY[a] (mOsm/L)	pH RANGE[b]
Dextrose Injection, USP	0	0	0	0	0	170	252	3.5–6.5
10% Dextrose Injection, USP	0	0	0	0	0	340	505	3.5–6.5
0.9% Sodium Chloride Injection, USP	154	0	0	154	0	0	308	4.5–7.0
Sodium Lactate Injection, USP (M/6 Sodium Lactate)	167	0	0	0	167	54	334	6.0–7.3
2.5% Dextrose & 0.45% Sodium Chloride Injection, USP	77	0	0	77	0	85	280	3.5–6.0
5% Dextrose & 0.2% Sodium Chloride Injection, USP	34	0	0	34	0	170	321	3.5–6.0
5% Dextrose & 0.33% Sodium Chloride Injection, USP	56	0	0	56	0	170	365	3.5–6.0
5% Dextrose & 0.45% Sodium Chloride Injection, USP	77	0	0	77	0	170	406	3.5–6.0
5% Dextrose & 0.9% Sodium Chloride Injection, USP	154	0	0	154	0	170	560	3.5–6.0
10% Dextrose & 0.9% Sodium Chloride Injection, USP	154	0	0	154	0	340	813	3.5–6.0
Ringer's Injection, USP	147.5	4	4.5	156	0	0	309	5.0–7.5
Lactated Ringer's Injection	130	4	3	109	28	9	273	6.0–7.5
5% Dextrose in Ringer's Injection	147.5	4	4.5	156	0	170	561	3.5–6.5
Lactated Ringer's with 5% Dextrose	130	4	3	109	28	180	525	4.0–6.5

[a] Normal physiological isotonicity range is approximately 280–310 mOsm/L. Administration of substantially hypotonic solutions may cause hemolysis, and administration of substantially hypertonic solutions may cause vein damage.
[b] pH ranges are USP for applicable solution, corporate specification for non-USP solutions.

■ *Hypertonic solutions.* Hypertonic solutions have a higher solute concentration than do the body's cells. When administered to a normally hydrated patient, they cause a fluid shift from the intracellular compartment into the extracellular compartment. Later, solutes will diffuse in the opposite direction.

■ *Hypotonic solutions.* Hypotonic solutions have a lower solute concentration than do the body's cells. When administered to a normally hydrated patient, they cause a fluid shift from the extracellular compartment into the intracellular compartment. Later, solutes will diffuse in the opposite direction.

(continued next page)

Clinical Note—*continued*
IV FLUID THERAPY

The type of IV fluid used depends upon the patient's needs. In the prehospital setting, isotonic fluids are usually used, because the patient's underlying electrolyte and hydration status is unknown (Figure 3–11●). However, once the patient arrives in the emergency department, blood electrolyte studies are used to guide fluid selection and administration. The IV fluids most frequently used in prehospital care include:

- *Lactated Ringer's.* Lactated Ringer's solution is an isotonic electrolyte solution that contains sodium chloride, potassium chloride, calcium chloride, and sodium lactate in water.
- *Normal saline.* Normal saline is an isotonic electrolyte solution that contains sodium chloride in water.

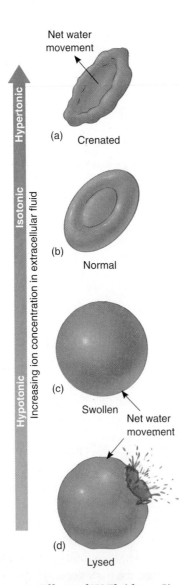

(a)   Crenated

(b)   Normal

(c)   Swollen

(d)   Lysed

Net water movement

Net water movement

Hypertonic / Isotonic / Hypotonic

Increasing ion concentration in extracellular fluid

● **FIGURE 3–11**   **Effects of IV Fluids on Circulating Red Blood Cells (Erythrocytes).** Hypertonic fluids cause cell shrinkage (crenation), while hypotonic fluids cause intracellular swelling that eventually leads to cell lysis.

- *5% dextrose in water (D_5W).* D_5W is a hypotonic glucose solution used to keep a vein patent and to dilute concentrated medications. While D_5W initially increases intravascular volume, glucose molecules rapidly diffuse across the vascular membrane and increase the amount of free water.

Both lactated Ringer's and normal saline are used for fluid replacement because of their immediate ability to expand the circulating fluid volume. However, due to the movement of electrolytes and water, two-thirds of either solution will be lost to the extravascular space within one hour (Figure 3–12●).

Dextrose-Containing Solutions
Several IV fluids contain dextrose (d-glucose) in varying concentrations. The most commonly used of these are 50% dextrose ($D_{50}W$) and 5% dextrose in water (D_5W). Other dextrose solutions include 25% dextrose ($D_{25}W$) and 10% dextrose and water ($D_{10}W$). Some dextrose-containing solutions will also contain electrolytes, usually sodium and chloride. Examples of these include:

- 2.5% dextrose and 0.45% sodium chloride ($D_{2.5}NS$)
- 5% dextrose and 0.20% sodium chloride (D_5NS)
- 5% dextrose and 0.33% sodium chloride (D_5NS)
- 5% dextrose and 0.45% sodium chloride (D_5NS)
- 5% dextrose and 0.9% sodium chloride (D_5NS)
- 10% dextrose and 0.9% sodium chloride ($D_{10}NS$)
- 5% dextrose in lactated Ringer's (D_5LR)

The high-concentration solutions, such as $D_{50}W$ and $D_{25}W$, are used for glucose replacement in documented hypoglycemia. $D_{10}W$ is used in patients such as chronic alcoholics who require calorie replacement in addition to water and electrolytes. D_5W and similar solutions are usually used for diluting intravenous medications and for conditions where an IV is started at a "to keep open" (TKO) or "keep vein open" (KVO) rate.

The solubility of dextrose is 1 gram per milliliter of water. Based on this property, dextrose-containing solutions are usually measured in weight-in-volume percentages. This system of measurement indicates the number of grams of dextrose in 100 mL of solution (water). A fully

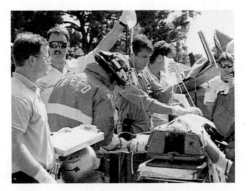

● **FIGURE 3–12**   **Emergency Fluid Resuscitation.** In the prehospital setting, it is best to administer just enough fluids to maintain the systolic pressure between 70–80 mmHg.

saturated solution contains 100 grams of dextrose per 100 mL of water and is considered a 100% solution ($D_{100}W$). $D_{50}W$, commonly used in prehospital care, contains 25 grams of glucose in 50 mL of water. Likewise, $D_{25}W$ contains 12.5 grams of glucose in 50 mL of water. $D_{25}W$ is preferred over $D_{50}W$ when administering intravenous dextrose to infants and children. $D_{25}W$ can be prepared by diluting 25 mL of $D_{50}W$ with 25 mL of sterile water for injection. This results in a solution that contains 12.5 grams of dextrose in 50 mL of water.

When considering glucose content, it is best to consider the amount of dextrose per mL of water. With a 50% dextrose solution, 1 mL of solution contains 0.5 grams of dextrose. Likewise, in a 25% dextrose solution, 1 mL of solution contains 0.25 grams of dextrose. A 100% dextrose solution would contain 1 gram/mL. A solution of 5% dextrose in water contains 0.05 grams of dextrose per mL.

Most intravenous fluids contain 5% dextrose in water or an electrolyte solution. When administered at a TKO or KVO rate, the amount of dextrose delivered to the patient is negligible. However, if IV solutions that contain 5% dextrose are used for volume replacement, then the amount of dextrose administered can be potentially dangerous for the patient. A 1000 mL bag of D_5W contains 50 grams of dextrose. This is equivalent to two prefilled syringes of $D_{50}W$. In multiple trauma, it is not uncommon to administer 2–3 liters of IV fluid in the prehospital setting. If D_5W were used, this would constitute a very high dextrose load. Infants and children are at increased risk of accidental overdose with dextrose. In documented hypoglycemia, the standard dose of dextrose is 0.5 grams per kilogram of body weight. A child who weighs 10 kilograms, for example, would receive 5 grams of dextrose for documented hypoglycemia (10 mL of $D_{50}W$ or 20 mL of $D_{25}W$). The amount of IV fluid administered to children who are dehydrated or in shock is 20 mL/kg of an isotonic crystalloid solution such as normal saline or lactated Ringer's. If D_5W were used by accident, 20 mL of fluid would provide 1 gram of dextrose per kilogram body weight. This is twice the recommended dose for documented hypoglycemia. Avoid using dextrose-containing solutions for any situation where volume replacement must be provided.

Colloids

Colloid solutions contain large proteins that cannot pass through the capillary membrane. Consequently, they remain in the intravascular compartment for a period of time. In addition, colloids have osmotic properties that attract water into the intravascular compartment. Thus, a small quantity of colloid can significantly increase intravascular volume. Common colloids include:

- *Plasma protein fraction (Plasmanate).* Plasmanate is a protein-containing colloid solution. Its principle protein, albumin, is suspended in a saline solvent.

- *Salt-poor albumin.* Salt-poor albumin contains only human albumin. Each gram of albumin will retain approximately 18 mL of water in the intravascular space.
- *Dextran.* Dextran is not a protein but a large sugar molecule with osmotic properties similar to albumins. It is supplied in two molecular weights: 40,000 and 70,000 Daltons. Dextran 40 has from two to two and a half times the colloidal osmotic pressure of albumin.
- *Hetastarch (Hespan).* Like Dextran, hetastarch is a sugar molecule with osmotic properties similar to those of proteins such as albumin.

Although colloids help to maintain intravascular volume, their use in the field is impractical. Their high cost, short shelf life, and specific storage requirements make them better suited to the hospital setting.

Blood

The most desirable fluid for replacement is whole blood. Unlike colloids and crystalloids, the hemoglobin available in blood carries oxygen. Blood, however, is a precious commodity and must be conserved so that it can be of benefit to the most people. Its use in the field is usually limited to aeromedical operations or mass-casualty incidents. O-negative blood does not contain any antigenic proteins and is thus considered the "universal donor" type.

OXYGEN-CARRYING SOLUTIONS

The problem with all IV fluids, except human packed red blood cells, is that they lack the capacity to carry oxygen. During times of stress, the cells of the body ultimately need oxygen and glucose.

Intense research is underway to develop a synthetic, or artificial, blood product that can carry oxygen. One of the first such solutions examined was the perfluorocarbons (PFCs). These are emulsions that carried oxygen and other gases. These gases are retained on the carrying molecule by surfactant; in this case, lecithin. However, initial research on the PFCs has been disappointing.

Many EMS systems and military operations now use another product, hemoglobin-based oxygen-carrying solutions (HBOCs). These contain actual hemoglobin that has been extracted from red blood cells, which is then washed, filtered, and chemically joined into a chain through a process called polymerization. Presently, there are two products on the market. The first, called Hemopure®, uses bovine (cow) blood as the hemoglobin source. It is being tested and used by the U.S. military, and has also been used in South Africa. It has a shelf life of nearly three years. The second product, PolyHeme®, uses expired human blood from blood banks as the hemoglobin source. PolyHeme® has been tested in EMS settings and found to be safe. Unlike standard IV fluids, HBOCs can provide the much-needed oxygen during emergencies. ■

Vesicular Transport

In **vesicular transport,** materials move into or out of the cell in vesicles, which are small membranous sacs that form at, or fuse with, the cell membrane. The two major categories of vesicular transport are *endocytosis* and *exocytosis*.

ENDOCYTOSIS.   The process called **endocytosis** (EN-dō-sīTŌ-sis; *endo-*, inside + *cyte*, cell) is the packaging of extracellular ma-

terials in a vesicle at the cell surface for import *into* the cell. Relatively large volumes of extracellular material may be involved. There are three major types of endocytosis: *receptor-mediated endocytosis*, *pinocytosis*, and *phagocytosis*. All three are active processes that require ATP or other sources of energy.

- **Receptor-mediated endocytosis** involves the formation of small vesicles at the membrane surface to import selected substances into the cell. This process produces vesicles that contain a specific target molecule in high concentrations. Receptor-mediated endocytosis begins when molecules in the extracellular fluid bind to receptors on the membrane surface (Figure 3–13●). The receptors bind to specific target molecules, called *ligands* (LĪ-gandz), such as a transport protein or hormone, and then cluster together on the membrane. The area of the membrane with the bound receptors forms a groove or pocket that pinches off to form a vesicle. Many important substances, such as cholesterol and iron ions (Fe^{2+}), are carried throughout the body attached to special transport proteins. These transport proteins are too large to pass through membrane channels but can enter cells through receptor-mediated endocytosis.

■ **Pinocytosis** (pi-nō-sī-TŌ-sis; *pinein*, to drink), or "cell drinking," is the formation of small vesicles filled with extracellular fluid. In this process, which is common to all cells, a deep groove or pocket forms in the cell membrane and then pinches off. Because no receptor proteins are involved, pinocytosis is not as selective a process as receptor-mediated endocytosis.

■ **Phagocytosis** (fag-ō-sī-TŌ-sis; *phagein*, to eat), or "cell eating," produces vesicles that contain solid objects that may be as large as the cell itself (Figure 3–14●). Cytoplasmic extensions called **pseudopodia** (soo-dō-PŌ-dē-ah; *pseudo-*, false + *podon*, foot) surround the object, and their membranes fuse to form a vesicle. This vesicle then fuses with many *lysosomes*, and the vesicle contents are broken down by the lysosomes' digestive enzymes. Most cells display pinocytosis, but phagocytosis is performed only by specialized cells that protect tissues by engulfing bacteria, cell debris, and other abnormal materials. Phagocytic cells will be considered in chapters that deal with blood cells (Chapter 11) and the immune response (Chapter 14).

■ The process called **exocytosis** (ek-sō-sī-TŌ-sis; *exo-*, outside) is the functional reverse of endocytosis. In exocytosis, a vesicle created inside the cell fuses with the cell membrane and discharges its contents into the extracellular environment. The ejected material may be a secretion such as a *hormone* (a compound that circulates in the blood and affects cells in other parts of the body), mucus, or waste products that remain from the recycling of damaged organelles (see 7 in Figure 3–14).

At any given moment, various transport mechanisms move materials into and out of the cell. These mechanisms are summarized in Table 3–4.

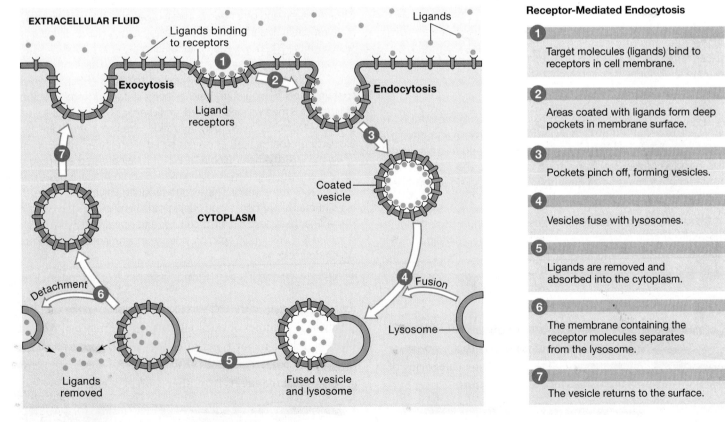

● **FIGURE 3–13   Receptor-Mediated Endocytosis.**

Receptor-Mediated Endocytosis

1. Target molecules (ligands) bind to receptors in cell membrane.

2. Areas coated with ligands form deep pockets in membrane surface.

3. Pockets pinch off, forming vesicles.

4. Vesicles fuse with lysosomes.

5. Ligands are removed and absorbed into the cytoplasm.

6. The membrane containing the receptor molecules separates from the lysosome.

7. The vesicle returns to the surface.

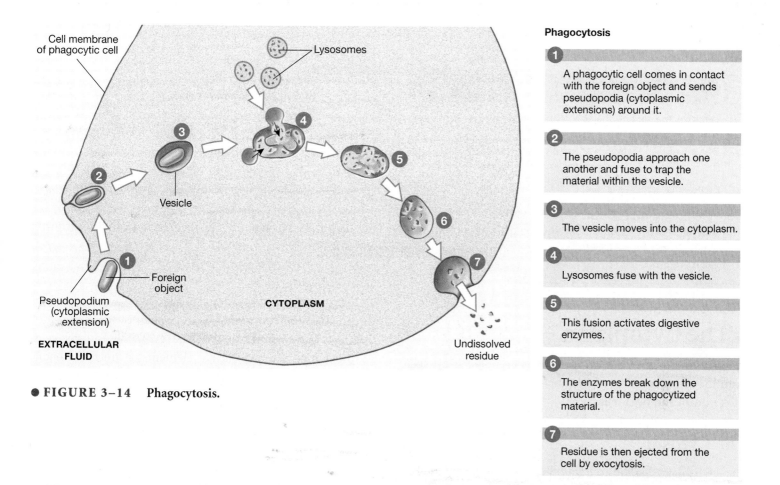

Cell membrane of phagocytic cell

Lysosomes

Vesicle

Foreign object

Pseudopodium (cytoplasmic extension)

EXTRACELLULAR FLUID

CYTOPLASM

Undissolved residue

● **FIGURE 3–14**   Phagocytosis.

Phagocytosis

1 A phagocytic cell comes in contact with the foreign object and sends pseudopodia (cytoplasmic extensions) around it.

2 The pseudopodia approach one another and fuse to trap the material within the vesicle.

3 The vesicle moves into the cytoplasm.

4 Lysosomes fuse with the vesicle.

5 This fusion activates digestive enzymes.

6 The enzymes break down the structure of the phagocytized material.

7 Residue is then ejected from the cell by exocytosis.

TABLE 3–4   *A Summary of the Mechanisms Involved in Movement Across Cell Membranes*

MECHANISM	PROCESS	FACTORS THAT AFFECT RATE OF MOVEMENT	SUBSTANCES INVOLVED
DIFFUSION	Molecular movement of solutes; direction determined by relative concentrations	Steepness of gradient, molecular size, electric charge, lipid solubility, temperature	Small inorganic ions, lipid-soluble materials (all cells)
Osmosis	Movement of water molecules toward solution that contains relatively higher solute concentration; requires selectively permeable membrane	Concentration gradient, opposing osmotic or hydrostatic pressure	Water only (all cells)
FILTRATION	Movement of water, usually with solute, by hydrostatic pressure; requires filtration membrane	Amount of pressure, size of pores in filtration membrane	Water and small ions (blood vessels)
CARRIER-MEDIATED TRANSPORT			
Facilitated diffusion	Carrier proteins passively transport solutes down a concentration gradient	Steepness of gradient, temperature, and availability of carrier proteins	Glucose and amino acids (all cells)
Active transport	Carrier proteins actively transport solutes regardless of any concentration gradients	Availability of carrier proteins, substrate, and ATP	Na^+, K^+, Ca^{2+}, Mg^{2+} (all cells); other solutes by specialized cells
VESICULAR TRANSPORT			
Endocytosis	Creation of membranous vesicles that contain fluid or solid material	Mechanism depends on substance being moved into cell; requires ATP	Fluids, nutrients (all cells); debris, pathogens (specialized cells)
Exocytosis	Fusion of vesicles that contain fluids and/or solids with the cell membrane	Mechanism depends on substance being carried; requires ATP	Fluids, debris (all cells)

→ CONCEPT CHECK QUESTIONS

1. What is the difference between active and passive transport processes?
2. During digestion in the stomach, the concentration of hydrogen (H⁺) ions rises to many times the concentration found in the cells lining the stomach. What type of transport process could produce this result?
3. When certain types of white blood cells encounter bacteria, they are able to engulf them and bring them into the cell. What is this process called?

Answers begin on p. 792.

■ The Cytoplasm

Cytoplasm is a general term for the material inside the cell, from the cell membrane to the nucleus. The cytoplasm contains cytosol and organelles.

The Cytosol

The **cytosol** is the intracellular fluid, which contains dissolved nutrients, ions, soluble and insoluble proteins, and waste products. It differs in composition from the extracellular fluid that surrounds most of the cells in the body in the following ways:

■ The cytosol contains a higher concentration of potassium ions and a lower sodium-ion concentration; whereas extracellular fluid contains a higher concentration of sodium ions and a lower potassium-ion concentration.

■ The cytosol contains a high concentration of dissolved proteins, many of them enzymes that regulate metabolic operations. These proteins give the cytosol a consistency that varies between that of thin maple syrup and almost-set gelatin.

■ The cytosol usually contains small quantities of carbohydrates and large reserves of amino acids and lipids. The carbohydrates are broken down to provide energy, and the amino acids are used to manufacture proteins. The lipids are used primarily as an energy source when carbohydrates are unavailable.

The cytosol may also contain insoluble materials known as **inclusions.** Examples include stored nutrients (such as glycogen granules in muscle and liver cells) and lipid droplets (in fat cells).

Organelles

Organelles (or-gan-ELZ; "little organs") are internal structures that perform specific functions essential to normal cell structure, maintenance, and metabolism (see Table 3–1). Membrane-enclosed organelles include the *nucleus, mitochondria, endoplasmic reticulum, Golgi apparatus, lysosomes,* and *peroxisomes.* The membrane isolates the organelle from the cytosol, so that the organelle can manufacture or store secretions, enzymes, or toxins that might otherwise damage the cell. The *cytoskeleton, microvilli, centrioles, cilia, flagella, ribosomes,* and *proteasomes* are nonmembranous organelles. Because they are not surrounded by membranes, their parts are in direct contact with the cytosol.

The Cytoskeleton

The **cytoskeleton** is an internal protein framework of various thread-like filaments and hollow tubules that gives the cytoplasm strength and flexibility (Figure 3–15●). In most cells, the most important cytoskeletal elements are microfilaments, intermediate filaments, and microtubules. Thick filaments are found only in muscle cells.

MICROFILAMENTS. The thinnest strands of the cytoskeleton are **microfilaments,** which are usually composed of the protein **actin.** In most cells, they form a dense layer just inside the cell membrane. Microfilaments attach the cell membrane to the underlying cytoplasm by forming connections with proteins of the cell membrane. In muscle cells, actin microfilaments interact with **thick filaments,** made of the protein **myosin,** to produce powerful contractions.

INTERMEDIATE FILAMENTS. These cytoskeletal filaments are intermediate in size between microfilaments and the thick filaments of muscle cells. Their protein composition varies among cell types. Intermediate filaments strengthen the cell and stabilize its position with respect to surrounding cells through specialized attachments to the cell membrane. Many cells contain specialized intermediate filaments with unique functions. For example, keratin fibers in the superficial layers of the skin are intermediate filaments that make these layers strong and able to resist stretching.

MICROTUBULES. All body cells contain **microtubules,** which are hollow tubes built from the globular protein **tubulin.** Microtubules form the primary components of the cytoskeleton, which gives the cell strength and rigidity, and anchors the positions of major organelles.

During cell division, microtubules form the *spindle apparatus,* which distributes the duplicated chromosomes to opposite ends of the dividing cell. This process will be considered in a later section.

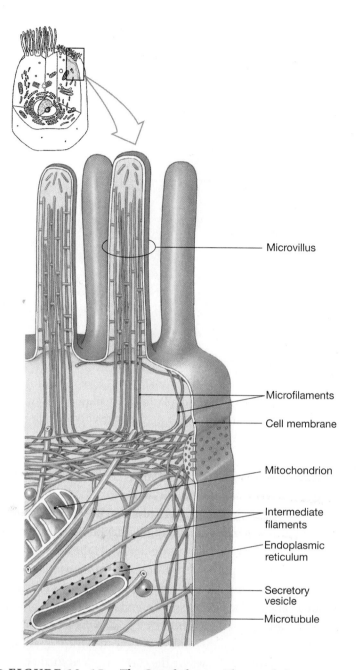

Centrioles, Cilia, and Flagella

In addition to functioning individually in the cytoskeleton, microtubules also interact to form more complex structures known as *centrioles, cilia,* and *flagella.*

CENTRIOLES. A **centriole** is a cylindrical structure composed of short microtubules (see Figure 3–2). All animal cells that are capable of dividing contain a pair of centrioles arranged perpendicular to each other. The centrioles produce the spindle fibers that move DNA strands during cell division. Mature red blood cells, skeletal muscle cells, cardiac muscle cells, and typical neurons lack centrioles; as a result, these cells cannot divide.

CILIA. Structures called **cilia** (SIL-ē-uh; singular *cilium*) are relatively long, slender extensions of the cell membrane (see Figure 3–2). They are supported internally by a cylindrical array of microtubules. Cilia undergo active movements that require energy from ATP. Their coordinated actions move fluids or secretions across the cell surface. For example, cilia that line the respiratory passageways beat in a synchronized manner to move sticky mucus and trapped dust particles toward the throat and away from delicate respiratory surfaces. If these cilia are damaged or immobilized by heavy smoking or a metabolic problem, the cleansing action is lost, and the irritants will no longer be removed. As a result, a chronic cough and respiratory infections develop.

FLAGELLA. Organelles called **flagella** (fla-JEL-uh; singular *flagellum,* whip) resemble cilia but are much longer. Flagella move a cell through the surrounding fluid, rather than moving the fluid past a stationary cell. Sperm cells are the only human cells that have a flagellum. If the flagella of sperm are paralyzed or otherwise abnormal, the individual will be sterile, because immobile sperm cannot perform fertilization.

Ribosomes

Ribosomes are organelles that manufacture proteins, using information provided by the DNA of the nucleus. Each ribosome consists of a small and large subunit composed of ribosomal RNA and protein. Ribosomes are found in all cells, but their number varies depending on the type of cell and its activities. For example, liver cells, which manufacture blood proteins, have many more ribosomes than do fat cells, which synthesize triglycerides.

There are two major types of ribosomes: free ribosomes and fixed ribosomes. **Free ribosomes** are scattered throughout the cytoplasm, and the proteins they manufacture enter the cytosol. **Fixed ribosomes** are attached to the *endoplasmic reticulum (ER),* a membranous organelle. Proteins manufactured by fixed ribosomes enter the endoplasmic reticulum, where they are modified and packaged for export.

— Microvillus

— Microfilaments

— Cell membrane

— Mitochondrion

— Intermediate filaments

— Endoplasmic reticulum

— Secretory vesicle

— Microtubule

● **FIGURE 13–15** **The Cytoskeleton.** The cytoskeleton provides strength and structural support for the cell and its organelles. Interactions between cytoskeletal components are also important in moving organelles and changing the shape of the cell.

Microvilli

Microvilli are small, finger-shaped projections of the cell membrane on the exposed surfaces of many cells (see Figure 3–2). An internal core of microfilaments supports the microvilli and connects them to the cytoskeleton (see Figure 3–15). Because they increase the surface area of the membrane, microvilli are common features of cells that are actively engaged in absorbing materials from the extracellular fluid, such as the cells of the digestive tract and kidneys.

Proteasomes

Whereas free ribosomes produce proteins within the cytoplasm, proteasomes remove them. **Proteasomes** are hollow, cylindrical organelles that contain an assortment of protein-breaking enzymes, or *proteases*. Proteasomes are responsible for removing and recycling damaged or denatured proteins and for breaking down abnormal proteins such as those produced within cells that are infected by viruses.

> ### → CONCEPT CHECK QUESTIONS
>
> 1. Cells that line the small intestine have numerous finger-like projections on their free surface. What are these structures, and what is their function?
> 2. How does the absence of centrioles affect a cell?
>
> *Answers begin on p. 792.*

The Endoplasmic Reticulum

The **endoplasmic reticulum** (en-dō-PLAZ-mik re-TIK-ū-lum; *reticulum*, a network), or **ER**, is a network of intracellular membranes connected to the membranous *nuclear envelope* that surrounds the nucleus (see Figure 3–2). The ER has four major functions:

1. *Synthesis.* Specialized regions of the ER manufacture proteins, carbohydrates, and lipids.
2. *Storage.* The ER can store synthesized molecules or materials absorbed from the cytosol without affecting other cellular operations.
3. *Transport.* Materials can be moved from place to place in the ER.
4. *Detoxification.* Drugs or toxins can be absorbed by the ER and neutralized by enzymes within it.

There are two types of endoplasmic reticulum, **smooth endoplasmic reticulum (SER)** and **rough endoplasmic reticulum (RER)** (Figure 3–16●). The term *smooth* refers to the fact that no ribosomes are associated with the SER. The SER is the site where lipids and carbohydrates are produced. The membranes of the RER contain fixed ribosomes, which gives the RER a beaded or rough appearance. The ribosomes participate in protein synthesis.

SER functions include: (1) the synthesis of the phospholipids and cholesterol needed for maintenance and growth of the cell membrane, ER, nuclear membrane, and Golgi apparatus in all cells; (2) the synthesis of steroid hormones, such as *testosterone* and *estrogen* (sex hormones) in cells of the reproductive organs; (3) the synthesis and storage of glycerides, especially triglyc-

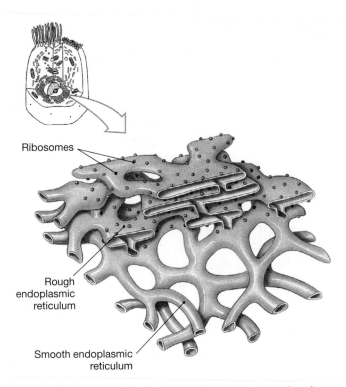

● FIGURE 3–16   The Endoplasmic Reticulum. This three-dimensional diagrammatic sketch shows the relationships between the rough and smooth endoplasmic reticula.

erides, in liver cells and fat cells; and (4) the synthesis and storage of glycogen in skeletal muscle and liver cells.

The **rough endoplasmic reticulum (RER)** functions as a combination workshop and shipping depot. The fixed ribosomes on its outer surface release newly-synthesized proteins into chamber-like spaces of the RER. Some proteins will remain in the RER and function as enzymes. Others are chemically modified and packaged into small membranous sacs that pinch off from the tips of the ER. The sacs, called *transport vesicles*, deliver the proteins to the Golgi apparatus, which is another membranous organelle, where they are processed further.

The amount of endoplasmic reticulum and the ratio of RER to SER vary with the type of cell and its activities. For example, pancreatic cells that manufacture digestive enzymes contain an extensive RER, and the SER is relatively small. The proportion is just the reverse in reproductive system cells that synthesize steroid hormones.

The Golgi Apparatus

The **Golgi** (GŌL-jē)**apparatus** consists of a set of five or six flattened, membranous discs. A single cell may contain several sets, each resembles a stack of dinner plates (see Figure 3–2). The major functions of the Golgi apparatus are: (1) the modification and packaging of secretions, such as hormones and enzymes; (2) the renewal or modification of the cell membrane; and (3) the packaging of special enzymes for use in the cytosol.

The various packaging roles of the Golgi apparatus are diagrammed in Figure 3–17a●. The synthesis of proteins and other substances occurs in the RER, and then transport vesicles move these products to the Golgi apparatus. Enzymes in the Golgi apparatus modify the newly-arrived molecules as other vesicles move them closer to the cell surface through succeeding membranes. Ultimately, the modified materials are repackaged in vesicles that leave the Golgi apparatus.

The Golgi apparatus creates three types of vesicles, each with a different fate. One type of vesicle, called *lysosomes*, contains digestive enzymes. These vesicles remain in the cytoplasm. A second type, **secretory vesicles,** contains secretions that will be discharged from the cell. Secretion occurs through exocytosis at the cell surface (Figure 3–17b●). A third type of vesicle, **membrane renewal vesicles,** fuses with the surface of the cell to add new lipids and proteins to the cell membrane. At the same time, other areas of the cell membrane are removed and recycled. Through such activities, the Golgi apparatus can change the properties of the cell membrane over time. For example, receptors can be added or removed, which makes the cell more or less sensitive to a particular stimulus.

Lysosomes

As just noted, **lysosomes** (LĪ-sō-sōmz; *lyso-*, breakdown + *soma*, body) are vesicles filled with digestive enzymes. Lysosomes perform cleanup and recycling functions within the cell. Their enzymes are activated when they fuse with the membranes of damaged organelles, such as mitochondria or fragments of the endoplasmic reticulum. Then the enzymes break down the lysosomal contents. Nutrients re-enter the cytosol through passive or active transport processes, and the remaining material is eliminated by exocytosis.

Lysosomes also function in defense against disease. Through endocytosis, immune system cells may engulf bacteria, fluids, and organic debris in their surroundings and isolate them within vesicles. ∞ p. 73 Lysosomes fuse with vesicles created in this way, and the digestive enzymes then break down the contents and release usable substances such as sugars or amino acids.

Lysosomes perform essential recycling functions inside the cell. For example, when muscle cells are inactive, lysosomes gradually break down their contractile proteins; if the cells become active once again, this destruction ends. However, in damaged or dead cells, lysosome membranes disintegrate, which releases active enzymes into the cytosol. These enzymes rapidly destroy the proteins and organelles of the cell, a process called **autolysis** (aw-TOL-i-sis; *auto-*, self). Because the breakdown of lysosomal membranes can destroy a cell, lysosomes have been called cellular "suicide packets." We do not know how lysosomal activities are controlled or why the enclosed enzymes do not digest the lysosomal membranes unless the cell is damaged.

Peroxisomes

Peroxisomes are smaller than lysosomes and carry a different group of enzymes. In contrast to lysosomes, which are produced at the Golgi apparatus, new peroxisomes arise from the growth

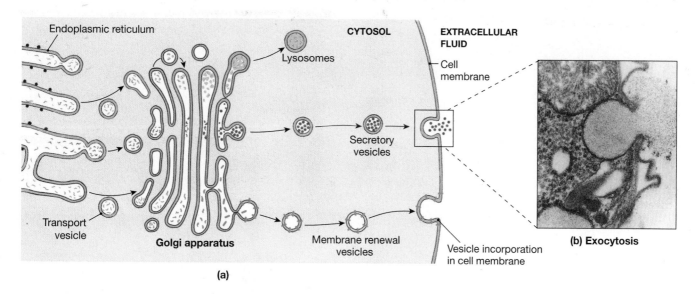

● **FIGURE 3–17   The Golgi Apparatus. (a)** Transport vesicles carry molecules manufactured in the endoplasmic reticulum to the Golgi apparatus, where these molecules are modified and transported through succeeding membranes toward the cell surface. At the membrane closest to the cell surface, three types of vesicles develop. Enzyme-filled lysosomes remain in the cytoplasm, secretory vesicles carry secretions from the Golgi apparatus to the cell surface, and membrane renewal vesicles add additional phospholipids and proteins to the cell membrane. **(b)** Exocytosis of secretions at the cell surface.

and subdivision of existing peroxisomes. Peroxisomes absorb and break down fatty acids and other organic compounds. As they do so, they generate hydrogen peroxide (H_2O_2), a potentially dangerous *free radical*. **Free radicals** are ions or molecules that contain unpaired electrons. They are highly reactive and enter additional reactions that can be destructive to vital compounds such as proteins. Other enzymes in peroxisomes break down hydrogen peroxide into oxygen and water, which protects the cell from the damaging effects of free radicals produced during catabolism. Peroxisomes are found in all cells, and are most abundant in metabolically active cells, such as liver cells.

Key Note

Cells respond directly to their environment and help maintain homeostasis at the cellular level. They can also change their internal structure and physiological functions over time.

Mitochondria

Mitochondria (mī-tō-KON-drē-uh; singular *mitochondrion; mitos,* thread + *chondrion,* granule) are small organelles that provide energy for the cell. The number of mitochondria in a particular cell varies with the cell's energy demands. Red blood cells, for example, have no mitochondria, whereas these organelles may account for 20 percent of the volume of an active liver cell.

Mitochondria are made up of an unusual double membrane. The outer membrane surrounds the entire organelle; the inner membrane contains numerous folds called *cristae* (Figure 3–18●). Cristae increase the surface area exposed to

the fluid contents, or *matrix,* of the mitochondria. Metabolic enzymes in the matrix catalyze energy-producing reactions.

Most of the chemical reactions that release energy occur in the mitochondria, but most of the cellular activities that require energy occur in the surrounding cytoplasm. Cells must, therefore, store energy in a form that can be moved from place to place. Energy is stored and transferred in the high-energy bond of ATP, as discussed in Chapter 2. ∞ p. 50 Living cells break the high-energy phosphate bond under controlled conditions, reconverting ATP to ADP and releasing energy for the cell's use.

MITOCHONDRIAL ENERGY PRODUCTION. Most cells generate ATP and other high-energy compounds through the breakdown of carbohydrates, especially glucose. Although most of the actual energy production occurs inside mitochondria, the first steps take place in the cytosol. In this reaction sequence, called *glycolysis,* six-carbon glucose molecules are broken down into three-carbon *pyruvic acid* molecules. These molecules are then absorbed by the mitochondria. If glucose or other carbohydrates are not available, mitochondria can absorb and utilize small carbon chains produced by the breakdown of proteins or lipids. So long as oxygen is present, these molecules will break down to carbon dioxide, which diffuses out of the cell, and hydrogen atoms, which participate in a series of energy-releasing steps that result in the enzymatic conversion of ADP to ATP. ∞ p. 51

Because the key reactions involved in mitochondrial activity consume oxygen, the process of mitochondrial energy production is known as **aerobic** (*aero-,* air + *bios,* life) **metabolism,** or *cellular respiration.* Aerobic metabolism in mitochondria pro-

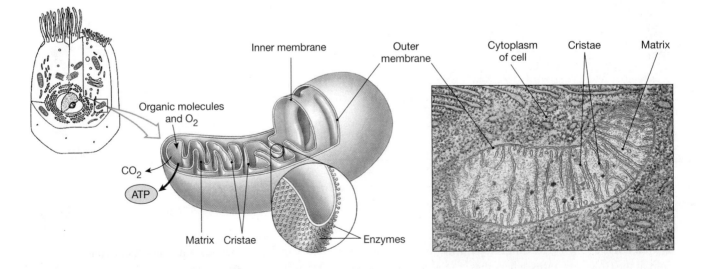

● **FIGURE 3–18   Mitochondria.** The three-dimensional organization of a typical mitochondrion and a color-enhanced TEM of a mitochondrion in section. (TEM × 46,332) Mitochondria absorb short carbon chains and oxygen and generate carbon dioxide and ATP.

duces about 95 percent of the energy a cell needs to stay alive. Aerobic metabolism is discussed in more detail in Chapters 7 and 17.

Several inheritable disorders result from abnormal mitochondrial activity. The mitochondria involved have defective enzymes, which reduces their ability to generate ATP. Cells throughout the body may be affected, but symptoms that involve muscle cells, nerve cells, and the light receptor cells in the eye are most common, because these cells have especially high energy demands.

 Key Note

Mitochondria provide most of the energy needed to keep your cells (and you) alive. They require oxygen and organic substrates and they generate carbon dioxide and ATP.

→ CONCEPT CHECK QUESTIONS

1. Why do certain cells in the ovaries and testes contain large amounts of smooth endoplasmic reticulum (SER)?
2. What does the presence of many mitochondria imply about a cell's energy requirements?

Answers begin on p. 792.

■ The Nucleus

The nucleus is usually the largest and most conspicuous structure in a cell. It is the control center for cellular operations. A single nucleus stores all the information needed to control the synthesis of the more than 400,000 different proteins in the human body. The nucleus determines both the structure of the cell and the functions it can perform by controlling which proteins are synthesized, under what circumstances, and in what amounts.

Most cells contain a single nucleus, but there are exceptions: skeletal muscle cells have many nuclei, and mature red blood cells have none. Figure 3–19● shows the structure of a typical nucleus. A **nuclear envelope** that consists of a double membrane surrounds the nucleus and separates its fluid contents—the *nucleoplasm*—from the cytosol. The nucleoplasm contains ions, enzymes, RNA and DNA nucleotides, proteins, small amounts of RNA, and DNA.

Chemical communication between the nucleus and the cytosol occurs through **nuclear pores.** These pores are large enough to permit the movement of ions and small molecules yet small enough to regulate the transport of proteins and RNA.

Most nuclei contain several **nucleoli** (noo-KLĒ-ō-lī; singular *nucleolus*). Nucleoli are organelles that synthesize **ribosomal RNA (rRNA)** and assemble the ribosomal subunits into functional ribosomes. For this reason, they are most prominent in cells that manufacture large amounts of proteins, such as muscle and liver cells.

Chromosome Structure

It is the DNA in the nucleus that stores instructions for protein synthesis, and this DNA is contained in **chromosomes** (*chroma*, color). The nuclei of human body cells contain 23 pairs of chromosomes. One member of each pair is derived from the mother and one from the father. The structure of a typical chromosome is shown in Figure 3–20●.

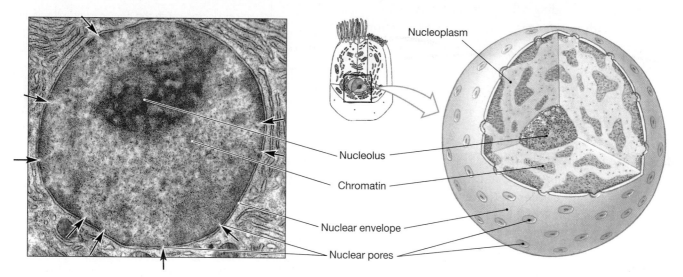

● **FIGURE 3–19   The Nucleus.** The electron micrograph and diagrammatic view show important nuclear structures. The arrows indicate the locations of nuclear pores. (TEM × 4828)

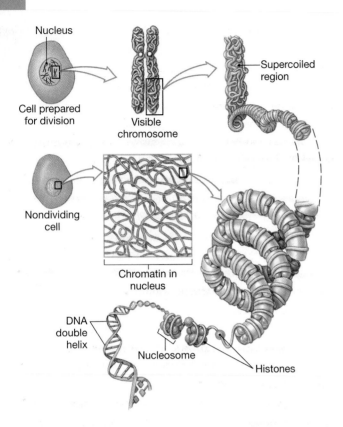

Nucleus

Cell prepared for division

Visible chromosome

Supercoiled region

Nondividing cell

Chromatin in nucleus

DNA double helix

Nucleosome

Histones

● **FIGURE 3–20** **Chromosome Structure.** DNA strands wound around histone proteins (at bottom) form coils that may be very tight or rather loose. In cells that are not dividing, the DNA is loosely-coiled, and forms a tangled network known as chromatin. When the coiling becomes tighter, as it does in preparation for cell division, the DNA becomes visible as distinct structures called chromosomes.

Each chromosome contains DNA strands wrapped around special proteins called *histones*. The tightness of DNA coiling determines whether the chromosome is long and thin or short and fat. Chromosomes in a dividing cell are very tightly coiled, and they can be seen clearly as separate structures in light or electron micrographs. In cells that are not dividing, the DNA is loosely coiled, and form a tangle of fine filaments known as **chromatin.**

 Key Note

The nucleus contains the genetic instructions needed to synthesize the proteins that determine cell structure and function. This information is stored in chromosomes, which consist of DNA and various proteins involved in controlling and accessing the genetic information.

The Genetic Code

Each protein molecule consists of a unique sequence of amino acids. ∞ p. 46 Any instructions for constructing a protein,

therefore, must include information about the amino acid sequence. This information is stored in the chemical structure of the DNA strands in the nucleus. The chemical "language" the cell uses is known as the **genetic code.** An understanding of the genetic code has enabled us to determine how cells build proteins and how various structural and functional traits, such as hair color or blood type, are inherited from generation to generation.

The basic structure of nucleic acids was described in Chapter 2. ∞ p. 49 A single DNA molecule consists of a pair of DNA strands held together by hydrogen bonding between complementary nitrogenous bases. Information is stored in the sequence of nitrogenous bases (adenine, A; thymine, T; cytosine, C; and guanine, G) along the length of the DNA strands.

The genetic code is called a *triplet code* because a sequence of three nitrogenous bases specifies the identity of a single amino acid. The DNA triplet adenine-cytosine-adenine (ACA), for example, codes for the amino acid cysteine.

A **gene** is the functional unit of heredity, and each gene consists of all the triplets needed to produce a specific protein. The number of triplets varies from gene to gene, depending on the size of the protein that will be produced. Each gene also contains special segments responsible for regulating its own activity. In effect these triplets say, "Do (or do not) read this message," "Message starts here," or "Message ends here." The "read me," "don't read me," and "start" signals form a special region of the DNA called the *promoter,* or *control segment,* at the start of each gene. Each gene ends with a "stop" signal.

 Clinical Note
DNA FINGERPRINTING

Every nucleated cell in the body carries a set of 46 chromosomes identical to the set formed at fertilization. Not all the DNA of these chromosomes codes for proteins, however, and long stretches of DNA have no known function. Some of the "useless" segments contain the same nucleotide sequence repeated over and over. The number of segments and the number of repetitions per segment vary from individual to individual. The chance that any two individuals, other than identical twins, will have the same pattern is less than one in nine billion. In other words, it is extremely unlikely that you will ever encounter someone else who has the same pattern of repeating nucleotide sequences as in your DNA.

The identification of individuals can, therefore, be made on the basis of DNA pattern analysis, just as it can on the basis of a fingerprint. Skin scrapings, blood, semen, hair, or other tissues can be used as a sample DNA source. Information from *DNA fingerprinting* is used to convict (and to acquit) persons accused of committing violent crimes, such as rape or murder. The science of molecular biology has, thus, become a useful addition to the crime-fighting toolbox. ■

Protein Synthesis

Each DNA molecule contains thousands of genes and, therefore, holds the information needed to synthesize thousands of proteins. These genes are normally tightly coiled and bound to histones, which prevent their activation and, in doing so, prevent the synthesis of proteins. Before a specific gene can be activated, enzymes must temporarily break the weak bonds between the gene's nitrogenous bases and remove the histone that guards the promoter at the start of each gene. Although the process of gene activation is only partially understood, more is known about protein synthesis. The process of **protein synthesis** is divided into *transcription*, which is the production of RNA from a single strand of DNA, and *translation*, which is the assembling of a protein by ribosomes, using the information carried by the RNA molecule. Transcription takes place within the nucleus, and translation occurs in the cytoplasm.

Transcription

Ribosomes, which are the organelles of protein synthesis, are located in the cytoplasm; whereas the genes are confined to the nucleus. This separation between the protein manufacturing site and the DNA's protein blueprint is overcome by the movement of a molecular messenger, a single strand of RNA known as **messenger RNA (mRNA)**. The process of mRNA formation is called **transcription** (Figure 3–21●). Transcription, which is the process of transcribing or "copying," is an appropriate term because the newly formed mRNA is a transcript (a copy) of the information contained in the gene.

Each DNA strand contains thousands of genes. Transcription begins when an enzyme, *RNA polymerase,* binds to the promoter of a gene (see Step 1 in Figure 3–21). This enzyme promotes the synthesis of an mRNA strand, using nucleotides complementary to those in the gene (Step 2). The nucleotides involved are those characteristic of RNA, not those of DNA; RNA polymerase may attach adenine, guanine, cytosine, or uracil (U), but never thymine. Thus, wherever an A occurs in the DNA strand, RNA polymerase will attach a U rather than a T. The mRNA strand, thus, contains a sequence of nitrogenous bases that are complementary to those of the gene. A sequence of three nitrogenous bases along the new mRNA strand represents a **codon** (KŌ-don) that is complementary

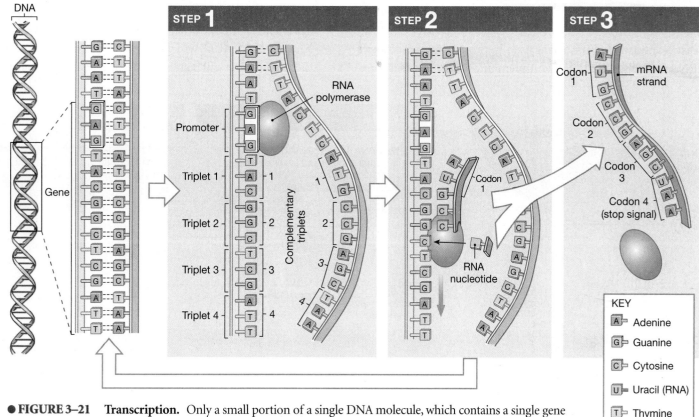

● **FIGURE 3–21　Transcription.** Only a small portion of a single DNA molecule, which contains a single gene available for transcription, is shown. **Step 1:** The two DNA strands separate, and RNA polymerase binds to the promoter of the gene. **Step 2:** The RNA polymerase moves from one nucleotide to another along the length of the gene. At each site, complementary RNA nucleotides form hydrogen bonds with the DNA nucleotides of the gene. The RNA polymerase then bonds the arriving nucleotides together into a strand of mRNA. **Step 3:** On reaching the stop signal at the end of the gene, the RNA polymerase and the mRNA strand detach, and the two DNA strands reattach.

to the corresponding triplet along the gene (Step 3). At the DNA "stop" signal, the enzyme and the mRNA strand detach, and the complementary DNA strands reassociate.

The mRNA formed in this way may be altered before it leaves the nucleus. For example, some regions (called *introns*) may be removed and the remaining segments (called *exons*) spliced together. This modification creates a shorter, functional mRNA strand that enters the cytoplasm through a nuclear pore. We now know that by removing different introns, a single gene can produce mRNAs that code for several different proteins. How such alterations are regulated remains a mystery.

Translation

Translation is the synthesis of a protein using the information provided by the sequence of codons along the mRNA strand. Every amino acid has at least one unique and specific codon; Table 3–5 includes several examples. During translation, the sequence of codons determines the sequence of amino acids in the protein.

Translation begins when the newly synthesized mRNA leaves the nucleus and binds with a ribosome in the cytoplasm. Molecules of **transfer RNA (tRNA)** then deliver amino acids that will be used by the ribosome to assemble a protein. There are more than 20 different types of transfer RNA, at least one for each amino acid used in protein synthesis. Each tRNA molecule contains a complementary triplet of nitrogenous bases, known as an **anticodon**, that will bind to a specific codon on the mRNA.

The translation process is illustrated in Figure 3–22●:

Step 1: Translation begins at the "start" codon of the mRNA strand, when it binds to the small ribosomal subunit and the first tRNA arrives. That tRNA carries a specific amino acid. The first codon of the mRNA strand always has the base sequence AUG, which codes for methionine. (This initial methionine will be removed from the finished protein.)

TABLE 3-5 *Examples of the Triplet Code*

DNA TRIPLET	mRNA CODON	tRNA ANTICODON	AMINO ACID (AND/OR INSTRUCTION)
AAA	UUU	AAA	Phenylalanine
AAT	UUA	AAU	Leucine
ACA	UGU	ACA	Cysteine
CAA	GUU	CAA	Valine
GGG	CCC	GGG	Proline
CGA	GCU	CGA	Alanine
TAC	AUG	UAC	Methionine; start codon
ATT	UAA	[none]	Stop codon

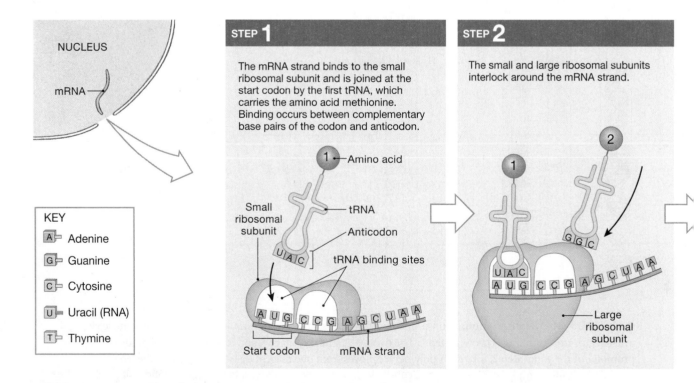

● **FIGURE 3-22** **Translation.** Once transcription completes, the mRNA diffuses into the cytoplasm and interacts with the ribosome.

Step 2: The small and large ribosomal units join together and enclose the mRNA.

Step 3: A second tRNA then arrives that carries a different amino acid and its anticodon binds to the second codon of the mRNA strand. Ribosomal enzymes now remove amino acid 1 from the first tRNA and attach it to amino acid 2 with a peptide bond. ∞ p. 47 The first tRNA then detaches from the ribosome and reenters the cytosol, where it can pick up another amino acid molecule and repeat the process. The ribosome now moves one codon farther along the length of the mRNA strand, and a third tRNA arrives, bearing amino acid 3.

Step 4: Ribosomal enzymes now remove amino acid 1 from the first tRNA and attach it to amino acid 2 with a peptide bond. The first tRNA then detaches from the ribosome and reenters the cytosol, where it can pick up another amino acid molecule and repeat the process. The ribosome now moves one codon farther along the length of the mRNA strand, and a third tRNA arrives that bears amino acid 3.

Step 5: Amino acids will continue to be added to the growing protein in this way until the ribosome reaches the "stop" codon. The ribosomal subunits then detach, which leaves an intact strand of mRNA and a completed polypeptide.

As you may recall from Chapter 2, a protein is a polypeptide that contains 100 or more amino acids. ∞ p. 47 Translation proceeds swiftly, and produces a typical protein (about 1000 amino acids) in around 20 seconds. The protein begins as a simple linear strand, but a more complex structure develops as it grows longer.

Key Note

Genes are the functional units of DNA that contain the instructions for making one or more proteins. The creation of specific proteins involves multiple enzymes and three types of RNA.

Clinical Note
RECOMBINANT DNA TECHNOLOGY

Many important medications are derived from animals. For example, insulin, which is an important hormone in glucose metabolism, was initially derived from swine or cattle. However, there are slight differences in the structure of human insulin and the structure of pork or beef insulin. The amino acid sequence in the animal insulin molecules differs slightly from human insulin. While animal insulin works well in humans, the body eventually begins to produce anti-insulin antibodies against the foreign part of the insulin molecule. Over time, humans on animal insulin require more and more insulin as their immune systems destroy increasing quantities of the animal insulin. This insulin resistance becomes a problem for long-term diabetics, as they require increasingly large doses of insulin.

(continued next page)

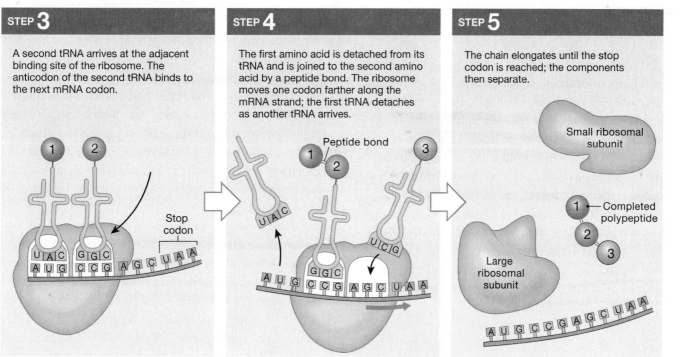

STEP 3

A second tRNA arrives at the adjacent binding site of the ribosome. The anticodon of the second tRNA binds to the next mRNA codon.

Stop codon

U A C   G G C
A U G   C C G   A G C U A A

STEP 4

The first amino acid is detached from its tRNA and is joined to the second amino acid by a peptide bond. The ribosome moves one codon farther along the mRNA strand; the first tRNA detaches as another tRNA arrives.

Peptide bond

U A C

U C G

G G C
A U G C C G A G C U A A

STEP 5

The chain elongates until the stop codon is reached; the components then separate.

Small ribosomal subunit

Completed polypeptide

Large ribosomal subunit

A U G C C G A G C U A A

However, scientists have developed a method of producing human insulin by a technique called *recombinant DNA technology*. With this technology, small fragments of human DNA that code for the production of insulin are inserted into specialized bacteria. These bacteria then begin producing insulin identical to human insulin. Diabetics can use this insulin without the risk of developing insulin resistance. Other medications produced by recombinant DNA technology include hepatitis B vaccines, "clot-busting" drugs (tPA) used in heart attacks, and others. ■

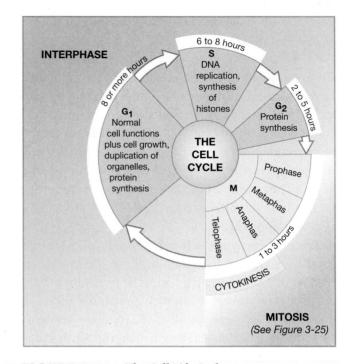

● **FIGURE 3–23** The Cell Life Cycle.

CONCEPT CHECK QUESTIONS

1. How does the nucleus control the cell's activities?
2. What process would be affected by the lack of the enzyme RNA polymerase?
3. During the process of transcription, a nucleotide was deleted from an mRNA sequence that coded for a protein. What effect would this deletion have on the amino acid sequence of the protein?

Answers begin on p. 792.

■ The Cell Life Cycle

During the time between fertilization and physical maturity, the number of cells that make up an individual increases from a single cell to roughly 75 trillion cells. This amazing increase in numbers occurs through a form of cellular reproduction called **cell division.** Even when development has been completed, cell division continues to be essential to survival because it replaces old and damaged cells.

For cell division to be successful, the genetic material in the nucleus must be duplicated accurately, and a complete copy must be distributed to each daughter cell. The duplication of the cell's genetic material is called *DNA replication,* and nuclear division is called **mitosis** (mī-TŌ-sis). Mitosis occurs during the division of **somatic** (*soma,* body) **cells,** which include the vast majority of the cells in the body. The production of sex cells—sperm and ova (eggs)—involves a different form of cell division; this process, called **meiosis** (mī-Ō-sis), is described in Chapter 19.

Figure 3–23● represents the life cycle of a typical cell. Most cells spend only a small part of their life cycle engaged in cell division, or mitosis. For most of their lives, cells are in **interphase,** which is an interval of time between cell divisions when they perform normal functions.

Interphase

Somatic cells differ in the length of time spent in interphase and their frequency of cell division. For example, stem cells divide repeatedly with very brief interphase periods. In contrast, mature skeletal muscle cells and most nerve cells never undergo mitosis or cell division. Certain other cells appear to be programmed not to divide but instead to self-destruct as a result of the activation of "suicide genes." The genetically controlled death of cells is called **apoptosis** (ap-op-TŌ-sis or ap-ō-TŌ-sis; *ptosis,* a falling away). Apoptosis is a key process in homeostasis. During an immune response, for example, some types of white blood cells activate genes in abnormal or infected cells that tell the cell to die.

A cell ready to divide first enters the G_1 phase. In this phase, the cell makes enough organelles and cytosol for two functional cells. These preparations may take hours, days, or weeks to complete, depending on the type of cell and the situation. For example, certain cells in the lining of the digestive tract divide every few days throughout life, whereas specialized cells in other tissues divide only under special circumstances, such as following an injury.

Once these preparations have been completed, the cell enters the S phase and replicates the DNA in its nucleus, a process that takes 6–8 hours. The goal of **DNA replication** is to copy the genetic information in the nucleus so that one set of chromosomes can be given to each of the two cells produced.

DNA replication starts when the complementary strands begin to separate and unwind (Figure 3–24●). Molecules of the enzyme *DNA polymerase* then bind to the exposed nitrogenous bases. As a result, complementary nucleotides present in the nucleoplasm attach to the exposed nitrogenous

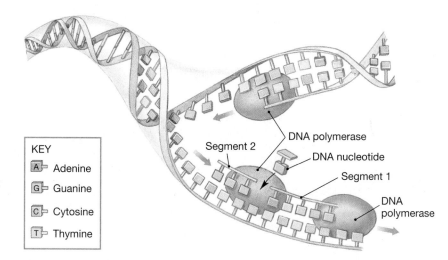

KEY

- $\boxed{A}$ Adenine
- $\boxed{G}$ Guanine
- $\boxed{C}$ Cytosine
- $\boxed{T}$ Thymine

● **FIGURE 3–24**   **DNA Replication.** In replication, the DNA strands unwind, and DNA polymerase begins attaching complementary DNA nucleotides along each strand. The complementary copy of one original strand is produced as a continuous strand. The other copy is produced as short segments, which are then spliced together. The end result is that two identical copies of the original DNA molecule are produced.

bases of the DNA strand and form a pair of identical DNA molecules. Shortly following DNA replication is a brief G_2 phase for protein synthesis and completing centriole replication. The cell then enters the M phase, and mitosis begins.

Mitosis

Mitosis is a process that separates and encloses the duplicated chromosomes of the original cell into two identical nuclei. Division of the cytoplasm to form two distinct cells involves a separate, but related, process known as **cytokinesis** (sī-tō-ki-NĒ-sis; *cyto-*, cell + *kinesis*, motion). Mitosis is divided into four stages: *prophase, metaphase, anaphase,* and *telophase* (Figure 3–25●).

Stage 1: Prophase

Prophase (PRŌ-fāz; *pro-*, before) begins when the chromosomes coil so tightly that they become visible as individual structures through a light microscope. As a result of DNA replication, there are now two copies of each chromosome. Each copy, called a **chromatid** (KRŌ-ma-tid), is connected to its duplicate copy at a single point, the **centromere** (SEN-trō-mēr).

As the chromosomes appear, the nucleoli disappear and the two pairs of centrioles move toward opposite poles of the nucleus. An array of microtubules, called **spindle fibers,** extends between the centriole pairs. Late in prophase, the nuclear envelope disappears and the chromatids become attached to the spindle fibers.

Stage 2: Metaphase

Metaphase (MET-a-fāz; *meta-*, after) begins as the chromatids move to a narrow central zone called the **metaphase plate.** Metaphase ends when all of the chromatids are aligned in the plane of the metaphase plate.

Stage 3: Anaphase

Anaphase (AN-a-fāz; *ana-,* apart) begins when the centromere of each chromatid pair splits and the chromatids separate. The two **daughter chromosomes** that result are now pulled toward opposite ends of the cell. Anaphase ends when the daughter chromosomes arrive near the centrioles at opposite ends of the cell.

Stage 4: Telophase

During **telophase** (TĒL-ō-fāz; *telos,* end), the cell prepares to return to interphase. The nuclear membranes form, the nuclei enlarge, and the chromosomes gradually uncoil. Once the chromosomes have relaxed and the fine filaments of chromatin become visible, nucleoli reappear and the nuclei resemble those of interphase cells.

🔒 Key Note

Mitosis is the separation of duplicated chromosomes into two identical sets and nuclei in the process of somatic cell division.

Cytokinesis

Telophase marks the end of mitosis proper, but the daughter cells have yet to complete their physical separation. **Cytokinesis,** which is the cytoplasmic division of the daughter cells, usually begins in late anaphase (see Figure 3–23). As the daughter chromosomes near the ends of the spindle fibers, the cytoplasm constricts along the plane of the metaphase plate, and forms a *cleavage furrow*. This process continues throughout telophase and is usually completed after a nuclear membrane has re-formed around each daughter nucleus. The completion of cytokinesis marks the end of cell division.

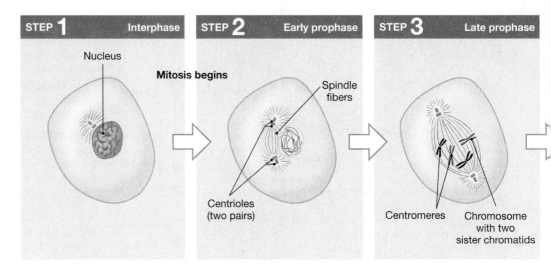

● **FIGURE 3–25**   Interphase, Mitosis, and Cytokinesis.

Cell Division and Cancer

When the rates of cell division and growth exceed the rate of cell death, a tissue begins to enlarge. A **tumor,** or *neoplasm,* is a mass or swelling produced by abnormal cell growth and division. In a **benign tumor,** the abnormal cells are confined within a connective-tissue capsule. Such a tumor is seldom life-threatening and can usually be surgically removed if its size or position disturbs tissue function.

Cells in a **malignant tumor** no longer respond to normal control mechanisms. These cells do not remain within a connective-tissue capsule but spread into surrounding tissues. The tumor of origin is called the *primary tumor* (or *primary neoplasm*), and the spreading process is called **invasion.** Malignant cells may also travel to distant tissues and organs and produce *secondary tumors.* This migration, called **metastasis** (me-TAS-ta-sis), is difficult to control.

Cancer is an illness that results from the effects of malignant cells. Such cancer cells lose their resemblance to normal cells. The secondary tumors they form are extremely active metabolically, and their presence stimulates the growth of blood vessels into the area. The increased blood supply provides additional nutrients to the cancer cells, which further accelerates tumor growth and metastasis.

As malignant tumors grow, organ function begins to deteriorate. The malignant cells may no longer perform their original functions, or they may perform normal functions in an abnormal way. Cancer cells do not use energy very efficiently, and they grow and multiply at the expense of healthy tissues, and compete for space and nutrients with normal cells. This competition accounts for the starved appearance of many patients in the late stages of cancer. Death may occur as a result of the compression of vital organs when nonfunctional cancer cells have killed or replaced the healthy cells in those organs, or when the cancer cells have starved normal tissues of essential nutrients.

Key Note

Cancer results from mutations that disrupt the control mechanism that regulates cell growth and division. Cancers most often begin where cells divide rapidly, because the more chromosomes are copied, the greater the chances of error.

CONCEPT CHECK QUESTIONS

1. What major events occur during interphase in cells preparing to undergo mitosis?
2. What are the four stages of mitosis?
3. What would happen if spindle fibers failed to form in a cell during mitosis?

Answers begin on p. 792.

■ Cell Diversity and Differentiation

All the somatic cells that comprise an individual have the same chromosomes and genes, yet liver cells, fat cells, and nerve cells are quite different from each other in appearance and function. These differences exist because, in each case, a different set of genes has been turned *off.* In other words, these cells differ because liver cells have one set of genes accessible for transcription and fat cells another. When a gene is deactivated, the cell loses the ability to make a particular protein and, thus, to perform any functions that involve the protein. As more genes are switched off, the cell's functions become more restricted or specialized. This specialization process is called **differentiation.**

Fertilization produces a single cell with all its genetic potential intact. A period of repeated cell divisions follows, and

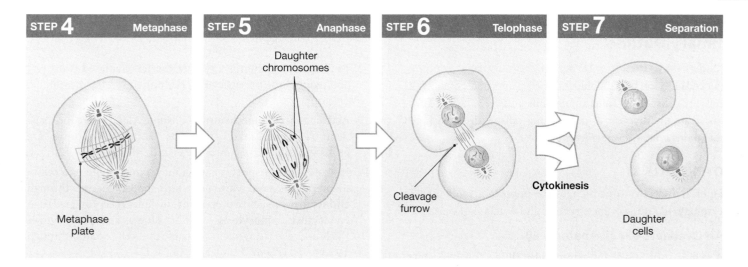

differentiation begins as the number of cells increases. Differentiation produces specialized cells with limited capabilities. These cells form organized collections known as *tissues,* each with different functional roles. The next chapter examines the structure and function of tissues, and the role of tissue interactions in the maintenance of homeostasis.

Chapter Review

Access more review material online at *www.prenhall.com/bledsoe.* There you will find quiz questions, labeling activities, animations, essay questions, and web links.

Key Terms

active transport 70
cells 59
chromosomes 81
cytoplasm 76
cytosol 76
diffusion 64
endocytosis 73
endoplasmic reticulum 78

exocytosis 74
gene 82
Golgi apparatus 78
mitochondria 80
mitosis 87
nucleus 60
organelles 76

osmosis 65
phagocytosis 74
protein synthesis 83
ribosome 77
translation 84
transcription 83
tumor 88

Related Clinical Terms

benign tumor A mass or swelling in which the cells usually remain within a connective-tissue capsule; rarely life-threatening.

cancer An illness characterized by gene mutations that lead to the formation of malignant tumors and metastasis.

carcinogen (kar-SIN-ō-jen) An environmental factor that stimulates the conversion of a normal cell to a cancer cell.

DNA fingerprinting A technique for identifying an individual on the basis of repeating nucleotide sequences in his or her DNA.

genetic engineering The research on and techniques for changing the genetic makeup (DNA) of an organism.

malignant tumor A mass or swelling in which the cells no longer respond to normal control mechanisms but divide rapidly and spread.

metastasis (me-TAS-ta-sis) The spread of malignant cells into distant tissues and organs, where secondary tumors subsequently develop.

normal saline A solution that approximates the normal osmotic concentration of extracellular fluids.

primary tumor *(primary neoplasm)* An abnormal mass where cancer cells first developed and the source of secondary tumors.

recombinant DNA DNA created by inserting (splicing) a specific gene from one organism into the DNA strand of another organism.

secondary tumor A colony of cancerous cells that result from metastasis.

tumor (neoplasm) A mass or swelling produced by abnormal cell growth and division.

Summary Outline

1. Modern **cell theory** incorporates several basic concepts: (1) **cells** are the building blocks of all plants and animals; (2) cells are the smallest functioning units of life; (3) cells are produced by the division of pre-existing cells; and (4) each cell maintains homeostasis. *(Figure 3–1)*

STUDYING CELLS 59

1. Light and electron microscopes are important tools used in **cytology,** the study of the structure and function of cells.

An Overview of Cell Anatomy 60

2. A cell is surrounded by **extracellular fluid.** The cell's outer boundary, the **cell membrane,** separates the **cytoplasm,** or cell contents, from the extracellular fluid. *(Figures 3–2 and 3–3; Table 3–1)*

THE CELL MEMBRANE 61

1. The functions of the cell membrane include: (1) physical isolation; (2) control of the exchange of materials with the cell's surroundings; (3) sensitivity; and (4) structural support.

Membrane Structure 63

2. The cell membrane, or plasma membrane, contains lipids, proteins, and carbohydrates. Its major components, lipid molecules, form a **phospholipid bilayer.** *(Figure 3–3)*

3. Membrane proteins may function as receptors, channels, carriers, enzymes, anchors, or identifiers. *(Table 3–2)*

Membrane Transport 64

4. Cell membranes are **selectively permeable.**

5. **Diffusion** is the net movement of material from an area where its concentration is relatively high to an area where its concentration is lower. Diffusion occurs until the **concentration gradient** is eliminated. *(Figure 3–4, Figure 3–5)*

6. **Osmosis** is the diffusion of water across a membrane in response to differences in concentration. The force of movement is **osmotic pressure.** *(Figure 3–6, Figure 3–7, Figure 3–8)*

Key Note 69

7. In **filtration,** hydrostatic pressure forces water across a membrane. If membrane pores are large enough, molecules of solute will be carried along with the water.

8. **Facilitated diffusion** is a type of **carrier-mediated transport** and requires the presence of carrier proteins in the membrane. *(Figure 3–9)*

9. **Active transport** mechanisms consume ATP and are independent of concentration gradients. Some **ion pumps** are **exchange pumps.** *(Figure 3–10, Figure 3–11, Figure 3–12)*

10. In **vesicular transport,** material moves into or out of a cell in membranous sacs. Movement into the cell occurs through **endocytosis,** an active process that includes **receptor-mediated endocytosis, pinocytosis** ("cell-drinking"), and **phagocytosis** ("cell-eating"). Movement out of the cell occurs through **exocytosis.** *(Figure 3–13, Figure 3–14; Table 3–4)*

THE CYTOPLASM 76

1. The cytoplasm surrounds the nucleus and contains a fluid **cytosol** and intracellular structures called *organelles.*

The Cytosol 76

2. The **cytosol** differs in composition from the extracellular fluid that surrounds most cells of the body.

Organelles 76

3. Membrane-enclosed **organelles** are surrounded by lipid membranes that isolate them from the cytosol. Membranous organelles include the endoplasmic reticulum, the nucleus, the Golgi apparatus, lysosomes, and mitochondria. *(Table 3–1)*

4. Nonmembranous organelles are always in contact with the cytosol. They include the cytoskeleton, microvilli, centrioles, cilia, flagella, proteasomes, and ribosomes. *(Table 3–1)*

5. The **cytoskeleton** gives the cytoplasm strength and flexibility. Its main components are **microfilaments, intermediate filaments,** and **microtubules.** *(Figure 3–15)*

6. **Microvilli** are small projections of the cell membrane that increase the surface area exposed to the extracellular environment.

7. **Centrioles** direct the movement of chromosomes during cell division.

8. **Cilia** beat rhythmically to move fluids or secretions across the cell surface.

9. **Flagella** move a cell through surrounding fluid rather than moving fluid past a stationary cell.

10. A **ribosome** is an intracellular factory that manufactures proteins. **Free ribosomes** are in the cytoplasm, and **fixed ribosomes** are attached to the endoplasmic reticulum.

11. **Proteasomes** remove and break down damaged or abnormal proteins.

12. The **endoplasmic reticulum (ER)** is a network of intracellular membranes. **Rough endoplasmic reticulum (RER)** contains ribosomes and is involved in protein synthesis. **Smooth endoplasmic reticulum (SER)** does not contain ribosomes; it is involved in lipid and carbohydrate synthesis. *(Figure 3–16)*

13. The **Golgi apparatus** forms **secretory vesicles** and new membrane components, and it packages *lysosomes*. Secretions are discharged from the cell by exocytosis. *(Figure 3–17)*

14. **Lysosomes** are vesicles filled with digestive enzymes. Their functions include ridding the cell of bacteria and debris.

Key Note 80

15. **Mitochondria** are responsible for 95 percent of the ATP production within a typical cell. The **matrix**, or fluid contents of a mitochondrion, lies inside **cristae**, or folds of an inner mitochondrial membrane. *(Figure 3–18)*

Key Note 81

THE NUCLEUS 81

1. The **nucleus** is the control center for cellular operations. It is surrounded by a **nuclear envelope**, through which it communicates with the cytosol by way of **nuclear pores**. *(Figure 3–19)*

Chromosome Structure 81

2. The nucleus controls the cell by directing the synthesis of specific proteins using information stored in the DNA of **chromosomes**. *(Figure 3–20)*

Key Note 82

The Genetic Code 82

3. The cell's information storage system, the **genetic code**, is called a *triplet code* because a sequence of three nitrogenous bases identifies a single amino acid. Each **gene** consists of all the triplets needed to produce a specific protein. *Table 3–5*

Protein Synthesis 83

4. **Protein synthesis** includes both *transcription*, which occurs in the nucleus, and *translation*, which occurs in the cytoplasm.

5. During **transcription**, a strand of **messenger RNA (mRNA)** is formed and carries protein-making instructions from the nucleus to the cytoplasm. *(Figure 3–21)*

6. During **translation**, a functional protein is constructed from the information contained in an mRNA strand. Each triplet of nitrogenous bases along the mRNA strand is a **codon**; the sequence of codons determines the sequence of amino acids in the protein. *(Figure 3–22)*

7. Molecules of **transfer RNA (tRNA)** bring amino acids to the ribosomes involved in translation.

Key Note 85

THE CELL LIFE CYCLE 86

1. **Cell division** is the reproduction of cells. **Apoptosis** is the genetically-controlled death of cells. **Mitosis** is the nuclear division of **somatic cells**. *(Figure 3–23)*

Interphase 86

2. Most somatic cells are in **interphase** most of the time. Cells preparing for mitosis undergo **DNA replication** in this phase. *(Figure 3–24)*

Mitosis 87

3. Mitosis proceeds in four stages: **prophase, metaphase, anaphase,** and **telophase**. *(Figure 3–25)*

Key Note 87

Cytokinesis 87

4. During **cytokinesis,** the cytoplasm divides, which produces two identical daughter cells.

Cell Division and Cancer 88

5. Abnormal cell growth and division forms **tumors** that are either **benign** (encapsulated) or **malignant** (able to invade other tissues). **Cancer** is a disease characterized by the presence of malignant tumors; over time, cancer cells tend to spread to new areas of the body.

Key Note 88

CELL DIVERSITY AND DIFFERENTIATION 88

1. **Differentiation** is the specialization that produces cells with limited capabilities. These specialized cells form organized collections called *tissues*, each of which has specific functional roles.

Review Questions

Level 1: Reviewing Facts and Terms

Match each item in column A with the most closely related item in column B. Place letters for answers in the spaces provided.

COLUMN A

____ 1. filtration

____ 2. osmosis

____ 3. hypotonic solution

____ 4. hypertonic solution

____ 5. isotonic solution

____ 6. facilitated diffusion

____ 7. carrier proteins

____ 8. vesicular transport

____ 9. cytosol

____ 10. cytoskeleton

____ 11. microvilli

____ 12. ribosomes

____ 13. mitochondria

____ 14. lysosomes

____ 15. nucleus

____ 16. chromosomes

____ 17. nucleoli

COLUMN B

a. water out of cell

b. passive carrier-mediated transport

c. endocytosis, exocytosis

d. movement of water

e. hydrostatic pressure

f. normal saline

g. ion pump

h. water into cell

i. manufacture proteins

j. digestive enzymes

k. internal protein framework

l. control center for cellular operations

m. intracellular fluid

n. DNA strands

o. cristae

p. synthesize components of ribosomes

q. increase cell surface area

18. The study of the structure and function of cells is called:
    (a) histology.
    (b) cytology.
    (c) physiology.
    (d) biology.

19. The proteins in the cell membranes may function as:
    (a) receptors and channels.
    (b) carriers and enzymes.
    (c) anchors and identifiers.
    (d) receptors, channels, carriers, enzymes, anchors, and identifiers.

20. All of the following membrane transport mechanisms are passive processes *except*:
    (a) diffusion.
    (b) facilitated diffusion.
    (c) vesicular transport.
    (d) filtration.

21. _____ ion concentrations are high in the extracellular fluids, and _____ ion concentrations are high in the cytoplasm.
    (a) Calcium, magnesium
    (b) Chloride, sodium
    (c) Potassium, sodium
    (d) Sodium, potassium

22. Structures that perform specific functions within the cell are:
    (a) organs.
    (b) organisms
    (c) organelles.
    (d) chromosomes.

23. The construction of a functional protein using the information provided by an mRNA strand is:
    (a) translation.
    (b) transcription.
    (c) replication.
    (d) gene activation.

24. The term *differentiation* refers to the:
    (a) loss of genes from cells.
    (b) acquisition of new functional capabilities by cells.
    (c) production of functionally specialized cells.
    (d) division of genes among different types of cells.

25. What are the four general functions of the cell membrane?

Level 2: Reviewing Concepts

29. Diffusion is important in body fluids because this process tends to:
    (a) increase local concentration gradients.
    (b) eliminate local concentration gradients.
    (c) move substances against their concentration gradients.
    (d) create concentration gradients.

30. When placed in a _____ solution, a cell will lose water through osmosis. The process results in the _____ of red blood cells.
    (a) hypotonic, crenation
    (b) hypertonic, crenation
    (c) isotonic, hemolysis
    (d) hypotonic, hemolysis

31. Suppose that a DNA segment has the following nucleotide sequence: CTC ATA CGA TTC AAG TTA. Which of the following nucleotide sequences would be found in a complementary mRNA strand?
    (a) GAG UAU GAU AAC UUG AAU
    (b) GAG TAT GCT AAG TTC AAT
    (c) GAG UAU GCU AAG UUC AAU
    (d) GUG UAU GGA UUG AAC GGU

Level 3: Critical Thinking and Clinical Applications

38. Experimental evidence shows that the transport of a certain molecule exhibits the following characteristics: (1) the molecule moves down its concentration gradient; (2) at concentrations above a given level, there is no increase in the rate of transport; and (3) cellular energy is not required for transport to occur. Which type of transport process is at work?

26. By what four major transport mechanisms do substances get into and out of cells?

27. What are the four major functions of the endoplasmic reticulum?

28. List the four stages of mitosis in their correct sequence.

32. How many amino acids are coded in the DNA segment in the previous question?
    (a) 18
    (b) 9
    (c) 6
    (d) 3

33. What are the similarities between facilitated diffusion and active transport? What are the differences?

34. How does the cytosol differ in composition from the extracellular fluid?

35. Differentiate between transcription and translation.

36. List the stages of mitosis and briefly describe the events that occur in each.

37. What is cytokinesis, and what role does it play in the cell cycle?

39. Two solutions, A and B, are separated by a selectively permeable barrier. Over a period of time, the level of fluid on side A increases. Which solution initially had the higher concentration of solute?

4 The Tissue Level of Organization

THE HUMAN BODY is extremely versatile and quite durable. However, it must operate within very close tolerances and maintain homeostasis. Some occupations, such as firefighting, expose the body to extreme stress, heat, and poisonous chemicals. Be- *cause of this, it is prudent to rotate and monitor fire personnel to ensure they have adequate time to replace fluids and rest before engaging in additional stressful work.*

Chapter Outline

Chapter Objectives

1. Identify the body's four major tissue types and describe their roles. (p. 96)
2. Discuss the types and functions of epithelial cells. (p. 96)
3. Describe the relationship between form and function for each epithelial type. (pp. 97–105)
4. Compare the structures and functions of the various types of connective tissues. (pp. 105–111)
5. Explain how epithelial and connective tissues combine to form four types of membranes and specify the functions of each. (pp. 112–113)
6. Describe the three types of muscle tissue and the special structural features of each. (pp. 113–115)
7. Discuss the basic structure and role of neural tissue. (p. 115)
8. Explain how tissues respond in a coordinated manner to maintain homeostasis. (p. 116)
9. Describe how aging affects the tissues of the body. (pp. 116–117)

Vocabulary Development

a- without; *avascular*
apo- from; *apocrine*
cardium heart; *pericardium*
chondros cartilage; *perichondrium*
dendron tree; *dendrites*
desmos ligament; *desmosome*
glia glue; *neuroglia*
histos tissue; *histology*

holos entire; *holocrine*
hyalos glass; *hyaline cartilage*
inter- between; *interstitial*
krinein to secrete; *exocrine*
lacus pool; *lacunae*
meros part; *merocrine*
neuro nerve; *neuron*
os bone; *osseous tissue*

peri- around; *perichondrium*
phagein to eat; *macrophage*
pleura rib; *pleural membrane*
pseudes false; *pseudostratified*
sistere to set; *interstitial*
soma body; *desmosome*
squama plate or scale; *squamous*
vas vessel; *vascular*

NO SINGLE CELL is able to perform the many functions of the human body. Instead, through differentiation, each cell specializes to perform a relatively restricted range of functions. Although there are trillions of individual cells in the human body, there are only about 200 different types of cells. These cell types combine to form **tissues,** which are collections of specialized cells and cell products that perform a limited number of functions. **Histology** (*histos,* tissue) is the study of tissues. Four basic *tissue types* exist: *epithelial tissue, connective tissue, muscle tissue,* and *neural tissue* (Figure 4–1●).

Key Note

Tissues are collections of cells and extracellular material that perform a specific but limited range of functions. There are four tissue types in varying combinations that form all of the structures of the human body: epithelial, connective, muscle, and neural tissue.

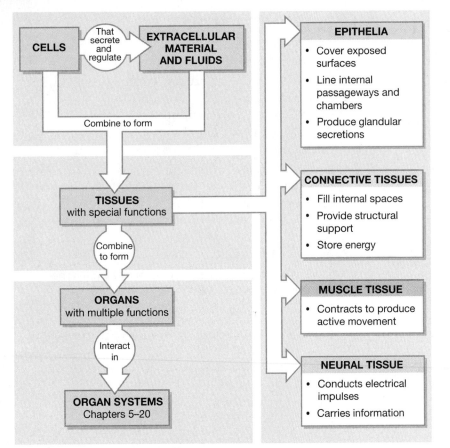

● **FIGURE 4–1** An Orientation to the Tissues of the Body.

■ Epithelial Tissue

Epithelial tissue includes epithelia and *glands*. **Epithelia** (ep-i-THĒ-lē-a; singular, *epithelium*) are layers of cells that cover internal or external surfaces. **Glands** are composed of secreting cells derived from epithelia. Important characteristics of epithelia include the following:

■ Cells that are bound closely together. In other tissue types, the cells are often widely separated by extracellular materials.

■ A free (apical) surface exposed to the environment or to some internal chamber or passageway.

■ Attachment to underlying connective tissue by a *basement membrane*.

■ The absence of blood vessels. Because of this **avascular** (ā-VAS-kū-lar; *a-*, without + *vas*, vessel) condition, epithelial cells must obtain nutrients across their attached surface from deeper tissues or across their exposed surfaces.

■ Epithelial cells that are damaged or lost at the exposed surface are continually replaced or regenerated.

Epithelia cover both external and internal body surfaces. In addition to covering the skin, epithelia line internal passageways that communicate with the outside world, such as the digestive, respi-

ratory, reproductive, and urinary tracts. These epithelia form selective barriers that separate the deep tissues of the body from the external environment.

Epithelia also line internal cavities and passageways, such as the chest cavity; the fluid-filled chambers in the brain, eye, and inner ear; and the inner surfaces of blood vessels and the heart. These epithelia prevent friction, regulate the fluid composition of internal cavities, and restrict communication between the blood and tissue fluids.

Functions of Epithelia

Epithelia perform four essential functions:

1. *Provide physical protection.* Epithelia protect exposed and internal surfaces from abrasion, dehydration, and destruction by chemical or biological agents. For example, as long as it remains intact, the epithelium of your skin resists impacts and scrapes, restricts water loss, and prevents invasion of underlying structures by bacteria.

2. *Control permeability.* Any substance that enters or leaves the body must cross an epithelium. Some epithelia are relatively impermeable; others are easily crossed by compounds as large as proteins.

3. *Provide sensation.* Specialized epithelial cells can detect changes in the environment and relay information about such changes to the nervous system. For example, touch receptors in the deepest layers of the epithelium of the skin respond by stimulating neighboring sensory nerves.

4. *Produce specialized secretions.* Epithelial cells that produce secretions are called **gland cells.** Individual gland cells are typically scattered among other cell types in an epithelium. In a **glandular epithelium,** most or all of the cells actively produce secretions. These secretions are classified according to where they are discharged:

■ **Exocrine** (*exo-*, outside + *krinein*, to secrete) secretions are discharged onto the surface of the epithelium. Examples include enzymes within the digestive tract, perspiration on the skin, and milk produced by mammary glands.

■ **Endocrine** (*endo-*, inside) secretions are released into the surrounding tissue fluid and blood. These secretions, called *hormones*, act as chemical messengers and regulate or coordinate the activities of other tissues, organs, and organ systems. (Hormones are discussed further in Chapter 10.) Endocrine secretions are produced in organs such as the pancreas, thyroid, and pituitary gland.

Intercellular Connections

To be effective in protecting other tissues, epithelial cells must remain firmly attached to the basement membrane and to one another to form a complete cover or lining. If an epithelium is damaged or the connections are broken, it is no longer an effective barrier. For example, when the epithelium of the skin is damaged by a burn or an abrasion, bacteria can enter underlying tissues and cause an infection. Undamaged epithelia form effective barriers because the epithelial cell membranes are held together by specialized transmembrane proteins called *cell adhesion molecules (CAMs)* and by a thin layer of *intercellular cement* (composed of a protein-polysaccharide

mixture). The CAMs bind to cytoskeletal filaments, to each other, and to extracellular materials, and form specialized attachment sites known as *cell junctions.* Three common intercellular connections are *tight junctions, gap junctions,* and *desmosomes* (Figure 4–2a●).

At a **tight junction,** the lipid layers of adjacent cell membranes are tightly bound together by interlocking membrane proteins (Figure 4–2b●). Basal to the tight junction is a continuous *adhesion belt* that encircles cells and binds them to their neighbors. The bands are connected to a network of actin filaments in the cytoskeleton. Tight junctions prevent the passage of water and solutes between cells. These junctions are common between epithelial cells exposed to harsh

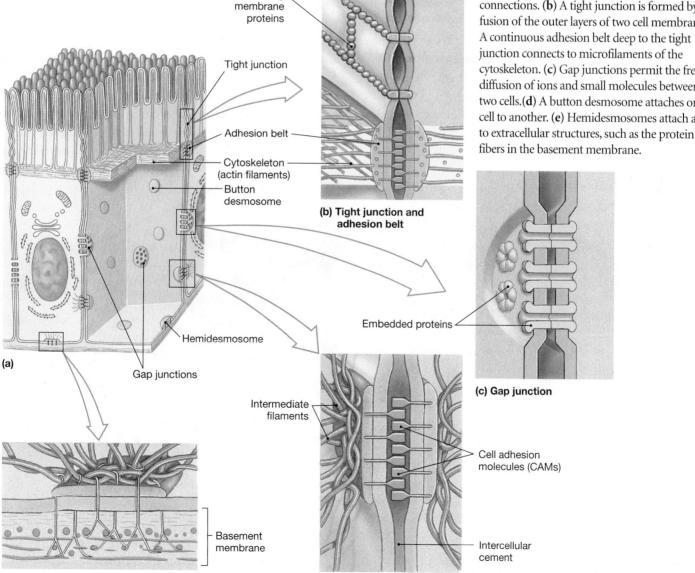

● **FIGURE 4–2** **Intercellular Connections.** (**a**) A diagrammatic view of an epithelial cell that shows the major types of intercellular connections. (**b**) A tight junction is formed by the fusion of the outer layers of two cell membranes. A continuous adhesion belt deep to the tight junction connects to microfilaments of the cytoskeleton. (**c**) Gap junctions permit the free diffusion of ions and small molecules between two cells.(**d**) A button desmosome attaches one cell to another. (**e**) Hemidesmosomes attach a cell to extracellular structures, such as the protein fibers in the basement membrane.

Interlocking membrane proteins

Tight junction

Adhesion belt

Cytoskeleton (actin filaments)

Button desmosome

(b) Tight junction and adhesion belt

Hemidesmosome

(a)

Gap junctions

Embedded proteins

(c) Gap junction

Intermediate filaments

Cell adhesion molecules (CAMs)

Basement membrane

Intercellular cement

(e) Hemidesmosome

(d) Button desmosome

chemicals or powerful enzymes. For example, tight junctions between epithelial cells that line the digestive tract keep digestive enzymes, stomach acids, or waste products from damaging underlying tissues.

Some epithelial functions require rapid intercellular communication. At a **gap junction,** two cells are held together by embedded membrane proteins (Figure 4–2c●). Because these are channel proteins, they form a narrow passageway that lets small molecules and ions pass from cell to cell. Gap junctions interconnect cells in some ciliated epithelia, but they are most abundant in cardiac muscle and smooth muscle tissue, where they are essential to the coordination of muscle contractions.

Most epithelial cells are subject to mechanical stresses—stretching, bending, twisting, or compression—so they must have durable interconnections. At a **desmosome** (DEZ-mō-sōm; *desmos*, ligament + *soma*, body), the cell membranes of two cells are locked together by intercellular cement and by membrane proteins connected to a network of intermediate filaments (Figure 4–2d●). Desmosomes that form a small disc are called *button desmosomes. Hemidesmosomes* resemble half of a button desmosome and attach a cell to the basement membrane (Figure 4–2e●). Desmosomes are abundant between cells in the superficial layers of the skin. As a result, damaged skin cells are usually lost in sheets rather than as individual cells. (That is why your skin peels rather than comes off as a powder after a sunburn.)

The Epithelial Surface

The *apical surface* of epithelial cells often have specialized structures that distinguish them from other body cells (Figure 4–3●). Many epithelia that line internal passageways have microvilli on their exposed surfaces. ∞ p. 77 Microvilli may vary in number from just a few to so many that they carpet the entire surface. They are especially abundant on epithelial surfaces where absorption and secretion take place, such as along portions of the digestive and urinary tracts. These epithelial cells specialize in the active and passive transport of materials across their cell membranes. ∞ p. 64 A cell with microvilli has at least 20 times the surface area of a cell without them; the greater the surface area of the cell membrane, the more transport proteins are exposed to the extracellular environment. Some epithelia contain cilia on their exposed surfaces. ∞ p. 77 A typical cell within a *ciliated epithelium* has roughly 250 cilia that beat in a coordinated fashion to move materials across the epithelial surface. For example, the ciliated epithelium that lines the respiratory tract (Figure 4–3b●) moves mucus-trapped irritants away from the lungs and toward the throat.

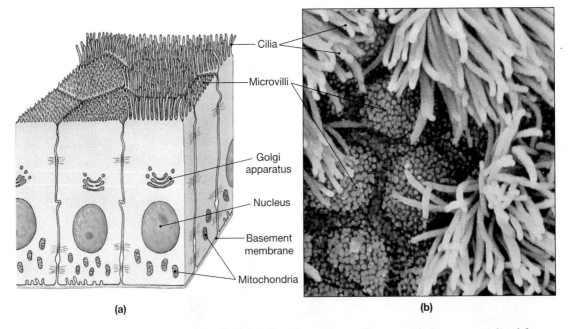

(a)   (b)

● **FIGURE 4–3 The Surfaces of Epithelial Cells.** The surfaces of most epithelia are specialized for specific functions. (**a**) In this diagram of a generalized epithelium, the free (apical) surface bears microvilli and cilia. Mitochondria are shown concentrated near the basal surface of the cells, where they likely provide energy for the cell's transport activities. (**b**) This SEM shows the surface of a ciliated epithelium that lines most of the respiratory tract. The small, bristly areas are microvilli on the exposed surfaces of mucus-producing cells that are scattered among the ciliated epithelial cells. (SEM × 13,469) *(Prof. P. Motta, Dept. of Anatomy, University La Sapienza)*

The Basement Membrane

Epithelial cells must not only adhere to one another but also must remain firmly connected to the rest of the body. This function is performed by the **basement membrane,** which lies between the epithelium and underlying connective tissues (see Figure 4–3a●). There are no cells within the basement membrane, which consists of a network of protein fibers. The epithelial cells adjacent to the basement membrane are firmly attached to these protein fibers by hemidesmosomes. In addition to providing strength and resisting distortion, the basement membrane also provides a barrier that restricts the movement of proteins and other large molecules from the underlying connective tissue into the epithelium.

Epithelial Renewal and Repair

An epithelium must continually repair and renew itself. Epithelial cells may survive for just a day or two, because they are lost or destroyed by exposure to disruptive enzymes, toxic chemicals, pathogenic microorganisms, or mechanical abrasion. The only way the epithelium can maintain its structure over time is through the continuous division of unspecialized cells known as **stem cells,** or *germinative cells.* These cells are found in the deepest layers of the epithelium, near the basement membrane.

Classifying Epithelia

There are many different specialized types of epithelia. Yet they can easily be classified according to the number of cell layers and the shape of the exposed cells. This classification scheme recognizes two types of layering—*simple* and *stratified*—and three cell shapes—*squamous, cuboidal,* and *columnar* (Table 4–1).

Cell Layers

A **simple epithelium** consists of a single layer of cells that covers the basement membrane. Simple epithelia are thin. A single layer of cells is fragile and cannot provide much mechanical protection, so simple epithelia are found only in protected areas inside the body. They line internal compartments and passageways, including the ventral body cavities, the heart chambers, and blood vessels.

Simple epithelia are characteristic of regions where secretion or absorption occurs, such as the lining of the digestive and urinary tracts, and the gas-exchange surfaces of the lungs. In such places, thinness is an advantage, because it reduces the diffusion time for materials that cross the epithelial barrier.

A **stratified epithelium** provides a greater degree of protection because it has several layers of cells above the basement membrane. Stratified epithelia are usually found in areas subject to mechanical or chemical stresses, such as the surface of the skin and the linings of the mouth and anus.

TABLE 4–1 *Classifying Epithelia*

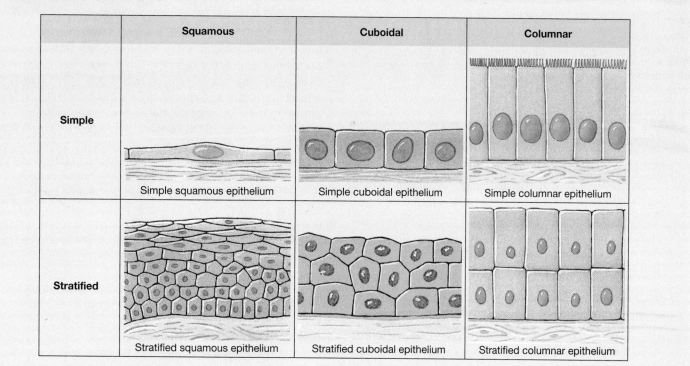

	Squamous	Cuboidal	Columnar
Simple	Simple squamous epithelium	Simple cuboidal epithelium	Simple columnar epithelium
Stratified	Stratified squamous epithelium	Stratified cuboidal epithelium	Stratified columnar epithelium

Cell Shape

In sectional view (perpendicular to the exposed surface and basement membrane), the cells at the surface of the epithelium usually have one of three basic shapes.

1. *Squamous.* In a **squamous epithelium** (SKWĀ-mus; *squama*, a plate or scale), the cells are thin and flat, and the nucleus occupies the thickest portion of each cell. Viewed from the surface, the cells look like fried eggs laid side by side.
2. *Cuboidal.* The cells of a **cuboidal epithelium** resemble little hexagonal boxes when seen in three dimensions, but in typical sectional view, they appear square. The nuclei lie near the center of each cell, and they form a neat row.
3. *Columnar.* In a **columnar epithelium** the cells are also hexagonal but taller and more slender. The nuclei are crowded into a narrow band close to the basement membrane, and the height of the epithelium is several times the distance between two nuclei.

The two basic epithelial arrangements (simple and stratified) and the three possible cell shapes (squamous, cuboidal, and columnar) enable one to describe almost every epithelium in the body. We will focus here on only a few major types of epithelia; additional examples will be encountered in later chapters.

Simple Squamous Epithelia

A **simple squamous epithelium** is found in protected regions where absorption takes place or where a slippery surface reduces friction (Figure 4–4a●). Examples are portions of the kidney tubules, the exchange surfaces of the lungs, the lining of ventral body cavities, and the linings of blood vessels and the inner surfaces of the heart.

Simple Cuboidal Epithelia

A **simple cuboidal epithelium** provides limited protection and occurs where secretion or absorption takes place (Figure 4–4b●). These functions are enhanced by larger cells, which have more room for the necessary organelles. Simple cuboidal epithelia secrete enzymes and buffers in the pancreas and salivary glands and line the ducts that discharge these secretions. Simple cuboidal epithelia also line portions of the kidney tubules involved in the production of urine.

Simple Columnar Epithelia

A **simple columnar epithelium** provides some protection and may also occur in areas of absorption or secretion (Figure 4–4c●). This type of epithelium lines the stomach, the intestinal tract, and many excretory ducts.

Pseudostratified Epithelia

Portions of the respiratory tract contain a columnar epithelium that includes a mixture of cell types. Because the nuclei are situated at varying distances from the surface, the epithelium has a layered appearance. But it is not a stratified epithelium, because all of the cells contact the basement membrane. Because it looks stratified but is not, it is known as a **pseudostratified columnar epithelium** (Figure 4–5a●). Epithelial cells of this tissue typically possess cilia. A ciliated pseudostratified columnar epithelium lines most of the nasal cavity, the trachea (windpipe) and bronchi, and portions of the male reproductive tract.

Transitional Epithelia

A **transitional epithelium** withstands considerable stretching. It lines the ureters and urinary bladder, where large changes in volume occur (Figure 4–5b●). In an empty urinary bladder, the epithelium seems to have many layers, and the outermost cells appear rounded or cuboidal. The multilayered appearance results from overcrowding. In a full urinary bladder, when the volume of urine has stretched the lining to its limits, the epithelium appears flattened, and more like a simple epithelium.

Stratified Squamous Epithelia

A **stratified squamous epithelium** is found where mechanical stresses are severe. The surface of the skin and the lining of the mouth, tongue, esophagus, and anus are good examples (Figure 4–5c●).

Clinical Note
CELLULAR ADAPTATION

Cells can change or adapt to their environment in order to protect themselves from injury. This process, called *cellular adaptation*, is a common and central aspect of many disease states. Cells can adapt by decreasing their size *(atrophy)*, increasing their size *(hypertrophy)*, increasing their numbers *(hyperplasia)*, or by reversibly replacing one mature cell type with another, less mature type *(metaplasia)*.

A good example of metaplasia is the response of bronchial (airway) tissues to prolonged exposure to cigarette smoke. In this process, normal columnar ciliated epithelial cells of the bronchial lining are replaced with stratified squamous epithelial cells. The newly formed replacement cells do not have cilia and do not secrete mucus. This results in a loss of two important pulmonary protective mechanisms. Bronchial metaplasia can often be reversed if the offending stimulus, usually cigarette smoke, is removed. When this occurs, normal columnar ciliated epithelial cells begin replacing the cells changed through the metaplastic process.

In addition to metaplasia, continued, prolonged exposure to cigarette smoke can result in the cancerous transformation of bronchial epithelial cells *(dysplasia)*. Bronchogenic (lung) cancers are a major health problem in industrialized countries. The mortality rate is high: lung cancers represent 32 percent of all cancer deaths. ■

SIMPLE SQUAMOUS EPITHELIUM

LOCATIONS: Epithelia lining ventral body cavities; lining of heart and blood vessels; portions of kidney tubules (thin sections of loops of Henle); inner lining of cornea; alveoli (air sacs) of lungs

FUNCTIONS: Reduces friction; controls vessel permeability; performs absorption and secretion

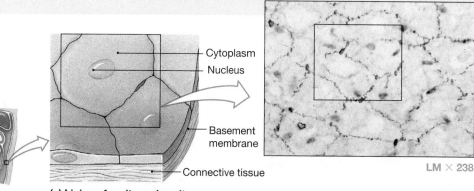

Cytoplasm
Nucleus
Basement membrane
Connective tissue

LM × 238

(a) Lining of peritoneal cavity

SIMPLE CUBOIDAL EPITHELIUM

LOCATIONS: Glands; ducts; portions of kidney tubules; thyroid gland

FUNCTIONS: Limited protection, secretion, absorption

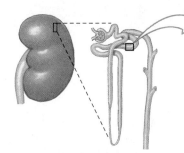

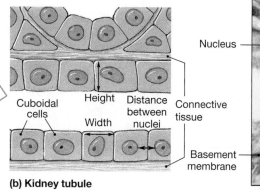

Cuboidal cells
Height
Width
Distance between nuclei
Connective tissue
Basement membrane

(b) Kidney tubule

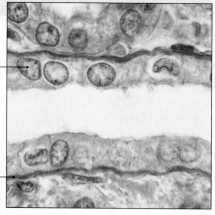

Nucleus

LM × 1350

SIMPLE COLUMNAR EPITHELIUM

LOCATIONS: Lining of stomach, intestine, gallbladder, uterine tubes, and collecting ducts of kidneys

FUNCTIONS: Protection, secretion, absorption

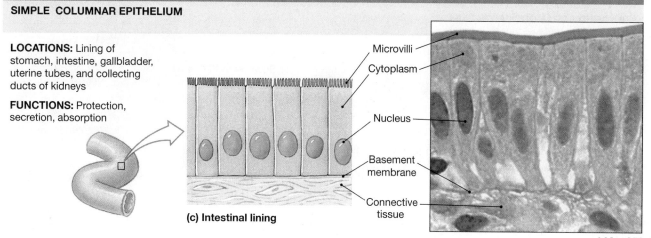

Microvilli
Cytoplasm
Nucleus
Basement membrane
Connective tissue

(c) Intestinal lining

LM × 350

● **FIGURE 4–4 Simple Epithelia.** (a) A simple squamous epithelium lines the peritoneal cavity. The three-dimensional drawing shows the epithelium in superficial and sectional views. (b) A simple cuboidal epithelium lines the kidney tubules. The diagrammatic view emphasizes the pertinent structural details for classifying this epithelium as cuboidal. (c) A simple columnar epithelium lines portions of the intestinal tract. In the diagrammatic view, note the relationships between the height and width of each cell; the relative size, shape, and location of nuclei; and the distance between adjacent nuclei. Compare with Figure 4–4b.

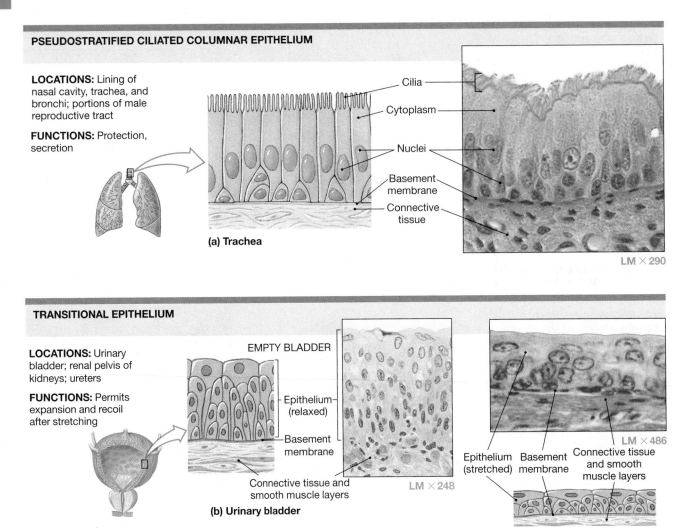

PSEUDOSTRATIFIED CILIATED COLUMNAR EPITHELIUM

LOCATIONS: Lining of nasal cavity, trachea, and bronchi; portions of male reproductive tract

FUNCTIONS: Protection, secretion

Cilia
Cytoplasm
Nuclei
Basement membrane
Connective tissue

(a) Trachea

LM × 290

TRANSITIONAL EPITHELIUM

LOCATIONS: Urinary bladder; renal pelvis of kidneys; ureters

FUNCTIONS: Permits expansion and recoil after stretching

EMPTY BLADDER

Epithelium (relaxed)
Basement membrane
Connective tissue and smooth muscle layers

LM × 248

LM × 486

Epithelium (stretched)   Basement membrane   Connective tissue and smooth muscle layers

(b) Urinary bladder

FULL BLADDER

STRATIFIED SQUAMOUS EPITHELIUM

LOCATIONS: Surface of skin; lining of mouth, throat, esophagus, rectum, anus, and vagina

FUNCTIONS: Provides physical protection against abrasion, pathogens, and chemical attack

Squamous superficial cells
Stem cells
Basement membrane
Connective tissue

(c) Surface of tongue

LM × 310

● **FIGURE 4–5 Stratified Epithelia.** (**a**) A pseudostratified ciliated columnar epithelium lines portions of the respiratory tract. Note that despite the uneven layering of the nuclei, all the cells contact the basement membrane. (**b**) A transitional epithelium lines the urinary bladder. At left is the relaxed epithelium in an empty bladder; in a full bladder (at right) the epithelium is stretched. (**c**) A stratified squamous epithelium covers the tongue.

Glandular Epithelia

Many epithelia contain gland cells that produce exocrine or endocrine secretions. Exocrine secretions are produced by exocrine glands, which discharge their products through a *duct*, or tube, onto some external or internal surface. Endocrine secretions (*hormones*) are produced by ductless glands and released into blood or tissue fluids.

Based on their structure, exocrine glands can be categorized as unicellular glands (called *goblet cells*) or as multicellular glands. The simplest multicellular exocrine gland is a *secretory sheet*, such as the epithelium of mucin-secreting cells that lines the stomach and protects it from its own acids and enzymes. Multicellular glands are further classified according to the branching pattern of the duct and the shape and branching pattern of the secretory portion of the gland. Additionally, exocrine glands can be classified according to *mode of secretion* or *type of secretion*.

Mode of Secretion

Each glandular epithelial cell releases its secretions by one of three mechanisms: (1) *merocrine secretion,* (2) *apocrine secretion,* or (3) *holocrine secretion* (Figure 4–6●).

In the most common mode of secretion, the product is released from secretory vesicles by exocytosis. ∞ p. 79 This mechanism, called **merocrine secretion** (MER-u-krin; *meros,* part + *krinein,* to secrete), is depicted in Figure 4–6a●. One

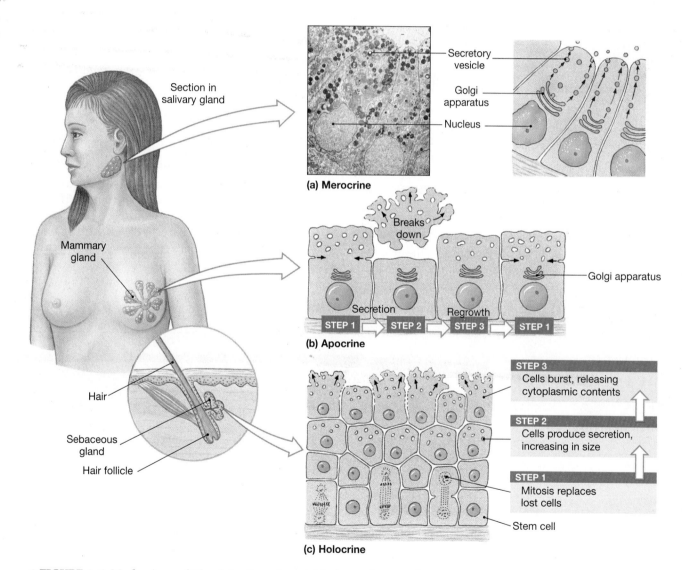

(a) Merocrine

(b) Apocrine

(c) Holocrine

●**FIGURE 4–6 Mechanisms of Glandular Secretion.** (**a**) In merocrine secretion, secretory vesicles are discharged at the free surface of the gland cell through exocytosis. (**b**) Apocrine secretion involves the loss of cytoplasm. Inclusions, secretory vesicles, and other cytoplasmic components are shed in the process. The gland cell then undergoes a period of growth and repair before releasing additional secretions. (**c**) Holocrine secretion occurs as superficial gland cells burst. Continued secretion involves the replacement of these cells through the mitotic divisions of underlying stem cells.

type of merocrine secretion, *mucin,* mixes with water to form **mucus,** which is an effective lubricant, a protective barrier, and a sticky trap for foreign particles and microorganisms. **Apocrine secretion** (AP-ō-krin; *apo-,* off) involves the loss of both cytoplasm and the secretory product (Figure 4–6b●). The outermost portion of the cytoplasm becomes packed with secretory vesicles before it is shed. Milk production in the mammary glands involves both merocrine and apocrine secretions. Whereas the processes of merocrine and apocrine secretion leave the cell intact and able to continue secreting, **holocrine secretion** (HOL-ō-krin; *holos,* entire) does not. Instead, the entire cell becomes packed with secretions and then bursts apart and dies (Figure 4–6c●). Sebaceous glands, which are associated with hair follicles, produce an oily hair coating by means of holocrine secretion.

Clinical Note
EXFOLIATIVE CYTOLOGY

Exfoliative cytology (eks-FŌ-lē-a-tiv; *ex-,* from + *folium,* leaf) is the study of cells shed or collected from epithelial surfaces. The cells are examined for a variety of reasons—for example, to check for cellular changes that indicate cancer or for genetic screening of a fetus. The cells are collected by sampling the fluids that cover the epithelia that line the respiratory, digestive, urinary, or reproductive tracts; by removing fluid from one of the ventral body cavities; or by removing cells from an epithelial surface.

A common example of exfoliative cytology is a *Pap test,* named after Dr. George Papanicolaou, who pioneered its use. The most familiar Pap test is that for cervical cancer, which involves scraping a small number of cells from the tip of the *cervix,* the portion of the uterus that projects into the vagina.

Amniocentesis is another important test that involves exfoliative cytology. In this procedure, exfoliated epithelial cells are collected from a sample of *amniotic fluid,* the fluid that surrounds and protects a developing fetus. Examination of these cells can determine whether the fetus has a genetic abnormality, such as *Down syndrome,* that affects the number or structure of chromosomes. ■

Type of Secretion

There are many kinds of exocrine secretions; all perform a variety of functions. Examples are enzymes that enter the digestive tract, perspiration on the skin, and milk produced by mammary glands.

Based on the type or types of secretions produced, exocrine glands can also be categorized as serous, mucous, or mixed. The secretions can have a variety of functions. *Serous glands* secrete a watery solution that contains enzymes. *Mucous glands* secrete mucins that form a thick, slippery mucus. *Mixed glands* contain more than one type of gland cell and may produce two different exocrine secretions, one serous and the other mucous.

Table 4–2 summarizes the classification of exocrine glands according to their mode of secretion and type of secretion.

→ CONCEPT CHECK QUESTIONS

1. You look at a tissue under a light microscope and see a simple squamous epithelium on the outer surface. Can it be a sample of the skin surface?
2. Secretory cells of sebaceous glands associated with hair follicles fill with secretions and then rupture, which releases their contents. What kind of secretion is this?
3. What physiological functions are enhanced by epithelial cells that bear microvilli and cilia?

Answers begin on p. 792.

TABLE 4–2 *A Classification of Exocrine Glands*

FEATURE	DESCRIPTION	EXAMPLES
MODE OF SECRETION		
Merocrine	Secretion occurs through exocytosis	Saliva from salivary glands; mucus in digestive and respiratory tracts; perspiration on the skin; milk in breasts
Apocrine	Secretion occurs through loss of cytoplasm that contains secretory product	Milk in breasts; viscous underarm perspiration
Holocrine	Secretion occurs through loss of entire cell that contains secretory product	Skin oils and waxy coating of hair (produced by sebaceous glands of the skin)
TYPE OF SECRETION		
Serous	Watery solution that contains enzymes	Secretions of parotid salivary gland
Mucous	Thick, slippery mucus	Secretions of sublingual salivary gland
Mixed	Produces more than one type of secretion	Secretions of submandibular salivary gland (serous and mucous)

■ Connective Tissues

Connective tissues are the most diverse tissues of the body. Bone, blood, and fat are familiar connective tissues that have very different functions and properties. All connective tissues have three basic components: (1) specialized cells, (2) protein fibers, and (3) a fluid known as **ground substance.** The extracellular protein fibers and ground substance form the **matrix** that surrounds the cells. Whereas epithelial tissue consists almost entirely of cells, the extracellular matrix accounts for most of the volume of connective tissues.

Connective tissues are distributed throughout the body but are never exposed to the outside environment. Many connective tissues are highly vascular (that is, they have many blood vessels) and contain receptors that provide pain, pressure, temperature, and other sensations. Connective tissue functions include the following:

■ *Support and protection.* The minerals and fibers produced by connective tissue cells establish a bony structural framework for the body, protect delicate organs, and surround and interconnect other tissue types.

■ *Transportation of materials.* Fluid connective tissues provide an efficient means of moving dissolved materials from one region of the body to another.

■ *Storage of energy reserves.* Fats are stored in connective tissue cells called *adipose cells* until needed.

■ *Defense of the body.* Specialized connective tissue cells respond to invasions by microorganisms through cell-to-cell interactions and the production of *antibodies.*

Classifying Connective Tissues

Based on the physical properties of their matrix, connective tissues are classified into three major types (Figure 4–7).

1. **Connective tissue proper** consists of many types of cells and fibers within a matrix that contains a syrupy ground substance. Examples are the tissue that underlies the skin, fatty tissue, and *tendons* and *ligaments*.
2. **Fluid connective tissues** have a distinctive population of cells suspended in a matrix of watery ground substance that contains dissolved proteins. The two fluid connective tissues are *blood* and *lymph*.
3. **Supporting connective tissues** have a less diverse cell population than connective tissue proper, and a matrix of dense ground substance and closely packed fibers. The body contains two supporting connective tissues: *cartilage* and *bone*. The fibrous matrix of bone is said to be calcified because it contains mineral deposits (primarily calcium salts) that give the bone strength and rigidity.

Connective Tissue Proper

Connective tissue proper contains a varied cell population, extracellular fibers, and a syrupy ground substance (Figure 4–8). Some cells of connective tissue proper are "permanent residents"; others are not always present because they leave to defend and repair areas of injured tissue.

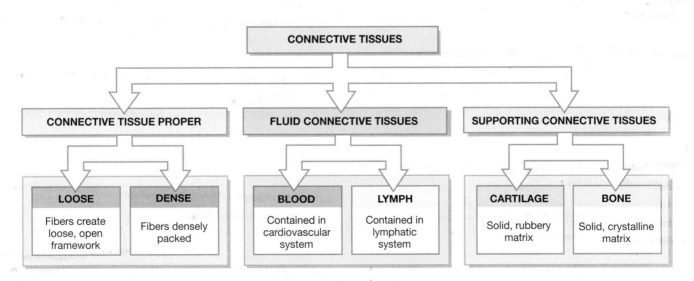

● **FIGURE 4–7** Major Types of Connective Tissue.

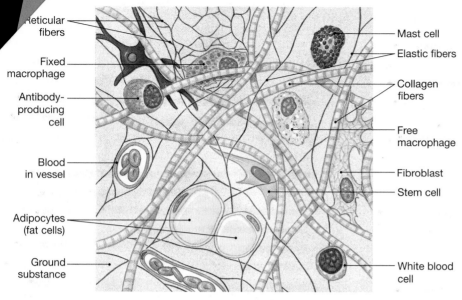

Reticular fibers
Fixed macrophage
Antibody-producing cell
Blood in vessel
Adipocytes (fat cells)
Ground substance

Mast cell
Elastic fibers
Collagen fibers
Free macrophage
Fibroblast
Stem cell
White blood cell

● **FIGURE 4–8 Cells and Fibers of Connective Tissue Proper.** This diagrammatic view shows the common cell types and fibers of connective tissue proper.

The Cell Population

Connective tissue proper includes the following major cell types:

- **Fibroblasts** (FĪ-brō-blasts) are the most abundant cells in connective tissue proper. These permanent residents are responsible for producing and maintaining the connective tissue fibers and ground substance.

- **Macrophages** (MAK-rō-fā-jez; *phagein,* to eat) are scattered throughout the matrix. These "big eater" cells engulf, or *phagocytize,* damaged cells or pathogens that enter the tissue; they also release chemicals that mobilize the immune system, and attract additional macrophages and other cells involved in tissue defense. ∞ p. 74 Macrophages that spend long periods of time in connective tissue are known as *fixed macrophages.* When an infection occurs, migrating macrophages (called *free macrophages*) are drawn to the affected area.

- **Fat cells**—also known as adipose cells, or **adipocytes** (AD-i-pō-sīts)—are permanent residents. A typical adipocyte contains such a large droplet of lipid that the nucleus and other organelles are squeezed to one side of the cell. The number of fat cells varies from one connective tissue to another, from one region of the body to another, and among individuals.

- **Mast cells** are small, mobile connective tissue cells often found near blood vessels. The cytoplasm of a mast cell is packed with vesicles filled with chemicals that are released to begin the body's defensive activities after an injury or infection (as discussed later in the chapter).

In addition to mast cells and free macrophages, both phagocytic and antibody-producing white blood cells may move through connective tissue proper. Their numbers increase markedly if the tissue is damaged, as does the production of **antibodies,** which are proteins that destroy invading microorganisms or foreign substances. *Stem cells* also respond to local injury by dividing to produce daughter cells that differentiate into fibroblasts, macrophages, or other connective tissue cells.

Connective Tissue Fibers

The three basic types of fibers—*collagen, elastic,* and *reticular*—are formed from protein subunits secreted by fibroblasts (see Figure 4–8●):

1. **Collagen fibers** are long, straight, and unbranched. These strong but flexible fibers are the most common fibers in connective tissue proper.
2. **Elastic fibers** contain the protein *elastin.* They are branched and wavy and after stretching will return to their original length.
3. **Reticular fibers** (*reticulum,* a network), the least common of the three, are thinner than collagen fibers and commonly form a branching, interwoven framework in various organs.

Ground Substance

Ground substance fills the spaces between cells and surrounds the connective tissue fibers (see Figure 4–8). In normal connective tissue proper, it is clear, colorless, and similar in consistency to maple syrup. This dense consistency slows the movement of bacteria and other pathogens, which makes them easier prey for phagocytes.

Clinical Note
MARFAN'S SYNDROME

Marfan's syndrome is an inherited condition caused by the production of an abnormal form of *fibrillin*, a carbohydrate-protein complex important to normal connective tissue strength and elasticity. Because most organs contain connective tissues, the effects of this defect are widespread. The most visible sign of Marfan's syndrome involves the skeleton; individuals with Marfan's syndrome are tall and have abnormally long arms, legs, and fingers. The most serious consequences involve the cardiovascular system. Roughly 90 percent of individuals with Marfan's syndrome have structural abnormalities in their cardiovascular systems. The most dangerous potential result is that the weakened connective tissues in the walls of major arteries, such as the aorta, may burst and cause a sudden, fatal loss of blood. ■

Clinical Note
BLUNT CHEST TRAUMA

The heart and great vessels are supported within the chest by strong connective tissues. These tissues allow limited movement of the heart and great vessels within the chest. The aortic arch is somewhat mobile, while the descending aorta is virtually immobile, particularly at the attachment of the fibrous *ligamentum arteriosum*. Vehicular trauma often involves a sudden and rapid deceleration, during which the heart continues to travel forward within the chest. This can place the aorta under great stress, and often results in tearing and rupture, typically at the point where the *ligamentum arteriosum* attaches. Traumatic thoracic aortic rupture has a very high mortality rate; 80–90 percent of victims die at the scene. ■

Connective tissue proper is categorized as either *loose connective tissues* or *dense connective tissues* on the basis of the relative proportions of cells, fibers, and ground substance. Loose connective tissues are the packing material of the body. These tissues fill spaces between organs, provide cushioning, and support epithelia. They also anchor blood vessels and nerves, store lipids, and provide a route for the diffusion of materials. Dense connective tissues are tough, strong, and durable. They resist tension and distortion and interconnect bones and muscles. Dense connective tissue also forms a thick fibrous layer, called a *capsule*, that surrounds internal organs (such as the liver, kidneys, and spleen) and encloses joint cavities.

Loose Connective Tissue

Loose connective tissue, or *areolar tissue* (*areola*, little space), is the least specialized connective tissue in the adult body (Figure 4–9a●). It contains all of the cells and fibers found in any connective tissue proper, in addition to an extensive blood supply.

Loose connective tissue forms a layer that separates the skin from underlying muscles, and provides both padding and a considerable amount of independent movement. Pinching the skin of the arm, for example, does not distort the underlying muscle. The ample blood supply in this tissue carries wandering cells to and from the tissue and provides for the metabolic needs (oxygen and nutrients) of nearby epithelial tissue.

Adipose Tissue

Adipose tissue, or fat, is a loose connective tissue that contains large numbers of fat cells, or adipocytes (Figure 4–9b●). The difference between loose connective tissue and adipose tissue is one of degree; a loose connective tissue is called adipose tissue when it becomes dominated by fat cells. Adipose tissue provides another source of padding and shock absorption for the body. It also provides insulation that slows heat loss through the skin, and it functions in energy storage.

Adipose tissue is common under the skin of the sides, buttocks, and breasts. It fills the bony sockets behind the eyes, surrounds the kidneys, and dominates extensive areas of loose connective tissue in the pericardial and peritoneal (abdominal) cavities.

Clinical Note
EMS AND OBESITY

Emergency Medical Services (EMS) is thought of as an intense, exciting, and physically challenging profession. In actuality, EMS is a fairly sedentary job punctuated by short periods of intense activity. Because of this, EMS providers are at risk for gaining weight. In addition, they also run the risk of physical injury if they do not remain physically fit.

As an EMS provider, you should carefully monitor your weight and maintain a healthy lifestyle. This should include some form of regular physical exercise, rational eating, and adequate rest and relaxation. You cannot depend on the job alone to provide the exercise required to maintain optimal fitness. If your agency does not have an ongoing fitness program, encourage them to start one. The result is a healthier, safer, and more satisfying career. ■

Dense Connective Tissues

Dense connective tissues consist mostly of collagen fibers; they may also be called *fibrous*, or *collagenous* (ko-LAJ-e-nus), tissues. The body has two types of dense connective tissues. In *dense regular connective tissue*, the collagen fibers are parallel to each other, packed tightly, and aligned with the forces applied to the tissue. **Tendons** are cords of dense regular connective tissue

LOOSE CONNECTIVE TISSUE

LOCATIONS: Beneath dermis of skin, digestive tract, respiratory and urinary tracts; between muscles; around blood vessels, nerves, and around joints

FUNCTIONS: Cushions organs; provides support but permits independent movement; phagocytic cells provide defense against pathogens

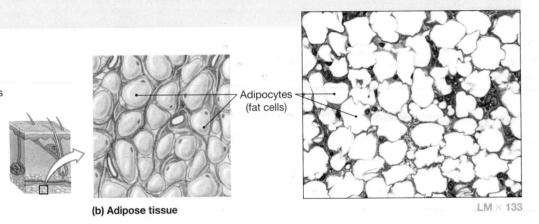

Collagen fibers

Mast cell

Fibroblasts

Fat cell

Elastic fibers

Macrophage

LM × 380

(a) Loose connective tissue

ADIPOSE TISSUE

LOCATIONS: Deep to the skin, especially at sides, buttocks, breasts; padding around eyes and kidneys

FUNCTIONS: Provides padding and cushions shocks; insulates (reduces heat loss); stores energy reserves

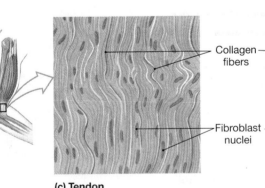

Adipocytes (fat cells)

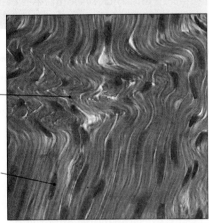

LM × 133

(b) Adipose tissue

DENSE CONNECTIVE TISSUES

LOCATIONS: Between skeletal muscles and skeleton (tendons); between bones (ligaments); covering skeletal muscles; capsules of internal organs

FUNCTIONS: Provides firm attachment; conducts pull of muscles; reduces friction between muscles; stabilizes relative positions of bones; helps prevent overexpansion of organs (such as the urinary bladder)

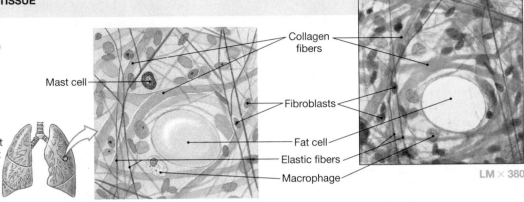

Collagen fibers

Fibroblast nuclei

LM × 440

(c) Tendon

● **FIGURE 4–9 Connective Tissue Proper: Loose and Dense Connective Tissues.** (a) All of the cells of connective tissue proper are found in loose connective tissue. (b) Adipose tissue is loose connective tissue dominated by adipocytes. In standard histological preparations, the tissue looks empty because the lipids in the fat cells dissolve in the alcohol used during tissue processing. (c) The dense regular connective tissue in a tendon largely consists of densely packed, parallel bundles of collagen fibers. The fibroblast nuclei are flattened between the bundles.

that attach skeletal muscles to bones (Figure 4–9c●). Their collagen fibers run along the length of the tendon and transfer the pull of the contracting muscle to the bone. **Ligaments** (LIG-aments) resemble tendons but connect one bone to another. Ligaments often contain elastic fibers as well as collagen fibers and thus can tolerate a modest amount of stretching. *Dense irregular connective tissue* contains an interwoven meshwork of collagen fibers (not shown). This structural pattern provides support to areas subjected to stresses from many directions and is what gives skin its strength.

Fluid Connective Tissues

Blood and **lymph** are connective tissues that contain distinctive collections of cells in a fluid matrix. Under normal conditions, the proteins dissolved in this watery matrix do not form large insoluble fibers. In blood, the watery matrix is called **plasma.**

A single cell type, the **red blood cell,** accounts for almost half the volume of blood. Red blood cells transport oxygen in the blood. Blood also contains small numbers of **white blood cells,** which are important components of the immune system, and **platelets,** which are cell fragments that function in blood clotting.

Together, plasma, lymph, and interstitial fluid constitute most of the extracellular fluid of the body. Plasma, which is confined to the blood vessels of the cardiovascular system, is kept in constant motion by contractions of the heart. When blood reaches tissues within thin-walled vessels called capillaries, filtration moves water and small solutes out of the capillaries and into the *interstitial fluid,* which surrounds the body's cells. Lymph forms as interstitial fluid enters small passageways, or *lymphatic vessels,* that eventually return it to the cardiovascular system. Along the way, cells of the immune system monitor the composition of the lymph and respond to signs of injury and infection.

Supporting Connective Tissues

Cartilage and bone are called supporting connective tissues because they provide a strong framework that supports the rest of the body. In these connective tissues the matrix contains numerous fibers and, in some cases, deposits of insoluble calcium salts.

Cartilage

The matrix of **cartilage** is a firm gel that contains embedded fibers. **Chondrocytes** (KON-drō-sīts), which are the only cells found within the matrix, live in small pockets known as *lacunae* (la-KOO-nē; *lacus,* pool). Because cartilage is avascular, chondrocytes must obtain nutrients and eliminate waste products by diffusion through the matrix. Blood vessels do not grow into cartilage because chondrocytes produce a chemical that discourages their formation. This lack of blood supply also limits the repair capabilities of cartilage. Structures of cartilage are covered and set apart from surrounding tissues by a **perichondrium** (per-i-KON-drē-um; *peri-,* around + *chondros,* cartilage), which contains an inner cellular layer and an outer fibrous layer.

TYPES OF CARTILAGE. The three major types of cartilage are *hyaline cartilage, elastic cartilage,* and *fibrocartilage* (Figure 4–10●):

1. **Hyaline cartilage** (HĪ-uh-lin; *hyalos,* glass) is the most common type of cartilage (Figure 4–10a●). The matrix contains closely-packed collagen fibers, which make hyaline cartilage tough but somewhat flexible. This type of cartilage connects the ribs to the sternum (breastbone), supports the conducting passageways of the respiratory tract, and covers opposing bone surfaces within joints.
2. **Elastic cartilage** contains numerous elastic fibers that make it extremely resilient and flexible (Figure 4–10b●). Elastic cartilage forms the external flap (the *auricle,* or *pinna*) of the outer ear, the epiglottis, and an airway to the middle ear (the *auditory tube*).
3. **Fibrocartilage** has little ground substance, and its matrix is dominated by collagen fibers (Figure 4–10c●). These fibers are densely interwoven, which makes this tissue extremely durable and tough. Pads of fibrocartilage lie between the vertebrae of the spinal column, between the pubic bones of the pelvis, and around or within a few joints and tendons. In these positions they resist compression, absorb shocks, and prevent damaging bone-to-bone contact. Cartilages heal poorly, and damaged fibrocartilage in joints such as the knee can interfere with normal movements.

Clinical Note
MENISCAL KNEE INJURIES

Sporting activities, particularly football, often result in knee injuries. A common knee injury is tearing or damaging one of the *menisci* (semilunar-shaped) cartilages. These cartilages, which contain hyaline, cushion the bones in the joint and prevent bone contact. The medial and lateral menisci are located on the tibial plateau. During stress, these cartilages can be torn or otherwise damaged. Because they are relatively avascular structures, they heal poorly and tend to be a source of ongoing problems for the patient. ■

HYALINE CARTILAGE

LOCATIONS: Between tips of ribs and bones of sternum; covering bone surfaces at synovial joints; supporting larynx (voice box), trachea, and bronchi; forming part of nasal septum

FUNCTIONS: Provides stiff but somewhat flexible support; reduces friction between bony surfaces

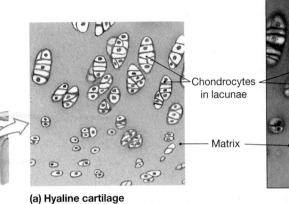
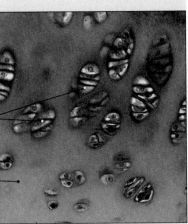

Chondrocytes in lacunae

Matrix

(a) Hyaline cartilage

LM × 500

ELASTIC CARTILAGE

LOCATIONS: Auricle of external ear; epiglottis; acoustic canal; cuneiform cartilages of larynx

FUNCTIONS: Provides support but tolerates distortion without damage and returns to original shape

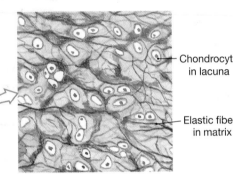
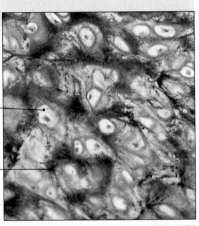

Chondrocyte in lacuna

Elastic fibers in matrix

(b) Elastic cartilage

LM × 358

FIBROCARTILAGE

LOCATIONS: Pads within knee joint; between pubic bones of pelvis; intervertebral discs separating vertebrae

FUNCTIONS: Resists compression; prevents bone-to-bone contact; limits relative movement

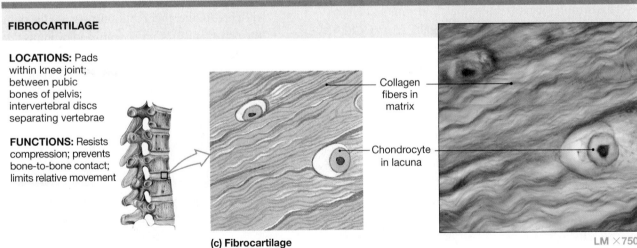

Collagen fibers in matrix

Chondrocyte in lacuna

(c) Fibrocartilage

LM ×750

● **FIGURE 4–10** **Types of Cartilage.** (a) Hyaline cartilage has a translucent matrix and lacks prominent fibers. (b) In elastic cartilage, the closely packed elastic fibers are visible between the chondrocytes. (c) In fibrocartilage, the collagen fibers are extremely dense, and the chondrocytes are relatively far apart.

Bone

Because the detailed histology of **bone,** or *osseous tissue* (OS-ē-us; *os,* bone), will be considered in Chapter 6, this discussion focuses on significant differences between cartilage and bone. The volume of ground substance in bone is very small. The matrix of bone consists mainly of hard calcium compounds and flexible collagen fibers. This combination gives bone truly remarkable properties, and makes it both strong and shatter-resistant. In its overall properties, bone can compete with the best steel-reinforced concrete.

The general organization of bone is shown in Figure 4–11●. Lacunae within the matrix contain bone cells, or **osteocytes** (OS-tē-ō-sīts; *os,* bone + *cyte,* cell). The lacunae surround the blood vessels that branch through the bony matrix. diffusion cannot occur through the bony matrix, os obtain nutrients through cytoplasmic extensions that blood vessels and other osteocytes. These extensions r through a branching network within the bony matrix called **canaliculi** (kan-a-LIK-ū-lē; little canals).

Except in joint cavities, where opposing surfaces are covered by hyaline cartilage, each bone is surrounded by a **periosteum** (per-ē-OS-tē-um), which is a covering made up of fibrous (outer) and cellular (inner) layers. Unlike cartilage, bone is constantly being remodeled throughout life, and complete repairs can be made even after severe damage has occurred. Table 4–3 compares cartilage and bone.

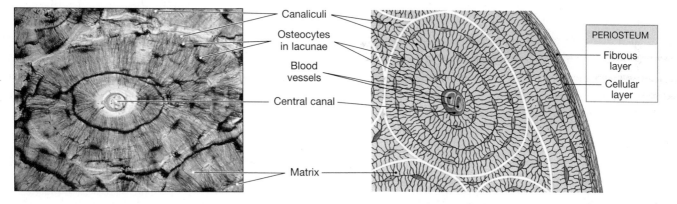

● **FIGURE 4–11 Bone.** The osteocytes in bone are usually organized in groups around a central space that contains blood vessels. In preparation for making the micrograph, a sample of bone was ground thin enough to become transparent. Bone dust filled the lacunae and the central canal, making them appear dark. (LM × 362)

TABLE 4–3 *A Comparison of Cartilage and Bone*

CHARACTERISTIC	CARTILAGE	BONE
STRUCTURAL FEATURES		
Cells	Chondrocytes in lacunae	Osteocytes in lacunae
Ground substance	Protein-polysaccharide gel and water	A small volume of liquid surrounds insoluble salts (calcium phosphate and calcium carbonate)
Fibers	Collagen, elastic, reticular fibers (proportions vary)	Collagen fibers predominate
Vascularity	None	Extensive
Covering	Perichondrium	Periosteum
Strength	Limited: bends easily but difficult to break	Strong: resists distortion until breaking point is reached
METABOLIC FEATURES		
Oxygen demands	Low	High
Nutrient delivery	By diffusion through matrix	By diffusion through cytoplasm and fluid in canaliculi
Repair capabilities	Limited	Extensive

■ Membranes

Some anatomical terms have more than one meaning, depending on the context. One such term is *membrane*. For example, at the cellular level, membranes are lipid bilayers that
restrict the passage of ions and other solutes. ∞ p. 61 At the
tissue level, membranes also form a barrier, such as the basement membranes that separate epithelia from connective tissues. At still another level, epithelia and connective tissues
combine to form membranes that cover and protect other
structures and tissues. The body has four such membranes:
mucous membranes, serous membranes, the *cutaneous membrane,* and *synovial membranes* (Figure 4–12●).

Mucous Membranes

Mucous membranes, or **mucosae** (mū-KŌ-sē), line cavities
that communicate with the exterior, including the digestive,
respiratory, reproductive, and urinary tracts (Figure 4–12a●).
The epithelial surfaces are kept moist at all times, typically by
mucous secretions or by exposure to fluids such as urine or semen. The connective tissue portion of a mucous membrane is
called the *lamina propria* (PRŌ-prē-uh).

Many mucous membranes are lined by simple epithelia
that perform absorptive or secretory functions, such as the
simple columnar epithelium of the digestive tract. However,
other types of epithelia may be involved. For example, a stratified squamous epithelium is part of the mucous membrane
of the mouth, and the mucous membrane along most of the
urinary tract has a transitional epithelium.

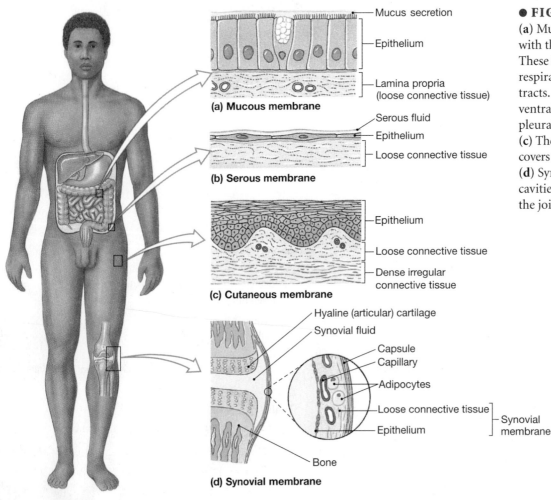

(a) Mucous membrane

(b) Serous membrane

(c) Cutaneous membrane

(d) Synovial membrane

● **FIGURE 4–12 Membranes.**
(**a**) Mucous membranes are coated
with the secretions of mucous glands.
These membranes line the digestive,
respiratory, urinary, and reproductive
tracts. (**b**) Serous membranes line the
ventral body cavities (the peritoneal,
pleural, and pericardial cavities).
(**c**) The cutaneous membrane, or skin,
covers the outer surface of the body.
(**d**) Synovial membranes line joint
cavities and produce the fluid within
the joint.

Serous Membranes

Serous membranes line the sealed, internal subdivisions of the ventral body cavity. There are three serous membranes; each consist of a simple epithelium supported by loose connective tissue (Figure 4–12b●). The **pleura** (PLOO-ra; *pleura*, rib) lines the pleural cavities and covers the lungs. The **peritoneum** (per-i-tō-NĒ-um; *peri*, around + *teinein*, to stretch) lines the peritoneal (abdominal) cavity and covers the surfaces of enclosed organs such as the liver and stomach. The **pericardium** (per-i-KAR-dē-um) lines the pericardial cavity and covers the heart.

A serous membrane has *parietal* and *visceral* portions that are in close contact at all times. ∞ p. 22 The parietal portion lines the inner surface of the cavity, and the visceral portion covers the outer surface of organs within the body cavity. For example, the visceral pericardium covers the heart, and the parietal pericardium lines the inner surfaces of the pericardial sac that surrounds the pericardial cavity. The primary function of any serous membrane is to minimize friction between the opposing parietal and visceral surfaces when an organ moves or changes shape. Friction is reduced by a watery, *serous fluid* formed by fluids that diffuse from underlying tissues.

The Cutaneous Membrane

The **cutaneous membrane,** or skin, covers the surface of the body (Figure 4–12c●). It consists of a stratified squamous epithelium and the underlying dense connective tissues. In contrast to serous or mucous membranes, the cutaneous membrane is thick, relatively waterproof, and usually dry. The skin is discussed in detail in Chapter 5.

Synovial Membranes

Bones contact one another at joints, or **articulations** (ar-tik-ū-LĀ-shuns). Joints that allow free movement are surrounded by a fibrous capsule and contain a joint cavity lined by a **synovial** (si-NO-vē-ul) **membrane** (Figure 4–12d●). Unlike the other three membranes, the synovial membrane consists primarily of loose connective tissue and an incomplete layer of epithelial tissue. In freely-movable joints, the bony surfaces do not come into direct contact with one another. If they did, impacts and abrasion would damage the opposing surfaces, and smooth movement would become almost impossible. Instead, the ends of the bones are covered with hyaline cartilage and separated by a viscous *synovial fluid* produced by fibroblasts in the connective tissue of the synovial membrane. The synovial fluid helps lubricate the joint and permits smooth movement.

> **CONCEPT CHECK QUESTIONS**

1. How does a cell membrane differ from a tissue-level membrane?
2. Serous membranes produce fluids. What is their function?
3. The lining of the nasal cavity is normally moist, contains numerous goblet cells, and rests on a layer of loose connective tissue. What type of membrane is this?

Answers begin on p. 792 .

■ Muscle Tissue

Muscle tissue is specialized for contraction. Muscle cell contraction involves interaction between filaments of *myosin* and *actin,* which are proteins found in the cytoskeletons of many cells. ∞ p. 76 In muscle cells, however, the filaments are more numerous and arranged so that their interaction produces a contraction of the entire cell.

There are three types of muscle tissue in the body— *skeletal, cardiac,* and *smooth muscle tissues* (Figure 4–13●). The contraction mechanism is the same in all of them, but the organization of their actin and myosin filaments differs. Because each type will be examined in later chapters, notably Chapter 7, this discussion will focus on general characteristics rather than specific details.

Skeletal Muscle Tissue

Skeletal muscle tissue contains very large, multinucleated cells (Figure 4–13a●). A large skeletal muscle cell may be 100 micrometers (μm; 1 μm = 1/25,000 in.) in diameter and 25 cm (10 in.) long. Because skeletal muscle cells are relatively long and slender, they are usually called *muscle fibers*. Skeletal muscle fibers are incapable of dividing, but new muscle fibers are produced through the divisions of stem cells in adult skeletal muscle tissue. As a result, at least partial repairs can occur after an injury.

Because the actin and myosin filaments are arranged in an organized pattern, skeletal muscle fibers appear to be marked by a series of bands known as *striations*. Skeletal muscle fibers will not usually contract unless stimulated by nerves. Because the nervous system provides voluntary control over its activities, skeletal muscle is described as *striated voluntary muscle.*

Cardiac Muscle Tissue

Cardiac muscle tissue is found only in the heart. Like skeletal muscle tissue, cardiac muscle tissue is striated, but each cardiac muscle cell is much smaller than a skeletal muscle fiber and usually has only a single nucleus (Figure 4–13b●). Cardiac

SKELETAL MUSCLE TISSUE

Cells are long, cylindrical, striated, and multinucleate.

LOCATIONS: Combined with connective tissues and neural tissue in skeletal muscles

FUNCTIONS: Moves or stabilizes the position of the skeleton; guards entrances and exits to the digestive, respiratory, and urinary tracts; generates heat; protects internal organs

Nuclei

Muscle fiber

Striations

(a) Skeletal muscle

LM × 180

CARDIAC MUSCLE TISSUE

Cells are short, branched, and striated, usually with a single nucleus; cells are interconnected by intercalated discs.

LOCATION: Heart

FUNCTIONS: Circulates blood; maintains blood (hydrostatic) pressure

Nucleus

Intercalated discs

Cardiac muscle cells

Striations

(b) Cardiac muscle

LM × 450

SMOOTH MUSCLE TISSUE

Cells are short, spindle-shaped, and nonstriated, with a single, central nucleus

LOCATIONS: Found in the walls of blood vessels and in digestive, respiratory, urinary, and reproductive organs

FUNCTIONS: Moves food, urine, and reproductive tract secretions; controls diameter of respiratory passageways; regulates diameter of blood vessels

Smooth muscle cell

Nucleus

(c) Smooth muscle

LM × 235

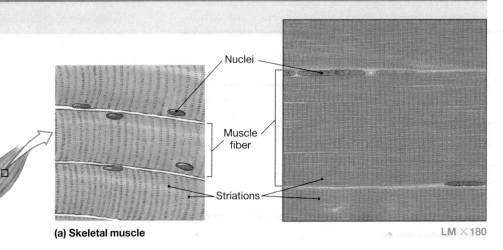
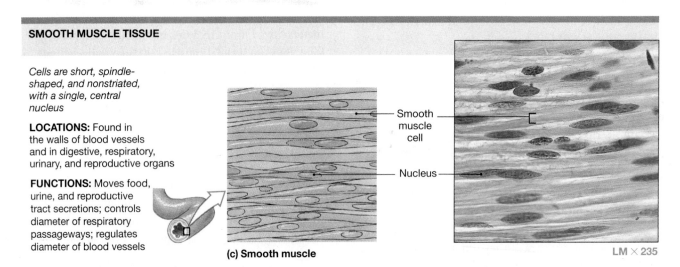

● **FIGURE 4–13 Muscle Tissue.** (**a**) Skeletal muscle fibers are large and have prominent striations (banding), multiple nuclei, and an unbranched arrangement. (**b**) Cardiac muscle cells differ from skeletal muscle fibers in three major ways: they are smaller, they branch, and they typically have a single, centrally placed nucleus. Like skeletal muscle fibers, cardiac muscle cells have striations. (**c**) Smooth muscle cells are small and spindle-shaped, have a central nucleus, and lack branches and striations.

muscle cells are interconnected at **intercalated** (in-TER-ka-lā-ted) **discs,** which are specialized attachment sites that contain gap junctions and desmosomes. Cardiac muscle cells branch, and form a network that efficiently conducts the force and stimulus for contraction from one area of the heart to another. Cardiac muscle tissue has a very limited ability to repair itself. Stem cells are lacking, and although some cardiac muscle cells do divide after an injury to the heart, the repairs are incomplete.

Cardiac muscle cells do not rely on nerve activity to start a contraction. Instead, specialized cells, called *pacemaker cells,* establish a regular rate of contraction. Although the nervous system can alter the rate of pacemaker activity, it does not provide voluntary control over individual cardiac muscle cells. Therefore, cardiac muscle is called *striated involuntary muscle.*

Smooth Muscle Tissue

Smooth muscle tissue is found in the walls of blood vessels; around hollow organs such as the urinary bladder; and in layers around the respiratory, circulatory, digestive, and reproductive tracts.

A smooth muscle cell is small and slender, and tapers to a point at each end; each smooth muscle cell has one nucleus (Figure 4–13c●). Unlike skeletal and cardiac muscle, the actin and myosin filaments in smooth muscle cells are scattered throughout the cytoplasm, so there are no striations. Smooth muscle cells can divide, so smooth muscle tissue can regenerate after injury.

Smooth muscle cells may contract on their own, or their contractions may be triggered by neural activity. The nervous system usually does not provide voluntary control over smooth muscle contractions, so smooth muscle is known as *nonstriated involuntary muscle.*

■ Neural Tissue

Neural tissue, which is also known as *nervous tissue* or *nerve tissue,* is specialized for the conduction of electrical impulses from one region of the body to another. Most neural tissue (98 percent) is concentrated in the brain and spinal cord, which are the control centers for the nervous system.

Neural tissue contains two basic types of cells: **neurons** (NOOR-onz; *neuro-,* nerve) and several different kinds of supporting cells, or **neuroglia** (noo-ROG-lē-uh or noo-rō-GLE-uh; *glia,* glue). Our conscious and unconscious thought processes reflect the communication among neurons. Neurons communicate through electrical events that affect their cell membranes. The neuroglia provide physical support for neural tissue, maintain the chemical composition of the tissue fluids, supply nutrients to neurons, and defend the tissue from infection.

The longest cells in your body are neurons, which reach up to a meter (39 in.) long. Most neurons cannot divide under normal circumstances, so they have a very limited ability to repair themselves after injury. A typical neuron has three main parts: (1) a **cell body** that contains a large nucleus, (2) numerous branching projections called **dendrites** (DEN-drīts; *dendron,* tree), and (3) one **axon** (Figure 4–14●). Dendrites receive information, typically from other neurons, and axons carry that information to other cells. Because axons tend to be very long and slender, they are also called *nerve fibers.* Each axon ends at *synaptic terminals,* where the neuron communicates with other cells. Chapter 8 considers the properties of neural tissue.

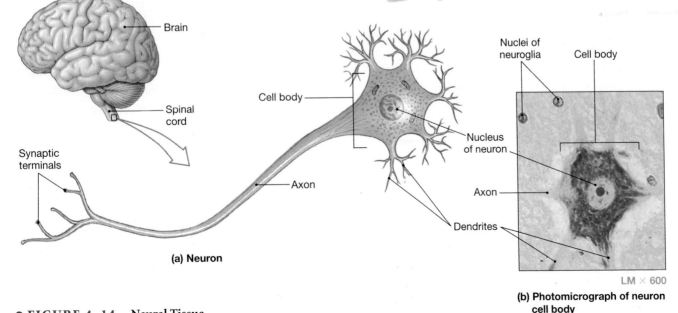

(a) Neuron

(b) Photomicrograph of neuron cell body

LM × 600

●**FIGURE 4–14** **Neural Tissue.**

■ Tissue Injuries and Repair

Tissues in the body are not independent of each other; they combine to form organs with diverse functions. Any injury affects several tissue types simultaneously, and these tissues must respond in a coordinated manner to restore homeostasis.

The restoration of homeostasis following a tissue injury involves two related processes: inflammation and regeneration. First, the area is isolated from neighboring healthy tissue while damaged cells, tissue components, and any dangerous microorganisms are cleaned up. This phase, which coordinates the activities of several different tissues, is called **inflammation**, or the *inflammatory response*. It produces several familiar indications, including swelling, warmth, redness, and pain. Inflammation can also result from an *infection*, which is the presence of pathogens (such as harmful bacteria) in tissues.

Inflammation can result from many other stimuli, including impact, abrasion, chemical irritation, and extreme temperatures (hot or cold). When any of these stimuli either kill cells, damage fibers, or injure tissues, they trigger the inflammatory response by stimulating connective tissue cells called *mast cells* (see p. 106). The mast cells release chemicals (*histamine and heparin*) that cause local blood vessels to *dilate* (enlarge in diameter) and become more permeable. The increased blood flow to the injured region makes it red and warm to the touch, and the diffusion of blood plasma causes the injured area to swell. The abnormal tissue conditions and the chemicals released by the mast cells also stimulate sensory nerve endings that produce the sensations of pain. These local circulatory changes increase the delivery of nutrients, oxygen, phagocytic white blood cells, and blood-clotting proteins, and they speed up the removal of waste products and toxins. Over a period of hours to days, this coordinated response generally succeeds in eliminating the inflammatory stimulus.

In the second phase that follows injury, the damaged tissues are replaced or repaired to restore normal function. This repair process is called **regeneration.** During regeneration, fibroblasts produce a dense network of collagen fibers known as *scar tissue* or *fibrous tissue*. Over time, scar tissue is usually remodeled and gradually assumes a more normal appearance.

Regeneration is more successful in some tissues than others: epithelia, connective tissues (except cartilage), and smooth muscle tissue usually regenerate well; other muscle tissues and neural tissue regenerate relatively poorly if at all. Because of different patterns of tissue organization, each organ has a different ability to regenerate after injury. Your skin, which is made up mostly of epithelia and connective tissues, regenerates rapidly. In contrast, damage to the heart is more serious, because, although its connective tissue can be repaired, most of the damaged cardiac muscle cells are replaced only by fibrous connective tissue. Such permanent replacement of normal tissues is called **fibrosis** (fī-BRŌ-sis). Fibrosis may occur in muscle and other tissues in response to injury, disease, or aging.

Inflammation and regeneration are controlled at the tissue level. The two phases overlap; isolation of the area of damaged tissue establishes a framework that guides the cells responsible for reconstruction, and repairs are under way well before cleanup operations have ended. Later chapters will examine inflammation (Chapter 14) and regeneration (Chapter 5) in more detail.

■ Tissues and Aging

Tissues change with age, and the speed and effectiveness of tissue repairs decrease. Repair and maintenance activities throughout the body slow down, and a combination of hormonal changes and alterations in lifestyle affect the structure and chemical composition of many tissues. Epithelia get thinner and connective tissues more fragile. Individuals bruise more easily and bones become brittle; joint pain and broken bones are common in the elderly. Cardiac muscle fibers and neurons cannot be replaced, and cumulative losses from even relatively minor damage can contribute to major health problems, such as cardiovascular disease or deterioration in mental function.

In later chapters we will consider the effects of aging on specific organs and systems. Some of these effects are genetically programmed. For example, as people age, their chondrocytes produce a slightly different form of the gelatinous compound that comprises the cartilage matrix. This difference in composition probably accounts for the thinner and less resilient cartilage of older people.

Other age-related changes in tissue structure have multiple causes. The age-related reduction in bone strength in women, a condition called *osteoporosis*, is often caused by a combination of inactivity, low dietary calcium intake, and a reduction in circulating estrogens (sex hormones). A program of exercise, calcium supplements, and hormonal replacement therapies can generally maintain normal bone structure for many years.

Aging and Cancer Incidence

Cancer rates increase with age, and roughly 25 percent of all Americans develop cancer at some point in their lives. It has been estimated that 70–80 percent of cancer cases result from chemical exposure, environmental factors, or some combination of the two, and 40 percent of these cancers are caused by cigarette smoke. Each year in the U.S., over 500,000 individuals die of cancer, which makes it second only to heart disease as a cause of death. Cancer development was discussed in Chapter 3. ∞ p. 88

Chapter Review

Access more review material online at *www.prenhall.com/bledsoe*. There you will find quiz questions, labeling activities, animations, essay questions, and web links.

Key Terms

basement membrane 99
blood 109
bone 111
cartilage 109
connective tissue 105
epithelium 96
fibroblasts 106

gap junction 98
gland cells 96
inflammation 116
lymph 109
macrophage 106
mucous membrane 112
muscle tissue 113

neural tissue 115
neuron 115
serous membrane 113
stem cells 99
tissue 95

Related Clinical Terms

adhesions Restrictive fibrous connections that can result from surgery, infection, or other injuries to serous membranes.

anaplasia (a-nuh-PLĀ-zē-uh) An irreversible change in the size and shape of tissue cells.

chemotherapy The administration of drugs that either kill cancerous tissues or prevent mitotic divisions.

dysplasia (dis-PLĀ-zē-uh) A change in the normal shape, size, and organization of tissue cells.

exfoliative cytology The study of cells shed or collected from epithelial surfaces.

liposuction A surgical procedure to remove unwanted adipose tissue by sucking it out through a tube.

metaplasia (me-tuh-PLĀ-zē-uh) A structural change in cells that alters the character of a tissue.

necrosis (ne-KRŌ-sis) Tissue destruction that occurs after cells have been injured or destroyed.

oncologists (on-KOL-o-jists) Physicians who specialize in identifying and treating cancers.

pathologists (pa-THOL-o-jists) Physicians who specialize in the study of disease processes.

pericarditis An inflammation of the pericardial lining that may lead to the accumulation of pericardial fluid (*pericardial effusion*).

peritonitis An inflammation of the peritoneum after infection or injury.

pleural effusion The accumulation of fluid within the pleural cavities as a result of chronic infection or inflammation of the pleura.

pleuritis (*pleurisy*) An inflammation of the pleural cavities.

regeneration The repair of injured tissues that follows inflammation.

remission A stage in which a tumor stops growing or becomes smaller; a major goal of cancer treatment.

Summary Outline

1. **Tissues** are collections of specialized cells and cell products that are organized to perform a relatively limited number of functions. The four **tissue types** are *epithelial tissue, connective tissue, muscle tissue,* and *neural tissue.* **Histology** is the study of tissues. *(Figure 4–1)*

   Key Note 95

EPITHELIAL TISSUE 96

1. An **epithelium** is an **avascular** layer of cells that forms a barrier that covers internal or external surfaces. **Glands** are secretory structures derived from epithelia.

   ### Functions of Epithelia 96

2. Epithelia provide physical protection, control permeability, provide sensations, and produce specialized secretions.

3. Gland cells are epithelial cells that produce secretions. **Exocrine** secretions are released onto body surfaces; **endocrine** secretions, known as *hormones,* are released by gland cells into the surrounding tissues.

   ### Intercellular Connections 97

4. The individual cells that make up tissues connect to one another or to extracellular protein fibers by means of *cell adhesion molecules (CAMs)* and intercellular cement to make specialized attachment sites called *cell junctions.* The three types are **tight junctions, gap junctions,** and **desmosomes.** *(Figure 4–2)*

   ### The Epithelial Surface 98

5. Many epithelial cells have microvilli. The coordinated beating of the cilia on a ciliated epithelium moves materials across the epithelial surface. *(Figure 4–3)*

   ### The Basement Membrane 99

6. The basal surface of each epithelium is connected to a noncellular **basement membrane.**

   ### Epithelial Renewal and Repair 99

7. Divisions by **stem cells,** or *germinative cells,* continually replace the short-lived epithelial cells.

   ### Classifying Epithelia 99

8. Epithelia are classified based on the number of cell layers and the shape of the exposed cells. *(Table 4–1)*

9. A **simple epithelium** has a single layer of cells that cover the basement membrane; a **stratified epithelium** has several layers. In a **squamous epithelium** the cells are thin and flat. Cells in a **cuboidal epithelium** resemble little hexagonal boxes; those in a **columnar epithelium** are taller and more slender. *(Figures 4–4, 4–5)*

   ### Glandular Epithelia 103

10. A glandular epithelial cell may release its secretions through *merocrine, apocrine,* or *holocrine* mechanisms. *(Figure 4–6)*

11. In **merocrine secretion** (the most common method of secretion), the product is released through exocytosis. **Apocrine secretion** involves the loss of both secretory product and cytoplasm. Unlike the first two methods, **holocrine secretion** destroys the cell, which becomes packed with secretions before it finally bursts.

12. Exocrine secretions may be serous (watery, usually contains enzymes), mucus (thick and slippery), or mixed (contains enzymes and lubricants). *(Table 4–2)*

CONNECTIVE TISSUES 105

1. All **connective** tissues have specialized cells and a **matrix,** which is composed of extracellular protein fibers and a **ground substance.**

2. Connective tissues are internal tissues with many important functions: establishing a structural framework; transporting fluids and dissolved materials; protecting delicate organs; supporting, surrounding, and interconnecting tissues; storing energy reserves; and defending the body from microorganisms.

   ### Classifying Connective Tissues 105

3. **Connective tissue proper** refers to connective tissues that contain varied cell populations and fiber types surrounded by a syrupy ground substance. *(Figure 4–7)*

4. **Fluid connective tissues** have a distinctive population of cells suspended in a watery ground substance that contains dissolved proteins. The two types are *blood* and *lymph.* *(Figure 4–7)*

5. **Supporting connective tissues** have a less diverse cell population than connective tissue proper and a dense matrix that contains closely-packed fibers. The two types of supporting connective tissues are cartilage and bone. *(Figure 4–7)*

   ### Connective Tissue Proper 105

6. Connective tissue proper contains fibers, a viscous ground substance, and a varied cell population.

7. Resident and migrating cells may include **fibroblasts, macrophages, fat cells, mast cells,** and various white blood cells. *(Figure 4–8)*

8. There are three types of fibers in connective tissue: **collagen fibers, reticular fibers,** and **elastic fibers.**

9. Connective tissue proper is classified as **loose** or **dense connective tissues.** Loose connective tissues include loose connective tissue (or *areolar tissue*) and **adipose tissue.** *(Figure 4–9a,b)*

10. Most of the volume in dense connective tissue consists of fibers. *Dense regular connective tissues* form **tendons** and **ligaments.** *(Figure 4–9c)*

   ### Fluid Connective Tissues 109

11. **Blood** and **lymph** are connective tissues that contain distinctive collections of cells in a fluid matrix.

12. Blood contains **red blood cells, white blood cells,** and **platelets;** the watery ground substance is called **plasma.**

13. Lymph forms as *interstitial fluid* enters the **lymphatic vessels,** which return lymph to the cardiovascular system.

Supporting Connective Tissues 109

14. Cartilage and bone are called supporting connective tissues because they support the rest of the body.

15. The matrix of **cartilage** consists of a firm gel and cells called **chondrocytes.** A fibrous **perichondrium** separates cartilage from surrounding tissues. The three types of cartilage are **hyaline cartilage, elastic cartilage,** and **fibrocartilage.** *(Figure 4–10)*

16. Chondrocytes rely on diffusion through the avascular matrix to obtain nutrients.

17. **Bone,** or *osseous tissue,* has a matrix that primarily consists of collagen fibers and calcium salts, which give it unique properties. *(Figure 4–11; Table 4–3)*

18. **Osteocytes** depend on diffusion through **canaliculi** for nutrient intake.

19. Each bone is surrounded by a **periosteum.**

MEMBRANES 112

1. Membranes form a barrier or an interface. Epithelia and connective tissues combine to form membranes that cover and protect other structures and tissues. There are four types of membranes: *mucous, serous, cutaneous,* and *synovial. (Figure 4–12)*

Mucous Membranes 112

2. **Mucous membranes** line cavities that communicate with the exterior. Their surfaces are normally moistened by mucous secretions.

Serous Membranes 113

3. **Serous membranes** line internal cavities and are delicate, moist, and very permeabie.

The Cutaneous Membrane 113

4. The **cutaneous membrane** covers the body surface. Unlike serous and mucous membranes, it is relatively thick, waterproof, and usually dry.

Synovial Membranes 113

5. **Synovial membranes,** which are located at joints (articulations), produce *synovial fluid* in joint cavities. Synovial fluid helps lubricate the joint and promotes smooth movement.

MUSCLE TISSUE 113

1. Muscle tissue is specialized for contraction. The three types of muscle tissue are *skeletal muscle, cardiac muscle,* and *smooth muscle. (Figure 4–13)*

Skeletal Muscle Tissue 113

2. **Skeletal muscle tissue** contains large cells, or *muscle fibers,* tied together by collagen and elastic fibers. Skeletal muscle fibers are multinucleated and have a striped appearance because of the organization of contractile proteins. Because we can control the contraction of skeletal muscle fibers through the nervous system, skeletal muscle is considered *striated voluntary muscle.*

Cardiac Muscle Tissue 113

3. **Cardiac muscle tissue** is found only in the heart. The nervous system does not provide voluntary control over cardiac muscle cells. Thus, cardiac muscle is *striated involuntary muscle.*

Smooth Muscle Tissue 115

4. **Smooth muscle tissue** is found in the walls of blood vessels, around hollow organs, and in layers around various tracts. It is classified as *nonstriated involuntary muscle.*

NEURAL TISSUE 115

1. **Neural tissue** is specialized to conduct electrical impulses that convey information from one area of the body to another.

2. Cells in neural tissue are either neurons or neuroglia. **Neurons** transmit information as electrical impulses in their cell membranes. Several kinds of **neuroglia** serve both supporting and defense functions. *(Figure 4–14)*

3. A typical neuron has a **cell body, dendrites,** and an **axon,** which ends at **synaptic terminals.**

TISSUE INJURIES AND REPAIR 116

1. Any injury affects several tissue types simultaneously, and they respond in a coordinated manner. Homeostasis is restored through two processes: *inflammation* and *regeneration.*

2. **Inflammation,** or the *inflammatory response,* isolates the injured area while damaged cells, tissue components, and any dangerous microorganisms are cleaned up.

3. **Regeneration** is the repair process that restores normal function.

TISSUES AND AGING 116

1. Tissues change with age. Repair and maintenance grow less efficient, and the structure and chemical composition of many tissues are altered.

Aging and Cancer Incidence 117

2. Cancer incidence increases with age; roughly three-quarters of all cases are caused by exposure to chemicals or environmental factors.

Review Questions

Level 1: Reviewing Facts and Terms

Match each item in column A with the most closely related item in column B. Place letters for answers in the spaces provided.

COLUMN A

____ 1. histology
____ 2. microvilli
____ 3. gap junction
____ 4. tight junction
____ 5. germinative cells
____ 6. destroys gland cell
____ 7. hormones
____ 8. adipocytes
____ 9. bone-to-bone attachment
____ 10. muscle-to-bone attachment
____ 11. skeletal muscle
____ 12. cardiac muscle

COLUMN B

a. repair and renewal
b. ligament
c. endocrine secretion
d. absorption and secretion
e. fat cells
f. holocrine secretion
g. study of tissues
h. tendon
i. intercellular connection
j. interlocking of membrane proteins
k. intercalated discs
l. striated, voluntary

13. The four basic tissue types found in the body are:
    (a) epithelia, connective, muscle, neural.
    (b) simple, cuboidal, squamous, stratified.
    (c) fibroblasts, adipocytes, melanocytes, mesenchymal.
    (d) lymphocytes, macrophages, microphages, adipocytes.

14. The most abundant connections between cells in the superficial layers of the skin are:
    (a) intermediate junctions.
    (b) gap junctions.
    (c) desmosomes.
    (d) tight junctions.

15. The three cell shapes that comprise epithelial tissue are:
    (a) simple, stratified, transitional.
    (b) simple, stratified, pseudostratified.
    (c) hexagonal, cuboidal, spherical.
    (d) cuboidal, squamous, columnar.

16. Mucous secretions that coat the passageways of the digestive and respiratory tracts result from _____ secretion.
    (a) apocrine
    (b) merocrine
    (c) holocrine
    (d) endocrine

17. The tissue that contains the fluid ground substance is:
    (a) epithelial.
    (b) neural.
    (c) muscle.
    (d) connective.

18. The three major types of cartilage in the body are:
    (a) collagen, reticular, elastic.
    (b) areolar, adipose, reticular.
    (c) hyaline, elastic, fibrocartilage.
    (d) keratin, reticular, elastic.

19. The primary function of serous membranes in the body is:
    (a) to minimize friction between opposing surfaces.
    (b) to line cavities that communicate with the exterior.
    (c) to perform absorptive and secretory functions.
    (d) to cover the surface of the body.

20. Large muscle fibers that are multinucleated, striated, and voluntary are found in _____ muscle tissue.
    (a) cardiac
    (b) skeletal
    (c) smooth
    (d) cardiac, skeletal, and smooth

21. Intercalated discs and pacemaker cells are characteristic of _____ muscle tissue.
    (a) smooth
    (b) cardiac
    (c) skeletal
    (d) smooth, cardiac, and skeletal

22. Dendrites, an axon, and a cell body are characteristics of cells found in _____ tissue.
    (a) neural
    (b) muscle
    (c) connective
    (d) epithelial

23. What are the four essential functions of epithelial tissue?

24. What three types of layering make epithelial tissue recognizable?

25. What three basic components are found in connective tissues?

26. Which fluid connective tissues and supporting connective tissues are found in the human body?

27. Which four kinds of membranes composed of epithelial and connective tissues cover and protect other structures and tissues in the body?

28. What two cell populations comprise neural tissue? What is the function of each?

Level 2: Reviewing Concepts

29. In body surfaces where mechanical stresses are severe, the dominant epithelium is _____ epithelium.
    (a) stratified squamous
    (b) simple cuboidal
    (c) simple columnar
    (d) stratified cuboidal

30. Why does holocrine secretion require continuous cell division?

31. What is the difference between an exocrine secretion and an endocrine secretion?

32. A significant structural feature in the digestive system is the presence of tight junctions located near the exposed surfaces of cells that line the digestive tract. Why are these junctions so important?

33. Why are infections always a serious threat after a severe burn or an abrasion?

34. What characteristics make the cutaneous membrane different from the serous and mucous membranes?

Level 3: Critical Thinking and Clinical Applications

35. A biology student loses the labels of two prepared slides she is studying. One is a slide of animal intestine, and the other is of animal esophagus. You volunteer to help her sort them out. How would you decide which slide is which?

36. You are asked to develop a scheme that can be used to identify the three different types of muscle tissue in two steps. What are the two steps?

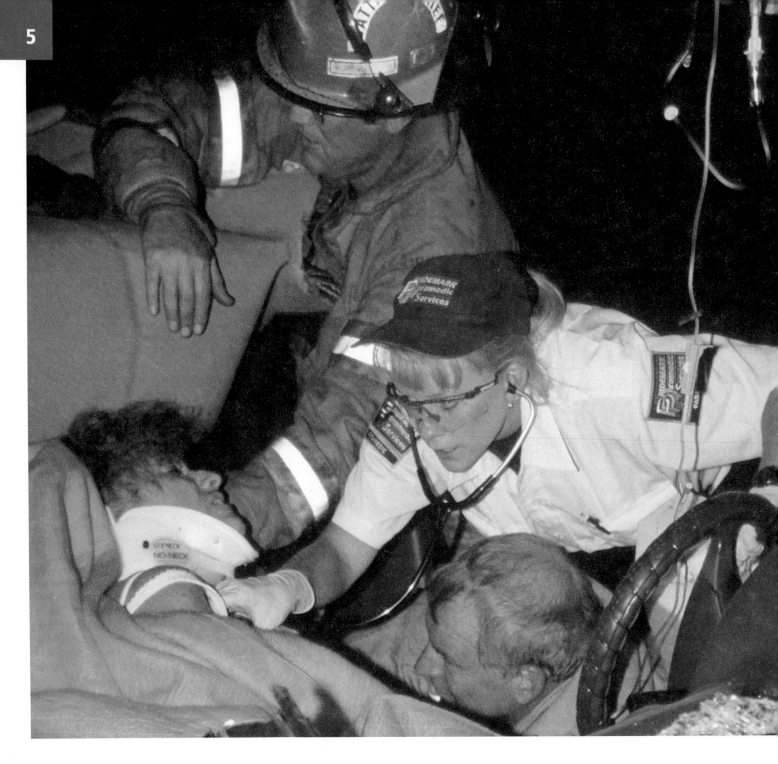

5 The Integumentary System

THE INTEGUMENTARY SYSTEM IS THE most visible organ system in the body. Because skin protects us from the environment, damage to the skin is a common emergency. Interruption in the integrity of the skin can provide a pathway for infectious substances to enter the body and cause infection. The initial pre-hospital care provided for skin injuries is a major factor in minimizing the chances of contamination and subsequent infection.

Chapter Outline

Chapter Objectives

1. Describe the general functions of the integumentary system. (pp. 123–124)
2. Describe the main structural features of the epidermis and explain their functional significance. (pp. 125–126)
3. Explain what accounts for individual differences in skin, such as skin color. (pp. 127–128)
4. Describe how the integumentary system helps regulate body temperature. (pp. 123, 137)
5. Discuss the effects of ultraviolet radiation on the skin and the role played by melanocytes. (pp. 126, 127, 128)
6. Discuss the functions of the skin's accessory structures. (pp. 134–138)
7. Describe the mechanisms that produce hair and that determine hair texture and color. (pp. 134–136)
8. Explain how the skin responds to injury and repairs itself. (pp. 138–140)
9. Summarize the effects of the aging process on the skin. (pp. 144–145)

Vocabulary Development

cornu horn; *stratum corneum*
cutis skin; *cutaneous*
derma skin; *dermis*
epi- above or over; *epidermis*
facere to make; *cornified*

germinare to start growing; *stratum germinativum*
keros horn; *keratin*
kyanos blue; *cyanosis*
luna moon; *lunula*

melas black; *melanin*
onyx nail; *eponychium*
papilla a nipple-shaped mound; *dermal papillae*

THE INTEGUMENTARY SYSTEM consists of the skin, hair, nails, and various glands. As the most visible organ system of the body, we devote a lot of time to improving its appearance. Washing your face and hands, brushing or trimming your hair, clipping your nails, showering, and applying deodorant are activities that modify the appearance or properties of the skin. And when something goes wrong with your skin, the effects are immediately apparent. You may notice a minor skin condition or blemish at once, whereas you may ignore more serious problems in other organ systems. Physicians, however, also pay attention to the skin because changes in its color, flexibility, or sensitivity may provide important clues about a disorder in another body system.

■ Integumentary Structure and Function

The **integumentary system,** or simply the **integument** (in-TEG-ū-ment), has two major components: the cutaneous membrane and the accessory structures. The **cutaneous membrane,** or *skin,* is an organ composed of the superficial epithelium, or **epidermis** (*epi-*, above), and the underlying connective tissues of the **dermis.** The **accessory structures** include hair, nails, and a variety of exocrine glands.

The general structure of the integument is shown in Figure 5–1●. Beneath the dermis, the loose connective tissue of the **subcutaneous layer,** or **hypodermis,** attaches the integument to deeper structures, such as muscles or bones. Although often not considered to be part of the integumentary system, we will include it here because its connective tissue fibers are interwoven with those of the dermis. As we will see, the five major functions of the integument are:

1. *Protection.* The skin covers and protects underlying tissues and organs from impacts, chemicals, and infections, and it prevents the loss of body fluids.
2. *Temperature maintenance.* The skin maintains normal body temperature by regulating heat exchange with the environment.
3. *Synthesis and storage of nutrients.* The epidermis synthesizes vitamin D_3, which is a steroid building block for a

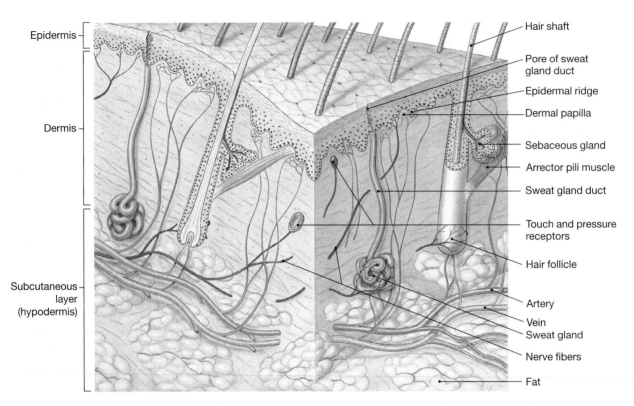

Epidermis

Dermis

Subcutaneous layer (hypodermis)

Hair shaft

Pore of sweat gland duct

Epidermal ridge

Dermal papilla

Sebaceous gland

Arrector pili muscle

Sweat gland duct

Touch and pressure receptors

Hair follicle

Artery

Vein

Sweat gland

Nerve fibers

Fat

● **FIGURE 5–1 The Components of the Integumentary System.** This diagrammatic section of skin shows the relationships among the major components of the integumentary system (with the exception of nails, shown in Figure 5–14).

hormone that aids calcium uptake. The dermis stores large reserves of lipids in adipose tissue.

4. *Sensory reception.*   Receptors in the integument detect touch, pressure, pain, and temperature stimuli and relay that information to the nervous system.

5. *Excretion and secretion.*   Integumentary glands excrete salts, water, and organic wastes. Additionally, specialized integumentary glands of the breasts secrete milk.

Each of these functions will be explored more fully as we discuss the individual components of the integument, beginning with the superficial layer of the skin: the epidermis.

The Epidermis

The epidermis consists of a stratified squamous epithelium of several different cell layers. ∞ p. 100 **Thick skin,** which is found on the palms of the hands and soles of the feet, contains five layers. Only four layers make up **thin skin,** which covers the rest of the body. Thin skin is about as thick as the wall of a plastic sandwich bag (about 0.08 mm thick), whereas thick skin is about as thick as a paper towel (0.5 mm). The words *thin* and *thick* refer to the relative thickness of the epidermis only, not to that of the integument as a whole.

Clinical Note
DERMATOLOGY

The integumentary system consists of the skin, hair, nails, and various glands. The largest and one of the most versatile organs of the body, it plays a major role in the maintenance of homeostasis. The study and treatment of diseases of the skin, hair, and nails is called *dermatology*, and physicians who specialize in the medical and surgical treatment of these disorders are called *dermatologists*. Dermatologists diagnose and treat skin, hair, and nail conditions and perform minor skin surgeries such as lesion removal. Physicians who perform major skin surgery are referred to as *plastic surgeons*. There are two general categories of plastic surgery: reconstructive and cosmetic (aesthetic). Reconstructive surgery corrects defects due to trauma, tumors, and disease, whereas cosmetic surgery is elective surgery that changes the physical appearance of a part of the body.

The integumentary system is essential to the body's continued maintenance of homeostasis. Its functions include:

■ *Protection.* The skin is a barrier that prevents the entry of microorganisms and other harmful substances. It also prevents the loss of water from the body to the environment.

■ *Temperature maintenance.* The integument plays a major role in temperature balance by regulating heat gain or heat loss to the environment.

■ *Nutrient storage.* The subcutaneous tissues contain a large reserve of lipids for use in metabolism and hormone production.

- *Sensory reception.* The skin contains a vast network of sensory fibers and nerves that detect touch, pressure, pain, and temperature.
- *Excretion and secretion.* The skin secretes water, salt, and organic wastes as the body strives to maintain homeostasis. The skin is important in the synthesis of important chemicals and hormones including the production of vitamin D.

The skin consists of the epidermis, the dermis, and the subcutaneous tissues. It is thickest on the lower back and thinnest on the genitalia. Various tensions on the skin cause it to be oriented into patterns called *cleavage lines* or *Langer's lines* (Figures 5–2● and 5–3●). Incisions and wounds perpendicular to cleavage lines tend to gape open; while incisions and wounds parallel to the cleavage lines tend to heal with a thin, less noticeable scar. ■

Layers of the Epidermis

Figure 5–4● shows the cell layers, or **strata** (singular *stratum*), in a section of thick skin. In order, from the basement membrane toward the free surface, are the *stratum germinativum*, three intermediate layers (the *stratum spinosum*, the *stratum granulosum*, and the *stratum lucidum*), and the *stratum corneum*.

STRATUM GERMINATIVUM. The deepest epidermal layer is called the **stratum germinativum** (STRA-tum jer-mi-na-TĒ-vum; *stratum*, layer + *germinare*, to start growing), or *stratum basale* (ba-SĀY-lē; *basis*, base). The cells of this layer are firmly attached to the basement membrane by hemidesmosomes. ∞ p. 99 The basement membrane separates the epidermis from the loose connective tissue of the adjacent dermis. The stratum germinativum forms **epidermal ridges,**

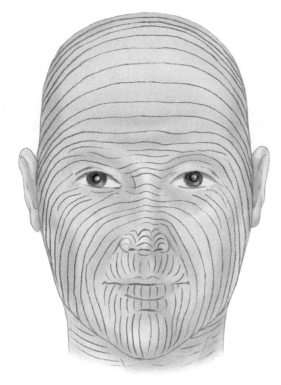

● **FIGURE 5–2 Cleavage Lines or Langer's Lines.** These lines represent natural tension patterns in the skin. If possible, surgeons try to make their incisions parallel to the cleavage lines so that the tension on the wound edges is kept to a minimum.

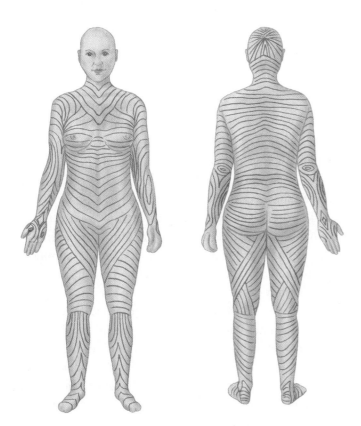

● **FIGURE 5–3 Cleavage Lines of the Face.**

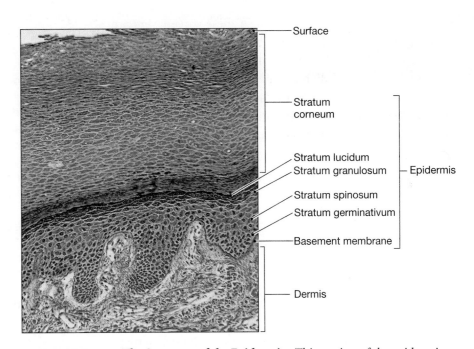

● **FIGURE 5–4 The Structure of the Epidermis.** This section of the epidermis in thick skin shows all five epidermal layers. (LM × 150)

which extend into the dermis and increase the area of contact between the two regions (see Figure 5–1). Dermal projections called *dermal papillae* (singular *papilla*; a nipple-shaped mound) extend upward between adjacent ridges. Because there are no blood vessels in the epidermis, epidermal cells must obtain nutrients delivered by dermal blood vessels. The combination of ridges and papillae increases the surface area for diffusion between the dermis and epidermis.

The contours of the skin surface follow the ridge patterns, which vary from small conical pegs (in thin skin) to the complex whorls on the thick skin of the palms and soles. The superficial ridges on the palms and soles (which overlie the dermal papillae) increase the surface area of the skin and increase friction, and ensure a secure grip. Ridge contours are genetically determined; those of each person are unique and do not change over the course of a lifetime. Fingerprints are ridge patterns on the tips of the fingers; they have been used to identify individuals in criminal investigations for over a century.

Large stem cells, or *germinative cells,* dominate the stratum germinativum, making it the layer where new cells are generated and begin to grow. The divisions of these cells replace cells that are lost or shed at the epithelial surface. The stratum germinativum also contains *melanocytes,* which are cells whose cytoplasmic processes extend between epithelial cells in this layer, and receptors that provide information about objects touching the skin. Melanocytes synthesize *melanin,* which is a brown, yellow-brown, or black pigment that colors the epidermis.

INTERMEDIATE STRATA. The cells in these three layers are progressively displaced from the basal layer as they become specialized to form the outer protective barrier of the skin. Each time a stem cell divides, one of the resulting daughter cells enters the next layer, the **stratum spinosum** (spiny layer), where it may continue to divide and add to the thickness of the epithelium. The **stratum granulosum** (grainy layer) consists of cells displaced from the stratum spinosum. The cells in this grainy layer have stopped dividing and begin making large amounts of the protein **keratin** (KER-a-tin; *keros,* horn). ∞ p. 45 Keratin is extremely durable and water-resistant. In humans, keratin not only coats the surface of the skin but also forms the basic structure of hair, calluses, and nails. In various other animals, it forms structures such as horns and hooves, feathers, and baleen plates (in the mouths of whales). In the thick skin of the palms and soles, a glassy **stratum lucidum** (clear layer) covers the stratum granulosum. The cells in this layer are flattened, densely packed, and filled with keratin.

STRATUM CORNEUM. The most superficial layer of the epidermis, the **stratum corneum** (KOR-nē-um; *cornu,* horn), consists of 15–30 layers of flattened and dead epithelial cells that have accumulated large amounts of keratin. Such cells are said to be **keratinized** (ker-A-tin-īzed), or **cornified** (KŌR-ni-fīd; *cornu,* horn + *facere,* to make). The dead cells in each layer of the stratum corneum remain tightly connected by desmosomes. ∞ p. 99 As a result, the keratinized cells of the stratum corneum are generally shed in large groups or sheets rather than individually.

It takes 2–4 weeks for a cell to move from the stratum germinativum to the stratum corneum. During this time, the cell is displaced from its oxygen and nutrient supply, becomes packed with keratin, and finally dies. The dead cells usually remain in the stratum corneum for an additional 2 weeks before they are shed or washed away. As superficial layers are lost, new layers arrive from the underlying strata. Thus, the deeper layers of the epithelium and underlying tissues remain protected by a barrier of dead, durable, and expendable cells. Normally, the surface of the stratum corneum is relatively dry, so it is unsuitable for the growth of many microorganisms.

Clinical Note
DRUG ADMINISTRATION THROUGH THE SKIN

Some drugs can be administered through the epidermis in a process called *transdermal medication administration.* Medications commonly administered by this route include nitroglycerin, female hormones, blood pressure medications, and narcotic pain relievers. Drug delivery systems have been developed that provide steady and prolonged delivery of the medication for as long as one week.

For transdermal administration, a predetermined amount of the drug is placed into an adhesive patch. These drugs must be lipid-soluble in order to penetrate the skin. Alternatively, the drug can be mixed with an inert solvent that is highly lipid-soluble. The resultant mixture is absorbed through the skin. The patch is applied to the skin, and the drug is slowly absorbed. Once the drug has penetrated the cell membranes in the *stratum corneum* and enters the underlying tissues, it is absorbed into the circulation. Transdermal administration provides significant patient convenience, because the drug is administered slowly and continuously over a long period, which helps minimize undesired side effects.

In the emergency setting, topical nitroglycerin preparations are among the drugs most frequently administered by the transdermal route. These can be rapidly applied to the patient's chest wall. Drug delivery begins promptly and continues at a steady rate. This minimizes many of the drug's unpleasant side effects (e.g., headache). ■

Skin Color

The color of your skin is caused by the interaction between (1) epidermal pigmentation and (2) the dermal blood supply.

PIGMENTATION. The epidermis contains variable amounts of two pigments: carotene and melanin. **Carotene** (KAR-uh-tēn) is an orange-yellow pigment that normally accumulates in epidermal cells. Carotene pigments are present in a variety of orange-colored vegetables, including carrots and squashes. Eating large amounts of carrots can actually cause the skin of light-skinned individuals to turn orange. The color change is less striking in the skin of darker individuals. Carotene can be converted to vitamin A, which is required for the normal maintenance of epithelial tissues and the synthesis of photoreceptor pigments in the eye. As noted earlier, **melanin** is a brown, yellow-brown, or black pigment produced by melanocytes.

Melanocytes (me-LAN-ō-sīts) manufacture and store melanin within intracellular vesicles. These vesicles are transferred to the epithelial cells of the stratum germinativum and stratum spinosum (Figure 5–5●). This transfer of pigmentation colors the entire epidermis. Melanocyte activity slowly increases in response to sunlight exposure, and peaks around

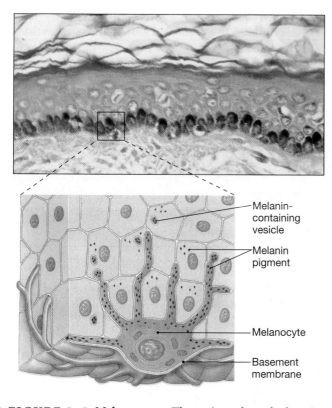

● **FIGURE 5–5 Melanocytes.** These views show the location and orientation of melanocytes in the deepest layer of the epidermis (stratum germinativum) of a dark-skinned person.

Melanin-containing vesicle

Melanin pigment

Melanocyte

Basement membrane

10 days after the initial exposure. *Freckles* are small, pigmented spots that appear on the skin of pale-skinned individuals. Freckles represent areas of greater-than-average melanin production. They tend to be most abundant on surfaces exposed to the sun, such as the face.

Sunlight contains significant amounts of **ultraviolet (UV) radiation.** A small amount of UV radiation is beneficial because it stimulates the synthesis of vitamin D_3 in the epidermis; this process is discussed in a later section. Too much ultraviolet radiation, however, produces immediate effects of mild or even serious burns. Melanin helps prevent skin damage by absorbing UV radiation before it reaches the deep layers of the epidermis and dermis. Within epidermal cells, melanin concentrates around the nuclear envelope and absorbs the UV radiation before it can damage nuclear DNA.

Despite the presence of melanin, long-term damage can result from repeated exposure to sunlight, even in darkly pigmented individuals. For example, alterations in the underlying connective tissues lead to premature wrinkling, and skin cancers can result from chromosomal damage in stem cells of the stratum germinativum or in melanocytes. One of the likely major consequences of the global depletion of the ozone layer in the upper atmosphere is a sharp increase in the rate of skin cancers, such as *malignant melanoma*. For this reason, limiting UV exposure through a combination of protective clothing and a sunblock with a sun protection factor (SPF) of at least 15 is recommended during outdoor activities. Individuals with fair skin are better off with an SPF of 20 to 30.

The ratio of melanocytes to stem cells ranges between 1:4 and 1:20, depending on the region of the body surveyed. The observed differences in skin color among individuals do not reflect the numbers of melanocytes but instead reflect the levels of melanin production. For example, in the inherited condition *albinism*, melanin is not produced by the melanocytes, even though these cells are of normal abundance and distribution. Individuals with this condition, known as *albinos*, have light-colored skin and hair.

Clinical Note
SKIN COLOR

Skin color is an important indicator of body function and varies from person to person. Changes in skin color are often evidence of a systemic disease. Normal skin color in light-skinned people is pink, which indicates adequate cardiorespiratory function and vascular integrity. In dark-skinned people, inspect the

(continued next page)

mucous membranes (such as the lips) to detect skin color changes. Paleness indicates decreased blood flow through the skin and results from anemia or conditions such as hypothermia or hypovolemia. A bluish discoloration of the skin is called *cyanosis* and is due to increased amounts of unoxygenated hemoglobin. The presence of cyanosis should alert care providers to an underlying problem that adversely affects oxygenation or perfusion. A yellowish discoloration of the skin, called *jaundice*, is due to the accumulation of bilirubin in the skin tissues. This is seen in liver diseases such as hepatitis and liver failure. A red coloration to the skin may be due to carbon monoxide poisoning. As carbon monoxide replaces oxygen on the hemoglobin molecules, the red color of the blood is enhanced, and this is evident in the skin. ■

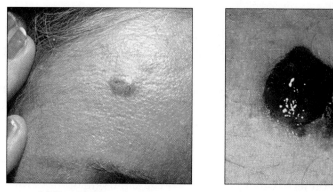

(a) Basal cell carcinoma   (b) Melanoma

● **FIGURE 5–6** Two Important Types of Skin Cancer.

DERMAL CIRCULATION. Blood with abundant oxygen is bright red, so blood vessels in the dermis normally give the skin a reddish tint that is most apparent in lightly pigmented individuals. When those vessels are dilated, as during inflammation, the red tones become much more pronounced. When the vessels are temporarily constricted, as when you are frightened, the skin becomes relatively pale. During a sustained reduction in circulatory supply, the blood in the skin loses oxygen and takes on a darker red tone. The skin then takes on a bluish coloration called **cyanosis** (sī-uh-NŌ-sis; *kyanos,* blue). In individuals of any skin color, cyanosis is most apparent in areas of thin skin, such as the lips, ears, or beneath the nails. It can be a response to extreme cold or a result of circulatory or respiratory disorders, such as heart failure or severe asthma.

The Epidermis and Vitamin D₃

Although strong sunlight can damage epithelial cells and deeper tissues, limited exposure to sunlight is very beneficial. When exposed to UV radiation, epidermal cells in the stratum spinosum and stratum germinativum convert a cholesterol-related steroid into **vitamin D₃.** This product is absorbed, modified, and released by the liver and then converted by the kidneys into *calcitriol,* which is a hormone essential for the absorption of calcium and phosphorus by the small intestine. An inadequate supply of vitamin D₃ leads to abnormal bone growth.

Skin Cancer

Skin cancers are the most common form of cancer. The most common *skin* cancer is **basal cell carcinoma** (Figure 5–6a●), which originates in the stratum germinativum

(basal) layer. Less common are **squamous cell carcinomas,** which involve more superficial layers of epidermal cells. Metastasis seldom occurs in either cancer, and most people survive these cancers. The usual treatment involves surgical removal of the tumor.

Compared with these common and seldom life-threatening cancers, **melanomas** (mel-a-NŌ-maz) (Figure 5–6b●) are extremely dangerous. A melanoma usually begins from a mole but may appear anywhere in the body. In this condition, cancerous melanocytes grow rapidly and metastasize through the lymphatic system. The outlook for long-term survival depends on when the condition is detected and treated. Avoiding exposure to UV radiation in sunlight (especially during the middle of the day) and using a sunblock (not a tanning oil) would largely prevent all three forms of cancer.

Key Note

The epidermis is a multi-layered, flexible, self-repairing barrier that prevents fluid loss, provides protection from UV radiation, produces vitamin D₃, and resists damage from abrasion, chemicals, and pathogens.

→ CONCEPT CHECK QUESTIONS

1. Excessive shedding of cells from the outer layer of skin in the scalp causes dandruff. What is the name of this epidermal layer?
2. Some criminals sand the tips of their fingers to avoid leaving recognizable fingerprints. Would this practice permanently remove fingerprints? Why or why not?
3. Why does exposure to sunlight or sunlamps darken the skin?

Answers begin on p. 792.

Clinical Note
SKIN LESIONS

A skin lesion is any disruption of normal skin tissue. The numerous types of skin lesions are usually classified as primary, and arise in previously normal skin (Figure 5–7●); secondary, which results from changes in primary lesions (Figure 5–8●); and vascular, which involves a blood vessel (Figure 5-9●).

Primary Lesions

- *Macule*—circumscribed change in skin color without elevation or depression of the surface; less than 1 cm in diameter. Examples: freckles, flat moles, measles.
- *Patch*—circumscribed change in skin color without elevation or depression of the surface; greater than 1 cm diameter. Examples: birthmarks, Mongolian spots, café-au-lait spots.
- *Papule*—solid, elevated area that varies in size; usually less than 1 cm in diameter. Examples: warts, elevated moles, insect bites.

- *Plaque*—solid, elevated area that varies in size; usually greater than 1 cm in diameter. Examples: psoriasis, seborrheic keratosis.
- *Wheal*—circumscribed, flat-topped, firm elevation with well-defined, palpable margin. Examples: allergic reactions, hives, insect stings.
- *Nodule*—small lesion in the dermal or subcutaneous tissue. Examples: skin cysts, fatty tumors (lipoma).
- *Tumor*—elevated, solid lesion; may be clearly demarcated; often greater than 2 cm in diameter. Examples: skin cancers, lipomas.
- *Vesicle*—elevated, circumscribed, superficial lesion that contains serous fluid; less than 1 cm in diameter. Examples: chicken pox (varicella), shingles (herpes zoster).
- *Bulla*—elevated, circumscribed, superficial lesion that contains serous fluid; less than 2 cm in diameter. Examples: blisters, second-degree burns.
- *Pustule*—elevated, superficial lesion; similar to a vesicle, but filled with purulent fluid (pus). Examples: acne pimple, spider bite.

● **FIGURE 5–7 Primary Skin Lesions.**

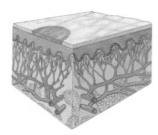

Macule – Flat spot, color varies from white to brown or from red to purple, diameter less than 1 cm

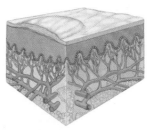

Plaque – Superficial papule, diameter more than 1 cm, rough texture

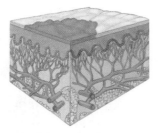

Patch – Irregular flat macule, diameter greater than 1 cm

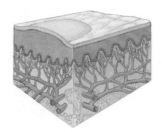

Wheal – Pink, irregular spot varying in size and shape

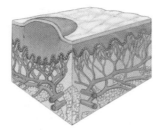

Papule – Elevated firm spot, color varies from brown to red or from pink to purplish red, diameter less than 1 cm

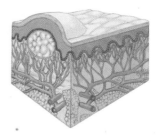

Nodule – Elevated firm spot, diameter 1–2 cm

(continued next page)

Clinical Note—*continued*
SKIN LESIONS

■ *Cyst*—elevated, circumscribed, encapsulated lesion; in dermis or subcutaneous tissue, filled with liquid or semisolid material. Examples: cystic acne, sebaceous cyst.

Secondary Lesions

■ *Fissure*—deep, linear crack or break from the epidermis that extends into the dermis; may be moist or dry. Examples: athlete's foot, cracks at corner of mouth.

■ *Erosion*—loss of part of the epidermis, depressed, moist, glistening; often follows rupture of a vesicle or bulla. Examples: varicella (after rupture), second-degree burn (after rupture).

■ *Ulcer*—loss of epidermis and dermis; concave; varies in size. Examples: decubitus ulcers (bed sores), stasis ulcers.

■ *Scar*—area of replacement fibrosis of the dermis that results from destruction of the dermis or subcutaneous layers. Examples: surgical scar, acne scars.

■ *Keloid*—sharply elevated, irregularly shaped, progressively enlarging scar; caused by excessive collagen; more common in blacks and Orientals. Example: large scar from minor trauma, such as ear piercing.

■ *Excoriation*—loss of the epidermis; linear, hollowed-out, crusted area. Examples: abrasions, scabies (mites).

■ *Scale*—desiccated, thin plates of cornified epithelial cells; often due to abnormal keratinization. Example: psoriasis.

■ *Crust*—results from drying of serum, blood, sebum, or interstitial fluid over the epidermis. Examples: seborrheic dermatitis, impetigo.

■ *Lichenification*—rough, thickened epidermis secondary to persistent rubbing, itching, or skin irritation; often involves flexor surface of extremity. Examples: chronic dermatitis, fungal infections.

■ *Atrophy*—thinning of the skin surface and loss of skin markings; skin appears translucent and paperlike. Examples: aged skin, striae.

Vascular Lesions

■ *Purpura*—reddish-purple patches due to blood in the dermis; greater than 0.5 cm in diameter. Example: idiopathic thrombocytopenia purpura (ITP).

Tumor – Elevated solid, diameter more than 2 cm, may be same color as skin

Pustule – Elevated area, diameter less than 1 cm, contains purulent fluid

Vesicle – Elevated area, diameter less than 1 cm, contains serous fluid

Cyst – Elevated, palpable area containing liquid or viscous matter

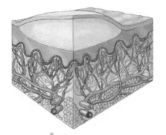

Bulla – Vesicle with diameter more than 1 cm

Telangiectasia – Red, threadlike line

● **FIGURE 5–7** *(continued)* **Primary Skin Lesions.**

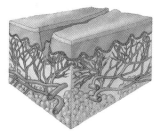

Fissure – Linear red crack ranging into dermis

Scar – Fibrous, depth varies, color ranges from white to red

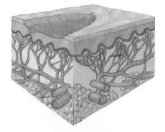

Erosion – Depression in epidermis, caused by tissue loss

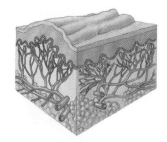

Keloid – Elevated scar, irregular shape, larger than original wound

Ulcer – Red or purplish depression ranging into dermis, caused by tissue loss

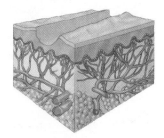

Excoriation – Linear, may be hollow or crusted, caused by loss of epidermis leaving dermis exposed

Scale – Elevated area of excessive exfoliation, varies in thickness, shape, and dryness, and ranges in color from white to silver or tan

Crust – Reddish, brown, black, tan, or yellowish dried blood, serum, or pus

Lichenification – Thickening and hardening of epidermis with emphasized lines in skin, resembles lichen

Atrophy – Skin surface thins and markings disappear, semitransparent parchment-like appearance

● **FIGURE 5–8 Secondary Skin Lesions.**

(continued next page)

■ *Petechiae*—reddish-purple patches due to blood in the dermis; less than 0.5 cm in diameter (often pinpoint). Examples: Rocky Mountain spotted fever, acute meningiococcemia.

■ *Ecchymosis*—reddish purple blotches due to leakage of blood from artery or vein; size varies. Example: bruising due to trauma.

■ *Spider angioma*—dilation of superficial skin arteriole with central, reddish, pulsating punctum with dilating legs. Examples: pregnancy, alcoholic liver disease.

■ *Venous star*—dilation of superficial skin vein; compressible punctum with dilating legs. Example: age-related lesion (usually on the head and neck).

■ *Capillary hemangioma*—irregular red spots due to capillary dilation; compressible. Examples: birthmarks in babies (often disappear by age 2 years).

■ *Telangiectasia*—fine, irregular red lines due to dilated superficial blood vessels. Examples: vascular inflammation, rosacea. ■

● **FIGURE 5–9** **Vascular Skin Lesions**

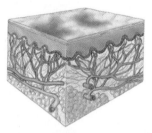

Purpura – Reddish-purple blotches, diameter more than 0.5 cm

Spider angioma – Reddish legs radiate from red spot

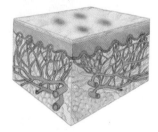

Petechiae – Reddish-purple spots, diameter less than 0.5 cm

Venous star – Bluish legs radiate from blue center

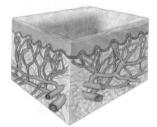

Ecchymoses – Reddish-purple blotch, size varies

Capillary hemangioma – Irregular red spots

The Dermis

The dermis lies beneath the epidermis. It has two major components: a superficial *papillary layer* and a deeper *reticular layer*.

Layers of the Dermis

The **papillary layer,** named after the dermal papillae, consists of loose connective tissue that supports and nourishes the epidermis. This region contains the capillaries and nerves that supply the surface of the skin.

The deeper **reticular layer** consists of an interwoven meshwork of dense, irregular connective tissue. Both *elastic fibers* and *collagen fibers* are present. The elastic fibers provide flexibility, and the collagen fibers limit that flexibility to prevent damage to the tissue. Bundles of collagen fibers blend into those of the papillary layer above, and blurs the boundary between these layers. Collagen fibers also extend into the subcutaneous layer below.

Other Dermal Components
In addition to protein-based elastic and collagen fibers, the dermis contains the mixed cell populations of connective tissue proper. ∞ p. 105 Epidermal accessory organs, such as hair follicles and sweat glands, extend into the dermis (see Figure 5–1).

Other organ systems communicate with the skin through their connections to the dermis. For example, both dermal layers contain a network of blood vessels (cardiovascular system), lymph vessels (lymphatic system), and nerve fibers (nervous system).

Blood vessels provide nutrients and oxygen and remove carbon dioxide and waste products. Both the blood vessels and the lymph vessels help local tissues defend and repair themselves after an injury or infection. The nerve fibers control blood flow, adjust gland secretion rates, and monitor sensory receptors in the dermis and the deeper layers of the epidermis. These receptors, which provide sensations of touch, pain, pressure, and temperature, will be described in Chapter 9.

Key Note
The dermis provides mechanical strength, flexibility, and protection for underlying tissues. It is highly vascular and contains a variety of sensory receptors that provide information about the external environment.

The Subcutaneous Layer

An extensive network of connective tissue fibers attaches the dermis to the subcutaneous layer (hypodermis). The boundary between the two is indistinct, and although the subcutaneous layer is not actually a part of the integument, it is important in stabilizing the position of the skin relative to underlying tissues, such as skeletal muscles or other organs, while permitting independent movement.

The subcutaneous layer consists of loose connective tissue with many fat cells. These adipose cells provide infants and small children with a layer of "baby fat," which helps them reduce heat loss. Subcutaneous fat also serves as an energy reserve and a shock absorber for the rough-and-tumble activities of our early years.

As we grow and mature, the distribution of subcutaneous fat changes. Beginning at puberty, men accumulate subcutaneous fat at the neck, upper arms, along the lower back, and over the buttocks; women do so in the breasts, buttocks, hips, and thighs. Both women and men, however, may accumulate distressing amounts of adipose tissue in the abdominal region, which produces a prominent "pot belly."

The hypodermis is quite elastic. Below its superficial region with its large blood vessels, the hypodermis contains no vital organs and few capillaries. The lack of vital organs makes *subcutaneous injection* a useful method for administering drugs using a *hypodermic needle*.

Clinical Note
MEDICAL CONDITIONS OF THE SKIN

Many diseases affect the skin. These can be localized to the skin or can involve other body organs and systems. Often, skin lesions are due to infections or problems elsewhere in the body.

Exanthems

Many of the viruses that infect humans cause characteristic skin rashes during the course of the infection. An exanthem is an acute, generalized skin eruption that can be caused by drugs, viruses, or certain bacteria. Usually, the eruption causes macules or papules, although some will have vesicles. Most of the childhood viral infections are characterized by recognizable viral exanthems (Figure 5–10●). These include: rubella (three-day, or German measles), rubeola (red measles), roseola, and varicella (chicken pox). Certain bacterial infections also result in the development of an exanthem. The most common of these, scarlet fever (scarletina), is caused by group A beta-hemolytic streptococcus infection, usually pharyngitis. The rash of scarlet fever

feels like sandpaper, and the face is flushed with an obvious pallor around the mouth.

Many of the viral exanthems are asymptomatic. Some, however, can cause itching and pain. Varicella causes vesicles that can actually shed the virus. Exanthems do not require specific treatment. Instead, treatment is directed at the underlying cause or at the symptoms that make the patient uncomfortable.

Herpes Zoster

Herpes zoster (shingles) is a disease characterized by the eruption of groups of vesicles along the dermatome of a sensory nerve. The virus that causes varicella (chicken pox) and herpes zoster is the varicella-zoster virus (VZV). After the initial infection with VZV, the patient will develop varicella. This usually occurs during childhood. During the course of the varicella infection, the VZV enters the ganglia for the sensory nerve and remains there for life. Later in life, the virus becomes reactivated and spreads along the sensory nerve that

(continued next page)

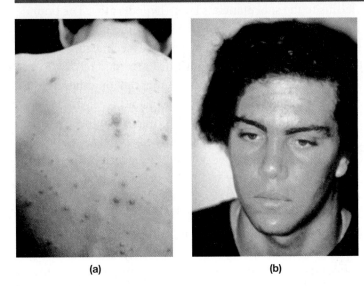

(a)   (b)

● **FIGURE 5–10** **Examples of Viral Exanthems Associated with Childhood Illness.** (**a**) varicella or chicken pox; (**b**) rubeola or red measles.

it infected. Stress, disease, and immunosuppression appear to be causes of virus reactivation, although in most cases the cause is idiopathic (cannot be determined). Initially, zoster causes pain that is often severe, followed by redness of the affected area. The redness and rash are limited to the dermatome of the affected nerve, and the rash does not cross the midline of the body. Zoster infections can also affect the cranial nerves or involve the eye. Eventually, the vesicles will break out. Although antiviral drugs do not eradicate the virus, they are effective to decrease the severity of the infection and the length of the outbreak, which may last for several weeks.

Staphylococcal Scalded Skin Syndrome

Staphylococcal scalded skin syndrome (SSSS) is most frequently seen in children less than 5 years of age. It is caused by infection with group II staphylococci, which produces a toxin that causes separation of the skin just below the granular layer of the epider-

mis. This results in sloughing of the skin in the affected areas. The toxin, an epidermolysin, is usually produced at a site other than the skin and is delivered to the skin via the circulatory system.

SSSS begins with fever, malaise, irritability, and runny nose, followed by generalized erythema (redness) with exquisite tenderness. The erythema spreads from the face and trunk to cover the entire body with the exception of the palms of the hands, soles of the feet, and mucous membranes. Within 48 hours, blisters and bullae may form and the pain becomes severe. As the blisters rupture, fluid is lost, which results in dehydration. In severe cases, the skin of the entire body may slough. Treatment of SSSS includes antibiotics to eradicate the underlying infection and replacement fluids if needed. The skin should be treated the same as with a severe burn. Children with a large area of skin involvement are best treated in a burn center.

Toxic Epidermal Necrolysis

Toxic epidermal necrolysis (TEN) is a serious and sometimes fatal drug reaction that is similar to staphylococcal scalded skin syndrome. It primarily affects adults. Antibiotics of the sulfonamide class, nonsteroidal anti-inflammatory drugs, and anticonvulsants seem to be the cause of most cases of TEN. The skin eruption is preceded by fever, malaise, anorexia, and inflammation of the eyelids and mucous membranes. The skin becomes reddened and is very tender, first in the axillae and groin, then extends over the body surface. Blisters and bullae form, and the entire epidermis may be shed. TEN causes full-thickness necrosis (death) of the epidermis with subepidermal blister formations. Treatment is identical as for burns, and severe cases are best managed in a burn unit. ■

Accessory Structures

Accessory structures include hair and hair follicles, sebaceous glands, sweat glands, and nails.

Hair and Hair Follicles

Hairs project above the surface of the skin almost everywhere except the sides and soles of the feet, the palms of the hands, the sides of the fingers and toes, the lips, and portions of the external genital organs. Hairs are nonliving structures produced in organs called **hair follicles.**

THE STRUCTURE OF HAIR AND HAIR FOLLICLES. Hair follicles project deep into the dermis and often extend into the underly-

ing subcutaneous layer (Figure 5–11a●). The walls of each follicle contain all the cell layers found in the epidermis. The epithelium at the base of a follicle forms a cap over the **hair papilla,** which is a peg of connective tissue that contains capillaries and nerves. Hair is formed by the repeated divisions of epithelial stem cells that surround the hair papilla. As the daughter cells are pushed toward the surface, the hair lengthens, and the cells undergo keratinization and die. The point at which this occurs is about halfway to the skin surface and marks the boundary between the **hair root** (the portion that anchors the hair into the skin) and the **hair shaft** (the part we see on the surface) (Figure 5–11b●). Each hair shaft consists of three layers of dead, keratinized cells (Figure 5–11c●). The surface layer, or **cuticle,** is made up of an overlapping shingle-like layer of cells. The un-

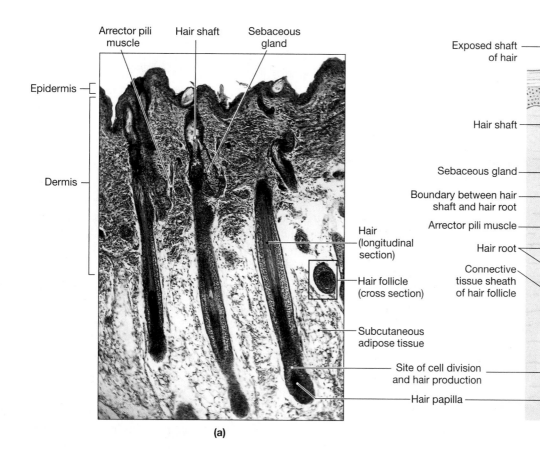

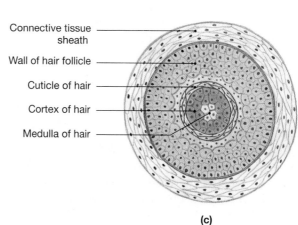

● **FIGURE 5–11 Hair Follicles.** (**a**) In this section of skin of the scalp, notice that the many hair follicles extend into the dermis and subcutaneous layer. (LM × 73) (**b**) This drawing shows a longitudinal section of a single hair follicle and hair. (**c**) This cross section through a hair follicle was taken at the boundary between the shaft and the root.

derlying layer is called the **cortex,** and the **medulla** makes up the core of the hair. The medulla contains a flexible *soft keratin;* the cortex and cuticle contain thick layers of *hard keratin,* which give the hair its stiffness.

Hairs grow and are shed according to a *hair growth cycle* based on the activity level of hair follicles. In general, a hair in the scalp grows for 2–5 years, at a rate of about 0.3 mm per day, and then its follicle may become inactive for a comparable period of time. When another growth cycle begins, the follicle produces a new hair, and the old hair gets pushed toward the surface to be shed. Variations in growth rate and in the length of the hair growth cycle account for individual differences in the length of uncut hair. Other differences in hair appearance result from the size of the follicles and the shapes of the hairs. For example, straight hairs are round in cross section, whereas curly ones are rather flattened.

FUNCTIONS OF HAIR. The 2.5 million hairs on the human body have important functions. The roughly 500,000 hairs on the head protect the scalp from UV light, help cushion a light blow to the head, and provide insulating benefits for the skull. The hairs that guard the entrances to the nostrils and external ear canals help prevent the entry of foreign particles, and eyelashes perform a similar function for the surface of the eye. A sensory nerve fiber is associated with the base of each hair follicle. As a result, you can feel the movement of the shaft of even a single hair. This sensitivity provides an early-warning system that may help prevent injury. For example, you may be able to swat a mosquito before it reaches the skin surface.

A bundle of smooth muscle cells forms the **arrector pili** (a-REK-tor PI-lē) muscle, which extends from the papillary dermis to the connective tissue sheath that surrounds each hair follicle (see Figure 5–11b). When stimulated, the arrector pili pulls on the follicle, forcing the hair to stand up. Contraction may be caused by emotional states (such as fear or rage) or a response to cold, which produce "goose bumps."

HAIR COLOR. Hair color reflects differences in the type and amount of pigment produced by melanocytes at the hair papilla. Different forms of melanin produce hair colors that can range from black to red. These pigment differences are genetically determined, but hormonal and environmental factors also influence the condition of your hair. As pigment production decreases with age, hair color lightens. White hair results from both a lack of pigment and the presence of air bubbles within the hair shaft. As the proportion of white hairs increases, the individual's hair color is described as gray. Because each hair is dead and inert, changes in coloration are gradual. Unless bleach is used, it is not possible for hair to "turn white overnight," as some horror stories would have us believe.

Clinical Note
THE HAIR IN DISEASE

The hair is a tactile sensory organ that covers the entire body, with the exception of the palms, soles, and parts of the genitalia. Like other body organs, the hair can be affected by disease. Changes in hair growth or distribution can be an aid in the diagnostic process.

Loss of hair occurs with aging. Normally, about 50 hairs are lost each day. Losses of more than 100 hairs per day (*alopecia*) may indicate underlying disease. Severe malnutrition and chemotherapy can cause hair loss. Chemotherapy, often used to treat cancer, inhibits cell growth. This is especially apparent in cells that are rapidly dividing. Cells in the hair follicle divide more rapidly than other body cells. Thus, hair loss often results from chemotherapy.

Hair growth and distribution are primarily controlled by masculine hormones (*androgens*). Diseases that cause an increase in androgens often result in excessive hair growth (*hirsutism*). These include certain tumors, polycystic ovarian disease, and hyperplasia of the adrenal glands. Both hair loss and excessive hair growth should be investigated to determine the underlying cause. ■

CONCEPT CHECK QUESTIONS

1. Describe the functions of the subcutaneous layer.
2. What happens when the arrector pili muscle contracts?
3. A person suffers a burn on the forearm that destroys the epidermis and the deep dermis. When the injury heals, would you expect to find hair growing again in the area of the injury?

Answers begin on p. 792.

Clinical Note
HAIR LOSS

Many people experience anxiety attacks when they find hairs clinging to their hairbrush instead of to their heads. On the average, about 50 hairs are lost from the head each day, but several factors can affect this rate. Sustained losses of over 100 hairs per day generally indicate a net loss of hair. Temporary increases in hair loss can result from drugs, dietary factors, radiation, high fever, stress, or hormonal factors related to pregnancy. In males, changes in the level of circulating sex hormones can affect the scalp, which causes a shift in production from normal hair to fine "peach fuzz" hairs, beginning at the temples and the crown of the head. This alteration is called *male pattern baldness*. Some cases of male pattern baldness respond to drug therapies, such as topical application of *minoxidil (Rogaine)*. ■

Sebaceous Glands

The integument contains two types of exocrine glands: *sebaceous glands* and *sweat glands*. **Sebaceous** (se-BĀ-shus) **glands,** or *oil glands*, are holocrine glands that discharge an oily lipid secretion into hair follicles or, in some cases, onto the skin (Figure 5–12●). The gland cells produce large quantities of lipids as they mature. The lipid is released through holocrine secretion, which is a process that involves the rupture and death of the cells. ∞ p. 104 The contraction of the arrector pili muscle that elevates the hair squeezes the sebaceous gland, which forces the oily secretions into the hair follicle and onto the surrounding skin. This secretion, called **sebum** (SĒ-bum), lubricates the hair and skin and inhibits the growth of bacteria. *Sebaceous follicles* are large sebaceous glands that discharge sebum directly onto the skin. They are located on the face, back, chest, nipples, and external genitalia.

Sebaceous glands are sensitive to changes in the concentrations of sex hormones, and their secretions accelerate at puberty. For this reason, individuals with large sebaceous glands may be especially prone to develop **acne** during adolescence. In acne, sebaceous ducts become blocked and secretions accumulate, causing inflammation and a raised "pimple." The trapped secretions provide a fertile environment for bacterial infection.

Sweat Glands

The skin contains two types of sweat glands, or *sudoriferous glands: apocrine sweat glands* and *merocrine sweat glands* (Figure 5–13●). The names refer to their mode of secretion, as discussed in Chapter 4 (see Table 4–2, p. 104).

APOCRINE SWEAT GLANDS. The sweat glands that secrete their products into hair follicles in the armpits, around the nipples, and in the groin are called **apocrine sweat glands.** The name *apocrine* was originally chosen because it was

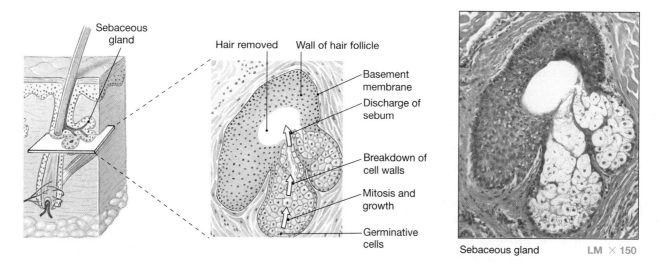

Sebaceous gland   LM × 150

● **FIGURE 5–12** The Structure of Sebaceous Glands and Their Relationship to Hair Follicles.

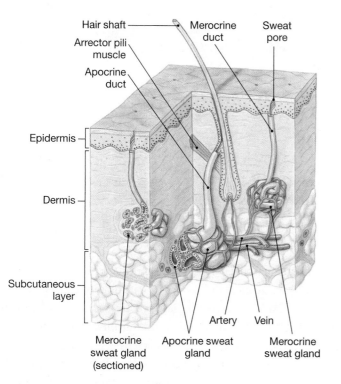

● **FIGURE 5–13 Sweat Glands.** Merocrine sweat glands secrete directly onto the skin, whereas apocrine sweat glands secrete into hair follicles.

thought these gland cells use an apocrine method of secretion. Even though we now know that they rely on merocrine secretion, the name has not changed. At puberty, these glands begin discharging a sticky, cloudy, and potentially odorous secretion. This sweat becomes odorous when bacteria break it down as a food source. In other mammals, this odor is an important form of communication; in our culture, whatever function it might have is masked by products such as deodorants. Other products, such as antiperspirants, contain astrin-

gent compounds that contract the skin and its sweat gland openings, thereby decreasing the quantity of both apocrine and merocrine secretions.

MEROCRINE SWEAT GLANDS. Coiled tubular glands that discharge their secretions directly onto the surface of the skin are **merocrine sweat glands,** or *eccrine* (EK-rin) *sweat glands.* They are far more numerous and widely distributed than apocrine glands. The skin of an adult contains 2–5 million eccrine glands. Palms and soles have the highest numbers; it has been estimated that the palm of the hand has about 500 glands per square centimeter (3000 per square inch).

The perspiration, or sweat, produced by merocrine glands is 99 percent water, but it also contains a mixture of electrolytes (chiefly sodium chloride), organic nutrients, and waste products such as urea. Sodium chloride gives sweat its salty taste. The primary function of merocrine gland activity and perspiration is to cool the surface of the skin and lower body temperature. When a person is sweating in the hot sun, all the merocrine glands are working together. The blood vessels beneath the epidermis are dilated and flushed with blood, the skin reddens in light-colored individuals, and the skin surface becomes warm and wet. As the moisture evaporates, the skin cools. If body temperature then falls below normal, perspiration ceases, blood flow to the skin is reduced, and the skin surface cools and dries, which releases little heat into the environment. The roles of the skin and negative feedback mechanisms in thermoregulation (temperature control) were considered in Chapter 1 (see ∞ p. 7); Chapter 17 will examine this process in greater detail.

Perspiration results in the excretion of water and electrolytes from the body. As a result, excessive perspiration to maintain normal body temperature can lead to problems. For example, when all of the merocrine sweat glands work at maximum, perspiration may exceed a gallon (about 4 liters) per hour, and dangerous fluid and electrolyte losses can occur. For this reason, marathoners and other endurance athletes must drink fluids at regular intervals.

Sweat also provides protection from environmental hazards. Sweat dilutes harmful chemicals in contact with the skin and flushes microorganisms from its surface. The presence of dermicidin, which is a small peptide molecule with antibiotic properties, provides additional protection from microorganisms.

The skin also contains other types of modified sweat glands with specialized secretions. For example, the mammary glands of the breasts are structurally related to apocrine sweat glands and secrete milk. Another example are the *ceruminous glands* in the passageway of the external ear. Their secretions combine with those of nearby sebaceous glands to form earwax.

Nails

Nails form on the dorsal surfaces of the fingers and toes, where they protect the exposed tips and help limit their distortion when they are subjected to mechanical stress—for example, when you run or grasp objects. The structure of a nail is shown in Figure 5–14●. The visible **nail body** consists of a dense mass of dead, keratinized cells. The nail body is recessed beneath the level of the surrounding epithelium. The body of the nail covers an area of epidermis called the **nail bed.** Nail production occurs at the **nail root,** which is an epithelial fold not visible from the surface. A portion of the stratum corneum of the fold extends over the exposed nail nearest the root, forming the **cuticle,** or *eponychium* (ep-ō-NIK-ē-um; *epi-,* over + *onyx,* nail). Underlying blood vessels give the nail its pink color, but near the root these vessels may be obscured, which leaves a pale crescent known as the **lunula** (LOO-nū-la; *luna,* moon).

→ **CONCEPT CHECK QUESTIONS**

1. What are the functions of sebaceous secretions?
2. Deodorants are used to mask the effects of secretions from which type of skin gland?

Answers begin on p. 792.

🔒 **Key Note**

The skin plays a major role in controlling body temperature. It acts as a radiator; the heat is delivered by the dermal circulation and removed primarily by the evaporation of sweat, or perspiration.

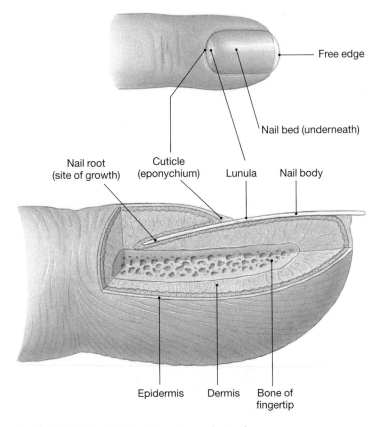

● **FIGURE 5–14** **The Structure of a Nail.**

■ Local Control of Homeostasis in the Integumentary System

The integumentary system can respond directly and independently to many local influences or stimuli. For example, when the skin is subjected to mechanical stresses, stem cells in the stratum germinativum divide more rapidly, and the thickness of the epithelium increases. That is why calluses form on your palms when you perform manual labor. A more dramatic example of local control can be seen after an injury to the skin.

Injury and Repair of the Skin

The skin can regenerate effectively even after considerable damage because stem cells are present in both its epithelial and connective tissue components. Divisions by these stem cells replace lost epidermal and dermal cells, respectively. This process can be slow, and when large surface areas are involved, infection and fluid loss complicate the situation. The relative speed and effectiveness of skin repair vary depending on the type of wound. A slender, straight cut, or *incision,* may heal relatively quickly compared with a scrape, or *abrasion,* which involves a much greater area.

Figure 5–15● illustrates the four stages in the regeneration of the skin after an injury. When damage extends through the epidermis and into the dermis, bleeding generally occurs. Mast cells in the dermis trigger an *inflammatory response* that will enhance blood flow to the surrounding region and attract phagocytes (Step 1). The blood clot, or **scab,** that forms at the surface temporarily restores the integrity of the epidermis and restricts the entry of additional microorganisms (Step 2).

STEP 1

Bleeding occurs at the site of injury immediately after the injury, and mast cells in the region trigger an inflammatory response.

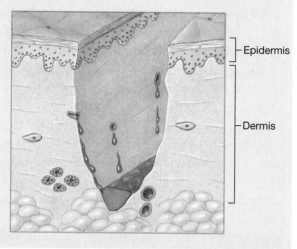

STEP 2

After several hours, a scab has formed and cells of the stratum germinativum are migrating along the edges of the wound. Phagocytic cells are removing debris, and more of these cells are arriving with the enhanced circulation in the area. Clotting around the edges of the affected area partially isolates the region.

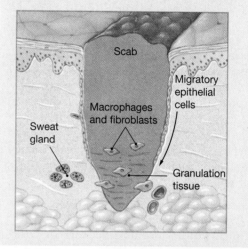

STEP 3

One week after the injury, the scab has been undermined by epidermal cells migrating over the meshwork produced by fibroblast activity. Phagocytic activity around the site has almost ended, and the fibrin clot is disintegrating.

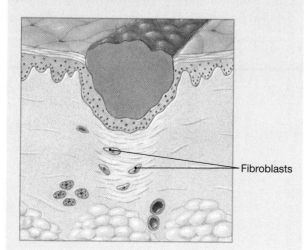

STEP 4

After several weeks, the scab has been shed, and the epidermis is complete. A shallow depression marks the injury site, but fibroblasts in the dermis continue to create scar tissue that will gradually elevate the overlying epidermis.

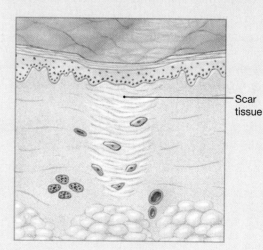

● **FIGURE 5–15** **Events in Skin Repair.**

Clinical Note
CHRONIC SKIN CONDITIONS

Two of the more common chronic skin conditions are eczema and psoriasis. *Eczema* is a general term that describes inflammatory skin diseases. Most cases of eczema are said to be *atopic*. Atopic eczema results from an abnormal response of the immune system and is characterized by dry, itchy patches of skin. In infants, eczema typically occurs on the forehead, cheeks, forearms, legs, scalp, and neck. In children and adults, eczema typically occurs on the face, neck, and the insides of the elbows, knees, and ankles. There is no cure but the condition can be controlled with moisturizing creams and topical corticosteroids.

Psoriasis is an inflammatory skin condition that is classically characterized by thickened, red areas of skin covered with silvery scales. The causes of psoriasis are not totally understood but are believed to result from an abnormal response by the immune system. Approximately 10–30 percent of people with psoriasis will develop psoriatic arthritis.

There are five types of psoriasis although approximately 80 percent of people with the disease will have the *plaque psoriasis* type. It occurs in raised and thickened patches of reddish skin called *plaques*, which are covered by silver-white scales. Plaques most often appear on the elbows, knees, scalp, chest, and lower back. However, they can appear anywhere on the body, including the genitals. Treatment involves the use of both topical and oral medications. Phototherapy is used in selected cases. ■

Most of the clot consists of an insoluble network of *fibrin,* which is a fibrous protein that forms from blood proteins during the clotting response. Cells of the stratum germinativum rapidly divide and begin to migrate along the sides of the wound to replace the missing epidermal cells. Meanwhile, macrophages and newly-arriving phagocytes patrol the damaged area of the dermis and clear away debris and pathogens.

If the wound covers an extensive area or involves a region covered by thin skin, dermal repairs must be under way before epithelial cells can cover the surface. Fiber-producing cells (*fibroblasts*) and connective tissue stem cells divide to produce mobile cells that invade the deeper areas of injury (see Step 2). Epithelial cells that line damaged blood vessels also begin to divide, and capillaries follow the fibroblasts, which provides a circulatory supply. The combination of blood clot, fibroblasts, and an extensive capillary network is called **granulation tissue** (see Step 2). Over time, the clot dissolves and the number of capillaries declines. Fibroblast activity formed an extensive meshwork of collagen fibers in the dermis (Step 3).

These repairs do not restore the integument to its original condition, however, because the dermis now contains an abnormally large number of collagen fibers and relatively few blood vessels. Severely damaged hair follicles, sebaceous or sweat glands, muscle cells, and nerves are seldom repaired, and they too are replaced by fibrous tissue. The formation of this rather inflexible, fibrous, noncellular **scar tissue** can be considered a practical limit to the healing process (Step 4).

The process of scar tissue formation is highly variable. For example, surgical procedures performed on a fetus do not leave scars. In some adults, most often those with dark skin, scar tissue formation may continue beyond the requirements of tissue repair. The result is a flattened mass of scar tissue that begins at the injury site and grows into the surrounding dermis. This thickened area of scar tissue, called a **keloid** (KĒ-loyd), is covered by a shiny, smooth epidermal surface. Keloids most commonly develop on the upper back, shoulders, anterior chest, and earlobes. They are harmless, and some aboriginal cultures intentionally produce keloids as a form of body decoration.

Burns

Burns are relatively common injuries that result from exposure of the skin to heat, radiation, electrical shock, or strong chemical agents. The severity of a burn depends on the depth of penetration and the total area affected. The most common classification of burns is based on the depth of penetration, as detailed in Table 5–1. The larger the area affected, the greater the impact on integumentary function.

TABLE 5–1 *A Common Classification of Burns*

CLASSIFICATION	DAMAGE REPORT	APPEARANCE AND SENSATION
FIRST-DEGREE BURN	*Killed:* superficial cells of epidermis *Injured:* deeper layers of epidermis, papillary dermis	Inflamed; tender
SECOND-DEGREE BURN	*Killed:* superficial and deeper cells of epidermis; dermis may be affected *Injured:* damage may extend into reticular layer of the dermis, but many accessory structures are unaffected	Blisters; very painful
THIRD-DEGREE BURN	*Killed:* all epidermal and dermal cells *Injured:* hypodermis and deeper tissues and organs	Charred; no sensation at all

Clinical Note
TRAUMATIC CONDITIONS OF THE SKIN

Traumatic disruption of the layers of the skin exposes the tissues underneath. This increases the likelihood of infection, loss of body fluids, and pain. Skin trauma can result from direct trauma (either sharp or blunt) or burns (either thermal or chemical).

Soft-Tissue Wounds

Trauma that results in the disruption of the layers of the skin is called a wound. There are several classifications of wounds. These include:

■ *Abrasion*—is a simple scrape or scratch where the outer layer of skin is damaged but not all of the layers are penetrated (Figure 5–16●).
■ *Laceration*—commonly called a cut, a laceration may be smooth or jagged (Figure 5–17●). Smooth wounds are usually caused by a sharp edge such as a razor blade or glass. Jagged wounds can be

due to injury from duller objects such as jagged metal. They can also occur as a result of blunt trauma such as a blow or fall.
■ *Puncture*—occurs when a sharp, pointed object penetrates the skin or other tissues (Figure 5–18●). Common causes include such items as nails, ice picks, splinters, or knives. There are two types of puncture wounds, penetrating and perforating. *Penetrating puncture wounds* can be shallow or deep and can injure underlying tissues and blood vessels. Penetrating puncture wounds carry an increased incidence of infection as foreign material may be carried into the wound during injury and remain there due to sealing of the surface. *Perforating puncture wounds* have both an entrance wound and an exit wound. Often, the exit wound is more serious than the entrance wound. An example of a penetrating puncture wound is a gunshot wound.
■ *Avulsion*—is an injury where a flap of skin and tissues are torn loose or pulled completely off (Figure 5–19●). Avulsions can cause serious tissue defects and subsequent scarring. A special type of avulsion is the *degloving avulsion*. This is a potentially devastating wound where the skin, usually of the hand, is stripped off like a glove (Figure 5–20●).

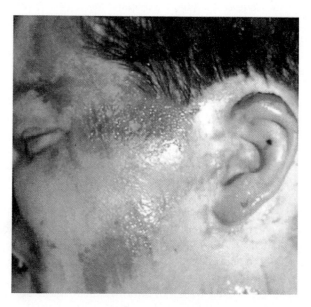

● **FIGURE 5–16 An Abrasion of the Left Side of the Face.**

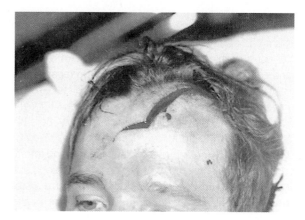

● **FIGURE 5–17 Jagged Laceration of the Forehead.**

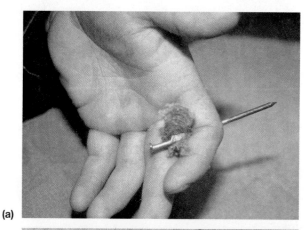
(a)

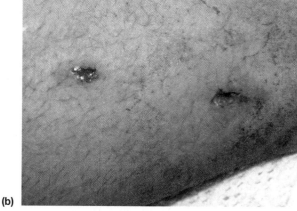
(b)

● **FIGURE 5–18 Examples of Puncture Wounds.** (a) Nail gun injury with embedded carpet fibers. (b) Gun shot wound of the left thigh—the entry wound is on the left and the exit wound is on the right. *(continued next page)*

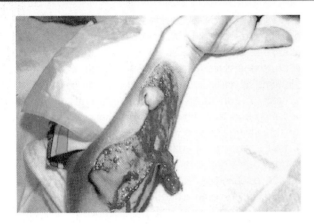

● **FIGURE 5–19** Avulsion of the Skin of the Forearm.

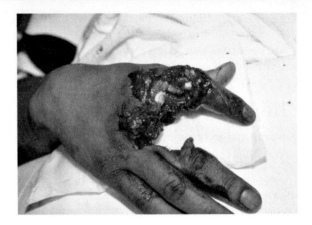

● **FIGURE 5–21** Partial Amputation of the Right Hand.

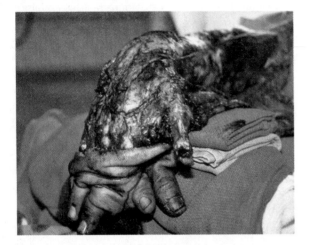

● **FIGURE 5–20** Degloving Injury.

■ *Amputation*—occurs when an extremity, or part of an extremity is completely cut through or torn off (Figure 5–21●). Amputations often are caused by machinery and can be life-threatening. Today, using microsurgical techniques, amputated body parts can be replanted if the condition of the tissues is satisfactory.

■ *Crush injury*—results when a body part, usually an extremity, becomes caught between heavy items such as parts of machinery. There is often massive damage to underlying blood vessels, nerves, and bone.

Soft-tissue injuries are commonly encountered in emergency care. They are often grotesque and can distract from other patient care activities. Fortunately, they are rarely life-threatening. In the pre-hospital setting, they are usually dressed to prevent further contamination, and definitive care, such as wound repair or surgery, is carried out at the hospital.

Limb Replantation

With the advent of microsurgical techniques, many body parts can be surgically reattached, or replanted, following amputation

(Figure 5–22●). Amputated body parts should be located at the emergency scene and should accompany the patient to the hospital if at all possible. Even if an amputated part is too damaged for replantation, the tissues can be used to help repair the wound.

The separated body part should be cleaned of any gross debris, exposed areas covered with a lightly moistened gauze, and then placed into a sealed plastic bag. The bag should be immersed into saline with a few pieces of ice added. Packing the body part in ice, as was the old practice, caused tissue damage at a cellular level, and adversely impacted limb survival. Many body parts have been successfully replanted including arms, hands, feet, legs, ears, the nose, and the penis. Patients with amputations where replantation is a possibility should be transported to medical centers with microsurgical replant capabilities.

The material placed directly on a wound is called a *dressing*. These are sterile and designed to control bleeding and protect the wound from contamination. A dressing is held in place by a *bandage*. Bandages can be made from gauze, elastic material, cloth, and many other materials. Sometimes, wounds require splinting. This is especially true in cases where there may be an associated fracture or where movement might cause the wound to open.

Injury and Repair

The skin is an amazingly resilient organ. It can regenerate effectively, even after considerable damage. This is due to stem cells in the epithelial and connective tissues. The rate of wound repair is related to several factors such as wound size, infection, fluid loss, vascular supply, and overall patient condition. Small lacerations may heal quickly while large avulsions and burns can take considerably more time.

Regeneration of a wound following injury involves several distinct stages. If the wound extends through the epidermis and into the dermis or connective tissue, bleeding usually occurs. The blood at the site of the wound will collect and clot, and eventually form a *scab*. This serves to temporarily restore the integrity of the epidermis and limits the entry of additional organisms. Next, the cells of the stratum germinativum, which is the deepest layer of the epidermis, begin to un-

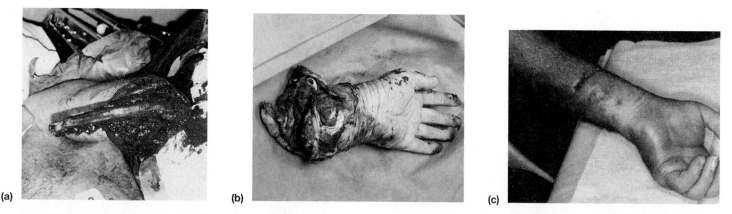

● **FIGURE 5–22 Amputation and Replantation.** (a) Complete amputation of the left forearm caused by tractor power takeoff device (PTO)—contraction of the forearm muscles exposes the radius and ulna. (b) The amputated hand was found and retrieved by EMS personnel. (c) Using microvascular surgical techniques, the hand was replanted.

dergo rapid division. As this occurs, they migrate along the sides of the wound, and attempt to replace the lost epidermal cells. Scavenger, phagocytic cells remove debris from the wound. Additional phagocytic cells are transported to the injury site via the circulatory system.

If the wound is extensive or deep, repairs to the dermis must be under way before epithelial cells can be laid down. Fibroblast and mesenchymal cells begin to divide, producing mobile cells that penetrate the deeper areas of the wound. Damaged blood vessels begin to repair themselves through division of endothelial cells. These cells follow the fibroblasts into the deeper areas of the wound, providing a blood supply. Together, fibroblasts, the blood clot, and the developing capillary network are called *granulation tissue*. Ultimately, the clot dissolves and the number of capillaries declines as the needs of the healing tissue decrease. Fibroblast activity leads to the formation of collagen fibers and *ground substance*.

The repaired wound is different from the original tissues. There is an increased number of collagen fibers and a decreased number of blood vessels that result in scar tissue. This scar tissue varies based upon the type of wound and the age and health of the patient.

Burns

Skin injury caused by heat, chemicals, electricity, and radiation is called a *burn*. Although burns usually affect the skin, they can also affect underlying tissues such as the subcutaneous tissues, muscle, blood vessels, nerves, and even bone (Figure 5–23●).

Burns are classified by depth and include the following:

■ *First-degree burns*—also called superficial burns, first-degree burns involve only the epidermis. They can be quite painful, as nerve fibers are uninjured. Example: sunburn (Figure 5–24●).

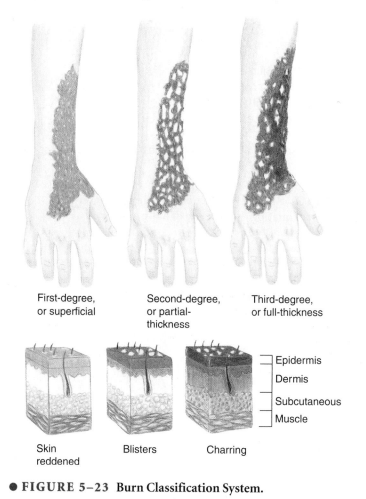

First-degree, or superficial

Second-degree, or partial-thickness

Third-degree, or full-thickness

Epidermis
Dermis
Subcutaneous
Muscle

Skin reddened

Blisters

Charring

● **FIGURE 5–23 Burn Classification System.**

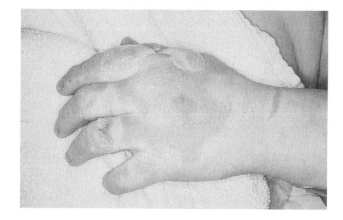

● **FIGURE 5–24 First-Degree Burn of the Left Hand.** (Note that some areas of second-degree burn are also present.)

(continued next page)

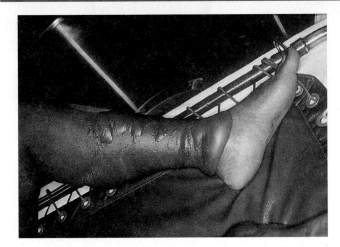

● **FIGURE 5–25** Second-Degree Burn of the Left Leg.

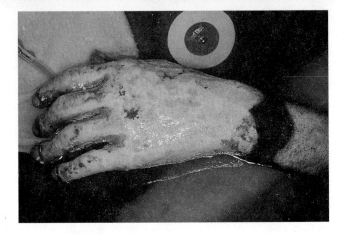

● **FIGURE 5–26** Third-Degree Burn of the Left Hand.

■ *Second-degree burns*—also called partial-thickness burns, the epidermis is burned through and the dermis is damaged (Figure 5–25●). There is usually intense pain and blistering. The blisters develop as plasma and interstitial fluids are released into the skin and elevate the top layer. Example: scald injury.

■ *Third-degree burns*—also called full-thickness burns, third-degree burns are characterized by damage to all layers of the skin (Figure 5–26●). The patient may not suffer as much pain as would be expected, because pain fibers in the skin may be destroyed by the injury. Third-degree burns are almost always accompanied by partial-thickness burns and some degree of pain is usually present. Third-degree burns usually require skin grafting and are very disfiguring.

Deep burns are sometimes seen with prolonged exposure to heat and with electrical and lightning injuries. In some cases, muscle tissue and bone will be involved. Although these are technically considered full-thickness burns, they are occasionally called fourth-degree burns to separate them from less severe injuries.

The severity of a burn is determined by consideration of the following factors:

■ Body regions burned
■ Depth of the burn
■ Extent of the burn
■ Agent or source of the burn
■ Age of the patient
■ Other associated illnesses and injuries

● **FIGURE 5–27** Rule of Nines, a System for Approximating the Severity of a Burn Injury.

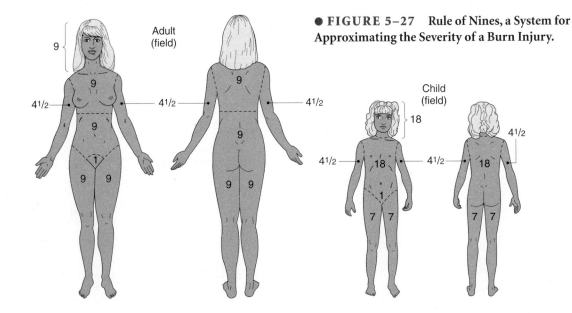

The agent or source of the burn can be significant. For example, electrical burns may cause only a small area of skin injury but cause massive injury to underlying tissues. Chemical burns are of special concern, because the chemical can remain on the skin and continue to burn for hours or even days.

The extent of a burn is important to determine. The amount of body surface area (BSA) involved can be quickly estimated by using the *rule of nines* (Figure 5–27●). In an adult, each of the following areas represents 9 percent of the BSA: head and neck, each upper extremity, chest, abdomen, upper back, lower back, the front of each lower extremity, and the back of each lower extremity. Together, these total 99 percent. The remaining 1 percent of BSA is assigned to the genital region. In children, the head is proportionally large compared to the body, and the body regions are adjusted accordingly. Another system of estimating the amount of BSA burned is the palmar method. The patient's palm equals about 1 percent of the patient's BSA. Considering this, the percentage of BSA burned can be quickly approximated. This system is particularly useful for smaller burns. ■

■ Aging and the Integumentary System

Aging affects all the components of the integumentary system. Major age-related changes include the following:

- *Skin injuries and infections become more common.* Such problems are more likely because the epidermis thins as stem cell activity declines.
- *The sensitivity of the immune system is reduced.* The number of macrophages and other immune system cells that reside in the skin decreases to around 50 percent of levels seen at maturity (roughly, age 21). This loss further encourages skin damage and infection.
- *Muscles become weaker, and bone strength decreases.* Such changes are related to reduced calcium and phosphate absorption due to a decline in vitamin D_3 production of around 75 percent.
- *Sensitivity to sun exposure increases.* Lesser amounts of melanin are produced because melanocyte activity declines. The skin of light-skinned individuals becomes very pale.
- *The skin becomes dry and often scaly.* Glandular activity declines, reducing sebum production and perspiration.
- *Hair thins and changes color.* Follicles stop functioning or they produce thinner, finer hairs. With decreased melanocyte activity, these hairs are gray or white.
- *Sagging and wrinkling of the skin occur.* The dermis becomes thinner, and the elastic fiber network decreases in size. The integument, therefore, becomes weaker and less resilient. These effects are most pronounced in areas exposed to the sun.
- *The ability to lose heat decreases.* The blood supply to the dermis is reduced at the same time that sweat glands become less active. This combination makes the elderly less able than younger people to lose body heat. As a result, overexertion or overexposure to high temperatures (such as a sauna or hot tub) can cause dangerously high body temperatures.
- *Skin repairs proceed relatively slowly.* For example, it takes 3–4 weeks to complete repairs to an uninfected blister site in a young adult. The same repairs could take 6–8 weeks at ages 65–75. Because repairs are slow, recurrent infections may occur.

→ **CONCEPT CHECK QUESTIONS**

1. Why can skin regenerate effectively even after considerable damage has occurred?
2. Older individuals do not tolerate summer heat as well as they did when they were young, and they are more prone to heat-related illnesses. What accounts for these changes?

Answers begin on p. 792.

■ Integration with Other Systems

Although the integumentary system can function independently, many of its activities are integrated with those of other organ systems. Figure 5–28● diagrams the major functional relationships.

→ **CONCEPT CHECK QUESTION**

1. How does vitamin D_3 production in the skin affect the functions of other organ systems in the body?

Answers begin on p. 792.

The Integumentary System in Perspective

For All Systems

Provides mechanical protection against environmental hazards

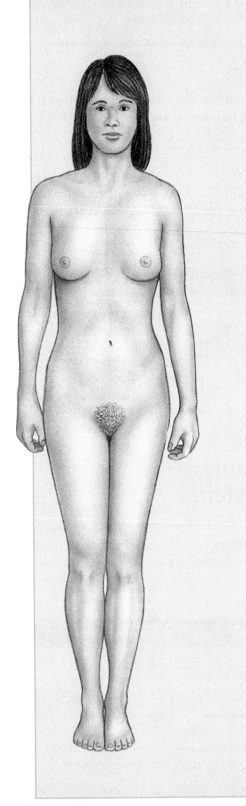

The Skeletal System

- Provides structural support
- Synthesizes vitamin D_3, essential for calcium and phosphorus absorption (bone maintenance and growth)

The Muscular System

- Contractions of skeletal muscle pull against skin of face, producing facial expressions important in communication
- Synthesizes vitamin D_3, essential for normal calcium absorption (calcium ions play an essential role in muscle contraction)

The Nervous System

- Controls blood flow and sweat gland activity for thermoregulation; stimulates contraction of arrector pili muscles to elevate hairs
- Receptors in dermis and deep epidermis provide sensations of touch, pressure, vibration, temperature, and pain

The Endocrine System

- Sex hormones stimulate sebaceous gland activity; male and female sex hormones influence hair growth, distribution of subcutaneous fat, and apocrine sweat gland activity; adrenal hormones alter dermal blood flow and help mobilize lipids from adipocytes
- Synthesizes vitamin D_3, precursor of calcitriol

The Cardiovascular System

- Provides oxygen and nutrients; delivers hormones and cells of immune system; carries away carbon dioxide, waste products, and toxins; provides heat to maintain normal skin temperature
- Stimulation by mast cells produces localized changes in blood flow and capillary permeability

The Lymphatic System

- Assists in defending the integument by providing additional macrophages and mobilizing lymphocytes
- Provides physical barriers that prevent pathogen entry; macrophages resist infection; mast cells trigger inflammation and initiate the immune reponse

The Respiratory System

- Provides oxygen and eliminates carbon dioxide
- Hairs guard entrance to nasal cavity

The Digestive System

- Provides nutrients for all cells and lipids for storage by adipocytes
- Synthesizes vitamin D_3, needed for absorption of calcium and phosphorus

The Urinary System

- Excretes waste products, maintains normal body fluid pH and ion composition
- Assists in elimination of water and solutes; keratinized epidermis limits fluid loss through skin

The Reproductive System

- Sex hormones affect hair distribution, adipose tissue distribution in subcutaneous layer, and mammary gland development
- Covers external genitalia; provides sensations that stimulate sexual behaviors; mammary gland secretions provide nourishment for newborn infant

● **FIGURE 5–28** Functional Relationships Between the Integumentary System and Other Systems.

Chapter Review

Access more review material online at *www.prenhall.com/bledsoe*. There you will find quiz questions, labeling activities, animations, essay questions, and web links.

Key Terms

cutaneous membrane 123
dermis 123
epidermis 123
hair 134

hair follicle 134
integument 123
keratin 126
melanin 127

nail 138
sebaceous glands 136
stratum germinativum 125
subcutaneous layer 123

Related Clinical Terms

acne A sebaceous gland inflammation caused by an accumulation of secretions.

basal cell carcinoma A cancer that originates in the stratum germinativum; the most common skin cancer. Roughly two-thirds of cases of this cancer appear in areas subjected to chronic UV exposure. Metastasis seldom occurs.

biopsy (BĪ-op-sē) The removal and examination of tissue from the body for the diagnosis of disease.

cavernous hemangioma (strawberry nevus) A mass of large blood vessels that can occur in the skin or other organs in the body; a "port wine stain" birthmark generally lasts a lifetime.

contact dermatitis Dermatitis that is generally caused by strong chemical irritants. It produces an itchy rash that may spread to other areas because scratching distributes the chemical agent; includes poison ivy.

cyanosis (sī-uh-NŌ-sis) Bluish skin color as a result of reduced oxygenation of the blood in superficial vessels.

dermatitis An inflammation of the skin that primarily involves the papillary region of the dermis.

dermatology The branch of medicine concerned with the diagnosis and treatment of diseases of the skin.

granulation tissue A combination of fibrin, fibroblasts, and capillaries that forms during tissue repair after inflammation.

keloid (KĒ-loyd) A thickened area of scar tissue covered by a shiny, smooth epidermal surface.

lesions (LĒ-zhuns) Changes in tissue structure caused by injury or disease.

male pattern baldness Hair loss in an adult male due to changes in levels of circulating sex hormones.

malignant melanoma (mel-uh-NŌ-muh) A skin cancer that originates in malignant melanocytes.

pruritis (proo-RĪ-tis) An irritating itching sensation, that is common in skin conditions.

psoriasis (sō-RĪ-uh-sis) A painless condition characterized by rapid stem cell divisions in the stratum germinativum of the scalp, elbows, palms, soles, groin, and nails. Affected areas appear dry and scaly.

sepsis (SEP-sis) A dangerous, widespread bacterial infection; the leading cause of death in burn patients.

squamous cell carcinoma A form of skin cancer less common than basal cell carcinoma, almost totally restricted to areas of sun-exposed skin. Metastasis seldom occurs.

ulcer A localized shedding of an epithelium.

xerosis (ze-RŌ-sis) "Dry skin," a common complaint of older persons and almost anyone who lives in an arid climate.

Summary Outline

INTEGUMENTARY STRUCTURE AND FUNCTION 123

1. The **integumentary system,** or **integument,** consists of the **cutaneous membrane,** which includes the **epidermis** and **dermis,** and the **accessory structures.** Beneath it lies the **subcutaneous layer** (or **hypodermis**). *(Figures 5–1 to 5–3)*

The Epidermis 124

2. **Thin skin** covers most of the body; heavily abraded body surfaces may be covered by **thick skin.**

3. Cell divisions by the stem cells that make up the **stratum germinativum** replace more superficial cells.

4. As epidermal cells age, they move up through the **stratum spinosum,** the **stratum granulosum,** the **stratum lucidum** (in thick skin), and the **stratum corneum.** In the process, they accumulate large amounts of **keratin.** Ultimately, the cells are shed or lost. *(Figure 5–4)*

5. **Epidermal ridges** interlock with the **dermal papillae** of the dermis. Together, they form superficial ridges on the palms and soles that improve the gripping ability of the hands and feet.

6. The color of the epidermis depends on two factors: blood supply and the concentrations of melanin and carotene. **Melanocytes** protect stem cells from **ultraviolet (UV) radiation.** *(Figure 5–5)*

7. Epidermal cells synthesize **vitamin D₃** when exposed to sunlight.

8. Skin cancer is the most common form of cancer. **Basal cell carcinoma** and **squamous cell carcinoma** are not as dangerous as **melanoma**. (*Figures 5–6 through 5–9*)

Key Note 128

The Dermis 132

9. The dermis consists of the **papillary layer** and the deeper **reticular layer.** (*Figure 5–10*)

10. The papillary layer of the dermis contains blood vessels, lymphatic vessels, and sensory nerves. This layer supports and nourishes the overlying epidermis. The reticular layer consists of a meshwork of collagen and elastic fibers oriented to resist tension in the skin.

11. Components of other organ systems (cardiovascular, lymphatic, and nervous) that communicate with the skin are in the dermis.

Key Note 133

The Subcutaneous Layer 133

12. The subcutaneous layer stabilizes the skin's position against underlying organs and tissues.

Accessory Structures 134

13. **Hairs** originate in complex organs called **hair follicles.** Each hair has a **shaft** composed of dead keratinized cells. Hairs have a central **medulla** of soft keratin surrounded by a **cortex** and an outer **cuticle** of hard keratin. (*Figure 5–11*)

14. Each **arrector pili** muscle can raise a single hair.

15. Our hairs grow and are shed according to the *hair growth cycle.* A single hair grows for 2–5 years and is then shed.

16. Typical **sebaceous glands** discharge waxy **sebum** into hair follicles. *Sebaceous follicles* are sebaceous glands that empty directly onto the skin. (*Figure 5–12*)

17. **Apocrine sweat glands** produce an odorous secretion; the more numerous **merocrine sweat glands** produce perspiration, which is a watery secretion. (*Figure 5–13*)

18. The **body** of a **nail** covers the **nail bed.** Nail production occurs at the **nail root.** (*Figure 5–14*)

Key Note 138

LOCAL CONTROL OF HOMEOSTASIS IN THE INTEGUMENTARY SYSTEM 138

Injury and Repair of the Skin 138

1. The skin can regenerate effectively even after considerable damage. (*Figures 5–15 through 5–27*)

2. Burns are relatively common injuries characterized by damage to layers of the epidermis and perhaps the dermis. (*Table 5–1*)

AGING AND THE INTEGUMENTARY SYSTEM 145

1. Aging affects all the components of the integumentary system.

INTEGRATION WITH OTHER SYSTEMS 146

1. Many activities of the integumentary system are integrated with those of other organ systems. (*Figure 5–28*)

Review Questions

Level 1: Reviewing Facts and Terms

Match each item in column A with the most closely related item in column B. Place letters for answers in the spaces provided.

COLUMN A

___ 1. cutaneous membrane

___ 2. carotene

___ 3. melanocytes

___ 4. epidermal layer that contains stem cells

___ 5. smooth muscle

___ 6. epidermal layer of flattened and dead cells

___ 7. bluish skin

___ 8. sebaceous glands

___ 9. merocrine (eccrine) glands

___ 10. vitamin D₃

COLUMN B

a. arrector pili

b. cyanosis

c. perspiration

d. sebum

e. stratum corneum

f. skin

g. orange-yellow pigment

h. bone growth

i. pigment cells

j. stratum germinativum

11. The two major components of the integument are:
    (a) the cutaneous membrane and the accessory structures.
    (b) the epidermis and the hypodermis.
    (c) the hair and the nails.
    (d) the dermis and the subcutaneous layer.

12. The fibrous protein that forms the basic structural component of hair and nails is:
    (a) collagen.
    (b) melanin.
    (c) elastin.
    (d) keratin.

13. The two types of exocrine glands in the skin are _____ and sweat glands.
    (a) merocrine
    (b) sebaceous
    (c) apocrine
    (d) eccrine

14. The following are all accessory structures of the integumentary system *except:*
    (a) nails.
    (b) hair.
    (c) dermal papillae.
    (d) sweat glands.

15. Sweat glands that communicate with hair follicles in the armpits and produce an odorous secretion are _____ glands.
    (a) apocrine
    (b) merocrine
    (c) sebaceous
    (d) apocrine, merocrine, and sebaceous

16. The reason older persons are more sensitive to sun exposure and more likely to get sunburned is that with age:
    (a) melanocyte activity declines.
    (b) vitamin D_3 production declines.
    (c) glandular activity declines.
    (d) skin thickness decreases.

17. Which two skin pigments are found in the epidermis?

18. Which two major layers constitute the dermis, and what components are found in each layer?

19. Which two groups of sweat glands are contained in the integument?

Level 2: Reviewing Concepts

20. During the transdermal administration of drugs, why are fat-soluble drugs more desirable than those that are water-soluble?

21. In our society, a tanned body is associated with good health. However, medical research constantly warns about the dangers of excessive exposure to the sun. What are the benefits of a tan?

22. In some cultures, women must be covered completely, except for their eyes, when they go outside. These women exhibit a high incidence of bone problems. Why?

23. Why is a subcutaneous injection with a hypodermic needle a useful method for administering drugs?

24. Why does skin sag and wrinkle as a person ages?

Level 3: Critical Thinking and Clinical Applications

25. A new mother notices that her six-month-old child has a yellow-orange complexion. Fearful that the child may have jaundice (a condition caused by bilirubin, a toxic yellow-orange pigment produced during the destruction of red blood cells), she takes him to her pediatrician. After examining the child, the pediatrician declares him perfectly healthy and advises the mother to watch the child's diet. Why?

26. Vanessa notices that even though her 80-year-old grandmother keeps her thermostat set at 80°F, she still wears a sweater in her house. When Vanessa asks her grandmother why, her grandmother tells her that she is cold. Vanessa can't understand this and asks you for an explanation. What would you tell her?

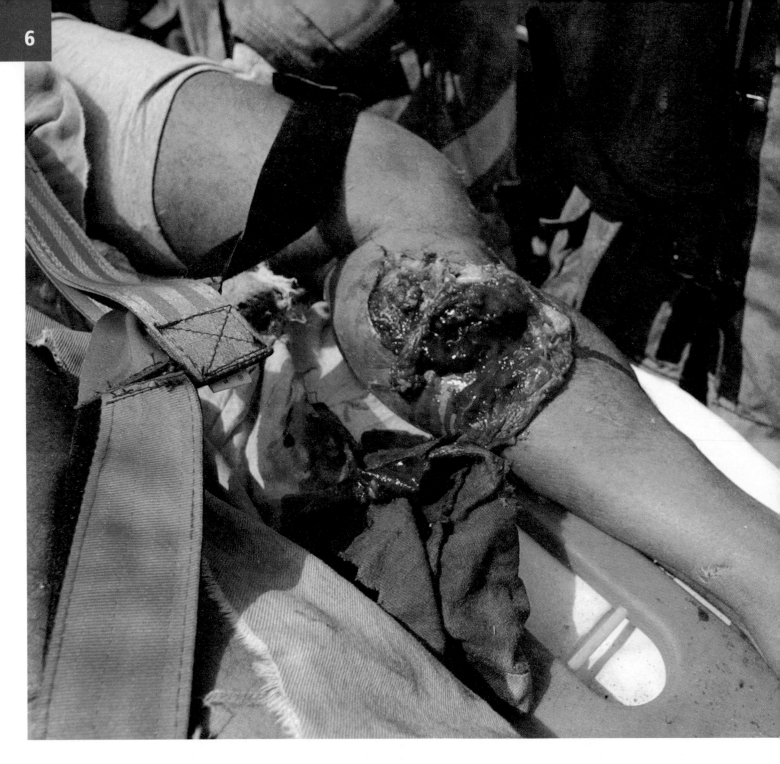

6 The Skeletal System

SKELETAL INJURIES CAN be quite grotesque and debilitating. The patient's chances of recovery are often dependent upon the prehospital care provided. EMS personnel must understand the anatomy and physiology of the skeletal system and understand that skeletal injuries are often associated with injuries to other important structures such as arteries and nerves. Basic life support skills, such as bandaging and splinting, are just as important as advanced life support skills, if not more so.

Chapter Outline

Chapter Objectives

1. Describe the functions of the skeletal system. (pp. 151–152)
2. Compare the structures and functions of compact and spongy bone. (pp. 153, 155–156)
3. Discuss bone growth and development, and account for variations in the internal structure of specific bones. (pp. 156–159)
4. Describe the remodeling and repair of the skeleton, and discuss homeostatic mechanisms responsible for regulating mineral deposition and turnover. (pp. 159–160)
5. Name the components and functions of the axial and appendicular skeletons. (p. 166)
6. Identify the bones of the skull. (pp. 166–175)
7. Discuss the differences in structure and function of the various vertebrae. (pp. 175–179)
8. Relate the structural differences between the pectoral and pelvic girdles to their various functional roles. (pp. 179–187)
9. Distinguish among different types of joints and link structural features to joint functions. (pp. 187–189)
10. Describe the dynamic movements of the skeleton and the structure of representative articulations. (pp. 189–194)
11. Explain the relationship between joint structure and mobility, using specific examples. (pp. 189–197)
12. Discuss the functional relationships between the skeletal system and other body systems. (pp. 197–199)

Vocabulary Development

ab- from; *abduction*
acetabulum a vinegar cup; *acetabulum of the hip joint*
ad- toward, to; *adduction*
amphi- on both sides; *amphiarthrosis*
arthros joint; *synarthrosis*
blast precursor; *osteoblast*
circum- around; *circumduction*
clast break; *osteoclast*
clavius clavicle; *clavicle*
concha shell; *middle concha*
corona crown; *coronoid fossa*
cranio- skull; *cranium*

cribrum sieve; *cribriform plate*
dens tooth; *dens*
dia- through; *diarthrosis*
duco to lead; *adduction*
e- out; *eversion*
gennan to produce; *osteogenesis*
gomphosis a bolting together; *gomphosis*
in- into; *inversion*
infra- beneath; *infraspinous fossa*
lacrimae tears; *lacrimal bones*
lamella thin plate; *lamellae of bone*
malleolus little hammer; *medial malleolus*
meniscus crescent; *menisci*

osteon bone; *osteocytes*
penia lacking; *osteopenia*
planta sole; *plantar*
porosus porous; *osteoporosis*
septum wall; *nasal septum*
stylos pillar; *styloid process*
supra- above; *supraspinous fossa*
sutura a sewing together; *suture*
teres cylindrical; *ligamentum teres*
trabecula wall; *trabeculae in spongy bone*
trochlea pulley; *trochlea*
vertere to turn; *inversion*

THE SKELETON HAS MANY FUNCTIONS, but the most obvious is supporting the weight of the body. This support is provided by bones, which are structures as strong as reinforced concrete but considerably lighter. Unlike concrete, bones can be remodeled and reshaped to meet changing metabolic demands and patterns of activity. Bones work with muscles to maintain body position and to produce controlled, precise movements. With the skeleton to pull against, contracting muscles can make us sit, stand, walk, or run.

The skeletal system includes the bones of the skeleton and the cartilages, joints, ligaments, and other connective tissues that stabilize or connect them. This system has <u>five primary</u> functions:

✱ **1.** *Support.* The skeletal system provides structural support for the entire body. Individual bones or groups of bones provide a framework for the attachment of soft tissues and organs.

✱ **2.** *Storage.* The calcium salts of bone represent a valuable mineral reserve that maintains normal concentrations of calcium and phosphate ions in body fluids. In addition, bones store energy reserves as lipids in areas filled with *yellow marrow.*

✱ **3.** *Blood cell production.* Red blood cells, white blood cells, and other blood elements are produced within the *red marrow,* which fills the internal cavities of many bones. The role of the bone marrow in blood cell formation will be discussed when we examine the cardiovascular and lymphatic systems (Chapters 11 and 14).

✱ **4.** *Protection.* Soft tissues and organs are often surrounded by skeletal elements. The ribs protect the heart and lungs, the skull encloses the brain, the vertebrae shield the spinal cord, and the pelvis cradles delicate digestive and reproductive organs.

✱ **5.** *Leverage.* Many bones function as levers that change the magnitude and direction of the forces generated by skeletal muscles. The resulting movements range from the delicate motion of a fingertip to powerful changes in the position of the entire body.

■ The Structure of Bone

Bone, or **osseous tissue,** is a supporting connective tissue that contains specialized cells and a matrix that consists of extracellular protein fibers and a ground substance. ∞ p. 111 The distinctive texture of bone results from the deposition of calcium salts within the matrix. Calcium phosphate, $Ca_3(PO_4)_2$, accounts for almost two-thirds of the weight of bone. The remaining third is dominated by collagen fibers; osteocytes and other cell types make up only around 2 percent of the mass of a bone.

✱ Macroscopic Features of Bone

The bones of the human skeleton have four general shapes: long, short, flat, and irregular (Figure 6–1●). **Long bones** are

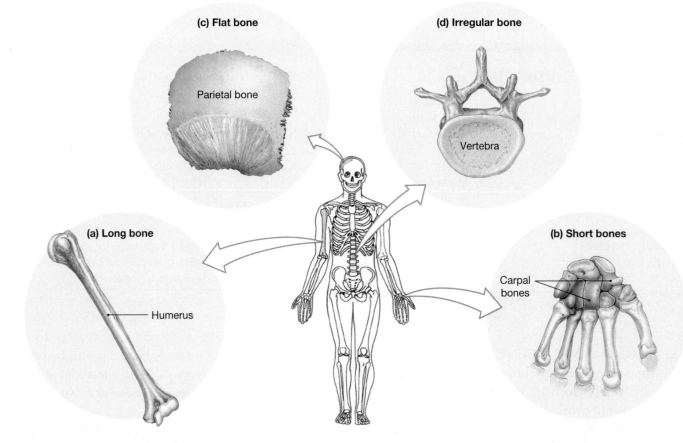

(c) Flat bone
Parietal bone

(d) Irregular bone
Vertebra

(a) Long bone
Humerus

(b) Short bones
Carpal bones

● **FIGURE 6–1 Shapes of Bones.**

longer than they are wide, whereas in **short bones** these dimensions are roughly equal. Examples of long bones are bones of the limbs, such as the bones of the arm (*humerus*) and thigh (*femur*). Short bones include the bones of the wrist (*carpal bones*) and ankles (*tarsal bones*). **Flat bones** are thin and relatively broad, such as the parietal bones of the skull, the ribs, and the shoulder blades (*scapulae*). **Irregular bones** have complex shapes that do not fit easily into any other category. An example is any of the vertebrae of the spinal column.

The typical features of a long bone such as the humerus are shown in Figure 6–2●. A long bone has a central shaft, or **diaphysis** (dī-AF-i-sis), that surrounds a central *marrow cavity* that contains **bone marrow,** which is a loose connective tissue. The expanded portions at each end, called **epiphyses** (ē-PIF-i-sēz), are covered by *articular cartilages*. Each epiphysis (ē-PIF-i-sis) of a long bone articulates with an adjacent

bone at a joint. As will be discussed shortly, growth in the length of an immature long bone occurs at the junctions between the epiphyses and the diaphysis.

The two types of bone tissue are visible in Figure 6–2. **Compact bone** (or dense bone) is relatively solid, whereas **spongy bone,** or *cancellous* (KAN-se-lus) *bone,* resembles a network of bony rods or struts separated by spaces. Both compact bone and spongy bone are present in the humerus; compact bone forms the diaphysis, and spongy bone fills the epiphyses.

The outer surface of a bone is covered by a **periosteum** (see Figure 6–2). The fibers of *tendons* and *ligaments* intermingle with those of the periosteum, and attach skeletal muscles to bones and one bone to another. The periosteum isolates the bone from surrounding tissues, provides a route for circulatory and nervous supplies, and participates in bone growth and repair. Within the bone, a cellular **endosteum** lines the marrow cavity and other inner surfaces. The endosteum is active during bone growth and whenever repair or remodeling is under way.

Microscopic Features of Bone

The general histology of bone was introduced in Chapter 4. Figure 6–3● presents the microscopic structure of bone in detail. Histologically, the periosteum consists of a fibrous outer layer and a cellular inner layer (Figure 6–3a●). Both compact bone and spongy bone contain bone cells, or **osteocytes** (OS-tē-ō-sīts; *osteon,* bone), in small pockets called **lacunae** (la-KOO-nē) (Figure 6–3b●). Lacunae are found between narrow sheets of calcified matrix that are known as **lamellae** (lah-MEL-lē; *lamella,* thin plate). Small channels, called **canaliculi** (ka-na-LIK-ū-lē), radiate through the matrix, and interconnect lacunae and link them to nearby blood vessels. The canaliculi contain cytoplasmic extensions of the osteocytes. Nutrients from the blood and waste products from osteocytes diffuse through the extracellular fluid that surrounds these cells as well as through their cytoplasmic extensions.

Compact and Spongy Bone

The basic functional unit of compact bone, the **osteon** (OS-tē-on), or *Haversian system,* is shown in Figure 6–3. Within an osteon, the osteocytes are arranged in concentric layers around a **central canal,** or *Haversian canal,* that contains one or more blood vessels. The lamellae are cylindrical, oriented parallel to the long axis of the central canal. **Perforating canals** provide passageways for linking the blood vessels of the central canals with those of the periosteum and the marrow cavity.

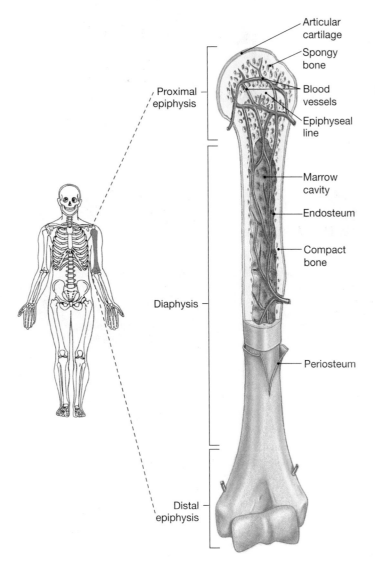

Articular cartilage
Spongy bone
Blood vessels
Epiphyseal line
Proximal epiphysis
Marrow cavity
Endosteum
Compact bone
Diaphysis
Periosteum
Distal epiphysis

● **FIGURE 6–2 The Structure of a Long Bone.**

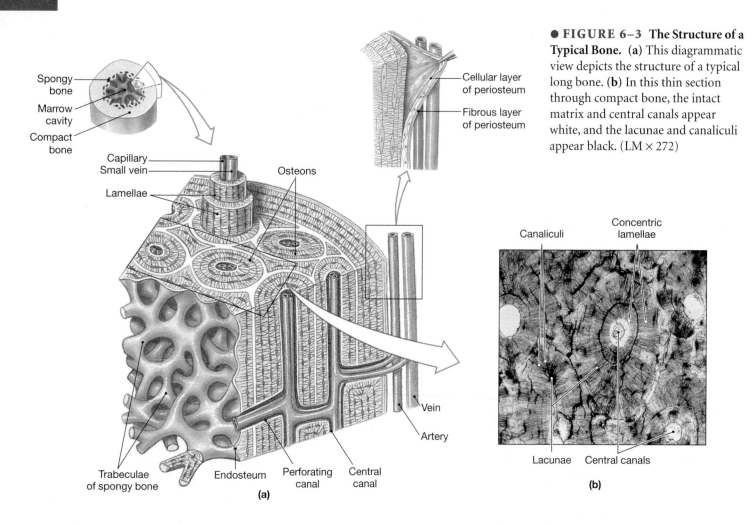

Spongy bone
Marrow cavity
Compact bone

Capillary
Small vein
Osteons
Lamellae

Cellular layer of periosteum
Fibrous layer of periosteum

Trabeculae of spongy bone
Endosteum
Perforating canal
Central canal

Vein
Artery

(a)

● **FIGURE 6–3** **The Structure of a Typical Bone.** **(a)** This diagrammatic view depicts the structure of a typical long bone. **(b)** In this thin section through compact bone, the intact matrix and central canals appear white, and the lacunae and canaliculi appear black. (LM × 272)

Canaliculi
Concentric lamellae

Lacunae
Central canals

(b)

Clinical Note
INTRAOSSEOUS NEEDLE PLACEMENT

The bones are highly vascular, living tissues. By taking advantage of this characteristic, we can place a needle into the medullary cavity of the bone in order to provide emergency fluids and medications. Recent literature has shown that the intraosseous (IO) route of medication administration in cardiac arrest is superior to endotracheal administration. The IO route should be considered for life-threatening cases, particularly cardiac arrest, when an IV cannot be placed.

The concept of placing a needle into the bone marrow was described as early as 1922. However, the technique was all but abandoned until the mid-1980s when it was re-introduced as an alternative fluid administration route in pediatric patients who needed fluids or medications emergently. Since then, the technique has become widespread and is now used in adults as well as children.

Large bones, such as the tibia and humerus, are extremely vascular and can accept large volumes of fluids and transfer them to the central circulation. The sternum is also used for IO administration in certain situations. The bone receives most of its blood supply through a nutrient artery, which enters the cortex of the bone and divides into ascending and descending branches. These branches further divide into arterioles and then capillaries. The capillaries drain into the medullary venous sinusoids throughout the medullary space of the bone (Figure 6–4●).

Specially designed IO needles are available to pierce the cortex of the bone (Figures 6–5● through 6–8●). Fluids and medications administered through a properly placed IO needle enter the medullary sinu-

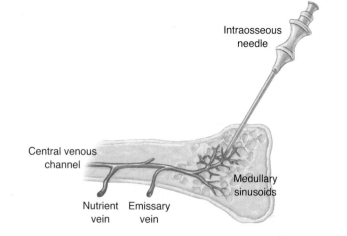

Intraosseous needle

Central venous channel

Medullary sinusoids

Nutrient vein
Emissary vein

● **FIGURE 6–4** **Intraosseous Needle Properly Placed in Marrow Cavity and Medullary Sinusoids.** Fluid primarily drains from the sinusoids and exits via the emissary vein or nutrient vein and enters the central venous circulation.

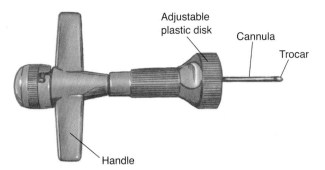
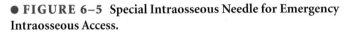

● **FIGURE 6–5** **Special Intraosseous Needle for Emergency Intraosseous Access.**

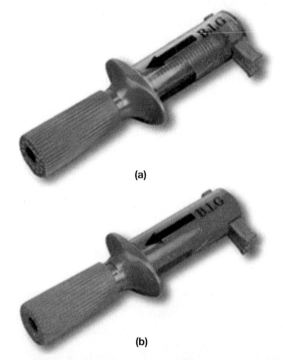

(a)

(b)

● **FIGURE 6–6** **Bone Injection Gun (BIG) IO Device.**
(a) Adult; **(b)** pediatric.

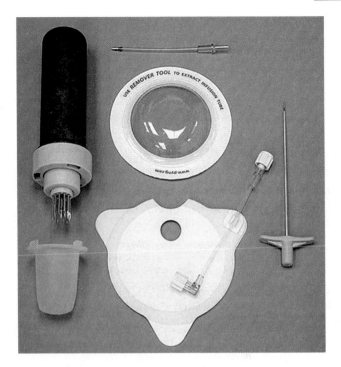

● **FIGURE 6–7** **F.A.S.T. 1 Sternal IO.**

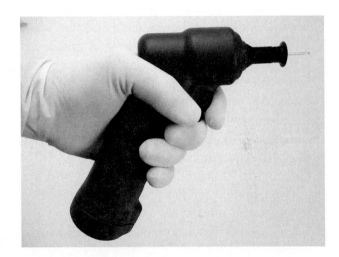

● **FIGURE 6–8** **EZ-IO Device.**

soids. The medullary cavity functions as a rigid, non-collapsible vein, even in the setting of cardiac arrest or shock. The medullary sinusoids act as a central venous channel that exits the bone as either nutrient veins or emissary veins. Fluids and medications administered through an IO needle enter the central circulation promptly, nearly as fast as through an intravenous line. Virtually all fluids and medications used in the prehospital setting can be administered by the IO route.

The IO route provides a viable alternative for critically ill or critically injured patients when medications or fluids are required. Current trauma practices call for limited fluid administration in trauma, so the expanding role of IO therapy appears to be in cardiac arrest when an IV cannot be placed. ■

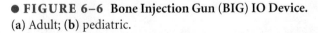

Spongy bone has a different lamellar arrangement and no osteons. Instead, the lamellae form rods or plates called **trabeculae** (tra-BEK-ū-lē; *trabecula,* wall). Frequent branchings of the thin trabeculae create an open network. Canaliculi that radiate from the lacunae of spongy bone end at the exposed surfaces of the trabeculae, where nutrients and wastes diffuse between the marrow and osteocytes.

A layer of compact bone covers bone surfaces everywhere except inside *joint capsules,* where articular cartilages protect opposing surfaces. Compact bone is usually found where stresses come from a limited range of directions. The limb bones, for example, are built to withstand forces applied at either end. Because osteons are parallel to the long axis of the shaft, a limb bone does not bend when a force (even a large one) is applied to either end. However, a much smaller force applied to the side of the shaft can break the bone.

In contrast, spongy bone is found where bones are not heavily stressed or where stresses arrive from many directions. For example, spongy bone is present at the epiphyses of long bones, where stresses are transferred across joints. Spongy bone is also much lighter than compact bone. This reduces the weight of the skeleton and makes it easier for muscles to move the bones. Finally, the trabecular network of spongy bone supports and protects the cells of red bone marrow, which are important sites of blood cell formation.

Cells in Bone

Although osteocytes are the most abundant cells in bone, other cell types are also present. These cells, called *osteoclasts* and *osteoblasts,* are associated with the endosteum that lines the inner cavities of both compact and spongy bone, and with the cellular layer of the periosteum. Three primary cell types occur in bone:

1. **Osteocytes** are mature bone cells. Osteocytes maintain normal bone structure by recycling the calcium salts in the bony matrix around themselves and by assisting in repairs.
2. **Osteoclasts** (OS-tē-ō-clasts; *clast,* break) are giant cells with 50 or more nuclei. Acids and enzymes secreted by osteoclasts dissolve the bony matrix and release the stored minerals through *osteolysis* (os-tē-OL-i-sis), or *resorption.* This process helps regulate calcium and phosphate concentrations in body fluids.
3. **Osteoblasts** (OS-tē-ō-blasts; *blast,* precursor) are the cells responsible for the production of new bone, a process called *osteogenesis* (os-tē-ō-JEN-e-sis; *gennan,* to produce). Osteoblasts produce new bone matrix and promote the deposition of calcium salts in the organic matrix. At any given moment, osteoclasts are removing matrix and osteoblasts are adding to it. When an osteoblast becomes completely surrounded by calcified matrix, it differentiates into an osteocyte.

→ CONCEPT CHECK QUESTIONS

1. How would the strength of a bone be affected if the ratio of collagen to calcium increased?
2. A sample of bone shows concentric layers that surround a central canal. Is it from the shaft or the end of a long bone?
3. If the activity of osteoclasts exceeds that of osteoblasts in a bone, how will the mass of the bone be affected?

Answers begin on p. 792.

■ Bone Formation and Growth

The growth of your skeleton determines the size and proportions of your body. Skeletal growth begins about six weeks after fertilization, when an embryo is about 12 mm (0.5 in.) long. (At this time, all skeletal elements are made of cartilage.) Bone growth continues through adolescence, and typically portions of the skeleton continue growing until roughly age 25. This section considers the process of osteogenesis (bone formation) and growth. The next section examines the maintenance and turnover of mineral reserves in the adult skeleton.

During development, cartilage or other connective tissues are replaced by bone. The process of replacing other tissues with bone is called **ossification.** (The process of **calcification,** which is the deposition of calcium salts, occurs during ossification, but it can also occur in tissues other than bone.) There are two major forms of ossification. In *intramembranous ossification,* bone develops within sheets or membranes of connective tissue. In *endochondral ossification,* bone replaces existing cartilage. Figure 6–9● shows some of the bones formed by these two processes in a 16-week-old fetus.

Intramembranous Ossification

Intramembranous (in-tra-MEM-bra-nus) **ossification** begins when osteoblasts differentiate within embryonic or fetal

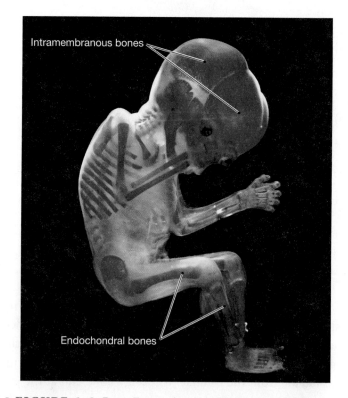

Intramembranous bones

Endochondral bones

● **FIGURE 6–9 Bone Formation in a 16–Week-Old Fetus.**

fibrous connective tissue. This type of ossification normally occurs in the deeper layers of the dermis. The osteoblasts differentiate from connective tissue stem cells after the organic components of the matrix secreted by the stem cells become calcified. The place where ossification first occurs is called an **ossification center.** As ossification proceeds and new bone branches outward, some osteoblasts become trapped inside bony pockets and change into osteocytes.

Bone growth is an active process, and osteoblasts require oxygen and a reliable supply of nutrients. Blood vessels begin to grow into the area to meet these demands and over time become trapped within the developing bone. At first, the intramembranous bone resembles spongy bone. Further remodeling around the trapped blood vessels can produce osteons typical of compact bone. The flat bones of the skull, the lower jaw (*mandible*), and the collarbones (*clavicles*) form this way.

Endochondral Ossification

Most of the bones of the skeleton are formed through **endochondral** (en-dō-KON-drul; *endo,* inside + *chondros,* cartilage) **ossification** of existing hyaline cartilage. The carti-

lages develop first; they are like miniature cartilage models of the future bone. By the time an embryo is six weeks old, the cartilage models of the future limb bones begin to be replaced by true bone. Steps in the growth and ossification of a limb bone are diagrammed in Figure 6–10●.

Step 1: Endochondral ossification starts as chondrocytes within the cartilage model enlarge and the surrounding matrix begins to calcify. The chondrocytes die because the diffusion of nutrients slows through the calcified matrix.

Step 2: Bone formation first occurs at the shaft surface. Blood vessels invade the perichondrium, and cells of its inner layer differentiate into osteoblasts that begin producing bone matrix. ∞ p. 109

Step 3: Blood vessels invade the inner region of the cartilage, and newly differentiated osteoblasts form spongy bone within the center of the shaft at a *primary center of ossification.* Bone development proceeds toward either end, and fills the shaft with spongy bone.

Step 4: As the bone enlarges, osteoclasts break down some of the spongy bone and create a marrow cavity. The cartilage model does not completely fill with bone because the

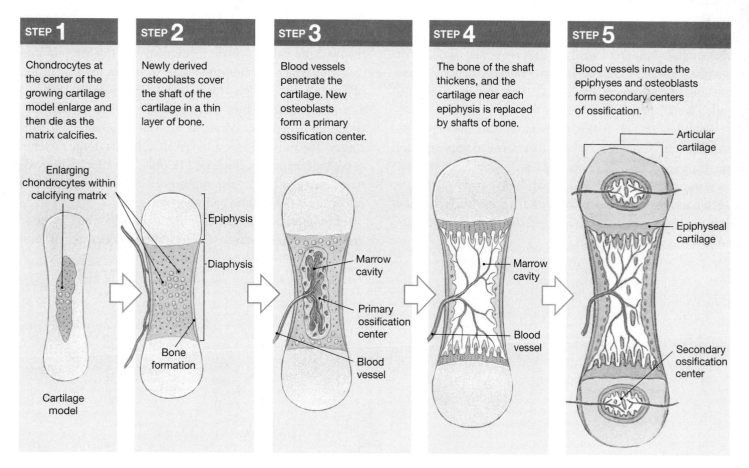

● **FIGURE 6–10   Endochondral Ossification.**

epiphyseal cartilages, or *epiphyseal plates,* on the ends continue to enlarge, and increase the length of the developing bone. Although osteoblasts from the shaft continuously invade the epiphyseal cartilages, the bone grows longer because new cartilage is continuously added in front of the advancing osteoblasts. This situation is like a pair of joggers, one in front of the other: as long as they run at the same speed, the one in back will never catch the one in front, no matter how far they travel.

Step 5: The centers of the epiphyses begin to calcify. As blood vessels and osteoblasts enter these areas, *secondary centers of ossification* form, and the epiphyses eventually become filled with spongy bone. A thin cap of the original cartilage model remains exposed to the joint cavity as the **articular cartilage.** At this stage, the bone of the shaft and the bone of each epiphysis are still separated by epiphyseal cartilage. As long as the rate of cartilage growth keeps pace with the rate of osteoblast invasion, the epiphyseal cartilage persists, and the bone continues to grow longer.

When sex hormone production increases at puberty, bone growth accelerates dramatically, and osteoblasts begin to produce bone faster than epiphyseal cartilage expands. As a result, the epiphyseal cartilages at each end of the bone get increasingly narrow, until they disappear. In adults, the former location of the epiphyseal cartilage is marked by a distinct *epiphyseal line* (see Figure 6–2) that remains evident in X-rays after epiphyseal growth has ended. The end of epiphyseal growth is called *epiphyseal closure.*

While the bone elongates, its diameter also enlarges at its outer surface. This enlargement process, called **appositional growth,** occurs as cells of the periosteum develop into osteoblasts and produce additional bony matrix (Figure 6–11●). As new bone is deposited on the outer surface of the shaft, the inner surface is eroded by osteoclasts, and the marrow cavity gradually enlarges.

Bone Growth and Body Proportions

The timing of epiphyseal closure varies from bone to bone and individual to individual. Ossification of the toes may be complete by age 11, whereas portions of the pelvis or the wrist may continue to enlarge until age 25. The epiphyseal cartilages in the arms and legs usually close by age 18 (women) or 20 (men). Differences in sex hormones account for variations in body size and proportions between men and women.

Requirements for Normal Bone Growth

Normal bone growth and maintenance cannot occur without a reliable source of minerals, especially calcium salts. During prenatal development these minerals are absorbed from the mother's bloodstream. The demands are so great that the maternal skeleton often loses bone mass during pregnancy. From infancy to adulthood, the diet must provide adequate amounts of calcium and phosphate, and the body must be able to absorb and transport these minerals to sites of bone formation.

Vitamin D_3 plays an important role in normal calcium metabolism. This vitamin can be obtained from dietary supplements or manufactured by epidermal cells exposed to UV radiation. ∞ p. 128 After vitamin D_3 has been processed in the liver, the kidneys convert a derivative of this vitamin into *calcitriol,* which is a hormone that stimulates the absorption of calcium and phosphate ions in the digestive tract. *Rickets* is a condition marked by a softening and bending of bones that occurs in growing children, as a result of vitamin D_3 deficiency. The reduced amounts of calcium salts in the skeleton cause the bones to become very flexible, and affected individuals develop a bowlegged appearance as the leg bones bend under the weight of the body.

Vitamin A and *vitamin C* are also essential for normal bone growth and maintenance. For example, a deficiency of vita-

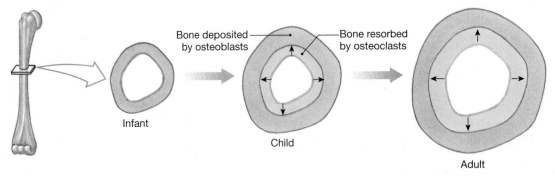

● **FIGURE 6–11** Appositional Bone Growth.

Bone deposited by osteoblasts
Bone resorbed by osteoclasts
Infant
Child
Adult

min C can cause *scurvy*. One of the primary features of this condition is a reduction in osteoblast activity that leads to weak and brittle bones. In addition to vitamins, various hormones (including growth hormone, thyroid hormones, sex hormones, and those involved in calcium metabolism) are essential to normal skeletal growth and development.

→ **CONCEPT CHECK QUESTIONS**

1. During intramembranous ossification, which type of tissue is replaced by bone?
2. How could X-rays of the femur be used to determine whether a person had reached full height?
3. In the Middle Ages, choirboys were sometimes castrated (had their testes removed) before puberty to prevent their voices from changing. How would castration have affected their height?
4. Why are pregnant women given calcium supplements and encouraged to drink milk even though their skeletons are fully formed?

Answers begin on p. 792.

■ Bone Remodeling and Homeostatic Mechanisms

Of the five major functions of the skeleton discussed earlier in this chapter, support and storage of minerals depend on the dynamic nature of bone. In adults, osteocytes in lacunae maintain the surrounding matrix, and continually remove and replace the surrounding calcium salts. But osteoclasts and osteoblasts also remain active, even after the epiphyseal cartilages have closed. Normally, their activities are balanced: as one osteon forms through the activity of osteoblasts, another is destroyed by osteoclasts. The turnover rate for bone is quite high, and in adults roughly 18 percent of the protein and mineral components are removed and replaced each year through the process of **remodeling.** Not every part of every bone is affected; there are regional and even local differences in the rate of turnover. For example, the spongy bone in the head of the femur may be replaced two or three times each year, whereas the compact bone along the shaft remains largely untouched.

The Role of Remodeling in Support

Regular mineral turnover gives each bone the ability to adapt to new stresses. Heavily stressed bones become thicker and stronger and develop more pronounced surface ridges; bones not subjected to ordinary stresses become thin and brittle.

Regular exercise is, thus, an important stimulus in maintaining normal bone structure.

Degenerative changes occur in the skeleton after even brief periods of inactivity. For example, using a crutch while wearing a cast takes the weight off the injured leg. After a few weeks, the unstressed leg will lose up to about a third of its bone mass. The bones rebuild just as quickly once they again carry their normal weight.

🔒 **Key Note**

What you don't use, you lose. The stresses applied to bones during exercise are essential to maintain bone strength and bone mass.

Homeostasis and Mineral Storage

The bones of the skeleton are more than just racks to hang muscles on. They are important mineral reservoirs—especially for calcium, which is the most abundant mineral in the human body. A typical human body contains 1–2 kg (2.2–4.4 lb) of calcium, 99 percent of which is deposited in the skeleton.

Calcium ions play an important role in many physiological processes, so calcium ion concentrations must be closely controlled. Even small variations from normal concentrations affect cellular operations, and larger changes can cause a clinical crisis. Neurons and muscle cells are particularly sensitive to changes in calcium ion concentration. If the calcium concentration in body fluids increases by 30 percent, neurons and muscle cells become relatively unresponsive. If calcium levels decrease by 35 percent, they become so excitable that convulsions may occur. A 50 percent reduction in calcium concentrations usually causes death. Such effects are relatively rare, however, because calcium ion concentrations are so closely regulated that daily fluctuations of more than 10 percent are very unusual.

The hormones *parathyroid hormone* (PTH) and *calcitriol* work together to elevate calcium levels in body fluids. Their actions are opposed by *calcitonin,* which depresses calcium levels in body fluids. These hormones and their regulation are discussed further in Chapter 10.

By providing a calcium reserve, the skeleton helps maintain calcium homeostasis in body fluids. This function can directly affect the shape and strength of the bones in the skeleton. When large numbers of calcium ions are mobilized, bones become weaker; when calcium salts are deposited, bones become more massive.

Injury and Repair

Despite its strength, bone cracks or even breaks if subjected to extreme loads, sudden impacts, or stresses from unusual directions. Every such crack or break in a bone constitutes a **fracture.** Fractures are classified according to many features, including their external appearance, the site of the fracture, and the nature of the break.

Bones will usually heal even after they have been severely damaged, so long as the blood supply remains and the cellular components of the endosteum and periosteum survive. Steps in the repair process, which may take from four months to well over a year following a fracture, are diagrammed in Figure 6–12●:

Step 1: In even a small fracture, many blood vessels are broken and extensive bleeding occurs. A large blood clot, called a **fracture hematoma** (*hemato-*, blood; + *tumere,* to swell), soon forms and closes off the injured blood vessels. Because the resulting lack of blood supply kills osteocytes, dead bone extends in either direction from the break.

Step 2: Cells of the periosteum and endosteum undergo mitosis, and the daughter cells migrate into the fracture zone. There they form localized thickenings—an **external callus** (*callum,* hard skin) and an **internal callus,** respectively. At the center of the external callus, cells differentiate into chondrocytes and produce hyaline cartilage.

Step 3: Osteoblasts replace the new central cartilage of the external callus with spongy bone. When this process is complete, the external and internal calluses form a continuous brace of spongy bone at the fracture site. The ends of the fracture are now held firmly in place and can withstand normal stresses from muscle contractions.

Step 4: The remodeling of spongy bone at the fracture site may continue over a period of four months to well over a year. When the remodeling is complete, the fragments of dead bone and the spongy bone of the calluses will be gone, and only living compact bone will remain. The repair may be "good as new," with no sign that a fracture occurred, but the bone may be slightly thicker than normal at the fracture site.

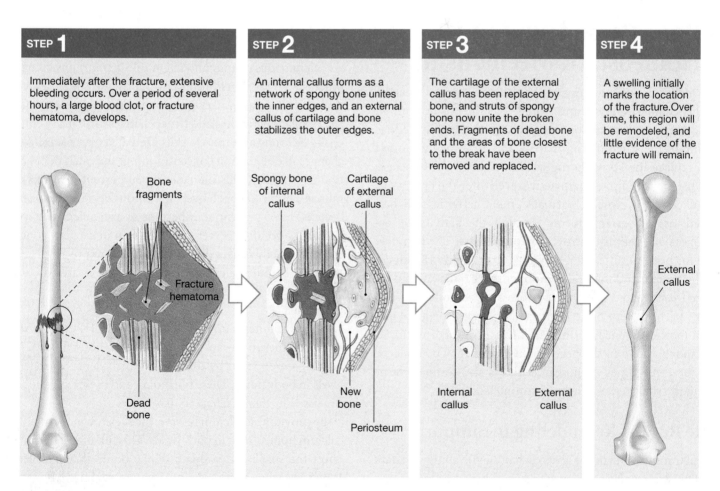

STEP 1

Immediately after the fracture, extensive bleeding occurs. Over a period of several hours, a large blood clot, or fracture hematoma, develops.

Bone fragments

Fracture hematoma

Dead bone

STEP 2

An internal callus forms as a network of spongy bone unites the inner edges, and an external callus of cartilage and bone stabilizes the outer edges.

Spongy bone of internal callus

Cartilage of external callus

New bone

Periosteum

STEP 3

The cartilage of the external callus has been replaced by bone, and struts of spongy bone now unite the broken ends. Fragments of dead bone and the areas of bone closest to the break have been removed and replaced.

Internal callus

External callus

STEP 4

A swelling initially marks the location of the fracture. Over time, this region will be remodeled, and little evidence of the fracture will remain.

External callus

● **FIGURE 6–12 Steps in the Repair of a Fracture.**

Clinical Note
SKELETAL INJURIES

There are four general types of skeletal and joint injuries: sprains, subluxations, dislocations, and fractures.

Sprains

The *sprain* is an injury that stretches or tears one or more ligaments within a joint. This tearing of ligaments weakens the joint. Stresses to a joint can extend the joint beyond its normal range of motion, which causes ligamentous injury (Figure 6–13●). The injury results in acute pain at the site, followed shortly by inflammation and swelling. Sprains are classified, or graded, according to their severity, using the following criteria:

■ *Grade I.* Minor and incomplete tear. The ligament is painful and tender, but there is no laxity. Swelling and ecchymosis are usually minimal. The joint is stable.

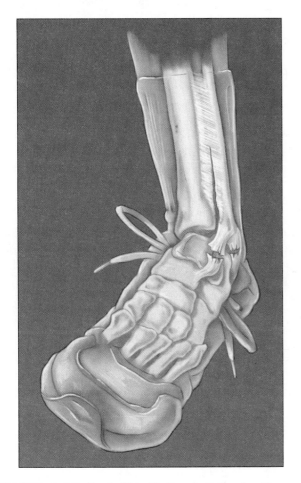

● **FIGURE 6–13 Grade II Ankle Sprain Following Common Inversion Injury.** The anterior talofibular ligament appears completely torn, while the posterior talofibular ligament is only partially torn. The vast majority of ankle sprains involve the lateral ligaments.

■ *Grade II.* Significant but incomplete tear. There is laxity, but also an endpoint beyond which no further opening of the joint occurs. Swelling and ecchymosis may be moderate to severe, and pain may range from moderate to severe. The joint is unstable but intact.

■ *Grade III.* Complete tear and total failure of the ligament or ligaments involved. No endpoint is felt when stress is applied to the ligament during examination of the joint. Pain and muscle spasm can often mask a grade III sprain, so the diagnosis is easily missed. Due to severe pain and spasm, the injury may be mistaken for a fracture. A repeat examination several days later can help confirm the diagnosis. The joint is unstable.

Subluxation

A *subluxation*, also called a *partial dislocation*, is a partial displacement of a bone end from its position within a joint capsule. It occurs as the joint separates under stress, which stretches the ligaments. A subluxation differs from a sprain in that it more significantly reduces the joint's integrity.

Dislocation

A *dislocation* is a complete displacement of bone ends from their normal position within a joint. The joint often fixes in an abnormal position with noticeable deformity (Figure 6–14●). This injury occurs when the bones of the joint move beyond their normal range of motion, usually with great force. It carries with it the danger of entrapping, compressing, or tearing nearby blood vessels and nerves. A dislocation should be suspected whenever a joint is deformed or does not move in a normal fashion.

Fracture

A *fracture* is an injury that interrupts the structural integrity of a bone. Most fractures are the result of significant trauma to a healthy bone. The bony cortex may be disrupted by many different forces including: a direct blow, angular (bending) forces, axial loading, twisting (torque) stress, or any combination of these.

Fractures also can occur in a bone that is diseased or otherwise abnormal. These pathological processes weaken the bone, making it susceptible to fracture by forces that would, under normal circumstances, not typically disrupt the cortex. These fractures are called pathological fractures and can result from relatively minor trauma. Examples of pathological fractures include fractures through lytic metastatic (cancerous) lesions, fractures through benign bone cysts, and vertebral compression fractures in patients with advanced osteoporosis. Vertebral compression fractures are the most common type of pathological fracture.

(continued next page)

Clinical Note—continued

SKELETAL INJURIES

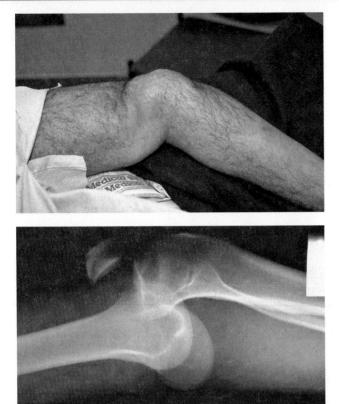

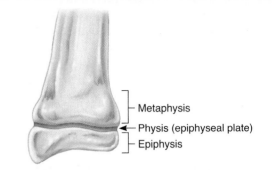
— Metaphysis
— Physis (epiphyseal plate)
— Epiphysis

● **FIGURE 6–15 Growth Plate in a Child's Long Bone.** This growth plate is also called the physis or epiphyseal plate. The portion of bone proximal to the physis is the metaphysics; the segment distal to the physis is the epiphysis.

is best for type I fractures and worst for type V fractures. The Salter-Harris classifications for growth-plate injuries are (Figure 6–16●):

- *Salter-Harris Type I.* The fracture line runs through the physis. There is usually little, if any, separation of the epiphysis from the rest of the bone. It is often difficult to see the fracture line on X-ray, as the line is hidden within the growth plate. A type I injury is the least severe epiphyseal fracture type.
- *Salter-Harris Type II.* The entire epiphysis and a portion of the metaphysis are broken off. The fracture line runs through the physis and into the metaphysis.
- *Salter-Harris Type III.* A portion of the epiphysis is broken off. The fracture line runs through the physis, into the epiphysis, and into the joint.
- *Salter-Harris Type IV.* A portion of the epiphysis and a portion of the metaphysis are broken off. The fracture line runs through the metaphysis, the physis, the epiphysis, and into the joint.
- *Salter-Harris Type V.* The epiphyseal plate is compressed, usually through an axial loading type force. These injuries are difficult to diagnose and are sometimes only evident retrospectively when a growth disturbance develops.

● **FIGURE 6–14 Anterior Dislocation of the Knee.** This rare injury poses a significant threat to blood vessels and nerves that transverse the knee. Immediate reduction is indicated.

GROWTH-PLATE INJURIES

Fractures can involve the epiphyseal growth plate in children. The cartilaginous epiphyseal plate, also called the physis, is readily injured because it is weaker than ossified bone or ligaments (Figure 6–15●). Damage to the epiphyseal plate during a child's growth may destroy all or part of the bone's ability to produce new bone, which results in stunted or deformed growth thereafter. The potential for a growth disturbance from a growth-plate injury is related to the number of years the child has yet to grow. Thus, the older the child, the less time remains for a deformity to develop. The *Salter-Harris system* is often used to classify growth-plate injuries. The potential for growth disturbance increases as the classification number increases. The prognosis

TYPES OF FRACTURES

There are many systems for classifying fractures. Generally, fractures are classified as either closed or open. In a *closed fracture*, the skin is not broken and there is no communication between the fracture site and the environment. In an *open fracture*, the skin is broken and a communication exists between the fracture site and the en-

● **FIGURE 6–16 Salter-Harris System of Classifying Growth Plate Injuries.** The likelihood of a permanent growth plate deformity increases as the classification number increases.

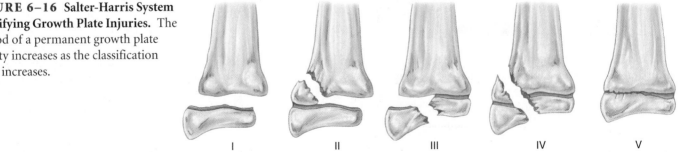

I   II   III   IV   V

vironment. The difference in these two classifications has significant emergency care considerations. Open fractures are usually taken to the operating room, where the fracture site is exposed, cleansed, and irrigated to prevent infection. Closed fractures can be splinted, placed in a cast, or otherwise immobilized (Figure 6–17●).

Fractures are also classified based upon the orientation of the fracture line as seen in radiographic (X-ray) studies (Figure 6–18●).

■ *Greenstick fracture.* Greenstick fractures are seen almost exclusively in children. In a greenstick fracture, one side of the bone is broken and the other side bent. This occurs due to the large amount of cartilage in the bones of children.

■ *Torus fracture.* Torus fractures occur almost exclusively in children. In a torus fracture, there is localized buckling or swelling (or torus) of the cortex, with little or no displacement of the bone itself.

■ *Transverse fracture.* A transverse fracture is a fracture line perpendicular to the long axis of the bone.

■ *Oblique fracture.* In an oblique fracture, the break extends obliquely to the long axis of the bone.

■ *Spiral fracture.* A spiral fracture, also called a torsion fracture, occurs when a twisting force is applied to a long bone.

■ *Comminuted fracture.* Comminuted fractures are those where the bone at the fracture site has multiple bone fragments.

■ *Segmental fracture.* A segmental fracture is an injury where there are multiple fracture sites along the axis of the bone, which leaves a free-floating segment of bone between the two fracture sites. Segmental fractures are often mistakenly called comminuted fractures.

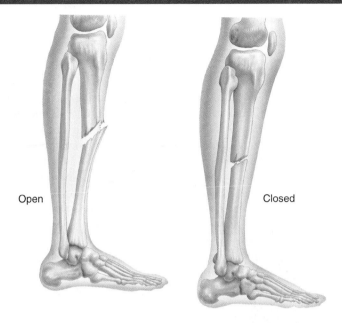

● **FIGURE 6–17 Open and Closed Fractures.** Open fractures, also called compound fractures, have a direct communication with the environment. Closed fractures, also called simple fractures, do not have any communication with the environment.

■ *Impacted fracture.* An impacted fracture is an injury where an axial loading force is applied to the bone, which drives the bone ends at the fracture site together.

● **FIGURE 6–18 Types of Fracture Based on the Appearance of the Fracture Line on Radiographs.**

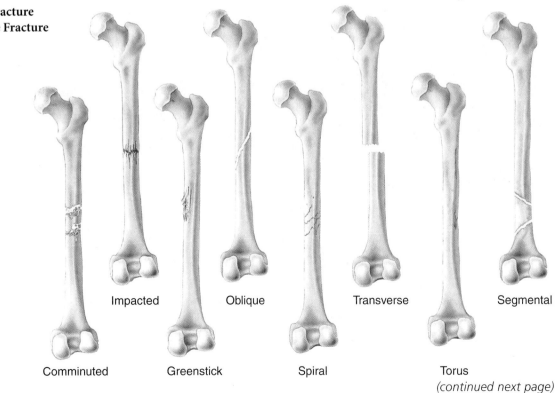

Impacted   Oblique   Transverse   Segmental

Comminuted   Greenstick   Spiral   Torus

(continued next page)

Clinical Note—continued
SKELETAL INJURIES

SPECIFIC FRACTURES
Several specific types of fractures are important to emergency medical care.

Upper Extremity Fractures
Upper extremity fractures are common. Falls on an outstretched arm can result in fractures of the wrist, elbow, humerus, and clavicle. In fact, the clavicle is the most frequently fractured bone in the body. A common forearm fracture is the *Colles' fracture* (Figure 6–19●). It usually results from a fall on an outstretched arm. The deformity resembles that of a dinner fork. Supracondylar fractures are fractures of the distal humerus, just above the elbow. Fractures that result from extension-type injuries usually cause posterior displacement of the distal segment. With flexion-type injuries, the distal fracture segment is usually displaced anteriorly. Flexion-type injuries occur much less frequently than extension-type injuries. Most supracondylar fractures occur in children and are associated with a number of complications, including nerve and vascular injuries due to the close proximity of these structures to the fracture site (Figure 6–20●). Supracondylar fractures that involve a growth plate can cause growth abnormalities that often result in permanent deformity. Because of the high complication rate, supracondylar fractures should be promptly evaluated and treated by an orthopaedic surgeon.

Hip Fractures
Although the femur is the largest bone in the body, fractures of the proximal femur, or hip, are frequently seen, especially in the elderly. The hip is a ball-and-socket joint that consists of the acetabulum and the proximal femur, two to three inches below the lesser trochanter. The incidence of hip fracture increases with age and doubles for every decade past the age of fifty. The incidence is two to three times higher in women than men, primarily due to decreased bone density secondary to osteoporosis. Hip fractures are usually classified as either intracapsular or extracapsular, depending on their location. With intracapsular fractures, the blood vessels that supply the femoral head are often compromised, which can lead to necrosis of the femoral head. The four types of intracapsular hip fractures are capital, subcapital, transcervical, or basicervical (Figure 6–21●). Subcapital fractures are by far the most common type of intracapsular hip fracture. There are three types of extracapsular hip fractures:

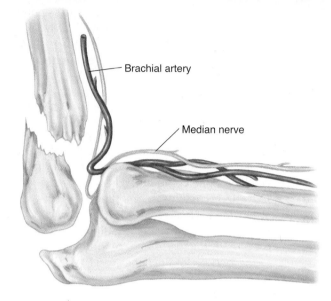

● **FIGURE 6–20 Supracondylar Fracture.** Supracondylar fractures have a high incidence of associated nervous and vascular tissue injury. In children, some supracondylar fractures can involve the growth plate, and possibly lead to permanent deformity or disability.

trochanteric, intertrochanteric, and subtrochanteric. Of these, intertrochanteric fractures are the most common. All hip fractures in ambulatory patients require open reduction and internal fixation in the operating room. Intracapsular fractures usually require replacement of the entire hip joint with a prosthetic hip and acetabulum. Most extracapsular fractures can be stabilized by placement of a surgical pin or nail to hold the bone segments together while healing. As a rule, early fixation of hip fractures (less than 72 hours) in the elderly reduces morbidity and mortality.

Facial Fractures
Fractures of the maxilla usually result from high-energy injuries. As a result, patients with facial fractures may also sustain other associated injuries such as spinal, chest, and abdominal injuries. There are several identifiable facial fracture patterns that were first described by LeFort. He developed a facial fracture classification system that bears his name (Figure 6–22●). A LeFort I fracture is limited to the maxilla at the level of the nares. In a LeFort I fracture, only the hard palate and upper teeth move with gentle palpation and mobilization. With a LeFort II injury, the triangular fracture line extends across the ridge of the cheeks and into the orbits. Mobilization of the fracture segment will move the nose but not the eyes. In a LeFort III fracture, the facial skeleton is separated from the skull. The entire face, including both orbits, shifts with palpation and gentle mobilization. This injury is also called a *cranial-facial disjunction*. Although not identified by LeFort, a LeFort IV facial fracture has been described. It is similar to a LeFort III fracture, but the fracture line extends upward into the frontal bones.

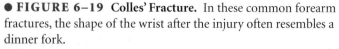

● **FIGURE 6–19 Colles' Fracture.** In these common forearm fractures, the shape of the wrist after the injury often resembles a dinner fork.

Intracapsular

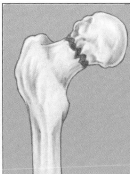

capital (uncommon)   subcapital (common)   trans- or midcervical (rare)   basicervical (uncommon)

Extracapsular

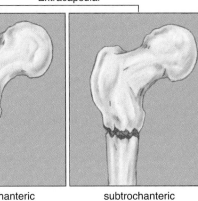

intertrochanteric   subtrochanteric

● **FIGURE 6–21 Types of Hip Fractures.** Subcapital and intertrochanteric are the most common.

LeFort I   **LeFort II**   **LeFort III**

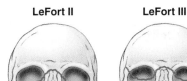

● **FIGURE 6–22 LeFort System of Classifying Facial Fractures.**

FRACTURE HEALING

A fracture heals in three phases: inflammatory, reparative, and remodeling. Each phase of healing gradually blends into the next.

Immediately following a fracture, microscopic blood vessels that cross the fracture line are severed, which interrupts blood supply to the injured bone ends. In the following days, these blood-deprived bone ends become necrotic. This triggers a classic inflammatory response and essential inflammatory cells migrate to the fracture site. Soon, granulation tissue begins to fill the affected area. Within the inflammatory response are cells capable of forming cartilage, collagen, and bone. Together, these three components form the bone callus that gradually surrounds the fractured bone ends and stabilizes them. There are both internal and external calluses. Over time, the external callus becomes harder as minerals, especially calcium, are laid down. Underneath the callus, the necrotic edges of bone at the fracture site are removed by osteoclasts, which are cells whose function is to resorb bone. Finally, following the reparative phase, the remodeling phase begins. Remodeling is the tendency of the bone to regain its original shape and contour. During this phase, the excess parts of the callus are removed and new bone is laid down along natural lines of stress. Remodeling can continue for years and is affected by such factors as the patient's age and health and the magnitude of the original fracture. Typically, within a year or so, all evidence that a fracture occurred has disappeared and the bone appears completely normal. Fractures heal most rapidly in young children and slowest in the elderly. A five-year-old child with an uncomplicated wrist fracture may only need to be in a cast for three weeks while an elderly patient with the same injury can expect to be in a cast for six weeks or more. ■

■ Aging and the Skeletal System

Bones become thinner and relatively weaker as a normal part of the aging process. Inadequate ossification is called **osteopenia** (os-tē-ō-PĒ-nē-uh; *penia*, lacking), and all of us become slightly osteopenic as we age. The reduction in bone mass begins between the ages of 30 and 40, as osteoblast activity begins to decline while osteoclast activity continues at normal levels. Once the reduction begins, women lose roughly eight percent of their skeletal mass every decade, whereas men's skeletons deteriorate at about three percent per decade. Not all parts of the skeleton are equally affected. Epiphyses, vertebrae, and the jaws lose more than their fair share, which results in fragile limbs, a reduction in height, and the loss of teeth.

Clinical Note
OSTEOPOROSIS

Osteoporosis (os-tē-ō-po-RĪ-sis; *porosus*, porous) is a condition that produces a reduction in bone mass great enough to compromise normal function. The difference between the "normal" osteopenia of aging and the clinical condition of osteoporosis is a matter of degree.

Sex hormones are important in maintaining normal rates of bone deposition. Among individuals over age 45, an estimated 29 percent of women and 18 percent of men have osteoporosis. In women, the increase in incidence after menopause has been linked to decreases in the production of estrogens (female sex hormones). Because men continue to produce androgens (male sex hormones) until relatively late in life, severe osteoporosis is less common in males under age 60 than in females in that same age group.

Because osteoporotic bones are more fragile, they break easily and do not repair well. Vertebrae may collapse, which distorts the vertebral articulations and puts pressure on spinal nerves. Therapies that boost estrogen levels in women, dietary changes that elevate calcium levels in the blood, and exercise that stresses bones and stimulates osteoblast activity appear to slow, but not completely prevent, the development of osteoporosis. ■

→ CONCEPT CHECK QUESTIONS

1. Why would you expect the arm bones of a weight lifter to be thicker and heavier than those of a jogger?
2. What is the difference between a simple fracture and a compound fracture?
3. Why is osteoporosis more common in women age 45 and over than in men of the same age?

Answers begin on p. 792.

■ An Overview of the Skeleton

Bone Markings (Surface Features)

Each bone in the human skeleton has not only a distinctive shape but also characteristic external and internal features. For example, elevations or projections form where tendons and ligaments attach and where adjacent bones articulate at joints. Depressions and openings indicate sites where blood vessels and nerves run alongside or penetrate the bone. These landmarks are called **bone markings,** or *surface features*. The most common terms used to describe bone markings are listed and illustrated in Table 6–1.

Skeletal Divisions

The skeletal system consists of 206 separate bones (Figure 6–23●) plus numerous associated cartilages. This system is divided into axial and appendicular divisions (Figure 6–24●). The **axial skeleton** forms the longitudinal axis of the body. This division's 80 bones can be subdivided into (1) the 22 bones of the **skull,** plus seven associated bones (six **auditory ossicles** and the **hyoid bone**); (2) the **thoracic cage** (*rib cage*), composed of 24 **ribs** and the **sternum;** and (3) the 26 bones of the **vertebral column.**

The **appendicular skeleton** includes the bones of the limbs and those of the **pectoral** and **pelvic girdles,** which attach the limbs to the trunk. All together there are 126 appendicular bones; 32 are associated with each upper limb, and 31 with each lower limb.

■ The Axial Division

The **axial skeleton** creates a framework that supports and protects organ systems in the brain and spinal cavities, and ventral body cavities. In addition, it provides an extensive surface area for the attachment of muscles that (1) adjust the positions of the head, neck, and trunk; (2) perform respiratory movements; and (3) stabilize or position elements of the appendicular skeleton.

The Skull

The bones of the skull protect the brain and support delicate sense organs involved with vision, hearing, balance, olfaction (smell), and gustation (taste). The skull is made up of 22 bones: eight form the **cranium,** and 14 are associated with the face. Seven additional bones are associated with the

TABLE 6-1   *An Introduction to the Surface Features of Bones*

GENERAL DESCRIPTION	ANATOMICAL TERM	DEFINITION
Elevations and projections (general)	Process	Any projection or bump
	Ramus	Extension of a bone that makes an angle with the rest of the structure
Processes formed where tendons or ligaments attach	Trochanter	Large, rough projection
	Tuberosity	Smaller, rough projection
	Tubercle	Small, rounded projection
	Crest	Prominent ridge
	Line	Low ridge
	Spine	Pointed process
Processes formed for articulation with adjacent bones	Head	Expanded articular end of an epiphysis, separated from the shaft by a neck
	Neck	Narrow connection between the epiphysis and the diaphysis
	Condyle	Smooth, rounded articular process
	Trochlea	Smooth, grooved articular process shaped like a pulley
	Facet	Small, flat articular surface
Depressions	Fossa	Shallow depression
	Sulcus	Narrow groove
Openings	Foramen	Rounded passageway for blood vessels or nerves
	Canal	Passageway through the substance of a bone
	Fissure	Elongate cleft
	Sinus	Chamber within a bone, normally filled with air

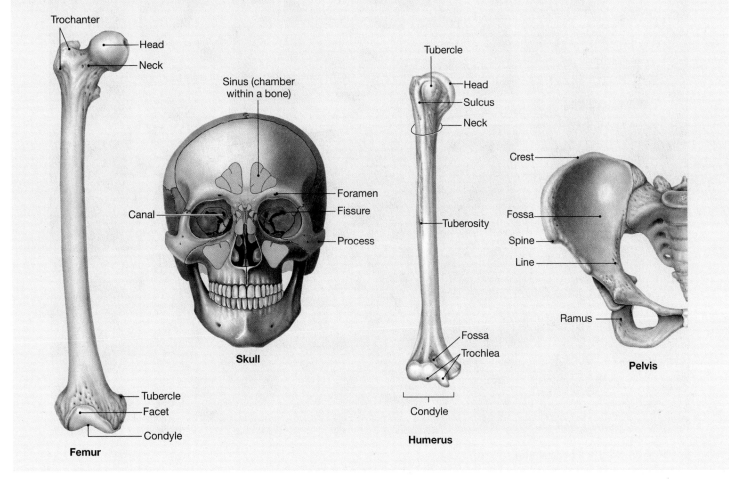

Femur — Trochanter, Head, Neck, Tubercle, Facet, Condyle

Skull — Sinus (chamber within a bone), Canal, Foramen, Fissure, Process

Humerus — Tubercle, Head, Sulcus, Neck, Tuberosity, Fossa, Trochlea, Condyle

Pelvis — Crest, Fossa, Spine, Line, Ramus

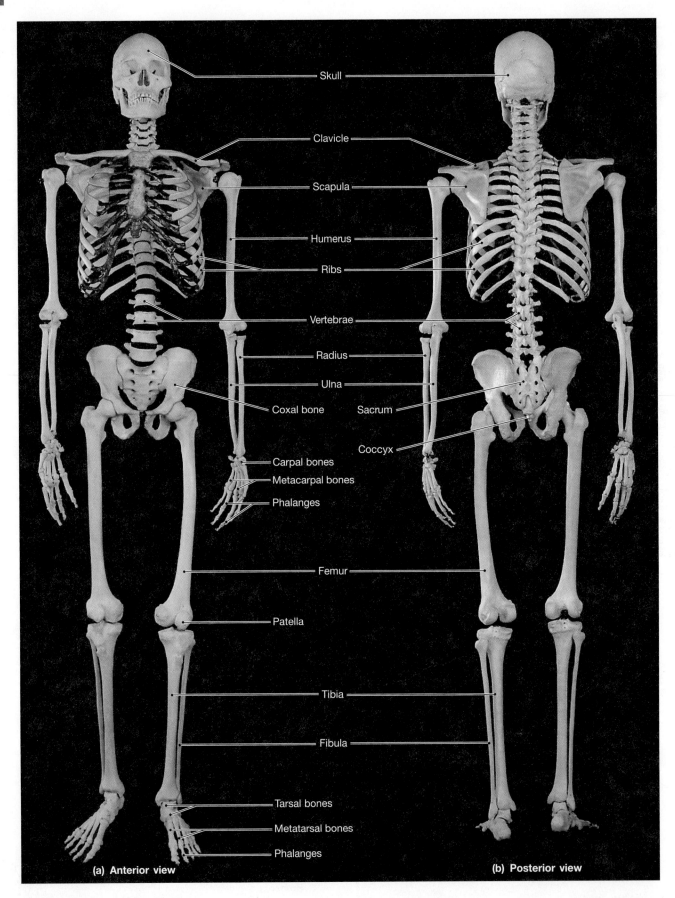

Skull

Clavicle

Scapula

Humerus

Ribs

Vertebrae

Radius

Ulna

Coxal bone

Sacrum

Coccyx

Carpal bones

Metacarpal bones

Phalanges

Femur

Patella

Tibia

Fibula

Tarsal bones

Metatarsal bones

Phalanges

(a) **Anterior view**

(b) **Posterior view**

● **FIGURE 6–23** The Skeleton.

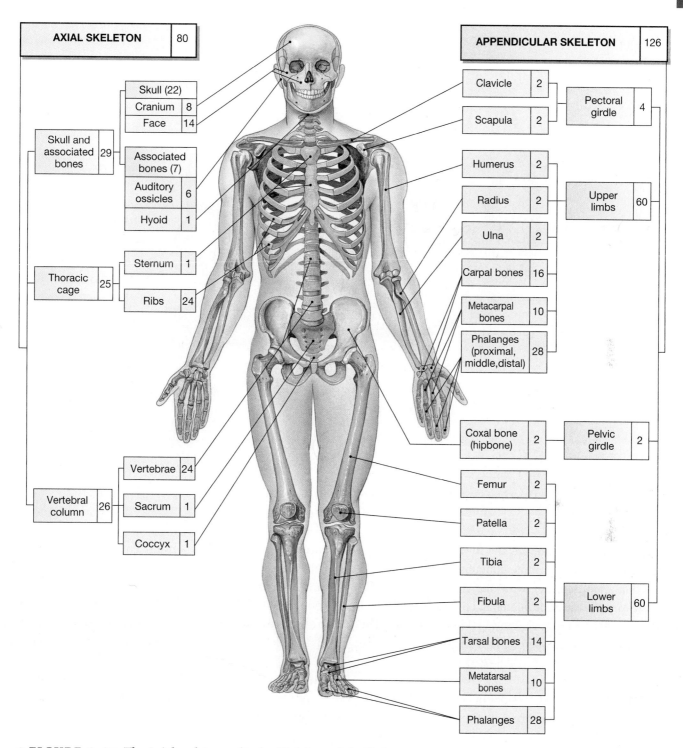

AXIAL SKELETON	80

Skull and associated bones	29

Skull (22)	
Cranium	8
Face	14

Associated bones (7)	
Auditory ossicles	6
Hyoid	1

Thoracic cage	25

Sternum	1
Ribs	24

Vertebral column	26

Vertebrae	24
Sacrum	1
Coccyx	1

APPENDICULAR SKELETON	126

Clavicle	2
Scapula	2

Pectoral girdle	4

Humerus	2
Radius	2
Ulna	2
Carpal bones	16
Metacarpal bones	10
Phalanges (proximal, middle, distal)	28

Upper limbs	60

Coxal bone (hipbone)	2

Pelvic girdle	2

Femur	2
Patella	2
Tibia	2
Fibula	2
Tarsal bones	14
Metatarsal bones	10
Phalanges	28

Lower limbs	60

● **FIGURE 6–24** The Axial and Appendicular Divisions of the Skeleton.

skull: six *auditory ossicles,* which are tiny bones involved in sound detection, are encased by the *temporal bones* of the cranium, and the *hyoid bone* is connected to the inferior surface of the skull by ligaments.

The cranium encloses the **cranial cavity,** which is a fluid-filled chamber that cushions and supports the brain. The outer surface of the cranium provides an extensive area for the attachment of muscles that move the eyes, jaws, and head.

The Bones of the Cranium

THE FRONTAL BONE. The **frontal bone** of the cranium forms the forehead and the roof of the **orbits,** the bony recesses that contain the eyes (Figures 6–25● and 6–26●). A **supraorbital foramen** is an opening that pierces the bony ridge above each orbit, and forms a passageway for blood vessels and nerves passing to or from the eyebrows and eyelids (see Figure 6–25). (Sometimes the ridge has a deep groove, called a *supraorbital*

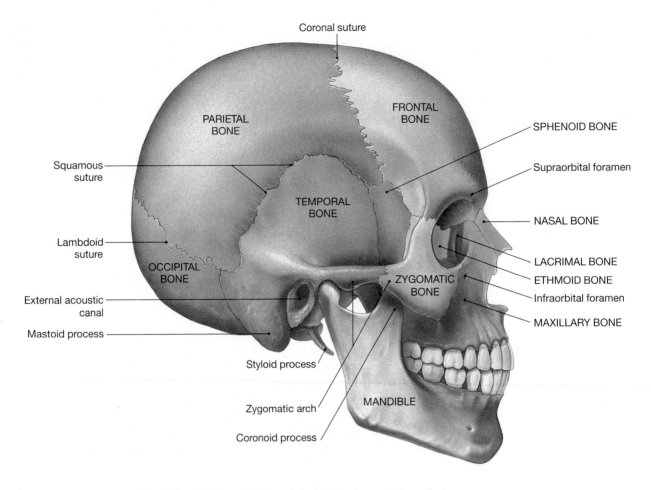

Coronal suture

PARIETAL BONE

FRONTAL BONE

SPHENOID BONE

Supraorbital foramen

Squamous suture

TEMPORAL BONE

NASAL BONE

Lambdoid suture

LACRIMAL BONE

OCCIPITAL BONE

ETHMOID BONE

ZYGOMATIC BONE

Infraorbital foramen

External acoustic canal

MAXILLARY BONE

Mastoid process

Styloid process

Zygomatic arch

MANDIBLE

Coronoid process

● **FIGURE 6–25 The Adult Skull, Part I.** The adult skull is shown in lateral view.

notch, rather than a foramen, but the function is the same.) Above the orbit, the frontal bone contains air-filled internal chambers that communicate with the nasal cavity. These **frontal sinuses** make the bone lighter and produce mucus that cleans and moistens the nasal cavities (Figure 6–27●). The **infraorbital foramen** is an opening for a major sensory nerve from the face (see Figures 6–25 and 6–26a).

THE PARIETAL BONES.   On both sides of the skull, a **parietal** (pa-RĪ-e-tal) **bone** is posterior to the frontal bone (see Figures 6–26a and 6–27). Together the parietal bones form the roof and the superior walls of the cranium. The parietal bones interlock along the **sagittal suture,** which extends along the midline of the cranium (see Figure 6–26a). Anteriorly, the two parietal bones articulate with the frontal bone along the **coronal suture** (see Figure 6–25).

THE OCCIPITAL BONE.   The **occipital bone** forms the posterior and inferior portions of the cranium (see Figures 6–25 and 6–26b). Along its superior margin, the occipital bone contacts the two parietal bones at the **lambdoid** (LAM-doyd)

suture. The **foramen magnum** connects the cranial cavity with the spinal cavity, which is enclosed by the vertebral column. The spinal cord passes through the foramen magnum to connect with the inferior portion of the brain. On either side of the foramen magnum are the **occipital condyles,** the sites of articulation between the skull and the vertebral column.

THE TEMPORAL BONES.   Below the parietal bones and contributing to the sides and base of the cranium are the **temporal bones.** The temporal bones contact the parietal bones along the **squamous** (SKWĀ-mus) **suture** on each side (see Figure 6–25).

The temporal bones display a number of distinctive anatomical landmarks. One of them, the **external acoustic canal,** leads to the **tympanum,** or *eardrum.* The eardrum separates the external acoustic canal from the *middle ear cavity,* which contains the *auditory ossicles,* or *ear bones.* The structure and function of the tympanum, middle ear cavity, and auditory ossicles will be considered in Chapter 9.

Anterior to the external acoustic canal is a transverse depression, the **mandibular fossa,** which marks the point of

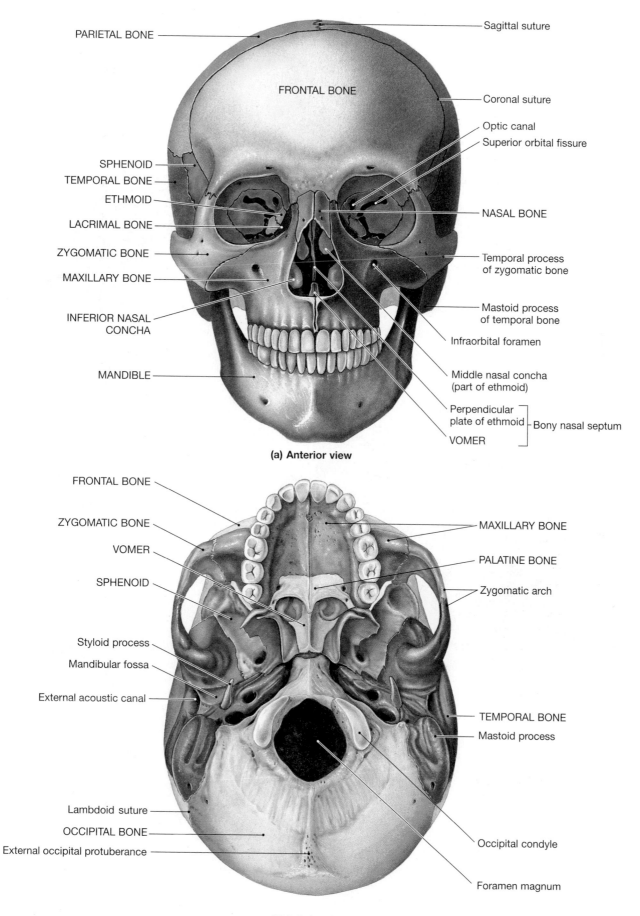

PARIETAL BONE

Sagittal suture

FRONTAL BONE

Coronal suture

Optic canal
Superior orbital fissure

SPHENOID
TEMPORAL BONE
ETHMOID
LACRIMAL BONE

NASAL BONE

ZYGOMATIC BONE

Temporal process
of zygomatic bone

MAXILLARY BONE

Mastoid process
of temporal bone

INFERIOR NASAL
CONCHA

Infraorbital foramen

MANDIBLE

Middle nasal concha
(part of ethmoid)

Perpendicular
plate of ethmoid
VOMER

Bony nasal septum

(a) Anterior view

FRONTAL BONE

ZYGOMATIC BONE

MAXILLARY BONE

VOMER

PALATINE BONE

SPHENOID

Zygomatic arch

Styloid process
Mandibular fossa
External acoustic canal

TEMPORAL BONE
Mastoid process

Lambdoid suture
OCCIPITAL BONE
External occipital protuberance

Occipital condyle

Foramen magnum

(b) Inferior view

● **FIGURE 6–26 The Adult Skull, Part II.**

171

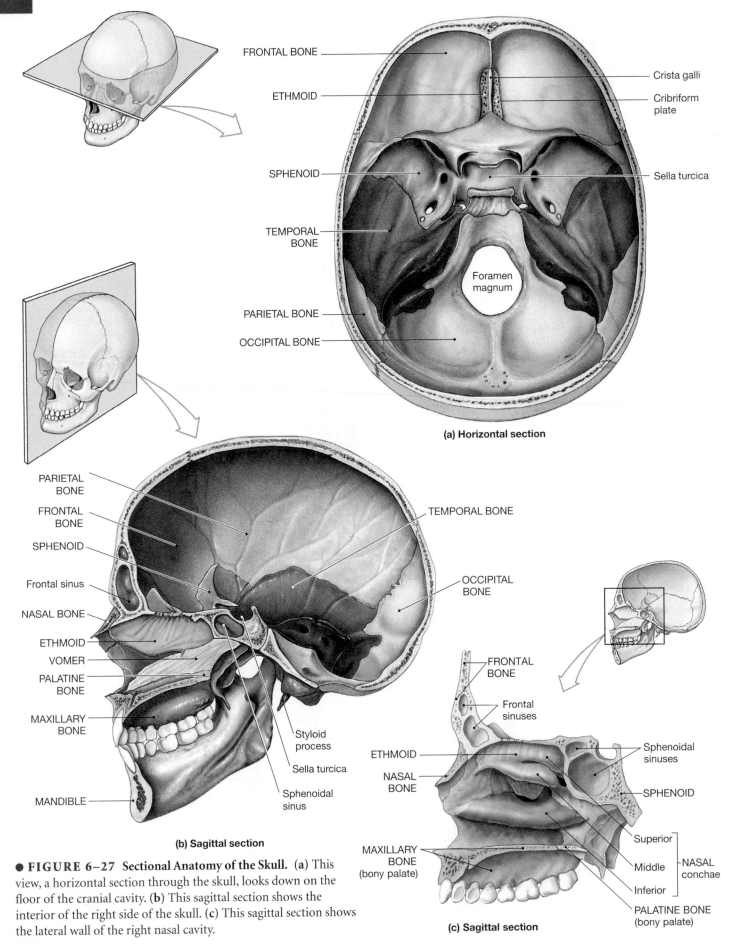

FRONTAL BONE

ETHMOID

Crista galli

Cribriform plate

SPHENOID

Sella turcica

TEMPORAL BONE

PARIETAL BONE

OCCIPITAL BONE

Foramen magnum

(a) Horizontal section

PARIETAL BONE

FRONTAL BONE

SPHENOID

Frontal sinus

NASAL BONE

ETHMOID

VOMER

PALATINE BONE

MAXILLARY BONE

MANDIBLE

TEMPORAL BONE

OCCIPITAL BONE

Styloid process

Sella turcica

Sphenoidal sinus

(b) Sagittal section

FRONTAL BONE

Frontal sinuses

ETHMOID

NASAL BONE

Sphenoidal sinuses

SPHENOID

Superior

Middle

Inferior

NASAL conchae

MAXILLARY BONE (bony palate)

PALATINE BONE (bony palate)

(c) Sagittal section

● **FIGURE 6–27 Sectional Anatomy of the Skull.** **(a)** This view, a horizontal section through the skull, looks down on the floor of the cranial cavity. **(b)** This sagittal section shows the interior of the right side of the skull. **(c)** This sagittal section shows the lateral wall of the right nasal cavity.

articulation with the lower jaw (mandible) (see Figure 6–26b). The prominent bulge just posterior and inferior to the entrance to the external acoustic canal is the **mastoid process,** which provides a site for the attachment of muscles that rotate or extend the head. Next to the base of the mastoid process is the long, sharp **styloid** (STĪ-loyd; *stylos*, pillar) **process.** The styloid process is attached to ligaments that support the hyoid bone and anchors muscles associated with the tongue and pharynx.

THE SPHENOID BONE. The **sphenoid** (SFĒ-noyd) **bone** forms part of the floor of the cranium (see Figure 6–26b). It also acts like a bridge: it unites the cranial and facial bones, and it braces the sides of the skull. The general shape of the sphenoid has been compared to that of a giant bat with wings extended; the wings can be seen most clearly on the superior surface (see Figure 6–27a). From the front (see Figure 6–26a) or side (see Figure 6–25), it is covered by other bones. Like the frontal bone, the sphenoid bone also contains a pair of sinuses, called **sphenoidal sinuses** (see Figures 6–27b,c).

The lateral "wings" of the sphenoid extend to either side from a central depression called the **sella turcica** (TUR-si-kuh) (Turk's saddle) (see Figure 6–27a). It encloses the pituitary gland, an endocrine organ that is connected to the inferior surface of the brain by a narrow stalk of neural tissue.

THE ETHMOID BONE. The **ethmoid bone** is anterior to the sphenoid bone. The ethmoid bone consists of two honeycombed masses of bone. It forms part of the cranial floor, contributes to the medial surfaces of the orbit of each eye, and forms the roof and sides of the nasal cavity (see Figures 6–26a and 6–27b). A prominent ridge, the **crista galli,** or "cock's comb," projects above the superior surface of the ethmoid (see Figure 6–26a). Holes in the **cribriform plate** (*cribrum*, sieve) permit passage of the olfactory nerves, which provide the sense of smell.

The lateral portions of the ethmoid bone contain the **ethmoidal sinuses,** which drain into the nasal cavity. Projections called the **superior** and **middle nasal conchae** (KONG-kē; *concha,* shell) extend into the nasal cavity toward the *nasal septum* (*septum,* wall), which divides the nasal cavity into left and right portions (see Figures 6–26a and 6–27b,c). The superior and middle nasal conchae, along with the inferior nasal conchae bones (discussed shortly), slow and break up the airflow through the nasal cavity. This deflection in airflow allows time for the air to become cleaned, moistened, and warmed before it reaches the delicate portions of the respiratory tract. It also directs air into contact with olfactory (smell) receptors in the superior portions of the nasal cavity. The **perpendicular plate** of the ethmoid bone extends inferiorly from the crista galli, then passes between the conchae to contribute to the nasal septum (see Figure 6–26a).

→ **CONCEPT CHECK QUESTIONS**

1. The mastoid and styloid processes are found on which skull bones?
2. What bone contains the depression called the *sella turcica?* What is located in the depression?
3. Which bone of the cranium articulates directly with the vertebral column?

Answers begin on p. 792.

The Bones of the Face

The facial bones protect and support the entrances to the digestive and respiratory tracts. They also provide sites for the attachment of muscles that control our facial expressions and help us manipulate food. Of the 14 facial bones, only the lower jaw, or mandible, is movable.

THE MAXILLARY BONES. The **maxillary** (MAK-si-ler-ē) **bones,** or *maxillae,* articulate with all other facial bones except the mandible. The maxillary bones form (1) the floor and medial portion of the rim of the orbit (see Figure 6–26a); (2) the walls of the nasal cavity; and (3) the anterior roof of the mouth, or *hard palate* (see Figure 6–27b). The maxillary bones contain large **maxillary sinuses,** which lighten the portion of the maxillary bones above the embedded teeth. Infections of the gums or teeth can sometimes spread into the maxillary sinuses, which increases pain and makes treatment more complicated.

THE PALATINE BONES. The paired **palatine bones** form the posterior surface of the *bony palate,* or *hard palate*—the "roof of the mouth" (see Figures 6–26b and 6–27b,c). The superior surfaces of the horizontal portion of each palatine bone contribute to the floor of the nasal cavity. The superior tip of the vertical portion of each palatine bone forms part of the floor of each orbit.

THE VOMER. The inferior margin of the **vomer** articulates with the paired palatine bones (see Figures 6–26b and 6–27b). The vomer supports a prominent partition that forms part of the *nasal septum,* along with the ethmoid bone (see Figures 6–26a and 6–27b).

THE ZYGOMATIC BONES. On each side of the skull, a **zygomatic** (zī-go-MA-tik) **bone** articulates with the frontal bone and the maxilla to complete the lateral wall of the orbit (see Figures 6–25 and 6–26a). Along its lateral margin, each zygomatic bone gives rise to a slender bony extension that curves laterally and posteriorly to meet a process from the temporal bone. Together these processes form the **zygomatic arch,** or *cheekbone.*

THE NASAL BONES. Forming the bridge of the nose midway between the orbits, the **nasal bones** articulate with the frontal bone and the maxillary bones (see Figures 6–25 and 6–26a).

THE LACRIMAL BONES. The **lacrimal** (*lacrimae,* tears) **bones** are located within the orbit on its medial surface. They articulate with the frontal, ethmoid, and maxillary bones (see Figures 6–25 and 6–26a).

THE INFERIOR NASAL CONCHAE. The paired **inferior nasal conchae** project from the lateral walls of the nasal cavity (see Figures 6–26a and 6–27c). Their shape helps slow airflow and deflects arriving air toward the olfactory (smell) receptors located near the upper portions of the nasal cavity.

THE NASAL COMPLEX. The **nasal complex** includes the bones that form the superior and lateral walls of the nasal cavities and the sinuses that drain into them. The ethmoid bone and vomer form the bony portion of the **nasal septum,** which separates the left and right portions of the nasal cavity (see Figure 6–26a). The frontal, sphenoid, ethmoid, palatine, and maxillary bones contain air-filled chambers collectively known as the **paranasal sinuses** (Figure 6–28●). (The tiny palatine sinuses, not shown, open into the sphenoid sinuses.) In addition to reducing the weight of the skull, the paranasal sinuses

help protect the respiratory system. The paranasal sinuses are connected to the nasal cavities and lined by a mucous membrane. The mucous secretions are released into the nasal cavities, and the ciliated epithelium passes the mucus back toward the throat, where it is eventually swallowed or expelled by coughing. Incoming air is humidified and warmed as it flows across this carpet of mucus. Foreign particles, such as dust and bacteria, become trapped in the sticky mucus and are swallowed or expelled. This mechanism helps protect more delicate portions of the respiratory tract.

THE MANDIBLE. The broad **mandible** is the bone of the lower jaw. It forms a broad, horizontal curve with vertical processes at either side. Each vertical process, or **ramus,** bears two processes. The more posterior **condylar process** ends at the *mandibular condyle,* a curved surface that articulates with the mandibular fossa of the temporal bone on that side. This articulation is quite mobile, and the disadvantage of such mobility is that the jaw can easily be dislocated. The anterior **coronoid** (kor-Ō-noyd) **process** (see Figure 6–25) is the attachment point for the *temporalis muscle,* a powerful muscle that closes the jaws.

The Hyoid Bone

The small, U-shaped **hyoid bone** is suspended below the skull (Figure 6–29●). Ligaments extend from the styloid processes of the temporal bones to the *lesser horns.* The hyoid (1) serves as a base for muscles associated with the *larynx* (voicebox), tongue, and pharynx and (2) supports and stabilizes the position of the larynx.

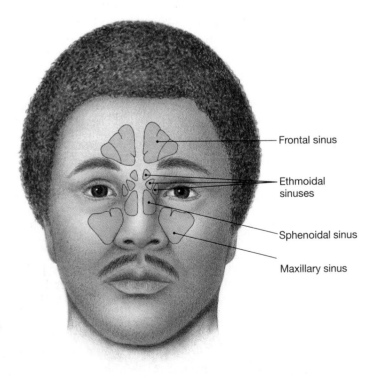

Frontal sinus

Ethmoidal sinuses

Sphenoidal sinus

Maxillary sinus

● **FIGURE 6–28 The Paranasal Sinuses.**

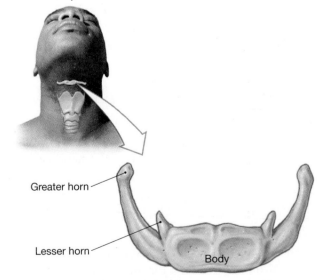

Greater horn

Lesser horn

Body

● **FIGURE 6–29 The Hyoid Bone.** The stylohyoid ligaments (not shown) connect the lesser horns of the hyoid to the styloid processes of the temporal bones.

1. During baseball practice, a ball hits Casey in the eye, which fractures the bones directly above and below the orbit. Which bones were broken?
2. What are the functions of the paranasal sinuses?
3. Why would a fracture of the coronoid process of the mandible make it difficult to close the mouth?
4. What signs would you expect to see in a person who suffers from a fractured hyoid bone?

Answers begin on p. 792.

The Skulls of Infants and Children

Many centers of ossification are involved in the formation of the skull. As a fetus develops, the individual centers begin to fuse. This fusion produces a smaller number of composite bones. For example, the sphenoid begins as 14 separate ossification centers but ends as just one bone. At birth, fusion is not yet complete, and there are two frontal bones, four occipital bones, and several sphenoid and temporal elements.

The developing skull organizes around the developing brain, and as the time of birth approaches, the brain enlarges rapidly. Although the bones of the skull are also growing, they fail to keep pace with the brain. At birth the cranial bones are connected by areas of fibrous connective tissue known as **fontanels** (fon-tah-NELZ). The fontanels, or "soft spots," are quite flexible and permit distortion of the skull without damage. During delivery such distortion occurs normally and eases the passage of the infant along the birth canal. Figure 6–30● shows the appearance of the skull at birth, including the prominent fontanels.

The Vertebral Column and Thoracic Cage

The rest of the axial skeleton consists of the thoracic cage, which we will discuss shortly, and the vertebral column, which we discuss next.

The **vertebral column,** or **spine,** consists of 26 bones: the 24 **vertebrae,** the *sacrum* (SĀ-krum), and the *coccyx* (KOK-siks) or *tailbone.* The vertebral column is subdivided on the basis of vertebral structure (Figure 6–31●). The **cervical region** of the vertebral column consists of the seven **cervical vertebrae** of the neck (abbreviated as C_1 to C_7). The cervical region begins at the articulation of C_1 with the occipital condyles of the skull and extends inferiorly to the articulation of C_7 with the first thoracic vertebra. The **thoracic region** consists of the 12 **thoracic vertebrae** (T_1 to T_{12}), each of which articulates with one or more pairs of ribs. The **lumbar region** contains the five **lumbar vertebrae** (L_1 to L_5). The first lumbar vertebra articulates with T_{12}, and the fifth lumbar vertebra articulates with the **sacrum.** The sacrum is a single bone formed by the fusion of the five embryonic vertebrae of the

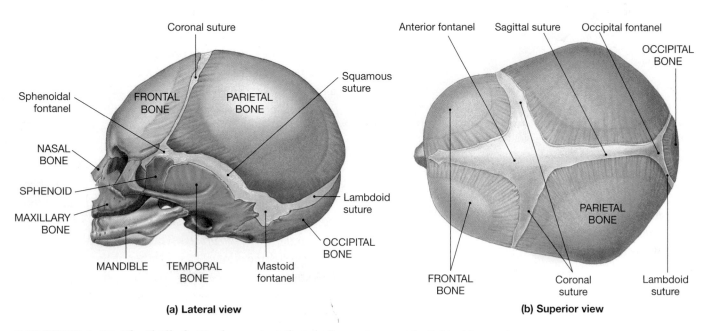

(a) Lateral view

(b) Superior view

● **FIGURE 6–30 The Skull of a Newborn.** An infant skull contains more individual bones than an adult skull. Many of these bones eventually fuse to create the adult skull. The flat bones of the skull are separated by areas of fibrous connective tissue called fontanels, which allow for cranial expansion and distortion during birth. By about age four these areas disappear, and skull growth is completed.

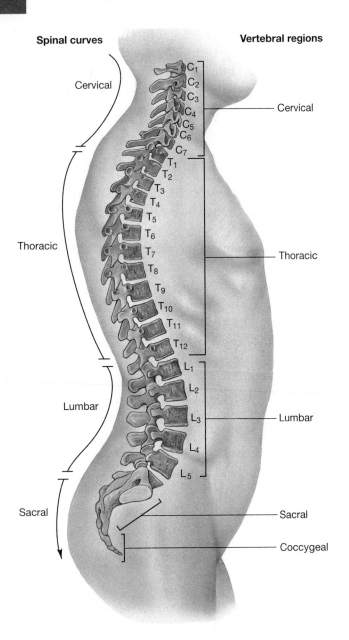

Spinal curves

Cervical

Thoracic

Lumbar

Sacral

Vertebral regions

Cervical

Thoracic

Lumbar

Sacral

Coccygeal

● **FIGURE 6–31 The Vertebral Column.** The major regions of the vertebral column and the four spinal curves are shown in this lateral view.

sacral region. The **coccygeal region** is made up of the small **coccyx,** which also consists of fused vertebrae. The total length of the adult vertebral column averages 71 cm (28 in.).

Spinal Curvature

The vertebral column is not straight and rigid. The lateral view of the spinal column in Figure 6–31 reveals four **spinal curves.** The *thoracic* and *sacral* curves are called **primary curves** because they appear late in fetal development, as the thoracic and abdominal organs enlarge. The *cervical* and *lumbar curves,*

known as **secondary curves,** do not appear until months after birth. The cervical curve develops as an infant learns to balance the head upright, and the lumbar curve develops with the ability to stand. When a person is standing, the body's weight must be transmitted through the spinal column to the pelvic girdle and ultimately to the legs. Yet most of the body weight lies in front of the spinal column. The secondary curves bring that weight in line with the body axis. All four spinal curves are fully developed by the time a child is 10 years old.

Several abnormal distortions of spinal curvature may appear during childhood and adolescence. Examples are *kyphosis* (kī-FŌ-sis; exaggerated thoracic curvature), *lordosis* (lor-DŌ-sis; exaggerated lumbar curvature), and *scoliosis* (skō-lē-Ō-sis; an abnormal lateral curvature).

Vertebral Anatomy

Figure 6–32● shows representative vertebrae from three different regions of the vertebral column. Features shared by all vertebrae include a *vertebral body,* a *vertebral arch,* and *articular processes.* The more massive, weight-bearing portion of a vertebra is called the **vertebral body.** The bony faces of the vertebral bodies usually do not contact one another because an **intervertebral disc** of fibrocartilage lies between them. Intervertebral discs are not found in the sacrum and coccyx, where the vertebrae have fused, or between the first and second cervical vertebrae.

The **vertebral arch** forms the posterior margin of each **vertebral foramen** (plural, *foramina*). Together, the vertebral foramina of successive vertebrae form the **vertebral canal,** which encloses the spinal cord. The vertebral arch has walls, called *pedicles* (PED-i-kulz), and a roof formed by flat layers called *laminae* (LAM-i-nē; singular, *lamina,* a thin plate). *Transverse processes* that project laterally or dorsolaterally from the pedicles serve as sites for muscle attachment. A *spinous process,* or spinal process, projects posteriorly from where the laminae fuse together. The spinous processes form the bumps that can be felt along the midline of your back.

The **articular processes** arise at the junction between the pedicles and laminae. Each side of a vertebra has a *superior* and *inferior articular process.* The articular processes of successive vertebrae contact one another at the **articular facets.** Gaps between the pedicles of successive vertebrae—the *intervertebral foramina*—permit the passage of nerves that run to or from the enclosed spinal cord.

Although all vertebrae have many similar characteristics, some regional structural differences reflect differences in function. The structural differences among the vertebrae are discussed next.

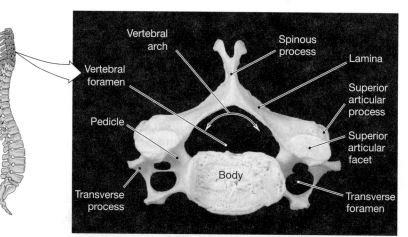

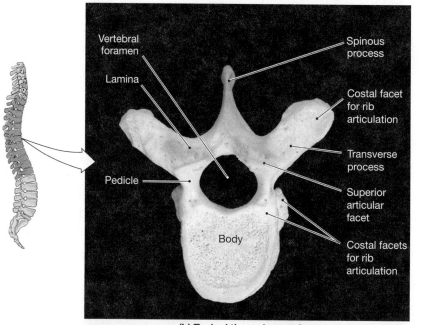

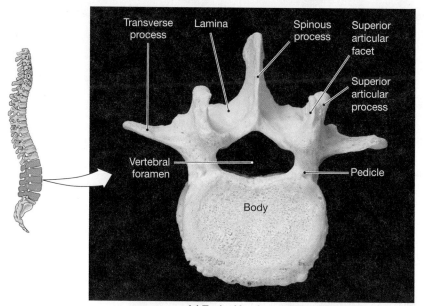

(a) Typical cervical vertebra

(b) Typical thoracic vertebra

(c) Typical lumbar vertebra

● **FIGURE 6–32 Typical Vertebrae of the Cervical, Thoracic, and Lumbar Regions.** Each vertebra is shown in superior view.

Clinical Note
COMPRESSION FRACTURES

Aging affects virtually every body system, and the skeletal system is no exception. Bone mass is normally lost with age in a process called *osteopenia*. However, some people will lose significantly more bone in a process called *osteoporosis*. In severe osteoporosis, over 50 percent of the bone mass can be lost. This is a particular problem in women following menopause where decreasing levels of female hormones (*estrogens*) result in loss of bone mass. This causes the bones to become weak and subject to fracture.

The weight-bearing parts of the spinal bones (*vertebral bodies*) are particularly vulnerable to osteoporosis. With advanced osteoporosis, relatively minor trauma, even as simple as rolling over in bed, can cause compression fractures. *Compression fractures* cause the vertebral body to collapse (much like crushing a soft drink can), which results in pain, limited movement, and a loss in body height. Elderly patients with multiple compression fractures may actually lose several inches of body height. ■

The Cervical Vertebrae

The seven cervical vertebrae extend from the head to the thorax. A typical cervical vertebra is illustrated in Figure 6–32a●. Notice that the body of the vertebra is not much larger than the size of the vertebral foramen. From the first thoracic vertebra to the sacrum, the diameter of the spinal cord decreases, and so does the size of the vertebral foramen. At the same time, the vertebral bodies gradually enlarge, because they must bear more weight.

Distinctive features of a typical cervical vertebra include (a) an oval, concave vertebral body; (b) a relatively large vertebral foramen; (c) a stumpy spinous process, usually with a notched tip; and (d) round **transverse foramina** within the transverse processes. These foramina protect important blood vessels that supply the brain.

The first two cervical vertebrae have unique characteristics that allow for specialized movements. The **atlas** (C_1) holds up the head, and articulates with the occipital condyles of the skull. It is named after Atlas, who, according to Greek myth, holds the weight of the world on his shoulders. The articulation between the occipital condyles and the atlas permits you to nod (as when indicating "yes"). The atlas, in turn, forms a pivot joint with the **axis** (C_2) through a projection on the axis called the **dens** (*denz*; tooth), or *odontoid process*. This articulation, which permits rotation (as when shaking your head to indicate "no"), is shown in Figure 6–33●.

The Thoracic Vertebrae

There are 12 thoracic vertebrae (Figure 6–32b●). Distinctive features of a thoracic vertebra include (1) a characteristic heart-shaped body that is more massive than that of a cervi-

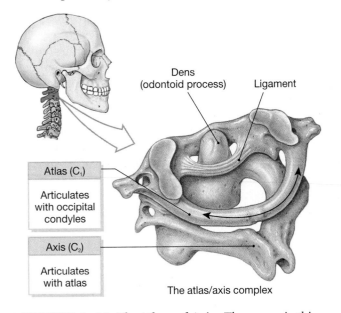

Dens
(odontoid process)   Ligament

| Atlas (C_1) |
| Articulates with occipital condyles |

| Axis (C_2) |
| Articulates with atlas |

The atlas/axis complex

● **FIGURE 6–33 The Atlas and Axis.** The arrows in this drawing indicate the direction of rotation at the articulation between the atlas (C_1) and the axis (C_2).

cal vertebra; (2) a large, slender spinous process that points inferiorly; and (3) *costal facets* on the body (and, in most cases, on the transverse processes) for articulating with the head of one or two pairs of ribs.

The Lumbar Vertebrae

The distinctive features of lumbar vertebrae (Figure 6–32c●) include (1) a vertebral body that is thicker and more oval than that of a thoracic vertebra; (2) a relatively massive, stumpy spinous process that projects posteriorly, and provides surface area for the attachment of the lower back muscles; and (3) blade-like transverse processes that lack articulations for ribs.

The lumbar vertebrae are the most massive and least mobile, for they support most of the body weight. As you increase the weight on the vertebrae, the intervertebral discs become increasingly important as shock absorbers. The lumbar discs, which are subjected to the most pressure, are the thickest of all. The articulations of lumbar vertebrae limit the stresses on the discs by restricting vertebral motion.

The Sacrum and Coccyx

The sacrum consists of the fused elements of five sacral vertebrae. It protects the reproductive, digestive, and excretory organs and attaches the axial skeleton to the appendicular skeleton by means of paired articulations with the pelvic girdle. The broad surface area of the sacrum provides an extensive area for the attachment of muscles, especially those responsible for leg movement. Figure 6–34● shows the posterior and anterior surfaces of the sacrum.

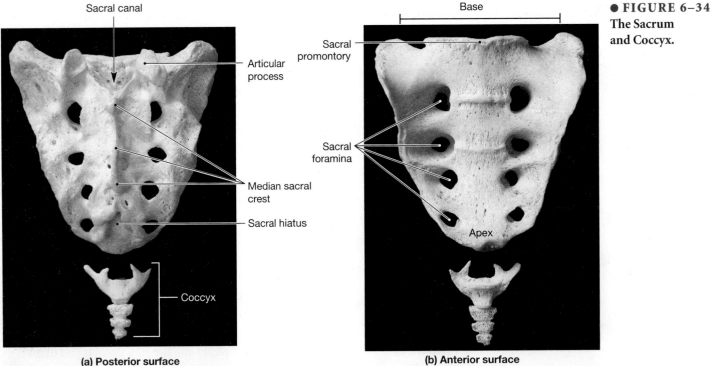

Sacral canal

Articular process

Median sacral crest

Sacral hiatus

Coccyx

(a) Posterior surface

Base

Sacral promontory

Sacral foramina

Apex

(b) Anterior surface

● **FIGURE 6–34**
The Sacrum and Coccyx.

Because the sacrum resembles a triangle, the narrow caudal portion is called the **apex,** and the broad superior surface is the **base.** The superior articular processes of the first sacral vertebra articulate with the last lumbar vertebra. The **sacral canal** is a passageway that begins between those processes and extends the length of the sacrum. Nerves and the membranes that line the vertebral canal in the spinal cord continue into the sacral canal. Its inferior end, the *sacral hiatus* (hī-Ā-tus), is covered by connective tissues. A prominent bulge at the anterior tip of the base, the **sacral promontory,** is an important landmark in females during pelvic examinations and during labor and delivery.

The sacral vertebrae begin fusing shortly after puberty and are typically completely fused at ages 25–30. Their fused spinal processes form a series of elevations along the *median sacral crest.* Four pairs of **sacral foramina** open on either side of the median sacral crest.

The coccyx provides an attachment site for a muscle that closes the anal opening. The fusion of the three to five (most often four) coccygeal vertebrae is not complete until late in adulthood. In elderly people, the coccyx may also fuse with the sacrum.

The Thoracic Cage

The skeleton of the chest, or thoracic cage, consists of the thoracic vertebrae, the ribs, and the sternum (Figure 6–35●). It provides bony support for the walls of the thoracic cavity. The ribs and the sternum form the *rib cage.* The thoracic cage protects the heart, lungs, and other internal organs and serves as a base for muscles involved in respiration.

Ribs, or *costal bones,* are elongate, flattened bones that originate on or between the thoracic vertebrae and end in the wall of the thoracic cavity. There are 12 pairs of ribs. The first seven pairs are called **true ribs.** These ribs reach the anterior body wall and are connected to the sternum by separate cartilaginous extensions, the **costal cartilages.** Ribs 8–12 are called the **false ribs** because they do not attach directly to the sternum. The costal cartilages of ribs 8–10 fuse together. This fused cartilage merges with the costal cartilage of rib 7 before it reaches the sternum. The last two pairs of ribs are called **floating ribs** because they have no connection with the sternum.

The adult **sternum,** or breastbone, has three parts. The broad, triangular **manubrium** (ma-NOO-brē-um) articulates with the clavicles of the appendicular skeleton and with the cartilages of the first pair of ribs. The *jugular notch* is the shallow indentation on the superior surface of the manubrium. The elongated **body** ends at the slender **xiphoid** (ZĪ-foyd) **process.** Ossification of the sternum begins at six to ten different centers, and fusion is not completed until at least age 25. The xiphoid process is usually the last of the sternal components to ossify and fuse. Impact or strong pressure can drive it into the liver, and cause severe damage. Cardiopulmonary resuscitation (CPR) training strongly emphasizes the proper positioning of the hand to reduce the chances of breaking the xiphoid process or ribs.

With their complex musculature, dual articulations at the vertebrae, and flexible connection to the sternum, the ribs are quite mobile. Because they are curved, their movements affect both the width and the depth of the thoracic cage, which increases or decreases its volume.

→ **CONCEPT CHECK QUESTIONS**

1. Joe suffered a hairline fracture at the base of the dens. Which bone is fractured, and where would you find it?
2. In adults, five large vertebrae fuse to form what single structure?
3. Why are the bodies of lumbar vertebrae so large?
4. What are the differences between true ribs and false ribs?
5. Improper administration of CPR (cardiopulmonary resuscitation) could result in a fracture of which bone?

Answers begin on p. 792.

■ The Appendicular Division

The appendicular skeleton includes the bones of the upper and lower limbs, and the supporting bones of the pectoral and pelvic girdles that connect the limbs to the trunk.

The Pectoral Girdle

Each upper limb articulates with the trunk at the pectoral girdle, or *shoulder girdle.* The pectoral girdle consists of two broad, flat **scapulae** (SKAP-ū-lē; singular, *scapula,* SKAP-ū-luh; or *shoulder blades*) and two slender, curved **clavicles** (KLAV-i-kulz; or *collarbones)* (see Figure 6–24). Each clavicle articulates with the manubrium of the sternum; these are the only direct connections between the pectoral girdle and the axial skeleton. Skeletal muscles support and position each scapula, which has no bony or ligamentous connections to the thoracic cage.

Movements of the clavicle and scapula position the shoulder joint and provide a base for arm movement. Once the shoulder joint is in position, muscles that originate on the pectoral girdle help to move the arm. The surfaces of the scapulae and clavicles are, therefore, extremely important as sites for muscle attachment.

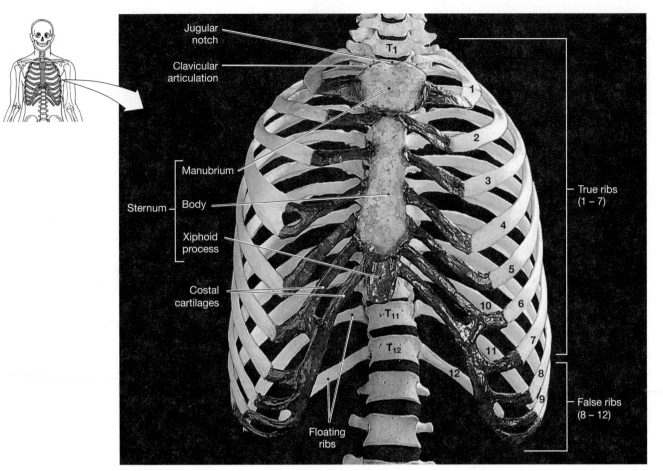

(a)

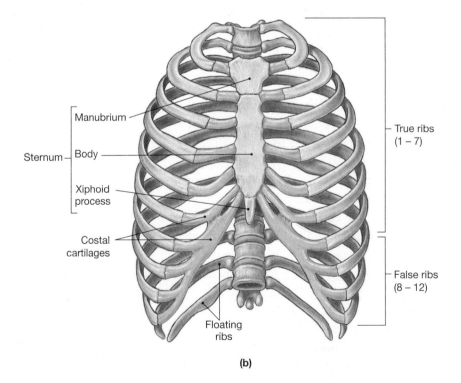

(b)

● **FIGURE 6–35 The Thoracic Cage.** Anterior views of the ribs, sternum, and costal cartilages are shown (**a**) in a photograph of a skeleton and (**b**) diagrammatically.

The Clavicle

The S-shaped clavicle, shown in Figure 6–36●, articulates with the manubrium of the sternum at its *sternal end* and with the *acromion* (a-KRŌ-mē-on), a process of the scapula, at its *acromial end*. The smooth superior surface of the clavicle lies just beneath the skin. The rough inferior surface of the acromial end is marked by prominent lines and tubercles, attachment sites for muscles and ligaments.

The clavicles are relatively small and fragile, so fractures are fairly common. For example, you can fracture a clavicle in a simple fall if you land on your hand with your arm outstretched. Fortunately, most fractures of the clavicle heal rapidly without a cast.

The Scapula

The anterior surface of the *body* of each scapula forms a broad triangle bounded by the **superior, medial,** and **lateral borders** (Figure 6–37●). Muscles that position the scapula attach along these edges. The head of the scapula at the intersection of the lateral and superior borders forms a broad process that supports the shallow, cup-shaped **glenoid cavity,** or *glenoid fossa* (FOS-sah). ∞ p. 167 At the glenoid cavity, the scapula articulates with the proximal end of the humerus to form the *shoulder joint.* The depression in the anterior surface of the body of the scapula is called the **subscapular fossa.** The *subscapularis muscle* attaches here and to the humerus, which is the proximal bone of the upper limb.

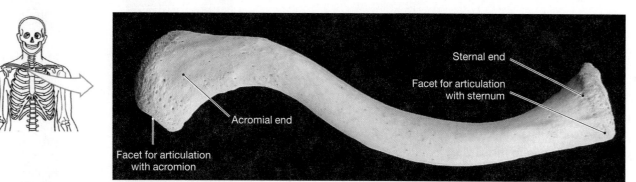

● **FIGURE 6–36 The Clavicle.** The right clavicle is shown in a superior view.

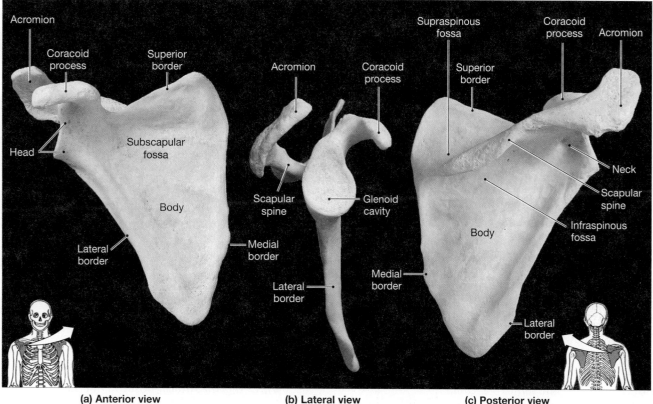

(a) Anterior view   (b) Lateral view   (c) Posterior view

● **FIGURE 6–37 The Scapula.** This photograph shows the major landmarks on the right scapula.

Figure 6–37b● shows a lateral view of the scapula and the two large processes that extend over the glenoid cavity. The smaller, anterior projection is the **coracoid** (KOR-uh-koyd) **process.** The **acromion** is the larger, posterior process. If you run your fingers along the superior surface of the shoulder joint, you will feel this process. The acromion articulates with the distal end of the clavicle.

The **scapular spine** divides the posterior surface of the scapula into two regions (Figure 6–37c●). The area superior to the spine is the **supraspinous fossa** (*supra-*, above); the *supraspinatus muscle* attaches here. The region below the spine is the **infraspinous fossa** (*infra-*, beneath), where the *infraspinatus muscle* attaches. Both muscles are also attached to the humerus.

The Upper Limb

The skeleton of each upper limb consists of the bones of the arm, forearm, wrist, and hand. Anatomically, the term *arm* refers only to the proximal portion of the upper limb (from shoulder to elbow), not to the entire limb. ∞ p. 18 The arm, or *brachium*, contains a single bone, the **humerus**, which extends from the scapula to the elbow.

The Humerus

At its proximal end, the round **head** of the humerus articulates with the scapula. The prominent **greater tubercle** of the humerus is a rounded projection near the lateral surface of the head (Figure 6–38●). It establishes the lateral contour of the shoulder. The **lesser tubercle** lies more anteriorly, separated from the greater tubercle by a deep *intertubercular groove*. Muscles are attached to both tubercles, and a large tendon runs along the groove. The *anatomical neck* lies between the tubercles and below the surface of the head. Distal to the tubercles, the narrow *surgical neck* corresponds to the region of growing bone, which is the *epiphyseal cartilage*. ∞ p. 157 The surgical neck earned its name by being a common fracture site.

The proximal shaft of the humerus is round in section. The elevated **deltoid tuberosity** that runs along the lateral border of the shaft is named after the *deltoid muscle,* which attaches to it.

Distally, the posterior surface of the shaft flattens, and the humerus expands to either side, forming a broad triangle. **Medial** and **lateral epicondyles** project to either side, which provides additional surface area for muscle attachment, and the smooth **condyle** dominates the inferior surface of the humerus. At the condyle, the humerus articulates with the bones of the forearm, the *radius* and *ulna.*

A low ridge crosses the condyle, and divides it into two distinct regions. The **trochlea** is the large medial portion shaped

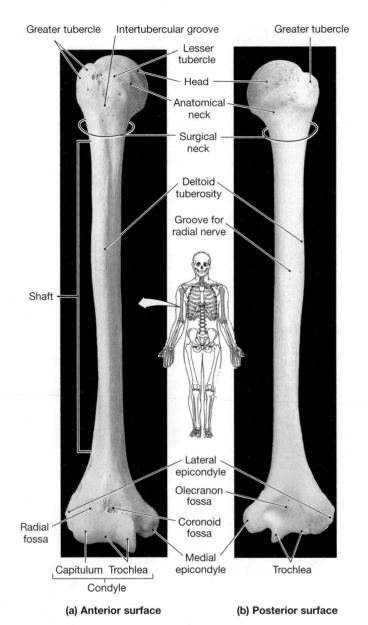

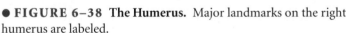

(a) Anterior surface   **(b) Posterior surface**

● **FIGURE 6–38 The Humerus.** Major landmarks on the right humerus are labeled.

like a spool or pulley (*trochlea,* a pulley). The trochlea extends from the base of the **coronoid** (*corona,* crown) **fossa** on the anterior surface to the **olecranon** (ō-LEK-ruh-non) **fossa** on the posterior surface. These depressions accept projections from the surface of the ulna as the elbow reaches its limits of motion. The **capitulum** forms the lateral region of the condyle. A shallow **radial fossa** proximal to the capitulum accommodates a small projection on the radius.

The Radius and Ulna

The **radius** and **ulna** are the bones of the forearm (Figure 6–39●). In the anatomical position, the radius lies along the lateral (thumb) side of the forearm while the ulna provides medial support of the forearm (Figure 6–39a).

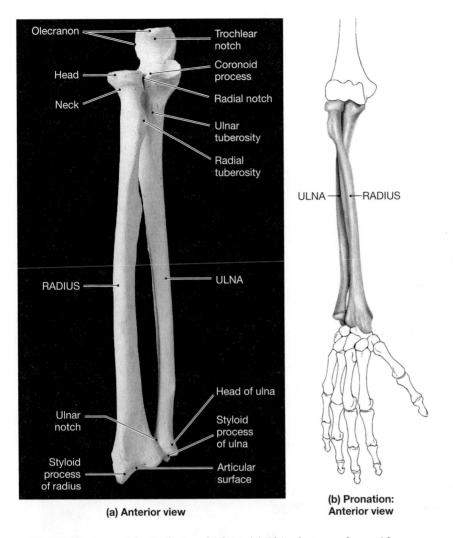

(a) Anterior view

ULNA — RADIUS

(b) Pronation:
Anterior view

● **FIGURE 6–39 The Radius and Ulna.** (a) This photograph provides an anterior view of the bones of the right forearm. (b) Notice the changes that occur during pronation.

The **olecranon** of the ulna is the point of the elbow. On its anterior surface, the **trochlear notch** articulates with the trochlea of the humerus at the elbow joint. The olecranon forms the superior lip of the notch, and the **coronoid process** forms its inferior lip. At the limit of *extension,* when the arm and forearm form a straight line, the olecranon swings into the olecranon fossa on the posterior surface of the humerus. At the limit of *flexion,* when the arm and forearm form a V, the coronoid process projects into the coronoid fossa on the anterior surface of the humerus. Lateral to the coronoid process, a smooth **radial notch** accommodates the head of the radius.

A fibrous sheet connects the lateral margin of the ulna to the radius along its length. Near the wrist, the ulnar shaft ends at a disc-shaped head whose posterior margin bears a short **styloid process.** The distal end of the ulna is separated from the wrist joint by a pad of cartilage, and only the large distal portion of the radius participates in the wrist joint. The styloid process of the

radius assists in the stabilization of the joint by preventing lateral movement of the bones of the wrist (*carpal bones*).

A narrow *neck* extends from the head of the radius to the **radial tuberosity,** which marks the attachment site of the *biceps brachii,* which is a large muscle on the anterior surface of the arm that flexes the forearm. The disc-shaped head of the radius articulates with the capitulum of the humerus at the elbow joint and with the ulna at the *radial notch.* This proximal articulation with the ulna allows the radius to roll across the ulna, and rotates the palm in a movement known as *pronation* (Figure 6–39b●). The reverse movement, which returns the forearm to the anatomical position, is called *supination.*

Bones of the Wrist and Hand

The wrist, palm, and fingers are supported by 27 bones (Figure 6–40●). The eight **carpal bones** of the wrist, or *carpus,* form two rows. There are four proximal carpal bones: (1) the *scaphoid bone,* (2) the *lunate bone,* (3) the *triquetrum bone,* and (4) the *pisiform* (PIS-i-form) *bone.* There are also four distal carpal bones: (1) the *trapezium,* (2) the *trapezoid bone,* (3) the *capitate bone,* and (4) the *hamate bone.* Joints between the carpal bones permit a limited degree of sliding and twisting.

Five **metacarpal** (met-uh-KAR-pul) **bones** articulate with the distal carpal bones and form the palm of the hand. The metacarpal bones in turn articulate with the finger bones, or **phalanges** (fa-LAN-jēz; singular, *phalanx*). Each hand has 14 phalangeal bones. Four of the fingers contain three phalanges each (proximal, middle, and distal), but the thumb or **pollex** (POL-eks) has only two phalanges (proximal and distal).

→ **CONCEPT CHECK QUESTIONS**

1. Why would a broken clavicle affect the mobility of the scapula?
2. The rounded projections on either side of the elbow are parts of which bone?
3. Which bone of the forearm is lateral in the anatomical position?

Answers begin on p. 792.

The Pelvic Girdle

The pelvic girdle articulates with the thigh bones (see Figure 6–24). Because of the stresses involved in weight bearing and locomotion, the bones of the pelvic girdle and lower limbs are

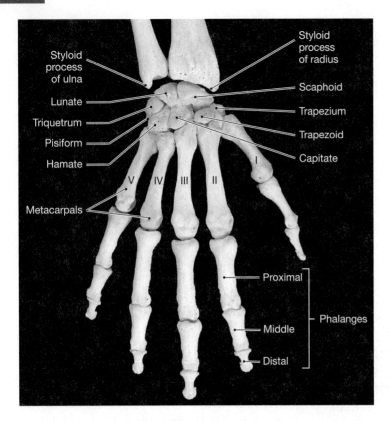

● FIGURE 6–40 Bones of the Wrist and Hand. A posterior view of the right hand is shown.

more massive than those of the pectoral girdle and upper limbs. The pelvic girdle is also much more firmly attached to the axial skeleton.

The pelvic girdle consists of two large hipbones, or **ossa coxae.** Each hipbone (os coxae or coxal bone) forms by the fusion of three bones: an **ilium** (IL-ē-um), an **ischium** (IS-kē-um), and a **pubis** (PŪ-bis) (Figure 6–41a●,b). Dorsally, the hipbones articulate with the sacrum at the **sacroiliac joint** (Figure 6–41c●). Ventrally, the hipbones are connected at the *pubic symphysis,* a fibrocartilage pad. At the hip joint on either side, the head of the femur (thighbone) articulates with the curved surface of the **acetabulum** (as-e-TAB-ū-lum; *acetabulum,* a vinegar cup).

The Coxal Bone (Hipbone)

The ilium is the most superior and largest component of the coxal bone (see Figure 6–41a,b). Above the acetabulum, the ilium forms a broad, curved surface that provides an extensive area for the attachment of muscles, tendons, and ligaments. The superior margin of the ilium, the **iliac crest,** marks the sites of attachments of both ligaments and muscles (see Figure 6–41c). Near the superior and posterior margin of the acetabulum, the ilium fuses with the ischium. The inferior surface of

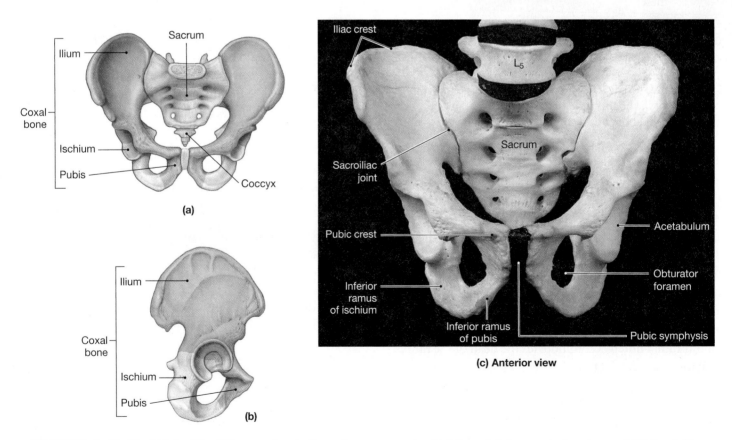

● FIGURE 6–41 The Pelvis. The different colors indicate the components of (**a**) the pelvis in an anterior view and (**b**) the right coxal bone in a lateral view. The photograph in (**c**) shows an anterior view of the pelvis of an adult male.

the ischium, the *ischial tuberosity* (not shown), supports the body's weight when sitting.

The fusion of a narrow branch of the ischium with a branch of the pubis completes the encirclement of the **obturator** (OB-tū-rā-tor) **foramen.** This space is closed by a sheet of collagen fibers whose inner and outer surfaces provide a base for the attachment of muscles and visceral structures.The anterior and medial surface of the pubis contains a roughened area that marks the **pubic symphysis,** an articulation with the pubis of the opposite side. The pubic symphysis limits movement between the two pubic bones.

The Pelvis

The **pelvis** consists of the two hipbones, the sacrum, and the coccyx (see Figure 6–41a). It is, thus, a composite structure that includes portions of both the appendicular and axial skeletons. An extensive network of ligaments connects the sacrum with the iliac crests, the ischia, and the pubic bones. Other ligaments tie the ilia to the posterior lumbar vertebrae. These interconnections increase the structural stability of the pelvis.

The shape of the pelvis of a female is somewhat different from that of a male (Figure 6–42●). Some of the differences result from variations in body size and muscle mass. For example, in females the pelvis is generally smoother, lighter in weight, and has less prominent markings. Other differences are adaptations for childbearing and are necessary for supporting the weight of the developing fetus and easing passage of the newborn through the pelvic outlet during delivery. The *pelvic outlet* is the inferior opening of the pelvis bounded by the coccyx, the ischia, and the pubic symphysis.

Compared to males, females have a relatively broad, low pelvis; a larger pelvic outlet; and a broader *pubic angle* (the angle between the pubic bones).

The Lower Limb

The skeleton of each lower limb consists of a *femur,* the thighbone; a *patella,* the kneecap; a *tibia* and a *fibula,* the bones of the leg; and the bones of the ankle and foot.

The Femur

The **femur,** or *thighbone,* is the longest and heaviest bone in the body (Figure 6–43●). The rounded epiphysis, or head, of the femur articulates with the pelvis at the acetabulum. The **greater** and **lesser trochanters** are large, rough projections that extend laterally from the juncture of the neck and the shaft. Both trochanters develop where large tendons attach to the femur. On the posterior surface of the femur, a prominent elevation, the **linea aspera,** marks the attachment of powerful muscles that pull the shaft of the femur toward the midline, a movement called *adduction* (*ad-,* toward + *duco,* to lead).

The proximal shaft of the femur is round in cross section. More distally, the shaft becomes more flattened and ends in two large **epicondyles** (**lateral** and **medial**). The inferior surfaces of the epicondyles form the **lateral** and **medial condyles.** The lateral and medial condyles form part of the knee joint.

The **patella** (*kneecap*) glides over the smooth anterior surface, or *patellar surface,* between the lateral and medial condyles. ∞ p. 169 The patella forms within the tendon of the *quadriceps femoris,* a group of muscles that straighten the knee.

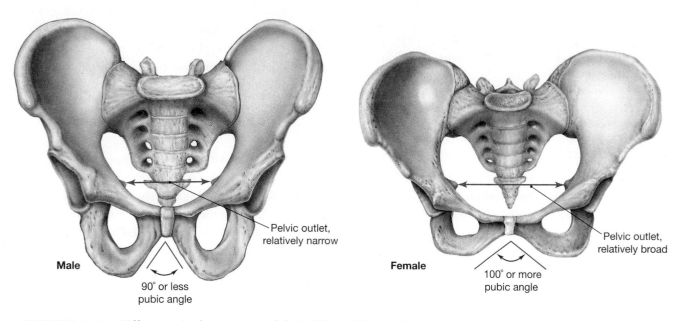

Male

Pelvic outlet, relatively narrow

90° or less pubic angle

Female

Pelvic outlet, relatively broad

100° or more pubic angle

● **FIGURE 6–42** **Differences in the Anatomy of the Pelvis in Males and Females.**

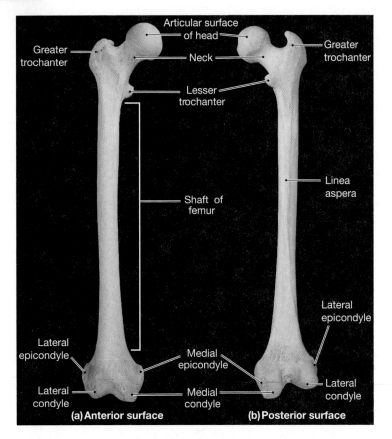

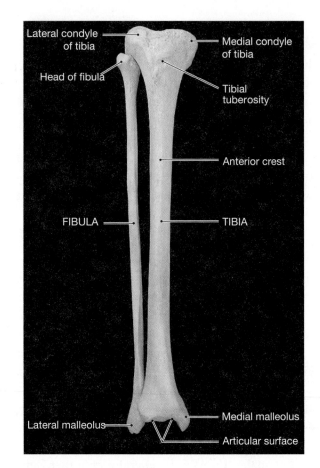

● FIGURE 6–43 The Femur. The labels indicate the various bone markings on the right femur.

● FIGURE 6–44 The Right Tibia and Fibula. These bones are shown in an anterior view.

Clinical Note
HIP FRACTURES

In the aged, simple falls can cause a *fractured* (broken) hip. These injuries are devastating; most patients require ambulance transport to the hospital. Most hip fractures require surgical repair. Fractures that occur low on the femoral neck can often be treated with a specialized nail that retains the native ball-and-socket joint. Fractures higher up the femoral neck usually require a prosthesis that replaces the native ball-and-socket joint. Hip fractures can be extremely debilitating and are one of the most common reasons for nursing home placement. ■

The Tibia and Fibula

The **tibia** (TIB-ē-uh), or *shinbone,* is the large medial bone of the leg (Figure 6–44●). The lateral and medial condyles of the femur articulate with the *lateral* and *medial condyles* of the **tibia.** The *patellar ligament* connects the patella to the **tibial tuberosity** just below the knee joint.

A projecting **anterior crest** extends almost the entire length of the anterior tibial surface. The tibia broadens at its distal end into a large process, the **medial malleolus** (ma-LĒ-o-lus; *malleolus,* hammer). The inferior surface of the tibia

forms a joint with the proximal bone of the ankle; the medial malleolus provides medial support for the ankle.

The slender **fibula** (FIB-ū-luh) parallels the lateral border of the tibia. The fibula articulates with the tibia inferior to the lateral condyle of the tibia. The fibula does not articulate with the femur or help transfer weight to the ankle and foot. However, it is an important surface for muscle attachment, and the distal **lateral malleolus** provides lateral stability to the ankle. A fibrous membrane that extends between the two bones helps stabilize their relative positions and provides additional surface area for muscle attachment.

Bones of the Ankle and Foot

The ankle, or *tarsus,* includes seven separate **tarsal bones** (Figure 6–45a●): (1) the *talus,* (2) the *calcaneus,* (3) the *navicular bone,* (4) the *cuboid bone,* and (5–7) the *first, second, and third cuneiform bones.* Only the proximal tarsal bone, the **talus,** articulates with the tibia and fibula. The talus then passes the body's weight to the ground through bones of the foot.

When you are standing normally, most of your weight is transmitted to the ground through the talus to the large **calcaneus** (kal-KĀ-nē-us), or *heel bone* (Figure 6–45b●). The

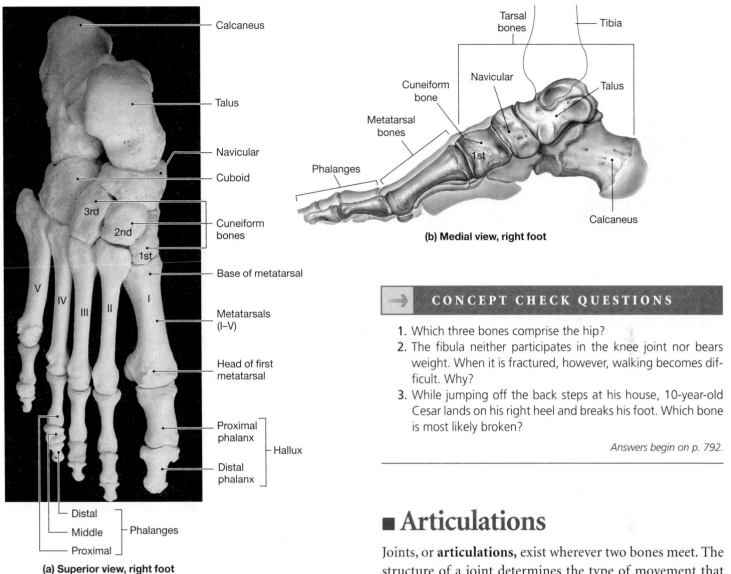

(b) Medial view, right foot

(a) Superior view, right foot

● **FIGURE 6–45 The Bones of the Ankle and Foot.** (a) This photograph shows the bones of the right foot in superior view. (b) This medial view shows the relative positions of the tarsal and metatarsal bones. Notice how the orientation of the tarsal bones conveys the weight of the body to the heel and the sole of the foot.

posterior projection of the calcaneus is the attachment site for the *calcaneal tendon,* or *Achilles tendon,* which arises from the calf muscles. These muscles raise the heel and depress the sole, as when you stand on tiptoes. The rest of the body weight is passed through the cuboid bone and cuneiform bones to the **metatarsal bones,** which support the sole of the foot.

The basic organizational pattern of the metatarsals and phalanges of the foot resembles that of the hand. The metatarsals are numbered by Roman numerals I to V from medial to lateral, and their distal ends form the ball of the foot. Like the thumb, the great toe (or **hallux**) has two phalanges; like the fingers, the other toes each contain three phalanges.

CONCEPT CHECK QUESTIONS

1. Which three bones comprise the hip?
2. The fibula neither participates in the knee joint nor bears weight. When it is fractured, however, walking becomes difficult. Why?
3. While jumping off the back steps at his house, 10-year-old Cesar lands on his right heel and breaks his foot. Which bone is most likely broken?

Answers begin on p. 792.

■ Articulations

Joints, or **articulations,** exist wherever two bones meet. The structure of a joint determines the type of movement that may occur. Each joint reflects a compromise between the need for strength and stability and the need for movement. When movement is not required, or when movement could actually be dangerous, joints can be very strong. For example, the sutures of the skull are such intricate and extensive joints that they lock the elements together as if they were a single bone. At other joints, movement is more important than strength. For example, the articulation at the shoulder permits a range of arm movement that is limited more by the surrounding muscles than by joint structure. The joint itself is relatively weak, and as a result shoulder injuries are rather common.

The Classification of Joints

Joints can be classified according to their structure or function. The structural classification is based on the anatomy of the joint. In this framework, joints are classified as **fibrous,** cartilaginous, or **synovial** (si-NŌ-vē-ul). The first two types reflect the type of connective tissue that binds them together. Such joints permit either no movement or slight movements. Synovial joints are surrounded by fibrous tissue, and the ends of bones are covered by cartilage that prevents bone-to-bone contact. Such joints permit free movement.

In a functional classification, joints are classified according to the range of motion they permit. An immovable joint is a **synarthrosis** (sin-ar-THRŌ-sis; *syn-*, together + *arthros*, joint); a slightly movable joint is an **amphiarthrosis** (am-fē-ar-THRŌ-sis; *amphi-*, on both sides); and a freely movable joint is a **diarthrosis** (dī-ar-THRŌ-sis; *dia-*, through), or *synovial joint.* Table 6–2 presents a functional classification of articulations, relates it to the structural classification scheme, and provides some examples.

Immovable Joints (Synarthroses)

At a synarthrosis, the bony edges are quite close together and may even interlock. A synarthrosis can be fibrous or cartilaginous. Two examples of fibrous immovable joints can be found in the skull. In a **suture** (*sutura*, a sewing together), the bones of the skull are interlocked and bound together by dense connective tissue. In a **gomphosis** (gom-FŌ-sis; *gomphosis*, a bolting together), a ligament binds each tooth in the mouth within a bony socket (*alveolus*).

A rigid, cartilaginous connection is called a **synchondrosis** (sin-kon-DRŌ-sis; *syn*, together + *chondros*, cartilage). The connection between the first pair of ribs and the sternum is a synchondrosis. Another example is the epiphyseal cartilage that connects the diaphysis and epiphysis in a growing long bone. ∞ p. 157

Slightly Movable Joints (Amphiarthroses)

An amphiarthrosis permits very limited movement, and the bones are usually farther apart than in a synarthrosis. Structurally, an amphiarthrosis can be fibrous or cartilaginous.

A **syndesmosis** (sin-dez-MŌ-sis; *desmos*, a band or ligament) is a fibrous joint connected by a ligament. The distal articulation between the two bones of the leg, the tibia and fibula, is an example. A **symphysis** is a cartilaginous joint because the bones are separated by a broad disc or pad of fibrocartilage. The articulations between the spinal vertebrae (at an *intervertebral disc*) and the anterior connection between the two pubic bones are examples of a symphysis.

Freely Movable Joints (Diarthroses)

Diarthroses, or **synovial joints,** permit a wide range of motion. The basic structure of a synovial joint was introduced in Chapter 4 in the discussion of synovial membranes. ∞ p. 113 Figure 6–46a● shows the structure of a representative synovial joint.

Synovial joints are typically found at the ends of long bones, such as those of the arms and legs. Under normal conditions the bony surfaces do not contact one another, for they are covered with special **articular cartilages.** The joint is surrounded by a fibrous **joint capsule,** or *articular capsule,* and the inner surfaces of the joint cavity are lined with a synovial membrane. **Synovial fluid** within the joint cavity provides lubrication that reduces the friction between the moving surfaces in the joint.

In some complex joints, additional padding lies between the opposing articular surfaces. An example of such shock-absorbing, fibrocartilage pads are the **menisci** (me-NIS-kē; *meniscus*, crescent) in the knee, shown in Figure 6–46b●. Also present in

---

TABLE 6–2 *A Functional Classification of Articulations*

FUNCTIONAL CATEGORY	STRUCTURAL CATEGORY	DESCRIPTION	EXAMPLE
Synarthrosis (no movement)	**Fibrous** Suture	Fibrous connections plus interlocked surfaces	Between the bones of the skull
	Gomphosis	Fibrous connections plus insertion in bony socket (alveolus)	Between the teeth and jaws
	Cartilaginous Synchondrosis	Interposition of cartilage plate	Epiphyseal cartilages
Amphiarthrosis (little movement)	**Fibrous** Syndesmosis	Ligamentous connection	Between the tibia and fibula
	Cartilaginous Symphysis	Connection by a fibrocartilage pad	Between right and left halves of pelvis; between adjacent vertebrae of spinal column
Diarthrosis (free movement)	**Synovial**	Complex joint bounded by joint capsule and contains synovial fluid	Numerous; subdivided by range of motion (see Figure 6–50)

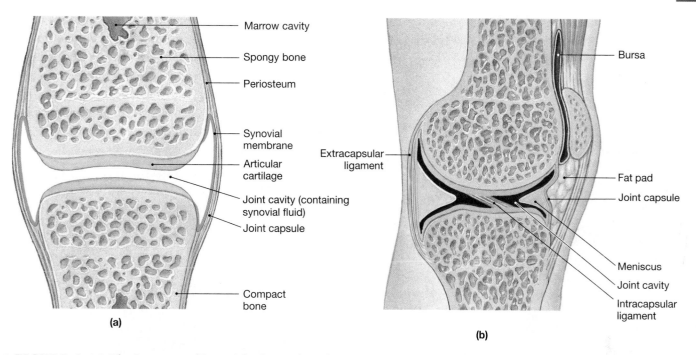

(a)

(b)

● **FIGURE 6–46 The Structure of Synovial Joints.** These diagrammatic sectional views show (**a**) a representative simple articulation and (**b**) the knee joint.

such joints are **fat pads,** which protect the articular cartilages and act as packing material. When the bones move, the fat pads fill in the spaces created as the joint cavity changes shape.

Clinical Note
RHEUMATISM AND ARTHRITIS

Rheumatism (ROO-muh-tizum) is a general term describing pain and stiffness that arises in the skeletal or muscular systems, or both. There are several major forms of rheumatism. Arthritis (ar-THRĪ-tis) includes all of the rheumatic diseases that affect synovial joints. Arthritis always involves damage to the articular cartilages, but the specific cause can vary. Arthritis can result from bacterial or viral infection, injury to the joint, metabolic problems, or severe physical stresses.

Osteoarthritis (os-tē-ō-ar-THRĪ-tis), also known as *degenerative arthritis*, or *degenerative joint disease (DJD)*, usually affects individuals age 60 or older. This disease can result from cumulative wear and tear at the joint surfaces or from genetic factors that affect collagen formation. In the U.S. population, 25 percent of women and 15 percent of men over age 60 show signs of this disease. Rheumatoid arthritis is an inflammatory condition that affects roughly 2.5 percent of the adult population. At least some cases result when the immune response mistakenly attacks the joint tissues. Allergies, bacteria, viruses, and genetic factors have all been proposed as contributing to or triggering the destructive inflammation.

Regular exercise, physical therapy, and drugs that reduce inflammation (such as aspirin) can slow the progress of the disease. Surgical procedures can realign or redesign the affected joint, and in extreme cases that involve the hip, knee, elbow, or shoulder, the defective joint can be replaced with an artificial one. ■

The joint capsule that surrounds the entire joint is continuous with the periostea of the articulating bones. In addition, **ligaments** that join bone to bone may be found outside or inside the joint capsule. Where a tendon or ligament rubs against other tissues, **bursae** (BUR-sē; singular *bursa,* a pouch)—small packets of connective tissue that contain synovial fluid—form to reduce friction and act as shock absorbers. Bursae are characteristic of many synovial joints and may also occur surrounding a tendon as a tubular sheath, covering a bone, or within other connective tissues exposed to friction or pressure.

Synovial Joints: Movement and Structure

Synovial joints are involved in all of our day-to-day movements. In discussions of motion at synovial joints, phrases such as "bend the leg" or "raise the arm" are not sufficiently precise. Anatomists use descriptive terms that have specific meanings.

Types of Movement

GLIDING. In **gliding,** two opposing surfaces slide past each other. Gliding occurs between the surfaces of articulating carpal bones and articulating tarsal bones and between the clavicles and sternum. The movement can occur in almost any direction, but the amount of movement is slight. Rotation is usually prevented by the joint capsule and associated ligaments.

ANGULAR MOTION. Examples of angular motion are *flexion, extension, adduction, abduction,* and *circumduction.* The description of each movement refers to an individual in the anatomical position.

Flexion (FLEK-shun) is movement in the anterior-posterior plane that reduces the angle between the articulating elements (Figure 6–47a●). **Extension** occurs in the same plane, but it in-

● **FIGURE 6–47 Angular Movements.** The red dots mark the locations of joints involved in the movements.

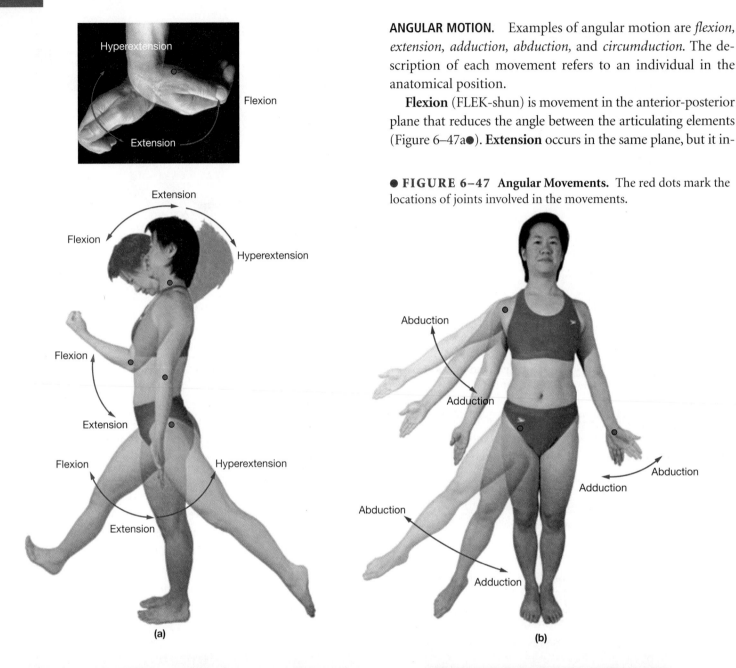

(a)

(b)

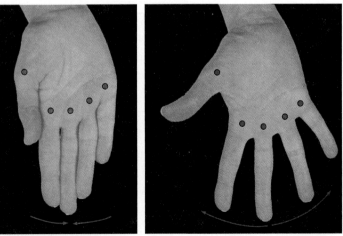

Adduction

Abduction

(c)

Circumduction

(d)

creases the angle between articulating elements. When you bring your head toward your chest, you flex the intervertebral joints of the neck. When you bend down to touch your toes, you flex the entire vertebral column. Extension reverses these movements.

Flexion at the shoulder joint or hip joint moves the limbs forward (*anteriorly*), whereas extension moves them back (*posteriorly*). Flexion of the wrist joint moves the hand forward, and extension moves it back. In each of these examples, extension can be continued past the anatomical position, in which case **hyperextension** occurs. You can also hyperextend the neck, a movement that enables you to gaze at the ceiling. Hyperextension of other joints is prevented by ligaments, bony processes, or soft tissues.

Abduction (*ab-*, from) is movement *away from the longitudinal axis of the body* in the frontal plane. For example, swinging the upper limb to the side is abduction of the limb (Figure 6–47b●). Moving it back to the anatomical position constitutes **adduction** (*ad,* to). Adduction of the wrist moves the heel of the hand toward the body, whereas abduction moves it farther away. Spreading the fingers or toes apart abducts them, because they move *away* from a central digit (finger or toe), as in Figure 6–47c●. Bringing them together is adduction. Abduction and adduction always refer to movements of the appendicular skeleton, not to those of the axial skeleton.

Circumduction (*circum,* around) is another type of angular motion. An example of circumduction is moving your arm in a loop, as when drawing a large circle on a chalkboard (Figure 6–47d●).

ROTATION.   Rotational movements are also described with reference to a figure in the anatomical position. **Rotation** involves turning around the longitudinal axis of the body or a limb. For example, you may rotate your head to look to one side or rotate your arm to screw in a lightbulb. Rotational movements are illustrated in Figure 6–48●.

The articulations between the radius and ulna permit the rotation of the distal end of the radius across the anterior surface of the ulna. This rotation moves the wrist and hand from palm-facing-front to palm-facing-back. This motion is called **pronation** (prō-NĀ-shun). The opposing movement, in which the palm is turned forward, is **supination** (soo-pi-NĀ-shun). ∞ p. 183

SPECIAL MOVEMENTS.   Some specific terms are used to describe unusual or special types of movement (Figure 6–49●).

■ **Inversion** (*in-*, into + *vertere*, to turn) is a twisting motion of the foot that turns the sole inward, which elevates the medial edge of the sole. The opposite movement is called **eversion** (ē-VER-zhun; *e-*, out).

■ **Dorsiflexion** is flexion of the ankle joint and elevation of the sole, as when you dig in your heel. **Plantar flexion** (*planta*, sole), the opposite movement, extends the ankle joint and elevates the heel, as when you stand on tiptoe.

■ **Opposition** is the movement of the thumb toward the palm or fingertips that enables you to grasp and hold an object.

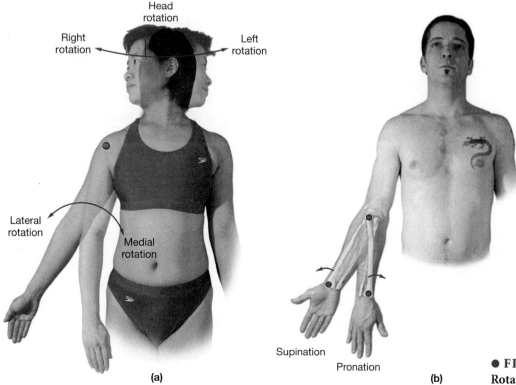

Head rotation

Right rotation

Left rotation

Lateral rotation

Medial rotation

(a)

Supination

Pronation

(b)

● **FIGURE 6–48**
Rotational Movements.

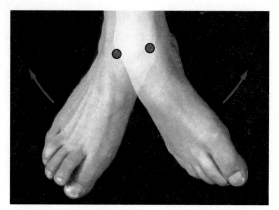
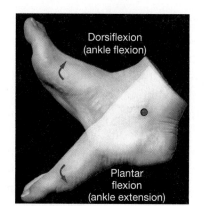
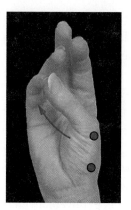

Eversion   Inversion

Opposition

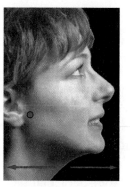

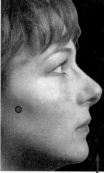

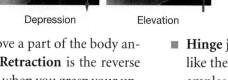

Retraction   Protraction   Depression   Elevation

● **FIGURE 6–49** Special Movements.

■ **Protraction** occurs when you move a part of the body anteriorly in the horizontal plane. **Retraction** is the reverse movement. You protract your jaw when you grasp your upper lip with your lower teeth, and you protract your clavicles when you cross your arms.

■ **Elevation** and **depression** occur when a structure moves in a superior or inferior direction, respectively. You depress your mandible when you open your mouth and elevate it as you close it.

A Structural Classification of Synovial Joints

Based on the shapes of the articulating surfaces, synovial joints can be described as *gliding, hinge, pivot, ellipsoidal, saddle,* or *ball-and-socket joints* (Figure 6–50●). Each type of joint permits a different type and range of motion:

■ **Gliding joints** have flattened or slightly curved faces (Figure 6–50a). The relatively flat articular surfaces slide across one another, but the amount of movement is very slight. Although rotation is theoretically possible at such a joint, ligaments usually prevent or restrict such movement. Gliding joints are found at the ends of the clavicles, between the carpal bones, between the tarsal bones, and between the articular facets of adjacent vertebrae.

■ **Hinge joints** permit angular movement in a single plane, like the opening and closing of a door (Figure 6–50b). Examples are the joint between the occipital bone and the atlas (in the axial skeleton), and the elbow, knee, ankle, and interphalangeal joints of the appendicular skeleton.

■ **Pivot joints** permit rotation only (Figure 6–50c). A pivot joint between the atlas and axis enables you to rotate your head to either side, and another between the head of the radius and the proximal shaft of the ulna permits pronation and supination of the palm.

■ In an **ellipsoidal joint** (or *condyloid joint*), an oval articular face nestles within a depression on the opposing surface (Figure 6–50d). Angular motion occurs in two planes, along or across the length of the oval. Ellipsoidal joints connect the radius with the proximal carpal bones, the phalanges of the fingers with the metacarpal bones, and the phalanges of the toes with the metatarsal bones.

■ **Saddle joints** have articular faces that fit together like a rider in a saddle (Figure 6–50e). Each face is concave on one axis and convex on the other, and the opposing faces nest together. This arrangement permits angular motion, including circumduction, but prevents rotation. The carpometacarpal joint at the base of the thumb is the best example of a saddle joint, and twiddling your thumbs will demonstrate the possible movements.

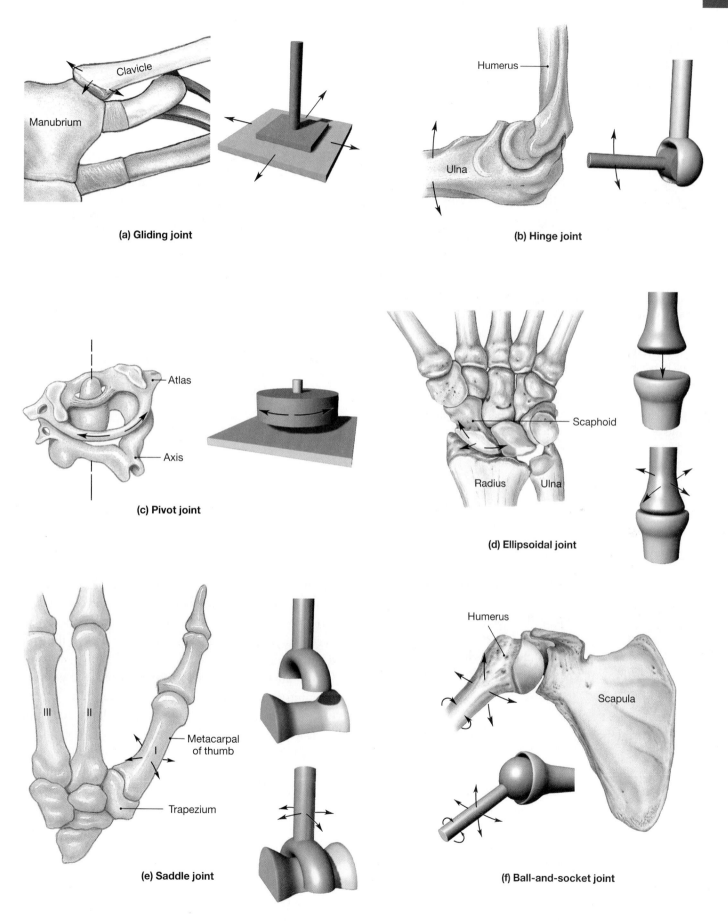

(a) Gliding joint

(b) Hinge joint

(c) Pivot joint

(d) Ellipsoidal joint

(e) Saddle joint

(f) Ball-and-socket joint

● **FIGURE 6–50 A Functional Classification of Synovial Joints.**

■ In a **ball-and-socket joint,** the round head of one bone rests within a cup-shaped depression in another bone (Figure 6–50f). All combinations of movements, including circumduction and rotation, can be performed at ball-and-socket joints. Examples are the shoulder and hip joints.

Key Note

A joint cannot be both highly mobile and very strong. The greater the mobility, the weaker the joint, because mobile joints rely on support from muscles and ligaments rather than solid bone-to-bone connections.

→ CONCEPT CHECK QUESTIONS

1. In a newborn, the large bones of the skull are joined by fibrous connective tissue. Which type of joint is this? These skull bones later grow, interlock, and form immovable joints. Which type of joints are these?
2. Give the proper term for each of the following types of motion: (a) moving the humerus away from the longitudinal axis of the body, (b) turning the palms so that they face forward, and (c) bending the elbow.
3. Which movements are associated with hinge joints?

Answers begin on p. 792.

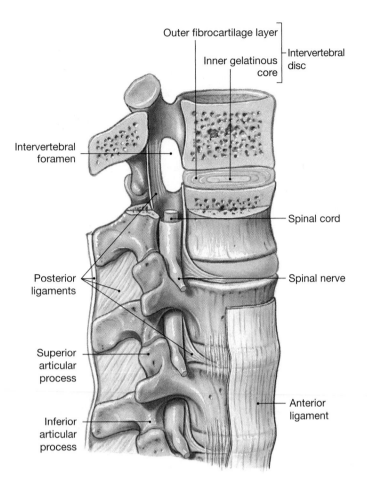

● **FIGURE 6–51 Intervertebral Articulations.**

Representative Articulations

This section describes examples of articulations that demonstrate important functional principles. We will first consider the *intervertebral articulations* of the axial skeleton. We will then proceed to a discussion of four *synovial articulations* of the appendicular skeleton: the shoulder and elbow of the upper limb, and the hip and knee of the lower limb.

Intervertebral Articulations

The vertebrae from the axis to the sacrum articulate with one another in two ways: (1) at gliding joints between the *superior* and *inferior articular processes,* and (2) at *symphyseal joints* between the vertebral bodies (Figure 6–51●). Articulations between the superior and inferior articular processes of adjacent vertebrae permit small movements that are associated with flexion and rotation of the vertebral column. Little gliding occurs between adjacent vertebral bodies.

Except for the first cervical vertebra, the vertebrae are separated and cushioned by pads called *intervertebral discs.* Each intervertebral disc consists of a tough outer layer of fibrocar-

tilage. The collagen fibers of that layer attach the discs to adjacent vertebrae. The fibrocartilage surrounds a soft, elastic, and gelatinous core, which gives intervertebral discs resiliency and enables them to act as shock absorbers, and compress and distort when stressed. This resiliency prevents bone-to-bone contact that might damage the vertebrae or jolt the spinal cord and brain.

Shortly after physical maturity is reached, the gelatinous mass within each disc begins to degenerate, and the "cushion" becomes less effective. Over the same period, the outer fibrocartilage loses its elasticity. If the stresses are sufficient, the inner mass may break through the surrounding fibrocartilage and protrude beyond the intervertebral space. This condition, called a *herniated disc,* further reduces disc function. The term *slipped disc* is often used to describe this problem, although the disc does not actually slip. The intervertebral discs also make a significant contribution to an individual's height; they account for roughly one-quarter of the length of the spinal column above the sacrum. As we grow older, the water content of each disc decreases; this loss accounts for the characteristic decrease in height with advancing age.

Clinical Note
DISLOCATIONS

The disruption or displacement of a joint due to trauma is a *dislocation.* In order for a joint to dislocate, the soft tissue of the joint capsule and ligaments must be stretched beyond the normal range of motion. Oftentimes, the ligaments are torn, which allows the bones of the joint to separate. Because of the associated soft-tissue damage, dislocations can cause paralysis of the affected limb as the nerves and arteries that lead to the extremity pass quite close to the joint and may be compressed or torn.

Dislocations are common in fingers, elbows, shoulders, hips, knees, and toes. Often, in addition to the dislocation, a fracture (broken bone) occurs at the time of injury.

The joints of the spine are at risk for dislocation, especially in high-energy accidents such as auto and motorcycle collisions, skiing injuries, and diving injuries. Spinal dislocations can be catastrophic as the spinal cord can be damaged when the dislocation occurs. ■

Articulations of the Upper Limb

The shoulder, elbow, and wrist are responsible for positioning the hand, which performs precise and controlled movements. The shoulder has great mobility, the elbow has great strength, and the wrist makes fine adjustments in the orientation of the palm and fingers.

THE SHOULDER JOINT. The shoulder joint permits the greatest range of motion of any joint in the body. Because it is also the most frequently dislocated joint, it provides an excellent demonstration of the principle that stability must be sacrificed to obtain mobility.

Figure 6–52● shows the ball-and-socket structure of the shoulder joint. The relatively loose joint capsule extends from the scapular neck to the humerus, and this oversized capsule permits an extensive range of motion. Bursae at the shoulder, like those at other joints, reduce friction where large muscles and tendons pass across the joint capsule. The bursae of the shoulder are especially large and numerous. Several bursae are associated with the capsule, the processes of the scapula, and large shoulder muscles. Inflammation of any of these bursae—a condition called *bursitis*—can restrict motion and produce pain.

The muscles that move the humerus do more to stabilize the shoulder joint than all its ligaments and capsular fibers combined. Powerful muscles that originate on the trunk, shoulder girdle, and humerus cover the anterior, superior, and posterior surfaces of the capsule. These muscles form the *rotator cuff,* which is a group of muscles that swing the arm through an impressive range of motion.

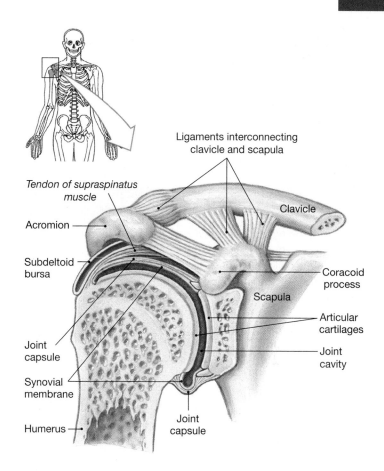

● **FIGURE 6–52 The Shoulder Joint.** The structure of the right shoulder joint is visible in this anterior view of a frontal section.

THE ELBOW JOINT. The elbow joint consists of two articulations: between the humerus and ulna, and between the humerus and radius (Figure 6–53●). The larger and stronger articulation is between the humerus and the ulna. This hinge joint provides stability and limits movement at the elbow joint.

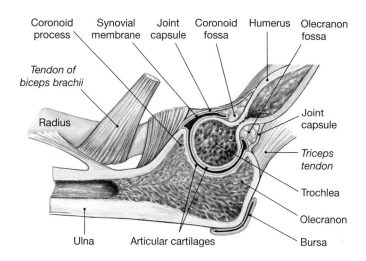

● **FIGURE 6–53 The Elbow Joint.** This longitudinal section reveals the anatomy of the right elbow joint.

The elbow joint is extremely stable because (1) the bony surfaces of the humerus and ulna interlock; (2) the joint capsule is very thick; and (3) the capsule is reinforced by stout ligaments. Nevertheless, the joint can be damaged by severe impacts or unusual stresses. When you fall on your hand with a partially flexed elbow, powerful contractions of the muscles that extend the elbow can break the ulna at the center of the trochlear notch.

→ **CONCEPT CHECK QUESTIONS**

1. Would a tennis player or a jogger be more likely to develop inflammation of the subdeltoid bursa? Why?
2. Daphne falls on her hands with her elbows slightly flexed. After the fall, she can't move her left arm at the elbow. If a fracture exists, which bone is most likely broken?

Answers begin on p. 792.

Articulations of the Lower Limb

The joints of the hip, ankle, and foot are sturdier than those at corresponding locations in the upper limb, and they have smaller ranges of motion. The knee has a range of motion comparable to that of the elbow, but it is subjected to much greater forces and, therefore, is less stable.

THE HIP JOINT. Figure 6–54● shows the structure of the hip joint, a ball-and-socket diarthrosis. The articulating surface of the acetabulum has a fibrocartilage pad along its edges, a fat pad covered by synovial membrane in its central portion, and a stout central ligament. This combination of coverings and membranes resists compression, absorbs shocks, and stretches and distorts without damage.

Compared with that of the shoulder, the joint capsule of the hip joint is denser and stronger. It extends from the lateral and inferior surfaces of the pelvic girdle to the femur and encloses both the femoral head and neck. This arrangement helps keep the head from moving away from the acetabulum. Three broad ligaments reinforce the joint capsule, while a fourth, the *ligament of the femoral head* (the *ligamentum teres*), is inside the acetabulum and attaches to the center of the femoral head. Additional stabilization comes from the bulk of the surrounding muscles.

The combination of an almost complete bony socket, a strong joint capsule, supporting ligaments, and muscular padding makes this an extremely stable joint. Fractures of the femoral neck or between the trochanters are actually more common than hip dislocations. Although flexion, extension, adduction, abduction, and rotation are permitted, the total range of motion is considerably less than that of the shoulder.

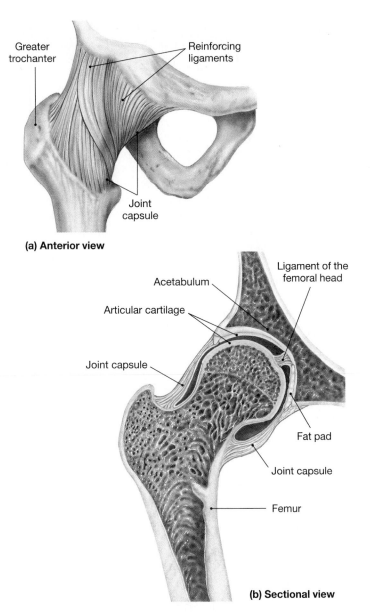

(a) Anterior view

(b) Sectional view

● **FIGURE 6–54 The Hip Joint. (a)** The hip joint is extremely strong and stable, in part because of the massive capsule and surrounding ligaments. **(b)** This sectional view of the right hip shows the structure of the joint.

Flexion is the most important normal hip movement, and range of flexion is primarily limited by the surrounding muscles. Other directions of movement are restricted by ligaments and the capsule.

THE KNEE JOINT. The hip joint passes weight to the femur, and at the knee joint the femur transfers the weight to the tibia. Although the knee functions as a hinge joint, the articulation is far more complex than that of the elbow or even the ankle. The rounded condyles of the femur roll across the top of the tibia, so the points of contact are constantly changing. Important features of the knee joint are shown in Figure 6–55●.

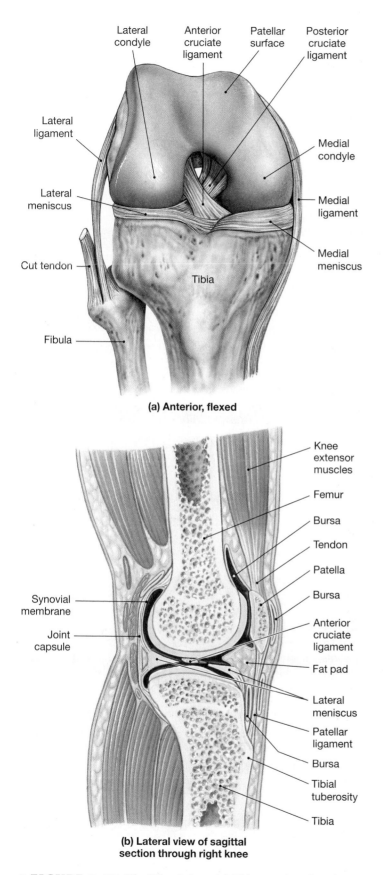

Lateral condyle
Anterior cruciate ligament
Patellar surface
Posterior cruciate ligament
Lateral ligament
Lateral meniscus
Lateral ligament
Cut tendon
Tibia
Fibula
Medial condyle
Medial ligament
Medial meniscus

(a) Anterior, flexed

Knee extensor muscles
Femur
Bursa
Tendon
Patella
Bursa
Anterior cruciate ligament
Synovial membrane
Joint capsule
Fat pad
Lateral meniscus
Patellar ligament
Bursa
Tibial tuberosity
Tibia

(b) Lateral view of sagittal section through right knee

● **FIGURE 6–55 The Knee Joint.** **(a)** This anterior view shows the right knee when flexed. **(b)** This sagittal section shows the internal anatomy of the extended right knee.

Structurally, the knee combines three separate articulations—two between the femur and tibia (medial condyle to medial condyle, and lateral condyle to lateral condyle) and one between the patella and the femur. There is no single unified joint capsule, nor is there a common synovial cavity. A pair of fibrocartilage pads, the **medial** and **lateral menisci,** lies between the femoral and tibial surfaces (see Figure 6–55a●). They act as cushions and conform to the shape of the articulating surfaces as the femur changes position. Prominent fat pads provide padding around the margins of the joint and assist the bursae in reducing friction between the patella and other tissues (see Figure 6–55b●).

Ligaments stabilize the anterior, posterior, medial, and lateral surfaces of this joint, and a complete dislocation of the knee is an extremely rare event. The tendon from the muscles responsible for extending the knee passes over the anterior surface of the joint. The patella lies within this tendon, and the **patellar ligament** continues its attachment on the anterior surface of the tibia. This ligament provides support to the front of the knee joint. Posterior ligaments between the femur and the heads of the tibia and fibula reinforce the back of the knee joint. The lateral and medial surfaces of the knee joint are reinforced by another pair of ligaments. These ligaments stabilize the joint at full extension.

Additional ligaments are found inside the joint capsule (see Figure 6–55b). Inside the joint a pair of ligaments, the *anterior cruciate* and *posterior cruciate,* cross each other as they attach the tibia to the femur (see Figure 6–55a). (The term *cruciate* is derived from the Latin word *crucialis,* which means a cross.) These ligaments limit the anterior and posterior movement of the femur.

> → **CONCEPT CHECK QUESTIONS**

1. Why is a complete dislocation of the knee joint an infrequent event?
2. What signs would you expect to see in an individual who has damaged the menisci of the knee joint?

Answers begin on p. 792.

■ Integration with Other Systems

Although bones may seem inert, you should now realize that they are quite dynamic structures. The entire skeletal system is intimately associated with other organ systems. For example,

The Skeletal System in Perspective

For All Systems

Provides mechanical support; stores energy reserves; stores calcium and phosphate reserves

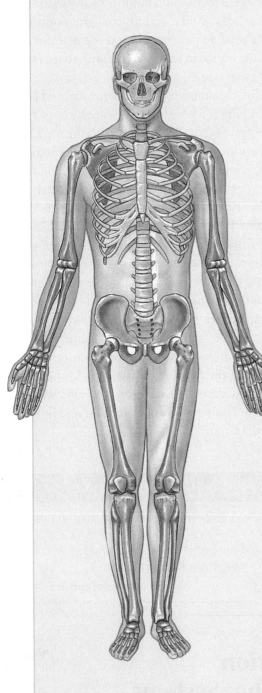

● **FIGURE 6–56** Functional Relationships Between the Skeletal System and Other Systems.

The Integumentary System

- Synthesizes vitamin D_3, essential for calcium and phosphorus absorption (bone maintenance and growth)
- Provides structural support

The Muscular System

- Stabilizes bone positions; tension in tendons stimulates bone growth and maintenance
- Provides calcium needed for normal muscle contraction; bones act as levers to produce body movements

The Nervous System

- Regulates bone position by controlling muscle contractions
- Provides calcium for neural function; protects brain, spinal cord; receptors at joints provide information about body position

The Endocrine System

- Skeletal growth regulated by growth hormone, thyroid hormones, and sex hormones; calcium mobilization regulated by parathyroid hormone and calcitonin
- Protects endocrine organs, especially in brain, chest, and pelvic cavity

The Cardiovascular System

- Provides oxygen, nutrients, hormones, blood cells; removes waste products and carbon dioxide
- Provides calcium needed for cardiac muscle contraction, blood cells produced in bone marrow

The Lymphatic System

- Lymphocytes assist in the defense and repair of bone following injuries
- Lymphocytes and other cells of the immune response are produced and stored in bone marrow

The Respiratory System

- Provides oxygen and eliminates carbon dioxide
- Movements of ribs important in breathing; axial skeleton surrounds and protects lungs

The Digestive System

- Provides nutrients, calcium, and phosphate
- Ribs protect portions of liver, stomach, and intestines

The Urinary System

- Conserves calcium and phosphate needed for bone growth; disposes of waste products
- Axial skeleton provides some protection for kidneys and ureters; pelvis protects urinary bladder and proximal urethra

The Reproductive System

- Sex hormones stimulate growth and maintenance of bones; surge of sex hormones at puberty causes acceleration of growth and closure of epiphyseal cartilages
- Pelvis protects reproductive organs of female, protects portion of ductus deferens and accessory glands in males

bones provide attachment sites for the muscular system, and they are extensively interconnected with the cardiovascular and lymphatic systems and largely under the physiological control of the endocrine system. These functional relationships are summarized in Figure 6–56●.

→ **CONCEPT CHECK QUESTIONS**

1. The bones of the skeletal system contain 99 percent of the calcium in the body. Which other organ systems depend on these reserves of calcium for normal functioning?

Answers begin on p. 792.

Chapter Review

Access more review material online at *www.prenhall.com/bledsoe*. There you will find quiz questions, labeling activities, animations, essay questions, and web links.

Key Terms

amphiarthrosis 188
appendicular skeleton 166
articulation 187
axial skeleton 166
bursa 189
compact bone 153
diaphysis 153
diarthrosis 188

epiphysis 153
fracture 160
ligament 189
marrow 153
meniscus 188
ossification 156
osteoblast 156

osteoclast 156
osteocyte 153
osteon 153
periosteum 153
spongy bone 153
synarthrosis 188
synovial fluid 188

Related Clinical Terms

ankylosis (ang-ki-LŌ-sis) An abnormal fusion between articulating bones in response to trauma and friction within a joint.

arthritis (ar-THRĪ-tis) Rheumatic diseases that affect synovial joints. Arthritis always involves damage to the articular cartilages, but the specific cause can vary. The diseases of arthritis are usually classified as either *degenerative* or *inflammatory*.

arthroscopic surgery The surgical modification of a joint using an *arthroscope* (a fiber-optic instrument used to view the inside of joint cavities).

bursitis Inflammation of a bursa that causes pain whenever the associated tendon or ligament moves.

carpal tunnel syndrome Inflammation of the tissues at the anterior wrist that causes compression of adjacent tendons and nerves. Symptoms are pain and a loss of wrist mobility.

fracture A crack or break in a bone.

gigantism A condition of extreme height that results from an overproduction of growth hormone before puberty.

herniated disc A condition caused by intervertebral disc compression severe enough

to rupture the outer fibrocartilage layer and release the inner soft, gelatinous core, which may protrude beyond the intervertebral space.

kyphosis (kī-FŌ-sis) An abnormal exaggeration of the thoracic spinal curve that produces a humpback appearance.

lordosis (lor-DŌ-sis) An abnormal lumbar curve of the spine that results in a swayback appearance.

luxation (luks-Ā-shun) A dislocation; a condition in which the articulating surfaces are forced out of position.

orthopedics (or-tho-PĒ-diks) A branch of surgery concerned with disorders of the bones and joints and their associated muscles, tendons, and ligaments.

osteomyelitis (os-tē-ō-mī-e-LĪ-tis) A painful infection in a bone, generally caused by bacteria.

osteopenia (os-tē-ō-PĒ-nē-uh) Inadequate ossification, which leads to thinner, weaker bones.

osteoporosis (os-tē-ō-po-RŌ-sis) A reduction in bone mass to a degree that compromises normal function.

rheumatism (ROO-muh-tizum) A general term that indicates pain and stiffness that arises in the skeletal system, the muscular system, or both.

rickets A childhood disorder that reduces the amount of calcium salts in the skeleton; typically characterized by a bowlegged appearance, because the leg bones bend under the body's weight.

scoliosis (skō-lē-Ō-sis) An abnormal lateral curvature of the spine.

scurvy A condition that involves weak, brittle bones as a result of a vitamin C deficiency.

spina bifida (SPĪ-nuh BI-fi-duh) A condition that results from the failure of the vertebral laminae to unite during development; commonly associated with developmental abnormalities of the brain and spinal cord.

sprain A condition in which a ligament is stretched to the point at which some of the collagen fibers are torn. The ligament remains functional, and the structure of the joint is not affected.

whiplash An injury that results when a sudden change in body position injures the cervical vertebrae.

Summary Outline

1. The skeletal system includes the bones of the skeleton and the cartilages, ligaments, and other connective tissues that stabilize or interconnect bones. Its functions include structural support, storage, blood cell production, protection, and leverage.

THE STRUCTURE OF BONE 152

1. **Bone,** or **osseous tissue,** is a supporting connective tissue with a solid *matrix.*

Macroscopic Features of Bone 152

2. General categories of bones are **long bones, short bones, flat bones,** and **irregular bones.** *(Figure 6–1)*

3. The features of a long bone include a **diaphysis,** two **epiphyses,** and a central *marrow cavity. (Figure 6–2)*

4. The two types of bone tissue are **compact** *(dense)* **bone** and **spongy** *(cancellous)* **bone.**

5. A bone is covered by a **periosteum** and lined with an **endosteum.**

Microscopic Features of Bone 153

6. Both types of bone tissue contain **osteocytes** in **lacunae.** Layers of calcified matrix are **lamellae,** interconnected by **canaliculi.** *(Figures 6–3 through 6–8)*

7. The basic functional unit of compact bone is the **osteon,** which contains osteocytes arranged around a **central canal.**

8. Spongy bone contains **trabeculae,** often in an open network.

9. Compact bone is located where stresses come from a limited range of directions; spongy bone is located where stresses are few or come from many different directions.

10. Cells other than osteocytes are also present in bone. **Osteoclasts** dissolve the bony matrix through the process of *osteolysis.* **Osteoblasts** synthesize the matrix in the process of *osteogenesis.*

BONE FORMATION AND GROWTH 156

1. **Ossification** is the process of converting other tissues to bone. *(Figure 6–9)*

Intramembranous Ossification 156

2. **Intramembranous ossification** begins when stem cells in connective tissue differentiate into osteoblasts and produce spongy or compact bone.

Endochondral Ossification 157

3. **Endochondral ossification** begins with the formation of a cartilage model of a bone that is gradually replaced by bone. *(Figure 6–10)*

4. Bone diameter increases through **appositional growth.** *(Figure 6–11)*

Bone Growth and Body Proportions 158

5. There are differences among bones and among individuals regarding the timing of epiphyseal closure.

Requirements for Normal Bone Growth 158

6. Normal osteogenesis requires a reliable source of minerals, vitamins, and hormones.

BONE REMODELING AND HOMEOSTATIC MECHANISMS 159

1. The organic and mineral components of bone are continuously recycled and renewed through the process of **remodeling.**

The Role of Remodeling in Support 159

2. The shapes and thicknesses of bones reflect the stresses applied to them. Mineral turnover enables bone to adapt to new stresses.

Key Note 159

Homeostasis and Mineral Storage 159

3. Calcium is the most abundant mineral in the human body; roughly 99 percent of it is located in the skeleton. The skeleton acts as a calcium reserve.

Injury and Repair 160

4. A **fracture** is a crack or break in a bone. Repair of a fracture involves the formation of a **fracture hematoma,** an **external callus,** and an **internal callus.** *(Figures 6–12 through 6–22)*

AGING AND THE SKELETAL SYSTEM 166

1. The effects of aging on the skeleton can include **osteopenia** and **osteoporosis.**

AN OVERVIEW OF THE SKELETON 166

Bone Markings (Surface Features) 166

1. **Bone markings** are surface features that can be used to describe and identify specific bones. *(Table 6–1)*

Skeletal Divisions 166

2. The **axial skeleton** can be subdivided into the **skull** and associated bones (including the **auditory ossicles,** or ear bones, and the **hyoid**); the **thoracic cage,** which is composed of the **ribs** and **sternum** *(rib cage)* and thoracic vertebrae; and the **vertebral column.** *(Figures 6–23, 6–24)*

3. The **appendicular skeleton** includes the upper and lower limbs and the **pectoral** and **pelvic girdles.**

THE AXIAL DIVISION 166

The Skull 166

1. The **cranium** encloses the **cranial cavity,** which encloses the brain.

2. The **frontal bone** forms the forehead and superior surface of each **orbit.** *(Figures 6–25, 6–26, 6–27)*

3. The **parietal bones** form the upper sides and roof of the cranium. *(Figures 6–25, 6–27)*

4. The **occipital bone** surrounds the **foramen magnum** and articulates with the sphenoid, temporal, and parietal bones to form the back of the cranium. *(Figures 6–25, 6–26, 6–27)*

5. The **temporal bones** help form the sides and base of the cranium and fuse with the parietal bones along the **squamous suture**. *(Figures 6–25, 6–26, 6–27)*

6. The **sphenoid bone** acts like a bridge that unites the cranial and facial bones. *(Figures 6–25, 6–26, 6–27)*

7. The **ethmoid bone** stabilizes the brain and forms the roof and sides of the nasal cavity. Its **cribriform plate** contains perforations for olfactory nerves, and the **perpendicular plate** forms part of the bony *nasal septum*. *(Figures 6–25, 6–26, 6–27)*

8. The left and right **maxillary bones**, or *maxillae*, articulate with all the other facial bones except the *mandible*. *(Figures 6–25, 6–26, 6–27)*

9. The **palatine bones** form the posterior portions of the hard palate and contribute to the walls of the nasal cavity and to the floor of each orbit. *(Figures 6–26, 6–27)*

10. The **vomer** forms the inferior portion of the bony nasal septum. *(Figures 6–26, 6–27)*

11. The **zygomatic bones** help complete the orbit and together with the temporal bones form the **zygomatic arch** *(cheekbone)*. *(Figures 6–25, 6–26)*

12. The **nasal bones** articulate with the frontal bone and the maxillary bones. *(Figures 6–25, 6–26, 6–27)*

13. The **lacrimal bones** are within the orbit on its medial surface. *(Figures 6–25, 6–26)*

14. The **inferior nasal conchae** inside the nasal cavity aid the **superior** and **middle nasal conchae** of the ethmoid bone in slowing incoming air. *(Figures 6–26a, 6–27c)*

15. The **nasal complex** includes the bones that form the superior and lateral walls of the nasal cavity and the sinuses that drain into them. The **nasal septum** divides the nasal cavities. Together the **frontal, sphenoidal, ethmoidal, palatine, and maxillary sinuses** make up the **paranasal sinuses**. *(Figures 6–26, 6–27, 6–28)*

16. The **mandible** is the bone of the lower jaw. *(Figures 6–25, 6–26, 6–27)*

17. The **hyoid bone** is suspended below the skull by ligaments from the styloid processes of the temporal bones. *(Figure 6–29)*

18. Fibrous tissue connections called **fontanels** permit the skulls of infants and children to continue growing. *(Figure 6–30)*

The Vertebral Column and Thoracic Cage 175

19. There are seven **cervical vertebrae**, 12 **thoracic vertebrae** (which articulate with ribs), and five **lumbar vertebrae** (the last articulates with the sacrum). The **sacrum** and **coccyx** consist of fused vertebrae. *(Figure 6–31)*

20. The spinal column has four **spinal curves**, which accommodate the unequal distribution of body weight and keep it in line with the body axis. *(Figure 6–31)*

21. A typical vertebra has a **body** and a **vertebral arch;** it articulates with other vertebrae at the **articular processes**. Adjacent vertebrae are separated by an **intervertebral disc.** *(Figure 6–32)*

22. Cervical vertebrae are distinguished by the oval body and **transverse foramina** on either side. *(Figure 6–32a)*

23. Thoracic vertebrae have distinctive heart-shaped bodies. *(Figure 6–32b)*

24. The lumbar vertebrae are the most massive, least mobile, and are subjected to the greatest strains. *(Figure 6–32c)*

25. The sacrum protects reproductive, digestive, and excretory organs. At its **apex,** the sacrum articulates with the coccyx. At its **base,** the sacrum articulates with the last lumbar vertebra. *(Figure 6–34)*

26. The skeleton of the chest, or thoracic cage, consists of the thoracic vertebrae, the ribs, and the sternum. The ribs and sternum form the rib cage. *(Figure 6–35)*

27. Ribs 1 to 7 are **true ribs.** Ribs 8 to 12 lack direct connections to the sternum and are called **false ribs;** they include two pairs of **floating ribs.** The medial end of each rib articulates with a thoracic vertebra. *(Figure 6–35)*

28. The sternum consists of a **manubrium,** a **body,** and a **xiphoid process.** *(Figure 6–35)*

THE APPENDICULAR DIVISION 179
The Pectoral Girdle 179

1. Each arm articulates with the trunk at the pectoral girdle, or shoulder girdle, which consists of the **scapulae** and **clavicles.** *(Figure 6–24)*

2. The clavicle and scapula position the shoulder joint, help move the arm, and provide a base for arm movement and muscle attachment. *(Figures 6–36, 6–37)*

3. Both the **coracoid process** and the **acromion** are attached to ligaments and tendons. The **scapular spine** crosses the posterior surface of the scapular body. *(Figure 6–37)*

The Upper Limb 182

4. The **humerus** articulates with the scapula at the shoulder joint. The **greater tubercle** and **lesser tubercle** of the humerus are important sites for muscle attachment. Other prominent landmarks include the **deltoid tuberosity,** the **medial** and **lateral epicondyles,** and the articular **condyle.** *(Figure 6–38)*

5. Distally, the humerus articulates with the radius and ulna. The medial **trochlea** extends from the **coronoid fossa** to the **olecranon fossa.** *(Figure 6–38)*

6. The **radius** and **ulna** are the bones of the forearm. The olecranon fossa accommodates the **olecranon process** during extension of the arm. The coronoid and radial fossae accommodate the **coronoid process** of the ulna. *(Figure 6–39)*

7. The bones of the wrist form two rows of **carpal bones.** The distal carpal bones articulate with the **metacarpal bones** of the palm. The metacarpal bones articulate with the proximal **phalanges,** or finger bones. Four of the fingers contain three phalanges; the **pollex,** or thumb, has only two. *(Figure 6–40)*

The Pelvic Girdle 183

8. The pelvic girdle consists of two **coxal bones.** *(Figures 6–24, 6–41, 6–42)*

9. The largest part of the coxal bone, the **ilium,** fuses with the **ischium,** which in turn fuses with the **pubis.** The **pubic symphysis** limits movement between the pubic bones.

10. The **pelvis** consists of the hipbones, the sacrum, and the coccyx.

The Lower Limb 185

11. The **femur,** or *thighbone,* is the longest bone in the body. It articulates with the **tibia** at the knee joint. A ligament from the **patella** (the *kneecap*) attaches at the **tibial tuberosity.** *(Figures 6–43, 6–44)*

12. Other tibial landmarks include the **anterior crest** and the **medial malleolus.** The **head** of the **fibula** articulates with the tibia below the knee, and the **lateral malleolus** stabilizes the ankle.

13. The ankle includes seven **tarsal bones;** only the **talus** articulates with the tibia and fibula. When we stand normally, most of our weight is transferred to the **calcaneus,** or *heel bone,* and the rest is passed on to the **metatarsal bones.** *(Figure 6–45)*

14. The basic organizational pattern of the metatarsals and phalanges of the foot resembles that of the hand.

ARTICULATIONS 187

The Classification of Joints 187

1. **Articulations** (joints) exist wherever two bones interact. Immovable joints are **synarthroses,** slightly-movable joints are **amphiarthroses,** and those that are freely-movable are called **diarthroses.** *(Table 6–2)*

2. Examples of synarthroses are a **suture,** a **gomphosis,** and a **synchondrosis.**

3. Examples of amphiarthroses are a **syndesmosis** and a **symphysis.**

4. The bony surfaces at diarthroses, or **synovial joints,** are covered by **articular cartilages,** lubricated by **synovial fluid,** and enclosed within a **joint capsule.** Other synovial structures include **menisci, fat pads, bursae,** and various **ligaments.** *(Figure 6–46)*

Synovial Joints: Movement and Structure 189

5. Important terms that describe dynamic motion at synovial joints are **flexion, extension, hyperextension, abduction, adduction, circumduction,** and **rotation.** *(Figures 6–47, 6–48)*

6. The bones in the forearm permit **pronation** and **supination.** *(Figure 6–48)*

7. Movements of the foot include **inversion** and **eversion.** The ankle undergoes **dorsiflexion** and **plantar flexion. Opposition** is the thumb movement that enables us to grasp and hold objects. *(Figure 6–49)*

8. **Protraction** involves moving a part of the body forward; **retraction** involves moving it back. **Depression** and **elevation** occur when we move a structure inferiorly and superiorly, respectively. *(Figure 6–49)*

9. Major types of synovial joints include gliding joints, hinge joints, pivot joints, ellipsoidal joints, saddle joints, and ball-and-socket joints. *(Figure 6–50)*

Key Note 194

Representative Articulations 194

10. The articular processes of adjacent vertebrae form gliding joints. *Symphyseal joints* connect adjacent vertebral bodies and are separated by pads called *intervertebral discs.* *(Figure 6–51)*

11. The shoulder joint is formed by the **glenoid cavity** and the head of the humerus. This joint is extremely mobile and, for that reason, it is also unstable and easily dislocated. *(Figure 6–52)*

12. Bursae at the shoulder joint reduce friction from muscles and tendons during movement. *(Figure 6–52)*

13. The elbow joint permits only flexion and extension. It is extremely stable because of extensive ligaments and the shapes of the articulating elements. *(Figure 6–53)*

14. The hip joint is formed by the union of the acetabulum with the head of the femur. This ball-and-socket diarthrosis permits flexion and extension, adduction and abduction, circumduction, and rotation. *(Figure 6–54)*

15. The knee joint is a complicated hinge joint. The joint permits flexion-extension and limited rotation. *(Figure 6–55)*

INTEGRATION WITH OTHER SYSTEMS 197

1. The skeletal system is dynamically associated with other systems. *(Figure 6–56)*

Review Questions

Level 1: Reviewing Facts and Terms

Match each item in column A with the most closely related item in column B. Place letters for answers in the spaces provided.

COLUMN A

_____ 1. osteocytes

_____ 2. diaphysis

_____ 3. auditory ossicles

_____ 4. cribriform plate

_____ 5. osteoblasts

_____ 6. C_1

_____ 7. C_2

_____ 8. hip and shoulder

_____ 9. patella

_____ 10. calcaneus

_____ 11. synarthrosis

_____ 12. moving the hand into a palm-front position

_____ 13. osteoclasts

_____ 14. raising the arm laterally

_____ 15. elbow and knee

COLUMN B

a. abduction

b. heel bone

c. ball-and-socket joints

d. bone-dissolving cells

e. hinge joints

f. axis

g. immovable joint

h. bone shaft

i. mature bone cells

j. bone-producing cells

k. atlas

l. olfactory nerves

m. ear bones

n. supination

o. kneecap

16. Skeletal bones store energy reserves as lipids in areas of:
    (a) red marrow.
    (b) yellow marrow.
    (c) the matrix of bone tissue.
    (d) the ground substance.

17. The two types of osseous tissue are _____ bone.
    (a) compact bone and spongy
    (b) dense bone and compact
    (c) spongy bone and cancellous
    (d) compact, spongy, dense, and cancellous

18. The basic functional units of mature compact bone are:
    (a) lacunae.
    (b) osteocytes.
    (c) osteons.
    (d) canaliculi.

19. The axial skeleton consists of the bones of the:
    (a) pectoral and pelvic girdles.
    (b) skull, thorax, and vertebral column.
    (c) arms, legs, hands, and feet.
    (d) limbs, pectoral girdle, and pelvic girdle.

20. The appendicular skeleton consists of the bones of the:
    (a) pectoral and pelvic girdles.
    (b) skull, thorax, and vertebral column.
    (c) arms, legs, hands, and feet.
    (d) limbs, pectoral girdle, and pelvic girdle.

21. Which of the following lists contains only bones of the cranium?
    (a) frontal, parietal, occipital, sphenoid
    (b) frontal, occipital, zygomatic, parietal
    (c) occipital, sphenoid, temporal, palatine
    (d) mandible, maxilla, nasal, zygomatic

22. Of the following bones, which one is unpaired?
    (a) vomer
    (b) maxilla
    (c) palatine
    (d) nasal

23. At the glenoid cavity, the scapula articulates with the proximal end of the:
    (a) humerus.
    (b) radius.
    (c) ulna.
    (d) femur.

24. While an individual is in the anatomical position, the ulna lies _____ to the radius.
    (a) medial
    (b) lateral
    (c) inferior
    (d) superior

25. Each coxal bone of the pelvic girdle consists of three fused bones: the
    (a) ulna, radius, and humerus.
    (b) ilium, ischium, and pubis.
    (c) femur, tibia, and fibula.
    (d) hamate, capitate, and trapezium.

26. Joints that are typically located at the end of long bones are:
    (a) synarthroses.
    (b) amphiarthroses.
    (c) diarthroses.
    (d) sutures.

27. The function of synovial fluid is:
    (a) to nourish chondrocytes.
    (b) to provide lubrication.
    (c) to absorb shock.
    (d) to nourish chondrocytes, provide lubrication, and absorb shock.

28. Abduction and adduction always refer to movements of the:
    (a) axial skeleton.
    (b) appendicular skeleton.
    (c) skull.
    (d) vertebral column.

29. Standing on tiptoe is an example of a movement called:
    (a) elevation.
    (b) dorsiflexion.
    (c) plantar flexion.
    (d) retraction.

30. What are the five primary functions of the skeletal system?

31. What is the primary difference between intramembranous ossification and endochondral ossification?

32. What unique characteristic of the hyoid bone makes it different from all the other bones in the body?

33. What two primary functions are performed by the thoracic cage?

34. Which two large scapular processes are associated with the shoulder joint?

Level 2: Reviewing Concepts

35. Why are stresses or impacts to the side of the shaft of a long bone more dangerous than stress applied along the long axis of the shaft?

36. During the growth of a long bone, how is the epiphysis forced farther from the shaft?

37. Why are ruptured intervertebral discs more common in lumbar vertebrae and dislocations and fractures more common in cervical vertebrae?

38. Why are clavicular injuries common?

39. What is the difference between the pelvic girdle and the pelvis?

40. How do articular cartilages differ from other cartilages in the body?

41. What is the significance of the fact that the pubic symphysis is a slightly movable joint?

Level 3: Critical Thinking and Clinical Applications

42. While playing on her swing set, 10-year-old Yasmin falls and breaks her right leg. At the emergency room, the physician tells Yasmin's parents that the proximal end of the tibia, where the epiphysis meets the diaphysis, is fractured. The fracture is properly set and eventually heals. During a routine physical when she is 18, Yasmin learns that her right leg is 2 cm shorter than her left, probably because of her accident. What might account for this difference?

43. Tess is diagnosed with a disease that affects the membranes that surround the brain. The physician tells Tess's family that the disease is caused by an airborne virus. Explain how this virus could have entered the cranium.

44. While working at an excavation, an archaeologist finds several small skull bones. She examines the frontal, parietal, and occipital bones and concludes that the skulls are those of children not yet one year old. How can she tell their ages from examining the bones?

45. Frank Fireman is fighting a fire in a building when part of the ceiling collapses and a beam strikes him on his left shoulder. He is rescued by his friends, but he has a great deal of pain in his shoulder and cannot move his arm properly, especially in the anterior direction. His clavicle is not broken, and his humerus is intact. What is the probable nature of Frank's injury?

46. Ed "turns over" his ankle while playing tennis. He experiences swelling and pain, but after examination, he is told that there are no torn ligaments and that the structure of the ankle is not affected. On the basis of the signs and symptoms and the examination results, what do you think happened to Ed's ankle?

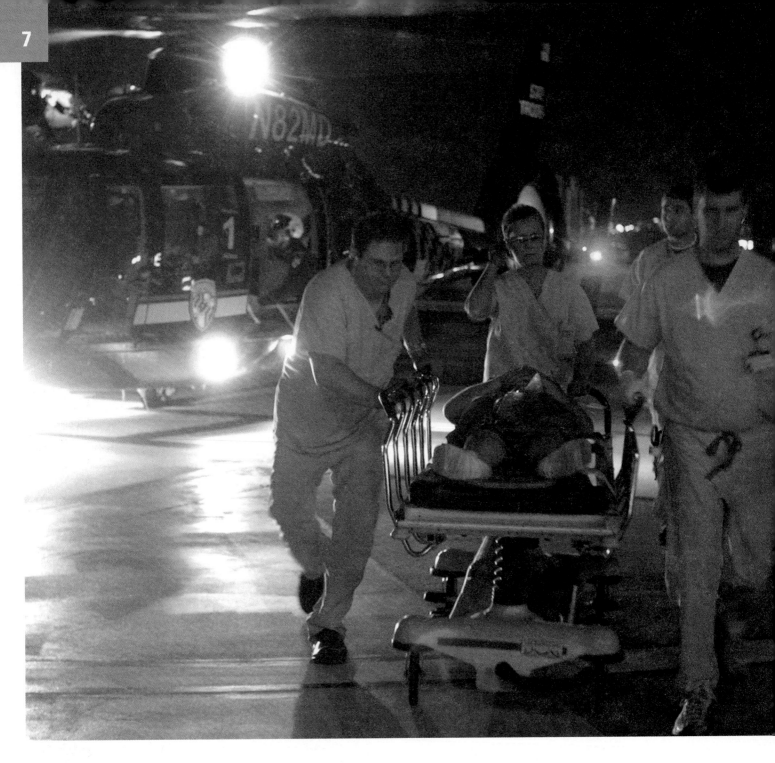

7 The Muscular System

MUCH OF PREHOSPITAL CARE is ensuring that patients are transported to the appropriate hospital. Research has shown that patient outcomes are uniformly better when patients are treated at facilities that routinely provide care for the illness or injury the patient is suffering. Thus, patients with severe trauma should be transported to a trauma center. Patients with burns should go to a burn center. Knowing the capabilities of your local health-care facilities and ensuring that patients are transported to the hospital with the most expertise in your patient's problem is almost as important as the care provided in the field.

Chapter Outline

Chapter Objectives

1. Describe the functions of skeletal muscle tissue. (p. 208)
2. Describe the organization of muscle at the tissue level. (pp. 208–209)
3. Identify the structural components of a sarcomere. (pp. 209–211)
4. Explain the key steps involved in the contraction of a skeletal muscle fiber. (pp. 212–217)
5. Compare the different types of muscle contractions. (pp. 218–221)
6. Describe the mechanisms by which muscles obtain and use energy to power contractions. (pp. 221–224)
7. Relate the types of muscle fibers to muscular performance. Distinguish between aerobic and anaerobic endurance, and explain their implications for muscular performance. (pp. 224–225)
8. Contrast skeletal, cardiac, and smooth muscles in terms of structure and function. (pp. 225–227)
9. Identify the main axial muscles of the body, along with their actions. Identify the main appendicular muscles of the body, along with their actions. (pp. 227–251)
10. Describe the effects of aging on muscle tissue. (pp. 251, 253)
11. Discuss the functional relationships between the muscular system and other organ systems. (pp. 252–253)

Vocabulary Development

aer air; *aerobic*
an not; *anaerobic*
bi two; *biceps*
caput head; *caput humeri*
clavius clavicle; *clavicle*
di two; *digastricus*
epi- on; *epimysium*
ergon work; *synergist*
fasciculus a bundle; *fascicle*

gaster stomach; *gastrocnemius*
hyper above; *hypertrophy*
iso- equal; *isometric*
kneme knee; *gastrocnemius*
lemma husk; *sarcolemma*
meros part; *sarcomere*
metron measure; *isometric*
mys muscle; *epimysium*

peri- around; *perimysium*
platys flat; *platysma*
sarkos flesh; *sarcolemma*
syn- together; *synergist*
tetanus convulsive tension; *tetanus*
tonos tension; *isotonic*
trope a turning; *tropomyosin*
-trophy nourishing; *atrophy*

IT IS HARD TO IMAGINE what life would be like without muscle tissue. We would be unable to sit, stand, walk, speak, or grasp objects. Blood would not circulate, because there would be no heartbeat to propel it through the vessels. The lungs could not rhythmically empty and fill, nor could food move through the digestive tract. Muscle tissue, one of the four primary tissue types, consists of elongated muscle cells that are highly specialized for contraction.

The three types of muscle tissue—*skeletal muscle, cardiac muscle,* and *smooth muscle*—were introduced in Chapter 4. ∞ p. 113

This chapter begins by discussing skeletal muscle tissue, the most abundant muscle tissue in the body. It is followed by an overview of the differences among skeletal, cardiac, and smooth muscle tissue. We will then proceed to a consideration of the muscular system.

■ Functions of Skeletal Muscle

Skeletal muscles are organs composed primarily of skeletal muscle tissue, but they also contain connective tissues, nerves, and blood vessels. These muscles are directly or indirectly attached to the bones of the skeleton. The muscular system includes approximately 700 skeletal muscles that perform the following functions:

1. *Produce movement of the skeleton.* Skeletal muscle contractions pull on tendons and thereby move the bones. These contractions may produce a simple motion, such as extending the arm, or the highly coordinated movements of swimming, skiing, or typing.

2. *Maintain posture and body position.* Continuous muscle contractions maintain body posture. Without this constant action, you could not sit upright without collapsing or stand without toppling over.

3. *Support soft tissues.* The abdominal wall and the floor of the pelvic cavity consist of layers of skeletal muscle. These muscles support the weight of visceral organs and shield internal tissues from injury.

4. *Guard entrances and exits.* Skeletal muscles encircle openings to the digestive and urinary tracts. These muscles provide voluntary control over swallowing, defecation, and urination.

5. *Maintain body temperature.* Muscle contractions require energy, and whenever energy is used in the body, some of it is converted to heat. The heat released by working muscles keeps body temperature in the range required for normal functioning.

■ The Anatomy of Skeletal Muscles

To understand how skeletal muscle contracts, we must study the structure of skeletal muscle. We begin with the organ-level structure of skeletal muscle before describing its cellular-level structure. In the following discussions we will often encounter the Greek words *sarkos* (flesh) and *mys* (muscle) as word roots in the names of the structural features of muscles and their components.

Gross Anatomy

Figure 7–1● illustrates the appearance and organization of a typical skeletal muscle. A skeletal muscle contains connective tissues, blood vessels, nerves, and skeletal muscle tissue. Each cell in skeletal muscle tissue is a single muscle *fiber*.

Connective Tissue Organization

Three layers of connective tissue are part of each muscle: the epimysium, the perimysium, and the endomysium (see Figure 7–1).

Surrounding the entire muscle is the **epimysium** (ep-i-MĪZ-ē-um; *epi-*, on + *mys*, muscle), which is a layer of collagen fibers that separates the muscle from surrounding tissues and organs.

The connective tissue fibers of the **perimysium** (per-i-MĪZ-ē-um; *peri-*, around) divide the skeletal muscle into bundles of muscle fibers called **fascicles** (FAS-i-klz; *fasciculus*, a bundle). In addition to collagen and elastic fibers, the perimysium contains blood vessels and nerves that supply the fascicles.

Within a fascicle, the **endomysium** (en-dō-MĪZ-ē-um; *endo-*, inside) surrounds each skeletal muscle fiber and ties

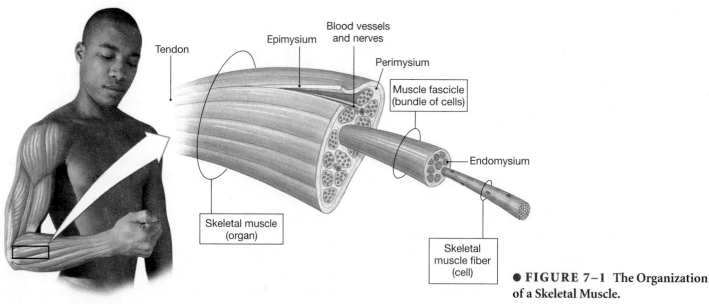

Tendon

Epimysium

Blood vessels and nerves

Perimysium

Muscle fascicle (bundle of cells)

Endomysium

Skeletal muscle (organ)

Skeletal muscle fiber (cell)

● **FIGURE 7–1 The Organization of a Skeletal Muscle.**

adjacent muscle fibers together. Stem cells scattered among the fibers help repair damaged muscle tissue.

At each end of the muscle, the collagen fibers of all three layers come together to form either a bundle known as a **tendon,** or a broad sheet called an *aponeurosis.* Tendons are bands of collagen fibers that attach skeletal muscles to bones, and aponeuroses connect different skeletal muscles. ∞ p. 106 The tendon fibers are interwoven into the periosteum of the bone, providing a firm attachment. Any contraction of the muscle exerts a pull on its tendon and in turn on the attached bone.

Blood Vessels and Nerves

The connective tissues of the epimysium and perimysium provide a passageway for the blood vessels and nerves that are necessary for the functioning of muscle fibers. Muscle contraction requires tremendous amounts of energy. An extensive network of blood vessels delivers the necessary oxygen and nutrients and carries away the metabolic wastes generated by active skeletal muscles.

Skeletal muscles contract only under stimulation from the central nervous system. *Axons* (nerve fibers) penetrate the epimysium, branch through the perimysium, and enter the endomysium to control individual muscle fibers. Skeletal muscles are often called *voluntary muscles* because we have voluntary control over their contractions. Many skeletal muscles may also be controlled at a subconscious level. For example, skeletal muscles involved with breathing, such as the *diaphragm,* usually work outside our conscious awareness.

Microanatomy

Skeletal muscle fibers are quite different from the "typical" cell described in Chapter 3. One obvious difference is their enormous size. A skeletal muscle fiber from a leg muscle, for example, could have a diameter of 100 μm and a length equal to that of the entire muscle (up to 60 cm, or 24 in.). In addition, each skeletal muscle fiber is *multinucleate,* and contains hundreds of nuclei just beneath the cell membrane. In the next sections we will examime the components of a typical skeletal muscle fiber.

The Sarcolemma and Transverse Tubules

The basic structure of a muscle fiber is depicted in Figure 7–2a●. The cell membrane, or **sarcolemma** (sar-kō-LEM-uh; *sarkos,* flesh + *lemma,* husk), of a muscle fiber surrounds the cytoplasm, or **sarcoplasm** (SAR-kō-plazm). Openings scattered across the surface of the sarcolemma lead into a network of narrow tubules called **transverse tubules,** or **T tubules.** Filled with extracellular fluid, the T tubules form passageways through the muscle fiber, like a series of tunnels through a mountain.

The T tubules play a major role in coordinating the contraction of the muscle fiber. A muscle fiber contraction occurs through the orderly interaction of both electrical and chemical events. Electrical impulses conducted by the sarcolemma trigger a contraction by altering the chemical environment everywhere inside the muscle fiber. The electrical impulses reach the cell's interior by traveling along the transverse tubules that extend deep into the sarcoplasm of the muscle fiber.

Myofibrils

Inside the muscle fiber, branches of T tubules encircle cylindrical structures called **myofibrils** (see Figure 7–2a). A myofibril is 1–2 μm in diameter and as long as the entire muscle fiber. Each muscle fiber contains hundreds to thousands of myofibrils. Myofibrils are bundles of thick and thin **myofilaments,** protein filaments that consist primarily of the proteins *actin* and *myosin* (Figure 7–2a–c●). Actin molecules are found in **thin filaments** and the myosin molecules in **thick filaments.** ∞ p. 76

Myofibrils are responsible for muscle fiber contraction. Because they are attached to the sarcolemma at each end of the cell, their contraction shortens the entire cell. Scattered among the myofibrils are mitochondria and granules of glycogen, a source of glucose. The breakdown of glucose and the activity of mitochondria provide the ATP needed to power muscular contractions.

The Sarcoplasmic Reticulum

Wherever a T tubule encircles a myofibril, the tubule is tightly bound to the membranes of the **sarcoplasmic reticulum (SR),** which is a specialized form of smooth endoplasmic reticulum (see Figure 7–2a). The sarcoplasmic reticulum forms a tubular network around each myofibril. On either side of a T tubule lie expanded chambers of the SR called *terminal cisternae* (singular: *cisterna*). As it encircles a myofibril, a transverse tubule lies sandwiched between a pair of terminal cisternae, and forms a *triad.*

The terminal cisternae contain high concentrations of calcium ions. The calcium ion concentration in the cytoplasm of all cells is kept very low. Most cells, including skeletal muscle fibers, pump calcium ions across their cell membranes and into the extracellular fluid. Skeletal muscle fibers, however, also actively transport calcium ions into the terminal cisternae of the sarcoplasmic reticulum. A muscle contraction begins when the stored calcium ions are released from the terminal cisternae into the sarcoplasm.

Sarcomeres

Myofilaments (thin and thick filaments) are organized into repeating functional units called **sarcomeres** (SAR-kō-mērz; *sarkos,* flesh + *meros,* part) (see Figure 7–2b). Each myofibril

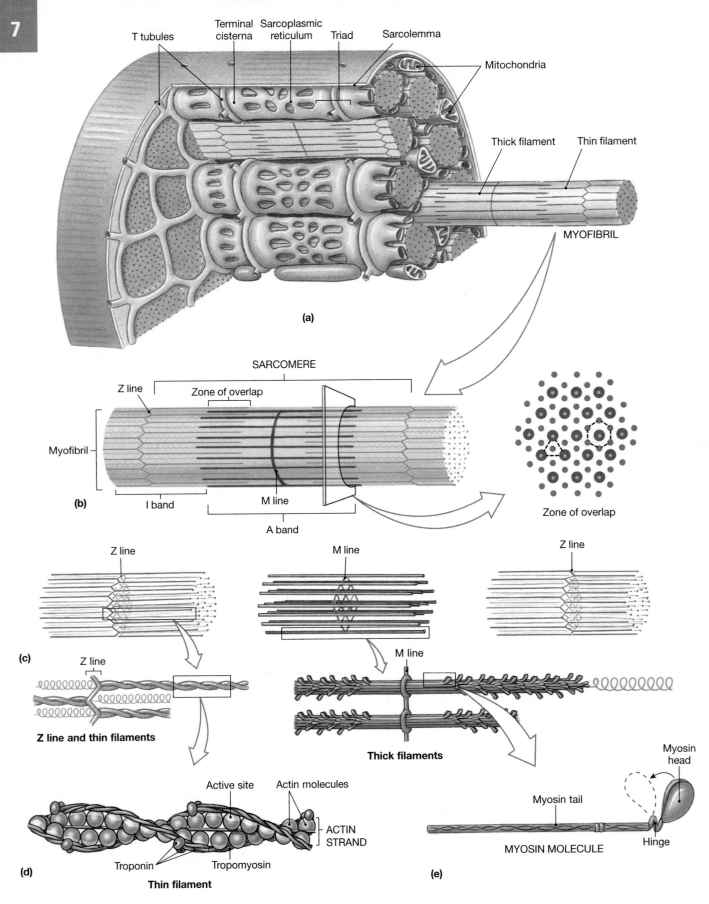

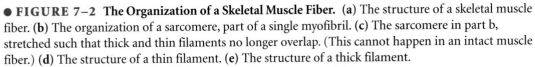

● **FIGURE 7–2 The Organization of a Skeletal Muscle Fiber.** (**a**) The structure of a skeletal muscle fiber. (**b**) The organization of a sarcomere, part of a single myofibril. (**c**) The sarcomere in part b, stretched such that thick and thin filaments no longer overlap. (This cannot happen in an intact muscle fiber.) (**d**) The structure of a thin filament. (**e**) The structure of a thick filament.

consists of approximately 10,000 sarcomeres arranged end to end. *The sarcomere is the smallest functional unit of the muscle fiber. Interactions between the thick and thin filaments of sarcomeres are responsible for muscle contraction.*

The arrangement of thick and thin filaments within a sarcomere produces a banded appearance. All of the myofibrils are arranged parallel to the long axis of the cell, and their sarcomeres lie side by side. As a result, the entire muscle fiber has a banded, or striated, appearance that corresponds to the bands of the individual sarcomeres (see Figure 7–2a).

Figure 7–2b diagrams the external and internal structure of an individual sarcomere. Each sarcomere has a resting length of about 2 μm. Neither type of filament spans the entire length of a sarcomere. The thick filaments lie in the center of the sarcomere. Thin filaments at either end of the sarcomere are attached to interconnecting proteins that make up the **Z lines,** which are the boundaries of each sarcomere. From the Z lines, the thin filaments extend toward the center of the sarcomere, and pass among the thick filaments in the *zone of overlap.* Strands of another protein extend from the Z lines to the ends of the thick filaments and keep both types of filaments in alignment. The **M line** is made up of proteins that connect the central portions of each thick filament to its neighbors. The relationships of Z lines and M lines are shown in Figure 7–2c.

Differences in the sizes and densities of thick and thin filaments account for the banded appearance of the sarcomere. The dark **A band** is the area that contains thick filaments. The light region between two successive A bands—including the Z line— is the **I band.** (It may help you to remember that in a light micrograph, the A band appears d**A**rk and the I band is l**I**ght.)

Clinical Note
RIGOR MORTIS

Dead human bodies that suddenly sit up in a morgue are a part of many urban legends. In actuality, this does not occur. Approximately six hours after death (the time varies depending on environmental temperature), the skeletal muscles have depleted all remaining glucose and ATP molecules. Waste products, primarily metabolic acids, accumulate. After the ATP is gone, the sarcoplasmic reticulum cannot remove calcium ions from the sarcoplasm. Then, myosin fibers cannot separate from actin fibers, and *rigor mortis*, which is a sustained contraction, sets in. The smaller muscles, usually those of the jaw, are affected first. Eventually, the entire body is affected. Finally, 12–24 hours later, lysosomal enzymes from the muscle cells break down the contracted myofilaments, and the muscles relax. ∎

Thin and Thick Filaments
Each thin filament consists of a twisted strand of actin molecules (Figure 7–2d●). Each actin molecule has an **active site**

capable of interacting with myosin. In a resting muscle, the active sites along the thin filaments are covered by strands of the protein **tropomyosin** (trō-pō-MĪ-ō-sin; *trope*, turning). The tropomyosin strands are held in position by molecules of **troponin** (TRŌ-pō-nin) that are bound to the actin strand.

Thick filaments are composed of myosin molecules, each with a *tail* and a globular *head* (Figures 7–2c,e●). The myosin molecules are oriented away from the center of the sarcomere, and the heads project outward. The myosin heads attach to actin molecules during a contraction. This interaction cannot occur unless the troponin changes position, which moves the tropomyosin and exposes the active sites.

Calcium is the "key" that "unlocks" the active sites and starts a contraction. When calcium ions bind to troponin, the protein changes shape, and swings the tropomyosin away from the active sites. Myosin-actin binding can then occur, and a contraction begins. The terminal cisternae of the sarcoplasmic reticulum are the source of the calcium that triggers muscle contraction.

Sliding Filaments and Cross-Bridges
When a sarcomere contracts, the I bands get smaller, the Z lines move closer together, and the zones of overlap get larger, but the width of the A bands does not change (Figure 7–3●). These observations make sense only if the thin filaments slide toward the center of the sarcomere, alongside the stationary thick filaments. This explanation for sarcomere contraction is called the *sliding filament theory.*

The mechanism responsible for sliding filaments involves the binding of the myosin heads of thick filaments to active sites on the thin filaments. When they connect thick filaments and thin filaments, the myosin heads are called **cross-bridges.** When a cross-bridge binds to an active site, it pivots toward the center of the sarcomere (see Figure 7–2e), and pulls the thin filament in that direction. The cross-bridge then detaches and returns to its original position, ready to repeat a cycle of "attach, pivot, detach, and return," like a person pulling in a rope one-handed.

Clinical Note
THE MUSCULAR DYSTROPHIES

Abnormalities in the genes that code for structural and functional proteins in muscle fibers are responsible for a number of inherited diseases collectively known as the *muscular dystrophies* (DIS-trō-fēz). These conditions, which cause progressive muscular weakness and deterioration, are the result of abnormalities in the sarcolemma or in the structure of internal proteins. The best-known example is *Duchenne's muscular dystrophy*, which typically develops in males ages three to seven years. ∎

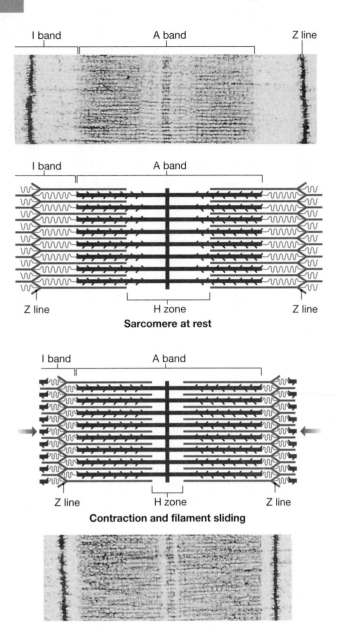

● **FIGURE 7–3 Changes in the Appearance of a Sarcomere During Contraction of a Skeletal Muscle Fiber.** During a contraction, the A band stays the same width, but the Z lines move closer together and the I band gets smaller.

CONCEPT CHECK QUESTIONS

1. How would severing the tendon attached to a muscle affect the muscle's ability to move a body part?
2. Why does skeletal muscle appear striated when viewed through a light microscope?
3. Where would you expect the greatest concentration of calcium ions in resting skeletal muscle to be?

Answers begin on p. 792.

Clinical Note
RHABDOMYOLYSIS

Patients who are unconscious or immobile for a prolonged time are at risk for developing rhabdomyolysis, which is a breakdown of muscle tissue where the patient has been lying. When this occurs, numerous substances, most notably myoglobin, muscle enzymes, and electrolytes, are released from the damaged muscle. Myoglobin concentrates in the urine (myoglobinuria), turning it dark reddish brown. In severe cases, the patient can develop kidney failure. ■

■ The Control of Muscle Fiber Contraction

Skeletal muscle fibers contract only under the control of the nervous system. The communication link between the nervous system and a skeletal muscle fiber occurs at a specialized intercellular connection known as a **neuromuscular junction** (Figure 7–4a●).

The Neuromuscular Junction

Each skeletal muscle fiber is controlled by a nerve cell called a *motor neuron.* A single axon of the neuron branches within the perimysium to form a number of fine branches, each of which ends at an expanded **synaptic terminal.** ∞ p. 115 Midway along the fiber's length, the synaptic terminal communicates with it as part of a neuromuscular junction. Figure 7–4b● depicts the key features of this structure.

The cytoplasm of the synaptic terminal contains mitochondria and vesicles filled with molecules of **acetylcholine** (as-ē-til-KŌ-lēn), or **ACh.** Acetylcholine is a *neurotransmitter,* which is a chemical released by a neuron to communicate with other cells. Neurotransmitters change the properties of another cell's membrane, in this case its permeability. The release of ACh from the synaptic terminal results in changes in the sarcolemma that trigger the contraction of the muscle fiber.

A narrow space, the **synaptic cleft,** separates the synaptic terminal from the sarcolemma. This portion of the membrane, which contains receptors that bind ACh, is known as the **motor end plate.** Both the synaptic cleft and the motor end plate contain the enzyme **acetylcholinesterase (AChE,** or *cholinesterase),* which breaks down molecules of ACh.

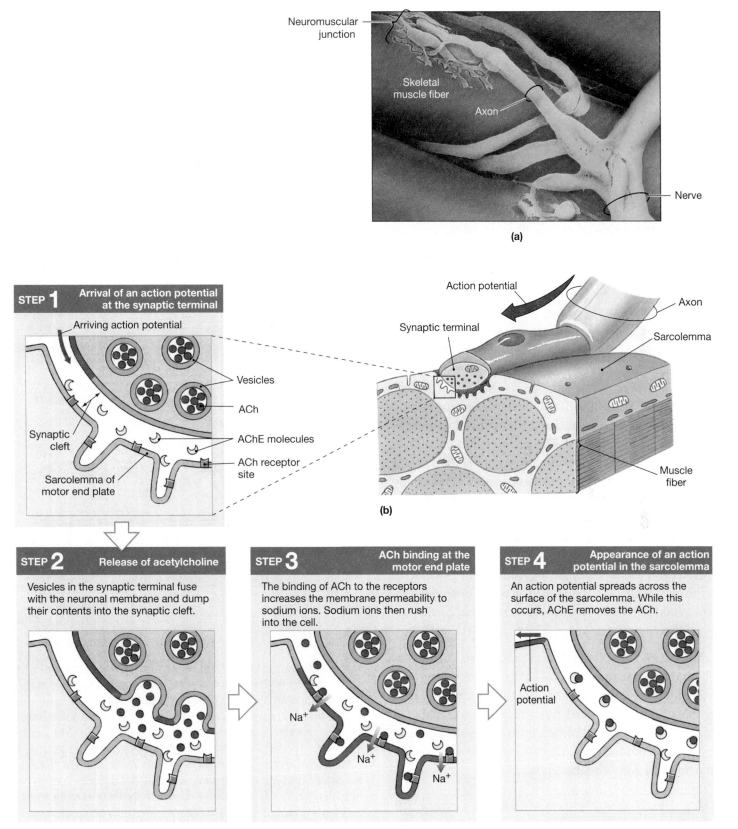

(a)

STEP 1 — Arrival of an action potential at the synaptic terminal

Arriving action potential

Vesicles

ACh

AChE molecules

ACh receptor site

Synaptic cleft

Sarcolemma of motor end plate

Action potential

Synaptic terminal

Axon

Sarcolemma

Muscle fiber

(b)

STEP 2 — Release of acetylcholine

Vesicles in the synaptic terminal fuse with the neuronal membrane and dump their contents into the synaptic cleft.

STEP 3 — ACh binding at the motor end plate

The binding of ACh to the receptors increases the membrane permeability to sodium ions. Sodium ions then rush into the cell.

Na$^+$

Na$^+$

Na$^+$

STEP 4 — Appearance of an action potential in the sarcolemma

An action potential spreads across the surface of the sarcolemma. While this occurs, AChE removes the ACh.

Action potential

(c)

● **FIGURE 7–4 The Structure and Function of the Neuromuscular Junction.** (a) This colorized SEM shows a neuromuscular junction. (b) An action potential (red arrow) carried by the nerve axon arrives at the synaptic terminal. (c) The chemical communication between the synaptic terminal and motor end plate occurs in the steps depicted here.

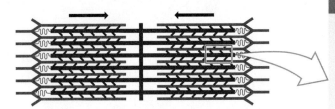

FIGURE 7–5 Molecular Events of the Contraction Process.

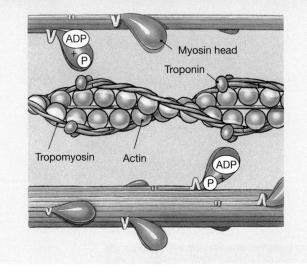

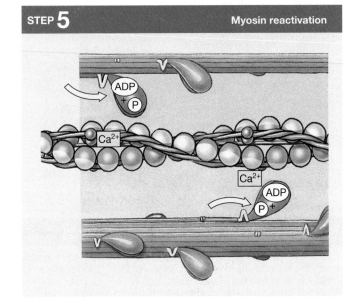

Neurons control skeletal muscle fibers by stimulating the production of an **action potential,** or electrical impulse, in the sarcolemma. In brief, events occur as follows (Figure 7–4c●):

Step 1: *An action potential arrives at the synaptic terminal.*

Step 2: *Acetylcholine releases when an action potential that is traveling along the axon of a motor neuron reaches the synaptic terminal,* vesicles in the synaptic terminal release acetylcholine into the synaptic cleft.

Step 3: *ACh binds at the motor end plate.* The ACh molecules diffuse across the synaptic cleft and bind to ACh receptors on the sarcolemma. This event changes the permeability of the membrane to sodium ions. The resulting sudden rush of sodium ions into the sarcoplasm produces an action potential in the sarcolemma. (We will examine the formation and conduction of action potentials more closely in Chapter 8.)

Step 4: *Appearance of an action potential at the sarcolemma.* The action potential spreads over the entire sarcolemma surface. It also travels down all of the transverse tubules toward the terminal cisternae that encircle the sarcomeres of the muscle fiber. The passage of an action potential triggers a sudden, massive release of calcium ions by the terminal cisternae.

As the calcium ion concentration rises, active sites are exposed on the thin filaments, cross-bridge interactions occur, and a contraction begins. Because all of the terminal cisternae in the muscle fiber are affected, this contraction is a combined effort that involves every sarcomere on every myofibril. While the contraction process gets under way, acetylcholine is being broken down by acetylcholinesterase.

Clinical Note
MYASTHENIA GRAVIS

Myasthenia gravis is an autoimmune disease characterized by muscle weakness and fatigue. It is particularly evident with repetitive use of voluntary muscles. Antibodies against acetylcholine receptors impair the function of the acetylcholine receptor at the neuromuscular junction, which causes varying degrees of motor weakness. The muscle weakness is most pronounced in the proximal muscles and is generally relieved by rest.

The signs and symptoms of myasthenia gravis can be quite varied. The first symptom is usually weakness of the eye muscles and drooping eyelids. The facial muscles are also often weak and can result in a peculiar smile known as the myasthenic snarl. As the illness progresses, the patient develops trouble swallowing and has difficulty holding his head upright. The weakness then spreads to the muscles of the trunk.

The diagnosis of myasthenia gravis can usually be made by performing a Tensilon test. Prior to administering Tensilon, a patient is given an apple to bite and chew. Then, a dose of the cholinesterase inhibitor *edrophonium chloride (Tensilon)* is administered. The patient is then asked to bite and chew the apple. A marked increase in muscle strength and the ability to eat the apple after administration of Tensilon helps establish the diagnosis. In myasthenia gravis, edropho-

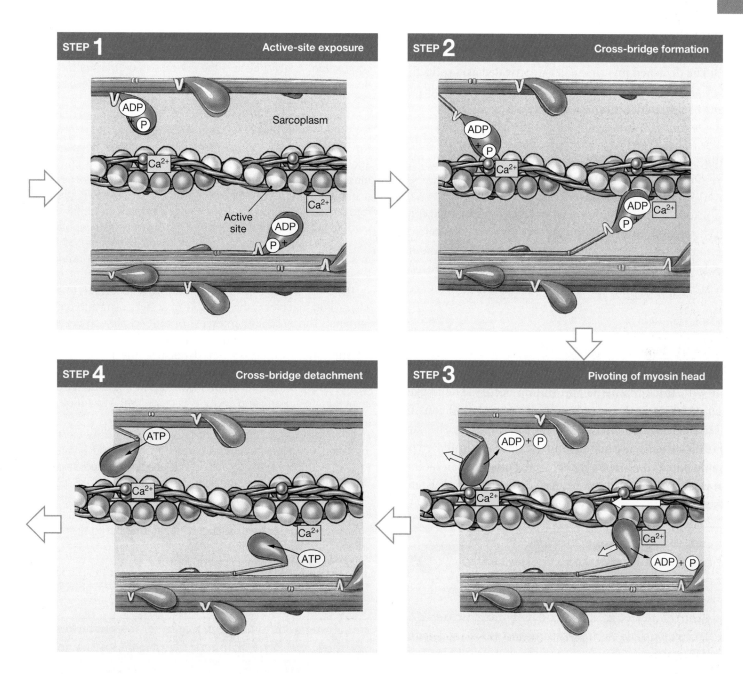

nium causes a dramatic difference in the patient's ability to eat the apple. Cholinesterase inhibitors, such as edrophonium or neostigmine, inhibit the enzyme cholinesterase. Cholinesterase breaks down acetylcholine at the neuromuscular junction. Inhibition of cholinesterase allows acetylcholine to accumulate in the neuromuscular junction and overcome the effects of the acetylcholine receptor antibodies. For chronic treatment of myasthenia gravis, long-acting cholinesterase agents can be used to prevent skeletal muscle weakness. ∎

The Contraction Cycle

In the resting sarcomere, each cross-bridge is bound to a molecule of ADP and a phosphate group (PO_4^{3-}), which are the products released by the breakdown of an ATP molecule. The

cross-bridge stores the energy released by the breakage of the high-energy bond. In effect, the resting cross-bridge is "primed" for a contraction, like a cocked pistol or a set mousetrap.

The contraction process involves the following five interlocking steps, which are shown schematically in Figure 7–5●:

Step 1: The active site is exposed following the binding of calcium ions (Ca^{2+}) to troponin.

Step 2: The myosin cross-bridge forms and attaches to the exposed active site on the thin filaments.

Step 3: The attached myosin head pivots toward the center of the sarcomere, and ADP and a phosphate group are released. This step uses the energy that was stored in the myosin molecule at rest.

Step 4: The cross-bridges detach when the myosin head binds another ATP molecule.

Step 5: The detached myosin head is reactivated as it splits the ATP and captures the released energy. The entire cycle can now be repeated, beginning with Step 2.

This cycle is broken when calcium ion concentrations return to normal resting levels, primarily through active transport into the sarcoplasmic reticulum. If a single action potential sweeps across the sarcolemma, calcium ion removal occurs very rapidly, and the contraction will be very brief. A sustained contraction will occur only if action potentials occur one after another, and calcium loss from the terminal cisternae continues.

Table 7–1 provides a summary of the contraction process, from ACh release to the end of the contraction.

Key Note

Skeletal muscle fibers shorten as thin filaments interact with thick filaments and sliding occurs. The trigger for contraction is the appearance of free calcium ions in the sarcoplasm; the calcium ions are released by the sarcoplasmic reticulum when the muscle fiber is stimulated by the associated motor neuron. Contraction is an active process; relaxation and the return to resting length is entirely passive.

→ CONCEPT CHECK QUESTIONS

1. How would a drug that interferes with cross-bridge formation affect muscle contraction?
2. What would you expect to happen to a resting skeletal muscle if the sarcolemma suddenly became very permeable to calcium ions?
3. What would happen to a muscle if the motor end plate did not produce acetylcholinesterase?

Answers begin on p. 792.

Clinical Note
NEUROMUSCULAR BLOCKADE

Certain emergency situations require chemically paralyzing a patient so that an endotracheal tube can be placed to assist or manage breathing. This procedure, referred to as *rapid sequence intubation (RSI),* involves the use of medications that act on the neuromuscular junction, the connection between the peripheral nerves and the skeletal muscle. Nerve impulses travel down the nerve and release a chemical neurotransmitter that stimulates (depolarizes) the associated skeletal muscle fibers, which results in contraction. Acetylcholine is the principle neurotransmitter in the neuromuscular junction. Blocking its action results in relaxation of skeletal (voluntary) muscles.

There are two ways to block the neuromuscular junction. One is by the administration of depolarizing agents that substitute for acetylcholine. Unlike acetylcholine, these depolarizing agents act over a prolonged time, which results in continued muscle depolarization and muscle paralysis. Because they have a stimulating effect, they often produce fasiculations (generalized, involuntary muscle twitching), especially in children, immediately after administration. The prototypical depolarizing neuromuscular blocker is succinylcholine (Anectine). It is the most commonly used neuromuscular blocker for rapid sequence intubation.

The other way to block the neuromuscular junction is to administer a drug that blocks the reuptake of acetylcholine into the nerve terminal. This produces an excess of acetylcholine in the neuromuscular junction, and inhibitis the stimulation of the muscle. Because these drugs do not depolarize the affected muscle fibers, they are referred to as *nondepolarizing blockers.* They do not cause muscle fasiculations. Drugs in this class are similar in action to curare and include vecuronium, atracurium, and pancuronium.

Chemically paralyzing a patient causes complete voluntary muscle relaxation, including the muscles of respiration. This allows caregivers to take control of the airway and provide unimpeded mechanical ventilation. Esophageal and stomach muscles, and therefore sphincter tone, also relax, which increases the risk of vomiting and aspiration. Because neuromuscular blocking agents do not affect mental status or pain sensation, patients should be sedated or put to sleep before administration of neuromuscular blockers. If pain is present or expected, an analgesic should be administered. ■

Clinical Note
INTERFERENCE WITH NEURAL CONTROL MECHANISMS

Any condition that interferes with the generation of an action potential in the sarcolemma will cause muscular paralysis. Two examples are worth noting.

Botulism results from the consumption of foods (often canned or smoked) contaminated with a toxin. The toxin, produced by the bacteria *Clostridium perfringens*, prevents the release of ACh at the synaptic terminals, which leads to a potentially fatal muscular paralysis.

The progressive muscular paralysis seen in *myasthenia gravis* results from the loss of ACh receptors at the motor end plate. The primary cause is a misguided attack on the ACh receptors by the immune system. Genetic factors play a role in predisposing individuals to develop this condition. ■

TABLE 7–1 *A Summary of the Steps Involved in Skeletal Muscle Contraction*

STEPS THAT START A CONTRACTION

1. At the neuromuscular junction, ACh released by the synaptic terminal binds to receptors on the sarcolemma.

2. The resulting change in the membrane potential of the muscle fiber leads to the production of an action potential that spreads across the entire surface of the muscle fiber and along the T tubules.

3. The sarcoplasmic reticulum (SR) releases stored calcium ions, which increases the calcium concentration of the sarcoplasm in and around the sarcomeres.

4. Calcium ions bind to troponin, which results in the movement of tropomyosin and the exposure of active sites on the thin (actin) filaments. Cross-bridges form when myosin heads bind to active sites.

5. The contraction begins as repeated cycles of cross-bridge binding, pivoting, and detachment occur, powered by the breakdown of ATP. These events produce filament sliding, and the muscle fiber shortens.

STEPS THAT END A CONTRACTION

6. Action potential generation ceases as ACh is broken down by acetylcholinesterase (AChE).

7. The SR reabsorbs calcium ions, and the concentration of calcium ions in the sarcoplasm declines.

8. When calcium ion concentrations approach normal resting levels, the troponin and tropomyosin molecules return to their normal positions. This change re-covers the active sites and prevents further cross-bridge interaction.

9. Without cross-bridge interactions, further sliding cannot take place and the contraction will end.

10. Muscle relaxation occurs, and the muscle returns passively to its resting length.

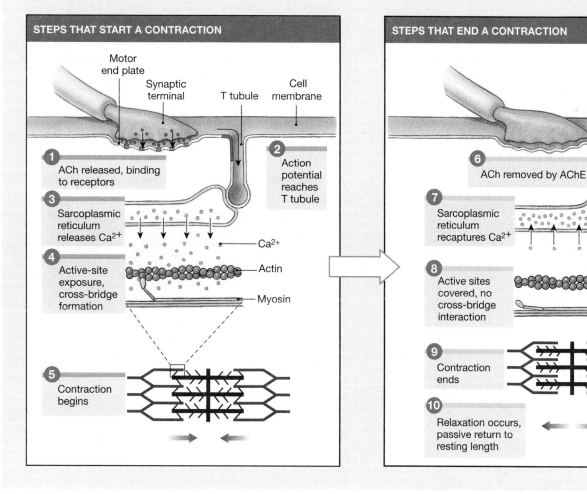

STEPS THAT START A CONTRACTION

Motor end plate
Synaptic terminal
T tubule
Cell membrane

1 ACh released, binding to receptors

2 Action potential reaches T tubule

3 Sarcoplasmic reticulum releases Ca²⁺

Ca²⁺

4 Active-site exposure, cross-bridge formation

Actin

Myosin

5 Contraction begins

STEPS THAT END A CONTRACTION

6 ACh removed by AChE

7 Sarcoplasmic reticulum recaptures Ca²⁺

8 Active sites covered, no cross-bridge interaction

9 Contraction ends

10 Relaxation occurs, passive return to resting length

■ Muscle Mechanics

Now that we are familiar with muscle contraction of individual muscle fibers, we can examine the performance of skeletal muscles. In this section we will consider the coordinated contractions of an entire population of muscle fibers.

The individual muscle cells in muscle tissue are surrounded and tied together by connective tissue. When muscle cells contract, they pull on collagen fibers, producing an active force called **tension.** Tension applied to an object tends to pull the object toward the source of the tension. However, before movement can occur, the applied tension must overcome the object's **resistance,** which is a passive force that opposes movement. The amount of resistance can depend on an object's weight and shape, friction, and other factors. In contrast, **compression**—a push applied to an object—tends to force the object away from the source of compression. Muscle cells can only contract (that is, shorten and generate tension); they cannot actively lengthen and generate compression.

The amount of tension produced by an individual muscle fiber depends solely on the number of pivoting cross-bridges it contains. There is no mechanism that regulates the amount of tension produced in that contraction by changing the number of contracting sarcomeres. The muscle fiber is either "on" (producing tension) or "off" (relaxed). Tension production does vary, however, depending on (1) the fiber's resting length at the time of stimulation, which determines the degree of overlap between thick and thin filaments, and (2) the frequency of stimulation, which determines the internal concentration of calcium ions and thus the amount bound to troponin molecules.

An entire skeletal muscle contracts when its component muscle fibers are stimulated. The amount of tension produced in the skeletal muscle *as a whole* is determined by (1) the frequency of neural stimulation and (2) the number of muscle fibers activated.

The Frequency of Muscle Fiber Stimulation

A **twitch** is a single stimulus-contraction-relaxation sequence in a muscle fiber. Its duration can be as brief as 7.5 msec, as in an eye muscle fiber, or up to 100 msec in fibers of the *soleus,* a small calf muscle. A *myogram* is a graph of tension development in a muscle during a twitch. Figure 7–6● is a myogram of the phases of a 40-msec twitch in the *gastrocnemius muscle,* a prominent calf muscle:

■ The **latent period** begins at stimulation and typically lasts about 2 msec. Over this period the action potential sweeps across the sarcolemma, and calcium ions are released by

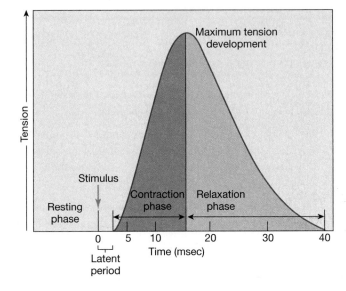

● **FIGURE 7–6 The Twitch and Development of Tension.** This myogram plots the time course of a single twitch in the gastrocnemius muscle.

the sarcoplasmic reticulum. No tension is produced by the muscle fiber because contraction has yet to begin.

■ In the **contraction phase,** tension rises to a peak. Throughout this period the cross-bridges interact with the active sites on the actin filaments. Maximum tension is reached roughly 15 msec after stimulation.

■ During the **relaxation phase,** muscle tension falls to resting levels as calcium levels drop, active sites are covered, and the number of cross-bridges declines. This phase lasts about 25 msec.

A single stimulation produces a single twitch, but twitches in a skeletal muscle do not accomplish anything useful. All normal activities involve sustained muscle contractions. Such contractions result from repeated stimulations.

Summation and Incomplete Tetanus

If a second stimulus arrives before the relaxation phase has ended, a second, more powerful contraction occurs. The addition of one twitch to another in this way is called **summation** (Figure 7–7a●). If the muscle is stimulated repeatedly such that it is never allowed to relax completely, the tension rises and peaks (Figure 7–7b●). A muscle that produces almost peak tension during rapid cycles of contraction and relaxation is said to be in **incomplete tetanus** (*tetanos,* convulsive tension). Virtually all normal muscular contractions involve incomplete tetanus of the participating muscle fibers.

Complete Tetanus

Complete tetanus occurs when the rate of stimulation is increased until the relaxation phase is completely eliminated,

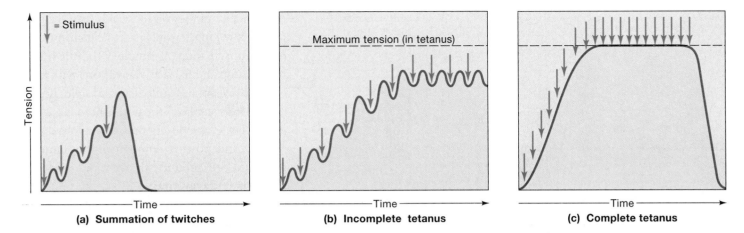

(a) Summation of twitches **(b) Incomplete tetanus** **(c) Complete tetanus**

● **FIGURE 7–7** **The Effects of Repeated Stimulations.** (**a**) During summation, tension rises when successive stimuli arrive before the relaxation phase has been completed. (**b**) Incomplete tetanus occurs if the rate of stimulation increases further. Tension production rises to a peak, and the periods of relaxation are very brief. (**c**) In complete tetanus, the frequency of stimulation is so high that the relaxation phase has been completely eliminated, and tension plateaus at the maximal level.

Clinical Note
THE DISEASE CALLED TETANUS

Children are often told to be careful around rusty nails. Parents should worry most not about the rust or the nail but about infection with a very common bacterium, *Clostridium tetani*. This bacterium can cause the disease called *tetanus*. Although they share a name, the disease tetanus has no relation to the normal muscle response to neural stimulation. The *Clostridium* bacteria, although found virtually everywhere, can thrive only in tissues that contain abnormally low amounts of oxygen. For this reason, a deep puncture wound, such as that from a nail, carries a much greater risk than a shallow, open cut that bleeds freely.

When active in body tissues, these bacteria release a powerful toxin that affects the central nervous system. Motor neurons, which control skeletal muscles throughout the body, are particularly sensitive to it. The toxin suppresses the mechanism that regulates motor neuron activity. The result is a sustained, powerful contraction of skeletal muscles throughout the body. The incubation period (the time between exposure and the onset of symptoms) is usually less than two weeks. The most common early complaints are headache, muscle stiffness, and difficulty in swallowing. Because it soon becomes difficult to open the mouth, this disease is also called *lockjaw*. Widespread muscle spasms usually develop within two to three days of the initial symptoms and continue for a week before subsiding. After two to four weeks, patients who survive recover with no aftereffects.

Although severe tetanus has a 40–60 percent mortality rate, immunization is effective in preventing the disease. Of the approximately 500,000 cases of tetanus worldwide each year, only about 100 occur in the United States, thanks to an effective immunization program. ("Tetanus shots," with booster shots every 10 years, are recommended.) Severe symptoms in unimmunized patients can be prevented by early administration of an antitoxin, usually *human tetanus immune globulin*. However, this treatment does not reduce symptoms that have already appeared. ■

which produces maximum tension (Figure 7–7c●). In complete tetanus, the action potentials are arriving so fast that the sarcoplasmic reticulum does not have time to reclaim calcium ions. The high calcium ion concentration in the cytoplasm prolongs the state of contraction, and makes it continuous.

The Number of Muscle Fibers Involved

We have a remarkable ability to control the amount of tension exerted by our skeletal muscles so that during a normal movement, our muscles contract smoothly, not jerkily. Such control is accomplished by controlling the number of stimulated muscle fibers in the skeletal muscle.

A typical skeletal muscle contains thousands of muscle fibers. Although some motor neurons control a single muscle fiber, most control hundreds or thousands of muscle fibers through multiple synaptic terminals. All the muscle fibers controlled by a single motor neuron constitute a **motor unit.**

The size of a motor unit indicates how fine the control of movement can be. In the muscles of the eye, where precise control is extremely important, a motor neuron may control two or three muscle fibers. We have much less precise control over our leg muscles, where up to 2000 muscle fibers may respond to stimulation by a single motor neuron. The muscle fibers of each motor unit are intermingled with those of other motor units (Figure 7–8●). This intermingling ensures that the direction of pull on a tendon does not change despite variations in the number of activated motor units. When you decide to perform a specific arm movement, specific groups of motor neurons within the spinal cord are stimulated. The

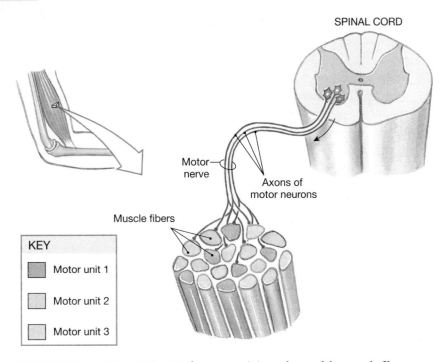

SPINAL CORD

Motor nerve

Axons of motor neurons

Muscle fibers

KEY

Motor unit 1

Motor unit 2

Motor unit 3

● **FIGURE 7–8   Motor Units.** Each motor unit is made up of the muscle fibers controlled by a single motor neuron. Muscle fibers of different motor units are intermingled, so the forces applied to the tendon remain roughly balanced regardless of which muscle groups are stimulated.

contraction begins with the activation of the smallest motor units in the stimulated muscle. Over time, the number of activated motor units gradually increases. The activation of more and more motor units is called **recruitment,** and the result is a smooth, steady increase in muscular tension.

Peak tension production occurs when all the motor units in the muscle contract in complete tetanus. Such contractions do not last long, however, because the muscle fibers soon use up their available energy supplies. During a sustained contraction, motor units are activated on a rotating basis, so that some are resting while others are contracting. As a result, when your muscles contract for sustained periods, they produce slightly less than maximal tension.

🔒 Key Note

All voluntary (intentional) movements involve the sustained contractions of skeletal muscle fibers in incomplete tetanus. The force exerted can be increased by increasing the frequency of action potentials or the number of stimulated motor units (recruitment).

Muscle Tone

Some of the motor units within any particular muscle are always active, even when the entire muscle is not contracting. Their contractions do not produce enough tension to cause

movement, but they do tense and firm the muscle. This resting tension in a skeletal muscle is called **muscle tone.** A muscle with little muscle tone is limp and flaccid, whereas one with moderate muscle tone is quite firm and solid. Resting muscle tone stabilizes the positions of bones and joints. For example, in muscles involved with balance and posture, enough motor units are stimulated to produce the tension needed to maintain body position.

A skeletal muscle that is not stimulated by a motor neuron on a regular basis will **atrophy** (AT-rō-fē; *a,* without + *-trophy,* nourishing): its muscle fibers will become smaller and weaker. Individuals paralyzed by spinal injuries or other damage to the nervous system gradually lose muscle tone and size in the areas affected. Even a temporary reduction in muscle use can lead to muscular atrophy; compare, for example, arm muscles after a cast has been worn to the same muscles in the other arm. Muscle atrophy is initially reversible, but dying muscle fibers are not replaced, and in extreme atrophy the functional losses are permanent. That is why physical therapy is so important for patients who are temporarily unable to move normally.

Isotonic and Isometric Contractions

Muscle contractions may be classified as isotonic or isometric based on the pattern of tension production and overall change in length. In an **isotonic** (*iso-,* equal + *tonos,* tension) **contraction,** tension rises and the skeletal muscle's length changes. Tension in the muscle remains at a constant level until relaxation occurs. Lifting an object off a desk, walking, and running involve isotonic contractions.

In an **isometric** (*metron,* measure) **contraction,** the muscle as a whole does not change length, and the tension produced never exceeds the resistance. Examples of isometric contractions are pushing against a closed door and trying to pick up a car. These examples are rather unusual, but many of the everyday reflexive muscle contractions that keep your body upright when you stand or sit involve isometric contractions of muscles that oppose the force of gravity.

Normal daily activities involve a combination of isotonic and isometric muscular contractions. As you sit reading this text, isometric contractions of postural muscles stabilize your vertebrae and maintain your upright position. When you next turn a page, the movements of your arm, forearm, hand, and fingers are produced by isotonic contractions.

Muscle Elongation

Recall that no active mechanism for muscle fiber elongation exists; contraction is active, but elongation is passive. After a contraction, a muscle fiber usually returns to its original length through a combination of *elastic forces,* the *movements of opposing muscles,* and *gravity.*

Elastic forces are generated when a muscle fiber contracts and tugs on the flexible extracellular fibers of the endomysium, perimysium, epimysium, and tendons. These fibers are also somewhat elastic, and their recoil gradually helps return the muscle fiber to its original resting length.

Much more rapid returns to resting length result from the contraction of opposing muscles. For example, contraction of the *biceps brachii* muscle on the anterior part of the arm flexes the elbow; contraction of the *triceps brachii* muscle on the posterior surface of the arm extends the elbow. When the biceps brachii contracts, the triceps brachii is stretched; when the biceps brachii relaxes, contraction of the triceps brachii extends the elbow and stretches the muscle fibers of the biceps brachii to their original length.

Gravity may also help lengthen a muscle after a contraction. For example, imagine the biceps brachii muscle fully contracted with the elbow pointing at the ground. When the muscle relaxes, gravity will pull the forearm down, and stretch the biceps brachii.

→ **CONCEPT CHECK QUESTIONS**

1. What factors are responsible for the amount of tension a skeletal muscle develops?
2. A motor unit from a skeletal muscle contains 1500 muscle fibers. Would this muscle be involved in fine, delicate movements or in powerful, gross movements? Explain.
3. Can a skeletal muscle contract without shortening? Explain.

Answers begin on p. 792.

■ The Energetics of Muscular Activity

Muscle contraction requires large amounts of energy. For example, an active skeletal muscle fiber may require some 600 trillion molecules of ATP each second, not including the energy needed to pump the calcium ions back into the sarcoplasmic reticulum. This enormous amount of energy is not available before a contraction begins in a resting muscle fiber. Instead, a resting muscle fiber contains only enough energy reserves to sustain a contraction until additional ATP can be generated. Throughout the rest of the contraction, the muscle fiber will generate ATP at roughly the same rate as it is used. This section discusses how muscle fibers meet the demand for ATP.

ATP and CP Reserves

The primary function of ATP is the transfer of energy from one location to another, not the long-term storage of energy. At rest, a skeletal muscle fiber produces more ATP than it needs. Under these conditions, ATP transfers energy to *creatine* (KRĒ-uh-tēn), which is a small molecule assembled from fragments of amino acids in muscle cells. As Figure 7–9a● shows, the energy transfer creates another high-energy compound, **creatine phosphate (CP):**

$$\text{ATP} + \text{creatine} \rightarrow \text{ADP} + \text{creatine phosphate}$$

During a contraction, each cross-bridge breaks down ATP, which produces ADP and a phosphate group. The energy stored in creatine phosphate is then used to "recharge" the ADP back to ATP through the reverse reaction:

$$\text{ATP} + \text{creatine phosphate} \rightarrow \text{ADP} + \text{creatine}$$

The enzyme that regulates this reaction is **creatine phosphokinase (CPK or CK).** When muscle cells are damaged, CPK leaks into the bloodstream. Thus, a high blood level of CPK usually indicates serious muscle (cardiac or skeletal muscle) damage.

A resting skeletal muscle fiber contains about six times as much creatine phosphate as ATP. But when a muscle fiber is in a sustained contraction, these energy reserves will be exhausted in about 15 seconds. The muscle fiber must then rely on other mechanisms to convert ADP to ATP.

ATP Generation

As you may recall from Chapter 3, cells in the body generate ATP through *aerobic* (requires oxygen) *metabolism* in mitochondria and through *glycolysis* in the cytoplasm. ∞ p. 80 Glycolysis is an **anaerobic** (does not require oxygen) process.

Aerobic Metabolism

Aerobic metabolism normally provides 95 percent of the ATP needed by a resting cell. In this process, mitochondria absorb oxygen, ADP, phosphate ions, and small organic substrate molecules from the surrounding cytoplasm. The organic substrates are carbon chains produced by the breakdown of carbohydrates, lipids, or proteins. The substrates enter the *TCA (tricarboxylic acid) cycle* (also known as the *citric acid cycle* or the

Krebs cycle) and are then completely disassembled by a series of chemical reactions. In brief, the carbon atoms and oxygen atoms are released as carbon dioxide (CO_2). The hydrogen atoms are shuttled to *respiratory enzymes* in the inner mitochondrial membrane where their electrons are removed. The protons and electrons thus produced ultimately recombine with oxygen to form water (H_2O). Along the way, large amounts of energy are released and used to make ATP. The aerobic metabolism of a common carbohydrate substrate, pyruvic acid, is quite efficient; for each pyruvic acid molecule broken down in the TCA cycle, the cell gains 17 ATP molecules.

Resting skeletal muscle fibers rely primarily on the aerobic metabolism of fatty acids to make ATP (see Figure 7–9a●). These fatty acids are absorbed from the circulation. When the muscle starts contracting, the mitochondria begin breaking down molecules of pyruvic acid instead of fatty acids. The pyruvic acid is provided through the process of glycolysis (discussed shortly).

The maximum rate of ATP generation within mitochondria is limited by the availability of oxygen. A sufficient supply of oxygen becomes a problem as the energy demands of the muscle fiber increase. Although oxygen consumption and energy production by mitochondria can increase to 40 times resting levels, the energy demands of the muscle fiber may increase by 120 times. Thus, at peak levels of exertion, mitochondrial activity provides only around one-third of the required ATP.

Glycolysis

Glycolysis is the breakdown of glucose to pyruvic acid in the cytoplasm of the cell. The ATP yield of glycolysis is much lower than that of aerobic metabolism; however, because it can proceed in the absence of oxygen, *glycolysis can continue to provide ATP when the availability of oxygen limits the rate of mitochondrial ATP production.*

The glucose broken down under these conditions is obtained from glycogen reserves in the sarcoplasm. Glycogen is a polysaccharide chain of glucose molecules. ∞ p. 42 Typical skeletal muscle fibers contain large glycogen reserves in the form of insoluble granules. When the muscle fiber begins to run short of ATP and CP, enzymes break the glycogen molecules apart, which releases glucose that can be used to generate more ATP.

Energy Use and the Level of Muscle Activity

In a resting skeletal muscle cell the demand for ATP is low. More than enough oxygen is available for the mitochondria to meet that demand and produce a surplus of ATP. The extra ATP is used to build up reserves of CP and glycogen (see Figure 7–9a). At moderate levels of activity, the demand for ATP

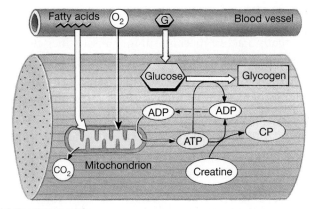

(a) Resting muscle

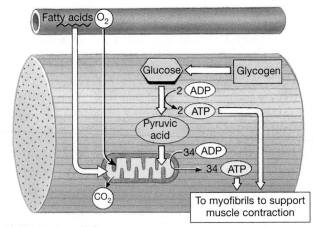

(b) Moderate activity

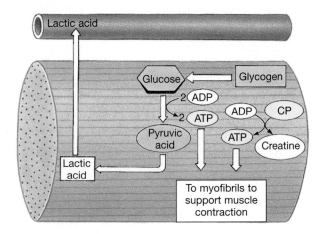

(c) Peak activity

● **FIGURE 7–9 Muscle Metabolism.** (a) A resting muscle generates ATP from the breakdown of fatty acids through aerobic metabolism. The ATP is used to build reserves of ATP, CP, and glycogen. (b) A muscle at a modest level of activity meets the ATP demands through the aerobic metabolism of fatty acids and glucose. (c) A muscle at peak activity meets the ATP demands primarily through glycolysis, which is an anaerobic process that leads to the production of lactic acid. Mitochondrial activity (not shown) now provides only about one-third of the ATP consumed.

increases (see Figure 7–9b●). As the rate of mitochondrial ATP production rises, so does the rate of oxygen consumption. So long as sufficient oxygen is available, the demand for ATP can be met by the mitochondria, and the amount of ATP provided by glycolysis remains relatively minor.

At periods of peak activity, oxygen cannot diffuse into the muscle fiber fast enough to enable the mitochondria to produce the required ATP. Mitochondrial activity can provide only about one-third the ATP needed, and glycolysis becomes the primary source of ATP (see Figure 7–9c●). The anaerobic process of glycolysis enables the cell to continue generating ATP when mitochondrial activity alone cannot meet the demand. However, this pathway has its drawbacks. For example, when glycolysis produces pyruvic acid faster than it can be used by the mitochondria, pyruvic acid levels rise in the sarcoplasm. Under these conditions, the pyruvic acid is converted to **lactic acid,** which is a related three-carbon molecule. The conversion of pyruvic acid to lactic acid poses a problem because lactic acid is an organic acid that in body fluids dissociates into a hydrogen ion and a *lactate ion*. The accumulation of hydrogen ions can lower the pH and alter the normal functioning of key enzymes. The muscle fiber then cannot continue to contract.

Glycolysis is also an inefficient way to generate ATP. Under anaerobic conditions, each glucose generates two pyruvic acids, which are converted to lactic acid. In return, the cell gains two ATP molecules. Had those two pyruvic acid molecules been catabolized aerobically in a mitochondrion, the cell would have gained an additional 34 ATP molecules.

Muscle Fatigue

A skeletal muscle fiber is said to be fatigued when it can no longer contract despite continued neural stimulation. **Muscle fatigue** is caused by the exhaustion of energy reserves or the buildup of lactic acid.

If the muscle contractions use ATP at or below the maximum rate of mitochondrial ATP generation, the muscle fiber can function aerobically. Under these conditions, fatigue will not occur until glycogen and other reserves such as lipids and amino acids are depleted. This type of fatigue affects the muscles of endurance athletes, such as marathon runners, after hours of exertion.

When a muscle produces a sudden, intense burst of activity, the ATP is provided by glycolysis. After a relatively short time (seconds to minutes), the rising lactic acid levels lower the tissue pH, and the muscle can no longer function normally. Athletes who run sprints, such as the 100-yard dash, suffer from this type of muscle fatigue.

The Recovery Period

When a muscle fiber contracts, conditions in the sarcoplasm are changed: energy reserves are consumed, heat is released, and lactic acid may be produced. During the **recovery period,** conditions within the muscle are returned to normal pre-exertion levels. The muscle's metabolic activity focuses on the removal of lactic acid and the replacement of intracellular energy reserves, and the body as a whole loses the heat generated during intense muscular contraction.

Lactic Acid Recycling

The reaction that converts pyruvic acid to lactic acid is freely reversible. During the recovery period, when oxygen is available, lactic acid can be recycled by converting it back to pyruvic acid. This pyruvic acid can then be used as a building block to synthesize glucose or by mitochondria to generate ATP. The ATP is used to convert creatine to creatine phosphate and to store the newly synthesized glucose as glycogen.

During the recovery period, the body's oxygen demand remains elevated above normal resting levels. The additional oxygen required during the recovery period to restore the normal pre-exertion levels is called the *oxygen debt*. Most of the extra oxygen is consumed by liver cells, because they produce ATP for the conversion of lactic acid back to glucose, and by muscle cells, as they restore their reserves of ATP, creatine phosphate, and glycogen. Other cells, including sweat gland cells, also increase their rate of oxygen use and ATP generation. While the oxygen debt is being repaid, breathing rate and depth are increased. That is why you continue to breathe heavily for a time after you stop exercising.

Key Note

Skeletal muscles at rest metabolize fatty acids and store glycogen. During light activity, muscles can generate ATP through the aerobic breakdown of carbohydrates, lipids, or amino acids. At peak levels of activity, most of the energy is provided by anaerobic reactions that generate lactic acid as a byproduct.

Heat Loss

Muscular activity generates substantial amounts of heat that warms the sarcoplasm, interstitial fluid, and circulating blood. Because muscle makes up a large portion of total body mass, muscle contractions play an important role in the maintenance of normal body temperature. Shivering, for example, can help keep you warm in a cold environment. But when skeletal muscles are contracting at peak levels, body temperature soon begins to climb. In response, blood flow to the skin increases, which promotes heat loss through mechanisms described in Chapters 1 and 5. ∞ p. 14, 136

■ Muscle Performance

Muscle performance can be considered in terms of **force,** which is the maximum amount of tension produced by a particular muscle or muscle group; and **endurance,** which is the amount of time over which the individual can perform a particular activity. Two major factors determine the performance capabilities of a particular skeletal muscle: the types of muscle fibers within the muscle, and physical conditioning or training.

Types of Skeletal Muscle Fibers

The human body contains two contrasting types of skeletal muscle fibers: fast (or fast-twitch) fibers and slow (or slow-twitch) fibers.

Fast Fibers

Most of the skeletal muscle fibers in the body are called **fast fibers** because they can contract in 0.01 second or less following stimulation. Fast fibers are large in diameter and contain densely packed myofibrils, large glycogen reserves, and relatively few mitochondria. The tension produced by a muscle fiber is directly proportional to the number of myofibrils, so fast-fiber muscles produce powerful contractions. However, because these contractions use ATP very quickly, their activity is primarily supported by glycolysis, and fast fibers fatigue rapidly.

Slow Fibers

Slow fibers are only about half the diameter of fast fibers, and they take three times as long to contract after stimulation; however, they can continue contracting for extended periods, long after a fast muscle would have become fatigued. Three specializations related to the availability and use of oxygen make this possible:

1. *Oxygen supply.* Slow muscle tissue contains a more extensive network of capillaries than does typical fast muscle tissue, so oxygen supply is dramatically increased.
2. *Oxygen storage.* Slow muscle fibers contain the red pigment **myoglobin** (MĪ-ō-glō-bin), which is a globular protein structurally related to hemoglobin, the oxy-

gen-carrying pigment found in blood. ∞ p. 47 Because myoglobin also binds oxygen molecules, resting slow muscle fibers contain oxygen reserves that can be mobilized during a contraction.
3. *Oxygen use.* Slow muscle fibers contain a relatively larger number of mitochondria than do fast muscle fibers.

The Distribution of Muscle Fibers and Muscle Performance

The percentages of fast and slow muscle fibers can vary considerably among skeletal muscles. Muscles dominated by fast fibers appear pale, and they are often called **white muscles.** Chicken breasts contain "white meat" because chickens use their wings for only brief intervals (as when fleeing from a predator), and the power for flight comes from the anaerobic process of glycolysis in the fast fibers of their breast muscles. The extensive blood vessels and myoglobin in slow muscle fibers give them a reddish color, and muscles dominated by slow fibers are, therefore, known as **red muscles.** Chickens walk around all day, and these movements are powered by aerobic metabolism in the slow muscle fibers of the "dark meat" of their legs.

Most human muscles contain a mixture of fiber types and, therefore, appear pink. However, there are no slow fibers in muscles of the eye and hand, where swift but brief contractions are required. Many back and calf muscles are dominated by slow fibers; these muscles contract almost continuously to maintain an upright posture. The percentage of fast versus slow fibers in each muscle is genetically determined, but the fatigue resistance of fast muscle fibers can be increased through athletic training.

Physical Conditioning

Physical conditioning and training schedules enable athletes to improve both power and endurance. In practice, the training schedule varies depending on whether the activity is primarily supported by aerobic or anaerobic energy production.

Anaerobic endurance is the length of time muscle contractions can be supported by glycolysis and existing energy reserves of ATP and CP. Examples of activities that require anaerobic endurance are a 50-yard dash or swim, a pole vault, and a weight-lifting competition. Such activities involve contractions of fast muscle fibers. Athletes training to develop anaerobic endurance perform frequent, brief, intense workouts. The net effect is an enlargement, or **hypertrophy** (hī-PER-trō-fē), of the stimulated muscles, as seen in champion weight lifters or bodybuilders. The number of muscle fibers does not change, but the muscle as a whole gets larger because each muscle fiber increases in diameter.

Aerobic endurance is the length of time a muscle can continue to contract while being supported by mitochondrial activities. Aerobic endurance is determined by the availability of substrates for aerobic metabolism from the breakdown of carbohydrates, lipids, or amino acids. Training to improve aerobic endurance usually involves sustained low levels of muscular activity. Examples are jogging, distance swimming, and other exercises that do not require peak tension production. Because glucose is a preferred energy source, endurance athletes often load up on carbohydrates ("carboload") on the day before an event. They may also consume glucose-rich "sports drinks" during an event.

Key Note

What you don't use, you lose. Muscle tone is an indication of the background level of activity in the motor units in skeletal muscles. When inactive for days or weeks, muscles become flaccid, and the muscle fibers break down their contractile proteins and grow smaller and weaker. If inactive for long periods, muscle fibers may be replaced by fibrous tissue.

→ CONCEPT CHECK QUESTIONS

1. Why would a sprinter experience muscle fatigue before a marathon runner would?
2. Which activity would be more likely to create an oxygen debt in an individual who regularly exercises: swimming laps or lifting weights?
3. Which type of muscle fibers would you expect to predominate in the large leg muscles of someone who excels at endurance activities such as cycling or long-distance running?

Answers begin on p. 792.

■ Cardiac and Smooth Muscle Tissues

Cardiac muscle tissue and smooth muscle tissue were first introduced in Chapter 4. Here we consider in greater detail the structural and functional properties of these types of muscle tissue.

Cardiac Muscle Tissue

Cardiac muscle cells are relatively small and usually have a single, centrally placed nucleus. Cardiac muscle tissue is found only in the heart.

Differences Between Cardiac Muscle and Skeletal Muscle

Like skeletal muscle fibers, cardiac muscle cells contain an orderly arrangement of myofibrils and are striated, but significant differences exist in their structure and function. The most obvious structural difference is that cardiac muscle cells are branched, and each cardiac cell contacts several others at specialized sites called **intercalated** (in-TER-ka-lā-ted) **discs** (Figure 7–10a●). These cellular connections contain gap junctions that allow the movement of ions and small molecules and the rapid passage of action potentials from cell to cell, which results in their simultaneous contraction. Because

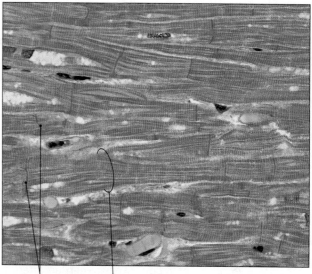

Intercalated   Cardiac   **(a)**
discs   muscle cell

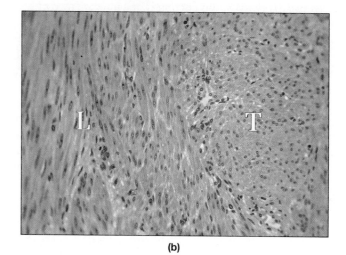

(b)

● **FIGURE 7–10   Cardiac and Smooth Muscle Tissues.** (**a**) Notice the striations and the intercalated discs in this LM of cardiac muscle tissue. (**b**) Many visceral organs contain layers or sheets of smooth muscle fibers. This LM view shows smooth muscle cells in longitudinal (L) and transverse (T) sections.

the myofibrils are also attached to the intercalated discs, the cells "pull together" quite efficiently.

Cardiac muscle and skeletal muscle also have the following important functional differences:

- Cardiac muscle tissue contracts without neural stimulation, a property called *automaticity*. The timing of contractions is normally determined by specialized cardiac muscle cells called **pacemaker cells.**
- Cardiac muscle cell contractions last roughly 10 times longer than those of skeletal muscle fibers.
- The properties of cardiac muscle cell membranes differ from those of skeletal muscle fibers. As a result, cardiac muscle tissue cannot undergo tetanus (sustained contraction). This property is important because a heart in tetany could not pump blood.
- An action potential not only triggers the release of calcium from the sarcoplasmic reticulum but also increases the permeability of the cell membrane to extracellular calcium ions.
- Cardiac muscle cells rely on aerobic metabolism for the energy needed to continue contracting. The sarcoplasm contains large numbers of mitochondria and abundant reserves of myoglobin (to store oxygen).

Clinical Note
LABORATORY TESTING IN HEART ATTACK

Diagnosing a heart attack is often difficult. A heart attack (*acute myocardial infarction*) occurs when an artery that supplies blood to the heart muscle is blocked. Initially, the area of heart muscle supplied by the artery will become injured (*myocardial ischemia*). If this continues, the affected area of heart muscle will die (*myocardial infarction*). Because of this, it is important for emergency personnel to recognize heart attacks, now referred to as *acute coronary syndrome*, so that treatment can be provided and blood flow restored before a significant mass of the heart muscle is permanently damaged.

Several diagnostic tools are routinely used to determine whether a heart attack has occurred. These include the electrocardiogram (ECG), X-rays, laboratory tests, and others. Following injury to the heart muscle, its chemical components are released into the circulation. Laboratory assays of these chemicals can aid in the diagnosis of heart attack. Commonly assayed chemicals include:

- *Creatine kinase (CK).* Creatine kinase (CK) is an enzyme found in muscle and brain tissues. There are subtle differences in CK structure (*isoenzymes),* depending on the source. CK-MM comes from skeletal muscle, CK-BB comes from brain tissue, and CH-MB comes from heart tissue. The amount of CK-MB in the blood does not normally exceed 2–4 percent of total CK values. Any elevation in the percentage of CK-MB indicates myocardial injury. The CK-MB level begins to increase within four to six hours of

injury, peaks at 18–24 hours, and remains elevated for three to four days.
- *Lactic dehydrogenase (LD or LDH).* LDH is found in heart muscle, skeletal muscle, liver, erythrocytes, kidney, and some types of tumors. It is increased in over 90 percent of myocardial infarctions. However, it can be increased in diseases of any of the above organs or hemolysis. There are five LDH isoenzymes:
  - LDH1: heart, erythrocytes, renal cortex
  - LDH2: reticuloendothelial system
  - LDH3: lung tissue
  - LDH4: placenta, kidney, pancreas
  - LDH5: skeletal muscle, liver
  Reversal of LDH1/LDH2 ratio is characteristic of an acute myocardial infarction; with an 80–85 percent sensitivity. LDH begins to rise within 24 hours of myocardial infarction, peaks in three days and returns to normal in eight to nine days.
- *Myoglobin.* Myoglobin is found in striated muscle and contains iron. It stores oxygen and gives muscle its red color. Damage to skeletal or cardiac muscle releases myoglobin into circulation. Myoglobin rapid assay kits are available that allow testing in the prehospital setting. Myoglobin rises fast (two hours) after myocardial infarction, peaks at six to eight hours and returns to normal in 20–36 hours.
- *Troponin I, T, and C.* The troponins are the contractile proteins of the myofibril. The cardiac isoforms are very specific for cardiac injury and are not present in serum from healthy people. Troponin I is the form frequently assessed. There is a new form (Troponin L) which may be detected earlier. Troponin rapid assay kits are available that allow testing in the prehospital setting. The troponins rise four to six hours after injury, peak in 12–16 hours, and remain elevated for up to 10 days.
- *B-natriuretic peptide (BNP).* BNP is a peptide found in the ventricles of the heart and increases when ventricular filling pressures are high. It can be used to detect congestive heart failure.

When myocardial ischemia is suspected, patients are usually admitted to the hospital and serial lab tests are performed, usually over a 24-hour period. The chance of detecting myocardial ischemia with a single sampling of CK-MB is only 34 percent. However, repeated sampling over 24 hours increases the accuracy to 90 percent or better. ■

Smooth Muscle Tissue

Smooth muscle cells are similar in size to cardiac muscle cells; they also contain a single, centrally located nucleus within each spindle-shaped cell (Figure 7–10b●). Smooth muscle tissue is found within almost every organ, and forms sheets, bundles, or sheaths around other tissues. In the skeletal, muscular, nervous, and endocrine systems, smooth muscles around blood vessels regulate blood flow through vital organs. In the digestive and urinary systems, rings of smooth muscles, called *sphincters*, regulate movement along internal passageways.

Differences Between Smooth Muscle and Other Muscle Tissues

STRUCTURAL DIFFERENCES. Actin and myosin are present in all three muscle types. However, the internal organization of a smooth muscle cell differs from that of skeletal or cardiac muscle cells in the following ways:

- Smooth muscle tissue lacks myofibrils, sarcomeres, or striations.
- In smooth muscle cells, the thick filaments are scattered throughout the sarcoplasm, and the thin filaments are anchored within the cytoplasm and to the sarcolemma.
- Adjacent smooth muscle cells are bound together at these anchoring sites, which thereby transmits the contractile forces throughout the tissue.

FUNCTIONAL DIFFERENCES. Smooth muscle tissue differs from other muscle types in several major ways:

- Calcium ions trigger contractions through a different mechanism than that found in other muscle types. Additionally, most of the calcium ions that trigger contractions enter the cell from the extracellular fluid.
- Smooth muscle cells are able to contract over a greater range of lengths than skeletal or cardiac muscle because the actin and myosin filaments are not rigidly organized. This property is important because layers of smooth muscle are found in the walls of organs that undergo large changes in volume, such as the urinary bladder and stomach.

- Many smooth muscle cells are not innervated by motor neurons, and the muscle cells contract either automatically (in response to *pacesetter cells*) or in response to environmental or hormonal stimulation. When smooth muscle fibers are innervated by motor neurons, the neurons involved are not under voluntary control.

Table 7–2 summarizes the structural and functional properties of skeletal, cardiac, and smooth muscle tissue.

→ CONCEPT CHECK QUESTIONS

1. How do intercalated discs enhance the functioning of cardiac muscle tissue?
2. Extracellular calcium ions are important for the contraction of what type(s) of muscle tissue?
3. Why can smooth muscle contract over a wider range of resting lengths than skeletal muscle?

Answers begin on p. 792.

■ Anatomy of the Muscular System

The **muscular system** includes all of the skeletal muscles, which can be controlled voluntarily (Figure 7–11●). The general appearance of each of the nearly 700 skeletal muscles provides clues to its primary function. Muscles involved with locomotion

---

TABLE 7-2 *A Comparison of Skeletal, Cardiac, and Smooth Muscle Tissues*

PROPERTY	SKELETAL MUSCLE FIBER	CARDIAC MUSCLE CELL	SMOOTH MUSCLE CELL
Fiber dimensions (diameter × length)	100 μm × up to 60 cm	10–20 μm × 50–100 μm	5–10 μm × 30–200 μm
Nuclei	Multiple, near sarcolemma	Usually single, centrally located	Single, centrally located
Filament organization	In sarcomeres along myofibrils	In sarcomeres along myofibrils	Scattered throughout sarcoplasm
Control mechanism	Neural, at single neuromuscular junction	Automaticity (pacemaker cells)	Automaticity (pacesetter cells), neural or hormonal control
Ca²⁺ source	Release from sarcoplasmic reticulum	Extracellular fluid and release from sarcoplasmic reticulum	Extracellular fluid and release from sarcoplasmic reticulum
Contraction	Rapid onset; tetanus can occur; rapid fatigue	Slower onset; tetanus cannot occur; resistant to fatigue	Slow onset; tetanus can occur; resistant to fatigue
Energy source	Aerobic metabolism at moderate levels of activity; glycolysis (anaerobic) during peak activity	Aerobic metabolism, usually lipid or carbohydrate substrates	Primarily aerobic metabolism

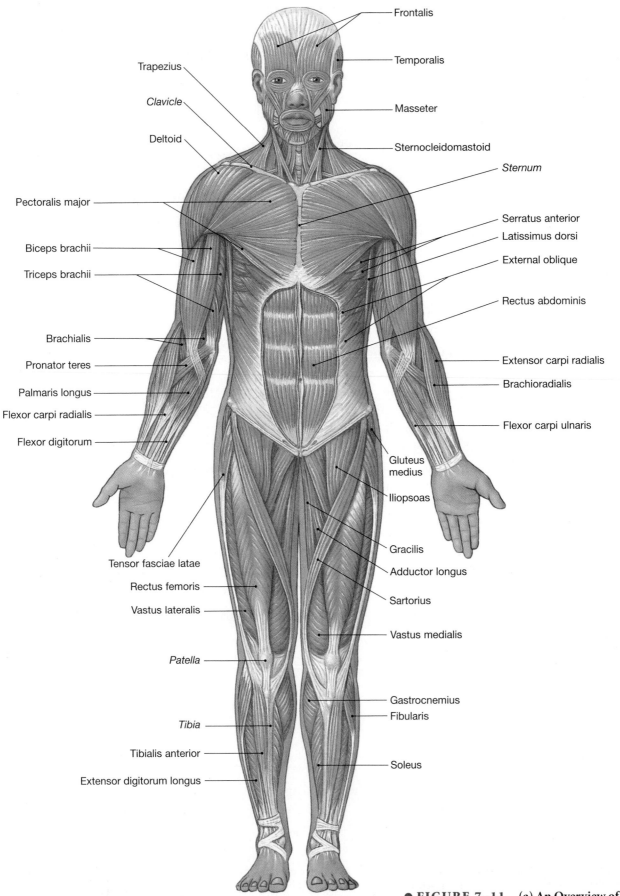

Frontalis

Temporalis

Trapezius

Masseter

Clavicle

Sternocleidomastoid

Deltoid

Sternum

Pectoralis major

Serratus anterior

Latissimus dorsi

Biceps brachii

External oblique

Triceps brachii

Rectus abdominis

Brachialis

Pronator teres

Extensor carpi radialis

Palmaris longus

Brachioradialis

Flexor carpi radialis

Flexor digitorum

Flexor carpi ulnaris

Gluteus
medius

Iliopsoas

Tensor fasciae latae

Rectus femoris

Gracilis

Adductor longus

Vastus lateralis

Sartorius

Patella

Vastus medialis

Tibia

Gastrocnemius

Fibularis

Tibialis anterior

Extensor digitorum longus

Soleus

(a)

● **FIGURE 7–11** **(a) An Overview of the Major
Skeletal Muscles.** Anterior view *(continued)*

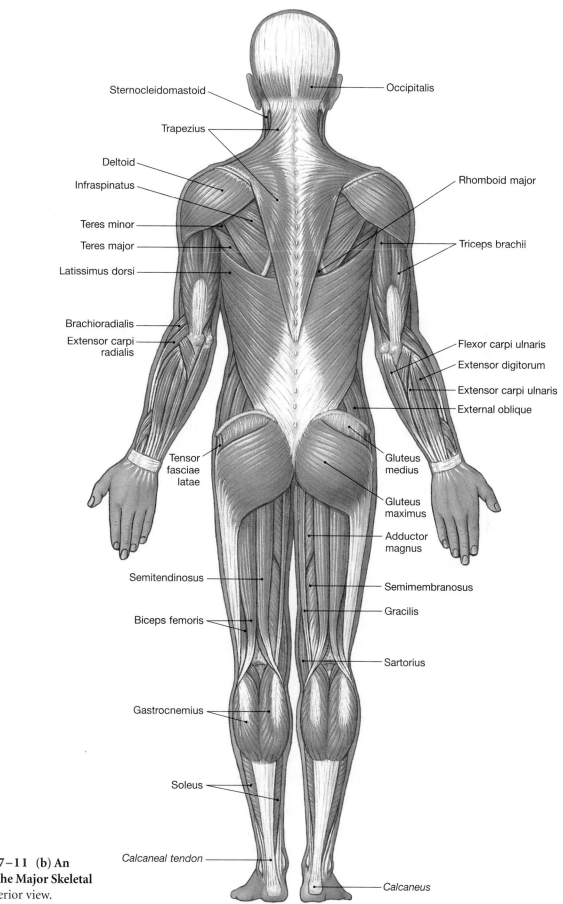

Sternocleidomastoid

Trapezius

Deltoid

Infraspinatus

Teres minor

Teres major

Latissimus dorsi

Brachioradialis

Extensor carpi
radialis

Occipitalis

Rhomboid major

Triceps brachii

Flexor carpi ulnaris

Extensor digitorum

Extensor carpi ulnaris

External oblique

Tensor
fasciae
latae

Gluteus
medius

Gluteus
maximus

Adductor
magnus

Semitendinosus

Semimembranosus

Biceps femoris

Gracilis

Sartorius

Gastrocnemius

Soleus

Calcaneal tendon

Calcaneus

● **FIGURE 7–11 (b) An
Overview of the Major Skeletal
Muscles.** Posterior view.

and posture work across joints, and produce movement of the skeleton. Those that support soft tissue form slings or sheets between relatively stable bony elements, whereas those that guard an entrance or exit completely encircle the opening.

Origins, Insertions, and Actions

Each muscle begins at an **origin,** ends at an **insertion,** and contracts to produce a specific **action.** In general, a muscle's origin remains stationary while the insertion moves. For example, the *gastrocnemius* muscle (in the calf) has its origin on the distal portion of the femur and inserts on the calcaneus. Its contraction pulls the insertion closer to the origin, which results in the action called *plantar flexion.* The determinations of origin and insertion are usually based on movement from the anatomical position.

Almost all skeletal muscles either originate or insert on the skeleton. When they contract, they may produce *flexion, extension, adduction, abduction, protraction, retraction, elevation, depression, rotation, circumduction, pronation, supination, inversion,* or *eversion.* (You may wish to review Figures 6-32 to 6–34, pp. 177–178.)

Actions of muscles may be described in two ways. The first describes muscle actions in terms of the bone affected. Accordingly, the *biceps brachii* muscle is said to perform "flexion of the forearm." The second method, which is increasingly used by specialists of human motion (*kinesiologists*), describes muscle action in terms of the joint involved. Thus, the biceps brachii muscle is said to perform "flexion at (or of) the elbow." We will primarily use the second method.

Muscles can also be described by their **primary actions:**

- A **prime mover,** or **agonist** (AG-o-nist), is a muscle whose contraction is chiefly responsible for producing a particular movement. The *biceps brachii* muscle is a prime mover that flexes the elbow.
- **Antagonists** (an-TAG-o-nists) are muscles whose actions oppose the movement produced by another muscle. An antagonist may also be a prime mover. For example, the *triceps brachii* muscle is a prime mover that extends the elbow. It is, therefore, an antagonist of the biceps brachii, and the biceps brachii is an antagonist of the triceps brachii. Agonists and antagonists are functional opposites—if one produces flexion, the other's primary action is extension.
- A **synergist** (*syn-*, together + *ergon,* work) is a muscle that helps a prime mover work efficiently. Synergists may either provide additional pull near the insertion or stabilize the point of origin. For example, the *deltoid muscle* acts to lift the arm away from the body (abduction). A smaller muscle,

the *supraspinatus muscle,* assists the deltoid in starting this movement. **Fixators** are synergists that stabilize the origin of a prime mover by preventing movement at another joint.

Names of Skeletal Muscles

The human body has approximately 700 skeletal muscles. You need not learn the name of every one of them, but you should become familiar with the most important ones. Fortunately, the names assigned to muscles provide clues to their identification. Table 7–3, which summarizes muscle terminology, can be a useful reference as you go through the rest of this chapter. (With the exception of the platysma and the diaphragm, the complete name of every muscle includes the word *muscle.* For simplicity, we have not included the word *muscle* in figures and tables.)

Some names, often those with Greek or Latin roots, refer to the orientation of the muscle fibers. For example, *rectus* means "straight," and *rectus muscles* are parallel muscles whose fibers generally run along the long axis of the body, as in the *rectus abdominis muscle.* In a few cases, a muscle is such a prominent feature that the regional name alone can identify it, such as the *temporalis muscle* of the head. Other muscles are named after structural features. For example, a biceps muscle has two tendons of origin (*bi-*, two + *caput,* head), whereas the *triceps* has three. Table 7–3 also lists names that reflect shape, length, or size, or whether a muscle is visible at the body surface (*externus, superficialis*) or lies beneath (*internus, profundus*). Superficial muscles that position or stabilize an organ are called *extrinsic muscles;* those that operate within an organ are called *intrinsic muscles.*

The first part of many names indicates the muscle's origin and the second part its insertion. The *sternohyoid muscle,* for example, originates at the sternum and inserts on the hyoid bone. Other names may also indicate the primary function of the muscle. For example, the *extensor carpi radialis muscle* is found along the radial (lateral) border of the forearm, and its contraction produces extension at the wrist (carpal) joint.

The separation of the skeletal system into axial and appendicular divisions provides a useful guideline for subdividing the muscular system as well:

- The **axial musculature** arises on the axial skeleton. It positions the head and spinal column and also moves the rib cage, and assists in the movements that make breathing possible. It does not play a role in movement or support of the pectoral or pelvic girdles or appendages. This category encompasses roughly 60 percent of the skeletal muscles in the body.
- The **appendicular musculature** stabilizes or moves components of the appendicular skeleton.

TABLE 7-3 *Muscle Terminology*

TERMS THAT INDICATE POSITION, DIRECTION, OR MUSCLE FIBER ORIENTATION	TERMS THAT INDICATE SPECIFIC REGIONS OF THE BODY*	TERMS THAT INDICATE STRUCTURAL CHARACTERISTICS OF THE MUSCLE	TERMS THAT INDICATE ACTIONS
Anterior (front)	Abdominis (abdomen)	**Origin**	**General**
Externus (superficial)	Anconeus (elbow)	Biceps (two heads)	Abductor
Extrinsic (outside)	Auricularis (auricle of ear)	Triceps (three heads)	Adductor
Inferioris (inferior)	Brachialis (brachium)	Quadriceps (four heads)	Depressor
Internus (deep, internal)	Capitis (head)		Extensor
Intrinsic (inside)	Carpi (wrist)	**Shape**	Flexor
Lateralis (lateral)	Cervicis (neck)	Deltoid (triangle)	Levator
Medialis/medius	Cleido/clavius (clavicle)	Orbicularis (circle)	Pronator
(medial, middle)	Coccygeus (coccyx)	Pectinate (comb-like)	Rotator
Obliquus (oblique)	Costalis (ribs)	Piriformis (pear-shaped)	Supinator
Posterior (back)	Cutaneous (skin)	Platys- (flat)	Tensor
Profundus (deep)	Femoris (femur)	Pyramidal (pyramid)	
Rectus (straight, parallel)	Genio- (chin)	Rhomboid	**Specific**
Superficialis (superficial)	Glosso/glossal (tongue)	Serratus (serrated)	Buccinator (trumpeter)
Superioris (superior)	Hallucis (great toe)	Splenius (bandage)	Risorius (laugher)
Transversus (transverse)	Ilio- (ilium)	Teres (long and round)	Sartorius (like a tailor)
	Inguinal (groin)	Trapezius (trapezoid)	
	Lumborum (lumbar region)		
	Nasalis (nose)	**Other Striking Features**	
	Nuchal (back of neck)	Alba (white)	
	Oculo- (eye)	Brevis (short)	
	Oris (mouth)	Gracilis (slender)	
	Palpebrae (eyelid)	Lata (wide)	
	Pollicis (thumb)	Latissimus (widest)	
	Popliteus (behind knee)	Longissimus (longest)	
	Psoas (loin)	Longus (long)	
	Radialis (radius)	Magnus (large)	
	Scapularis (scapula)	Major (larger)	
	Temporalis (temples)	Maximus (largest)	
	Thoracis (thoracic region)	Minimus (smallest)	
	Tibialis (tibia)	Minor (smaller)	
	Ulnaris (ulna)	-tendinosus (tendinous)	
	Uro- (urinary)	Vastus (great)	

*For other regional terms, refer to Figure 1–6, p. 16, which shows anatomical landmarks.

→ **CONCEPT CHECK QUESTIONS**

1. Which type of muscle would you expect to find guarding the opening between the stomach and the small intestine?
2. Which muscle is the antagonist of the biceps brachii?
3. What does the name *flexor carpi radialis* tell you about this muscle?

Answers begin on p. 792.

The Axial Muscles

The axial muscles fall into four logical groups based on location, function, or both:

1. *The muscles of the head and neck.* These muscles include the muscles responsible for facial expression, chewing, and swallowing.

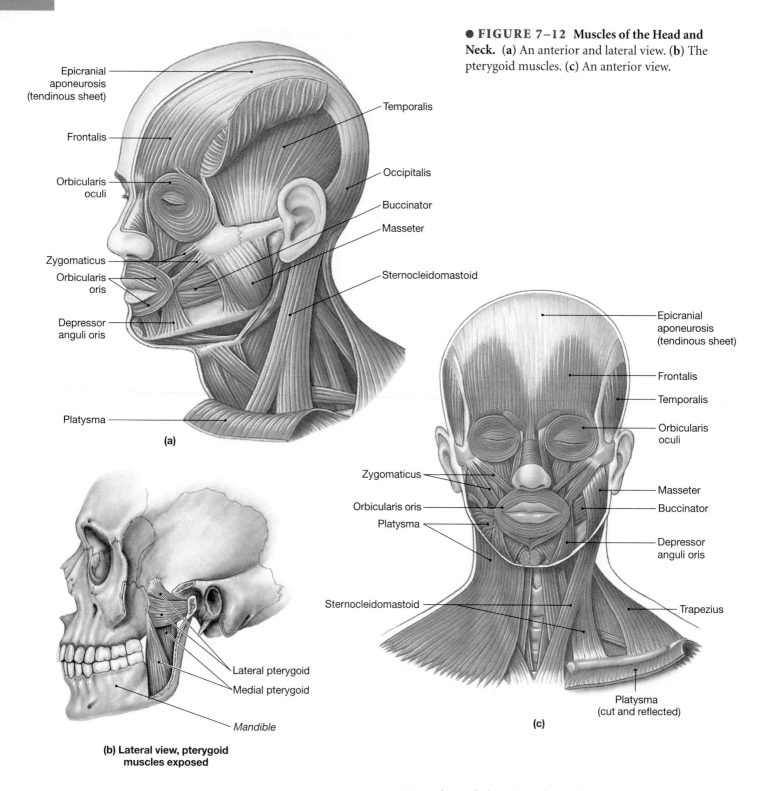

● **FIGURE 7–12** Muscles of the Head and Neck. (**a**) An anterior and lateral view. (**b**) The pterygoid muscles. (**c**) An anterior view.

Epicranial aponeurosis (tendinous sheet)

Frontalis

Orbicularis oculi

Zygomaticus

Orbicularis oris

Depressor anguli oris

Platysma

Temporalis

Occipitalis

Buccinator

Masseter

Sternocleidomastoid

(a)

Epicranial aponeurosis (tendinous sheet)

Frontalis

Temporalis

Orbicularis oculi

Masseter

Buccinator

Depressor anguli oris

Trapezius

Zygomaticus

Orbicularis oris

Platysma

Sternocleidomastoid

Platysma (cut and reflected)

(c)

Lateral pterygoid

Medial pterygoid

Mandible

(b) Lateral view, pterygoid muscles exposed

2. *The muscles of the spine.* This group includes flexors and extensors of the head, neck, and spinal column.

3. *The muscles of the trunk.* The *oblique* and *rectus* muscles form the muscular walls of the thoracic and abdominopelvic cavities.

4. *The muscles of the pelvic floor.* These muscles extend between the sacrum and pelvic girdle and form the muscular *perineum,* which closes the pelvic outlet.

Muscles of the Head and Neck

The muscles of the head and neck are shown in Figures 7–12● and 7–13● and detailed in Table 7–4. The muscles of the face originate on the surface of the skull and insert into the dermis of the skin. When they contract, the skin moves. For example, the **frontalis** muscle of the forehead raises the eyebrows and pulls on the skin of the scalp. The largest group of facial muscles is associated with the mouth. The **orbicularis oris** con-

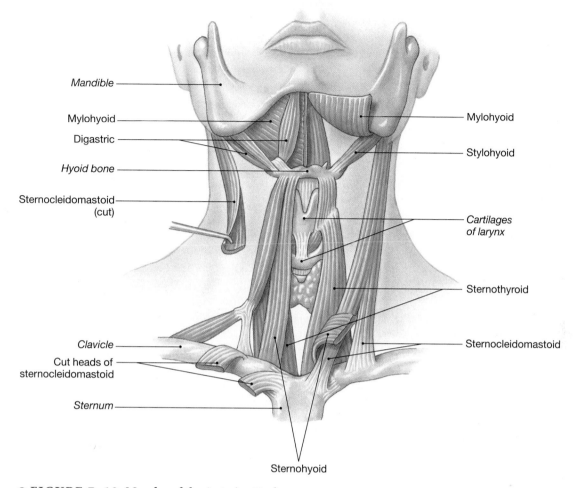

Mandible

Mylohyoid

Digastric

Hyoid bone

Sternocleidomastoid
(cut)

Mylohyoid

Stylohyoid

*Cartilages
of larynx*

Sternothyroid

Sternocleidomastoid

Clavicle

Cut heads of
sternocleidomastoid

Sternum

Sternohyoid

● **FIGURE 7–13** **Muscles of the Anterior Neck.**

stricts the opening, and other muscles move the lips or the corners of the mouth. The **buccinator** (BUK-si-nā-tor), which is one of the muscles associated with the mouth, compresses the cheeks, as when pursing the lips and blowing forcefully. (*Buccinator* translates as "trumpet player.") During chewing, contraction and relaxation of the buccinators move food back across the teeth from the space inside the cheeks. The chewing motions are primarily produced by contractions of the **masseter,** assisted by the **temporalis** and the **pterygoid** muscles used in various combinations. In infants, the buccinator produces suction for suckling at the breast.

Smaller groups of muscles control movements of the eyebrows and eyelids, the scalp, the nose, and the external ear. The **epicranium** (ep-i-KRĀ-nē-um), or *scalp,* contains two muscles—the frontalis and the **occipitalis.** These muscles are separated by an *aponeurosis,* or tendinous sheet, called the **epicranial aponeurosis.** The **platysma** (pla-TIZ-muh; *platys,* flat) covers the ventral surface of the neck, and ex-

tends from the base of the neck to the mandible and the corners of the mouth.

The muscles of the neck control the position of the larynx, depress the mandible, tense the floor of the mouth, and provide a stable foundation for muscles of the tongue and pharynx (see Figure 7–13). These muscles include the following:

■ The **digastric,** which has two bellies (*di-,* two + *gaster,* stomach), opens the mouth by depressing the mandible.

■ The broad, flat **mylohyoid** provides a muscular floor to the mouth and supports the tongue.

■ The **stylohyoid** forms a muscular connection between the hyoid bone and the styloid process of the skull.

■ The **sternocleidomastoid** (ster-nō-klī-dō-MAS-toyd) extends from the clavicles and the sternum to the mastoid region of the skull. It can rotate the head or flex the neck.

TABLE 7-4 *Muscles of the Head and Neck*

REGION/MUSCLE	ORIGIN	INSERTION	ACTION
MOUTH			
Buccinator	Maxillary bone and mandible	Blends into fibers of orbicularis oris	Compresses cheeks
Orbicularis oris	Maxillary bone and mandible	Lips	Compresses, purses lips
Depressor anguli oris	Anterolateral surface of mandible	Skin at angle of mouth	Depresses corner of mouth
Zygomaticus	Zygomatic bone	Angle of mouth; upper lip	Draws corner of mouth back and up
EYE			
Orbicularis oculi	Medial margin of orbit	Skin around eyelids	Closes eye
SCALP			
Frontalis	Epicranial aponeurosis	Skin of eyebrow and bridge of nose	Raises eyebrows, wrinkles forehead
Occipitalis	Occipital bone	Epicranial aponeurosis	Tenses and retracts scalp
LOWER JAW			
Masseter	Zygomatic arch	Lateral surface of mandible	Elevates mandible
Temporalis	Along temporal lines of skull	Coronoid process of mandible	Elevates mandible
Pterygoids	Inferior processes of sphenoid	Medial surface of mandible	Elevate, protract, and/or move mandible to either side
NECK			
Platysma	From cartilage of second rib to acromion of scapula	Mandible and skin of cheek	Tenses skin of neck, depresses mandible
Digastric	Mastoid region of temporal bone and inferior surface of mandible	Hyoid bone	Depresses mandible and/or elevates larynx
Mylohyoid	Medial surface of mandible	Median connective tissue band that runs to hyoid bone	Elevates floor of mouth and hyoid, and/or depresses mandible
Sternohyoid	Clavicle and sternum	Hyoid bone	Depresses hyoid bone and larynx
Sternothyroid	Dorsal surface of sternum and 1st rib	Thyroid cartilage of larynx	As above
Stylohyoid	Styloid process of temporal bone	Hyoid bone	Elevates larynx
Sternocleidomastoid	Superior margins of sternum and clavicle	Mastoid region of skull	Both sides together flex the neck; alone one side bends head toward shoulder and turns face to opposite side

Clinical Note
SPASMODIC TORTICOLLIS

Spasmodic torticollis is an intermittent or continuous spasm of the muscles of the neck, most commonly the *sternocleidomastoid* and *trapezius* muscles. It is usually more pronounced on one side, which results in turning or tipping of the head toward the affected muscles. Torticollis is involuntary and cannot be inhibited. It tends to be worse when the patient sits, stands, or walks. Torticollis affects women twice as often as men.

Torticollis should not be confused with a cervical muscle strain or spasm. Torticollis belongs to a class of diseases referred to as *focal dystonias*. Dystonias are an abnormal state of muscle tone. Focal dystonias affect a single area of the body. They occur more frequently in adults than in children, remain stable, and rarely spread to other body parts. Spasmodic torticollis is the most common focal dystonia. It is always important to exclude an extrapyramidal system reaction as a cause of spasmodic torticollis. Many of the antipsychotic drugs can cause focal dystonias, including torticollis, especially when administered in higher doses.

A simple neck muscle spasm, generally referred to as a "crick in the neck," often results from overuse, awkward positioning, or sleeping in an unusual position. The spasm can be quite intense and painful and can last for several days. The initial "injury" can slightly tear or stretch the affected muscle and cause pain and spasm. Most cases respond to local ice or heat, immobilization, and anti-inflammatory drugs. In severe cases, narcotics may be required for pain control, and intravenous diazepam (Valium) may be needed to help alleviate the muscle spasm. ■

→ CONCEPT CHECK QUESTIONS

1. If you were contracting and relaxing your masseter muscle, what would you probably be doing?
2. Which facial muscle would you expect to be well developed in a trumpet player?

Answers begin on p. 792.

Muscles of the Spine

The muscles of the spine are covered by more superficial back muscles, such as the trapezius and latissimus dorsi (see Figure 7–11b). The most superior of the spinal muscles are the posterior neck muscles: the superficial **splenius capitis** and the deeper **semispinalis capitis** (Figure 7–14● and Table 7–5). When the left and right pairs of these muscles contract together, they assist each other in extending the head. When both contract on one side, they assist in tilting the head. Because of its more lateral insertion, contraction of the splenius capitis also acts to rotate the head. The *spinal extensors*, or **erector spinae**, act to maintain an erect spinal column and head. Moving laterally from the spine, these muscles can be subdivided into **spinalis, longissimus,** and **iliocostalis** divisions. In the lower lumbar and sacral regions, the border between the longissimus and iliocostalis muscles is indistinct, and they are sometimes known as the *sacrospinalis* muscles. When contracting together, these muscles extend the spinal column. When only the muscles on one side contract, the spine is bent laterally (lateral flexion). Deep to the spinalis muscles, smaller muscles interconnect and stabilize the vertebrae.

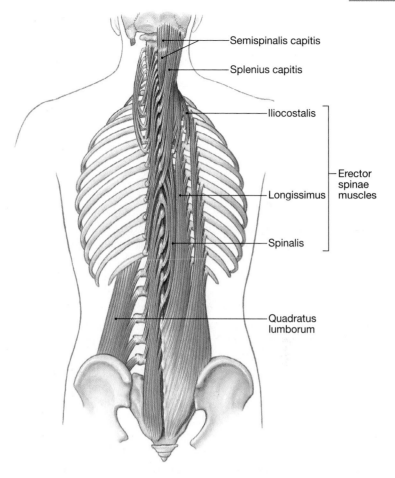

● **FIGURE 7–14** Muscles of the Spine.

TABLE 7–5 *Muscles of the Spine*

REGION/MUSCLE	ORIGIN	INSERTION	ACTION
SPINAL EXTENSORS			
Splenius capitis	Spinous processes of lower cervical and upper thoracic vertebrae	Mastoid process, base of the skull, and upper cervical vertebrae	The two sides act together to extend the neck; either alone rotates and laterally flexes head to that side
Semispinalis capitis	Spinous processes of lower cervical and upper thoracic vertebrae	Base of skull, upper cervical vertebrae	The two sides act together to extend the neck; either alone laterally flexes head to that side
Spinalis group	Spinous processes and transverse processes of cervical, thoracic, and upper lumbar vertebrae	Base of skull and spinous processes of cervical and upper thoracic vertebrae	The two sides act together to extend vertebral column; either alone extends neck and laterally flexes head or rotates vertebral column to that side
Longissimus group	Processes of lower cervical, thoracic, and upper lumbar vertebrae	Mastoid processes of temporal bone, transverse processes of cervical vertebrae and inferior surfaces of ribs	The two sides act together to extend vertebral column; either alone rotates and laterally flexes head or vertebral column to that side
Iliocostalis group	Superior borders of ribs and iliac crest	Transverse processes of cervical vertebrae and inferior surfaces of ribs	Extends vertebral column or laterally to that side; moves ribs
SPINAL FLEXOR			
Quadratus lumborum	Iliac crest	Last rib and transverse processes of lumbar vertebrae	Together they depress ribs, flex vertebral column; one side acting alone produces lateral flexion

Clinical Note
MUSCULOSKELETAL BACK DISORDERS

Back pain is one of the most common complaints encountered in modern emergency medical practice. In fact, low back pain is secondary only to the common cold as a cause of missed time from work. It is also the primary cause of reduced work capacity. Between 60 and 90 percent of the population will experience back pain in their lifetime. EMS is a physically demanding occupation (Figure 7–15●). EMS personnel are particularly vulnerable to back injury and should take precautions to minimize the chances of injury. This includes proper lifting techniques and requesting assistance when needed (Figure 7–16●).

There are numerous causes of back pain, which range from a simple muscle strain to a ruptured aortic aneurysm. Back strains usually occur when an abnormal or exaggerated movement stretches or tears a muscle or group of muscles in the back (Figure 7–17●). Carrying heavy loads can also stress or tear back muscles and result in a strain. Back strains, especially those that involve a group of muscles, initially causes localized bleeding at the injury site. This is followed by the formation of a hematoma. Hematoma formation is often accompanied by pain. Often, the affected muscles, and other muscles in the area, will spasm. This causes increased irritation of the previously injured muscle and increased pain. Occasionally, the spasm can be so severe that pressure is placed on nerve roots that exit the spine at each level. Irritation of the nerve roots can cause pain along the distribution of the affected spinal nerve.

There are three layers of muscles along the spine, collectively referred to as the *paraspinous muscles.* The superficial layer, which consists of the *spinalis*, the *longissimus*, and the *iliocostalis*, is palpable during physical examination. Tenderness or spasm can usually be palpated. Occasionally, spasm is limited to one side and an obvious lateral curvature of the spine can be seen.

A lumbar muscle strain can be quite painful. It is not uncommon for these patients to be in such severe pain that they must be medicated before they can lie down on the ambulance stretcher for transport. The definitive treatment of a lumbar strain is to rest the affected muscles and minimize inflammation. Initially, this should include the application of ice and rest. Later, moist heat can be applied to the

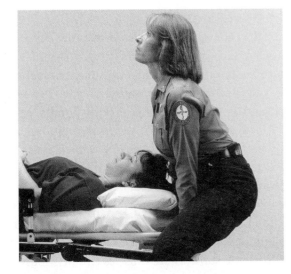

● **FIGURE 7–16 Proper Lifting Technique.** By lifting with the legs, the large paraspinous muscles of the back are not overstressed.

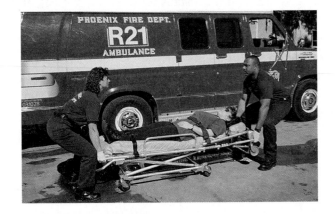

● **FIGURE 7–17 Lifting in Unison.** Back strains in EMS personnel usually occur when a team member moves awkwardly during a lift. Twisting, turning, or other movements can place stress on the large back muscles and result in back strain.

affected muscles. Anti-inflammatory medications are also helpful. Moderate to severe pain may require a short course of narcotic pain medicine. Muscle relaxant medications are somewhat controversial. Their effectiveness in the treatment of lumbar muscle strains varies significantly from patient to patient. Most of the muscle relaxants have sedating side effects that may help a patient to sleep. The application of physical therapy modalities, osteopathic manipulation, or chiropractic adjustment can sometimes help to alleviate acute pain and shorten the recovery period. ■

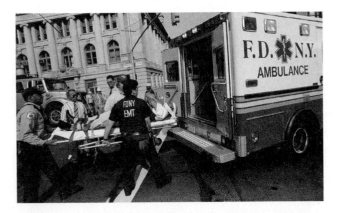

● **FIGURE 7–15 Physical Demands of EMS.** Lifting and moving patients and equipment can cause low back injury if not performed correctly.

The Axial Muscles of the Trunk

The *oblique muscles* and the *rectus muscles* form the muscular walls of the thoracic and abdominopelvic cavities between the first thoracic vertebra and the pelvis. In the thoracic area, these muscles are partitioned by the ribs, but over the abdominal surface they form broad muscular sheets (Figure 7–18● and Table 7–6). The oblique muscles can

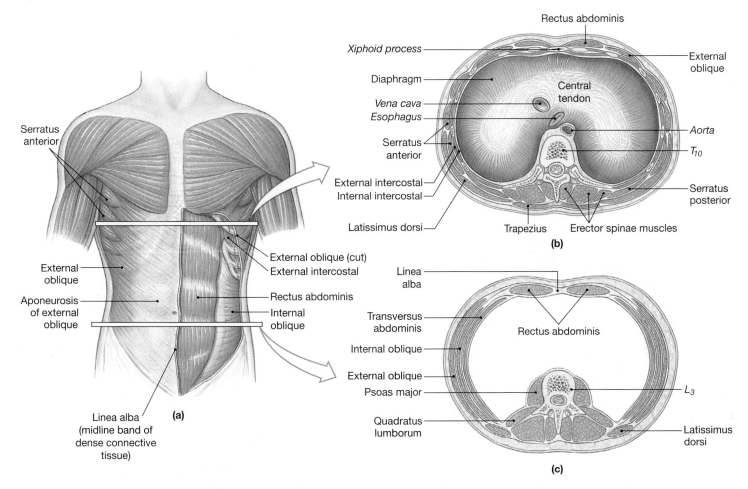

● **FIGURE 7–18** **Oblique and Rectus Muscles and the Diaphragm.** (**a**) An anterior view. (**b**) A sectional view at the level of the diaphragm. (**c**) A sectional view at the level of the umbilicus.

TABLE 7–6	*Axial Muscles of the Trunk*		
REGION/MUSCLE	**ORIGIN**	**INSERTION**	**ACTION**
THORACIC REGION			
External intercostals	Inferior border of each rib	Superior border of next rib	Elevate ribs
Internal intercostals	Superior border of each rib	Inferior border of the preceding rib	Depress ribs
Diaphragm	Xiphoid process, cartilages of ribs 4–10, and anterior surfaces of lumbar vertebrae	Central tendinous sheet	Contraction expands thoracic cavity, compresses abdominopelvic cavity
ABDOMINAL REGION			
External oblique	Lower eight ribs	Linea alba and iliac crest	Compresses abdomen, depresses ribs, flexes or laterally flexes vertebral column
Internal oblique	Iliac crest and adjacent connective tissues	Lower ribs, xiphoid of sternum, and linea alba	As above
Transversus abdominis	Cartilages of lower ribs, iliac crest, and adjacent connective tissues	Linea alba and pubis	Compresses abdomen
Rectus abdominis	Superior surface of pubis around symphysis	Inferior surfaces of costal cartilages (ribs 5–7) and xiphoid process	Depresses ribs, flexes vertebral column

compress underlying structures or rotate the spinal column, depending on whether one or both sides are contracting. The rectus muscles are important flexors of the spinal column; they oppose the erector spinae.

The axial muscles of the trunk include (1) the **external** and **internal intercostals,** (2) the muscular **diaphragm** that separates the thoracic and abdominopelvic cavities, (3) the **external** and **internal obliques,** (4) the **transversus abdominis,** (5) the **rectus abdominis,** and (6) the muscles that form the floor of the pelvic cavity.

Clinical Note
HERNIAS

Contraction of the abdominal muscles can significantly increase the pressure within the abdominal cavity. During forceful exercise or lifting, the pressures within the abdominopelvic cavity can increase even more. If a weakness exists in the wall of the abdominal cavity, this increased pressure can force a portion of a visceral organ into the weakened area. The presence of a visceral organ in a weakened area of the abdominal wall is referred to as a *hernia*. Hernias can develop virtually anywhere in the abdominal wall. However, the inguinal region is particularly vulnerable to hernia development.

During development of the male, the testes descend from the abdominal cavity into the scrotum. They pass through the abdominal wall at the inguinal canals and carry remnants of the abdominal wall and membranes with them. In the adult male, the spermatic ducts and associated blood vessels penetrate the abdominal wall at the inguinal canals on their way to the testicles. The inguinal canal is weaker than other portions of the muscular abdominal wall. With age and increased intra-abdominal pressure, the inguinal canal slowly enlarges. On occasion, the inguinal canal enlarges and allows some of the abdominal contents, usually a portion of the small intestine, to enter. The presence of abdominal viscera in the inguinal canal is referred to as an *inguinal hernia*. Usually, the affected small intestine is forced into the canal when pressure within the abdomen increases; it returns to the abdomen when pressure is relieved. If the affected portion of the hernial site becomes trapped, and cannot return to the abdominal cavity, the hernia is said to be *incarcerated*. If blood supply to the hernia is compromised, the hernia is said to be *strangulated*. Emergency surgery is necessary to remove the hernia sac from the inguinal canal before bowel ischemia and necrosis occurs (Figure 7–19●).

In addition to inguinal hernias, several other types of hernias can occur. Patients who have had abdominal surgery may have a weakness in the abdominal wall along the scar of the incision. When intra-abdominal pressure increases, a visceral organ can be forced through this weakness to form an *incisional hernia*. This situation is more common in obese patients. Often, a piece of surgical mesh must be placed in the abdominal wall in order to pre-

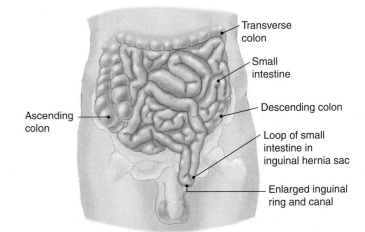

● **FIGURE 7–19 Inguinal Hernia.** An inguinal hernia occurs when a loop of abdominal viscera, usually the small intestine, enters a weakened and dilated inguinal canal. In severe cases, the loop of small intestine can be entrapped within the canal, often cutting off the blood supply. This situation, referred to as a strangulated inguinal hernia, is a surgical emergency.

vent hernia formation. Females tend to develop *femoral hernias*, which develop in the femoral ring underneath the inguinal ligament. The incidence of femoral hernias is much lower than that of inguinal hernias.

A weakness in the diaphragm can result in some of the abdominal contents being forced into the chest cavity. A *hiatal hernia* is common. In these, the proximal portion of the stomach is forced upward through the esophageal hiatus (the point where the esophagus enters the abdomen). Hiatal hernias are usually managed with diet, weight loss, and medication. Surgery is rarely indicated. ■

→ CONCEPT CHECK QUESTIONS

1. Damage to the external intercostal muscles would interfere with what important process?
2. If someone were to hit you in your rectus abdominis muscle, how would your body position change?

Answers begin on p. 792.

Muscles of the Pelvic Floor

The floor of the pelvic cavity is called the **perineum** (Figure 7–20● and Table 7–7). It is formed by a broad sheet of muscles that connects the sacrum and coccyx to the ischium and pubis. These muscles support the organs of the pelvic cavity, flex the coccyx, and control the movement of materials through the urethra and anus.

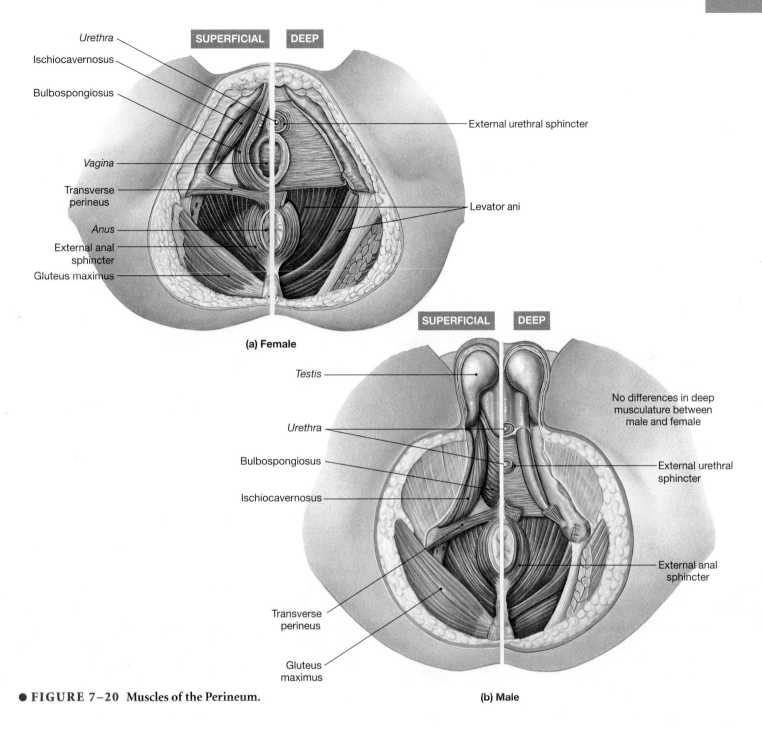

Urethra
Ischiocavernosus
Bulbospongiosus

SUPERFICIAL   DEEP

External urethral sphincter

Vagina

Transverse
perineus

Anus

External anal
sphincter

Gluteus maximus

Levator ani

(a) Female

Testis

SUPERFICIAL   DEEP

No differences in deep
musculature between
male and female

Urethra

Bulbospongiosus

Ischiocavernosus

External urethral
sphincter

External anal
sphincter

Transverse
perineus

Gluteus
maximus

(b) Male

● **FIGURE 7–20 Muscles of the Perineum.**

The Appendicular Muscles

The appendicular musculature includes (1) the muscles of the shoulders and upper limbs and (2) the muscles of the pelvic girdle and lower limbs. Because the functions and required ranges of motion are very different, few similarities exist between the two groups. In addition to increasing the mobility of the upper limb, the muscular connections between the pectoral girdle and the axial skeleton must act as shock absorbers. For example, people can still perform delicate hand movements while jogging because the muscular connections between the axial and appendicular skeleton smooth out the bounces in their stride. In contrast, the pelvic girdle has evolved to transfer weight from the axial to the appendicular skeleton. A muscular connection would reduce the efficiency of the transfer, and the emphasis is on sheer power rather than mobility.

TABLE 7-7 *Muscles of the Perineum*

MUSCLE	ORIGIN	INSERTION	ACTION
BULBOSPONGIOSUS			
Males	Base of penis; fibers cross over urethra	Midline and central tendon of perineum	Compresses base and stiffens penis; ejects urine or semen
Females	Base of clitoris; fibers run on either side of urethral and vaginal openings	Central tendon of perineum	Compresses and stiffens clitoris; narrows vaginal opening
Ischiocavernosus	Inferior medial surface of ischium	Symphysis pubis anterior to base of penis or clitoris	Compresses and stiffens penis or clitoris
Transverse perineus	Inferior medial surface of ischium	Central tendon of perineum	Stabilizes central tendon of perineum
EXTERNAL URETHRAL SPHINCTER			
Males	Inferior medial surfaces of ischium and pubis	Midline at base of penis; inner fibers encircle urethra	Closes urethra, compresses prostate and bulbourethral glands
Females	As above	Midline; inner fibers encircle urethra	Closes urethra, compresses vagina and greater vestibular glands
EXTERNAL ANAL SPHINCTER	By tendon from coccyx	Encircles anal opening	Closes anal opening
LEVATOR ANI	Ischial spine, pubis	Coccyx	Tenses floor of pelvis, supports pelvic organs, flexes coccyx, elevates and retracts anus

Muscles of the Shoulders and Upper Limbs

The large, superficial **trapezius** muscles cover the back and portions of the neck, and reach to the base of the skull. These muscles form a broad diamond (Figure 7–21● and Table 7–8). Its actions are quite varied because specific regions can be made to contract independently. The **rhomboid** muscles and the **levator scapulae** are covered by the trapezius. Both originate on vertebrae and insert on the scapula. Contraction of the rhomboids adducts the scapula, and pulls it toward the center of the back. The levator scapulae elevates the scapula, as when you shrug your shoulders.

On the chest, the **serratus anterior** originates along the anterior surfaces of several ribs (see Figure 7–21) and inserts along the vertebral border of the scapula. When the serratus anterior contracts, it pulls the shoulder anteriorly. The **pectoralis minor** attaches to the coracoid process of the scapula. When it contracts, it depresses and protracts the scapula.

TABLE 7-8 *Muscles of the Shoulder*

MUSCLE	ORIGIN	INSERTION	ACTION
Levator scapulae	Transverse processes of first four cervical vertebrae	Vertebral border of scapula	Elevates scapula
Pectoralis minor	Anterior surfaces of ribs 3–5	Coracoid process of scapula	Depresses and protracts shoulder; rotates scapula laterally (downward); elevates ribs if scapula is stationary
Rhomboid muscles	Spinous processes of lower cervical and upper thoracic vertebrae	Vertebral border of scapula	Adducts and rotates scapula laterally (downward)
Serratus anterior	Anterior and superior margins of ribs 1–9	Anterior surface of vertebral border of scapula	Protracts shoulder, abducts and medially rotates scapula (upward)
Subclavius	First rib	Clavicle	Depresses and protracts shoulder
Trapezius	Occipital bone and spinous processes of thoracic vertebrae	Clavicle and scapula (acromion and scapular spine)	Depends on active region and state of other muscles; may elevate, adduct, depress, or rotate scapula and/or elevate clavicle; can also extend or hyperextend neck

● FIGURE 7–21 Muscles of the Shoulder.

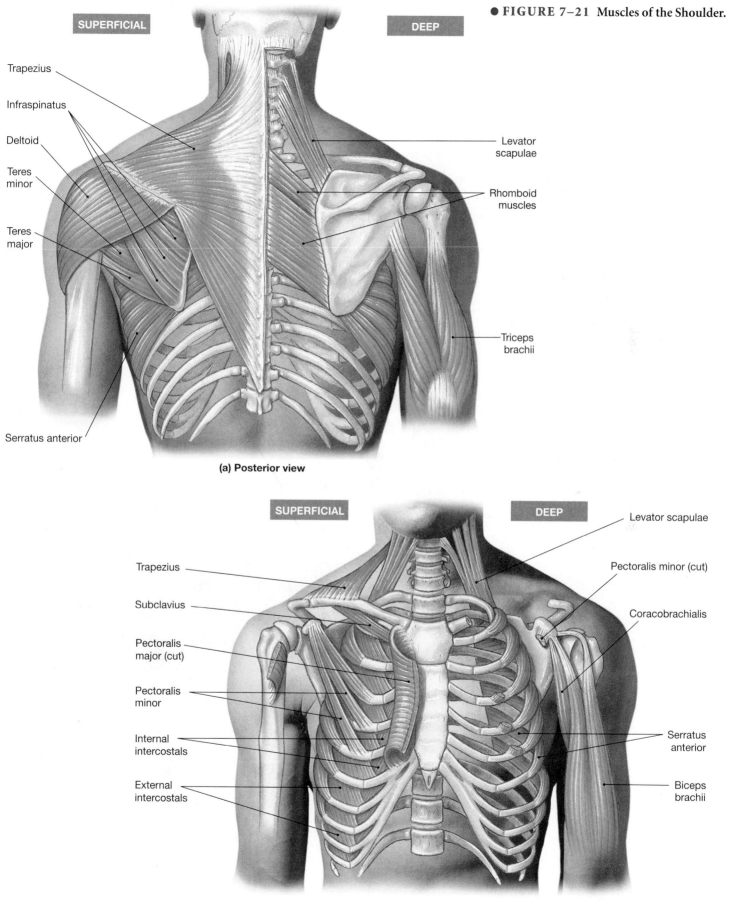

SUPERFICIAL

DEEP

Trapezius

Infraspinatus

Deltoid

Teres minor

Teres major

Serratus anterior

Levator scapulae

Rhomboid muscles

Triceps brachii

(a) Posterior view

SUPERFICIAL

DEEP

Trapezius

Subclavius

Pectoralis major (cut)

Pectoralis minor

Internal intercostals

External intercostals

Levator scapulae

Pectoralis minor (cut)

Coracobrachialis

Serratus anterior

Biceps brachii

(b) Anterior view

MUSCLES THAT MOVE THE ARM. The muscles that move the arm (Figure 7–22● and Table 7–9) are easiest to remember when grouped by primary actions:

■ The **deltoid** is the major abductor of the arm, and the **supraspinatus** assists at the start of this movement.

■ The **subscapularis, teres major, infraspinatus,** and **teres minor** rotate the arm.

● **FIGURE 7–22** Muscles That Move the Arm.

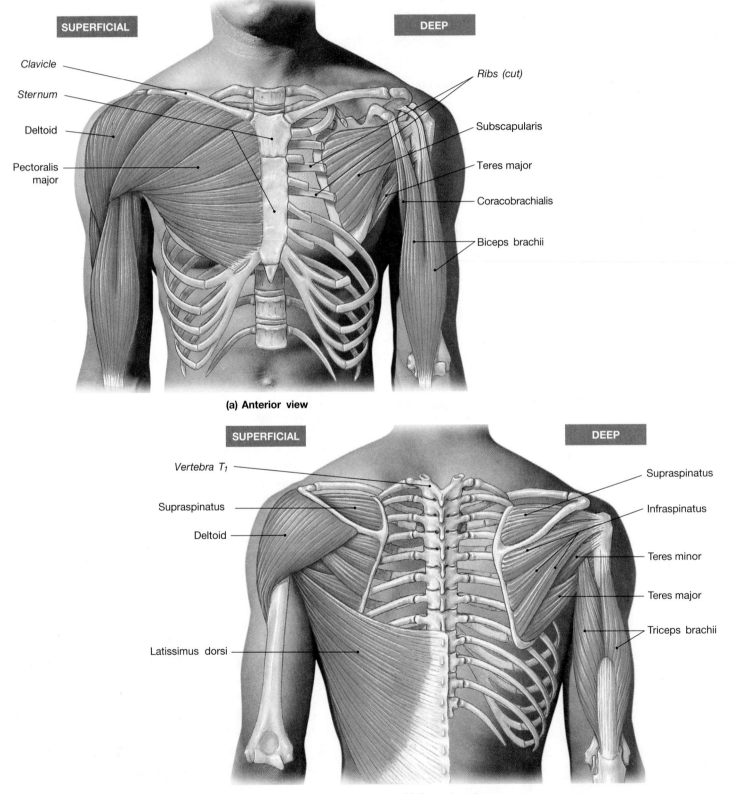

SUPERFICIAL   DEEP

Clavicle
Sternum
Deltoid
Pectoralis major

Ribs (cut)
Subscapularis
Teres major
Coracobrachialis
Biceps brachii

(a) Anterior view

SUPERFICIAL   DEEP

Vertebra T_1
Supraspinatus
Deltoid
Latissimus dorsi

Supraspinatus
Infraspinatus
Teres minor
Teres major
Triceps brachii

(b) Posterior view

TABLE 7-9 *Muscles That Move the Arm*

MUSCLE	ORIGIN	INSERTION	ACTION
Coracobrachialis	Coracoid process	Medial margin of shaft of humerus	Adduction and flexion at shoulder
Deltoid	Clavicle and scapula (acromion and adjacent scapular spine)	Deltoid tuberosity of humerus	Abduction at shoulder
Latissimus dorsi	Spinous processes of lower thoracic vertebrae, ribs, and lumbar vertebrae	Intertubercular groove of humerus	Extension, adduction, and medial rotation at shoulder
Pectoralis major	Cartilages of ribs 2–6, body of sternum, and clavicle	Greater tubercle of humerus	Flexion, adduction, and medial rotation at shoulder
Supraspinatus*	Supraspinous fossa of scapula	Greater tubercle of humerus	Abduction at shoulder
Infraspinatus*	Infraspinous fossa of scapula	Greater tubercle of humerus	Lateral rotation at shoulder
Subscapularis*	Subscapular fossa of scapula	Lesser tubercle of humerus	Medial rotation at shoulder
Teres minor*	Lateral border of scapula	Greater tubercle of humerus	Lateral rotation at shoulder
Teres major	Inferior angle of scapula	Intertubercular groove of humerus	Adduction and medial rotation at shoulder

*Rotator cuff muscles

Clinical Note
INTRAMUSCULAR DRUG INJECTION

Injecting medications into muscle tissue (*intramuscular administration, IM*) is a safe, effective, simple, and relatively painless method of drug administration. The medication is absorbed into the network of blood vessels within the muscle and subsequently enters the circulatory system. The onset of action of medications administered by this route is typically 10–15 minutes. Drug absorption is steady and usually predictable and may continue for hours to days, depending upon the medication injected. In addition, drugs administered by this route, unlike those administered by the oral route, do not have to pass through the liver before arriving at their site of action.

The muscles most often used for IM injection are the deltoid muscle of the upper arm and the gluteus muscle in the buttock. These muscles are easy to access and large enough to handle large volumes of medication. Care must be taken to avoid nearby neurovascular structures. Typically, a 1–1/2 inch needle is attached to a syringe that contains the medication. The needle is inserted through the skin and subcutaneous tissue into the middle of the muscle. The plunger on the syringe is pulled back to ensure that a blood vessel has not been inadvertently entered. Then, the plunger is depressed and the drug is deposited into the muscle tissue. The medication is then absorbed into the blood vessels and continues until all of the medication is gone. Second to the oral route, IM injection is the most frequently used method of drug administration. ■

The **pectoralis major,** which extends between the chest and the greater tubercle of the humerus, produces flexion at the shoulder joint. The **latissimus dorsi,** which extends between the thoracic vertebrae and the intertubercular groove of the humerus, produces extension. The two muscles also work together to produce adduction and rotation of the humerus.

These muscles provide substantial support for the shoulder joint. The tendons of the supraspinatus, infraspinatus, subscapularis, and teres minor blend with and support the capsular fibers that enclose the shoulder joint. They are the muscles of the *rotator cuff*, which is a common site of sports injuries. Sports that involve throwing a ball, such as a baseball or football, place considerable strain on the muscles of the rotator cuff, which can lead to a *muscle strain* (a tear or break in the muscle), *bursitis*, and other painful injuries.

MUSCLES THAT MOVE THE FOREARM AND WRIST. Although most of the muscles that insert on the forearm and wrist (Figure 7–23● and Table 7–10. originate on the humerus, there are two notable exceptions. The **biceps brachii** and the *long head* tendon of the **triceps brachii** originate on the scapula and insert on the bones of the forearm. Although their contractions can have a secondary effect on the shoulder, their primary actions are at the elbow. The triceps brachii

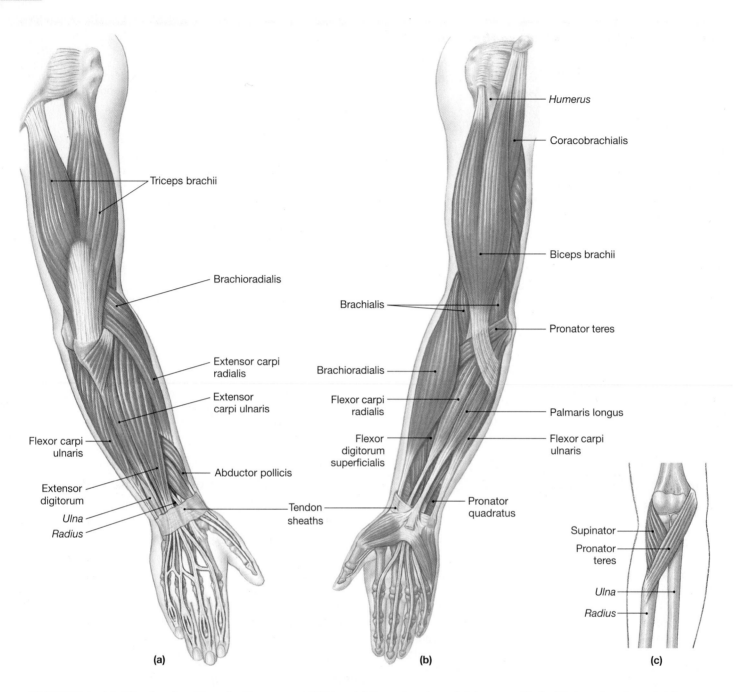

● FIGURE 7–23 Muscles that Move the Forearm and Wrist. (a) Posterior view of right upper limb. (b) Anterior view of right upper limb. (c) Anterior view of the muscles of pronation and supination when the limb is supinated.

extends the elbow when, for example, you do pushups. The biceps brachii both flexes the elbow and supinates the forearm. With the forearm pronated (palm facing back), the biceps brachii cannot function effectively. As a result, you are strongest when you flex your elbow with a supinated forearm; the biceps brachii then makes a prominent bulge.

Other important muscles include the following:

■ The **brachialis** and **brachioradialis** also flex the elbow, opposed by the triceps brachii.

■ The **flexor carpi ulnaris,** the **flexor carpi radialis,** and the **palmaris longus** are superficial muscles that work together to produce flexion of the wrist. Because they originate on opposite sides of the humerus, the flexor carpi radialis flexes and abducts the wrist, whereas the flexor carpi ulnaris flexes and adducts the wrist.

■ The **extensor carpi radialis** muscles and the **extensor carpi ulnaris** have a similar relationship; the former produces extension and abduction at the wrist, the latter extension and adduction.

TABLE 7–10 *Muscles That Move the Forearm, Wrist, and Hand*

MUSCLE	ORIGIN	INSERTION	ACTION
PRIMARY ACTION AT THE ELBOW			
Flexors			
Biceps brachii	*Short head* from the coracoid process and *long head* from the supraglenoid tubercle (both on the scapula)	Tuberosity of radius	Flexion at shoulder and elbow; supination
Brachialis	Anterior, distal surface of humerus	Tuberosity of ulna	Flexion at elbow
Brachioradialis	Lateral epicondyle of humerus	Styloid process of radius	As above
Extensor			
Triceps brachii	Superior, posterior, and lateral margins of humerus, and the scapula	Olecranon of ulna	Extension at elbow
Pronators/Supinator			
Pronator quadratus	Medial surface of distal portion of ulna	Anterior and lateral surface of distal portion of radius	Pronation
Pronator teres	Medial epicondyle of humerus and coronoid process of ulna	Distal lateral surface of radius	As above
Supinator	Lateral epicondyle of humerus and ulna	Anterior and lateral surface of radius distal to the radial tuberosity	Supination
PRIMARY ACTION AT THE WRIST			
Flexors			
Flexor carpi radialis	Medial epicondyle of humerus	Bases of second and third metacarpal bones	Flexion and abduction at wrist
Flexor carpi ulnaris	Medial epicondyle of humerus and adjacent surfaces of ulna	Pisiform bone, hamate bone, and base of fifth metacarpal bone	Flexion and adduction at wrist
Palmaris longus	Medial epicondyle of humerus	A tendinous sheet on the palm	Flexion at wrist
Extensors			
Extensor carpi radialis	Distal lateral surface and lateral epicondyle of humerus	Bases of second and third metacarpal bones	Extension and abduction at wrist
Extensor carpi ulnaris	Lateral epicondyle of humerus and adjacent surface of ulna	Base of fifth metacarpal bone	Extension and adduction at wrist
ACTION AT THE HAND			
Extensor digitorum	Lateral epicondyle of humerus	Posterior surfaces of the phalanges	Extension at finger joints and wrist
Flexor digitorum	Medial epicondyle of humerus; anterior surfaces of ulna and radius; medial and posterior surfaces of ulna	Distal phalanges	Flexion at finger joints and wrist

■ The **pronators** and the **supinator** rotate the radius at its proximal and distal articulations with the ulna without either flexing or extending the elbow.

MUSCLES THAT MOVE THE HAND AND FINGERS. The muscles of the forearm flex and extend the finger joints (see Table 7–10). These muscles end before reaching the hand, and only their tendons cross the wrist. These are relatively large muscles, and keeping them clear of the joints ensures maximum mobility at both the wrist and hand. The tendons that cross the dorsal and ventral surfaces of the wrist are held in place by *tendon sheaths,* which are wide tubular bursae that reduce friction. ∞ p. 189

Inflammation of tendon sheaths can restrict movement and irritate the *median nerve,* which is a nerve that innervates the palm of the hand. Chronic pain, which is often associated with weakness in the hand muscles, is the result. This condition is known as *carpal tunnel syndrome.* Fine control of the hand involves small *intrinsic muscles,* which originate on the carpal and metacarpal bones. No muscles originate on the phalanges, and only tendons extend across the distal joints of the fingers.

→ **CONCEPT CHECK QUESTIONS**

1. Which muscle do you use to shrug your shoulders?
2. Sometimes baseball pitchers suffer from rotator cuff injuries. Which muscles are involved in this type of injury?
3. Injury to the flexor carpi ulnaris would impair which two movements?

Answers begin on p. 792.

Muscles of the Pelvis and Lower Limbs

The muscles of the pelvis and the lower limb can be divided into three functional groups: (1) muscles that work across the hip joint to move the thigh; (2) muscles that work across the knee joint to move the leg; and (3) muscles that work across the various joints of the foot to move the ankles, feet, and toes.

MUSCLES THAT MOVE THE THIGH. The muscles that move the thigh are detailed in Figure 7–24● and Table 7–11.

■ **Gluteal muscles** cover the lateral surfaces of the ilia (Figure 7–24a). The **gluteus maximus** is the largest and most posterior of the gluteal muscles, which produce extension, rotation, and abduction at the hip.

TABLE 7–11   *Muscles That Move the Thigh*

GROUP/MUSCLE	ORIGIN	INSERTION	ACTION
GLUTEAL GROUP			
Gluteus maximus	Iliac crest of ilium, sacrum, and coccyx	Iliotibial tract and gluteal tuberosity of femur	Extension and lateral rotation at hip
Gluteus medius	Anterior iliac crest and lateral surface of ilium	Greater trochanter of femur	Abduction and medial rotation at hip
Gluteus minimus	Lateral surface of ilium	As above	As above
Tensor fasciae latae	Iliac crest and surface of ilium between anterior iliac spines	Iliotibial tract	Flexion, abduction, and medial rotation at hip; tenses fascia lata, which laterally supports the knee
ADDUCTOR GROUP			
Adductor brevis	Inferior ramus of pubis	Linea aspera of femur	Adduction, flexion, and medial rotation at hip
Adductor longus	Inferior ramus of pubis anterior to adductor brevis	As above	As above
Adductor magnus	Inferior ramus of pubis posterior to adductor brevis	As above	Adduction at hip joint; superior portion produces flexion; inferior portion produces extension
Pectineus	Superior ramus of pubis	Inferior to lesser trochanter of femur	Adduction, flexion, and medial rotation at hip joint
Gracilis	Inferior ramus of pubis	Medial surface of tibia inferior to medial condyle	Flexion at knee; adduction and medial rotation at hip
ILIOPSOAS GROUP			
Iliacus	Medial surface of ilium	Femur distal to lesser trochanter; tendon fused with that of psoas major	Flexion at hip and/or lumbar intervertebral joints
Psoas major	Anterior surfaces and transverse processes of T_{12} and lumbar vertebrae	Lesser trochanter in company with iliacus	As above

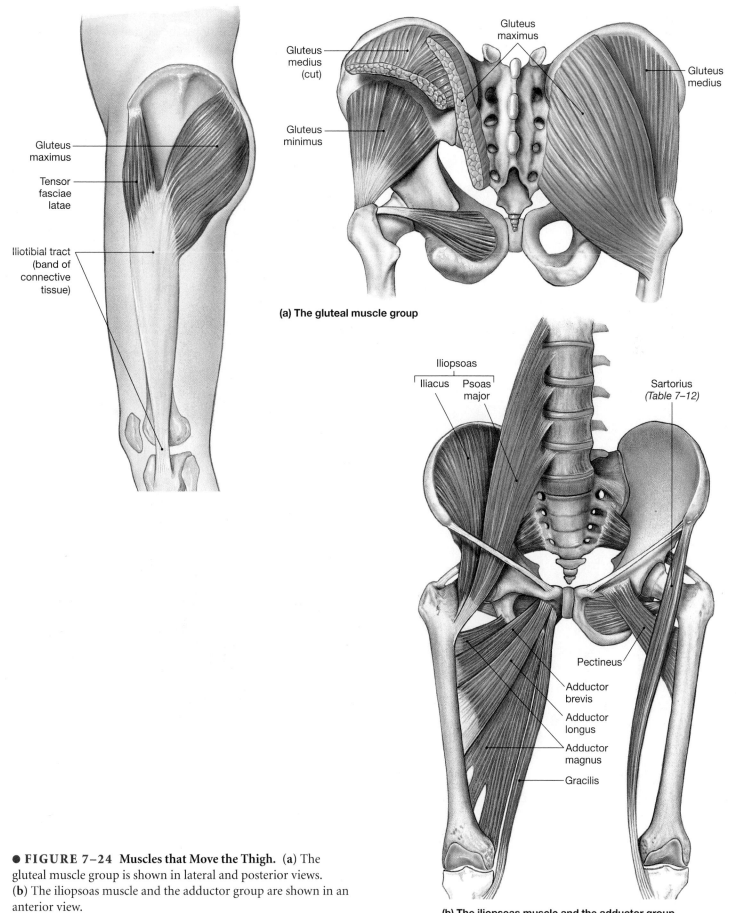

(a) The gluteal muscle group

(b) The iliopsoas muscle and the adductor group

● **FIGURE 7–24 Muscles that Move the Thigh.** (a) The gluteal muscle group is shown in lateral and posterior views. (b) The iliopsoas muscle and the adductor group are shown in an anterior view.

- The adductors of the thigh include the **adductor magnus,** the **adductor brevis,** the **adductor longus,** the **pectineus** (pek-TI-nē-us), and the **gracilis** (GRAS-i-lis) (Figure 7–24b). When an athlete suffers a *pulled groin,* the problem is a *strain*—a muscle tear or break—in one of these adductor muscles.

- The largest hip flexor is the **iliopsoas** (il-ē-ō-SŌ-us) muscle. The iliopsoas is really two muscles, the **psoas major** and the **iliacus** (il-Ē-ah-kus), that share a common insertion at the lesser trochanter of the femur.

MUSCLES THAT MOVE THE LEG. The pattern of muscle distribution in the lower limb is like that in the upper limb: extensors are found along the anterior and lateral surfaces of the limb, and flexors lie along the posterior and medial surfaces.

- The four flexors of the knee include three muscles collectively known as the *hamstrings*—the **biceps femoris** (FEMor-is), the **semimembranosus** (sem-ē-mem-bra-NŌ-sus), and the **semitendinosus** (sem-ē-ten-di-NŌ-sus)—and the **sartorius** (sar-TŌR-ē-us) (Figure 7–25a●). A *pulled hamstring* is a relatively common sports injury caused by a strain that affects one of the hamstring muscles.

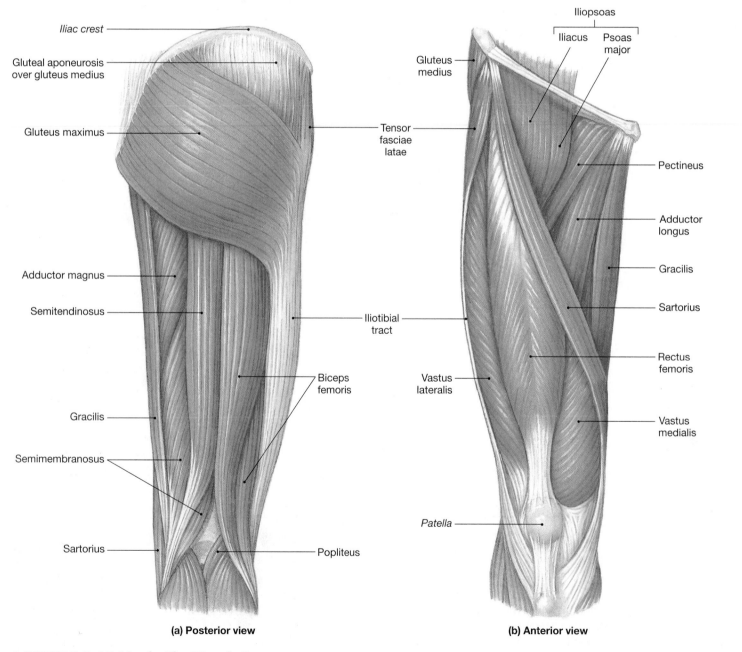

(a) Posterior view

(b) Anterior view

● **FIGURE 7–25** Muscles That Move the Leg.

TABLE 7-12 *Muscles That Move the Leg*

MUSCLE	ORIGIN	INSERTION	ACTION
FLEXORS			
Biceps femoris*	Ischial tuberosity and linea aspera of femur	Head of fibula, lateral condyle of tibia	Flexion at knee, extension and lateral rotation at hip
Semimembranosus*	Ischial tuberosity	Posterior surface of medial condyle of tibia	Flexion at knee; extension and medial rotation at hip
Semitendinosus*	As above	Proximal medial surface of tibia	As above
Sartorius	Anterior superior spine of ilium	Medial surface of tibia near tibial tuberosity	Flexion at knee; flexion and lateral rotation at hip
Popliteus	Lateral condyle of femur	Posterior surface of proximal tibial shaft	Rotates tibia medially (or rotates femur laterally); flexion at knee
EXTENSORS			
Rectus femoris	Anterior inferior iliac spine and superior acetabular rim of ilium	Tibial tuberosity by way of patellar ligament	Extension at knee, flexion at hip
Vastus intermedius	Anterior and lateral surface of femur along linea aspera	As above	Extension at knee
Vastus lateralis	Anterior and inferior to greater trochanter of femur and along linea aspera	As above	As above
VASTUS MEDIALIS	Entire length of linea aspera of femur	As above	As above

*Hamstring muscles

- Collectively the *knee extensors* are known as the **quadriceps femoris.** The three **vastus** muscles and the **rectus femoris** insert on the patella, which is attached to the tibial tuberosity by the patellar ligament (Figure 7–25b●). (Because the vastus intermedius lies under the other quadriceps femoris muscles, it is not visible in Figure 7–25.)
- When you stand, a slight lateral rotation of the tibia can lock the knee in the extended position. This enables you to stand for long periods with minimal muscular effort, but the locked knee cannot be flexed. The **popliteus** (pop-LI-tē-us) muscle medially rotates the tibia back into its normal position, which unlocks the joint.

The muscles that move the leg are detailed in Table 7–12.

MUSCLES THAT MOVE THE FOOT AND TOES. Muscles that move the foot and toes are shown in Figure 7–26● and detailed in Table 7–13 Most of the muscles that move the ankle produce the plantar flexion involved with walking and running movements.

- The large **gastrocnemius** (gas-trok-NĒ-mē-us; *gaster,* stomach + *kneme,* knee) of the calf is assisted by the underlying **soleus** (SŌ-lē-us) muscle. These muscles share a common tendon, the **calcaneal tendon,** or *Achilles tendon.*
- A pair of deep **fibularis** muscles, or *peroneus* muscles, produces eversion of the foot as well as plantar flexion at the ankle.
- Inversion of the foot is caused by contraction of the **tibialis** (tib-ē-A-lis) muscles; the large **tibialis anterior** dorsiflexes the ankle and opposes the gastrocnemius.

Important digital muscles originate on the surface of the tibia, the fibula, or both, and their tendons pass through tendon sheaths at the ankle joint. Several smaller intrinsic muscles originate on the tarsal and metatarsal bones; their contractions move the toes.

CONCEPT CHECK QUESTIONS

1. You often hear of athletes who suffer from a "pulled hamstring." To what does this phrase refer?
2. How would you expect a torn calcaneal tendon to affect movement of the foot?

Answers begin on p. 792.

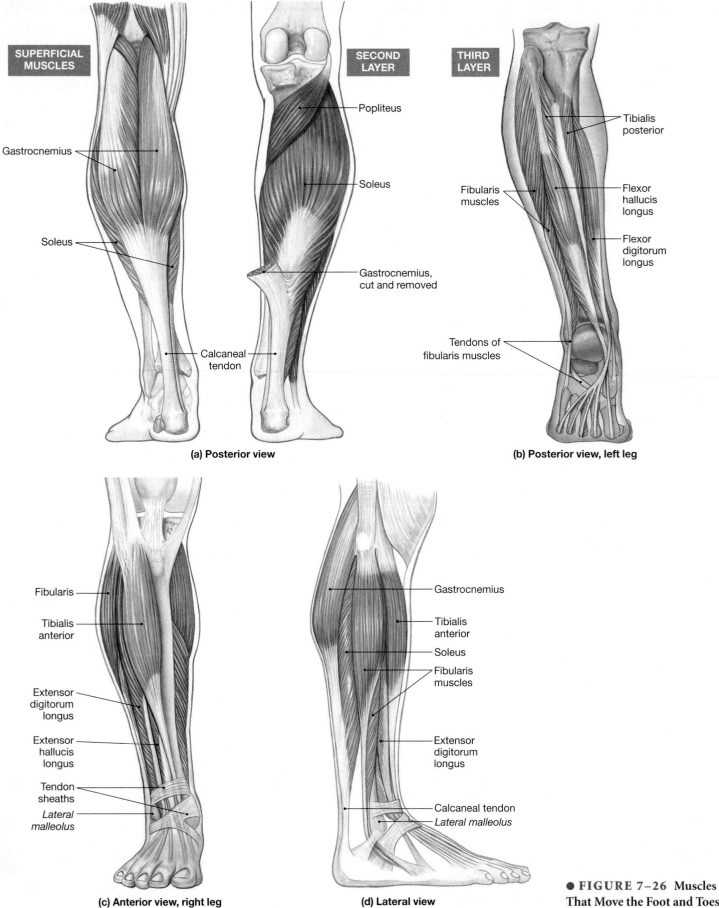

SUPERFICIAL MUSCLES

Gastrocnemius

Soleus

Calcaneal tendon

(a) Posterior view

SECOND LAYER

Popliteus

Soleus

Gastrocnemius, cut and removed

THIRD LAYER

Tibialis posterior

Fibularis muscles

Flexor hallucis longus

Flexor digitorum longus

Tendons of fibularis muscles

(b) Posterior view, left leg

Fibularis

Tibialis anterior

Extensor digitorum longus

Extensor hallucis longus

Tendon sheaths

Lateral malleolus

(c) Anterior view, right leg

Gastrocnemius

Tibialis anterior

Soleus

Fibularis muscles

Extensor digitorum longus

Calcaneal tendon

Lateral malleolus

(d) Lateral view

● **FIGURE 7–26** Muscles That Move the Foot and Toes.

TABLE 7-13 *Muscles That Move the Foot and Toes*

MUSCLE	ORIGIN	INSERTION	ACTION
Dorsiflexor			
Tibialis anterior	Lateral condyle and proximal shaft of tibia	Base of first metatarsal bone	Dorsiflexion at ankle; inversion of foot
Plantar flexors			
Gastrocnemius	Femoral condyles	Calcaneus by way of calcaneal tendon	Plantar flexion at ankle; inversion and adduction of foot; flexion at knee
Fibularis	Fibula and lateral condyle of tibia	Bases of first and fifth metatarsal bones	Eversion of foot and plantar flexion at ankle
Soleus	Head and proximal shaft of fibula, and adjacent shaft of tibia	Calcaneus by way of calcaneal tendon	Plantar flexion at ankle; adduction of foot
Tibialis posterior	Connective tissue membrane and adjacent shafts of tibia and fibula	Tarsal and metatarsal bones	Adduction and inversion of foot; plantar flexion at ankle
ACTION AT THE TOES **Flexors**			
Flexor digitorum longus	Posterior and medial surface of tibia	Inferior surface of phalanges, toes 2–5	Flexion at joints of toes 2–5
Flexor hallucis longus	Posterior surface of fibula	Inferior surface, distal phalanx of great toe	Flexion at joints of great toe
Extensors			
Extensor digitorum longus	Lateral condyle of tibia, anterior surface of fibula	Superior surfaces of phalanges, toes 2–5	Extension at joints of toes 2–5
Extensor hallucis longus	Anterior surface of fibula	Superior surface, distal phalanx of great toe	Extension at joints of great toe

Clinical Note
FIBROMYALGIA

Fibromyalgia is a chronic inflammatory disorder of the muscular system. It is classified by the American Academy of Rheumatology as the presence of 11 of 18 specific tender points, nonrestorative sleep, muscle stiffness, and generalized aching pain, with symptoms present for more than three months' duration. Furthermore, the pains and stiffness cannot be explained by other mechanisms. Fibromyalgia can be quite debilitating. It commonly affects women under 40 years of age and is almost always associated with chronic fatigue. Many of the problems associated with fibromyalgia can be attributed to other conditions, such as depression. However, the presence of the tender points is the diagnostic key to fibromyalgia. ■

■ Aging and the Muscular System

As the body ages, a general reduction in the size and the power of all muscle tissues occurs. The effects of aging on the muscular system can be summarized as follows:

1. *Skeletal muscle fibers become smaller in diameter.* The reduction in size reflects a decrease in the number of myofibrils. The overall effects are a reduction in muscle strength and endurance and a tendency to fatigue rapidly. Because cardiovascular performance also decreases with age, blood flow to active muscles does not increase with exercise as rapidly as it does in younger people.

2. *Skeletal muscles become less elastic.* Aging skeletal muscles develop increasing amounts of fibrous connective tissue, a process called *fibrosis*. Fibrosis makes the muscle less flexible, and the collagen fibers can restrict movement and circulation.

3. *The tolerance for exercise decreases.* A lower tolerance for exercise as age increases results in part from the tendency for rapid fatigue and in part from the reduction in thermoregulatory ability described in Chapters 1 and 5. ∞ pp. 14, 124 Individuals over age 65 cannot eliminate muscle-generated heat as effectively as younger people, which leads to overheating.

4. *The ability to recover from muscular injuries decreases.* When an injury occurs, repair capabilities are limited, and scar tissue formation is the usual result.

The Muscular System in Perspective

For All Systems
Generates heat that maintains normal body temperature

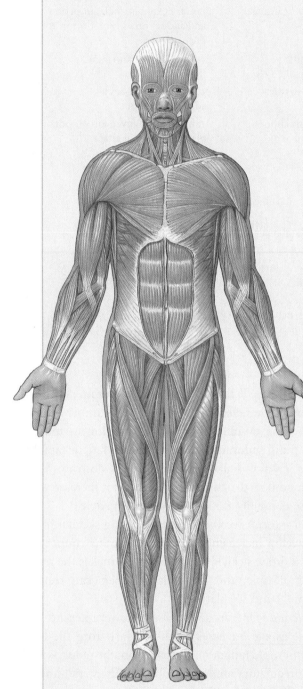

The Integumentary System

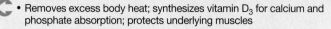

- Removes excess body heat; synthesizes vitamin D_3 for calcium and phosphate absorption; protects underlying muscles
- Skeletal muscles pulling on skin of face produce facial expressions

The Skeletal System

- Maintains normal calcium and phosphate levels in body fluids; supports skeletal muscles; provides sites of attachment
- Provides movement and support; stresses exerted by tendons maintain bone mass; stabilizes bones and joints

The Nervous System

- Controls skeletal muscle contractions; adjusts activities of respiratory and cardiovascular systems during periods of muscular activity
- Muscle spindles monitor body position; facial muscles express emotion; muscles of the larynx, tongue, lips and cheeks permit speech

The Endocrine System

- Hormones adjust muscle metabolism and growth; parathyroid hormone and calcitonin regulate calcium and phosphate ion concentrations
- Skeletal muscles provide protection for some endocrine organs

The Cardiovascular System

- Delivers oxygen and nutrients; removes carbon dioxide, lactic acid, and heat
- Skeletal muscle contractions assist in moving blood through veins; protects deep blood vessels

The Lymphatic System

- Defends skeletal muscles against infection and assists in tissue repairs after injury
- Protects superficial lymph nodes and the lymphatic vessels in the abdominopelvic cavity

The Respiratory System

- Provides oxygen and eliminates carbon dioxide
- Muscles generate carbon dioxide; control entrances to respiratory tract, fill and empty lungs, control airflow through larynx, and produce sounds

The Digestive System

- Provides nutrients; liver regulates blood glucose and fatty acid levels and removes lactic acid from circulation
- Protects and supports soft tissues in abdominal cavity; controls entrances to and exits from digestive tract

The Urinary System

- Removes waste products of protein metabolism; assists in regulation of calcium and phosphate concentrations
- External sphincter controls urination by constricting urethra

The Reproductive System

- Reproductive hormones accelerate skeletal muscle growth
- Contractions of skeletal muscles eject semen from male reproductive tract; muscle contractions during sex act produce pleasurable sensations

● **FIGURE 7–27** Functional Relationships Between the Muscular System and Other Systems.

The *rate* of decline in muscular performance is the same in all people, regardless of their exercise patterns or lifestyle. Therefore, to be in good shape late in life, an individual must be in *very* good shape early in life. Regular exercise helps control body weight, strengthens bones, and generally improves the quality of life at all ages. Extremely demanding exercise is not as important as regular exercise. In fact, extreme exercise in the elderly can damage tendons, bones, and joints. Although it has obvious effects on the quality of life, there is no clear evidence that exercise prolongs life expectancy.

→ **CONCEPT CHECK QUESTION**

1. What major structural change occurs in skeletal muscle fibers as we age, and what effects does that change have on muscle performance?

Answers begin on p. 792.

■ Integration with Other Systems

To operate at maximum efficiency, the muscular system must be supported by many other systems. The changes that occur during exercise provide a good example of such interactions. As noted previously, active muscles consume oxygen and generate carbon dioxide and heat. Responses of other organ systems include the following:

- *Cardiovascular system.* Blood vessels in active muscles and in the skin dilate, and heart rate increases. These adjustments speed up the delivery of oxygen and the removal of carbon dioxide at the muscle and bring heat to the skin for radiation into the environment.
- *Respiratory system.* The rate and depth of respiration increase during exercise. Air moves into and out of the lungs more quickly, and keeps pace with the increased rate of blood flow through the lungs.
- *Integumentary system.* Blood vessels dilate, and sweat gland secretion increases. This combination helps promote evaporation at the skin surface and removes the excess heat generated by muscular activity.
- *Nervous and endocrine systems.* These systems direct the responses of other organ systems by controlling heart rate, respiratory rate, and sweat gland activity.

Even when the body is at rest, the muscular system is interacting with other organ systems. Figure 7–27● summarizes the range of interactions between the muscular system and other systems of the body.

→ **CONCEPT CHECK QUESTION**

1. Which organ systems are involved in the recovery period following the contraction of skeletal muscles?

Answers begin on p. 792.

Chapter Review

Access more review material online at *www.prenhall.com/bledsoe*. There you will find quiz questions, labeling activities, animations, essay questions, and web links.

Key Terms

anaerobic 221
complete tetanus 218
cross-bridges 211
glycolysis 222
insertion 230
isometric contraction 220
isotonic contraction 220

lactic acid 223
motor unit 219
myofilaments 209
myoglobin 224
neuromuscular junction 212
origin 230

prime mover 230
sarcomere 209
sarcoplasmic reticulum 209
synergist 230
tendon 209
transverse tubules 209

Related Clinical Terms

botulism A disease characterized by severe, potentially fatal paralysis of skeletal muscles, that results from the consumption of a bacterial toxin.

carpal tunnel syndrome Inflammation of the sheath that surrounds the flexor tendons of the palm and leads to nerve compression and pain.

compartment syndrome Ischemia (defined shortly) that results from accumulated blood and fluid trapped within limb muscle compartments formed by partitions of dense connective tissue.

fibrosis A process in which a tissue is replaced by fibrous connective tissue. Fibrosis makes muscles weaker and less flexible.

hernia A condition involving an organ or a body part that protrudes through an abnormal opening in the wall of a body cavity.

intramuscular (IM) injection The administration of a drug by injecting it into the mass of a large skeletal muscle.

ischemia (is-KĒ-mē-uh) A deficiency of blood ("blood starvation") in a body part due to compression of regional blood vessels.

muscle cramps Prolonged, involuntary, painful muscular contractions.

muscular dystrophies (DIS-trō-fēz) A varied collection of inherited diseases that produce progressive muscle weakness and deterioration.

myalgia (mī-AL-jē-uh) Muscular pain; a common symptom of a wide variety of conditions and infections.

myasthenia gravis (mī-as-THĒ-nē-uh GRA-vis) A general muscular weakness that results from a reduction in the number of ACh receptors on the motor end plate.

myoma A benign tumor of muscle tissue.

myositis (mi-ō-SĪ-tis) Inflammation of muscle tissue.

polio A viral disease in which the destruction of motor neurons produces paralysis and atrophy of motor units.

rigor mortis A state following death during which muscles are locked in the contracted position, which makes the body extremely stiff.

sarcoma A malignant tumor of mesoderm-derived tissue (muscle, bone, or other connective tissue).

strains Tears or breaks in muscles.

tendinitis Inflammation of the connective tissue that surrounds a tendon.

tetanus A disease caused by a bacterial toxin that results in sustained, powerful contractions of skeletal muscles throughout the body.

Summary Outline

1. The three types of muscle tissue are *skeletal muscle, cardiac muscle,* and *smooth muscle.* The muscular system includes all of the body's skeletal muscles, which can be controlled voluntarily.

FUNCTIONS OF SKELETAL MUSCLE 208

1. **Skeletal muscles** attach to bones directly or indirectly and perform the following functions: (1) produce movement of the skeleton, (2) maintain posture and body position, (3) support soft tissues, (4) guard entrances and exits, and (5) maintain body temperature.

THE ANATOMY OF SKELETAL MUSCLES 208

Gross Anatomy 208

1. Each muscle fiber is surrounded by an **endomysium.** Bundles of muscle fibers are sheathed by a **perimysium,** and the entire muscle is covered by an **epimysium.** At the end of the muscle is a **tendon.** (*Figure 7–1*)

Microanatomy 209

2. A muscle cell has a **sarcolemma** (cell membrane), **sarcoplasm** (cytoplasm), and a **sarcoplasmic reticulum,** similar to the smooth endoplasmic reticulum of other cells. **Transverse tubules (T tubules)** and **myofibrils** aid in contraction. Filaments in a myofibril are organized into repeating functional units called **sarcomeres.** (*Figure 7–2a–c*)

3. **Myofilaments** consist of **thin filaments** (*actin*) and **thick filaments** (*myosin*). (*Figure 7–2d,e*)

4. The spatial relationships between the thick and thin filaments change as the muscle contracts and shortens. The **Z lines** move closer together as the thin filaments slide past the thick filaments. (*Figure 7–3*)

5. The contraction process involves **active sites** on thin filaments and **cross-bridges** of the thick filaments. For each cross-bridge, sliding involves repeated cycles of "attach, pivot, detach, and return." At rest, the necessary interactions are prevented by **tropomyosin** and **troponin** proteins on the thin filaments.

THE CONTROL OF MUSCLE FIBER CONTRACTION 212

1. Neural control of muscle function involves a link between electrical activity in the sarcolemma and the initiation of a contraction.

The Neuromuscular Junction 212

2. Each skeletal muscle fiber is controlled by a neuron at a **neuromuscular junction;** the junction includes the **synaptic terminal,** the **synaptic cleft,** and the **motor end plate. Acetylcholine (ACh)** and **acetylcholinesterase (AChE)** play roles in the chemical communication between the synaptic terminal and muscle fiber. (*Figure 7–4a,b*)

3. When an **action potential** arrives at the synaptic terminal, acetylcholine is released into the synaptic cleft. The binding of ACh to receptors on the motor end plate leads to the generation of an action potential in the sarcolemma. The passage of an action potential along a transverse tubule triggers the release of calcium ions from the *terminal cisternae* of the sarcoplasmic reticulum. (*Figure 7–4c*)

The Contraction Cycle 215

4. A contraction involves a repeated cycle of "attach, pivot, detach, and return." It begins when calcium ions are released by the sar-

coplasmic reticulum. The calcium ions bind to troponin, which changes position and moves tropomyosin away from the active sites of actin. Cross-bridge binding of myosin heads to actin can then occur. After binding, each myosin head pivots at its base, and pulls the actin filament toward the center of the sarcomere. *(Figure 7–5)*

5. A summary of the contraction process, from ACh release to the end of the contraction, is shown in *Table 7–1*.

Key Note 216

MUSCLE MECHANICS 218

1. The amount of tension produced by a muscle fiber depends on the number of cross-bridges formed.

2. Both the number of activated muscle fibers and their rate of stimulation control the tension developed by an entire skeletal muscle.

The Frequency of Muscle Fiber Stimulation 218

3. A muscle fiber **twitch** is a cycle of contraction and relaxation produced by a single stimulus. *(Figure 7–6)*

4. Repeated stimulation before the relaxation phase ends can result in the addition of twitches (known as **summation**). The result can be either **incomplete tetanus** (in which tension peaks because the muscle is never allowed to relax completely) or **complete tetanus** (in which the relaxation phase is completely eliminated). *(Figure 7–7)*

The Number of Muscle Fibers Involved 219

5. The number and size of a muscle's **motor units** indicate how precisely the muscle's movements are controlled. *(Figure 7–8)*

6. Muscle tension is increased by increasing the number of motor units involved—a process called **recruitment.**

Key Note 220

7. Resting **muscle tone** stabilizes bones and joints. Inadequate stimulation causes muscles to undergo **atrophy.**

Isotonic and Isometric Contractions 220

8. Normal activities usually include both **isotonic contractions** (in which the tension in a muscle remains constant as the muscle shortens) and **isometric contractions** (in which the muscle's tension rises but the length of the muscle remains constant).

Muscle Elongation 221

9. Contraction is an active process, but elongation of a muscle fiber is passive. Elongation can result from elastic forces, the contraction of opposing muscles, or the effects of gravity.

THE ENERGETICS OF MUSCULAR ACTIVITY 221

1. Muscle contractions require large amounts of energy from ATP.

ATP and CP Reserves 221

2. ATP is an energy-transfer molecule, not an energy-storage molecule. **Creatine phosphate (CP)** can release stored energy to convert ADP to ATP. A resting muscle cell contains many times more CP than ATP. *(Figure 7–9a)*

ATP Generation 221

3. At rest or moderate levels of activity, aerobic metabolism in mitochondria can provide most of the ATP required to support muscle contractions.

4. When a muscle fiber runs short of ATP and CP, enzymes can break down glycogen molecules to release glucose that can be broken down by **glycolysis.** *(Figure 7–9b)*

5. At peak levels of activity the cell relies heavily on the **anaerobic** process of glycolysis to generate ATP, because the mitochondria cannot obtain enough oxygen to meet the existing ATP demands. *(Figure 7–9c)*

Muscle Fatigue 223

6. **Muscle fatigue** occurs when a muscle can no longer contract, because of pH changes due to the buildup of **lactic acid,** a lack of energy, or other problems.

The Recovery Period 223

7. The **recovery period** begins immediately after a period of muscle activity and continues until conditions inside the muscle have returned to pre-exertion levels. The *oxygen debt* created during exercise is the amount of oxygen used during the recovery period to restore normal conditions.

Key Note 223

MUSCLE PERFORMANCE 224

1. Muscle performance can be considered in terms of **force** (the maximum amount of tension produced by a particular muscle or muscle group) and **endurance** (the duration of muscular activity).

Types of Skeletal Muscle Fibers 224

2. The two types of human skeletal muscle fibers are **fast fibers** and **slow fibers.**

3. Fast fibers are large in diameter, contain densely packed myofibrils, large reserves of glycogen, and few mitochondria. They produce rapid and powerful contractions of relatively short duration.

4. Slow fibers are smaller in diameter and take three times as long to contract after stimulation. Specializations such as an extensive capillary supply, abundant mitochondria, and high concentrations of **myoglobin** enable them to contract for long periods of time.

Physical Conditioning 224

5. **Anaerobic endurance** is the time over which a muscle can support sustained, powerful contractions through anaerobic mechanisms. Training to develop anaerobic endurance can lead to **hypertrophy** (enlargement) of the stimulated muscles.

6. **Aerobic endurance** is the time over which a muscle can continue to contract while supported by mitochondrial activities.

Key Note 225

CARDIAC AND SMOOTH MUSCLE TISSUES 225

Cardiac Muscle Tissue 225

1. Cardiac muscle cells differ from skeletal muscle fibers in that cardiac muscle cells are smaller, typically have a single central nucleus,

have a greater reliance on aerobic metabolism when contracting at peak levels, and have **intercalated discs.** (*Figure 7–10a; Table 7–2*)

2. Cardiac muscle cells have *automaticity* and do not require neural stimulation to contract. Their contractions last longer than those of skeletal muscles, and cardiac muscle cannot undergo tetanus.

Smooth Muscle Tissue 226

3. Smooth muscle is nonstriated, involuntary muscle tissue that can contract over a greater range of lengths than skeletal muscle cells. (*Figure 7–10b; Table 7–2*)

4. Many smooth muscle fibers lack direct connections to motor neurons; those that are innervated are not under voluntary control.

ANATOMY OF THE MUSCULAR SYSTEM 227

1. The **muscular system** includes approximately 700 skeletal muscles, which can be voluntarily controlled. (*Figure 7–11*)

Origins, Insertions, and Actions 230

2. Each muscle can be identified by its **origin, insertion,** and **primary action.** A muscle can be classified by its primary action as a **prime mover,** or **agonist;** as a **synergist;** or as an **antagonist.**

Names of Skeletal Muscles 230

3. The names of muscles often provide clues to their location, orientation, or function. (*Table 7–3*)

4. The **axial musculature** arises on the axial skeleton; it positions the head and spinal column and moves the rib cage. The **appendicular musculature** stabilizes or moves components of the appendicular skeleton.

The Axial Muscles 231

5. The axial muscles fall into four groups based on location and/or function: muscles of (a) the head and neck, (b) the spine, (c) the trunk, and (d) the pelvic floor.

6. The muscles of the head include the **frontalis, orbicularis oris, buccinator, masseter, temporalis,** and **pterygoids.** (*Figure 7–12; Table 7–4*)

7. The muscles of the neck include the **platysma, digastric, mylohyoid, stylohyoid,** and **sternocleidomastoid.** (*Figures 7–12, 7–13; Table 7–4*)

8. The **splenius capitis** and **semispinalis capitis** are the most superior muscles of the spine. The extensor muscles of the spine, or **erector spinae,** can be classified into the **spinalis, longissimus,** and **iliocostalis** groups. In the lower lumbar and sacral regions, the longissimus and iliocostalis are sometimes called the *sacrospinalis* muscles. (*Figures 7–14 through 7–17; Table 7–5*)

9. The muscles of the trunk include the **oblique** and **rectus** muscles. The thoracic region muscles include the **intercostal** and **transversus** muscles. The **diaphragm** is also important to respiration. (*Figure 7–18; Table 7–6*)

10. The muscular floor of the pelvic cavity is called the **perineum.** These muscles support the organs of the pelvic cavity and control the movement of materials through the urethra and anus. (*Figures 7–19, 7–20; Table 7–7*)

The Appendicular Muscles 239

11. Together, the **trapezius** and the sternocleidomastoid affect the position of the shoulder, head, and neck. Other muscles that insert on the scapula include the **rhomboid,** the **levator scapulae,** the **serratus anterior,** and the **pectoralis minor.** (*Figure 7–21; Table 7–8*)

12. The **deltoid** and the **supraspinatus** produce abduction of the arm at the shoulder. The **subscapularis, teres major, infraspinatus,** and **teres minor** rotate the arm at the shoulder. (*Figure 7–21; Table 7–9*)

13. The **pectoralis major** flexes the shoulder joint, and the **latissimus dorsi** extends it. Both of these muscles adduct and rotate the arm at the shoulder joint. (*Figure 7–21; Table 7–9*)

14. The primary actions of the **biceps brachii** and the **triceps brachii** (long head) affect the elbow. The **brachialis** and **brachioradialis** flex the elbow. The **flexor carpi ulnaris,** the **flexor carpi radialis,** and the **palmaris longus** cooperate to flex the wrist. They are opposed by the **extensor carpi radialis** and the **extensor carpi ulnaris.** The **pronator** muscles pronate the forearm, opposed by the **supinator** and the biceps brachii. (*Figures 7–22, 7–23; Table 7–10*)

15. **Gluteal muscles** cover the lateral surfaces of the ilia. They produce extension, abduction, and rotation at the hip. (*Figure 7–24a; Table 7–11*)

16. Adductors of the thigh work across the hip joint; these muscles include the **adductor magnus, adductor brevis, adductor longus, pectineus,** and **gracilis.** (*Figure 7–24b; Table 7–11*)

17. The **psoas major** and the **iliacus** merge to form the **iliopsoas** muscle, which is a powerful flexor of the hip. (*Figure 7–24b; Table 7–11*)

18. The flexors of the knee include the hamstrings (**biceps femoris, semimembranosus,** and **semitendinosus**) and **sartorius.** The **popliteus** aids flexion by unlocking the knee. (*Figure 7–25; Table 7–12*)

19. The *knee extensors* are known as the **quadriceps femoris.** This group includes the three **vastus** muscles and the **rectus femoris.** (*Figure 7–25; Table 7–12*)

20. The **gastrocnemius** and **soleus** muscles produce plantar flexion. A pair of **fibularis** muscles produces eversion as well as plantar flexion. The **tibialis anterior** performs dorsiflexion. (*Figure 7–26; Table 7–13*)

21. Control of the phalanges is provided by muscles that originate at the tarsal bones and at the metatarsal bones. (*Table 7–13*)

AGING AND THE MUSCULAR SYSTEM 251

1. The aging process reduces the size, elasticity, and power of all muscle tissues. Both exercise tolerance and the ability to recover from muscular injuries decrease.

INTEGRATION WITH OTHER SYSTEMS 253

1. To operate at maximum efficiency, the muscular system must be supported by many other organ systems. Even at rest, it interacts extensively with other systems. (*Figure 7–27*)

Review Questions

Level 1: Reviewing Facts and Terms

Match each item in column A with the most closely related item in column B. Place letters for answers in the spaces provided.

COLUMN A

____ 1. epimysium

____ 2. fascicle

____ 3. endomysium

____ 4. motor end plate

____ 5. transverse tubule

____ 6. actin

____ 7. myosin

____ 8. extensor of the knee

____ 9. sarcomeres

____ 10. tropomyosin

____ 11. recruitment

____ 12. muscle tone

____ 13. white muscles

____ 14. flexor of the leg

____ 15. red muscles

____ 16. hypertrophy

COLUMN B

a. resting tension

b. contractile units

c. thin filaments

d. surrounds muscle fiber

e. enlargement

f. surrounds muscle

g. slow fibers

h. thick filaments

i. muscle bundle

j. hamstring muscles

k. covers active sites

l. conducts action potentials

m. fast fibers

n. quadriceps muscles

o. multiple motor units

p. binds ACh

17. A skeletal muscle contains:
   (a) connective tissues.
   (b) blood vessels and nerves.
   (c) skeletal muscle tissue.
   (d) connective tissues, blood vessels and nerves, and skeletal muscle tissue.

18. The type of contraction in which the tension rises but the resistance does not move is called:
   (a) a wave summation.   (c) an isotonic contraction.
   (b) a twitch.   (d) an isometric contraction.

19. What are the five functions of skeletal muscle?

20. What five interlocking steps are involved in the contraction process?

21. What forms of energy reserves are found in resting skeletal muscle cells?

23. What is the functional difference between the axial musculature and the appendicular musculature?

Level 2: Reviewing Concepts

24. Areas of the body where no slow fibers are found include the:
   (a) back and calf muscles.
   (b) eye and hand.
   (c) chest and abdomen.
   (d) back and calf muscles, eye and hand, and chest and abdomen.

25. Describe the basic sequence of events that occurs at a neuromuscular junction.

26. Why is the multinucleate condition important in skeletal muscle fibers?

27. The muscles of the spine include many dorsal extensors but few ventral flexors. Why?

28. What specific structural characteristic makes voluntary control of urination and defecation possible?

29. What types of movements are affected when the hamstrings are injured?

Level 3: Critical Thinking and Clinical Applications

30. Many potent insecticides contain toxins called *organophosphates*, which interfere with the action of the enzyme acetylcholinesterase. Terry is using an insecticide that contains organophosphates and is very careless. He does not use gloves or a mask, so he absorbs some of the chemical through his skin and inhales a large amount as well. What signs would you expect to observe in Terry as a result of organophosphate poisoning?

31. The time of a murder victim's death is commonly estimated by the flexibility of the body. Explain why this is possible.

32. Makani is interested in building up his thigh muscles, specifically the quadriceps group. What exercises would you recommend to help him accomplish his goal?

8 The Nervous System

AS AN EMS PROVIDER *you can never tell where your next call will be or the nature of the call. Being a good EMT or paramedic means that you have the capacity to adapt to* whatever environment you find yourself in. It also means that you are good at "thinking on the run" and solving problems quickly to ensure your patients receive the best care available.

Chapter Outline

Chapter Objectives

1. Describe the two major anatomical divisions of the nervous system and their general functions. (pp. 260–261)
2. Distinguish between neurons and neuroglia on the basis of their structure and function. (pp. 261–266)
3. Discuss the events that generate action potentials in the membranes of nerve cells. (pp. 266–270)
4. Explain the mechanism of nerve impulse transmission at the synapse. (pp. 270–274)
5. Describe the three meningeal layers that surround the central nervous system. (pp. 274–276)
6. Discuss the structure and functions of the spinal cord. (pp. 278–282)
7. Name the major regions of the brain and describe their functions. (pp. 282–284)
8. Locate the motor, sensory, and association areas of the cerebral cortex and discuss their functions. (pp. 284–294)
9. Identify the cranial nerves and relate each pair of cranial nerves to its principal functions. (pp. 294–297)
10. Relate the distribution pattern of spinal nerves to the regions they innervate. (pp. 297–298)
11. Describe the components of a reflex arc. (pp. 298–303)
12. Identify the principal sensory and motor pathways. (pp. 303–305)
13. Compare and contrast the functions and structures of the sympathetic and parasympathetic divisions. (pp. 306–313)
14. Summarize the effects of aging on the nervous system. (p. 314)
15. Discuss the interrelationships between the nervous system and other organ systems. (p. 319)

Vocabulary Development

a- without; *aphasia*
af to; *afferent*
arachne spider; *arachnoid membrane*
astro- star; *astrocyte*
ataxia a lack of order; *ataxia*
axon axis; *axon*
cauda tail; *cauda equina*
cephalo- head; *diencephalon*
chiasm a crossing; *optic chiasm*
choroid a vascular coat; *choroid plexus*
colliculus a small hill; *superior colliculus*
commissura a joining together; *commissure*
cortex rind; *neural cortex*

cyte cell; *astrocyte*
dia through; *diencephalon*
dura hard; *dura mater*
equus horse; *cauda equina*
ef-, ex- from; *efferent*
ferre to carry; *afferent*
ganglio knot; *ganglion*
glia glue; *neuroglia*
hypo- below; *hypothalamus*
inter- between; *interneurons*
lexis diction; *dyslexia*
limbus a border; *limbic system*
mamilla a little breast; *mamillary bodies*

mater mother; *dura mater*
meninx membrane; *meninges*
meso- middle; *mesencephalon*
neuro- nerve; *neuron*
nigra black; *substantia nigra*
oligo- few; *oligodendrocytes*
phasia speech; *aphasia*
pia delicate; *pia mater*
plexus a network; *choroid plexus*
saltare to leap; *saltatory*
syn- together; *synapse*
vagus wandering; *vagus nerve*
vas vessel; *vasomotor*

TWO ORGAN SYSTEMS, the *nervous system* and the *endocrine system,* coordinate organ system activities to maintain homeostasis in response to changing environmental conditions. The nervous system responds relatively swiftly but briefly to stimuli, whereas endocrine responses develop more slowly but last much longer. Thus, the nervous system adjusts body position and moves your eyes across this page, while the endocrine system adjusts the entire body's daily rate of energy use and directs such long-term processes as growth and maturation. (We will consider the endocrine system in Chapter 10.)

The nervous system is the most complex organ system. As you read these words and think about them, at the involuntary level your nervous system is also monitoring the external environment and your internal systems and issuing commands as needed to maintain homeostasis. In a few hours—at mealtime or while you are sleeping—the pattern of nervous system activity will be very different. The change from one pattern of activity to another can be almost instantaneous because neural function relies on electrical events that proceed at great speed. This chapter examines the structure and function of the nervous system, from its cells through their organization into two major divisions: the *central nervous system* and the *peripheral nervous system.* ∞ p. 9

■ The Nervous System

The **nervous system** (1) monitors the internal and external environments, (2) integrates sensory information, and (3) coordinates voluntary and involuntary responses of many other organ systems. These functions are performed by cells called neurons, which are supported and protected by surrounding cells collectively termed *neuroglia.*

The nervous system has two major anatomical subdivisions. The **central nervous system (CNS),** which consists of the *brain* and the *spinal cord,* integrates and coordinates the processing of sensory data and the transmission of motor commands. The CNS is also the seat of higher functions, such as intelligence, memory, and emotion. All communication between the CNS and the rest of the body occurs over the **peripheral nervous system (PNS),** which includes all the neural tissue *outside* the CNS.

The functional relationships of the two major anatomical subdivisions of the nervous system are detailed in Figure 8–1●. Sensory information detected outside the nervous system—by structures called *receptors*—is transmitted by the **afferent division** (*af-,* to + *ferre,* to carry) of the PNS to sites in the CNS, where the information is processed. The CNS then sends motor commands by means of the **efferent division** (*ef-,* from) of the PNS to muscles, glands, and adipose tissue, which are called *effectors.*

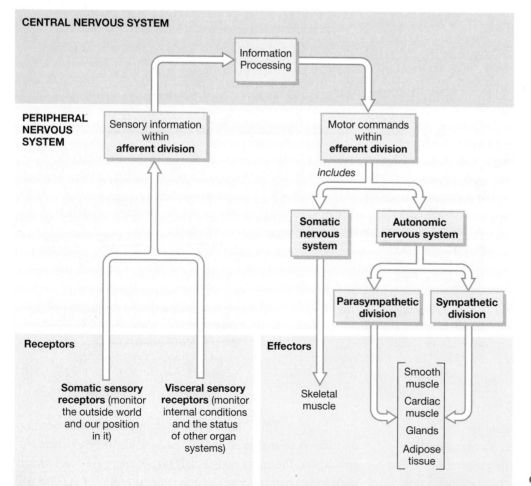

● **FIGURE 8–1 A Functional Overview of the Nervous System.**

The efferent division of the PNS is subdivided into the **somatic nervous system (SNS),** which provides control over skeletal muscle contractions, and the **autonomic nervous system (ANS)** or *visceral motor system,* which provides automatic involuntary regulation of smooth muscle, cardiac muscle, glandular secretions, and adipose tissue. The ANS includes a *sympathetic division* and a *parasympathetic division,* which commonly have opposite effects. For example, activity of the sympathetic division accelerates the heart rate, whereas the parasympathetic division slows the heart rate.

■ Cellular Organization in Neural Tissue

The nervous system includes all the neural tissue in the body. Neural tissue, introduced in Chapter 4, consists of two kinds of cells, *neurons* and *neuroglia.* ∞ p. 115 **Neurons** (*neuro-,* nerve) are the basic units of the nervous system. All neural functions involve the communication of neurons with one another and with other cells. The **neuroglia** (noo-RŌG-lē-uh

or noo-rō-GLĒ-uh; *glia,* glue) regulate the environment around the neurons, provide a supporting framework for neural tissue, and act as phagocytes. Although they are much smaller cells, neuroglia (also called *glial cells*) far outnumber neurons. Unlike most neurons, most glial cells retain the ability to divide.

Neurons

The General Structure of Neurons

A "representative" neuron has (1) a cell body; (2) several branching, sensitive **dendrites,** which receive incoming signals; and (3) an elongate **axon,** which carries outgoing signals toward (4) one or more **synaptic terminals** (Figure 8–2●). At each synaptic terminal, the neuron communicates with another cell. Neurons can have a variety of shapes; Figure 8–2 shows a *multipolar neuron,* the most common type of neuron in the CNS.

The cell body of a typical neuron contains a large, round nucleus with a prominent nucleolus. Most neurons lack centrioles, which are organelles involved in the movement of chromosomes during mitosis. ∞ p. 77 As a result, typical CNS neurons cannot divide, so they cannot be replaced if lost to injury or disease. Although neural stem cells persist in the adult nervous system, they are typically inactive except in the nose, where the regeneration of olfactory (smell) receptors maintains our sense of smell, and in the *hippocampus,* which is a portion of the brain involved with memory storage. The mechanisms that trigger neural stem cell activity are now being investigated, with the goal of preventing or reversing neuron loss due to trauma, disease, or aging.

The cell body also contains organelles that provide energy and synthesize organic compounds. The numerous mitochondria, free and fixed ribosomes, and membranes of the rough endoplasmic reticulum (RER) give the cytoplasm a coarse, grainy

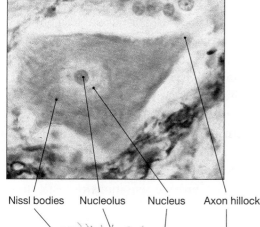

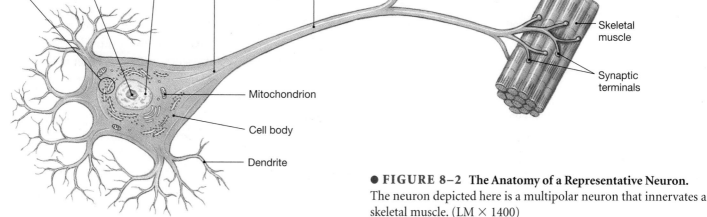

Nissl bodies   Nucleolus   Nucleus   Axon hillock   Axon   Collateral   Skeletal muscle   Synaptic terminals   Mitochondrion   Cell body   Dendrite

● **FIGURE 8–2 The Anatomy of a Representative Neuron.** The neuron depicted here is a multipolar neuron that innervates a skeletal muscle. (LM × 1400)

appearance. Clusters of rough ER and free ribosomes, known as **Nissl bodies,** give a gray color to areas that contain neuron cell bodies and account for the color of *gray matter* seen in brain and spinal cord dissections.

Projecting from the cell body are a variable number of dendrites and a single large axon. The cell membrane of the dendrites and cell body is sensitive to chemical, mechanical, or electrical stimulation. In a process described later, such stimulation often leads to the generation of an electrical impulse, or *action potential,* that travels along the axon. Action potentials begin at a thickened region of the cell body called the *axon hillock.* The axon may branch along its length, producing branches called *collaterals.* Synaptic terminals are found at the tips of each branch. A synaptic terminal is part of a **synapse,** which is a site where a neuron communicates with another cell.

Structural Classification of Neurons

The billions of neurons in the nervous system are variable in form. Based on the relationship of the dendrites to the cell body and axon, neurons are classified into three types (Figure 8–3●):

1. A **multipolar neuron** has two or more dendrites and a single axon (Figure 8–3a●). These are the most common neurons in the CNS. All of the motor neurons that control skeletal muscles are multipolar.

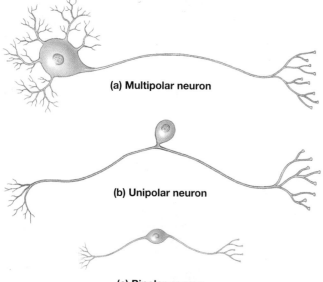

(a) Multipolar neuron

(b) Unipolar neuron

(c) Bipolar neuron

● **FIGURE 8–3   A Structural Classification of Neurons.** The neurons are not drawn to scale; bipolar neurons are many times smaller than typical unipolar and multipolar neurons.

2. In a **unipolar neuron,** the dendrites and axon are continuous, and the cell body lies off to one side (Figure 8–3b●). In a unipolar neuron, the action potential begins at the base of the dendrites, and the rest of the process is considered an axon. Most sensory neurons of the peripheral nervous system are unipolar.
3. **Bipolar neurons** have two processes—one dendrite and one axon—with the cell body between them (Figure 8–3c●). Bipolar neurons are rare but occur in special sense organs, where they relay information about sight, smell, or hearing from receptor cells to other neurons.

Functional Classification of Neurons

Neurons are sorted into three functional groups: (1) *sensory neurons,* (2) *motor neurons,* and (3) *interneurons.*

SENSORY NEURONS.   The approximately 10 million **sensory neurons** in the human body form the afferent division of the PNS. Sensory neurons receive information from *sensory receptors* that monitor the external and internal environments and then relay the information to other neurons in the CNS (spinal cord or brain). The receptor may be a dendrite of a sensory neuron or specialized cells of other tissues that communicate with the sensory neuron.

Receptors may be categorized according to the information they detect. Two types of **somatic sensory receptors** detect information about the outside world or our physical position within it. (1) **External receptors** provide information about the external environment in the form of touch, temperature, and pressure sensations and the more complex senses of sight, smell, hearing, and touch; and (2) **proprioceptors** (prō-prē-ō-SEP-torz; *proprius,* one's own + *capio,* to take) monitor the position and movement of skeletal muscles and joints. **Visceral receptors,** or **internal receptors,** monitor the activities of the digestive, respiratory, cardiovascular, urinary, and reproductive systems and provide sensations of taste, deep pressure, and pain.

MOTOR NEURONS.   The half million **motor neurons** of the efferent division carry instructions from the CNS to other tissues, organs, or organ systems. The peripheral targets are called *effectors* because they respond by *doing* something. For example, a skeletal muscle is an effector that contracts upon neural stimulation. Neurons in the two efferent divisions of the PNS target separate classes of effectors. The **somatic motor neurons** of the somatic nervous system innervate skeletal muscles, whereas the **visceral motor neurons** of the autonomic nervous system innervate all other effectors, including cardiac muscle, smooth muscle, glands, and adipose tissue.

INTERNEURONS.   The 20 billion **interneurons,** or *association neurons,* are located entirely within the brain and the spinal cord. Interneurons, as the name implies (*inter-,* between), interconnect other neurons. They are responsible for the distribution of sensory information and the coordination of motor activity. The more complex the response to a given stimulus, the greater the number of interneurons involved. Interneurons also play a role in all higher functions, such as memory, planning, and learning.

Neuroglia

Neuroglia are found in both the CNS and PNS, but the CNS has a greater variety of glial cells. There are four types of neuroglial cells in the central nervous system (Figure 8–4●):

● **FIGURE 8–4 Neuroglia in the CNS.** This diagrammatic view of neural tissue in the CNS depicts the relationships between neuroglia and neurons.

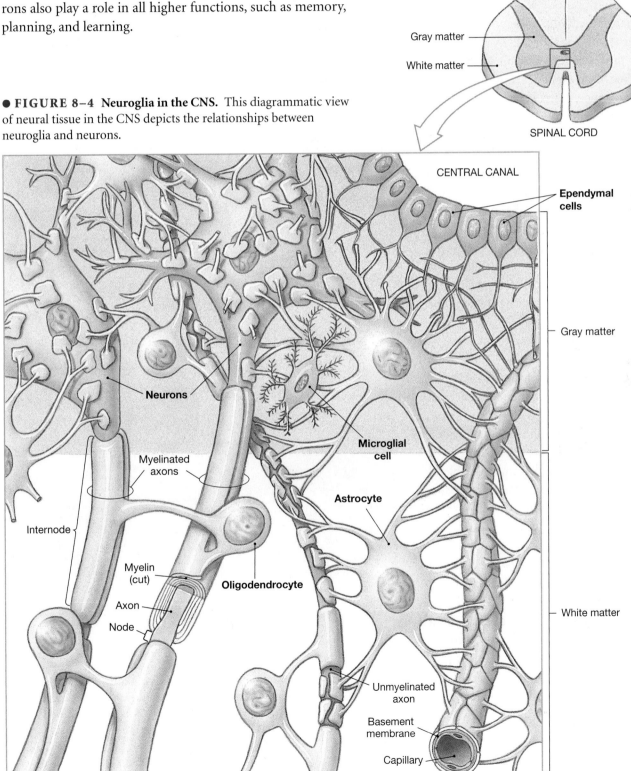

1. **Astrocytes** (AS-trō-sīts; *astro-,* star + *cyte,* cell) are the largest and most numerous neuroglia. Astrocytes secrete chemicals vital to the maintenance of the *blood-brain barrier,* which isolates the CNS from the general circulation. The secretions cause the capillaries of the CNS to become impermeable to many compounds that could interfere with neuron function. Astrocytes also create a structural framework for CNS neurons and perform repairs in damaged neural tissues.

2. **Oligodendrocytes** (ol-i-gō-DEN-drō-sīts; *oligo-,* few) have fewer processes (cytoplasmic extensions) than astrocytes. Their thin, expanded tips wrap around axons, which creates a membranous sheath of insulation made of **myelin** (MĪ-e-lin). Each oligodendrocyte myelinates short segments of several axons, so many oligodendrocytes are needed to coat an entire axon with myelin. Such an axon is said to be **myelinated.** The gaps between adjacent cell processes are called *nodes,* or the *nodes of Ranvier* (rahn-vē-Ā). The areas covered in myelin are called *internodes.* Myelin increases the speed at which an action potential travels along the axon. Not every axon in the CNS is myelinated, and those without a myelin coating are said to be **unmyelinated.** Myelin is lipid-rich, and on dissection, areas of the CNS that contain myelinated axons appear glossy white. These areas constitute the **white matter** of the CNS, whereas areas of **gray matter** are dominated by neuron cell bodies.

3. **Microglia** (mī-KRŌG-lē-uh) are the smallest and rarest of the neuroglia in the CNS. Microglia are phagocytic cells derived from white blood cells that migrated into the CNS as the nervous system formed. They perform protective functions such as engulfing cellular waste and pathogens.

4. **Ependymal** (ep-EN-di-mul) **cells** line both the *central canal* of the spinal cord and the chambers *(ventricles)* of the brain, which are cavities in the CNS that are filled with cerebrospinal fluid (CSF). This lining of epithelial cells is called the **ependyma** (ep-EN-di-muh). In some regions of the brain, the ependyma produces CSF, and the cilia on ependymal cells in other locations help circulate this fluid within and around the CNS.

There are two types of neuroglia in the PNS. **Satellite cells** surround and support neuron cell bodies in the peripheral nervous system, much as astrocytes do in the CNS. The other glial cells in the PNS are **Schwann cells** (Figure 8–5a●).

Schwann cells cover every axon outside the CNS. Wherever a Schwann cell covers an axon, the outer surface of the Schwann cell is called the *neurilemma* (noor-i-LEM-uh). Whereas an oligodendrocyte in the CNS may myelinate portions of several adjacent axons, a Schwann cell can myelinate only one segment of a single axon (Figure 8–5a). However, a Schwann cell can enclose portions of several different unmyelinated axons (Figure 8–5b●).

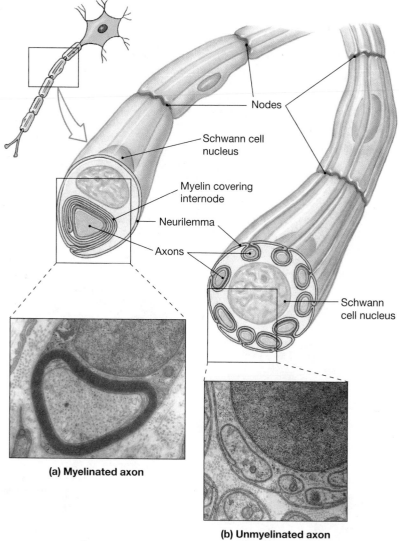

(a) Myelinated axon

(b) Unmyelinated axon

● **FIGURE 8–5** **Schwann Cells and Peripheral Axons.** (a) A myelinated axon in the PNS is covered by several Schwann cells, each of which forms a myelin sheath around a portion of the axon. This arrangement differs from the way myelin forms in the CNS; compare with Figure 8–4. (b) A single Schwann cell can encircle several unmyelinated axons. Every axon in the PNS is completely enclosed by Schwann cells. (TEMs × 14,048)

Clinical Note
DEMYELINATION DISORDERS

Demyelination is the progressive destruction of myelin sheaths, accompanied by inflammation, axon damage, and scarring of neural tissue. The result is a gradual loss of sensation and motor control that leaves affected regions numb and paralyzed. In one demyelination disorder, *multiple sclerosis* (skler-Ō-sis; *sklerosis*, hardness), or *MS,* axons in the optic nerve, brain, and/or spinal cord are affected. Common signs and symptoms of MS include partial loss of vision and problems with speech, balance, and general motor coordination. Other important demyelination disorders include *heavy metal poisoning, diphtheria,* and *Guillain-Barré syndrome.* ■

Key Note

Neurons perform all of the communication, information processing, and control functions of the nervous system. Neuroglia outnumber neurons and have functions essential to preserving the physical and biochemical structure of neural tissue and the survival of neurons.

Anatomical Organization of Neurons

Neuron cell bodies and their axons are not randomly scattered in the CNS and PNS. Instead, they are organized into masses or bundles that have distinct anatomical boundaries and are identified by specific terms (Figure 8–6●). We will use these terms again, so a brief overview here may prove helpful.

In the PNS:

■ Neuron cell bodies (gray matter) are located in **ganglia.**

■ The white matter of the PNS contains axons bundled together in **nerves;** *spinal nerves* are connected to the spinal cord, and *cranial nerves* are connected to the brain. Both sensory and motor axons may be present in the same nerve.

In the CNS:

■ A collection of neuron cell bodies with a common function is called a **center.** A center with a discrete boundary is called a **nucleus.** Portions of the brain surface are covered by a thick layer of gray matter called **neural cortex** (*cortex,* rind). The term *higher centers* refers to the most complex integration centers, nuclei, and cortical areas in the brain.

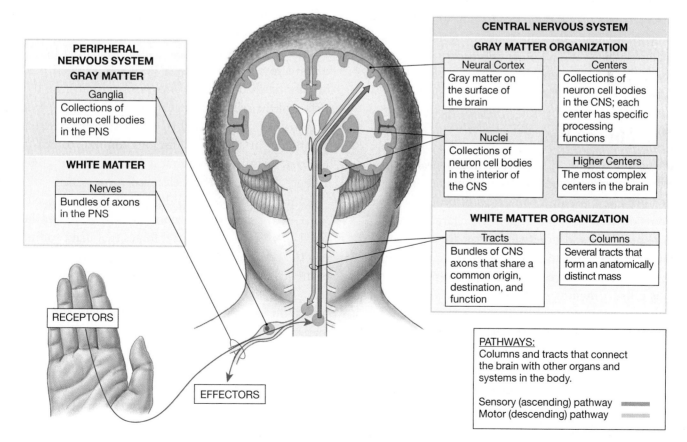

● **FIGURE 8–6   The Anatomical Organization of the Nervous System.**

- The white matter of the CNS contains bundles of axons that share common origins, destinations, and functions. These bundles are called **tracts.** Tracts in the spinal cord form larger groups called **columns.**

- **Pathways** link the centers of the brain with the rest of the body. For example, **sensory** (*ascending*) **pathways** distribute information from sensory receptors to processing centers in the brain, and **motor** (*descending*) **pathways** begin at CNS centers concerned with motor activity and end at the skeletal muscles they control.

→ **CONCEPT CHECK QUESTIONS**

1. What would be the effect of damage to the afferent division of the PNS?
2. Examination of a tissue sample reveals unipolar neurons. Are these more likely to be sensory neurons or motor neurons?
3. Which type of glial cell would occur in greater-than-normal numbers in the brain tissue of a person with a CNS infection?

Answers begin on p. 792.

■ Neuron Function

The sensory, integrative, and motor functions of the nervous system are dynamic and ever changing. All of the communications between neurons and other cells occur through their membrane surfaces. These membrane changes are electrical events that proceed at great speed.

The Membrane Potential

A characteristic feature of a living cell is a *polarized* cell membrane. An undisturbed cell has a cell membrane that is polarized because it separates an excess of positive charges on the outside from an excess of negative charges on the inside. When positive and negative charges are held apart, a *potential difference* is said to exist between them. Because the charges are separated by a cell membrane, this potential difference is called a **membrane potential,** or *transmembrane potential.*

The unit of measurement of potential difference is the *volt* (V). Most cars, for example, have 12 V batteries. The membrane potential of cells is much smaller and is usually reported in *millivolts* (mV, thousandths of a volt). The membrane potential of an undisturbed cell is known as its **resting potential.** The resting potential of a neuron is –70 mV; the minus sign indicates that the inside of the cell membrane contains an excess of negative charges as compared with the outside.

Factors Responsible for the Membrane Potential

In addition to an imbalance of electrical charges, the intracellular and extracellular fluids differ markedly in ionic composition. For example, the extracellular fluid contains relatively high concentrations of sodium ions (Na^+) and chloride ions (Cl^-) whereas the intracellular fluid contains high concentrations of potassium ions (K^+) and negatively charged proteins (Pr^-).

The selective permeability of the cell membrane maintains these differences between the intracellular and extracellular fluids. The proteins within the cytoplasm are too large to cross the membrane, and the ions can enter or leave the cell only with the aid of membrane channel and/or carrier proteins. ∞ p. 63 There are many different types of membrane channels; some are always open (*leak channels*), whereas others (*gated channels*) open or close under specific circumstances, such as a change in voltage.

Both passive and active processes act across the cell membrane to determine the membrane potential. The passive forces are chemical and electrical. Chemical concentration gradients move potassium ions out of the cell and sodium ions into the cell (through separate leak channels). However, because it is easier for potassium ions to diffuse through a potassium channel than for sodium ions to diffuse through sodium channels, potassium ions diffuse out of the cell faster than sodium ions enter the cell. The passive movement of sodium and potassium ions is also influenced by electrical forces across the membrane. Positively charged potassium ions are repelled by the overall positive charge on the outer surface of the cell membrane, whereas the positively charged sodium ions are attracted to the negatively charged inner membrane surface. Potassium ions continue to exit the cell, however, because its chemical concentration gradient is stronger than the repelling electrical force.

To maintain a potential difference across the cell membrane, active processes are needed both to counter the combined chemical and electrical forces that drive sodium ions into the cell and to maintain the potassium concentration gradient. The resting potential remains stable over time because of the actions of a carrier protein, the sodium–potassium exchange pump. ∞ p. 70 This ion pump exchanges three intracellular sodium ions for two extracellular potassium ions. At the normal resting potential of –70 mV sodium ions are ejected as fast as they enter the cell. The cell, therefore, undergoes a net loss of positive charges, and as a result, the interior of the cell membrane contains an excess of negative charges, primarily from negatively charged proteins. Figure 8–7● presents a diagrammatic view of the cell membrane at the resting potential.

● **FIGURE 8–7 The Cell Membrane at the Resting Potential.** The resting potential is the membrane potential of an undisturbed cell. The action of the ATP-driven sodium-potassium exchange pump maintains the concentration gradients of sodium and potassium, as well as the imbalance of electric charges, across the cell membrane of an undisturbed cell.

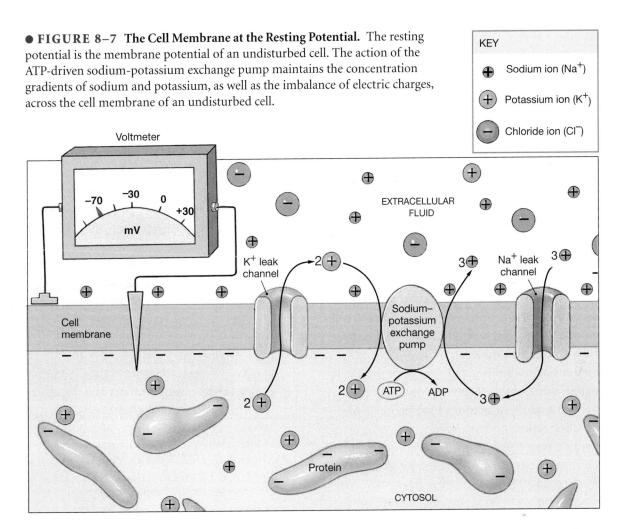

Key Note

A transmembrane potential exists across the cell membrane. It is there because (1) the cytosol differs from extracellular fluid in its chemical and ionic composition and (2) the cell membrane is selectively permeable. The transmembrane potential can change from moment to moment, as the cell membrane changes its permeability in response to chemical or physical stimuli.

Changes in the Membrane Potential

Any stimulus that (1) alters membrane permeability to sodium or potassium or (2) alters the activity of the exchange pump will disturb the resting potential of a cell. Examples of stimuli that can affect membrane potential include exposure to specific chemicals, mechanical pressure, changes in temperature, or shifts in the extracellular ion concentrations. Any change in the resting potential can have an immediate effect on the cell. For example, permeability changes in the sarcolemma of a skeletal muscle fiber trigger a contraction. ∞ p. 216

In most cases, a stimulus opens gated ion channels that are closed when the cell membrane is at its resting potential. The opening of these channels accelerates the movement of ions across the cell membrane, and this movement changes the membrane potential. For example, the opening of gated sodium channels accelerates the entry of sodium into the cell. As the number of positively charged ions on the inner surface of the cell membrane increases, the membrane potential shifts toward 0 mV. A shift in this direction is called a **depolarization** of the membrane. A stimulus that opens gated potassium ion channels will shift the membrane potential away from 0 mV, because additional potassium ions will leave the cell. Such a change, which may take the membrane potential from –70 mV to –80 mV is called a **hyperpolarization.**

Information transfer between neurons and other cells involves graded potentials and action potentials. **Graded potentials** are changes in the membrane potential that cannot spread far from the site of stimulation, and thereby affects only a limited portion of the cell membrane. For example, if a chemical stimulus applied to the cell membrane of a neuron opens gated sodium ion channels at a single site, the sodium ions that enter the cell will depolarize the membrane at that location. Attracted to surrounding negative ions, the sodium ions move along the inner surface of the membrane in all directions. The degree of depolarization decreases with distance from the point of entry, however, because the cytosol resists ion movement and because some sodium ions are lost as they recross the membrane through leak channels.

Graded potentials occur in the membranes of all cells in response to environmental stimuli. They often trigger specific cell functions. For example, a graded potential in the membrane of a gland cell may trigger secretion. However, graded potentials affect too small an area to have an effect on the activities of such large cells as skeletal muscle fibers or neurons. In these cells, graded potentials can influence operations in distant portions of the cell only if they lead to the production of an *action potential*, which is an electrical signal that affects the surface of the entire membrane.

An **action potential** is a propagated change in the membrane potential of the entire cell membrane. Only skeletal muscle fibers and the axons of neurons have *excitable membranes* that conduct action potentials. In a skeletal muscle fiber, the action potential begins at the neuromuscular junction and travels along the entire membrane surface, including the T tubules. ∞ p. 214 The resulting ion movements trigger a contraction. In an axon, an action potential usually begins near the axon hillock and travels along the length of the axon toward the synaptic terminals, where its arrival activates the synapses.

Action potentials are generated by the opening and closing of gated sodium and potassium channels in response to a graded potential (local depolarization). This depolarization acts like pressure on the trigger of a gun. A gun fires only after a certain minimum pressure has been applied to the trigger. It does not matter whether the pressure builds gradually or is exerted suddenly—when the pressure reaches a critical point, the gun will fire. Whenever the gun fires, the forces that were applied to the trigger have no effect on the speed of the bullet leaving the gun. In an axon, the graded potential is the pressure on the trigger, and the action potential is the firing of the gun. An action potential will not appear unless the membrane depolarizes sufficiently to a level known as the **threshold.**

Every stimulus (whether minor or extreme) that brings the membrane to threshold will generate an identical action potential. This is called the **all-or-none principle:** a given stimulus either triggers a typical action potential, or it does not produce one at all.

The Generation of an Action Potential

An action potential begins when the cell membrane at the axon hillock depolarizes to threshold. The steps involved in the generation of an action potential, beginning with a graded depolarization to threshold (from –70 to –60 mV) and ending with a return to the resting potential (–70 mV) are illustrated in Figure 8–8●.

From the moment the voltage-regulated sodium channels open at threshold until *repolarization* (the return to the resting potential) is complete, the membrane cannot respond normally to further stimulation. This is the **refractory period.** The refractory period limits the rate at which action potentials can be generated in an excitable membrane. (The maximum rate of action potential generation is 500–1000 per second.)

Clinical Note
NEUROTOXINS IN SEAFOOD

Potentially deadly forms of poisoning result from eating seafood that contains *neurotoxins*, which are poisons that primarily affect neurons. Several neurotoxins, such as *tetrodotoxin (TTX)* from puffer fish, *saxitoxin (SSX)* from bivalve shellfish, and *ciguatoxin (CTX)* from carnivorous, tropical coral reef fish prevent sodium channels from opening. Motor neurons cannot function under these conditions, and death may result from paralysis of the respiratory muscles. ■

Propagation of an Action Potential

An action potential initially involves a relatively small portion of the total membrane surface. But unlike graded potentials, which diminish rapidly with distance, action potentials go on to affect the entire membrane surface. The basic mechanisms of action potential propagation along unmyelinated and myelinated axons are shown in Figure 8–9●.

At a given site, for a brief moment at the peak of the action potential, the inside of the cell membrane contains an excess of positive ions. Because opposite charges attract, these ions immediately begin spreading along the inner surface of the membrane, drawn to the surrounding negative charges. This *local current* depolarizes adjacent portions of the membrane, and when threshold is reached, action potentials occur at these locations (Figure 8–9a●). Each time a local current develops, the action potential moves forward, but not backward, because the previous segment of the axon is still in the refractory period.

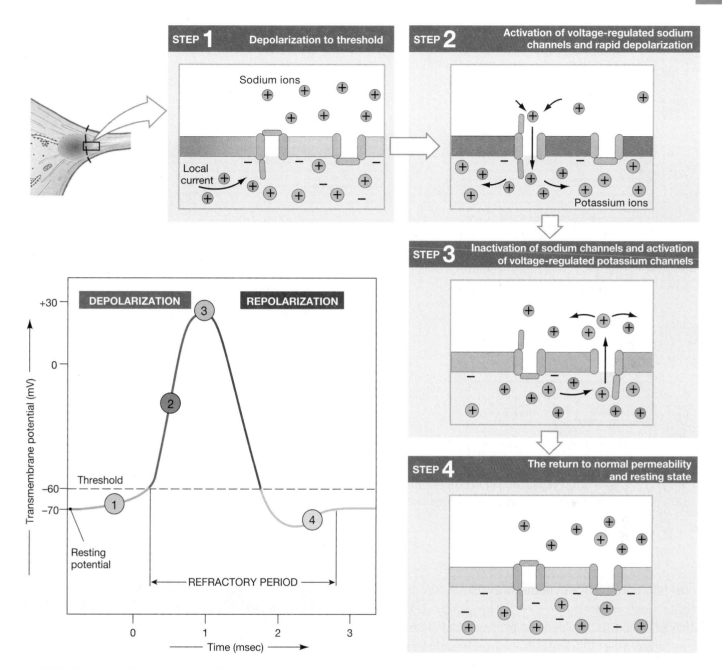

● **FIGURE 8–8 The Generation of an Action Potential.** A sufficiently strong depolarizing stimulus will bring the membrane potential to threshold and trigger an action potential. The numbers on the graph correspond to the steps illustrated at the right.

The process continues in a chain reaction that soon reaches the most distant portions of the cell membrane. This form of action potential transmission is known as **continuous propagation.** You might compare continuous propagation to a person walking toward some destination by taking small "baby steps"; progress is made, but slowly. Continuous propagation occurs along unmyelinated axons at a speed of about 1 meter per second (2 mph).

In a myelinated fiber, the axon is wrapped in layers of myelin. This wrapping is complete except at the nodes, where adjacent glial cells contact one another. Between the nodes, the lipids of the myelin sheath block the flow of ions across the membrane. As a result, continuous propagation cannot occur. Instead, when an action potential occurs at the axon hillock, the local current skips the internode and depolarizes the closest node to threshold (Figure 8–9b●). Thus, the action

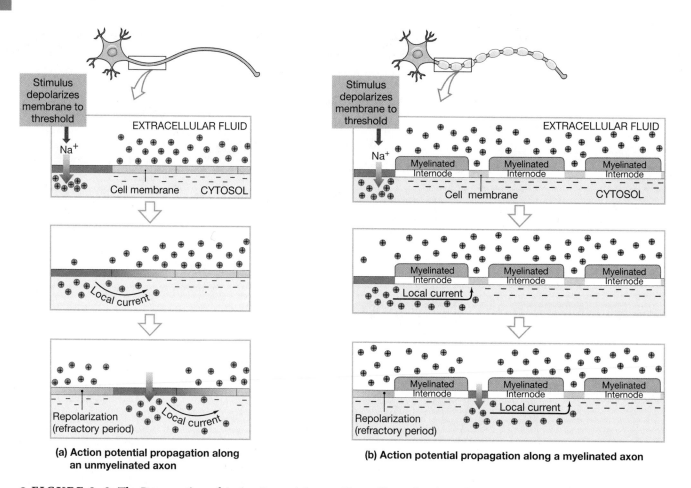

(a) Action potential propagation along an unmyelinated axon

(b) Action potential propagation along a myelinated axon

● **FIGURE 8–9** **The Propagation of Action Potentials over Unmyelinated and Myelinated Axons.** (**a**) Continuous propagation occurs along unmyelinated axons. (**b**) Saltatory propagation occurs along myelinated axons.

potential jumps from node to node rather than proceeding in a series of small steps (Figure 8–9b). This process is called **saltatory propagation,** which takes its name from *saltare*, the Latin word meaning "to leap." Saltatory propagation carries nerve impulses along an axon at speeds that range from 18–140 meters per second (40–300 mph). This faster process might be compared to a person jumping over puddles on their way to their destination.

Key Note

"Information" travels within the nervous system primarily in the form of propagated electrical signals known as action potentials. The most important information, including vision and balance sensations and the motor commands to skeletal muscles, is carried by myelinated axons.

→ **CONCEPT CHECK QUESTIONS**

1. How would a chemical that blocks the sodium channels in a neuron's cell membrane affect the neuron's ability to depolarize?
2. Two axons are tested for propagation velocities. One carries action potentials at 50 meters per second, the other at one meter per second. Which axon is myelinated?

Answers begin on p. 792.

■ Neural Communication

In the nervous system, information moves from one location to another in the form of action potentials along axons. These electrical events are also known as **nerve impulses.** At the end of an axon, the arrival of an action potential results in the transfer of information to another neuron or to an effector cell. The information transfer occurs through the release of chemicals called **neurotransmitters** from the synaptic terminal.

When one neuron communicates with another, the synapse may occur on a dendrite, on the cell body, or along the length of the axon. Synapses between a neuron and another cell type are called **neuroeffector junctions.** As we saw in Chapter 7, a neuron communicates with a muscle cell at a *neuromuscular junction.* ∞ p. 214 At a *neuroglandular junction,* a neuron controls or regulates the activity of a secretory cell.

Structure of a Synapse

Communication between neurons and other cells occurs in only one direction across a synapse. At a synapse between two neurons, the impulse passes from the **synaptic knob** of the **presynaptic neuron** to the **postsynaptic neuron** (Figure 8–10●). (The synaptic terminal of a presynaptic neuron at such synapses is called a *synaptic knob* because of its rounded shape.) The opposing cell membranes are separated by a narrow space called the **synaptic cleft.**

Each synaptic terminal contains mitochondria, synaptic vesicles, and endoplasmic reticulum. Every synaptic vesicle contains several thousand molecules of a specific neurotrans-

mitter and, upon stimulation, many of these vesicles release their contents into the synaptic cleft. The neurotransmitter then diffuses across the synaptic cleft and binds to receptors on the postsynaptic membrane.

Synaptic Function and Neurotransmitters

There are many different neurotransmitters. The neurotransmitter **acetylcholine,** or **ACh,** is released at **cholinergic synapses.** Cholinergic synapses are widespread inside and outside of the CNS; the neuromuscular junction described in Chapter 7 is one example. Figure 8–11● shows the major events that occur at a cholinergic synapse after an action potential arrives at the presynaptic neuron:

Step 1: *The arrival of an action potential at the synaptic knob.* The arriving action potential depolarizes the presynaptic membrane of the synaptic knob.

Step 2: *The release of the neurotransmitter ACh.* Depolarization of the presynaptic membrane causes the brief opening of calcium channels, which allows extracellular calcium ions to enter the synaptic knob. Their arrival triggers the exocytosis of the synaptic vesicles and the release of ACh. The release of ACh stops very soon thereafter because the calcium ions are rapidly removed from the cytoplasm by active transport mechanisms.

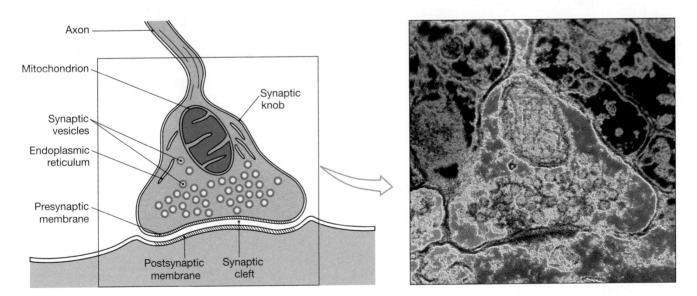

● **FIGURE 8–10 The Structure of a Typical Synapse.** A synapse between two neurons is shown diagrammatically (at left) and in a color-enhanced micrograph (at right, TEM × 222,000).

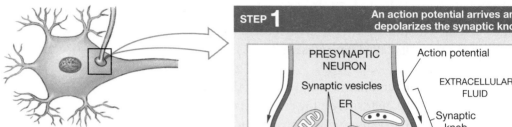

● **FIGURE 8–11** The Events at a Cholinergic Synapse.

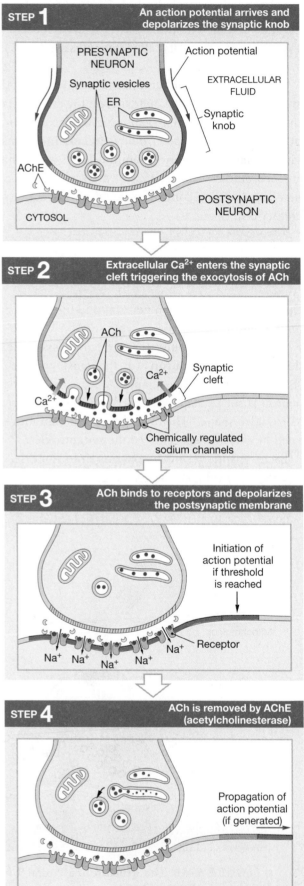

Step 3: *The binding of ACh and the depolarization of the postsynaptic membrane.* The binding of ACh to sodium channels causes them to open and allows sodium ions to enter. If the resulting depolarization of the postsynaptic membrane reaches threshold, an action potential is produced.

Step 4: *The removal of ACh by AChE.* The effects on the postsynaptic membrane are temporary because the synaptic cleft and postsynaptic membrane contain the enzyme acetylcholinesterase (AChE). ∽ p. 212 The AChE removes ACh by breaking it into acetate and choline.

Table 8–1 summarizes the sequence of events that occurs at a cholinergic synapse.

TABLE 8-1 *The Sequence of Events at a Typical Cholinergic Synapse*

Step 1:
- An arriving action potential depolarizes the synaptic knob and the presynaptic membrane.

Step 2:
- Calcium ions enter the cytoplasm of the synaptic knob.
- ACh release occurs through exocytosis of neurotransmitter vesicles.

Step 3:
- ACh diffuses across the synaptic cleft and binds to receptors on the postsynaptic membrane.
- Sodium channels on the postsynaptic surface are activated, which produces a graded depolarization.
- ACh release stops because calcium ions are removed from the cytoplasm of the synaptic knob.

Step 4:
- The depolarization ends as ACh is broken down into acetate and choline by AChE.
- The synaptic knob reabsorbs choline from the synaptic cleft and uses it to resynthesize ACh.

Clinical Note
ORGANOPHOSPHATE POISONING

Organophosphates are a class of insecticides used widely in agriculture. Examples of organophosphates include Diazinon, orthene, malathion, parathion, and others. In addition, organophosphates have been used as chemical warfare agents since World War II. Most recently, the organophosphate sarin was used in the terrorist attack on a Japanese subway in 1995. The potency of organophosphate compounds varies significantly.

The principle action of the organophosphates is deactivation of the enzyme acetylcholinesterase (cholinesterase) in the nervous system. This leads to the accumulation of acetylcholine at nerve synapses and at the neuromuscular junctions, which results in overstimulation of the acetylcholine receptors. This is followed by paralysis of cholinergic synaptic transmission.

Organophosphates bind irreversibly to cholinesterase, thus inactivating the enzyme. Organophosphates can enter the body through the skin or through the respiratory tract. Acute systemic organophosphate poisoning results in a variety of CNS, muscarinic, nicotinic, and somatic motor manifestations.

CNS symptoms include anxiety, restlessness, tremor, headache, dizziness, mental confusion, and seizures. Muscarinic receptor stimulation causes increased salivation, lacrimation, sweating, urinary incontinence, diarrhea, gastrointestinal system distress, vomiting, and bradycardia. Nicotinic receptor stimulation causes pallor, dilated pupils (mydriasis), and hypertension. Nicotinic stimulation at the neuromuscular junction also causes muscle fasciculations, cramps, and muscle weakness. This can progress to paralysis and a loss of reflexes.

The diagnosis of organophosphate poisoning is made by recognition of the signs and symptoms. Treatment is directed at support of essential functions such as maintenance of the airway and support of respirations. Initially, the drug atropine sulfate is administered. Atropine competes with acetylcholine for the acetylcholine receptors. In severe poisonings, exceedingly large doses of atropine may be required. This helps to reverse muscarinic and CNS symptoms. The antidote for organophosphate poisoning is *pralidoxime (2-PAM)*, which restores cholinesterase activity by regenerating cholinesterase. Pralidoxime also appears to prevent toxicity by detoxifying the remaining organophosphate molecules. ■

Another common neurotransmitter, **norepinephrine** (nor-ep-i-NEF-rin), or **NE,** is important in the brain and in portions of the autonomic nervous system. It is *noradrenaline,* and synapses releasing NE are described as **adrenergic.** The neurotransmitters **dopamine** (DŌ-puh-mēn), **gamma aminobutyric** (GAM-ma a-MĒ-nō-bū-TĒR-ik) **acid** (also known as **GABA**), and **serotonin** (ser-o-TŌ-nin) function in the CNS. There are at least 50 other neurotransmitters whose functions are not well understood. In addition, two gases are now known to be important neurotransmitters: *nitric oxide (NO)* and *carbon monoxide (CO).*

The neurotransmitters released at a synapse may have excitatory or inhibitory effects. Both ACh and NE usually have an excitatory effect due to the depolarization of postsynaptic neurons. Like ACh, NE's effect is temporary; it is broken down by an enzyme called *monoamine oxidase*. The effects of dopamine, GABA, and serotonin are usually inhibitory due to the hyperpolarization of postsynaptic neurons.

Whether or not an action potential appears in the postsynaptic neuron depends on the balance between the depolarizing and hyperpolarizing stimuli that arrives at any moment. For example, suppose that a neuron will generate an action potential if it receives 10 depolarizing stimuli. That could mean 10 active synapses if all release excitatory neurotransmitters. But if, at the same moment, 10 other synapses release inhibitory neurotransmitters, the excitatory and inhibitory effects will cancel one another out, and no action potential will develop. The activity of a neuron thus depends on the balance between excitation and inhibition. The interactions are extremely complex—synapses at the cell body and dendrites may involve tens of thousands of other neurons, some release excitatory neurotransmitters and others release inhibitory neurotransmitters.

Clinical Note
NEUROTRANSMITTERS

No actual physical connection exists between two nerve cells or between a nerve cell and the organ it innervates. Instead, there is a space or *synapse*, between nerve cells. Specialized chemicals called *neurotransmitters* conduct the nervous impulse between nerve cells or between a nerve cell and its target organ.

The two neurotransmitters of the autonomic nervous system are *acetylcholine* and *norepinephrine.* Acetylcholine is utilized in the preganglionic nerves of the sympathetic nervous system and in both the preganglionic and postganglionic nerves of the parasympathetic nervous system. Norepinephrine is the postganglionic neurotransmitter of the sympathetic nervous system. Synapses that use acetylcholine as the neurotransmitter are *cholinergic synapses*. Synapses that use norepinephrine as the neurotransmitter are *adrenergic synapses.* ■

Neuronal Pools

As noted earlier, the human body has about 10 million sensory neurons, 20 billion interneurons, and one-half million motor neurons. These individual neurons represent the simplest level of organization within the CNS. However, the integration of sensory and motor information to produce complex responses requires groups of interneurons that act together. A **neuronal pool** is a group of interconnected interneurons with specific functions. Each neuronal pool has a limited number of input sources and output destinations, and each may contain excitatory and inhibitory neurons. The output of one pool may stimulate or depress the activity of other pools, or it may exert direct control over motor neurons or peripheral effectors.

Neurons and neuronal pools communicate with one another in several patterns, or neural circuits. The two simplest circuit patterns are *divergence* and *convergence*. In **divergence,** information spreads from one neuron to several neurons, or from one neuronal pool to multiple neuronal pools (Figure 8–12a●). Considerable divergence occurs when sensory neurons bring information into the CNS, because the sensory information is distributed to neuronal pools throughout the spinal cord and brain. For example, visual information that arrives from the eyes reaches your conscious awareness at the same time it is carried to areas of the brain that control posture and balance at the subconscious level.

Divergence is also involved when you step on a sharp object; that action stimulates sensory neurons that distribute information to a number of neuronal pools. As a result, you might react in several ways—withdraw your foot, shift your weight, move your arms, feel the pain, and shout "Ouch!"—all at the same time.

In **convergence** (Figure 8–12b●), several neurons synapse on a single postsynaptic neuron. Convergence makes possible both voluntary and involuntary control of some body processes. For example, right now the movements of your diaphragm and ribs are being involuntarily, or subconsciously, controlled by respiratory centers in the brain. These centers activate or inhibit motor neurons in the spinal cord that control the respiratory muscles. But those same movements can

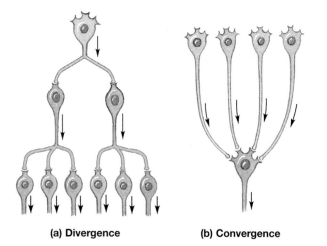

(a) Divergence   (b) Convergence

● **FIGURE 8–12 Two Common Types of Neuronal Pools.**
(**a**) Divergence is a mechanism for spreading stimulation to multiple neurons or neuronal pools in the CNS. (**b**) Convergence is a mechanism for providing input to a single neuron from multiple sources.

be controlled voluntarily, or consciously, as when you take a deep breath and hold it. Two neuronal pools are involved, and both synapse on the same motor neurons.

Key Note

At a chemical synapse a synaptic terminal releases a neurotransmitter that binds to the postsynaptic cell membrane. The result is a temporary, localized change in the permeability or function of the postsynaptic cell. This change may have broader effects on the cell, depending on the nature and number of the stimulated receptors. Many drugs affect the nervous system by stimulating receptors that otherwise respond only to neurotransmitters. These drugs can have complex effects on perception, motor control, and emotional states.

→ CONCEPT CHECK QUESTIONS

1. A neurotransmitter causes potassium channels (but not sodium channels) to open. What effect would this neurotransmitter produce at the postsynaptic membrane?
2. How would synapse function be affected if the uptake of calcium were blocked at the presynaptic membrane of a cholinergic synapse?
3. What type of neural circuit permits both conscious and subconscious control of the same motor neurons?

Answers begin on p. 792.

■ The Central Nervous System

The central nervous system consists of the spinal cord and brain. These masses of neural tissue are extremely delicate and must be protected against shocks, infections, and other dangers. In addition to glial cells within the neural tissue, the CNS is protected by a series of covering layers, the *meninges.*

The Meninges

Neural tissue has a very high metabolic rate and requires abundant nutrients and a constant supply of oxygen. At the same time, the CNS must be isolated from a variety of compounds in the blood that could interfere with its complex operations. Delicate neural tissues must also be defended against damaging contact with the surrounding bones. The **meninges** (me-NIN-jēz), which are three layers of specialized membranes that surround the brain and spinal cord, provide physical stability and shock absorption to CNS tissue (Figure 8–13●). Blood vessels branching within these layers also deliver needed

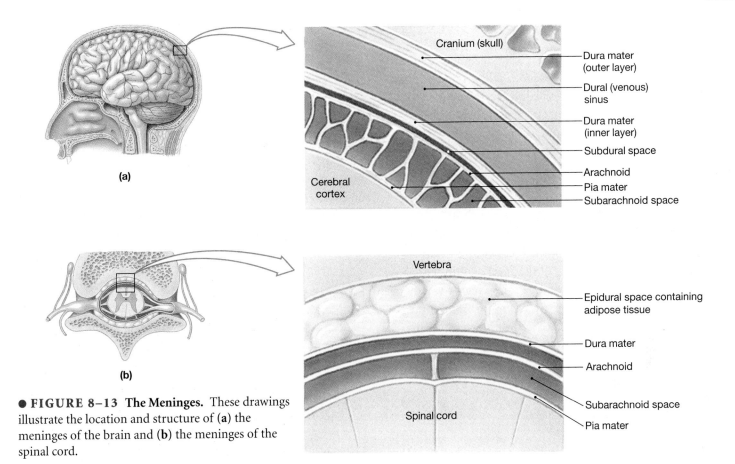

(a)

(b)

● **FIGURE 8–13 The Meninges.** These drawings illustrate the location and structure of (**a**) the meninges of the brain and (**b**) the meninges of the spinal cord.

oxygen and nutrients. At the foramen magnum of the skull, the *cranial meninges* covering the brain are continuous with the *spinal meninges* that surround the spinal cord. The three meningeal layers are the *dura mater,* the *arachnoid,* and the *pia mater.* The meninges cover cranial and spinal nerves as they penetrate the skull or pass through the intervertebral foramina, and become continuous with the connective tissues that surround the peripheral nerves.

The Dura Mater

The tough, fibrous **dura mater** (DOO-ruh MĀ-ter; *dura,* hard + *mater,* mother) forms the outermost covering of the central nervous system. The dura mater that surrounds the brain consists of two fibrous layers. The outer layer is fused to the periosteum of the skull, and the two layers are separated by a slender gap that contains tissue fluids and blood vessels (Figure 8–13a●).

At several locations, the inner layer of the dura mater extends deep into the cranial cavity, and forms folded membranous sheets called *dural folds.* The dural folds act like seat belts to hold the brain in position. Large collecting veins known as *dural sinuses* lie between the two layers of a dural fold.

In the spinal cord, the outer layer of the dura mater is not fused to bone. Between the dura mater of the spinal cord and the walls of the vertebral canal lies the **epidural space,** which contains loose connective tissue, blood vessels, and adipose tissue (Figure 8–13b●). Injecting an anesthetic into the epidural space produces a temporary sensory and motor paralysis known as an *epidural block.* This technique has the advantage of affecting only the spinal nerves in the immediate area of the injection. Epidural blocks in the lower lumbar or sacral regions may be used to control pain during childbirth.

> ### *Clinical Note*
> **BURR HOLES**
>
> Bleeding in the epidural space, called an *epidural hematoma,* usually results from laceration of an artery during blunt head trauma. This causes rapid swelling and compression of the brain. Life-saving emergency treatment may require drilling *burr holes* through the skull over the hematoma to relieve the pressure. Although burr holes are usually performed by neurosurgeons, the emergency physician may have to place them if a neurosurgeon is unavailable. ■

The Arachnoid

A narrow **subdural space** separates the inner surface of the dura mater from the second meningeal layer, the **arachnoid** (a-RAK-noyd; *arachne,* spider). This intervening space contains a small quantity of lymphatic fluid, which reduces friction between the opposing surfaces. The arachnoid is a layer of squamous cells; deep to this epithelial layer lies the **subarachnoid space,** which contains a delicate web of collagen and elastic fibers. The subarachnoid space is filled with *cerebrospinal fluid,* which acts as a shock absorber and transports dissolved gases, nutrients, chemical messengers, and waste products.

The Pia Mater

The subarachnoid space separates the arachnoid from the innermost meningeal layer, the pia mater (*pia,* delicate + *mater,* mother). The pia mater is bound firmly to the underlying neural tissue. The blood vessels that service the brain and spinal cord run along the surface of this layer, within the subarachnoid space. The pia mater of the brain is highly vascular, and large vessels branch over the surface of the brain, supplying the superficial areas of neural cortex. This extensive circu-

latory supply is extremely important, for the brain has a very high rate of metabolism; at rest, the 1.4 kg (3.1 lb) brain uses as much oxygen as 28 kg (61.6 lb) of skeletal muscle.

Clinical Note
EPIDURAL AND SUBDURAL DAMAGE

A severe head injury may damage cerebral blood vessels and cause bleeding into the cranial cavity. If blood leaks out between the dura mater and the cranium, the condition is known as an *epidural hemorrhage.* The flow of blood into the lower layer of the dura mater and subdural space is called a *subdural hemorrhage.* These are serious conditions because the blood that enters these spaces compresses and distorts the relatively soft tissues of the brain. The signs and symptoms vary depending on whether an artery or a vein is damaged. Because arterial blood pressure is higher, bleeding from an artery can cause more rapid and severe distortion of neural tissue. ■

Clinical Note
NERVOUS SYSTEM INFECTIONS

Infections of the nervous system can be life threatening. The most common nervous system infection is meningitis. Its causes include viruses, bacteria, fungi, parasites, and prions. Less common CNS infections include encephalitis and brain abscess. CNS infections are classified as acute, subacute, or chronic depending upon the duration of the disease.

Meningitis

Meningitis is an infection of the meninges. Bacterial meningitis is a medical emergency that begins when the causative organisms enter the subarachnoid space. The three principle organisms are *Streptococcal pneumoniae, Haemophilus influenza (type b),* and *Neissiera meningitidis.* They enter the subarachnoid space, usually through the upper airway, and trigger infection and inflammation. The signs and symptoms of meningitis are fever, headache, altered mental status, photophobia, and stiffness of the neck (meningismus). With meningismus, passive flexion of the neck causes flexion of the hips and knees. This finding, referred to as *Brudzinski's sign,* is due to inflammation of the meninges. *Kernig's sign,* an inability to extend the legs when the knees are flexed, is also due to meningeal inflammation. Brudzinski's sign and Kernig's sign are seen in approximately 50 percent of patients with bacterial meningitis. The diagnosis of meningitis is based on the history, physical examination, and lumbar puncture (spinal tap).

A *lumbar puncture* is an important diagnostic procedure that obtains *cerebrospinal fluid (CSF)* for laboratory analysis. The puncture is usually made between the third and fourth lumbar vertebrae, well below the terminal end of the spinal cord (Figure 8–14●). The patient is placed in a seated or a lateral recumbent position. The skin is cleansed

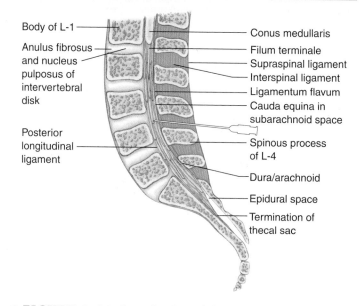

● **FIGURE 8–14 Cross Section of the Distal Spine and Spinal Cord That Shows Proper Positioning of a Spinal Needle During Lumbar Puncture.** Note that the puncture site is several segments below the terminal end of the spinal cord.

and a sterile spinal needle is slowly inserted through the skin of the lower back into the subarachnoid space. Often, a "pop" can be felt as the needle penetrates the tough dura mater. Once the needle is properly placed, the CSF pressure is measured. Then, approximately 3–4 milliliters of CSF are removed. Spinal fluid is normally clear and does not contain cells. In meningitis, the spinal fluid becomes cloudy due to the presence of white blood cells (*pleocytosis*) and bacteria.

Bacterial meningitis can be rapidly fatal. As soon as the diagnosis is suspected, empiric antibiotics are administered. In adults, *N. meningitidis* and *S. pneumoniae* are the most common organisms. The incidence of *H. influenzae* has steadily declined with the introduction of a vaccine. Infection with *N. meningitidis*, which is referred to as meningiococcemia, is particularly severe. Emergency personnel who have been in close contact with a patient who has meningiococcemia may require prophylactic antibiotics.

VIRAL MENINGITIS

The signs and symptoms of viral meningitis are typically less severe than those of bacterial forms. The CSF fluid will typically contain white blood cells, but an infectious organism cannot be seen or cultured. Because of this, viral meningitis is often called aseptic meningitis. Viral meningitis often occurs in winter, and outbreaks are not uncommon. Treatment is supportive.

UNUSUAL FORMS OF MENINGITIS

Although viral and bacterial meningitis are by far the most common, some other types of meningitis can be life threatening. The incidence of some infections is increasing due to increasing numbers of patients with incompetent immune systems. These include AIDS patients, patients who have received an organ transplant, and patients who are undergoing chemotherapy treatment for cancer. Often, the causative agent is not infectious and may actually be a part of the body's normal flora, but in patients with immunosuppression from drugs or disease, it can cause infection. Because of this, these types of meningitis are referred to as *opportunistic infections*.

Fungal meningitis is uncommon and, when present, tends to be chronic. Treatment is difficult and often requires long periods of therapy with antifungal drugs. It is most commonly seen in AIDS patients or in others who are immunocompromised.

Tubercular (TB) meningitis is being seen with increasing frequency, primarily in AIDS patients. The symptoms include headache, low-grade fever, nausea, vomiting, irritability, difficulty sleeping, and fatigue. As the disease progresses, confusion, stiff neck, behavioral changes, and seizures can occur. Despite the severity, the recovery rate approaches 90 percent if therapy is initiated in time.

Occasionally, meningitis can result from infection with free-living amoebas, particularly Naegleria. These organisms are found in both fresh and brackish water including that from lakes, swimming pools, hot springs, and heating and air conditioning units. Infection occurs when persons swimming in, or exposed to, the water inhale the organism. It invades the olfactory nervous tissue from the nasal cavity. After an incubation period of 2–15 days, the patient develops high fever, severe headache, nausea, vomiting, and meningismus. Rapid progression to seizures and coma occurs, and most patients die within a week. Only four survivors have been reported.

PARASITIC BRAIN INFECTIONS

Cestodes are flatworms commonly referred to as tapeworms. The most commonly encountered member of this group is the pork tapeworm (Taenia solium). It is a significant problem in developing countries and is occasionally encountered in the U.S., especially in immigrants and visitors from Central America and the Middle East.

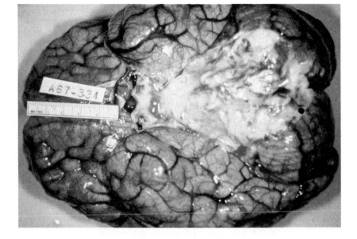

● **FIGURE 8–15** **Large Brain Abscess Secondary to Intravenous Injection of Narcotics.** Bacteria are introduced to the brain, which results in abscess formation.

The larval stage of *T. solium* can cause clinical disease referred to as *cysticercosis*, which can be serious and often fatal. *Taenia* cysts can enter the subcutaneous tissue, the eye, the brain, and the heart and cause seizures and hydrocephalus. Cysticercosis should be considered a possible cause of new-onset seizures in recent immigrants from Central America. It can be identified through CT scanning of the brain and by isolation of cysts in the stool. Treatment with antiparasitic medications is usually effective if administered in time.

BRAIN ABSCESS

Brain abscesses are localized collections of pus within the parenchyma of the brain or spinal cord (Figure 8–15●). They tend to occur most frequently in men between the ages of 30 and 40 years. They can be caused by open trauma or neurosurgical procedures, by infection that has spread from the middle ear or nose, or by spreading from abscesses elsewhere in the body. AIDS patients are particularly susceptible to brain abscesses caused by the protozoan *Toxoplasma gondii (toxoplasmosis)*. Treatment of brain abscess generally requires surgical drainage.

ENCEPHALITIS

Encephalitis is an acute infection, usually of viral origin, with CNS involvement. Most cases of encephalitis are due to viruses carried by mosquitoes. Outbreaks of encephalitis usually occur in summer. Causative viruses include: *Eastern equine encephalitis*, *Western equine encephalitis*, *St. Louis encephalitis*, and *California encephalitis*. Encephalitis can also occur as a complication of viral diseases that usually affect other body systems, such as rabies or mononucleosis. Encephalitis is also seen in AIDS patients. Common causative viruses in this population include *herpes virus* and *cytomegalovirus (CMV)*. The most common signs and symptoms of encephalitis are headache and fever. Delirium, confusion, seizures, and unconsciousness can occur in severe cases. Treatment for epidemic encephalitis is usually supportive. Opportunistic viral infections in AIDS patients are often treated with antiviral drugs.

(continued next page)

PRION DISEASES

Until recently, scientists thought the smallest particle capable of transmitting disease was a virus. However, researchers have demonstrated that a protein alone is capable of transmitting disease. Protein particles, referred to as prions, have been identified as the causative agent of a group of diseases called *spongiform encephalopathies*. These diseases cause the brain to develop large vacuoles in the cortex and cerebellum, which adversely affect central nervous system (CNS) function. It appears that traditional sterilization methods do not inactivate prions and they can remain communicable for long periods of time.

Previously, prion diseases have been rare. However, with the outbreak of mad cow disease in Europe, considerable attention has been focused on them. Mad cow disease, also known as *bovine spongiform encephalopathy (BSE)*, initially appeared in the United Kingdom and has spread to parts of Europe. BSE apparently is transmitted through cattle feed. In the cattle industry, it is not uncommon for cattle feed to contain the byproducts of other cattle. Apparently, parts of infected cattle were unknowingly ground up and mixed in cattle feed and fed to cattle in commercial feedlots. Some cattle that ate the tainted feed developed BSE.

Several human diseases are known to be caused by prions, the most common being *Creutzfeldt-Jakob Disease (CJD)*. Prior to the outbreak of BSE, the incidence of CJD was only one person per million per year. However, the incidence has increased since the outbreak of BSE. Now, there appears to be a direct link between BSE and a form of CJD.

Early symptoms of CJD include failing memory, changes in behavior, lack of coordination, and visual disturbances. These are followed by a rapid, progressive dementia, involuntary jerking movements, and progressive motor dysfunction. The duration of CJD from the onset of symptoms to death is usually less than a year. Death is typically caused by secondary pneumonia. There is no treatment or cure for CJD.

Two interesting prion diseases are *fatal familial insomnia (FFI)* and *kuru*. FFI presents with untreatable insomnia, autonomic nervous system dysfunction, and eventually death. It is associated with severe, selective atrophy of the thalamus. Kuru occurs only in isolated tribes in the Fore highlands of New Guinea. In these tribes, the brains of dead relatives were removed and eaten as part of a religious ritual that honors the dead relative. Tribal members ground the brain up into a pale, gray soup, heated it, and ate it. The disease clinically resembles CJD, and some scientists have speculated that the brain tissue was highly infectious, with the causative prion being transmitted through ingestion. An effort by the New Guinea government to curtail cannibalism has caused the incidence of kuru to rapidly decline, and the disease is now almost unknown. ■

The Spinal Cord

The spinal cord serves as the major highway for the passage of sensory impulses to the brain and of motor impulses from the brain. In addition, the spinal cord integrates information on its own and controls *spinal reflexes*, which are automatic motor responses ranging from withdrawal from pain to complex reflex patterns involved in sitting, standing, walking, and running.

Gross Anatomy

The spinal cord is approximately 45 cm (18 in.) long (Figure 8–16●); it has a maximum width of roughly 14 mm (0.55 in.). With two exceptions, the diameter of the cord decreases as it extends toward the sacral region. The two exceptions are regions concerned with the sensory and motor control of the limbs. The *cervical enlargement* supplies nerves to the shoulder girdles and upper limbs, and the *lumbar enlargement* provides innervation to the pelvis and lower limbs. Below the lumbar enlargement, the spinal cord becomes tapered and conical. A slender strand of fibrous tissue extends from the inferior tip of the spinal cord to the coccyx, and serves as an anchor that prevents upward movement.

The spinal cord has a **central canal,** which is a narrow internal passageway filled with cerebrospinal fluid (Figure 8–16b●).

The posterior surface of the spinal cord has a shallow groove, the *posterior median sulcus,* and the anterior surface has a deeper groove, the *anterior median fissure.*

The entire spinal cord consists of 31 segments, each identified by a letter and number (see Figure 8–16a●). Every spinal segment is associated with a pair of **dorsal root ganglia** (singular: ganglion), which contain the cell bodies of sensory neurons (see Figure 8–16b). The **dorsal roots,** which contain the axons of these neurons, bring sensory information to the spinal cord. A pair of **ventral roots** contains the axons of CNS motor neurons that control muscles and glands. On either side, the dorsal and ventral roots from each segment leave the vertebral column between adjacent vertebrae at the *intervertebral foramen.*

Distal to each dorsal root ganglion, the sensory (dorsal) and motor (ventral) roots are bound together into a single *spinal nerve.* All spinal nerves are classified as *mixed nerves,* because they contain both sensory and motor fibers. The spinal nerves on either side form outside the vertebral canal, where the ventral and dorsal roots unite. ∞ p. 176

In adults, the spinal cord extends only to the level of the first or second lumbar vertebrae. When seen in gross dissection, the long ventral and dorsal roots of spinal nerves inferior to the tip of the spinal cord, plus the cord's thread-like exten-

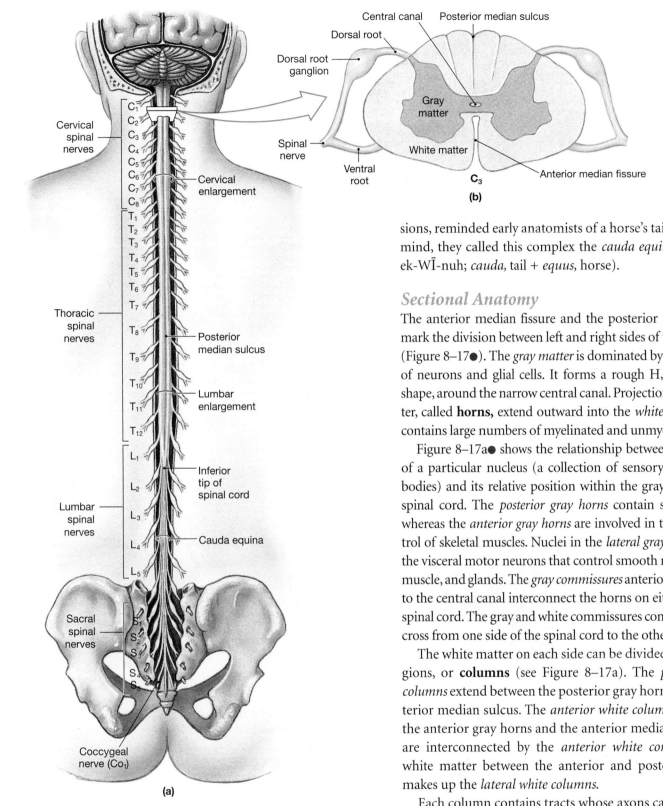

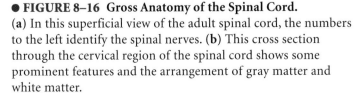

● **FIGURE 8–16 Gross Anatomy of the Spinal Cord.**
(**a**) In this superficial view of the adult spinal cord, the numbers to the left identify the spinal nerves. (**b**) This cross section through the cervical region of the spinal cord shows some prominent features and the arrangement of gray matter and white matter.

sions, reminded early anatomists of a horse's tail. With that in mind, they called this complex the *cauda equina* (KAW-duh ek-WĪ-nuh; *cauda,* tail + *equus,* horse).

Sectional Anatomy

The anterior median fissure and the posterior median sulcus mark the division between left and right sides of the spinal cord (Figure 8–17●). The *gray matter* is dominated by the cell bodies of neurons and glial cells. It forms a rough H, or a butterfly shape, around the narrow central canal. Projections of gray matter, called **horns,** extend outward into the *white matter,* which contains large numbers of myelinated and unmyelinated axons.

Figure 8–17a● shows the relationship between the function of a particular nucleus (a collection of sensory or motor cell bodies) and its relative position within the gray matter of the spinal cord. The *posterior gray horns* contain sensory nuclei, whereas the *anterior gray horns* are involved in the motor control of skeletal muscles. Nuclei in the *lateral gray horns* contain the visceral motor neurons that control smooth muscle, cardiac muscle, and glands. The *gray commissures* anterior and posterior to the central canal interconnect the horns on either side of the spinal cord. The gray and white commissures contain axons that cross from one side of the spinal cord to the other.

The white matter on each side can be divided into three regions, or **columns** (see Figure 8–17a). The *posterior white columns* extend between the posterior gray horns and the posterior median sulcus. The *anterior white columns* lie between the anterior gray horns and the anterior median fissure; they are interconnected by the *anterior white commissure.* The white matter between the anterior and posterior columns makes up the *lateral white columns.*

Each column contains tracts whose axons carry either sensory data or motor commands. Small tracts carry sensory or motor signals between segments of the spinal cord, and larger tracts connect the spinal cord with the brain. **Ascending tracts** carry sensory information toward the brain, and **descending tracts** convey motor commands into the spinal cord.

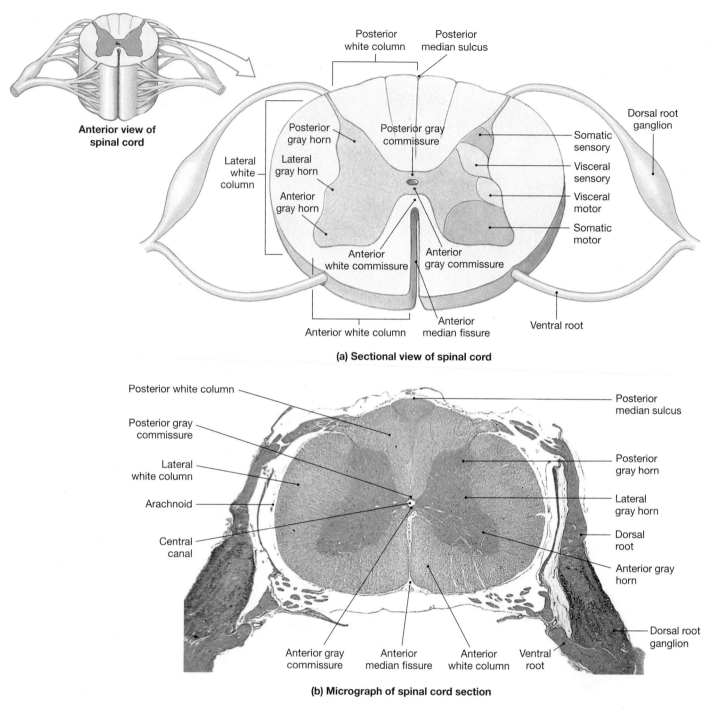

Anterior view of
spinal cord

Posterior
white column

Posterior
median sulcus

Posterior
gray horn

Posterior gray
commissure

Somatic
sensory

Lateral
white
column

Lateral
gray horn

Visceral
sensory

Dorsal root
ganglion

Anterior
gray horn

Visceral
motor

Somatic
motor

Anterior
white commissure

Anterior
gray commissure

Anterior white column

Anterior
median fissure

Ventral root

(a) Sectional view of spinal cord

Posterior white column

Posterior
median sulcus

Posterior gray
commissure

Posterior
gray horn

Lateral
white column

Lateral
gray horn

Arachnoid

Dorsal
root

Central
canal

Anterior gray
horn

Anterior gray
commissure

Anterior
median fissure

Anterior
white column

Ventral
root

Dorsal root
ganglion

(b) Micrograph of spinal cord section

● **FIGURE 8–17 Sectional Anatomy of the Spinal Cord. (a)** The left half of this diagrammatic sectional view labels important anatomical landmarks and the major regions of white matter and gray matter. The right half indicates the functional organization of the gray matter in the posterior, lateral, and anterior gray horns. **(b)** This micrograph of a section through the spinal cord shows the major landmarks diagrammed in (a).

Clinical Note
SPINAL-CORD INJURIES

The spinal cord is well-protected by the vertebral bodies (Figure 8–18●). Spinal-cord injuries usually are due to vertebral injuries that usually result from the extremes of normal motion such as flexion, extension, rotation, and lateral bending. Damaging mechanisms also include forces transmitted along the axis of the spine: *axial loading* and *distraction*. Finally, spinal injury can result either directly from blunt or

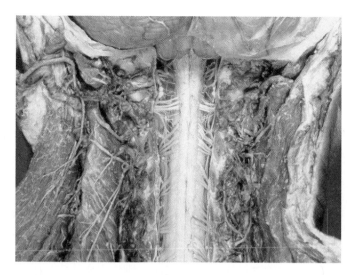

● **FIGURE 8–18 Detailed Dissection of the Proximal Spinal Cord That Illustrates Essential Structures and Spinal Nerve Roots.**

- *Anterior-cord syndrome.* Results from compression of the anterior part of the spinal cord, often from hyperflexion of the cervical spine. It causes complete paralysis below the lesion with loss of pain and temperature sensation. Vibratory and light touch ability are spared in this injury.
- *Central-cord syndrome.* Results from disruption of blood supply to the spinal cord, often due to hyperextension injuries. It causes quadriparesis that is greater in the upper extremities than in the lower extremities. There may be some loss of pain and temperature sensation.
- *Brown-Séquard syndrome.* Results from transection of half of the spinal cord or from unilateral cord compression. It causes spastic paresis and loss of position and vibratory sensation on the affected side and loss of temperature and pain sensation on the opposite side.

penetrating trauma or indirectly when an expanding mass (edema or hematoma) compresses the cord or a disruption of blood supply damages it (Figure 8–19●). It is possible to have spinal-cord injury without any spinal-column injury.

The bones, ligaments, and joints of the vertebral column may be damaged by hyperflexion or hyperextension. This most frequently occurs in the cervical and lumbar regions. These motions can cause the vertebral column to fracture or dislocate. With fractures, bone fragments can damage the spinal cord. With dislocations, abnormal movement of the vertebrae can compress or even shear the spinal cord.

Spinal-cord injury varies in severity; its classifications include the following:

- *Cord concussion.* A temporary interruption in cord-mediated functions.
- *Cord contusion.* Bruising of the neural tissue that results in swelling and temporary loss of cord-mediated functions.
- *Cord compression.* Results from pressure on the cord which causes ischemia. Surgery to decompress the cord is often required to prevent permanent damage.
- *Cord lacerations.* Tearing of the neural tissues from bone fragments or shearing forces. Slight damage may be reversible, but permanent damage can occur if spinal tracts are disrupted.
- *Complete transection.* Severing of the spinal cord that causes permanent loss of function.
- *Incomplete transection.* Part of the spinal cord is severed, but some tracts are spared.
- *Cord hemorrhage.* Bleeding into the neural tissues secondary to blood vessel damage following injury.

Several syndromes can develop with spinal-cord injury. These result from partial cord injuries where some spinal tracts are destroyed while others are spared. These syndromes include:

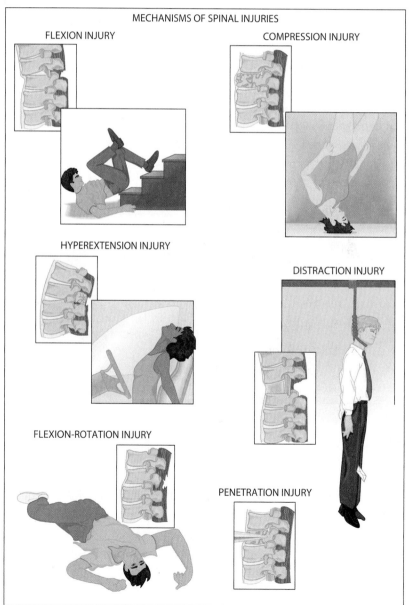

● **FIGURE 8–19 Mechanisms Associated with Cervical Spine, Vertebral, and Spinal-Cord Injury.**   *(continued next page)*

Clinical Note—continued
SPINAL-CORD INJURIES

■ *Cauda equina syndrome.* Results from injury to the cauda equina from lumbar vertebral fractures of herniated discs. It can cause weakness and sensory loss in the lower extremities, sciatica, and possible bowel or bladder dysfunction.

■ *Spinal shock.* Results from partial or complete spinal-cord injury at T_6 or above. It causes loss of reflexes, sensation, and flaccid paralysis below the level of the lesion. In addition, there is a flaccid bladder, loss of rectal sphincter tone, bradycardia, and hypotension.

As central nervous system tissues cannot regenerate, spinal-cord injuries are usually permanent. Occasionally, if treated in time, the injury can be minimized by the administration of extremely high doses of corticosteroids. This appears to limit the inflammation associated with the injury and prevent secondary injury from swelling.

Significant research is directed at the biochemical control of nerve growth and regeneration. In addition, electronic devices and computers are being used to stimulate specific muscles and muscle groups. This has allowed persons with spinal cord injuries to walk short distances. ■

 Key Note

The spinal cord has a narrow central canal surrounded by gray matter that contains sensory and motor nuclei. Sensory nuclei are dorsal; motor nuclei are ventral. The gray matter is covered by a thick layer of white matter that consists of ascending and descending axons. These axons are organized in columns that contain axon bundles with specific functions. Because the spinal cord is so highly organized, it is often possible to predict the results of injuries to localized areas.

→ CONCEPT CHECK QUESTIONS

1. Damage to which root of a spinal nerve would interfere with motor function?
2. A man with polio has lost the use of his leg muscles. In what area of the spinal cord would you expect to locate virus-infected motor neurons?
3. Why are spinal nerves also called mixed nerves?

Answers begin on p. 792.

The Brain

The brain is far more complex than the spinal cord, and its responses to stimuli are more versatile. The brain contains roughly 35 billion neurons organized into hundreds of neuronal pools; it has a complex three-dimensional structure and performs a bewildering array of functions. All of our dreams, passions, plans, and memories are the result of brain activity.

The adult human brain contains almost 98 percent of the neural tissue in the body. A "typical" adult brain weighs 1.4 kg (3 lb) and has a volume of 1200 cc (71 in.^{3}). There is considerable individual variation, and the brains of males are generally about 10 percent larger than those of females, because of differences in average body size. There is no correlation between brain size and intelligence. Individuals with the smallest brains (750 cc) or largest brains (2100 cc) are functionally normal.

Major Divisions of the Brain

The adult brain has six major regions: (1) the *cerebrum*, (2) the *diencephalon*, (3) the *midbrain*, (4) the *pons*, (5) the *medulla oblongata*, and (6) the *cerebellum*. Major landmarks are indicated in Figure 8–20●.

Viewed from the superior surface, the **cerebrum** (SER-e-brum or se-RĒ-brum) can be divided into large, paired **cerebral hemispheres** (Figure 8–20a●). Conscious thoughts, sensations, intellectual functions, memory storage and retrieval, and complex movements originate in the cerebrum. The hollow **diencephalon** (dī-en-SEF-a-lon; *dia-*, through + *cephalo*, head) is connected to the cerebrum (Figure 8–20c●). Its largest portion is the **thalamus** (THAL-a-mus), which contains relay and processing centers for sensory information. A narrow stalk connects the **hypothalamus** (*hypo-*, below) to the *pituitary gland*. The hypothalamus contains centers involved with emotions, autonomic function, and hormone production. The pituitary gland, which is the primary link between the nervous and endocrine systems, is discussed in Chapter 10. The **epithalamus** contains another endocrine structure, the *pineal gland*.

The **brain stem** contains three major regions of the brain: the midbrain, pons, and medulla oblongata. The brain stem contains important processing centers and relay stations for information headed to or from the cerebrum or cerebellum. Nuclei in the **midbrain,** or *mesencephalon* (mez-en-SEF-a-lon; *meso-*, middle), process visual and auditory information and generate involuntary motor responses. This region also

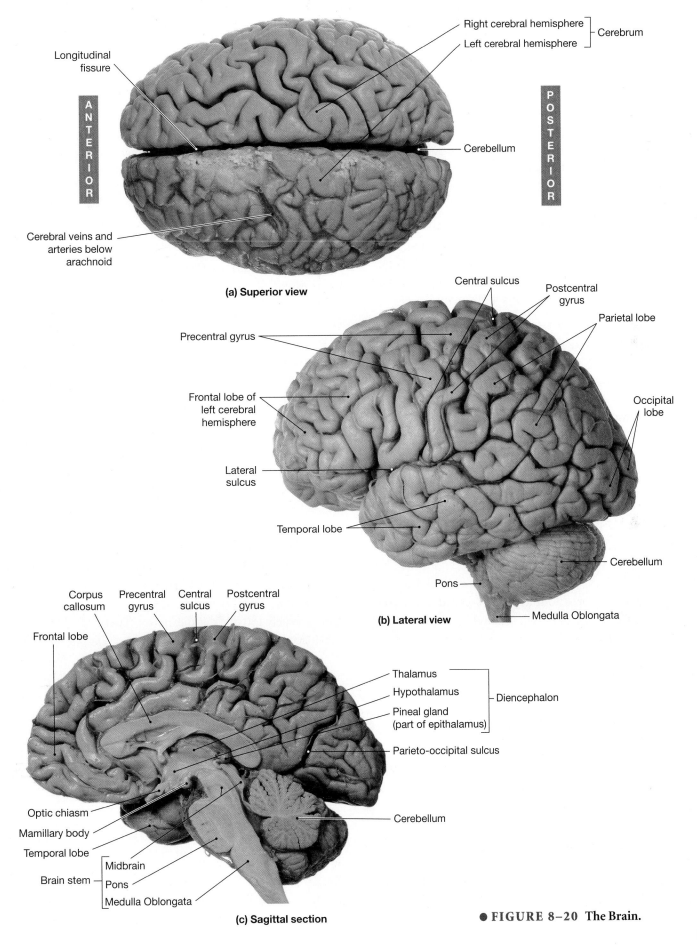

Right cerebral hemisphere ⎤
Left cerebral hemisphere ⎦ Cerebrum

Longitudinal fissure

ANTERIOR

POSTERIOR

Cerebellum

Cerebral veins and arteries below arachnoid

(a) Superior view

Central sulcus

Postcentral gyrus

Parietal lobe

Precentral gyrus

Frontal lobe of left cerebral hemisphere

Occipital lobe

Lateral sulcus

Temporal lobe

Cerebellum

Pons

Medulla Oblongata

(b) Lateral view

Corpus callosum   Precentral gyrus   Central sulcus   Postcentral gyrus

Frontal lobe

Thalamus

Hypothalamus

Diencephalon

Pineal gland (part of epithalamus)

Parieto-occipital sulcus

Optic chiasm

Mamillary body

Temporal lobe

Cerebellum

Brain stem   Midbrain

Pons

Medulla Oblongata

(c) Sagittal section

● **FIGURE 8–20** **The Brain.**

contains centers that help maintain consciousness. The term *pons* refers to a bridge, and the **pons** of the brain connects the cerebellum to the brain stem. In addition to tracts and relay centers, this region of the brain also contains nuclei involved in somatic and visceral motor control. The pons is also connected to the **medulla oblongata,** which is the segment of the brain that is attached to the spinal cord. The medulla oblongata relays sensory information to the thalamus and other brain stem centers; it also contains major centers that regulate autonomic function, such as heart rate, blood pressure, respiration, and digestive activities.

The large cerebral hemispheres and the smaller hemispheres of the **cerebellum** (ser-e-BEL-um) almost completely cover the brain stem. The cerebellum adjusts voluntary and involuntary motor activities on the basis of sensory information and stored memories of previous movements.

CONCEPT CHECK QUESTIONS

1. Describe one major function of each of the six regions of the brain.
2. The pituitary gland links the nervous and endocrine systems. To which portion of the diencephalon is it attached?

Answers begin on p. 792.

The Ventricles of the Brain

As previously noted, the brain and spinal cord contain internal cavities filled with cerebrospinal fluid and lined by ependymal cells. ∞ p. 264 The brain has a central passageway that expands to form four chambers called **ventricles** (VEN-tri-kls) (Figure 8–21●). Each cerebral hemisphere contains a large **lateral ventricle.** There is no direct connection between the lateral ventricles, but an opening, the *interventricular foramen,* allows each of them to communicate with the **third ventricle** in the diencephalon. Instead of a ventricle, the midbrain has a slender canal known as the *mesencephalic aqueduct (cerebral aqueduct),* which connects the third ventricle with the **fourth ventricle** in the pons and upper portion of the medulla oblongata. Within the medulla oblongata the fourth ventricle narrows and becomes continuous with the central canal of the spinal cord.

Cerebrospinal Fluid

Completely surrounding and bathing the exposed surfaces of the CNS and cushioning delicate neural structures is **cerebrospinal fluid,** or **CSF.** It also provides support, as the brain essentially floats in the cerebrospinal fluid. A human brain weighs about 1400 g (3.1 lb) in air but only about 50 g (1.76 oz) when supported by cerebrospinal fluid. Finally, the CSF transports nutrients, chemical messengers, and waste products. Except at the *choroid plexus,* where CSF is produced, the ependymal lining is freely permeable, and CSF is in constant chemical communication with the interstitial fluid of the CNS. Because free exchange occurs between the interstitial fluid and CSF, changes in CNS function may produce changes in the composition of CSF. Samples of CSF can be obtained through a *lumbar puncture,* or *spinal tap,* which provides useful clinical information about CNS injury, infection, or disease.

Cerebrospinal fluid is produced at the **choroid plexus** (*choroid,* a vascular coat + *plexus,* a network), a network of permeable capillaries that extends into each of the four ventricles (Figure 8–22a●). The capillaries of the choroid plexus are covered by large ependymal cells that secrete CSF at a rate of about 500 mL/day. The total volume of CSF at any given moment is approximately 150 mL; this means that the entire volume of CSF is replaced roughly every eight hours. Despite this rapid

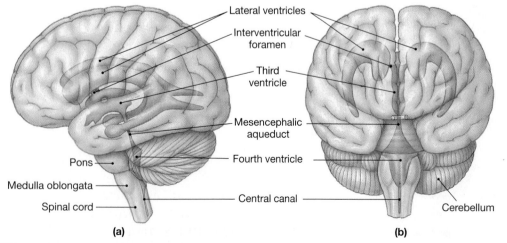

Lateral ventricles
Interventricular foramen
Third ventricle
Mesencephalic aqueduct
Fourth ventricle
Pons
Medulla oblongata
Spinal cord
Central canal
Cerebellum

(a)

(b)

● **FIGURE 8–21 The Ventricles of the Brain.** These drawings of the brain as if it were transparent show the orientation and extent of the ventricles in (**a**) a lateral view and (**b**) an anterior view.

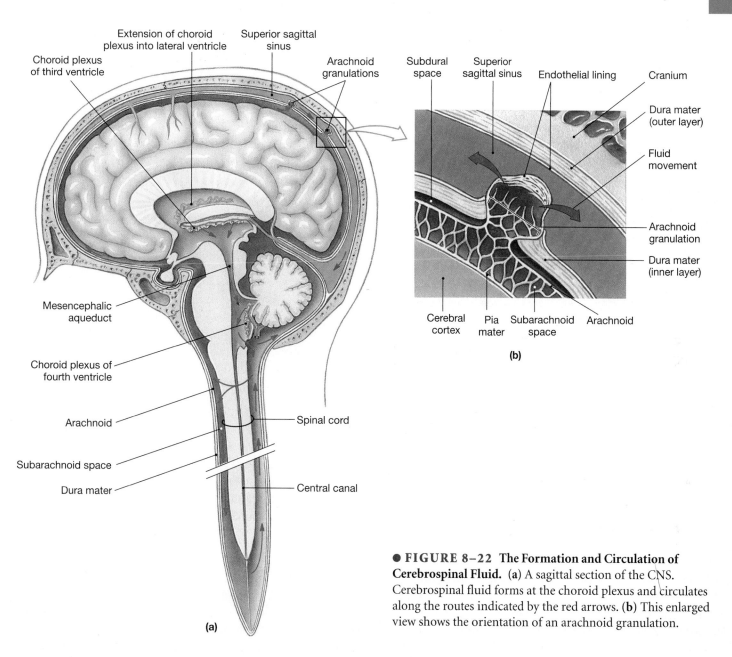

Extension of choroid
plexus into lateral ventricle

Superior sagittal
sinus

Choroid plexus
of third ventricle

Arachnoid
granulations

Subdural
space

Superior
sagittal sinus

Endothelial lining

Cranium

Dura mater
(outer layer)

Fluid
movement

Arachnoid
granulation

Dura mater
(inner layer)

Cerebral
cortex

Pia
mater

Subarachnoid
space

Arachnoid

(b)

Mesencephalic
aqueduct

Choroid plexus of
fourth ventricle

Arachnoid

Spinal cord

Subarachnoid space

Dura mater

Central canal

(a)

● **FIGURE 8–22** **The Formation and Circulation of Cerebrospinal Fluid.** (**a**) A sagittal section of the CNS. Cerebrospinal fluid forms at the choroid plexus and circulates along the routes indicated by the red arrows. (**b**) This enlarged view shows the orientation of an arachnoid granulation.

turnover, the composition of CSF is closely regulated, and the rate of removal normally keeps pace with the rate of production. If it does not, a variety of clinical problems may appear.

CSF circulates between the different ventricles, passes along the central canal, and enters the subarachnoid space (see Figure 8–22a). Once inside the subarachnoid space, the CSF circulates around the spinal cord and cauda equina and across the surfaces of the brain. Between the cerebral hemispheres, slender extensions of the arachnoid penetrate the inner layer of the dura mater. Clusters of these extensions form **arachnoid granulations,** which project into the *superior sagittal sinus,* a large cerebral vein (Figure 8–22b●). Diffusion across the arachnoid granulations returns excess cerebrospinal fluid to the venous circulation.

The Cerebrum

The cerebrum, the largest region of the brain, is the site where conscious thought and intellectual functions originate. Much of the cerebrum is involved in receiving somatic sensory information and then exerting voluntary or involuntary control over somatic motor neurons. In general, we are aware of these events. However, most sensory processing and all visceral motor (*autonomic*) control occur elsewhere in the brain, usually outside our conscious awareness.

The cerebrum includes gray matter and white matter. Gray matter is found in a superficial layer of neural cortex and in deeper basal nuclei. The central white matter, composed of myelinated axons, lies beneath the neural cortex and surrounds the basal nuclei.

Clinical Note
HYDROCHEPHALUS

Hydrocephalus is an abnormal accumulation of cerebrospinal fluid (CSF) within the ventricles of the brain. It results from an imbalance between the amount of CSF that is produced and the rate at which it is absorbed. As the CSF builds up, it causes the ventricles to enlarge and results in increased intracranial pressure (Figure 8–23●).

Hydrocephalus may be congenital or acquired. The causes of congenital hydrocephalus are complex and believed to be an interaction between environmental and genetic factors. Acquired hydrocephalus may result from intraventricular hemorrhage, meningitis, head trauma, tumors, and cysts. Hydrocephalus is believed to occur in about two out of 1000 births. The incidences of adult-onset hydrocephalus and acquired hydrocephalus are not known.

The symptoms of hydrocephalus vary with age, disease progression, and individual differences in tolerance to CSF. In infancy, the most obvious indication of hydrocephalus is often a rapid increase in head circumference. In older children and adults, symptoms may include headache accompanied by vomiting, nausea, *papilledema* (swelling of the optic disk, which is part of the optic nerve), downward deviation of the eyes (called *sunsetting*), problems with balance, poor coordination, gait disturbance, urinary incontinence, slowing or loss of development (in children), lethargy, drowsiness, irritability, or other changes in personality or cognition, including memory loss.

Hydrocephalus is diagnosed through clinical neurological evaluation and by using cranial imaging techniques such as ultrasonography, computer tomography (CT), magnetic resonance imaging (MRI), or pressure-monitoring techniques.

There is no known way to prevent or cure hydrocephalus. The most effective treatment is surgical insertion of a shunt that drains the fluid from the brain, which thus decreases intracranial pressures. ■

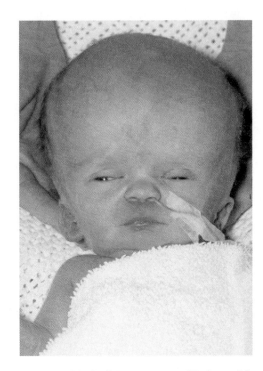

● **FIGURE 8–23** **Typical Appearance of Infant with Hydrocephalus.**

temporal lobe, which overlaps the **insula** (IN-sū-luh), an "island" of cortex that is otherwise hidden (see Figure 8–24). The **parietal lobe** extends between the central sulcus and the **parieto-occipital sulcus** (see Figure 8–20c). What remains is the **occipital lobe.**

In each lobe, some regions are concerned with sensory information and others with motor commands. Additionally, each hemisphere receives sensory information from, and sends motor commands to, the opposite side of the body. This results in the left cerebral hemisphere controlling the right side of the body and the right cerebral hemisphere controlling the left side. This crossing over has no known functional significance.

MOTOR AND SENSORY AREAS OF THE CORTEX. Figure 8–24 labels the major motor and sensory regions of the cerebral cortex. The central sulcus separates the motor and sensory portions of the cortex. The **precentral gyrus** of the frontal lobe forms the anterior margin of the central sulcus, and its surface is the **primary motor cortex.** Neurons of the primary motor cortex direct voluntary movements by controlling somatic motor neurons in the brain stem and spinal cord.

The **postcentral gyrus** of the parietal lobe forms the posterior margin of the central sulcus, and its surface contains the **primary sensory cortex.** Neurons in this region receive somatic sensory information from touch, pressure, pain, and temperature receptors. We are consciously aware of these sen-

STRUCTURE OF THE CEREBRAL HEMISPHERES. A thick blanket of neural cortex known as the cerebral cortex covers the superior and lateral surfaces of the cerebrum (Figure 8–24●). This outer surface forms a series of elevated ridges, or *gyri* (JĪ-rī; singular, *gyrus*), separated by shallow depressions, called *sulci* (SUL-s), or by deeper grooves, called *fissures.* Gyri increase the surface area of the cerebrum and, thus, the number of neurons in the cortex; the total surface area of the cerebral hemispheres is roughly equivalent to 2200 cm² (2.5 ft²) of flat surface.

The two cerebral hemispheres are separated by a deep **longitudinal fissure** (see Figure 8–20a). Each hemisphere can be divided into well-defined regions, or **lobes,** named after the overlying bones of the skull (see Figures 8–20b and 8–24). Extending laterally from the longitudinal fissure is a deep groove, the **central sulcus.** Anterior to this is the **frontal lobe,** bordered inferiorly by the **lateral sulcus** (see Figure 8–20b). The cortex inferior to the lateral sulcus is the

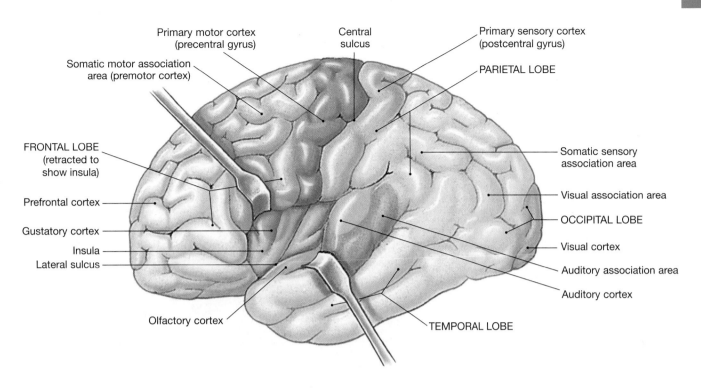

● **FIGURE 8–24** **The Surface of the Cerebral Hemispheres.** Major anatomical landmarks on the surface of the left cerebral hemisphere are shown. The colored areas represent various motor, sensory, and association areas of the cerebral cortex.

sations because brain stem nuclei relay sensory information to the primary sensory cortex.

Sensations of sight, taste, sound, and smell arrive at other portions of the cerebral cortex. The **visual cortex** of the occipital lobe receives visual information, the **gustatory cortex** of the frontal lobe receives taste sensations, and the **auditory cortex** and **olfactory cortex** of the temporal lobe receive information about hearing and smell, respectively.

ASSOCIATION AREAS. The sensory and motor regions of the cortex are connected to nearby **association areas,** regions that interpret incoming data or coordinate a motor response. The **somatic sensory association area** monitors activity in the primary sensory cortex. This area allows you to recognize a touch as light as the arrival of a mosquito on your arm. The special senses of smell, sight, and hearing involve separate areas of sensory cortex, and each has its own association area. The **somatic motor association area,** or **premotor cortex,** is responsible for coordinating learned movements. When you perform a voluntary movement, such as picking up a glass or scanning these lines of type, instructions are relayed to the primary motor cortex by the premotor cortex.

The functional distinctions between the motor and sensory association areas are most evident after localized brain damage has occurred. For example, someone with damage to the premotor cortex might understand written letters and words but be unable to read due to an inability to track along the lines on a printed page. In contrast, someone with a damaged **visual association area** can scan the lines of a printed page but cannot figure out what the letters mean.

CORTICAL CONNECTIONS. The various regions of the cerebral cortex are interconnected by the white matter that lies beneath the cerebral cortex. Axons of different lengths interconnect gyri within a single cerebral hemisphere and link the two hemispheres across the **corpus callosum** (see Figure 8–20c). Other bundles of axons link the cerebral cortex with the diencephalon, brain stem, cerebellum, and spinal cord.

CEREBRAL PROCESSING CENTERS. "Higher-order" integrative centers receive information from many different association areas. These integrative centers control extremely complex motor activities and perform complicated analytical functions. Even though integrative centers may be found in both cerebral hemispheres, many are lateralized—restricted to either the left or the right hemisphere (Figure 8–25●). Examples of integrative centers that are lateralized are those concerned with complex processes such as speech, writing, mathematical computation, and understanding spatial relationships.

■ *The General Interpretive Area.* The **general interpretive area,** or *Wernicke's area,* receives information from all the sensory association areas. This region plays an essential role

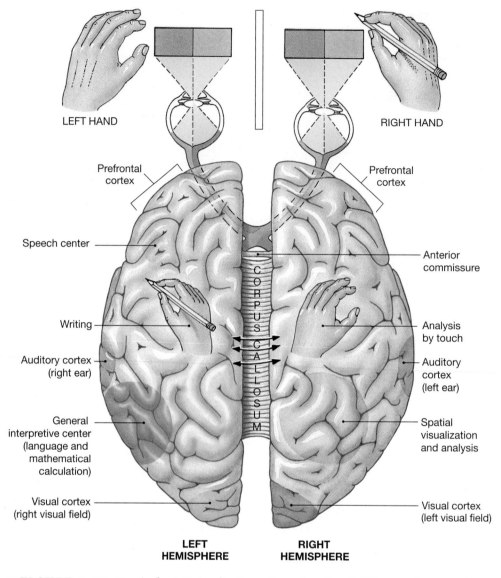

LEFT HAND

RIGHT HAND

Prefrontal cortex

Prefrontal cortex

Speech center

Anterior commissure

Writing

Analysis by touch

Auditory cortex (right ear)

Auditory cortex (left ear)

General interpretive center (language and mathematical calculation)

Spatial visualization and analysis

Visual cortex (right visual field)

Visual cortex (left visual field)

CORPUS CALLOSUM

LEFT HEMISPHERE

RIGHT HEMISPHERE

● **FIGURE 8–25 Hemispheric Lateralization.** Some functional differences between the left and right cerebral hemispheres are depicted.

in your personality by integrating sensory information and coordinating access to complex visual and auditory memories. This center is present in only one hemisphere, usually the left. Damage to this area affects the ability to interpret what is read or heard, even though the words are understood as individual entities. For example, an individual might understand the meaning of the spoken words "sit" and "here," because word recognition occurs in the auditory association areas, but be totally bewildered by the instruction "sit here."

■ *The Speech Center.* Some of the neurons in the general interpretive area connect to the **speech center** *(Broca's area).* This center lies along the edge of the premotor cortex in the same hemisphere as the general interpretive area. The speech center regulates the patterns of breathing and vocalization required for normal speech. A person with a damaged speech center can make sounds but not words.

The motor commands issued by the speech center are adjusted by feedback from the auditory association area. Damage there can cause a variety of speech-related problems. Some affected individuals have difficulty speaking, even though they know exactly which words to use; others talk constantly but use all the wrong words.

■ *The Prefrontal Cortex.* The **prefrontal cortex** of the frontal lobe (see Figure 8–25) coordinates information from the association areas of the entire cortex. In doing so, it performs such abstract intellectual functions as predicting the future consequences of events or actions. Damage to this area leads to problems in estimating time relationships between events. Questions such as "How long ago did this happen?" or "What happened first?" become difficult to answer. The prefrontal cortex also has connections with other cortical areas and with other portions of the brain. Feelings of frustration, tension, and anxiety are generated at the prefrontal cortex as it interprets ongoing events and predicts future situations or consequences. If the connections between the prefrontal cortex and other brain regions are severed, the tensions, frustrations, and anxieties are removed. Early in the 1900s this rather drastic procedure, called a *prefrontal lobotomy,* was used to "cure" a variety of mental illnesses, especially those associated with violent or antisocial behavior.

Clinical Note
APHASIA AND DYSLEXIA

Aphasia (*a-,* without + *phasia,* speech) is a disorder that affects the ability to speak or read. *Global aphasia* results from extensive damage to the general interpretive area or to the associated sensory tracts. Affected individuals cannot speak, read, understand, or interpret the speech of others. Global aphasia often accompanies a severe stroke or tumor that affects a large area of cortex, including the speech and language areas. Recovery is possible when the condition results from *edema* (an abnormal accumulation of fluid) or hemorrhage, but the process often takes months. Lesser degrees of aphasia often follow minor strokes with no initial period of global aphasia. Such individuals can understand spoken and written words and may recover completely.

Dyslexia (*lexis* diction) is a disorder that affects the comprehension and use of words. Developmental dyslexia affects children; estimates indicate that up to 15 percent of children in the U.S. suffer from some degree of dyslexia. These children have difficulty reading and writing, although their other intellectual functions may be normal or above normal. Their writing looks uneven and unorganized; letters are typically written in the wrong order (*dig* becomes *gid*) or reversed (*E* becomes Ǝ). Recent evidence suggests that at least some forms of dyslexia result from problems in processing and sorting visual information. ∎

HEMISPHERIC LATERALIZATION. As shown in Figure 8–25 each of the two cerebral hemispheres is responsible for specific functions that are not ordinarily performed by the opposite hemisphere. This specialization has been called *hemispheric lateralization.* In most people, the left hemisphere contains the general interpretive and speech centers and is responsible for language-based skills (reading, writing, and speaking). In addition, the premotor cortex involved in the control of hand movements is larger on the left side in right-handed individuals than in those who are left-handed. The left hemisphere is also important in performing analytical tasks, such as mathematical calculations and logical decision making. For these reasons, the left hemisphere has been called the *dominant hemisphere,* or the *categorical hemisphere.*

The right cerebral hemisphere analyzes sensory information and relates the body to the sensory environment. Interpretive centers in this hemisphere enable you to identify familiar objects by touch, smell, taste, or feel. For example, the right hemisphere plays a dominant role in recognizing faces and in understanding three-dimensional relationships. It is also important in analyzing the emotional context of a conversation—for instance, distinguishing between the threat "Get lost" and the question "Get lost?"

Interestingly, there may be a link between handedness and sensory/spatial abilities. An unusually high percentage of musicians and artists are left-handed; the complex motor activities performed by these individuals are directed by the primary motor cortex and association areas on the right hemisphere, near the association areas involved with spatial visualization and emotions.

Hemispheric lateralization does not mean that the two hemispheres function independently of each other. As previously noted, the white fibers of the corpus callosum link the two hemispheres, including their sensory information and motor commands. The corpus callosum alone contains over 200 million axons, and carry an estimated 4 billion impulses per second!

THE ELECTROENCEPHALOGRAM. The primary sensory cortex and the primary motor cortex have been mapped by direct stimulation in patients undergoing brain surgery. The functions of other regions of the cerebrum can be revealed by the behavioral changes that follow localized injuries or strokes, and the activities of specific regions can be examined by noninvasive techniques such as a PET scan or sequential MRI scans. ∞ p. 23

The electrical activity of the brain is commonly monitored to assess brain activity. Neural function depends on electrical events within the cell membrane. The brain contains billions of nerve cells, and their activity generates an electrical field that can be measured by placing electrodes on the brain or on the outer surface of the skull. The electrical activity changes constantly as nuclei and cortical areas are stimulated or quiet down. An **electroencephalogram (EEG)** is a printed record of this electrical activity over time. The electrical patterns are called **brain waves,** which can be correlated with the individual's level of consciousness. Electroencephalograms can also provide useful diagnostic information regarding brain disorders. Four types of brain wave patterns are shown in Figure 8–26●.

MEMORY. What was the topic of the last sentence you read? What is your Social Security number? How do you open a screw-top jar? How do you throw a Frisbee? Answering these questions involves accessing memories, which are stored bits of information gathered through prior experience. Answering the

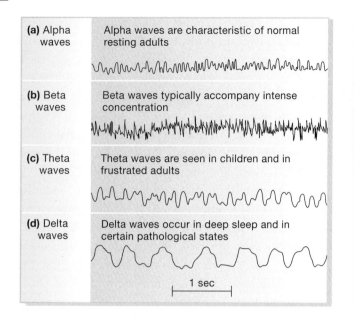

(a) Alpha waves	Alpha waves are characteristic of normal resting adults
(b) Beta waves	Beta waves typically accompany intense concentration
(c) Theta waves	Theta waves are seen in children and in frustrated adults
(d) Delta waves	Delta waves occur in deep sleep and in certain pathological states

1 sec

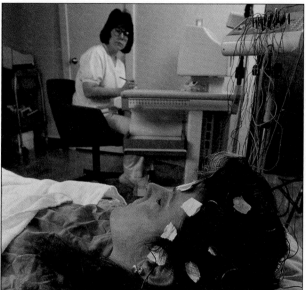

● **FIGURE 8–26 Brain Waves.**

first two questions involves **fact memories,** which are specific bits of information. Answering the last two questions involves **skill memories,** which are learned motor behaviors. With repetition, skill memories become incorporated at the unconscious level. Examples include the complex motor patterns involved in skiing or playing the violin. Skill memories related to programmed behaviors, such as eating, are stored in appropriate portions of the brain stem. Complex skill memories involve an interplay between the cerebellum and the cerebral cortex.

Memories are often classified according to duration. **Short-term memories,** or *primary memories,* do not last long, but while they persist the information can be recalled immediately. Primary memories contain small bits of information, such as a person's name or a telephone number. Repeating a phone number or other bit of information reinforces the original short-term memory and helps ensure its conversion to a long-term memory. **Long-term memories** remain for much longer periods, in some cases for an entire lifetime. The conversion from short-term to long-term memory is called *memory consolidation.* Some long-term memories fade with time and may require considerable effort to recall. Other long-term memories seem to be part of consciousness, such as your name or the contours of your own body.

Most long-term memories are stored in the cerebral cortex. Conscious motor and sensory memories are referred to the appropriate association areas. For example, visual memories are stored in the visual association area, and memories of voluntary motor activity are kept in the premotor cortex. Special portions of the occipital and temporal lobes retain the memories of faces, voices, and words.

Amnesia refers to the loss of memory from disease or trauma. The type of memory loss depends on the specific regions of the brain affected. For example, damage to the auditory association areas may make it difficult to remember sounds. Damage to thalamic and limbic structures, especially the *hippocampus,* will affect memory storage and consolidation.

The Basal Nuclei

While your cerebral cortex is consciously active, other centers of your cerebrum, diencephalon, and brain stem are processing sensory information and issuing motor commands at a subconscious level. Many of these activities outside our conscious awareness are directed by the basal nuclei, or *cerebral nuclei.* The **basal nuclei** are masses of gray matter that lie beneath the lateral ventricles and within the central white matter of each cerebral hemisphere (Figure 8–27●). The **caudate nucleus** has a massive head and slender, curving tail that follows the curve of the lateral ventricle. Inferior to the head of the caudate nucleus is the **lentiform** (*lens-shaped*) **nucleus,** which consists of a medial **globus pallidus** (GLŌ-bus PAL-i-dus; pale globe) and a lateral **putamen** (pū-TA-men). Together, the caudate and lentiform nuclei are also called the *corpus striatum* (striated body). Inferior to the caudate and lentiform nuclei is another nucleus, the **amygdaloid** (ah-MIG-da-loyd; *amygdale,* almond) **body.** It is a component of the *limbic system* and is discussed in the next section.

The basal nuclei function in the subconscious control of skeletal muscle tone and the coordination of learned movement patterns. These nuclei do not start a movement—that decision is a voluntary one—but once a movement is under way,

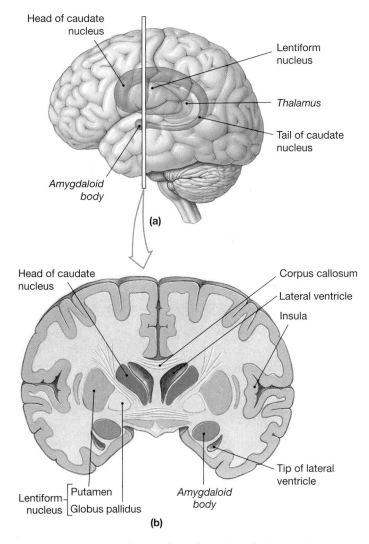

Clinical Note
SEIZURES

A *seizure* is an episode of abnormal neurological function caused by an abnormal electrical discharge of brain neurons. The seizure is the clinical event experienced by the patient following the abnormal electrical discharge. *Epilepsy* is a clinical syndrome in which an individual is subject to recurrent seizures. The occurrence of one or more seizures indicates an abnormal function of cerebral neurons. Where a cause for attacks in patients who are otherwise normal cannot be determined, seizures are referred to as *primary,* or *idiopathic.* Seizures that result from some identifiable condition, such as brain tumor or brain trauma, are referred to as *secondary seizures.*

Seizures are common and are a frequent reason that EMS is summoned. Most are self-limited and last less than a minute. Seizures can be categorized as generalized or partial, depending upon the part of the brain involved. Generalized seizures involve the entire cerebral cortex, and usually begin with a loss of consciousness. This may be the only symptom, or it may be followed by motor activity. *Generalized tonic-clonic seizures* are the most familiar and dramatic seizure type. Often referred to as *grand mal seizures,* generalized tonic-clonic seizures cause the patient to become stiff and fall to the ground. This is followed by alternating contraction and relaxation of the skeletal muscles. Urinary and fecal incontinence is common. A seizure usually lasts 6–90 seconds, and as it ends, the patient is left flaccid and unconscious. The patient may be confused for up to an hour following the attack (*postictal confusion*). Fatigue is common and may last for hours after the event.

Absence seizures, also called *petit mal seizures,* are very brief, and often last only a few seconds. Absence seizures are generalized seizures in which the patient suddenly loses consciousness without any loss of postural tone. These attacks end abruptly and the patients resume their activities. Often the patient and bystanders are unaware that anything has happened.

Focal seizures are due to electrical discharges that are limited to a portion of the cerebral cortex and are often secondary seizures that result from a lesion in the brain. The effects of focal seizure depend upon the part of the brain involved. If the seizure's focus is in the motor cortex, tonic or clonic muscle contractions that involve a single extremity may result. Sensory hallucinations suggest a focus in the sensory cortex. Visual symptoms, such as flashing lights, suggest a focus in the occipital lobe. Focal seizures isolated in the temporal lobe can cause altered thinking or behavior. Commonly referred to as psychomotor seizures, they can be mistakenly diagnosed as psychiatric disease.

Most seizure disorders can be controlled with medication. Modern anticonvulsant medications have minimal side effects and are highly effective. In fact, most seizures seen in the emergency setting occur because patients fail to take their medication or take it improperly. ■

●**FIGURE 8–27 The Basal Nuclei.** The relative positions of the basal nuclei can be seen in (**a**) a lateral view of a transparent brain and (**b**) this frontal section.

the basal nuclei provide the general pattern and rhythm. For example, during a simple walk the basal nuclei control the cycles of arm and thigh movements that occur between the time you decide to "start" walking and the time you give the "stop" order.

The Limbic System

The **limbic system** (LIM-bik; *limbus,* a border) includes the olfactory cortex, several basal nuclei, gyri, and tracts along the border between the cerebrum and diencephalon (Figure 8–28●). This system is a functional grouping rather than an anatomical one. The functions of the limbic system include (1) establishing emotional states and related behavioral drives; (2) linking the conscious, intellectual functions of the cerebral cortex with the unconscious and autonomic functions of the brain stem; and (3) long-term memory storage and retrieval. Whereas the sensory cortex,

motor cortex, and association areas of the cerebral cortex enable you to perform complex tasks, it is largely the limbic system that makes you *want* to do them.

The amygdaloid bodies link the limbic system, the cerebrum, and various sensory systems. These nuclei play a role in

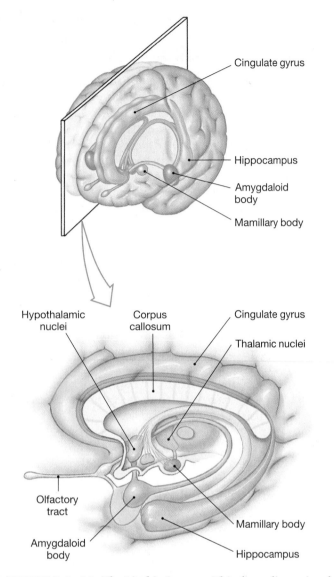

→ CONCEPT CHECK QUESTIONS

1. How would decreased diffusion across the arachnoid granulations affect the volume of cerebrospinal fluid in the ventricles?
2. Mary suffers a head injury that damages her primary motor cortex. Where is this area located?
3. What senses would be affected by damage to the temporal lobes of the cerebrum?

Answers begin on p. 792.

● FIGURE 8–28 The Limbic System. This three-dimensional reconstruction of the limbic system shows the relationships among the system's major components.

the regulation of heart rate, in the control of the "fight or flight" response, and in linking emotions with specific memories. The hippocampus is important in learning and in the storage of long-term memories. Damage to the hippocampus that occurs in Alzheimer's disease interferes with memory storage and retrieval.

The limbic system also includes hypothalamic centers that control (1) emotional states, such as rage, fear, and sexual arousal, and (2) reflex movements that can be consciously activated. For example, the limbic system includes the *mamillary bodies* (MAM-i-lar-ē; *mamilla*, a little breast) of the hypothalamus. These nuclei process olfactory sensations and control reflex movements associated with eating, such as chewing, licking, and swallowing.

The Diencephalon

The diencephalon (Figure 8–29●) contains switching and relay centers that integrate conscious and unconscious sensory information and motor commands. It surrounds the third ventricle and consists of the *epithalamus, thalamus,* and *hypothalamus.*

THE EPITHALAMUS. The epithalamus lies superior to the third ventricle where it forms the roof of the diencephalon. The anterior portion contains an extensive area of choroid plexus. The posterior portion contains the **pineal gland** (Figure 8–29b●), which is an endocrine structure that secretes the hormone *melatonin.* Among other functions, melatonin is important in regulating day–night cycles.

THE THALAMUS. The left thalamus and right thalamus are separated by the third ventricle, and each contains a rounded mass of thalamic nuclei. The thalamus (see Figures 8–20c and 8–29) is the final relay point for all ascending sensory information, other than olfactory, that will reach our conscious awareness. It acts as a filter, and passes on to the primary sensory cortex only a small portion of the arriving sensory information. The rest is relayed to the basal nuclei and centers in the brain stem. The thalamus also plays a role in the coordination of voluntary and involuntary motor commands.

THE HYPOTHALAMUS. The hypothalamus lies inferior to the third ventricle (see Figure 8–20c). The hypothalamus contains important control and integrative centers in addition to those associated with the limbic system. Its diverse functions include (1) the subconscious control of skeletal muscle contractions associated with rage, pleasure, pain, and sexual arousal; (2) adjusting the activities of autonomic centers in the pons and medulla oblongata (such as heart rate, blood pressure, respiration, and digestive functions); (3) coordinating activities of the nervous and endocrine systems; (4) secreting a variety of hormones, including *antidiuretic hormone (ADH)* and *oxytocin;* (5) producing the behavioral "drives" involved in hunger and thirst; (6) coordinating voluntary and autonomic functions;

(7) regulating normal body temperature; and (8) coordinating the daily cycles of activity.

The Midbrain

The midbrain (see Figure 8–29) contains various nuclei and bundles of ascending and descending nerve fibers. It includes two pairs of sensory nuclei, or *colliculi* (ko-LIK-ū-lī; singular: *colliculus* a small hill), involved in the processing of visual and auditory sensations. The *superior colliculi* control the reflex movements of the eyes, head, and neck in response to visual stimuli, such as a blinding flash of light. The *inferior colliculi* control reflex movements of the head, neck, and trunk in response to auditory stimuli, such as a loud noise. The midbrain also contains motor nuclei for two of the cranial nerves (N III, IV) involved in the control of eye movements. Descending bundles of nerve fibers on the ventrolateral surface of the midbrain make up the **cerebral peduncles** (*peduncles*, little feet). Some of the descending fibers go to the cerebellum by way of the pons, and others carry voluntary motor commands from the primary motor cortex of each cerebral hemisphere.

The midbrain is also headquarters to one of the most important brain stem components, the **reticular formation,** which regulates many involuntary functions. The reticular formation is a network of interconnected nuclei that extends the length of the brain stem. The reticular formation of the midbrain contains the *reticular activating system (RAS).* The output of this system directly affects the activity of the cerebral cortex. When the RAS is inactive, so are we; when the RAS is stimulated, so is our state of attention or wakefulness.

The maintenance of muscle tone and posture is controlled by midbrain nuclei that integrate information from the cerebrum and cerebellum and issue the appropriate involuntary motor commands. Other midbrain nuclei play an important role in regulating the motor output of the basal nuclei. For example, the *substantia nigra* (NĪ-gruh; black) inhibit the activity of the basal nuclei by releasing the neurotransmitter dopamine. ∞ p. 273 If the substantia nigra are damaged or the neurons secrete less dopamine, the basal nuclei become more active. The result is a gradual increase in muscle tone and the appearance of symptoms characteristic of *Parkinson's disease.* Persons with Parkinson's disease have difficulty starting voluntary movements because opposing muscle groups do not relax—they must be overpowered. Once a movement is under way, every aspect must be voluntarily controlled through intense effort and concentration.

The Pons

The pons (see Figure 8–29a●) links the cerebellum with the midbrain, diencephalon, cerebrum, and spinal cord. One group of nuclei within the pons includes the sensory and motor nuclei for four of the cranial nerves (N V–VIII).

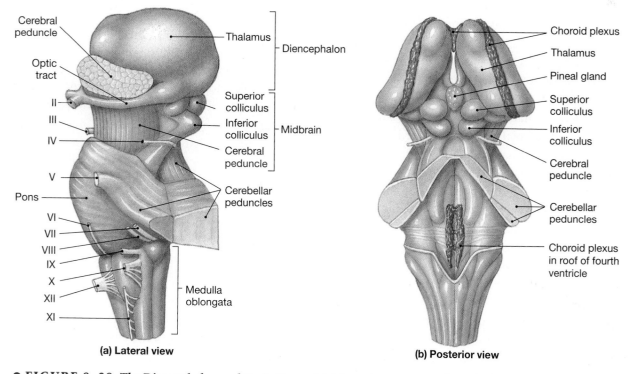

(a) Lateral view

(b) Posterior view

● **FIGURE 8–29 The Diencephalon and Brain Stem.** **(a)** A lateral view, as seen from the left side. The Roman numerals refer to the cranial nerves. **(b)** A posterior view.

Other nuclei are concerned with the involuntary control of the pace and depth of respiration. Tracts that pass through the pons link the cerebellum with the brain stem, cerebrum, and spinal cord.

The Cerebellum

The cerebellum (see Figure 8–20b,c) is an automatic processing center. Its two important functions are (1) adjusting the postural muscles of the body to maintain balance and (2) programming and fine-tuning movements controlled at the conscious and subconscious levels. These functions are performed indirectly by regulating activity along motor pathways at the cerebral cortex, basal nuclei, and brain stem. The cerebellum compares the motor commands with proprioceptive information (position sense) and performs adjustments needed to make the movement smooth. The tracts that link the cerebellum with these different regions are the **cerebellar peduncles** (see Figure 8–29). Like the cerebrum, the cerebellum is composed of white matter covered by a layer of neural cortex called the *cerebellar cortex*.

The cerebellum can be permanently damaged by trauma or stroke or temporarily affected by drugs such as alcohol. These alterations can produce *ataxia* (a-TAK-sē-uh; *ataxia*, a lack of order), which is a disturbance in balance.

The Medulla Oblongata

The medulla oblongata (see Figure 8–29) connects the brain with the spinal cord. All communication between the brain and spinal cord involves tracts that ascend or descend through the medulla oblongata. These tracts often synapse in the medulla oblongata at sensory or motor nuclei that act as relay stations and processing centers. In addition to these nuclei, the medulla oblongata contains sensory and motor nuclei associated with five of the cranial nerves (N VIII–XII).

The portion of the reticular system within the medulla oblongata contains nuclei and centers that regulate vital autonomic functions. These *reflex centers* receive inputs from cranial nerves, the cerebral cortex, and the brain stem, and their output controls or adjusts the activities of the cardiovascular and respiratory systems. The **cardiovascular centers** adjust heart rate, the strength of cardiac contractions, and the flow of blood through peripheral tissues. In terms of function, the cardiovascular centers are subdivided into a *cardiac center* that regulates the heart rate and a *vasomotor center* that controls peripheral blood flow. The **respiratory rhythmicity centers** set the basic pace for respiratory movements, and their activity is adjusted by the *respiratory centers* of the pons.

Key Note

The brain is a large, delicate mass of neural tissue that contains internal passageways and chambers filled with cerebrospinal fluid. Each of the six major regions of the brain has specific functions. As you ascend from the medulla oblongata, which connects to the spinal cord, to the cerebrum, those functions become more complex and variable. Conscious thought and intelligence are provided by the neural cortex of the cerebral hemispheres.

CONCEPT CHECK QUESTIONS

1. The thalamus acts as a relay point for all but what type of sensory information?
2. Which area of the diencephalon is stimulated by changes in body temperature?
3. The medulla oblongata is one of the smallest sections of the brain. Why can damage there cause death, whereas similar damage in the cerebrum might go unnoticed?

Answers begin on p. 792.

■ The Peripheral Nervous System

The peripheral nervous system (PNS) is the link between the neurons of the central nervous system (CNS) and the rest of the body; all sensory information and motor commands are carried by axons of the PNS (see Figure 8–1, p. 260). These axons, bundled together and wrapped in connective tissue, form **peripheral nerves,** or simply nerves. Cranial nerves originate from the brain, and spinal nerves connect to the spinal cord. The PNS also contains both the cell bodies and the axons of sensory neurons and motor neurons of the autonomic nervous system. The cell bodies are clustered together in masses called **ganglia** (singular: *ganglion*) (see Figure 8–6, p. 265).

The Cranial Nerves

Twelve pairs of **cranial nerves** connect to the brain (Figure 8–30●). Each cranial nerve has a name related to its appearance or function; each also has a designation that consists of the letter N (for "nerve") and a Roman numeral (for its position along the longitudinal axis of the brain). For example, N I refers to the first pair of cranial nerves, the olfactory nerves.

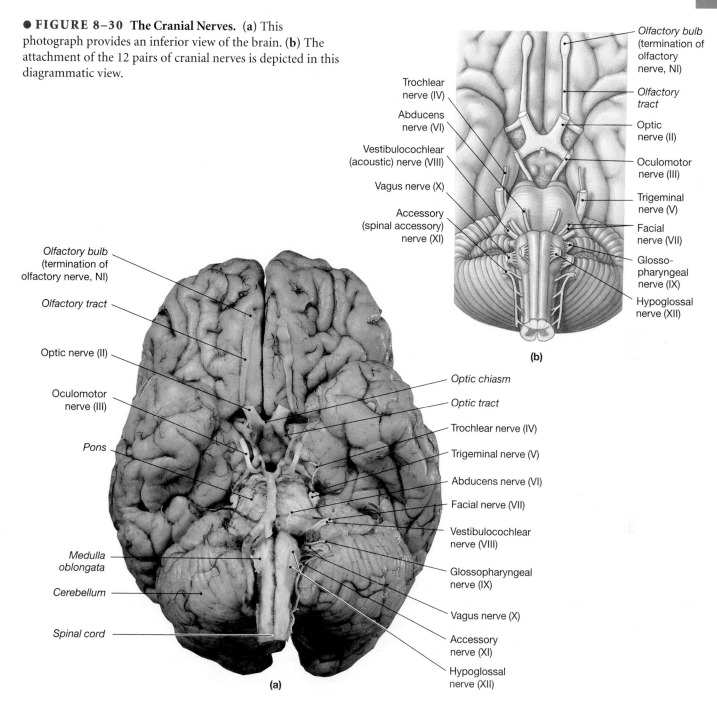

● **FIGURE 8–30 The Cranial Nerves. (a)** This photograph provides an inferior view of the brain. **(b)** The attachment of the 12 pairs of cranial nerves is depicted in this diagrammatic view.

Olfactory bulb (termination of olfactory nerve, NI)

Olfactory tract

Trochlear nerve (IV)

Abducens nerve (VI)

Vestibulocochlear (acoustic) nerve (VIII)

Vagus nerve (X)

Accessory (spinal accessory) nerve (XI)

Olfactory bulb (termination of olfactory nerve, NI)

Olfactory tract

Optic nerve (II)

Oculomotor nerve (III)

Trigeminal nerve (V)

Facial nerve (VII)

Glosso-pharyngeal nerve (IX)

Hypoglossal nerve (XII)

(b)

Olfactory bulb (termination of olfactory nerve, NI)

Olfactory tract

Optic nerve (II)

Oculomotor nerve (III)

Pons

Medulla oblongata

Cerebellum

Spinal cord

Optic chiasm

Optic tract

Trochlear nerve (IV)

Trigeminal nerve (V)

Abducens nerve (VI)

Facial nerve (VII)

Vestibulocochlear nerve (VIII)

Glossopharyngeal nerve (IX)

Vagus nerve (X)

Accessory nerve (XI)

Hypoglossal nerve (XII)

(a)

Distribution and Function of Cranial Nerves

Functionally, each nerve can be classified as primarily sensory, primarily motor, or mixed (sensory and motor). Many cranial nerves, however, have secondary functions. For example, several cranial nerves (N III, VII, IX, and X) also carry autonomic fibers to PNS ganglia, just as spinal nerves deliver them to ganglia along the spinal cord. Next we consider the distribution and functions of the cranial nerves.

Few people are able to remember the names, numbers, and functions of the cranial nerves without some effort. Many people use mnemonic phrases, such as Oh, Once One Takes The Anatomy Final, Very Good Vacations Are Heavenly, in which the first letter of each word represents the names of the cranial nerves.

THE OLFACTORY NERVES (N I). The first pair of cranial nerves, the **olfactory nerves,** are the only cranial nerves attached to the cerebrum. (The rest start or end within nuclei of the diencephalon or brain stem.) These nerves carry special sensory information responsible for the sense of smell. The olfactory nerves originate in the epithelium of the upper nasal cavity and penetrate the cribriform plate of the ethmoid bone to synapse in the olfactory bulbs of the brain. From the olfactory bulbs, the axons of postsynaptic neurons travel within the olfactory tracts to the olfactory centers of the brain.

THE OPTIC NERVES (N II).   The **optic nerves** carry visual information from the eyes. After passing through the **optic foramina** of the orbits, these nerves intersect at the **optic chiasm** ("crossing") (Figure 8–30a●) before they continue as the *optic tracts* to nuclei of the left and right thalamus.

THE OCULOMOTOR NERVES (N III).   The midbrain contains the motor nuclei that control the third and fourth cranial nerves. Each **oculomotor nerve** innervates four of the six muscles that move an eyeball (the superior, medial, and inferior rectus muscles and the inferior oblique muscle). These nerves also carry autonomic fibers to intrinsic eye muscles that control the amount of light that enters the eye and the shape of the lens.

THE TROCHLEAR NERVES (N IV).   The **trochlear** (TRŌK-lē-ar; *trochlea*, a pulley) **nerves,** which are the smallest of the cranial nerves, innervate the superior oblique muscles of the eyes. The motor nuclei that control these nerves lie in the midbrain. The name *trochlear* refers to the pulley-shaped, ligamentous sling through which the tendon of the superior oblique muscle passes to reach its attachment on the eyeball (see Figure 9-9a, p. 336).

THE TRIGEMINAL NERVES (N V).   The pons contains the nuclei associated with cranial nerve V. The **trigeminal** (trī-JEM-i-nal) **nerves** are the largest of the cranial nerves. These nerves provide sensory information from the head and face and motor control over the chewing muscles, such as the temporalis and masseter. The trigeminal has three major branches. The *ophthalmic branch* provides sensory information from the orbit of the eye, the nasal cavity and sinuses, and the skin of the forehead, eyebrows, eyelids, and nose. The *maxillary branch* provides sensory information from the lower eyelid, upper lip, cheek, nose, upper gums and teeth, palate, and portions of the pharynx. The *mandibular branch,* which is the largest of the three, provides sensory information from the skin of the temples, the lower gums and teeth, the salivary glands, and the anterior portions of the tongue. It also provides motor control over the chewing muscles (the temporalis, masseter, and pterygoid muscles). ⟷ p. 232

THE ABDUCENS NERVES (N VI).   The **abducens** (ab-DŪ-senz) **nerves** innervate only the lateral rectus, the sixth of the extrinsic eye muscles. The nuclei of the abducens nerves are in the pons. The nerves emerge at the border between the pons and the medulla oblongata and reach the orbit of the eye

along with the oculomotor and trochlear nerves. The name *abducens* is based on the action of this nerve's innervated muscle, which abducts the eyeball, and causes it to rotate laterally, away from the midline of the body.

THE FACIAL NERVES (N VII).   The **facial nerves** are mixed nerves of the face whose sensory and motor roots emerge from the side of the pons. The sensory fibers monitor proprioceptors in the facial muscles, provide deep pressure sensations over the face, and provide taste information from receptors along the anterior two-thirds of the tongue. The motor fibers produce facial expressions by controlling the superficial muscles of the scalp and face and muscles near the ear. These nerves also carry autonomic fibers that result in control of the tear glands and salivary glands.

THE VESTIBULOCOCHLEAR NERVES (N VIII).   The **vestibulocochlear nerves,** also called acoustic nerves, monitor the sensory receptors of the inner ear. The pons and medulla oblongata contain nuclei associated with these nerves. Each vestibulocochlear nerve has two components: (1) a **vestibular nerve** (*vestibulum*, a cavity), which originates at the *vestibule* (the portion of the inner ear concerned with balance sensations) and conveys information on position, movement, and balance; and (2) the **cochlear** (KOK-lē-ar; *cochlea*, snail shell) **nerve,** which monitors the receptors of the cochlea (the portion of the inner ear responsible for the sense of hearing).

THE GLOSSOPHARYNGEAL NERVES (N IX).   The **glossopharyngeal** (glos-ō-fah-RIN-jē-al; *glossus*, tongue) **nerves** are mixed nerves that innervate the tongue and pharynx. The associated sensory and motor nuclei are in the medulla oblongata. The sensory portion of this nerve provides taste sensations from the posterior third of the tongue and monitors blood pressure and dissolved gas concentrations in major blood vessels. The motor portion controls the pharyngeal muscles involved in swallowing. These nerves also carry autonomic fibers that result in control of the parotid salivary glands.

THE VAGUS NERVES (N X).   The **vagus** (VĀ-gus; *vagus*, wandering) **nerves** provide sensory information from the ear canals, the diaphragm, and taste receptors in the pharynx, and from visceral receptors along the esophagus, respiratory tract, and abdominal organs as far away as the last portions of the large intestine. The associated sensory and motor nuclei of the vagus are located in the medulla oblongata. The sensory information provided by N X is vital to the autonomic control of visceral function, but we are not consciously aware of these sensations because they are seldom relayed to the cerebral cortex. The motor components of the vagus nerves control skeletal muscles of the soft palate, pharynx, and esophagus and affect cardiac muscle, smooth muscle, and glands of the esophagus, stomach, intestines, and gallbladder.

THE ACCESSORY NERVES (N XI). The **accessory nerves,** sometimes called the *spinal accessory nerves,* are motor nerves that innervate structures in the neck and back. These nerves differ from other cranial nerves in that some of their motor fibers originate in the lateral gray horns of the first five cervical segments of the spinal cord, as well as in the medulla oblongata. The internal branch joins the vagus nerve and innervates the voluntary swallowing muscles of the soft palate and pharynx, as well as the laryngeal muscles that control the vocal cords and produce speech. The external branch controls the sternocleidomastoid and trapezius muscles associated with the pectoral girdle. ∞ pp. 233, 240

THE HYPOGLOSSAL NERVES (N XII). The **hypoglossal** (hī-pō-GLOS-al) **nerves** provide voluntary control over the skeletal muscles of the tongue. The nuclei for these motor nerves are located in the medulla oblongata.

The distribution and functions of the cranial nerves are summarized in Table 8–2.

Key Note

There are 12 pairs of cranial nerves. They are responsible for the special senses of smell, sight, and hearing/balance, and for control over the muscles of the eye, jaw, face, and tongue, and superficial muscles of the neck, back, and shoulders. The cranial nerves also provide sensory information from the face, neck, and upper chest and autonomic innervation to organs in the thoracic and abdominopelvic cavities.

CONCEPT CHECK QUESTIONS

1. What signs would you associate with damage to the abducens nerve (N VI)?
2. John is having trouble moving his tongue. His physician tells him it is due to pressure on a cranial nerve. Which cranial nerve is involved?

Answers begin on p. 792.

The Spinal Nerves

The 31 pairs of **spinal nerves** are grouped according to the region of the vertebral column from which they originate

TABLE 8-2 *The Cranial Nerves*

CRANIAL NERVES (NUMBER)	PRIMARY FUNCTION	INNERVATION
Olfactory (I)	Special sensory	Olfactory epithelium
Optic (II)	Special sensory	Retina of eye
Oculomotor (III)	Motor	Inferior, medial, superior rectus; inferior oblique; and intrinsic muscles of eye
Trochlear (IV)	Motor	Superior oblique muscle of eye
Trigeminal (V)	Mixed	*Sensory:* orbital structures, nasal cavity, skin of forehead, eyelids, eyebrows, nose, lips, gums and teeth, cheek, palate, pharynx, and tongue
		Motor: chewing (temporalis, masseter, pterygoids) muscles
Abducens (VI)	Motor	Lateral rectus muscle of eye
Facial (VII)	Mixed	*Sensory:* taste receptors on the anterior 2/3 of tongue
		Motor: muscles of facial expression, lacrimal (tear) gland, and submandibular and sublingual salivary glands
Vestibulocochlear (Acoustic) (VIII)	Special sensory	Cochlea (receptors for hearing) Vestibule (receptors for motion and balance)
Glossopharyngeal (IX)	Mixed	*Sensory:* posterior 1/3 of tongue; pharynx and palate (part); receptors for blood pressure, pH, oxygen, and carbon dioxide concentrations
		Motor: pharyngeal muscles, parotid salivary gland
Vagus (X)	Mixed	*Sensory:* pharynx; auricle and external acoustic canal; diaphragm; visceral organs in thoracic and abdominopelvic cavities
		Motor: palatal and pharyngeal muscles and visceral organs in thoracic and abdominopelvic cavities
Accessory (Spinal Accessory) (XI)	Motor	Voluntary muscles of palate, pharynx, and larynx (with vagus nerve); sternocleidomastoid and trapezius muscles
Hypoglossal (XII)	Motor	Tongue muscles

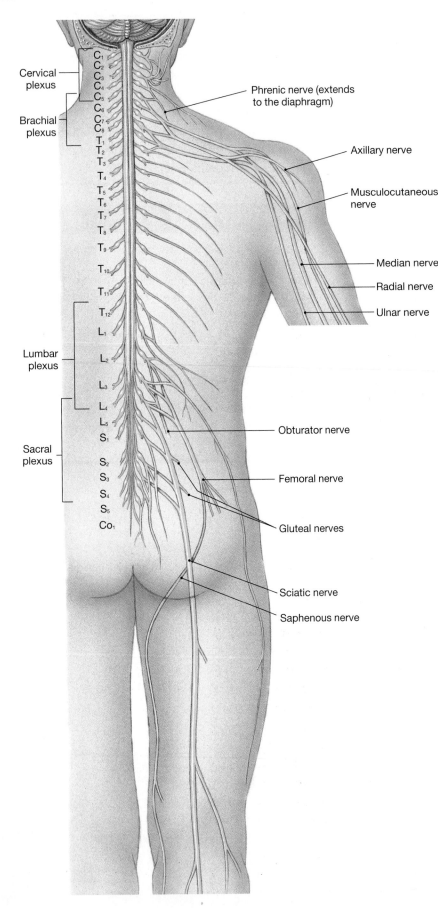

Cervical plexus

Brachial plexus

C₁
C₂
C₃
C₄
C₅
C₆
C₇
C₈
T₁
T₂
T₃
T₄
T₅
T₆
T₇
T₈
T₉
T₁₀
T₁₁
T₁₂

Phrenic nerve (extends to the diaphragm)

Axillary nerve

Musculocutaneous nerve

Median nerve

Radial nerve

Ulnar nerve

Lumbar plexus

L₁
L₂
L₃
L₄
L₅
S₁

Sacral plexus

S₂
S₃
S₄
S₅
Co₁

Obturator nerve

Femoral nerve

Gluteal nerves

Sciatic nerve

Saphenous nerve

● **FIGURE 8–31** **Peripheral Nerves and Nerve Plexuses.**

(Figure 8–31●). They include eight pairs of cervical nerves (C_1–C_8), 12 pairs of thoracic nerves (T_1–T_{12}), five pairs of lumbar nerves (L_1–L_5), five pairs of sacral nerves (S_1–S_5), and one pair of coccygeal nerves (Co_1). Each pair of spinal nerves monitors a specific region of the body surface known as a **dermatome** (Figure 8–32●). Dermatomes are clinically important because damage or infection of a spinal nerve or of dorsal root ganglia produces a characteristic loss of sensation in the corresponding region of the skin. For example, *shingles* is a virus that infects dorsal root ganglia and causes a painful rash whose distribution corresponds to that of the affected sensory nerves.

Nerve Plexuses

During development, skeletal muscles commonly fuse, and form larger muscles innervated by nerve trunks that contain axons derived from several spinal nerves. These compound nerve trunks originate at networks called **nerve plexuses.** The four plexuses and the major peripheral nerves are shown in Figure 8–31.

The **cervical plexus** innervates the muscles of the neck and extends into the thoracic cavity to control the diaphragm. The **brachial plexus** innervates the shoulder girdle and upper limb. The **lumbar plexus** and the **sacral plexus** supply the pelvic girdle and lower limb. These plexuses are sometimes designated the *lumbosacral plexus.* Table 8–3 lists the spinal nerve plexuses and describes the distribution of some of the major nerves.

The nerves that arise at the nerve plexuses contain sensory as well as motor fibers. *Peripheral nerve palsies,* also known as *peripheral neuropathies,* are characterized by regional losses of sensory and motor function as the result of nerve trauma or compression. You have experienced a mild, temporary palsy if your arm or leg has ever "fallen asleep."

Reflexes

The central and peripheral nervous systems can be studied separately, but they function

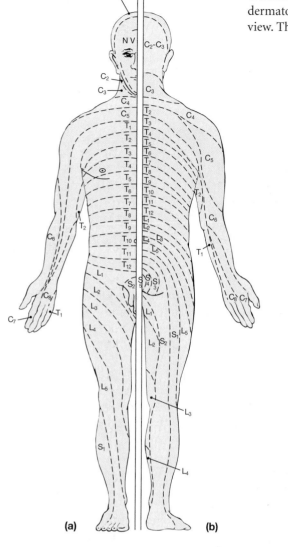

(a)   (b)

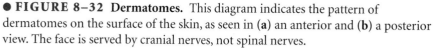

● **FIGURE 8–32 Dermatomes.** This diagram indicates the pattern of dermatomes on the surface of the skin, as seen in (**a**) an anterior and (**b**) a posterior view. The face is served by cranial nerves, not spinal nerves.

together. To consider the ways the CNS and PNS interact, we begin with simple reflex responses to stimulation. A **reflex** is an automatic motor response to a specific stimulus. Reflexes help preserve homeostasis by making rapid adjustments in the function of organs or organ systems. The response shows little variability—when a particular reflex is activated, it usually produces the same motor response.

Simple Reflexes

A reflex involves sensory fibers that deliver information from peripheral receptors to the CNS and motor fibers that carry motor commands to peripheral effectors. The "wiring" of a single reflex is called a **reflex arc.** Figure 8–33● diagrams the five steps involved in the action of a reflex arc: (1) the arrival of a stimulus and activation of a receptor, (2) the activation of a sensory neuron, (3) information processing by an interneuron, (4) the activation of a motor neuron, and (5) the response by an effector (muscle or gland).

A reflex response usually removes or opposes the original stimulus. In Figure 8–33●, the contracting muscle pulls the hand away from the painful stimulus. This reflex arc is, therefore, an example of *negative feedback.* ∞ p. 7 By opposing potentially harmful changes in the internal or external environment, reflexes play an important role in maintaining homeostasis.

TABLE 8-3 *Nerve Plexuses and Major Nerves*

PLEXUS	MAJOR NERVE	DISTRIBUTION
Cervical Plexus (C$_1$–C$_5$)	Phrenic nerve	Diaphragm
	Other branches	Muscles of the neck; skin of upper chest, neck, and ears
Brachial Plexus (C$_5$–T$_1$)	Axillary nerve	Deltoid and teres minor muscles; skin of shoulder
	Musculocutaneous nerve	Flexor muscles of the arm and forearm; skin on lateral surface of forearm
	Median nerve	Flexor muscles of forearm and hand; skin over lateral surface of hand
	Radial nerve	Extensor muscles of the arm, forearm, and hand; skin over posterolateral surface of the arm
	Ulnar nerve	Flexor muscles of forearm and small digital muscles; skin over medial surface of hand
Lumbosacral Plexus **Lumbar Plexus (T$_{12}$–L$_4$)**	Femoral nerve	Flexors and adductors of hip, extensors of knee; skin over medial surfaces of thigh, leg, and foot
	Obturator nerve	Adductors of hip; skin over medial surface of thigh
	Saphenous nerve	Skin over medial surface of leg
Sacral Plexus (L$_4$–S$_4$)	Gluteal nerve	Adductors and extensors of hip; skin over posterior surface of thigh
	Sciatic nerve	Flexors of knee and ankle, flexors and extensors of toes; skin over anterior and posterior surfaces of leg and foot

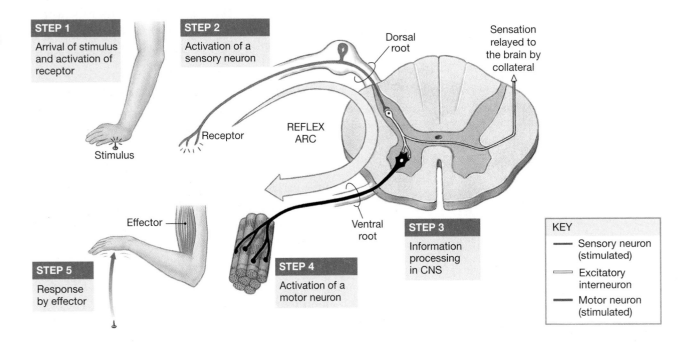

● **FIGURE 8-33 The Components of a Reflex Arc.** A simple reflex arc, such as the withdrawal reflex shown here, consists of a sensory neuron, an interneuron, and a motor neuron.

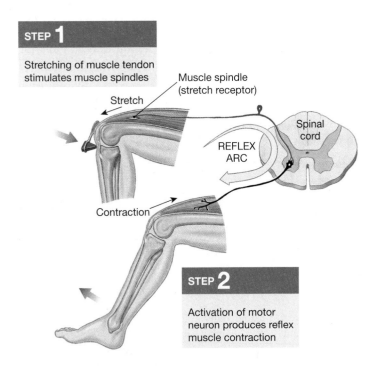

● **FIGURE 8-34 A Stretch Reflex.** The patellar reflex is a stretch reflex controlled by stretch receptors (muscle spindles) in the muscles that straighten the knee. When a reflex hammer strikes the patellar ligament, the muscle spindles are stretched. This stretching results in a sudden increase in the activity of the sensory neurons, which synapse on motor neurons in the spinal cord. The activation of spinal motor neurons produces an immediate muscle contraction and a reflexive kick.

In the simplest reflex arc, a sensory neuron synapses directly on a motor neuron, which performs the information-processing function. Such a reflex is called a **monosynaptic reflex.** Because there is only one synapse, monosynaptic reflexes control the most rapid, stereotyped motor responses of the nervous system. The best-known example is the stretch reflex.

The **stretch reflex** provides automatic regulation of skeletal muscle length. The sensory receptors in the stretch reflex are called **muscle spindles,** which are bundles of small, specialized skeletal muscle fibers scattered throughout skeletal muscles. The stimulus (increasing muscle length) activates a sensory neuron that triggers an immediate motor response (contraction of the stretched muscle) that counteracts the stimulus.

Stretch reflexes are important to maintain normal posture and balance and to make automatic adjustments in muscle tone. Physicians can use the sensitivity of the stretch reflex to test the general condition of the spinal cord, peripheral nerves, and muscles. For example, in the **knee jerk reflex** (or *patellar reflex*), a sharp rap on the patellar ligament stretches muscle spindles in the quadriceps muscles (see Figure 8-34●). With so brief a stimulus, the reflexive contraction occurs unopposed and produces a noticeable kick. If this contraction shortens the muscle spindles to below their original resting lengths, the sensory nerve endings are compressed, the sensory neuron is inhibited, and the leg drops back.

Complex Reflexes

Many spinal reflexes have at least one interneuron between the sensory (afferent) neuron and the motor (efferent) neuron (see Figure 8–33). Because there are more synapses, such **polysynaptic reflexes** include a longer delay between stimulus and response. But they can produce far more involved responses because the interneurons can control several muscle groups simultaneously.

Withdrawal reflexes move stimulated parts of the body away from a source of stimulation. The strongest withdrawal reflexes are triggered by painful stimuli, but these reflexes are also initiated by the stimulation of touch or pressure receptors. A **flexor reflex** is a withdrawal reflex affecting the muscles of a limb. If you grab an unexpectedly hot pan on the stove, a dramatic flexor reflex will occur (Figure 8–35●). When the pain receptors in your hand are stimulated, the sensory neurons activate interneurons in the spinal cord that stimulate motor neurons in the anterior gray horns. The result is a contraction of flexor muscles that yanks your hand away from the stove.

When a specific muscle contracts, opposing (antagonistic) muscles are stretched. The flexor muscles that bend the elbow, for example, are opposed by extensor muscles, which straighten it out. A potential conflict exists here: contraction of a flexor muscle should trigger in the extensors a stretch reflex that would cause them to contract, and oppose the movement that is under way. Interneurons in the spinal cord prevent such competition through **reciprocal inhibition.** When one set of motor neurons is stimulated, those controlling antagonistic muscles are inhibited.

Integration and Control of Spinal Reflexes

Although reflexes are automatic, higher centers in the brain influence these responses by stimulating or inhibiting the interneurons and motor neurons involved. The sensitivity of a reflex can, thus, be modified. For example, a voluntary effort to pull apart clasped hands elevates the general state of stimulation along the spinal cord, which leads to an enhancement of all spinal reflexes.

Other descending fibers have an inhibitory effect on spinal reflexes. Stroking an infant's foot on the side of the sole produces a fanning of the toes known as the **Babinski sign,** or *positive Babinski reflex.* This response disappears as descending inhibitory synapses develop, so in adults the same stimulus produces a curling of the toes, called a **plantar reflex,** or *negative Babinski reflex,* after about a one-second delay. If either the higher centers or the descending tracts are damaged, the Babinski sign will reappear. As a result, this reflex is often tested if CNS injury is suspected.

Spinal reflexes produce consistent, stereotyped motor patterns that are triggered by specific external stimuli. However, the same motor patterns can also be activated as needed by higher centers in the brain. The use of pre-existing motor patterns allows a relatively small number of descending fibers to control complex motor functions. For example, the motor patterns for walking, running, and jumping are directed primarily by neuronal pools in the spinal cord. The descending pathways from the brain facilitate, inhibit, or fine-tune the established patterns.

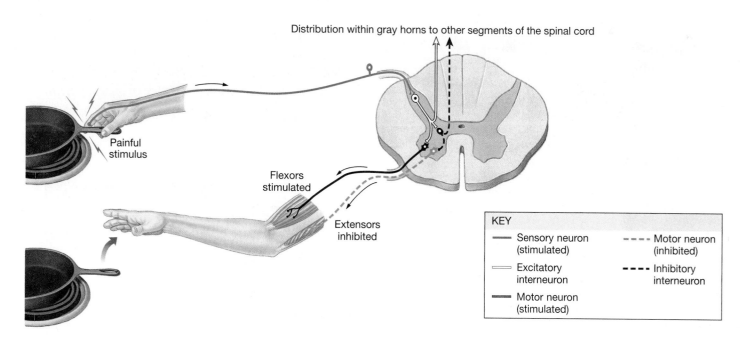

Distribution within gray horns to other segments of the spinal cord

Painful stimulus

Flexors stimulated

Extensors inhibited

KEY
— Sensory neuron (stimulated)
▭ Excitatory interneuron
— Motor neuron (stimulated)
--- Motor neuron (inhibited)
- - - Inhibitory interneuron

● **FIGURE 8–35**   **The Flexor Reflex, a Type of Withdrawal Reflex.**

Clinical Note
REFLEXES

Reflex testing can be a useful physical examination tool. A *reflex* is an automatic motor response triggered by a particular stimulus. A *reflex arc* includes a receptor, a sensory neuron, a motor neuron, and an effector. In simple reflex arcs, a sensory neuron interacts directly with a motor neuron. These simple monosynaptic reflexes are quite rapid. They protect parts of the body from injury by withdrawing the affected part when it is exposed to a noxious stimulus, such as heat. The best example of this is the stretch reflex, which provides automatic regulation of skeletal muscle length.

Stretch reflexes can be evaluated by tapping on a part of the muscle with a reflex hammer. The brief stimulus of a hammer tap results in a noticeable contraction in the affected muscle. This response is considered the reflex and is generally graded on a scale of 0–4. Table 8–4 illustrates the reflex grading scale. Reflexes on one side of the body are usually compared to their corresponding reflexes on the other side of the body. Asymmetry of the reflexes may indicate an abnormality. Testing reflexes provides information about the corresponding spinal segment. The four most commonly examined reflexes are the ankle jerk, biceps reflex, patellar reflex, and abdominal reflex. The ankle jerk reflex evaluates sacral spinal segments S_1 and S_2. The biceps reflex evaluates spinal nerves C_5 and C_6. The patellar reflex examines spinal nerves L_2, L_3, and L_4. The abdominal reflex is a contraction of the abdominal muscles that moves the umbilicus (navel) toward the stimulus. The portion of the abdomen above the umbilicus corresponds to thoracic spinal segments T_8, T_9, and T_{10}, while the portion of the abdomen below the umbilicus corresponds to the thoracic spinal segments T_{10}, T_{11}, and T_{12} (Figure 8–36●).

A decreased reflex (hyporeflexia), or an absent reflex (areflexia), may be due to temporary or permanent damage to skele-

TABLE 8–4	*Reflex Scale*
GRADE	DESCRIPTIONS
0	No response
+	Diminished, below normal
++	Average, normal
+++	Brisker than normal
++++	Hyperactive, associated with clonus

tal muscles, dorsal or ventral nerve roots, spinal nerves, the spinal cord, or the brain. An increased reflex (hyperreflexia) usually results from diseases that affect higher centers or descending tracts. In certain patients, a tap on a tendon with a reflex hammer can cause a sustained contraction or several successive contractions, referred to as clonus.

In a pronounced type of hyperreflexia that occurs following severe spinal injury, the motor neurons of the spinal cord lose contact with higher centers in the brain. Initially following the injury is a period of areflexia known as spinal shock. When the reflexes return, they respond in an exaggerated fashion, even to mild stimuli.

Stroking the bottom of the feet along the lateral aspect of the sole should cause plantar flexion of the toes (Figure 8–37●). If the big toe dorsiflexes and the other toes fan out, this indicates a central nervous system lesion. Known as a *Babinski response*, this reflex should be assessed in all critically ill or critically injured patients. ■

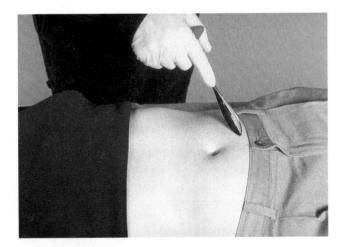

● **FIGURE 8–36 Abdominal Reflex.** Gently stroking the skin of the abdomen should cause contraction of the underlying muscles, and move the umbilicus toward the location of the stimulus.

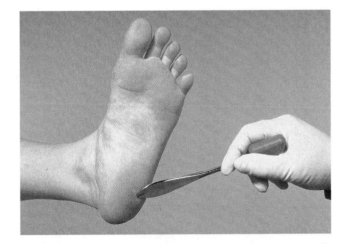

● **FIGURE 8–37 Plantar Reflex.** Stroking the lateral aspect of the plantar surface of the foot should cause plantar flexion of the toes. Dorsiflexion of the great toe and fanning of the other toes following stimulation is a positive Babinski reflex, which suggests problems with higher centers in the brain.

Key Note

Reflexes are rapid, automatic responses to stimuli that "buy time" for the planning and execution of more complex responses that are often consciously directed.

→ **CONCEPT CHECK QUESTIONS**

1. Which common reflex do physicians use to test the general condition of the spinal cord, peripheral nerves, and muscles?
2. Why can polysynaptic reflexes produce more involved responses than can monosynaptic reflexes?
3. After suffering an injury to his back, Tom exhibits a positive Babinski reflex. What does this reaction imply about Tom's injury?

Answers begin on p. 792.

Sensory and Motor Pathways

The communication among the CNS, the PNS, and organs and organ systems occurs over pathways, nerve tracts, and nuclei that relay sensory and motor information. ∞ p. 265 The major sensory (ascending) and motor (descending) tracts of the spinal cord are named with regard to the destinations of the axons. If the name of a tract begins with *spino-*, the tract starts in the spinal cord and ends in the brain, and it therefore carries sensory information. If the name of a tract ends in *-spinal,* its axons start in the higher centers and end in the spinal cord, and bear motor commands. The rest of the tract's name indicates the associated nucleus or cortical area of the brain.

Table 8–5 lists some examples of sensory and motor pathways, and their functions.

Sensory Pathways

Sensory receptors monitor conditions in the body or the external environment. A **sensation,** the information gathered by a sensory receptor, arrives in the form of action potentials in an afferent (sensory) fiber. Most of the processing of arriving sensations occurs in centers along the sensory pathways in the spinal cord or brain stem; only about one percent of the information provided by afferent fibers reaches the cerebral cortex and our conscious awareness. For example, we usually do not feel the clothes we wear or hear the hum of our car's engine.

THE POSTERIOR COLUMN PATHWAY.   One example of an ascending sensory pathway is the **posterior column pathway** (Figure 8–38●). It sends highly localized ("fine") touch, pressure, vibration, and proprioceptive (position) sensations to the cerebral cortex. In the process, the information is relayed from one neuron to another.

Sensations travel along the axon of a sensory neuron, and reach the CNS through the dorsal roots of spinal nerves. Once inside the spinal cord, the axons ascend within the posterior column pathway to synapse in a sensory nucleus of the medulla oblongata. The axons of the neurons in this nucleus (the second neuron in this pathway) cross over to the opposite side of the brain stem before continuing to the thalamus. The location of the synapse in the thalamus depends on the region of the body involved. The thalamic (in this case, third) neuron then relays the information to an appropriate region of the primary sensory cortex.

The sensations arrive organized such that sensory information from the toes reaches one end of the primary sensory cortex, and information from the head reaches the other. As a result, the sensory cortex contains a miniature map of the body surface. That map is distorted because the

---

TABLE 8–5   *Sensory and Motor Pathways*

PATHWAY	FUNCTION
SENSORY	
Posterior column pathway	Delivers highly localized sensations of fine touch, pressure, vibration, and proprioception to the primary sensory cortex
Spinothalamic pathway	Delivers poorly localized sensations of touch, pressure, pain, and temperature to the primary sensory cortex
Spinocerebellar pathway	Delivers proprioceptive information concerning the positions of muscles, bones, and joints to the cerebellar cortex
MOTOR	
Corticospinal pathway	Provides conscious control of skeletal muscles throughout the body
Medial and lateral pathways	Provides subconscious regulation of skeletal muscle tone, controls reflexive skeletal muscle responses to equilibrium sensations and to sudden or strong visual and auditory stimuli

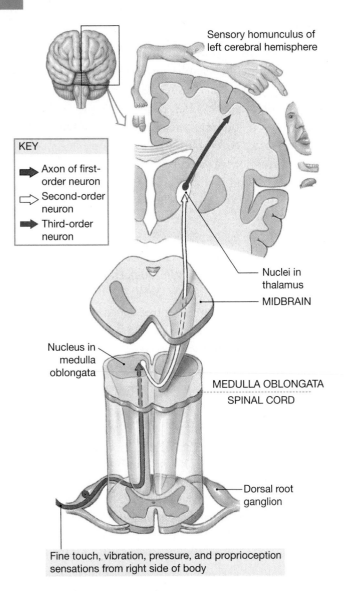

Sensory homunculus of left cerebral hemisphere

KEY

➡ Axon of first-order neuron
⇨ Second-order neuron
➡ Third-order neuron

Nuclei in thalamus

MIDBRAIN

Nucleus in medulla oblongata

MEDULLA OBLONGATA
SPINAL CORD

Dorsal root ganglion

Fine touch, vibration, pressure, and proprioception sensations from right side of body

● **FIGURE 8–38 The Posterior Column Pathway.** The posterior column pathway delivers fine touch, vibration, and proprioception information to the primary sensory cortex of the cerebral hemisphere on the opposite side of the body. (For clarity, this figure shows only the pathway for sensations that originate on the right side of the body.)

area of sensory cortex devoted to a particular region is proportional not to its size, but to the number of sensory receptors it contains. In other words, it takes many more cortical neurons to process sensory information that arrives from the tongue, which has tens of thousands of taste and touch receptors, than it does to analyze sensations that originate on the back, where touch receptors are few and far between.

Motor Pathways

In response to information provided by sensory systems, the CNS issues motor commands that are distributed by the

somatic nervous system (SNS) and the autonomic nervous system (ANS) of the efferent division of the PNS (see Figure 8–1). The SNS, which is under voluntary control, issues somatic motor commands that direct the contractions of skeletal muscles. The motor commands of the ANS, which are issued outside our conscious awareness, control the smooth and cardiac muscles, glands, and fat cells.

Three motor pathways provide control over skeletal muscles: the corticospinal pathway, the medial pathway, and the lateral pathway. The corticospinal pathway provides conscious, voluntary control over skeletal muscles; whereas the medial and lateral pathways exert more indirect, subconscious control. Table 8–5 lists some examples and functions of these motor pathways. We begin our examination of motor pathways with the corticospinal pathway.

THE CORTICOSPINAL PATHWAY. The **corticospinal pathway,** sometimes called the *pyramidal system,* provides conscious, voluntary control of skeletal muscles. Figure 8–39● shows the motor pathway that provides voluntary control over the right side of the body. As was the case for the sensory map on the primary sensory cortex, the proportions of the neurons of the primary motor cortex reflect the number of motor units present in that portion of the body. For example, the grossly oversized hands provide an indication of how many different motor units are involved in writing, grasping, and manipulating objects in our environment.

The corticospinal pathway begins at triangular-shaped *pyramidal cells* of the cerebral cortex. The axons of these upper motor neurons extend into the brain stem and spinal cord, where they synapse on lower, somatic motor neurons (see Figure 8–39). All axons of the corticospinal tracts eventually cross over to reach motor neurons on the opposite side of the body. As a result, the left side of the body is controlled by the right cerebral hemisphere, and the right side is controlled by the left cerebral hemisphere.

THE MEDIAL AND LATERAL PATHWAYS. The **medial and lateral pathways** provide subconscious, involuntary control of muscle tone and movements of the neck, trunk, and limbs. They also coordinate learned movement patterns and other voluntary motor activities (see Table 8–5). Together, these pathways were known as the *extrapyramidal system* because it was thought that they operated independently of and parallel to the *pyramidal system* (the corticospinal pathway). It is now known that the control of the body's motor functions is integrated among all three motor pathways.

The components of the medial and lateral pathways are spread throughout the brain. These components include nu-

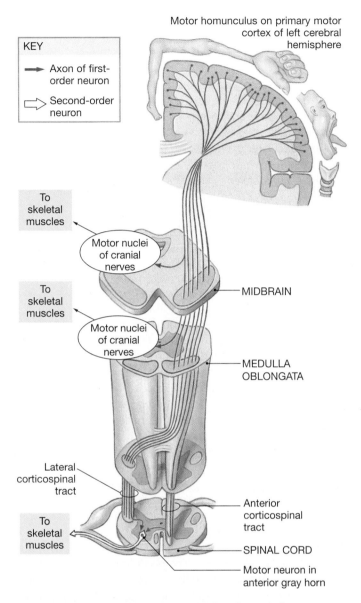

Motor homunculus on primary motor cortex of left cerebral hemisphere

KEY

→ Axon of first-order neuron

⇨ Second-order neuron

To skeletal muscles

Motor nuclei of cranial nerves

MIDBRAIN

To skeletal muscles

Motor nuclei of cranial nerves

MEDULLA OBLONGATA

Lateral corticospinal tract

To skeletal muscles

Anterior corticospinal tract

SPINAL CORD

Motor neuron in anterior gray horn

● **FIGURE 8–39 The Corticospinal Pathway.** The corticospinal pathway originates at the primary motor cortex. Axons of the pyramidal cells of the primary motor cortex descend to reach motor nuclei in the brain stem and spinal cord. Most of the fibers cross over in the medulla oblongata before descending into the spinal cord as the corticospinal tracts.

clei in the brain stem (midbrain, pons, and medulla oblongata), relay stations in the thalamus, the basal nuclei of the cerebrum, and the cerebellum. Output from the basal nuclei and cerebellum exerts the highest level of control. For example, their output can (1) stimulate or inhibit other nuclei of these pathways or (2) stimulate or inhibit the activities of pyramidal cells in the primary motor cortex. Additionally, axons from the upper motor neurons in the medial and lateral pathways synapse on the same motor neurons innervated by the corticospinal pathway.

Clinical Note
EXTRAPYRAMIDAL MOTOR SYNDROMES

Several of the drugs used in emergency medicine can cause side effects that involve the *extrapyramidal system (EPS)*. The drugs most frequently implicated are those used in the treatment of nausea and vomiting and in the treatment of acute psychotic disorders. Most belong to a class of medications called *phenothiazines*. These drugs block the neurotransmitter *dopamine* in the brain thus causing EPS.

EPS side effects are similar to the effects of Parkinson's disease but are reversible. They are usually seen in the first few days of treatment and can be misdiagnosed as anxiety. EPS signs and symptoms include muscle spasms of the neck, face, tongue, and back (*dystonia*); a sensation of restlessness (*akathisia*); a shuffling gait; rigidity of the extremities; and drooling. These symptoms can be quite disconcerting for patients and their families.

Treatment usually results in prompt, and often dramatic, reversal of symptoms. Drugs used in the treatment of Parkinson's disease (benztropine) or antihistamines (diphenhydramine) are highly effective. ■

Clinical Note
CEREBRAL PALSY

The term *cerebral palsy* refers to a number of disorders that affect voluntary motor performance (including speech, movement, and posture) that appear during infancy or childhood and persist throughout life. The cause may be trauma associated with premature or unusually stressful birth; maternal exposure to drugs, including alcohol; or a genetic defect that causes improper development of the motor pathways. Problems with labor and delivery result from the compression or interruption of placental circulation or oxygen supplies. If the oxygen concentration of fetal blood declines significantly for as little as 5–10 minutes, CNS function can be permanently impaired. The cerebral cortex, cerebellum, basal nuclei, hippocampus, and thalamus are likely targets, and there are abnormalities in motor skills, posture and balance, memory, speech, and learning abilities. ■

→ **CONCEPT CHECK QUESTIONS**

1. As a result of pressure on her spinal cord, Jill cannot feel touch or pressure on her legs. What sensory pathway is being compressed?
2. The primary motor cortex of the right cerebral hemisphere controls motor function on which side of the body?
3. An injury to the superior portion of the motor cortex would affect which part of the body?

Answers begin on p. 792.

Clinical Note
HEADACHE

Headache is a common complaint and usually a benign symptom. However, it can be associated with serious disease processes such as meningitis, brain tumor, and uncontrolled hypertension. The several *benign primary headache syndromes* include migraine headache, cluster headaches, and tension headaches.

Migraine Headache

Migraine headache is a common benign primary headache syndrome. It usually begins in the early teenage years and is more common in women than men. Migraine headaches appear to be caused by a particular trigger that sets a chain of events into motion. This complex chain of events includes neurological, vascular, hormonal, and neurotransmitter components. The phases of migraine headache are: (1) initial trigger phase initiated by external factors; (2) an aura with inhibition of neuronal activity in the cortex and a reduction in blood flow; (3) release of chemical substances that affect the blood vessels including serotonin and histamine; and (4) activation of fibers in the fifth cranial nerve (trigeminal nerve), which causes dilation of the dural arteries. These vascular changes cause the typical migraine headache.

Migraine headaches are usually classified as classic migraine or common migraine. In classic migraine, the headache is often preceded by an aura that is due to a slowly expanding area of reduced blood flow. These auras vary from patient to patient but commonly include blurred vision and flashing lights (scotoma). Some patients may report a strange taste or smell. The aura always occurs before the headache begins. In common migraine, the headache develops without a preceding aura.

Migraine headache generally develops slowly and lasts from four to 72 hours. It is typically located on one side of the head and pulsates. Physical activity usually worsens migraine headache, and causes the patient to lie motionless. Nausea and vomiting are very common. Sensitivity to light (photophobia) and sound (phonophobia) commonly accompany migraine headache. In rare cases (complex migraine), migraines can cause total blindness, weakness or paralysis of one side of the body, and speech difficulties.

Migraine headaches are a common reason people seek emergency care. Due to a better understanding of the mechanism of migraine, several medications have been developed that will actually abort the headache if administered in time. These medications are most effective if administered as soon as possible following the onset of the aura or the headache. Once the headache has developed, medications for nausea and vomiting and for pain are often required.

Cluster Headache

Cluster headache is characterized by very severe, unilateral pain in the orbit, forehead, or temple. These headaches usually occur in men, with onset typically after age 20. The pain of cluster headache is usually so severe that patients cannot lie still. Generally, they pace and are very restless. Often there will be tearing of the eye or redness of the conjunctiva on the affected side. The headaches usually last from 15 to 180 minutes and generally occur in "clusters" that occur daily on the same side of the face for several weeks. Oxygen is an effective treatment in up to 70 percent of patients, and some of the medications developed for migraine headaches also have proven effective in cluster headaches.

Tension Headache

Tension headache is a common type of headache that occurs in 40–60 percent of the population. The average age of onset is from 25 to 30 years. Tension headache is usually located on both sides of the head and is nonpulsating. Unlike migraine headaches, tension headaches are not associated with nausea and vomiting and are not worsened by physical exertion. Treatment is aimed at alleviating the symptoms. ■

■ The Autonomic Nervous System

Using the pathways already discussed, the body can respond to sensory information and exert voluntary control over the activities of skeletal muscles. Yet our conscious sensations, plans, and responses represent only a tiny fraction of the activities of the nervous system. In practical terms, conscious activities have little to do with our immediate or long-term survival, and the adjustments made by the **autonomic nervous system (ANS)** are much more important. Without the ANS, a simple night's sleep would be a life-threatening event.

As parts of the efferent division of the PNS, both the ANS and the somatic nervous system (SNS) carry motor commands to peripheral effectors. ∞ p. 261 However, clear anatomical differences exist between the SNS and ANS (Figure 8–40●). In the SNS, lower motor neurons exert direct control over skeletal muscles (Figure 8–40a●). In the ANS, a second motor neuron always separates the CNS and the peripheral effector (Figure 8–40b●). The ANS motor neurons in the CNS, known as **preganglionic neurons,** send their axons, called *preganglionic fibers*, to autonomic ganglia outside the CNS. In these ganglia, the axons of preganglionic neurons synapse on **ganglionic neurons.** The axons, or **postganglionic fibers,** of these neurons leave the ganglia and innervate cardiac muscle, smooth muscles, glands, and fat cells (adipocytes).

The ANS consists of two divisions: the sympathetic division and the parasympathetic division. Preganglionic fibers from the thoracic and lumbar spinal segments synapse in ganglia near the spinal cord; these axons and ganglia are part of

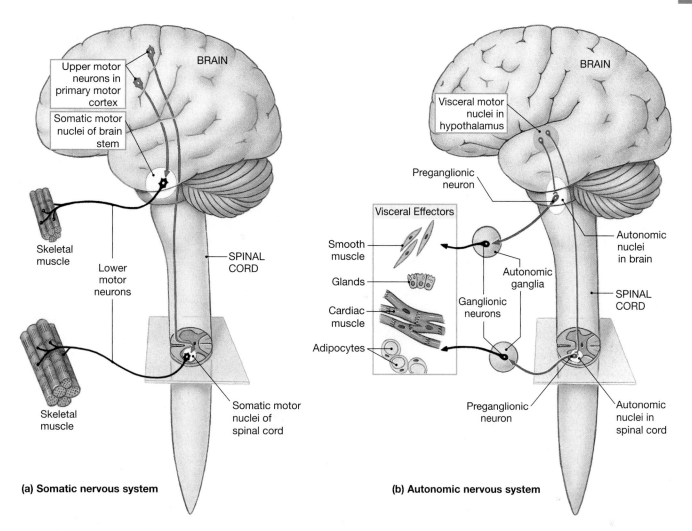

(a) Somatic nervous system

(b) Autonomic nervous system

● **FIGURE 8–40 The Somatic and Autonomic Nervous Systems.** (a) In the SNS, a motor neuron in the CNS has direct control over skeletal muscle fibers. (b) In the ANS, visceral motor neurons in the hypothalamus synapse on preganglionic neurons. The preganglionic neurons synapse on ganglionic neurons that innervate effectors, such as smooth and cardiac muscles, glands, and fat cells.

the **sympathetic division** of the ANS. The sympathetic division is often called the "fight or flight" system because it usually stimulates tissue metabolism, increases alertness, and prepares the body to deal with emergencies.

Preganglionic fibers that originate in the brain and the sacral spinal segments synapse on neurons of terminal ganglia located near the target organ or *intramural ganglia* (*murus,* wall) embedded within the tissues of visceral organs. These components are part of the **parasympathetic division** of the ANS, often regarded as the "rest and repose" or "rest and digest" system because it conserves energy and promotes sedentary activities, such as digestion.

The sympathetic and parasympathetic divisions affect target organs through the controlled release of specific neurotransmitters by the postganglionic fibers. Whether the result is a stimulation or an inhibition of activity depends on the response of the membrane receptor to the presence of the neurotransmitter. Some general patterns are worth noting:

- All preganglionic autonomic fibers are cholinergic: they release acetylcholine (ACh) at their synaptic terminals. ∞ p. 271 The effects are always excitatory.
- Postganglionic parasympathetic fibers are also cholinergic, but the effects are excitatory or inhibitory, depending on the nature of the target cell receptor.
- Most postganglionic sympathetic fibers release norepinephrine (NE). Neurons that release NE are called *adrenergic.* ∞ p. 273 The effects of NE are usually excitatory.

Key Note

The autonomic nervous system operates largely outside of our conscious awareness. It has two divisions: a sympathetic division concerned with increasing alertness, metabolic rate, and muscular abilities; and a parasympathetic division concerned with reducing metabolic rate and promoting visceral activities such as digestion.

The Sympathetic Division

The sympathetic division of the ANS (Figure 8–41●) consists of the following components:

■ *Preganglionic neurons located between segments T_1 and L_2 of the spinal cord.* These neurons are situated in the lateral gray horns, and their short axons enter the ventral roots of these segments.

■ *Ganglionic neurons located in ganglia near the vertebral column.* Two types of sympathetic ganglia exist. Paired sympathetic chain ganglia on either side of the vertebral column contain neurons that control effectors in the body wall and inside the thoracic cavity. Unpaired collateral ganglia, anterior to the vertebral column, contain ganglionic neurons that innervate tissues and organs in the abdominopelvic cavity.

■ *The adrenal medullae.* The center of each adrenal gland, an area known as the **adrenal medulla,** is a modified ganglion. Its ganglionic neurons have very short axons.

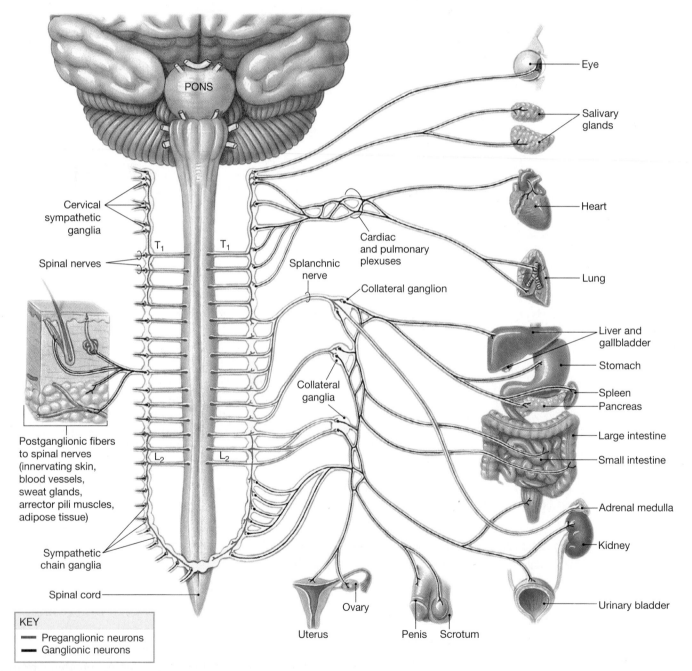

● **FIGURE 8–41 The Sympathetic Division.** The distribution of sympathetic fibers is the same on both sides of the body. For clarity, the innervation of somatic structures is shown to the left and the innervation of visceral structures to the right.

Organization of the Sympathetic Division

THE SYMPATHETIC CHAIN. From spinal segments T_1 to L_2, sympathetic preganglionic fibers join the ventral root of each spinal nerve. All of these fibers then exit the spinal nerve to enter the sympathetic chain ganglia (see Figure 8–41). For motor commands to the body wall, a synapse occurs at the chain ganglia, and then the postganglionic fibers return to the spinal nerve for distribution. For the thoracic cavity, a synapse also occurs at the chain ganglia, but the postganglionic fibers then form nerves that go directly to their targets (see Figure 8–41).

THE COLLATERAL GANGLIA. The abdominopelvic tissues and organs receive sympathetic innervation over preganglionic fibers from lower thoracic and upper lumbar segments that pass through the sympathetic chain without synapsing and instead synapse within three unpaired **collateral ganglia** (see Figure 8–41). The nerves traveling to the collateral ganglia are known as *splanchnic nerves*. The postganglionic fibers leaving the collateral ganglia innervate organs throughout the abdominopelvic cavity.

THE ADRENAL MEDULLAE. Preganglionic fibers entering each adrenal gland proceed to its central region where they synapse on modified neurons that perform an endocrine function. When stimulated, these cells release the neurotransmitters norepinephrine (NE) and epinephrine (E) into surrounding capillaries, which carry them throughout the body. In general, the effects of these neurotransmitters resemble those produced by the stimulation of sympathetic postganglionic fibers.

General Functions of the Sympathetic Division

The sympathetic division stimulates tissue metabolism, increases alertness, and prepares the individual for sudden, intense physical activity. Sympathetic innervation distributed by the spinal nerves stimulates sweat gland activity and arrector pili muscles (which produces "goose bumps"), reduces circulation to the skin and body wall, accelerates blood flow to skeletal muscles, releases stored lipids from adipose tissue, and dilates the pupils. The activation of the sympathetic nerves to the thoracic cavity accelerates the heart rate, increases the force of cardiac contractions, and dilates the respiratory passageways. The postganglionic fibers from the collateral ganglia reduce the blood flow to, and energy use by, visceral organs that are not important to short-term survival (such as the digestive tract), and they stimulate the release of stored energy reserves. The release of NE and E by the adrenal medullae broadens the effects of sympathetic activation to cells not innervated by sympathetic postganglionic fibers and also makes the effects last much longer than those produced by direct sympathetic innervation.

Clinical Note
SYMPATHETIC NERVOUS SYSTEM

Sympathetic stimulation ultimately results in the release of norepinephrine from postganglionic sympathetic nerves. The norepinephrine crosses the synaptic cleft and interacts with adrenergic receptors on the postsynaptic nerves or target organ. Shortly thereafter, the norepinephrine is either taken up whole by the presynaptic neuron for reuse or broken down by enzyme systems within the synapse. The two most common of these enzyme systems are *monamine oxidase (MAO)* and *catechol-O-methyltransferase (COMT)* (Figure 8–42●).

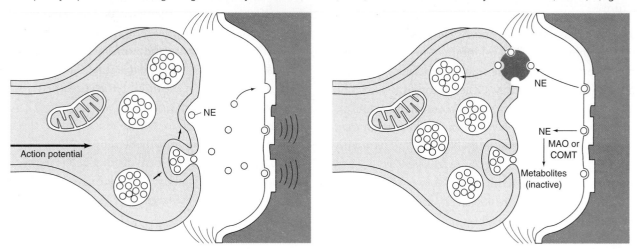

● **FIGURE 8–42 Norepinephrine Neurotransmission.** Norepinephrine is the neurotransmitter of the postganglionic sympathetic nervous system. It is taken up into the presynaptic membrane and broken down by the enzyme monamine oxidase or catechol-O-methyltransferase.

(continued next page)

Clinical Note—continued
SYMPATHETIC NERVOUS SYSTEM

Sympathetic stimulation also results in the release of epinephrine and norepinephrine from the adrenal medulla. These hormones interact with other adrenergic receptors on the membranes of target organs. This action effectively amplifies the magnitude of the sympathetic response.

The two known types of sympathetic receptors are the adrenergic receptors and the dopaminergic receptors. The adrenergic receptors are generally divided into four types. These four receptors are designated alpha 1 (α_1), alpha 2 (α_2), beta 1 (β_1), and beta 2 (β_2). The α_1 receptors cause peripheral vasoconstriction, mild bronchoconstriction, and stimulation of metabolism. The α_2 receptors are found on the presynaptic surfaces of sympathetic neuroeffector junctions. Stimulation of α_2 receptors is inhibitory. These receptors serve to prevent overrelease of norepinephrine in the synapse. When the level of norepinephrine in the synapse gets high enough, the α_2 receptors are stimulated and norepinephrine release is inhibited. Stimulation of the β_1 receptors causes an increase in heart rate, cardiac contractile force, and cardiac automaticity and conduction. Stimulation of β_2 receptors causes vasodilation and bronchodilation.

Dopaminergic receptors—although not fully understood—evidently cause dilation of the renal, coronary, and cerebral arteries. This helps maintain circulation to critical organs during periods of intense stress. Table 8–6 describes the chief locations and primary actions of each receptor. ■

TABLE 8–6 *Primary Actions and Locations of Adrenergic and Dopaminergic Receptors*

RECEPTOR	RESPONSE TO STIMULATION	LOCATION
Alpha 1 (α_1)	Constriction	Arterioles
	Constriction	Veins
	Mydriasis	Eye
	Ejaculation	Penis
Alpha 2 (α_2)	Inhibition of presynaptic terminals*	
Beta 1 (β_1)	Increased heart rate	Heart
	Increased conductivity	
	Increased automaticity	
	Increased contractility	
	Renin release	Kidney
Beta 2 (β_2)	Bronchodilation	Lungs
	Dilation	Arterioles
	Inhibition of contractions	Uterus
	Tremors	Skeletal muscle
Dopaminergic	Vasodilation (increased blood flow)	Kidney, heart, brain

*Stimulation of α_2 adrenergic receptors inhibits the continued release of norepinephrine from the presynaptic terminal. It is a feedback mechanism that limits the adrenergic response at the synapse. These receptors have no other identified peripheral effects.

The Parasympathetic Division

The parasympathetic division of the ANS includes the following structures:

- *Preganglionic neurons in the brain stem and in sacral segments of the spinal cord.* The midbrain, pons, and medulla oblongata contain autonomic nuclei associated with cranial nerves III, VII, IX, and X. Other autonomic nuclei lie in the lateral gray horns of spinal cord segments S_2 to S_4.
- *Ganglionic neurons in peripheral ganglia within or adjacent to the target organs.* Preganglionic fibers of the parasympathetic division do not diverge as extensively as those of the sympathetic division. Thus, the effects of parasympathetic stimulation are more localized and specific than those of the sympathetic division.

Organization of the Parasympathetic Division

Figure 8–43● diagrams the pattern of parasympathetic innervation. Preganglionic fibers leaving the brain travel within cranial nerves III (oculomotor), VII (facial), IX (glossopharyngeal), and X (vagus). These fibers synapse in terminal ganglia located in peripheral tissues, and short postganglionic fibers then continue to their targets. The vagus nerves provide preganglionic parasympathetic innervation to ganglia in organs of the thoracic and abdominopelvic cavities as distant as the last segments of the large intestine. The vagus nerves provide roughly 75 percent of all parasympathetic outflow and innervate most of those organs.

Preganglionic fibers in the sacral segments of the spinal cord carry the sacral parasympathetic output. They do not join the spinal nerves but instead form distinct **pelvic nerves,** which innervate intramural ganglia in the kidney and urinary bladder, the last segments of the large intestine, and the sex organs.

General Functions of the Parasympathetic Division

Among other things, the parasympathetic division constricts the pupils, increases secretions by the digestive glands, increases

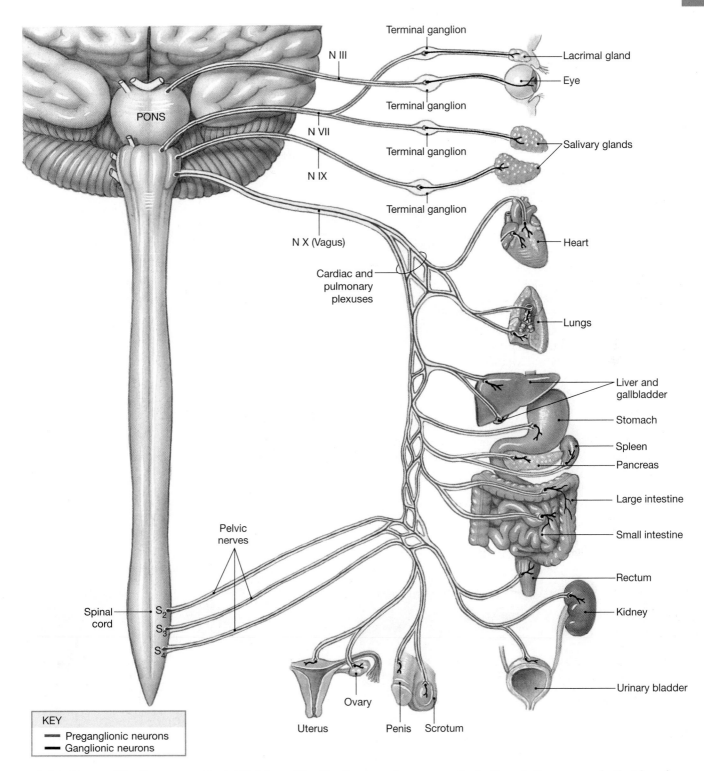

● **FIGURE 8–43 The Parasympathetic Division.** The distribution of parasympathetic fibers is the same on both sides of the body.

smooth muscle activity of the digestive tract, stimulates defecation and urination, constricts respiratory passageways, and reduces heart rate and the force of cardiac contractions. These functions center on relaxation, food processing, and energy absorption. Stimulation of the parasympathetic division leads to a general increase in the nutrient content of the blood. Cells throughout the body respond to this increase by absorbing nutrients and using them to support growth and the storage of energy reserves. The effects of parasympathetic stimulation are usually brief and are restricted to specific organs and sites.

Clinical Note
PARASYMPATHETIC NERVOUS SYSTEM

Acetylcholine, which is present in the neuromuscular junction, is the neurotransmitter for the somatic nervous system as well as for the parasympathetic nervous system. Acetylcholine is very short-lived. Within a fraction of a second after its release, it is deactivated by another chemical called *acetylcholinesterase*. Acetylcholinesterase, commonly called *cholinesterase*, breaks acetylcholine into *acetic acid* and *choline*. These two substances are taken back up by the presynaptic neuron and recycled for future use (Figure 8–44●).

The parasympathetic nervous system has two types of acetylcholine receptors, *nicotinic* and *muscarinic*. Understanding these receptors will significantly aid in understanding the function of many emergency medications. $Nicotinic_N$ (neuron) receptors are found in all autonomic ganglia, both parasympathetic and sympathetic, where acetylcholine serves as the neurotransmitter. $Nicotinic_M$ (muscle) receptors are found on the neuromuscular junction and initiate muscle contraction as part of somatic nervous system function. *Muscarinic receptors* are found in many organs throughout the body and are primarily responsible for promoting the parasympathetic response. Table 8–7 summarizes the locations and actions of muscarinic receptors. Because both nicotinic and muscarinic receptors are specific for acetylcholine, they are collectively referred to as cholinergic receptors. ■

TABLE 8-7 *Location and Effect of Muscarinic Receptors*

ORGAN	FUNCTIONS	LOCATION
Heart	Decreased heart rate	Sinoatrial node
	Decreased conduction rate	Atrioventricular node
Arterioles	Dilation	Coronary
	Dilation	Skin and mucosa
	Dilation	Cerebral
GI tract	Relaxed	Sphincters
	Increased	Motility
	Increased salivation	Salivary glands
	Increased secretion	Exocrine glands
Lungs	Bronchoconstriction	Bronchiole smooth muscle
	Increased mucus production	Bronchial glands
Gallbladder	Contraction	
Urinary bladder	Relaxation	Urinary sphincter
	Contraction	Detrusor muscle
Liver	Glycogen synthesis	
Lacrimal glands	Secretion (increased tearing)	Eye
Eye	Contraction for near vision	Ciliary muscle
	Constriction	Pupil
Penis	Erection	

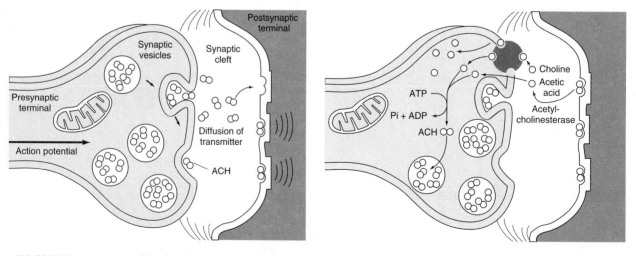

● **FIGURE 8–44 Acetylcholine Neurotransmission.** Acetylcholine is the neurotransmitter of the parasympathetic nervous system, the ganglia of the sympathetic nervous system, and the somatic nervous system. It is promptly degraded into acetic acid and choline by the enzyme acetylcholinesterase and those components are taken up by the presynaptic membrane.

Relationships Between the Sympathetic and Parasympathetic Divisions

The sympathetic division has widespread effects, and reaches visceral and somatic structures throughout the body; whereas the parasympathetic division innervates only visceral structures either serviced by cranial nerves or lying within the abdominopelvic cavity. Although some organs are innervated by one division or the other, most vital organs receive **dual innervation**—that is, instructions from both autonomic divisions. Where dual innervation exists, the two divisions often have opposing effects. Table 8–8 provides examples of the effects of either single or dual innervation on selected organs.

→ **CONCEPT CHECK QUESTIONS**

1. While out for a brisk walk, Megan is suddenly confronted by an angry dog. Which division of the ANS is responsible for the physiological changes that occur as she turns and runs from the animal?
2. Why is the parasympathetic division of the ANS sometimes referred to as "the anabolic system"?
3. What effect would loss of sympathetic stimulation have on the flow of air to the lungs?
4. What physiological changes would you expect to observe in a patient who is about to undergo a root canal procedure and who is quite anxious about it?

Answers begin on p. 792.

TABLE 8–8 *The Effects of the Sympathetic and Parasympathetic Divisions of the ANS on Various Body Structures*

STRUCTURE	SYMPATHETIC EFFECTS	PARASYMPATHETIC EFFECTS
EYE	Dilation of pupil Focusing for distance vision	Constriction of pupil Focusing for near vision
Tear Glands	None (not innervated)	Secretion
SKIN		
Sweat glands	Increases secretion	None (not innervated)
Arrector pili muscles	Contraction, erection of hairs	None (not innervated)
CARDIOVASCULAR SYSTEM		
Blood vessels	Vasoconstriction and vasodilatior	None (not innervated)
Heart	Increases heart rate, force of contraction, and blood pressure	Decreases heart rate, force of contraction, and blood pressure
ADRENAL GLANDS	Secretion of epinephrine and norepinephrine by adrenal medullae	None (not innervated)
RESPIRATORY SYSTEM		
Airways	Increases diameter	Decreases diameter
Respiratory rate	Increases rate	Decreases rate
DIGESTIVE SYSTEM		
General level of activity	Decreases activity	Increases activity
Liver	Glycogen breakdown, glucose synthesis and release	Glycogen synthesis
SKELETAL MUSCLES	Increases force of contraction, glycogen breakdown	None (not innervated)
ADIPOSE TISSUE	Lipid breakdown, fatty acid release	None (not innervated)
URINARY SYSTEM		
Kidneys	Decreases urine production	Increases urine production
Urinary bladder	Constricts sphincter, relaxes urinary bladder	Tenses urinary bladder, relaxes sphincter to eliminate urine
REPRODUCTIVE SYSTEM	Increased glandular secretions; ejaculation in males	Erection of penis (males) or clitoris (females)

■ Aging and the Nervous System

Age-related anatomical and physiological changes in the nervous system begin shortly after maturity (probably by age 30) and accumulate over time. Although an estimated 85 percent of individuals above age 65 lead relatively normal lives, noticeable changes in mental performance and CNS functioning occur. Common age-related anatomical changes include the following:

■ *A reduction in brain size and weight.* This change results primarily from a decrease in the volume of the cerebral cortex. The brains of elderly individuals have narrower gyri and wider sulci than those of young persons.

■ *A reduction in the number of neurons.* Brain shrinkage has been linked to a loss of cortical neurons. Evidence indicates that the loss of neurons does not occur (at least to the same degree) in brain stem nuclei.

■ *A decrease in blood flow to the brain.* With age, the gradual accumulation of fatty deposits in the walls of blood vessels reduces the rate of blood flow through arteries. (This process, called *arteriosclerosis*, affects arteries throughout the body; it is discussed further in Chapter 13.) The reduction in blood flow may not cause a cerebral crisis, but it does increase the chances that the individual will suffer a stroke.

■ *Changes in synaptic organization of the brain.* The number of dendritic branchings and interconnections appears to decrease. Synaptic connections are lost, and neurotransmitter production declines.

■ *Intracellular and extracellular changes in CNS neurons.* Many neurons in the brain accumulate abnormal intracellular deposits (pigments or abnormal proteins) that have no apparent function. These changes appear to occur in all aging brains, but when present in excess they seem to be associated with clinical abnormalities.

These anatomical changes are linked to impaired neural function. Memory consolidation—the conversion of short-term memory to long-term memory—often becomes more difficult. Other memories, especially those of the recent past, also become more difficult to access. The sensory systems of the elderly (notably hearing, balance, vision, smell, and taste) become less acute. Light must be brighter, sounds louder, and smells stronger before they are perceived. Reaction times are slowed, and reflexes—even some withdrawal reflexes—weaken or disappear. The precision of motor control decreases, so it takes longer to perform a given motor pattern than it did 20 years earlier. For roughly 85 percent of the elderly, these changes do not

---

Clinical Note
ALZHEIMER'S DISEASE

The most common age-related incapacitating condition of the central nervous system is *Alzheimer's disease*, which is a progressive disorder characterized by the loss of higher cerebral functions. It is the most common cause of *senile dementia*, or *senility*. The first symptoms may appear at 50 to 60 years of age, although the disease occasionally affects younger individuals. The effects of Alzheimer's disease are widespread; an estimated two million people in the U.S., including roughly 15 percent of those over age 65, have some form of the condition, and it causes approximately 100,000 deaths each year. Moreover, the condition can have devastating emotional effects on the patient's immediate family.

In its characteristic form, Alzheimer's disease produces a gradual deterioration of mental organization. The afflicted individual loses memories, verbal and reading skills, and emotional control. As memory losses continue to accumulate, problems become more severe. The affected person may forget relatives, a home address, or how to use the telephone. The loss of memory affects both intellectual and motor abilities, and a patient with severe Alzheimer's disease has difficulty performing even the simplest motor tasks. There is no cure for Alzheimer's disease, but a few medications and supplements slow its progress in many patients. ■

---

interfere with their abilities to function. But for as yet unknown reasons, many individuals become incapacitated by progressive CNS changes. The most common of such incapacitating conditions is *Alzheimer's disease* (see the Clinical Note above).

→ CONCEPT CHECK QUESTION

1. What is the major cause of age-related brain shrinkage in the CNS?

Answers begin on p. 792.

Clinical Note
CEREBROVASCULAR ACCIDENT (STROKE)

Cerebrovascular accident (CVA), also called *stroke*, is a general term that describes injury or death of brain tissue. CVA usually results from an interruption of blood supply to the affected area of the brain. The term *brain attack* is used because it compares the physiology of a stroke with that of a heart attack. Strokes are the third most common cause of death and the leading cause of disability.

A stroke results from any disease process that interrupts the blood flow to parts of the brain. The two general categories of stroke are ischemic and hemorrhagic. Ischemic strokes are due to blockage of a blood vessel that supplies a part of the brain. Hemorrhagic strokes occur due to rupture of a blood vessel that supplies a part of the brain (Figure 8–45●). Approximately 80–85 percent of strokes are ischemic, whereas 15–20 percent are hemorrhagic.

Ischemic strokes can be divided into two general categories: thrombotic and embolic. In *thrombotic strokes* a clot forms at the site of the blockage. Often, the artery is narrowed or irregular due to atherosclerosis, which promotes clot formation. In embolic strokes, the blood clot forms elsewhere in the body, usually in the heart or great vessels, travels through the circulatory system, and ultimately lodges in a cerebral artery, which causes occlusion. Unlike thrombotic strokes, embolic strokes tend to occur in blood vessels that are relatively free of disease or narrowing.

Hemorrhagic strokes tend to occur in younger patients than do ischemic strokes. Most hemorrhagic strokes occur within the substance of the brain (intracerebral hemorrhage) (Figure 8–46●). Some hemorrhagic strokes, however, will cause bleeding into the subarachnoid space (subarachnoid hemorrhage). In *subarachnoid hemorrhage*, the sudden release of blood under high pressure and the subsequent rise in intracranial pressure (ICP) causes direct cellular injury. Most subarachnoid hemorrhages are due to rupture of *berry aneurysms*. Berry aneurysms are sac-like dilations of blood vessels that supply a part of the brain. They are usually congenital and produce weakened areas of the blood vessel. The highest incidence of berry-aneurysm rupture or bleeding occurs in patients from 20 to 50 years of age.

The signs and symptoms of a stroke depend upon the part or parts of the brain affected. Blockage of one of the smaller cerebral vessels may cause minimal, if any, symptoms. Blockage of a larger cerebral vessel can cause catastrophic symptoms and even death. Blood supply to the brain is derived from two sources: the anterior and posterior circulation. The *anterior circulation* is supplied by the carotid arteries and provides blood to 80 percent of the brain. The *posterior circulation* receives blood from the vertebral arteries. Although the posterior circulation provides only 20 percent of the brain's blood, it supplies critical structures such as

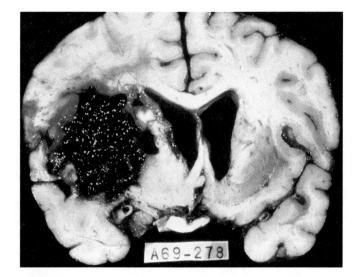

● **FIGURE 8–46** Hemorrhagic Stroke That Involves the Tissues That Adjoin the Lateral Ventricle (Periventricular Hemorrhage).

the brain stem, which is essential for movement, sensation, and normal consciousness.

Because each hemisphere of the brain controls the opposite side of the body, weakness and sensory loss will be noted on the side of the body opposite the stroke. If the dominant hemisphere of the brain is affected, there may be a loss of the ability to speak (*aphasia*). In right-handed patients and in up to 80 percent of left-handed patients, the left hemisphere is the dominant hemisphere.

Diagnosis of stroke is based on the history, physical examination, and computed tomography (CT). With CT, hemorrhagic strokes are readily identified. Ischemic strokes are more difficult to see in the acute phase. As time progresses, scar tissue replaces the parts of the brain affected, and ischemic lesions become more visible.

Treatment of strokes has changed significantly over the last decade. If treated promptly (usually within three hours), patients with ischemic strokes may be candidates for thrombolytic therapy. A thrombolytic agent, such as *tissue plasminogen activator (tPA)*, can be administered and may dissolve the clot causing the ischemia. If blood flow can be restored to the affected parts of the brain in time, then the patient may not suffer any permanent injury. Because of this, the concept of "brain attack" has been promoted so that the public will recognize the signs and symptoms of stroke early and get the patient to an appropriate treatment facility.

Transient Ischemic Attacks

Some patients will develop stroke-like symptoms that spontaneously resolve. This is referred to as a *transient ischemic attack (TIA)* and is often a precursor to a full-blown stroke. In most cases of TIA, the symptoms resolve in a few hours, although some may last for 24 hours or more. The occurrence of TIAs prompts a detailed investigation to determine the cause. ■

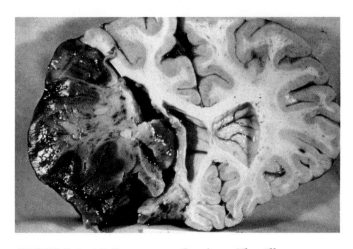

● **FIGURE 8–45** Postmortem Specimen That Illustrates Massive Hemorrhagic Stroke That Affects the Vast Majority of One Hemisphere of the Brain.

Clinical Note
NEUROLOGIC TRAUMA

Although well protected by bony structures, the brain and spinal cord are susceptible to injury. Injuries are the leading cause of death in persons less than 45 years of age, with approximately half being due to head trauma. *Traumatic brain injury (TBI)* can be devastating, and cause permanent disability. Young male adults are at greatest risk of neurological trauma, although children and the elderly are at increased risk due to underlying anatomical and physiological factors.

Neurological trauma is usually classified as penetrating or blunt. *Blunt trauma* is more common and results from motor vehicle collisions (MVCs), assaults, and sporting injuries. *Penetrating trauma* is most often due to gunshot or stab wounds. Neurological trauma can also be described as primary or secondary. In *primary injury*, neuronal damage occurs immediately at impact. In *secondary injury,* neuronal damage occurs from minutes to days after the event and results from indirect causes such as brain swelling, lack of oxygen, or inadequate perfusion.

Blunt Trauma

Blunt trauma can cause focal injuries or diffuse injuries, depending upon the location of the force. *Focal injuries* occur at a specific location in the brain, whereas *diffuse injuries* are generalized. Head trauma can cause injury immediately under the point of impact (*coup injury*) or on the opposite side of the brain (*contrecoup injury*). Contrecoup injury results from the forceful movement of the brain away from the impact, which causes it to impact the interior of the skull opposite the injury. Contrecoup injuries can occur with trauma to the front or back of the head, as well as on either side (Figure 8–47●).

DIFFUSE INJURIES
During head impact, a shearing, tearing, or stretching force is applied to the nerve fibers and causes damage to the axons. Referred to as

diffuse axonal injury (DAI), the damage can range from mild to severe. A concussion is a mild to moderate form of DAI and the most common result of blunt head trauma. In a concussion, there is neuronal dysfunction without underlying anatomical damage. This is characterized by a transient episode of confusion, disorientation, or event amnesia, followed by a rapid return to normal. A brief loss of consciousness can occur. Usually there is no permanent neurological impairment.

Bruising of brain tissue following injury causes moderate DAI. If the cerebral cortex or the reticular activating system is affected, the patient may be rendered unconscious. This injury is more severe than a mild concussion and can cause both short- and long-term signs and symptoms. These include immediate unconsciousness, followed by persistent confusion, inability to concentrate, disorientation, and amnesia.

Severe DAI is caused by mechanical disruption of many axons in both cerebral hemispheres with extension into the brain stem. This injury usually results in coma and causes an increase in intracranial pressure (ICP). Many patients do not survive severe DAI, and those who do survive usually have some degree of permanent neurological impairment.

FOCAL INJURIES
Focal injuries occur at a specific location in the brain and include contusions and intracranial hemorrhages. A *cerebral contusion* is due to capillary bleeding into the substance of the brain at the location of the impact. It can cause confusion and other types of neurological deficits depending upon the location involved. For example, injury to the frontal lobe can cause personality changes.

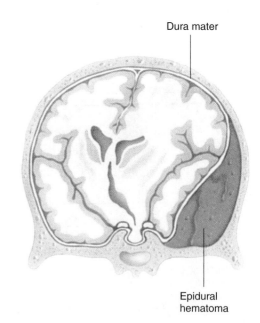

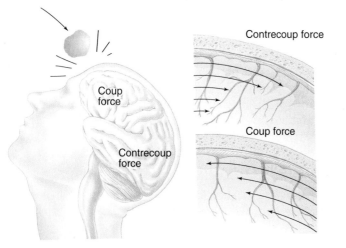

● **FIGURE 8–47 Coup and Contrecoup Injuries After Blunt Trauma to the Head.**

● **FIGURE 8–48 Epidural Hematoma.** The bleeding usually results from arterial bleeding and can develop rapidly.

Blunt trauma can cause bleeding at several locations within the brain. Bleeding between the dura mater and the interior surface of the skull is an *epidural hematoma* (Figure 8–48●). This usually is caused by damage to the middle meningeal arteries from a skull fracture. Because the bleeding is arterial, epidural hematomas develop quickly and can cause herniation of the brain stem if not treated expeditiously. The classic history of an epidural hematoma is for the patient to experience an immediate loss of consciousness after blunt head trauma. The patient then awakens and has a *lucent period* before again falling unconscious as the hematoma expands. However, the classic syndrome occurs in only about 20 percent of patients with epidural hematoma. Treatment is neurosurgical evacuation of the hematoma as soon as possible. Rarely, when neurosurgical care is unavailable, emergency physicians may be required to place burr holes in the skull above the hematoma to prevent brain stem herniation.

Bleeding beneath the dura mater and within the subarachnoid space is a *subdural hematoma.* This type of bleeding occurs very slowly and is usually due to rupture of venous vessels, such as the bridging veins of the dural sinus. Subdural hematoma usually causes few initial signs and symptoms unless the hematoma is large (Figure 8–49●). Treatment of subdural hematoma is based on its severity. Small hematomas may be monitored with the blood being resorbed over a few days. Large hematomas may require surgical decompression. Associated brain injury is common.

Intracerebral hemorrhage (ICH) results from a ruptured blood vessel within the substance of the brain. Blood loss is usually minimal but particularly damaging. It causes inflammation and swelling of the brain. The signs and symptoms of ICH, which are similar to

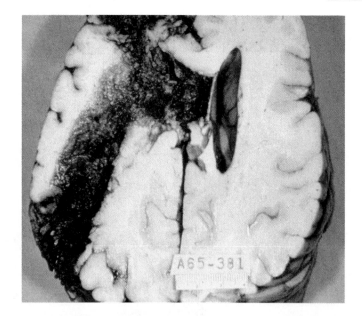

● **FIGURE 8–50** **Postmortem Specimen That Shows Massive Missile Tract Through the Substance of the Brain After a High-Velocity, High-Energy Gunshot.**

those of a stroke, depend upon the portion of the brain involved. Treatment is directed at controlling ICP in order to minimize secondary injury.

PENETRATING TRAUMA

Penetrating trauma to the brain or spinal cord is potentially devastating. Penetrating injuries are usually due to gunshots or stab wounds or to sharp objects encountered in an MVC. Often, as the penetrating object enters the head, it fractures the skull and carries bone fragments into the substance of the brain. Penetrating injuries can also cause bleeding if a blood vessel is struck during entry. As the brain is exposed to the environment, the possibility of infection exists.

The velocity and energy associated with the penetrating object are major factors in the type of injury encountered. Stab wounds are generally low-velocity and low-energy. Their damage is limited primarily to the point of entry. Small-caliber bullets may not have enough energy to both enter and exit the skull. Instead, after the bullet enters the skull, it strikes the interior surface of the skull on the opposite side and ricochets throughout the brain, which often causes significant tissue damage. High-energy, large-caliber bullets usually have enough energy to penetrate and exit the skull. Because of the tremendous energy associated with these missiles, damage to the brain is significant, and the exit wound destroys a significant part of the skull (Figure 8–50●). Increased ICP is usually not a concern in penetrating trauma due to the open wound. Penetrating injuries are often fatal. Those who survive often have serious permanent neurological damage. ■

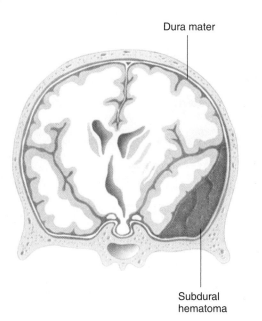

Dura mater

Subdural hematoma

● **FIGURE 8–49** **Subdural Hematoma.** The bleeding is usually venous and develops much slower than in epidural hematomas.

The Nervous System in Perspective

For All Systems

Monitors pressure, pain, and temperature; adjusts tissue blood flow patterns

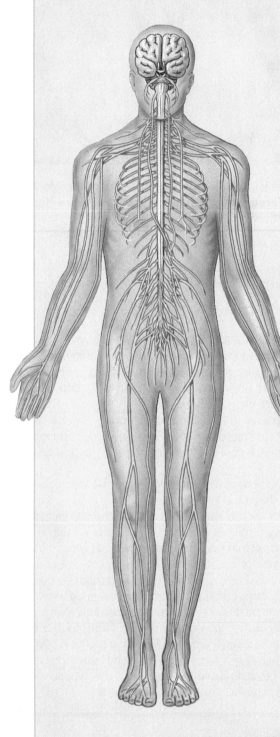

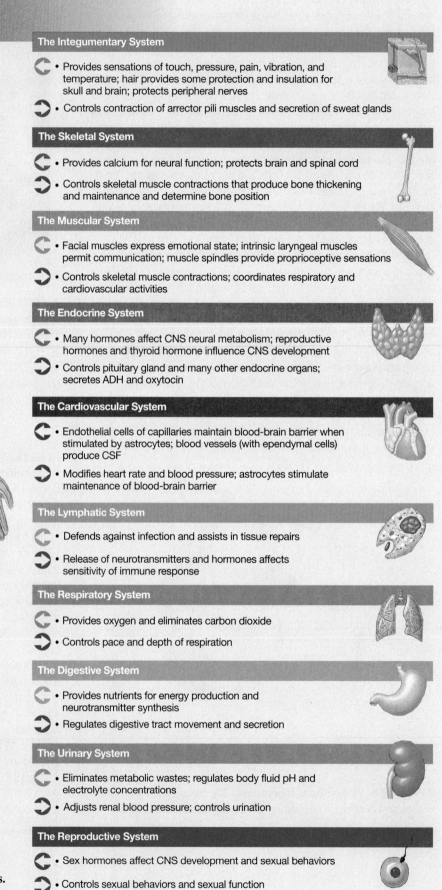

The Integumentary System

- Provides sensations of touch, pressure, pain, vibration, and temperature; hair provides some protection and insulation for skull and brain; protects peripheral nerves
- Controls contraction of arrector pili muscles and secretion of sweat glands

The Skeletal System

- Provides calcium for neural function; protects brain and spinal cord
- Controls skeletal muscle contractions that produce bone thickening and maintenance and determine bone position

The Muscular System

- Facial muscles express emotional state; intrinsic laryngeal muscles permit communication; muscle spindles provide proprioceptive sensations
- Controls skeletal muscle contractions; coordinates respiratory and cardiovascular activities

The Endocrine System

- Many hormones affect CNS neural metabolism; reproductive hormones and thyroid hormone influence CNS development
- Controls pituitary gland and many other endocrine organs; secretes ADH and oxytocin

The Cardiovascular System

- Endothelial cells of capillaries maintain blood-brain barrier when stimulated by astrocytes; blood vessels (with ependymal cells) produce CSF
- Modifies heart rate and blood pressure; astrocytes stimulate maintenance of blood-brain barrier

The Lymphatic System

- Defends against infection and assists in tissue repairs
- Release of neurotransmitters and hormones affects sensitivity of immune response

The Respiratory System

- Provides oxygen and eliminates carbon dioxide
- Controls pace and depth of respiration

The Digestive System

- Provides nutrients for energy production and neurotransmitter synthesis
- Regulates digestive tract movement and secretion

The Urinary System

- Eliminates metabolic wastes; regulates body fluid pH and electrolyte concentrations
- Adjusts renal blood pressure; controls urination

The Reproductive System

- Sex hormones affect CNS development and sexual behaviors
- Controls sexual behaviors and sexual function

● **FIGURE 8–51 Functional Relationships Between the Nervous System and Other Systems.**

■ Integration with Other Systems

The relationships between the nervous system and other organ systems are depicted in Figure 8–51●. Many of these interactions will be explored in detail in later chapters.

→ **CONCEPT CHECK QUESTION**

1. How are the nervous system and the integumentary system functionally related?

Answers begin on p. 792.

Chapter Review

Access more review material online at *www.prenhall.com/bledsoe*. There you will find quiz questions, labeling activities, animations, essay questions, and web links.

Key Terms

action potential 268
autonomic nervous system (ANS) 261
axon 261
cerebellum 284
cerebrospinal fluid 284
cerebrum 282
cranial nerves 294
diencephalon 282
dual innervation 313
ganglion/ganglia 265
hypothalamus 282

limbic system 291
medulla oblongata 284
membrane potential 266
meninges 274
midbrain 282
myelin 264
nerve plexus 298
neuroglia 261
neurotransmitter 271
parasympathetic division 307
polysynaptic reflex 301

pons 284
postganglionic fiber 306
preganglionic neuron 306
reflex 299
somatic nervous system (SNS) 261
spinal nerves 297
sympathetic division 307
synapse 262
thalamus 282

Related Clinical Terms

amyotrophic lateral sclerosis (ALS) A progressive, degenerative disorder that affects motor neurons of the spinal cord, brain stem, and cerebral hemispheres; commonly known as Lou Gehrig's disease.

aphasia A disorder that affects the ability to speak or read.

ataxia A disturbance of balance that in severe cases leaves the individual unable to stand without assistance. It is caused by problems that affect the cerebellum.

cerebrovascular accident (CVA), or *stroke* A condition in which the blood supply to a portion of the brain is blocked off.

dementia A stable, chronic state of consciousness characterized by deficits in memory, spatial orientation, language, or personality.

demyelination The destruction of the myelin sheaths around axons in the CNS and PNS.

diphtheria (dif-THĒ-rē-uh) A disease that results from a bacterial infection of the respiratory tract. Among other effects, the bacterial toxins damage Schwann cells and cause PNS demyelination.

epidural block The injection of anesthetic into the epidural space of the spinal cord to eliminate sensory and motor innervation by spinal nerves in the area of injection.

Hansen's disease *(leprosy)* A bacterial infection that begins in sensory nerves of the skin and gradually progresses to a motor paralysis of the same regions.

Huntington's disease An inherited disease marked by a progressive deterioration of mental abilities and by motor disturbances.

meningitis Inflammation of the meninges that involves the spinal cord (*spinal meningitis*) and/or brain (*cerebral meningitis*); generally caused by bacterial or viral pathogens.

multiple sclerosis (skler-Ō-sis) (MS) A disease marked by recurrent incidents of demyelination that affect axons in the optic nerve, brain, and/or spinal cord.

myelography A diagnostic procedure in which a radiopaque dye is introduced into the cerebrospinal fluid to obtain an X-ray image of the spinal cord and cauda equina.

neurology The branch of medicine that deals with the study of the nervous system and its disorders.

neurotoxin A compound that disrupts normal nervous system function by interfering with the generation or propagation of action potentials. Examples include *tetrodotoxin (TTX), saxitoxin (STX), paralytic shellfish poisoning (PSP),* and *ciguatoxin (CTX).*

paraplegia Paralysis that involves a loss of motor control of the lower (but not the upper) limbs.

paresthesia An abnormal tingling sensation, usually described as "pins and needles," that accompanies the return of sensation after a temporary palsy.

Parkinson's disease A condition characterized by a pronounced increase in muscle tone, which results from the excitation of neurons in the basal nuclei.

quadriplegia Paralysis that involves the loss of sensation and motor control of the upper and lower limbs.

sciatica (sī-AT-i-kuh) The painful result of compression of the roots of the sciatic nerve.

seizure A temporary disorder of cerebral function, accompanied by abnormal movements, unusual sensations, and/or inappropriate behavior.

shingles A condition caused by the infection of neurons in dorsal root ganglia by the virus *Herpes* varicella-zoster. The primary symptom is a painful rash along the sensory distribution of the affected spinal nerves.

spinal shock A period of depressed sensory and motor function that follows any severe injury to the spinal cord.

Summary Outline

1. Two organ systems, the nervous and endocrine systems, coordinate organ system activity. The nervous system provides swift but brief responses to stimuli; the endocrine system adjusts metabolic operations and directs long-term changes.

THE NERVOUS SYSTEM 260

1. The **nervous system** includes all the neural tissue in the body. Its major anatomical divisions include the **central nervous system (CNS)** (the brain and spinal cord) and the **peripheral nervous system (PNS)** (all of the neural tissue outside the CNS).

2. Functionally, the PNS can be divided into an **afferent division,** which brings sensory information to the CNS, and an **efferent division,** which carries motor commands to muscles, glands, and fat cells. The efferent division includes the **somatic nervous system (SNS)** (voluntary control over skeletal muscle contractions) and the **autonomic nervous system (ANS)** (automatic, involuntary regulation of smooth muscle, cardiac muscle, glandular activity, and adipose tissue). *(Figure 8–1)*

CELLULAR ORGANIZATION IN NEURAL TISSUE 261

1. There are two types of cells in neural tissue: **neurons,** which are responsible for information transfer and processing, and **neuroglia,** or *glial cells,* which provide a supporting framework and act as phagocytes.

Neurons 261

2. **Sensory neurons** form the afferent division of the PNS and deliver information to the CNS. **Motor neurons** stimulate or modify the activity of a peripheral tissue, organ, or organ system. **Interneurons (association neurons)** may be located between sensory and motor neurons; they analyze sensory inputs and coordinate motor outputs.

3. A typical neuron has a cell body, an **axon,** and several branching **dendrites,** and synaptic terminals. *(Figure 8–2)*

4. Neurons may be described as **unipolar, bipolar,** or **multipolar.** *(Figure 8–3)*

Neuroglia 263

5. The four types of neuroglia in the CNS are (1) **astrocytes,** which are the largest and most numerous; (2) **oligodendrocytes,** which are responsible for the **myelination** of CNS axons; (3) **microglia,** phagocytic cells derived from white blood cells; and (4) **ependymal cells,** with functions related to the cerebrospinal fluid (CSF). *(Figure 8–4)*

6. Nerve cell bodies in the PNS are clustered into **ganglia** (singular: *ganglion*). Their axons are covered by myelin wrappings of **Schwann cells.** *(Figure 8–5)*

Key Note 265

Anatomical Organization of Neurons 265

7. In the CNS, a collection of neuron cell bodies that share a particular function is called a **center.** A center with a discrete anatomical boundary is called a **nucleus.** Portions of the brain surface are covered by a thick layer of gray matter called the **neural cortex.** The white matter of the CNS contains bundles of axons, or **tracts,** that share common origins, destinations, and functions. Tracts in the spinal cord form larger groups, called *columns.* *(Figure 8–6)*

8. **Sensory** *(ascending)* **pathways** carry information from peripheral sensory receptors to processing centers in the brain; **motor** *(descending)* **pathways** extend from CNS centers concerned with motor control to the associated skeletal muscles. *(Figure 8–6)*

NEURON FUNCTION 266

The Membrane Potential 266

1. The **resting potential** (or **membrane potential**) of an undisturbed nerve cell results from a balance between the rates of sodium ion entry and potassium ion loss achieved by the sodium-potassium exchange pump. Any stimulus that affects this balance will alter the resting potential of the cell. *(Figure 8–7)*

2. An **action potential** appears when the membrane depolarizes to a level known as the **threshold.** The steps involved include the opening of sodium channels and membrane depolarization, the closing of sodium channels and opening of potassium channels, and the return to normal permeability. *(Figure 8–8)*

Key Note 267

Propagation of an Action Potential 268

3. In **continuous propagation,** an action potential spreads along the entire excitable membrane surface in a series of small steps. During **saltatory propagation,** the action potential appears to

leap from node to node, and skips the intervening membrane surface. *(Figure 8–9)*

Key Note 270

NEURAL COMMUNICATION 270

1. A **synapse** is a site where intercellular communication occurs through the release of chemicals called **neurotransmitters.** A synapse where neurons communicate with other cell types is a **neuroeffector junction.**

Structure of a Synapse 271

2. Neural communication moves from the **presynaptic neuron** to the **postsynaptic neuron** across the **synaptic cleft.** *(Figure 8–10)*

Synaptic Function and Neurotransmitters 271

3. **Cholinergic synapses** release the neurotransmitter **acetylcholine (ACh).** ACh is broken down in the synaptic cleft by the enzyme **acetylcholinesterase (AChE).** *(Figure 8–11; Table 8–1)*

Neuronal Pools 273

4. The roughly 20 billion interneurons are organized into **neuronal pools** (groups of interconnected neurons with specific functions). **Divergence** is the spread of information from one neuron to several neurons or from one neuronal pool to several pools. In **convergence,** several neurons synapse on the same postsynaptic neuron. *(Figure 8–12)*

Key Note 274

THE CENTRAL NERVOUS SYSTEM 274

1. The CNS is made up of the *spinal cord* and *brain.*

The Meninges 274

2. Special covering membranes, the **meninges,** protect and support the spinal cord and the delicate brain. The *cranial meninges* (dura mater, arachnoid, and pia mater) are continuous with those of the spinal cord, the *spinal meninges. (Figure 8–13)*

3. The **dura mater** covers the brain and spinal cord. The **epidural space** separates the spinal dura mater from the walls of the vertebral canal. The subarachnoid space of the arachnoid layer contains **cerebrospinal fluid** (the CSF), which acts as a shock absorber and a diffusion medium for dissolved gases, nutrients, chemical messengers, and waste products. The **pia mater** is bound to the underlying neural tissue. *(Figures 8–14, 8–15)*

The Spinal Cord 278

4. In addition to relaying information to and from the brain, the spinal cord integrates and processes information on its own.

5. The spinal cord has 31 segments, each associated with a pair of **dorsal root ganglia** and their **dorsal roots,** and a pair of **ventral roots.** *(Figures 8–16, 8–17)*

6. The white matter contains myelinated and unmyelinated axons; the gray matter contains cell bodies of neurons and glial cells. The projections of gray matter toward the outer surface of the spinal cord are called **horns.** *(Figures 8–16 through 8–19)*

Key Note 282

The Brain 282

7. There are six regions in the adult brain: cerebrum, diencephalon, midbrain, pons, medulla oblongata, and cerebellum. *(Figure 8–20)*

The Ventricles of the Brain 284

8. The central passageway of the brain expands to form four chambers called **ventricles.** Cerebrospinal fluid continuously circulates from the ventricles and central canal of the spinal cord into the subarachnoid space of the meninges that surround the CNS. *(Figures 8–21, 8–22, 8–23)*

9. Conscious thought, intellectual functions, memory, and complex involuntary motor patterns originate in the **cerebrum.** *(Figure 8–21)*

10. The cortical surface of the cerebrum contains **gyri** (elevated ridges) separated by **sulci** (shallow depressions) or deeper grooves **(fissures).** The **longitudinal fissure** separates the two **cerebral hemispheres.** The **central sulcus** marks the boundary between the **frontal lobe** and the **parietal lobe.** Other sulci form the boundaries of the **temporal lobe** and the **occipital lobe.** *(Figures 8–20; 8–24)*

11. Each cerebral hemisphere receives sensory information and generates motor commands that concern the opposite side of the body. The **primary motor cortex** of the **precentral gyrus** directs voluntary movements. The **primary sensory cortex** of the **postcentral gyrus** receives somatic sensory information from touch, pressure, pain, and temperature receptors. **Association areas,** such as the **visual association area** and **premotor cortex** (motor association area), control our ability to understand sensory information and coordinate a motor response. *(Figure 8–24)*

12. The left hemisphere is usually the *categorical hemisphere,* which contains the general interpretive and speech centers and is responsible for language-based skills. The right hemisphere, or *representational hemisphere,* is concerned with spatial relationships and analyses. *(Figure 8–25)*

13. An electroencephalogram (EEG) is a printed record of brain waves. *(Figure 8–26)*

14. The basal nuclei lie within the central white matter and aid in the coordination of learned movement patterns and other somatic motor activities. *(Figure 8–27)*

15. The limbic system includes the *hippocampus,* which is involved in memory and learning, and the *mamillary bodies,* which control reflex movements associated with eating. The functions of the limbic system involve emotional states and related behavioral drives. *(Figure 8–28)*

16. The diencephalon provides the switching and relay centers needed to integrate the conscious and unconscious sensory and motor pathways. It is made up of the epithalamus, which contains the *pineal gland* and *choroid plexus* (a vascular network that produces cerebrospinal fluid), the thalamus, and the hypothalamus. *(Figure 8–29)*

17. The thalamus is the final relay point for ascending sensory information. Only a small portion of the arriving sensory information is passed to the cerebral cortex; the rest is relayed to the basal nuclei and centers in the brain stem. *(Figures 8–20c; 8–29)*

18. The hypothalamus contains important control and integrative centers. It can produce emotions and behavioral drives, coordinate activities of the nervous and endocrine systems, secrete hormones, coordinate voluntary and autonomic functions, and regulate body temperature.

19. Three regions make up the brain stem. (1) The midbrain processes visual and auditory information and generates involuntary somatic motor responses. (2) The pons connects the cerebellum to the brain stem and is involved with somatic and visceral motor control. (3) The spinal cord connects to the brain at the medulla oblongata, which relays sensory information and regulates autonomic functions. *(Figures 8–20c; 8–29)*

20. The cerebellum oversees the body's postural muscles and programs and fine-tunes voluntary and involuntary movements. The cerebellar peduncles are tracts that link the cerebellum with the brain stem, cerebrum, and spinal cord. *(Figures 8–20; 8–29)*

21. The medulla oblongata connects the brain to the spinal cord. Its nuclei relay information from the spinal cord and brain stem to the cerebral cortex. Its reflex centers, including the **cardiovascular centers** and the **respiratory rhythmicity centers,** control or adjust the activities of one or more peripheral systems. *(Figure 8–29)*

Key Note 294

THE PERIPHERAL NERVOUS SYSTEM 294

1. The **peripheral nervous system (PNS)** links the central nervous system (CNS) with the rest of the body; all sensory information and motor commands are carried by axons of the PNS. The sensory and motor axons are bundled together into **peripheral nerves,** or nerves, and clusters of cell bodies, or *ganglia.*

2. The PNS includes cranial nerves and spinal nerves.

The Cranial Nerves 294

3. There are 12 pairs of **cranial nerves,** which connect to the brain, not to the spinal cord. *(Figure 8–30; Table 8–2)*

4. The **olfactory nerves (N I)** carry sensory information responsible for the sense of smell.

5. The **optic nerves (N II)** carry visual information from special sensory receptors in the eyes.

6. The **oculomotor nerves (N III)** are the primary sources of innervation for four of the six muscles that move the eyeball.

7. The **trochlear nerves (N IV),** the smallest cranial nerves, innervate the superior oblique muscles of the eyes.

8. The **trigeminal nerves (N V),** the largest cranial nerves, are mixed nerves with ophthalmic, maxillary, and mandibular branches.

9. The **abducens nerves (N VI)** innervate the sixth extrinsic eye muscle, the lateral rectus.

10. The **facial nerves (N VII)** are mixed nerves that control muscles of the scalp and face. They provide pressure sensations over the face and receive taste information from the tongue.

11. The **vestibulocochlear nerves (N VIII)** contain the vestibular nerves, which monitor sensations of balance, position, and movement; and the cochlear nerves, which monitor hearing receptors.

12. The **glossopharyngeal nerves (N IX)** are mixed nerves that innervate the tongue and pharynx and control swallowing.

13. The **vagus nerves (N X)** are mixed nerves that are vital to the autonomic control of visceral function and have an variety of motor components.

14. The **accessory nerves (N XI)** have a internal branch, which innervates voluntary swallowing muscles of the soft palate and pharynx; and an external branch, which controls muscles associated with the pectoral girdle.

15. The **hypoglossal nerves (N XII)** provide voluntary control over tongue movements.

Key Note 297

The Spinal Nerves 297

16. There are 31 pairs of **spinal nerves:** eight cervical, 12 thoracic, five lumbar, five sacral, and one coccygeal. Each pair monitors a region of the body surface known as a **dermatome.** *(Figures 8–31, 8–32)*

Nerve Plexuses 298

17. A **nerve plexus** is a complex, interwoven network of nerves. The four large plexuses are the **cervical plexus,** the **brachial plexus,** the **lumbar plexus,** and the **sacral plexus.** The latter two can be united into a *lumbosacral plexus. (Figure 8–31; Table 8–3)*

Reflexes 298

18. A **reflex** is an automatic involuntary motor response to a specific stimulus.

19. A **reflex arc** is the "wiring" of a single reflex. There are five steps involved in the action of a reflex arc: (1) arrival of a stimulus and activation of a receptor, (2) activation of a sensory neuron, (3) information processing, (4) activation of a motor neuron, and (5) response by an effector. *(Figure 8–33)*

20. A **monosynaptic reflex** is the simplest reflex arc, in which a sensory neuron synapses directly on a motor neuron that acts as the processing center. A **stretch reflex** is a monosynaptic reflex that automatically regulates skeletal muscle length and muscle tone. The sensory receptors involved are **muscle spindles.** *(Figure 8–34)*

21. **Polysynaptic reflexes,** which have at least one interneuron between the sensory afferent neuron and the motor efferent neuron, have a longer delay between stimulus and response than a monosynaptic synapse. Polysynaptic reflexes can also produce more involved responses. A **flexor reflex** is a withdrawal reflex that affects the muscles of a limb. *(Figures 8–35, 8–36, 8–37; Table 8–4)*

22. The brain can facilitate or inhibit reflex motor patterns based in the spinal cord.

Key Note 303

Sensory and Motor Pathways 303

23. The essential communication between the CNS and PNS occurs over pathways that relay sensory information and motor commands. *(Table 8–5)*

24. A **sensation** arrives in the form of an action potential in an afferent fiber. The **posterior column pathway** carries fine touch, pressure, and proprioceptive sensations. The axons ascend within this pathway and synapse with neurons in the medulla oblongata. These axons then cross over and travel on to the thalamus. The thalamus sorts the sensations according to the region of the body involved and projects them to specific regions of the primary sensory cortex. *(Figure 8–38; Table 8–5)*

25. The **corticospinal pathway** provides conscious skeletal muscle control. The **medial** and **lateral pathways** generally exert subconscious control over skeletal muscles. *(Figure 8–39; Table 8–5)*

THE AUTONOMIC NERVOUS SYSTEM 306

1. The autonomic nervous system (ANS) coordinates cardiovascular, respiratory, digestive, excretory, and reproductive functions.

2. **Preganglionic neurons** in the CNS send axons to synapse on **ganglionic neurons** in **autonomic ganglia** outside the CNS. The axons of the ganglionic neurons (postganglionic fibers) innervate cardiac muscle, smooth muscles, glands, and adipose tissues. *(Figure 8–40)*

3. Preganglionic fibers from the thoracic and lumbar segments form the **sympathetic division** ("fight or flight" system) of the

ANS. Preganglionic fibers leaving the brain and sacral segments form the **parasympathetic division** ("rest and repose" or "rest and digest" system).

Key Note 307

The Sympathetic Division 308

4. The sympathetic division consists of preganglionic neurons between segments T_1 and L_2, ganglionic neurons in ganglia near the vertebral column, and specialized neurons in the adrenal gland. Sympathetic ganglia are paired **sympathetic chain ganglia** or unpaired **collateral ganglia.** *(Figures 8–41, 8–42; Tables 8–6, 8–7)*

5. Preganglionic fibers entering the adrenal glands synapse within the **adrenal medullae.** During sympathetic activation these endocrine organs secrete epinephrine and norepinephrine into the bloodstream.

6. In crises, the entire division responds, and produces increased alertness, a feeling of energy and euphoria, increased cardiovascular and respiratory activity, and elevation in muscle tone.

The Parasympathetic Division 310

7. The parasympathetic division includes preganglionic neurons in the brain stem and sacral segments of the spinal cord, and ganglionic neurons in peripheral ganglia located within or next to target organs. Preganglionic fibers leaving the sacral segments form **pelvic nerves.** *(Figures 8–43, 8–44)*

8. The effects produced by the parasympathetic division center on relaxation, food processing, and energy absorption; they are usually brief and restricted to specific sites.

Relationships Between the Sympathetic and Parasympathetic Divisions 313

9. The sympathetic division has widespread effects, and reaches visceral and somatic structures throughout the body. The parasympathetic division innervates only visceral structures either serviced by cranial nerves or lying within the abdominopelvic cavity. Organs with **dual innervation** receive instructions from both divisions. *(Table 8–8)*

AGING AND THE NERVOUS SYSTEM 314

1. Age-related changes in the nervous system include (1) a reduction of brain size and weight, (2) a reduction of the number of neurons, (3) decreased blood flow to the brain, (4) changes in synaptic organization of the brain, and (5) intracellular and extracellular changes in CNS neurons. *(Figures 8–45 through 8–50)*

INTEGRATION WITH OTHER SYSTEMS 319

1. The nervous system monitors pressure, pain, and temperature and adjusts tissue blood flow for all systems. *(Figure 8–51)*

Review Questions

Level 1: Reviewing Facts and Terms

Match each item in column A with the most closely related item in column B. Place letters for answers in the spaces provided.

COLUMN A

____ 1. neuroglia

____ 2. autonomic nervous system

____ 3. sensory neurons

____ 4. dual innervation

____ 5. ganglia

____ 6. oligodendrocytes

____ 7. ascending tracts

____ 8. descending tracts

____ 9. saltatory propagation

____ 10. continuous propagation

____ 11. dura mater

____ 12. monosynaptic reflex

____ 13. sympathetic division

____ 14. cerebellum

____ 15. somatic nervous system

____ 16. hypothalamus

____ 17. medulla oblongata

____ 18. choroid plexus

____ 19. parasympathetic division

____ 20. motor neurons

COLUMN B

a. cover CNS axons with myelin

b. carry sensory information to the brain

c. occurs along unmyelinated axons

d. outermost covering of brain and spinal cord

e. production of CSF

f. supporting cells

g. controls smooth and cardiac muscle, glands, and fat cells

h. occurs along myelinated axons

i. link between nervous and endocrine systems

j. carry motor commands to spinal cord

k. efferent division of the PNS

l. controls contractions of skeletal muscles

m. masses of neuron cell bodies

n. connects the brain to the spinal cord

o. stretch reflex

p. afferent division of the PNS

q. maintains muscle tone and posture

r. "rest and repose"

s. opposing effects

t. "fight or flight"

21. Regulation by the nervous system provides:
    (a) relatively slow but long-lasting responses to stimuli.
    (b) swift, long-lasting responses to stimuli.
    (c) swift but brief responses to stimuli.
    (d) relatively slow, short-lived responses to stimuli.

22. All the motor neurons that control skeletal muscles are:
    (a) multipolar neurons.
    (b) myelinated bipolar neurons.
    (c) unipolar, unmyelinated sensory neurons.
    (d) proprioceptors.

23. Depolarization of a neuron cell membrane will shift the membrane potential toward:
    (a) 0 mV.
    (b) –70 mV.
    (c) –90 mV.
    (d) 0 mV, –70 mV, and –90 mV.

24. The structural and functional link between the cerebral hemispheres and the components of the brain stem is the:
    (a) neural cortex.
    (b) medulla oblongata.
    (c) midbrain.
    (d) diencephalon.

25. The ventricles in the brain are filled with:
    (a) blood.
    (b) cerebrospinal fluid.
    (c) air.
    (d) neural tissue.

26. Reading, writing, and speaking are dependent on processing in the:
    (a) right cerebral hemisphere.
    (b) left cerebral hemisphere.
    (c) prefrontal cortex.
    (d) postcentral gyrus.

27. Establishment of emotional states and related behavioral drives are functions of the:
    (a) limbic system.
    (b) pineal gland.
    (c) mamillary bodies.
    (d) thalamus.

28. The final relay point for ascending sensory information that will be projected to the primary sensory cortex is the:
    (a) hypothalamus.
    (b) thalamus.
    (c) spinal cord.
    (d) medulla oblongata.

29. Spinal nerves are called mixed nerves because they:
    (a) are associated with a pair of dorsal root ganglia.
    (b) exit at intervertebral foramina.
    (c) contain sensory and motor fibers.
    (d) are associated with a pair of dorsal and ventral roots.

30. There is always a synapse between the CNS and the peripheral effector in:
    (a) the ANS.
    (b) the SNS.
    (c) a reflex arc.
    (d) the ANS, SNS, and a reflex arc.

Level 2: Reviewing Concepts

35. A graded potential:
    (a) decreases with distance from the point of stimulation.
    (b) spreads passively because of local currents.
    (c) may involve either depolarization or hyperpolarization.
    (d) a, b, and c are correct.

36. The loss of positive ions from the interior of a neuron produces:
    (a) depolarization.
    (b) threshold.
    (c) hyperpolarization.
    (d) an action potential.

Level 3: Critical Thinking and Clinical Applications

41. If neurons in the central nervous system lack centrioles and are unable to divide, how can a person develop brain cancer?

42. A police officer has just stopped Bill on suspicion of driving while intoxicated. The officer asks Bill to walk the yellow line on the road and then asks him to place the tip of his index finger on the tip of his nose. How would these activities indicate Bill's level of sobriety? Which part of the brain is being tested by these activities?

43. In some severe cases of stomach ulcers, the branches of the vagus nerve (N X) that lead to the stomach are surgically severed. How might this procedure control the ulcers?

31. Approximately 75 percent of parasympathetic outflow is provided by the:
    (a) pelvic nerves.
    (b) sciatic nerve.
    (c) glossopharyngeal nerves.
    (d) vagus nerve.

32. State the all-or-none principle of action potentials.

33. Using the mnemonic device "Oh, Once One Takes The Anatomy Final, Very Good Vacations Are Heavenly," list the 12 pairs of cranial nerves and their functions.

34. How does the emergence of sympathetic fibers from the spinal cord differ from the emergence of parasympathetic fibers?

37. What would happen if the ventral root of a spinal nerve was damaged or transected?

38. Which major part of the brain is associated with respiratory and cardiac activity?

39. Why is response time in a monosynaptic reflex much faster than response time in a polysynaptic reflex?

40. Compare the general effects of the sympathetic and parasympathetic divisions of the ANS.

44. Improper use of crutches can produce a condition known as crutch paralysis, which is characterized by a lack of response by the extensor muscles of the arm and a condition known as wrist drop. Which nerve is involved?

45. While playing football, Ramon is tackled hard and suffers an injury to his left leg. As he tries to get up, he finds that he cannot flex his left hip or extend the knee. Which nerve is damaged, and how would this damage affect sensory perception in his left leg?

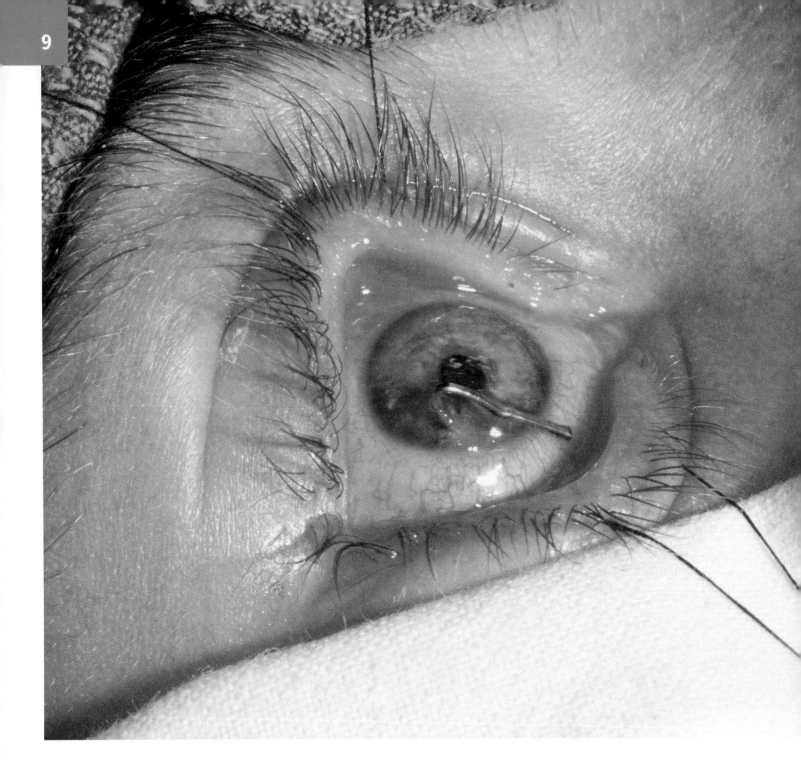

The General and Special Senses

IT HAS BEEN often said that the eyes are the windows to the soul. People have a natural reluctance to touch the eyes. However, in prehospital care, EMS personnel must understand the anatomy and physiology of the eye and be prepared to provide the necessary treatment. While eye injuries may not be life threatening, they can be life altering. Like many emergencies, the care provided for eye injuries in the prehospital setting significantly impacts the patient's subsequent quality of life.

Chapter Outline

Chapter Objectives

1. Distinguish between the general senses and the special senses. (pp. 328, 333)
2. Identify the receptors for the general senses and describe how they function. (pp. 328–333)
3. Describe the receptors and processes involved in the sense of smell. (pp. 333–334)
4. Discuss the receptors and processes involved in the sense of taste. (pp. 334–336)
5. Identify the parts of the eye and their functions. (pp. 336–348)
6. Explain how we are able to see objects and distinguish colors. (pp. 345–349)
7. Discuss how the central nervous system processes information related to vision. (pp. 327–328, 347–349)
8. Discuss the receptors and processes involved in the sense of equilibrium. (pp. 351, 353–356)
9. Describe the parts of the ear and their roles in the process of hearing. (pp. 356–360)
10. Describe the effects of aging on smell, taste, vision, and hearing. (pp. 360–361)

Vocabulary Development

akousis hearing; *acoustic*
baro- pressure; *baroreceptors*
circa about; *circadian*
circum- around; *circumvallate papillae*
cochlea snail shell; *cochlea*
dies day; *circadian*
emmetro- proper measure; *emmetropia*
incus anvil; *incus* (auditory ossicle)
iris colored circle; *iris*

labyrinthos network of canals; *labyrinth*
lacrima tear; *lacrimal gland*
lithos a stone; *otolith*
macula spot; *macula lutea*
malleus a hammer; *malleus* (auditory ossicle)
myein to shut; *myopia*
noceo hurt; *nociceptor*
olfacere to smell; *olfaction*

ops eye; *myopia*
oto- ear; *otolith*
presbys old man; *presbyopia*
skleros hard; *sclera*
stapes stirrup; *stapes* (auditory ossicle)
tectum roof; *tectorial membrane*
tympanon drum; *tympanum*
vallum wall; *circumvallate papillae*
vitreus glassy; *vitreous body*

OUR KNOWLEDGE OF THE WORLD around us is limited to those characteristics that stimulate our sensory receptors. Although we may not realize it, our picture of the environment is incomplete. Colors invisible to us guide insects to flowers, and sounds and smells we cannot detect provide important information to dolphins, dogs, and cats about their surroundings. Moreover, our senses are sometimes deceptive: in phantom limb pain, a person "feels" pain in a missing limb, and during an epileptic seizure an individual may experience sights, sounds, or smells that have no physical basis.

All sensory information is picked up by *sensory receptors*, specialized cells or cell processes that monitor conditions inside or outside the body. The simplest receptors are the dendrites of sensory neurons. The branching tips of these dendrites are called **free nerve endings.** Free nerve endings are sensitive to many types of stimuli. For example, a given free nerve ending in the skin may provide the sensation of pain in response to crushing, heat, or a cut. Other receptors are especially sensitive to one kind of stimulus. For example, a touch receptor is very sensitive to pressure but relatively insensitive to chemical stimuli; a taste receptor is sensitive to dissolved chemicals but insensitive to pressure. The most complex receptors, such as the visual receptors of the eye, are protected by accessory cells and layers of connective tissue. Not only are these receptor cells specialized to detect light, but also they are seldom exposed to any stimulus *except* light.

The area monitored by a single receptor cell is its *receptive field* (Figure 9–1●). Whenever a sufficiently strong stimulus arrives in the receptive field, the CNS receives the information "stimulus arriving at receptor X." The larger the receptive field, the poorer your ability to localize a stimulus. For example, a touch receptor on the general body surface with a receptive field 7 cm (2.5 in.) in diameter provides less precise information than receptors on the tongue or fingertips, which have receptive fields less than a millimeter in diameter.

All sensory information arrives at the CNS in the form of action potentials in a sensory (afferent) fiber. In general, the stronger the stimulus, the higher the frequency of action potentials. The arriving information is called a **sensation.** When sensory information arrives at the CNS, it is routed according to the location and nature of the stimulus. For example, touch, pressure, pain, temperature, and taste sensations arrive at the primary sensory cortex; visual, auditory, and olfactory information reaches the visual, auditory, and olfactory regions of the cortex, respectively. The conscious awareness of a sensation is called a *perception.*

The CNS interprets the nature of sensory information entirely on the basis of the area of the brain stimulated; it cannot tell the difference between a "true" sensation and a "false" one. For instance, when rubbing your eyes, you may "see" flashes of light. Although the stimulus is mechanical rather than visual, any activity along the optic nerve is projected to the visual cortex and experienced as a visual perception.

Adaptation is a reduction in sensitivity in the presence of a constant stimulus. Familiar examples are stepping into a hot bath or jumping into a cold lake; within moments neither temperature seems as extreme as it did initially. Adaptation reduces the amount of information that arrives at the cerebral cortex. Most sensory information is routed to centers along the spinal cord or brain stem, and potentially triggers such involuntary reflexes as the withdrawal reflexes. ∞ p. 301 Only about 1 percent of the information provided by afferent fibers reaches the cerebral cortex and our conscious awareness.

Output from higher centers, however, can increase receptor sensitivity or facilitate transmission along a sensory pathway. For example, the *reticular activating system* in the midbrain helps focus attention and, thus, heightens or reduces awareness of arriving sensations. ∞ p. 293 This adjustment of sensitivity can occur under conscious or unconscious direction. When we "listen carefully," our sensitivity to and awareness of auditory stimuli increase. The reverse occurs when we enter a noisy factory or walk along a crowded city street, as we automatically "tune out" the high level of background noise.

The **general senses** include temperature, pain, touch, pressure, vibration, and **proprioception** (body position). The receptors for the general senses are scattered throughout the body. The **special senses** are smell (**olfaction**), taste (**gustation**), vision, balance (**equilibrium**), and hearing. The receptors for the five special senses are concentrated within specific structures—the sense organs. This chapter explores both the general senses and the special senses.

Key Note

Stimulation of a receptor produces action potentials along the axon of a sensory neuron. The frequency or pattern of action potentials contains information about the stimulus. Your perception of the nature of that stimulus depends on the path it takes inside the CNS and the region of the cerebral cortex it stimulates.

■ The General Senses

Receptors for the general senses are scattered throughout the body and are relatively simple in structure. These receptors are classified according to the nature of the stimulus that excites them. Important receptor classes include receptors sensitive to pain (*nociceptors*); to temperature (*thermoreceptors*); to physical distortion that results from touch, pressure, and body position (*mechanoreceptors*); and to chemical stimuli (*chemoreceptors*).

Pain

Pain receptors, or **nociceptors** (nō-sē-SEP-tōrz; *noceo*, hurt), are free nerve endings. They are especially common in the superficial portions of the skin, in joint capsules, within the periostea that cover bones, and around blood vessel walls.

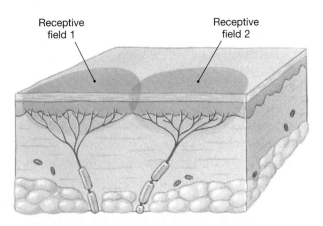

● **FIGURE 9–1 Receptors and Receptive Fields.** Each receptor cell monitors a specific area known as a receptive field.

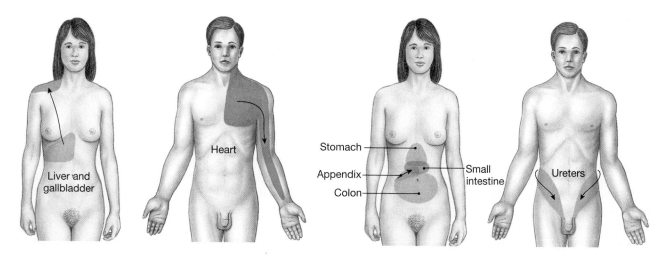

● **FIGURE 9–2 Referred Pain.** In referred pain, sensations that originate in visceral organs are perceived as pain in other body regions innervated by the same spinal nerves. Each region of perceived pain is labeled according to the organ at which the pain originates.

Other deep tissues and most visceral organs contain few nociceptors. Pain receptors have large receptive fields (see Figure 9–1) and, as a result, it is often difficult to determine the exact source of a painful sensation.

Nociceptors may be sensitive to (1) extremes of temperature, (2) mechanical damage, or (3) dissolved chemicals, such as those released by injured cells. Very strong stimuli by any of these sources may excite a nociceptor. For that reason, people who describe very painful sensations—whether caused by heat, a deep cut, or inflammatory chemicals—use a similar descriptive term, such as "burning."

Once pain receptors in a region are stimulated, two types of axons carry the painful sensations. Myelinated fibers carry very localized sensations of **fast pain** (or *prickling pain*), such as that caused by an injection or a deep cut. These sensations reach the CNS very quickly, where they often trigger somatic reflexes. They are also relayed to the primary sensory cortex and so receive conscious attention. Slower, unmyelinated fibers carry sensations of **slow pain,** or *burning and aching pain.* Unlike fast pain sensations, slow pain sensations enable you to identify only the general area involved.

Pain sensations from visceral organs are often perceived as originating at the body surface, generally in those regions innervated by the same spinal nerves. The perception of pain coming from parts of the body that are not actually stimulated is called **referred pain.** The precise mechanism responsible for referred pain is not yet clear, but several clinical examples are shown in Figure 9–2●. Cardiac pain, for example, is often perceived as originating in the skin of the upper chest and left arm.

Pain receptors continue to respond as long as the painful stimulus remains. However, the perception of the pain can decrease over time because of the inhibition of centers in the thalamus, reticular formation, lower brain stem, and spinal cord.

Temperature

Temperature receptors, or **thermoreceptors,** are free nerve endings located in the dermis, in skeletal muscles, in the liver, and in the hypothalamus. Cold receptors are three or four times as numerous as warm receptors. There are no known structural differences between warm and cold thermoreceptors.

Temperature sensations are relayed along the same pathways that carry pain sensations. They are sent to the reticular formation, the thalamus, and (to a lesser extent) the primary sensory cortex. Thermoreceptors are very active when the temperature is changing, but they quickly adapt to a stable temperature. When you enter an air-conditioned classroom on a hot summer day or a warm lecture hall on a brisk fall evening, the temperature seems extreme at first, but you quickly become comfortable as adaptation occurs.

Touch, Pressure, and Position

Mechanoreceptors are sensitive to stimuli such as stretching, compression, or twisting. Distortion of the receptor's cell membrane in response to these stimuli causes mechanically regulated ion channels to open or close. There are three classes of mechanoreceptors: (1) *tactile receptors* (touch), (2) *baroreceptors* (pressure), and (3) *proprioceptors* (position).

Tactile Receptors

Tactile receptors provide sensations of touch, pressure, and vibration. The distinctions between these sensations are hazy, for a touch also represents a pressure, and a vibration consists of an oscillating touch/pressure stimulus. **Fine touch and pressure receptors** provide detailed information about a source of stimulation, including its exact location, shape, size, texture, and movement. **Crude touch and pressure**

Clinical Note
PAIN

Pain is an important symptom in medicine. In fact, over 60 percent of people who present to hospital emergency departments do so because of pain. Despite this, inadequate analgesia continues to be a problem in emergency care, especially in children. Emergency pain management should include pain relief when possible and management of associated anxiety. This must be provided while continuously monitoring the patient's airway, vital signs, and mental status (Figure 9–3●).

The pathophysiology of the pain response is very complex. There are both peripheral and central mediators of pain. The peripheral pain system is activated when nociceptors and free nerve endings register the original noxious stimulus in the peripheral tissues and transmit it to the central nervous system. Several neurotransmitters are involved in the pain response including the excitatory amino acid glutamate and the neuropeptides *neurokinin-A, calcitonin-gene related peptide,* and *substance P.* Pain signals are integrated in the dorsal horn of the spinal cord. These are relayed to higher centers in the brain including the hypothalamus, thalamus, and the limbic and reticular activating systems. These centers integrate and process pain information, which allows the detection of and perception of pain. Interpretation, identification, and localization of pain also occur at these sites.

Because pain is subjective, it is often difficult to assess. In the emergency setting, it is common to use a pain scale to determine the sever-ity of a patient's pain. The most popular pain scale asks patients to rate their pain on a numeric scale that ranges from 0 to 10; with 0 indicates no pain and 10 indicates the worst possible pain. Pain scale ratings can be monitored to determine the effectiveness of medications and other treatments.

Analgesics are medications that help to alleviate pain. They may work on peripheral pain mediators, central pain mediators, or both. Nonsteroidal anti-inflammatory (NSAID) agents are commonly used for mild to moderate pain. NSAIDs primarily act on peripheral pain mediators and include aspirin, ibuprofen, naproxen, ketoprofen, and many others. These drugs decrease levels of inflammatory mediators, such as prostaglandins, generated at the site of tissue injury. Because they act peripherally, they do not cause sedation or respiratory depression and they do not interfere with bowel or bladder function. Acetaminophen (Tylenol) is also a peripherally acting analgesic. However, unlike the NSAIDs, it does not have anti-inflammatory properties and does not affect platelet aggregation (as does aspirin). Most peripherally acting analgesics must be administered orally or by topical application. The exception is ketorolac (Toradol), which is available for intramuscular or intravenous injection.

Moderate to severe pain usually requires opioid analgesics. Opioid analgesics include morphine, codeine, hydromorphone, meperidine, fentanyl, and others. They are most effective when they are administered parenterally (outside of the digestive system). Common routes of opioid injection are subcutaneous, intramuscular, and intravenous. Several opioids referred to as endorphins and enkephalins occur naturally in the body. These substances serve as natural painkillers. The opiate medications act on the same receptors as the endorphins and enkephalins. Opiate receptors' different shapes influence the fit of the corresponding opiate/opioid molecules. The principle opioid receptors, referred to as mu (μ) receptors, produce analgesia, euphoria, and respiratory depression. The mu receptors can be further classified as mu-1 and mu-2. Mu-1 (μ1) receptors cause analgesia, while mu-2 (μ2) receptors produce constipation, euphoria, physical dependence, and respiratory depression. Several other receptors involved in the central pain response include the delta, sigma, kappa, and epsilon receptors. The delta (δ) receptors also produce analgesia. The sigma (ζ) receptors stimulate respiratory and vasomotor activity as well as hallucinations and dysphoria. The kappa (κ) receptors influence spinal analgesia, sedation, and pupillary constriction. Finally, the epsilon (ε) receptors produce analgesia. Morphine and the other opiate derivatives have an affinity for the mu and kappa receptors.

Peripherally acting and centrally acting analgesics are often mixed to provide highly effective pain control. Usually, these are a combination of an opioid and acetaminophen or an opioid and ibuprofen. In addition, long-term pain control can be provided with delayed-release medications or skin patches that deliver a standard amount of opioid analgesia over 72 hours. Patient comfort and pain control are important aspects of emergency care. Emergency personnel should assess the severity of a patient's pain and expeditiously provide adequate analgesia when possible. ■

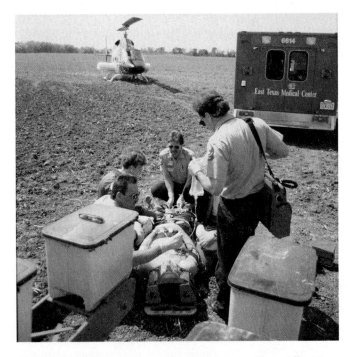

● **FIGURE 9–3 Pain Control in Emergency Care.** All patients should be assessed for pain. When possible, analgesics should be administered, especially for injuries, such as fractures, that may be worsened by ambulance transport.

receptors provide poor localization and little additional information about the stimulus.

Tactile receptors range in complexity from free nerve endings to specialized sensory complexes with accessory cells and supporting structures. Figure 9–4● depicts six types of tactile receptors in the skin:

1. Free nerve endings sensitive to touch and pressure are situated between epidermal cells. No structural differences have been found between these receptors and the free nerve endings that provide temperature or pain sensations.

2. The **root hair plexus** is made up of free nerve endings that are stimulated by hair displacement. They monitor distortions and movements across the body surface.

3. **Tactile discs,** or *Merkel's* (MER-kelz) *discs,* are fine touch and pressure receptors. Hairless skin contains large epithelial cells (*Merkel cells*) in its deepest epidermal layer. The dendrites of a single sensory neuron make close contact with a group of Merkel cells. When compressed,

Merkel cells release chemicals that stimulate the neuron.

4. **Tactile corpuscles,** or *Meissner's* (MĪS-nerz) *corpuscles,* are sensitive to fine touch and pressure and to low frequency vibration. They are abundant in the eyelids, lips, fingertips, nipples, and external genitalia.

5. **Lamellated** (LAM-e-lāt-ed) **corpuscles,** or *pacinian* (pa-SIN-ē-an) *corpuscles,* are large receptors sensitive to deep pressure and to pulsing or high-frequency vibrations. They are common in the skin of the fingers, breasts, and external genitalia. They are also present in joint capsules, mesenteries, the pancreas, and the walls of the urethra and urinary bladder.

6. **Ruffini** (roo-FĒ-nē) **corpuscles** are sensitive to pressure and distortion of the skin, but they are located in the deepest layer of the dermis.

Tactile sensations travel through the posterior column and spinothalamic pathways. ∞ p. 303 Sensitivity to tactile sensations can be altered by infection, disease, and damage to sensory neurons or pathways. The locations of tactile responses

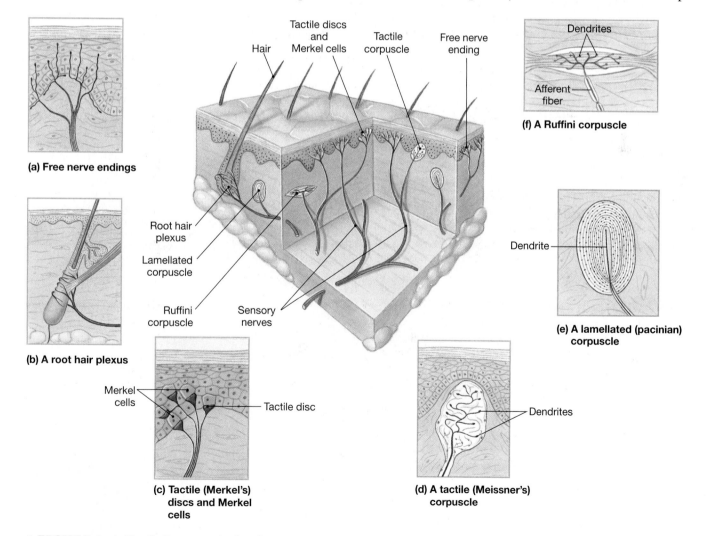

(a) Free nerve endings

(b) A root hair plexus

Root hair plexus

Lamellated corpuscle

Ruffini corpuscle

Sensory nerves

Hair

Tactile discs and Merkel cells

Tactile corpuscle

Free nerve ending

Dendrites

Afferent fiber

(f) A Ruffini corpuscle

Dendrite

(e) A lamellated (pacinian) corpuscle

Merkel cells

Tactile disc

(c) Tactile (Merkel's) discs and Merkel cells

Dendrites

(d) A tactile (Meissner's) corpuscle

● **FIGURE 9–4** Tactile Receptors in the Skin.

may have diagnostic significance. For example, sensory loss along the boundary of a dermatome can help identify the affected spinal nerve or nerves. ∞ p. 297

Baroreceptors

Baroreceptors (bar-ō-rē-SEP-tōrz; *baro-*, pressure) provide information essential to the regulation of autonomic activities by monitoring changes in pressure. These receptors consist of free nerve endings that branch within the elastic tissues in the wall of a distensible organ, such as a blood vessel or a portion of the respiratory, digestive, or urinary tract. When the pressure changes, the elastic walls of these vessels or tracts expand or recoil. This movement distorts the dendritic branches and alters the rate of action potential generation. Although baroreceptors respond immediately to a change in pressure, they also adapt rapidly; the result is that the output along the afferent fibers returns to "normal" gradually.

Figure 9–5● illustrates the locations of some baroreceptors and summarizes their functions in autonomic activities. Baroreceptors monitor blood pressure in the walls of major blood vessels, including the carotid artery (at the *carotid sinus*) and the aorta (at the *aortic sinus*). The information plays a major role in regulating cardiac function and adjusting blood flow to vital tissues. Baroreceptors in the lungs monitor the degree of lung expansion. This information is relayed to the respiratory rhythmicity centers in the brain, which set the pace of respira-

tion. ∞ p. 294 Baroreceptors in the digestive and urinary tracts trigger various visceral reflexes, including those of urination, movement of materials along the digestive tract, and defecation.

Proprioceptors

Proprioceptors monitor the position of joints, the tension in tendons and ligaments, and the state of muscular contraction. *Free nerve endings* in joint capsules detect pressure, tension, and movement at the joint. *Golgi tendon organs* lie between a skeletal muscle and its tendon and monitor the strain on a tendon during muscle contraction. *Muscle spindles* monitor the length of a skeletal muscle and trigger stretch reflexes. ∞ p. 300 Proprioceptors do not adapt to constant stimulation, and each receptor continuously sends information to the CNS. Most of this information is processed subconsciously; only a small proportion of it reaches your conscious awareness. Your sense of body position results from the integration of information from these three types of proprioceptors with information from the receptors of the inner ear.

Chemical Detection

In general, **chemoreceptors** respond only to water-soluble and lipid-soluble substances that are dissolved in the surrounding fluid. Adaptation usually occurs over a few seconds following stimulation. Except for the special senses of

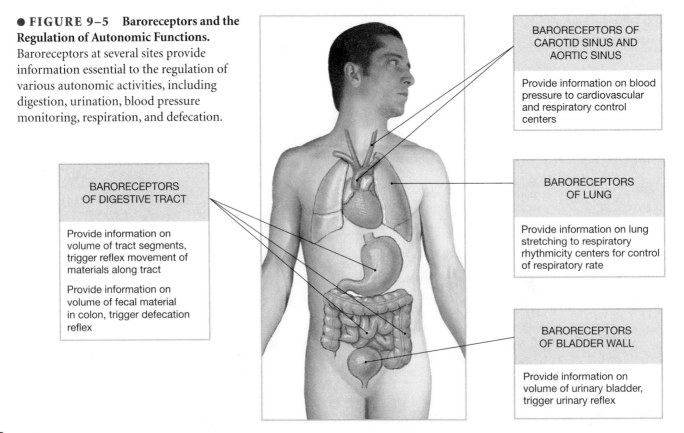

● **FIGURE 9–5 Baroreceptors and the Regulation of Autonomic Functions.** Baroreceptors at several sites provide information essential to the regulation of various autonomic activities, including digestion, urination, blood pressure monitoring, respiration, and defecation.

BARORECEPTORS OF DIGESTIVE TRACT

Provide information on volume of tract segments, trigger reflex movement of materials along tract

Provide information on volume of fecal material in colon, trigger defecation reflex

BARORECEPTORS OF CAROTID SINUS AND AORTIC SINUS

Provide information on blood pressure to cardiovascular and respiratory control centers

BARORECEPTORS OF LUNG

Provide information on lung stretching to respiratory rhythmicity centers for control of respiratory rate

BARORECEPTORS OF BLADDER WALL

Provide information on volume of urinary bladder, trigger urinary reflex

taste and smell, there are no well-defined chemosensory pathways in the brain or spinal cord. The chemoreceptors of the general senses send their information to brain stem centers that participate in the autonomic control of respiratory and cardiovascular functions. The locations of important chemoreceptors are shown in Figure 9–6●. Neurons within the respiratory centers of the brain respond to the concentrations of hydrogen ions (pH) and carbon dioxide molecules in the cerebrospinal fluid. Chemoreceptors are also located in the **carotid bodies,** near the origin of the internal carotid arteries on each side of the neck, and in the **aortic bodies,** between the major branches of the aortic arch. These receptors monitor the pH and the carbon dioxide and oxygen concentrations of arterial blood. The afferent fibers leaving the carotid and aortic bodies reach the respiratory centers by traveling along the glossopharyngeal (N IX) and vagus (N X) cranial nerves.

CONCEPT CHECK QUESTIONS

1. Receptor A has a circular receptive field with a diameter of 2.5 cm. Receptor B has a circular receptive field 7.0 cm in diameter. Which receptor provides more precise sensory information?
2. When the nociceptors in your hand are stimulated, what sensation do you perceive?
3. What would happen to you if the information from proprioceptors in your legs was blocked from reaching the CNS?

Answers begin on p. 792.

■ The Special Senses

Next we turn our attention to the five *special senses*: smell, taste, vision, equilibrium, and hearing. Although the sense organs involved are structurally more complex than those of the general senses, the same basic principles of receptor function apply. The information these receptors provide is distributed to specific areas of the cerebral cortex and to the brain stem.

■ Smell

The sense of smell, or *olfaction*, is provided by paired **olfactory organs** (Figure 9–7●). These organs are located in the nasal cavity on either side of the nasal septum just inferior to the cribriform plate of the ethmoid bone. Each olfactory organ consists of an **olfactory epithelium,** which contains the **olfactory receptor cells,** supporting cells, and regenerative *basal cells* (stem cells). The underlying layer of loose connective tissue contains large **olfactory glands** whose secretions absorb water and form a pigmented mucus that covers the epithelium. The mucus is produced in a continuous stream that passes across the surface of the olfactory organ, preventing the buildup of potentially dangerous or overpowering stimuli and keeping the area moist and free from dust or other debris.

When you draw air in through your nose, the air swirls within the nasal cavity. A normal, relaxed inhalation carries a small sample (about 2 percent) of the inhaled air to

●**FIGURE 9–6 Locations and Functions of Chemoreceptors.** Chemoreceptors are located in the CNS (on the ventrolateral surfaces of the medulla oblongata) and in the aortic and carotid bodies. These receptors are involved in the autonomic regulation of respiratory and cardiovascular function.

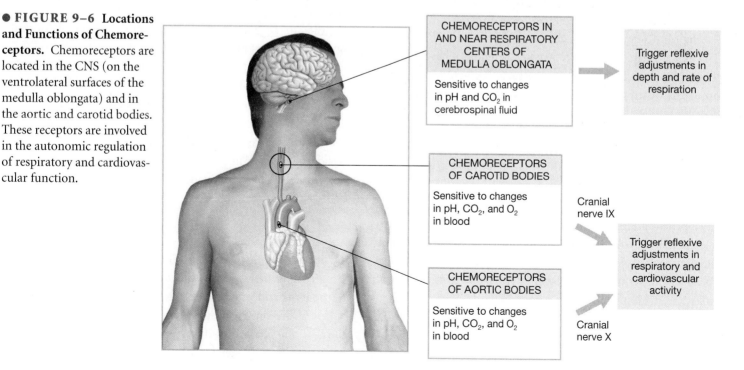

CHEMORECEPTORS IN AND NEAR RESPIRATORY CENTERS OF MEDULLA OBLONGATA

Sensitive to changes in pH and CO_2 in cerebrospinal fluid

Trigger reflexive adjustments in depth and rate of respiration

CHEMORECEPTORS OF CAROTID BODIES

Sensitive to changes in pH, CO_2, and O_2 in blood

Cranial nerve IX

Trigger reflexive adjustments in respiratory and cardiovascular activity

CHEMORECEPTORS OF AORTIC BODIES

Sensitive to changes in pH, CO_2, and O_2 in blood

Cranial nerve X

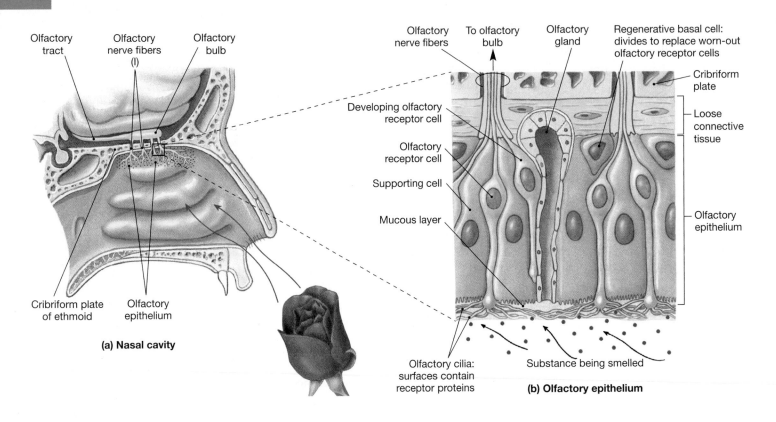

● **FIGURE 9–7 The Olfactory Organs.** (a) The structure and location of the olfactory organ on the left side of the nasal septum is shown. (b) An olfactory receptor is a modified neuron with multiple cilia that extend from its free surface.

the olfactory organs. Repeated sniffing increases the flow of air across the olfactory epithelium, and intensifies the stimulation of the receptors. Once airborne compounds have reached the olfactory organs, water-soluble and lipid-soluble chemicals must diffuse into the mucus before they can stimulate the olfactory receptors.

The olfactory receptor cells are highly modified neurons. The exposed tip of each receptor cell provides a base for cilia that extend into the surrounding mucus. Olfactory reception occurs as dissolved chemicals interact with receptors, called *odorant binding proteins*, on the surfaces of the cilia. *Odorants* are chemicals that stimulate olfactory receptors. The binding of an odorant changes the permeability of the receptor membrane, and produces action potentials. This information is relayed to the central nervous system, which interprets the smell on the basis of the particular pattern of receptor activity.

Approximately 10–20 million olfactory receptor cells are packed into an area of roughly 5 cm². If we take into account the surface area of the exposed cilia, the actual sensory area approaches that of the entire body surface. Nevertheless, our olfactory sensitivities cannot compare with those of other vertebrates such as dogs, cats, or fishes. A German shepherd who sniffs for smuggled drugs or explosives has an olfactory receptor surface 72 times greater than that of its human handler.

The Olfactory Pathways

The axons leaving the olfactory epithelium collect into 20 or more bundles that penetrate the cribriform plate of the ethmoid bone to reach the **olfactory bulbs,** where the first synapse occurs. Axons leaving each olfactory bulb travel along the olfactory tract to reach the olfactory cortex of the cerebrum, the hypothalamus, and portions of the limbic system.

Olfactory stimuli are the only type of sensory information that reaches the cerebral cortex without first synapsing in the thalamus. The extensive limbic and hypothalamic connections help explain the profound emotional and behavioral responses that certain smells can produce. The perfume industry, which understands the practical implications of these connections, spends billions of dollars to develop odors that trigger sexual responses.

■ Taste

Taste receptors, or *gustatory* (GUS-ta-tor-ē) *receptors*, are distributed over the surface of the tongue and adjacent portions of the pharynx and larynx (Figure 9–8●). The most important taste receptors are on the tongue; by the time we reach adulthood, those on the pharynx and larynx have decreased in importance and abundance. Taste receptors and specialized epithelial cells

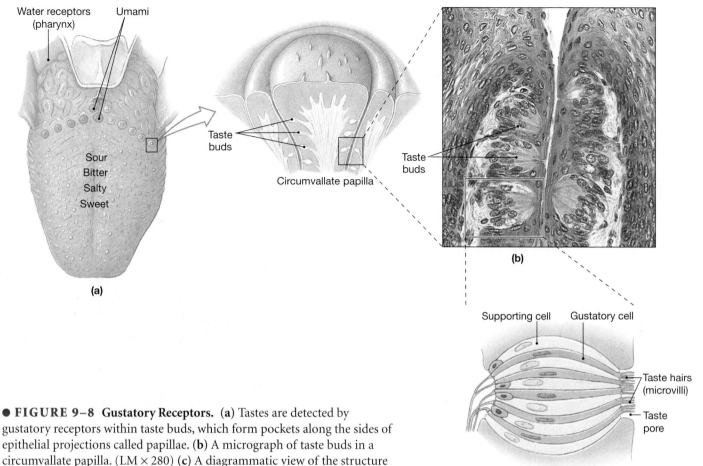

● **FIGURE 9–8 Gustatory Receptors.** (a) Tastes are detected by gustatory receptors within taste buds, which form pockets along the sides of epithelial projections called papillae. (b) A micrograph of taste buds in a circumvallate papilla. (LM × 280) (c) A diagrammatic view of the structure of a taste bud that shows receptor (gustatory) cells and supporting cells.

form sensory structures called **taste buds.** The taste buds are particularly well protected from the excessive mechanical stress due to chewing, for they lie along the sides of epithelial projections called **papillae** (pa-PIL-lē). The greatest numbers of taste buds are associated with the large *circumvallate papillae,* which form a V that points toward the attached base of the tongue.

Each taste bud contains slender sensory receptors known as **gustatory cells** and supporting cells. Each gustatory cell extends slender microvilli, sometimes called *taste hairs*, into the surrounding fluids through a narrow opening, the **taste pore.** The mechanism behind gustatory reception seems to parallel that of olfaction. Dissolved chemicals that contact the taste hairs stimulate a change in the membrane potential of the taste cell, which leads to action potentials in the sensory neuron.

You are probably already familiar with the four **primary taste sensations:** sweet, salty, sour, and bitter. There is some evidence for differences in sensitivity to tastes along the long axis of the tongue, with greatest sensitivity to salty-sweet anteriorly and to sour-bitter posteriorly (Figure 9–8a●). However, there are no differences in the structure of the taste buds, and taste buds in all portions of the tongue provide all four primary taste sensations.

Two additional tastes, *umami* and *water*, have been discovered in humans. **Umami** (oo-MAH-mē) is a pleasant taste that is characteristic of beef broth, chicken broth, and Parmesan cheese. Most people say water has no flavor, yet **water receptors** are present, especially in the pharynx. Their sensory output is processed in the hypothalamus and affects several systems involved in water balance and the regulation of blood volume.

The threshold for receptor stimulation varies for each of the primary taste sensations, and the taste receptors respond most readily to unpleasant rather than pleasant stimuli. For example, we are much more sensitive to acids, which taste sour, than to either sweet or salty chemicals, and we are more sensitive to bitter compounds than to acids. This sensitivity has survival value, because acids can damage the mucous membranes of the mouth and pharynx, and many biological toxins have an extremely bitter taste.

The Taste Pathways

Taste buds are monitored by the facial (N VII), glossopharyngeal (N IX), and vagus (N X) cranial nerves. The sensory afferent fibers of these different nerves synapse within a nucleus in the medulla oblongata, and the axons of the postsynaptic neurons synapse in the thalamus. There the neurons join axons that carry sensory information on touch, pressure, and proprioception. The information is then projected to the appropriate portions of the primary sensory cortex.

A conscious perception of taste is produced as the information received from the taste buds is correlated with other sensory data. Information about the texture of food, along with taste-related sensations such as "peppery" or "spicy hot," is provided by sensory afferents in the trigeminal nerve (V). In addition, information from olfactory receptors plays an overwhelming role in taste perception. You are several thousand times more sensitive to "tastes" when your olfactory organs are fully functional. By contrast, if you have a cold and your nose is stuffed up, airborne molecules cannot reach your olfactory receptors, so meals taste dull and unappealing (even though your taste buds are responding normally).

Key Note

Olfactory information is routed directly to the cerebrum, and olfactory stimuli have powerful effects on mood and behavior. Gustatory sensations are strongest and clearest when integrated with olfactory sensations.

→ CONCEPT CHECK QUESTIONS

1. How does sniffing repeatedly help to identify faint odors?
2. If you completely dry the surface of your tongue and then place salt or sugar crystals on it, you cannot taste them. Why not?

Answers begin on p. 792.

■ Vision

We rely more on vision than on any other special sense. Our visual receptors are contained in the eyes, elaborate structures that enable us to detect not only light but also detailed images. We will begin our discussion of these complex organs by considering the *accessory structures* of the eye, which provide protection, lubrication, and support.

The Accessory Structures of the Eye

The **accessory structures** of the eye include the (1) eyelids and associated exocrine glands and (2) the superficial ep-

ithelium of the eye (Figure 9–9a●); (3) structures associated with the production, secretion, and removal of tears (Figure 9–9b●); and (4) the extrinsic eye muscles (Figure 9–10●).

The **eyelids,** or **palpebrae** (pal-PĒ-brē), are a continuation of the skin. They act like windshield wipers: their blinking movements keep the surface of the eye lubricated and free from dust and debris. They can also close firmly to protect the delicate surface of the eye. The upper and lower eyelids are connected at the **medial canthus** (KAN-thus) and the **lateral canthus** (see Figure 9–9a). The eyelashes are very robust hairs that help prevent foreign particles such as dirt and insects from contacting the surface of the eye.

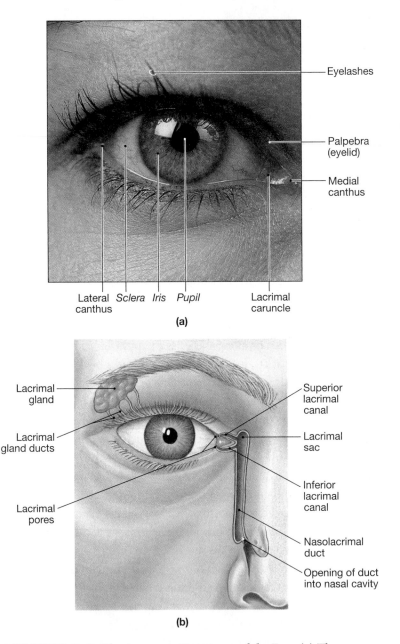

● **FIGURE 9–9 The Accessory Structures of the Eye.** (a) The gross and superficial anatomies of the accessory structures. (b) The details of the lacrimal apparatus.

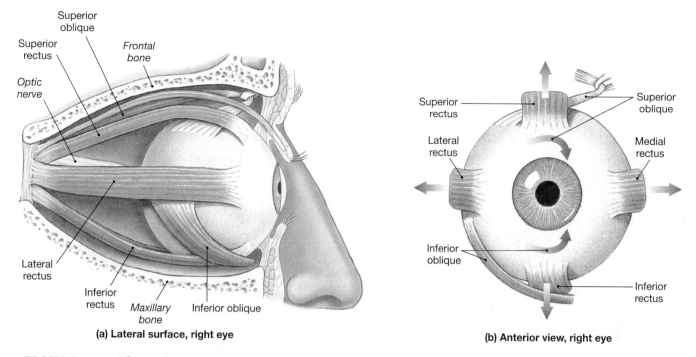

(a) Lateral surface, right eye

(b) Anterior view, right eye

● **FIGURE 9–10** The Extrinsic Eye Muscles.

Several types of exocrine glands protect the eye and its accessory structures. Large sebaceous glands are associated with the eyelashes, as they are with other hairs and hair follicles. ∞ p. 136 Along the inner margins of the eyelids, modified sebaceous glands *(tarsal glands)* secrete a lipid-rich substance that keeps the eyelids from sticking together. At the medial canthus, the **lacrimal caruncle** (KAR-ung-kul), a soft mass of tissue, contains glands that produce thick secretions that contribute to the gritty deposits occasionally found after a night's sleep. These various glands sometimes become infected by bacteria. An infection in a sebaceous gland of one of the eyelashes, in a tarsal gland, or in one of the sweat glands between the eyelash follicles produces a painful localized swelling known as a **sty.**

The epithelium that covers the inner surfaces of the eyelids and the outer surface of the eye is a mucous membrane called the **conjunctiva** (kon-junk-TĪ-vuh) (Figure 9–11●). The conjunctiva extends to the edges of the **cornea** (KOR-nē-uh), which is a transparent part of the outer layer of the eye. The cornea is covered by a delicate *corneal epithelium,* which is continuous with the conjunctiva. The conjunctiva contains many free nerve endings and is very sensitive. The painful condition of **conjunctivitis,** or pinkeye, results from damage to and irritation of the conjunctival surface. The most obvious sign, redness, is due to the dilation of the blood vessels beneath the conjunctival epithelium.

A constant flow of tears keeps the surface of the eyeball moist and clean. Tears reduce friction, remove debris, prevent bacterial infection, and provide nutrients and oxygen to the conjunctival epithelium. The **lacrimal apparatus** produces, distributes, and removes tears (see Figure 9–9b). Superior and lateral to the eyeball is the **lacrimal gland,** or *tear gland,* which has a dozen or more ducts that empty into the pocket between the eyelid and the eye. This gland nestles within a depression in the frontal bone, just inside the orbit. The lacrimal gland normally provides the key ingredients and most of the volume of the tears. Its watery, slightly alkaline secretions also contain *lysozyme,* an enzyme that attacks bacteria. The mixture of secretions from the lacrimal glands, accessory glands, and tarsal glands forms a superficial "oil slick" that assists in lubrication and slows evaporation.

Blinking sweeps tears across the surface of the eye to the medial canthus. Two small pores direct the tears into the **lacrimal canals,** which are passageways that end at the lacrimal sac (see Figure 9–9b). From this sac, the **nasolacrimal duct** carries the tears to the nasal cavity.

Six **extrinsic eye muscles,** or *oculomotor* (ok-ū-lō-MŌ-ter) *muscles,* originate on the surface of the orbit and control the position of the eye (Figure 9–10 and Table 9–1). These muscles are the **inferior rectus, medial rectus, superior rectus, lateral rectus, inferior oblique,** and **superior oblique.**

The Eye

The eyes are sophisticated visual instruments—more versatile and adaptable than the most expensive cameras, yet compact and durable. Each eye is roughly spherical, has a diameter of

TABLE 9-1 *The Extrinsic Eye Muscles (Figure 9–10)*

MUSCLE	ORIGIN	INSERTION	ACTION	INNERVATION
Inferior rectus	Sphenoid bone around optic canal	Inferior, medial surface of eyeball	Eye looks down	Oculomotor nerve (N III)
Medial rectus	As above	Medial surface of eyeball	Eye looks medially	As above
Superior rectus	As above	Superior surface of eyeball	Eye looks up	As above
Lateral rectus	As above	Lateral surface of eyeball	Eye looks laterally	Abducens nerve (N VI)
Inferior oblique	Maxillary bone at anterior portion of orbit	Inferior, lateral surface of eyeball	Eye rolls, looks up and laterally	Oculomotor nerve (N III)
Superior oblique	Sphenoid bone around optic canal	Superior, lateral surface of eyeball	Eye rolls, looks down and laterally	Trochlear nerve (N IV)

nearly 2.5 cm (1 in.), and weighs around 8 g (0.28 oz). The eyeball shares space within the orbit with the extrinsic eye muscles, the lacrimal gland, and the various cranial nerves and blood vessels that service the eye and adjacent areas of the orbit and face. A mass of *orbital fat* cushions and insulates the eye.

The eyeball is hollow (see Figure 9–11●); its interior can be divided into two cavities: the posterior cavity and the anterior cavity (Figure 9–11b●). The large **posterior cavity** is also called the *vitreous chamber* because it contains the gelatinous *vitreous body*. The smaller **anterior cavity** is subdivided into the *anterior chamber* and the *posterior chamber* (Figure 9–11c●). The shape of the eye is stabilized in part by the vitreous body and the *aqueous humor*, which is a clear fluid that fills the anterior cavity. The wall of the eye contains three distinct layers, or *tunics* (see Figure 9–11b): an outer *fibrous tunic*, an intermediate *vascular tunic*, and an inner *neural tunic*.

The Fibrous Tunic

The **fibrous tunic,** the outermost layer of the eye, consists of the *sclera* (SKLER-uh) and the *cornea*. The fibrous tunic (1) provides mechanical support and some degree of physical protection, (2) serves as an attachment site for the extrinsic eye muscles, and (3) assists in the focusing process. The **sclera,** or "white of the eye," is a layer of dense fibrous connective tissue that contains both collagen and elastic fibers (see Figure 9–11c). It is thickest over the posterior surface of the eye and thinnest over the anterior surface. The six extrinsic eye muscles insert on the sclera.

The surface of the sclera contains small blood vessels and nerves that penetrate the sclera to reach internal structures. On the anterior surface of the eye, however, these blood vessels lie under the conjunctiva. Because this network of small capillaries does not carry enough blood to lend an obvious color to the sclera, the white color of the collagen fibers is visible.

The transparent **cornea** is continuous with the sclera, but the collagen fibers of the cornea are organized into a series of layers that does not interfere with the passage of light. The cornea has no blood vessels, and its epithelial cells obtain their oxygen and nutrients from the tears that flow across their surfaces. Because the cornea has only a limited ability to repair itself, corneal injuries must be treated immediately to prevent serious vision losses. Restoration of vision after corneal scarring usually requires replacement of the cornea through a corneal transplant. Such transplants can be performed between unrelated individuals because there are no blood vessels to carry white blood cells, which attack foreign tissues, into the area.

The Vascular Tunic

The **vascular tunic** contains numerous blood vessels, lymphatic vessels, and the *intrinsic eye muscles*. The functions of this layer include (1) providing a route for blood vessels and lymphatic vessels that supply tissues of the eye, (2) regulating the amount of light entering the eye, (3) secreting and reabsorbing the *aqueous humor* that circulates within the eye, and (4) controlling the shape of the lens, which is an essential part of the focusing process.

The vascular tunic includes the *iris*, the *ciliary body*, and the *choroid* (see Figure 9–11b). Visible through the transparent cornea, the **iris** contains blood vessels, pigment cells, loose connective tissue, and two layers of intrinsic smooth muscle fibers. When these *pupillary muscles* contract, they change the diameter of the central opening, or **pupil,** of the iris (Figure 9–12●).

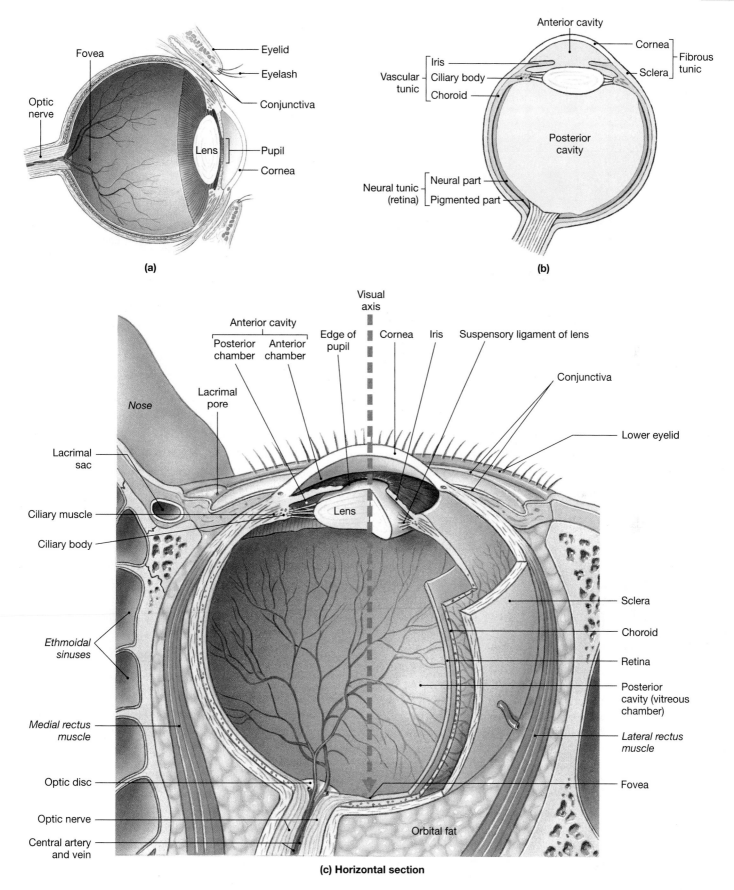

(a)

Fovea
Eyelid
Eyelash
Conjunctiva
Optic nerve
Lens
Pupil
Cornea

(b)

Anterior cavity
Cornea
Iris
Fibrous tunic
Vascular tunic
Ciliary body
Sclera
Choroid
Posterior cavity
Neural tunic (retina)
Neural part
Pigmented part

(c) Horizontal section

Visual axis
Anterior cavity
Posterior chamber
Anterior chamber
Edge of pupil
Cornea
Iris
Suspensory ligament of lens
Lacrimal pore
Conjunctiva
Nose
Lower eyelid
Lacrimal sac
Ciliary muscle
Ciliary body
Lens
Sclera
Choroid
Retina
Ethmoidal sinuses
Posterior cavity (vitreous chamber)
Medial rectus muscle
Lateral rectus muscle
Optic disc
Fovea
Optic nerve
Central artery and vein
Orbital fat

● **FIGURE 9–11** **The Sectional Anatomy of the Eye.** (**a**) This sagittal section bisects the left eye. (**b**) This horizontal section reveals the three layers, or tunics, of the right eye. (**c**) This horizontal section of the right eye shows various landmarks and features, including the path light takes (the "visual axis").

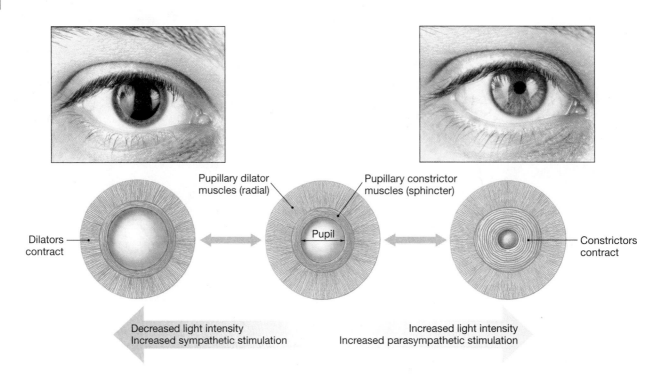

Pupillary dilator muscles (radial)

Pupillary constrictor muscles (sphincter)

Pupil

Dilators contract

Constrictors contract

Decreased light intensity
Increased sympathetic stimulation

Increased light intensity
Increased parasympathetic stimulation

● **FIGURE 9–12 The Pupillary Muscles.** Two sets of intrinsic smooth muscle within the iris control the diameter of the pupil. In bright light, the concentric muscle fibers contract, which constricts the pupil and reduces the level of incoming light. In dim light, the radial muscle fibers contract and the pupil enlarges, or dilates, to allow the entry of additional light.

One layer of these muscles forms concentric circles around the pupil; their contraction decreases, or constricts, the diameter of the pupil. The second layer of muscles extends radially away from the edge of the pupil; their contraction enlarges, or dilates, the pupil. Dilation and constriction are controlled by the autonomic nervous system in response to changes in light intensity. Parasympathetic activation in response to bright light causes the pupils to constrict, and sympathetic activation in response to dim light causes the pupils to dilate.

Clinical Note
ANISOCORIA

Anisocoria, or unequal pupils, is always a concern in the emergency patient. Pupillary constriction is controlled by the *oculomotor nerve (CN III)*. An expanding lesion in the brain can compress the nerve and cause pupillary dilation on the affected side. In a trauma patient, this may indicate an intracranial injury. In a medical patient, anisocoria may indicate intracranial bleeding or tumor. Because of this, the presence of anisocoria requires further investigation.

Anisocoria can be a normal finding. In fact, 20 percent of the general population has some degree of anisocoria. The difference is typically 1 millimeter or less and is not accompanied by other findings (such as drooping eyelids or extraocular muscle paralysis). Anisocoria may vary from day to day in the same person. Also, many medications, such as decongestants, can cause anisocoria. If necessary, old photographs, such as those on a driver's license, can help determine whether anisocoria is old or new. ■

Eye color is determined by (1) the number of melanocytes in the iris and (2) the presence of melanin granules in the pigmented epithelium on the posterior surface of the iris. (This pigmented epithelium is part of the neural tunic.) When the iris contains no melanocytes, light passes through the iris and bounces off the pigmented epithelium; the eye then appears blue. The irises of green, brown, and black eyes have increased numbers of melanocytes. The eyes of albino humans are very pale gray or blue gray.

Along its outer edge, the iris attaches to the anterior portion of the **ciliary body,** most of which consists of the *ciliary muscle,* a ring of smooth muscle that projects into the interior of the eye (see Figure 9–11c). The ciliary body begins at the junction between the cornea and sclera and extends to the scalloped border that also marks the anterior edge of the neural tunic. Posterior to the iris, the surface of the ciliary body is thrown into folds called *ciliary processes.* The **suspensory ligaments** of the lens attach to these processes. The connective tissue fibers of these ligaments hold the lens so that light passing through the pupil passes through the center of the lens along the visual axis. The choroid is a layer that separates the fibrous and neural tunics posterior to the ciliary body (see Figure 9–11c). The choroid contains a capillary network that delivers oxygen and nutrients to the neural tunic.

The Neural Tunic

The **neural tunic,** or **retina,** is the innermost layer of the eye. It consists of a thin outer pigment layer called the *pigmented part* and a thick inner layer called the *neural part* (see Figure

9–11b). The pigmented part absorbs light after it passes through the neural part. The neural part contains (1) the photoreceptors that respond to light, (2) supporting cells and neurons that perform preliminary processing and integration of visual information, and (3) blood vessels that supply tissues that line the posterior cavity. The two layers of the retina are normally very close together but not tightly interconnected.

The pigmented part continues over the ciliary body and iris. The neural part forms a cup that establishes the posterior and lateral boundaries of the posterior cavity.

ORGANIZATION OF THE RETINA. The retina contains several layers of cells (Figure 9–13a●). The outermost layer, closest to the wall of the pigmented part of the retina, contains the

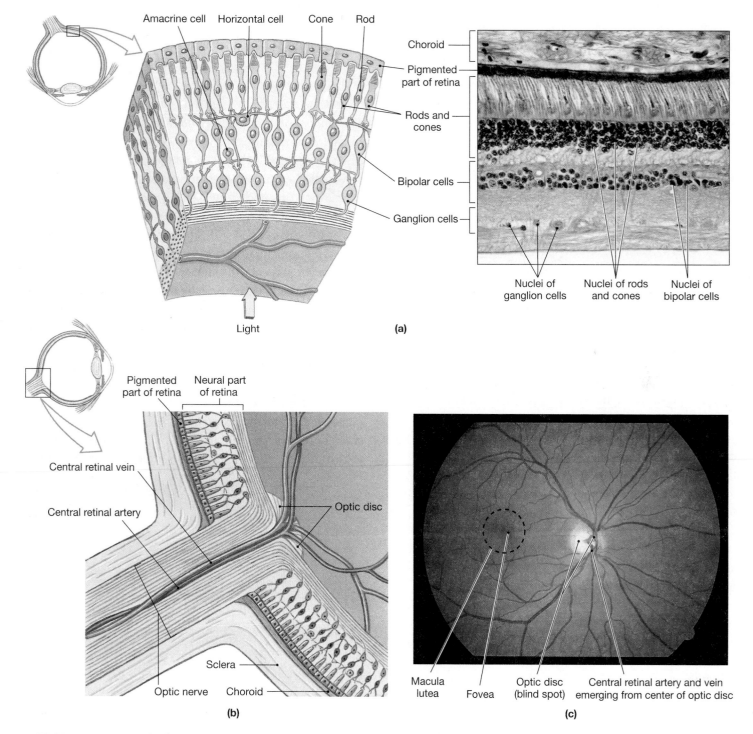

● FIGURE 9–13 Retinal Organization. (a) The cellular organization of the retina is shown in a drawing and a micrograph. Note that the photoreceptors are closer to the choroid than to the posterior cavity. (LM × 290) **(b)** This diagrammatic section passes through the optic disc. **(c)** This photograph shows the retina as seen through the pupil of the right eye.

photoreceptors, the cells that detect light. The eye has two types of photoreceptors: rods and cones. **Rods** do not discriminate among colors of light. These very light-sensitive receptors enable us to see in dimly lit rooms, at twilight, or in pale moonlight. **Cones** provide us with color vision. Three types of cones are present, and their stimulation in various combinations provides the perception of different colors. Cones give us sharper, clearer images, but they require brighter light than do rods. When you watch a sunset, you can notice your vision shifting from cone-based vision (a clear image in full color) to rod-based vision (a less distinct image in black and white).

Rods and cones are not evenly distributed across the retina. If you think of the retina as a cup, approximately 125 million rods are found on the sides, and roughly 6 million cones dominate the bottom. Most of these cones are concentrated in the area where the visual image arrives after passing through the cornea and lens. This region is the **macula lutea** (MAK-ū-luh LOO-tē-uh; yellow spot) (Figure 9–13c●). The highest concentration of cones is found in the center of the macula lutea in an area called the **fovea** (FŌ-vē-uh; shallow depression), or *fovea centralis.* The fovea is the center of color vision and the site of sharpest vision. When you look directly at an object, its image falls on this portion of the retina. An imaginary line drawn from the center of that object through the center of the lens to the fovea establishes the **visual axis** of the eye (see Figure 9–11).

You are probably already aware of the visual consequences of this distribution. During the day, when there is enough light to stimulate the cones, you see a very clear image. In very dim light, however, cones cannot function. When you try to stare at a dim star, for example, you are unable to see it. But if you look a little to one side rather than directly at the star, you can see it quite clearly. Shifting your gaze moves the image of the star from the fovea, where it does not provide enough light to stimulate the cones, to the sides of the retina, where it stimulates the more sensitive rods.

The rods and cones synapse with roughly 6 million **bipolar cells** (see Figure 9–13a). Bipolar cells in turn synapse within the layer of **ganglion cells** adjacent to the posterior cavity. The axons of the ganglion cells deliver the sensory information to the brain. *Horizontal cells* and *amacrine* (AM-a-krin) cells can regulate communication between photoreceptors and ganglion cells, and adjust the sensitivity of the retina. The effect is comparable to adjusting the contrast on a television. These cells play an important role in the eye's adjustment to dim or brightly lit environments.

THE OPTIC DISC.   Axons from an estimated 1 million ganglion cells converge on the **optic disc,** which is a circular region just medial to the fovea. The optic disc is the origin of the optic nerve (N II) (Figure 9–13b●). From this point, the axons turn,

penetrate the wall of the eye, and proceed toward the diencephalon. Blood vessels that supply the retina pass through the center of the optic nerve and emerge on the surface of the optic disc (see Figure 9–13b,c). The optic disc has no photoreceptors or other retinal structures. Because light that strikes this area goes unnoticed, it is commonly called the **blind spot.** You do not notice a blank spot in your visual field because involuntary eye movements keep the visual image moving and allow your brain to fill in the missing information. A simple activity, shown in Figure 9–14●, can demonstrate the presence and location of the blind spot.

The Chambers of the Eye

The ciliary body and lens divide the interior of the eye into a small anterior cavity and a larger posterior cavity, or vitreous chamber (Figure 9–15●). The anterior cavity is further subdivided into the **anterior chamber,** which extends from the cornea to the iris, and the **posterior chamber,** which is between the iris and the ciliary body and lens. The anterior and posterior chambers are filled with **aqueous humor.** This fluid circulates within the anterior cavity, and passes from the posterior to the anterior chamber through the pupil. The posterior cavity also contains aqueous humor, but most of this cavity is filled with a clear gelatinous substance known as the **vitreous body,** or *vitreous humor.* The vitreous body helps maintain the shape of the eye and also holds the retina against the choroid.

AQUEOUS HUMOR.   Aqueous humor is secreted into the posterior chamber by epithelial cells of the ciliary processes (see Figure 9–15). Pressure exerted by this fluid helps maintain the shape of the eye, and the circulation of aqueous humor transports nutrients and wastes. In the anterior chamber near the

● **FIGURE 9–14 A Demonstration of the Presence of a Blind Spot.** Close your left eye and stare at the cross with your right eye, and keep the cross in the center of your field of vision. Begin with the page a few inches away from your eye and gradually increase the distance. The dot will disappear when its image falls on the blind spot at the optic disc. To check the blind spot in your left eye, close your right eye, stare at the dot, and repeat this sequence.

● **FIGURE 9–15** **Eye Chambers and the Circulation of Aqueous Humor.** The lens is suspended between the vitreous chamber and the posterior chamber. Its position is maintained by suspensory ligaments, which attach the lens to the ciliary body. Aqueous humor secreted at the ciliary body circulates through the posterior and anterior chambers and is reabsorbed after passing along the canal of Schlemm.

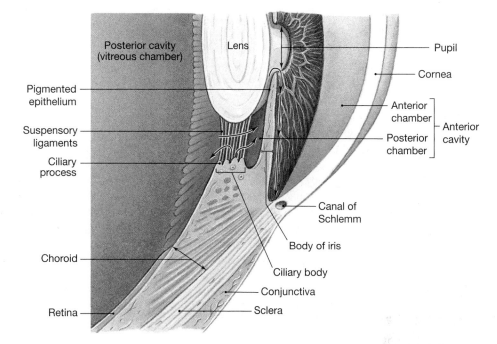

Posterior cavity (vitreous chamber)

Lens

Pupil

Cornea

Pigmented epithelium

Anterior chamber — Anterior cavity

Suspensory ligaments

Posterior chamber

Ciliary process

Canal of Schlemm

Body of iris

Choroid

Ciliary body

Conjunctiva

Retina

Sclera

edge of the iris, the aqueous humor enters a passageway, known as the *canal of Schlemm,* that empties into veins in the sclera and returns this fluid to the venous system.

Interference with the normal circulation and reabsorption of aqueous humor leads to an elevation in pressure inside the eye. If this condition, called **glaucoma,** is left untreated, it can eventually produce blindness by distorting the retina and the optic disc.

Clinical Note
HYPHEMA

Blunt trauma to the eye can rupture one of the blood vessels within the iris, and cause blood to enter the anterior chamber of the eye. Blood in the anterior chamber, called a *hyphema,* is a serious emergency (Figure 9–16●). The amount of blood in the chamber can range from microscopic to the so-called eight-ball hyphema, where blood fills the entire anterior chamber. Up to one-third of patients will suffer a rebleed within 3–5 days, and these rebleeds are often worse than the original bleed. Complications of hyphema include reduced vision, secondary glaucoma, and staining of the cornea.

Treatment of hyphema includes use of medications that reduce pressure within the anterior chamber and anti-inflammatory medicines. Some patients require hospitalization and complete bed rest for up to 5 days to prevent clot dislodgement and rebleeding. Patients who suffer a hyphema have the best outcome when the injury is promptly recognized and ophthalmologic care is given. ■

The Lens

The **lens** lies posterior to the cornea and is held in place by suspensory ligaments that extend from the ciliary body of the

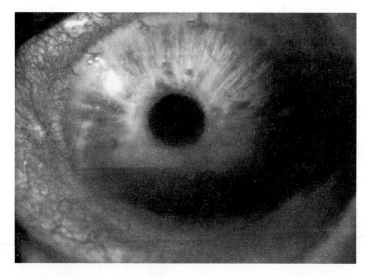

● **FIGURE 9–16** **Hyphema.** Hyphema is the presence of blood in the anterior chamber. Notice the blood level in chamber. Hyphemas can be sight-threatening emergencies.

choroid. The primary function of the lens is to focus the visual image on the photoreceptors. The lens does so by changing its shape.

THE STRUCTURE OF THE LENS. The transparent lens consists of concentric layers of cells wrapped in a dense fibrous capsule. The cells that comprise the interior of the lens lack organelles and are filled with transparent proteins. The capsule contains many elastic fibers that, in the absence of any outside force, contract and make the lens spherical. However, tension in the suspensory ligaments can overpower their contraction and pull the lens into a flattened oval.

Clinical Note
CATARACTS

The transparency of the lens depends on a precise combination of structural and biochemical characteristics. When that balance is disturbed, the lens loses its transparency, which is a condition known as a *cataract*. Cataracts can result from drug reactions, injuries, or radiation, but *senile cataracts* are the most common form. Over time, the lens becomes less elastic, takes on a yellowish hue, and eventually begins to lose its transparency. As the lens becomes opaque, or "cloudy," the individual needs brighter and brighter reading lights, and visual clarity begins to fade. If the lens becomes completely opaque, the person will be functionally blind, even though the photoreceptors are normal. Surgical procedures involve removing the lens, either intact or in pieces after it has been shattered with high-frequency sound. The missing lens is then replaced by an artificial substitute, and vision is fine-tuned with glasses or contact lenses. ■

LIGHT REFRACTION AND ACCOMMODATION. The eye is often compared to a camera. To provide useful information, the lens of the eye, like a camera lens, must focus the arriving image. To say that an image is "in focus" means that the rays of light that arrive from an object strike the sensitive surface of the retina (or of photographic film) so as to form a sharp miniature image of the original. If the rays are not perfectly focused, the image will be blurry. In the eye, focusing normally occurs in two steps, as light passes through first the cornea and then the lens.

Light is bent, or *refracted*, when it passes from one medium to a medium with a different density. In the human eye, the greatest amount of refraction occurs when light passes from the air into the cornea, which has a density close to that of water. As the light enters the relatively dense lens, the lens provides the extra refraction needed to focus the light rays from an object toward a specific **focal point**—the point at which the light rays converge (Figure 9–17a●). The distance between

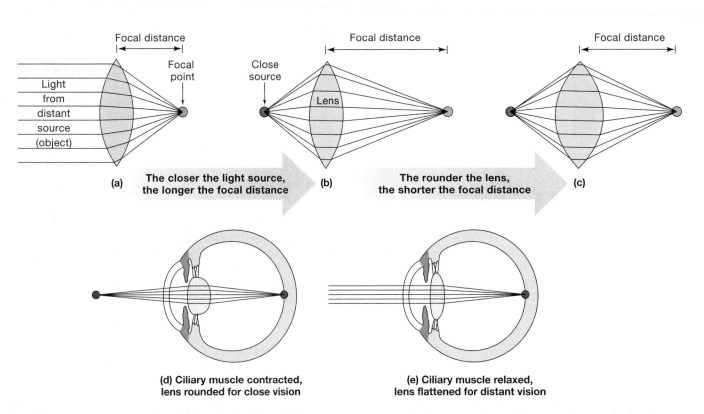

● **FIGURE 9–17** **Focal Point, Focal Distance, and Visual Accommodation.** A lens refracts light toward a specific focal point. The distance from the center of the lens to that point is the focal distance of the lens. (**a**) Light arrives from a distant source in parallel waves, and the lens is able to focus them over a short focal distance. Light from a nearby source, however, is still spreading out from its source when it strikes the lens; the result is a longer focal distance. (**b**) Lens shape also affects focal distance: the rounder the lens, the shorter the focal distance. (**c**) For the eye to form a sharp image, the focal distance must equal the distance between the center of the lens and the retina. The lens compensates for variations in the distance between the eye and the object in view by changing its shape. This process is called accommodation. (**d**) For close objects, the ciliary muscle contracts, and the suspensory ligaments allow the lens to round up. (**e**) For distant objects, the ciliary muscle relaxes, and the ligaments pull against the margins of the lens and flatten it.

the center of the lens and the focal point is the *focal distance.* This distance is determined by two factors: the distance of the object from the lens and the shape of the lens. The closer the object, the longer the focal distance (see Figure 9–17a,b●); the rounder the lens, the more refraction occurs, and the shorter is the focal distance (Figure 9–17b,c●). In the eye, the lens changes shape to keep the focal distance constant, thereby keeping the image focused on the retina.

Accommodation is the process of focusing an image on the retina by changing the shape of the lens (Figure 9–17d,e●). During accommodation, the lens either becomes rounder (to focus the image of a nearby object on the retina) or flattens (to focus the image of a distant object on the retina).

The lens is held in place by the suspensory ligaments that originate at the ciliary body. Smooth muscle fibers in the ciliary body encircle the lens and act like sphincter muscles. As you view a nearby object, the ciliary muscle contracts, and the ciliary body moves toward the lens (Figure 9–17d). This movement reduces the tension in the suspensory ligaments, and the elastic capsule pulls the lens into a more spherical shape. When you view a distant object, the ciliary muscle relaxes, the suspensory ligaments pull at the circumference of the lens, and the lens becomes flatter (Figure 9–17e).

IMAGE FORMATION. The image of an object that reaches the retina is a miniature image of the original, but it is upside down and backward. This makes sense if an object in view can be treated as a large number of individual light sources. Figure 9–18a● shows why an image formed on the retina is upside down. Light from the top of the pole lands at the bottom of the retina, and light from the bottom of the pole hits the top of the retina. Figure 9–18b● illustrates why an image formed on the retina is backward. Light from the left side of the fence falls on the right side of the retina, and light from the right side of the fence falls on the left side of the retina. The brain compensates for both aspects of image reversal without our conscious awareness.

Key Note

Light passes through the conjunctiva and cornea, crosses the anterior cavity to reach the lens, transits the lens, crosses the posterior chamber, and then penetrates the neural tissue of the retina before reaching and stimulating the photoreceptors. Cones are most abundant at the fovea and macula lutea, and they provide high-resolution color vision in brightly lit environments. Rods dominate the peripheral areas of the retina, and they provide relatively low-resolution black and white vision in dimly lit environments.

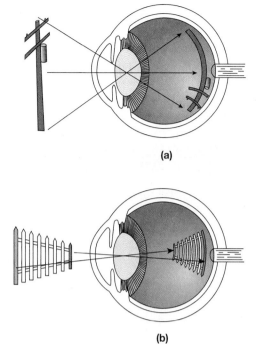

● **FIGURE 9–18 Image Formation.** Light from each portion of an object is focused on a different part of the retina. The resulting image arrives (**a**) upside down and (**b**) backward.

Visual Physiology

The rods and cones of the retina are called **photoreceptors** because they detect *photons*, which are basic units of visible light. Light is a form of radiant energy that travels in waves with a characteristic wavelength (distance between wave

Clinical Note
VISUAL ACUITY

How well the eye discriminates small details is determined by measuring the patient's visual acuity. Visual acuity testing is standardized and involves placing the patient exactly 20 feet from a Snellen eye chart. Eyes are individually tested by having the patient read the smallest line on the chart while convering the opposite eye. Then both eyes are tested together. Visual acuity is recorded as a fraction in which the numerator (top number) indicates the distance of the patient from the chart and the denominator (bottom number) indicates the distance at which the normal eye can read the line. A visual acuity of 20/20 or better is considered normal. Patients who cannot read any letter on the chart should be asked to count the number of fingers the examiner holds up. If the patient can accomplish this, then *CF* (counts fingers) is recorded. Patients who cannot count fingers should be examined to determine whether they are able to detect light. If they can detect light, then *LO* (light only) is recorded. ■

peaks). Our eyes are sensitive to wavelengths that make up the spectrum of **visible light** (700–400 nm). This spectrum, seen in a rainbow, can be remembered by the acronym ROY G. BIV (*Red, Orange, Yellow, Green, Blue, Indigo, Violet*). Color de-

pends on the wavelength of the light. Photons of red light have the longest wavelength and carry the least energy. Photons from the violet portion of the spectrum have the shortest wavelength and carry the most energy.

Clinical Note
ACCOMMODATION PROBLEMS

In the normal eye, when the ciliary muscles are relaxed and the lens is flattened, a distant image will be focused on the retinal surface (Figure 9–19a●), which is a condition called *emmetropia (emmetro-, proper measure)*. However, irregularities in the shape of the lens or cornea can affect refraction and the clarity of the resulting visual image. This condition, called *astigmatism*, can usually be corrected by glasses or special contact lenses.

Figure 9–19 also diagrams two common accommodation problems. If the eyeball is too deep, the image of a distant object will form in front of the retina, and the image will be blurry and out of focus (Figure 9–19b●). Vision at close range will be normal, because the lens will be able to round up enough to focus the image on the retina. As a result, such individuals are said to be "nearsighted." Their condition is more formally termed *myopia (myein, to shut + ops, eye)*. Myopia can be corrected by placing a diverging lens in front of the eye (Figure 9–19c●).

If the eyeball is too shallow, *hyperopia* results (Figure 9–19d●). The ciliary muscles must contract to focus even a distant object on the retina, and at close range the lens cannot pro-

vide enough refraction. These individuals are said to be "farsighted" because they can see distant objects most clearly. Older individuals become farsighted as their lenses lose elasticity; this form of hyperopia is called *presbyopia (presbys, old man)*. Hyperopia can be treated by placing a converging lens in front of the eye (Figure 9–19e●).

Variable success at correcting myopia and hyperopia has been achieved by surgically reshaping the cornea to alter its refractive powers. In *radial keratotomy*, several corneal incisions flatten the surface. Some scarring and localized corneal weakness may occur, and only two-thirds of patients are satisfied with the results. In a newer procedure called *photorefractive keratectomy (PRK)*, a computer-guided laser shapes the cornea to exact specifications. Tissue is removed only to a depth of 10–20 μm more than about 10 percent of the cornea's thickness. The entire procedure can be done in less than a minute. Each year, an estimated 100,000 people undergo the PRK procedure in the U.S. Corneal scarring is rare; however, many still need reading glasses, and both immediate and long-term visual problems can occur. ■

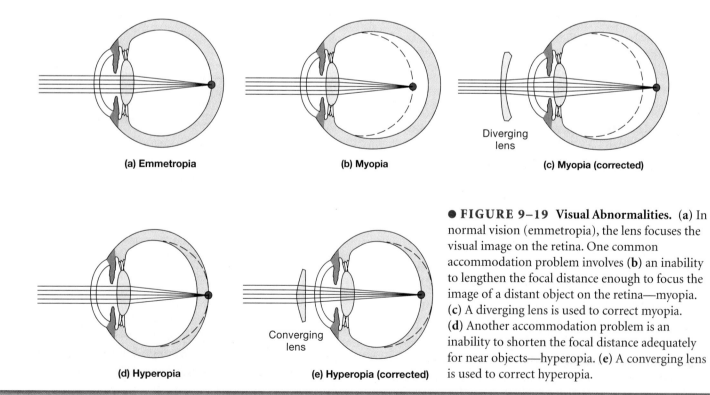

(a) Emmetropia   **(b) Myopia**   **(c) Myopia (corrected)**

(d) Hyperopia   **(e) Hyperopia (corrected)**

● **FIGURE 9–19 Visual Abnormalities.** (**a**) In normal vision (emmetropia), the lens focuses the visual image on the retina. One common accommodation problem involves (**b**) an inability to lengthen the focal distance enough to focus the image of a distant object on the retina—myopia. (**c**) A diverging lens is used to correct myopia. (**d**) Another accommodation problem is an inability to shorten the focal distance adequately for near objects—hyperopia. (**e**) A converging lens is used to correct hyperopia.

Rods and Cones

Rods provide the CNS with information about the presence or absence of photons, without regard to wavelength. As a result, they do not discriminate among colors of light. They are very sensitive, however, and enable us to see in dim conditions.

Cones provide information about the wavelength of photons. Because cones are less sensitive than rods, they function only in relatively bright light. We have three types of cones: *blue cones*, *green cones*, and *red cones*. Each type of cone contains pigments sensitive to blue, green, or red wavelengths of light; their stimulation in various combinations accounts for our perception of colors.

Persons unable to distinguish certain colors have a form of *color blindness*. The standard tests for color vision involve picking numbers or letters out of a complex image, such as the one in Figure 9–20●. Color blindness occurs because one or more classes of cones are absent or nonfunctional. In the most common condition, the red cones are missing and the individual cannot distinguish red light from green light. Ten percent of men have some color blindness, whereas the incidence among women is only around 0.67 percent. Total color blindness is extremely rare; only 1 person in 300,000 has no cone pigments of any kind.

Photoreceptor Structure

Figure 9–21a● compares the structure of rods and cones. The *outer segment* of a photoreceptor contains hundreds to thousands of flattened membranous discs. The names *rod* and *cone* refer to the outer segment's shape. The *inner segment* of a pho-

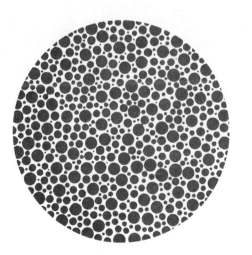

● **FIGURE 9–20 A Standard Test for Color Vision.**
Individuals who lack one or more populations of cones are unable to distinguish the patterned image (the number 12).

toreceptor contains typical cellular organelles and forms synapses with other cells. In the dark, each photoreceptor continually releases neurotransmitters. The arrival of a photon initiates a chain of events that alters the membrane potential of the photoreceptor and changes the rate of neurotransmitter release.

The discs of the outer segment in both rods and cones contain special organic compounds called **visual pigments.** The absorption of photons by visual pigments is the first key step in the process of *photoreception*, which is the detection of light. The visual pigments are derived from the compound **rhodopsin** (rō-DOP-sin). Rhodopsin consists of a protein, **opsin,** bound to the pigment **retinal** (RET-i-nal) (Figure 9–21b●). Retinal is synthesized from **vitamin A.** Retinal is identical in both rods and cones, but a different form of opsin is found in the rods and in each of the three types of cones (red, blue, and green).

Photoreception

Photoreception begins when a photon strikes a rhodopsin molecule in the outer segment of a photoreceptor. When the photon is absorbed, a change in the shape of the retinal component activates opsin, which starts a chain of enzymatic events that alters the rate of neurotransmitter release. This change is the signal that light has struck a photoreceptor at that particular location on the retina.

Shortly after the retina changes shape, the rhodopsin molecule begins to break down into retinal and opsin, which is a process known as *bleaching* (Figure 9–22●). The retina must be converted to its former shape before it can recombine with opsin. This conversion requires energy in the form of ATP, and it takes time. Bleaching contributes to the lingering visual impression that you have after a camera flash goes off. After an intense exposure to light, a photoreceptor cannot respond to further stimulation until its rhodopsin molecules have been regenerated. As a result, a "ghost" image remains on the retina.

The Visual Pathway

The visual pathway begins at the photoreceptors and ends at the visual cortex of the cerebral hemispheres. In other sensory pathways we have examined, at most one synapse lies between a receptor and a sensory neuron that delivers information to the CNS. In the visual pathway, the message must cross two synapses (photoreceptor to bipolar cell, and bipolar cell to ganglion cell) before it moves toward the

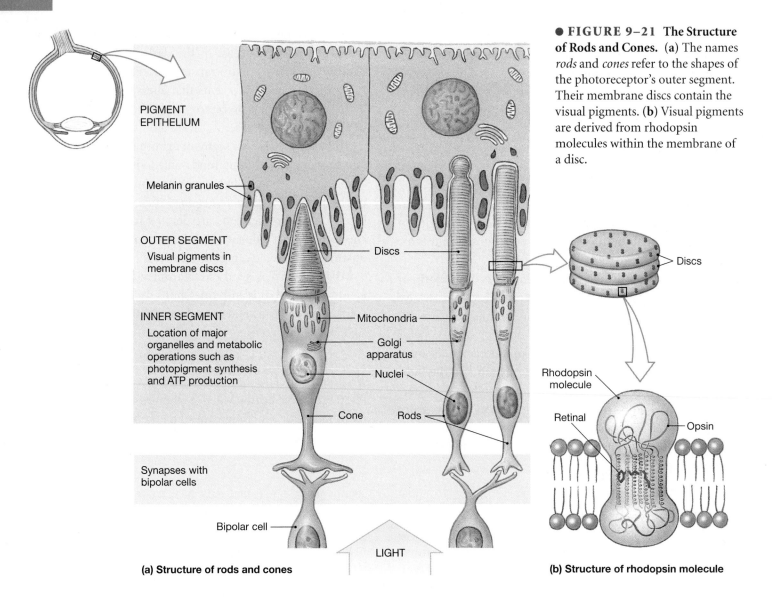

PIGMENT EPITHELIUM

Melanin granules

OUTER SEGMENT
Visual pigments in membrane discs

Discs

INNER SEGMENT
Location of major organelles and metabolic operations such as photopigment synthesis and ATP production

Mitochondria

Golgi apparatus

Nuclei

Cone   Rods

Synapses with bipolar cells

Bipolar cell

LIGHT

Discs

Rhodopsin molecule

Retinal

Opsin

(a) Structure of rods and cones

(b) Structure of rhodopsin molecule

brain. Axons from the entire population of ganglion cells converge on the optic disc, penetrate the wall of the eye, and proceed toward the diencephalon as the optic nerve (N II). The two optic nerves (one from each eye) meet at the optic chiasm (Figure 9–23●). From this point, approximately half of the fibers of each optic nerve proceed within the optic tracts toward the thalamic nucleus on the same side of the brain, while the other half cross over to reach the thalamic nucleus on the opposite side. The thalamic nuclei act as switching and processing centers that relay visual information to reflex centers in the brain stem as well as to the cerebral cortex. The visual information received by the *superior colliculi* (midbrain nuclei in the brain stem) controls constriction or dilation of the pupil and reflexes that control eye movement. ∞ p. 293

The sensation of vision arises from the integration of information that arrives at the visual cortex of the cerebrum.

Clinical Note
NIGHT BLINDNESS

The visual pigments of the photoreceptors are synthesized from vitamin A. The body contains vitamin A reserves sufficient for several months, and a significant amount is stored in the cells of the pigmented part of the retina. If dietary sources are inadequate, these reserves are gradually exhausted, and the amount of visual pigment in the photoreceptors begins to drop. Daylight vision is affected, but in daytime the light is usually bright enough to stimulate any remaining visual pigments in the densely packed cone population. As a result, the problem first becomes apparent at night, when the dim light proves insufficient to activate the rods. This condition, known as *night blindness,* can be treated by administration of vitamin A. The body can convert the carotene pigments in many vegetables to vitamin A. Carrots are a particularly good source of carotene, which explains the old adage that carrots are good for your eyes. ■

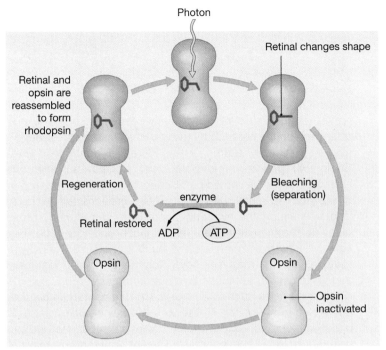

● **FIGURE 9–22** **The Bleaching and Regeneration of Visual Pigments.**

The visual cortex of each occipital lobe contains a sensory map of the entire field of vision. As with the primary sensory cortex, the map does not faithfully duplicate the relative areas within the sensory field. For example, the area assigned to the fovea covers about 35 times the surface it would cover if the map were proportionally accurate.

Many centers in the brain stem receive visual information from the thalamic nuclei or over collateral branches from the optic tracts. For example, some collaterals that bypass the thalamic nuclei synapse in the hypothalamus. Visual inputs there and at the pineal gland establish a daily pattern of activity that is tied to the day-night cycle. This *circadian* (*circa*, about + *dies*, day) *rhythm* affects your metabolic rate, endocrine function, blood pressure, digestive activities, awake-sleep cycle, and other processes.

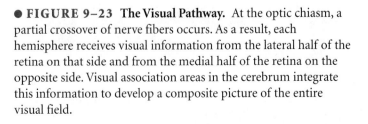

● **FIGURE 9–23** **The Visual Pathway.** At the optic chiasm, a partial crossover of nerve fibers occurs. As a result, each hemisphere receives visual information from the lateral half of the retina on that side and from the medial half of the retina on the opposite side. Visual association areas in the cerebrum integrate this information to develop a composite picture of the entire visual field.

Clinical Note
EYE EMERGENCIES

Vision is one of our most important senses. The eyes are the principle sensory organs of vision, and emergencies that involve the eye can threaten sight. Because of this, detailed evaluation and treatment of emergent eye illnesses and emergencies is essential. Two types of doctors specialize in treating eye disorders: ophthalmologists and optometrists. *Ophthalmologists* are physicians (M.D. or D.O.) who have graduated from medical school and completed a residency in ophthalmology. They specialize in the medical and surgical management of eye disorders. Optometrists are doctors of optometry (O.D.) who have completed four years of optometry school. They primarily perform refractive examinations and prescribe glasses and contact lenses. Optometrists do not perform surgery and, in many states, do not prescribe medication.

Conjunctival Injuries

Acute eye pain or a red eye are the most common initial complaints of patients with an ocular emergency. Injuries to the eye are common. One of the most striking eye injuries is a subconjunctival hemorrhage. Trauma can cause the fragile blood vessels within the conjunctiva to rupture. The bleeding is evident as it occurs over the white of the eye. In addition to trauma, sneezing, coughing, vomiting, and straining can cause a subconjunctival hemorrhage. High blood pressure also can cause a conjunctival hemorrhage. In many cases, a specific cause cannot be identified. Subconjunctival hemorrhages are painless, do not affect vision, generally do not require treatment, and resolve completely within a week or two. Abrasions of the conjunctival membranes, which are also common, heal completely within two to three days (Figure 9–24●).

Corneal Injuries

Trauma to the cornea can cause an abrasion. In addition to trauma, corneal abrasions can develop from contact lens wear.

● **FIGURE 9–24 Conjunctival Abrasion.** A conjunctival foreign body and abrasion overlie a large subconjunctival hemorrhage.

Pain, redness, tearing, and light sensitivity usually accompany a corneal abrasion. Assessment of corneal abrasions is often difficult because of the patient's discomfort. Often, these patients feel as though a foreign body is embedded in the eye, and they often worsen the abrasion by trying to remove the perceived foreign body. Usually, a drop or two of topical ophthalmic anesthetic will provide rapid pain relief so that an adequate examination can be carried out. Under magnification, a defect in the cornea can usually be seen. Corneal abrasions can be visualized by staining the eye with fluorescein. The injured cornea takes up the fluorescein, and examination under an ultraviolet light will clearly demonstrate the injury. Corneal abrasions usually heal within a matter of days. Treatment includes analgesics and placement of antibiotic drops or ointment.

Small particles, such as dust or metal fragments, can become embedded in the cornea. These corneal foreign bodies cause an underlying abrasion. Most foreign bodies are superficial and can be removed in the emergency department. A ring of rust may develop around metallic foreign bodies that are embedded more than 24 hours. The rust must be removed or it will permanently scar the cornea. Following removal of a corneal foreign body, it is important to evert the eyelid to make sure a second foreign body is not present.

Blunt Eye Trauma

Blunt trauma to the eye can cause swelling of the lids and the periorbital tissues. A direct blow to the eye can result in a *hyphema,* which is bleeding into the anterior chamber. A hyphema is a serious injury that can result in permanent blindness and should always be evaluated by an ophthalmologist. Hyphemas can cause increased pressures within the eye and permanent injury. The patient's head should be elevated to help decrease intraocular pressure. Spontaneous rebleeding is not uncommon with hyphemas.

Blunt trauma to the eye can sometimes result in a blowout fracture, which is a fracture of the wall or walls of the orbit. Most frequently, the inferior wall of the orbit is fractured into the maxillary sinus. Occasionally, the inferior rectus muscle may be entrapped in the fracture, preventing the eye from moving superiorly (looking up). The patient may report double vision on upward gaze. Blowout fractures require surgical treatment (Figure 9–25●).

Penetrating Eye Injury

Any injury that penetrates the globe or ruptures the globe is extremely serious. Common causes include BB pellets, lawn mower projectiles, particles from hammering, grinding injuries, knife wounds, and gunshot wounds. Any penetrating injury has the potential for entering the eye. If a ruptured globe is suspected, a protective shield should be immediately placed over the affected eye, and the patient should be kept calm to prevent exacerbating the injury (Figure 9–26●).

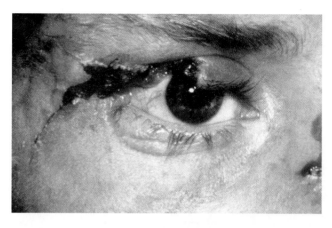

● **FIGURE 9–25 Trauma to the Right Eye.** Significant trauma, such as this upper lid laceration, necessitates a detailed examination for other injuries such as a blowout fracture.

Chemical Injuries

Chemical injuries to the eye are common. The severity of a chemical injury is directly related to the chemical agent involved. To help remove the offending agent immediately following a chemical injury, the eye should be irrigated with copious amounts of water for 10 minutes (if the chemical was an alkali, irrigation is carried out longer until pH of tears returns to normal). Following this, a detailed examination of the affected eye, including possible fluorescein staining, should be carried out to determine the severity of tissue injury.

Ultraviolet Keratitis

Ultraviolet keratitis is severe pain, tearing, light sensitivity, and foreign-body sensation that occurs from 6 to 12 hours after ocular exposure to a welding arc, tanning lights, or bright snow. Often, the

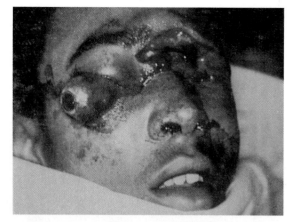

● **FIGURE 9–26 Blunt Facial Trauma That Results in Enucleation of the Right Eye.** The globe should be carefully protected and the patient transported to a facility with ophthalmological surgery capabilities.

patient is awakened with severe eye pain and tearing. This injury can be extremely painful but responds readily to topical anesthetics. It usually resolves in 24 to 48 hours.

Acute Glaucoma

Failure of the aqueous humor to enter the canal of Schlemm leads to glaucoma. Although drainage is impaired, the production of aqueous humor continues, and the intraocular pressure begins to rise. As this progresses, the soft tissues within the eye become distorted. *Acute angle-closure glaucoma* is a serious medical emergency. The patient often complains of cloudy vision, eye ache, headache, and frequently nausea and vomiting. Usually, there is no history of glaucoma. Acute angle-closure glaucoma requires hospitalization and treatment with medications that decrease intraocular pressure. ■

■ Equilibrium and Hearing

The special senses of equilibrium and hearing are provided by the *inner ear*, which is a receptor complex located in the temporal bone of the skull. ∞ p. 170 *Equilibrium* informs us of the position of the body in space by monitoring gravity, linear acceleration, and rotation; *hearing* enables us to detect and interpret sound waves. The basic receptor mechanism for these senses is the same. The receptors—*hair cells*—are simple mechanoreceptors. The complex structure of the inner ear and the different arrangements of accessory structures permit hair cells to respond to different stimuli and, thus, to provide the input for both senses.

The Anatomy of the Ear

The ear is divided into three anatomical regions: the *external ear*, the *middle ear*, and the *inner ear* (Figure 9–27●). The external ear—the visible portion of the ear—collects and directs sound

waves toward the middle ear, which is a chamber located in a thickened portion of the temporal bone. Structures of the middle ear collect and amplify sound waves and transmit them to an appropriate portion of the inner ear. The inner ear contains the sensory organs for hearing and equilibrium.

The External Ear

The **external ear** includes the fleshy **auricle,** or *pinna*, which surrounds the entrance to the **external acoustic canal,** or *ear canal*. The auricle, which is supported by elastic cartilage, protects the opening of the canal. It also provides directional sensitivity to the ear: sounds coming from behind the head are partially blocked by the auricle; sounds coming from the side are collected and channeled into the external auditory canal. (When you "cup" your ear with your hand to hear a faint sound more clearly, you are exaggerating this effect.) **Ceruminous** (se-ROO-mi-nus) **glands** along the external

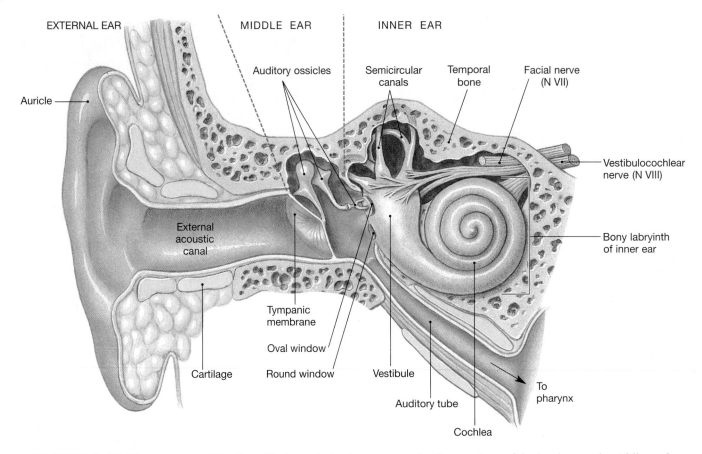

EXTERNAL EAR   MIDDLE EAR   INNER EAR

Auditory ossicles   Semicircular canals   Temporal bone   Facial nerve (N VII)

Auricle

Vestibulocochlear nerve (N VIII)

External acoustic canal

Bony labryinth of inner ear

Tympanic membrane

Oval window

Cartilage   Round window   Vestibule   To pharynx

Auditory tube

Cochlea

● **FIGURE 9–27 The Anatomy of the Ear.** The boundaries that separate the three regions of the ear (external, middle, and inner) are roughly marked by the dashed lines.

acoustic canal secrete a waxy material (*cerumen*) that helps prevent the entry of foreign objects and insects, as do many small, outwardly projecting hairs. The waxy cerumen also slows the growth of microorganisms in the ear canal and reduces the likelihood of infection. The external acoustic canal ends at the **tympanic membrane,** also called the *tympanum* or *eardrum.* The tympanic membrane is a thin sheet that separates the external ear from the middle ear (see Figure 9–27●).

The Middle Ear

The **middle ear,** or *tympanic cavity,* is an air-filled chamber separated from the external acoustic canal by the tympanum. The middle ear communicates with the superior portion of the pharynx, which is a region known as the *nasopharynx,* and with *air cells* in the mastoid process of the temporal bone. The connection with the nasopharynx is the **auditory tube,** also called the *pharyngotympanic tube* or the *Eustachian tube* (see Figure 9–28●). The auditory tube enables the equalization of pressure on either side of the eardrum. Unfortunately, it can also allow microorganisms to travel from the nasopharynx into the tympanic cavity, and lead to an unpleasant middle ear infection known as *otitis media.*

THE AUDITORY OSSICLES.   The middle ear contains three tiny ear bones, which are collectively called **auditory ossicles.** The ear bones connect the tympanum with the receptor complex of the inner ear (Figure 9–28●). The three auditory ossicles are the malleus, the incus, and the stapes. The **malleus** (*malleus,* hammer) attaches at three points to the interior surface of the tympanum. The middle bone—the **incus** (*incus,* anvil)—attaches the malleus to the inner bone, the **stapes** (*stapes,* stirrup). The base of the stapes almost completely fills the *oval window,* a small opening in the bone that encloses the inner ear.

The in-and-out vibrations of the tympanic membrane convert arriving sound energy into mechanical movements of the auditory ossicles. The ossicles act as levers that conduct the vibrations to the inner ear. The tympanum is larger and heavier than the delicate membrane that spans the oval window, so the amount of movement increases markedly from tympanum to oval window.

This magnification in movement enables us to hear very faint sounds. It can also be a problem, however, when we are exposed to very loud noises. In the middle ear, two small muscles protect the eardrum and ossicles from violent movements under noisy conditions. The *tensor tympani* (TEN-sor tim-PAN-ē) *muscle*

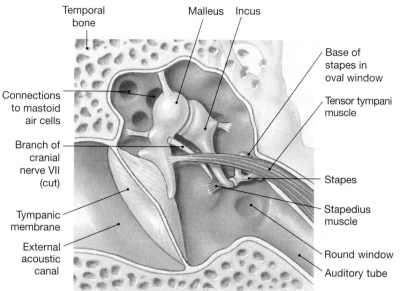

Temporal bone

Malleus   Incus

Base of stapes in oval window

Tensor tympani muscle

Connections to mastoid air cells

Branch of cranial nerve VII (cut)

Stapes

Tympanic membrane

Stapedius muscle

External acoustic canal

Round window

Auditory tube

● **FIGURE 9–28   The Structure of the Middle Ear.**

pulls on the malleus, which increases the stiffness of the tympanum and reduces the amount of possible movement. The *stapedius* (sta-PĒ-dē-us) *muscle* pulls on the stapes, which thereby reduces its movement at the oval window.

The Inner Ear

The senses of equilibrium and hearing are provided by the receptors within the **inner ear.** These receptors are protected by the **bony labyrinth** (Figure 9–29a●); its outer walls are fused with the surrounding temporal bone (see Figure 9–27). The bony labyrinth surrounds and protects the **membranous labyrinth** (*labyrinthos*, network of canals), a collection of tubes and chambers that follow the contours of the surrounding bony labyrinth and are filled with a fluid called **endolymph** (EN-dō-limf). Between the bony and membranous labyrinths flows another fluid, the **perilymph** (PER-i-limf) (Figure 9–29b●). The receptors lie within the membranous labyrinth.

The bony labyrinth can be subdivided into three parts (Figure 9–29a):

1. *Vestibule.* The **vestibule** (VES-ti-būl) includes a pair of membranous sacs, the **saccule** (SAK-ūl) and the **utricle** (Ū-tri-kul). Receptors in these sacs provide sensations of gravity and linear acceleration.

2. *Semicircular canals.* The **semicircular canals** enclose slender *semicircular ducts.* Receptors in the semicircular ducts are stimulated by rotation of the head. The combination of vestibule and semicircular canals is called the **vestibular complex,** because the fluid-filled chambers within the vestibule are continuous with those of the semicircular canals.

3. *Cochlea.* The bony, spiral-shaped **cochlea** (KOK-lē-uh; *cochlea*, snail shell) contains the **cochlear duct** of the membranous labyrinth. Receptors in the cochlear duct provide the sense of hearing. The cochlear duct is sandwiched between a pair of perilymph-filled chambers, and the entire complex is coiled around a central bony hub.

The bony labyrinth's walls are dense bone everywhere except at two small areas near the base of the cochlea. The round window is an opening in the bone of the cochlea. A thin membrane spans the opening and separates perilymph in the cochlea from the air in the middle ear. The membrane that spans the oval window is firmly attached to the base of the stapes. When a sound vibrates the tympanic membrane, the movements are conducted over the malleus and incus to the stapes. Movement of the stapes ultimately leads to the stimulation of receptors in the cochlear duct, and we hear the sound.

RECEPTOR FUNCTION IN THE INNER EAR.   The receptors of the inner ear are called **hair cells** (Figure 9–29c●). Regardless of location, they are always surrounded by supporting cells and monitored by the dendrites of sensory neurons. Each hair cell communicates with a sensory neuron by continually releasing small quantities of neurotransmitter. The free surface of this receptor supports 80–100 long microvilli called *stereocilia.* Hair cells do not actively move these processes. Instead, when some external force causes the stereocilia to move, their movement distorts the cell surface and alters its rate of neurotransmitter release. Displacement of the stereocilia in one direction stimulates the hair cells (and increases neurotransmitter release); displacement in the opposite direction inhibits the hair cells (and decreases neurotransmitter release).

Equilibrium

There are two aspects of equilibrium: (1) **dynamic equilibrium,** which aids us in maintaining our balance when the head and body are moved suddenly, and (2) **static equilibrium,** which maintains our posture and stability when the body is motionless. All equilibrium sensations are provided by hair cells of the vestibular complex. The semicircular ducts, which monitor dynamic equilibrium, provide information about rotational movements of the head. For example, when you turn your head to the left, receptors in the semicircular ducts tell you how rapid the movement is and in which direction. The saccule and the utricle, which monitor static equilibrium, provide information about your position with respect to gravity. If you

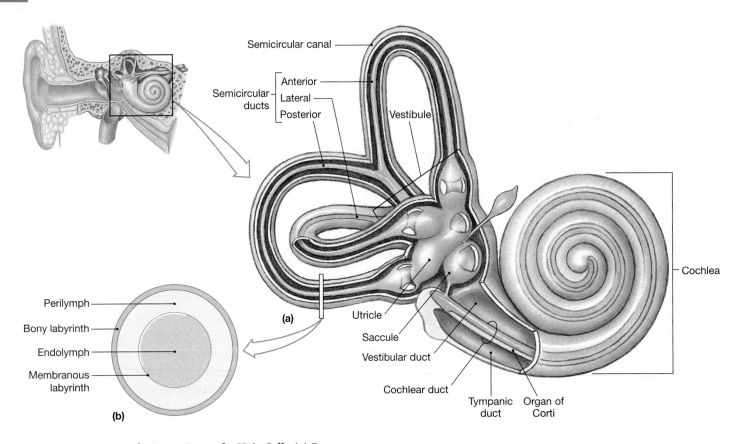

(a)

Semicircular canal

Semicircular ducts
- Anterior
- Lateral
- Posterior

Vestibule

Cochlea

Utricle

Saccule

Vestibular duct

Cochlear duct

Tympanic duct

Organ of Corti

Perilymph

Bony labyrinth

Endolymph

Membranous labyrinth

(b)

● **FIGURE 9–29 The Inner Ear and a Hair Cell.** (**a**) Part of this anterior view of the bony labyrinth has been removed to show the outline of the enclosed membranous labyrinth. (**b**) A section through one of the semicircular canals, which shows the relationship between the bony and membranous labyrinths and the locations of the perilymph and endolymph. (**c**) A representative hair cell (receptor) from the vestibular complex. Bending the stereocilia in one direction results in stimulating its sensory neuron. Displacement in the opposite direction results in inhibition of its sensory neuron.

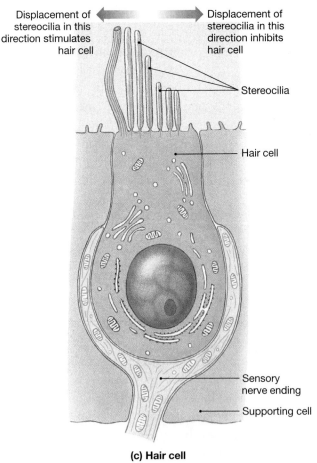

Displacement of stereocilia in this direction stimulates hair cell

Displacement of stereocilia in this direction inhibits hair cell

Stereocilia

Hair cell

Sensory nerve ending

Supporting cell

(c) Hair cell

stand with your head tilted to one side, receptors in the saccule and utricle will report the angle involved and whether your head is tilting forward or backward. These receptors are also stimulated by sudden changes in velocity. For example, when your car accelerates, the saccular and utricular receptors give you the sensation of increasing speed.

The Semicircular Ducts: Rotational Motion

Sensory receptors in the semicircular ducts respond to rotational movements of the head. Figure 9–29a shows the **anterior, posterior,** and **lateral semicircular ducts** and their continuity with the utricle. Each semicircular duct contains a swollen region, the **ampulla,** which contains the sensory receptors (Figure 9–30a●). Hair cells attached to the wall of the ampulla form a raised structure known as a *crista* (Figure 9–30b●).

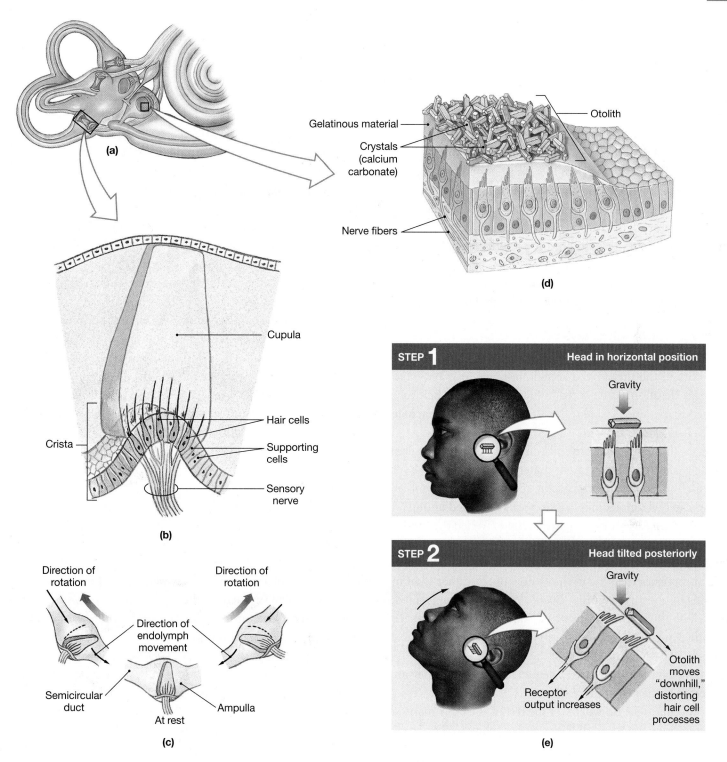

● FIGURE 9–30 The Vestibular Complex. (a) This anterior view of the right semicircular ducts, utricle, and saccule shows the location of the sensory receptors. **(b)** A cross section through the ampulla of a semicircular duct reveals the relationship of hair cells to the cupula and the crista. **(c)** Endolymph movement along the axis of the semicircular duct moves the cupula and stimulates the hair cells. **(d)** This enlargement reveals the structure of a macula. **(e)** This diagram depicts the function of an otolith when the head is level and tilted back.

The stereocilia of the hair cells are embedded in a gelatinous structure called the *cupula* (KŪ-pū-luh), which nearly fills the ampulla. When the head rotates in the plane of the semicircular duct, movement of the endolymph pushes against this structure and stimulates the hair cells (Figure 9–30c●).

Each semicircular duct responds to one of three possible rotational movements. To distort the cupula and stimulate the receptors, endolymph must flow along the axis of the duct; such flow will occur only when there is rotation in that plane. A horizontal rotation, as in shaking the head "no," stimulates the hair cells of the lateral semicircular duct. Nodding "yes" excites receptors of the anterior duct, and tilting the head from side to side activates receptors in the posterior duct. The three planes monitored by the semicircular ducts correspond to the three dimensions in the world around us, and they provide accurate information about even the most complex movements.

The Vestibule: Gravity and Linear Acceleration

Receptors in the utricle and saccule respond to gravity and linear acceleration. As depicted in Figure 9–30a, the hair cells of the utricle and saccule are clustered in oval **maculae** (MAK-ū-lē; *macula*, spot). As in the ampullae, the hair cell processes in the maculae are embedded in a gelatinous mass, but the surface of this gelatinous material contains a thin layer of densely packed calcium carbonate crystals. The complex as a whole (gelatinous mass and crystals) is called an *otolith* (*oto-*, ear + *lithos*, a stone) (Figure 9–30d●). When the head is in the normal, upright position, the otolith crystals sit atop the macula. Their weight presses down on the macular surface, and pushes the sensory hairs downward rather than to one side or another (Figure 9–30e●). When the head is tilted, the pull of gravity on the otolith crystals shifts their weight to the side, which distorts the sensory hairs. The change in receptor activity tells the CNS that the head is no longer level.

Otolith crystals are relatively dense and heavy, and they are connected to the rest of the body only by the sensory processes of the macular cells. So whenever the rest of the body makes a sudden movement, the otolith crystals lag behind. When an elevator starts downward, for example, we are immediately aware of it because the otolith crystals no longer push so forcefully against the surfaces of the receptor cells. Once they catch up and the elevator has reached a constant speed, we are no longer aware of any movement until the elevator brakes to a halt. As the body slows down, the otolith crystals press harder against the hair cells and we "feel" the force of gravity increase.

A similar mechanism accounts for our perception of linear acceleration in a car that speeds up suddenly. The otoliths lag behind, which distorts the sensory hairs and changes the activity in the sensory neurons. A comparable otolith movement occurs when the chin is raised and gravity pulls the otoliths backward. On the basis of visual information, the brain decides whether the arriving sensations indicate acceleration or a change in head position.

Pathways for Equilibrium Sensations

Hair cells of the vestibule and of the semicircular canals are monitored by sensory neurons whose fibers form the **vestibular branch** of the vestibulocochlear nerve (N VIII). These fibers synapse on neurons in the *vestibular nuclei* located at the boundary between the pons and the medulla oblongata. The two vestibular nuclei (1) integrate sensory information that arrives from each side of the head; (2) relay information to the cerebellum; (3) relay information to the cerebral cortex, which provides a conscious sense of position and movement; and (4) send commands to motor nuclei in the brain stem and in the spinal cord. These reflexive motor commands are distributed to the motor nuclei for cranial nerves involved with eye, head, and neck movements (N III, IV, VI, and XI). Descending instructions along the *vestibulospinal tracts* of the spinal cord adjust peripheral muscle tone to complement the reflexive movements of the head or neck.

Hearing

The receptors of the cochlear duct provide us with a sense of hearing that enables us to detect the quietest whisper, yet remain functional in a crowded, noisy room. The receptors responsible for auditory sensations are hair cells similar to those of the vestibular complex. However, both their placement within the cochlear duct and the organization of the surrounding accessory structures shield them from stimuli other than sound. In conveying vibrations from the tympanic membrane to the oval window, the auditory ossicles convert sound energy (pressure waves) in air to pressure pulses in the perilymph of the cochlea. These pressure pulses stimulate hair cells along the cochlear spiral. The *frequency* (pitch) of the perceived sound is determined by *which part* of the cochlear duct is stimulated. The *intensity* (volume) of the perceived sound is determined by *how many* hair cells at that location are stimulated.

The Cochlear Duct

In sectional view, the cochlear duct, or *scala media*, lies between a pair of chambers that contain perilymph: the **vestibular duct** (*scala vestibuli*) and the **tympanic duct** (*scala tympani*) (Figure 9–31a●). The vestibular and tympanic ducts are interconnected at the tip of the cochlear spiral. The outer surfaces of these ducts are encased by the bony labyrinth every-

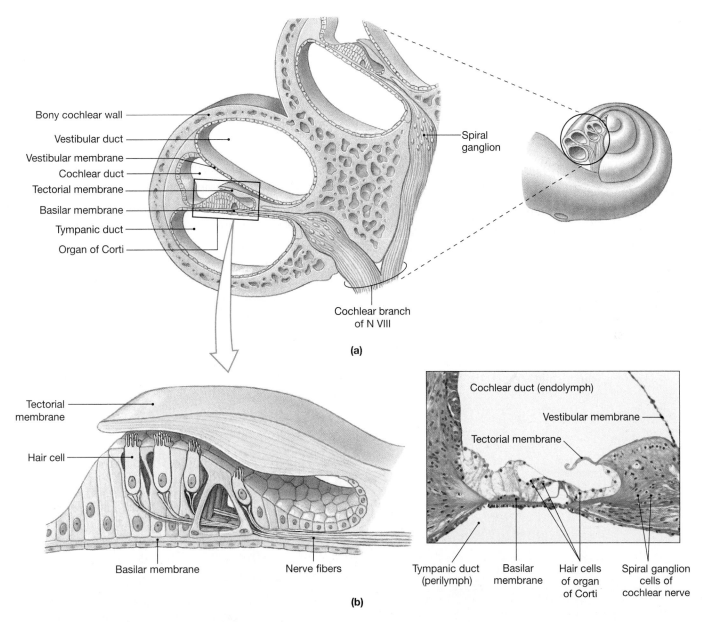

Bony cochlear wall

Vestibular duct

Vestibular membrane

Cochlear duct

Tectorial membrane

Basilar membrane

Tympanic duct

Organ of Corti

Spiral ganglion

Cochlear branch of N VIII

(a)

Tectorial membrane

Hair cell

Basilar membrane

Nerve fibers

Cochlear duct (endolymph)

Vestibular membrane

Tectorial membrane

Tympanic duct (perilymph)

Basilar membrane

Hair cells of organ of Corti

Spiral ganglion cells of cochlear nerve

(b)

● **FIGURE 9–31 The Cochlea and Organ of Corti.** (**a**) A section of the cochlea reveals its internal structures, including the organ of Corti. (**b**) The drawing shows the three-dimensional structure of the tectorial membrane and hair cell complex of the organ of Corti; the photomicrograph shows the actual structures. (LM × 1233)

where except at the oval window (the base of the vestibular duct) and the round window (the base of the tympanic duct).

THE ORGAN OF CORTI.   The hair cells of the cochlear duct are located in the **organ of Corti** (Figure 9–31b●). This sensory structure sits above the **basilar membrane,** which separates the cochlear duct from the underlying tympanic duct. The hair cells are arranged in a series of longitudinal rows, with their stereocilia in contact with the overlying **tectorial membrane** (tek-TOR-ē-al; *tectum,* roof). This membrane is firmly attached to the inner wall of the cochlear duct. When a portion of the basilar membrane bounces up and down in response to pressure waves in the

perilymph, the stereocilia of the hair cells are distorted as they are pushed up against the tectorial membrane.

The Hearing Process

Hearing is the detection of sound, which consists of waves of pressure that are conducted through a medium such as air or water. Physicists use the term **cycles** rather than waves, and the number of cycles per second (cps)—or **hertz (Hz)**—represents the **frequency** of the sound. What we perceive as the **pitch** of a sound (how high or low it is) is our sensory response to its frequency. A sound of high frequency (high pitch) might have a frequency of 15,000 Hz or more; a sound of low frequency (low pitch) could have a frequency of 100 Hz or less. The amount of energy, or power, of a sound determines its *intensity*, or volume. Intensity is reported in **decibels** (DES-i-belz). Some examples of different sounds and their intensities include a soft whisper (30 decibels), a refrigerator (50 decibels), a gas lawn mower (90 decibels), a chain saw (100 decibels), and a jet plane (140 decibels).

Hearing can be divided into six basic steps, diagrammed in Figure 9–32●.

Step 1: *Sound waves arrive at the tympanic membrane.* Sound waves enter the external acoustic canal and travel toward the tympanic membrane. Sound waves that approach the side of the head have direct access to the tympanic membrane on that side, whereas sounds that arrive from another direction must bend around corners or pass through the auricle or other body tissues.

Step 2: *Movement of the tympanic membrane causes displacement of the auditory ossicles.* The tympanic membrane provides the surface for sound collection. It vibrates to sound waves with frequencies between approximately 20 and 20,000 Hz (in a young child). When the tympanic membrane vibrates, so do the malleus and (through their articulations) the incus and stapes.

Step 3: *The movement of the stapes at the oval window establishes pressure waves in the perilymph of the vestibular duct.* When the stapes moves, it applies pressure to the perilymph of the vestibular duct. Because the rest of the cochlea is sheathed in bone, pressure applied at the oval window can be relieved only at the round window. When the stapes moves inward, the membrane that spans the round window bulges outward. As the stapes moves in and out, and vibrates at the frequency of the sound at the tympanic membrane, it creates pressure waves within the perilymph.

Step 4: *The pressure waves distort the basilar membrane on their way to the round window of the tympanic duct.* These pressure waves cause movement in the basilar membrane. The basilar membrane does not have the same structure throughout its length. Near the oval window, it is narrow and stiff; at its terminal end, it is wider and more flexible. As a result, the location of maximum stimulation varies with the frequency of the sound. High-frequency sounds vibrate

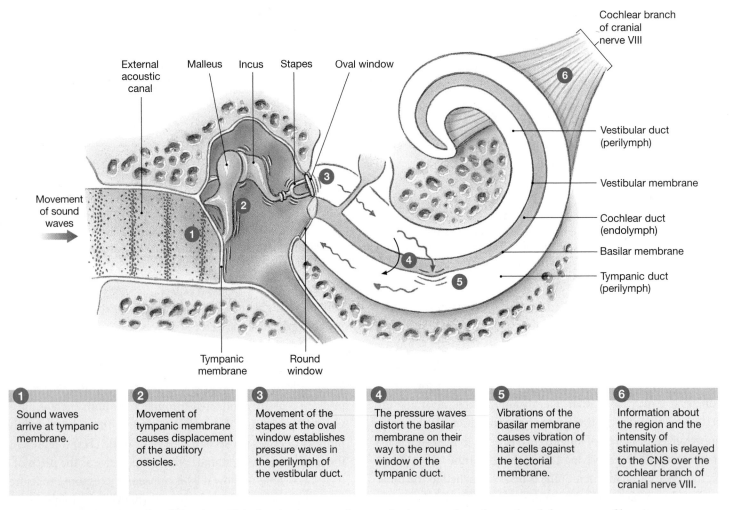

1	**2**	**3**	**4**	**5**	**6**
Sound waves arrive at tympanic membrane.	Movement of tympanic membrane causes displacement of the auditory ossicles.	Movement of the stapes at the oval window establishes pressure waves in the perilymph of the vestibular duct.	The pressure waves distort the basilar membrane on their way to the round window of tympanic duct.	Vibrations of the basilar membrane causes vibration of hair cells against the tectorial membrane.	Information about the region and the intensity of stimulation is relayed to the CNS over the cochlear branch of cranial nerve VIII.

● **FIGURE 9–32 Sound and Hearing.** This drawing presents the steps in the reception of sound and the process of hearing.

the basilar membrane near the oval window. The lower the frequency of the sound, the farther from the oval window is the area of maximum distortion. The actual *amount* of movement at a given location depends on the amount of force applied by the stapes. The louder the sound, the more the basilar membrane moves.

Step 5: *Vibration of the basilar membrane causes vibration of hair cells against the tectorial membrane.* The vibration of the affected region of the basilar membrane moves hair cells against the tectorial membrane. The displacement of the hair cells results in the release of neurotransmitters and the stimulation of sensory neurons. The hair cells are arranged in several rows; a very soft sound may stimulate only a few hair cells in a portion of one row. As the volume of a sound increases, not only do these hair cells become more active, but also additional hair cells—at first in the same row and then in adjacent rows—are stimulated as well. The number of hair cells that respond in a given region of the organ of Corti thus provides information on the intensity of the sound.

Step 6: *Information about the region and intensity of stimulation is relayed to the CNS over the cochlear branch of N VIII.* The cell bodies of the sensory neurons that monitor the cochlear hair cells are located at the center of the bony cochlea (see Figure 9–32a●) in the *spiral ganglion*. From there, the information is carried by the cochlear branch of cranial nerve VIII to the cochlear nuclei of the medulla oblongata for distribution to other centers in the brain.

Auditory Pathways

Hair cell stimulation activates sensory neurons whose cell bodies are in the nearby spiral ganglion. Their afferent fibers (axons) form the **cochlear branch** of the vestibulocochlear nerve (N VIII) (Figure 9–33●). These axons enter the medulla oblongata and synapse at the *cochlear nucleus*. From there, the information crosses to the opposite side of the brain and ascends to the *inferior colliculus* of the midbrain. This processing center coordinates a number of responses to acoustic stimuli, including auditory reflexes that involve skeletal muscles of the head, face, and trunk. For example, these reflexes automatically change the position of your head in response to a sudden loud noise.

Before reaching the cerebral cortex and your conscious awareness, ascending auditory sensations synapse in the thalamus. Thalamic fibers then deliver the information to the auditory cortex of the temporal lobe. In effect, the auditory cortex contains a map of the organ of Corti. High-frequency sounds activate one portion of the cortex, and low-frequency sounds affect another. If the auditory cortex is damaged, the individual will respond to sounds and have

normal acoustic reflexes, but sound interpretation and pattern recognition will be difficult or impossible. Damage to the adjacent association area leaves the ability to detect the tones and patterns unaffected but produces an inability to comprehend their meaning.

Auditory Sensitivity

Our hearing abilities are remarkable, but it is difficult to assess the absolute sensitivity of the system. The range from the softest audible sound to the loudest tolerable blast represents a trillionfold increase in power. The receptor mechanism is so sensitive that if we were to remove the stapes, we could, in theory, hear air molecules that bounce off the oval window. We never utilize the full potential of this system, because body movements and our internal organs produce squeaks, groans, thumps, and other sounds that are tuned out by adaptation. When other environmental noises fade away, the level of adaptation drops and the system becomes increasingly sensitive. If we relax in a quiet room, our heartbeat seems to get louder and louder as the auditory system adjusts to the lower level of background noise.

Clinical Note
HEARING LOSS

Hearing occurs by air conduction and bone conduction. Hearing loss is a common problem, and affects 5–10 percent of the general population. Problems in the external auditory canal and the middle ear cause *conductive hearing losses*, while problems in the inner ear or *vestibulocochlear nerve (CN VIII)* cause *sensorineural hearing loss*.

Conductive hearing loss results from blockage of the external auditory canal, damage to the tympanic membrane, disruption of the auditory ossicles, or from fluid or scarring within the middle ear. Sensorineural hearing loss is primarily due to damage to the hair cells of the organ of Corti. Causes include intense noise, infections, ototoxic drugs, fracture of the temporal bone, Ménière's disease, and aging.

The type of hearing loss can be differentiated by comparing the threshold of hearing by air conduction to that of bone conduction (*Rinne's test*). For this, a tuning fork is struck and placed near the ear. Then the tuning fork is struck and the stem placed on the mastoid process. Normally, air conduction is louder than bone conduction. When bone conduction is louder than air conduction, a conductive loss is suspected. With a sensorineural hearing loss, both are reduced.

Sensorineural loss can be detected with *Weber's test*. For this, a tuning fork is struck and the stem placed on the head in the midline. The tone should be heard equally in both ears. With unilateral conductive hearing loss, the tone is perceived in the affected ear. With unilateral sensorineural loss, the tone is perceived in the unaffected ear. ■

Key Note

Balance and hearing rely on the same basic types of sensory receptors (hair cells). The nature of the stimulus that stimulates a particular group of hair cells depends on the structure of the associated sense organ. In the semicircular ducts, the stimulus is fluid movement caused by head rotation in the horizontal, sagittal, or frontal planes. In the utricle and saccule, the stimulus is gravity-induced shifts in the position of attached otoliths. In the cochlea, the stimulus is movement of the basilar membrane by pressure waves.

→ **CONCEPT CHECK QUESTIONS**

1. If the round window were not able to bulge out with increased pressure in the perilymph, how would sound perception be affected?
2. How would the loss of stereocilia from the hair cells of the organ of Corti affect hearing?

Answers begin on p. 792.

■ Aging and the Senses

The general lack of replacement of neurons leads to an inevitable decline in sensory function with age. Although increases in stimulus strength can compensate for part of this functional decline, the loss of axons necessary for conducting sensory action potentials cannot compensate to the same degree. Next we consider the toll of aging on various special senses.

Smell

Unlike populations of other neurons, the population of olfactory receptor cells is regularly replaced by the division of stem cells in the olfactory epithelium. Despite this process, the total number of receptors declines with age, and the remaining receptors be-

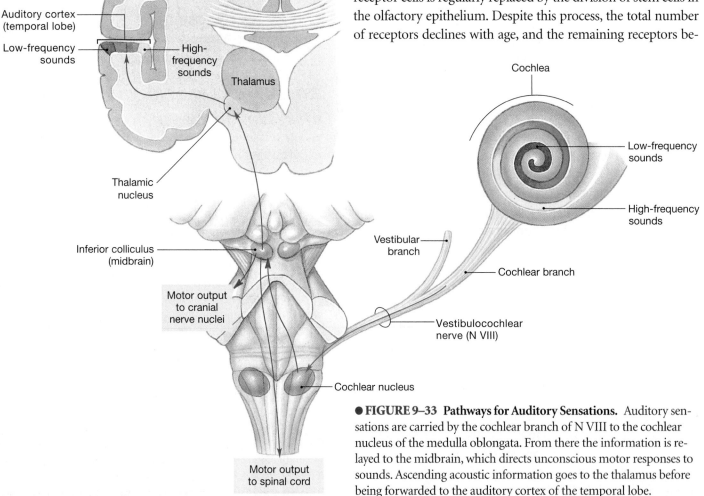

● **FIGURE 9–33 Pathways for Auditory Sensations.** Auditory sensations are carried by the cochlear branch of N VIII to the cochlear nucleus of the medulla oblongata. From there the information is relayed to the midbrain, which directs unconscious motor responses to sounds. Ascending acoustic information goes to the thalamus before being forwarded to the auditory cortex of the temporal lobe.

come less sensitive. As a result, elderly individuals have difficulty detecting odors in low concentrations. This drop in the number of receptors explains "Grandmother's" tendency to apply too much perfume and why "Grandfather's" aftershave lotion seems so overpowering. They must apply more to be able to smell it.

Taste

Tasting ability declines with age due to the thinning of mucous membranes and a reduction in the number and sensitivity of taste buds. We begin life with more than 10,000 taste buds, but that number begins declining dramatically by age 50. The sensory loss becomes especially significant because aging individuals also experience a decline in the number of olfactory receptors. As a result, many of the elderly find that their food tastes bland and unappetizing. Children, however, find the same food too spicy.

Vision

Various disorders of vision are associated with normal aging; the most common involve the lens and the neural part of the retina. With age, the lens loses its elasticity and stiffens. As a result, seeing objects up close becomes more difficult, and older individuals become farsighted—a condition called *presbyopia*. ∞ p. 346 For example, the inner limit of clear vision, known as the *near point of vision*, changes from 7–9 cm in children to 15–20 cm in young adults and typically reaches 83 cm by age 60. As noted earlier, the most common cause of the development of a cataract (the loss of transparency in the lens) is advancing age. Such cataracts are called *senile*

cataracts. ∞ p. 344 In addition to changes in the near point of vision and some changes in lens transparency, a gradual loss of rods occurs with age. This reduction explains why individuals over age 60 need almost twice as much light for reading than individuals at age 40.

Another contributor to the loss of vision with age is *macular degeneration*, which is the leading cause of blindness in persons over 50. This condition is typically associated with the growth and proliferation of blood vessels in the retina. The leakage of blood from these abnormal vessels causes retinal scarring and a loss of photoreceptors. The vascular proliferation begins in the macula lutea, which is the area of the retina correlated with acute vision. Color vision is affected as the cones deteriorate.

Hearing

Hearing is generally affected less by aging than are the other senses. However, because the tympanic membrane loses some of its elasticity, it becomes more difficult to hear high-pitched sounds. The progressive loss of hearing that occurs with aging is called *presbycusis* (prez-bē-KŪ-sis; *presbys*, old man + *akousis*, hearing).

> ### → CONCEPT CHECK QUESTIONS
>
> 1. How can a given food be both too spicy for a child and too bland for an elderly individual?
> 2. Explain why we have an increasingly difficult time seeing close objects as we age.
>
> *Answers begin on p. 792.*

Chapter Review

Access more review material online at *www.prenhall.com/bledsoe*. There you will find quiz questions, labeling activities, animations, essay questions, and web links.

Key Terms

accommodation 345	**iris** 338	**proprioception** 328
cochlea 353	**macula** 356	**pupil** 338
fovea 342	**nociceptors** 328	**retina** 340
gustation 328	**olfaction** 328	**sclera** 338

Related Clinical Terms

analgesic A drug that relieves pain without eliminating sensitivity to other stimuli, such as touch or pressure.

anesthesia A total or partial loss of sensation.

cataract Opacity (loss of transparency) of the lens.

color blindness A condition in which a person is unable to distinguish certain colors.

conductive deafness Deafness that results from conditions in the outer or middle ear that block the transfer of vibrations from the tympanic membrane to the oval window.

glaucoma A condition characterized by increased fluid pressure within the eye due to the impaired reabsorption of aqueous humor; can result in blindness.

hyperopia, or *farsightedness* A condition in which nearby objects are blurry but distant objects are clear.

Ménière's disease A condition in which high fluid pressures rupture the walls of the membranous labyrinth, which results in acute vertigo (an inappropriate sense of motion) and inappropriate auditory sensations.

myopia, or *nearsightedness* A condition in which vision at close range is normal but distant objects appear blurry.

nerve deafness Deafness that results from problems within the cochlea or along the auditory pathway.

nystagmus Abnormal eye movements that may appear after the brain stem or inner ear is damaged.

opthalmology (of-thal-MOL-o-jē) The study of the eye and its diseases.

presbyopia A type of hyperopia that develops with age as the lens becomes less elastic.

retinitis pigmentosa A group of inherited retinopathies (see the next term) characterized by the progressive deterioration of photoreceptors, which eventually results in blindness.

retinopathy (ret-i-NOP-ah-thē) A disease or disorder of the retina.

scotomas (skō-TŌ-muhz) Abnormal blind spots in the field of vision (that is, those not caused by the optic disc).

strabismus Deviations in the alignment of the eyes to each other; one or both eyes turn inward or outward.

Summary Outline

1. The **general senses** are temperature, pain, touch, pressure, vibration, and proprioception; receptors for these sensations are distributed throughout the body. Receptors for the **special senses** (smell, taste, vision, balance, and hearing) are located in specialized areas or in sense organs.

2. A *sensory receptor* is a specialized cell that, when stimulated, sends a sensation to the CNS. The simplest receptors are **free nerve endings;** the most complex have specialized accessory structures that isolate the receptors from all but a specific type of stimulus.

3. Each receptor cell monitors a specific *receptive field. (Figure 9–1)*

4. Sensory information is relayed in the form of action potentials in a sensory (afferent) fiber. In general, the larger the stimulus, the greater is the frequency of action potentials. The CNS interprets the nature of the arriving sensory information on the basis of the area of the brain stimulated.

5. **Adaptation**—a reduction in sensitivity in the presence of a constant stimulus—involves changes in receptor sensitivity or inhibition along sensory pathways.

Key Note 328

THE GENERAL SENSES 328

Pain 328

1. **Nociceptors** respond to a variety of stimuli usually associated with tissue damage. The two types of these painful sensations are **fast pain,** or *prickling pain,* and **slow pain,** or *burning and aching pain.*

2. The perception of pain in parts of the body that are not actually stimulated is called **referred pain.** *(Figures 9–2, 9–3)*

Temperature 329

3. **Thermoreceptors** respond to changes in temperature.

Touch, Pressure, and Position 329

4. **Mechanoreceptors** respond to physical distortion of, contact with, or pressure on their cell membranes; **tactile receptors** respond to touch, pressure, and vibration; **baroreceptors** respond to pressure changes in the walls of blood vessels, the digestive and urinary tracts, and the lungs; and **proprioceptors** respond to positions of joints and muscles.

5. **Fine touch and pressure receptors** provide detailed information about a source of stimulation; **crude touch and pressure receptors** are poorly localized. Important tactile receptors include the *root hair plexus, tactile discs, tactile corpuscles, lamellated corpuscles,* and *Ruffini corpuscles. (Figure 9–4)*

6. Baroreceptors in the walls of major arteries and veins respond to changes in blood pressure, and those along the digestive tract help coordinate reflex activities of digestion. *(Figure 9–5)*

7. Proprioceptors monitor the position of joints, tension in tendons and ligaments, and the state of muscular contraction. Proprioceptors include Golgi tendon organs and muscle spindles.

Chemical Detection 332

8. In general, **chemoreceptors** respond to water-soluble and lipid-soluble substances dissolved in the surrounding fluid. They monitor the chemical composition of body fluids. *(Figure 9–6)*

THE SPECIAL SENSES 333

SMELL 333

1. The **olfactory organs** consist of an **olfactory epithelium** that contains **olfactory receptor cells** (neurons sensitive to chemicals dissolved in the overlying mucus), supporting cells, and *basal* (stem) *cells.* Their surfaces are coated with the secretions of the **olfactory glands.** *(Figure 9–7)*

2. The olfactory receptors are modified neurons.

The Olfactory Pathways 334

3. The olfactory system has extensive limbic and hypothalamic connections.

TASTE 334

1. **Taste (gustatory) receptors** are clustered in **taste buds;** each taste bud contains **gustatory cells,** which extend *taste hairs* through a narrow **taste pore.** *(Figure 9–8)*

2. Taste buds are associated with **papillae,** which are epithelial projections on the superior surface of the tongue. *(Figure 9–8)*

3. The **primary taste sensations** are sweet, salty, sour, and bitter; umami and water receptors are also present. *(Figure 9–8)*

The Taste Pathways 335

4. The taste buds are monitored by cranial nerves that synapse within a nucleus of the medulla oblongata.

Key Note 336

VISION 336

The Accessory Structures of the Eye 336

1. The **accessory structures** of the eye include the eyelids and associated exocrine glands, the superficial epithelium of the eye, structures associated with the production and removal of tears, and the extrinsic eye muscles.

2. An epithelium called the **conjunctiva** covers most of the exposed surface of the eye except the transparent **cornea.**

3. The secretions of the **lacrimal gland** bathe the conjunctiva; these secretions contain *lysozyme* (an enzyme that attacks bacteria). Tears reach the nasal cavity after passing through the **lacrimal canals,** the **lacrimal sac,** and the **nasolacrimal duct.** *(Figure 9–9)*

4. Six **extrinsic eye muscles** control external eye movements: the **inferior** and **superior rectus,** the **lateral** and **medial rectus,** and the **superior** and **inferior obliques.** *(Figure 9–10; Table 9–1)*

The Eye 337

5. The eye has three layers: an outer fibrous tunic, a vascular tunic, and an inner neural tunic. Most of the ocular surface is covered by the **sclera** (a dense fibrous connective tissue), which is continuous with the cornea, both of which are part of the **fibrous tunic.** *(Figure 9–11)*

6. The **vascular tunic** includes the **iris,** the **ciliary body,** and the **choroid.** The iris forms the boundary between the eye's anterior and posterior chambers. The iris regulates the amount of light that enters the eye. The ciliary body contains the *ciliary muscle* and the *ciliary processes,* which attach to the **suspensory ligaments** of the **lens.** *(Figures 9–11, 9–12)*

7. The **neural tunic** consists of an outer *pigmented part* and an inner *neural part*; the latter contains visual receptors and associated neurons. *(Figures 9–11, 9–13)*

8. From the photoreceptors, the information is relayed to **bipolar cells,** then to **ganglion cells,** and to the brain by the optic nerve. Horizontal cells and amacrine cells modify the signals passed between other retinal components. *(Figure 9–14)*

9. The ciliary body and lens divide the interior of the eye into a large **posterior cavity** and a smaller **anterior cavity.** The anterior cavity is subdivided into the **anterior chamber,** which extends from the cornea to the iris, and a **posterior chamber** between the iris and the ciliary body and lens. The posterior cavity contains the *vitreous body,* a gelatinous mass that helps stabilize the shape of the eye and supports the retina. *(Figures 9–15, 9–16)*

10. **Aqueous humor** circulates within the eye and reenters the circulation after diffusing through the walls of the anterior chamber and into veins of the sclera through the *canal of Schlemm.* *(Figure 9–15)*

11. The lens, held in place by the suspensory ligaments, focuses a visual image on the retinal receptors. Light is refracted (bent) when it passes through the cornea and lens. During **accommodation,** the shape of the lens changes to focus an image on the retina. *(Figures 9–17, 9–18, 9–19)*

Key Note 345

Visual Physiology 345

12. Light is radiated in waves with a characteristic wavelength. A *photon* is a single energy packet of visible light. The two types of **photoreceptors** (visual receptors of the retina) are **rods** and **cones.** Rods respond to almost any photon, regardless of its energy content; cones have characteristic ranges of sensitivity. Many cones are densely packed within the **fovea** (the central portion of the **macula lutea**), the site of sharpest vision. *(Figure 9–21)*

13. Each photoreceptor contains an outer segment with membranous **discs** that contain **visual pigments.** Light absorption occurs in the visual pigments, which are derivatives of **rhodopsin** (**opsin** plus the pigment **retinal,** which is synthesized from **vitamin A**). A photoreceptor responds to light by changing its rate of neurotransmitter release and thereby altering the activity of a bipolar cell. *(Figures 9–21, 9–22)*

The Visual Pathway 347

14. The message is relayed from photoreceptors to bipolar cells to ganglion cells within the retina. The axons of ganglion cells converge at the optic disc and leave the eye as the optic nerve. A partial crossover occurs at the optic chiasm before the information reaches a nucleus in the thalamus on each side of the brain. From these nuclei, visual information is relayed to the visual cortex of the occipital lobe, which contains a sensory map of the field of vision. *(Figures 9–23 through 9–26)*

EQUILIBRIUM AND HEARING 351

1. The senses of equilibrium (**dynamic equilibrium** and **static equilibrium**) and hearing are provided by the receptors of the **inner ear.** Its chambers and canals contain the fluid **endolymph.** The **bony labyrinth** surrounds and protects the **membranous labyrinth,** and the space between them contains the fluid **perilymph.** The bony labyrinth consists of the **vestibule,** the **semicircular canals** (receptors in the vestibule and semicircular canals provide the sense of equilibrium), and the **cochlea** (these receptors provide the sense of hearing). The structures and air spaces of the **external ear** and **middle ear** help capture and transmit sound to the cochlea. *(Figures 9–27, 9–28, 9–29)*

The Anatomy of the Ear 351

2. The external ear includes the **auricle** (*pinna*), which surrounds the entrance to the **external acoustic canal,** which ends at the **tympanic membrane (eardrum).** *(Figures 9–27, 9–28)*

3. The middle ear is connected to the nasopharynx by the **auditory tube** (*pharyngotympanic tube* or *Eustachian tube*). The middle ear encloses and protects the **auditory ossicles,** which connect the tympanic membrane with the receptor complex of the inner ear. *(Figures 9–27, 9–28)*

4. The vestibule includes a pair of membranous sacs, the **saccule** and **utricle,** whose receptors provide sensations of gravity and linear acceleration. The semicircular canals contain the **semicircular**

ducts, whose receptors provide sensations of rotation. The cochlea contains the **cochlear duct,** which is an elongated portion of the membranous labyrinth. *(Figure 9–29)*

5. The basic receptors of the inner ear are **hair cells,** whose surfaces support *stereocilia*. Hair cells provide information about the direction and strength of mechanical stimuli. *(Figure 9–29)*

Equilibrium 353

6. The **anterior, posterior,** and **lateral semicircular ducts** are attached to the utricle. Each semicircular duct contains an **ampulla** with sensory receptors. There the stereocilia contact the *cupula,* a gelatinous mass that is distorted when endolymph flows along the axis of the duct. *(Figures 9–29, 9–30a,b,c)*

7. In the saccule and utricle, hair cells cluster within **maculae,** where their cilia contact the *otolith* (densely packed mineral crystals in a gelatinous mass). When the head tilts, the otoliths shift, and the resulting distortion in the sensory hairs signals the CNS. *(Figure 9–30d,e)*

8. The vestibular receptors activate sensory neurons whose axons form the **vestibular branch** of the vestibulocochlear nerve (N VIII).

Hearing 356

9. Sound waves travel toward the tympanic membrane, which vibrates; the auditory ossicles amplify and conduct the vibrations to the inner ear. Movement at the oval window applies pressure to the perilymph of the **vestibular duct.** *(Figures 9–31, 9–32)*

10. Pressure waves distort the **basilar membrane** and push the hair cells of the **organ of Corti** against the **tectorial membrane.** *(Figure 9–32)*

11. The sensory neurons are located in the **spiral ganglion** of the cochlea. Afferent fibers of sensory neurons form the **cochlear branch** of the vestibulocochlear nerve (N VIII), and synapse at the cochlear nucleus. *(Figure 9–33)*

Key Note 360

AGING AND THE SENSES 360

1. As part of the aging process, there are (1) gradual reductions in smell and taste sensitivity, (2) a tendency toward *presbyopia* and cataract formation in the eyes, and (3) a progressive loss of hearing (*presbycusis*).

Review Questions

Level 1: Reviewing Facts and Terms

Match each item in column A with the most closely related item in column B. Place letters for answers in the spaces provided.

COLUMN A
___ 1. myopia
___ 2. fibrous tunic
___ 3. nociceptors
___ 4. proprioceptors
___ 5. cones
___ 6. accommodation
___ 7. tympanic membrane
___ 8. thermoreceptors
___ 9. rods
___ 10. olfaction
___ 11. fovea
___ 12. hyperopia
___ 13. maculae
___ 14. semicircular ducts

COLUMN B
a. pain receptors
b. free nerve endings
c. sclera and cornea
d. rotational movements
e. provide information on joint position
f. color vision
g. site of sharpest vision
h. active in dim light
i. eardrum
j. change in lens shape to focus retinal image
k. nearsighted
l. farsighted
m. smell
n. gravity and acceleration receptors

15. Regardless of the nature of a stimulus, sensory information must be sent to the central nervous system in the form of:
    (a) dendritic processes.
    (b) action potentials.
    (c) neurotransmitter molecules.
    (d) generator potentials.

16. A reduction in sensitivity in the presence of constant stimulus is called:
    (a) transduction. (c) line labeling.
    (b) sensory coding. (d) adaptation.

17. Mechanoreceptors that detect pressure changes in the walls of blood vessels and in portions of the digestive, reproductive, and urinary tracts are:
    (a) tactile receptors. (c) proprioceptors.
    (b) baroreceptors. (d) free nerve endings.

18. Examples of proprioceptors that monitor the position of joints and the state of muscular contraction are:
    (a) lamellated and Meissner's corpuscles.
    (b) carotid and aortic sinuses.
    (c) Merkel's discs and Ruffini corpuscles.
    (d) Golgi tendon organs and muscle spindles.

19. When chemicals dissolve in the nasal cavity, they stimulate:
    (a) gustatory cells.   (c) rod cells.
    (b) olfactory hairs.   (d) tactile receptors.

20. Taste receptors are also known as:
    (a) tactile discs.   (c) hair cells.
    (b) gustatory receptors.   (d) olfactory receptors.

21. The purpose of tears produced by the lacrimal apparatus is to:
    (a) keep conjunctival surfaces moist and clean.
    (b) reduce friction and remove debris from the eye.
    (c) provide nutrients and oxygen to the conjunctival epithelium.
    (d) a, b, and c are correct.

22. The thickened gel-like substance that helps support the structure of the eyeball is the:
    (a) vitreous body.   (c) cupula.
    (b) aqueous humor.   (d) perilymph.

23. The retina is considered to be a component of the:
    (a) vascular tunic.   (c) neural tunic.
    (b) fibrous tunic.   (d) vascular, fibrous, and neural tunics.

24. At sunset your visual system adapts to _____ vision.
    (a) fovea   (c) macular
    (b) rod-based   (d) cone-based

25. The malleus, incus, and stapes are the tiny ear bones located in the:
    (a) outer ear.   (c) inner ear.
    (b) middle ear.   (d) membranous labyrinth.

26. Receptors in the saccule and utricle provide sensations of:
    (a) balance and equilibrium.
    (b) hearing.
    (c) vibration.
    (d) gravity and linear acceleration.

27. The organ of Corti is located within the _____ of the inner ear.
    (a) utricle   (c) vestibule
    (b) bony labyrinth   (d) cochlea

28. What three types of mechanoreceptors respond to stretching, compression, twisting, or other distortions of the cell membrane?

29. Identify six types of tactile receptors found in the skin and their sensitivities.

30. (a) What structures make up the fibrous tunic of the eye?
    (b) What are the functions of the fibrous tunic?

31. What structures are parts of the vascular tunic of the eye?

32. What six basic steps are involved in the process of hearing?

Level 2: Reviewing Concepts

33. The CNS interprets sensory information entirely on the basis of the:
    (a) strength of the action potential.
    (b) number of generator potentials.
    (c) area of brain stimulated.
    (d) a, b, and c are correct.

34. If the auditory cortex is damaged, the individual will respond to sounds and have normal acoustic reflexes, but:
    (a) the sounds may produce nerve deafness.
    (b) the auditory ossicle may be immobilized.
    (c) sound interpretation and pattern recognition may be impossible.
    (d) normal transfer of vibration to the oval window is inhibited.

35. Distinguish between the general senses and the special senses in the human body.

36. In what form does the CNS receive a stimulus detected by a sensory receptor?

37. Why are olfactory sensations long-lasting and an important part of our memories and emotions?

38. Jane makes an appointment with the optometrist for a vision test. Her test results are reported as 20/15. What does this test result mean? Is a rating of 20/20 better or worse?

Level 3: Critical Thinking and Clinical Applications

39. You are at a park watching some deer 35 feet away from you. Your friend taps you on the shoulder to ask a question. As you turn to look at your friend, who is standing 2 feet away, what changes will occur regarding your eyes?

40. After attending a Fourth of July fireworks extravaganza, Millie finds it difficult to hear normal conversation, and her ears keep "ringing." What is causing her hearing problems?

41. After riding the express elevator from the twentieth floor to the ground floor, for a few seconds you still feel as if you are descending, even though you have obviously come to a stop. Why?

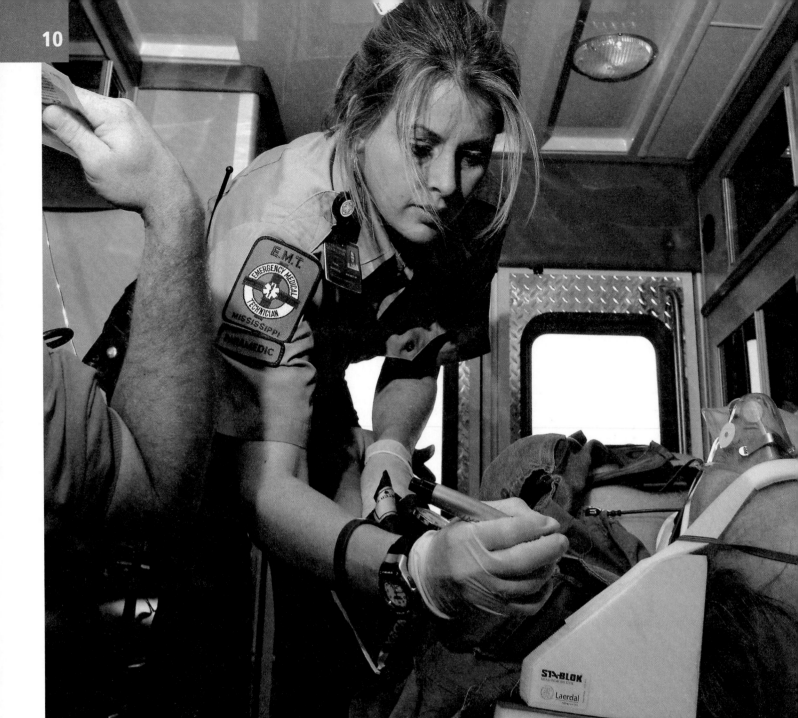

10 The Endocrine System

WHILE A GREAT DEAL of prehospital care occurs on scene, it is important to remember that patient care must continue until the patient is delivered to the hospital. It is essential that EMTs routinely and periodically re-evaluate the patient under their care. Re-evaluation is important in ensuring that all emergent problems have been addressed, but it is also important in determining whether the patient is improving or worsening and whether or not the patient is responding to the care provided.

Chapter Outline

Chapter Objectives

1. Compare the similarities between the endocrine and nervous systems. (pp. 367–368)
2. Compare the major chemical classes of hormones. (pp. 368–369)
3. Explain the general mechanisms of hormonal action. (pp. 369–372)
4. Describe how endocrine organs are controlled. (pp. 372–373)
5. Discuss the location, hormones, and functions of the following endocrine glands and tissues: pituitary gland, thyroid gland, parathyroid glands, adrenal glands, pineal gland, pancreas, kidneys, heart, thymus gland, testes, ovaries, and adipose tissues. (pp. 373–396)
6. Explain how hormones interact to produce coordinated physiological responses. (p. 396)
7. Identify the hormones that are especially important to normal growth and discuss their roles. (pp. 396–397)
8. Explain how the endocrine system responds to stress. (pp. 397–398)
9. Discuss the results of abnormal hormone production or abnormal responses. (pp. 398–399)
10. Discuss the functional relationships between the endocrine system and other body systems. (pp. 400–401)

Vocabulary Development

ad- to or toward; *adrenal*
andros man; *androgen*
angeion vessel; *angiotensin*
corpus body; *corpus luteum*
diabetes to pass through; *diabetes*
diourein to urinate; *diuresis*
erythros red; *erythropoietin*
infundibulum funnel; *infundibulum*

insipidus tasteless; *diabetes insipidus*
krinein to secrete; *endocrine*
lac milk; *prolactin*
mellitum honey; *diabetes mellitus*
natrium sodium; *natriuretic*
ouresis making water; *polyuria*
oxy- quick; *oxytocin*
para beyond; *parathyroid*

poiesis making; *erythropoietin*
pro- before; *prolactin*
renes kidneys; *adrenal*
synergia working together; *synergistic*
teinein to stretch; *angiotensin*
thyreos an oblong shield; *thyroid*
tokos childbirth; *oxytocin*
tropos turning; *gonadotropins*

TO MAINTAIN HOMEOSTASIS, every cell in the body must communicate with its neighbors and with cells and tissues in distant portions of the body. Most of this communication involves the release and receipt of chemical messages. Each living cell is continually "talking" to its neighbors by releasing chemicals into the extracellular fluid. These chemicals tell cells what their neighbors are doing at any given moment, and the result is the coordination of tissue function at the local level.

Cellular communication over greater distances is coordinated by the nervous and endocrine systems. The nervous system functions somewhat like a telephone company, and carries specific "messages" from one location to another inside the body. The source and the destination are quite specific, and the effects are short-lived. This form of communication is ideal for crisis management; if you are in danger of being hit by a speeding bus, the nervous system can coordinate and direct your leap to safety.

Many life processes, however, require long-term cellular communication. This type of regulation is provided by the endocrine system, which uses chemical messengers called *hormones* to relay information and instructions between cells. In such communication, hormones are like addressed letters, and the cardiovascular system is the postal service. Each hormone released into and distributed by the circulation has specific *target cells* that will respond to its presence. These target cells possess the receptors required for binding and "reading" the hormonal message when it arrives. But hormones are really like bulk mail advertisements—cells throughout the body are exposed to them whether or not they have the necessary receptors. At any moment, each individual cell can respond to only a few of the hormones present. The other hormones are ignored, because the cell lacks the receptors needed to read the messages they contain.

Because target cells can be anywhere in the body, a single hormone can alter the metabolic activities of multiple tissues and organs simultaneously. The effects may be slow to appear, but they often persist for days. This persistence makes hormones effective in coordinating cell, tissue, and organ activities on a sustained, long-term basis. For example, circulating hormones keep body water content and levels of electrolytes and organic nutrients within normal limits 24 hours a day throughout our entire lives.

While the effects of a single hormone persist, a cell may receive additional instructions from other hormones. The result is a further modification in cellular operations. Gradual changes in the quantities and identities of circulating hormones can produce complex changes in physical structure and physiological capabilities. Examples are the processes of embryological and fetal development, growth, and puberty.

When viewed from a broad perspective, the differences between the nervous and endocrine systems seem relatively clear. In fact, these broad organizational and functional distinctions are the basis for treating them as two separate systems. Yet when considered in detail, the two systems are organized along parallel lines. For example:

- Both systems rely on the release of chemicals that bind to specific receptors on target cells.
- Both systems share various chemical messengers; for example, norepinephrine and epinephrine are called *hormones* when released into the bloodstream and *neurotransmitters* when released across synapses.
- Both systems are primarily regulated by negative feedback control mechanisms.
- Both systems coordinate and regulate the activities of other cells, tissues, organs, and systems and maintain homeostasis.

This chapter introduces the components and functions of the endocrine system and explores the interactions between the nervous and endocrine systems.

■ An Overview of the Endocrine System

The endocrine system includes all of the endocrine cells and tissues of the body. As noted in Chapter 4, **endocrine cells** are glandular secretory cells that release their secretions into the extracellular fluid. This feature distinguishes them from *exocrine cells*, which secrete onto epithelial surfaces. ∞ p. 96 The chemicals released by endocrine cells may affect adjacent cells only—as is the case of *cytokines* (or *local hormones*) such as *prostaglandins*—or they may affect cells throughout the body. We define **hormones** as chemical messengers that are released in one tissue and transported by the bloodstream to reach target cells in other tissues.

The tissues and organs of the endocrine system, and some of the major hormones they produce, are introduced in Figure 10–1●. Some of these organs, such as the pituitary gland, have endocrine secretion as a primary function. Others, such as the pancreas, have many other functions besides endocrine secretion; chapters on other organ systems consider such endocrine organs in more detail.

The Structure of Hormones

Based on their chemical structures, hormones can be divided into the following three groups:

1. *Amino acid derivatives.* Some hormones are relatively small molecules that are structurally similar to amino acids. (Amino acids, the building blocks of proteins, were introduced in Chapter 2.) ∞ p. 45 This group includes *epinephrine, norepinephrine,* the *thyroid hormones,* and *melatonin.*

2. *Peptide hormones.* **Peptide hormones** consist of chains of amino acids. These molecules range from short amino acid chains, such as *ADH* and *oxytocin,* to small proteins such as *growth hormone* and *prolactin.* This is the largest class of hormones and includes all of the hormones secreted by the hypothalamus, pituitary gland, heart, kidneys, thymus, digestive tract, and pancreas.

3. *Lipid derivatives.* There are two classes of lipid-based hormones: steroid hormones and eicosanoids (ī-KŌ-sa-noydz). **Steroid hormones** are lipids that are derived from (and, thus, are structurally similar to) cholesterol, which is a lipid introduced in Chapter 2. ∞ p. 45 Steroid hor-

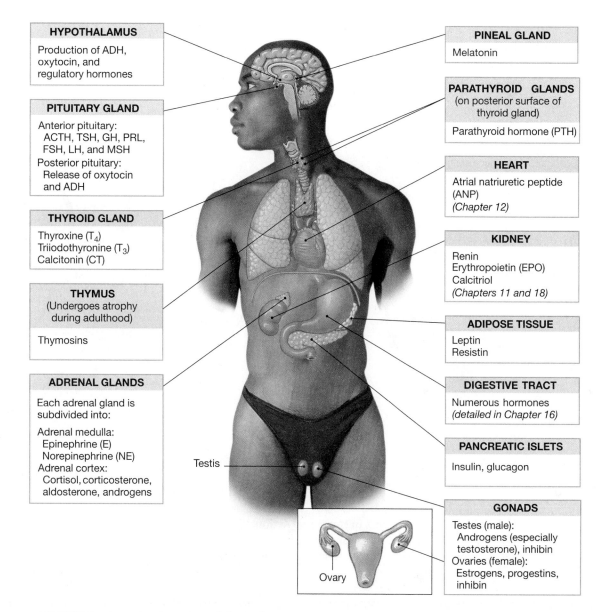

HYPOTHALAMUS
Production of ADH, oxytocin, and regulatory hormones

PITUITARY GLAND
Anterior pituitary:
  ACTH, TSH, GH, PRL, FSH, LH, and MSH
Posterior pituitary:
  Release of oxytocin and ADH

THYROID GLAND
Thyroxine (T$_4$)
Triiodothyronine (T$_3$)
Calcitonin (CT)

THYMUS
(Undergoes atrophy during adulthood)

Thymosins

ADRENAL GLANDS
Each adrenal gland is subdivided into:

Adrenal medulla:
  Epinephrine (E)
  Norepinephrine (NE)
Adrenal cortex:
  Cortisol, corticosterone, aldosterone, androgens

Testis

Ovary

PINEAL GLAND
Melatonin

PARATHYROID GLANDS
(on posterior surface of thyroid gland)
Parathyroid hormone (PTH)

HEART
Atrial natriuretic peptide (ANP)
(Chapter 12)

KIDNEY
Renin
Erythropoietin (EPO)
Calcitriol
(Chapters 11 and 18)

ADIPOSE TISSUE
Leptin
Resistin

DIGESTIVE TRACT
Numerous hormones
(detailed in Chapter 16)

PANCREATIC ISLETS
Insulin, glucagon

GONADS
Testes (male):
  Androgens (especially testosterone), inhibin
Ovaries (female):
  Estrogens, progestins, inhibin

● **FIGURE 10–1 An Overview of the Endocrine System.**

mones are released by the reproductive organs and the adrenal glands. Insoluble in water, steroid hormones are bound to specific transport proteins in blood. *Eicosanoids* are fatty acid-based compounds derived from the 20-carbon fatty acid *arachidonic* (a-rak-i-DON-ik) *acid.* Eicosanoids, which include the **prostaglandins,** coordinate local cellular activities and affect enzymatic processes in extracellular fluids, including blood clotting.

The Mechanisms of Hormonal Action

All cellular structures and functions are determined by proteins. Structural proteins determine the general shape and internal structure of a cell, and enzymes direct the cell's

metabolism. Hormones alter cellular operations by changing the *identities, activities, locations,* or *quantities* of important enzymes and structural proteins in various **target cells.** A target cell's sensitivity is determined by the presence or absence of a specific **receptor** with which a given hormone interacts (Figure 10–2●). Hormone receptors are located either on the cell membrane or inside the cell.

Hormonal Action at the Cell Membrane
The receptors for epinephrine, norepinephrine, peptide hormones, and eicosanoids are in the cell membranes of their respective target cells (Figure 10–3a●). Because epinephrine, norepinephrine, and the peptide hormones are not lipid-soluble, they cannot diffuse through a cell membrane, and they

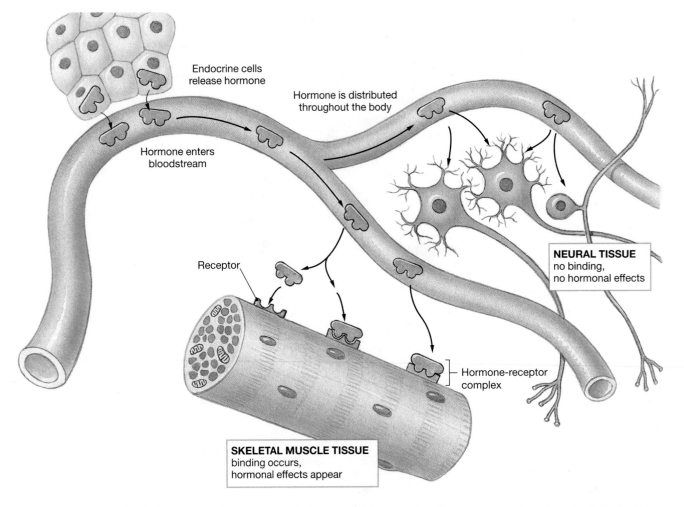

Endocrine cells
release hormone

Hormone is distributed
throughout the body

Hormone enters
bloodstream

Receptor

NEURAL TISSUE
no binding,
no hormonal effects

Hormone-receptor
complex

SKELETAL MUSCLE TISSUE
binding occurs,
hormonal effects appear

● **FIGURE 10–2 The Role of Target Cell Receptors in Hormonal Action.** For a hormone to affect a target cell, that cell must have receptors that can bind the hormone and initiate a change in cellular activity. The hormone in this illustration affects skeletal muscle tissue but not neural tissue because only the muscle tissue has the appropriate receptors.

bind to receptor proteins on the *outer* surface of the cell membrane. Eicosanoids, which *are* lipid-soluble, diffuse across the cell membrane and bind to receptor proteins on the *inner* surface of the cell membrane.

Hormones that bind to cell membrane receptors do not have direct effects on the target cell. Instead, when these hormones bind to an appropriate receptor, they are considered **first messengers** that then trigger the appearance of a **second messenger** in the cytoplasm. The link between the first messenger and the second messenger usually involves a **G protein,** which is an enzyme complex that is coupled to a membrane receptor. The G protein is activated when a hormone binds to the hormone's receptor at the membrane surface. The second messenger may function as an enzyme activator or inhibitor, but the net result is a change in the cell's metabolic activities.

One of the most important second messengers is **cyclic-AMP (cAMP)** (see Figure 10–3a). Its appearance depends on an activated G protein, which activates an enzyme called

adenylate cyclase. In turn, adenylate cyclase converts ATP to a ring-shaped molecule of cAMP. Cyclic-AMP activates *kinase* enzymes, which attach a high-energy phosphate group (PO_4^{3-}) to another molecule in a process called *phosphorylation.*

The effect on the target cell depends on the nature of the proteins affected. The phosphorylation of membrane proteins can open ion channels, and in the cytoplasm many enzymes can be activated only by phosphorylation. As a result, a single hormone can have one effect in one target tissue and quite different effects in other target tissues. The effects of cAMP are usually very short-lived, because another enzyme in the cell, *phosphodiesterase (PDE),* quickly breaks down cAMP. In a few instances, the activation of a G protein can *lower* the concentration of cAMP within the cell by stimulating PDE activity. The decline in cAMP has an inhibitory effect on the cell because without phosphorylation key enzymes remain inactive.

Although cyclic-AMP is one of the most common second messengers, there are many others. Important examples are

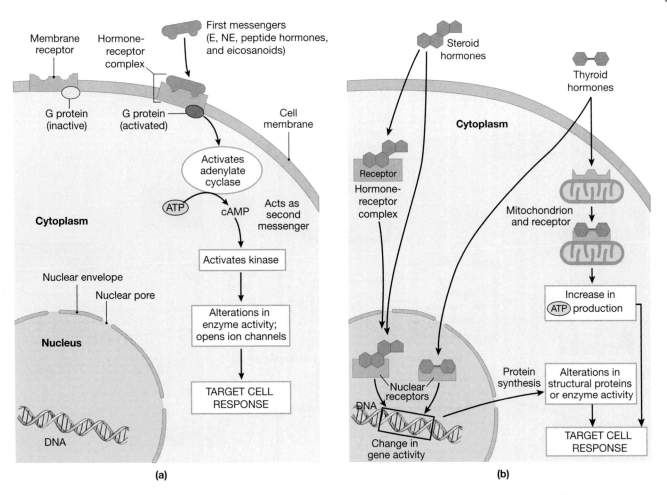

● **FIGURE 10–3 Mechanisms of Hormone Action.** (**a**) Nonsteroidal hormones, such as epinephrine (E), norepinephrine (NE), peptide hormones, and eicosanoids, bind to membrane receptors and activate G proteins. They exert their effects on target cells through a second messenger, such as cAMP, which alters the activity of enzymes present in the cell. (**b**) Both steroid hormones and thyroid hormones pass directly through target cell membranes. Steroid hormones bind to receptors in the cytoplasm or nucleus. Thyroid hormones either proceed directly to receptors in the nucleus or bind to receptors on mitochondria in the cytoplasm. In the nucleus, both steroid and thyroid hormone-receptor complexes directly affect gene activity and protein synthesis. Thyroid hormones also increase the rate of ATP production in the cell.

calcium ions and *cyclic-GMP*, a derivative of the high-energy compound *guanosine triphosphate (GTP)*.

Hormone Interaction with Intracellular Receptors

Steroid hormones and thyroid hormones cross the cell membrane before binding to receptors inside the cell (Figure 10–3b●). Steroid hormones diffuse rapidly through the lipid portion of the cell membrane and bind to receptors in the cytoplasm or nucleus. The resulting *hormone-receptor complex* then activates or inactivates specific genes in the nucleus. By this mechanism, steroid hormones can alter the rate of mRNA transcription, and thereby change the structure or function of the cell. The steroid hormone testosterone, for example, stimulates the production of enzymes and proteins in skeletal muscle fibers, and increases muscle size and strength.

Thyroid hormones cross the cell membrane either by diffusion or by a transport mechanism. Once within the cell, thyroid hormones bind to receptors within the nucleus or on mitochondria. The hormone-receptor complexes in the nucleus activate specific genes or change the rate of mRNA transcription. The result is an increase in metabolic activity due to changes in the nature or number of enzymes in the cytoplasm. Thyroid hormones bound to mitochondria increase the mitochondrial rates of ATP production.

The Secretion and Distribution of Hormones

Hormone release occurs where capillaries are abundant, and the hormones quickly enter the bloodstream for distribution throughout the body. Within the blood, hormones may

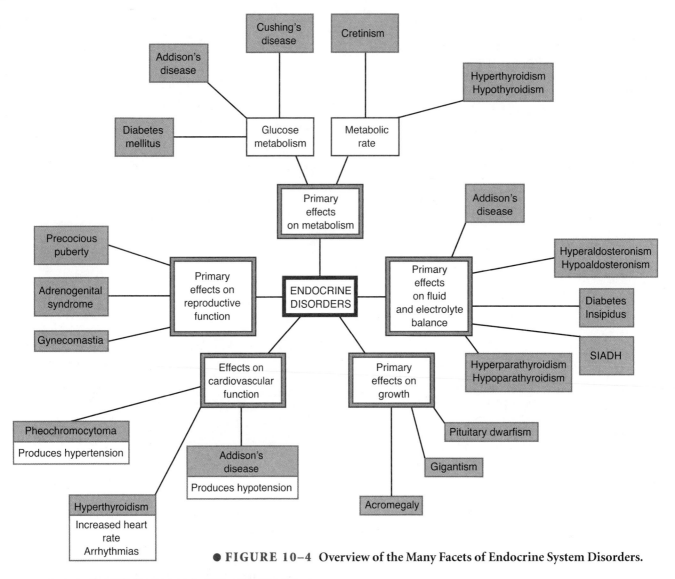

● **FIGURE 10–4 Overview of the Many Facets of Endocrine System Disorders.**

Clinical Note
ENDOCRINE EMERGENCIES

The symptoms of endocrine emergencies (Figure 10–4●) can usually be assigned to one of two basic categories: symptoms of underproduction (inadequate hormonal effects) or symptoms of overproduction (excessive hormonal effects). The observed symptoms may reflect either abnormal hormone production (hyposecretion or hypersecretion) or abnormal cellular sensitivity. These conditions are interesting because they highlight the significance of normally "silent" hormonal contributions. ■

circulate freely or attach to special transport proteins. A freely circulating hormone remains functional for less than one hour and sometimes for as little as two minutes. Free hormones are inactivated when (1) they diffuse out of the bloodstream and bind to receptors on target cells, (2) they are absorbed and broken down by certain liver or kidney cells, or (3) they are broken down by enzymes in the plasma or interstitial fluids.

Steroid hormones and thyroid hormones remain in circulation much longer because almost all become attached to special transport proteins. For each hormone an equilibrium occurs between the bound hormones and the small number that remain in a free state. As the free hormones are removed, they are replaced by the release of bound hormones.

Key Note

Hormones coordinate cell, tissue, and organ activities on a sustained basis. They circulate in the extracellular fluid and bind to specific receptors on or in target cells. They then modify cellular activities by altering membrane permeability, activating or inactivating key enzymes, or changing genetic activity.

The Control of Endocrine Activity

Endocrine activity—specifically, hormonal secretion—is controlled by negative feedback mechanisms. In this case, a stimulus triggers the production of a hormone whose direct or indirect effects reduce the intensity of the stimulus. ∞ p. 14 In the simplest case, endocrine activity may be controlled by *humoral* ("liquid") *stimuli,* which are changes in the composition of the extracellular fluid. Consider, for example, the control of blood calcium levels by two hormones, *parathyroid hormone* and *calcitonin.* When calcium levels in the blood decline, parathyroid hormone is released, and the responses of target cells elevate blood calcium levels. When calcium levels in the blood rise, calcitonin is released, and the responses of target cells lower blood calcium levels. Endocrine activity may also be controlled by *hormonal stimuli,* which are changes in the levels of circulating hormones. Such control may involve one or more intermediary steps and two or more hormones. Finally, endocrine control can also occur through *neural stimulation* that results from the arrival of a neurotransmitter at a neuroglandular junction. An important example of endocrine control involves the activity of the hypothalamus.

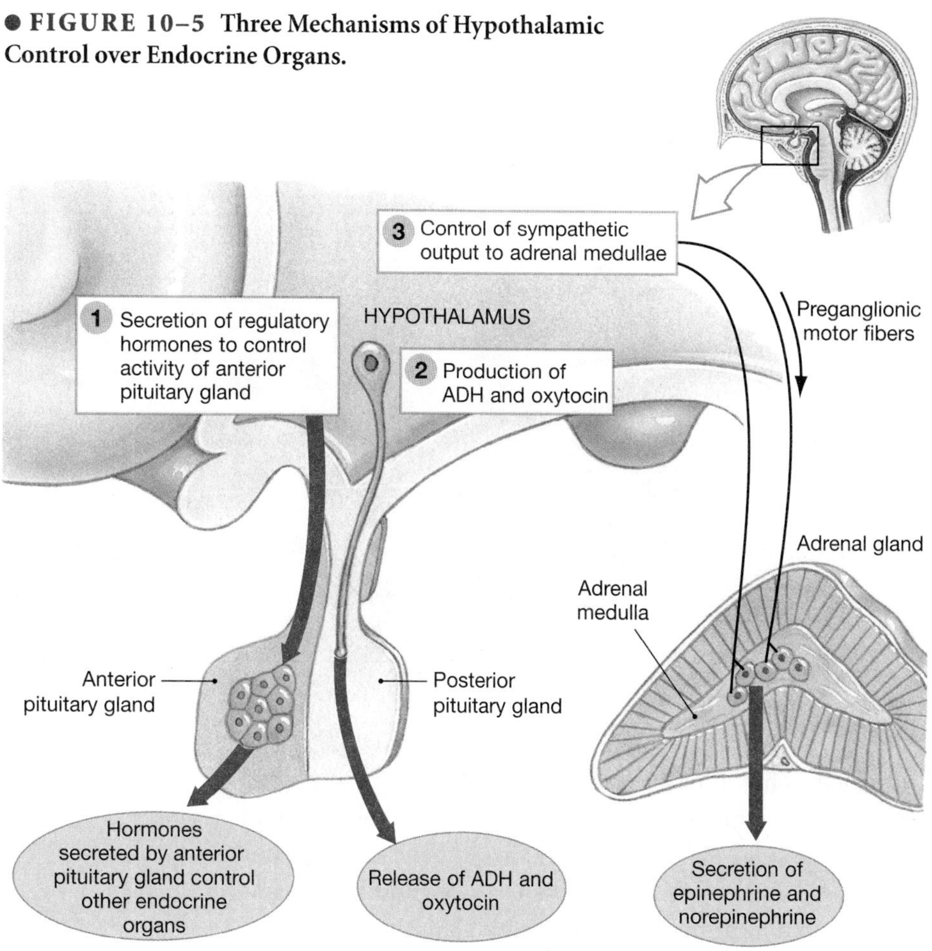

The Hypothalamus and Endocrine Regulation

The hypothalamus provides the highest level of endocrine control by acting as an important link between the nervous and endocrine systems. Coordinating centers in the hypothalamus regulate the activities of the nervous and endocrine systems in three ways (Figure 10–5●):

1. The hypothalamus secretes **regulatory hormones,** which are special hormones that control the activity of endocrine cells in the anterior pituitary gland. There are two classes of regulatory hormones: **releasing hormones (RH)** stimulate the production of one or more hormones in the anterior pituitary; whereas **inhibiting hormones (IH)** prevent the synthesis and secretion of pituitary hormones. The hormones released by the anterior pituitary gland control other endocrine glands.

2. The hypothalamus acts as an endocrine organ by synthesizing two hormones, ADH and oxytocin, which are then released into the circulation at the posterior pituitary gland.

3. The hypothalamus contains autonomic nervous system centers that control the endocrine cells of the adrenal

● **FIGURE 10–5** Three Mechanisms of Hypothalamic Control over Endocrine Organs.

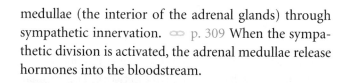

medullae (the interior of the adrenal glands) through sympathetic innervation. ∞ p. 309 When the sympathetic division is activated, the adrenal medullae release hormones into the bloodstream.

→ **CONCEPT CHECK QUESTIONS**

1. What is the primary factor that determines each cell's sensitivities to hormones?
2. How would the presence of a molecule that blocks adenylate cyclase affect the activity of a hormone that produces its cellular effects through cAMP?
3. Why is cAMP described as a second messenger?
4. What are the three types of stimuli that control hormone secretion?

Answers begin on p. 792.

■ The Pituitary Gland

The **pituitary gland,** or **hypophysis** (hī-POF-i-sis), secretes nine different hormones. All are peptides or small proteins that bind to membrane receptors, and all use cAMP as a second

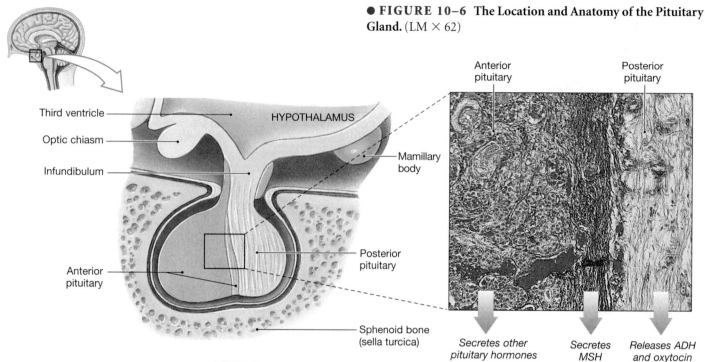

● **FIGURE 10–6** **The Location and Anatomy of the Pituitary Gland.** (LM × 62)

messenger. The pituitary gland is a small, oval gland nestled within the *sella turcica*, which is a depression in the sphenoid bone of the skull (Figure 10–6●). It hangs beneath the hypothalamus, connected by a slender stalk, the **infundibulum** (in-fun-DIB-ū-lum; funnel). The pituitary gland has a complex structure, including distinct anterior and posterior regions.

The Anterior Pituitary Gland

The **anterior pituitary gland** contains endocrine cells surrounded by an extensive capillary network. This capillary network, which provides entry into the circulatory system for the hormones secreted by endocrine cells of the anterior pituitary, is part of the *hypophyseal portal system.*

The Hypophyseal Portal System

As previously noted, regulatory hormones produced by the hypothalamus control the activities of the anterior pituitary. These hormones, released by hypothalamic neurons near the attachment of the infundibulum, enter a network of highly permeable capillaries. Before leaving the hypothalamus, this capillary network unites to form a series of slightly larger vessels that descend to the anterior pituitary before forming a second capillary network (Figure 10–7●).

The circulatory arrangement illustrated in Figure 10–7, in which blood flows from one capillary bed to another, is very unusual. Typically, blood flows from the heart through increasingly smaller arteries to a capillary network and then returns to the heart through increasingly larger veins. Blood

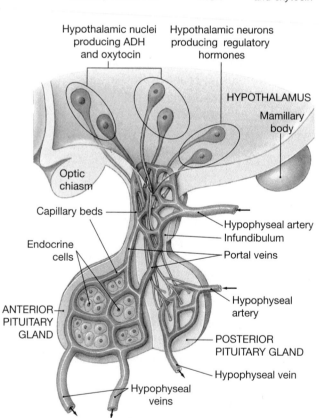

● **FIGURE 10–7** **The Hypophyseal Portal System and the Blood Supply to the Pituitary Gland.**

vessels that link two capillary networks—including the vessels between the hypothalamus and the anterior pituitary—are called *portal vessels;* in this case, they have the structure of veins. The entire complex is termed a **portal system.**

● **FIGURE 10–8** Negative Feedback Control of Endocrine Secretion. (a) In the typical pattern of regulation (in which multiple endocrine organs are involved), the hypothalamus produces a releasing hormone (RH) to stimulate hormone production by other glands; control occurs via negative feedback. (b) Shown here are two variations on the theme outlined in part (a). For the regulation of prolactin (PRL) production by the anterior pituitary (at left), the hypothalamus produces both a releasing factor (PRF) and an inhibiting hormone (PIH); when one is stimulated, the other is inhibited. In the regulation of growth hormone (GH) production by the anterior pituitary (at right), whenever GH-RH release is inhibited, GH-IH release is stimulated.

Portal systems ensure that all of the blood that enters the portal vessels reaches certain target cells before returning to the general circulation. Portal systems are named after their destinations, so this particular network is called the **hypophyseal** (hī-pō-FI-sē-al) **portal system.**

Hypothalamic Control of the Anterior Pituitary

An endocrine cell in the anterior pituitary may be controlled by releasing hormones (RH), inhibiting hormones (IH), or some combination of the two. The regulatory hormones released at the hypothalamus are transported directly to the anterior pituitary by the hypophyseal portal system.

The rate of regulatory hormone secretion by the hypothalamus is regulated through negative feedback mechanisms. The basic regulatory patterns are diagrammed in Figure 10–8●; these will be referenced in the following description of pituitary hormones. Many of these regulatory hormones are called *tropic hormones* (*tropé*, a turning) because they "turn on" other endocrine glands or support the functions of other organs.

Hormones of the Anterior Pituitary

The anterior pituitary gland produces seven hormones. The first four described in the following list regulate the production of hormones by other endocrine glands.

1. **Thyroid-stimulating hormone (TSH),** or *thyrotropin*, targets the thyroid gland and triggers the release of thyroid hormones. TSH is released in response to *thyrotropin-releasing hormone (TRH)* from the hypothalamus. As circulating concentrations of thyroid hormones rise, the rates of TRH and TSH production decline (Figure 10–8a●).

2. **Adrenocorticotropic hormone (ACTH)** stimulates the release of steroid hormones by the *adrenal cortex*, the

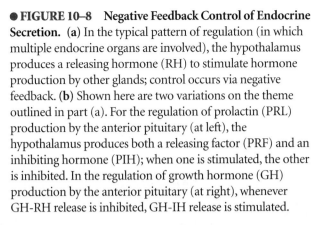

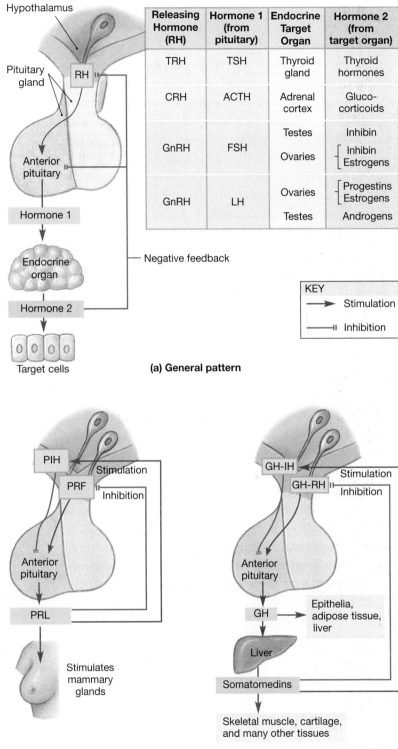

Releasing Hormone (RH)	Hormone 1 (from pituitary)	Endocrine Target Organ	Hormone 2 (from target organ)
TRH	TSH	Thyroid gland	Thyroid hormones
CRH	ACTH	Adrenal cortex	Gluco-corticoids
GnRH	FSH	Testes	Inhibin
		Ovaries	Inhibin Estrogens
GnRH	LH	Ovaries	Progestins Estrogens
		Testes	Androgens

(a) General pattern

KEY
→ Stimulation
—⊪ Inhibition

(b) Pattern variations

outer portion of the adrenal glands. ACTH specifically targets cells that produce hormones called *glucocorticoids* (gloo-kō-KOR-ti-koydz), which affect glucose metabolism. ACTH release occurs under the stimulation of *corticotropin-releasing hormone (CRH)* from the hypothalamus. A rise in glucocorticoid levels causes a decline in the production of ACTH and CRH. This type of negative feedback control is comparable to that for TSH (see Figure 10–8a).

A group of hormones called **gonadotropins** (gō-nad-ō-TRŌ-pinz) regulates the activities of the male and female sex organs, or *gonads*. The production of gonadotropins is stimulated by *gonadotropin-releasing hormone (GnRH)* from the hypothalamus. An abnormally low production of gonadotropins produces *hypogonadism*. Children with this condition will not undergo sexual maturation, and adults with hypogonadism cannot produce functional sperm or ova. The anterior pituitary produces two gonadotropins: follicle-stimulating hormone (FSH) and luteinizing hormone (LH).

3. **Follicle-stimulating hormone (FSH)** promotes follicle (and egg) development in females, and it stimulates the secretion of estrogens—steroid hormones produced by ovarian cells. In males, FSH production supports sperm production in the testes. A peptide hormone called *inhibin,* which is released by the cells of the testes and ovaries, inhibits the release of FSH and GnRH through a negative feedback control mechanism comparable to that for TSH (see Figure 10–8a).

4. **Luteinizing** (LOO-tē-in-ī-zing) **hormone (LH)** induces *ovulation,* which is the production of reproductive cells in females. It also promotes the secretion by the ovaries of estrogens and **progestins** (such as *progesterone*), which prepare the body for possible pregnancy. In males, LH is sometimes called *interstitial cell-stimulating hormone (ICSH)* because it stimulates the interstitial cells of the testes to produce sex hormones. These sex hormones are called **androgens** (AN-drō-jenz; *andros,* man); the most important is *testosterone*. GnRH production is inhibited by estrogens, progestins, and androgens through a negative feedback control mechanism comparable to that of TSH (see Figure 10–8a).

5. **Prolactin** (prō-LAK-tin; *pro-,* before + *lac,* milk), or **PRL,** works with other hormones to stimulate mammary gland development. In pregnancy and during the period of nursing following delivery, PRL also stimulates the production of milk by the mammary glands. Prolactin's effects in the human male are poorly understood, but it may help regulate androgen production. The regulation of prolactin release involves interactions between prolactin-releasing (PRF) and prolactin-inhibiting (PIH) hormones from the hypothalamus. The regulatory pattern is diagrammed in Figure 10–8b●.

6. **Growth hormone (GH),** also called *human growth hormone (hGH)* or *somatotropin (soma,* body), stimulates

cell growth and replication by accelerating the rate of protein synthesis. Although virtually every tissue responds to some degree, skeletal muscle cells and chondrocytes (cartilage cells) are particularly sensitive to levels of growth hormone.

The stimulation of growth by GH involves two mechanisms. The indirect, primary mechanism is best understood. Liver cells respond to the presence of growth hormone by synthesizing and releasing **somatomedins,** or *insulin-like growth factors (IGFs),* which are peptide hormones that bind to receptor sites on a variety of cell membranes. Somatomedins increase the rates at which amino acids are taken up and incorporated into new proteins. These effects develop almost immediately after GH release occurs, and they are particularly important after a meal, when the blood contains high concentrations of glucose and amino acids.

The direct actions of GH usually do not appear until after blood glucose and amino acid concentrations have returned to normal levels. In epithelia and connective tissues, GH stimulates stem cell divisions and the differentiation of daughter cells. GH also has metabolic effects in adipose tissue and in the liver. In adipose tissue, it stimulates the breakdown of stored fats and the release of fatty acids into the blood. In turn, many tissues stop breaking down glucose and start breaking down fatty acids to generate ATP. This process is termed a *glucose-sparing effect.* In the liver, GH stimulates the breakdown of glycogen reserves and the release of glucose into the circulation. Thus, GH plays a role in mobilizing energy reserves.

The production of GH is regulated by *growth hormone-releasing hormone (GH-RH)* and *growth hormone-inhibiting hormone (GH-IH)* from the hypothalamus. Somatomedins stimulate GH-IH and inhibit GH-RH. This regulatory mechanism is summarized in Figure 10–8b.

7. **Melanocyte-stimulating hormone (MSH)** stimulates the melanocytes in the skin, and increases their production of melanin. MSH is important in the control of skin and hair pigmentation in fish, amphibians, reptiles, and many mammals other than primates. The MSH-producing cells of the pituitary gland in adult humans are virtually nonfunctional, and the circulating blood usually does not contain MSH. However, the human pituitary secretes MSH (1) during fetal development, (2) in very young children, (3) in pregnant women, and (4) in some diseases. The functions of MSH under these circumstances are not known. The administration of a synthetic form of MSH causes darkening of the skin, so MSH has been suggested as a means of obtaining a "sunless tan."

The Posterior Pituitary Gland

The **posterior pituitary gland** contains axons from two different groups of neurons located within the hypothalamus. One group manufactures antidiuretic hormone (ADH) and the other oxytocin. These products are transported within axons along the infundibulum to the posterior pituitary, as indicated in Figure 10–7.

Antidiuretic hormone (ADH) is released in response to such stimuli as a rise in the concentration of electrolytes in the blood (an increased osmotic pressure) or a fall in blood volume or pressure. The primary function of ADH is to decrease the amount of water lost in the urine. With losses minimized, any water absorbed from the digestive tract will be retained, reducing the concentration of electrolytes. ADH also causes *vasoconstriction,* which is a constriction of peripheral blood vessels that helps increase blood pressure. ADH release is inhibited by alcohol, which explains the increased fluid excretion that follows the consumption of alcoholic beverages.

In women, **oxytocin** (*oxy-,* quick + *tokos,* childbirth) stimulates smooth muscle contractions in the wall of the uterus during labor and delivery and in special contractile cells associated with the mammary glands. Until the final stages of pregnancy, the uterine muscles are insensitive to oxytocin, but they become more sensitive as the time of delivery approaches. The stimulation of uterine muscles by oxytocin helps maintain and complete normal labor and childbirth (discussed in Chapter 20). After delivery, oxytocin also stimulates the contraction of special cells that surround the secretory cells and ducts of the mammary glands. In the "milk let-down" reflex, oxytocin secreted in response to suckling triggers the release of milk from the breasts.

Although oxytocin's functions in sexual activity remain uncertain, circulating oxytocin levels are known to rise during sexual arousal and peak at orgasm in both sexes. In men, oxytocin stimulates smooth muscle contraction in the walls of the sperm duct and prostate gland. These actions may be important in *emission,* which is the ejection of prostatic secretions, sperm, and the secretions of other glands into the male reproductive tract before ejaculation. In women, oxytocin released during intercourse may stimulate smooth muscle contractions in the uterus and vagina that promote the transport of sperm toward the uterine tubes.

Figure 10–9● and Table 10–1 summarize important information concerning the hormones produced by the pituitary gland.

TABLE 10–1 *The Pituitary Hormones*

REGION	HORMONE	TARGET	HORMONAL EFFECTS
Anterior pituitary	Thyroid-stimulating hormone (TSH)	Thyroid gland	Secretion of thyroid hormones
	Adrenocorticotropic hormone (ACTH)	Adrenal cortex	Glucocorticoid secretion (cortisol, corticosterone)
	Gonadotropins:		
	Follicle-stimulating hormone (FSH)	Follicle cells of ovaries	Estrogen secretion, follicle development
		Sustentacular cells of testes	Sperm maturation
	Luteinizing hormone (LH)	Follicle cells of ovaries	Ovulation, formation of corpus luteum, and progesterone secretion
		Interstitial cells of testes	Testosterone secretion
	Prolactin (PRL)	Mammary glands	Production of milk
	Growth hormone (GH)	All cells	Growth, protein synthesis, lipid mobilization and catabolism
	Melanocyte-stimulating hormone (MSH)	Melanocytes of skin	Increased melanin synthesis in epidermis
Posterior pituitary	Antidiuretic hormone (ADH)	Kidneys	Reabsorption of water, elevation of blood volume and pressure
	Oxytocin	Uterus, mammary glands (females)	Labor contractions, milk ejection
		Sperm duct and prostate gland (males)	Contractions of sperm duct and prostate gland

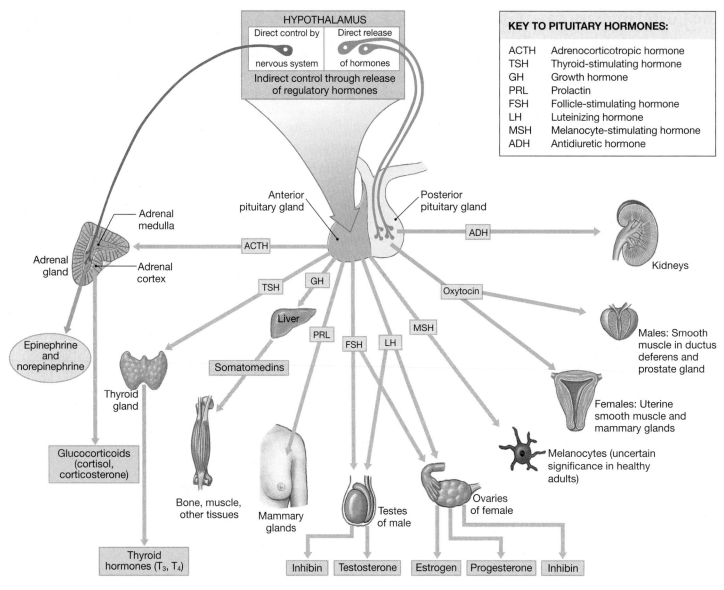

● FIGURE 10–9 Pituitary Hormones and Their Targets.

Clinical Note
VASOPRESSIN

Antidiuretic hormone (ADH), also called *vasopressin,* is one of two hormones secreted by the posterior pituitary. It decreases the amount of water lost through the kidney and causes constriction of peripheral blood vessels (vasoconstriction). Both mechanisms serve to increase the blood pressure.

When vasopressin is given in unnaturally high doses, much higher than those needed for its antidiuretic hormone effects, its vasoconstrictive properties are enhanced. Because of this, vasopressin can be used to treat certain types of cardiac arrest and gastrointestinal bleeding (particularly bleeding esophageal varices).

The side effects include nausea, intestinal cramps, the urge to defecate, bronchial constriction, and pallor of the skin. In women, it can also cause uterine contractions.

Studies have shown that circulating levels of natural vasopressin in patients who receive CPR are higher in those who survive than in those who do not. This appears to result from increased blood flow to vital organs, including the heart and brain. Because vasopressin increases blood flow to vital organs, it appears to be a suitable alternative to epinephrine (adrenalin) in treating certain types of cardiac arrest. ■

Clinical Note
DIABETES INSIPIDUS

Diabetes (*diabetes*, to pass through) occurs in several forms, all characterized by excessive urine production (*polyuria*). Although diabetes can be caused by physical damage to the kidneys, most forms are the result of endocrine abnormalities. The two most important forms are diabetes mellitus and diabetes insipidus. Diabetes mellitus is described on pp. 391–393.

Diabetes insipidus (*insipidus*, tasteless) develops when the posterior pituitary no longer releases adequate amounts of ADH or the kidneys fail to respond to ADH. Water conservation at the kidneys is impaired, and excessive amounts of water are lost in the urine. As a result, the individual is constantly thirsty, a condition known as *polydipsia* (*dipsa*, thirst), but the fluids consumed are not retained by the body. Mild cases may not require treatment, so long as fluid and electrolyte intake keep pace with urinary losses. In severe cases, fluid losses can reach 10 liters per day, and a fatal dehydration will occur unless treatment is provided. ■

Key Note

The hypothalamus produces regulatory factors that adjust the activities of the anterior pituitary gland, which produces seven hormones. Most of these hormones control other endocrine organs, including the thyroid gland, adrenal gland, and gonads. It also produces growth hormone, which stimulates cell growth and protein synthesis. The posterior pituitary gland releases two hormones produced in the hypothalamus. ADH restricts water loss and promotes thirst, and oxytocin stimulates smooth muscle contractions in the mammary glands and uterus (in females) and the prostate gland (in males).

→ CONCEPT CHECK QUESTIONS

1. If a person became dehydrated, how would the level of ADH released by the posterior pituitary change?
2. A blood sample shows elevated levels of somatomedins. Which pituitary hormone would you expect to be elevated as well?
3. What effect would elevated levels of cortisol, a hormone from the adrenal gland, have on the pituitary secretion of ACTH?

Answers begin on p. 792.

■ The Thyroid Gland

The **thyroid gland** lies anterior to the trachea and just inferior to the **thyroid** ("shield-shaped") **cartilage**, which forms most of the anterior surface of the larynx (Figure 10–10a●). The two lobes of the thyroid gland are united by a slender connection, the *isthmus* (IS-mus). An extensive blood supply gives the thyroid gland a deep red color.

Thyroid Follicles and Thyroid Hormones

The thyroid gland contains numerous **thyroid follicles,** which are spheres lined by a simple cuboidal epithelium (Figure 10–10b●). The cavity within each follicle contains a viscous *colloid*, which is a fluid that contains large amounts of suspended proteins and thyroid hormones. A network of capillaries surrounds each follicle, and delivers nutrients and regulatory hormones to the glandular cells and picks up their secretory products and metabolic wastes.

Thyroid hormones are manufactured by the follicular epithelial cells and stored within the follicle cavities. Under TSH stimulation from the anterior pituitary, the epithelial cells remove hormones from the follicle cavities and release them into the circulation. However, almost all of the released thyroid hormones are unavailable because they become attached to plasma proteins in the bloodstream. Only the remaining unbound thyroid hormones, which are a small percentage of the total released, are free to diffuse into target cells in body tissues. As the concentration of unbound hormone molecules decreases, the plasma proteins release additional bound hormone. The bound thyroid hormones are a substantial reserve; in fact, the bloodstream normally contains more than a week's supply of thyroid hormones.

The thyroid hormones are derived from molecules of the amino acid tyrosine to which iodine atoms have been attached. The hormone **thyroxine** (thī-ROKS-ēn) contains four atoms of iodine; it is also known as *tetraiodothyronine* (tet-ra-ī-ō-dō-THĪ-rō-nēn), or T_4. Thyroxine accounts for roughly 90 percent of all thyroid secretions. **Triiodothyronine,** or T_3, is a related, more potent molecule that contains three iodine atoms.

Thyroid hormones readily cross cell membranes, and they affect almost every cell in the body. Inside a cell, they bind to receptor sites on mitochondria and in the nucleus (see Figure 10–3b). The binding of thyroid hormones to mitochondria increases the rate of ATP production. Thyroid hormone-receptor complexes in the nucleus activate gene coding for the synthesis of enzymes involved in glycolysis and energy production, resulting in an increase in cellular rates of metabolism and oxygen consumption. Because the cell consumes more energy, and energy use is measured in *calories*, the effect

● **FIGURE 10–10 The Thyroid Gland.** (a) This drawing shows the location and gross anatomy of the thyroid gland. (b) This micrograph reveals histological details, including the thyroid follicles. (LM × 211)

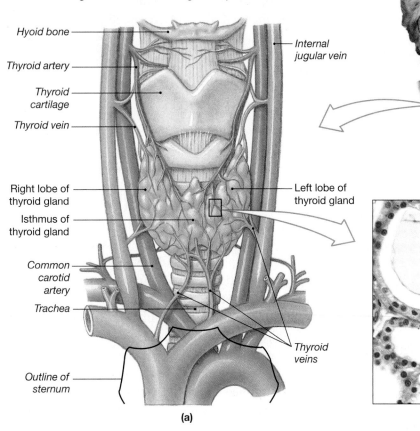

is called the **calorigenic effect** of thyroid hormones. When the metabolic rate increases, more heat is generated and body temperature rises. In growing children, thyroid hormones are essential to normal development of the skeletal, muscular, and nervous systems.

Normal production of thyroid hormones establishes the background rates of cellular metabolism. These hormones exert their primary effects on active tissues and organs, including skeletal muscles, the liver, the heart, and the kidneys. Overproduction or underproduction of thyroid hormones can, therefore, cause very serious metabolic problems. In many parts of the world, inadequate dietary iodine intake leads to an inability to synthesize thyroid hormones. Under these conditions, TSH stimulation continues, and the thyroid follicles become distended with nonfunctional secretions. The result is an enlarged thyroid gland, or *goiter*. Goiters vary in size, and a large goiter can interfere with breathing and swallowing. This is seldom a problem in the U.S. because the typical American diet provides roughly three times the minimum daily requirement of iodine, thanks to the addition of iodine to table salt ("iodized salt").

The C Cells of the Thyroid Gland: Calcitonin

C cells, or *parafollicular cells,* are endocrine cells sandwiched between the follicle cells and their basement membrane (see Figure 10–10b). C cells produce the hormone **calcitonin (CT).** Calcitonin helps regulate calcium ion concentrations in body fluids. The control of calcitonin secretion is independent of the hypothalamus or pituitary gland. As Figure 10–11● illustrates, C cells release calcitonin when the calcium ion concentration of the blood rises above normal. The target organs are the bones and the kidneys. Calcitonin inhibits osteoclasts (which slows the release of calcium from bone) and stimulates calcium excretion at the kidneys. The resulting reduction in calcium ion concentrations eliminates the stimulus and "turns off" the C cells.

Calcitonin is most important during childhood, when it stimulates active bone growth and calcium deposition in the skeleton. It also acts to reduce the loss of bone mass during prolonged starvation and during late pregnancy, when the maternal skeleton competes with the developing fetus for absorbed calcium ions. The role of calcitonin in healthy, nonpregnant adults is unclear.

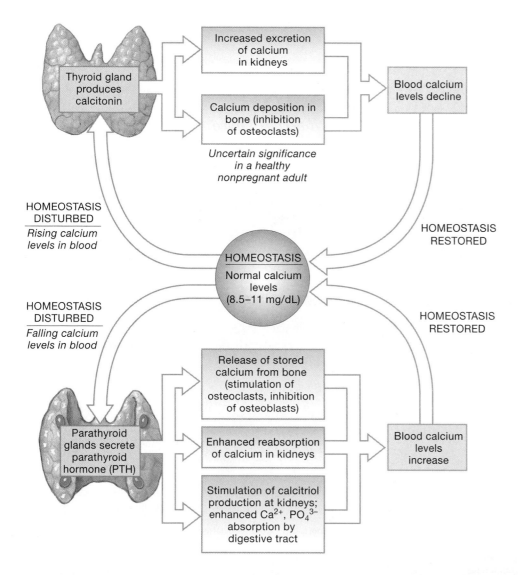

● **FIGURE 10–11** The Homeostatic Regulation of Calcium Ion Concentrations.

We have seen the importance of calcium ions in controlling muscle cell and nerve cell activities. ∞ pp. 215, 271 Calcium ion concentrations also affect the sodium permeabilities of excitable membranes. At high calcium ion concentrations, sodium permeability decreases, and membranes become less responsive. Such problems are prevented by the secretion of calcitonin under appropriate conditions. However, under normal conditions, calcium ion levels seldom rise enough to trigger calcitonin secretion. Most homeostatic adjustments function to prevent lower than normal calcium ion concentrations. Low calcium concentrations are dangerous because sodium permeabilities then increase, and muscle cells and neurons become extremely excitable. If calcium levels fall too far, convulsions or muscular spasms can result. Such disastrous events are prevented by the actions of the parathyroid glands.

Clinical Note
DISORDERS OF THE THYROID GLAND

Disorders of the thyroid gland are typically chronic conditions. However, you may see patients with acute complications of thyroid disorders. The most common of these are:

- *Hyperthyroidism.* The presence of excess thyroid hormones in the blood.
- *Thyrotoxic crisis.* A condition that reflects prolonged exposure of body organs to excess thyroid hormones, with resultant changes in structure and function. Thyrotoxicosis is generally caused by Graves' disease.
- *Hypothyroidism.* The presence of inadequate thyroid hormones in the blood.
- *Myxedema.* A condition that reflects long-term exposure to inadequate levels of thyroid hormones, with resultant changes in structure and function.

The risk factors for developing thyroid dysfunction include:

- Female sex
- Older age

- Personal history:
  - Goiter
  - Previous surgery or radiotherapy that affects the thyroid gland
  - Other autoimmune diseases, including diabetes mellitus, vitiligo, pernicious anemia, and leukotrichia
  - Use of lithium or compounds that contain iodine
- Family history:
  - Thyroid disease
  - Pernicious anemia
  - Diabetes mellitus
  - Primary adrenal insufficiency

The thyroid gland is controlled by the anterior pituitary gland through the release of thyroid-stimulating hormone (TSH). The anterior pituitary, in turn, is controlled by the hypothalamus through the release of thyrotropin-releasing hormone (TRH). An increased level of TSH results in increased thyroid function, while a decreased

(continued next page)

level results in a decline. A negative feedback loop exists between the thyroid gland and the anterior pituitary gland and hypothalamus (Figure 10–12●). Table 10–2 compares common signs and symptoms of hypothyroidism and hyperthyroidism.

Graves' Disease

Excess circulating thyroid hormones result from Graves' disease. Roughly 15 percent of Graves' patients have a close relative with the disease, which suggests a strong hereditary role in predisposition to the disorder. In addition, Graves' disease is about six times more common in women than in men, with onset typically in young adulthood (20s and 30s).

Graves' disease has an autoimmune origin. Autoantibodies are generated that stimulate thyroid tissue to produce excessive amounts of thyroid hormones (Figure 10–13●). The resultant changes in organ function are responses to either excess thyroid hormones or to the autoantibodies themselves.

The signs and symptoms of Graves' disease include agitation, emotional lability, insomnia, poor heat tolerance, weight loss de-

TABLE 10–2 *Common Signs and Symptoms of Thyroid Disease*	
HYPOTHYROIDISM	**HYPERTHYROIDISM**
Fatigue	Fatigue
Weight gain	Weight loss
Cold intolerance	Heat intolerance
Skin dry	Skin moist (hyperhidrous)
Hair dryness and/or loss	Hair fine and silky
Depression	Nervousness
Dementia	Insomnia
	Tremor
Muscle cramps and myalgia	Muscle weakness
	Dyspnea
Bradycardia	Tachycardia
	Palpations
Constipation	Hyperdefecation
Infertility	
Edema	
Menstrual irregularity (hypermenorrhea common)	Menstrual irregularity (hypermenorrhea common)

Source: "American Thyroid Association Guidelines for Detection of Thyroid Dysfunction." *Archives of Internal Medicine* 160 (2000): 1573–75.

spite increased appetite, weakness, dyspnea, and tachycardia or new-onset atrial fibrillation in the absence of a cardiac history. Nervous system symptoms tend to be more common in younger adults, whereas serious cardiovascular symptoms tend to predominate in older individuals. Prolonged exposure of orbital tissues to the pathological thyroid-stimulating autoantibodies can cause exophthalmos (protrusion of the eyeballs), whereas interaction of autoantibodies with thyroid tissue often produces diffuse goiter (a generally enlarged thyroid gland).

Cardiac dysfunction is probably the most likely context in which an emergency call may arise from thyrotoxicosis, usually caused by Graves' disease. Use of β-adrenergic blockers such as propranolol may temporarily reduce cardiac stress, but make sure the patient does not have heart failure or asthma before considering use. Glucocorticoid therapy (namely, dexamethasone) is sometimes helpful in quickly reducing the level of circulating T_4.

Thyrotoxic Crisis (Thyroid Storm)

Thyrotoxic crisis, or thyroid storm, is a life-threatening emergency that can be fatal within as few as 48 hours if untreated. It is usually associated with severe physiological stress (e.g., trauma, infec-

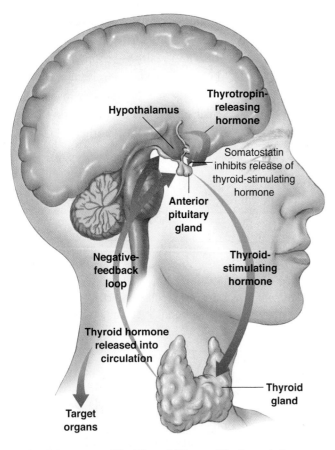

● **FIGURE 10–12 The Thyroid Loop.** The hypothalamus controls the anterior pituitary, which in turn controls the thyroid gland. Negative feedback loops prevent the oversecretion of hormones.

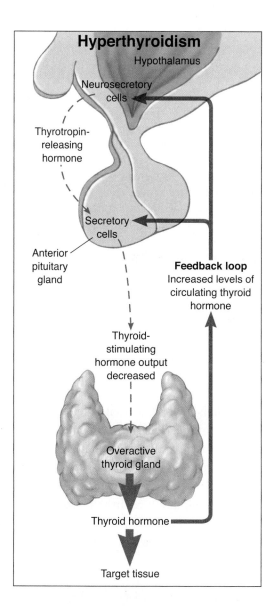

● **FIGURE 10–13 Graves' Disease.** Hyperthyroidism results in an increase in the levels of circulating thyroid hormone.

tion), less often with psychological stress. You may also encounter thyroid storm secondary to overdose of thyroid hormone medication in a hypothyroid individual.

The mechanisms that underlie thyrotoxic crisis are poorly understood. An acute increase in the levels of thyroid hormones does not appear to be the cause. More likely, thyroid storm is caused by a shift of thyroid hormone in the blood from the protein-bound (biologically inactive) to the free (biologically active) state. This significantly increases the amount of active hormone in the circulation, and thus stimulates the thyroid gland.

The signs and symptoms associated with thyrotoxic crisis reflect the patient's extreme hypermetabolic state and increased activity of the sympathetic nervous system. The syndrome is characterized by high fever (106°F/41°C or higher), irritability, delirium or coma, tachycardia, hypotension, vomiting, and diarrhea. A less severe presentation may also occur, with slight fever and marked lethargy.

In the presence of the signs and symptoms of thyrotoxic crisis, field management is largely focused on supportive care: oxygenation, ven-

tilatory assistance, fluid resuscitation, and cardiac monitoring. Glucocorticoids and β-adrenergic blockers may be helpful, especially if transport times are long. Transport should be expedited for definitive therapy that blocks the high blood levels of thyroid hormones.

Hypothyroidism and Myxedema

Hypothyroidism can be congenital or acquired and can affect both sexes. The recent increase in incidence of hypothyroidism in middle-aged women may reflect better diagnostics, a true rise in incidence, or both. Advanced myxedema in middle-aged and elderly individuals is the condition you are most likely to see in the emergency setting.

Hypothyroidism creates a low metabolic state, and early signs reflect poor organ function and poor response to challenges such as exercise or infection (Figure 10–14●). Over time, untreated severe hypothyroidism causes the additional sign of myxedema, which is a thickening of connective tissue in the skin and other tissues, including

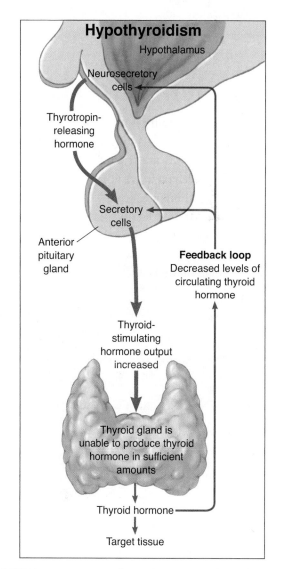

● **FIGURE 10–14 Hypothyroidism.** Hypothyroidism results from a decrease in the levels of circulating thyroid hormones.

(continued next page)

the heart. Patients with myxedema may progress into a hypothermic, stuporous state called *myxedema coma*, which can be fatal if respiratory depression occurs. Triggers for progression to myxedema coma include infection, trauma, a cold environment, or exposure to central nervous system depressants such as alcohol or certain drugs.

Early signs of hypothyroidism may be subtle. Symptoms may be as slight as fatigue and slowed mental function attributed falsely to aging. Typically, patients with hypothyroidism or myxedema show lethargy, cold intolerance, constipation, decreased mental function, or decreased appetite with increased weight. In addition, the relaxation stage of deep tendon reflexes (DTRs) is slowed. The classic appearance of myxedema is an unemotional, puffy-faced, and pale individual with thinned hair, enlarged tongue, and cool skin that looks and feels like dough. Myxedema coma may be difficult to identify. Note if the history is consistent with hypothyroidism and look for the physical appear-

ance of myxedema. Other signs include profound hypothermia (temperatures as low as 75°F/24°C are not uncommon), low amplitude bradycardia, and carbon dioxide retention.

These signs and symptoms may alert you to the possible presence of myxedema. Keep in mind that heart failure due to the combination of age, atherosclerosis, and myxedematous enlargement is not uncommon, so focus on maintaining the ABCs and closely monitoring cardiac and pulmonary status. Most patients with myxedema coma require intubation and ventilatory assistance. Active rewarming is contraindicated due to the risk of cardiac dysrhythmias and cardiovascular collapse secondary to vasodilatation. Although IV access is important, limit fluids because fluid and electrolyte imbalance is common and cardiac function is compromised. IV therapy should be guided by appropriate laboratory studies. Expedite transport to an appropriate facility for definitive treatment. ■

■ The Parathyroid Glands

Two tiny pairs of **parathyroid glands** are embedded in the posterior surfaces of the thyroid gland (Figure 10–15a●). The cells of the two adjacent glands are separated by connective tissue fibers that surround each parathyroid gland (Figure 10–15b●). At least two cell populations are found in the parathyroid gland. **Chief cells** produce parathyroid hormone; the functions of the other cell type are unknown.

Like the C cells of the thyroid, the chief cells monitor the concentration of circulating calcium ions. When the calcium

concentration falls below normal, the chief cells secrete **parathyroid hormone (PTH),** or *parathormone* (see Figure 10–11). Although parathyroid hormone acts on the same target organs as calcitonin, it produces the opposite effects. PTH stimulates osteoclasts, inhibits the bone-building functions of osteoblasts, and reduces urinary excretion of calcium ions. PTH also stimulates the kidneys to form and secrete *calcitriol*, which promotes the absorption of Ca^{2+} and PO_4^{3-} by the digestive tract.

Information concerning the hormones of the thyroid and parathyroid glands is summarized in Table 10–3.

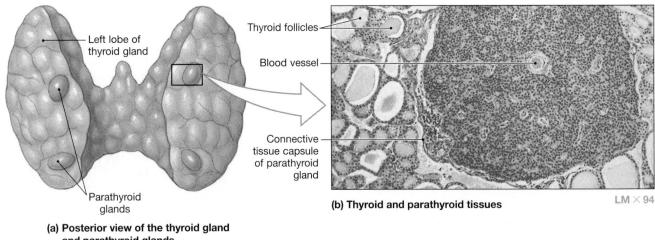

Thyroid follicles

Blood vessel

Left lobe of thyroid gland

Connective tissue capsule of parathyroid gland

Parathyroid glands

LM × 94

(a) Posterior view of the thyroid gland and parathyroid glands

(b) Thyroid and parathyroid tissues

● **FIGURE 10–15 The Parathyroid Glands. (a)** The parathyroid glands lie embedded in the posterior surface of the thyroid lobes. **(b)** This photomicrograph shows the histologic features of parathyroid and thyroid tissues.

TABLE 10–3 *Hormones of the Thyroid Gland and Parathyroid Glands*

GLAND/CELLS	HORMONE(S)	TARGETS	HORMONAL EFFECTS
THYROID **Follicular epithelium**	Thyroxine (T_4), triiodothyronine (T_3)	Most cells	Increased energy utilization, oxygen consumption, growth, and development
C cells	Calcitonin (CT)	Bone, kidneys	Decreased calcium concentrations in body fluids (see Figure 10–11)
PARATHYROIDS **Chief cells**	Parathyroid hormone (PTH)	Bone, kidneys	Increased calcium concentrations in body fluids (see Figure 10–11)

Key Note

The thyroid gland produces (1) hormones that adjust tissue metabolic rates and (2) a hormone that usually plays a minor role in calcium ion homeostasis by opposing the action of parathyroid hormone.

CONCEPT CHECK QUESTIONS

1. What clinical signs would you expect to see in an individual whose diet lacks iodine?
2. Why do signs of reduced thyroid hormone concentration not appear until about one week after a person's thyroid gland is removed?
3. Removal of the parathyroid glands would result in decreased blood concentrations of what important mineral?

Answers begin on p. 792.

■ The Adrenal Glands

A yellow, pyramid-shaped **adrenal gland,** or *suprarenal* (soo-pra-RĒ-nal; *supra-,* above + *renes,* kidneys) *gland,* sits on the superior border of each kidney (Figure 10–16a●). Each adrenal gland has two parts: an outer *adrenal cortex* and an inner *adrenal medulla* (Figure 10–16b●).

The Adrenal Cortex

The yellowish color of the **adrenal cortex** is due to the presence of stored lipids, especially cholesterol and various fatty acids. The adrenal cortex produces more than two dozen steroid hormones, collectively called *adrenocortical steroids,* or simply **corticosteroids**. In the bloodstream, these hormones are bound to transport proteins. These hormones are vital; if the adrenal glands are destroyed or removed, the individual will die unless corticosteroids are administered. Overproduction or underproduction of any of the corticosteroids will have severe consequences because these hormones affect metabolism in many different tissues.

Corticosteroids

The adrenal cortex contains three distinct regions or zones. Each zone synthesizes specific steroid hormones; the outer zone produces *mineralocorticoids,* the middle zone produces *glucocorticoids,* and the inner zone produces *androgens* (Figure 10–16c●).

MINERALOCORTICOIDS (MCS). The **mineralocorticoids** affect the electrolyte composition of body fluids. **Aldosterone** (al-DOS-ter-ōn), which is the principal MC, stimulates the conservation of sodium ions and the elimination of potassium ions by targeting cells that regulate the ionic composition of excreted fluids.

Specifically, aldosterone causes the retention of sodium by preventing the loss of sodium ions in urine, sweat, saliva, and digestive secretions. The retention of sodium ions is accompanied by a loss of potassium ions. Secondarily, the reabsorption of sodium ions results in the osmotic reabsorption of water at the kidneys, sweat glands, salivary glands, and pancreas. Aldosterone also increases the sensitivity of salt receptors in the tongue, which results in greater interest in consuming salty foods.

Aldosterone secretion occurs in response to a drop in blood sodium content, blood volume, or blood pressure, or to a rise in blood potassium levels. Aldosterone release also occurs in response to the hormone *angiotensin II* (*angeion,* vessel + *teinein,* to stretch). This hormone will be discussed later in the chapter.

GLUCOCORTICOIDS (GCS). The steroid hormones collectively known as **glucocorticoids** affect glucose metabolism. **Cortisol**

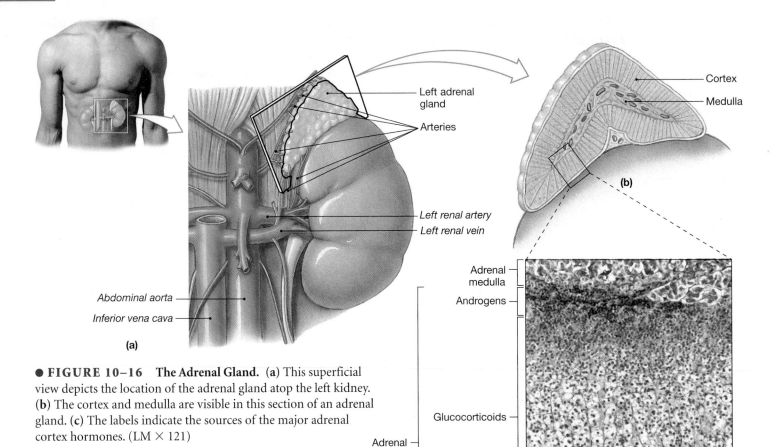

Left adrenal gland

Arteries

Cortex

Medulla

(b)

Left renal artery
Left renal vein

Abdominal aorta
Inferior vena cava

(a)

Adrenal medulla

Androgens

Glucocorticoids

Adrenal cortex

Mineralocorticoids

Capsule

(c)

● **FIGURE 10–16**   **The Adrenal Gland.** (a) This superficial view depicts the location of the adrenal gland atop the left kidney. (b) The cortex and medulla are visible in this section of an adrenal gland. (c) The labels indicate the sources of the major adrenal cortex hormones. (LM × 121)

(KOR-ti-sol; also called *hydrocortisone*), **corticosterone** (kor-ti-KOS-te-rōn), and **cortisone** are the three most important glucocorticoids.

Glucocorticoid secretion occurs under ACTH stimulation and is regulated by negative feedback (see Figure 10–8a). These hormones accelerate the rates of glucose synthesis and glycogen formation, especially within the liver. Simultaneously, adipose tissue responds by releasing fatty acids into the blood, and other tissues begin to break down fatty acids instead of glucose. This *glucose-sparing effect* results in an increase in blood glucose levels. Glucocorticoids also have *anti-inflammatory* effects: they suppress the activities of white blood cells and other components of the immune system. "Steroid creams" are often used to control irritating allergic rashes, such as those produced by poison ivy, and injections of glucocorticoids may be used to control more severe allergic reactions. Because they slow wound healing and suppress immune defenses against infectious organisms, topical steroids are used to treat superficial rashes but are never applied to open wounds.

ANDROGENS.   The adrenal cortex in both sexes produces small quantities of **androgens,** which are the sex hormones produced in large quantities by the testes in males. Once in the blood-

stream, some of the androgens are converted to estrogens, which are the dominant sex hormone in females. When secreted in normal amounts, neither androgens nor estrogens affect sexual characteristics, so the importance of the small adrenal production of androgens in both sexes remains unclear.

The Adrenal Medulla

The **adrenal medulla** has a reddish brown coloration partly because of the many blood vessels within it. It contains large, rounded cells similar to those found in other sympathetic ganglia, and these cells are innervated by preganglionic sympathetic fibers. The secretory activities of the adrenal medullae are controlled by the sympathetic division of the ANS. ∞ p. 308

The adrenal medulla contains two populations of secretory cells, one that produces **epinephrine** (**E,** or *adrenaline*) and the other **norepinephrine** (**NE,** or *noradrenaline*). These hormones are continuously released at a low rate, but sympathetic stimulation accelerates the rate of discharge dramatically.

Epinephrine makes up 75–80 percent of the secretions from the medulla; the rest is norepinephrine. Receptors for epinephrine and norepinephrine are found on skeletal muscle fibers, adipocytes, liver cells, and cardiac muscle fibers. In skeletal muscles, adrenal medulla secretions trigger a mobilization of glycogen reserves and accelerate the breakdown of glucose to provide ATP; this combination results in increased muscular power and endurance. In adipose tissue, stored fats are broken down to fatty acids, and in the liver, glycogen molecules are converted to glucose. The fatty acids and glucose are then released into the circulation for use by peripheral tissues. The heart responds to adrenal medulla hormones with an increase in the rate and force of cardiac contractions.

The metabolic changes that follow epinephrine and norepinephrine release peak 30 seconds after adrenal stimulation and linger for several minutes thereafter. Thus, the effects produced by stimulation of the adrenal medullae outlast the other results of sympathetic activation.

The characteristics of the adrenal hormones are summarized in Table 10–4.

TABLE 10-4 *The Adrenal Hormones*

REGION	HORMONE	TARGET	EFFECTS
Adrenal cortex	Mineralocorticoids, primarily aldosterone	Kidneys	Increased reabsorption of sodium ions and water by the kidneys; accelerates urinary loss of potassium ions
	Glucocorticoids: cortisol (hydrocortisone), corticosterone, cortisone	Most cells	Release of amino acids from skeletal muscles and lipids from adipose tissues; promotes liver formation of glycogen and glucose; promotes peripheral use of lipids; anti-inflammatory effects
	Androgens		Uncertain significance under normal conditions
Adrenal medulla	Epinephrine (E, adrenaline), norepinephrine (NE, noradrenaline)	Most cells	Increased cardiac activity, blood pressure, glycogen breakdown, and blood glucose levels; release of lipids by adipose tissue (see Table 8–8, p. 313)

Clinical Note
DISORDERS OF THE ADRENAL GLANDS

Two disorders of the adrenal cortex, Cushing's syndrome and Addison's disease, can play a part in medical emergencies or complicate responses to trauma. Cushing's syndrome is caused by excessive adrenocortical activity, while Addison's disease is caused by deficient adrenocortical activity.

Hyperadrenalism (Cushing's Syndrome)

Cushing's syndrome is a relatively common disorder of the adrenal glands. It usually affects middle-aged persons and is more common in women than in men. It results from excess glucocorticoids, primarily cortisol, which can be due to abnormalities in the anterior pituitary gland or in the adrenal cortex. It also can be due to treatment with glucocorticoids, such as prednisone. Check the history to note any recent steroid treatment for nonendocrine conditions such as cancer or rheumatologic disorders.

Long-term exposure to excess glucocorticoids produces numerous changes. Metabolically, cortisol is an antagonist to insulin. Gluconeogenesis is prominent, with profound protein catabolism. The body's handling of fats is altered; because of this, atherosclerosis and hypercholesterolemia are common. Over time, diabetes mellitus also may develop. Cortisol's mineralocorticoid activity causes sodium retention and increased blood volume. Increased vascular sensitivity to catecholamines occurs, and this may also contribute to hypertension. Potassium loss through the kidneys may cause hypokalemia. Cortisol's anti-inflammatory and immunosuppressive properties predispose the patient to infection.

Regardless of its cause, the presenting signs and symptoms of hyperadrenalism are the same. The earliest sign is weight gain, particularly through the trunk of the body, face, and neck. A "moon-faced" appearance often develops (Figure 10–17●). The accumulation of fat on the upper back is occasionally referred to as a buffalo hump. Skin changes are also very common and may be an early clue to potential problems. These include the skin's thinning to an almost transparent appearance, a tendency to bruise easily, delayed healing from even minor wounds, and development of facial hair among women (hirsutism). Mood swings and impaired memory or concentration are also common.

Although you probably will not see patients with acute hyperadrenal crisis, you are likely to encounter patients with signs

(continued next page)

and symptoms of Cushing's syndrome. Remember that these patients have a higher incidence of cardiovascular disease, including hypertension and stroke. Pay particular attention to skin preparation when starting IV lines because of the patient's fragile skin and susceptibility to infection. Note any observations indicative of Cushing's syndrome in your report and relay them to hospital staff.

Adrenal Insufficiency (Addison's Disease)

Addison's disease is due to cortical destruction. Addison's has become less common as its former leading causes, such as tuberculosis, have come under control. Currently, over 90 percent of Addison's disease cases are due to autoimmune disease. As with Graves' disease, which is another autoimmune disorder, heredity plays a prominent role in an individual's predisposition for Addison's disease. In fact, patients with Addison's are more likely than average to have other autoimmune disorders, including Graves' disease.

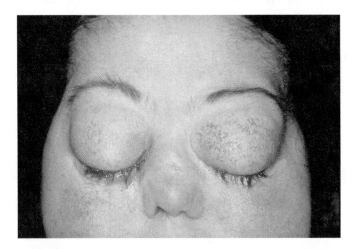

● **FIGURE 10–17 Facial Feature's of Cushing's Syndrome.**

Destruction of the adrenal cortex results in minimal production of all three classes of hormones: glucocorticoids, mineralocorticoids, and androgens. Low mineralocorticoid activity is key to the changes of Addison's, which causes major disturbances in water and electrolyte balance. Increased sodium excretion in urine results in low blood volume, and potassium retention can cause hyperkalemia and ECG changes. Many cases of adrenal insufficiency are due to therapy with steroids such as prednisone, which can completely suppress normal adrenal function. Sudden cessation of the drug may trigger symptoms of Addison's disease or even an Addisonian crisis, with cardiovascular collapse.

Addison's disease is characterized by changes related to low corticosteroid activity: progressive weakness, fatigue, decreased appetite, and weight loss. Hyperpigmentation of the skin and mucous membranes, particularly in sun-exposed areas, is also characteristic.

Acute stresses, such as infection or trauma, may tip Addison's patients into a metabolic failure called Addisonian crisis, which is characterized by profound hypotension and electrolyte imbalances. Many patients will have gastrointestinal problems such as vomiting or diarrhea, which will exacerbate electrolyte imbalances, low blood volume, and hypotension and increase the potential for cardiac dysrhythmias. Be alert for this potentially life-threatening emergency, and include it in your list of possible causes of unexplained cardiovascular collapse, particularly if the history suggests primary Addison's disease or Addison's disease secondary to drug therapy.

The patient may reveal the presence of Addison's disease during the history, or the disease's signs and symptoms may lead you to suspect its presence. Focus emergency management on maintaining the ABCs and on closely monitoring cardiac and oxygenation status as well as blood-glucose level. Hypoglycemia poses its own threat. Assess blood-glucose levels and administer 25–50 grams of 50 percent dextrose to patients with blood glucose levels less than 50 mg/dL or those with altered mental status. Obtain a baseline 12-lead ECG to check for dysrhythmias related to electrolyte imbalance. Be aggressive in fluid resuscitation. Follow your local protocol or contact medical direction for specific orders based on your patient's presentation. Expedite transport to an appropriate facility for definitive treatment. ■

■ The Pineal Gland

The **pineal gland** lies in the posterior portion of the roof of the third ventricle. ∞ p. 292 It contains neurons, glial cells, and secretory cells that synthesize the hormone **melatonin** (mel-a-TŌ-nin). Branches of the axons of neurons that make up the visual pathways enter the pineal gland and affect the rate of melatonin production, which is lowest during daylight hours and highest at night.

Several functions have been suggested for melatonin in humans:

- *Inhibition of reproductive function.* In some mammals, melatonin slows the maturation of sperm, ova, and reproductive organs. The significance of this effect remains unclear, but circumstantial evidence suggests that melatonin may play a role in the timing of human sexual maturation. Melatonin levels in the blood decline at puberty, and pineal tumors that eliminate melatonin production cause premature puberty in young children.
- *Antioxidant activity.* Melatonin is a very effective antioxidant that may protect CNS neurons from *free radicals*, such as nitric oxide (NO) or hydrogen peroxide (H_2O_2)

that may be generated in active neural tissue. (Free radicals are highly reactive atoms or molecules that contain unpaired electrons in their outer electron shell.) ∞ p. 273

- *Establishment of day-night cycles of activity.* Because of the cyclical nature of its rate of secretion, the pineal gland may also be involved in maintaining basic *circadian rhythms*—daily changes in physiological processes that follow a regular day-night pattern. Increased melatonin secretion in darkness has been suggested as a primary cause of *seasonal affective disorder (SAD)*. This condition, characterized by changes in mood, eating habits, and sleeping patterns, can develop during the winter in high latitudes, where sunshine is scarce or lacking.

Key Note

The adrenal glands produce hormones that adjust metabolic activities at specific sites, which affects either the pattern of nutrient utilization, mineral ion balance, or the rate of energy consumption by active tissues.

CONCEPT CHECK QUESTIONS

1. What effect would elevated cortisol levels have on blood glucose levels?
2. Increased amounts of light would inhibit the production of which hormone?

Answers begin on p. 792.

■ The Pancreas

The **pancreas** lies in the J-shaped loop between the stomach and proximal portion of the small intestine (Figure 10–18●). It is a slender, pale organ with a nodular (lumpy) consistency, and it contains both exocrine and endocrine cells. The pancreas is primarily a digestive organ whose exocrine cells make digestive enzymes. The **exocrine pancreas** is discussed further in Chapter 16. The endocrine cells of the pancreas produce two hormones, *glucagon* and *insulin*.

Cells of the **endocrine pancreas** form clusters known as **pancreatic islets,** or the *islets of Langerhans* (LAN-ger-hanz). The islets are scattered among the exocrine cells (Figure 10–18b●) and account for only about 1 percent of all pancreatic cells. Each islet contains several cell types. The two most important are **alpha cells,** which produce the hormone **glucagon** (GLOO-ka-gon), and **beta cells,** which secrete **insulin** (IN-suh-lin). Glucagon and insulin regulate blood glucose concentrations in much the same way parathyroid hormone and calcitonin control blood calcium levels.

Regulation of Blood Glucose Concentrations

Figure 10–19● diagrams the mechanism of hormonal regulation of blood glucose levels. Glucose is the preferred energy source for most cells in the body, and under normal

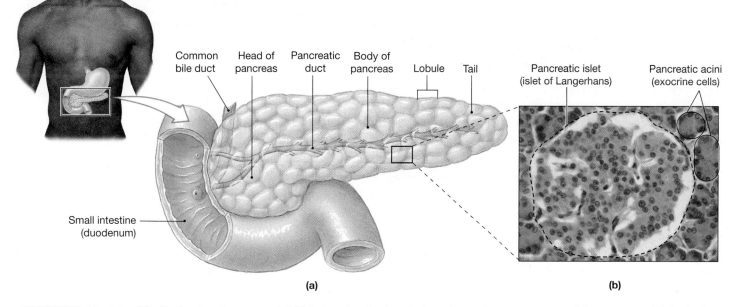

● FIGURE 10–18   The Endocrine Pancreas. (a) This drawing depicts the location and gross anatomy of the pancreas. (b) Each pancreatic islet is surrounded by exocrine cells. (LM × 276)

conditions it is the only energy source for neurons. When blood glucose concentrations rise above normal homeostatic levels, the beta cells release insulin, which then stimulates glucose transport and utilization in its target cells. The cell membranes of almost all cells in the body contain insulin receptors; the only exceptions are (1) neurons and red blood cells, which cannot metabolize nutrients other than glucose; (2) epithelial cells of the kidney tubules, where glucose is reabsorbed; and (3) epithelial cells of the intestinal lining, where glucose is obtained from the diet. When glucose is abundant, all cells use it as an energy source and stop breaking down amino acids and lipids.

ATP generated by the breakdown of glucose molecules is used to build proteins and to increase energy reserves, and most cells increase their rates of protein synthesis in response to insulin. A secondary effect is an increase in the rate of amino acid transport across cell membranes. Insulin also stimulates fat cells to increase their rates of triglyceride (fat) synthesis and storage. In the liver and in skeletal muscle fibers, insulin also accelerates the formation of glycogen. In summary, when glucose is abundant, insulin secretion by beta cells stimulates glucose utilization to support growth and to establish glycogen and fat reserves (see Figure 10–19).

When glucose concentrations fall below normal homeostatic levels, insulin secretion is suppressed, as is glucose transport into its target cells. These cells now shift over to other energy sources, such as fatty acids. At the same time, the alpha cells release glucagon, and energy reserves are mobilized. Skeletal muscles and liver cells break down glycogen into glucose (for energy in the muscle or for release by the liver), adipose tissue releases fatty acids for use by other tissues, and proteins are broken down into their component amino acids (see Figure 10–19). The liver takes in the amino acids and converts them to glucose that can be released into the circulation. As a result, blood glucose concentrations rise toward normal levels. The interplay between insulin and glucagon both stabilizes blood glucose levels and prevents competition between neural tissue and other tissues for limited glucose supplies.

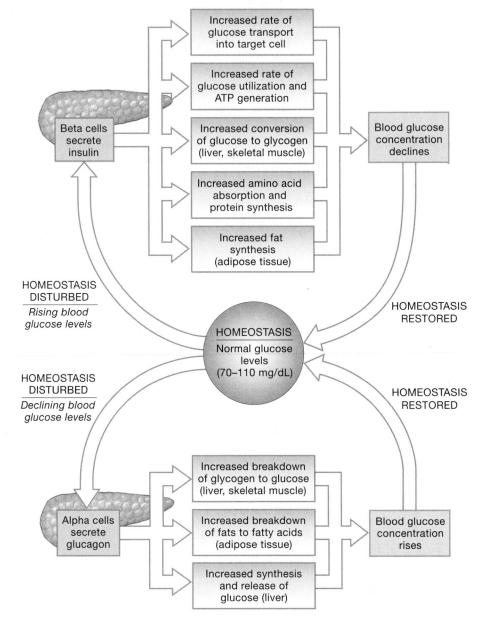

● **FIGURE 10–19** The Hormonal Regulation of Blood Glucose Concentrations.

Pancreatic alpha and beta cells are sensitive to blood glucose concentrations, and the secretion of glucagon and insulin can occur without endocrine or nervous system instructions. Yet because the islet cells are very sensitive to variations in blood glucose levels, any hormone that affects blood glucose concentrations will indirectly affect the production of insulin and glucagon. Insulin and glucagon production are also influenced by autonomic activity. Parasympathetic stimulation enhances insulin release, whereas sympathetic stimulation inhibits it. Sympathetic stimulation promotes glucagon release.

Clinical Note
DIABETES MELLITUS

The disease diabetes mellitus is marked by inadequate insulin activity in the body. Insulin is critical to maintaining normal blood glucose levels. Glucose is important for all cells, but it is especially critical for brain cells. In fact, glucose is the only substance that brain cells can readily and efficiently use as an energy source. In addition, insulin enables the body to store energy as glycogen, protein, and fat.

Diabetes mellitus, or sugar diabetes, is a common and ancient serious disease. Over 8 million Americans have been diagnosed with diabetes, and U.S. health experts believe nearly the same number of the populace may be living with undiagnosed diabetes. The disease was named in ancient times by Greek physicians who noted that affected persons produced large volumes of urine that attracted bees and other insects, hence *diabetes* (meaning "to siphon," or "to pass through") for excessive urine production and *mellitus* (meaning "honey sweet") for the presence of sugar in the urine.

Diabetes mellitus is typically categorized as Type I diabetes or Type II diabetes. In addition, diabetes can also be classified as primary or secondary. Secondary diabetes is due to another cause and is uncommon; it affects only about 1 percent of diabetics. Causes include drugs, infections, genetic disorders, other endocrine disorders, and immune-mediated diseases. Gestational diabetes, which occurs only during pregnancy, affects about 4 percent of pregnancies. Careful control of blood glucose levels in gestational diabetes is essential in preventing fetal cardiac and nervous system abnormalities. Untreated gestational diabetes is a major cause of large birth weight infants (macrosomia). Gestational diabetes must be treated with insulin, as oral hypoglycemic medications cross the placental barrier and adversely affect the developing fetus.

Type I Diabetes Mellitus

Type I diabetes mellitus is characterized by the destruction of the beta (β) cells of the pancreas, which usually leads to absolute insulin deficiency. Type I diabetes is commonly called juvenile-onset diabetes because of the average age at diagnosis. The term *insulin-dependent diabetes mellitus (IDDM)* also is used because patients require regular insulin injections to maintain glucose homeostasis. This type of diabetes is less common than Type II diabetes, but it is more serious. Diabetes is regularly among the ten leading causes of death in the U.S. and Type I diabetes accounts for most diabetes-related deaths.

Heredity is important in determining who will be predisposed to develop Type I diabetes. Although the cause of Type I diabetes is often unclear, viral infection, production of autoantibodies directed against beta cells, and genetically determined early deterioration of beta cells are all possible causes. Regardless, the immediate cause of the disease is destruction of beta cells in the islets of Langerhans in the pancreas.

In untreated Type I diabetes, blood-glucose levels rise because, without adequate insulin, cells cannot take up the circulating glucose. Hyperglycemia in the range of 300 to 500 milligrams per deciliter is not uncommon. As glucose spills into the urine, large amounts of water are lost through osmotic diuresis. Catabolism of fat becomes significant as the body switches to fatty acids as the primary energy source. Overall, this pathophysiology accounts for the constant thirst (polydipsia), excessive urination (polyuria), ravenous appetite (polyphagia), weakness, and weight loss associated with untreated Type I diabetes. Ketosis can occur as the result of fat catabolism, and it may proceed to frank diabetic ketoacidosis, a medical emergency that you will encounter in the field.

Treatment of Type I diabetes requires the administration of insulin. Tight control of blood glucose levels minimizes many of the long-term complications of diabetes mellitus. To achieve this, insulin dosing must be individualized based upon the patient's activities of daily living. Frequent dosing of short- and medium-acting insulin provides tighter control of blood-glucose levels than a single daily dose of long-term or mixed insulin. Selected patients can benefit from an insulin pump that slowly administers insulin over a 24-hour period as programmed based on the patient's daily activities.

Type II Diabetes Mellitus

Type II diabetes mellitus, also called adult-onset diabetes or non-insulin-dependent diabetes (NIDDM), is associated with a moderate decline in insulin production accompanied by a markedly deficient response to the insulin present in the body. Some Type II patients will predominantly have insulin resistance with relative insulin deficiency, while others will have a predominantly secretory defect with insulin resistance.

Heredity may also play a role in predisposition to Type II diabetes. In addition, obese persons are more likely to develop Type II diabetes, and obesity probably plays a role in development of the disease. Increased weight (and increased size of fat cells) causes a relative deficiency in the number of insulin receptors per cell, which makes fat cells less responsive to insulin. This type of diabetes is far more common than Type I diabetes, and accounts for about 90 percent of cases of diabetes mellitus. It is also less serious. The major risk factors for Type II diabetes mellitus are:

- Family history of diabetes (i.e., parents or siblings with diabetes)
- Obesity (≥ 20 percent over desired body weight or body mass index ≥ 27 kg/m²)
- Race or ethnicity with high risk of diabetes (e.g., African American, Hispanic American, Native American, Asian American, Pacific Islander)
- Age ≥ 45 years
- Previously identified impaired fasting glucose or impaired glucose tolerance
- Hypertension (≥ 140/90 mmHg)
- Hyperlipidemia (HDL cholesterol level ≤ 35 mg/dL [0.90 mmol/L] or triglyceride level ≥ 250 mg/dL [2.82 mmol/L] or both)
- History of gestational diabetes or delivery of baby over 9 lb (4.1 kg)

(continued next page)

Untreated Type II diabetes typically presents with a lower level of hyperglycemia and fewer major signs of metabolic disruption than Type I diabetes. For instance, limited glucose use is usually sufficient to keep the body from switching to fats as the primary energy source. Thus, diabetic ketoacidosis (DKA) is uncommon in these patients. However, a complication called *nonketotic hyperosmolar coma* can occur, and you may see it as a medical emergency.

Medical treatment of Type II diabetes is less intensive than for Type I diabetes. Initial therapy often consists of dietary change and increased exercise in an attempt to improve body weight. If nonpharmacological therapy is insufficient to bring blood glucose levels down to the normal range, oral hypoglycemic agents may be prescribed. These drugs stimulate insulin secretion by beta cells and promote an increase in the number of insulin receptors per cell. In some cases, however, control may eventually require use of insulin.

Emergency Complications of Diabetes

Some of the complications of diabetes mellitus are medical emergencies and require prompt intervention. These include diabetic hypoglycemia (insulin shock), ketoacidosis (diabetic coma), and nonketotic hyperosmolar coma.

HYPOGLYCEMIA (INSULIN SHOCK)

Hypoglycemia, or low blood glucose, is a medical emergency. It can occur when a patient takes too much insulin, eats too little to match an insulin dose, or physically overexerts and uses almost all of the available blood glucose. As the duration of hypoglycemia lengthens, the risk increases that brain cells will be permanently damaged or killed due to lack of glucose. Although brain cells can adapt to use fats as an energy source, this adaptation requires hours to develop, and the switch to fat-based metabolism cannot correct any damage already incurred. This is why every second counts in treating hypoglycemia.

Hypoglycemia, or insulin shock, reflects high insulin and low blood-glucose levels. Regardless of the reason for low blood sugar, insulin causes almost all remaining blood glucose to be taken up by cells. Because of the high level of insulin, glucagon may be ineffective in raising blood-glucose levels. In prolonged fasts, almost half the glucose normally produced through gluconeogenesis is of renal origin. This activity is stimulated by epinephrine. Diabetic patients with kidney failure may be predisposed to hypoglycemia because of a lack of renal gluconeogenesis.

The signs and symptoms of hypoglycemia are many and varied. Altered mental status is the most important. As blood-glucose levels fall, the patient may display inappropriate anger (even rage) or display a bizarre behavior. Sometimes the patient may be placed in police custody for such behavior or be involved in an automobile collision. Physical signs may include diaphoresis and tachycardia. If the blood glucose falls to a critically low level, the patient may have a hypoglycemic seizure or become comatose. In contrast to diabetic ketoacidosis, hypoglycemia can

develop quickly. A clear change in mental status can occur without warning. Always consider hypoglycemia when encountering a patient with bizarre behavior. Additionally, hypoglycemia can cause symptoms that resemble a CVA (hypoglycemic hemiparesis), which is why blood sugar evaluation is mandatory prior to treating for stroke.

In suspected cases of hypoglycemia, perform the initial assessment quickly. Look for a medical identification device, such as a MedicAlert bracelet. If possible, determine blood-glucose level. Because of the urgency of this emergency, most paramedic units must be able to perform this task or to rush a blood sample along with the patient. If the blood-glucose level is less that 60 mg/dL, 50 percent dextrose solution should be administered intravenously (Figure 10–20●). If the patient is conscious and able to swallow, glucose administration may be provided orally with orange juice, sugared soft drinks, or commercially available glucose pastes.

When an IV cannot be started, hypoglycemic patients may improve following the administration of glucagon, which can be administered intramuscularly. This is a much slower process and will work only if adequate stores of glycogen are available. Glucagon must be reconstituted immediately prior to administration.

DIABETIC KETOACIDOSIS (DIABETIC COMA)

Diabetic ketoacidosis (DKA) is a serious, potentially life-threatening complication associated with Type I diabetes. It occurs when profound insulin deficiency is coupled with increased glucagon and stress hormone activity. It may occur as the initial presentation of severe diabetes, as a result of patient noncompliance with insulin injections, or as the result of physiologic stress, such as surgery, a myocardial infarction, or serious infection.

Diabetic ketoacidosis reflects amplification of the same physiological mechanisms as ketosis. In the initial phase of diabetic ketoacidosis, profound hyperglycemia develops because of lack of insulin. Body cells cannot take in glucose for normal metabolic

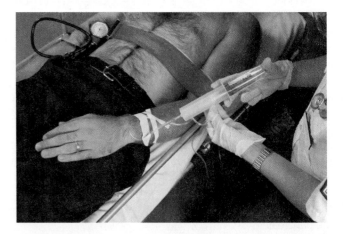

● **FIGURE 10–20** **Intravenous Glucose Administration.**
Patients with confirmed or suspected hypoglycemia must receive glucose before brain cells are destroyed.

processes. Gluconeogenesis, which is the compensatory mechanism for low glucose levels within cells, only contributes more blood glucose. The consequent loss of glucose in the urine, accompanied by loss of water through osmotic diuresis, produces significant dehydration.

As the body switches to fat-based metabolism, the blood levels of ketones rise. The ketone load accounts for the observed acidosis. By the time the characteristic decrease in pH from about 7.4 to about 6.9 has occurred, the patient is within hours of death if left untreated. The onset of clinically obvious diabetic ketoacidosis is slow, and lasts from 12 to 24 hours. In the initial phase, signs of osmotic diuresis appear, including increased urine production and dry, warm skin and mucous membranes. The individual often has excessive hunger and thirst coupled with a progressive sense of general malaise. Volume depletion induces tachycardia and feelings of physical weakness.

As ketoacidosis develops, a rapid deep breathing pattern termed Kussmaul's respirations appears. This major compensatory mechanism for acidosis helps expel carbon dioxide (CO_2) from the body. The breath itself may have a fruity or acetone-like smell as some blood acetone is expelled through the lungs. The blood profile includes not only hyperglycemia and acidic pH but also multiple electrolyte abnormalities. Low bicarbonate levels (HCO_3^-) reflect loss of acid-base buffer via Kussmaul's respirations. Low potassium levels may be found secondary to diuresis, with marked hypokalemia that increases the risk for cardiac dysrhythmias or death. Over time, mental function declines and frank coma may occur. A fever is not characteristic of ketoacidosis. If present, it is a sign of infection.

The treatment for a patient suffering from diabetic ketoacidosis is essentially the same as for any other patient who has mental impairment or is unconscious. The sweet, fruity odor of ketones occasionally can be detected in the breath. If possible, the blood glucose level should be determined. It is not uncommon for patients with ketoacidosis to have blood-glucose levels well in excess of 300 mg/dL. Prehospital treatment of DKA includes airway maintenance and fluid resuscitation to counteract dehydration. Often, the DKA patient will require several liters of an isotonic fluid. Definitive treatment includes insulin administration and correction of electrolyte deficiencies.

NONKETOTIC HYPEROSMOLAR COMA

Nonketotic hyperosmolar coma (NKHC) is a serious complication associated with Type II diabetes. Typically, both insulin and glucagon activity are present. NKHC develops when two conditions occur: sustained hyperglycemia causes osmotic diuresis sufficient to produce marked dehydration, and water intake is inadequate to replace lost fluids. Renal or peritoneal dialysis, high-osmolarity feeding supplements, infection, and certain drugs also can be associated with development of NKHC.

As sustained hyperglycemia develops, glucose spills into the urine, which causes osmotic diuresis and resultant dehydration. The level of hyperglycemia is often much higher than is seen in diabetic ketoacidosis (up to 1000 mg/dL). However, insulin activity in patients with NKHC is usually sufficient to prevent significant production of ketone bodies. Inadequate fluid replacement results in characteristic signs and symptoms.

The mortality rate for NKHC coma is higher than for ketoacidosis: it ranges from 40 to 70 percent. The higher mortality rate may be due to the lack of early signs and symptoms that would bring patients with ketoacidosis to the attention of family or health-care professionals.

The onset of NKHC is even slower than that of ketoacidosis; development often occurs over several days. Early signs include increased urination and increased thirst. Subsequent volume depletion can result in orthostatic hypotension when the patient gets out of bed, along with other signs such as dry skin and mucous membranes, as well as tachycardia. The patient may become lethargic, confused, or enter frank coma. Kussmaul's respirations are rarely seen because of the lack of ketoacidosis.

Prehospital treatment of the patient suffering from NKHC is essentially the same as of any other patient who has mental impairment or is unconscious. Distinguishing diabetic ketoacidosis from NKHC is often difficult in the field. Therefore, the prehospital treatment of both emergencies is identical, and transportation should be expedited. ■

🔒 Key Note

The pancreatic islets release insulin and glucagon. Insulin is released when blood glucose levels rise, and it stimulates glucose transport into, and utilization by, peripheral tissues. Glucagon is released when blood glucose levels decline, and it stimulates glycogen breakdown, glucose synthesis, and fatty acid release.

■ The Endocrine Tissues of Other Organ Systems

Many organs that are part of other body systems have secondary endocrine functions. Examples include the intestines (digestive system), the kidneys (urinary system), the heart (cardiovascular system), the thymus (lymphatic system), and the gonads—the testes in the male and ovaries in the female (reproductive system). Recent discoveries have led to the realization that the body's adipose tissue has important endocrine functions, which are also considered in this section.

The Intestines

The intestines, which process and absorb nutrients, release a variety of hormones that coordinate the activities of the digestive system. Although the pace of digestive activities can be affected by the autonomic nervous system, most digestive processes are controlled locally. The hormones involved will be discussed in Chapter 16.

The Kidneys

The kidneys release the steroid hormone *calcitriol,* the peptide hormone *erythropoietin,* and the enzyme *renin.* Calcitriol is important to calcium ion homeostasis. Erythropoietin and renin are involved in the regulation of blood pressure and blood volume.

Calcitriol is secreted by the kidneys in response to the presence of parathyroid hormone (PTH). Its synthesis depends on the availability of vitamin D_3, which may be synthesized in the skin or absorbed from the diet. Vitamin D_3 is absorbed by the liver and converted to an intermediary product that is released into the circulation and absorbed by the kidneys. Calcitriol stimulates the absorption of calcium and phosphate ions across the intestinal lining of the digestive tract.

Erythropoietin (e-rith-rō-POY-ē-tin; *erythros,* red + *poiesis,* making), or **EPO,** is released by the kidneys in response to low oxygen levels in kidney tissues. EPO stimulates the production of red blood cells by the bone marrow. The increase in the number of red blood cells elevates blood volume. Because these cells transport oxygen, their increased abundance improves oxygen delivery to peripheral tissues. EPO will be considered in greater detail when we discuss the formation of blood cells in Chapter 11.

Renin (RĒ-nin) is released by specialized kidney cells in response to a decline in blood volume, blood pressure, or both. Once in the bloodstream, renin starts an enzymatic chain reaction, known as the *renin-angiotensin system,* that leads to the formation of the hormone **angiotensin II.** Angiotensin II stimulates the production of aldosterone (by the adrenal cortex) and ADH (in the posterior pituitary gland). This combination restricts salt and water loss at the kidneys. Angiotensin II also stimulates thirst and elevates blood pressure. (Because renin plays a leading role in the formation of angiotensin II, many physiological and endocrinological references consider renin to be a hormone.) The *renin-angiotensin system* will be detailed in Chapters 13 and 18.

The Heart

The endocrine cells in the heart are cardiac muscle cells in the walls of the *right atrium,* which is the chamber that receives blood from the largest veins. If blood volume becomes too great, these cardiac muscle cells are excessively stretched, which stimulates them to release the hormone *atrial natriuretic peptide* (*ANP*) (nā-trē-ū-RET-ik; *natrium,* sodium + *ouresis,* making water). In general, the effects of ANP oppose those of angiotensin II: ANP promotes the loss of sodium ions and water at the kidneys and inhibits renin release and the secretion of ADH and aldosterone. ANP secretion results in a reduction in both blood volume and pressure. We will consider the actions of this hormone further when we discuss the control of blood pressure and volume in Chapter 13.

Clinical Note
NATRIURETIC PEPTIDES

Natriuretic peptides are a group of naturally occurring substances that counteract the effects of the renin-angiotensin system. They cause vasodilation, stimulate the kidneys to increase sodium excretion (natriuresis), and water loss. They appear to be effective in the management of congestive heart failure because they decrease preload and promote sodium and water loss.

Three types of natriuretic peptides have been identified:

- *Atrial natriuretic peptide (ANP).* ANP is produced in the atria. The identification of ANP was the first indication that the heart also has some endocrine function.
- *Brain natriuretic peptide (BNP).* BNP is synthesized in the ventricles but named BNP as it was first identified in the porcine brain.
- *C-type natriuretic peptide (CNP).* CNP is produced in the brain.

Both ANP and BNP are released in response to atrial and ventricular stretch, respectively. They cause vasodilation, inhibition of aldosterone secretion from the adrenal cortex, and inhibition of renin in the kidney. Both ANP and BNP will cause natriuresis and a reduction in intravascular volume. These effects are amplified by antagonism of antidiuretic hormone (ADH). The physiologic effects of CNP are different from those of ANP and BNP. CNP has a hypotensive effect, but no direct diuretic or natriuretic actions.

In the emergency setting, BNP can be measured in the blood. An elevated BNP is suggestive of congestive heart failure. The higher the level of BNP the more severe is the heart failure. ■

The Thymus

The **thymus** is located in the *mediastinum,* usually just posterior to the sternum. In a newborn infant, the thymus is relatively enormous, and often extends from the base of the neck to the superior border of the heart. As the child grows, the thymus continues to enlarge slowly, and reaches its maximum size just before puberty, at a weight of about 40 g (1.4 oz). After puberty, it gradually diminishes in size; by age 50, the thymus may weigh less than 12 g (0.4 oz).

The thymus produces several hormones collectively known as the **thymosins** (THĪ-mō-sinz), which play a key role in the development and maintenance of normal immune defenses. It has been suggested that the gradual decrease in the size and secretory abilities of the thymus may make the elderly more susceptible to disease. The structure of the thymus and the functions of the thymosins will be considered in more detail in Chapter 14.

The Gonads

The Testes

In males, the **interstitial cells** of the testes produce the steroid hormones known as androgens, of which **testosterone** (tes-TOS-ter-ōn) is the most important. Testosterone promotes the production of functional sperm, maintains the secretory glands of the male reproductive tract, and determines secondary sex characteristics such as the distribution of facial hair and body fat. Testosterone also affects metabolic operations throughout the body. It stimulates protein synthesis and muscle growth, and it produces aggressive behavioral responses. During embryonic development, the production of testosterone affects the development of male reproductive ducts, external genitalia, and CNS structures, including hypothalamic nuclei that will later affect sexual behaviors.

Sustentacular cells in the testes support the formation of functional sperm. Under FSH stimulation, these cells secrete the hormone **inhibin,** which inhibits the secretion of FSH by the anterior pituitary. Throughout adult life, these two hormones interact to maintain sperm production at normal levels.

The Ovaries

In the ovaries, female sex cells *(ova)* develop in specialized structures called **follicles,** under stimulation by FSH. Follicle cells that surround the ova produce **estrogens** (ES-trō-jenz), which are steroid hormones that support the maturation of the eggs and stimulate the growth of the lining of the uterus. Under FSH stimulation, follicle cells secrete inhibin, which suppresses FSH release through a negative feedback mechanism comparable to that in males. After ovulation has occurred, the follicular cells reorganize into a **corpus luteum.** The cells of the corpus luteum then begin to release a mixture of estrogens and progestins, especially **progesterone** (prō-JES-ter-ōn). Progesterone accelerates the movement of fertilized eggs along the uterine tubes and prepares the uterus for the arrival of a developing embryo. In combination with other hormones, it also causes an enlargement of the mammary glands.

The production of androgens, estrogens, and progestins is controlled by regulatory hormones released by the anterior pituitary gland. During pregnancy, the placenta functions as an endocrine organ, and works with the ovaries and the pituitary gland to promote normal fetal development and delivery. (Table 10–5 summarizes the characteristics of the reproductive hormones.)

Clinical Note
ESTROGEN REPLACEMENT THERAPY

Estrogen protects women against heart disease and vaginal atrophy and conserves calcium and phosphorus, which helps prevent the bone loss of osteoporosis. Estrogen levels begin to slowly fall when a woman enters menopause, and although the incidence of heart disease in women is significantly lower than in men, women quickly catch up with men once menopause occurs.

Estrogen replacement therapy (ERT) has become common practice. In addition to preventing heart disease and osteoporosis, estrogen helps to minimize some of the uncomfortable side effects of menopause including "hot flashes" and vaginal atrophy. Most patients receiving ERT take estrogen on a cyclical basis unless they have had a hysterectomy, in which case they usually take it daily. ■

Adipose Tissue

Adipose tissue is a type of loose connective tissue introduced in Chapter 4. ∞ p. 107 Adipose tissue is known to produce two peptide hormones: *leptin* and *resistin*. **Leptin,** which is secreted by adipose tissue throughout the body, has several functions, including the negative feedback control of appetite. When you eat, adipose tissue absorbs glucose and lipids and synthesizes

STRUCTURE/CELLS	HORMONE	PRIMARY TARGET	EFFECTS
TABLE 10–5 *Hormones of the Reproductive System*			
TESTES			
Interstitial cells	Androgens	Most cells	Support functional maturation of sperm, protein synthesis in skeletal muscles, male secondary sex characteristics, and associated behaviors
Sustentacular cells	Inhibin	Anterior pituitary	Inhibits secretion of FSH
OVARIES			
Follicular cells	Estrogens	Most cells	Support follicle maturation, female secondary sex characteristics, and associated behaviors
	Inhibin	Anterior pituitary	Inhibits secretion of FSH
Corpus luteum	Progestins	Uterus, mammary glands	Prepare uterus for implantation; prepare mammary glands for secretory functions

triglycerides (fats) for storage. At the same time, it releases leptin into the bloodstream. Leptin binds to neurons in the hypothalamus involved with emotion and appetite control. The result is a sense of satiation and the suppression of appetite.

Leptin also enhances GnRH and gonadotropin synthesis. This effect explains why (1) thin girls commonly enter puberty relatively late, (2) an increase in body fat content can improve fertility, and (3) women stop menstruating when their body fat content becomes very low.

Resistin reduces insulin sensitivity throughout the body; it has been proposed as the "missing link" between obesity and Type II diabetes mellitus. (See the Clinical Note "Diabetes Mellitus" on p. 391.)

→ CONCEPT CHECK QUESTIONS

1. Which pancreatic hormone causes skeletal muscle and liver cells to convert glucose to glycogen?
2. What effect would increased levels of glucagon have on the amount of glycogen stored in the liver?
3. What are the primary targets and effects of the hormones calcitriol and erythropoietin?

Answers begin on p. 792.

■ Patterns of Hormonal Interaction

Even though hormones are usually studied individually, extracellular fluids contain a mixture of hormones whose concentrations change daily and even hourly. When a cell receives instructions from two different hormones at the same time, four outcomes are possible:

■ The two hormones may have **antagonistic** (opposing) **effects,** as is the case for parathyroid hormone and calcitonin, or insulin and glucagon.

■ The two hormones may have additive effects, in which case the net result is greater than the effect that each would produce acting alone. In some cases, the net result is greater than the *sum* of their individual effects. An example of such a **synergistic** (sin-er-JIS-tik; *syn,* together + *ergon,* work) **effect** is the glucose-sparing action of GH and glucocorticoids. ∞ pp. 376, 385

■ Hormones can have a **permissive effect** on other hormones. In such cases, one hormone must be present if a second hormone is to produce its effects. Thus, epinephrine, for example, has no apparent effect on energy consumption unless thyroid hormones are also present in normal concentrations.

■ Hormones may also produce different but complementary results in a given tissue or organ. These **integrative effects** are important in coordinating the activities of diverse physiological systems. The differing effects of calcitriol and parathyroid hormone on tissues involved in calcium metabolism are an example.

The next few sections will discuss how hormones interact to control normal growth, reactions to stress, alterations of behavior, and the effects of aging. More detailed discussions can be found in chapters on cardiovascular function, metabolism, excretion, and reproduction.

Hormones and Growth

Normal growth requires the cooperation of several endocrine organs. Six hormones—growth hormone, thyroid hormones, insulin, parathyroid hormone, calcitriol, and reproductive hormones—are especially important, although many others have secondary effects on growth rates and patterns:

■ *Growth hormone (GH).*   The effects of GH on protein synthesis and cellular growth are most apparent in children, in whom GH supports muscular and skeletal development. In adults, GH helps maintain normal blood glucose concentrations and mobilizes lipid reserves stored in adipose tissues. It is not the primary hormone involved, however, and an adult with a growth hormone deficiency but normal levels of thyroxine, insulin, and glucocorticoids will have no physiological problems. Undersecretion or oversecretion of GH can lead to *pituitary dwarfism* or *gigantism,* respectively.

■ *Thyroid hormones.*   Normal growth also requires appropriate levels of thyroid hormones. If these hormones are absent for the first year after birth, the nervous system fails to develop normally, which produces mental retardation. If thyroxine concentrations decline later in life but before puberty, normal skeletal development will not continue.

■ *Insulin.*   Growing cells need adequate supplies of energy and nutrients. Without insulin, the passage of glucose and amino acids across cell membranes is drastically reduced or eliminated.

■ *Parathyroid hormone (PTH) and calcitriol.*   Parathyroid hormone and calcitriol promote the absorption of calcium for building bone. Without adequate levels of both hormones, bones can enlarge but will be poorly mineralized, weak, and flexible. For example, in *rickets,* which is a condition that typically results from inadequate production of calcitriol in growing children, the limb bones are so weak that they bend under the body's weight. ∞ p. 158

■ *Reproductive hormones.*   Both the activity of osteoblasts in key locations and the growth of specific cell populations are affected by the presence or absence of sex hormones (androgens in males, estrogens in females). The targets differ for androgens and estrogens, and the differential growth induced by these hormonal changes accounts for sex-related differences in skeletal proportions and secondary sex characteristics.

Hormones and Stress

Any condition—whether physical or emotional—that threatens homeostasis is a form of **stress.** Stresses produced by the action of *stressors* may be (1) physical, such as illness or injury; (2) emotional, such as depression or anxiety; (3) environmental, such as extreme heat or cold; or (4) metabolic, such as acute starvation. Many stresses are opposed by specific homeostatic adjustments. For example, a decline in body temperature leads to shivering or changes in the pattern of blood flow, which act to restore normal body temperature.

In addition, the body has a general response to stress that can occur while other, more specific, responses are under way. *All stress-causing factors produce the same basic pattern of hormonal and physiological adjustments.* These responses are part of the **general adaptation syndrome (GAS),** also known as the **stress response.** The GAS has three basic phases: the *alarm phase,* the *resistance phase,* and the *exhaustion phase* (Figure 10–21●).

The **alarm phase** is an immediate response to the stress; it is under the direction of the sympathetic division of the autonomic nervous system. During this phase, energy reserves (mainly in the form of glucose) are mobilized, and the body prepares for any physical activities needed to eliminate or escape from the source of the stress. Epinephrine is the dominant hormone of the alarm phase, and its secretion accompanies the sympathetic activation that produces the "fight or flight" response. ∞ p. 307 Even though the effects of epinephrine are most apparent during the alarm phase, other hormones play supporting roles. For example, the reduction of water losses that result from ADH production and aldosterone secretion can be very important if the stress involves a loss of blood.

The temporary adjustments of the alarm phase often remove or overcome the stress. But some stresses, including starvation, acute illness, or severe anxiety, can persist for hours, days, or even weeks. If stress lasts longer than a few hours, the individual enters the **resistance phase** of the GAS. Glucocorticoids (GCs) are the dominant hormones of the resistance phase. Epinephrine, GH, and thyroid hormones are also involved. Energy demands in the resistance phase remain higher than normal due to the combined effects of these hormones.

Neural tissue has a high demand for energy, and if blood glucose concentrations fall too far, neural function deteriorates. The endocrine secretions of the resistance phase coordinate three integrated actions to maintain adequate levels of glucose in the blood: (1) the mobilization of lipid and protein reserves,

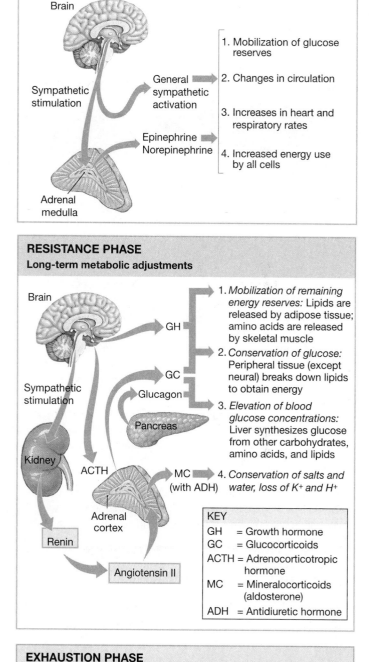

ALARM PHASE (Fight or flight)
Immediate short-term response to crises

1. Mobilization of glucose reserves
2. Changes in circulation
3. Increases in heart and respiratory rates
4. Increased energy use by all cells

RESISTANCE PHASE
Long-term metabolic adjustments

1. *Mobilization of remaining energy reserves:* Lipids are released by adipose tissue; amino acids are released by skeletal muscle
2. *Conservation of glucose:* Peripheral tissue (except neural) breaks down lipids to obtain energy
3. *Elevation of blood glucose concentrations:* Liver synthesizes glucose from other carbohydrates, amino acids, and lipids
4. *Conservation of salts and water, loss of K+ and H+*

KEY
GH = Growth hormone
GC = Glucocorticoids
ACTH = Adrenocorticotropic hormone
MC = Mineralocorticoids (aldosterone)
ADH = Antidiuretic hormone

EXHAUSTION PHASE
Collapse of vital systems

Causes may include:
- Exhaustion of lipid reserves
- Inability to produce glucocorticoids
- Failure of electrolyte balance
- Cumulative structural or functional damage to vital organs

● **FIGURE 10–21** The General Adaptation Syndrome.

(2) the conservation of glucose for neural tissues, and (3) the synthesis and release of glucose by the liver. In addition, blood volume is maintained by the actions of ADH and aldosterone.

When the resistance phase ends, homeostatic regulation breaks down, and the **exhaustion phase** begins. Unless corrective actions are taken almost immediately, the ensuing failure of one or more organ systems will prove fatal. Mineral imbalances contribute to the existing problems with major

systems. The production of aldosterone throughout the resistance phase results in a conservation of sodium ions at the expense of potassium ions. As the body's potassium content declines, a variety of cells—notably neurons and muscle fibers—begin to malfunction. Although a single cause, such as heart failure, may be listed as the cause of death, the underlying problem is the inability to support the endocrine and metabolic adjustments of the resistance phase.

Clinical Note
THE PHYSIOLOGIC RESPONSE TO STRESS

EMS is an inherently stressful profession. *Stress* can be defined as any condition within the body that threatens *homeostasis*. The word *stress* also refers to a "hardship or strain," or a "physical or emotional response to a stimulus." A person's reactions to stress are individual and varied. They are affected by previous exposure to the *stressor* (a stimulus that causes stress), the perception of the event, general life events, and personal coping skills.

Adapting to stress is a dynamic and evolving process. The body has a general physiological response to stress. In fact, all stressors produce the same basic pattern of hormonal and physiological adjustments. The stress response is called the *general adaptation syndrome* (*GAS*) and consists of three basic phases: *alarm, resistance,* and *exhaustion* (see Figure 10–21). At the end comes a period of rest and recovery.

- *Stage I: Alarm.* The alarm phase is the "fight-or-flight" response. It occurs when the body physically and rapidly prepares to defend itself against a threat, whether real or imagined. Hormones begin to flood the body via the sympathetic nervous system under the control of the hypothalamus. *Epinephrine* and *norepinephrine* from the adrenal medulla increase the heart rate and blood pressure, dilate the pupils, increase the blood sugar level, slow digestion, and dilate the bronchial tree. In addition, the pituitary gland

begins releasing *adrenocorticotropic (ACTH) hormones* that stimulate the adrenal cortex.

- *Stage II: Resistance.* If the stress lasts longer than a few hours, the individual will enter the resistance phase. The glucocorticoid hormones dominate this stage, although other hormones are involved. Energy demands remain higher because of increased production of the glucocorticoids, epinephrine, growth hormone, and thyroid hormone. These serve to maintain an elevated level of blood glucose, which facilitates the mobilization of lipid and protein, the conservation of glucose for the brain and neural tissues, and the synthesis and release of glucose by the liver. In this stage, the individual beings to cope with the stress. Over time, the individual may become desensitized or adapted to stressors. Late in this stage, physiological parameters, such as blood pressure and pulse rate, may return to normal.

- *Stage III: Exhaustion.* When the resistance phase ends, the homeostatic regulatory mechanisms break down and the exhaustion phase begins. Prolonged exposure to the same stressors leads to exhaustion of an individual's ability to resist and adapt. Resistance to all stressors declines, and susceptibility to physical and psychological ailments increases. A period of rest and recovery is necessary for a healthy outcome. Unless corrective actions are taken, organ system failure will begin. ■

Hormones and Behavior

The hypothalamus regulates many endocrine functions, and its neurons monitor the levels of many circulating hormones. Other portions of the brain that affect how we act, or behave, are also quite sensitive to hormonal stimulation.

The clearest demonstrations of the behavioral effects of specific hormones involve individuals whose endocrine glands are oversecreting or undersecreting. But even normal changes in circulating hormone levels can cause behavioral changes. In *precocious* (premature) *puberty*, sex hormones are produced at an inappropriate time, perhaps as early as age 5 or 6. An affected child not only begins to develop adult secondary sex characteristics but also undergoes significant behavioral changes. The "nice little kid" disappears, and the child becomes aggressive and assertive due to the effects of sex hormones on CNS function.

Thus, behaviors that in normal teenagers are usually attributed to environmental stimuli, such as peer pressure, can have a physiological basis as well. In adults, changes in the mixture of hormones that reach the CNS can have significant effects on intellectual capabilities, memory, learning, and emotional states.

Hormones and Aging

The endocrine system undergoes relatively few functional changes with age. The most dramatic exception is the decline in the concentration of reproductive hormones. Effects of these hormonal changes on the skeletal system were noted in Chapter 6 ∞ p. 166; further discussion can be found in Chapter 20.

Blood and tissue concentrations of many other hormones, including TSH, thyroid hormones, ADH, PTH, prolactin, and glucocorticoids, remain unchanged with increasing age. Although circulating hormone levels may remain within normal limits, some endocrine tissues become less responsive to stimulation. For example, in elderly individuals, less GH and insulin are secreted after a carbohydrate-rich meal. The reduction in levels of GH and other tropic hormones affects tissues throughout the body; these hormonal effects are associated with the reductions in bone density and muscle mass noted in earlier chapters.

Finally, it should be noted that age-related changes in peripheral tissues may make them less responsive to some hormones. This loss of sensitivity has been documented for glucocorticoids and ADH.

Clinical Note
ENDOCRINE DISORDERS

Some endocrine disorders, such as diabetes mellitus and hypothyroidism, are common. Others are rare. Most endocrine disorders begin with an increase or decrease in production of a specific hormone. Some conditions will result from under- or overproduction of several hormones from the same gland. Many of the endocrine disorders result in physical findings that make them fairly easily to recognize (Figure 10–22●). Some are more common in Third World countries, because many of these patients cannot afford or have never been offered the necessary treatment. ■

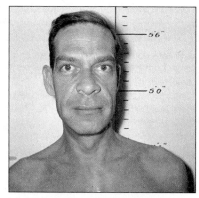

(a) *Acromegaly* results from the overproduction of growth hormone after puberty, when most of the epiphyseal cartilages have fused. Bone shapes change, and cartilaginous areas of the skeleton enlarge. Notice the broad facial features and the enlarged lower jaw.

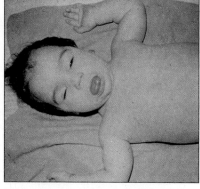

(b) *Cretinism or congenital hypothyroidism* results from thyroid hormone insufficiency in infancy.

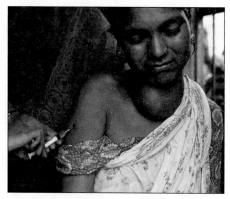

(c) An enlarged thyroid gland, or *goiter,* can be associated with thyroid hyposecretion due to iodine insufficiency in adults.

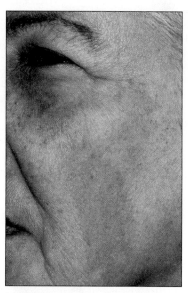

(d) *Addison's disease* is caused by hyposecretion of corticosteroids, especially glucocorticoids. Pigment changes result from stimulation of melanocytes by ACTH, which is structurally similar to MSH.

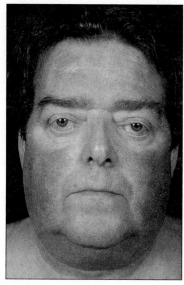

(e) *Cushing's disease* is caused by hypersecretion of glucocorticoids. Lipid reserves are mobilized, and adipose tissue accumulates in the cheeks and at the base of the neck.

● **FIGURE 10–22** Endocrine Abnormalities.

The Endocrine System in Perspective

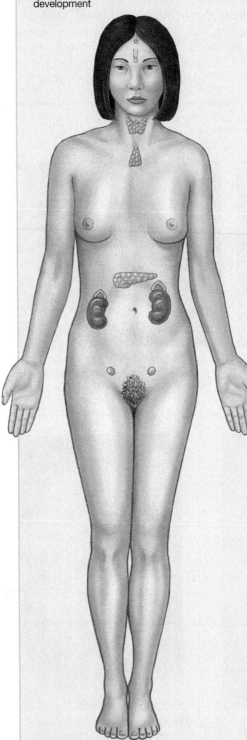

For All Systems

Adjusts metabolic rates and substrate utilization; regulates growth and development

The Integumentary System

↺ • Protects superficial endocrine organs; epidermis synthesizes vitamin D_3

↻ • Sex hormones stimulate sebaceous gland activity, influence hair growth, fat distribution, and apocrine sweat gland activity; PRL stimulates development of mammary glands; adrenal hormones alter dermal blood flow, stimulate release of lipids from adipocytes; MSH stimulates melanocyte activity

The Skeletal System

↺ • Protects endocrine organs, especially in brain, chest, and pelvic cavity

↻ • Skeletal growth regulated by several hormones; calcium mobilization regulated by parathyroid hormone and calcitonin; sex hormones speed growth and closure of epiphyseal cartilages at puberty and help maintain bone mass in adults

The Muscular System

↺ • Skeletal muscles provide protection for some endocrine organs

↻ • Hormones adjust muscle metabolism, energy production, and growth; regulate calcium and phosphate levels in body fluids; speed skeletal muscle growth

The Nervous System

↺ • Hypothalamic hormones directly control pituitary secretions and indirectly control secretions of other endocrine organs; controls adrenal medullae; secretes ADH and oxytocin

↻ • Several hormones affect neural metabolism; hormones help regulate fluid and electrolyte balance; reproductive hormones influence CNS development and behaviors

The Cardiovascular System

↺ • Circulatory system distributes hormones throughout the body; heart secretes ANP

↻ • Erythropoietin regulates production of RBCs; several hormones elevate blood pressure; epinephrine elevates heart rate and contraction force

The Lymphatic System

↺ • Lymphocytes provide defense against infection and, with other WBCs, assist in repair after injury

↻ • Glucocorticoids have anti-inflammatory effects; thymosins stimulate development of lymphocytes; many hormones affect immune function

The Respiratory System

↺ • Provides oxygen and eliminates carbon dioxide generated by endocrine cells

↻ • Epinephrine and norepinephrine stimulate respiratory activity and dilate respiratory passageways

The Digestive System

↺ • Provides nutrients to endocrine cells; endocrine cells of pancreas secrete insulin and glucagon; liver produces angiotensinogen

↻ • E and NE stimulate constriction of sphincters and depress activity along digestive tract; digestive tract hormones coordinate secretory activities along tract

The Urinary System

↺ • Kidney cells (1) release renin and erythropoietin when local blood pressure declines and (2) produce calcitriol

↻ • Aldosterone, ADH, and ANP adjust rates of fluid and electrolyte reabsorption in kidneys

The Reproductive System

↺ • Steroid sex hormones and inhibin suppress secretory activities in hypothalamus and pituitary

↻ • Hypothalamic regulatory hormones and pituitary hormones regulate sexual development and function; oxytocin stimulates uterine and mammary gland smooth muscle contractions

● **FIGURE 10–23** Functional Relationships Between the Endocrine System and Other Systems.

■ Integration with Other Systems

The relationships between the endocrine system and other organ systems are summarized in Figure 10–23●. This overview excludes all the hormones associated with the digestive system and the control of digestive functions. These hormones will be discussed in Chapter 16.

→ **CONCEPT CHECK QUESTIONS**

1. What type of hormonal interaction occurs when insulin lowers the level of glucose in the blood and glucagon causes glucose levels to rise?
2. The lack of which hormones inhibits skeletal formation?

Answers begin on p. 792.

Chapter Review

Access more review material online at *www.prenhall.com/bledsoe*. There you will find quiz questions, labeling activities, animations, essay questions, and web links.

Key Terms

adrenal cortex 385
adrenal medulla 386
endocrine cell 368
first messenger 370
general adaptation syndrome 397

glucagon 389
hormone 368
hypophyseal portal system 375
hypophysis 373
insulin 389

pancreas 389
peptide hormone 368
pituitary gland 373
second messenger 370
steroid hormone 368

Related Clinical Terms

Addison's disease A condition caused by the hyposecretion of glucocorticoids and mineralocorticoids; characterized by an inability to mobilize energy reserves and maintain normal blood glucose levels.

cretinism (KRĒ-tin-ism) A condition caused by hypothyroidism at birth or in infancy; marked by inadequate skeletal and nervous development and a metabolic rate as much as 40 percent below normal levels.

Cushing's disease A condition caused by the hypersecretion of glucocorticoids; characterized by the excessive breakdown of lipid reserves and proteins, and relocation of lipids.

diabetes insipidus A disorder that develops either when the posterior pituitary no longer releases adequate amounts of ADH or when the kidneys cannot respond to ADH.

diabetes mellitus (MEL-ĭ-tus) A disorder characterized by glucose concentrations high enough to overwhelm the kidneys' reabsorption capabilities.

diabetic retinopathy, nephropathy, neuropathy Disorders of the retina, kidneys, and peripheral nerves, respectively, related to diabetes mellitus; these conditions most often afflict middle-aged or older diabetics.

endocrinology (EN-do-kri-NOL-o-jē) The study of hormones and hormone-secreting tissues and glands and their roles in physiological and disease processes in the body.

general adaptation syndrome (GAS) The pattern of hormonal and physiological adjustments with which the body responds to all forms of stress.

glycosuria (glī-ko-SOO-rē-a) The presence of glucose in the urine.

goiter An abnormal enlargement of the thyroid gland.

hyperglycemia Abnormally high glucose levels in the blood.

hypoglycemia Abnormally low glucose levels in the blood.

myxedema In adults, the effects of hyposecretion of thyroid hormones, including subcutaneous swelling, hair loss, dry skin, low body temperature, muscle weakness, and slowed reflexes.

polyuria The production of excessive amounts of urine; a clinical sign of diabetes.

thyrotoxicosis A condition caused by the oversecretion of thyroid hormones (*hyperthyroidism*). Signs and symptoms include increases in metabolic rate, blood pressure, and heart rate; excitability and emotional instability; and lowered energy reserves.

Type I diabetes mellitus, or *insulin-dependent diabetes mellitus (IDDM)*, or *juvenile-onset diabetes* A type of diabetes mellitus; the primary cause is inadequate insulin production by the beta cells of the pancreatic islets.

Type II diabetes mellitus, or *non-insulin-dependent diabetes mellitus (NIDDM)*, or *maturity-onset diabetes* A type of diabetes mellitus in which insulin levels are normal or elevated but peripheral tissues no longer respond to insulin normally.

Summary Outline

1. In general, the nervous system performs short-term "crisis management," whereas the endocrine system regulates longer-term, ongoing metabolic processes. Endocrine cells release **hormones,** which are chemicals that alter the metabolic activities of many different tissues and organs. *(Figure 10–1)*

AN OVERVIEW OF THE ENDOCRINE SYSTEM 368

The Structure of Hormones 368

1. Hormones can be divided into three groups based on chemical structure: amino acid derivatives, peptide hormones, and lipid derivatives.

2. *Amino acid derivatives* are structurally similar to amino acids; they include *epinephrine, norepinephrine, thyroid hormones,* and *melatonin.*

3. **Peptide hormones** are chains of amino acids.

4. There are two classes of *lipid derivatives*: **steroid hormones,** which are lipids that are structurally similar to cholesterol; and *eicosanoids,* which are fatty acid-based hormones that include **prostaglandins.**

The Mechanisms of Hormonal Action 369

5. Hormones exert their effects by modifying the activities of **target cells** (peripheral cells that are sensitive to that particular hormone). *(Figure 10–2)*

6. Receptors for amino acid-derived hormones, peptide hormones, and fatty acid-derived hormones are located on the cell membranes of target cells; in this case, the hormone acts as a **first messenger** that causes the formation of a **second messenger** in the cytoplasm. Thyroid and steroid hormones cross the cell membrane and bind to receptors in the cytoplasm or nucleus. Thyroid hormones also bind to mitochondria, where they increase the rate of ATP production. *(Figure 10–3)*

The Secretion and Distribution of Hormones 371

7. Hormones may circulate freely or bind to transport proteins. Free hormones are rapidly removed from the bloodstream. *(Figure 10–4)*

Key Note 372

The Control of Endocrine Activity 372

8. The most direct patterns of endocrine control involve negative feedback on the endocrine cells that result from changes in the extracellular fluid.

9. The most complex endocrine responses involve the hypothalamus, which regulates the activities of the nervous and endocrine systems by three mechanisms: (1) it acts as an endocrine organ by releasing hormones into the circulation; (2) it secretes **regulatory hormones** that control the activities of endocrine cells in the pituitary gland; and (3) its autonomic centers exert direct neural control over the endocrine cells of the adrenal medullae. *(Figure 10–5)*

THE PITUITARY GLAND 373

1. The **pituitary gland** (*hypophysis*) releases nine important peptide hormones; all bind to membrane receptors, and most use cyclic-AMP as a second messenger. *(Figure 10–6)*

The Anterior Pituitary Gland 374

2. Hypothalamic neurons release regulatory factors into the surrounding interstitial fluids, which then enter highly permeable capillaries.

3. The **hypophyseal portal system** ensures that all of the blood that enters the *portal vessels* will reach target cells in the anterior pituitary before returning to the general circulation. *(Figure 10–7)*

4. The rate of regulatory hormone secretion by the hypothalamus is regulated through negative feedback mechanisms. *(Figure 10–8)*

5. The seven hormones of the **anterior pituitary gland** are: (1) **thyroid-stimulating hormone (TSH),** which triggers the release of thyroid hormones; (2) **adrenocorticotropic hormone (ACTH),** which stimulates the release of **glucocorticoids** by the adrenal gland; (3) **follicle-stimulating hormone (FSH),** which stimulates estrogen secretion and egg development in females and sperm production in males; (4) **luteinizing hormone (LH),** which causes ovulation and **progestin** production in females and **androgen** production in males; (5) **prolactin (PRL),** which stimulates the development of the mammary glands and the production of milk; (6) **growth hormone (GH),** which stimulates cell growth and replication by triggering the release of **somatomedins** from liver cells; and (7) **melanocyte-stimulating hormone (MSH),** which stimulates melanocytes to produce melanin (but is not normally secreted by nonpregnant human adults).

The Posterior Pituitary Gland 377

6. The **posterior pituitary gland** contains the axons of hypothalamic neurons that manufacture **antidiuretic hormone (ADH)** and **oxytocin.** ADH decreases the amount of water lost at the kidneys. In females, oxytocin stimulates smooth muscle cells in the uterus and contractile cells in the mammary glands. In males, it stimulates sperm duct and prostate gland smooth muscle contractions. *(Figure 10–9; Table 10–1)*

Key Note 379

THE THYROID GLAND 379

1. The **thyroid gland** lies near the **thyroid cartilage** of the larynx and consists of two lobes. *(Figure 10–10)*

Thyroid Follicles and Thyroid Hormones 379

2. The thyroid gland contains numerous **thyroid follicles.** Thyroid follicles release several hormones, including **thyroxine (TX or T_4)** and **triiodothyronine (T_3).** *(Tables 10–2, 10–3)*

3. Thyroid hormones exert a **calorigenic effect,** which enables us to adapt to cold temperatures.

The C Cells of the Thyroid Gland: Calcitonin 380

4. The **C cells** of the follicles produce **calcitonin (CT),** which helps lower calcium ion concentrations in body fluids. *(Figures 10–12, 10–13, 10–14; Table 10–3)*

THE PARATHYROID GLANDS 384

1. Four **parathyroid glands** are embedded in the posterior surface of the thyroid gland. The **chief cells** of the parathyroid produce **parathyroid hormone (PTH)** in response to lower than normal concentrations of calcium ions. Chief cells and the C cells of the thyroid gland maintain calcium ion levels within relatively narrow limits. *(Figures 10–11, 10–15; Table 10–3)*

Key Note 385

THE ADRENAL GLANDS 385

1. A single **adrenal gland** lies along the superior border of each kidney. Each gland, which is surrounded by a fibrous capsule, can be subdivided into the superficial adrenal cortex and the inner adrenal medulla. *(Figure 10–16)*

The Adrenal Cortex 385

2. The **adrenal cortex** manufactures steroid hormones called *adrenocortical steroids* (**corticosteroids**). The cortex produces (1) **glucocorticoids**—notably, **cortisol, corticosterone,** and **cortisone**—which, in response to ACTH, affect glucose metabolism; (2) **mineralocorticoids**—principally **aldosterone**—which, in response to *angiotensin II*, restricts sodium and water losses at the kidneys, sweat glands, digestive tract, and salivary glands; and (3) androgens of uncertain significance. *(Figure 10–16; Table 10–4)*

The Adrenal Medulla 386

3. The **adrenal medulla** produces **epinephrine** and **norepinephrine.** *(Figures 10–16, 10–17; Table 10–4)*

THE PINEAL GLAND 388

1. The **pineal gland** synthesizes **melatonin.** Melatonin appears to: (1) slow the maturation of sperm, eggs, and reproductive organs; (2) protect neural tissue from free radicals; and (3) establish daily circadian rhythms.

Key Note 389

Regulation of Blood Glucose Concentrations 389

THE PANCREAS 389

1. The **pancreas** contains both exocrine and endocrine cells. The **exocrine pancreas** secretes an enzyme-rich fluid that functions in the digestive tract. Cells of the **endocrine pancreas** form clusters called **pancreatic islets** (*islets of Langerhans*) that contain **alpha cells** (which produce the hormone **glucagon**) and **beta cells** (which secrete **insulin**). *(Figure 10–18)*

2. Insulin lowers blood glucose by increasing the rate of glucose uptake and utilization; glucagon raises blood glucose by increasing the rates of glycogen breakdown and glucose synthesis in the liver. *(Figures 10–19, 10–20)*

Key Note 393

THE ENDOCRINE TISSUES OF OTHER ORGAN SYSTEMS 393

The Intestines 393

1. The intestines release hormones that coordinate digestive activities.

The Kidneys 394

2. Endocrine cells in the kidneys produce hormones important in calcium metabolism and in the maintenance of blood volume and blood pressure.

3. **Calcitriol** stimulates calcium and phosphate ion absorption along the digestive tract.

4. **Erythropoietin (EPO)** stimulates red blood cell production by the bone marrow.

5. **Renin** activity leads to the formation of **angiotensin II,** the hormone that stimulates the production of aldosterone in the adrenal cortex.

The Heart 394

6. Specialized muscle cells in the heart produce **atrial natriuretic peptide (ANP)** when blood pressure or blood volume becomes excessive.

The Thymus 394

7. The **thymus** produces several hormones called **thymosins,** which play a role in developing and maintaining normal immunological defenses.

The Gonads 395

8. The **interstitial cells** of the paired testes in males produce androgens and **inhibin.** The androgen testosterone is the most important sex hormone in males. *(Table 10–5)*

9. In females, ova (eggs) develop in **follicles;** follicle cells that surround the eggs produce **estrogens** and inhibin. After ovulation, the cells reorganize into a **corpus luteum** that releases a mixture of estrogens and progestins, especially **progesterone.** If pregnancy occurs, the placenta functions as an endocrine organ. *(Table 10–5)*

Adipose Tissue 395

10. Adipose tissue secretes **leptin** (which participates in negative feedback control of appetite) and **resistin** (which reduces insulin sensitivity).

PATTERNS OF HORMONAL INTERACTION 396

1. The endocrine system functions as an integrated unit and hormones often interact. These interactions may have: (1) **antagonistic** (opposing) effects; (2) **synergistic** (additive) effects; (3) **permissive** effects; or (4) **integrative** effects, in which hormones produce different but complementary results.

Hormones and Growth 396

2. Normal growth requires the cooperation of several endocrine organs. Six hormones are especially important: growth hormone, thyroid hormones, insulin, parathyroid hormone, calcitriol, and reproductive hormones.

Hormones and Stress 397

3. Any condition that threatens homeostasis is a **stress.** Our bodies respond to stress through the **general adaptation syndrome (GAS).** The GAS is divided into three phases: (1) the **alarm phase,** which is under the direction of the sympathetic division of the ANS; (2) the **resistance phase,** which is dominated by glucocorticoids; and (3) the **exhaustion phase,** which is the eventual breakdown of homeostatic regulation and failure of one or more organ systems. *(Figures 10–21, 10–22)*

Hormones and Behavior 398

4. Many hormones affect the functional state of the nervous system, and produce changes in mood, emotional states, and various behaviors.

Hormones and Aging 398

5. The endocrine system undergoes relatively few functional changes with advancing age. The most dramatic endocrine change is a decline in the concentration of reproductive hormones.

INTEGRATION WITH OTHER SYSTEMS 401

1. The endocrine system affects all organ systems by adjusting metabolic rates and regulating growth and development. *(Figure 10–23)*

Review Questions

Level 1: Reviewing Facts and Terms

Match each item in column A with the most closely related item in column B. Place letters for answers in the spaces provided.

COLUMN A
____ 1. thyroid gland
____ 2. pineal gland
____ 3. polyuria
____ 4. parathyroid gland
____ 5. thymus gland
____ 6. adrenal cortex
____ 7. heart
____ 8. endocrine pancreas
____ 9. gonadotropins
____ 10. hypothalamus
____ 11. pituitary gland
____ 12. growth hormone

COLUMN B
a. isets of Langerhans
b. atrophies by adulthood
c. atrial natriuretic peptide
d. cell growth
e. melatonin
f. hypophysis
g. excessive urine production
h. calcitonin
i. secretes regulatory hormones
j. FSH and LH
k. secretes androgens, mineralocorticoids, and glucocorticoids
l. stimulated by low calcium levels

13. Adrenocorticotropic hormone (ACTH) stimulates the release of:
    (a) thyroid hormones by the hypothalamus.
    (b) gonadotropins by the adrenal glands.
    (c) somatotropins by the hypothalamus.
    (d) steroid hormones by the adrenal glands.

14. FSH production in males supports:
    (a) maturation of sperm by stimulating sustentacular cells.
    (b) development of muscles and strength.
    (c) production of male sex hormones.
    (d) increased desire for sexual activity.

15. The hormone that induces ovulation in women and promotes the ovarian secretion of progesterone is:
    (a) interstitial cell-stimulating hormone.
    (b) estradiol.
    (c) luteinizing hormone.
    (d) prolactin.

16. The two hormones released by the posterior pituitary are:
    (a) somatotropin and gonadotropin.
    (b) estrogen and progesterone.
    (c) growth hormone and prolactin.
    (d) antidiuretic hormone and oxytocin.

17. The primary function of antidiuretic hormone (ADH) is to:
    (a) increase the amount of water lost at the kidneys.
    (b) decrease the amount of water lost at the kidneys.
    (c) dilate peripheral blood vessels to decrease blood pressure.
    (d) increase absorption along the digestive tract.

18. The element required for normal thyroid function is:
    (a) magnesium.
    (b) calcium.
    (c) potassium.
    (d) iodine.

19. Reduced fluid losses in the urine due to retention of sodium ions and water is a result of the action of:
    (a) insulin.
    (b) calcitonin.
    (c) aldosterone.
    (d) cortisone.

20. The adrenal medullae produce the hormones:
    (a) cortisol and cortisone.
    (b) epinephrine and norepinephrine.
    (c) corticosterone and testosterone.
    (d) androgens and progesterone.

21. What seven hormones are released by the anterior pituitary gland?

22. What effects do calcitonin and parathyroid hormone have on blood calcium levels?

23. (a) What three phases of the general adaptation syndrome (GAS) constitute the body's response to stress? (b) What endocrine secretions play dominant roles in the alarm and resistance phases?

Level 2: Reviewing Concepts

24. What is the primary difference in the way the nervous and endocrine systems communicate with their target cells?

25. How can a hormone modify the activities of its target cells?

26. What possible results occur when a cell receives instructions from two different hormones at the same time?

27. How would blocking the activity of phosphodiesterase affect a cell that responds to hormonal stimulation by the cAMP second messenger system?

Level 3: Critical Thinking and Clinical Applications

28. Roger has been suffering from extreme thirst; he drinks numerous glasses of water every day and urinates a great deal. Name two disorders that could produce these symptoms. What test could a clinician perform to determine which disorder is present?

29. Julie is pregnant and is not receiving any prenatal care. She has a poor diet that consists mostly of fast food. She drinks no milk, and prefers colas instead. How will this situation affect Julie's level of parathyroid hormone?

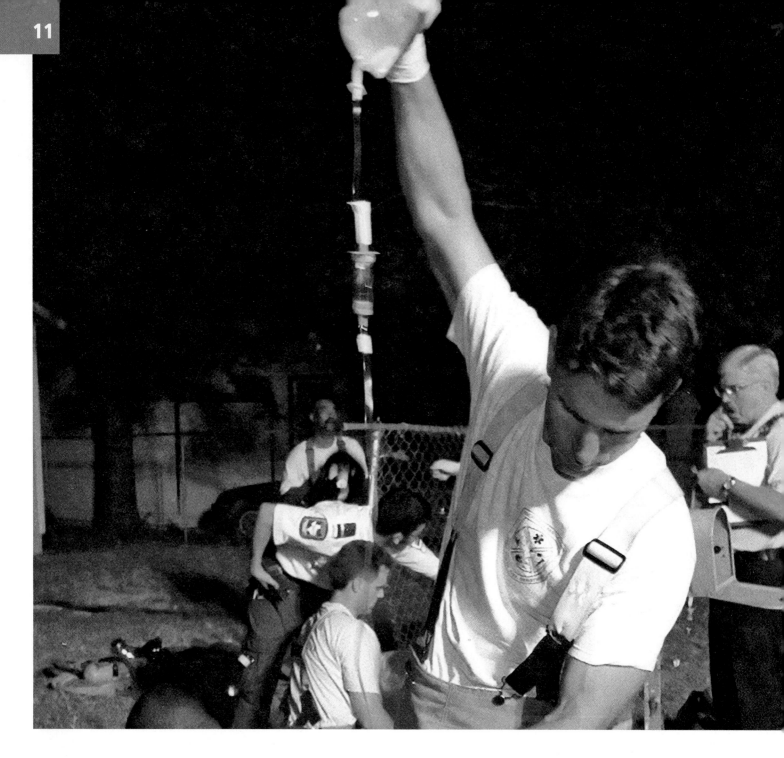

11 The Cardiovascular System: Blood

THE LOSS OF circulating blood volume (hypovolemia) is a common problem in trauma. An adequate blood volume and an adequate amount of hemoglobin molecules are necessary to ensure oxygen transport to the peripheral tissues. The ability to administer blood products in the prehospital setting is extremely limited. Current trauma practice calls for the administration of small volumes of salt-containing solutions in order to maintain a low level of perfusion.

Chapter Outline

Chapter Objectives

1. Describe the important components and major functions of blood. (p. 408)
2. Discuss the composition and functions of plasma. (pp. 408, 410)
3. Describe the origins and production of the formed elements in blood. (pp. 410–411)
4. Discuss the characteristics and functions of red blood cells. (pp. 411–412)
5. Explain the factors that determine a person's blood type, and why blood types are important. (pp. 417–419)
6. Categorize the various white blood cells on the basis of their structures and functions. (pp. 419–424)
7. Describe the mechanisms that reduce blood loss after an injury. (pp. 425–428)

Vocabulary Development

agglutinins gluing; *agglutinization*
embolos plug; *embolus*
erythros red; *erythrocytes*
haima blood; *hemostasis*
hypo- below; *hypoxia*
karyon nucleus; *megakaryocyte*

leukos white; *leukocyte*
megas big; *megakaryocyte*
myelos marrow; *myeloid*
-osis condition; *leukocytosis*
ox- presence of oxygen; *hypoxia*
penia poverty; *leukopenia*

poiesis making; *hemopoiesis*
punctura a piercing; *venipuncture*
stasis halt; *hemostasis*
thrombos clot; *thrombocytes*
vena vein; *venipuncture*

THE LIVING BODY is in constant chemical communication with its external environment. Nutrients are absorbed through the lining of the digestive tract, gases move across the thin epithelium of the lungs, and wastes are excreted in the feces and urine. Even though these chemical exchanges occur at specialized sites, they affect every cell, tissue, and organ in a matter of moments because all parts of the body are linked by the **cardiovascular system,** which is an internal transport network.

The cardiovascular system can be compared to the cooling system of a car. Both systems have a circulating fluid (blood versus water), a pump (the heart versus a water pump), and flexible tubing to carry the fluid (the blood vessels versus radiator hoses). Although the cardiovascular system is far more complicated and versatile, both mechanical and biological systems can malfunction from fluid losses, pump failures, or damaged tubing.

Small embryos do not need cardiovascular systems because diffusion across their exposed surfaces can exchange materials rapidly enough to meet their demands. By the time an embryo has reached a few millimeters in length, however, developing tissues consume oxygen and nutrients and generate waste products faster than they can be provided or removed through simple diffusion. At that stage, the cardiovascular system must

begin functioning to provide a rapid-transport system for oxygen, nutrients, and waste products. It is the first organ system to become fully operational: the heart begins beating by the end of the third week of embryonic life, when most other systems have barely begun to develop. When the heart starts beating, blood begins circulating. The embryo can now make more efficient use of the nutrients obtained from the maternal bloodstream, and its size doubles in the next week.

This chapter considers the nature of the circulating blood. Chapter 12 focuses on the structure and function of the heart, and Chapter 13 examines the organization of blood vessels and the integrated functions of the cardiovascular system. Chapter 14 considers the lymphatic system, which is a defense system intimately connected to the cardiovascular system.

■ The Functions of Blood

The circulating fluid of the body is **blood,** which is a specialized connective tissue that contains cells suspended in a fluid matrix. p. 109 Blood has five major functions:

1. *Transportation of dissolved gases, nutrients, hormones, and metabolic wastes.* Blood carries oxygen from the lungs to

the tissues and carbon dioxide from the tissues to the lungs. It distributes nutrients that are either absorbed at the digestive tract or released from storage in adipose tissue or in the liver. Blood carries hormones from endocrine glands toward their target cells. It also absorbs the wastes produced by active cells and carries these wastes to the kidneys for excretion.

2. *Regulation of the pH and ion composition of interstitial fluids throughout the body.* Diffusion between interstitial fluids and blood eliminates local deficiencies or excesses of ions such as calcium or potassium. Blood also absorbs and neutralizes the acids generated by active tissues, such as lactic acid produced by skeletal muscle contractions.

3. *Restriction of fluid losses at injury sites.* Blood contains enzymes and factors that respond to breaks in vessel walls by initiating the process of *blood clotting.* The resulting blood clot acts as a temporary patch that prevents further reductions in blood volume.

4. *Defense against toxins and pathogens.* Blood transports white blood cells, which are specialized cells that migrate into body tissues to fight infections or remove debris. Blood also delivers *antibodies,* which are special proteins that attack invading organisms or foreign compounds.

5. *Stabilization of body temperature.* Blood absorbs the heat generated by active skeletal muscles and redistributes it to other tissues. When body temperature is high, blood is directed to the skin surface, where heat is lost to the environment. When body temperature is too low, the flow of warm blood is restricted to crucial structures— to the brain and to other temperature-sensitive organs.

■ The Composition of Blood

Blood has a unique composition (Figure 11–1●). It is a fluid connective tissue that consists of a matrix called *plasma* (PLAZ-muh) and several formed elements.

Plasma contains dissolved proteins rather than the network of insoluble fibers like those in loose connective tissue or cartilage. Because these proteins are in solution, plasma is slightly denser than water. **Formed elements** are blood cells and cell fragments (platelets) suspended in the plasma. **Red blood cells (RBCs),** or **erythrocytes** (e-RITH-rō-sīts; *erythros,* red), transport oxygen and carbon dioxide. The less numerous **white blood cells (WBCs),** or **leukocytes** (LOO-kō-sīts; *leukos,* white), function as part of the body's defense mechanisms. **Platelets** are small, membrane-enclosed packets of cytoplasm that contain enzymes and factors important to blood clotting.

Together, the plasma and formed elements constitute **whole blood.** The volume of whole blood varies from 5 to 6 liters (5.3–6.4 quarts) in the cardiovascular system of an adult man, and from 4 to 5 liters (4.2–5.3 quarts) in that of an adult woman. Whole blood components may be separated, or **fractionated,** for analytical or clinical purposes.

Blood Collection and Analysis

Fresh whole blood is usually collected from a superficial vein, such as the median cubital vein on the anterior surface of the elbow (Figure 11–1a). This procedure is called **venipuncture** (VĒN-i-punk-chur; *vena,* vein + *punctura,* a piercing). It is a common sampling technique because superficial veins are easy to locate, the walls of veins are thinner than those of arteries of comparable size, and blood pressure in the venous system is relatively low, so the puncture wound seals quickly. The most common clinical procedures examine venous blood.

Blood from peripheral capillaries can be obtained by puncturing the tip of a finger, an ear lobe, or (in infants) the great toe or heel of the foot. A small drop of capillary blood can be used to prepare a *blood smear,* which is a thin film of blood on a microscope slide. The blood smear is then stained with special dyes to show different types of formed elements.

An **arterial puncture,** or "arterial stick," may be required for evaluating the efficiency of gas exchange at the lungs. Samples are usually drawn from the radial artery at the wrist or the brachial artery at the elbow. Whole blood from any of these sources has the same basic physical characteristics:

- *Temperature.* The temperature of blood is roughly 38°C (100.4°F), which is slightly above normal body temperature.
- *Viscosity.* Blood is five times as viscous as water—that is, five times stickier, more cohesive, and resistant to flow than water. The high viscosity results from interactions among the dissolved proteins, formed elements, and water molecules in plasma.
- *pH.* Blood is slightly alkaline, with a pH between 7.35 and 7.45 (average: 7.4). ∞ p. 39

■ Plasma

Plasma makes up approximately 55 percent of the volume of whole blood, and water accounts for 92 percent of plasma volume (Figure 11–1b). Together, plasma and interstitial fluid account for most of the volume of extracellular fluid (ECF) in the body.

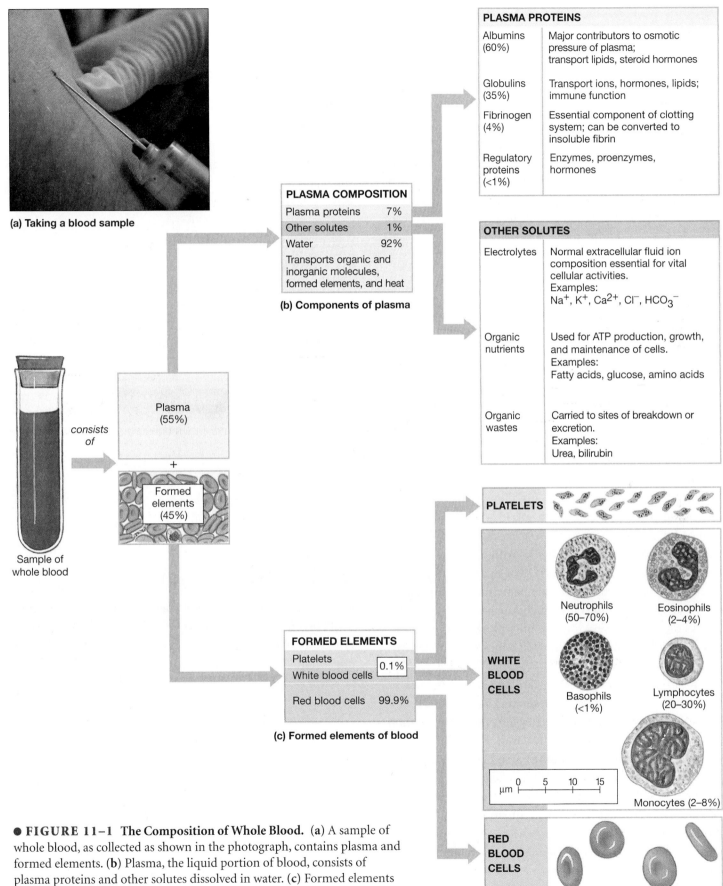

(a) Taking a blood sample

PLASMA PROTEINS

Albumins (60%)	Major contributors to osmotic pressure of plasma; transport lipids, steroid hormones
Globulins (35%)	Transport ions, hormones, lipids; immune function
Fibrinogen (4%)	Essential component of clotting system; can be converted to insoluble fibrin
Regulatory proteins (<1%)	Enzymes, proenzymes, hormones

PLASMA COMPOSITION

Plasma proteins	7%
Other solutes	1%
Water	92%

Transports organic and inorganic molecules, formed elements, and heat

(b) Components of plasma

OTHER SOLUTES

Electrolytes	Normal extracellular fluid ion composition essential for vital cellular activities. Examples: Na^+, K^+, Ca^{2+}, Cl^-, HCO_3^-
Organic nutrients	Used for ATP production, growth, and maintenance of cells. Examples: Fatty acids, glucose, amino acids
Organic wastes	Carried to sites of breakdown or excretion. Examples: Urea, bilirubin

Plasma (55%)

consists of

+

Formed elements (45%)

Sample of whole blood

PLATELETS

WHITE BLOOD CELLS

Neutrophils (50–70%)   Eosinophils (2–4%)

Basophils (<1%)   Lymphocytes (20–30%)

μm 0 5 10 15

Monocytes (2–8%)

FORMED ELEMENTS

Platelets	0.1%
White blood cells	
Red blood cells	99.9%

(c) Formed elements of blood

RED BLOOD CELLS

● **FIGURE 11–1 The Composition of Whole Blood.** (**a**) A sample of whole blood, as collected as shown in the photograph, contains plasma and formed elements. (**b**) Plasma, the liquid portion of blood, consists of plasma proteins and other solutes dissolved in water. (**c**) Formed elements include red blood cells, white blood cells, and platelets.

In many respects, the composition of plasma resembles that of interstitial fluid (IF). The concentrations of the major plasma ions (electrolytes), for example, are similar to those of the IF and differ from those inside cells. The similarity is understandable because a constant exchange of water, ions, and small solutes occurs between the interstitial fluid and plasma across the walls of capillaries (considered in Chapter 13). The primary differences between plasma and interstitial fluid involve the concentrations of dissolved proteins and the levels of respiratory gases (oxygen and carbon dioxide). The protein concentrations differ because plasma proteins are too large to cross the walls of capillaries. The differences in the levels of respiratory gases are due to the respiratory activities of cells (as discussed in Chapter 15).

Plasma Proteins

Plasma contains considerable quantities of dissolved proteins. On average, each 100 mL of plasma contains about 7 g of protein (see Figure 11–1b), almost five times the concentration in interstitial fluid. The large size of most blood proteins prevents them from crossing capillary walls, so they remain confined to the circulatory system. The three primary types of plasma proteins are *albumins* (al-BŪ-minz), *globulins* (GLOB-û-linz), and *fibrinogen* (fī-BRIN-ō-jen).

Albumins constitute roughly 60 percent of all plasma proteins. As the most abundant proteins, they are major contributors to the osmotic pressure of the plasma. **Globulins,** which account for another 35 percent of plasma proteins, include antibodies and transport proteins. **Antibodies,** also called **immunoglobulins** (i-mū-nō-GLOB-ū-linz), attack foreign proteins and pathogens. **Transport proteins** bind small ions, hormones, or compounds that might otherwise be lost at the kidneys or that have very low solubility in water. One example is thyroid-binding globulin, which binds and transports thyroid hormones. Regulates calcium or body weight

Both albumins and globulins can bind to lipids, such as triglycerides, fatty acids, or cholesterol. These lipids are not themselves water soluble, but the protein-lipid combination readily dissolves in plasma. In this way, the cardiovascular system transports insoluble lipids to peripheral tissues. Globulins involved in lipid transport are called *lipoproteins* (LĪ-pō-prō-tēnz).

The third type of plasma protein, **fibrinogen,** functions in blood clotting. Under certain conditions, fibrinogen molecules interact and convert to form large, insoluble strands of **fibrin** (FĪ-brin). These fibers provide the basic framework for a blood clot. If steps are not taken to prevent clotting in a plasma sample, fibrinogen will convert to fibrin. The fluid left after the clotting proteins are removed is known as **serum.**

The liver synthesizes more than 90 percent of the plasma proteins including all albumins and fibrinogen, and most of the globulins. Antibodies (immunoglobulins) are produced by *plasma cells* of the lymphatic system. Because the liver is the primary source of plasma proteins, liver disorders can alter the composition and functional properties of the blood. For example, some forms of liver disease can lead to uncontrolled bleeding that results from the inadequate synthesis of fibrinogen and other plasma proteins involved in the clotting response.

Key Note

Approximately half of the volume of whole blood consists of cells and cell products. Plasma resembles interstitial fluid but it contains a unique mixture of proteins not found in other extracellular fluids.

CONCEPT CHECK QUESTIONS

1. Why is venipuncture a common technique for obtaining a blood sample?
2. What would be the effects of a decrease in the amount of plasma proteins?

Answers begin on p. 792.

■ Formed Elements

The most abundant formed elements are red blood cells and white blood cells, but platelets—small noncellular packets of cytoplasm that function in the clotting response—are also an important formed element (Figure 11–1c).

The Production of Formed Elements

Formed elements are produced through the process of **hemopoiesis** (hēm-ō-poy-Ē-sis), or *hematopoiesis*. Embryonic blood cells appear in the bloodstream during the third week of development. These cells divide repeatedly, and increase in number. The vessels of the yolk sac, which is an embryonic membrane, are the primary sites of blood formation for the first eight weeks of development. As other organ systems develop, some of the embryonic blood cells move out of the bloodstream and into the liver, spleen, thymus, and bone marrow. These embryonic cells become *pluripotent stem cells,* or **hemocytoblasts,** which subsequently divide to produce *myeloid stem cells* and *lymphoid stem cells.* Both populations of these stem cells retain the ability to divide, and their daughter cells go on to produce specific types of blood cells. We will consider the fates of the myeloid and lymphoid stem cells as we discuss the formation of each type of formed element.

The liver and spleen are the primary sites of hemopoiesis from the second to fifth month of development, but as the skeleton enlarges, the bone marrow becomes an increasingly important site. In adults, red bone marrow is the only site of red blood cell production and the primary site of white blood cell formation as well.

Red Blood Cells

Red blood cells (RBCs) contain the pigment *hemoglobin,* which binds and transports oxygen and carbon dioxide. Red blood cells are the most abundant blood cells, accounting for 99.9 percent of the formed elements.

The number of RBCs in the blood of a normal individual staggers the imagination. A standard blood test reports the number of RBCs per microliter (μL) of whole blood as the *red blood cell count.* In adult males, 1 microliter, or 1 cubic millimeter (mm³), of whole blood contains roughly 5.4 million erythrocytes; in adult females, 1 microliter contains about 4.8 million. A single drop of whole blood contains some 260 million RBCs. Roughly one-third of the 75 trillion cells in the human body are red blood cells.

The **hematocrit** (he-MA-tō-krit) is the percentage of whole blood volume occupied by cellular elements. The hematocrit is measured after a blood sample has been spun in a centrifuge so that all the formed elements come out of suspension. In adult males, it averages 46 percent (range: 40–54); in adult females, 42 percent (range: 37–47). The difference in hematocrit between males and females reflects the fact that androgens (male hormones) stimulate red blood cell production, whereas estrogens (female hormones) do not. Because whole blood contains roughly 1000 red blood cells for each white blood cell, the hematocrit closely approximates the volume of RBCs. For this reason, hematocrit values are often reported as the *volume of packed red cells* (*VPRC*), or simply the *packed cell volume* (*PCV*).

Many conditions can affect the hematocrit. The hematocrit increases, for example, during dehydration (owing to a reduction in plasma volume) or after *erythropoietin* (*EPO*) stimulation. ∞ p. 394 The hematocrit decreases as a result of internal bleeding or problems with RBC formation. As a result, the hematocrit alone does not provide specific diagnostic information. However, a change in hematocrit is an indication that more specific tests are needed. (Some of those tests are discussed on p. 416.)

The Structure of RBCs

Red blood cells are specialized to transport oxygen and carbon dioxide within the bloodstream. As Figure 11–2● shows, each RBC is a biconcave disc with a thin central region and a thick outer margin. This unusual shape has two important effects on RBC function: (1) it gives each RBC a relatively large surface area to volume ratio that increases the rate of diffusion between its cytoplasm and the surrounding plasma; and (2) it enables RBCs to bend and flex to squeeze through narrow capillaries.

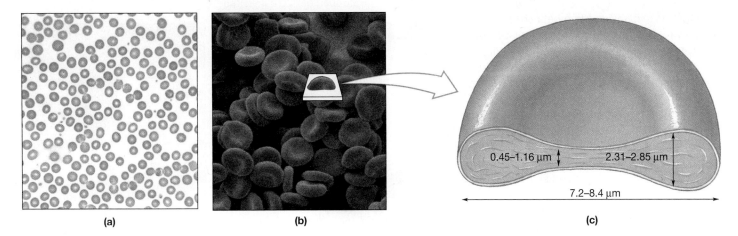

(a)　(b)　(c)

● **FIGURE 11–2 The Anatomy of Red Blood Cells.** (a) When viewed in a standard blood smear, red blood cells appear as two-dimensional objects because they are flattened against the surface of the slide. (b) A scanning electron micrograph of red blood cells reveals their three-dimensional structure. (c) The average dimensions of a mature red blood cell are indicated in this sectional view.

During their formation, RBCs lose most of their organelles, including mitochondria, ribosomes, and nuclei. Without a nucleus or ribosomes, RBCs can neither undergo cell division nor synthesize structural proteins or enzymes. Without mitochondria, they can obtain energy only through anaerobic metabolism, and rely on glucose obtained from the surrounding plasma. This characteristic makes RBCs relatively inefficient in terms of energy use, but it ensures that any oxygen they absorb will be carried to peripheral tissues, not "stolen" by mitochondria in the cytoplasm.

Hemoglobin Structure and Function

A mature red blood cell consists of a cell membrane that encloses a mass of transport proteins. Molecules of **hemoglobin** (HĒ-mō-glō-bin) (**Hb**) account for over 95 percent of an RBC's intracellular proteins. Hemoglobin is responsible for the cell's ability to transport oxygen and carbon dioxide.

Two pairs of globular proteins (each pair composed of slightly different polypeptide chains) combine to form a single hemoglobin molecule. ∞ p. 47 Each of the four subunits contains a single molecule of an organic pigment called **heme.** Each heme molecule holds an iron ion in such a way that it can interact with an oxygen molecule (O_2). The iron-oxygen inter-

action is very weak, and the two can easily separate. RBCs that contain hemoglobin with bound oxygen give blood a bright red color. The RBCs give blood a dark red, almost burgundy, color when oxygen is not bound to hemoglobin.

The amount of oxygen bound in each RBC depends on the conditions in the surrounding plasma. When oxygen is abundant in the plasma, hemoglobin molecules gain oxygen until all the heme molecules are occupied. As plasma oxygen levels decline, plasma carbon dioxide levels usually rise. Under these conditions, hemoglobin molecules release their oxygen reserves, and the globin portion of each hemoglobin molecule begins to bind carbon dioxide molecules in a process that is just as reversible as the binding of oxygen to heme.

As red blood cells circulate, they are exposed to varying concentrations of oxygen and carbon dioxide. At the lungs, where diffusion brings oxygen into the plasma and removes carbon dioxide, the hemoglobin molecules in red blood cells respond by absorbing oxygen and releasing carbon dioxide. In peripheral tissues, the situation is reversed; active cells consume oxygen and produce carbon dioxide. As blood flows through these areas, oxygen diffuses out of the plasma, and carbon dioxide diffuses in. Under these conditions, hemoglobin releases its bound oxygen and binds carbon dioxide.

Clinical Note
SICKLE CELL DISEASE

Sickle cell disease is an inherited disorder caused by abnormal hemoglobin. Hemoglobin molecules consist of four hemoglobin chains. Most adults have *hemoglobin A*, which contains two alpha chains and two beta chains ($\alpha_2\beta_2$). In sickle cell disease, the beta chains are replaced with an abnormal hemoglobin chain called *hemoglobin S*. Sickle cell disease is passed from parent to child in an autosomal recessive pattern. That is, if a child receives hemoglobin S from one parent and a normal beta hemoglobin chain from the other parent, the child has sickle-cell trait ($\alpha_2\beta S$). But, if the child receives an S chain from each parent, the child has sickle cell disease ($\alpha_2 S_2$).

Persons with sickle cell trait rarely have problems. However, persons with sickle cell disease have significant problems. When oxygenated, hemoglobin S functions normally. However, when deoxygenated, hemoglobin S polymerizes with the red blood cell (RBC), which causes the classic sickle (crescent) shape. Sickled RBCs increase the viscosity of the blood, which leads to sludging and obstruction of small blood vessels. Eventually, the cells be-

come irreversibly sickled. Sickled RBCs are rapidly hemolyzed, resulting in an RBC life span of 10–20 days, compared with the normal RBC life span of 120 days (Figure 11–3●). ■

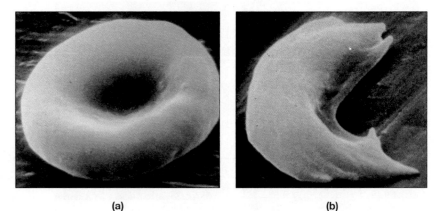

(a)   (b)

● **FIGURE 11–3 Sickling in Red Blood Cells.** (a) When fully oxygenated, the red blood cells of an individual with sickle-cell trait have the same appearance as normal RBCs. (b) At lower oxygen concentrations the RBCs change shape and become more rigid and sharply curved ("sickled"). (SEM × 67,500)

Normal activity levels can be sustained only when tissue oxygen levels are kept within normal limits. The blood of a person who has a low hematocrit, or whose RBCs have a reduced hemoglobin content, has a reduced oxygen-carrying capacity. This condition is called **anemia.** Anemia causes a variety of symptoms, including premature muscle fatigue, weakness, and a general lack of energy.

RBC Life Span and Circulation

RBCs are exposed to severe physical stresses. A single round-trip through the circulatory system usually takes less than a minute, but in that time a red blood cell is forced along vessels where it bounces off the walls, collides with other red blood cells, and is squeezed through tiny capillaries. With all this mechanical battering and no repair mechanisms, an RBC has a relatively short life span—only about 120 days. The continuous elimination of RBCs usually goes unnoticed because new ones enter the circulation at a comparable rate. About 1 percent of the circulating RBCs are replaced each day, and approximately 3 million new RBCs enter the circulation each second!

HEMOGLOBIN RECYCLING. As red blood cells age or are damaged, some of them rupture. When this occurs, the hemoglobin breaks down in the blood, and the individual polypeptide chains are filtered from the blood by the kidneys and lost in the urine. When large numbers of RBCs break down in the circulation, the urine can turn reddish or brown, which is a condition called **hemoglobinuria.**

Fortunately, only about 10 percent of RBCs survive long enough to rupture, or **hemolyze** (HĒ-mō-līz), within the bloodstream. Instead, phagocytic cells (macrophages) in the liver, spleen, and bone marrow usually recognize and engulf RBCs before they undergo *hemolysis,* in the process recycling hemoglobin and other components of RBCs. (Phagocytosis and phagocytic cells were introduced in Chapter 3; additional details are given in Chapter 14.) ∞ p. 74

The recycling of hemoglobin and turnover of red blood cells is shown in Figure 11–4●. Once a red blood cell has been engulfed and broken down by a macrophage, each component of a hemoglobin molecule has a different fate:

1. The four globular proteins of each hemoglobin molecule are disassembled into their component amino acids. These amino acids are either metabolized by the cell or released into the circulation for use by other cells.
2. Each heme molecule is stripped of its iron and converted to **biliverdin** (bil-i-VER-din), an organic compound with a green color. (Bad bruises commonly appear greenish because biliverdin forms in the blood-filled tissues.)

Biliverdin is then converted to **bilirubin** (bil-i-ROO-bin), which is an orange-yellow pigment, and released into the circulation. Liver cells absorb the bilirubin and normally release it into the small intestine within the bile. If, however, the bile ducts are blocked (by gallstones, for example), bilirubin then diffuses into peripheral tissues, giving them a yellow color that is most apparent in the skin and the sclera of the eyes. The condition characterized by yellow skin and eyes is called **jaundice** (JAWN-dis). Bilirubin that reaches the large intestine is converted to related pigment molecules, some of which are absorbed into the bloodstream and excreted into urine. The yellow color of urine and the brown color of feces are due to these bilirubin-derived pigments.

3. Iron extracted from heme molecules may be stored in the macrophage or released into the bloodstream, where it binds to **transferrin** (tranz-FER-in), a plasma transport protein. Red blood cells developing in the bone marrow absorb amino acids and transferrins from the circulation and use them in the synthesis of new hemoglobin molecules. Excess transferrins are removed in the liver and spleen, where the iron is stored in special protein-iron complexes.

In summary, most of the components of an individual red blood cell are recycled following hemolysis or phagocytosis. The entire system is remarkably efficient; although roughly 26 mg of iron are incorporated into hemoglobin molecules each day, a dietary supply of 1–2 mg can keep pace with the incidental losses that occur in the feces and urine.

Any impairment in iron uptake or metabolism can cause serious clinical problems because RBC formation will be affected. Women are especially dependent on a normal dietary supply of iron because their iron reserves are smaller than those of men. The body of a normal man contains around 3.5 g of iron in the ionic form Fe^{2+}. Of that amount, 2.5 g is bound to the hemoglobin of circulating red blood cells, and the rest is stored in the liver and bone marrow. In women, total body iron content averages 2.4 g, with roughly 1.9 g incorporated into red blood cells. Thus, a woman's iron reserves consist of only 0.5 g, half that of a typical man. If dietary supplies of iron are inadequate, hemoglobin production slows, and symptoms of *iron deficiency anemia* appear. The accumulation of too much iron in the liver and in cardiac muscle tissue can also cause problems. Excessive iron deposition in cardiac muscle cells has recently been linked to heart disease.

Red Blood Cell Formation

Red blood cell formation, or **erythropoiesis** (e-rith-rō-poy-Ē-sis), occurs in *red bone marrow,* or **myeloid tissue** (MĪ-e-loyd;

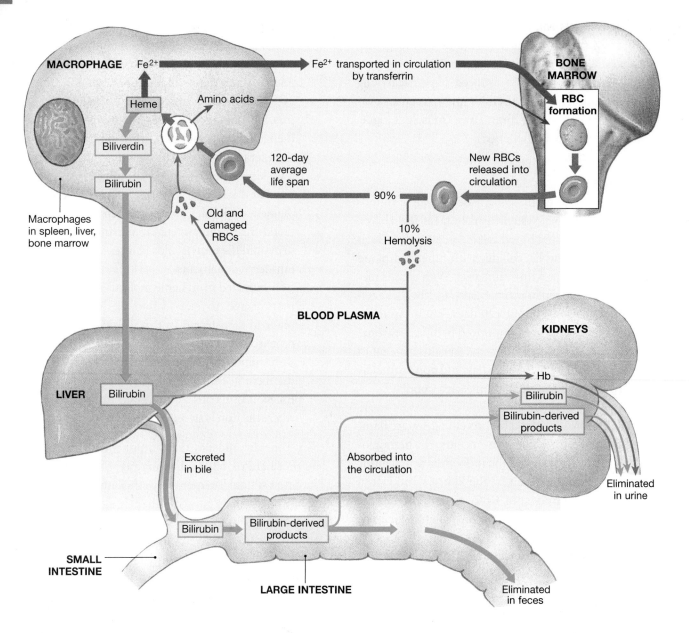

● **FIGURE 11–4** **Hemoglobin Recycling.** Macrophages in the spleen, liver, and bone marrow phagocytize and break down old and damaged RBCs and release the components of the hemoglobin molecules they contain. The breakdown of the protein portion of hemoglobin releases amino acids into the bloodstream for use by other cells, including RBCs developing in the red bone marrow. Iron from the heme units is either used in the formation of new hemoglobin or stored in the liver, spleen, and bone marrow. After removal of the iron, the remaining portions of the heme units are converted to bilirubin. Bilirubin in the circulation is absorbed by the liver and excreted within bile, which is released into the intestine. Bilirubin-derived products are excreted in urine or feces.

myelos, marrow). This tissue is located in the vertebrae, sternum, ribs, scapulae, pelvis, and proximal limb bones. Other marrow areas contain a fatty tissue known as *yellow bone marrow.* Under extreme stimulation, such as a severe and sustained blood loss, areas of yellow marrow can convert to red marrow, increasing the rate of RBC formation.

STAGES OF RBC MATURATION. Specialists in blood formation and function—**hematologists** (hē-ma-TOL-o-jists)—give spe-

cific names to key stages in the maturation of the formed elements. Like all formed elements, RBCs result from the divisions of hemocytoblasts (pluripotent stem cells) in red bone marrow (Figure 11–5●). In giving rise to the cells that ultimately become RBCs, the hemocytoblasts produce **myeloid stem cells,** some of which proceed through a series of stages in their development to mature erythrocytes (see left side of Figure 11–5).

Erythroblasts are very immature red blood cells that actively synthesize hemoglobin. After roughly four days of

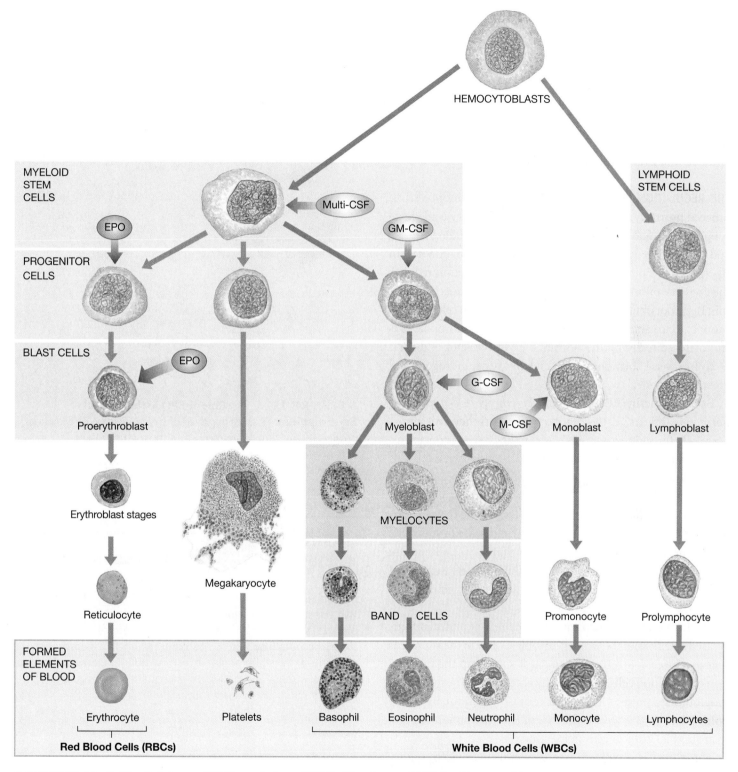

● **FIGURE 11–5** **The Origins and Differentiation of RBCs, Platelets, and WBCs.** Hemocytoblast divisions give rise to (1) lymphoid stem cells, which produce the various lymphocytes; and (2) myeloid stem cells, which produce cells that ultimately become red blood cells, platelets, and the other types of white blood cells. Hematologists use specific terms for the various key stages the formed elements pass through on the way to maturity. The targets of EPO and four colony-stimulating factors (CSFs) are also indicated.

differentiation, each erythroblast sheds its nucleus and becomes a **reticulocyte** (re-TIK-ū-lō-sīt). After two or three more days in the bone marrow synthesizing proteins, reticulocytes enter the bloodstream. At this time they can still be detected in a blood smear with stains that specifically combine with RNA. Normally, reticulocytes account for about 0.8 percent of the circulating erythrocytes. After 24 hours in circulation, the reticulocytes complete their maturation and become indistinguishable from other mature RBCs.

THE REGULATION OF ERYTHROPOIESIS. For erythropoiesis to proceed normally, the red bone marrow must receive adequate supplies of amino acids, iron, and vitamins (including B_{12}, B_6, and folic acid) required for protein synthesis. We obtain **vitamin B_{12}** from dairy products and meat, and its absorption requires the presence of *intrinsic factor* produced in the stomach. If vitamin B_{12} is not obtained from the diet, normal stem cell divisions cannot occur, and *pernicious anemia* results. Erythropoiesis is stimulated directly by the hormone erythropoietin and indirectly by several hormones, including thyroxine, androgens, and growth hormone.

Erythropoietin (EPO), also called *erythropoiesis-stimulating hormone,* appears in the plasma when peripheral tissues—especially the kidneys—are exposed to low oxygen concentrations (Figure 11–6●). The state of low tissue oxygen levels is called **hypoxia** (hī-POKS-ē-uh; *hypo-,* below + *ox-,* presence of oxygen). EPO is released (1) during anemia, (2) when blood flow to the kidneys declines, (3) when the oxygen content of air in the lungs declines (owing to disease or high altitude), and (4) when the respiratory surfaces of the lungs are damaged. Once in the bloodstream, EPO travels to areas of red bone marrow, where it stimulates stem cells and developing RBCs.

Erythropoietin has two major effects: (1) it stimulates increased cell division rates in erythroblasts and in the stem cells that produce erythroblasts; and (2) it speeds up the maturation of red blood cells, primarily by accelerating the rate of hemoglobin synthesis. Under maximum EPO stimulation, the bone marrow can increase the rate of RBC formation tenfold, to around 30 million cells per second.

This ability is important to a person recovering from a severe blood loss. However, if EPO is administered to a healthy person, the hematocrit may rise to 65 or more, and the resulting increase in blood viscosity increases the workload on the heart, which can lead to sudden death from heart failure. Similar risks result from *blood doping,* in which athletes reinfuse packed RBCs removed at an earlier date for the purpose of improving performance by increasing oxygen delivery to muscles.

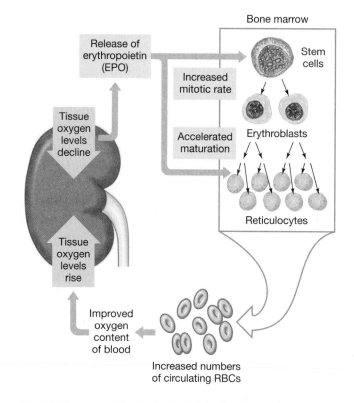

● **FIGURE 11–6**   **The Role of EPO in the Control of Erythropoiesis.** Tissues deprived of oxygen release EPO, which accelerates division of stem cells and the maturation of erythroblasts. More red blood cells then enter the circulation and improve the delivery of oxygen to peripheral tissues.

Clinical Note
RAPID BLOOD TESTS

In an emergency it is important to quickly determine baseline information about a patient's red blood cells (RBCs). The two most common rapid tests are the *hematocrit* and *hemoglobin* (H&H). The hematocrit determines the percentage of the blood occupied by RBCs. The hemoglobin test determines the concentration of hemoglobin in the RBC. Together, the H&H provides a quick evaluation of the quantity and quality of the patient's RBCs. Other tests routinely performed on RBCs are described in Table 11–1. ■

Key Note

Red blood cells (RBCs) are the most numerous cells in the body. They remain in circulation for approximately four months before being recycled; several million are produced each second. The hemoglobin inside RBCs transports oxygen from the lungs to peripheral tissues; it also carries carbon dioxide from those tissues to the lungs.

TABLE 11-1 *Common RBC Tests and Related Terminology*

TEST	DETERMINES	TERMS ASSOCIATED WITH ABNORMAL VALUES	
		ELEVATED	DEPRESSED
HEMATOCRIT (HCT)	Percentage of formed elements in whole blood. Normal = 37–54%	Polycythemia (may result from erythrocytosis or leukocytosis)	Anemia
COMPLETE BLOOD COUNT (CBC) RBC count	Number of RBCs per mL of whole blood. Normal = 42–63 million/µL	Erythrocytosis/polycythemia	Anemia
Hemoglobin concentration (Hb)	Concentration of hemoglobin in blood. Normal = 12–18 g/dL		Anemia
Reticulocyte count (Retic.)	Circulating percentage of reticulocytes. Normal = 0.8%	Reticulocytosis	
Mean corpuscular volume (MCV)	Average volume of a single RBC. Normal = 82–101 µm³ (normocytic)	Macrocytic	Microcytic
Mean corpuscular hemoglobin concentration (MCHC)	Average amount of Hb in one RBC. Normal = 27–34 pg/µL (normochromic)	Hyperchromic	Hypochromic

CONCEPT CHECK QUESTIONS

1. What effect does dehydration have on an individual's hematocrit?
2. How is the level of bilirubin in the blood affected by diseases (such as hepatitis and cirrhosis) that damage the liver?
3. What effect does a reduction in oxygen supply to the kidneys have on levels of erythropoietin in the blood?

Answers begin on p. 792.

RBCs and Blood Types

Antigens are substances (most often proteins) that can trigger an *immune response,* which is a defense mechanism that protects the body from infection. The membranes of all your cells contain surface antigens, which are substances that your immune defenses recognize as "normal." In other words, your immune system ignores these substances rather than attacking them as "foreign."

Red blood cells have at least 50 kinds of surface antigens. Which antigens your RBCs have is genetically determined. The presence or absence of three antigens on the surface of RBCs—**A, B,** and **Rh**—determines your **blood type.**

All of the red blood cells of a given individual have the same pattern of surface antigens. **Type A** blood has antigen A only, **Type B** has antigen B only, **Type AB** has both A and B, and **Type O** has neither A nor B (Figure 11–7a●). The average values for the U.S. population are Type O, 46 percent; Type A, 40 percent; Type B, 10 percent; and Type AB, 4 percent. Variations in these values differ by ethnic group and by region (Table 11–2).

TABLE 11-2 *The Distribution of Blood Types in Selected Populations*

POPULATION	PERCENTAGE WITH EACH BLOOD TYPE				
	O	A	B	AB	Rh⁺
U.S. (AVERAGE)	46	40	10	4	85
African American	49	27	20	4	95
Caucasian	45	40	11	4	85
Chinese American	42	27	25	6	100
Filipino American	44	22	29	6	100
Hawaiian	46	46	5	3	100
Japanese American	31	39	21	10	100
Korean American	32	28	30	10	100
NATIVE NORTH AMERICAN	79	16	4	< 1	100
NATIVE SOUTH AMERICAN	100	0	0	0	100
AUSTRALIAN ABORIGINE	44	56	0	0	100

The term **Rh positive** (Rh⁺) indicates the presence of the Rh antigen on the surface of RBCs. The absence of this antigen is indicated as **Rh negative** (Rh⁻). When an individual's complete blood type is recorded, the term *Rh* is usually omitted, and the data are reported as O negative (O⁻), A positive (A⁺), and so on.

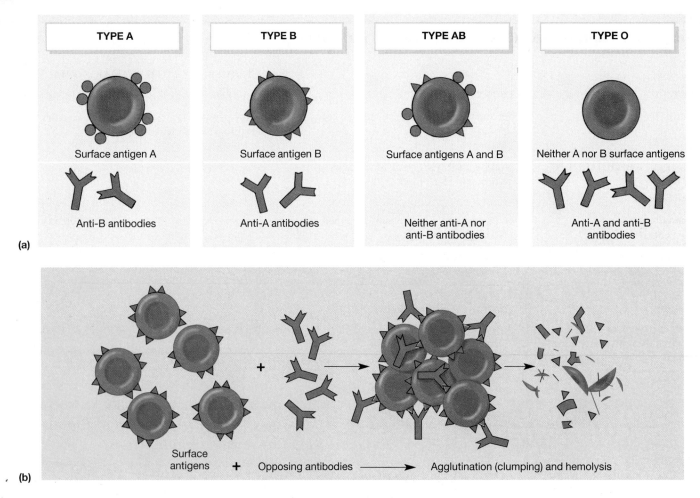

TYPE A	TYPE B	TYPE AB	TYPE O
Surface antigen A	Surface antigen B	Surface antigens A and B	Neither A nor B surface antigens
Anti-B antibodies	Anti-A antibodies	Neither anti-A nor anti-B antibodies	Anti-A and anti-B antibodies

(a)

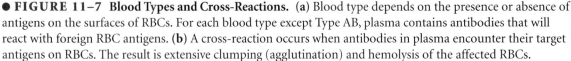

Surface antigens + Opposing antibodies ⟶ Agglutination (clumping) and hemolysis

(b)

● **FIGURE 11–7 Blood Types and Cross-Reactions.** (a) Blood type depends on the presence or absence of antigens on the surfaces of RBCs. For each blood type except Type AB, plasma contains antibodies that will react with foreign RBC antigens. (b) A cross-reaction occurs when antibodies in plasma encounter their target antigens on RBCs. The result is extensive clumping (agglutination) and hemolysis of the affected RBCs.

ANTIBODIES AND CROSS-REACTIONS. We noted previously that your immune system ignores the surface antigens—also called *agglutinogens* (a-gloo-TIN-ō-jenz)—on your own RBCs. However, your plasma contains antibodies, or *agglutinins* (a-GLOO-ti-ninz), that will attack surface antigens on RBCs of a different blood type (see Figure 11–7a). Thus, the plasma of individuals with Type A blood contains circulating anti-B antibodies, which will attack Type B surface antigens, and the plasma of Type B individuals contains anti-A antibodies, which will attack Type A surface antigens. Similarly, Type AB individuals lack antibodies against either A or B surface antigens, whereas the plasma of individuals with Type O blood contains both anti-A and anti-B antibodies.

The presence of these antibodies is why, before blood is transfused, the blood types of donor and recipient are identified. If an individual receives blood of a different blood type, antibodies in the recipient's plasma meet their specific anti-gen on the donated RBCs, and a **cross-reaction** occurs (Figure 11–7b●). Initially the binding of antigens and antibodies causes the foreign RBCs to clump together—a process called **agglutination** (a-gloo-ti-NĀ-shun). Subsequently, the RBCs may break up, or *hemolyze.* Clumps and fragments of RBCs under attack from antibodies form drifting masses that can plug small vessels in the kidneys, lungs, heart, or brain, damaging or destroying tissues. Such cross-reactions, or *transfusion reactions,* can be avoided by ensuring that the blood types of donor and recipient are **compatible.**

In practice, the surface antigens on the donor's cells are more important in determining compatibility than are the antibodies in the donor's plasma. Unless large volumes of whole blood or plasma are transferred, cross-reactions between the donor's plasma and the recipient's blood cells will fail to produce significant agglutination. Packed RBCs, with a minimal amount of plasma, are commonly transfused. Even when whole blood is transfused, the plasma is diluted through mixing with the recipient's relatively large plasma volume.

Unlike the case for Type A and Type B individuals, the plasma of an Rh-negative individual does not normally contain anti-Rh antibodies. These antibodies are present only if the individual has been *sensitized* by previous exposure to Rh-positive RBCs. Such exposure can occur accidentally, during a transfusion, but it can also occur in a normal pregnancy when an Rh-negative mother carries an Rh-positive fetus.

Clinical Note
EMERGENCY TRANSFUSIONS

Blood is a precious commodity and must be utilized so that it can provide the most benefit for the greatest number of people. A unit of whole blood consists of approximately 450 milliliters of blood with about 65 grams of hemoglobin. In current medical practice, it is uncommon to administer whole blood. Instead, the blood is separated into its various elements, and those elements are administered as needed. Elements derived from whole blood include the red blood cells (*packed red blood cells*), *platelets, granulocytes,* and plasma (*fresh frozen plasma*). Patients receive only the blood elements that they need. A patient who has sustained hemorrhage, for instance, will receive packed RBCs and crystalloid fluids, while a patient with a platelet disorder will receive platelet transfusion.

In order to prevent a *hemolytic transfusion reaction,* blood must be tested for compatibility between the donor and the recipient. Blood products are tested for the major antigens (AB and Rh). In addition, they are tested for minor antigen compatibility. The analysis and selection of blood products for transfusion is a time-consuming process called *typing* and *crossmatching*. In a critical emergency, there may not be time to wait for blood typing and crossmatching. In these situations, Type O-negative (O⁻) blood can be safely administered. Type O⁻ blood does not contain the A, B, or Rh antigen. Thus, it will not induce a hemolytic reaction in a patient with a different blood type. Because of this, persons with Type O⁻ blood are referred to as *universal donors.*

Persons who have type AB blood are referred to as *universal recipients.* They have both the A and B antigens on their RBCs and thus lack circulating antibodies against both A and B. In an emergency, they can receive any type of blood, as they already have the antibodies.

In an emergency setting, if time permits, it is preferred to wait until the blood is fully typed and crossmatched. If there is not adequate time for this, the patient can be given *type-specific blood*. Type-specific blood has been tested for the major (ABO and Rh) antigens. It has not been tested for the minor antigens. Finally, as described above, in critical situations, the patient may be given un-crossmatched Type O⁻ blood until typing and crossmatching have been completed.

Platelets do not contain any minor blood antigens. Platelets should be of the same ABO and Rh type. They are rapidly destroyed by the body and must be frequently administered. ■

CONCEPT CHECK QUESTIONS

1. Which blood types can be transfused into a person with Type AB blood?
2. Why can't a person with Type A blood receive blood from a person with Type B blood?

Answers begin on p. 792.

White Blood Cells

White blood cells, also known as WBCs or **leukocytes,** can be distinguished from RBCs by their larger size and by the presence of a nucleus and other organelles. WBCs also lack hemoglobin. White blood cells help defend the body against invasion by pathogens, and they remove toxins, wastes, and abnormal or damaged cells. Traditionally, WBCs have been divided into two groups on the basis of their appearance after staining: (1) *granulocytes* (with abundant stained granules) and (2) *agranulocytes* (with few if any stained granules). This categorization is convenient but somewhat misleading because the "granules" in granulocytes are actually secretory vesicles and lysosomes, and the "agranulocytes" also contain vesicles and lysosomes—they are just smaller and more difficult to see with the light microscope.

Clinical Note
HEMOLYTIC DISEASE OF THE NEWBORN

Genes that control the presence or absence of any surface antigen in the membrane of an RBC are provided by both parents, so a child's blood type can differ from that of either parent. During pregnancy, when fetal and maternal circulatory systems are closely intertwined, certain kinds of the mother's antibodies may cross the placenta, and attack and destroy fetal RBCs. The resulting potentially fatal condition is called *hemolytic disease of the newborn (HDN)*.

The sensitization that causes this condition usually occurs during delivery, when bleeding at the placenta and uterus exposes an Rh-negative mother to an Rh-positive fetus's Rh antigens. This event can trigger the production of anti-Rh antibodies in the mother. Because these antibodies are not produced in significant amounts until after delivery, the first Rh-positive infant is not affected. However, a sensitized Rh-negative mother will respond to a second Rh-positive fetus by producing massive amounts of anti-Rh antibodies. These antibodies attack and hemolyze the fetal RBCs, producing a dangerous anemia, which increases the fetal demand for blood cells. The resulting RBCs leave the bone marrow and enter the circulation before completing their development. Because these immature RBCs are erythroblasts, HDN is also known as *erythroblastosis fetalis* (e-rith-rō-blas-TŌ-sis fe-TAL-is). ■

Clinical Note
TESTING FOR BLOOD COMPATIBILITY

Testing for blood compatibility normally involves two steps: a determination of blood type and a *crossmatch test.* Although some 50 RBC surface antigens are known, the standard test for blood type categorizes a blood sample on the basis of the three most likely to produce dangerous cross-reactions. The test involves mixing drops of blood with solutions containing anti-A, anti-B, and anti-Rh antibodies and noting any cross-reactions. For example, if the RBCs clump together when exposed to anti-A and anti-B, the individual has Type AB blood. If no reactions occur, the person must be Type O. The presence or absence of the Rh antigen is also noted, and the individual is classified as Rh-positive or Rh-negative. In the most common type—Type O-positive (O^+)—the RBCs lack surface antigens A and B, but they do have the Rh antigen. Standard blood typing of both donor and recipient can be completed in a matter of minutes.

In an emergency (such as a severe gunshot wound), a patient may require 5 *liters* or more of blood before the damage can be repaired. Under these circumstances, Type O blood can be safely administered to a victim of any blood type because Type O RBCs lack A and B surface antigens. Because their blood cells are unlikely to produce severe cross-reactions in a recipient, Type O (especially O^-) individuals are sometimes called *universal donors.* Type AB individuals are called *universal recipients* because they lack anti-A or anti-B antibodies, which would attack donated RBCs. This term is used less frequently because reliable blood supplies and quick compatibility testing typically allow the administration of Type AB blood to Type AB recipients. With at least 48 other possible antigens on the cell surface, however, cross-reactions can occur, even to Type O blood. Whenever time and facilities permit, further testing is performed to ensure complete compatibility. *Crossmatch testing* involves exposing the donor's RBCs to a sample of the recipient's plasma under controlled conditions. This procedure reveals the presence of significant cross-reactions that involve other antigens and antibodies. Another way to avoid compatibility problems is to replace lost blood with synthetic blood substitutes, which do not contain surface antigens that can trigger a cross-reaction.

Other applications of blood compatibility—for example, paternity tests and crime detection—stem from the fact that blood groups are inherited. Results from such tests cannot prove that a particular individual is the criminal or the father involved, but it can prove that he is not involved. For example, a man with Type AB blood cannot be the father of an infant with Type O blood. ■

Clinical Note
BLOOD TRANSFUSION REACTIONS

Human red blood cell membranes contain hundreds of different antigenic structures. Great care is taken to ensure that blood products match the intended recipient before transfusion is undertaken. All blood products are tested to determine the ABO blood type and for the presence of the Rh antigen, as well as for several minor antigens. Blood that has been tested and deemed suitable for transfusion to a patient is said to be crossmatched. This means that the blood product in question has the lowest possible likelihood of causing a transfusion reaction.

Blood transfusion reactions can develop quite rapidly and are sometimes fatal. Transfusion reactions are usually classified as immune or nonimmune. Immune reactions are antibody-mediated reactions directed against a blood component. Nonimmune reactions include such complications as circulatory overload or transmission of an infectious agent.

A common type of blood transfusion reaction is the hemolytic reaction, in which red blood cell antibodies lyse (rupture) red cells within the circulatory system. This allows the contents of the cell to spill into the circulatory system. Hemoglobin can collect in the urine and adversely affect renal function. Patients with hemolytic reactions often complain of flushing, pain at the infusion site, chest or back pain, restlessness, anxiety, nausea, or diarrhea. Fever and chills are common findings. Shock and renal failure also can occur. Treatment of a hemolytic transfusion reaction should begin with termination and removal of the product being transfused, which should be returned to the laboratory to determine the cause of the reaction. Fluids should be administered to maintain adequate renal output. On occasion, it may be necessary to administer a diuretic such as furosemide (Lasix) or an osmotic diuretic such as mannitol (Osmotrol) in order to maintain renal function and prevent fluid overload. Intravenous antihistamines and corticosteroids are sometimes required. Analgesics and anti-inflammatory agents may be required for the treatment of pain as well as fever and chills.

Nonimmune transfusion reactions are managed by treating the underlying problem. Circulatory overload is treated with fluid restriction and diuretics as needed. Infections are treated based on the type of infection present. ■

WBC Circulation and Movement

Unlike RBCs, WBCs circulate for only a short portion of their life span. White blood cells migrate through the loose and dense connective tissues of the body; they use the bloodstream to travel from one organ to another and for rapid transportation to areas of invasion or injury. WBCs are sensitive to the chemical signs of damage to surrounding tissues. When problems are detected, these cells leave the bloodstream and enter the damaged area.

Circulating WBCs have four characteristics:

1. They are capable of *amoeboid movement.* Amoeboid movement is a gliding motion accomplished by the flow

Typical WBCs in the circulating blood are shown in Figure 11–8●. *Neutrophils, eosinophils,* and *basophils* are granulocytes; *monocytes* and *lymphocytes* are agranulocytes. Even though a microliter of blood typically contains 6000–9000 WBCs, circulating WBCs represent only a small fraction of the total population. Most of the WBCs in the body are located in connective tissue proper or in organs of the lymphatic system.

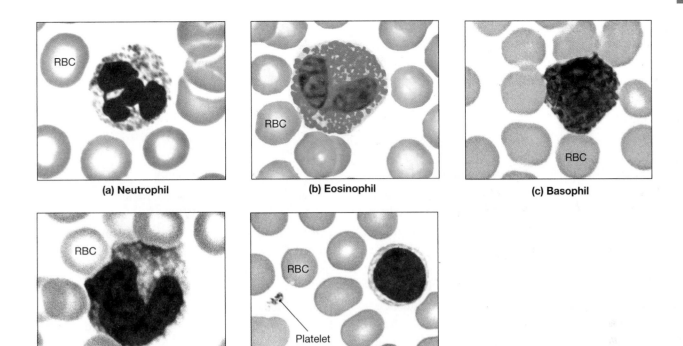

(a) Neutrophil

(b) Eosinophil

(c) Basophil

(d) Monocyte

(e) Lymphocyte

RBC

RBC

RBC

RBC

RBC

Platelet

● **FIGURE 11–8** White Blood Cells. (LMs × 1500)

of cytoplasm into slender cellular processes extended out from the cell. This mobility allows WBCs to move along the walls of blood vessels and, when outside the bloodstream, through surrounding tissues.

2. They can migrate out of the bloodstream. WBCs can enter surrounding tissue by squeezing between adjacent epithelial cells in the capillary wall. This process is called *diapedesis* (dī-a-pe-DĒ-sis).

3. They are attracted to specific chemical stimuli. This characteristic, called **positive chemotaxis** (kē-mō-TAK-sis), guides WBCs to invading pathogens, damaged tissues, and other active WBCs.

4. Neutrophils, eosinophils, and monocytes are capable of *phagocytosis.* ∞ p. 74 These cells can engulf pathogens, cell debris, or other materials. Neutrophils and eosinophils are sometimes called *microphages* to distinguish them from the larger macrophages in connective tissues. Macrophages are monocytes that have moved out of the bloodstream and become actively phagocytic.

Types of WBCs

Neutrophils, eosinophils, basophils, and monocytes contribute to the body's *nonspecific defenses.* These defenses respond to a variety of stimuli but always in the same way—they do not discriminate between one type of threat and another. Lymphocytes, in contrast, are the cells responsible for *specific defenses:* the body's ability to attack invading pathogens or foreign proteins *on a specific, individual basis.* The interactions between WBCs, and the relationships between specific and nonspecific defenses, will be discussed in Chapter 14.

Neutrophils

Fifty to 70 percent of the circulating white blood cells are **neutrophils** (NOO-trō-filz). This name reflects the fact that their granules are chemically neutral and, thus, are difficult to stain with either acidic or basic dyes. A mature neutrophil has a very dense, contorted nucleus with two to five lobes that resemble beads on a string (Figure 11–8a).

Neutrophils are usually the first WBCs to arrive at an injury site. They are very active phagocytes, specializing in attacking and digesting bacteria. Most neutrophils have a short life span (about 10 hours). After engulfing one to two dozen bacteria, a neutrophil dies, but its breakdown releases chemicals that attract other neutrophils to the site. A mixture of dead neutrophils, cellular debris, and other waste products form the *pus* associated with infected wounds.

Eosinophils

Eosinophils (ē-ō-SIN-ō-filz) were so named because their granules stain darkly with the red dye eosin (Figure 11–8b). They usually represent 2–4 percent of circulating WBCs and are similar in size to neutrophils. Their deep red granules and a two-lobed nucleus make them easy to identify. Eosinophils attack objects that are coated with antibodies. Although they will engulf antibody-marked bacteria, protozoa, or cellular debris, their primary mode of attack is the exocytosis of toxic compounds, including nitric oxide and cytotoxic enzymes. Their numbers increase dramatically during a parasitic infection or an allergic reaction.

Basophils

Basophils (BĀ-sō-filz) have numerous granules that stain darkly with basic dyes. In a standard blood smear, the granules are a deep purple or blue (Figure 11–8c). These cells are somewhat smaller than neutrophils or eosinophils and are relatively rare: they account for less than 1 percent of the circulating WBC population. Basophils migrate to sites of injury and cross the capillary wall to accumulate within damaged tissues, where they discharge their granules into the interstitial fluids. The granules contain the chemicals *heparin,* which prevents blood clotting, and *histamine.* The release of histamine by basophils enhances the local inflammation initiated by mast cells within damaged connective tissues. ∞ p. 116 Other chemicals released by stimulated basophils attract eosinophils and other basophils to the area.

Monocytes

Monocytes (MON-ō-sīts) are nearly twice the size of a typical erythrocyte (Figure 11–8d). The nucleus is large and commonly oval or shaped like a kidney bean. Monocytes normally account for 2–8 percent of circulating WBCs. They remain in circulation for only about 24 hours before entering peripheral tissues to become tissue macrophages. These migrating monocytes are called *free macrophages* to distinguish them from the immobile *fixed macrophages* present in many connective tissues. ∞ p. 106 They are aggressive phagocytes and often attempt to engulf items as large as or larger than themselves. Active monocytes release chemicals that attract and stimulate neutrophils, additional monocytes, and other phagocytes and that draw fibroblasts to the region. The fibroblasts then begin to produce scar tissue, which walls off the injured area.

Lymphocytes

Typical **lymphocytes** (LIM-fō-sīts) are slightly larger than RBCs and contain a relatively large nucleus surrounded by a thin halo of cytoplasm (Figure 11–8e). Lymphocytes account for 20–30 percent of the WBC population in blood. Lymphocytes are continuously migrating from the bloodstream, through peripheral tissues, and back to the bloodstream. Circulating lymphocytes represent a very small fraction of the entire WBC population, for at any moment most lymphocytes are in other connective tissues and in organs of the lym-

phatic system. They also protect the body and its tissues, but they do not rely on phagocytosis. Some kinds of lymphocytes attack foreign cells and abnormal cells of the body; other kinds secrete antibodies into the circulation. The antibodies can attack foreign cells or proteins in distant parts of the body.

The Differential Count and Changes in WBC Abundance

A variety of disorders, including infections, inflammation, and allergic reactions, cause characteristic changes in the circulating populations of WBCs. By examining a stained blood smear, we can obtain a **differential count** of the WBC population. The values reported indicate the number of each type of cell in a sample of 100 WBCs.

The normal range for the differential count for each WBC type is included in Table 11–3. The term **leukopenia** (loo-kō-PĒ-ne-uh; *penia,* poverty) indicates reduced numbers of WBCs. **Leukocytosis** (loo-kō-sī-TO-sis) refers to excessive numbers of WBCs. Extreme leukocytosis (WBC counts of 100,000/μL or more) usually indicates the presence of some form of **leukemia** (loo-KĒ-mē-uh), which is a cancer of blood-forming tissues. Only some of the many types of leukemia are characterized by leukocytosis; other indications are the presence of abnormal or immature WBCs. Treatment helps in some cases; unless treated, all are fatal.

White Blood Cell Formation

As previously noted, hemocytoblasts—the cells from which all the formed elements develop—are present in red bone marrow. Hemocytoblasts produce (1) lymphoid stem cells, which give rise to lymphocytes; and (2) myeloid stem cells, which give rise to all the other types of formed elements (see Figure 11–5, p. 415). Basophils, eosinophils, and neutrophils complete their development in myeloid (marrow) tissue; monocytes begin their differentiation in the bone marrow, enter the circulation, and complete their development when they become free macrophages in peripheral tissues. Each of these cell types goes through a characteristic series of developmental stages.

Many of the lymphoid stem cells responsible for the production of lymphocytes migrate from the bone marrow to peripheral **lymphoid tissues,** including the thymus, spleen, and lymph nodes. As a result, lymphocytes are produced in these organs as well as in the bone marrow. The process of lymphocyte production is called **lymphopoiesis.**

Various hormones are involved in the regulation of white blood cell populations. The WBCs other than lymphocytes are regulated by hormones called *colony-stimulating factors* (*CSFs*). Four CSFs have been identified, and each targets sin-

TABLE 11-3   *A Review of the Formed Elements of the Blood*

	CELL	ABUNDANCE (AVERAGE PER μL)	FUNCTIONS	REMARKS
	RED BLOOD CELLS	5.2 million (range: 4.4–6.0 million)	Transport oxygen from lungs to tissues, and carbon dioxide from tissues to lungs	Remain in bloodstream; 120-day life expectancy; amino acids and iron recycled; produced in bone marrow
	WHITE BLOOD CELLS	7000 (range: 6000–9000)		
	Neutrophils	4150 (range: 1800–7300) Differential count: 50%–70%	Phagocytic: Engulf pathogens or debris in tissues, release cytotoxic enzymes and chemicals	Move into tissues after several hours; survive minutes to days, depending on tissue activity; produced in bone marrow
	Eosinophils	165 (range: 0–700) Differential count: 2%–4%	Attack antibody-labeled materials through release of cytotoxic enzymes and/or phagocytosis	Move into tissues after several hours; survive minutes to days, depending on tissue activity; produced in bone marrow
	Basophils	44 (range: 0–150) Differential count: <1%	Enter damaged tissues and release histamine and other chemicals that promote inflammation	Survival time unknown; assist mast cells of tissues in producing inflammation; produced in bone marrow
	Monocytes	456 (range: 200–950) Differential count: 2%–8%	Enter tissues to become macrophages; engulf pathogens or debris	Move into tissues after 1–2 days; survive months or longer; primarily produced in bone marrow
	Lymphocytes	2185 (range: 1500–4000) Differential count: 20%–30%	Cells of lymphatic system; provide defense against specific pathogens or toxins	Survive months to decades; circulate from blood to tissues and back; produced in bone marrow and lymphoid tissues
	PLATELETS	350,000 (range: 150,000–500,000)	Hemostasis: Clump together and stick to vessel wall (platelet phase); initiate coagulation phase	Remain in circulation or in vascular organs; survive 7–12 days; produced by megakaryocytes in bone marrow

gle stem cell lines or groups of stem cell lines (see Figure 11–5, p. 415). Prior to an individual's maturity, hormones (thymosins) produced by the thymus gland promote the differentiation of one type of lymphocyte, the *T cells*. (In adults, the production of lymphocytes is regulated by exposure to antigens such as foreign proteins, cells, and toxins. We will describe the control mechanisms in Chapter 14.)

Many aspects of the body's defenses depend on the effects of lymphocyte hormones on the production of other WBCs.

For example, active macrophages release chemicals that make lymphocytes more sensitive to antigens and accelerate the development of specific immunity. In turn, active lymphocytes release CSFs, which reinforce nonspecific defenses by stimulating the production of other WBCs. Because of their potential clinical importance, immune system hormones are being studied intensively. Several CSFs are now produced by genetic engineering; one of them is used to stimulate the production of neutrophils in patients undergoing cancer chemotherapy.

Key Note

White blood cells (WBCs) are usually outnumbered by RBCs by a ratio of 1000:1. WBCs are responsible for defending the body against infection, foreign cells, or toxins, and for assisting in the cleanup and repair of damaged tissues. The most numerous are neutrophils, which engulf bacteria; and lymphocytes, which are responsible for the specific defenses of the immune response.

■ Platelets

Platelets (PLĀT-lets) are another component of the formed elements. In nonmammalian vertebrates, platelets are nucleated cells called **thrombocytes** (THROM-bō-sīts; *thrombos,* clot). Because in humans these formed elements are cell fragments rather than individual cells, the term *platelet* is preferred when referring to our blood.

Bone marrow contains enormous cells with large nuclei called **megakaryocytes** (meg-a-KAR-ē-ō-sīts; *megas,* big +

karyon, nucleus + *-cyte,* cell) (see Figure 11–5). Megakaryocytes continuously shed cytoplasm in small membrane-enclosed packets. These packets are the platelets that enter the bloodstream. Platelets initiate the clotting process and help close injured blood vessels. They are a major participant in a vascular *clotting system,* which is detailed in the next section.

Platelets are continuously replaced. An individual platelet circulates for 9–12 days before being removed by phagocytes. Each microliter of circulating blood contains 150,000–500,000 platelets; 350,000/μL is the average concentration. An abnormally low platelet count (or less) is known as *thrombocytopenia* (throm-bō-sī-tō-PĒ-nē-uh), and this condition usually results from excessive platelet destruction or inadequate platelet production. Clinical signs include bleeding along the digestive tract, within the skin, and occasionally inside the CNS.

In *thrombocytosis* (throm-bō-sī-TŌ-sis), platelet counts can exceed 1,000,000/μL. Thrombocytosis usually results from accelerated platelet formation in response to infection, inflammation, or cancer.

Clinical Note
BLOOD BANKING AND TRANSFUSION

Each year in the U.S. approximately 13.9 million units of whole blood are donated and are transfused into approximately 4.5 million patients (Figure 11–9●). Typically, each donated unit of blood, referred to as *whole blood,* is separated, or *fractionated,* into multiple components, and each component is generally transfused into different individuals according to their specific needs. The components of whole blood include red blood cells, plasma, platelets, white blood cells, and plasma derivatives.

Red Blood Cells

Red blood cells (RBCs) are the most recognizable component of whole blood. The RBCs, also called *erythrocytes,* contain hemoglobin, which transports oxygen. With an average life of approximately 120 days, RBCs are continuously being produced and broken down by the body and are ultimately removed by the spleen. RBC transfusions, often called *packed cells,* are used in the treatment of acute blood loss and anemia.

Plasma

The plasma is the liquid portion of the blood in which the red cells, white cells, and platelets are suspended. It is a protein-salt solution that is 90 percent water. Plasma contains *albumin,* which is the principle protein in the blood. In addition, it contains *fibrino-*

gen and other proteins that are part of the clotting system, and *globulins,* which include the antibodies that aid in fighting infection. Plasma is administered to patients with bleeding disorders.

Cryoprecipitated AHF is the part of the plasma that contains certain clotting factors, including Factor VIII, fibrinogen, von Willebrand

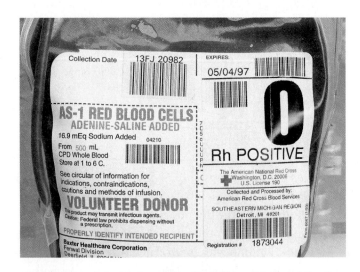

● **FIGURE 11–9 Modern Blood Banking.** Modern blood banking techniques have made transfusion a safe procedure. Liquid blood can remain in storage for up to 42 days before it must be discarded.

factor, and Factor XIII. Cryoprecipitated AHF is removed from plasma by freezing and then slowly thawing the plasma. Used only when specific factor concentrates are unavailable, it helps to control bleeding in individuals with hemophilia and von Willebrand's disease.

Platelets

Platelets, also called *thrombocytes,* are small blood components that aid in the clotting process by adhering to the lining of blood vessels. Platelets are manufactured in the bone marrow and remain viable in the circulation for an average of 9 to 10 days before being removed by the spleen. Units of platelets for transfusion are prepared by using a centrifuge to separate the platelets from plasma. Platelets can also be obtained through *apheresis,* a procedure in which blood is drawn from a patient, the platelets are removed, and the other blood components are returned to the patient. Platelets are used to treat thrombocytopenia or conditions that involve an abnormality in platelet function.

White Blood Cells

White blood cells (WBCs), also called *leukocytes,* are responsible for protecting the body from invasion by foreign substances such as bacteria, viruses, fungi, and parasites. The majority of WBCs are produced in the bone marrow. There are several types of WBCs, which have been previously detailed in the text. WBCs must be used within 24 hours after collection. They are most frequently used for infections that are unresponsive to antibiotic therapy. However, the effectiveness of WBC transfusions is still not clear.

Plasma Derivatives

Plasma derivatives are concentrates of specific plasma proteins prepared from pools of many donors through a process known as *fractionation.* They are usually heat-treated or washed with a solvent-detergent that kills certain viruses including HIV and hepatitis B and C. Plasma derivatives include:

- Factor VIII concentrate
- Factor IX concentrate
- Anti-inhibitor coagulation complex (AICC)
- Albumin
- Immune globulins (including Rh immune globulin)
- Antithrombin III concentrate
- Alpha 1-proteinase inhibitor complex

Blood transfusions may be either allogenic or autologous. In *allogenic* blood transfusions, blood is transfused to someone other than the donor. In *autologous* transfusions, the blood donor and transfusion recipient are the same individual. The most common autologous donation is the preoperative donation of blood for transfusion back to the donor during elective surgery. Blood can be stored in its liquid form for up to 42 days. Preoperative autologous donation is not allowed within 72 hours of surgery due to the time necessary to recover from donation. ■

■ Hemostasis

Hemostasis (*haima,* blood + *stasis,* halt), which is the process that halts bleeding, prevents the loss of blood through the walls of damaged vessels. In restricting blood loss, this process also establishes a framework for tissue repairs. Hemostasis is a process that consists of three overlapping phases: (1) the vascular phase, (2) the platelet phase, and (3) the coagulation phase.

Phase 1: The walls of blood vessels contain smooth muscle and an inner lining of simple squamous epithelium known as an **endothelium** (en-dō-THĒ-lē-um). Cutting the wall of a blood vessel triggers a contraction in the smooth muscle fibers in the vessel wall that decreases the vessel's diameter. This local contraction, called a *vascular spasm,* can slow or even stop the loss of blood through the wall of a small vessel. The vascular spasm lasts about 30 minutes, a period that constitutes the **vascular phase** of hemostasis. Changes also occur in the vessel's endothelium; the membranes of endothelial cells at the injury site become "sticky," and in small capillaries the cells may stick together and block the opening completely.

Phase 2: The platelets begin to attach to the sticky endothelial surfaces and exposed collagen fibers within 15 seconds of the injury. These attachments mark the start of the **platelet phase** of hemostasis. As more platelets arrive, they begin sticking to one another as well. This process forms a *platelet plug,* which is a mass that may close the break in the vessel wall if the damage is not severe or occurs in a small blood vessel.

Phase 3: The vascular and platelet phases begin within a few seconds after the injury. The **coagulation** (cō-ag-ū-LĀ-shun) **phase** does not start until 30 seconds or more

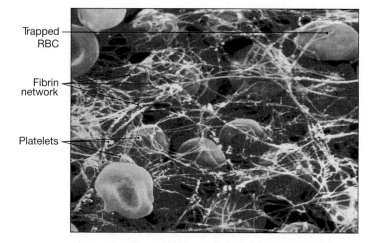

● **FIGURE 11–10 The Structure of a Blood Clot.** This colorized scanning electron micrograph shows the fibrin network that forms the framework of a clot. (SEM × 4120) Red blood cells trapped in those fibers add to the mass of the blood clot and make it red. Platelets that stick to the fibrin strands gradually contract, which shrinks the clot and tightly packs the RBCs within it.

after the vessel has been damaged. **Coagulation,** or *blood clotting,* involves a complex sequence of steps that leads to the conversion of circulating fibrinogen into the insoluble protein fibrin. As the fibrin network grows, blood cells and additional platelets are trapped within the fibrous tangle, forming a **blood clot** that effectively seals off the damaged portion of the vessel (Figure 11–10●).

The Clotting Process

Normal blood clotting cannot occur unless the plasma contains the necessary **clotting factors,** which include calcium ions and 11 different plasma proteins. These proteins are converted from inactive *proenzymes* to active enzymes that direct essential reactions in the clotting response. Most of the circulating clotting proteins are synthesized by the liver.

During the coagulation phase, the clotting proteins interact in sequence, such that one protein is converted into an enzyme that activates a second protein and so on, in a chain reaction, or *cascade.* Figure 11–11● provides an overview of the cascades involved in the *extrinsic, intrinsic,* and *common pathways,* which result in the formation of a blood clot. The extrinsic pathway begins outside the bloodstream, in the vessel wall; the intrinsic pathway begins inside the bloodstream. These pathways join at the common pathway through the activation of Factor X, which is a clotting protein produced by the liver.

The Extrinsic, Intrinsic, and Common Pathways

When a blood vessel is damaged, both the extrinsic and intrinsic pathways are activated. Clotting usually begins in 15 seconds and is initiated by the shorter and faster extrinsic pathway. The slower, intrinsic pathway reinforces the initial clot, and makes it larger and more effective.

The **extrinsic pathway** begins with the release of a lipoprotein called **tissue factor** by damaged endothelial cells or peripheral tissues. The greater the damage, the more tissue factor is released and the faster clotting occurs. Tissue factor then combines with calcium ions and another clotting protein (Factor VII) to form an enzyme capable of activating Factor X.

The **intrinsic pathway** begins with the activation of proenzymes exposed to collagen fibers at the injury site. This pathway proceeds with the assistance of a *platelet factor* released by aggregating platelets. Platelets also release a variety of other factors that speed up the reactions of the intrinsic pathway. After a series of linked reactions, activated clotting

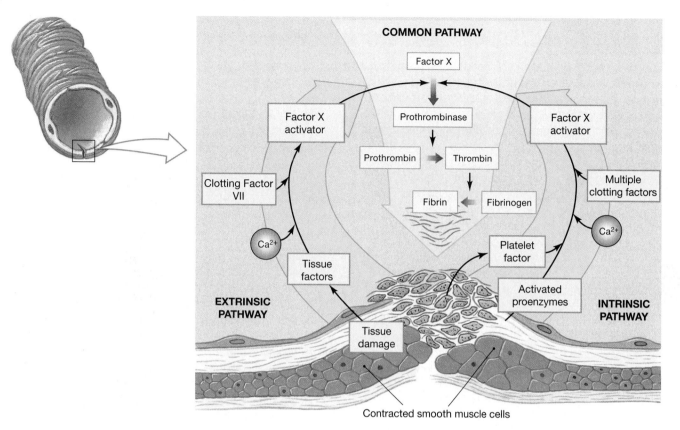

COMMON PATHWAY

Factor X

Factor X activator

Prothrombinase

Factor X activator

Clotting Factor VII

Prothrombin → Thrombin

Multiple clotting factors

Ca²⁺

Fibrin ← Fibrinogen

Ca²⁺

Tissue factors

Platelet factor

EXTRINSIC PATHWAY

Activated proenzymes

INTRINSIC PATHWAY

Tissue damage

Contracted smooth muscle cells

THE COAGULATION PHASE

● **FIGURE 11–11** Events in the Coagulation Phase of Hemostasis.

proteins combine to form an enzyme capable of activating Factor X.

The **common pathway** begins when enzymes from either the extrinsic or the intrinsic pathway activate Factor X, forming the enzyme prothrombinase. Prothrombinase converts the clotting protein **prothrombin** into the enzyme **thrombin** (THROM-bin). Thrombin then completes the clotting process by converting fibrinogen to fibrin.

The thrombin formed in the common pathway also acts to increase the rate of clot formation. Thrombin stimulates the formation of tissue factor and the release of platelet factor by platelets. The presence of these factors further stimulates both the extrinsic and intrinsic pathways. This positive feedback loop accelerates the clotting process, and speed is important in reducing blood loss after a severe injury. ∞ p. 15

Calcium ions and a single vitamin, **vitamin K,** affect almost every aspect of the clotting process. Because all three pathways (intrinsic, extrinsic, and common) require the presence of calcium ions, any disorder that lowers plasma calcium ion concentrations will also impair blood clotting. Because adequate amounts of vitamin K must be present for the liver to synthesize four of the clotting factors (including prothrombin), a vi-

tamin K deficiency leads to the breakdown of the common pathway and inactivates the clotting system. Roughly half of the vitamin K the body needs is absorbed directly from the diet, and the rest is synthesized by intestinal bacteria.

Clot Retraction and Removal

Once the fibrin network has appeared, platelets and red blood cells stick to the fibrin strands. The platelets then contract, pulling the torn edges of the wound closer together as the entire clot begins to undergo **clot retraction.** This process reduces the size of the damaged area, making it easier for the fibroblasts, smooth muscle cells, and endothelial cells in the area to make the necessary repairs.

As repairs proceed, the clot gradually dissolves. This process, called **fibrinolysis** (fi-bri-NOL-i-sis), begins with the activation of the plasma protein **plasminogen** (plaz-MIN-ō-jen) by thrombin and **tissue plasminogen activator,** or **t-PA,** which is released by damaged tissues. The activation of plasminogen produces the enzyme **plasmin** (PLAZ-min), which begins digesting the fibrin strands and breaking down the clot.

Clinical Note
HEPARIN AND FIBRINOLYTIC THERAPY

The localized formation of a blood clot (thrombosis) is a normal part of the body's repair and healing response. A physiological thrombus serves to limit hemorrhage that results from microscopic or macroscopic vascular injury. Physiological thrombus is counterbalanced by physiological anticoagulation and physiological fibrinolysis. In the normal setting, a physiological thrombus is confined to the immediate area of injury and does not obstruct blood flow to critical areas. Certain pathological conditions can lead to thrombus formation. In addition, under pathological conditions, a thrombus can expand into otherwise normal blood vessels, and obstruct blood flow to critical tissues. An abnormal thrombus can occur anywhere in the body but is particularly problematic when it causes acute coronary syndrome, deep vein thrombosis, pulmonary embolism, acute nonhemorrhagic stroke, or blockage of peripheral arteries.

A thrombus in a coronary artery interrupts blood supply to a portion of the myocardium, which results in acute coronary syndrome. Likewise, the formation of a thrombus within the vessels of the brain can cause a stroke. The time from blockage of the vessel until irreversible tissue injury occurs is short. In the setting of acute myocardial infarction, treatment must be provided in six hours or less in order to be most effective. When fibrinolytics are administered to treat nonhemorrhagic strokes, they generally must be administered within three hours.

Therapy for abnormal thrombosis uses drugs that can be divided into three categories. The first is fibrinolytic agents that break down the offending thrombus (*thrombolysis*). Another is anticoagulants, which are agents that help prevent thrombus formation. Finally, antiplatelet drugs, which inhibit platelet aggregation, are helpful in preventing clot formation. All three categories play essential roles in emergency medicine.

■ *Fibrinolytic therapy.* Fibrinolytic agents are able to dissolve preformed arterial and venous thrombi and restore blood flow to the affected tissues. They are derived from bacteria (streptokinase) or from recombinant DNA technology (tissue plasminogen activator, tPA). The effects of fibrinolytic therapy impact the entire body. Because of this, complications, most notably bleeding, can occur. Patients must be carefully screened before receiving fibrinolytic therapy to ensure that the possible benefits exceed the risks.

■ *Anticoagulation therapy.* Anticoagulants are drugs that inhibit clot formation. They do not affect a clot that has already formed. These drugs are administered to patients at increased risk of clot formation or following removal of a clot. Heparin and warfarin (Coumadin) are the most frequently used. Heparin must be administered by injection, while warfarin must be administered orally. Recently, newer forms of heparin have been created that are much easier to administer and do not require frequent blood testing. These agents, called *low-molecular-weight heparins (LMWHs),* include enoxaparin (Fragmin), dalteparin (Lovenox), and ardeparin (Fraxiparine). These drugs can be administered once or twice daily, and allow selected patients to be treated in the outpatient setting.

■ *Antiplatelet therapy.* In addition to anticoagulants and fibrinolytics, several other agents are beneficial in preventing thrombosis. Aspirin, which is a mainstay of medical therapy, inhibits platelet function and aggregation; it is inexpensive and highly effective. A new class of antiplatelet agents is the adenosine diphosphate (ADP) inhibitors. These agents are potent inhibitors of platelet aggregation through a mechanism that differs from aspirin. Drugs in this class include: abciximab (Reopro), eptifibatide (Integrilin), and tirofibin (Aggrastat). These agents are being used with increasing frequency in the treatment of acute coronary syndrome.

Other thrombotic disorders, including deep venous thrombosis, pulmonary embolism, and peripheral arterial occlusion, benefit from these new therapies. With the development of safer and easier-to-administer medications, the use of fibrinolytics and anticoagulants in prehospital care is increasing. ■

Key Note

Platelets are involved in the coordination of hemostasis (blood clotting). When platelets are activated by abnormal changes in their local environment, they release clotting factors and other chemicals. Hemostasis is a complex cascade that establishes a fibrous patch that can subsequently be remodeled and then removed as the damaged area is repaired.

→ CONCEPT CHECK QUESTIONS

1. If a sample of bone marrow has fewer than normal numbers of megakaryocytes, what body process would you expect to be impaired as a result?

2. Two alternate pathways of interacting clotting proteins lead to coagulation, or blood clotting. How is each pathway initiated?

3. How do inadequate levels of vitamin K affect blood clotting (coagulation)?

Answers begin on p. 792.

Clinical Note
ABNORMAL HEMOSTASIS

Hemostasis involves a complex chain of events, and any disorder that affects any individual clotting factor can disrupt the entire process. As a result, many clinical conditions affect the clotting process.

Excessive Coagulation

If the clotting process is inadequately controlled, clots will form in the circulation rather than at an injury site. These blood clots do not stick to the wall of the vessel but drift downstream until plasmin digests them or they become stuck in a small blood vessel. A drifting blood clot is a type of **embolus** (EM-bo-lus; *embolos,* plug), an abnormal mass within the bloodstream. (Other common emboli are drifting air bubbles and fat globules.) When an embolus becomes lodged in a blood vessel, it blocks circulation to the area downstream, killing the affected tissues. The condition that results from vascular blockage by an embolus is called *embolism.*

An *embolus* in the arterial system can become lodged in the capillaries in the brain, causing a tissue-damaging event known as a *stroke.* An embolus that forms in the venous system is likely to become lodged in the capillaries in the lung, causing a condition known as *pulmonary embolism.*

A *thrombus*—a blood clot attached to a vessel wall—begins to form when platelets stick to the wall of an intact blood vessel. Often the platelets are attracted to roughened areas called *plaques,* where endothelial and smooth muscle cells contain large quantities of lipids. The blood clot gradually enlarges, projecting into the vessel's channel (or *lumen*) and reducing its diameter. Eventually the vessel may be completely blocked, or a large chunk of the clot can break off, creating an equally dangerous embolus.

Treatment of these conditions must be prompt to prevent irreparable damage to tissues whose vessels have been restricted or blocked by emboli or thrombi. The clots can be surgically removed, or they can be attacked by administering anticoagulants—such as the enzymes *streptokinase* (strep-tō-KĪ-nās) or *urokinase* (ū-rō-KĪ-nās), which convert plasminogen to plasmin— or by administering *tissue plasminogen activator* (*tPA*), which stimulates plasmin formation.

Inadequate Coagulation

Hemophilia (hē-mō-FĒL-ē-uh) is one of many inherited disorders characterized by the inadequate production of clotting factors. The incidence of this condition in the general population is about 1 in 10,000, with males accounting for 80–90 percent of those affected. In hemophilia, the production of a single clotting factor (most often Factor VIII, an essential component of the intrinsic clotting pathway) is inadequate; the severity of the condition depends on the degree of underproduction. In severe cases, extensive bleeding accompanies the slightest mechanical stresses, and hemorrhages occur spontaneously at joints and around muscles.

Transfusions of clotting factors can reduce or control the effects of hemophilia, but plasma samples from many individuals must be pooled (combined) to obtain adequate amounts of clotting factors. This procedure makes treatment very expensive and increases the risk of blood-borne infections such as hepatitis or HIV. Gene-splicing techniques have been used to manufacture Factor VIII. As methods are developed to synthesize other clotting factors, treatment of the various forms of hemophilia will become safer and cheaper. ■

Clinical Note
FIBRINOLYTIC THERAPY: THE ERA OF REPERFUSION

Fibrinolytic agents, also called *thrombolytic agents,* are drugs that dissolve arterial and venous blood clots and restore blood flow to anoxic tissues. Fibrinolytic therapy has become an essential component of emergency care in the twenty-first century (Figure 11–12●). Several life-threatening conditions result from the formation of blood clots in the vascular system. These can be localized clots (*thrombi*) or clots that move through the circulatory system (*emboli*). Conditions that result from blood clot formation

● **FIGURE 11–12 Battery-Powered ECG Monitor/ Defibrillator.** The development of the battery-powered 12-lead ECG monitor/defibrillator now makes it possible to obtain quality 12-lead ECGs in the field and screen patients for fibrinolytic therapy.

(continued next page)

Clinical Note—continued

FIBRINOLYTIC THERAPY: THE ERA OF REPERFUSION

include acute coronary syndrome, acute ischemic stroke, pulmonary embolus, deep venous thrombosis, and peripheral arterial occlusion.

Acute Coronary Syndrome

Acute coronary syndrome results from the formation of a clot in one of the coronary arteries. The coronary arteries provide blood supply to the heart, and blockage of one of these vessels can cause death of myocardial tissue (acute myocardial infarction). The larger the affected vessel, the more severe is the resultant tissue damage. To be effective, fibrinolytic therapy should be started within six hours after the onset of symptoms. Because of this, it is important to determine, if possible, when the patient's symptoms began. In rural areas and in areas subject to delays in getting patients to a hospital with invasive cardiology capabilities, paramedics in the field may have to initiate fibrinolytic therapy.

Acute Ischemic Stroke

An acute ischemic stroke is the presence of a blood clot in the brain circulation that interrupts blood flow to the part of the brain supplied by the affected artery. For many years, there was no effective therapy for strokes. Now, in selected patients, administration of a fibrinolytic agent can dissolve the offending clot and restore blood flow to the brain. The results can be striking. The best outcome occurs when fibrinolytic therapy is started within three hours of the onset of symptoms.

Pulmonary Embolus

An acute pulmonary embolus is the presence of a blood clot in the pulmonary circulation. The clot develops in another part of the body, usually the large veins of the leg and pelvis, and travels through the circulatory system until it lodges in the lungs. This interrupts blood flow to the affected lung tissue, reduces pulmonary venous return, and can markedly decrease oxygenation of the blood and inhibit the removal of metabolic waste products. Fibrinolytic therapy is usually limited to patients with massive emboli who have experienced symptoms for 48 hours or less and who show evidence of hemodynamic compromise.

Deep Venous Thrombosis

Deep venous thrombosis is the formation of a clot within one of the larger veins, usually in the lower extremity. These patients are at risk of the clot breaking loose and traveling to the lungs or brain, where it can be fatal. Fibrinolytic therapy is limited to large clots that pose an immediate threat to the patient's life.

Peripheral Arterial Occlusion

The formation of a blood clot in a peripheral artery can adversely affect the tissues supplied by that artery. Thrombolytic agents can be administered directly into the affected artery or systemically.

Like clot formation, clot dissolution is an important body function. The process of clot dissolution is referred to as *fibrinolysis.* It begins with activation of the plasma protein *plasminogen* by *tissue plasminogen activator (tPA).* Damaged tissues release tPA. The activation of plasminogen produces the enzyme *plasmin,* which begins to digest the fibrin strands and thus erode the clot foundation. Fibrinolytic agents cause the conversion of plasminogen to plasmin. *Alteplase (recombinant tPA, Activase)* is human tPA derived through recombinant DNA technology. Because it is chemically identical to natural tPA, the risks of an allergic reaction to the drug are minimized. Older fibrinolytic agents were derived from beta-hemolytic streptococcal bacteria. The most common drug of this group is *streptokinase (Streptase).* It promotes fibrinolysis by activating the conversion of plasminogen to plasmin. Plasmin in turn degrades fibrin, fibrinogen, and other procoagulant proteins. Because streptokinase is derived from bacteria, the risk of developing a serious allergic reaction is much higher. *Anistreplase (Eminase)* is a newer product that is derived from streptococcal bacteria. It appears to have less antigenicity than streptokinase and is much easier to administer.

Fibrinolytic medications are very potent. In addition to dissolving the clot, they cause significant blood thinning that places the patient at increased risk of hemorrhage (Figure 11–13●). In fact, hemorrhage is the most common significant risk factor. Serious hemorrhage should be treated with transfusion of red blood cells to replace blood loss. Plasma and platelet administration may be indicated based upon laboratory evaluation of the patient's clotting system. ■

● **FIGURE 11–13 Monitoring for Hemorrhage.** It is essential that emergency personnel constantly monitor patients who have recently received a fibrinolytic agent for signs and symptoms of hemorrhage.

Chapter Review

Access more review material online at *www.prenhall.com/bledsoe*. There you will find quiz questions, labeling activities, animations, essay questions, and web links.

Key Terms

blood 407
cardiovascular system 407
coagulation 426
embolus 429
erythrocyte 408

fibrin 410
fibrinolysis 427
hematocrit 411
hemoglobin 412
hemopoiesis 410

hemostasis 415
leukocyte 408
plasma 408
platelets 408
serum 410

Related Clinical Terms

anemia (a-NĒ-mē-uh) A condition in which the oxygen-carrying capacity of blood is reduced due to low hematocrit or low blood hemoglobin concentrations.

embolism A condition in which a drifting blood clot (embolus) becomes stuck in a blood vessel, and blocks circulation to the area downstream.

erythrocytosis (e-rith-rō-sī-TŌ-sis) A polycythemia that involves red blood cells only.

hematocrit (he-MAT-ō-krit) The percentage of whole blood volume occupied by cellular elements.

hematology (HĒM-ah-tol-o-jē) The study of blood and its disorders.

hematuria (hē-ma-TOO-rē-uh) The presence of red blood cells in the urine.

hemolytic disease of the newborn (HDN) A condition in which fetal red blood cells are destroyed by maternal antibodies.

hemophilia (hē-mō-FĒL-ē-uh) Inherited disorders characterized by inadequate production of clotting factors.

heterologous marrow transplant The transplantation of bone marrow from one individual to another to replace bone marrow destroyed during cancer therapy.

hypoxia (hī-POKS-ē-uh) Low tissue oxygen levels.

jaundice (JAWN-dis) A condition characterized by yellow skin and eyes, which results from abnormally high levels of bilirubin in the plasma.

leukemia (loo-KĒ-mē-uh) A cancer of blood-forming tissues characterized by extremely elevated levels of circulating white blood cells; includes both *myeloid leukemia* (abnormal granulocytes or other cells of the bone marrow) and *lymphoid leukemia* (abnormal lymphocytes).

leukopenia (loo-kō-PĒ-ne-uh) Inadequate numbers of white blood cells in the bloodstream.

normovolemic (nor-mō-vō-LĒ-mik) Having a normal blood volume.

phlebotomy (fle-BOT-o-mē) The process of withdrawing blood from a vein.

polycythemia (po-lē-sī-THĒ-me-uh) An elevated hematocrit accompanied by a normal blood volume.

septicemia A dangerous condition in which bacteria and bacterial toxins are distributed throughout the body in the bloodstream.

sickle cell anemia An anemia that results from the production of an abnormal form of hemoglobin; causes red blood cells to become sickle-shaped at low oxygen levels.

thalassemia An inherited blood disorder that results from the production of an abnormal form of hemoglobin.

thrombus A blood clot attached to the internal surface of a blood vessel.

transfusion A procedure in which blood components are given to someone to restore blood volume or to remedy some deficiency in blood composition.

venipuncture (VĒN-i-punk-chur) The puncturing of a vein for any purpose, including the withdrawal of blood or the administration of medication.

Summary Outline

1. The **cardiovascular system** provides a mechanism for the rapid transport of nutrients, waste products, respiratory gases, and cells within the body.

THE FUNCTIONS OF BLOOD 407

1. **Blood** is a specialized fluid connective tissue. Its functions include (1) transporting dissolved gases, nutrients, hormones, and metabolic wastes; (2) regulating the pH and electrolyte composition of the interstitial fluids; (3) restricting fluid losses through damaged vessels; (4) defending against pathogens and toxins; and (5) regulating body temperature by absorbing and redistributing heat.

THE COMPOSITION OF BLOOD 408

Blood Collection and Analysis 408

1. Blood contains **plasma, red blood cells (RBCs), white blood cells (WBCs),** and **platelets.** The plasma and formed elements constitute whole blood, which can be **fractionated** for analytical or clinical purposes. *(Figure 11–1a)*

PLASMA 408

1. Plasma accounts for about 55 percent of the volume of blood; roughly 92 percent of plasma is water. *(Figure 11–1b)*

2. Compared with interstitial fluid, plasma has a higher dissolved oxygen concentration and more dissolved proteins. The three classes of plasma proteins are *albumins, globulins,* and *fibrinogen.*

Plasma Proteins 410

3. **Albumins** constitute about 60 percent of plasma proteins. **Globulins** constitute roughly 35 percent of plasma proteins; they include **antibodies (immunoglobulins),** which attack foreign proteins and pathogens, and **transport proteins,** which bind ions, hormones, and other compounds. In the clotting reaction, **fibrinogen** molecules are converted to **fibrin.** The removal of clotting proteins from plasma leaves a fluid called **serum.**

Key Note 410

FORMED ELEMENTS 410

The Production of Formed Elements 410

1. **Hemopoiesis** is the process by which all formed elements are produced. **Stem cells** called **hemocytoblasts** divide to form all three types of formed elements.

Red Blood Cells 411

2. Red blood cells **(RBCs),** or **erythrocytes,** account for slightly less than half of the blood volume and 99.9 percent of the formed elements. The **hematocrit** is the percentage of whole blood volume occupied by cellular elements. *(Figure 11–1c; Table 11–1)*

3. RBCs transport oxygen and carbon dioxide within the bloodstream. They are highly specialized cells with a large surface area to volume ratio. They lack many organelles and usually degenerate after 120 days in the bloodstream. *(Figures 11–2, 11–3)*

4. Molecules of **hemoglobin (Hb)** account for over 95 percent of RBC proteins. Hemoglobin is a globular protein formed from four subunits. Each subunit contains a single molecule of **heme** and can reversibly bind an oxygen molecule. Hemoglobin from damaged or dead RBCs is recycled by phagocytes. *(Figure 11–4)*

5. **Erythropoiesis,** the formation of RBCs, occurs mainly in the **red bone marrow (myeloid tissue)** in adults. RBC formation increases under stimulation by **erythropoietin (EPO),** or erythropoiesis-stimulating hormone, which occurs when peripheral tissues are exposed to low oxygen concentrations. Stages in RBC development include **erythroblasts** and **reticulocytes.** *(Figures 11–5, 11–6)*

Key Note 416

6. **Blood type** is determined by the presence or absence of three specific **surface antigens** *(agglutinogens)* in the cell membranes of RBCs: antigens **A, B,** and **Rh.** Antibodies *(agglutinins)* in the plasma of individuals of some blood types can react with surface antigens on the RBCs of different blood types. Anti-Rh antibodies are synthesized only after an Rh-negative individual becomes sensitized to the Rh surface antigen. *(Figure 11–7; Table 11–2)*

White Blood Cells 419

7. White blood cells (WBCs), or **leukocytes,** defend the body against pathogens and remove toxins, wastes, and abnormal or damaged cells. *(Figure 11–8).*

8. Leukocytes exhibit *diapedesis* (the ability to move through vessel walls) and **chemotaxis** (an attraction to specific chemicals).

9. *Granulocytes* (granular leukocytes) are *neutrophils, eosinophils,* and *basophils.* Fifty to 70 percent of circulating WBCs are **neutrophils,** which are highly mobile phagocytes. The much less common **eosinophils** are phagocytes attracted to foreign substances that have reacted with circulating antibodies. The

relatively rare **basophils** migrate to damaged tissues and release histamine, and aid the inflammation response. *(Figure 11–8)*

10. *Agranulocytes* (agranular leukocytes) are *monocytes* and *lymphocytes*. **Monocytes** that migrate into peripheral tissues become free macrophages. Most **lymphocytes** are in the tissues and organs of the **lymphatic system,** where they function in the body's specific defenses. Different classes of lymphocytes attack foreign cells directly, produce antibodies, and destroy abnormal body cells. *(Figure 11–8)*

11. Granulocytes and monocytes are produced by **myeloid stem cells** in the bone marrow. **Lymphoid stem cells** responsible for **lymphopoiesis** (production of lymphocytes) also originate in the bone marrow, but many migrate to lymphoid tissues. *(Figure 11–5)*

12. Factors that regulate lymphocyte maturation are not completely understood. *Colony-stimulating factors (CSFs)* are hormones involved in regulating other WBC populations.

Key Note 424

PLATELETS 424

13. **Megakaryocytes** in the bone marrow release packets of cytoplasm (platelets) into the circulating blood. Platelets are essential to the clotting process. *(Figure 11–5)*

HEMOSTASIS 425

1. **Hemostasis** prevents the loss of blood through the walls of damaged vessels.

2. The initial step of hemostasis, the **vascular phase,** is a period of local contraction of vessel walls that results from a vascular spasm at the injury site. The **platelet phase** follows as platelets stick to damaged surfaces. The third phase of hemostasis is the **coagulation phase.**

The Clotting Process 426

3. The coagulation phase occurs as factors released by endothelial cells or peripheral tissues (**extrinsic pathway**) and platelets (**intrinsic pathway**) interact with **clotting factors** to form a **blood clot.** *(Figures 11–10 through 11–13)*

Clot Retraction and Removal 427

4. During **clot retraction,** platelets contract, pulling the torn edges of a damaged blood vessel closer together. During **fibrinolysis,** the clot gradually dissolves through the action of **plasmin,** which is the activated form of circulating **plasminogen.**

Key Note 428

Review Questions

Level 1: Reviewing Facts and Terms

Match each item in column A with the most closely related item in column B. Place letters for answers in the spaces provided.

COLUMN A

____ 1. interstitial fluid
____ 2. hemopoiesis
____ 3. stem cells
____ 4. hypoxia
____ 5. surface antigens
____ 6. antibodies
____ 7. diapedesis
____ 8. leukopenia
____ 9. agranulocyte
____ 10. leukocytosis
____ 11. granulocyte
____ 12. thrombocytopenia

COLUMN B

a. hemocytoblasts
b. abundant WBCs
c. agglutinogens
d. neutrophil
e. WBC migration
f. extracellular fluid
g. monocyte
h. few WBCs
i. low platelet count
j. agglutinins
k. blood cell formation
l. low oxygen concentration

13. The formed elements of the blood include:
    (a) plasma, fibrin, serum.
    (b) albumins, globulins, fibrinogen.
    (c) WBCs, RBCs, platelets.
    (d) a, b, and c are correct.

14. Blood temperature is approximately _____, and the blood pH averages _____.
    (a) 98.6°F, 7.0
    (b) 104°F, 7.8
    (c) 100.4°F, 7.4
    (d) 96.8°F, 7.0

15. Plasma contributes approximately _____ percent of the volume of whole blood, and water accounts for _____ percent of the plasma volume.
    (a) 55, 92
    (b) 25, 55
    (c) 92, 55
    (d) 35, 72

16. When the clotting proteins are removed from plasma, _____ remains.
    (a) fibrinogen
    (b) fibrin
    (c) serum
    (d) heme

17. In an adult, the only site of red blood cell production, and the primary site of white blood cell formation, is the:
    (a) liver.
    (b) spleen.
    (c) thymus.
    (d) red bone marrow.

18. The most numerous WBCs found in a differential count of a "normal" individual are:
    (a) neutrophils.
    (b) basophils.
    (c) lymphocytes.
    (d) monocytes.

19. Stem cells responsible for the process of lymphopoiesis are located in the:
    (a) thymus and spleen.
    (b) lymph nodes.
    (c) red bone marrow.
    (d) a, b, and c are correct.

20. The first step in the process of hemostasis is:
    (a) coagulation.
    (b) the platelet phase.
    (c) fibrinolysis.
    (d) vascular spasm.

21. The complex sequence of steps that leads to the conversion of fibrinogen to fibrin is called:
    (a) fibrinolysis.
    (b) clotting.
    (c) retraction.
    (d) the platelet phase.

22. What five major functions are performed by the blood?

23. What three primary classes of plasma proteins are found in the blood? What is the major function of each?

24. What type(s) of antibodies does the plasma contain for each of the following blood types?
    (a) Type A
    (b) Type B
    (c) Type AB
    (d) Type O

25. What three processes facilitate the movement of WBCs to areas of invasion or injury?

26. What contribution from the intrinsic and extrinsic pathways is necessary for the common pathway to begin?

27. Distinguish between an embolus and a thrombus.

Level 2: Reviewing Concepts

28. Dehydration would cause:
    (a) an increase in the hematocrit.
    (b) a decrease in the hematocrit.
    (c) no effect in the hematocrit.
    (d) an increase in plasma volume.

29. Erythropoietin directly stimulates RBC formation by:
    (a) increasing rates of mitotic divisions in erythroblasts.
    (b) speeding up the maturation of red blood cells.
    (c) accelerating the rate of hemoglobin synthesis.
    (d) a, b, and c are correct.

30. A person with Type A blood has:
    (a) anti-A antibodies in the plasma.
    (b) B antigens in the plasma.
    (c) A antigens on red blood cells.
    (d) anti-B antibodies on red blood cells.

31. Hemolytic disease of the newborn can result if:
    (a) the mother is Rh-positive and the father is Rh-negative.
    (b) both the father and the mother are Rh-negative.
    (c) both the father and the mother are Rh-positive.
    (d) an Rh-negative woman carries an Rh-positive fetus.

32. How do red blood cells differ from typical body cells?

Level 3: Critical Thinking and Clinical Applications

33. Which of the formed elements would you expect to increase after you have donated a pint of blood?

34. Why do many individuals with advanced kidney disease become anemic?

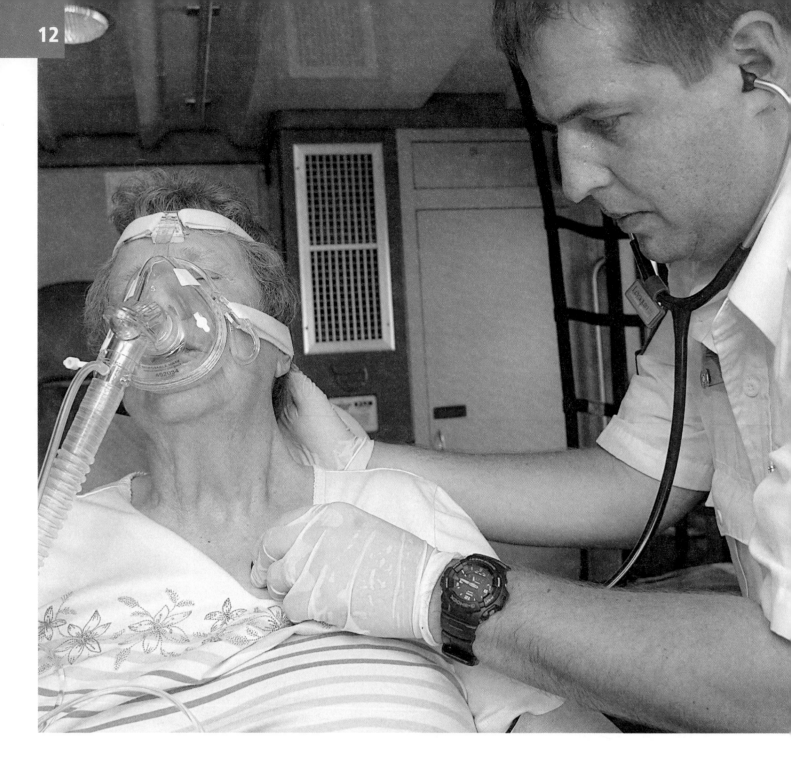

12 The Cardiovascular System: The Heart

EMS, AS WE KNOW IT TODAY, was developed primarily to treat heart attacks and heart-related problems. As our society ages, we will be seeing more and more patients with heart disease and other chronic diseases. Newer treatments, such as continuous positive airway pressure (CPAP) as shown here, *are extremely effective in the treatment of congestive heart failure (CHF) and acute pulmonary edema. CPAP, coupled with judicious medication therapy, can prevent many CHF patients from being intubated and placed on mechanical ventilators.*

Chapter Outline

Chapter Objectives

1. Describe the location and general features of the heart. (p. 437)
2. Identify the layers of the heart wall. (pp. 438–441)
3. Trace the flow of blood through the heart, identifying the major blood vessels, chambers, and heart valves. (pp. 442–446)
4. Describe the differences in the action potentials and twitch contractions of skeletal muscle fibers and cardiac muscle cells. (pp. 450–451)
5. Describe the components and functions of the conducting system of the heart. (pp. 452–457)
6. Explain the events of the cardiac cycle and relate the heart sounds to specific events in this cycle. (pp. 457–458)
7. Define stroke volume and cardiac output and describe the factors that influence each. (pp. 458–462)

Vocabulary Development

anastomosis outlet; *anastomoses*
atrion hall; *atrium*
auris ear; *auricle*
bi- two; *bicuspid*
bradys slow; *bradycardia*
cuspis point; *bicuspid valve*

diastole expansion; *diastole*
-gram record; *electrocardiogram*
luna moon; *semilunar valve*
mitre a bishop's hat; *mitral valve*
papilla nipple-shaped elevation; *papillary muscles*

semi- half; *semilunar valve*
septum wall; *interatrial septum*
systole a drawing together; *systole*
tachys swift; *tachycardia*
tri- three; *tricuspid valve*
ventricle little belly; *ventricle*

OUR BODY CELLS rely on the surrounding interstitial fluid for oxygen, nutrients, and waste disposal. Conditions in the interstitial fluid are kept stable through continuous exchange between the peripheral tissues and circulating blood. If the blood stops moving, its oxygen and nutrient supplies are quickly exhausted, its capacity to absorb wastes is soon saturated, and neither hormones nor white blood cells can reach their intended targets. Thus, all cardiovascular functions ultimately depend on the heart. This muscular organ beats approximately 100,000 times each day and pumps roughly 8000 liters of blood—enough to fill forty 55-gallon drums or nearly 8500 quart-sized milk cartons.

We begin this chapter by examining the structural features that enable the heart to perform so reliably. We will then consider the physiological mechanisms that regulate the activities of the heart to meet the body's ever-changing needs.

■ The Heart's Place in the Circulatory System

Blood flows through a network of blood vessels that extends between the heart and peripheral tissues. Blood vessels are subdivided into a **pulmonary circuit,** which carries blood to and from the exchange surfaces of the lungs, and a **systemic circuit,** which transports blood to and from the rest of the body (Figure 12–1●). Each circuit begins and ends at the heart, and blood travels through these circuits in sequence. Thus, blood returning to the heart from the systemic circuit must complete the pulmonary circuit before re-entering the systemic circuit.

Arteries, or *efferent* vessels, carry blood away from the heart; **veins,** or *afferent* vessels, return blood to the heart. **Capillaries** are small, thin-walled vessels between the smallest arteries and the smallest veins. The thin walls of capillaries permit the exchange of nutrients, dissolved gases, and waste products between the blood and surrounding tissues.

Despite its impressive workload, the heart is a small organ, roughly the size of a clenched fist. The heart contains four chambers, two associated with each circuit. The **right atrium** (A-trē-um; hall; plural, *atria*) receives blood from the systemic circuit, and the **right ventricle** (VEN-tri-kl; "little belly") discharges blood into the pulmonary circuit. The **left atrium** collects blood from the pulmonary circuit, and the **left ventricle** ejects it into the systemic circuit. When the heart beats, the two atria contract first, then the ventricles. The two ventricles contract at the same time and eject equal volumes of blood.

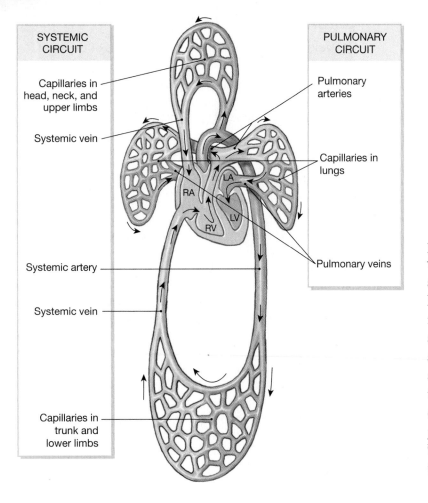

SYSTEMIC CIRCUIT

Capillaries in head, neck, and upper limbs

Systemic vein

Systemic artery

Systemic vein

Capillaries in trunk and lower limbs

PULMONARY CIRCUIT

Pulmonary arteries

Capillaries in lungs

Pulmonary veins

RA · LA · RV · LV

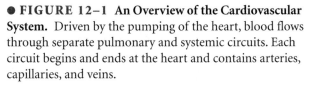

● **FIGURE 12–1 An Overview of the Cardiovascular System.** Driven by the pumping of the heart, blood flows through separate pulmonary and systemic circuits. Each circuit begins and ends at the heart and contains arteries, capillaries, and veins.

The pericardium can be subdivided into the visceral pericardium and the parietal pericardium. The **visceral pericardium,** or **epicardium,** covers the outer surface of the heart, whereas the **parietal pericardium** lines the inner surface of the *pericardial sac,* which surrounds the heart (see Figure 12–2b). The pericardial sac consists of a dense network of collagen fibers that stabilizes the positions of the pericardium, heart, and associated vessels in the mediastinum. The space between the parietal and visceral surfaces is the pericardial cavity. It normally contains a small quantity of *pericardial fluid* secreted by the pericardial membranes. The fluid acts as a lubricant and reduces friction between the opposing surfaces as the heart beats.

■ The Anatomy and Organization of the Heart

The heart lies near the anterior chest wall, directly behind the sternum (Figure 12–2a●). It is enclosed by the *mediastinum,* which is the connective tissue mass that divides the thoracic cavity into two pleural cavities (see Figure 1–11c, p. 21) and also contains the thymus, esophagus, and trachea.

The heart is surrounded by the **pericardial** (per-i-KAR-dē-al) **cavity.** The lining of the pericardial cavity is a serous membrane called the **pericardium.** ∞ p. 113 To visualize the relationship between the heart and the pericardial cavity, imagine pushing your fist toward the center of a large balloon (Figure 12–2b●). The balloon represents the pericardium, and your fist represents the heart. Your wrist, where the balloon folds back on itself, corresponds to the **base** of the heart (see Figure 12–2a). The air space inside the balloon corresponds to the pericardial cavity.

The Surface Anatomy of the Heart

Several external features of the heart can be used to identify its four chambers (Figure 12–3●). The two atria have relatively thin muscular walls and are highly expandable, so when an atrium is not filled with blood, its outer portion deflates into a lumpy, wrinkled flap called an **auricle** (AW-ri-kl; *auris,* ear). The **coronary sulcus,** which is a deep groove usually filled with substantial amounts of fat, marks the border between the atria and the ventricles. Shallower depressions—the **anterior interventricular sulcus** and **posterior interventricular sulcus**—mark the boundary between the left and right ventricles. In addition to fat, the sulci (SUL-sī) also contain the major arteries and veins that supply blood to the cardiac muscle.

The great veins and arteries of the circulatory system are connected to the superior end of the heart at the base. The inferior, pointed tip of the heart is the **apex** (Ā-peks) (see Figure 12–2a). A typical heart measures approximately 12.5 cm (5 in.) from the attached base to the apex.

The heart sits at an angle to the longitudinal axis of the body. It is also rotated slightly toward the left, so the anterior

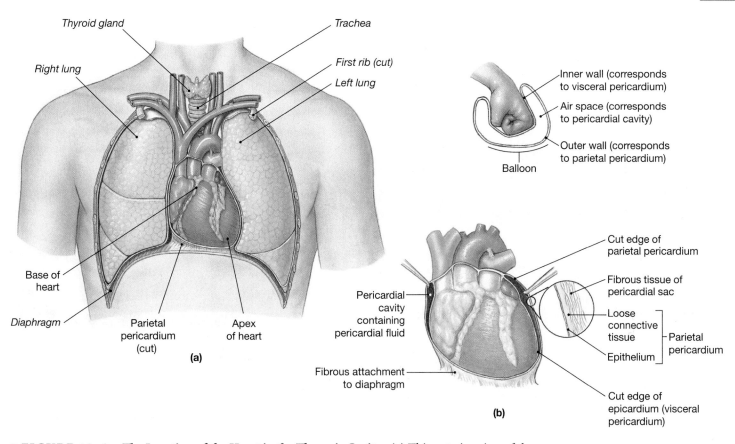

● FIGURE 12–2   The Location of the Heart in the Thoracic Cavity. (a) This anterior view of the open chest cavity shows the position of the heart and major vessels relative to the lungs. (b) The pericardial cavity that surrounds the heart is formed by the visceral epicardium and the parietal epicardium. The relationship between the heart and the pericardial cavity can be likened to a fist pushed into a balloon.

surface primarily consists of the right atrium and right ventricle (see Figure 12–3a). The wall of the left ventricle forms much of the posterior surface between the base and the apex of the heart (see Figure 12–3b).

The Heart Wall

The wall of the heart contains three distinct layers: the epicardium (visceral pericardium), the myocardium, and the endocardium (Figure 12–4a●). The **epicardium** covers the outer surface of the heart. This serous membrane consists of an exposed epithelium and an underlying layer of loose connective tissue that is attached to the myocardium. The **myocardium,** or muscular wall of the heart, contains cardiac muscle tissue, blood vessels, and nerves. The cardiac muscle tissue of the myocardium forms concentric layers that wrap around the atria and spiral into the walls of the ventricles (Figure 12–4b●). This arrangement results in squeezing and twisting contractions that increase the pumping efficiency of

the heart. The heart's inner surfaces, including the heart valves, are covered by the **endocardium** (en-dō-KAR-dē-um), which is a simple squamous epithelium that is continuous with the endothelium (epithelial lining) of the attached blood vessels.

Cardiac Muscle Cells

Typical cardiac muscle cells are shown in Figure 12–4c,d●. These cells are smaller than skeletal muscle fibers and contain a single, centrally located nucleus. Like skeletal muscle fibers, each cardiac muscle cell contains myofibrils, and contraction involves the shortening of individual sarcomeres. Because cardiac muscle cells are almost totally dependent on aerobic metabolism to obtain the energy needed to continue contracting, they have many mitochondria and abundant reserves of myoglobin (to store oxygen). Energy reserves are stored as glycogen and lipids.

Each cardiac muscle cell is in contact with several others at specialized sites known as **intercalated** (in-TER-ka-lā-ted)

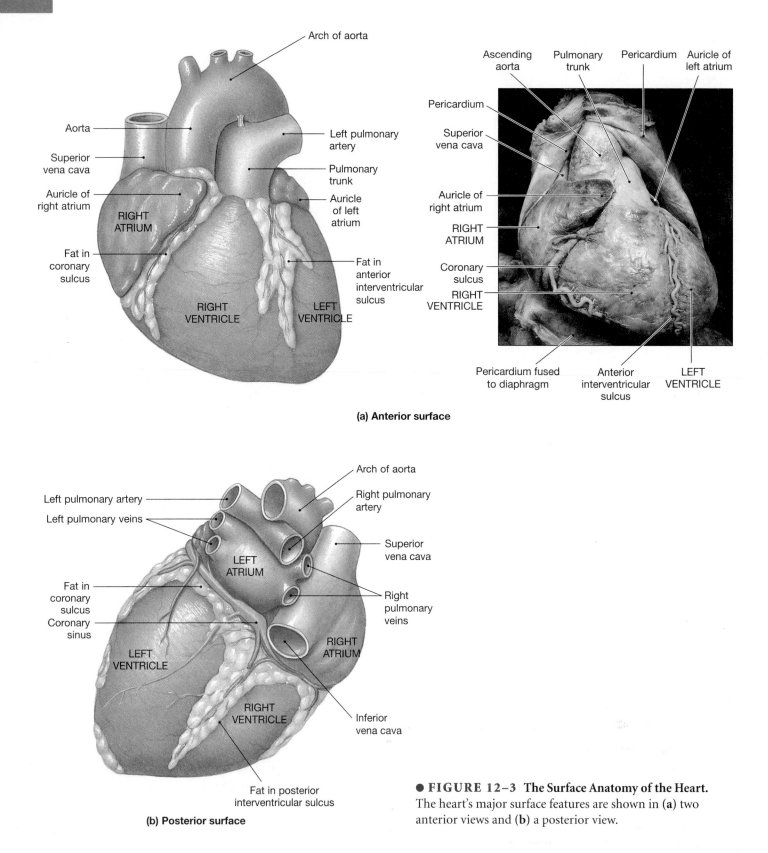

(a) Anterior surface

(b) Posterior surface

● **FIGURE 12–3** **The Surface Anatomy of the Heart.**
The heart's major surface features are shown in (**a**) two
anterior views and (**b**) a posterior view.

discs. ∞ p. 113 At an intercalated disc, the interlocking
membranes of adjacent cells are held together by desmosomes
and linked by gap junctions. ∞ p. 98 The desmosomes help
convey the force of contraction from cell to cell and increase

their efficiency as they "pull together" during a contraction.
The gap junctions provide for the movement of ions and
small molecules and enable action potentials to travel rapidly
from cell to cell.

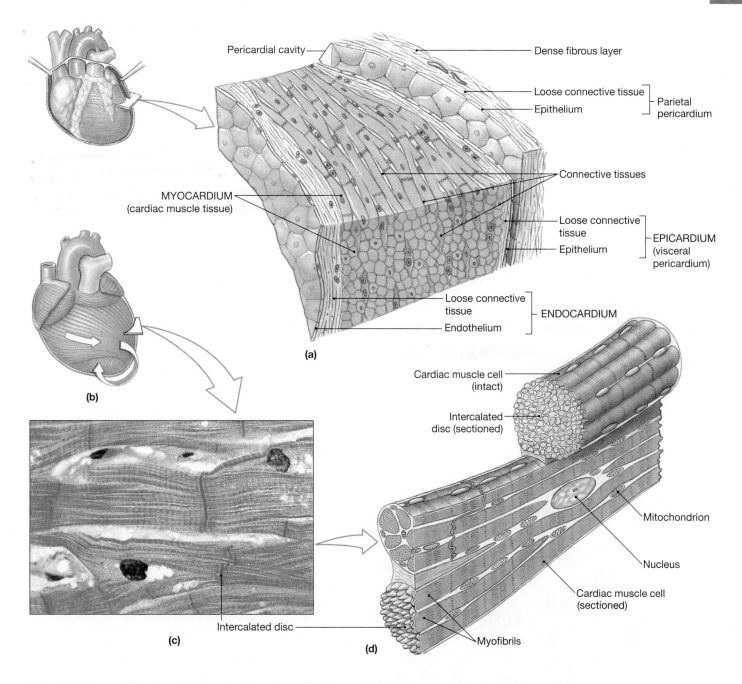

● FIGURE 12–4 **The Heart Wall and Cardiac Muscle Tissue.** (a) This diagrammatic section through the heart wall illustrates the relative positions of the epicardium, myocardium, and endocardium. (b) Cardiac muscle tissue forms concentric layers that wrap around the atria and spiral within the walls of the ventricles. (c, d) These sectional and diagrammatic views of cardiac muscle tissue show the characteristic single central nucleus, branching interconnections, and intercalated discs of cardiac muscle cells. (LM × 575)

Connective Tissue in the Heart

The connective tissues of the heart include abundant collagen and elastic fibers that wrap around each cardiac muscle cell and tie together adjacent cells. These fibers: (1) provide support for cardiac muscle fibers, blood vessels, and nerves of the myocardium; (2) add strength and prevent overexpansion of the heart; and (3) help the heart return to normal shape after contractions. Connective tissue also forms the *fibrous skeleton* of the heart, which is discussed in a later section.

Clinical Note
CHAGAS' DISEASE

In many parts of Central and South America, heart disease is the number one cause of death. But, unlike the U.S. and Canada, the type of heart disease seen in those areas is called *Chagas' disease* and is due to infection by the parasite *Trypanosoma cruzi*. Thought to infect over 16 million persons, it is spread through the bite of an insect known as the "kissing bug." The initial signs and symptoms are mild. However, ten to twenty years later, the patient develops an enlarged heart and dysrhythmias and eventually dies from heart failure. ■

Internal Anatomy and Organization

The four chambers of the heart are shown in sectional view in Figure 12–5●. The two atria are separated by the **interatrial septum** (*septum*, wall), and the two ventricles are divided by the **interventricular septum.** Each atrium opens into the ventricle on the same side through an **atrioventricular (AV) valve,** which is comprised of folds of fibrous tissue that ensure a one-way flow of blood from the atria into the ventricles.

The right atrium receives blood from the systemic circuit through two large veins, the superior vena cava (VĒ-na KĀ-vuh) and the inferior vena cava. The **superior vena cava** delivers blood from the head, neck, upper limbs, and chest. The **inferior vena cava** carries blood from the rest of the trunk, the viscera, and the lower limbs. The *cardiac veins* of the heart return venous blood to the **coronary sinus,** which opens into the right atrium slightly below the connection with the inferior vena cava.

A small depression called the *fossa ovalis* (see Figure 12–5) persists where an oval opening—the *foramen ovale*—penetrated the interatrial septum from the fifth week of embryonic development until birth. The foramen ovale (see Figure 13–29, p. 501) allowed blood to flow from the right atrium to the left atrium while the lungs were developing; at birth, the foramen ovale closes, and after 48 hours

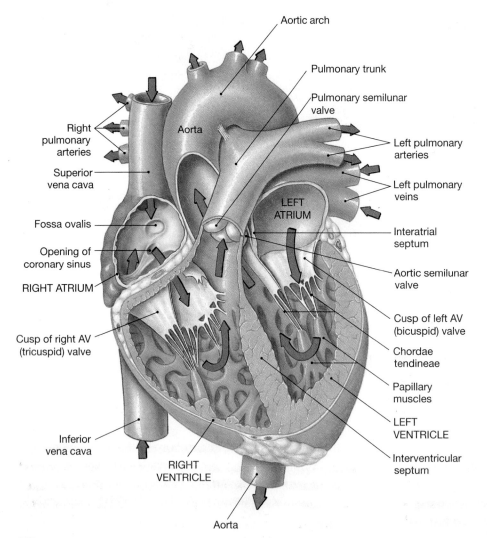

● **FIGURE 12–5 The Sectional Anatomy of the Heart.** The heart's major internal landmarks and the path of blood flow through the atria and ventricles are shown in this diagrammatic frontal section. Red arrows represent oxygenated blood, and blue arrows represent deoxygenated blood.

it is permanently sealed. Occasionally, the foramen ovale remains open even after birth. Under these conditions, contractions of the left atrium push blood back into the pulmonary circuit. This leads to heart enlargement and eventual heart failure and death if the condition is not surgically corrected.

Blood travels from the right atrium into the right ventricle through a broad opening bounded by three flaps of fibrous tissue. These flaps, or **cusps,** are part of the **right atrioventricular (AV) valve,** also known as the **tricuspid** (trī-KUS-pid; *tri-,* three + *cuspis,* point) **valve** (see Figure 12–5). Each cusp is braced by connective tissue fibers called **chordae tendineae** (KOR-dē TEN-di-nē-ē; "tendinous cords"). These fibers are connected to **papillary** (PAP-i-ler-ē) **muscles,** cone-shaped projections on the inner surface of the ventricle. The contraction of these muscles tenses the chordae tendineae, limiting the movement of the cusps and preventing the backflow of blood into the right atrium.

Blood leaving the right ventricle flows into the **pulmonary trunk,** which is the start of the pulmonary circuit. The **pulmonary semilunar** (*semi-,* half + *luna,* moon; a crescent or half-moon shape) **valve** (*pulmonary valve*) guards the entrance to this efferent trunk. Within the pulmonary trunk, blood flows into the **left** and **right pulmonary arteries.** These vessels branch repeatedly in the lungs before supplying the capillaries where gas exchange occurs. From these respiratory capillaries, oxygenated blood moves into the **left** and **right pulmonary veins,** which deliver it to the left atrium.

Like the right atrium, the left atrium has an external auricle and a valve, which is the **left atrioventricular (AV) valve,** or **bicuspid** (bī-KUS-pid) **valve.** As the name *bicuspid* implies, the left AV valve contains a pair, not a trio, of cusps. Clinicians often call this valve the **mitral** (MĪ-tral; *mitre,* a bishop's hat) **valve.**

The internal organization of the left ventricle resembles that of the right ventricle. A pair of papillary muscles braces the chordae tendineae that insert on the bicuspid valve. Blood leaving the left ventricle passes through the **aortic semilunar valve** (*aortic valve*) and into the **aorta,** which is the start of the systemic circuit.

Structural Differences Between the Left and Right Ventricles

The function of either atrium is to collect blood returning to the heart and deliver that blood to the attached ventricle. Thus, the demands on the two atria are very similar, and they look almost identical. But the demands on the right and left ventricles are very different, as the significant structural differences between the two suggest.

The lungs are close to the heart, and the pulmonary arteries and veins are relatively short and wide. Thus, the right ventricle normally does not need to push very hard to propel blood through the pulmonary circuit. The wall of the right ventricle is relatively thin (see Figure 12–5), and in sectional view it resembles a pouch attached to the massive wall of the left ventricle. When the right ventricle contracts, it acts like a bellows pump and squeezes the blood against the left ventricle and then out through the pulmonary semilunar valve. This mechanism moves blood very efficiently with minimal effort, but it develops relatively low pressures.

A comparable pumping arrangement would not be suitable for the left ventricle, because propelling blood through the systemic circuit takes six to seven times more force than pumping blood around the pulmonary circuit. The left ventricle has an extremely thick muscular wall that is round in cross section. When this ventricle contracts, two things happen: (1) the distance between the heart's base and apex decreases, and (2) the diameter of the ventricular chamber decreases. (If you can imagine the effects of simultaneously squeezing and rolling up the end of a toothpaste tube, you get the idea.) As the powerful left ventricle contracts, it also bulges into the right ventricular cavity, an action that helps force blood out of the right ventricle. An individual whose right ventricular musculature has been severely damaged may survive because the contraction of the left ventricle helps push blood through the pulmonary circuit.

The Heart Valves

Details of the structure and function of the heart valves are depicted in Figure 12–6●.

THE ATRIOVENTRICULAR VALVES. The atrioventricular valves prevent the backflow of blood from the ventricles into the atria. The chordae tendineae and papillary muscles play important roles in the normal function of the AV valves. When the ventricles are relaxed (Figure 12–6a), the chordae tendineae are loose and the AV valve offers no resistance to the flow of blood from atrium to ventricle. When the ventricles contract (Figure 12–6b), blood moving back toward the atrium swings the cusps together, which closes the valves. During ventricular contraction, tension in the papillary muscles and chordae tendineae keeps the cusps from swinging into the atrium. This action prevents the backflow, or **regurgitation,** of blood into the atrium each time the ventricle contracts. A small amount of regurgitation often occurs, even in normal individuals. The swirling blood creates a soft but distinctive sound called a *heart murmur.*

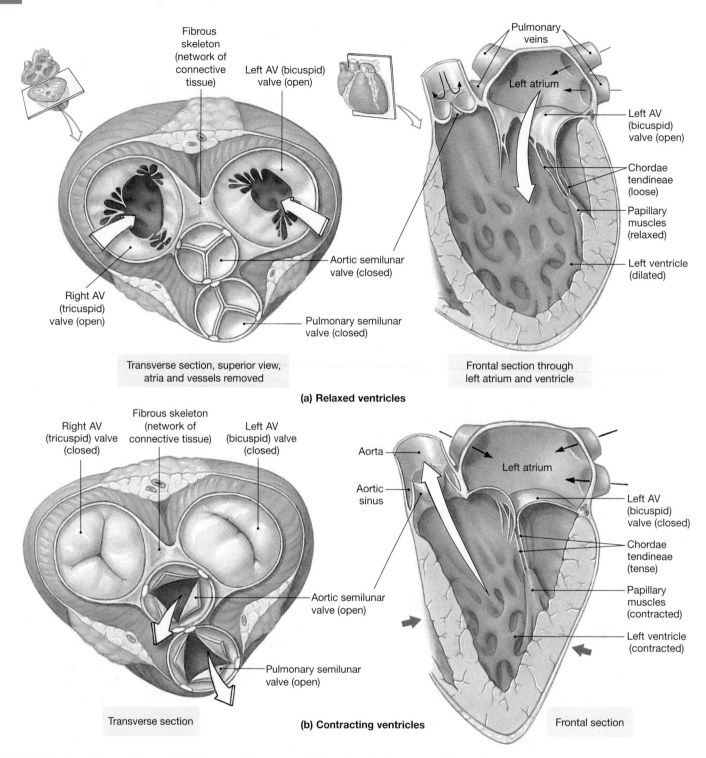

Fibrous skeleton (network of connective tissue)

Left AV (bicuspid) valve (open)

Right AV (tricuspid) valve (open)

Aortic semilunar valve (closed)

Pulmonary semilunar valve (closed)

Transverse section, superior view, atria and vessels removed

Pulmonary veins

Left atrium

Left AV (bicuspid) valve (open)

Chordae tendineae (loose)

Papillary muscles (relaxed)

Left ventricle (dilated)

Frontal section through left atrium and ventricle

(a) Relaxed ventricles

Right AV (tricuspid) valve (closed)

Fibrous skeleton (network of connective tissue)

Left AV (bicuspid) valve (closed)

Aortic semilunar valve (open)

Pulmonary semilunar valve (open)

Transverse section

Aorta

Aortic sinus

Left atrium

Left AV (bicuspid) valve (closed)

Chordae tendineae (tense)

Papillary muscles (contracted)

Left ventricle (contracted)

Frontal section

(b) Contracting ventricles

● **FIGURE 12–6 The Valves of the Heart.** White arrows indicate blood flow into and out of the ventricle; black arrows, blood flow into an atrium; and, green arrows, ventricular contraction. (**a**) When the ventricles are relaxed, the AV valves are open, and the semilunar valves are closed. The chordae tendineae are slack, and the papillary muscles are relaxed. (**b**) When the ventricles contract, the AV valves are closed, and the semilunar valves are open. Notice in the frontal section that the chordae tendineae and papillary muscles prevent backflow through the left AV valve.

Clinical Note
MITRAL VALVE PROLAPSE

Minor abnormalities in valve shape are relatively common. For example, an estimated 10 percent of normal individuals ages 14–30 have some degree of *mitral valve prolapse*. In this condition, the cusps of the left AV valve (the mitral valve) do not close properly. The problem may stem from abnormally long (or short) chordae tendineae or malfunctioning papillary muscles. Because the valve does not work perfectly, some regurgitation occurs during the contraction of the left ventricle. Most individuals with mitral valve prolapse are completely asymptomatic; they live normal, healthy lives unaware of any circulatory malfunction. ∎

THE SEMILUNAR VALVES. The pulmonary and aortic semilunar valves prevent the backflow of blood from the pulmonary trunk and aorta into the right and left ventricles, respectively. The semilunar valves do not require muscular bracing as the AV valves do, because the arterial walls do not contract and the relative positions of the cusps are stable. When the semilunar valves close, the three symmetrical cusps in each valve support one another like the legs of a tripod, which prevents the movement of blood back into the ventricles (see Figure 12–6a●).

Sac-like expansions of the base of the ascending aorta occur next to each cusp of the aortic semilunar valve. These sacs, called **aortic sinuses** (see Figure 12–6b●), prevent the cusps from sticking to the wall of the aorta when the valve opens. The *right* and *left coronary arteries* originate at the aortic sinuses.

Clinical Note
VALVULAR HEART DISEASE

Prior to the latter half of the twentieth century, valvular heart disease was the most common type of heart disease encountered. Many valvular heart problems began following rheumatic fever. Rheumatic fever can cause the body's immune system to mistakenly manufacture antibodies against the leaflets of the valves, and cause vegetative growths. These growths prevent the valves from properly closing, which results in leaking that reduces pumping efficiency and eventually leads to congestive heart failure. Since the introduction of antibiotics, the incidence of rheumatic fever has dropped markedly. ∎

The Fibrous Skeleton of the Heart

The **fibrous skeleton** of the heart consists of dense bands of tough, elastic connective tissue that encircle the bases of the large blood vessels that carry blood away from the heart (*pulmonary trunk* and *aorta*) and each of the heart valves (see Figure 12–6). The fibrous skeleton stabilizes the position of the heart valves and also physically isolates the atrial muscle tissue from the ventricular muscle tissue. This isolation is important to normal heart function because it means that the timing of ventricular contraction relative to atrial contraction can be precisely controlled.

Key Note

The heart has four chambers, two associated with the pulmonary circuit (right atrium and right ventricle) and two with the systemic circuit (left atrium and left ventricle). The left ventricle has a greater workload and is much more massive than the right ventricle, but the two chambers pump equal amounts of blood. AV valves prevent backflow from the ventricles into the atria, and semilunar valves prevent backflow from the aortic and pulmonary trunks into the ventricles.

The Blood Supply to the Heart

The heart works continuously, so cardiac muscle cells require reliable supplies of oxygen and nutrients. The **coronary circulation** (Figure 12–7●) supplies blood to the muscle tissue of the heart. During exertion, the oxygen demand rises considerably, and blood flow to the heart may increase to up to nine times that of resting levels.

The left and right **coronary arteries** originate at the base of the aorta (Figure 12–7a) at the aortic sinuses. Blood pressure here is the highest of any place in the systemic circuit. This high pressure ensures a continuous flow of blood to meet the demands of active cardiac muscle. The right coronary artery (RCA) supplies blood to the right atrium and to portions of both ventricles. The left coronary artery (LCA) supplies blood to the left ventricle, left atrium, and interventricular septum.

Each coronary artery gives rise to two branches. The right coronary artery forms the *marginal* and *posterior interventricular (descending)* branches, and the left coronary artery forms the *circumflex* and *anterior interventricular (descending)* branches. Small tributaries from these branches of the left and right coronary arteries form interconnections called **anastomoses** (a-nas-tō-MŌ-sēz; *anastomosis*, outlet). Because the arteries are interconnected in this way, alternate

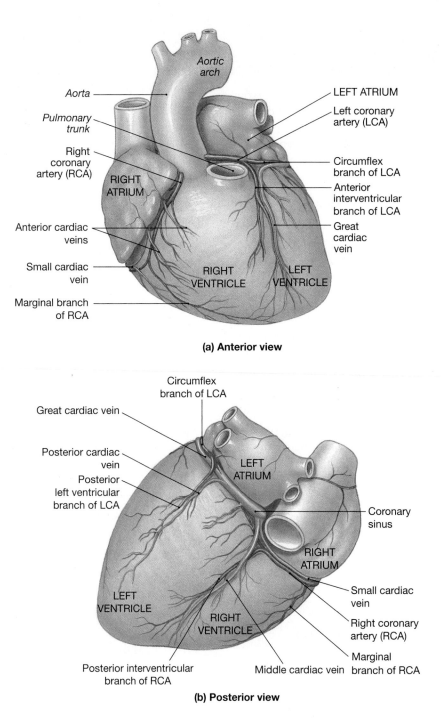

(a) Anterior view

(b) Posterior view

● **FIGURE 12–7 The Coronary Circulation.** The blood vessels that constitute the coronary circulation are labeled in (**a**) anterior and (**b**) posterior views.

pathways exist for the blood supply to reach cardiac muscle. The **great** and **middle cardiac veins** carry blood away from the coronary capillaries. They drain into the **coronary sinus,** which is a large, thin-walled vein in the posterior portion of the coronary sulcus. The coronary sinus opens into the right atrium near the base of the inferior vena cava (see Figure 12–5).

An area of dead tissue caused by an interruption in blood flow is called an **infarct.** In a **myocardial** (mī-ō-KAR-dē-al) **infarction (MI),** or *heart attack,* the coronary circulation becomes blocked, and cardiac muscle cells die from a lack of oxygen. Heart attacks most often result from severe *coronary artery disease,* which is a condition characterized by the buildup of fatty deposits in the walls of the coronary arteries.

Clinical Note
CORONARY ARTERY DISEASE

The initial biological process in the progression of coronary artery disease is the development of microscopic *atherosclerosis*. Atherosclerosis is a form of *arteriosclerosis* where soft deposits of intra-arterial fat and fibrin cause thickening and hardening of the arterial wall. The initial occurrence in this process is an endothelial injury, followed by a series of pathological events that lead to the development of a fatty streak on the lining of the vessel. These lesions can be found in the walls of most people's arteries, even young children's. They are reversible with treatment that lowers *low-density lipoprotein (LDL)* levels. Once formed, however, fatty streaks produce toxic oxygen radicals and cause inflammatory changes that result in progressive damage to the arterial wall.

The next stage of *atherogenesis* is the incorporation of fibrous tissue and damaged smooth muscle cells into the area that forms a *fibrous plaque* or *fibroadenoma*. This causes further endothelial dysfunction, necrosis of underlying vessel tissue, and narrowing of the vessel lumen. As the plaque continues to develop, it can ulcerate or rupture because of the mechanical shear forces and continued necrosis of the vessel wall. Platelets aggregate and adhere to the vessel surface, which simultaneously activates the coagulation cascade. In severe cases, a thrombus (blood clot) can form, completely obstructing the vessel lumen and resulting in tissue ischemia and infarction.

The rate of progression of atherosclerosis can be significantly influenced by specific conditions and behaviors referred to as *risk factors* including age, gender, tobacco use, diabetes, obesity, hypertension, and many others. Atherosclerosis is a disease spectrum with various events that occur along the continuum. Cardiac arrest and myocardial infarction (heart attack) are, in the vast majority of cases, end points in a decades-long evolution of atherosclerotic arterial disease (Figure 12–8●). The rate of progression of atherosclerosis is the primary determinant of the age at which myocardial infarction or sudden death occurs. Control or elimination of risk factors can usually be achieved by establishing positive health attitudes and behaviors, especially in the young. Significant modifications in risk factors can occur by exer-

cise, smoking cessation, dietary modifications, control of hypertension, and use of medications to reduce lipid levels, blood pressure, and platelet aggregation. Atherosclerotic heart disease is a disease that must be identified and controlled in order to assure long-term health.

Acute Coronary Syndrome

Our knowledge and treatment of acute myocardial infarction (AMI) has evolved dramatically over the past decade. We now recognize that AMI and unstable angina are part of a spectrum of clinical disease referred to as *acute coronary syndrome*. Most patients with acute coronary syndrome have some level of underlying atherosclerotic coronary artery disease. Often, an atheromatous plaque will rupture and block the artery, or a thrombus (blood clot) will form and block blood flow through the affected vessel. The degree and duration of the occlusion determine the type of infarction. The severity of the infarction can be reduced if collateral blood vessels supply some blood to the tissues distal to the obstruction.

Initially following coronary artery occlusion, the myocardial tissue supplied by the affected artery will become ischemic (Figures 12–9●, 12–10●, and 12–11●). If blood flow is restored in time, myocardial ischemia will resolve without permanent muscle injury; however, if myocardial ischemia continues, then actual tissue injury will begin. Tissue injury may or may not be reversed with the restoration of blood flow. Finally, if blood flow is interrupted long enough, then irreversible tissue damage (myocardial infarction) will occur (Figure 12–12●).

The signs and symptoms of acute coronary syndrome are due to coronary artery occlusion and related muscle injury and death. They include chest pain, difficulty breathing, sweating (diaphoresis), nausea, vomiting, and dizziness.

Half of all patients who die of AMI do so early, usually before they reach a hospital. Most of these patients develop *dysrhythmias*, which are irregularities in their heart's electrical activity, some of which can

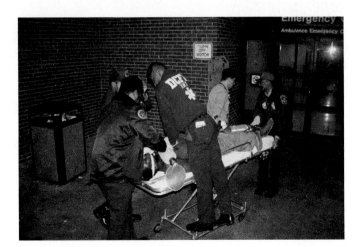

● **FIGURE 12–8 Cardiac Arrest.** Acute myocardial infarction and cardiac arrest are often the end points of atherosclerotic arterial disease over a number of decades.

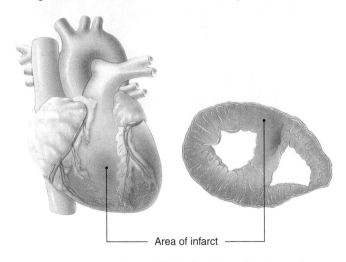

Area of infarct

● **FIGURE 12–9 Myocardial Infarction.** Blockage of a coronary artery results in injury and death to the myocardial tissue supplied by that artery. *(continued next page)*

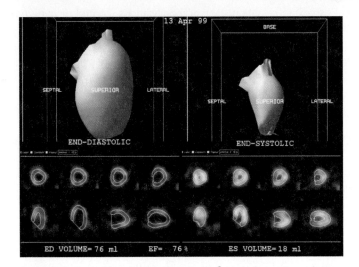

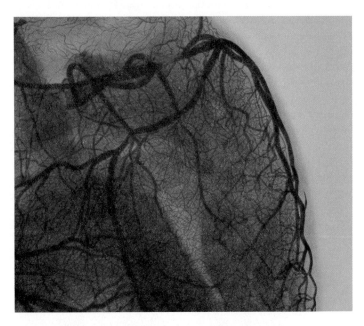

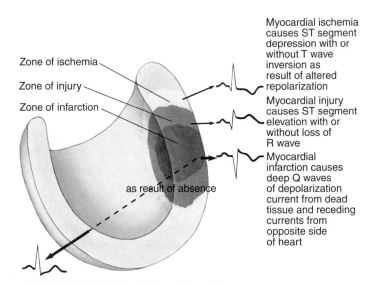

● **FIGURE 12–10 Nuclear Medicine Scan of the Heart That Illustrates Good Ventricular Function.** Nuclear medicine technology allows detailed examination of cardiac perfusion and function.

● **FIGURE 12–11 Coronary Arteriography.**

● **FIGURE 12–12 Evolving Myocardial Infarction.** The center of the lesion is the zone of infarction where tissue death has actually occurred. Adjacent to that is the zone of injury where some recovery is possible. At the periphery of the lesion is the zone of ischemia where actual tissue injury has not yet occurred.

be immediately fatal. Common dysrhythmias associated with myocardial ischemia and sudden death are *ventricular fibrillation* or *ventricular tachycardia*. In ventricular fibrillation, no organized electrical activity occurs, and the myocardial muscle mass simply quivers, or fibrillates (Figure 12–13●). Thus, no myocardial contraction takes place. Ventricular fibrillation can be effectively treated by defibrillation, which is the rapid application of an electrical countershock (Figure 12–14●). Defibrillation stops the irregular electrical activity of the heart and allows the normal pacemaker of the heart to resume

control. Defibrillator technology has evolved significantly. Now, portable defibrillators are completely automated and readily available. Persons who learn CPR and basic life support are now trained in the use of an automated defibrillator.

Reperfusion

One of the most significant advances in emergency medicine over the last decade has been the development of reperfusion therapy, which is the process of restoring blood flow through an occluded artery. This can be achieved with medications, surgery, or both. Initially limited to acute myocardial infarction, reperfusion therapy is being used now in the treatment of selected stroke patients and patients with large blood clots in the lungs (pulmonary emboli).

The 12-lead electrocardiogram (ECG) is a major tool in diagnosing acute coronary syndrome. It is also used to stratify, or triage, patients into one of three groups for treatment:

1. ST segment elevation or new left bundle branch block (LBBB)
2. ST segment depression (> 1 millimeter) or T wave inversion
3. Nondiagnostic or normal ECG

Patients who have signs and symptoms consistent with acute coronary syndrome and have demonstrable high-risk ECG changes (ST segment elevation or new left bundle branch block) should be con-

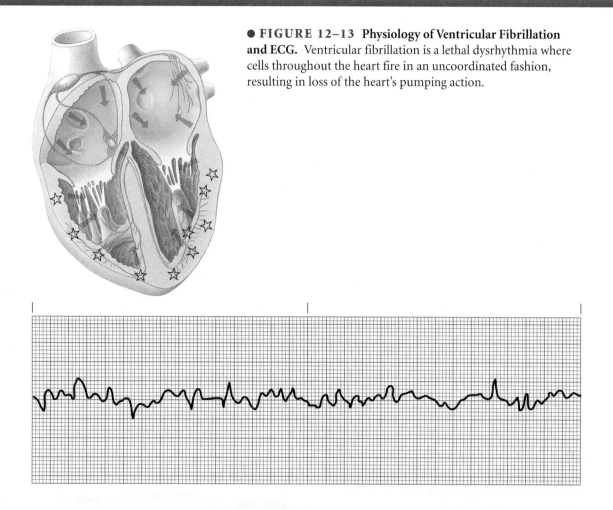

● **FIGURE 12–13 Physiology of Ventricular Fibrillation and ECG.** Ventricular fibrillation is a lethal dysrhythmia where cells throughout the heart fire in an uncoordinated fashion, resulting in loss of the heart's pumping action.

sidered candidates for thrombolytic therapy. Patients who meet the screening criteria and whose symptoms have been present for six hours or less should receive thrombolytic therapy. The thrombolytic agent varies depending upon local recommendations. Regardless of the agent, the patient should be monitored for the development of adverse effects such as intracranial hemorrhage. In addition to thrombolytic agents, patients usually will receive aspirin, heparin (blood thinner), beta blockers, and nitroglycerin. The treatment is based upon the assessment of the patient's overall condition.

Patients who have signs and symptoms consistent with acute coronary syndrome and demonstrable moderate-risk ECG changes (ST segment depression or T wave inversion) should be considered as strongly suspicious for myocardial ischemia. These patients should be treated with heparin, aspirin, beta blockers, and nitroglycerin. Some patients in this category might benefit from glycoprotein IIb/IIIa inhibitors. These agents provide benefits beyond those of aspirin and heparin. A commonly used drug from this classification is abciximab (ReoPro). Aggrastat and Integrilin are also commonly used in the non-catheterization setting.

Finally, patients who have signs and symptoms consistent with acute coronary syndrome but do not have demonstrable ECG

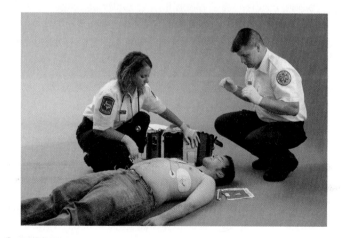

● **FIGURE 12–14 EMS-Provided Defibrillation.** Defibrillation can be life saving in the treatment of sudden death and acute coronary syndrome when complicated by ventricular fibrillation or ventricular tachycardia.

(continued next page)

Clinical Note—continued
CORONARY ARTERY DISEASE

---

changes or have a normal ECG should receive aspirin and other therapy as appropriate. The concept of risk-stratification in acute coronary syndrome is designed to ensure that patients receive the best possible therapy as quickly as possible.

Revascularization

Following thrombolytic therapy, or when thrombolytic therapy is not indicated or is ineffective, patients should be referred for coronary angiography and possible interventional therapy. In many cases, emergent cardiac catheterization is superior to thrombolytic therapy. It allows visualization of the coronary artery anatomy including the degree and magnitude of vessel blockage.

While in the cardiac catheterization lab for coronary angiography, patients with atherosclerotic lesions and occlusions suitable for angioplasty can be treated. The process, called *percutaneous transluminal coronary angioplasty (PTCA),* involves introducing a small catheter into the affected artery. An uninflated balloon catheter is placed at the site of the lesion, or lesions, and then inflated. This compresses the atheromatous plaque and increases the size of the vessel lumen. This procedure may need to be repeated several times until the artery remains open. If the vessel will not remain adequately open following balloon angioplasty, a metal stent can be placed at the site of the lesion. The metal stent is compressed for placement. Once placed, the stent is released and expands to hold back the plaque, which restores patency to the blood vessel. Other technologies, such as intraluminal lasers and rotary files that break down inflammatory and atheromatous plaques, are used occasionally.

Some patients have severe and diffuse atherosclerotic coronary artery disease. Their coronary angiograms reveal multiple regions of obstruction that make angioplasty impossible. These patients are evaluated for *coronary artery bypass grafting (CABG).* CABG is a major procedure that requires that the sternum be split, the heart stopped (cardioplegia), and the patient placed on a bypass pump for the duration of the surgery. The medial aspect of the leg is opened and a part of the saphenous vein is harvested for the bypass grafts. The grafts are prepared and sewn into the aorta, just above the aortic valve. The distal end is then attached to the diseased coronary artery distal to the blockage. Some patients require multiple grafts. It is not uncommon to perform four or five grafts in one operation. In younger patients, arteries in the chest, such as the internal mammary artery, are used to provide blood to one of the new grafts. New technologies are making the surgical procedure less invasive, which shortens the patient's hospitalization and recovery period. In "keyhole surgery" a small camera and the necessary surgical instruments are introduced through several small incisions in the chest. This operation is referred to as "off pump" surgery because the entire procedure is performed while the heart remains beating. This significantly lessens the patient's pain and decreases the hospital stay because the patient has several small incisions instead of two large ones. ■

→ **CONCEPT CHECK QUESTIONS**

1. Damage to the semilunar valves on the right side of the heart would interfere with blood flow to which vessel?
2. What prevents the AV valves from opening back into the atria?
3. Why is the left ventricle more muscular than the right ventricle?

Answers begin on p. 792.

■ The Heartbeat

In a single **heartbeat,** the entire heart—atria and ventricles—contracts in a coordinated manner so that blood flows in the correct direction at the proper time. Each time the heart beats, the contractions of individual cardiac muscle cells in the atria and ventricles must occur in a specific sequence. Two types of cardiac muscle cells are involved in a normal heartbeat: (1) *contractile cells,* which produce the powerful contractions that propel blood; and (2) specialized noncontractile muscle cells of the *conducting system,* which control and coordinate the activities of the contractile cells.

Contractile Cells

Contractile cells form the bulk of the heart's muscle tissue (about 99 percent of all cardiac muscle cells). In both cardiac muscle cells and skeletal muscle fibers, an action potential leads to the appearance of Ca^{2+} among the myofibrils, and its binding to troponin on the thin filaments begins a contraction. However, skeletal and cardiac muscle cells differ in terms of the duration of action potentials, the source of Ca^{2+}, and the duration of the resulting contraction. Figure 12–15a● shows the sequence of events in an action potential in a cardiac muscle cell, and Figure 12–15b● compares the action potentials and a twitch contraction in skeletal muscle and cardiac muscle.

An action potential in a ventricular cardiac muscle cell begins when the membrane is brought to threshold by a stimulus (typically by the excitation of an adjacent muscle cell). The action potential then proceeds in three steps (see Figure 12–15a):

Step 1: *Rapid depolarization.* At threshold, voltage-regulated sodium channels open, and the influx of sodium ions

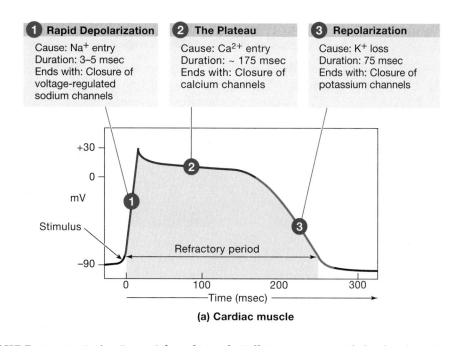

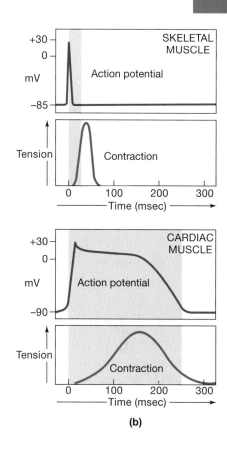

1 Rapid Depolarization
Cause: Na$^+$ entry
Duration: 3–5 msec
Ends with: Closure of voltage-regulated sodium channels

2 The Plateau
Cause: Ca^{2+} entry
Duration: ~ 175 msec
Ends with: Closure of calcium channels

3 Repolarization
Cause: K$^+$ loss
Duration: 75 msec
Ends with: Closure of potassium channels

(a) Cardiac muscle

(b)

● **FIGURE 12–15 Action Potentials and Muscle Cell Contraction in Skeletal and Cardiac Muscle. (a)** This graph indicates the events of an action potential in a ventricular cardiac muscle cell. **(b)** In skeletal muscle fibers, an action potential is relatively brief and ends as the related twitch contraction begins. In cardiac muscle cells, both the action potential and twitch contraction are prolonged. The shaded areas indicate the duration of the refractory period.

rapidly depolarizes the sarcolemma (cell membrane). The sodium channels close when the transmembrane potential reaches +30 mV.

Step 2: *The plateau.* As the cell now begins to actively pump sodium ions out, voltage-regulated calcium channels open and *extracellular* calcium ions enter the sarcoplasm. The calcium channels remain open for a relatively long period, and the entering positive charges (Ca^{2+}) roughly balance the loss of Na$^+$ from the cell. The extracellular calcium ions (1) delay repolarization (their positive charges maintain a transmembrane potential near 0 mV—the *plateau*) and (2) initiate contraction. Their increased concentration within the cell also triggers the release of Ca^{2+} from reserves in the sarcoplasmic reticulum (SR), which continues the contraction.

Step 3: *Repolarization.* As the calcium channels begin to close, potassium channels open and potassium ions (K$^+$) rush out of the cell. The net result is a repolarization that restores the resting potential.

In a skeletal muscle fiber, the 10-msec (millisecond) action potential of a rapid depolarization is immediately followed by a rapid repolarization. This brief action potential ends as the related twitch contraction begins (see Figure 12–15b). The twitch contraction is short and ends as the SR

reclaims the Ca^{2+} it released. The action potential is prolonged in a cardiac muscle cell because calcium ions continue to enter the cell throughout the plateau. As a result, the period of muscle contraction continues until the plateau ends. As the calcium channels close, the intracellular calcium ions are absorbed by the SR or are pumped out of the cell, and the muscle cell relaxes.

The complete depolarization-repolarization process in a cardiac muscle cell lasts 250–300 msec, 25–30 times as long as an action potential in a skeletal muscle fiber (see Figure 12–15b). Until the membrane repolarizes, it cannot respond to further stimulation, so the refractory period of a cardiac muscle cell membrane is relatively long. Thus, a normal cardiac muscle cell is limited to a maximum rate of about 200 contractions per minute.

In skeletal muscle fibers, the refractory period ends before the muscle fiber develops peak tension and relaxes. As a result, twitches can build on one another until tension reaches a sustained peak; this state is called *tetanus*. ∞ p. 218 In cardiac muscle cells, the refractory period continues until relaxation is under way. A summation of twitches is, therefore, not possible, and tetanic contractions cannot occur in a normal cardiac muscle cell, regardless of the frequency and intensity of stimulation. This feature is absolutely vital because a heart in tetany could not pump blood.

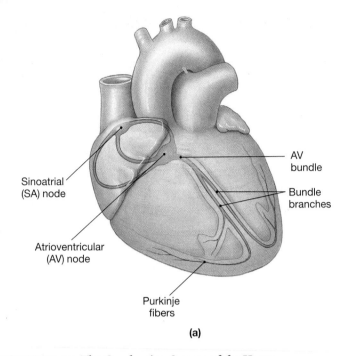

(a)

● **FIGURE 12–16 The Conducting System of the Heart.**
(**a**) The locations of the elements of the conducting system and pathways of electrical impulses are shown in this diagrammatic view of the heart. (**b**) The stimulus for contraction moves through the heart in a predictable sequence of events (steps 1–5 as diagrammed).

The Conducting System

In contrast to skeletal muscle, cardiac muscle tissue contracts on its own in the absence of neural or hormonal stimulation. (This property, called *automaticity,* or *autorhythmicity,* also characterizes some types of smooth muscle tissue discussed in Chapter 7. ∞ p. 226

In the heart's normal pattern of activity, each contraction follows a precise sequence: the atria contract first, followed by the ventricles. Cardiac contractions are coordinated by the heart's **conducting system,** a network of specialized cardiac muscle cells that initiates and distributes electrical impulses (Figure 12–16a●). The network is made up of two types of cardiac muscle cells that do not contract: (1) **nodal cells,** which are responsible for establishing the rate of cardiac contraction and are located at the *sinoatrial (SA)* and *atrioventricular (AV) nodes;* and (2) **conducting cells,** which distribute the contractile stimulus to the general myocardium. Major sites of conducting cells include the *AV bundle,* the *bundle branches,* and the *Purkinje fibers.*

Nodal cells are unusual because their cell membranes depolarize spontaneously and generate action potentials at regular intervals. Nodal cells are electrically coupled to one

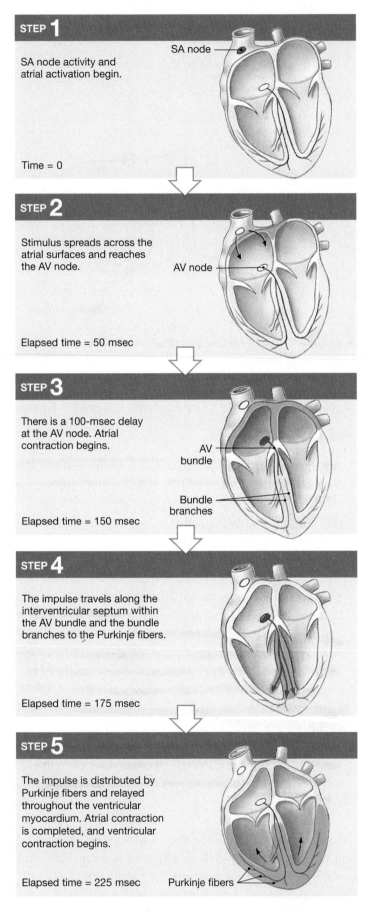

STEP 1

SA node activity and atrial activation begin.

SA node

Time = 0

STEP 2

Stimulus spreads across the atrial surfaces and reaches the AV node.

AV node

Elapsed time = 50 msec

STEP 3

There is a 100-msec delay at the AV node. Atrial contraction begins.

AV bundle

Bundle branches

Elapsed time = 150 msec

STEP 4

The impulse travels along the interventricular septum within the AV bundle and the bundle branches to the Purkinje fibers.

Elapsed time = 175 msec

STEP 5

The impulse is distributed by Purkinje fibers and relayed throughout the ventricular myocardium. Atrial contraction is completed, and ventricular contraction begins.

Elapsed time = 225 msec

Purkinje fibers

(b)

another, to conducting cells, and to normal cardiac muscle cells. As a result, when an action potential is initiated in a nodal cell, it sweeps through the conducting system, reaching all of the cardiac muscle tissue and causing a coordinated contraction. In this way, nodal cells determine the heart rate.

Not all nodal cells depolarize at the same rate, and the normal rate of contraction is established by **pacemaker cells,** which are the nodal cells that reach threshold first. These pacemaker cells are located in the **sinoatrial** (sī-no-Ā-trē-al) **node (SA node),** or *cardiac pacemaker,* which is a tissue mass embedded in the posterior wall of the right atrium near the entrance of the superior vena cava (see Figure 12–16a). Pacemaker cells depolarize rapidly and spontaneously, and generate 70–80 action potentials per minute. This results in a heart rate of 70–80 beats per minute (bpm).

After the stimulus for a contraction is generated at the SA node, it must be distributed so that (1) the atria contract together, before the ventricles; and (2) the ventricles contract together, in a wave that begins at the apex and spreads toward the base. When the ventricles contract in this way, blood is pushed toward the base of the heart into the aorta and pulmonary trunk.

The cells of the SA node are electrically connected to those of the larger **atrioventricular** (ā-trē-ō-ven-TRIK-ū-lar) **node (AV node)** by conducting cells in the atrial walls (see Figure 12–16a). Although the AV nodal cells also depolarize spontaneously, they generate only 40–60 action potentials per minute. Under normal circumstances, before an AV cell depolarizes to threshold spontaneously, it is stimulated by an action potential generated by the SA node. However, if the AV node does not receive this action potential, it will then become the pacemaker of the heart and establish a heart rate of 40–60 beats per minute.

The AV node is located in the floor of the right atrium near the opening of the coronary sinus. From there the action potentials travel to the **AV bundle,** which is also known as the *bundle of His* (pronounced *hiss*). This bundle of conducting cells extends along the interventricular septum before dividing into **left** and **right bundle branches,** which radiate across the inner surfaces of the left and right ventricles. At this point, specialized **Purkinje** (pur-KIN-jē) **fibers** (*Purkinje cells*) convey the impulses to the contractile cells of the ventricular myocardium.

As shown in Figure 12–16b●, it takes an action potential roughly 50 msec to travel from the SA node to the AV node over the conducting pathways. Along the way, the conducting cells pass the contractile stimulus to cardiac muscle cells of the right and left atria. The action potential then spreads across the atrial surfaces through cell-to-cell contact. The stimulus affects only the atria because the fibrous skeleton (see Figure 12–16b) electrically isolates the atria from the ventricles everywhere except at the AV bundle.

At the AV node, the impulse slows down, and another 100 msec pass before it reaches the AV bundle. This delay is important because the atria must be contracting, and blood must be moving, before the ventricles are stimulated. Once the impulse enters the AV bundle, it flashes down the interventricular septum, along the bundle branches, and into the ventricular myocardium along the Purkinje fibers. Within another 75 msec, the stimulus to begin a contraction has reached all of the ventricular muscle cells.

A number of clinical problems result from deviations from normal pacemaker function, which produces a heart rate that averages 70–80 bpm. **Bradycardia** (brād-ē-KAR-dē-uh; *bradys,* slow) is a condition in which the heart rate is slower than normal (less than 60 bpm). **Tachycardia** (tak-ē-KAR-dē-uh; *tachys,* swift) is a faster than normal heart rate (100 bpm or more). In some cases, an abnormal conducting cell or ventricular muscle cell may begin generating action potentials so rapidly that they override those of the SA or AV node. The origin of such abnormal signals is called an **ectopic** (ek-TOP-ik; out of place) **pacemaker.** The action potentials may completely bypass the conducting system and disrupt the timing of ventricular contractions. Such conditions are commonly diagnosed with the aid of an *electrocardiogram,* which we will discuss next.

Clinical Note
DEFIBRILLATION

Defibrillation is used to treat life-threatening dysrhythmias such as ventricular fibrillation and nonperfusing ventricular tachycardia. Defibrillation is the delivery of sufficient electrical energy to the heart to depolarize the myocardial muscle mass and at the same time causes minimal electrical injury to the heart. This allows a pacemaker cell within the heart to resume its rhythmic firing, and restores a normal electrical pattern. A shock will not terminate the dysrhythmia if the energy or current is too low. Functional and structural damage to the heart can occur if the energy or current are too high.

(continued next page)

Clinical Note—continued

DEFIBRILLATION

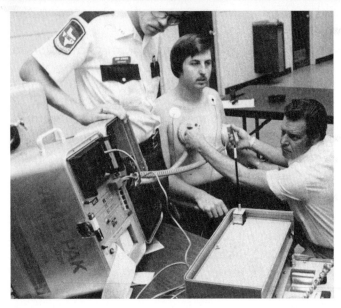

● **FIGURE 12–17 Early Portable Defibrillator.** Defibrillator and EMS technology has evolved significantly. This 1975 photo shows a portable defibrillator/monitor that weighed nearly 40 pounds. The large device at the bottom of the screen is a Motorola Apcor radio modified to operate as a mobile telephone.

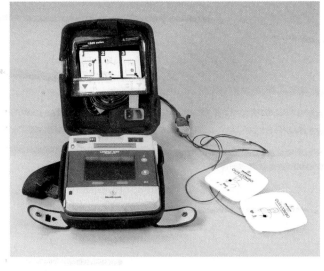

● **FIGURE 12–18 Modern Defibrillator.** Defibrillators are now compact and completely automated as illustrated by the device shown here.

When first developed, defibrillators were extremely large units that operated on standard electrical current. These older units used alternating current (AC) and were limited to hospitals. In the 1970s, portable defibrillators that could be used in the prehospital setting were developed. These units utilized direct current (DC) but were extremely heavy and bulky, often weighing 40 pounds or more (Figure 12–17●). Battery technology was limited and the unit required recharging immediately after use. Now, defibrillators are available that are approximately the size of a notebook. Many are completely automated, which allows use by nonmedical personnel (Figure 12–18●).

Technology has also allowed the development of a defibrillator unit that can be permanently implanted in patients deemed at risk for life-threatening dysrhythmias. These devices, referred to as *implanted automatic cardioverter-defibrillators (IACDs),* have become the treatment of choice for sudden cardiac death. They have reduced mortality from 30 to 45 percent per year to less than 2 percent per year. The IACD consists of a pulse generator, a lead system with both sensing and shocking electrodes, integrated circuitry to analyze the cardiac rhythm and trigger defibrillation, and a power supply. The life span of these units is approximately eight years depending upon the frequency of discharge.

Today, there are three general types of external defibrillators: automated external defibrillators (AEDs), semi-automated external defibrillators (SAEDs), and manual defibrillators. Automated units simply require application of the electrodes and turning on the device. Semiautomated units require that the operator press a button to deliver

the electrical shock as directed by the device. These extremely automated units are designed for use by first responders and nonmedical personnel. Most airlines now carry automated external defibrillators on flights longer than three hours. Soon, all large commercial passenger aircraft will carry these devices. The manual units are designed for use by trained medical personnel. They usually have several other features including continuous cardiac monitoring, synchronized cardioversion, hard-copy recording, and external cardiac pacing. Newer units allow recording and monitoring of a 12-lead ECG. They also contain a diagnostic module that provides a computerized interpretation of the ECG tracing.

The ability to defibrillate a fibrillating heart depends primarily on the electrical current delivered to the myocardium. All modern defibrillators use direct current (DC) and store the energy in a capacitor that discharges when triggered by the operator. Factors that impede the flow of energy to the myocardium include resistance to current flow by the chest wall (transthoracic impedance) and internal loss of energy within the defibrillator. The current delivered to the myocardium is a function of the energy delivered by the defibrillator, the electrical impedance within the device and chest, and the duration of current flow:

$$\text{Energy} = \text{current} \times \text{impedance} \times \text{duration}$$

$$(\text{Joules} = \text{amperes} \times \text{ohms} \times \text{seconds})$$

An increase in either current or duration will increase the energy. An increase in transthoracic impedance (resistance to current flow) will reduce the delivered current.

Modern defibrillators deliver energy, or current, in waveforms. The energy levels vary with the type of device and type of wave-

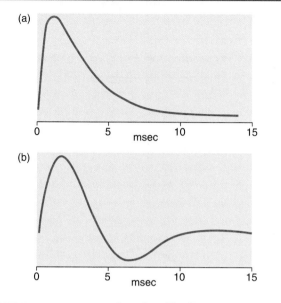

(a)

(b)

● **FIGURE 12–19 Examples of Defibrillator Waveforms.** The top waveform is monophasic, the lower waveform is biphasic. With a biphasic waveform, current travels in a positive direction for the first part of the energy delivery and then in a negative direction for the remainder.

form it uses. Modern defibrillators use two types of waveforms: monophasic and biphasic (Figure 12–19●). For years the standard DC waveform was monophasic, and delivered current in one direction only. Biphasic waveforms, in contrast, deliver current that flows in a positive direction for a specific duration and then flows in a negative direction for the remaining millisecond of the electrical discharge. A biphasic waveform has been demonstrated to be superior in patients who have implantable cardioverter-defibrillators. However, research has not determined whether a biphasic waveform provides any advantage in external defibrillators. The recommended energy settings for defibrillation with a monophasic-type machine start at 200 Joules, then increase to 300 Joules with the second shock, and increase again to 360 Joules with the third and subsequent shocks. The optimum energy for biphasic-type machines has not been clearly identified.

Many of the manual defibrillator/monitors can provide synchronized cardioversion. This technology, monitors the patient's ECG, electronically marks the QRS complex, and triggers the shock at precisely the proper time in the cardiac cycle to depolarize the heart. Because the shock targets the vulnerable period, the amount of energy needed to depolarize the myocardium is markedly less. ■

→ **CONCEPT CHECK QUESTIONS**

1. Cardiac muscle does not undergo tetanus as skeletal muscle does. How does this characteristic affect the functioning of the heart?
2. If the cells of the SA node were not functioning, how would the heart rate be affected?
3. Why is it important for the impulses from the atria to be delayed at the AV node before passing into the ventricles?

Answers begin on p. 792.

The Electrocardiogram

The electrical events that occur in the heart are powerful enough to be detected by electrodes on the body surface. A recording of these events is an **electrocardiogram** (ē-lek-trō-KAR-dē-ō-gram), also called an **ECG** or **EKG.** Each time the heart beats, a wave of depolarization radiates through the atria, reaches the AV node, travels down the interventricular septum to the apex, turns, and spreads through the ventricular myocardium toward the base.

By comparing the information obtained from electrodes placed at different locations, a clinician can monitor the electrical activity of the heart and check the performance of specific nodal, conducting, and contractile components. When, for example, a portion of the heart has been damaged, the affected muscle cells will no longer conduct action potentials, so an ECG will reveal an abnormal pattern of impulse conduction.

The appearance of the ECG tracing varies with the placement of the monitoring electrodes, or *leads*. The important features of an electrocardiogram as analyzed with the leads in one of the standard configurations are shown in Figure 12–20●:

■ A small **P wave** accompanies the depolarization of the atria. The atria begin contracting around 100 msec after the start of the P wave.

■ A **QRS complex** appears as the ventricles depolarize. This portion of the electrical signal is relatively strong because the mass of the ventricular muscle is much larger than that of the atria. The ventricles begin contracting shortly after the peak of the R wave.

■ A smaller **T wave** indicates ventricular repolarization. Atrial repolarization is not apparent because it occurs while the ventricles are depolarizing, and the electrical events there are masked by the QRS complex.

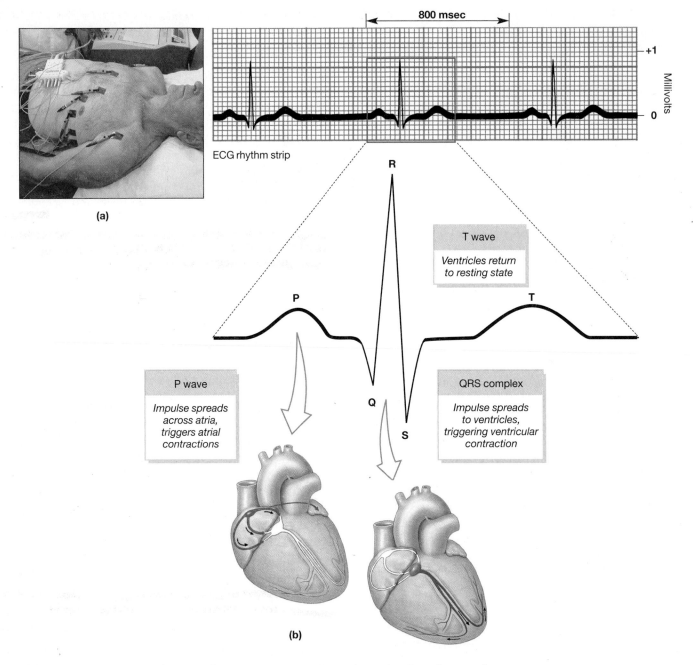

● **FIGURE 12–20 An Electrocardiogram.** An ECG printout is a strip of graph paper that contains a continuous record of the electrical events in the heart as monitored by electrodes attached to the body surface. **(a)** The placement of electrodes affects the size and shape of the waves recorded. The electrode placement for a standard ECG is shown here. **(b)** The enlarged section indicates the major components of the ECG and the events that correspond to them.

Analyzing an ECG involves measuring the size of the voltage changes and determining the temporal relationships of the various components. Attention usually focuses on the amount of depolarization that occurs during the P wave and the QRS complex. For example, a smaller than normal electrical signal can mean that the mass of the heart muscle has decreased, whereas excessively strong depolarizations can mean that the heart muscle has become enlarged.

Electrocardiogram analysis is useful in detecting and diagnosing **cardiac arrhythmias** (ā-RITH-mē-az), abnormal patterns of cardiac activity. Momentary arrhythmias are not inherently dangerous, and about 5 percent of the normal population experiences a few abnormal heartbeats each day. Clinical problems appear when the arrhythmias reduce the heart's pumping efficiency. Serious arrhythmias can indicate damage to the myocardium, injuries to the pacemaker or conduction pathways, exposure to drugs, or variations in the electrolyte composition of the extracellular fluids.

Clinical Note
EMERGENCY CARDIAC CARE

Emergency cardiac care (ECC) is a comprehensive system designed to deal with sudden, often life-threatening events that affect the cardiovascular, cerebrovascular, and respiratory systems. ECC specifically includes:

1. Recognition of the early warning signs of a heart attack and stroke, efforts to prevent complications, reassurance of the victim, activation of the EMS system, and prompt availability of monitoring equipment.
2. Provision of immediate basic life support (BLS) and cardiopulmonary resuscitation (CPR) at the scene when needed.
3. Provision of advanced cardiac life support (ACLS) at the scene as quickly as possible to defibrillate if necessary and to stabilize the victim prior to transport.
4. Transfer of the stabilized victim to a hospital where definitive cardiac care can be provided.

The key link in the ECC system is the layperson who recognizes the medical emergency, summons EMS, and provides initial CPR and other BLS measures.

Advanced prehospital care and paramedic practice were first developed to treat life-threatening cardiac emergencies. Now, in addition to ECC, paramedics treat all types of medical and traumatic emergencies and provide effective on-scene stabilization. ■

Key Note

The heart rate is normally established by the cells of the SA node, but that rate can be modified by autonomic activity, hormones, and other factors. From the SA node the stimulus is conducted to the AV node, the AV bundle, the bundle branches, and Purkinje fibers before reaching the ventricular muscle cells. The electrical events associated with the heartbeat can be monitored in an electrocardiogram (ECG).

The Cardiac Cycle

The period between the start of one heartbeat and the start of the next is a single **cardiac cycle** (Figure 12–21●). The cardiac cycle, therefore, includes both a period of contraction and one of relaxation. For any one chamber in the heart, the cardiac cycle can be divided into two phases. During contraction, or **systole** (SIS-tō-lē), the chamber squeezes blood into an adjacent chamber or into an arterial trunk. Systole is followed by the second phase—one of relaxation, or **diastole** (dī-AS-tō-lē)—when the chamber fills with blood and prepares for the start of the next cardiac cycle.

Fluids tend to move from an area of higher pressure to one of lower pressure. During the cardiac cycle, the pressure within each chamber rises during systole and falls during diastole. An increase in pressure in one chamber causes the blood to flow to another chamber (or vessel) where the pressure is lower. The atrioventricular and semilunar valves ensure that blood flows in one direction only during the cardiac cycle.

The correct pressure relationships among the heart chambers are maintained by the careful timing of contractions. The elaborate pacemaking and conduction systems normally provide the required interval of time between atrial systole and ventricular systole. If the atria and ventricles were to contract simultaneously, blood could not leave the atria because the AV valves would be closed. In the normal heart, atrial systole and atrial diastole are slightly out of phase with ventricular systole and diastole. Figure 12–21 shows the duration and timing of systole and diastole for a heart rate of 75 bpm.

The cardiac cycle begins with atrial systole. At the start of a cardiac cycle, the ventricles are partially filled with blood. During atrial systole, the atria contract and the ventricles become completely filled with blood (Figure 12–21a). As atrial systole ends (Figure 12–21b), atrial diastole and ventricular systole begin. As pressures in the ventricles rise above those in the atria, the AV valves swing shut (Figure 12–21c). But blood cannot begin moving into the arterial trunks until ventricular pressures exceed the arterial pressures. At this point, the blood pushes open the semilunar valves and flows into the aorta and pulmonary trunk (Figure 12–21d). This blood flow continues for the duration of ventricular systole.

When ventricular diastole begins (Figure 12–21e), ventricular pressures decline rapidly. As they fall below the pressures of the arterial trunks, the semilunar valves close. Ventricular pressures continue to drop; as they fall below atrial pressures, the AV valves open and blood flows from the atria into the ventricles. Both atria and ventricles are now in diastole; blood now flows from the major veins through the relaxed atria and into the ventricles (Figure 12–21f). By the time atrial systole marks the start of another cardiac cycle, the ventricles are roughly 70 percent filled. The relatively minor contribution atrial systole makes to ventricular volume explains why individuals can survive quite normally when their atria have been so severely damaged that they can no longer function. In contrast, damage to one or both ventricles can leave the heart unable to maintain adequate cardiac output. A condition of *heart failure* then exists.

Heart Sounds

Physicians use an instrument called a **stethoscope** to listen for four **heart sounds**. When you listen to your own heart, you

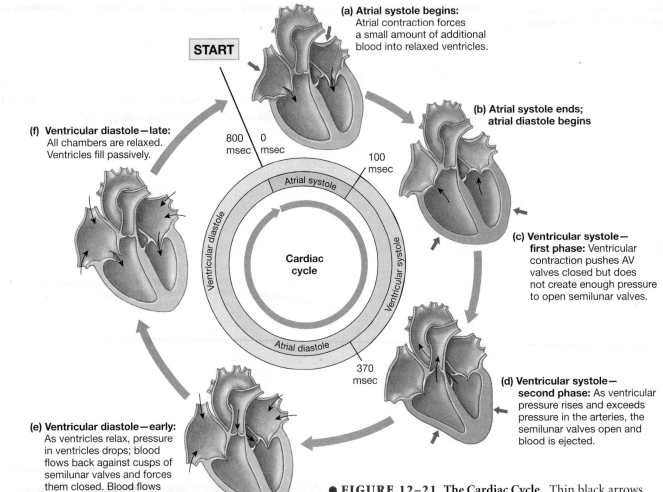

(a) Atrial systole begins: Atrial contraction forces a small amount of additional blood into relaxed ventricles.

START

(b) Atrial systole ends; atrial diastole begins

(c) Ventricular systole—first phase: Ventricular contraction pushes AV valves closed but does not create enough pressure to open semilunar valves.

(d) Ventricular systole—second phase: As ventricular pressure rises and exceeds pressure in the arteries, the semilunar valves open and blood is ejected.

(e) Ventricular diastole—early: As ventricles relax, pressure in ventricles drops; blood flows back against cusps of semilunar valves and forces them closed. Blood flows into the relaxed atria.

(f) Ventricular diastole—late: All chambers are relaxed. Ventricles fill passively.

800 msec / 0 msec

100 msec

370 msec

Atrial systole

Atrial diastole

Ventricular diastole

Ventricular systole

Cardiac cycle

● **FIGURE 12–21 The Cardiac Cycle.** Thin black arrows indicate blood flow, and green arrows indicate contractions.

hear the familiar "lubb-dupp" that accompanies each heartbeat. These two sounds accompany the actions of the heart valves. The *first heart sound* ("lubb") is produced as the AV valves close and the semilunar valves open. It marks the start of ventricular systole and lasts a little longer than the second sound. The *second heart sound*, "dupp," occurs at the beginning of ventricular diastole, when the semilunar valves close.

Third and *fourth heart sounds* may be audible as well, but they are usually very faint and are seldom detectable in healthy adults. These sounds are associated with atrial contraction and blood flowing into the ventricles rather than with valve action.

> **CONCEPT CHECK QUESTIONS**
>
> 1. Is the heart always pumping blood when pressure in the left ventricle is rising? Explain.
> 2. What causes the "lubb-dupp" sounds of the heart that are heard with a stethoscope?
>
> *Answers begin on p. 792.*

■ Heart Dynamics

The term *heart dynamics* refers to the movements and forces generated during cardiac contractions. Each time the heart beats, the two ventricles eject equal amounts of blood. The amount ejected by a ventricle during a single beat is the **stroke volume (SV).** The stroke volume can vary from beat to beat, so physicians are often more interested in the **cardiac output (CO),** or the amount of blood pumped by each ventricle in one minute. Cardiac output provides an indication of the blood flow through peripheral tissues; without adequate blood flow, homeostasis cannot be maintained.

Cardiac output can be calculated by multiplying the average stroke volume by the heart rate (HR):

$$\underset{\substack{\text{cardiac output} \\ \text{(mL/min)}}}{\textbf{CO}} = \underset{\substack{\text{stroke volume} \\ \text{(mL/beat)}}}{\textbf{SV}} \times \underset{\substack{\text{heart rate} \\ \text{(beats/min)}}}{\textbf{HR}}$$

For example, if the average stroke volume is 80 mL per beat and the heart rate is 70 bpm, the cardiac output will be:

$$CO = 80 \text{ mL/beat} \times 70 \text{ bpm}$$
$$= 5600 \text{ mL/min (5.6 liters per minute)}$$

This value represents a cardiac output equivalent to the total volume of blood of an average adult every minute. Cardiac output is highly variable, however, because a normal heart can increase both its rate of contraction and its stroke volume.

Clinical Note
BLUNT MYOCARDIAL INJURY

Blunt trauma to the myocardium can result in injuries that range from contusions to concussion. In a myocardial contusion, blunt trauma to the anterior chest transmits energy to the underlying heart. The most common cause of blunt myocardial injury is high-speed motor-vehicle collisions, often with crushing of the steering wheel. This can result in focal myocardial edema, interstitial hemorrhage, and occasionally subendocardial hemorrhage. The portions of the heart most frequently affected are the anterior right ventricular wall, the anterior interventricular septum, and the anterior/apical aspect of the left ventricle.

A myocardial contusion usually causes chest pain similar to that seen in angina pectoris. There is often tachycardia associated with rhythm and conduction disturbances. Severe injuries can cause a reduction in cardiac output. Myocardial contusions can cause abnormalities in the ECG such as ST segment elevation in the leads over the injured myocardium. In some cases, an elevation of cardiac enzymes is often seen. Most patients who sustain a myocardial contusion have complete recovery within three to six weeks.

Commotio Cordis

A *myocardial concussion,* often called *commotio cordis,* is a blow to the myocardium where no obvious gross or microscopic injury can be found. However, the patient can develop a life-threatening dysrhythmia, such as ventricular fibrillation, and die. Myocardial concussion usually results from a direct force to the chest, often a seemingly minor blow during sporting activities. The sports most frequently associated with *commotio cordis* are karate, hockey, softball, and baseball. Young male athletes aged 5–18 years are particularly at risk for this catastrophe. Death is usually instantaneous. In animal models, *commotio cordis* has been reproduced. The patient develops ventricular fibrillation immediately upon impact. The chest impact must fall within the vulnerable period of the cardiac cycle (during repolarization just before peaking of the T wave) to induce ventricular fibrillation. Successful resuscitation has been documented with prompt defibrillation, cardiopulmonary resuscitation, or application of a precordial thump. *Commotio cordis* is an extremely rare cause of death. The personal, physiological, and cardiovascular benefits of athletics far outweigh the risks. When possible, protective pads and similar devices should be used by those who are at increased risk. ∎

When necessary, stroke volume in a normal heart can almost double, and the heart rate can increase by 250 percent. When both increase together, the cardiac output can increase by 500 to 700 percent, or up to 30 liters per minute.

Factors That Control Cardiac Output

Cardiac output is precisely regulated so that peripheral tissues receive an adequate blood supply under a variety of conditions. The major factors that regulate cardiac output often affect both heart rate and stroke volume simultaneously. These primary factors include *blood volume reflexes, autonomic innervation,* and *hormones.* Secondary factors include the concentration of ions in the extracellular fluid and body temperature (see the Clinical Note "Extracellular Ions, Temperature, and Cardiac Output" on page 461).

Blood Volume Reflexes

Cardiac muscle contraction is an active process, but relaxation is entirely passive. The force needed to return cardiac muscle to its precontracted length is provided by the blood pouring into the heart, aided by the elasticity of the fibrous skeleton. As a result, there is a direct relationship between the amount of blood that enters the heart and the amount of blood ejected during the next contraction.

Two heart reflexes respond to changes in blood volume. One of these occurs in the right atrium and affects heart rate. The other is a ventricular reflex that affects stroke volume.

The **atrial reflex** *(Bainbridge reflex)* produces adjustments in heart rate in response to an increase in the **venous return,** which is the flow of venous blood to the heart. The entry of blood stimulates stretch receptors in the right atrial walls, and triggers a reflexive increase in heart rate through increased sympathetic activity. As a result of sympathetic stimulation, the cells of the SA node depolarize faster and heart rate increases.

The amount of blood pumped out of a ventricle each heartbeat (the stroke volume) depends not only on venous return but also on the **filling time**—the duration of ventricular diastole, when blood can flow into the ventricles. Filling time depends primarily on the heart rate: the faster the heart rate, the shorter the available filling time. Venous return changes in response to alterations in cardiac output, peripheral circulation, and other factors that affect the rate of blood flow through the venae cavae.

Over the range of normal activities, the greater the volume of blood that enters the ventricles, the more powerful the contraction (and the more pumping force is produced), and the greater the stroke volume. In a resting individual, venous return is relatively low and the walls are not stretched much, so the ventricles develop little power, and stroke volume is low. If venous return suddenly increases (that is, more blood flows into the heart), the myocardium stretches farther, the ventricles produce greater force upon contraction, and stroke volume increases. This general rule of "more in = more out" is often called the **Frank-Starling principle,** in honor of the physiologists who first demonstrated the relationship. The major effect of the Frank-Starling principle is that the output of blood from both ventricles is balanced under a variety of conditions.

Autonomic Innervation

As previously noted, the pacemaker cells of the SA node establish the basic heart rate, but this rate can be modified by the autonomic nervous system (ANS). As shown in Figure 12–22●, both the sympathetic and parasympathetic divisions of the ANS innervate the heart. Postganglionic sympathetic fibers extend from neuron cell bodies located in the cervical and upper thoracic ganglia. The vagus nerves (N X) carry parasympathetic preganglionic fibers to small ganglia near the heart. Both ANS divisions innervate the SA and AV nodes as well as the atrial and ventricular cardiac muscle cells.

AUTONOMIC EFFECTS ON HEART RATE. Autonomic effects on heart rate primarily reflect the responses of the SA node to acetylcholine (ACh) and to norepinephrine (NE). Acetylcholine released by parasympathetic motor neurons results in a lowering of the heart rate. Norepinephrine released by sympathetic neurons increases the heart rate. A more sustained rise in heart rate follows the release of epinephrine (E) and norepinephrine by the adrenal medullae during sympathetic activation.

AUTONOMIC EFFECTS ON STROKE VOLUME. The ANS also affects stroke volume by altering the force of myocardial contractions through the release of NE, E, and ACh:

● **FIGURE 12–22 Autonomic Innervation of the Heart.**

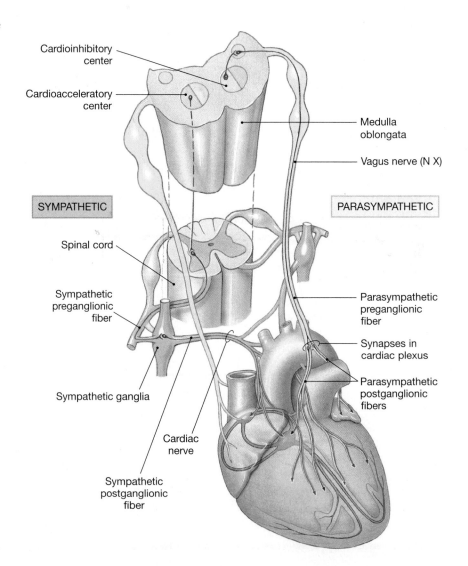

- *The Effects of NE and E.* The sympathetic release of NE at synapses in the myocardium and the release of NE and E by the adrenal medullae stimulate cardiac muscle cell metabolism and increase the force and degree of contraction. The result is an increase in stroke volume.

- *The Effects of ACh.* The primary effect of parasympathetic ACh release is inhibition, which results in a decrease in the force of cardiac contractions. Because parasympathetic innervation of the ventricles is relatively limited, the greatest reduction in contractile force occurs in the atria.

Both autonomic divisions are normally active at a steady background level, and release ACh and NE both at the nodes and into the myocardium. Thus, cutting the vagus nerves increases the heart rate, and sympathetic blocking agents slow the heart rate. Through dual innervation and adjustments in autonomic tone, the ANS can make very delicate adjustments in cardiovascular function.

THE COORDINATION OF AUTONOMIC ACTIVITY. The *cardiac centers* of the medulla oblongata contain the autonomic

headquarters for cardiac control. p. 294 The stimulation of the **cardioacceleratory center** activates sympathetic motor neurons; the nearby **cardioinhibitory center** controls the parasympathetic motor neurons (see Figure 12–22). Information concerning the status of the cardiovascular system arrives at the cardiac centers over sensory fibers from the vagus nerves and from the sympathetic nerves of the cardiac plexus.

The cardiac centers respond to changes in blood pressure and in the arterial concentrations of dissolved oxygen and carbon dioxide. These properties are monitored by baroreceptors and chemoreceptors innervated by the glossopharyngeal (N IX) and vagus nerves. A decline in blood pressure or oxygen concentrations or an increase in carbon dioxide levels usually indicates that the oxygen demands of peripheral tissues have increased. The cardiac centers then call for an increase in the cardiac output, and the heart works harder.

In addition to making automatic adjustments in response to sensory information, the cardiac centers can be influenced by higher centers, especially centers in the hypothalamus. As a result, changes in emotional state (such as rage, fear, or arousal) have an immediate effect on heart rate.

Clinical Note
THE DENERVATED HEART

Cardiac transplantation is now a fairly common procedure; many transplants last 15–20 years with proper care. It is important to remember that the transplanted heart is a denervated heart. That is, when the heart is removed from the donor, the autonomic nerves that control the heart are severed and cannot be reattached upon transplant into the recipient. Thus, there is no nervous control of the transplanted heart although there is evidence that some nervous control develops over time in a small percentage of patients.

In the transplanted heart, the resting heart rate is controlled by the heart's intrinsic pacemaker system and the autonomic nervous system through hormonal influence. This delicate balance keeps the heart functioning at the appropriate rate. Also, the modification of these systems is responsible for the change in heart rate in the appropriate direction. It has been shown that the parasympathetic nervous system is the major controlling factor in the resting state. Early after transplantation, the rate must be supported by administered drugs (chronotropes) or pacing. Later, circulating endogenous catecholamines support the resting heart rate. In the transplanted denervated heart, the resting rate is faster than normal due to the loss of parasympathetic inhibition.

The transplanted heart responds to exercise as expected by increasing its rate. However, the rate of increase is delayed just as the decrease of heart rate. The mechanism implicated in the increase of heart rate is the release of catecholamines from the peripheral nervous system. ∎

Clinical Note
EXTRACELLULAR IONS, TEMPERATURE, AND CARDIAC OUTPUT

Changes in extracellular calcium ion concentrations primarily affect the strength and duration of cardiac contractions and, thus, stroke volume. If such calcium concentrations are elevated (*hypercalcemia*), cardiac muscle cells become extremely excitable, and their contractions become powerful and prolonged. In extreme cases, the heart goes into an extended state of contraction that is usually fatal. When calcium levels are abnormally low (*hypocalcemia*), the contractions become very weak and may cease altogether.

Abnormal extracellular potassium ion concentrations alter the resting potential at the SA node and primarily change the heart rate. When potassium concentrations are high (*hyperkalemia*), cardiac contractions become weak and irregular. When the extracellular concentration of potassium is abnormally low (*hypokalemia*), heart rate is reduced.

Temperature changes affect metabolic operations throughout the body. For example, a lowered body temperature slows the rate of depolarization at the SA node, lowers the heart rate, and reduces the strength of cardiac contractions. An elevated body temperature accelerates the heart rate and the contractile force—one reason why your heart seems to race and pound when you have a fever. ∎

Hormones

As previously noted, epinephrine and norepinephrine secreted by the adrenal medullae act to increase both heart rate and the force of contraction. Thyroid hormones and glucagon also act to increase the force of contraction. Before synthetic drugs were available, glucagon was widely used to stimulate heart function.

Various drugs that affect the contractility of the heart have been developed for use in cardiac emergencies and to treat heart disease. Those that act by increasing the Ca^{2+} concentration within cardiac muscle cells, such as *digitalis* and related drugs, result in an increase in the force of contractions. Many drugs used to treat hypertension (high blood pressure) have a negative effect on contractility.

Key Note

Cardiac output is the amount of blood pumped by the left ventricle each minute. It is adjusted on a moment-to-moment basis by the ANS, and in response to circulating hormones, changes in blood volume, and alterations in venous return. Most healthy people can increase cardiac output by 300–500 percent.

→ CONCEPT CHECK QUESTIONS

1. What effect does stimulation of the acetylcholine receptors of the heart have on cardiac output?
2. What effect does increased venous return have on stroke volume?
3. How would increased sympathetic stimulation of the heart affect stroke volume?

Answers begin on p. 792.

Chapter Review

Access more review material online at *www.prenhall.com/bledsoe*. There you will find quiz questions, labeling activities, animations, essay questions, and web links.

Key Terms

atrioventricular valve 442
atrium 437
cardiac cycle 457
cardiac output 458
diastole 457

electrocardiogram (ECG, EKG) 455
endocardium 439
epicardium 439
intercalated disc 439
myocardium 439

pericardium 438
Purkinje fibers 453
sinoatrial node 453
systole 457
ventricle 437

Related Clinical Terms

angina pectoris (an-JĪ-nuh PEK-tor-is) Severe chest pain that results from temporary ischemia whenever the heart's workload increases (as during exertion or stress).

balloon angioplasty A technique for reducing the size of a coronary plaque by compressing it against the arterial walls using a catheter with an inflatable collar.

bradycardia (brād-ē-KAR-dē-uh) A slower than normal heart rate.

cardiac arrhythmias (ā-RITH-mē-az) Abnormal patterns of cardiac electrical activity that indicate abnormal contractions.

cardiac tamponade A condition that results from pericardial irritation and inflam-

mation; fluid collects in the pericardial sac and restricts cardiac output.

cardiology (kar-dē-OL-o-jē) The study of the heart, its functions, and diseases.

carditis (kar-DĪ-tis) Inflammation of the heart.

coronary arteriography The production of an X-ray image of coronary circulation after the introduction of a radiopaque dye into one of the coronary arteries through a catheter; the resulting image is a *coronary angiogram*.

coronary artery bypass graft (CABG) The routing of blood around an obstructed coronary artery (or one of its branches) by a vessel transplanted from another part of the body.

coronary artery disease (CAD) A disorder that results from the obstruction of coronary circulation.

coronary ischemia A deficiency of blood supply to the heart due to restricted circulation; may cause cardiac tissue damage and a reduction in cardiac efficiency.

coronary thrombosis The presence of a thrombus (clot) in a coronary artery; may cause circulatory blockage that results in a heart attack.

defibrillator A device used to eliminate atrial or ventricular fibrillation (uncoordinated contractions) and restore normal cardiac rhythm.

echocardiography Ultrasound analysis of the heart and of the blood flow through the great vessels.

electrocardiogram (ē-lek-trō-KAR-dē-ō-gram) (**ECG** or **EKG**) A recording of the electrical activities of the heart over time.

heart block An impairment of conduction in the heart in which damage to conduction pathways (due to mechanical distortion, ischemia, infection, or inflammation) disrupts the heart's normal rhythm.

heart failure A condition in which the heart weakens and peripheral tissues suffer from oxygen and nutrient deprivation.

myocardial (mī-ō-KAR-dē-al) **infarction** (**MI**) A condition in which blockage of the coronary circulation causes cardiac muscle cells to die from oxygen starvation; also called a *heart attack*.

pericarditis Inflammation of the pericardium.

rheumatic heart disease (**RHD**) A disorder in which the heart valves become thickened and stiffen into a partially closed position and reduces the efficiency of the heart.

tachycardia (tak-ē-KAR-dē-uh) A faster than normal heart rate.

valvular heart disease (**VHD**) A disorder caused by abnormal functioning of one of the cardiac valves; its severity depends on the degree of damage and the valve involved.

Summary Outline

THE HEART'S PLACE IN THE CIRCULATORY SYSTEM 437

1. The circulatory system can be subdivided into the **pulmonary circuit** (which carries blood to and from the lungs) and the **systemic circuit** (which transports blood to and from the rest of the body). **Arteries** carry blood away from the heart; **veins** return blood to the heart. **Capillaries** are tiny vessels between the smallest arteries and smallest veins. (*Figure 12–1*)

2. The heart has four chambers: the **right atrium, right ventricle, left atrium,** and **left ventricle.**

THE ANATOMY AND ORGANIZATION OF THE HEART 438

1. The heart is surrounded by the **pericardial cavity** (lined by the **pericardium**). The **visceral pericardium (epicardium)** covers the heart's outer surface, whereas the **parietal pericardium** lines the inner surface of the *pericardial sac,* which surrounds the heart. (*Figure 12–2*)

The Surface Anatomy of the Heart 438

2. The **coronary sulcus,** a deep groove, marks the boundary between the atria and ventricles. (*Figure 12–3*)

The Heart Wall 439

3. The bulk of the heart consists of the muscular **myocardium.** The **endocardium** lines the inner surfaces of the heart. (*Figure 12–4a,b*)

4. **Cardiac muscle cells** are interconnected by **intercalated discs,** which convey the force of contraction from cell to cell and conduct action potentials. (*Figure 12–4c,d*)

Internal Anatomy and Organization 442

5. The atria are separated by the **interatrial septum,** and the ventricles are divided by the **interventricular septum.** The right atrium receives blood from the systemic circuit via two large veins, the **superior vena cava** and **inferior vena cava.** (*Figure 12–5*)

6. Blood flows from the right atrium into the right ventricle through the **right atrioventricular (AV) valve (tricuspid valve).** This opening is bounded by three **cusps** of fibrous tissue braced by the tendinous **chordae tendineae,** which are connected to **papillary muscles.**

7. Blood leaving the right ventricle enters the **pulmonary trunk** after passing through the **pulmonary semilunar valve.** The pulmonary trunk divides to form the **left** and **right pulmonary arteries.** The **left** and **right pulmonary veins** return oxygenated blood to the left atrium. Blood leaving the left atrium flows into the left ventricle through the **left atrioventricular (AV) valve (bicuspid valve** or **mitral valve).** Blood leaving the left ventricle passes through the **aortic semilunar valve** and into the systemic circuit through the **aorta.** (*Figure 12–5*)

8. Anatomical differences between the ventricles reflect the functional demands on them. The wall of the right ventricle is relatively thin, whereas the left ventricle has a massive muscular wall.

9. Valves normally permit blood flow in only one direction, which prevents the **regurgitation** (backflow) of blood. (*Figure 12–6*)

10. The **fibrous skeleton** supports the heart's contractile cells and valves. (*Figure 12–6*)

Key Note 445

The Blood Supply to the Heart 445

11. The **coronary circulation** meets the high oxygen and nutrient demands of cardiac muscle cells. The coronary arteries originate at the base of the aorta. Arterial **anastomoses**—interconnections between arteries—ensure a constant blood supply. The **great** and **middle cardiac veins** carry blood from the coronary capillaries to the **coronary sinus.** (*Figures 12–7 through 12–14*)

THE HEARTBEAT 450

1. Two general classes of cardiac cells are involved in the normal heartbeat: *contractile cells* and cells of the conducting system.

Contractile Cells 450

2. Cardiac muscle cells have a long refractory period, so rapid stimulation produces isolated contractions (twitches) rather than tetanic contractions. (*Figure 12–15*)

The Conducting System 452

3. The conducting system includes **nodal cells** and **conducting cells.** The conducting system initiates and distributes electrical impulses within the heart. Nodal cells establish the rate of cardiac contraction; **pacemaker cells** are nodal cells that reach

threshold first. Conducting cells distribute the contractile stimulus to the general myocardium.

4. Unlike skeletal muscle, cardiac muscle contracts without neural or hormonal stimulation. Pacemaker cells in the **sinoatrial (SA) node** *(cardiac pacemaker)* normally establish the rate of contraction. From the SA node, impulses travel to the **atrioventricular (AV) node** and then to the **AV bundle,** which divides into **bundle branches.** From there, **Purkinje fibers** convey the impulses to the ventricular myocardium. *(Figures 12–16 through 12–19)*

The Electrocardiogram 455

5. A recording of electrical activities in the heart is an **electrocardiogram (ECG** or **EKG).** Important landmarks of an ECG include the **P wave** (atrial depolarization), **QRS complex** (ventricular depolarization), and **T wave** (ventricular repolarization). *(Figure 12–20)*

Key Note 457

The Cardiac Cycle 457

6. A **cardiac cycle** consists of **systole** (contraction) followed by **diastole** (relaxation). Both ventricles contract at the same time, and they eject equal volumes of blood. *(Figure 12–21)*

7. The closing of the heart valves and the rushing of blood through the heart cause characteristic **heart sounds.**

HEART DYNAMICS 458

1. *Heart dynamics* refers to the movements and forces generated during contractions. The amount of blood ejected by a ventricle during a single beat is the **stroke volume (SV);** the amount of blood pumped by a ventricle each minute is the **cardiac output (CO).**

Factors That Control Cardiac Output 459

2. The major factors that affect cardiac output are blood volume reflexes, autonomic innervation, and hormones.

3. Blood volume reflexes are stimulated by changes in **venous return,** the amount of blood that enters the heart. The **atrial reflex** accelerates the heart rate when entering blood stretches the walls of the right atrium. Ventricular contractions become more powerful and increase stroke volume when the ventricular walls are stretched (the **Frank-Starling principle**).

4. The basic heart rate is established by pacemaker cells, but it can be modified by the ANS. *(Figure 12–22)*

5. Acetylcholine (ACh) released by parasympathetic motor neurons lowers the heart rate and stroke volume. Norepinephrine (NE) released by sympathetic neurons increases the heart rate and stroke volume.

6. Epinephrine (E) and norepinephrine, which are hormones released by the adrenal medullae during sympathetic activation, increase both heart rate and stroke volume. Thyroid hormones and glucagon also act to increase cardiac output.

7. The **cardioacceleratory center** in the medulla oblongata activates sympathetic neurons; the **cardioinhibitory center** governs the activities of the parasympathetic neurons. These cardiac centers receive inputs from higher centers and from receptors that monitor blood pressure and the levels of dissolved gases.

Key Note 462

Review Questions

Level 1: Reviewing Facts and Terms

Match each item in column A with the most closely related item in column B. Place letters for answers in the spaces provided.

COLUMN A

____ 1. epicardium

____ 2. right AV valve

____ 3. left AV valve

____ 4. anastomoses

____ 5. myocardial infarction

____ 6. SA node

____ 7. systole

____ 8. diastole

____ 9. cardiac output

____ 10. HR slower than usual

____ 11. HR faster than normal

____ 12. atrial reflex

COLUMN B

a. heart attack

b. cardiac pacemaker

c. tachycardia

d. $SV \times HR$

e. tricuspid valve

f. bradycardia

g. mitral valve

h. interconnections between arteries

i. visceral pericardium

j. increased venous return

k. contractions of heart chambers

l. relaxation of heart chambers

13. The blood supply to the muscles of the heart is provided by the:
    (a) systemic circulation.
    (b) pulmonary circulation.
    (c) coronary circulation.
    (d) coronary portal system.

14. The autonomic centers for cardiac function are located in:
    (a) myocardial tissue of the heart.
    (b) cardiac centers of the medulla oblongata.
    (c) the cerebral cortex.
    (d) a, b, and c are correct.

15. The simple squamous epithelium that covers the valves of the heart constitutes the:
    (a) epicardium.
    (b) endocardium.
    (c) myocardium.
    (d) fibrous skeleton.

16. The structure that permits blood flow from the right atrium to the left atrium while the lungs are developing is the:
    (a) foramen ovale.
    (b) interatrial septum.
    (c) coronary sinus.
    (d) fossa ovalis.

17. Blood leaves the left ventricle by passing through the:
    (a) aortic semilunar valve.
    (b) pulmonary semilunar valve.
    (c) mitral valve.
    (d) tricuspid valve.

18. The QRS complex of the ECG is produced when the:
    (a) atria depolarize.
    (b) ventricles depolarize.
    (c) ventricles repolarize.
    (d) atria repolarize.

19. During diastole, the chambers of the heart:
    (a) relax and fill with blood.
    (b) contract and push blood into an adjacent chamber.
    (c) undergo a sharp increase in pressure.
    (d) reach a pressure of approximately 120 mmHg.

20. What roles do the chordae tendineae and papillary muscles have in the normal function of the AV valves?

21. What are the principal heart valves, and what is the function of each?

22. Trace the normal path of an electrical impulse through the conducting system of the heart.

23. (a) What is the cardiac cycle? (b) What phases and events are necessary to complete the cardiac cycle?

Level 2: Reviewing Concepts

24. Tetanic muscle contractions cannot occur in a normal cardiac muscle cell because:
    (a) cardiac muscle tissue contracts on its own.
    (b) there is no neural or hormonal stimulation.
    (c) the refractory period lasts until the muscle cell relaxes.
    (d) the refractory period ends before the muscle cell reaches peak tension.

25. The amount of blood forced out of the heart depends on the:
    (a) degree of stretching at the end of ventricular diastole.
    (b) contractility of the ventricle.
    (c) amount of pressure required to eject blood.
    (d) a, b, and c are correct.

26. Cardiac output cannot increase indefinitely because:
    (a) available filling time becomes shorter as the heart rate increases.
    (b) cardiovascular centers adjust the heart rate.
    (c) the rate of spontaneous depolarization decreases.
    (d) the ion concentrations that surround pacemaker cell membranes decrease.

27. Describe the relationships of the four chambers of the heart to the pulmonary and systemic circuits.

28. What are the source and significance of the heart sounds?

29. (a) What effect does sympathetic stimulation have on the heart? (b) What effect does parasympathetic stimulation have on the heart?

Level 3: Critical Thinking and Clinical Applications

30. A patient's ECG tracing shows a consistent pattern of two P waves followed by a normal QRS complex and T wave. What is the cause of this abnormal wave pattern?

31. Karen is taking the medication *verapamil*, which is a drug that blocks the calcium channels in cardiac muscle cells. What effect would you expect this medication to have on Karen's stroke volume?

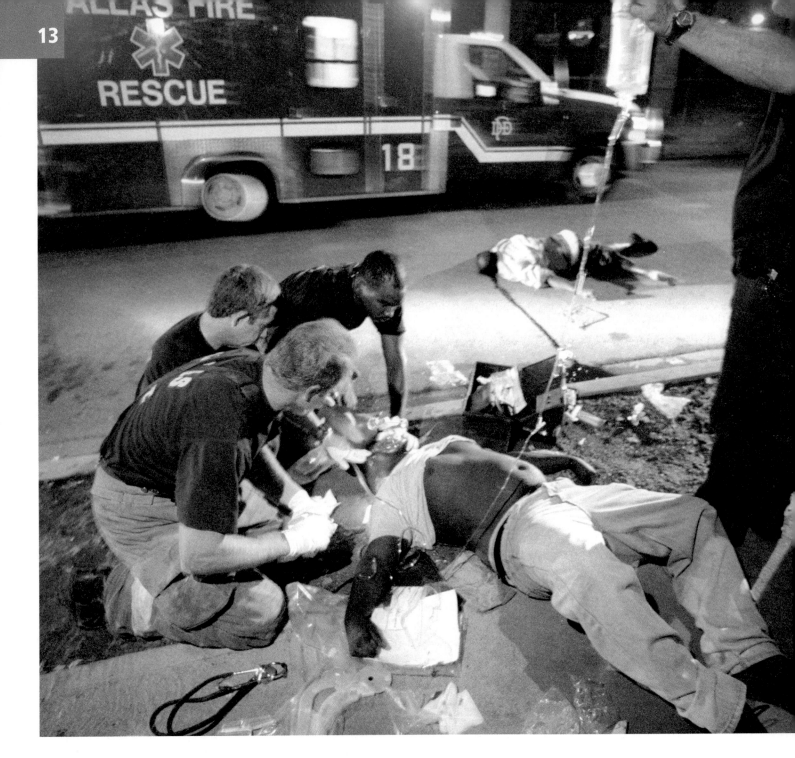

13 The Cardiovascular System: Blood Vessels and Circulation

THERE ARE THREE categories of trauma death. In the first group (approximately 50% of all trauma deaths) the injured person typically dies within minutes of injury, which is usually vascular or neurological and generally not survivable. The second group (approximately 30% of all trauma deaths) usually dies within hours of injury. Death in this group is largely due to hypovolemia and hypoxia. The third group (approximately 20% of trauma deaths) usually dies within days of injury often due to multiple organ failure and other complications. The current trend in EMS is to concentrate on the second group—those most likely to benefit from prehospital care.

Chapter Outline

Chapter Objectives

1. Distinguish among the types of blood vessels on the basis of their structure and function. (pp. 408–472)
2. Explain the mechanisms that regulate blood flow through arteries, capillaries, and veins. (pp. 472–473)
3. Discuss the mechanisms and various pressures involved in the movement of fluids between capillaries and interstitial spaces. (pp. 473–476)
4. Describe the factors that influence blood pressure and the mechanisms that regulate blood pressure. (pp. 476–478)
5. Describe how central and local control mechanisms interact to regulate blood flow and pressure in tissues. (pp. 478–481)
6. Explain how the activities of the cardiac, vasomotor, and respiratory centers are coordinated to control blood flow through tissues. (pp. 481–482)
7. Explain how the cardiovascular system responds to the demands of exercise and hemorrhage. (pp. 482–483)
8. Identify the major arteries and veins and the areas they serve. (pp. 484–500)
9. Describe the age-related changes that occur in the cardiovascular system. (pp. 500–501)
10. Discuss the structural and functional interactions between the cardiovascular system and other organ systems. (p. 503)

Vocabulary Development

baro- pressure; *baroreceptors*
capillaris hair-like; *capillary*
manometer device for measuring pressure; *sphygmomanometer*

porta gate; *portal vein*
pulmo- lung; *pulmonary*
pulsus stroke; *pulse*
saphenes prominent; *saphenous vein*

skleros hard; *arteriosclerosis*
sphygmos pulse; *sphygmomanometer*
vaso- vessel; *vasoconstriction*

THE PREVIOUS TWO CHAPTERS examined the composition of blood and the structure and function of the heart, whose pumping action keeps blood in motion. Here we will consider the vessels that carry blood to peripheral tissues and the nature of the exchange that occurs between the blood and interstitial fluids of the body.

Blood leaves the heart by way of the pulmonary trunk and aorta, each with a diameter of around 2.5 cm (1 in.). These vessels branch repeatedly and form the major **arteries** that distribute blood to body organs. Within these organs further branching occurs, which creates several hundred million tiny arteries, or **arterioles** (ar-TĒR-e-ōls). The arterioles provide blood to more than 10 billion **capillaries.** These capillaries, barely the diameter of a single red blood cell, form extensive, branching networks. If all the capillaries in an average adult's body were placed end to end, they would circle the globe in an unbroken chain 25,000 miles long.

The vital functions of the cardiovascular system occur at the capillary level: *chemical and gaseous exchange between the blood and interstitial fluid takes place across capillary walls.* Tissue cells rely on capillary diffusion to obtain nutrients and oxygen and to remove metabolic wastes such as carbon dioxide and urea.

Blood that flows out of a capillary network first enters the **venules** (VEN-ūls), which are the smallest vessels of the venous system. These slender vessels subsequently merge to form small **veins.** Blood then passes through medium-sized and large veins before reaching the venae cavae (in the systemic circuit) or the pulmonary veins (in the pulmonary circuit) (see Figure 12–1, p. 438).

This chapter first examines the structure of arteries, capillaries, and veins. Then it discusses their functions and the basic principles of cardiovascular regulation. The final section of the chapter describes the distribution of major blood vessels of the body.

■ The Structure of Blood Vessels

The walls of arteries and veins contain three distinct layers (Figure 13–1●):

1. The **tunica interna** (in-TER-nuh), or *tunica intima*, is the innermost layer of a blood vessel. It includes the endothelial lining of the vessel and an underlying layer of connective tissue dominated by elastic fibers.

2. The **tunica media,** the middle layer, contains smooth muscle tissue in a framework of collagen and elastic fibers. When these smooth muscles contract, vessel diameter decreases; when they relax, vessel diameter increases.

3. The outer **tunica externa** (eks-TER-nuh), or *tunica adventitia* (ad-ven-TISH-a), forms a sheath of connective tissue around the vessel. Its collagen fibers may intertwine with those of adjacent tissues, stabilizing and anchoring the blood vessel.

Arteries and veins often lie side by side in a narrow band of connective tissue, as shown in Figure 13–1. Such a sectional view clearly shows the greater wall thickness that is character-istic of arteries. The thicker tunica media of an artery contains more elastic fibers and smooth muscle than a vein. The elastic components enable arterial vessels to resist the pressure generated by the heart as it forces blood into the arterial network. The smooth muscle provides a means of actively controlling vessel diameter. Arterial smooth muscle is under the control of the sympathetic division of the autonomic nervous system. When stimulated, these muscles in the vessel wall contract and the artery constricts in a process called **vasoconstriction.** Relaxation increases the diameter of the artery and its central opening, or *lumen*, in a process called **vasodilation.**

Arteries

In traveling from the heart to the capillaries, blood passes through elastic arteries, muscular arteries, and arterioles (Figure 13–2●).

Elastic arteries are large, extremely resilient vessels with diameters of up to 2.5 cm (1 in.). Some examples are the pulmonary trunk and aorta, and their major arterial branches. The walls of elastic arteries contain a tunica media dominated by elastic fibers rather than smooth muscle cells. As a

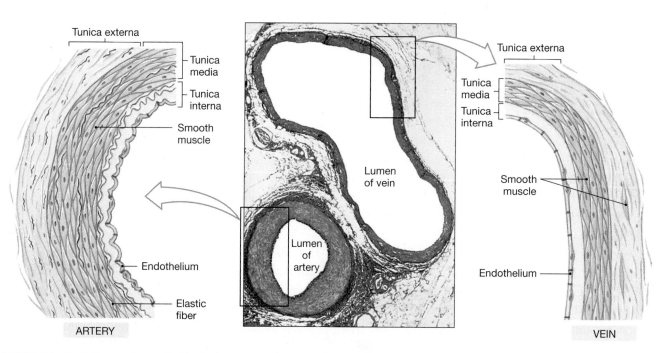

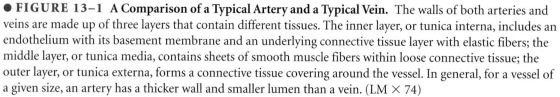

● **FIGURE 13–1 A Comparison of a Typical Artery and a Typical Vein.** The walls of both arteries and veins are made up of three layers that contain different tissues. The inner layer, or tunica interna, includes an endothelium with its basement membrane and an underlying connective tissue layer with elastic fibers; the middle layer, or tunica media, contains sheets of smooth muscle fibers within loose connective tissue; the outer layer, or tunica externa, forms a connective tissue covering around the vessel. In general, for a vessel of a given size, an artery has a thicker wall and smaller lumen than a vein. (LM × 74)

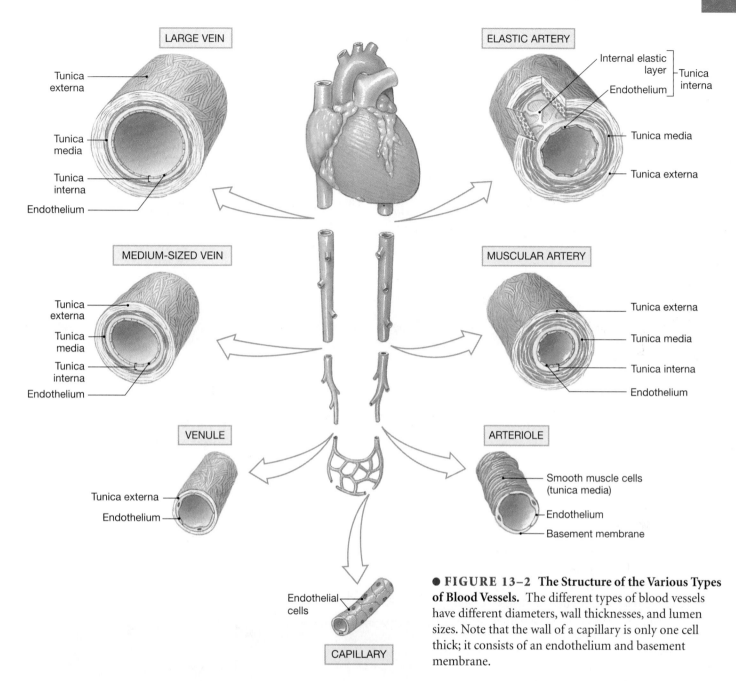

● FIGURE 13–2 The Structure of the Various Types of Blood Vessels. The different types of blood vessels have different diameters, wall thicknesses, and lumen sizes. Note that the wall of a capillary is only one cell thick; it consists of an endothelium and basement membrane.

result, elastic arteries are able to absorb the pressure changes that occur during the cardiac cycle. During ventricular systole, blood pressure rises quickly as additional blood is pushed into the systemic circuit. Over this period, the elastic arteries stretch, and their diameter increases. During ventricular diastole, arterial blood pressure declines, and the elastic fibers recoil to their original dimensions. The net result is that the expansion of the arteries dampens the rise in pressure during ventricular systole, and the ensuing arterial recoil slows the decline in pressure during ventricular diastole. If the arteries were solid pipes rather than elastic tubes, pressures would rise much higher during systole and would fall much lower during diastole.

Muscular arteries, also known as *medium-sized arteries* or *distribution arteries,* distribute blood to skeletal muscles and internal organs. A typical muscular artery has a diameter of approximately 0.4 cm (0.15 in.). The external carotid arteries of the neck are one example. The thick tunica media in a muscular artery contains more smooth muscle and fewer elastic fibers than an elastic artery (see Figure 13–2).

Arterioles, with an internal diameter of about 30 μm, are much smaller than muscular arteries. The tunica media of an arteriole consists of one to two layers of smooth muscle cells. These muscle layers enable muscular arteries and arterioles to change the diameter of the lumen, thereby altering blood pressure and the rate of flow through the dependent tissues.

Capillaries

Capillaries are the *only* blood vessels whose walls permit exchange between the blood and the surrounding interstitial fluid. Because capillary walls are relatively thin, diffusion distances are small, and exchange can occur quickly. In addition, the small diameter of capillaries slows blood flow and allows sufficient time for the diffusion or active transport of materials across the capillary walls.

A typical capillary consists of a single layer of endothelial cells inside a basement membrane (see Figure 13–2). Neither a tunica externa nor a tunica media is present. The average diameter of a capillary is 8 μm, very close to that of a red blood cell. In most regions of the body, the endothelium forms a complete lining, and most substances enter or leave the capillary by diffusing across endothelial cells or through gaps between adjacent endothelial cells. Water, small solutes, and lipid-soluble materials can easily diffuse into the surrounding interstitial fluid. In other areas (notably, the choroid plexus of the brain, the hypothalamus, and filtration sites in the kidneys), small pores in the endothelial cells also permit the passage of relatively large molecules, including proteins.

Clinical Note
ARTERIOSCLEROSIS

Arteriosclerosis (ar-tē-rē-ō-skle-RŌ-sis; *skleros,* hard) is a thickening and toughening of arterial walls. Complications related to arteriosclerosis account for roughly half of all deaths in the U.S. The effects of arteriosclerosis are varied; for example, arteriosclerosis of coronary vessels is responsible for *coronary artery disease (CAD),* and arteriosclerosis of arteries that supply the brain can lead to strokes. ∞ p. 445 There are two major forms of arteriosclerosis: *focal calcification* and *atherosclerosis.*

In *focal calcification,* degenerating smooth muscle in the tunica media is replaced by calcium deposits. It occurs as part of the aging process, in association with atherosclerosis, and as a complication of diabetes mellitus, which is an endocrine disorder. ∞ p. 391

Atherosclerosis (ath-er-ō-skle-RŌ-sis; *athero-,* a pasty deposit) is the formation of lipid deposits in the tunica media associated with damage to the endothelial lining. Many factors may contribute to the development of atherosclerosis. One major factor is lipid levels in the blood. Atherosclerosis tends to develop in individuals whose blood contains elevated levels of plasma lipids, specifically cholesterol. Circulating cholesterol is transported to peripheral tissues in protein-lipid complexes called *lipoproteins.* (The various types of lipoproteins are discussed in Chapter 17.) When cholesterol-rich lipoproteins remain in circulation for an extended period, circulating monocytes begin removing them from the bloodstream. Eventually the monocytes, now filled with lipid droplets, attach themselves to the endothelia of blood vessels. These cells then release growth factors that stimulate the divisions of smooth muscle cells near the tunica interna. The vessel wall thickens and stiffens.

Other monocytes then invade the area and migrate between the endothelial cells. As these changes occur, the monocytes, smooth muscle fibers, and endothelial cells begin phagocytizing lipids as well. The result is a *plaque,* which is a fatty mass of tissue that projects into the lumen of the vessel. At this point, the plaque has a relatively simple structure, and evidence indicates that the process can be reversed if appropriate dietary adjustments are made.

If the conditions persist, the endothelial cells become swollen with lipids, and gaps appear in the endothelial lining. Platelets now begin sticking to the exposed collagen fibers, and platelet adhesion and aggregation lead to the formation of a localized blood clot that further restricts blood flow through the vessel. The structure of the plaque is now relatively complex (Figure 13–3●); plaque growth can be halted, but the structural changes are permanent.

Elderly individuals, especially elderly men, are most likely to develop atherosclerotic plaques. In addition to age, sex, and high blood cholesterol levels, other important risk factors include high blood pressure, cigarette smoking, diabetes mellitus, obesity, and stress. ■

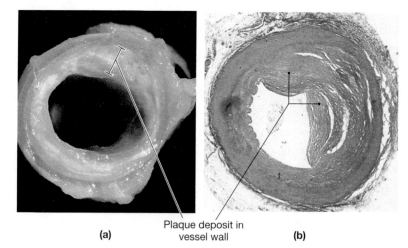

Plaque deposit in
vessel wall

(a) **(b)**

● **FIGURE 13–3 A Plaque Within an Artery.** **(a)** This photograph shows a section of a coronary artery narrowed by plaque formation. **(b)** This photomicrograph provides a cross-sectional view of a large plaque.

Capillary Beds

Capillaries do not function as individual units but as part of an interconnected network called a **capillary bed** (Figure 13–4●). A single arteriole usually gives rise to dozens of capillaries, which in turn collect into several *venules,* the smallest vessels of the venous system. The entrance to each capillary is guarded by a **precapillary sphincter,** which is a band of smooth muscle. Contraction of the smooth muscle fibers narrows the diameter of the capillary entrance and reduces the flow of blood. The relaxation of the sphincter dilates the opening and allows blood to enter the capillary more rapidly.

Although blood usually flows from arterioles to venules at a constant rate, the blood flow within any single capillary can be quite variable. Each precapillary sphincter undergoes cycles of activity, alternately contracting and relaxing perhaps a dozen times each minute. As a result of this cyclical change—called **vasomotion** (*vaso-,* vessel)—the blood flow within any one capillary is intermittent rather than a steady and constant stream. The net effect is that blood may reach the venules by one route now and by a quite different route later. This process is controlled at the tissue level, as smooth muscle fibers respond to local changes in the concentrations of chemicals and dissolved gases in the interstitial fluid. For example, when dissolved oxygen levels decline within a tissue, the capillary sphincters relax, and blood flow to the area increases. Such control at the tissue level is called *autoregulation.*

Sometimes blood vessels provide alternate routes for blood flow by forming an **anastomosis** (a-nas-tō-MŌ-sis; "outlet"; plural, *anastomoses*), which is the joining of two tubes. Under certain conditions, blood completely bypasses a capillary bed through an **arteriovenous** (ar-tēr-ē-ō-VĒ-nus) **anastomosis,** a vessel that connects an arteriole to a venule (see Figure 13–4a). In other cases, a single capillary bed is supplied by an **arterial anastomosis,** in which more than one artery fuses before giving rise to arterioles. An arterial anastomosis in effect provides an insurance policy for capillary beds: if one artery is compressed or blocked, the others can continue to deliver blood to the capillary bed, and the dependent tissues will not be damaged. Arterial anastomoses occur in the brain, in the coronary circulation, and in many other sites as well. ∞ p. 445

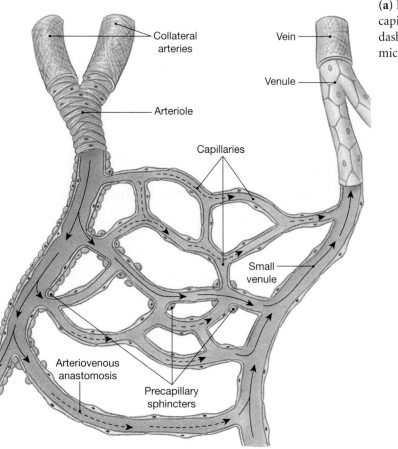

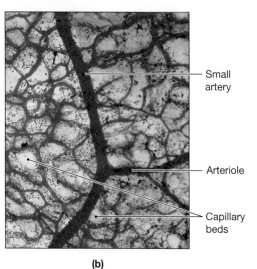

● **FIGURE 13–4** **The Organization of a Capillary Bed.** **(a)** In this diagram, which shows the basic features of a typical capillary bed, solid arrows indicate consistent blood flow and dashed arrows indicate variable blood flow. **(b)** This micrograph shows several capillary beds.

(a)

(b)

Veins

Veins collect blood from all tissues and organs and return it to the heart. Veins are classified on the basis of their internal diameters. The smallest—**venules**—resemble expanded capillaries, and venules with diameters smaller than 50 µm lack a tunica media altogether (see Figure 13–2). **Medium-sized veins** range from 2 to 9 mm in diameter, comparable in size to muscular arteries. In these veins, the tunica media contains several smooth muscle layers, and the relatively thick tunica externa has longitudinal bundles of elastic and collagen fibers. **Large veins** include the two venae cavae and their tributaries in the abdominopelvic and thoracic cavities. In these vessels, the thin tunica media is surrounded by a thick tunica externa composed of elastic and collagenous fibers.

Veins have relatively thin walls because they need not withstand much pressure. In venules and medium-sized veins, the pressure is so low that it cannot overcome the force of gravity. In the limbs, medium-sized veins contain **valves,** which are folds of endothelium that function like the valves in the heart, and prevent the backflow of blood (Figure 13–5●). As long as

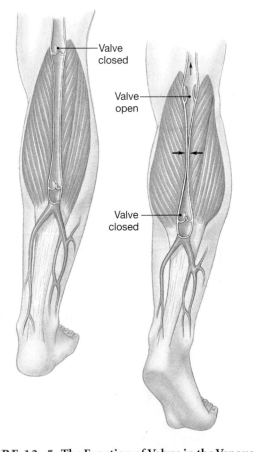

● **FIGURE 13–5 The Function of Valves in the Venous System.** Valves in the walls of medium-sized veins prevent the backflow of blood. The compression of veins by the contraction of adjacent skeletal muscles helps maintain venous blood flow.

the valves function normally, any body movement that compresses a vein will push blood toward the heart and improve *venous return.* ∞ p. 459 If the walls of the veins near the valves weaken or become stretched and distorted, the valves may not work properly. Blood then pools in the veins, which become distended. The effects range from mild discomfort and a cosmetic problem (as in superficial *varicose veins* in the thighs and legs) to painful distortion of adjacent tissues (as in *hemorrhoids,* which are swollen veins in the lining of the anal canal).

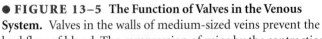

> ## CONCEPT CHECK QUESTIONS
>
> 1. A histological cross section shows several small, thin-walled vessels with very little smooth muscle tissue in the tunica media. Which type of vessels are these?
> 2. How would the relaxation of precapillary sphincters affect the blood flow through a tissue?
> 3. Why are valves found in veins of the limbs but not in arteries?
>
> *Answers begin on p. 792.*

■ Circulatory Physiology

The primary function of the components of the cardiovascular system (the blood, heart, and blood vessels) is to maintain an adequate blood flow through the capillaries in all the tissues of the body. The capillaries are the sites of nutrient and waste exchange for the body's cells. Under normal circumstances, blood flow is equal to cardiac output. When cardiac output goes up, so does blood flow through capillary beds; when cardiac output declines, blood flow is reduced. Next we consider how the flow of blood through the capillaries also depends on two additional factors—*pressure* and *resistance.*

Factors That Affect Blood Flow

Pressure and resistance both affect blood flow to the tissues, but they have opposing effects. Blood flow and pressure are directly related: when pressure increases, flow increases. Blood flow and resistance are inversely related: when resistance increases, flow decreases. To keep blood moving in the body, the heart must generate enough pressure to overcome the resistance to blood flow in the pulmonary and systemic circuits.

Pressure

Liquids, including blood, cannot be compressed. As a result, a force exerted against a liquid generates a fluid (or hydrostatic) pressure that is conducted in all directions. If a pressure difference exists, a liquid will flow from an area of higher pressure toward an area of lower pressure. The flow rate is directly proportional to the pressure difference: the greater the difference in pressure, the faster the flow.

The largest pressure difference, or *pressure gradient*, is found in the systemic circuit between the base of the aorta (where blood leaves the left ventricle) and the entrance to the right atrium (where blood returns to the heart). This pressure difference, called the *circulatory pressure*, averages about 100 mmHg (millimeters of mercury, a standard unit of pressure). This relatively high circulatory pressure is needed primarily to force blood through the arterioles and into the capillaries.

Circulatory pressure is divided into three components: (1) *arterial pressure* (routinely measured on a person's arm and commonly referred to as **blood pressure**), (2) *capillary pressure*, and (3) *venous pressure*. Each component of circulatory pressure will be discussed in some detail shortly.

Resistance

A *resistance* is any force that opposes movement. In the cardiovascular system, resistance opposes the movement of blood. For blood to flow, the circulatory pressure must be great enough to overcome the *total peripheral resistance*, which is the resistance of the entire cardiovascular system. The greatest pressure difference within the cardiovascular system—about 65 mm Hg—occurs in the arterial network because of the high resistance of the arterioles. The resistance of the arterial system is termed **peripheral resistance.** Sources of peripheral resistance include *vascular resistance, viscosity,* and *turbulence.*

VASCULAR RESISTANCE. The largest component of peripheral resistance is **vascular resistance,** which is the resistance of the blood vessels to blood flow. *The most important factor in vascular resistance is friction between the blood and the vessel walls.* The amount of friction depends on the length of the vessel and its diameter. Friction increases with increasing vessel length (longer vessels have a larger surface area in contact with the blood) and with decreasing vessel diameter (in small-diameter vessels, nearly all the blood is slowed down by friction with the walls). Vessel length cannot ordinarily be changed, so vascular resistance is controlled by changing the diameter of blood vessels—through the contraction or relaxation of smooth muscle in the vessel walls.

Most of the vascular resistance occurs in the arterioles, which are extremely muscular. The walls of an arteriole with a 30–μm internal diameter, for example, can have a 20–μm-thick layer of smooth muscle. Local, neural, and hormonal stimuli that stimulate or inhibit this smooth muscle tissue can adjust the diameters of these vessels, and a small change in diameter can produce a very large change in resistance.

VISCOSITY. The property called **viscosity** is the resistance to flow that results from interactions among molecules and sus-

pended materials in a liquid. Liquids of low viscosity, such as water, flow at low pressures; whereas thick, syrupy liquids such as molasses flow only under higher pressures. Because whole blood contains plasma proteins and suspended blood cells, it has a viscosity about five times that of water.

Under normal conditions, the viscosity of blood remains stable. But disorders that affect the hematocrit or the plasma protein content can change blood viscosity and increase or decrease peripheral resistance. For example, in **anemia,** the hematocrit is reduced due to inadequate production of hemoglobin, RBCs, or both. As a result, both the oxygen-carrying capacity of the blood and blood viscosity are reduced. A reduction in blood viscosity can also result from protein deficiency diseases, in which the liver cannot synthesize normal amounts of plasma proteins.

TURBULENCE. Blood usually flows through a vessel smoothly, with the slowest flow near the walls, and the fastest flow at the center of the vessel. High flow rates, irregular surfaces caused by injury or disease processes, or sudden changes in vessel diameter upset this smooth flow and create eddies and swirls. This phenomenon, called **turbulence,** slows flow and increases resistance.

Turbulence normally occurs when blood flows between the heart's chambers and from the heart into the aorta and pulmonary trunk. In addition to increasing resistance, this turbulence generates the *third* and *fourth heart sounds* often heard through a stethoscope. Turbulent blood flow across damaged or misaligned heart valves produces the sound of *heart murmurs.* ∞ p. 443

The Interplay Between Pressure and Resistance

Neural and hormonal control mechanisms regulate blood pressure, keeping it relatively stable. Adjustments in the peripheral resistance of vessels supplying specific organs allow the rate of blood flow to be precisely controlled. An example is the increase in blood flow to skeletal muscles during exercise, discussed in Chapter 7. ∞ p. 253 That increase occurs because there is a drop in the peripheral resistance of the arteries supplying active muscles. Of the three sources of resistance, only vascular resistance can be adjusted by the nervous or endocrine system to regulate blood flow. Viscosity and turbulence, which affect peripheral resistance, are normally constant.

Cardiovascular Pressures Within the Systemic Circuit

Pressure varies along the path of blood flow within the systemic circuit. Systemic pressures are highest in the aorta, and peak at around 120 mmHg; they are lowest at the venae cavae

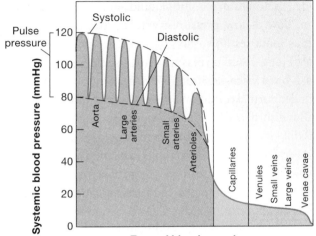

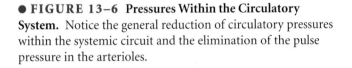

Type of blood vessel

● **FIGURE 13–6 Pressures Within the Circulatory System.** Notice the general reduction of circulatory pressures within the systemic circuit and the elimination of the pulse pressure in the arterioles.

Blood Pressure (Arterial Pressure)

As Figure 13–6 shows, the pressure in large and small arteries fluctuates; it rises during ventricular systole and falls during ventricular diastole. **Systolic pressure** is the peak blood pressure measured during ventricular systole, and **diastolic pressure** is the minimum blood pressure at the end of ventricular diastole. In recording blood pressure, we separate systolic and diastolic pressures by a slash mark, as in "120/80" (read as "one twenty over eighty"). A *pulse* is a rhythmic pressure oscillation that accompanies each heartbeat. The difference between the systolic and diastolic pressures is the **pulse pressure** (*pulsus*, stroke).

Pulse pressure lessens as the distance from the heart increases. As already discussed, the average pressure declines as a result of friction between the blood and the vessel walls. The pulse pressure fades because arteries are elastic tubes rather than solid pipes. Much as a balloon's walls expand upon the entry of air, the elasticity of the arteries allows them to expand with blood during systole. When diastole begins and blood pressures fall, the arteries recoil to their original dimensions. Because the aortic semilunar valve prevents the return of blood to the heart, arterial recoil adds an extra push to the flow of blood. The magnitude of this phe-

nomenon, called *elastic rebound,* is greatest near the heart and drops in succeeding arterial sections.

By the time blood reaches a precapillary sphincter, there are no pressure oscillations, and the blood pressure remains steady at about 35 mmHg (see Figure 13–6). Along the length of a typical capillary, blood pressure gradually falls from about 35 mmHg to roughly 18 mmHg, which is the pressure at the start of the venous system.

Clinical Note
HYPERTENSION

A consistent elevation of the blood pressure is called *hypertension*. Approximately 50 million people in the U.S. have hypertension. The diagnosis is made when the average of two or more diastolic blood pressure measurements taken on two different occasions is 90 mmHg or higher or when the average of two or more systolic blood pressure measurements taken on two different occasions is 140 mmHg or higher.

The blood pressure is a function of cardiac output and peripheral vascular resistance. An increase in cardiac output, peripheral vascular resistance, or both will increase the blood pressure. A specific cause of hypertension has not been determined. However, both genetic and environmental factors appear responsible for its development.

Chronic hypertension damages the walls of systemic blood vessels. Prolonged vasoconstriction and high pressures within these vessels, particularly the arteries and arterioles, cause the vessels to thicken and strengthen to withstand the stress.

Treatment of hypertension depends on the severity. Exercise, dietary changes, weight control, and medications can keep blood pressure under control and reduce the long-term complications. ■

Capillary Pressures and Capillary Exchange

The pressure of blood within a capillary bed, or **capillary pressure,** pushes against the capillary walls, just as it does in the arteries. But unlike other blood vessels, capillary walls are quite permeable to small ions, nutrients, organic wastes, dissolved gases, and water. Most of these materials are reabsorbed by the capillaries, but about 3.6 liters (0.95 gallons) of water and solutes flow through peripheral tissues each day and enter the **lymphatic vessels** of the lymphatic system, which empty into the bloodstream (Figure 13–7●).

This continuous movement and exchange of water and solutes from the capillaries through the body tissues and back into the bloodstream plays an important role in homeostasis. Capillary exchange has four important functions: (1) maintaining constant communication between plasma and interstitial fluid; (2) speeding the distribution

and average about 2 mmHg (Figure 13–6●). In this section we first consider pressures in the arterial network; then we examine how pressure in the capillaries drives the exchange of substances between the bloodstream and body tissues before considering pressures in the venous system.

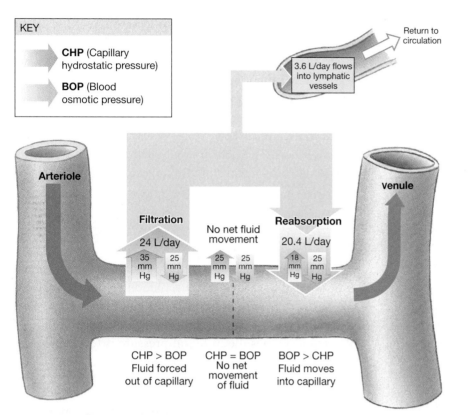

KEY

CHP (Capillary hydrostatic pressure)

BOP (Blood osmotic pressure)

3.6 L/day flows into lymphatic vessels

Return to circulation

Arteriole

venule

Filtration

24 L/day

No net fluid movement

Reabsorption

20.4 L/day

| 35 mm Hg | 25 mm Hg | 25 mm Hg | 25 mm Hg | 18 mm Hg | 25 mm Hg |

CHP > BOP Fluid forced out of capillary

CHP = BOP No net movement of fluid

BOP > CHP Fluid moves into capillary

● **FIGURE 13–7** **Forces That Act Across Capillary Walls.** At the arteriolar end of the capillary, capillary hydrostatic pressure (CHP) is greater than blood osmotic pressure (BOP), so fluid moves out of the capillary (filtration). Near the venule, CHP is lower than BOP, so fluid moves into the capillary (reabsorption).

of nutrients, hormones, and dissolved gases throughout tissues; (3) assisting the movement of insoluble lipids and tissue proteins that cannot cross capillary walls; and (4) flushing bacterial toxins and other chemical stimuli to lymphoid tissues and organs that function in providing immunity to disease.

DYNAMICS OF CAPILLARY EXCHANGE. The movement of materials across capillary walls occurs by diffusion, filtration, and osmosis. Chapter 3 described these processes, so only a brief overview is provided here. ∞ p. 64 Solute molecules tend to diffuse across the capillary lining, driven by their individual concentration gradients. Water-soluble materials—including ions and small organic molecules such as glucose, amino acids, or urea—diffuse through small spaces between adjacent endothelial cells. Larger water-soluble molecules, such as plasma proteins, cannot normally leave the bloodstream. Lipid-soluble materials—including steroids, fatty acids, and dissolved gases—diffuse across the endothelial lining, and pass through the membrane lipids.

Water molecules will move when driven by either hydrostatic (fluid) pressure or osmotic pressure. ∞ p. 66 In filtration, hydrostatic pressure pushes water molecules from an area of high pressure to an area of lower pressure. At a capillary, the hydrostatic pressure, or *capillary hydrostatic pressure* (*CHP*), is greatest at the arteriolar end (35 mmHg) and least at the venous end (18 mmHg). The tendency for water and small solutes to move out of the blood is, therefore, greatest at the start of a capillary, where the CHP is highest, and declines along the length of the capillary as CHP falls.

Osmosis, in contrast, is the movement of water across a selectively permeable membrane that separates two solutions with different solute concentrations. Water moves into the solution with the higher solute concentration, and the force of this water movement is called osmotic pressure. Because blood contains more dissolved proteins than does interstitial fluid, its osmotic pressure is higher and water tends to move from the interstitial fluid into the blood. ∞ p. 66 The osmotic pressure of blood (25 mmHg) is constant along the length of the capillaries. Thus, capillary hydrostatic pressure (CHP) tends to push water out of the capillary, whereas blood osmotic pressure (BOP) forces tends to reabsorb, or pull water back in (see Figure 13–7).

Venous Pressure

Although pressure at the start of the venous system is only about one-tenth that at the start of the arterial system (see Figure 13–6), the blood must still travel through a vascular network as complex as the arterial system before returning to the heart. However, venous pressures are low, and the veins offer little resistance. As a result, once blood enters the venous system, pressure declines very slowly.

As blood travels through the venous system toward the heart, the veins become larger, resistance drops further, and the flow rate increases. Pressures at the entrance to the right atrium fluctuate, but they average approximately 2 mmHg. This means that the driving force pushing blood through the venous system is a mere 16 mmHg (18 mmHg in the venules – 2 mmHg in the venae cavae = 16 mmHg) as compared to the 65 mmHg pressure acting along the arterial system (100 mmHg at the aorta – 35 mmHg at the capillaries = 65 mmHg).

Clinical Note
CAPILLARY PHYSIOLOGY

The exchange of fluid through the capillary membranes is an important physiological process. Normally, capillaries are able to maintain normal fluid volume distribution between the plasma and the interstitial fluid. Normal capillary pressure at the arterial end of the capillaries is 15–25 mmHg greater than at the venous end. Because of this difference, fluid "filters" out of the capillaries at the arterial end and is reabsorbed back into the capillaries at the venous end. This results in a small amount of fluid "flowing" through the tissues from the arterial ends of the capillaries to the venous ends.

Forces at the arterial end of the capillary move fluid outward. These include the *capillary blood pressure* and the *interstitial fluid colloid osmotic pressure*. At the same time, *plasma colloid osmotic pressure* tends to move fluid inward. At the arterial end, the outward pressures, called the *net filtration pressure,* exceed the inward pressures, which results in loss of fluid from the capillaries to the interstitial space.

At the venous end of the capillary, *plasma colloid osmotic pressure* tends to move fluid inward. Likewise, *capillary blood pressure* and *interstitial fluid colloid osmotic pressure* tend to move fluids outward. Here, the inward pressure, called the *reabsorption pressure,* exceeds the outward pressure and causes fluid to move into the capillaries.

In hemorrhage or dehydration, the loss in blood volume causes a drop in blood pressure. Also, as water is lost from the plasma, the blood becomes more concentrated. This results in increased plasma colloid osmotic pressure that pulls fluid from the interstitial space into the intravascular space. The loss of fluid in the interstitial spaces causes a decrease in *skin turgor,* a characteristic finding of dehydration.

An increase in circulating fluid volume increases the capillary blood pressure at the arterial end. Plasma colloids are diluted by the excess fluid, which decreases the plasma colloid osmotic pressure. This causes an accumulation of fluid in the interstitial spaces, commonly called *edema*. ■

When you are lying down, a pressure gradient of 16 mmHg is sufficient to maintain venous flow. But when you are standing, the venous blood in regions below the heart must overcome gravity as it ascends within the inferior vena cava. Two factors help overcome gravity and propel venous blood toward the heart:

1. **Muscular compression.** The contractions of skeletal muscles near a vein compress it and help push blood toward the heart. The valves in medium-sized veins ensure that blood flow occurs in one direction only (see Figure 13–5, p. 472).

2. *The* **respiratory pump.** As you inhale, decreased pressure in the thoracic cavity draws air into the lungs. This drop in pressure causes the inferior vena cava and right atrium to expand and fill with blood, which increases venous return. During exhalation, the increased pressure that forces air out of the lungs compresses the venae cavae, pushing blood into the right atrium.

During exercise, both factors cooperate to increase venous return and push cardiac output to maximal levels. However, when an individual stands at attention, with knees locked and leg muscles immobile, these mechanisms are impaired, and the reduction in venous return leads to a fall in cardiac output. The blood supply to the brain is in turn reduced, sometimes enough to cause *fainting*, which is a temporary loss of consciousness. The person then collapses; once the body is horizontal, both venous return and cardiac output return to normal.

Key Note

Blood flow is the goal, and total peripheral blood flow is equal to cardiac output. Blood pressure is needed to overcome friction and elastic forces and sustain blood flow. If blood pressure is too low, vessels collapse, blood flow stops, and tissues die; if blood pressure is too high, vessel walls stiffen and capillary beds may rupture.

→ CONCEPT CHECK QUESTIONS

1. In a normal individual, where would you expect the cardiovascular pressure to be greater, in the aorta or in the inferior vena cava? Explain.
2. Explain why Sally begins to feel light-headed and then faints while standing in the hot sun.

Answers begin on p. 792.

■ Cardiovascular Regulation

Homeostatic mechanisms regulate cardiovascular activity to ensure that tissue blood flow, also called *tissue perfusion,* meets the demand for oxygen and nutrients. The three variable factors that influence tissue blood flow are (1) cardiac output, (2) peripheral resistance, and (3) blood pressure. We discussed cardiac output in Chapter 12 (p. 458) and considered peripheral resistance and pressure earlier in this chapter.

Most cells are relatively close to capillaries. When a group of cells becomes active, the circulation to that region must increase to deliver the oxygen and nutrients they need and to carry away the waste products and carbon dioxide they generate. The goal of cardiovascular regulation is to ensure that these blood flow changes occur (1) at an appropriate time, (2) in the right area, and (3) without drastically altering blood pressure and blood flow to vital organs.

Clinical Note
VITAL SIGNS

Vital signs are outward indicators of what is going on inside the body. They include blood pressure (BP), heart rate, respiratory rate, and temperature. These measurements provide a great deal of information about a patient's condition and should be measured frequently.

The *pulse rate* is a measure of the number of times the heart beats per minute. The pulse is best measured where an artery lies close to the body's surface and crosses over a bone. Common locations for pulse determination are illustrated in Figure 13–8a●. To measure the pulse rate, locate a suitable site and palpate the pulse. Then, count the number of pulse waves that occur in one minute. Alternatively, you can count the number of pulse waves in 30 seconds and multiply by two. In order to ensure accuracy, it is not recommended that intervals less than 30 seconds be measured.

The *respiratory rate* is a measure of the number of times a person breathes in one minute. To measure the respiratory rate, have the patient assume a comfortable position. While the patient is at rest, count the number of respirations that occur in one minute. People will alter their respiratory pattern if they feel it is being watched, so it is best to distract the patient. Skilled personnel will often take a patient's wrist and measure the pulse. Then while still holding the wrist, they will count the respiratory rate while the patient thinks that the pulse is being measured.

The *blood pressure* is a function of the amount of blood being pumped by the heart (cardiac output) and the peripheral vascular resistance. To determine the blood pressure, a *blood pressure cuff (sphygmomanometer)* is placed on the arm 3–4 centimeters above the elbow (Figure 13–8b●). A stethoscope is placed over the brachial artery. The blood pressure cuff is inflated to approximately 30 mmHg above the point where it occludes the brachial artery and stops the flow of blood. Then, the pressure in the cuff is slowly released. When the pressure in the cuff falls below systolic pressure, blood flow resumes in the artery and produces *Korotkoff sounds*. As the pressure falls, the vessel remains open longer and the sounds in the artery change. When the cuff pressure falls below the diastolic pressure, blood flow becomes continuous and the Korotkoff sounds disappear completely.

The *temperature* is not often measured in EMS. However, electronic thermometers have made this a simple procedure. The temperature is measured in the mouth (under the tongue), rectally, or through the ear (tympanic membrane). Normal body temperature is 37°C (98.6°F). ∎

● **FIGURE 13–8 Checking the Pulse and Blood Pressure.** (**a**) Several pressure points can be used to monitor the pulse or control peripheral bleeding. (**b**) A sphygmomanometer and a stethoscope are used to check an individual's blood pressure.

Temporal artery

External carotid artery

Facial artery

Brachial artery

Radial artery

Femoral artery

Popliteal artery

Posterial tibial artery

Dorsalis pedis artery

(a)

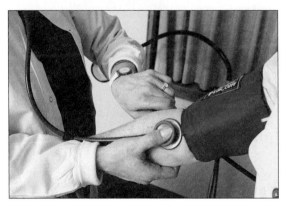

(b)

The mechanisms involved in the regulation of cardiovascular function include the following (Figure 13–9●):

- *Autoregulation.* Changes in tissue conditions act directly on precapillary sphincters to alter peripheral resistance, producing local changes in the pattern of blood flow within capillary beds. Such autoregulation causes immediate, localized homeostatic adjustments. If autoregulation is unable to normalize tissue conditions, neural and endocrine mechanisms are activated.
- *Neural mechanisms.* Neural mechanisms respond to changes in arterial pressure or blood gas levels at specific sites. When those changes occur, the autonomic nervous system adjusts cardiac output and peripheral resistance to maintain adequate blood flow.
- *Endocrine mechanisms.* The endocrine system releases hormones that enhance short-term adjustments and direct long-term changes in cardiovascular performance.

Short-term responses adjust cardiac output and peripheral resistance to stabilize blood pressure and blood flow to tissues. Long-term adjustments involve alterations in blood volume that affect cardiac output and the transport of oxygen and carbon dioxide to and from active tissues.

Autoregulation of Blood Flow

Under normal resting conditions, cardiac output remains stable, and precapillary sphincters control local blood flow by automatically adjusting peripheral resistance in individual tissues. The smooth muscles of precapillary sphincters respond automatically to certain alterations in the local environment, such as changes in oxygen and carbon dioxide levels or the presence of specific chemicals. For example, when oxygen is abundant, the smooth muscle cells in the sphincters contract and slow the flow of blood. As the cells take up and use O_2 for aerobic respiration, tissue oxygen supplies dwindle, carbon dioxide levels rise, and pH falls. These changes signal the smooth muscle cells in the precapillary sphincters to relax, and blood flow increases. Similarly, the release of nitric oxide (NO) by capillary endothelial cells stimulated by high shear forces along the capillary walls or the presence of histamine at an injury site during inflammation triggers the relaxation of precapillary sphincters, which increases blood flow.

Factors that promote the dilation of precapillary sphincters are called **vasodilators,** and those that stimulate the constriction of precapillary sphincters are called **vasoconstrictors.** Together, such factors control blood flow in a single capillary bed. When present in high concentrations, these factors also affect arterioles

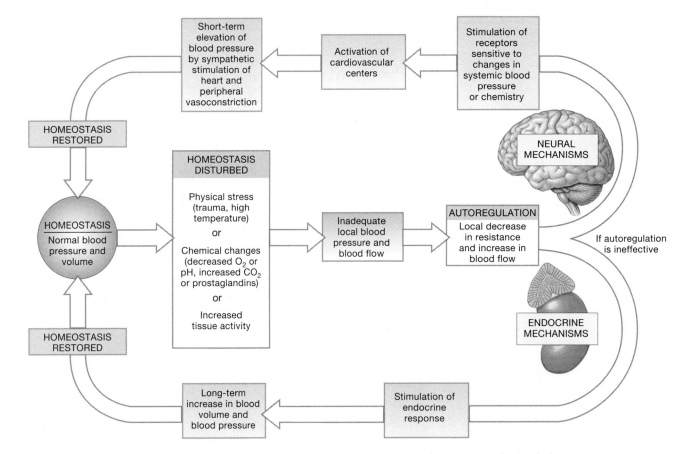

● **FIGURE 13–9** Local, Neural, and Endocrine Adjustments That Maintain Blood Pressure and Blood Flow.

and increase or decrease blood flow to all the capillary beds in a given region. Such an event often triggers a neural response, because significant changes in blood flow to one region of the body have an immediate effect on circulation to other regions.

Neural Control of Blood Pressure and Blood Flow

The nervous system adjusts cardiac output and peripheral resistance to maintain adequate blood flow to vital tissues and organs. The *cardiac centers* and *vasomotor centers* of the medulla oblongata are the *cardiovascular (CV) centers* responsible for these regulatory activities. As noted in Chapter 12, each cardiac center includes a *cardioacceleratory center,* which increases cardiac output through sympathetic innervation, and a *cardioinhibitory center,* which reduces cardiac output through parasympathetic innervation. ∞ p. 460

The vasomotor centers of the medulla oblongata primarily control the diameters of the arterioles through sympathetic innervation. Inhibition of the vasomotor centers leads to vasodilation (the dilation of arterioles), which reduces peripheral resistance. Stimulation of the vasomotor centers causes vasoconstriction (the constriction of peripheral arterioles) and, with very strong stimulation, **venoconstriction** (the constriction of peripheral veins), both of which increase peripheral resistance.

The cardiovascular centers detect changes in tissue demand by monitoring arterial blood, especially blood pressure, pH, and dissolved gas concentrations. *Baroreceptor reflexes* respond to changes in blood pressure, and *chemoreceptor reflexes* respond to changes in chemical composition. These reflexes are regulated through negative feedback: the stimulation of a receptor by an abnormal condition leads to a response that counteracts the stimulus and restores normal conditions.

Baroreceptor Reflexes

Baroreceptors monitor the degree of stretch in the walls of expandable organs. ∞ p. 329 The baroreceptors involved in cardiovascular regulation are located (1) in the **aortic sinuses,** which are pockets in the walls of the aorta adjacent to the heart (see Figure 12–6b; p. 444); (2) in the walls of the **carotid sinuses,** which are expanded chambers near the bases of the *internal carotid arteries* of the neck; and (3) in the wall of the right atrium. These receptors initiate the **baroreceptor reflexes** (*baro-,* pressure), which are autonomic reflexes that adjust cardiac output and peripheral resistance to maintain normal arterial pressures.

Aortic baroreceptors monitor blood pressure within the ascending aorta. There the *aortic reflex* adjusts blood pressure in response to changes in pressure. The purpose of the aortic reflex is to maintain adequate blood pressure and blood flow

through the systemic circuit. Carotid sinus baroreceptors respond to changes in blood pressure at the carotid sinuses. These receptors trigger reflexes that maintain adequate blood flow to the brain. The blood flow to the brain must remain constant, and the *carotid sinus reflex* is extremely sensitive. The baroreceptor reflexes triggered by changes in blood pressure at these locations are diagrammed in Figure 13–10●.

When blood pressure climbs (Figure 13–10a), increased output from the baroreceptors travels to the cardiovascular centers of the medulla oblongata, where it inhibits the cardioacceleratory center, stimulates the cardioinhibitory center, and inhibits the vasomotor centers. Two effects are produced: (1) under the command of the cardioinhibitory center, the vagus nerves release acetylcholine (ACh), which reduces the rate and strength of cardiac contractions, decreasing cardiac output; and (2) the inhibition of the vasomotor centers leads to dilation of peripheral arterioles throughout the body. This combination of reduced cardiac output and decreased peripheral resistance then lowers blood pressure.

If blood pressure falls below normal (Figure 13–10b), inhibition of baroreceptor output produces the opposite effects: (1) an increase in cardiac output and (2) widespread peripheral vasoconstriction. Reduced baroreceptor activity stimulates the cardioacceleratory center, inhibits the cardioinhibitory center, and stimulates the vasomotor centers. The cardioacceleratory center stimulates sympathetic neurons that innervate the sinoatrial (SA) node, atrioventricular (AV) node, and general myocardium. This stimulation increases heart rate and stroke volume, and leads to an immediate increase in cardiac output. Vasomotor activity, also carried by sympathetic motor neurons, produces a rapid vasoconstriction, which increases peripheral resistance. These adjustments—increased cardiac output and increased peripheral resistance—work together to elevate blood pressure.

Atrial baroreceptors monitor blood pressure at the end of the systemic circuit—at the venae cavae and the right atrium. The atrial reflex responds to the stretching of the wall of the right atrium. ∞ p. 459 Under normal circumstances, the heart pumps blood into the aorta at the same rate at which it arrives at the right atrium. A rise in blood pressure at the atrium means that blood is arriving at the heart faster than it is being pumped out. The atrial baroreceptors correct the situation by stimulating the cardioacceleratory center, increasing cardiac output until the backlog of venous blood is removed. Atrial pressure then returns to normal.

Chemoreceptor Reflexes

The **chemoreceptor reflexes** respond to changes in carbon dioxide, oxygen, or pH in blood and cerebrospinal fluid (Figure 13–11●). The chemoreceptors involved are sensory neurons

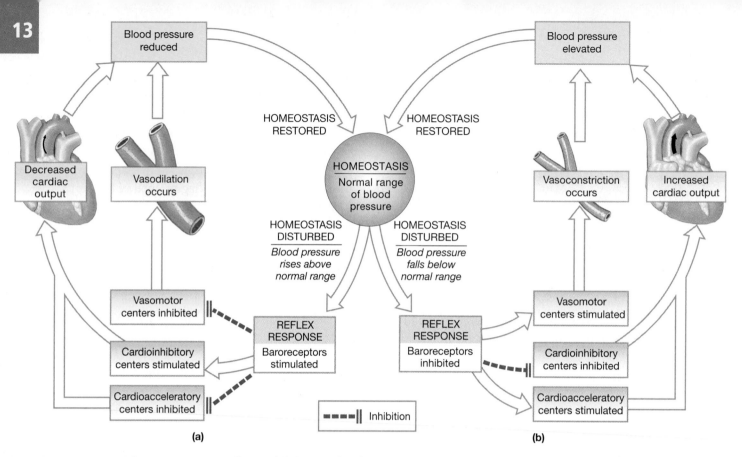

● **FIGURE 13–10** The Baroreceptor Reflexes of the Carotid and Aortic Sinuses.

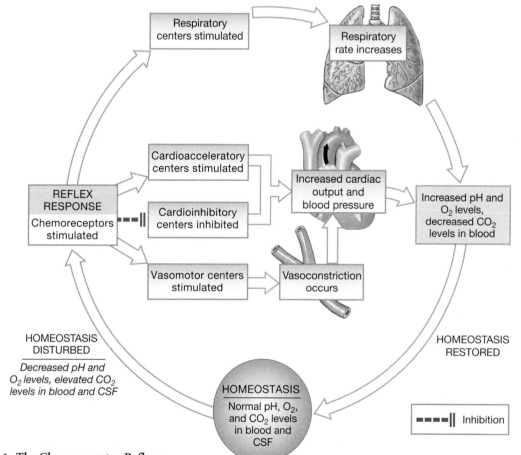

● **FIGURE 13–11** The Chemoreceptor Reflexes.

● **FIGURE 13–12 The Hormonal Regulation of Blood Pressure and Blood Volume.** The negative feedback mechanisms that compensate for (**a**) decreased blood pressure and volume and (**b**) increased blood pressure and volume.

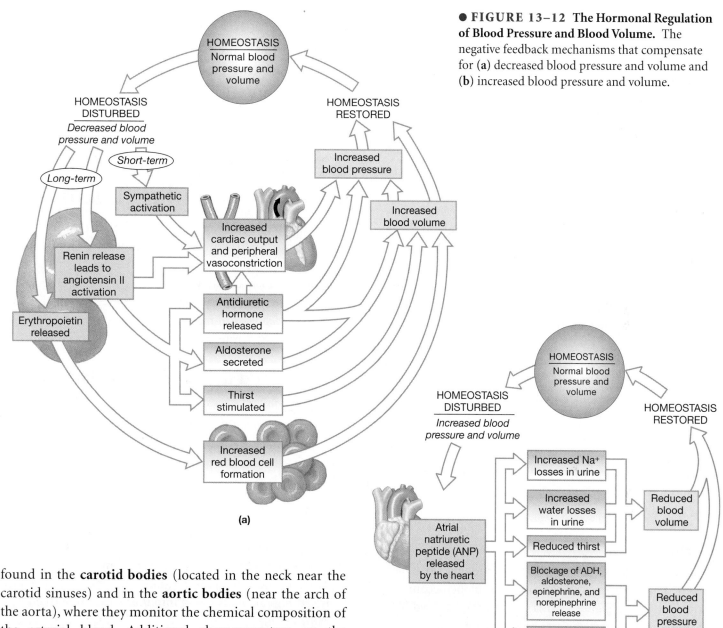

(a)

(b)

found in the **carotid bodies** (located in the neck near the carotid sinuses) and in the **aortic bodies** (near the arch of the aorta), where they monitor the chemical composition of the arterial blood. Additional chemoreceptors on the medulla oblongata monitor the composition of the cerebrospinal fluid (CSF).

Chemoreceptors are activated by a drop in pH or in plasma O_2 or by a rise in CO_2. Any of these changes leads to a stimulation of the cardioacceleratory and vasomotor centers. This elevates arterial pressure and increases blood flow through peripheral tissues. Chemoreceptor output also affects the respiratory centers in the medulla oblongata. As a result, a rise in blood flow and blood pressure is associated with an elevated respiratory rate. The coordination of cardiovascular and respiratory activity is vital, because accelerating tissue blood flow is useful only if the circulating blood contains adequate oxygen. In addition, a rise in the respiratory rate accelerates venous return through the action of the respiratory pump. ∞ p. 475

Hormones and Cardiovascular Regulation

The endocrine system provides both short-term and long-term regulation of cardiovascular performance. In the short term, epinephrine (E) and norepinephrine (NE) from the adrenal medullae stimulate cardiac output and peripheral vasoconstriction. Long-term cardiovascular regulation involves the participation of other hormones—antidiuretic hormone (ADH), angiotensin II, erythropoietin (EPO), and atrial natriuretic peptide (ANP), introduced in Chapter 10. ∞ pp. 377, 394 The roles of these hormones in the regulation of blood pressure and blood volume are diagrammed in Figure 13–12●.

Antidiuretic Hormone

Antidiuretic hormone (ADH) is released at the posterior pituitary gland in response to a decrease in blood volume, an increase in the osmotic concentration of the plasma, or in response to the presence of angiotensin II (Figure 13–12a). The immediate result is peripheral vasoconstriction that elevates blood pressure. ADH also has a water-conserving effect on the kidneys, and thus prevents a reduction in blood volume.

Angiotensin II

Angiotensin II is formed in the blood following the release of the enzyme renin by specialized kidney cells in response to a fall in blood pressure (see Figure 13–12a). Renin starts a chain reaction that ultimately converts an inactive plasma protein, *angiotensinogen,* to the hormone angiotensin II. Angiotensin II stimulates cardiac output and triggers arteriole constriction, which in turn elevates systemic blood pressure almost immediately. It also stimulates the secretion of ADH by the pituitary gland and aldosterone by the adrenal cortex. The effects of these hormones are complementary: ADH stimulates water conservation at the kidneys, and aldosterone stimulates the reabsorption of sodium ions and water from the urine. In addition, angiotensin II stimulates thirst, and the presence of ADH and aldosterone ensures that the additional water consumed will be retained and elevate blood volume.

Erythropoietin

Erythropoietin (EPO) is released by the kidneys when blood pressure falls or the oxygen content of the blood becomes abnormally low (see Figure 13–12a). EPO stimulates red blood cell production, which elevates blood volume and improves the oxygen-carrying capacity of the blood.

Atrial Natriuretic Peptide

In contrast to the three hormones just described, atrial natriuretic peptide (ANP) release is stimulated by *increased* blood pressure (Figure 13–12b). ANP is produced by specialized cardiac muscle cells in the atrial walls when they are stretched by excessive venous return. ANP reduces blood volume and blood pressure by (1) increasing the loss of sodium ions and water at the kidneys; (2) promoting water losses by increasing the volume of urine produced; (3) reducing thirst; (4) blocking the release of ADH, aldosterone, E, and NE; and (5) stimulating peripheral vasodilation. As blood volume and blood pressure decline, tension on the atrial walls is removed, and ANP production ceases.

Key Note

Cardiac output cannot increase indefinitely, and blood flow to active versus inactive tissues must be differentially controlled. This is accomplished by a combination of autoregulation, neural regulation, and hormone release.

■ Patterns of Cardiovascular Response

In our day-to-day lives, the components of the cardiovascular system—the blood, heart, and blood vessels—operate in an integrated way. In this section we will examine this system's adaptability in maintaining homeostasis in response to two common stresses: exercise and blood loss. We will also consider the physiological mechanisms involved in shock, which is an important cardiovascular disorder.

Exercise and the Cardiovascular System

At rest, cardiac output averages about 5.8 liters per minute. During exercise, both cardiac output and the pattern of blood distribution change markedly. As exercise begins, several interrelated changes take place:

- *Extensive vasodilation occurs* as the rate of oxygen consumption in skeletal muscles increases. Peripheral resistance drops, blood flow through the capillaries increases, and blood enters the venous system at an accelerated rate.
- *The venous return increases* as skeletal muscle contractions squeeze blood along the peripheral veins, and an increased breathing rate pulls blood into the venae cavae (the respiratory pump).
- *Cardiac output rises* as a result of the increased venous return. This increase occurs in direct response to ventricular stretching (the Frank-Starling principle) and in a reflexive response to atrial stretching (the atrial reflex). ∞ p. 459 The increased cardiac output keeps pace with the elevated demand, and arterial pressures are maintained despite the drop in peripheral resistance.

This regulation by venous feedback gradually increases cardiac output to about double resting levels. Over this range, which is typical of light exercise, the pattern of blood distribution remains relatively unchanged.

At higher levels of exertion, other physiological adjustments occur as the cardiac and vasomotor centers activate the sympathetic nervous system. Cardiac output increases (up to 20–25 liters per minute), blood pressure increases, and blood flow is severely restricted to "nonessential" organs (such as those of the digestive system) in order to increase the blood flow to active skeletal muscles. When you exercise at maximal levels, your blood essentially races among your skeletal muscles, lungs, and heart.

The Cardiovascular Response to Hemorrhage

In Chapter 11 we considered the local circulatory reaction to a break in the wall of a blood vessel. p. 425 When the clotting response fails to prevent a significant blood loss and blood pressure falls, the entire cardiovascular system begins making adjustments. The immediate goal is to maintain adequate blood pressure and peripheral blood flow; the long-term goal is to restore normal blood volume (see Figure 13–12).

The Short-Term Elevation of Blood Pressure

Short-term responses appear almost as soon as blood pressure starts to decline, when the carotid and aortic reflexes increase cardiac output and cause peripheral vasoconstriction. When you donate blood at a blood bank, the amount collected is usually 500 mL, roughly 10 percent of your total blood volume. Such a loss initially causes a drop in cardiac output, but the vasomotor centers quickly improve venous return and restore cardiac output to normal levels by mobilizing the **venous reserve** (large reservoirs of slowly moving venous blood in the liver, bone marrow, and skin) through venoconstriction (see p. 479). This venous compensation can restore normal arterial pressures and peripheral blood flow after losses of 15–20 percent of total blood volume. With a more substantial blood loss, cardiac output is maintained by increasing the heart rate, often to 180–200 bpm. Sympathetic activation assists by constricting the muscular arteries and arterioles, which elevates blood pressure.

Sympathetic activation also causes the secretion of E and NE by the adrenal medullae. At the same time, ADH is released by the posterior pituitary gland, and the fall in blood pressure in the kidneys causes the release of renin, and initiates the activation of angiotensin II. Both E and NE increase cardiac output, and in combination with ADH and angiotensin II, they cause a powerful vasoconstriction that elevates blood pressure and improves peripheral blood flow.

The Long-Term Restoration of Blood Volume

After serious hemorrhaging, several days may pass before blood volume returns to normal. When short-term responses are unable to maintain normal cardiac output and blood pressure, the decline in capillary blood pressure triggers a recall of fluids from the interstitial spaces. p. 475 Over this period, ADH

Clinical Note
SHOCK

Shock (hypoperfusion) is a state of inadequate tissue perfusion and is often the end product of many disease and injury processes. During shock, body tissues are deprived of oxygen and essential nutrients. At the same time, carbon dioxide and other waste products accumulate. Regardless of the cause, shock results from one of three physiological problems. These include an inadequate pump (cardiogenic shock), inadequate fluid (hypovolemic shock), or an inadequate container (neurogenic shock). All of these can result in a fall in blood pressure and, ultimately, inadequate tissue perfusion.

Cardiogenic shock occurs when the heart fails as an effective pump. This is a complication of heart attack or congestive heart failure. *Hypovolemic shock* results from the loss of blood or plasma from the intravascular space. It is seen following severe hemorrhage, burns, crush injuries, and severe nausea and vomiting. *Neurogenic shock* occurs when autonomic control of peripheral blood vessels is lost, resulting in marked dilation and a marked decrease in peripheral vascular resistance. This can be seen following a spinal-cord injury where nerve fibers are severed. It can also be seen in severe allergic reactions (*anaphylactic shock*) where histamine and other chemical mediators cause marked peripheral vascular dilation and loss of plasma from the capillaries. Thus, anaphylactic shock involves problems with both fluid loss and peripheral vascular dilation. Another form of neurogenic shock, *septic shock,* occurs following a severe bacterial infection, whereby bacteria release potent *endotoxins* that cause marked peripheral vasodilation that results in shock.

The signs and symptoms of shock, which can often be subtle, include altered mental status; pale, cool, clammy skin; nausea and vomiting; increased pulse rate; increased respiratory rate; and a fall in blood pressure. A fall in blood pressure is considered a late sign. Shock is best treated when it is promptly identified and emergency interventions are provided before the blood pressure falls. ■

and aldosterone promote fluid retention and reabsorption at the kidneys, which prevents further reductions in blood volume. Thirst increases, and additional water is absorbed across the digestive tract. This intake of fluid elevates blood volume and ultimately replaces the interstitial fluids "borrowed" at the capillaries. Erythropoietin targets the bone marrow, stimulating the maturation of red blood cells, which increases blood volume and improves oxygen delivery to peripheral tissues.

CONCEPT CHECK QUESTIONS

1. How would slightly compressing the common carotid artery affect your heart rate?
2. Why does blood pressure increase during exercise?
3. What effect would vasoconstriction of the renal artery have on blood pressure and blood volume?

Answers begin on p. 792.

■ The Blood Vessels

As previously discussed, the circulatory system is divided into the **pulmonary circuit** and the **systemic circuit.** ∞ p. 437 The pulmonary circuit is composed of arteries and veins that transport blood between the heart and the lungs; this circuit begins at the right ventricle and ends at the left atrium. The systemic circuit is composed of arteries that transport oxygenated blood and nutrients to all other organs and tissues and veins that return deoxygenated blood to the heart; this circuit begins at the left ventricle and ends at the right atrium. Figure 13–13● summarizes the primary circulatory routes within the pulmonary and systemic circuits.

The distribution of arteries and veins on the left and right sides of the body is usually identical—except near the heart, where large vessels connect to the atria or ventricles. For example, the distribution of the *left* and *right subclavian, axillary, brachial,* and *radial arteries* parallels the *left* and *right subclavian, axillary, brachial,* and *radial veins.* In addition, a single vessel may undergo several name changes as it crosses specific anatomical boundaries. For example, the *external iliac artery* becomes the *femoral artery* as it leaves the trunk and enters the thigh.

The Pulmonary Circulation

Blood that enters the right atrium has just returned from the peripheral capillary beds, where oxygen was released and carbon dioxide was absorbed. After traveling through the right atrium and ventricle, blood enters the **pulmonary trunk** (*pulmo-*, lung), the start of the pulmonary circuit (Figure 13–14●). In this circuit, carbon dioxide is released, oxygen stores are replenished, and the oxygenated blood is returned to the heart for distribution in the systemic circuit.

The arteries of the pulmonary circuit differ from those of the systemic circuit in that they carry deoxygenated blood. (For this reason, color-coded diagrams usually show the pulmonary arteries in blue, the same color as systemic veins.) As the pulmonary trunk curves over the superior border of the heart, it gives rise to the **left** and **right pulmonary arteries.** These large arteries enter the lungs before branching repeatedly, giving rise to smaller and smaller arteries. The smallest branches, the pulmonary arterioles, provide blood to capil-

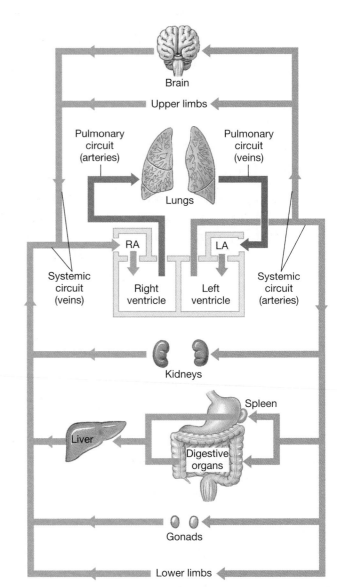

● **FIGURE 13–13 An Overview of the Pattern of Circulation.** RA stands for right atrium; LA for left atrium.

lary networks that surround small air pockets, or **alveoli** (al-VĒ-ō-lī; *alveolus,* sac). The walls of alveoli are thin enough for gas exchange to occur between the capillary blood and inspired air, a process detailed in Chapter 15. As oxygenated blood leaves the alveolar capillaries, it enters venules, which in turn unite to form larger vessels that lead to the **pulmonary veins.** These four veins (two from each lung) empty into the left atrium, which completes the pulmonary circuit.

Clinical Note
ACUTE PULMONARY EMBOLISM

Acute pulmonary embolism (PE) occurs when a blood clot or other particle lodges in a pulmonary artery and blocks blood flow through that vessel. Pulmonary emboli may be composed of air, fat, amniotic fluid, or blood clots. Factors that predispose a patient to blood clots include prolonged immobilization, thrombophlebitis (inflammation and clots in a vein), use of certain medications, and atrial fibrillation. Long periods of travel in airplanes and automobiles increase an individual's chances of developing a PE.

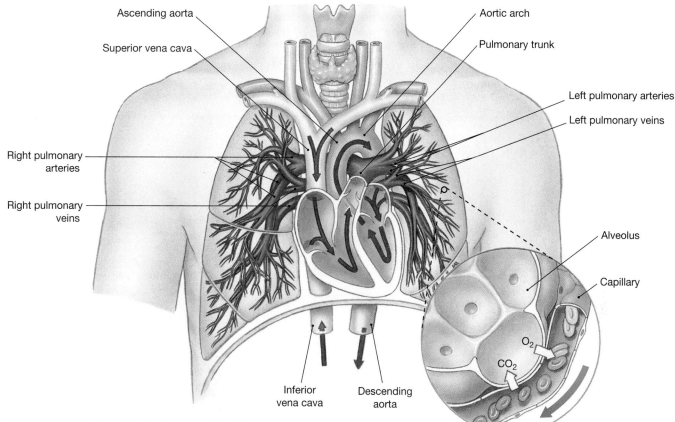

● **FIGURE 13–14 The Pulmonary Circuit.**

When a PE blocks the blood flow through a pulmonary vessel, the right heart must pump against increased resistance, which in turn increases pulmonary capillary pressure (Figure 13–15●). The area of the lung supplied by the occluded vessel then stops functioning, and gas exchange decreases. The larger the volume of the lung affected, the more significant the problems with gas exchange.

The signs and symptoms of pulmonary embolism depend upon the size of the obstruction. The patient who suffers from an acute pulmonary embolism may report a sudden onset of severe and unexplained dyspnea that may or may not be associated with chest pain. The signs and symptoms of an acute pulmonary embolism can be subtle and easily overlooked. The diagnosis can be made using several imaging techniques. A frequently used test is the ventilation/perfusion scan (V/Q scan), for which the patient inhales radioactive material and also has radioactive material injected into his or her venous system. Two sets of images of the patient's lungs are obtained,

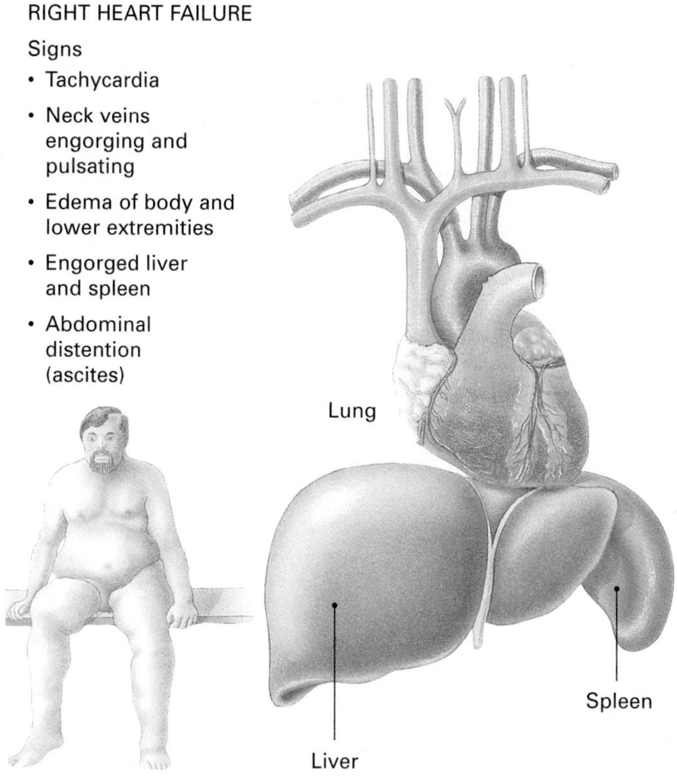

RIGHT HEART FAILURE

Signs

- Tachycardia
- Neck veins engorging and pulsating
- Edema of body and lower extremities
- Engorged liver and spleen
- Abdominal distention (ascites)

Lung

Liver

Spleen

● **FIGURE 13–15 Congestive Heart Failure.** Massive acute pulmonary embolism can cause a sudden increase in pulmonary vascular resistance, resulting in right heart failure. This initially causes tachycardia and distended neck veins. If it continues, it can cause accumulation of fluid in the liver (hepatomegaly), spleen (splenomegaly), and in the peritoneal cavity (ascites). Eventually, edema of the extremities develops.

one that shows pulmonary ventilation through the respiratory tree and the other that shows pulmonary perfusion in the pulmonary circulatory system. Areas of the lung that are ventilated but not perfused indicate a PE. The gold standard for the diagnosis of PE is a pulmonary arteriogram that will show any blockage in the pulmonary circulation. Pulmonary angiograms are higher risk procedures than V/Q scans and require specialized facilities.

The treatment of a PE varies depending upon its size. With small to moderate emboli, the patient will usually receive a blood thinner and be closely monitored in the hospital. Large emboli, as evidenced by marked hypoxia and low blood pressure, may require thrombolytic therapy. In most severe cases, the affected pulmonary artery is catheterized and a thrombolytic agent is injected directly into the affected vessel. ■

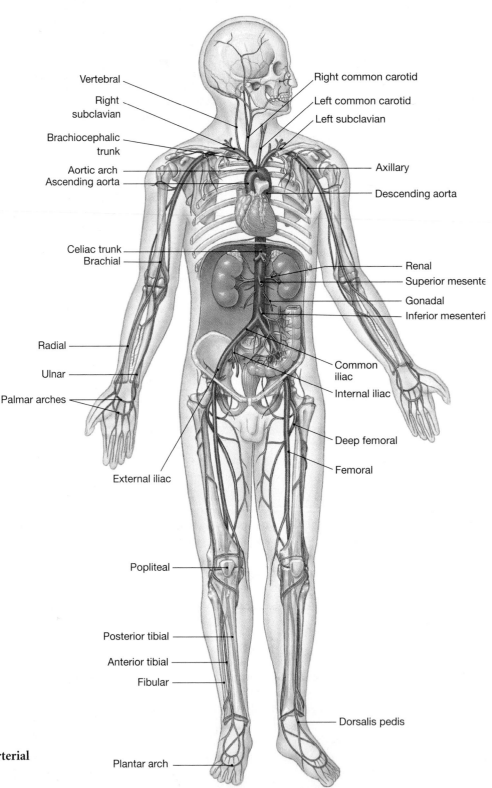

Vertebral
Right subclavian
Brachiocephalic trunk
Aortic arch
Ascending aorta
Celiac trunk
Brachial
Radial
Ulnar
Palmar arches
External iliac
Popliteal
Posterior tibial
Anterior tibial
Fibular
Plantar arch

Right common carotid
Left common carotid
Left subclavian
Axillary
Descending aorta
Renal
Superior mesente
Gonadal
Inferior mesenteri
Common iliac
Internal iliac
Deep femoral
Femoral
Dorsalis pedis

● **FIGURE 13–16** **An Overview of the Arterial System.**

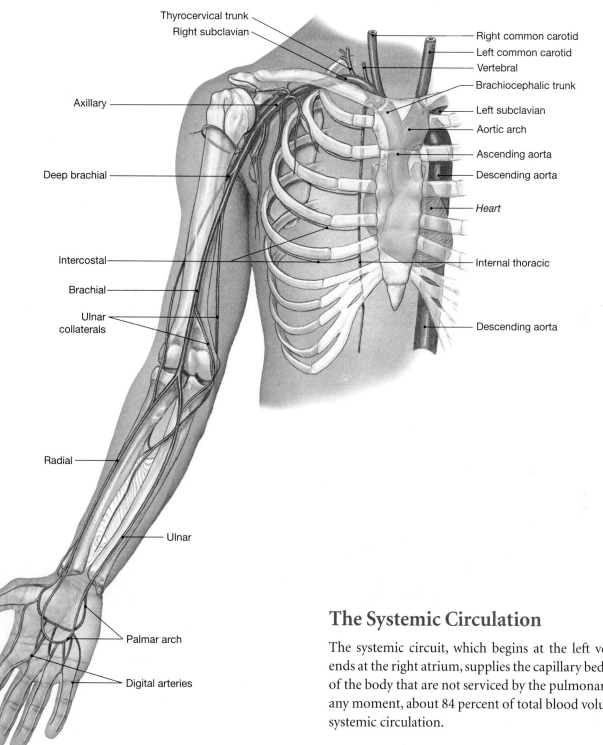

Thyrocervical trunk

Right subclavian

Axillary

Deep brachial

Intercostal

Brachial

Ulnar collaterals

Radial

Ulnar

Palmar arch

Digital arteries

Right common carotid

Left common carotid

Vertebral

Brachiocephalic trunk

Left subclavian

Aortic arch

Ascending aorta

Descending aorta

Heart

Internal thoracic

Descending aorta

● **FIGURE 13–17 Arteries of the Chest and Upper Limb.**

The Systemic Circulation

The systemic circuit, which begins at the left ventricle and ends at the right atrium, supplies the capillary beds in all parts of the body that are not serviced by the pulmonary circuit. At any moment, about 84 percent of total blood volume is in the systemic circulation.

Systemic Arteries

Figure 13–16● provides an overview of the locations of the major systemic arteries. The first systemic vessel and largest artery is the aorta. The **ascending aorta** begins at the aortic semilunar valve of the left ventricle, and the *left* and *right coronary arteries* originate near its base (see Figure 12–7, p. 446). The **aortic arch** curves across the superior surface of the heart and connects the ascending aorta with the **descending aorta** (Figure 13–17●).

Clinical Note
ANEURYSM

An *aneurysm* is a bulging in the weakened wall of a blood vessel. Aneurysms are at risk for rupture, and rupture of aneurysms in the brain, aorta, or heart can be rapidly fatal. The several types of aneurysms include:

- Atherosclerotic
- Dissecting
- Infectious
- Congenital
- Traumatic

Most aneurysms result from atherosclerosis and involve the aorta because the blood pressure there is higher than at any other vessel in the body (Figures 13–18● and 13–19●). An aneurysm usually occurs gradually. Eventually, blood surges into the aortic wall through a tear in the aortic *tunica intima*. Infectious aneurysms are most commonly associated with syphilis and are rare in the U.S. Congenital aneurysms can occur with several disease states, such as Marfan's syndrome, which is a hereditary disease that affects the connective tissue. Aortic aneurysm occurs in people with Marfan's syndrome because it involves the connective tissue within the vessel wall. Those affected may experience sudden death, usually from spontaneous rupture of the aorta, often at a fairly young age. Aortic aneurysms can occur in the chest or in the abdomen. Thoracic aortic aneurysms usually involve the arch of the aorta. They cause symptoms by eroding into nearby structures. Rupture of a thoracic aortic aneurysm is usually a catastrophic event.

Abdominal Aortic Aneurysm

Abdominal aortic aneurysm usually results from atherosclerosis and occurs most frequently in the aorta, below the renal arteries and above the bifurcation of the common iliac arteries (Figure 13–20●). It is ten times more common in men than in women and most prevalent between ages 60 and 70. Many abdominal aortic aneurysms are detected by physical exam or by diagnostic imaging (X-rays, ultrasound). Nondissecting aneurysms less than 5 centimeters in diameter are usually monitored. Larger aneurysms, or those that are rapidly expanding, require surgical repair.

The signs and symptoms of an abdominal aneurysm include:

- Abdominal pain
- Back and flank pain
- Hypotension
- Urge to defecate, caused by the retroperitoneal leakage of blood

Rupture or leaking of an aortic aneurysm is a surgical emergency. Patients suspected of having this condition should be promptly taken to surgery for repair and grafting. Massive transfusion of blood in these cases is common.

● **FIGURE 13–18 Normal Artery.** Post-mortem specimen of a large artery, opened to reveal minimal atherosclerotic changes.

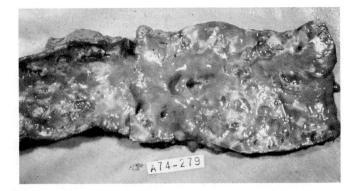

● **FIGURE 13–19 Atherosclerotic Changes.** Marked atherosclerosis in a medium-sized artery that causes turbulent blood flow and marked blood flow resistance.

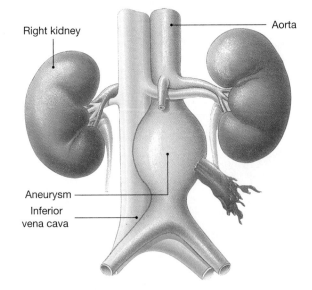

● **FIGURE 13–20 Abdominal Aortic Aneurysm.** Most abdominal aortic aneurysms occur below the level of the branches of the renal arteries and extend down to the bifurcation of the iliac arteries. Rupture can cause rapid exsanguination that will be rapidly fatal without emergency surgical repair.

Dissecting Aneurysm

Some aneurysms will begin to dissect within the wall of the blood vessel. Most dissecting aortic aneurysms result from degenerative changes in the smooth muscle and elastic tissue of the aortic tunica media that can result in hematoma and, subsequently, aneurysm. The original tear often is due to *cystic medial necrosis,* which is a degenerative disease of connective tissue commonly associated with hypertension and to a certain extent, aging. Predisposing factors include hypertension, which is present in 75–85 percent of

cases. It occurs more frequently in patients older than 40–50, although it can occur in younger individuals, especially pregnant women. A tendency for this disease also runs in families.

Of dissecting aortic aneurysms, 67 percent involve the ascending aorta. Once dissection has started, it can extend to all of the abdominal aorta as well as its branches, including the coronary arteries, aortic valve, subclavian arteries, and carotid arteries. The aneurysm can rupture at any time, usually into the pericardial or pleural cavity, which often causes death. ■

ARTERIES OF THE AORTIC ARCH. Three elastic arteries—the **brachiocephalic** (brā-kē-ō-se-FAL-ik) **trunk,** the **left common carotid,** and the **left subclavian** (sub-CLĀ-vē-an)— originate along the aortic arch and deliver blood to the head, neck, shoulders, and upper limbs (see Figures 13–17 and 13–18). The brachiocephalic trunk ascends for a short distance before branching to form the **right common carotid artery** and the **right subclavian artery.** Note that we have

only one brachiocephalic trunk and that the left common carotid and left subclavian arteries arise separately from the aortic arch. In terms of their peripheral distribution, however, the vessels on the left side are mirror images of those on the right side. Because most of the major arteries are paired, with one artery of each pair on either side of the body, the descriptions that follow will not use the terms *right* and *left* (Figure 13–21●).

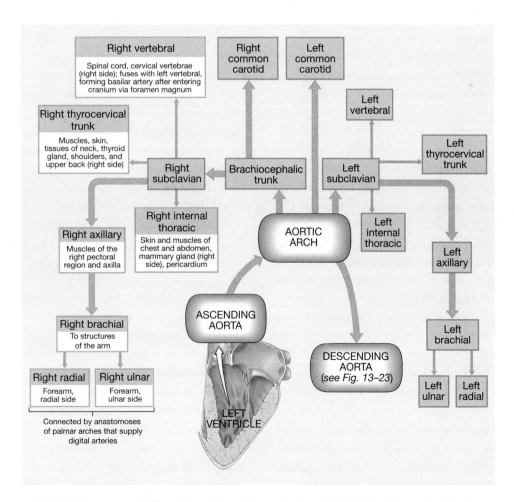

● **FIGURE 13–21** **A Flow Chart That Shows the Arterial Distribution to the Head, Chest, and Upper Limbs.**

THE SUBCLAVIAN ARTERIES. The subclavian arteries supply blood to the arms, chest wall, shoulders, back, and central nervous system. Before a subclavian artery leaves the thoracic cavity, it gives rise to an **internal thoracic artery,** which supplies the pericardium and anterior wall of the chest; a **vertebral artery,** which supplies the brain and spinal cord; and the **thyrocervical trunk,** which supplies muscles and other tissues of the neck, shoulder, and upper back (see Figure 13–17).

After passing the first rib, the subclavian gets a new name: the **axillary artery** (see Figure 13–17). This artery crosses the axilla (armpit) to enter the arm, where its name changes again and becomes the **brachial artery.** The brachial artery provides blood to the arm before it branches to create the **radial artery** and **ulnar artery** of the forearm. These arteries connect at the palm to form anastomoses—the superficial and deep *palmar arches*—from which the *digital arteries* originate.

THE CAROTID ARTERY AND THE BLOOD SUPPLY TO THE BRAIN. The common carotid arteries ascend deep in the tissues of the neck. Each common carotid artery divides into an **external carotid** and an **internal carotid artery** (Figure 13–22a●). The external carotid artery can usually be located by pressing gently along either side of the windpipe (trachea) until a strong pulse is felt. The external carotid arteries supply blood to the pharynx, esophagus, larynx, and face. The internal carotid arteries enter the skull to deliver blood to the brain and the eyes.

The brain is extremely sensitive to changes in blood supply. An interruption of blood flow for several seconds will produce unconsciousness, and after four minutes permanent neural damage may result. Such circulatory crises are rare, however, because blood reaches the brain by two routes: through the vertebral arteries, and through the internal carotid arteries.

The vertebral arteries ascend within the transverse foramina of the cervical vertebrae, and penetrate the skull at the foramen magnum. Inside the cranium, they fuse to form a large **basilar artery,** which continues along the ventral surface of the brain. This artery gives rise to the vessels shown in Figure 13–22b●.

Normally, the internal carotid arteries supply the arteries of the anterior half of the cerebrum, and the rest of the brain receives blood from the vertebral arteries. But this circulatory pattern can easily change, because the internal carotids and the basilar artery are interconnected in the **cerebral arterial circle,** or *circle of Willis,* which is a ring-shaped anastomosis that encircles the infundibulum (stalk) of the pituitary gland. With this arrangement, the brain can receive blood from either the carotid or the vertebral arteries, and the chances for a serious interruption of circulation are reduced.

THE DESCENDING AORTA. The **descending aorta** is continuous with the aortic arch. The diaphragm divides the descending aorta into a superior **thoracic aorta** and an inferior **abdominal aorta** (Figure 13–23●). The thoracic aorta travels within the mediastinum, providing blood to the intercostal arteries, which carry blood to the vertebral column area and the body wall. The thoracic aorta also gives rise to small arteries that end in capillary beds in the esophagus, pericardium, and other mediastinal structures. Near the diaphragm, the **phrenic** (FREN-ik) **arteries** deliver blood to the muscular diaphragm, which separates the thoracic and abdominopelvic cavities.

The abdominal aorta delivers blood to all the abdominopelvic organs and structures (see Figure 13–23). The **celiac** (SĒ-lē-ak) **trunk, superior mesenteric** (mez-en-TER-ik) **artery,** and **inferior mesenteric artery** arise on the anterior surface of the abdominal aorta and branch in the connective tissues of the mesenteries. These three vessels provide blood to all of the digestive organs in the abdominopelvic cavity. The celiac trunk divides into three branches that deliver blood to the liver, gallbladder, stomach, and spleen. The superior mesenteric artery supplies the pancreas, small intestine, and most of the large intestine. The inferior mesenteric delivers blood to the last portion of the large intestine and rectum.

Paired **gonadal** (gō-NAD-al) **arteries** originate between the superior and inferior mesenteric arteries. In males they are called *testicular arteries;* in females, *ovarian arteries.* The **suprarenal arteries** and **renal arteries** arise along the lateral surface of the abdominal aorta and travel behind the peritoneal lining to reach the adrenal glands and kidneys. Small **lumbar arteries** begin on the posterior surface of the aorta and supply the spinal cord and the abdominal wall.

Near the level of vertebra L4 the abdominal aorta divides to form a pair of muscular arteries. These **common iliac** (IL-ē-ak) **arteries** carry blood to the pelvis and lower limbs (see Figure 13–16). As it travels along the inner surface of the ilium, each common iliac divides to form an **internal iliac artery,** which supplies smaller arteries of the pelvis, and an **external iliac artery,** which enters the lower limb.

Once in the thigh, the external iliac artery branches, forming the **femoral artery** and the **deep femoral artery.** When it reaches the back of the knee, the femoral artery becomes the **popliteal artery,** which almost immediately branches to form the **anterior tibial, posterior tibial,** and **fibular arteries.** At the ankle, the anterior tibial artery becomes the *dorsalis pedis artery,* and the posterior tibial artery divides in two. These three arteries are connected by two anastomoses. The arrangement produces a *dorsal arch* on the top of the foot and a *plantar arch* on the bottom.

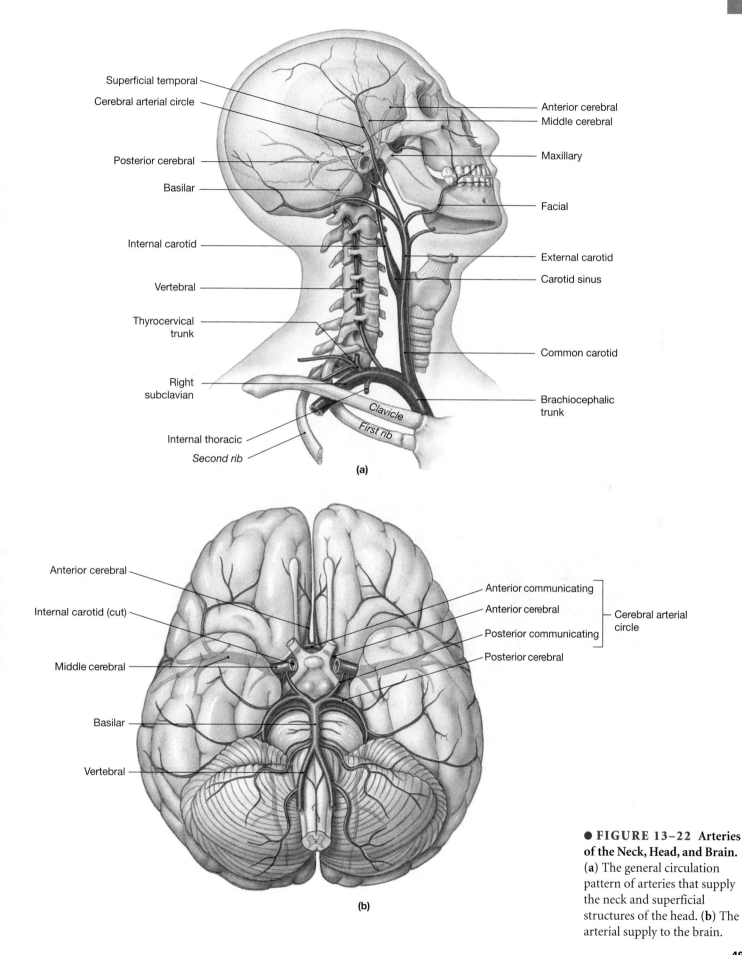

Superficial temporal

Cerebral arterial circle

Posterior cerebral

Basilar

Internal carotid

Vertebral

Thyrocervical trunk

Right subclavian

Internal thoracic

Second rib

Anterior cerebral

Middle cerebral

Maxillary

Facial

External carotid

Carotid sinus

Common carotid

Brachiocephalic trunk

Clavicle

First rib

(a)

Anterior cerebral

Internal carotid (cut)

Middle cerebral

Basilar

Vertebral

Anterior communicating

Anterior cerebral

Posterior communicating

Posterior cerebral

Cerebral arterial circle

(b)

● **FIGURE 13–22 Arteries of the Neck, Head, and Brain.** (**a**) The general circulation pattern of arteries that supply the neck and superficial structures of the head. (**b**) The arterial supply to the brain.

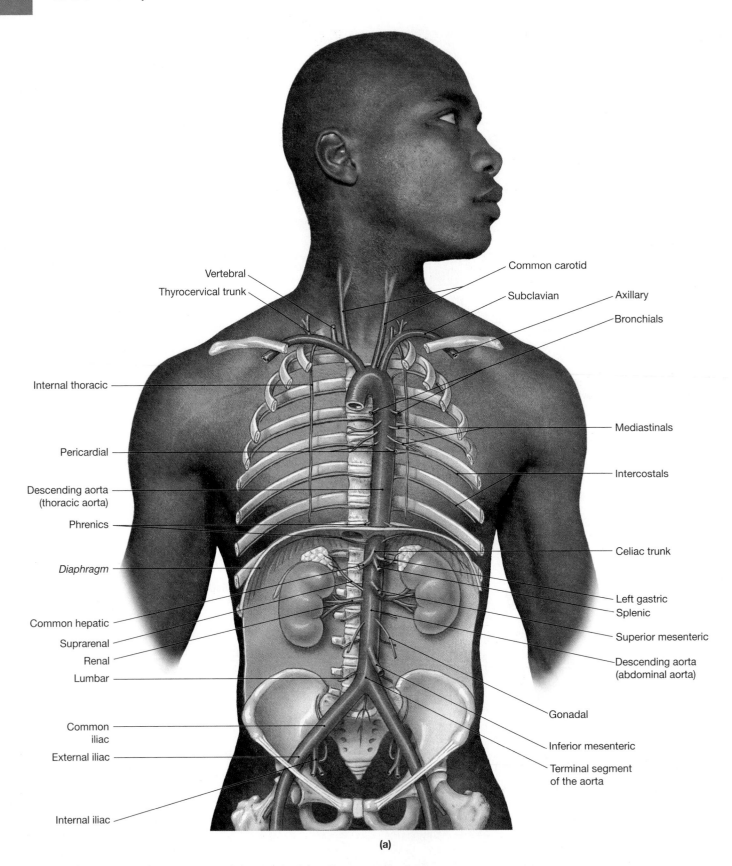

Vertebral

Thyrocervical trunk

Common carotid

Subclavian

Axillary

Bronchials

Internal thoracic

Mediastinals

Pericardial

Intercostals

Descending aorta
(thoracic aorta)

Phrenics

Celiac trunk

Diaphragm

Left gastric

Splenic

Common hepatic

Superior mesenteric

Suprarenal

Descending aorta
(abdominal aorta)

Renal

Lumbar

Gonadal

Common
iliac

Inferior mesenteric

External iliac

Terminal segment
of the aorta

Internal iliac

(a)

● **FIGURE 13–23 Major Arteries of the Trunk. (a)** A diagrammatic view.

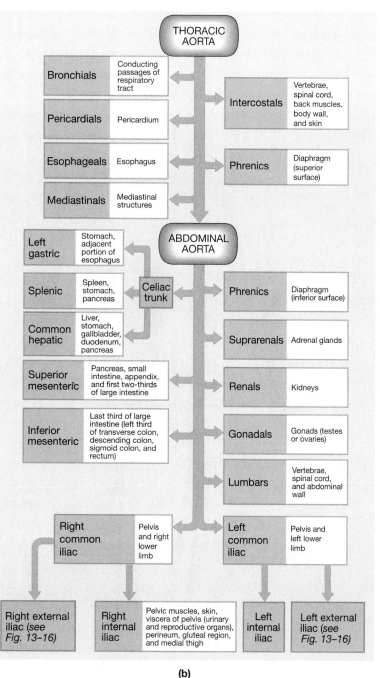

● FIGURE 13–23 (continued) (b) A flow chart.

Clinical Note
VASCULITIS

Vasculitis is an inflammation of a blood vessel. Most vasculitis stems from a variety of rheumatic diseases and syndromes. The inflammatory process is usually segmental, and inflammation within the tunica media of a muscular artery tends to destroy the internal elastic lamina. Necrosis and hypertrophy (enlarging) of the vessel occur, and the vessel wall has a high likelihood of breaching, leaking fibrin and red blood cells into the surrounding tissue. This potentially can lead to partial or total vascular occlusion and subsequent necrosis.

Raynaud's phenomenon is a type of vasculitis that is characterized by episodic ischemia of the fingers or toes as evidenced by digital blanching, cyanosis, and pain. Cold weather or emotional upset have been identified as triggers. Treatment of Raynaud's involves reassurance and having the patient dress warmly and wear gloves. In severe cases, medications that dilate the blood vessels (calcium-channel blockers) are used. ■

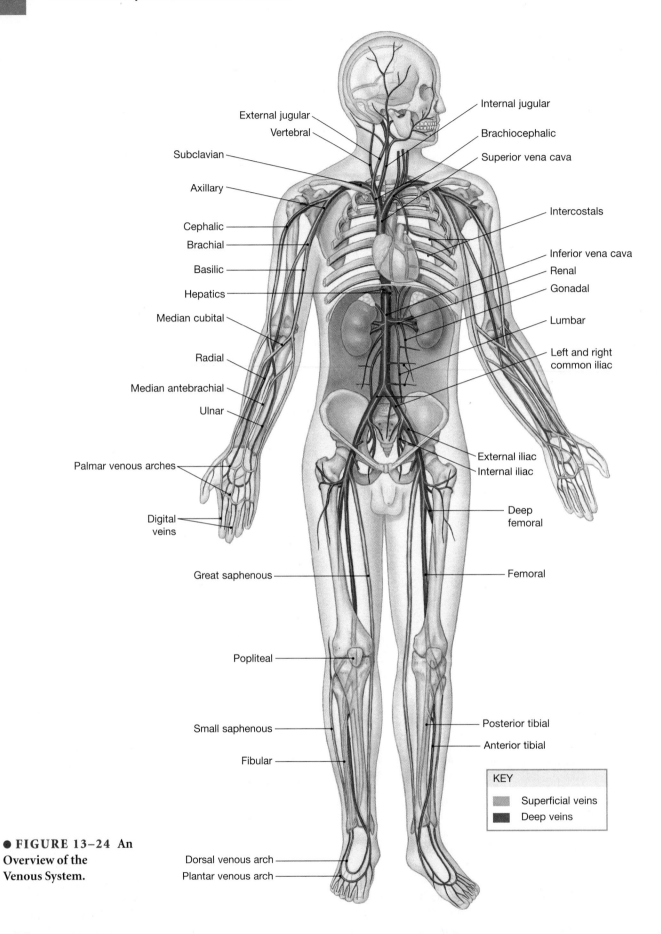

External jugular
Vertebral
Subclavian
Axillary
Cephalic
Brachial
Basilic
Hepatics
Median cubital
Radial
Median antebrachial
Ulnar
Palmar venous arches
Digital veins

Internal jugular
Brachiocephalic
Superior vena cava
Intercostals
Inferior vena cava
Renal
Gonadal
Lumbar
Left and right common iliac
External iliac
Internal iliac
Deep femoral
Femoral

Great saphenous
Popliteal
Small saphenous
Fibular
Posterior tibial
Anterior tibial

Dorsal venous arch
Plantar venous arch

KEY

Superficial veins
Deep veins

● **FIGURE 13–24** An Overview of the Venous System.

Clinical Note
ACUTE ARTERIAL OCCLUSION

An acute arterial occlusion is the sudden occlusion of arterial blood flow due to trauma, thrombosis, tumor, embolus, or idiopathic means. Emboli are probably the most common cause of acute arterial occlusion. They can arise from within a chamber of the heart (mural emboli), as from a thrombus in the left ventricle, from an atrial thrombus secondary to atrial fibrillation, or from a thrombus caused by abdominal aortic atherosclerosis. Arterial occlusions most commonly involve vessels in the abdomen or lower extremities.

Acute arterial occlusions are usually treated by an embolectomy, which is the surgical removal of the clot. In this procedure, a balloon catheter or similar device is inserted in the artery distal to the obstruction. The balloon is inflated, the catheter is withdrawn, and the embolus is removed from the vessel lumen. ∎

Systemic Veins

Blood from each of the tissues and organs of the body returns to the heart by means of a venous network that drains into the right atrium through the superior and inferior venae cavae.

Figure 13–24● illustrates the major vessels of the venous system. Complementary arteries and veins often run side by side, and in many cases they have comparable names. For example, the axillary arteries run alongside the axillary veins. In addition, arteries and veins often travel in the company of peripheral nerves that have the same names and innervate the same structures.

One significant difference between the arterial and venous systems concerns the distribution of major veins in the neck and limbs. Arteries in these areas are located deep beneath the skin, protected by bones and surrounding soft tissues. In contrast, the neck and limbs usually have two sets of peripheral veins, one superficial and the other deep. This dual venous drainage helps control body temperature. In a hot environment, venous blood flows in superficial veins, through which heat loss can occur; in a cold environment, blood is routed to the deep veins to minimize heat loss.

THE SUPERIOR VENA CAVA. The **superior vena cava** (SVC) receives blood from two regions: the head and neck (Figure 13–25●), and the upper limbs, shoulders, and chest (Figure 13–26●).

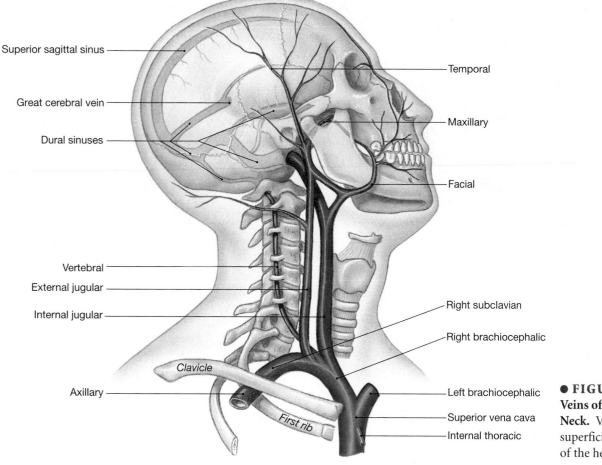

● **FIGURE 13–25 Major Veins of the Head and Neck.** Veins that drain superficial and deep portions of the head and neck.

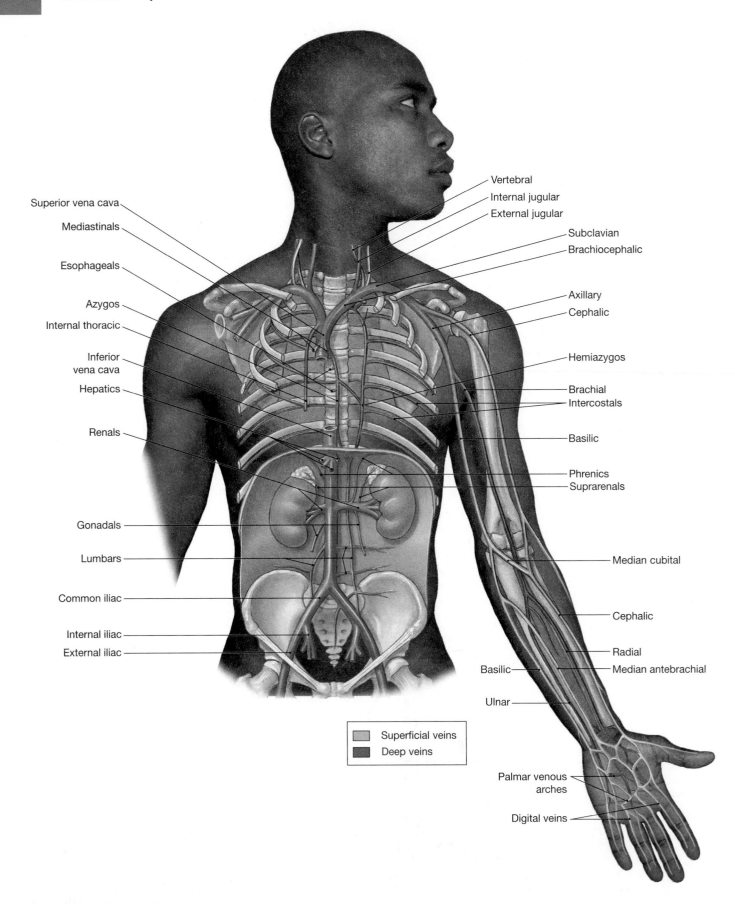

Superior vena cava

Mediastinals

Esophageals

Azygos

Internal thoracic

Inferior vena cava

Hepatics

Renals

Gonadals

Lumbars

Common iliac

Internal iliac

External iliac

Vertebral

Internal jugular

External jugular

Subclavian

Brachiocephalic

Axillary

Cephalic

Hemiazygos

Brachial

Intercostals

Basilic

Phrenics

Suprarenals

Median cubital

Cephalic

Radial

Median antebrachial

Basilic

Ulnar

Palmar venous arches

Digital veins

Superficial veins

Deep veins

● **FIGURE 13–26** **The Venous Drainage of the Abdomen and Chest.**

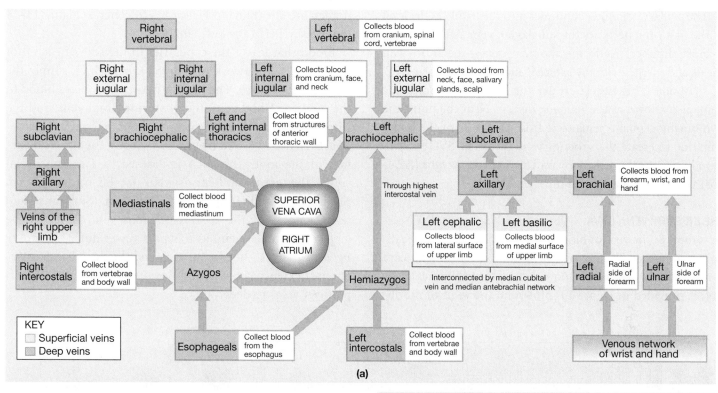

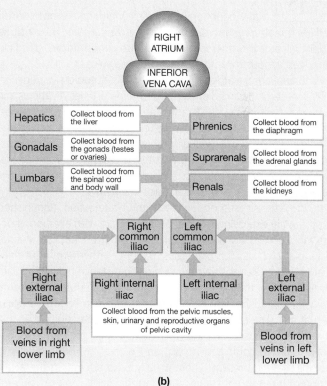

● **FIGURE 13–27** **A Flow Chart of the Circulation to the Superior and Inferior Venae Cavae. (a)** The tributaries of the SVC. **(b)** The tributaries of the IVC.

VENOUS RETURN FROM THE HEAD AND NECK. Small veins in the neural tissue of the brain empty into a network of thin-walled channels called the **dural sinuses.** ∞ p. 275 The largest, the **superior sagittal sinus,** is located within the fold of dura mater that lies between the cerebral hemispheres. Most of the blood that leaves the brain passes through one of the dural sinuses and leaves the skull in one of the **internal jugular veins,** which descend parallel to the common carotid artery in the neck. The more superficial **external jugular veins** collect blood from the superficial structures of the head and neck. These veins travel just beneath the skin, and a *jugular venous pulse (JVP)* can sometimes be detected at the base of the neck. **Vertebral veins** drain the cervical spinal cord and the posterior surface of the skull and descend within the transverse foramina of the cervical vertebrae alongside the vertebral arteries.

VENOUS RETURN FROM THE UPPER LIMBS AND CHEST. The major veins of the upper body are illustrated in Figure 13–26, and a flow chart that indicates the venous tributaries of the superior vena cava is provided in (Figure 13–27a●). A venous network in the palms collects blood from the digital veins. These vessels drain into the **cephalic vein** and the **basilic vein.** The superficial **median cubital vein** passes from the cephalic vein, medially and at an oblique angle, to connect to the basilic vein. (The median cubital is the vein from which

venous blood samples are typically collected.) The deeper veins of the forearm consist of a **radial vein** and an **ulnar vein.** After crossing the elbow, these veins fuse to form the **brachial vein.** As the brachial vein continues toward the trunk, it joins the basilic vein before entering the axilla as the **axillary vein.** The cephalic vein drains into the axillary vein at the shoulder.

The axillary vein then continues into the trunk; at the level of the first rib it becomes the **subclavian vein.** After traveling a short distance inside the thoracic cavity, the subclavian meets and merges with the external and internal jugular veins of that side. This fusion creates the large **brachiocephalic vein,** also known as the *innominate vein.* Near the heart, the two brachiocephalic veins (one from each side of the body) combine to create the superior vena cava. The SVC receives blood from the thoracic body wall through the **azygos** (AZ-i-gos) **vein** before it arrives at the right atrium.

THE INFERIOR VENA CAVA. The **inferior vena cava (IVC)** collects most of the venous blood from organs inferior to the diaphragm. (A small amount reaches the superior vena cava through the azygos vein.) A flow chart of the tributaries of the IVC is provided in Figure 13–27b●, and the veins of the abdomen are illustrated in Figure 13–26. Refer to Figure 13–24 to see the veins of the lower limbs.

Blood that leaves the capillaries in the sole of each foot collects into a network of *plantar veins*, which supply the *plantar venous arch*. The plantar network provides blood to the **anterior tibial vein,** the **posterior tibial vein,** and the **fibular vein,** which are the deep veins of the leg. A *dorsal venous arch* drains blood from capillaries on the superior surface of the foot. This arch is drained by two superficial veins, the **great saphenous vein** (sa-FĒ-nus; *saphenes,* prominent) and the **small saphenous vein.** (Surgeons often use segments of the great saphenous vein, which is the largest superficial vein, as a bypass vessel during *coronary bypass surgery.*) The plantar arch and the dorsal arch interconnect extensively, so blood flow can easily shift from superficial veins to deep veins.

Clinical Note
PERIPHERAL VASCULAR CONDITIONS

Many peripheral vascular conditions are not considered life threatening but may require prehospital care. They include peripheral arterial atherosclerotic disease, intermittent claudication, deep venous thrombosis, and varicose veins.

Peripheral arterial atherosclerotic disease is a progressive degenerative disease of the medium-sized and large arteries. It affects the aorta and its branches, the brachial and femoral peripheral arteries, and the cerebral arteries. For reasons unknown, it does not affect the coronary arteries. It is a gradual, progressive disease that is often associated with diabetes mellitus. In extreme cases, significant arterial insufficiency may lead to ulcers and gangrene. Occlusion of the peripheral arteries causes chronic and acute ischemia.

In the chronic setting, intermittent claudication (diminished blood flow in exercising muscle) produces pain with exertion. It occurs most commonly with the calf, but can affect any leg muscle. Rest initially relieves this pain. Most patients with intermittent claudication report being able to walk a given distance before leg pain develops. After they rest and the pain resolves, they can resume walking, but as the disease progresses, the pain begins to occur even at rest. The extremity usually appears normal, but pulses will be reduced or absent. As the ischemia worsens, the extremity becomes painful, cold, and numb, and ulceration, gangrene, and necrosis may be present. There usually is no edema. In the acute setting of intermittent claudication, arterial occlusion from an embolus, aneurysm, or thrombosis occurs. The patient experiences a sudden onset of pain, coldness, numbness, and pallor. Pulses are absent distal to the occlusion. Acute occlusion may cause severe ischemia with motor and sensory deficits. Edema is not present.

Deep venous thrombosis is a blood clot in a vein. It most commonly occurs in the larger veins of the thigh and calf, although the subclavian vein is sometimes involved. Predisposing factors include a recent history of trauma, inactivity, pregnancy, or varicose veins. The patient frequently complains of gradually increasing pain and calf tenderness. Often the leg and foot are swollen because of occluded venous drainage. Leg elevation may alleviate the signs and symptoms. In some cases, the patient may be asymptomatic. Gentle palpation of the calf and thigh may reveal tenderness and, on occasion, cord-like clotted veins. Dorsiflexion of the foot may cause *Homan's sign,* which is discomfort behind the knee. This is associated with deep venous thrombosis. The skin may be warm and red. Deep venous thrombosis has traditionally been treated with a blood thinner (heparin) and hospitalization with the affected leg elevated. Heparin must be administered by continuous intravenous infusion, and the patient's coagulation profile must be checked frequently to ensure proper dosing. In many cases, patients are hospitalized at complete bed rest for 7 to 10 days. With the development of low-molecular-weight heparin (Lovenox), which can be administered by periodic subcutaneous injections, patients with deep venous thrombosis can be treated as outpatients. Patients are instructed to self-administer the drug at home and keep the affected extremity elevated.

Varicose veins are dilated superficial veins, usually in the lower extremities. Predisposing factors include pregnancy, obesity, and genetics. Signs and symptoms include the visible distention of the leg veins, lower leg swelling and discomfort (especially at the end of the day), and skin color and texture changes in the legs and ankles. If the condition is chronic, venous stasis ulcers can develop. Although venous stasis ulcers can rupture, direct pressure usually can control the bleeding, which occasionally is significant. Venous stasis ulcers are usually slow healing and require proper care to prevent additional complications. Treatment of varicose veins primarily involves use of support stockings. Laser therapy and injection of the veins with a sclerosing agent will help improve the legs' cosmetic appearance. ■

Behind the knee, the small saphenous, tibial, and fibular veins unite to form the **popliteal vein.** When the popliteal vein reaches the femur, it becomes the **femoral vein.** Before penetrating the abdominal wall, the great saphenous and **deep femoral** veins join the femoral vein. The femoral vein penetrates the body wall and emerges into the pelvic cavity as the **external iliac vein.** As the external iliac travels across the inner surface of the ilium, it is joined by the **internal iliac vein,** which drains the pelvic organs. The resulting **common iliac vein** then meets its counterpart from the opposite side to form the IVC.

Like the aorta, the IVC lies posterior to the abdominopelvic cavity. As it ascends to the heart, it collects blood from several lumbar veins. In addition, the IVC receives blood from the *gonadal, renal, suprarenal, phrenic,* and *hepatic veins* before reaching the right atrium (see Figure 13–26).

THE HEPATIC PORTAL SYSTEM. You may have noticed that the list of veins did not include any names that refer to digestive organs other than the liver. Instead of traveling directly to the inferior vena cava, blood leaving the capillaries supplied by the celiac, superior, and inferior mesenteric arteries flows to the liver through the **hepatic portal system** (*porta,* a gate). A blood vessel that connects two capillary beds is called a *portal vessel,* and the network formed is called a *portal system.* ∞ p. 374

Blood in the hepatic portal vessels is quite different in composition from that in other systemic veins because it contains substances absorbed by the digestive tract, including high concentrations of glucose and amino acids, various wastes, and an occasional toxin. Because a portal system carries blood from one capillary bed to another, the blood within it does not immediately mix with blood in the general circulation. Instead, the hepatic portal system delivers blood that contains these compounds to the liver, where liver cells absorb them for storage, metabolic conversion, or excretion. In the process, the liver regulates the concentrations of nutrients in the circulating blood.

Figure 13–28● shows the anatomy of the hepatic portal system. The system begins in the capillaries of the digestive organs. Blood from capillaries along the lower portion of the large intestine enters the **inferior mesenteric vein.** On their way toward the liver, veins from the spleen, the lateral border of the stomach,

● **FIGURE 13–28**
The Hepatic Portal System.

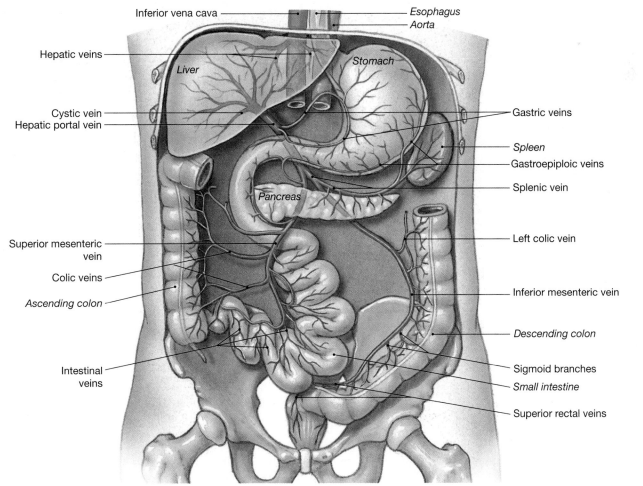

and the pancreas fuse with the inferior mesenteric, and form the **splenic vein.** The **superior mesenteric vein** also drains the lateral border of the stomach, through an anastomosis with one of the branches of the splenic vein. In addition, the superior mesenteric collects blood from the entire small intestine and two-thirds of the large intestine. The **hepatic portal vein** forms through the fusion of the superior mesenteric and splenic veins. Of the two, the superior mesenteric normally contributes the greater volume of blood and most of the nutrients. As it proceeds, the hepatic portal vein receives blood from the **gastric veins,** which drain the medial border of the stomach, and the **cystic vein** from the gallbladder. The hepatic portal system ends where the hepatic portal vein empties into the liver capillaries.

After passing through the liver capillaries, blood collects in the hepatic veins, which empty into the inferior vena cava. Because blood goes to the liver before returning to the heart, the composition of the blood in the systemic circulation remains relatively stable, regardless of the digestive activities under way.

→ CONCEPT CHECK QUESTIONS

1. Blockage of which branch of the aortic arch would interfere with blood flow to the left arm?
2. Why would compression of one of the common carotid arteries cause a person to lose consciousness?
3. Grace ruptures her celiac trunk in an automobile accident. Which organs would be affected most directly by this injury?
4. Describe the general distribution of major arteries and veins in the neck and limbs. What functional advantage does this provide?

Answers begin on p. 792.

Fetal Circulation

The fetal and adult circulatory systems have significant differences that reflect different sources of respiratory and nutritional support. The embryonic lungs are collapsed and nonfunctional, and the embryonic digestive tract has nothing to digest. All of the embryo's nutritional and respiratory needs are provided by diffusion across the *placenta,* which is a structure within the uterine wall where the maternal and fetal circulatory systems are in close contact. Circulation in a full-term (nine-month-old) fetus is diagrammed in Figure 13–29a●.

Placental Blood Supply

The fetus's blood reaches the placenta through a pair of **umbilical arteries,** which arise from the internal iliac arteries before entering the umbilical cord. At the placenta, the blood gives up CO_2 and wastes and picks up oxygen and nutrients. Blood that returns from the placenta flows through a single **umbilical vein** before reaching the developing liver. Some of the blood flows through capillary networks within the liver; the rest bypasses the liver capillaries and reaches the inferior vena cava within the **ductus venosus.** When the placental connection is broken at birth, blood flow through the umbilical vessels ceases, and they soon degenerate.

Circulation in the Heart and Great Vessels

One of the most interesting aspects of circulatory development reflects the differences between the life of an embryo or fetus and that of an infant. Throughout embryonic and fetal life, the lungs are collapsed; yet after delivery, the newborn infant must be able to extract oxygen from inspired air rather than across the placenta.

Although the interatrial and interventricular septa of the heart develop early in fetal life, the interatrial partition remains functionally incomplete until birth. The interatrial opening, or **foramen ovale,** is associated with an elongate flap that acts as a valve. Blood can flow freely from the right atrium to the left atrium, but any backflow will close the valve and isolate the two chambers. Thus, blood can enter the heart at the right atrium and bypass the pulmonary circuit. A second short-circuit exists between the pulmonary and aortic trunks. This connection, the **ductus arteriosus,** consists of a short, muscular vessel.

With the lungs collapsed, the capillaries are compressed and little blood flows through the lungs. During diastole, blood enters the right atrium and flows into the right ventricle, but it also passes into the left atrium through the foramen ovale. About 25 percent of the blood that arrives at the right atrium bypasses the pulmonary circuit in this way. In addition, over 90 percent of the blood that leaves the right ventricle passes through the ductus arteriosus and enters the systemic circuit rather than continuing to the lungs.

Circulatory Changes at Birth

At birth, dramatic changes occur. When an infant takes its first breath, the lungs expand, and so do the pulmonary vessels. Within a few seconds, the smooth muscles in the ductus arteriosus contract, isolating the pulmonary and aortic trunks, and blood begins flowing through the pulmonary circuit. As pressures rise in the left atrium, the valvular flap closes the foramen ovale (Figure 13–29b●). In adults, the interatrial septum bears a shallow depression, the *fossa ovalis,* that marks the site of the foramen ovale (see Figure 12–5, p. 442). The remnants of the ductus arteriosus persist as a fibrous cord, which is the *ligamentum arteriosum.*

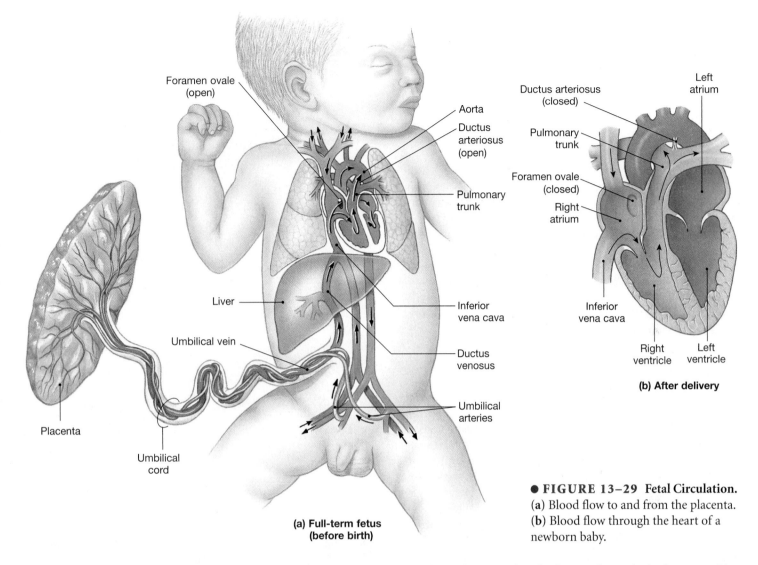

● FIGURE 13–29 Fetal Circulation.
(**a**) Blood flow to and from the placenta.
(**b**) Blood flow through the heart of a newborn baby.

If the proper vascular changes do not occur at birth or shortly thereafter, problems will eventually develop because the heart will have to work too hard to provide adequate amounts of oxygen to the systemic circuit. Treatment may involve surgical closure of the foramen ovale, the ductus arteriosus, or both. Other forms of congenital heart defects result from abnormal cardiac development or inappropriate connections between the heart and major arteries and veins.

■ Aging and the Cardiovascular System

The capabilities of the cardiovascular system gradually decline with age. Major changes affect all parts of the cardiovascular system: blood, heart, and vessels.

In the blood, age-related changes may include (1) decreased hematocrit; (2) the constriction or blockage of peripheral veins by the formation of a *thrombus* (stationary blood clot),

which can become detached, pass through the heart, and become wedged in a small artery (most often in the lungs, causing *pulmonary embolism*); and (3) the pooling of blood in the veins of the legs because valves are not working effectively.

In the heart, age-related changes include (1) a reduction in maximum cardiac output; (2) changes in the activities of the nodal and conducting cells; (3) a reduction in the elasticity of the fibrous skeleton; (4) progressive atherosclerosis, which can restrict coronary circulation; and (5) the replacement of damaged cardiac muscle cells by scar tissue.

In blood vessels, age-related changes are often related to arteriosclerosis, which is a thickening and toughening of arterial walls. For example, (1) the inelastic walls of arteries become less tolerant of sudden pressure increases, which can lead to a localized dilation, or *aneurysm*, whose rupture may cause a stroke, myocardial infarction, or massive blood loss (depending on the vessel); (2) calcium salts can be deposited on weakened vascular walls, increasing the risk of a stroke or myocardial infarction; and (3) thrombi can form at atherosclerotic plaques.

Clinical Note
VASCULAR TRAUMA

Injury to the vascular system can interrupt blood flow to the part of the body that is supplied by the artery. Also, some fractures and dislocations can impinge upon nearby arteries, obstructing blood flow; the blood flow often returns when the affected extremity is returned to its anatomical position. Any vascular structure is subject to injury. Laceration of a moderate-sized artery or vein can cause life-threatening hemorrhage. If the vessel is cut cleanly, however, as with a knife, the muscles in the wall of the vessel contract, which constricts the vessel lumen and retracts the end of the severed vessel into the nearby soft tissue. As the muscle is drawn back from the wound, it thickens and further restricts the vessel lumen, which thus restricts blood flow. This, in turn, reduces the rate of blood loss and assists the body's clotting mechanisms. Clean lacerations and amputations therefore do not bleed profusely, but if the vessel is not cleanly lacerated, muscle contraction may actually open the wound, increasing blood flow and loss (Figure 13–30●).

Traumatic Aneurysm or Rupture of the Aorta

Aortic aneurysm and rupture are extremely life-threatening injuries that result from either blunt or penetrating trauma. The aorta is most commonly injured by blunt trauma and carries an overall mortality of 85 to 95 percent. It is responsible for 15 percent of all thoracic trauma deaths. Aneurysm and rupture are usually associated with high-speed automobile crashes (most commonly lateral impact) and, in some cases, with high falls. Unlike myocardial rupture, a significant proportion of these victims, possibly as high as 20 percent, will survive the initial insult and aneurysm. Some 30 percent of those initial survivors will die in six hours if not treated; this increases to about 50 percent at 24 hours, and to just under 70 percent by the end of the first week. It is this subset of patients who survive the initial impact and are alive at the scene whom you can most benefit by recognizing their potential injury and then rapidly extricating, packaging, and transporting them to a trauma center.

The aorta is a large, high-pressure vessel that provides outflow from the left ventricle for distribution to the body. It is relatively fixed at three points as it passes through the thoracic cavity: the aortic annulus, where the aorta joins the heart; the aortic isthmus, where it is joined by the ligamentum arteriosum; and the diaphragm, where it exits the chest. Because of this, the aorta experiences shear forces secondary to severe deceleration of the chest. Traumatic dissecting aneurysm occurs most commonly to the descending aorta and, infrequently, to the ascending aorta. With severe deceleration, shear forces separate the layers of the artery, specifically the interior surface (the *tunica intima)* from the muscle layer (the *tunica media)*. This allows blood to enter the area between the layers, and because the blood is under great pressure, it begins to dissect the aortic lining like a bulging inner tube (Figure 13–31●). The aorta is likely to rupture if it is not surgically repaired.

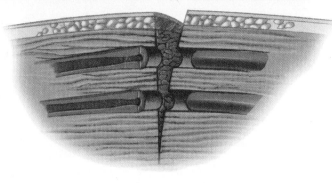

a. A clean lateral cut permits the vessel to retract and thicken its wall.

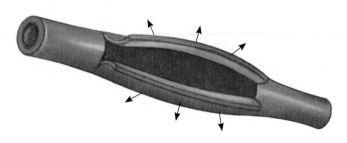

b. A longitudinal cut to the vessel causes the wound to open.

●**FIGURE 13–30 Arterial Laceration. (a)** Clean, transverse lacerations cause spasm and contraction of the vessel which slows bleeding. **(b)** In longitudinal lacerations, the muscle fibers can actually hold the wound open, which worsens blood loss.

The patient with aortic rupture will be severely hypotensive, quickly lose all vital signs, and die unless taken immediately to surgery. Dissecting aortic aneurysm progresses more slowly, though the aneurysm may rupture at any moment. The patient will probably have a history of a high fall or severe auto impact and deceleration. Lateral impact is an especially high risk factor for aortic aneurysm. The patient may complain of severe tearing chest pain that may radiate to the back. The patient may have a pulse deficit between the left and right upper extremities and/or reduced pulse strength in the lower extremities. Blood pressure may be high (hypertension) due to stretching of sympathetic nerve fibers in the aorta near the ligamentum arteriosum, or the pressure may be low due to leakage and hypovolemia. Turbulence as the blood exits the heart and passes the disrupted blood vessel wall may create a harsh systolic murmur.

Other Vascular Injuries

Other thoracic vascular structures that can sustain injury during chest trauma are the pulmonary arteries and vena cava. Their in-

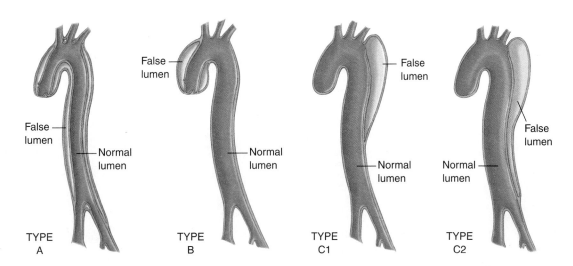

● **FIGURE 13–31 Different Types of Thoracic Aortic Aneurysm.** **(a)** The dissecting aneurysm begins near the aortic valve in the ascending aorta and extends throughout the aorta down to the external iliac arteries. **(b).** Dissecting aneurysm limited to the ascending aorta (as seen in Marfan's syndrome). **(c1).** Dissecting thoracic aortic aneurysm that originates distal to the left subclavian artery. The localized nature of this lesion makes it readily accessible for surgical excision.**(c2)** Dissecting aneurysm that arises distal to the left subclavian artery but extends into the abdominal aorta. Only partial excision is possible.

jury, and the resulting hemorrhage, may cause significant hemothorax, and possibly lead to hypotension and respiratory insufficiency. The blood may also flow into the mediastinum and compress the great vessels, esophagus, and heart. Penetrating trauma is the primary cause of injury to the pulmonary arteries and vena cava. The patient with pulmonary artery or vena cava injuries will likely have a penetrating wound to the central chest, or elsewhere with a likelihood of central chest involvement. These injuries initially present with the signs and symptoms of hypovolemia and shock and then present with the signs and symptoms of hemothorax or hemomediastinum as those conditions develop. ■

■ Integration with Other Systems

The cardiovascular system is both structurally and functionally linked to all other systems. Figure 13–32● summarizes the physiological relationships between the cardiovascular system and other organ systems. The most extensive communication occurs between the cardiovascular system and the lymphatic system. Not only are the two systems physically interconnected, but also cells of the lymphatic system use the cardiovascular system as a highway to move from one part of the body to another. Chapter 14 examines the lymphatic system in detail.

→ CONCEPT CHECK QUESTIONS

1. A blood sample taken from the umbilical cord has high concentrations of oxygen and nutrients and low concentrations of carbon dioxide and waste products. Is this sample from an umbilical artery or from the umbilical vein?
2. The cardiovascular system is most closely interconnected with what other organ system?

Answers begin on p. 792.

The Cardiovascular System in Perspective

For All Systems

Delivers oxygen, hormones, nutrients, and WBCs; removes carbon dioxide and metabolic wastes; transfers heat

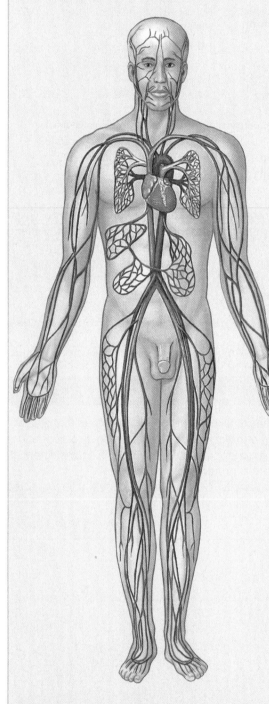

The Integumentary System

- Stimulation of mast cells produces localized changes in blood flow and capillary permeability
- Delivers immune system cells to injury sites; clotting response seals breaks in skin surface; carries away toxins from sites of infection; provides heat

The Skeletal System

- Provides calcium needed for normal cardiac muscle contraction; protects blood cells developing in bone marrow
- Provides calcium and phosphate for bone deposition; delivers EPO to bone marrow, parathyroid hormone, and calcitonin to osteoblasts and osteoclasts

The Muscular System

- Skeletal muscle contractions assist in moving blood through veins; protects superficial blood vessels, especially in neck and limbs
- Delivers oxygen and nutrients, removes carbon dioxide, lactic acid, and heat during skeletal muscle activity

The Nervous System

- Controls patterns of circulation in peripheral tissues; modifies heart rate and regulates blood pressure; releases ADH
- Endothelial cells maintain blood-brain barrier; help generate CSF

The Endocrine System

- Erythropoietin regulates production of RBCs; several hormones elevate blood pressure; epinephrine stimulates cardiac muscle, elevating heart rate and contractile force
- Distributes hormones throughout the body; heart secretes ANP

The Lymphatic System

- Defends against pathogens or toxins in blood; fights infections of cardiovascular organs; returns tissue fluid to circulation
- Distributes WBCs; carries antibodies that attack pathogens; clotting response assists in restricting spread of pathogens; granulocytes and lymphocytes produced in bone marrow

The Respiratory System

- Provides oxygen to cardiovascular organs and removes carbon dioxide
- RBCs transport oxygen and carbon dioxide between lungs and peripheral tissues

The Digestive System

- Provides nutrients to cardiovascular organs; absorbs water and ions essential to maintenance of normal blood volume
- Distributes digestive tract hormones; carries nutrients, water, and ions away from sites of absorption; delivers nutrients and toxins to liver

The Urinary System

- Releases renin to elevate blood pressure and erythropoietin to accelerate red blood cell production
- Delivers blood to capillaries, where filtration occurs; accepts fluids and solutes reabsorbed during urine production

The Reproductive System

- Sex hormones maintain healthy vessels, estrogen slows development of atherosclerosis
- Distributes reproductive hormones; provides nutrients, oxygen, and waste removal for developing fetus; local blood pressure changes responsible for physical changes during sexual arousal

● **FIGURE 13–32 Functional Relationships Between the Cardiovascular System and Other Systems.**

Chapter Review

Access more review material online at *www.prenhall.com/bledsoe*. There you will find quiz questions, labeling activities, animations, essay questions, and web links.

Key Terms

anastomosis 471
arteriole 467
artery 467
blood pressure 473
capillary 467
hepatic portal system 499

peripheral resistance 473
pulmonary circuit 484
pulse pressure 474
respiratory pump 474
systemic circuit 484

valve 472
vasoconstriction 468
vasodilation 468
vein 472
venule 467

Related Clinical Terms

aneurysm (AN-ū-rizm) A bulge in the weakened wall of a blood vessel, generally an artery.
arteriosclerosis (ar-tē-rē-ō-skle-RŌ-sis) A condition characterized by the thickening and toughening of arterial walls.
atherosclerosis (ath-er-ō-skle-RŌ-sis) A type of arteriosclerosis characterized by changes in the endothelial lining and the formation of a plaque.
edema (e-DĒ-muh) An abnormal accumulation of fluid in peripheral tissues.
hypertension Abnormally high blood pressure; usually defined in adults as blood pressure higher than 140/90.
hypervolemic (hī-per-vō-LĒ-mik) Having an excessive blood volume.

hypotension Blood pressure so low that circulation to vital organs may be impaired.
hypovolemic (hī-pō-vō-LE-mik) Having a low blood volume.
orthostatic hypotension Low blood pressure upon standing, often accompanied by dizziness or fainting; results from a failure of the regulatory mechanisms that increase blood pressure to maintain adequate blood flow to the brain.
phlebitis Inflammation of a vein.
pulmonary embolism Circulatory blockage caused by the trapping of an embolus (often a detached thrombus) in a pulmonary artery.

shock An acute circulatory crisis marked by hypotension and inadequate peripheral blood flow.
sphygmomanometer A device that measures blood pressure using an inflatable cuff placed around a limb.
thrombus A stationary blood clot within a blood vessel.
varicose (VAR-i-kōs) **veins** Sagging, swollen veins distorted by gravity and by the failure of the venous valves.

Summary Outline

1. Blood flows through a network of arteries, veins, and capillaries. The vital functions of the cardiovascular system depend on events at the capillary level: all chemical and gaseous exchange between the blood and interstitial fluid takes place across **capillary** walls.

THE STRUCTURE OF BLOOD VESSELS 468

1. Arteries and veins form an internal distribution system, propelled by the heart. **Arteries** branch repeatedly and decrease in size until they become **arterioles;** from the arterioles, blood enters the capillary networks. Blood that flows from the capillaries enters small **venules** before entering larger **veins.**

2. The walls of arteries and veins contain three layers: the **tunica interna, tunica media,** and outermost **tunica externa.** *(Figure 13–1)*

Arteries 468

3. The walls of arteries are usually thicker than the walls of veins. The arterial system includes the large **elastic arteries,** medium-sized **muscular arteries,** and smaller arterioles. As blood proceeds toward the capillaries, the number of vessels increases, but the diameter of the individual vessels decreases and the walls become thinner. *(Figure 13–2)*

Capillaries 470

4. Capillaries are the only blood vessels whose walls permit exchange between blood and interstitial fluid.

5. Capillaries form interconnected networks called **capillary beds.** A **precapillary sphincter** (a band of smooth muscle) adjusts blood flow into each capillary. Blood flow in a capillary changes as **vasomotion** occurs. *(Figure 13–4)*

Veins 472

6. Venules collect blood from capillaries and merge into **medium-sized veins** and then **large veins.** The arterial system is a high-pressure system; pressure in veins is much lower. **Valves** in these vessels prevent backflow of blood. *(Figure 13–5)*

CIRCULATORY PHYSIOLOGY 472

Factors That Affect Blood Flow 472

1. Flow is proportional to the difference in pressure; blood will flow from an area of higher pressure to one of relatively lower pressure.

2. For circulation to occur, the *circulatory pressure* must be greater than the *total peripheral resistance* (the resistance of the entire circulatory system). For blood to flow into peripheral capillaries, **blood pressure** (arterial pressure) must be greater than the **peripheral resistance** (the resistance of the arterial system). Neural and hormonal control mechanisms regulate blood pressure.

3. The most important determinant of peripheral resistance is the diameter of arterioles.

Cardiovascular Pressures Within the Systemic Circuit 473

4. The high arterial pressures overcome peripheral resistance and maintain blood flow through peripheral tissues. **Capillary pressures** are normally low; small changes in capillary pressure determine the rate of fluid movement into or out of the bloodstream. Venous pressure, which is normally low, determines venous return and affects cardiac output and peripheral blood flow.

5. Arterial pressure rises in ventricular systole and falls in ventricular diastole. The difference between the **systolic** and **diastolic pressures** is **pulse pressure.** *(Figures 13–6, 13–8)*

6. At the capillaries, solute molecules diffuse across the capillary lining, and water-soluble materials diffuse through small spaces between endothelial cells. Water will move when driven by either hydrostatic pressure or osmotic pressure. The direction of water movement is determined by the balance between these two opposing pressures. *(Figure 13–7)*

7. Valves, **muscular compression,** and the **respiratory pump** help the relatively low venous pressures propel blood toward the heart. *(Figure 13–5)*

Key Note 476

CARDIOVASCULAR REGULATION 476

1. Homeostatic mechanisms ensure that tissue blood flow (*tissue perfusion*) delivers adequate oxygen and nutrients.

2. Blood flow varies with cardiac output, peripheral resistance, and blood pressure.

3. Autoregulation, neural mechanisms, and endocrine mechanisms influence the coordinated regulation of cardiovascular function. Autoregulation involves local factors that change the pattern of blood flow within capillary beds in response to chemical changes in interstitial fluids. Central nervous system mechanisms respond to changes in arterial pressure or blood gas levels. Hormones can assist in short-term adjustments (changes in cardiac output and peripheral resistance) and long-term adjustments (changes in blood volume that affect cardiac output and gas transport). *(Figure 13–9)*

Autoregulation of Blood Flow 478

4. Peripheral resistance is adjusted at the tissues by the dilation or constriction of precapillary sphincters.

Neural Control of Blood Pressure and Blood Flow 479

5. **Baroreceptor reflexes** are autonomic reflexes that adjust cardiac output and peripheral resistance to maintain normal arterial pressures. Baroreceptor populations include the **aortic** and **carotid sinuses** and *atrial baroreceptors.* *(Figure 13–10)*

6. **Chemoreceptor reflexes** respond to changes in the oxygen or carbon dioxide levels in the blood and cerebrospinal fluid. Sympathetic activation leads to stimulation of the *cardioacceleratory* and *vasomotor centers*; parasympathetic activation stimulates the *cardioinhibitory center.* *(Figure 13–11)*

Hormones and Cardiovascular Regulation 481

7. The endocrine system provides short-term regulation of cardiac output and peripheral resistance with epinephrine and norepinephrine from the adrenal medullae. Hormones involved in long-term regulation of blood pressure and volume are antidiuretic hormone (ADH), angiotensin II, erythropoietin (EPO), and atrial natriuretic peptide (ANP). *(Figure 13–12)*

8. ADH and angiotensin II also promote peripheral vasoconstriction in addition to their other functions. ADH and aldosterone promote water and electrolyte retention and stimulate thirst. EPO stimulates red blood cell production. ANP encourages sodium loss and fluid loss, reduces blood pressure, inhibits thirst, and lowers peripheral resistance.

Key Note 482

PATTERNS OF CARDIOVASCULAR RESPONSE 482

Exercise and the Cardiovascular System 482

1. During exercise, blood flow to skeletal muscles increases at the expense of circulation to nonessential organs, and cardiac output rises. Cardiovascular performance improves with training. Athletes have larger stroke volumes, slower resting heart rates, and greater cardiac reserves than do nonathletes.

The Cardiovascular Response to Hemorrhage 483

2. Blood loss causes an increase in cardiac output, mobilization of venous reserves, peripheral vasoconstriction, and the liberation of hormones that promote fluid retention and the manufacture of red blood cells.

THE BLOOD VESSELS 484

1. The peripheral distributions of arteries and veins are usually identical on both sides of the body except near the heart.

The Pulmonary Circulation 484

2. The **pulmonary circuit** includes the **pulmonary trunk,** the **left** and **right pulmonary arteries,** and the **pulmonary veins,** which empty into the left atrium. *(Figures 13–13, 13–14, and 13–15)*

The Systemic Circulation 487

3. In the **systemic circuit,** the **ascending aorta** gives rise to the coronary circulation. The **aortic arch** communicates with the **descending aorta.** *(Figures 13–16 through 13–23)*

4. Arteries in the neck and limbs are deep beneath the skin; in contrast, there are usually two sets of peripheral veins in those sites, one superficial and one deep. This dual-venous drainage is important for controlling body temperature.

5. The **superior vena cava (SVC)** receives blood from the head, neck, chest, shoulders, and arms. *(Figures 13–24 through 13–27)*

6. The **inferior vena cava (IVC)** collects most of the venous blood from organs inferior to the diaphragm. *(Figure 13–27)*

7. The **hepatic portal system** directs blood from the other digestive organs to the liver before the blood returns to the heart. *(Figure 13–28)*

Fetal Circulation 500

8. The placenta receives blood from the two **umbilical arteries.** Blood returns to the fetus through the umbilical vein, which delivers blood to the **ductus venosus** in the liver. *(Figure 13–29a)*

9. Prior to delivery, blood bypasses the pulmonary circuit by flowing (1) from the right atrium into the left atrium through the **foramen ovale,** and (2) from the pulmonary trunk into the aortic arch via the **ductus arteriosus.** *(Figure 13–29b)*

AGING AND THE CARDIOVASCULAR SYSTEM 501

1. Age-related changes in the blood can include (1) decreased hematocrit, (2) constriction or blockage of peripheral veins by a *thrombus* (stationary blood clot), and (3) the pooling of blood in the veins of the legs because the valves are not working effectively. *(Figures 13–30 and 13–31)*

2. Age-related changes in the heart include (1) a reduction in maximum cardiac output, (2) changes in the activities of the nodal and conducting cells, (3) a reduction in the elasticity of the fibrous skeleton, (4) progressive **atherosclerosis** that can restrict coronary circulation, and (5) replacement of damaged cardiac muscle cells by scar tissue.

3. Age-related changes in blood vessels, often related to **arteriosclerosis,** include (1) reduced tolerance of inelastic walls of arteries to sudden pressure increases, which can lead to an aneurysm; (2) the deposition of calcium salts on weakened vascular walls, which increases the risk of stroke or infarction; and (3) the formation of thrombi at atherosclerotic plaques.

INTEGRATION WITH OTHER SYSTEMS 503

1. The cardiovascular system delivers oxygen, nutrients, and hormones to all the body systems. *(Figure 13–32)*

Review Questions

Level 1: Reviewing Facts and Terms

Match each item in column A with the most closely related item in column B. Place letters for answers in the spaces provided.

COLUMN A
____ 1. diastolic pressure
____ 2. arterioles
____ 3. hepatic vein
____ 4. renal vein
____ 5. aorta
____ 6. precapillary sphincter
____ 7. medulla oblongata
____ 8. internal iliac artery
____ 9. external iliac artery
____ 10. baroreceptors
____ 11. systolic pressure
____ 12. saphenous vein

COLUMN B
a. drains the liver
b. largest superficial vein in body
c. carotid sinus
d. minimum blood pressure
e. blood supply to leg
f. blood supply to pelvis
g. peak blood pressure
h. vasomotion
i. largest artery in body
j. drains the kidney
k. vasomotor center
l. smallest arterial vessels

13. Blood vessels that carry blood away from the heart are called:
   (a) veins.
   (b) arterioles.
   (c) venules.
   (d) arteries.

14. The layer of the arteriole wall that provides the properties of contractility and elasticity is the tunica:
   (a) adventitia.
   (b) media.
   (c) interna.
   (d) externa.

15. The two-way exchange of substances between blood and body cells occurs only through:
    (a) arterioles.
    (b) capillaries.
    (c) venules.
    (d) a, b, and c are correct.

16. The blood vessels that collect blood from all tissues and organs and return it to the heart are the:
    (a) veins.
    (b) arteries.
    (c) capillaries.
    (d) arterioles.

17. Blood is compartmentalized within the veins because of the presence of:
    (a) venous reservoirs.   (c) clots.
    (b) muscular walls.   (d) valves.

18. The most important factor in vascular resistance is:
    (a) the viscosity of the blood.
    (b) friction between the blood and the vessel walls.
    (c) turbulence due to irregular surfaces of blood vessels.
    (d) the length of the blood vessels.

19. In a blood pressure reading of 120/80, the 120 represents _____ and the 80 represents _____.
    (a) diastolic pressure; systolic pressure
    (b) pulse pressure; mean arterial pressure
    (c) systolic pressure; diastolic pressure
    (d) mean arterial pressure; pulse pressure

20. Hydrostatic pressure forces water _____ a solution; osmotic pressure forces water _____ a solution.
    (a) into; out of   (c) out of; out of
    (b) out of; into   (d) a, b, and c are correct.

21. The two factors that assist the relatively low venous pressures in propelling blood toward the heart are:
    (a) ventricular systole and valve closure.
    (b) gravity and vasomotion.
    (c) muscular compression and the respiratory pump.
    (d) atrial and ventricular contractions.

22. The arteries of the pulmonary circuit differ from those of the systemic circuit in that they carry:
    (a) oxygen and nutrients.
    (b) deoxygenated blood.
    (c) oxygenated blood.
    (d) oxygen, carbon dioxide, and nutrients.

23. The two arteries formed by the division of the brachiocephalic trunk are the:
    (a) aorta and internal carotid.
    (b) axillary and brachial.
    (c) external and internal carotid.
    (d) common carotid and subclavian.

24. The unpaired arteries that supply blood to the visceral organs include the:
    (a) suprarenal, renal, and lumbar arteries.
    (b) iliac, gonadal, and femoral arteries.
    (c) celiac trunk and superior and inferior mesenteric arteries.
    (d) a, b, and c are correct.

25. The artery generally used to feel the pulse at the wrist is the _____ artery.
    (a) ulnar
    (b) radial
    (c) fibular
    (d) dorsalis

26. The vein that drains the dural sinuses of the brain is the _____ vein.
    (a) cephalic
    (b) great saphenous
    (c) internal jugular
    (d) superior vena cava

27. The vein that collects most of the venous blood from below the diaphragm is the:
    (a) superior vena cava.
    (b) great saphenous vein.
    (c) inferior vena cava.
    (d) azygos vein.

28. (a) What are the primary forces that cause fluid to move out of a capillary and into the interstitial fluid at its arterial end? (b) What are the primary forces that cause fluid to move into a capillary from the interstitial fluid at its venous end?

29. What two effects result when the baroreceptor response to elevated blood pressure is triggered?

30. What factors affect the activity of chemoreceptors in the carotid and aortic bodies?

31. What circulatory changes occur at birth?

32. What age-related changes take place in the blood, heart, and blood vessels?

Level 2: Reviewing Concepts

33. When dehydration occurs:
    (a) water reabsorption at the kidneys accelerates.
    (b) fluids are reabsorbed from the interstitial fluid.
    (c) blood osmotic pressure increases.
    (d) a, b, and c are correct.

34. Increased CO_2 levels in tissues would promote:
    (a) constriction of precapillary sphincters.
    (b) an increase in the pH of the blood.
    (c) dilation of precapillary sphincters.
    (d) a decrease of blood flow to tissues.

35. Elevated levels of the hormones ADH and angiotensin II will produce increased:
    (a) peripheral vasodilation.
    (b) peripheral vasoconstriction.
    (c) peripheral blood flow.
    (d) venous return.

36. Relate the anatomical differences between arteries and veins to their functions.

37. Why do capillaries permit the diffusion of materials, whereas arteries and veins do not?

38. Why is blood flow to the brain relatively continuous and constant?

39. An accident victim displays the following signs and symptoms: hypotension; pale, cool, moist skin; and confusion and disorientation. Identify her condition, and explain why these signs and symptoms occur. If you took her pulse, what would you find?

Level 3: Critical Thinking and Clinical Applications

40. Bob is sitting outside on a warm day and is sweating profusely. His friend Mary wants to practice taking blood pressures, and he agrees to play patient. Mary finds that Bob's blood pressure is elevated, even though he is resting and has lost fluid from sweating. (She reasons that fluid loss should lower blood volume and, thus, blood pressure.) Mary asks you why Bob's blood pressure is high instead of low. What should you tell her?

41. People with allergies frequently take antihistamines and decongestants to relieve their symptoms. The medication's label warns that the medication should not be taken by individuals being treated for high blood pressure. Why?

42. Gina awakens suddenly to the sound of her alarm clock. Realizing she is late for class, she jumps to her feet, feels lightheaded, and falls back on her bed. What probably caused this to happen? Why doesn't this always happen?

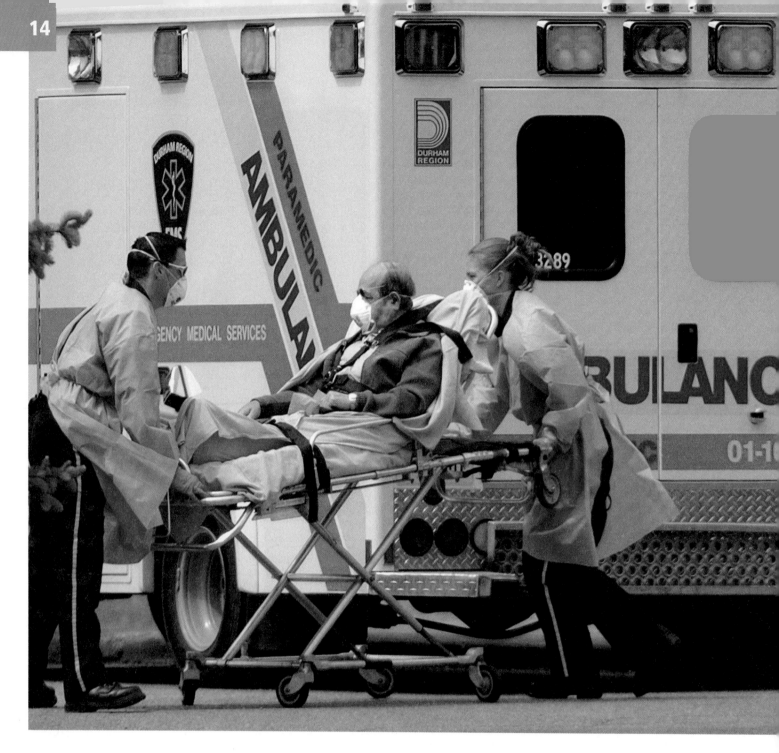

14 The Lymphatic System and Immunity

THE IMMUNE SYSTEM plays a major role in fighting infection. However, in this age of transcontinental air travel and extremely mobile society, infectious diseases are now spreading globally. This was never more evident than the outbreak of severe acute respiratory syndrome (SARS) that struck Toronto and southern Ontario. This new virus infected several EMS workers and resulted in a significant part of the workforce being quarantined. It required that EMS personnel begin significant isolation measures to protect the rest of the population. It is important to remember that the next pandemic may be as close as the next airliner that lands at Los Angeles International Airport or any other large transportation hub.

Chapter Outline

Chapter Objectives

1. Identify the major components of the lymphatic system and explain their functions. (p. 512)

2. Discuss the importance of lymphocytes and describe where they are found in the body. (pp. 513–519)

3. List the body's nonspecific defenses and explain how each functions. (pp. 519–522)

4. Define specific resistance and identify the forms and properties of immunity. (pp. 522–525)

5. Distinguish between cell-mediated immunity and antibody-mediated (humoral) immunity. (p. 526)

6. Discuss the different types of T cells and the role played by each in the immune response. (pp. 526–528)

7. Describe the structure of antibody molecules and explain how they function. (pp. 528–531)

8. Describe the primary and secondary immune responses to antigen exposure. (pp. 532–533)

9. Relate allergic reactions and autoimmune disorders to immune mechanisms. (pp. 534–537)

10. Describe the changes in the immune system that occur with aging. (p. 537)

11. Discuss the structural and functional interactions between the lymphatic system and other body systems. (pp. 537–540)

Vocabulary Development

anamnesis a memory; *anamnestic response*
apo- away; *apoptosis*
chemo- chemistry; *chemotaxis*
dia- through; *diapedesis*
-gen to produce; *pyrogen*

humor a liquid; *humoral immunity*
immunis safe; *immune*
inflammare to set on fire; *inflammation*
lympha water; *lymph*
nodulus little knot; *nodule*

pathos disease; *pathogen*
pedesis a leaping; *diapedesis*
ptosis a falling; *apoptosis*
pyr fire; *pyrogen*
taxis arrangement; *chemotaxis*

WE DO NOT LIVE IN A HARMLESS WORLD. Bumps, cuts, scrapes, chemical and thermal burns, extreme cold, and ultraviolet radiation are just a few of the hazards in the physical environment. Making matters worse, our world also contains an assortment of viruses, bacteria, fungi, and parasites capable of not only surviving but also thriving inside our bodies—and potentially causing us great harm. These organisms, called **pathogens** (*pathos*, disease + *-gen*, to produce), are responsible for many human diseases.

Each pathogen has a different mode of life and interacts with the body in a characteristic way. For example, most of the time viruses exist within cells, which they often eventually destroy. (Viruses lack a cellular structure; they consist only of nucleic acid and protein, and can reproduce themselves only within a living cell.) Many bacteria multiply in the interstitial fluids, and some of the largest parasites, such as roundworms, burrow through internal organs.

Many different organs and systems work together to keep us alive and healthy. In this ongoing struggle, the **lymphatic system** plays a central role.

This chapter begins by examining the organization of the lymphatic system. We then consider how the lymphatic system interacts with cells and tissues of other systems to defend the body against infection and disease.

■ Organization of the Lymphatic System

One of the least familiar organ systems, the lymphatic system includes the following four components:

1. *Vessels.* A network of **lymphatic vessels,** often called **lymphatics,** begins in peripheral tissues and ends at connections to the venous system.
2. *Fluid.* A fluid called **lymph** flows through the lymphatic vessels. Lymph resembles plasma but contains a much lower concentration of suspended proteins.
3. *Lymphocytes.* **Lymphocytes** are specialized cells that perform an array of specific functions in defending the body.
4. *Lymphoid tissues and organs.* **Lymphoid tissues** are collections of loose connective tissue and lymphocytes in structures called *lymphoid nodules;* an example is the tonsils. **Lymphoid organs** are more complex structures that contain large numbers of lymphocytes and are connected to lymphatic vessels; examples include the lymph nodes, spleen, and thymus.

Figure 14–1● provides an overview of the components of the lymphatic system.

Functions of the Lymphatic System

The lymphatic system has the following primary functions:

■ *The production, maintenance, and distribution of lymphocytes.* Lymphocytes are produced and stored within lymphoid organs, such as the spleen, thymus, and bone marrow. These cells are vital to the body's ability to resist or overcome infection and disease. Lymphocytes respond to the presence of (1) invading pathogens, such as bacteria or viruses; (2) abnormal body cells, such as virus-infected cells or cancer cells; and (3) foreign proteins, such as the toxins produced by some bacteria. Lymphocytes attempt to eliminate these threats or render them harmless through a combination of physical and chemical actions.

Lymphocytes respond to specific threats, such as a bacterial invasion of a tissue, by mounting a defense against that specific type of bacterium. Such a *specific defense* of the body is known as an **immune response. Immunity** is the body's ability to resist infection and disease through the activation of specific defenses.

■ *The return of fluid and solutes from peripheral tissues to the blood.* The return of tissue fluids through the lymphatic system maintains normal blood volume and eliminates local variations in the composition of the interstitial fluid. The volume of flow is considerable—roughly 3.6 liters per

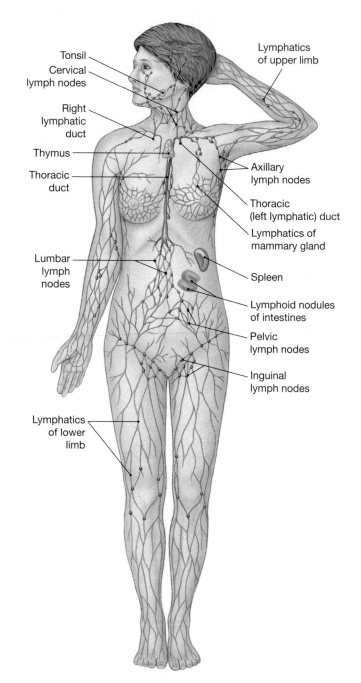

● **FIGURE 14–1 The Components of the Lymphatic System.**

day—and a break in a major lymphatic vessel can cause a rapid and potentially fatal decline in blood volume.

■ *The distribution of hormones, nutrients, and waste products from their tissues of origin to the general circulation.* Substances unable to enter the bloodstream directly may do so by way of lymphatic vessels. For example, lipids absorbed by the digestive tract do not often enter the bloodstream through capillaries. They reach the bloodstream only after they have traveled along lymphatic vessels (a process described further in Chapter 16).

Lymphatic Vessels

Lymphatic vessels, or lymphatics, carry lymph from peripheral tissues to the venous system. The smallest lymphatic vessels—called **lymphatic capillaries**—begin as blind pockets in peripheral tissues (Figure 14–2a●). Lymphatic capillaries are lined by an endothelium (simple squamous epithelium). The lack of a basement membrane around the endothelium permits fluids and solutes, as well as viruses, bacteria, and cell debris, to flow into a lymphatic capillary. The overlapping

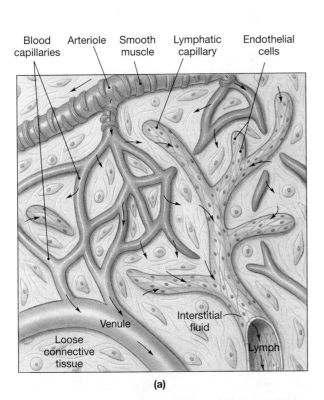

(a)

(b)

● **FIGURE 14–2 Lymphatic Capillaries.** (a) This drawing provides a three-dimensional representation of the association of blood capillaries, tissue, interstitial fluid, and lymphatic capillaries. Arrows indicate the directions of interstitial fluid and lymph movement. (b) Like some veins, some lymphatic vessels contain valves. (LM × 43)

arrangement of the endothelial cells acts as a one-way valve to prevent backflow of fluid into the intercellular spaces.

From lymphatic capillaries, lymph flows into larger lymphatic vessels that lead toward the trunk of the body. The walls of these lymphatics contain layers comparable to those of veins and, like veins, such lymphatic vessels contain valves (Figure 14–2b●). Pressures within the lymphatic system are extremely low, so the valves are essential to maintaining normal lymph flow.

The lymphatic vessels ultimately empty into two large collecting structures called *lymphatic ducts* (Figure 14–3●). The **thoracic duct** collects lymph from the lower abdomen, pelvis, and lower limbs, and from the left half of the head, neck, and chest. It empties its collected lymph into the venous system near the junction between the left internal jugular vein and the left subclavian vein. The smaller **right lymphatic duct,** which ends at a comparable location on the right side, delivers lymph from the right side of the body above the diaphragm. It empties into the right subclavian vein. Blockage of lymphatic drainage from a limb can cause the limb to swell due to the accumulation of interstitial fluid. This condition is called *lymphedema.*

Lymphocytes

As noted in Chapter 11, lymphocytes account for roughly 25 percent of the circulating white blood cell population. ∞ p. 422 But circulating lymphocytes constitute only a small fraction of the total lymphocyte population. At any given moment, most of the approximately 1 trillion (10^{12}) lymphocytes—with a combined weight of over a kilogram (1 kg = 2.2 lb)—are found within lymphoid organs or other tissues. The bloodstream provides a rapid transport system for lymphocytes that move from one site to another.

Types of Circulating Lymphocytes

The blood contains three classes of lymphocytes: *T cells, B cells,* and *NK cells.* Each lymphocyte class has distinctive functions.

T CELLS. Approximately 80 percent of circulating lymphocytes are **T** (**t**hymus-dependent) **cells.** *Cytotoxic T cells* directly attack foreign cells or body cells infected by viruses. These lymphocytes are the primary providers of *cell-mediated immunity,* or *cellular immunity. Helper T cells* stimulate the activities of both T cells and B cells. *Suppressor T cells* inhibit both T cells and B cells. Helper T cells and suppressor T cells are also called **regulatory T cells.**

B CELLS. **B** (**b**one marrow-derived) **cells** make up 10–15 percent of circulating lymphocytes. B cells can differentiate into

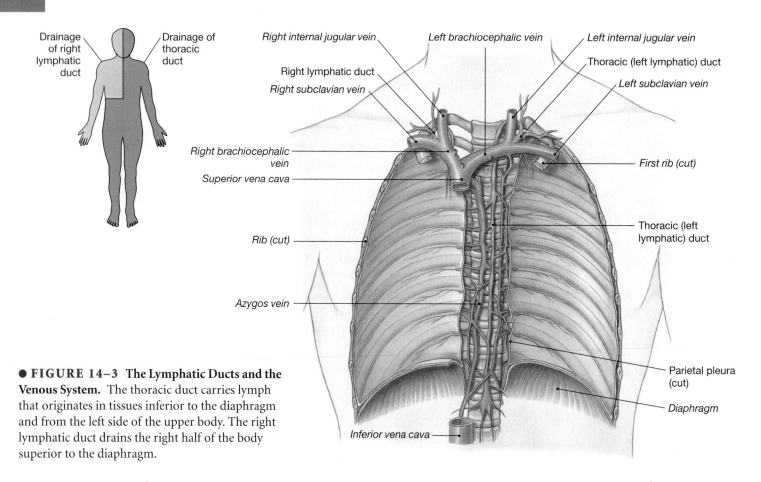

Drainage of right lymphatic duct

Drainage of thoracic duct

Right internal jugular vein

Left brachiocephalic vein

Left internal jugular vein

Right lymphatic duct

Thoracic (left lymphatic) duct

Right subclavian vein

Left subclavian vein

Right brachiocephalic vein

First rib (cut)

Superior vena cava

Thoracic (left lymphatic) duct

Rib (cut)

Azygos vein

Parietal pleura (cut)

Diaphragm

Inferior vena cava

● **FIGURE 14–3** **The Lymphatic Ducts and the Venous System.** The thoracic duct carries lymph that originates in tissues inferior to the diaphragm and from the left side of the upper body. The right lymphatic duct drains the right half of the body superior to the diaphragm.

plasma cells, which produce and secrete **antibodies**—soluble proteins that are also called **immunoglobulins.** ∞ p. 410 Antibodies bind to specific chemical targets called **antigens,** which are usually pathogens, parts or products of pathogens, or other foreign compounds. Formation of an antigen-antibody complex starts a chain of events that lead to the destruction of the target compound or organism. B cells are said to be responsible for *antibody-mediated immunity,* which is also known as *humoral* ("liquid") *immunity* because antibodies function in body fluids.

NK CELLS. The remaining 5–10 percent of circulating lymphocytes are **NK** (**n**atural **k**iller) **cells.** These lymphocytes attack foreign cells, normal cells infected with viruses, and cancer cells that appear in normal tissues. Their continual monitoring of peripheral tissues is known as *immunological surveillance.*

The Origin and Circulation of Lymphocytes

Lymphocytes in the blood, bone marrow, spleen, thymus, and peripheral lymphoid tissues are visitors, not residents. Lymphocytes move throughout the body; after they enter a tissue, they wander through it and then enter a blood vessel or lymphatic vessel for transport to another site. In general, lymphocytes have relatively long life spans. Roughly 80 percent survive for four

years, and some last 20 years or more. Throughout life, normal lymphocyte populations are maintained through the divisions of stem cells in the bone marrow and lymphoid tissues.

Lymphocyte production and development, or **lymphopoiesis** (lim-fō-poy-Ē-sis), involves the bone marrow and thymus (Figure 14–4●). As each B cell and T cell develop, they gain the ability to respond to the presence of a specific antigen; similarly, NK cells gain the ability to recognize abnormal cells.

Hemocytoblasts in the bone marrow produce lymphoid stem cells with two distinct fates. One group remains in the bone marrow and generates B cells and functional NK cells (Figure 14–4a). The second group of lymphoid stem cells migrates to the thymus. Under the influence of thymic hormones (collectively known as *thymosins*), these cells divide repeatedly, and produce large numbers of T cells (Figure 14–4b). As they mature, all three types of lymphocytes enter the bloodstream and migrate to peripheral tissues (Figure 14–4c), including lymphoid tissues and organs, such as the spleen. As these lymphocyte populations migrate through peripheral tissues, they retain the ability to divide and produce daughter cells of the same type. For example, a dividing B cell produces other B cells, not T cells or NK cells. As we will see, the ability of a specific type of lymphocyte to increase in number is important to the success of the immune response.

● **FIGURE 14–4** **The Origins of Lymphocytes.**
(**a**) Hemocytoblast divisions produce lymphoid stem cells with two fates. One group remains in the bone marrow and produces daughter cells that mature into B cells and NK cells. (**b**) The second group migrates to the thymus, where subsequent divisions produce daughter cells that mature into T cells. (**c**) Mature B cells, NK cells, and mature T cells are transported in the bloodstream to temporary sites within peripheral tissues.

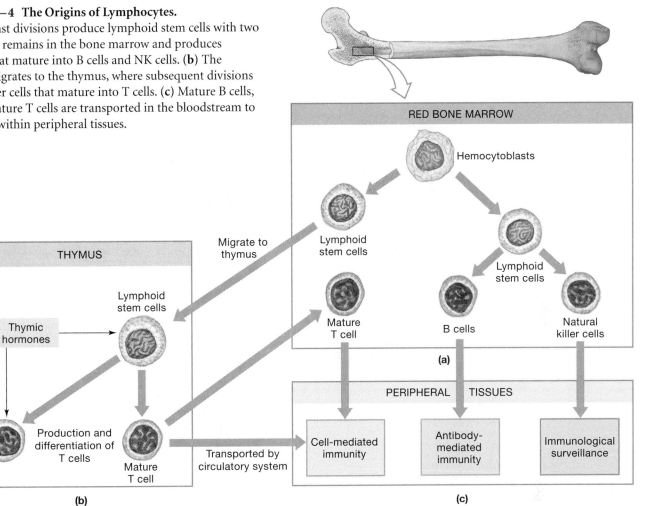

Lymphoid Nodules

Lymphoid tissues are composed of loose connective tissue and lymphocytes. **Lymphoid nodules** are masses of lymphoid tissue that are not surrounded by a fibrous capsule. As a result, their size can increase or decrease, depending on the number of lymphocytes present at any given moment. Large lymphoid nodules often contain a pale central region, called a *germinal center,* where lymphocytes actively divide.

Lymphoid nodules are found beneath the epithelia that line various organs of the respiratory, digestive, and urinary systems. All these systems are open to the external environment and, therefore, provide a route of entry into the body for potentially harmful organisms and toxins.

Because our food usually contains foreign proteins and often contains bacteria, lymphoid tissue and nodules associated with the digestive tract play a particularly important role in the defense of the body. The **tonsils,** which are large lymphoid nodules in the walls of the pharynx, guard the entrance to the digestive and respiratory tracts (Figure 14–5●). Five tonsils are usually present: a single *pharyngeal tonsil,* or *adenoids;* a pair of *palatine tonsils;* and a pair of *lingual tonsils.* Clusters of lymphoid nodules, or *Peyer's patches,* also lie beneath the epithelial lining of the intestines, and fused lymphoid nodules dominate the walls of the *appendix,* which is a blind pouch located near the junction of the small and large intestines.

The lymphocytes in a lymphoid nodule are not always able to destroy bacterial or viral invaders, and if pathogens become established in a lymphoid nodule, an inflammatory response to the infection develops. Two examples are probably familiar to you: *tonsillitis,* which is inflammation of one of the tonsils (usually the pharyngeal tonsil); and *appendicitis,* which is inflammation of the lymphoid nodules in the appendix.

Lymphoid Organs

Lymphoid organs are separated from surrounding tissues by a fibrous connective-tissue capsule. Important lymphoid organs include the *lymph nodes,* the *thymus,* and the *spleen.*

Lymph Nodes

Lymph nodes are small, oval lymphoid organs covered by a fibrous capsule and are 1–25 mm (up to 1 in.) in diameter (Figure 14–6●). *Afferent lymphatics* deliver lymph to a lymph node, and

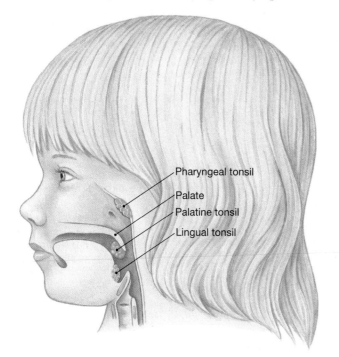

● **FIGURE 14–5 The Tonsils.** The tonsils are lymphoid nodules in the wall of the pharynx. A single pharyngeal tonsil (the adenoids) lies above the paired palatine and lingual tonsils. Each tonsil contains germinal centers where lymphocyte divisions occur.

efferent lymphatics carry the lymph onward, toward the venous system. Lymph nodes filter and purify the lymph before it reaches the venous system. As lymph flows through a lymph node, at least 99 percent of the antigens present in the arriving lymph are removed. As the antigens are detected and removed, T cells and B cells are stimulated, initiating an immune response. Lymph nodes are located in regions where they can detect and eliminate harmful "intruders" before they reach vital organs of the body (see Figure 14–1, p. 512).

Clinical Note
"SWOLLEN GLANDS"

Lymph nodes are often called *lymph glands,* and "swollen glands" usually accompany tissue inflammation or infection. Chronic or excessive enlargement of lymph nodes, a sign called *lymphadenopathy* (lim-fad-e-NOP-a-thē), may occur in response to bacterial or viral infections, endocrine disorders, or cancer.

Because lymphatic capillaries offer little resistance to the passage of cancer cells, cancer cells often spread along the lymphatics and become trapped in lymph nodes. Thus, an analysis of swollen lymph nodes can provide information on the nature and distribution of cancer and aid in the selection of appropriate therapies. *Lymphomas*—cancers that arise from lymphocytes or lymphoid stem cells—are an important group of lymphatic system cancers. ■

The Thymus

The **thymus** is a pink gland that lies in the mediastinum posterior to the sternum (Figure 14–7a●). It is the site of T cell production and maturation. The thymus reaches its greatest size (relative to body size) in the first year or two after birth and its maximum absolute size during puberty, when it weighs between 30 and 40 g (1.06 to 1.41 oz). Thereafter, the thymus gradually decreases in size.

The thymus has two lobes, each divided into *lobules* by fibrous partitions, or *septae* (*septum,* a wall) (Figure 14–7b●). Each lobule consists of a densely packed outer cortex and a paler, central *medulla* (Figure 14–7c●). The cortex contains clusters of lymphocytes surrounded by other cells that secrete the hormones collectively known as *thymosins*. Thymosins stimu-

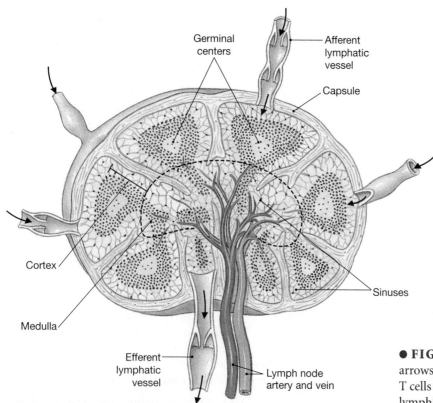

● **FIGURE 14–6   The Structure of a Lymph Node.** The arrows indicate the direction of lymph flow. Mature B cells and T cells in the cortex and medulla remove antigens from the lymph and initiate immune responses.

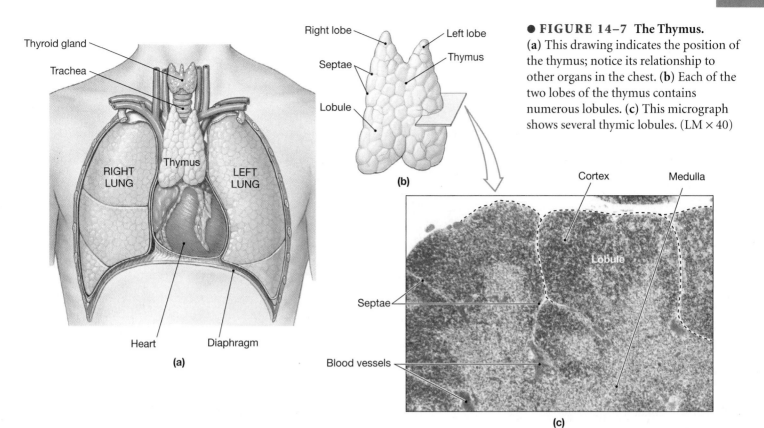

● FIGURE 14–7 The Thymus.
(**a**) This drawing indicates the position of the thymus; notice its relationship to other organs in the chest. (**b**) Each of the two lobes of the thymus contains numerous lobules. (**c**) This micrograph shows several thymic lobules. (LM × 40)

late lymphocyte stem cell divisions and T cell maturation. After the T cells migrate into the medulla, they leave the thymus in one of the blood vessels in that region.

The Spleen

The adult **spleen** contains the largest collection of lymphoid tissue in the body. It is about 12 cm (5 in.) long and can weigh about 160 g (5.6 oz). A major function of the spleen parallels that of the lymph nodes, except that it filters blood instead of lymph. It removes abnormal blood cells and components, and it initiates the responses of B cells and T cells to antigens in the circulating blood. In addition, the spleen stores iron from recycled red blood cells. ∞ p. 413

As indicated in Figure 14–8a●, the spleen is wedged between the stomach, the left kidney, and the muscular diaphragm. The spleen normally has a deep red color because of the blood it contains (Figure 14–8b●). The cellular components of the spleen are arranged into areas of *red pulp*, which contain large quantities of red blood cells, and areas of *white pulp* that resemble lymphoid nodules (Figure 14–8c●).

After the splenic artery enters the spleen, it branches outward toward the capsule into many smaller arteries that are surrounded by white pulp. Capillaries then discharge the blood into a network of reticular fibers that makes up the red pulp. Blood from the red pulp enters *venous sinusoids* (small vessels lined by macrophages) and then flows into small veins and the

splenic vein. As blood passes through the spleen, macrophages identify and engulf any damaged or infected cells. The presence of lymphocytes nearby ensures that any microorganisms or other abnormal antigens stimulate an immune response.

Clinical Note
INJURY TO THE SPLEEN

An impact to the left side of the abdomen can distort or damage the spleen. Such injuries are known risks of contact sports, such as football or hockey, and other athletic activities, such as skiing or sledding. However, because even a seemingly minor blow to the side may rupture the capsule, the results can be serious internal bleeding and eventual circulatory shock. The spleen can also be damaged by infection, inflammation, or invasion by cancer cells.

Because the spleen is relatively fragile, it is very difficult to repair surgically. (Sutures usually tear out before they have been tensed enough to stop the bleeding.) The trend in management of splenic injuries is nonoperative or conservative management. Typically, patients with stable vital signs, stable hemoglobin levels over 12–48 hours, minimal transfusion requirements (two units or less), suitable findings on computed tomography (CT) scans, and patients younger than 55 years are candidates for conservative therapy (Figure 14–9●). If the patient is unstable, or if there is a treatment for a severely ruptured spleen, complete removal, a process called a *splenectomy* (sple-NEK-to-mē), may be indicated. A person without a spleen survives without difficulty but has a greater risk for bacterial infections than does an individual with a functional spleen. ■

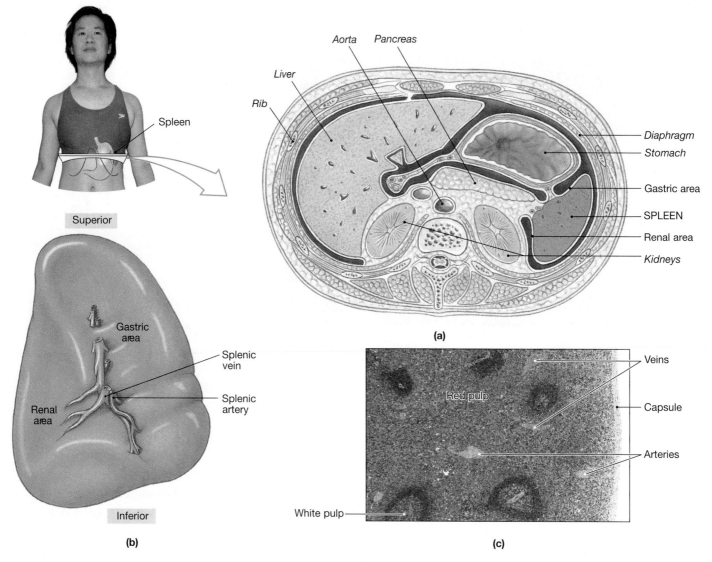

Spleen

Superior

Gastric area

Splenic vein

Splenic artery

Renal area

Inferior

(b)

Aorta Pancreas

Liver

Rib

Diaphragm

Stomach

Gastric area

SPLEEN

Renal area

Kidneys

(a)

Veins

Red pulp

Capsule

Arteries

White pulp

(c)

● **FIGURE 14–8   The Spleen. (a)** The shape of the spleen roughly conforms to the shapes of adjacent organs. This transverse section through the trunk shows the typical position of the spleen within the abdominopelvic cavity. **(b)** This external view of the intact spleen includes some important anatomical landmarks. **(c)** This photomicrograph shows the internal structure of the spleen. The white pulp appears blue because the nuclei of lymphocytes stain very darkly. Red pulp contains an abundance of red blood cells.

> ### CONCEPT CHECK QUESTIONS
>
> 1. How would blockage of the thoracic duct affect the circulation of lymph?
> 2. If the thymus gland failed to produce thymic hormones, which population of lymphocytes would be affected?
> 3. Why do lymph nodes enlarge during some infections?
>
> *Answers begin on p. 792.*

The Roles of the Lymphatic System in Body Defenses

The human body has many defense mechanisms, but they can be sorted into two categories:

1. **Nonspecific defenses** do not distinguish between one threat and another. These defenses, which are present at birth, include *physical barriers, phagocytic cells, immunological surveillance, interferons, complement, inflammation,* and *fever*—all of which are discussed shortly. They provide the body with a defensive capability known as **nonspecific resistance.**

2. **Specific defenses** protect against particular threats. For example, a specific defense may oppose infection by one type of bacterium but ignore other bacteria and all viruses. Many specific defenses develop after birth as a result of exposure to environmental hazards or infectious agents. Specific defenses are dependent on the activities of lymphocytes. The body's specific defenses

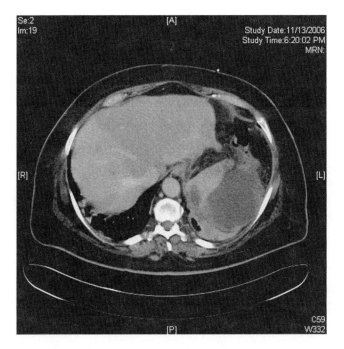

● **FIGURE 14–9 Splenic Trauma.** The spleen is a solid organ and vulnerable to the effects of blunt trauma.

produce a state of protection known as immunity, or **specific resistance.**

Nonspecific and specific resistance does not function in an either-or fashion; both are necessary to provide adequate resistance to infection and disease.

■ Nonspecific Defenses

Nonspecific defenses deny the entry, or limit the spread within the body, of microorganisms or other environmental agents (Figure 14–10●). Their response is the same regardless of the type of invading agent.

Physical Barriers

Physical barriers keep hazardous organisms and materials outside the body. To cause infection or disease, a foreign (antigenic) compound or pathogen must enter body tissues, which requires crossing an epithelium. The epithelial covering of the skin has a keratin coating, multiple layers, and a network of desmosomes that locks adjacent cells together. ∞ p. 124 These barriers provide very effective protection for underlying tissues.

The exterior surface of the body has several other defenses. The hairs found in most areas provide some protection against mechanical abrasion (especially on the scalp), and they often prevent hazardous materials or insects from contacting the skin's surface. The epidermal surface also receives the secretions of sebaceous glands and sweat glands. These secretions flush the surface and wash away microorganisms and chemical agents. The secretions also contain microbe-killing chemicals, destructive enzymes (*lysozymes*), and antibodies.

The epithelia that line the digestive, respiratory, urinary, and reproductive tracts are more delicate, but they are equally well defended. Mucus bathes most surfaces of the digestive tract, and the stomach contains a powerful acid that can destroy many potential pathogens. Mucus is swept across the lining of the respiratory tract, urine flushes the urinary passageways, and glandular secretions flush structures in the reproductive tract. Special enzymes, antibodies, and an acidic pH add to the effectiveness of these secretions.

Phagocytes

Phagocytes in peripheral tissues remove cellular debris and respond to invasion by foreign compounds or pathogens. These cells represent the "first line" of cellular defense, and often attack and remove microorganisms before lymphocytes become aware of their presence. Two general classes of phagocytic cells are found in the human body: *microphages* and *macrophages.*

Microphages are the neutrophils and eosinophils that normally circulate in the blood. These phagocytic cells leave the bloodstream and enter peripheral tissues subjected to injury or infection. As noted in Chapter 11, neutrophils are abundant, mobile, and quick to phagocytize cellular debris or invading bacteria. ∞ p. 421 The less abundant eosinophils target foreign compounds or pathogens that have been coated with antibodies.

The body also contains several types of **macrophages**—large, actively phagocytic cells derived from circulating monocytes. Almost every tissue in the body shelters resident (fixed) or visiting (free) macrophages. The relatively diffuse collection of phagocytic cells throughout the body is called the **monocyte-macrophage system,** or the *reticuloendothelial system.* In some organs, fixed macrophages have special names. For example, *microglia* are macrophages in the central nervous system, and *Kupffer* (KOOP-fer) *cells* are macrophages found in and around blood channels in the liver.

All phagocytic cells function in much the same way, although the targets of phagocytosis may differ from one cell type to another. Mobile macrophages and microphages also share a number of other functional characteristics in addition to phagocytosis. All can move through capillary walls by squeezing between adjacent endothelial cells, a process known as *diapedesis (dia,* through + *pedesis,* a leaping). They may also be attracted to or repelled by chemicals in the surrounding fluids, which is a phenomenon called **chemotaxis** (*chemo-,* chemistry

+ *taxis,* arrangement). They are particularly sensitive to chemicals released by other body cells or by pathogens.

Immunological Surveillance

Our immune defenses generally ignore normal cells in the body's tissues, but abnormal cells are attacked and destroyed. The constant monitoring of normal tissues is called **immunological surveillance,** and it primarily involves the lymphocytes known as NK (natural killer) cells. The cell membrane of an abnormal cell generally contains antigens that are not found on the membranes of normal cells. NK cells recognize an abnormal cell by detecting the presence of those antigens. NK cells are much less selective about their targets than are other lymphocytes. When they encounter such antigens on a bacterium, a cancer cell, or a cell infected with viruses, NK cells secrete proteins called *perforins,* which kill the abnormal cell by creating large pores in its cell membrane.

NK cells respond much more rapidly than T cells or B cells upon contact with an abnormal cell. Killing the abnormal cells can slow the spread of a bacterial or viral infection, and it may eliminate cancer cells before they spread to other tissues. Unfortunately, some cancer cells avoid detection, a process called *immunological escape.* Once immunological escape has occurred, cancer cells can multiply and spread without interference by NK cells.

Interferons

Interferons (in-ter-FĒR-onz) are small proteins released by activated lymphocytes, macrophages, and tissue cells infected with viruses. Normal cells exposed to interferon molecules respond by producing antiviral proteins that interfere with viral replication inside the cell. In addition to slowing the spread of viral infections, interferons stimulate the activities of macrophages and NK cells. Interferons are examples of **cytokines** (SĪ-tō-kīnz),

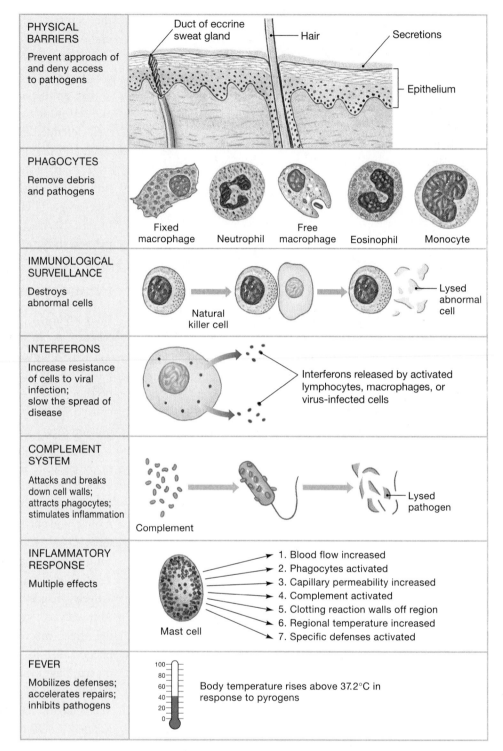

● **FIGURE 14–10 The Body's Nonspecific Defenses.** Nonspecific defenses deny pathogens access to the body or destroy them without distinguishing among specific types.

which are chemical messengers that are released by tissue cells and coordinate local activities. Most cytokines act only within one tissue, but those released by cellular defenders also act as hormones; they affect the activities of cells and tissues throughout the body. Their role in the regulation of specific defenses will be discussed later in the chapter.

Complement

Plasma contains 11 special *complement proteins* that constitute the **complement system.** The term *complement* refers to the fact that this system complements the actions of antibodies. Complement proteins interact with one another in chain reactions similar to those of the clotting system. The reaction begins when a particular complement protein binds either to an antibody molecule attached to a bacterial cell wall or directly to bacterial cell walls. The bound complement protein then interacts with a series of other complement proteins. Complement activation is known to (1) attract phagocytes, (2) stimulate phagocytosis, (3) destroy cell membranes, and (4) promote inflammation.

Inflammation

Inflammation, or the *inflammatory response,* is a localized tissue response to injury. ∞ p. 116 Inflammation produces local swelling, redness, heat, and pain. Inflammation can be produced by any stimulus that kills cells or damages loose connective tissue. *Mast cells* within the affected tissue play a pivotal role in this process. ∞ p. 106

The purposes of inflammation include the following:

- To perform a temporary repair at the injury site and prevent the access of additional pathogens.
- To slow the spread of pathogens from the injury site.
- To mobilize a wide range of defenses that can overcome the pathogens and aid permanent tissue repair. The repair process is called *regeneration.*

Figure 14–11● summarizes the events that occur during inflammation of the skin. (Comparable events will occur in almost any tissue subjected to physical damage or to infection.) When stimulated by mechanical stress or chemical changes in the local environment, mast cells release chemicals, including *histamine* and *heparin,* into the interstitial fluid. These chemicals initiate the process of inflammation. The released histamine increases capillary permeability and accelerates blood flow through the area. Increased vessel permeability allows clotting factors and complement proteins to leave the bloodstream. The combination of abnormal tissue conditions and chemicals released by mast cells stimulates local sensory neurons, producing sensations of pain. The increased blood flow reddens the area and elevates local temperature, increasing the rate of enzymatic reactions and accelerating the activity of phagocytes. Clotting does not occur at the actual site of injury, due to the presence of heparin. However, a clot soon forms around the damaged area. Additionally, the release of histamine stimulates other

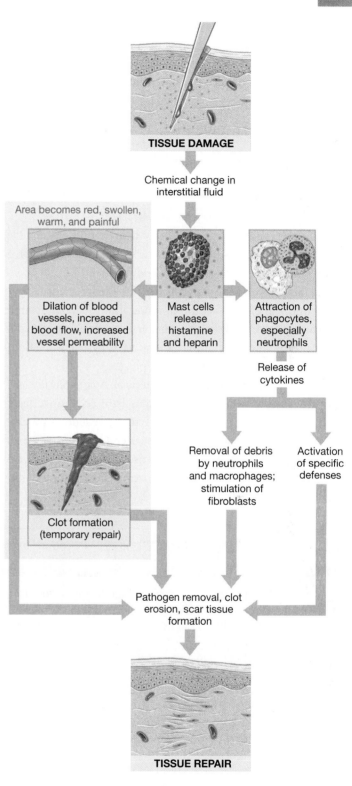

● **FIGURE 14–11 Events in Inflammation.**

series of events that activate specific defenses and further pave the way for the repair of the injured tissue.

After an injury, tissue conditions generally become more abnormal before they begin to improve. **Necrosis** (ne-KRŌ-sis) is the tissue destruction that occurs after cells have been injured or destroyed. The process begins several hours after the original injury and is caused by lysosomal enzymes. Lysosomes break down by autolysis, which releases digestive enzymes that first destroy the injured cells and then attack surrounding tissues. ∞ p. 79 As local inflammation continues, debris and dead and dying cells accumulate at the injury site. This thick fluid mixture is known as **pus.** An accumulation of pus in an enclosed tissue space is called an *abscess.*

Fever

Fever is the maintenance of a body temperature greater than 37.2°C (99°F). The hypothalamus, which contains nuclei that regulate body temperature, acts as the body's thermostat. ∞ p. 292 Circulating proteins called **pyrogens** (PĪ-rō-jenz; *pyr,* fire + *-gen,* to produce) in effect reset the thermostat in the hypothalamus and cause a rise in body temperature. Pathogens, bacterial toxins, and antigen-antibody complexes may act as pyrogens or stimulate the release of pyrogens by macrophages.

Within limits, a fever may be beneficial. A rise in body temperature increases the rate of metabolism; cells move faster (enhancing phagocytosis) and enzymatic reactions proceed more quickly. However, high fevers (over 42°C, or 107°F) can damage many physiological systems. Such high fevers can cause CNS problems, including nausea, disorientation, hallucinations, or convulsions.

Clinical Note
FEVER

It is important to remember that fever is a normal response to infectious diseases. Fever is defined as a temperature 1° or more above the normal. Mild or short-term elevations are common with minor infections. High fevers, with temperatures of 103°F and above, can signal a potentially dangerous infection. Fever, although frightening for parents, can actually be a good thing. It is important to remember that fever is part of the body's way of fighting infections. Thus, not all fevers need to be treated. High fever can make a child uncomfortable and can aggravate problems such as dehydration. Some children who develop high fevers will have febrile seizures. While these are scary for the parents, they are usually self-limited and cause no permanent damage.

If the child is prone to develop febrile seizures or is extremely uncomfortable, fever may be treated with antipyretic medications such as ibuprofen or acetaminophen. Aspirin should never be administered. Sponge baths with tepid water (never with alcohol) can also help lower the child's temperature. With sponging, be sure not to overcool the child; overcooling can cause shivering and shut down the body's cooling mechanisms. The treatment of fever in the prehospital setting is controversial. Always follow local protocols in this regard. ■

CONCEPT CHECK QUESTIONS

1. What types of cells would be affected by a decrease in the monocyte-forming cells in the bone marrow?
2. A rise in the level of interferon in the body suggests what kind of infection?
3. What effects do pyrogens have in the body?
4. Describe how fever can aid the body in fighting infection.

Answers begin on p. 792.

■ Specific Defenses: The Immune Response

Specific resistance, or immunity, is provided by the coordinated activities of T cells and B cells, which respond to the presence of *specific* antigens. *T cells* provide a defense against abnormal cells and pathogens inside living cells; this process is called **cell-mediated immunity,** or *cellular immunity.* B cells provide a defense against antigens and pathogens in body fluids. This process is called **antibody-mediated immunity,** or *humoral immunity.* Regardless of whether T cells or B cells are involved, immunity can be categorized into several types according to when and how it arises in the body (Figure 14–12●).

Types of Immunity

Immunity is either innate or acquired. **Innate immunity** is genetically determined; it is present at birth and is independent of previous exposure to the antigen involved. For example, people do not get the same diseases as goldfish. By contrast, **acquired immunity** is not present at birth but instead arises by active or passive means.

Active immunity appears after exposure to an antigen as a consequence of the immune response. Even though the immune system is capable of defending against an enormous number of antigens, the appropriate defenses are mobilized only after an individual's lymphocytes encounter a particular antigen. Active immunity may develop as a result of natural exposure to an antigen in the environment (naturally acquired active immunity) or from deliberate exposure to an antigen (induced active immunity).

Naturally acquired immunity normally begins to develop after birth, and it is continually enhanced as an individual encounters "new" pathogens or other antigens. This process might be likened to vocabulary development—a child begins with a few basic common words and learns new ones as needed.

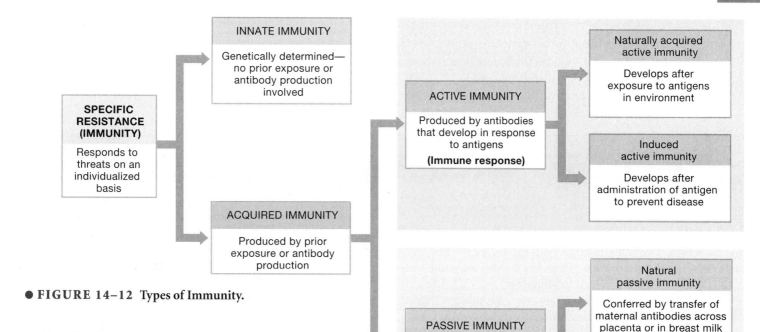

● **FIGURE 14–12** Types of Immunity.

In *induced active immunity,* antibody production is stimulated under controlled conditions so that the individual will be able to overcome natural exposure to the same type of pathogen some time in the future. This is the basic principle behind *immunization,* or *vaccination,* to prevent disease. A **vaccine** is a preparation designed to induce an immune response. It contains either a dead or an inactive pathogen or antigens derived from that pathogen. Vaccines are given orally or by intramuscular or subcutaneous injection.

Passive immunity is produced by the transfer of antibodies to an individual from some other source. *Naturally acquired passive immunity* results when antibodies produced by a mother protect her baby against infections during gestation (by crossing the placenta) or in early infancy (through breast milk). In *induced passive immunity,* antibodies are administered to fight infection or prevent disease after exposure to the pathogen. For example, antibodies that attack the rabies virus are injected into a person recently bitten by a rabid animal.

Properties of Immunity

Immunity has four general properties:

1. *Specificity.* A specific defense is activated by a specific antigen, and the immune response targets only that particular antigen. This process is known as **antigen recognition. Specificity** occurs because the cell membrane of each T cell and B cell has receptors that will bind only one specific antigen and ignore all other antigens. Each lymphocyte will inactivate or destroy that specific antigen only, without affecting other antigens or normal tissues.

2. *Versatility.* In the course of a normal lifetime, an individual encounters an enormous number of antigens—perhaps tens of thousands. Your immune system cannot anticipate which antigens it will encounter, so it must be ready to confront *any* antigen at *any* time. The immune system achieves **versatility** by producing millions of different lymphocyte populations, each with different antigen receptors, and through variability in the structure of synthesized antibodies. In this way, the immune system can produce appropriate and specific responses to each antigen when exposure does occur.

3. *Memory.* The immune system "remembers" antigens that it encounters. As a result of immunologic **memory,** the immune response to a second exposure to an antigen is stronger and lasts longer than the response to the first exposure. During the initial response to an antigen, lymphocytes sensitive to its presence undergo repeated cell divisions. Two kinds of cells are produced: some that attack the antigen and others that remain inactive unless they are exposed to the same antigen at a later date. These latter cells—**memory cells**—enable the immune system to "remember" previously encountered antigens and launch a faster, stronger response if one of them ever appears again.

4. *Tolerance.* Although the immune response targets foreign cells and compounds, it generally ignores normal tissues and their antigens. **Tolerance** is said to exist when the immune system does not respond to such antigens. Any B cells or T cells undergoing differenti-

ation (in the bone marrow and thymus, respectively) that react to normal body antigens are destroyed. As a result, normal B cells and T cells will ignore normal (or *self*) antigens and will attack foreign (or *nonself*) antigens.

Clinical Note
IMMUNOLOGY

Overview

The immune system is the principal body system involved in fighting disease. Much of immune system function occurs in the lymphatic system, which consists of the lymphatic vessels, the lymph, and the lymph organs. Its primary functions are the production, maintenance, and distribution of lymphocytes (white blood cells); the return of fluid and solutes from the peripheral tissues to the blood; and the distribution of hormones, nutrients, and waste products from their tissues of origin to the general circulation. Immunity is one of the body's major defense mechanisms. It is provided by the coordinated activities of T cells and B cells, which respond to the presence of specific antigens. T cells provide cellular immunity, while B cells provide humoral immunity.

Physicians who specialize in the immune system disorders are referred to as *allergists* or *allergists/immunologists.* They usually train in general internal medicine or pediatrics and take additional training in allergy and immunology. An immunologist is a scientist (Ph.D., M.D., or D.O.) who specializes in the study of the immune system. A closely related medical discipline is infectious disease. Because infectious disease processes almost always involve the body's immune system, infectious disease specialists often treat immune system disorders. Like an allergist/immunologist, an infectious disease specialist first trains in general internal medicine or pediatrics and then serves a fellowship in infectious disease.

Immunity

The ability of the body's defenses to combat infection with specific responses is called *immunity.* The immune response is a complex cascade of events that occurs following activation by an invading substance, or pathogen. Its goal is the destruction or inactivation of pathogens, abnormal cells, or foreign molecules such as toxins. The body can accomplish this through two mechanisms, cellular immunity and humoral immunity. Cellular immunity involves a direct attack on the foreign substance by specialized cells of the immune system. These cells physically engulf and deactivate or destroy the offending agent. Humoral immunity is much more complicated. Humoral immunity is basically a chemical attack on the invading substance. The principal chemical agents of this attack are *antibodies,* also called *immunoglobulins (Igs).* Antibodies are a unique class of chemicals that are manufactured by specialized cells of the immune system called B cells. There are five different classes of antibodies: IgA, IgD, IgE, IgG, and IgM.

The humoral immune response begins with exposure of the body to an antigen. An *antigen* is any substance capable of inducing an immune response. Most antigens are proteins. Following exposure to an antigen, antibodies are released from cells of the immune system and attach themselves to the invading substance to facilitate its removal from the body by other cells of the immune system.

The immune system responds differently to a particular antigen depending upon if the body has or has not been exposed previously to that antigen. The initial response to an antigen is called the *primary response.* Following exposure to a new antigen, both the cellular and humoral components of the immune system require several days to respond. Generalized antibodies (IgG and IgM) are first released to help fight the antigen.

At the same time, other components of the immune system begin to develop antibodies specific for the antigen. These cells also develop a memory of the particular antigen. If the body is exposed to the same antigen again, the immune system responds much faster. This is called the *secondary response.* As a part of the secondary response, antibodies specific for the offending antigen are released. Antigen-specific antibodies are much more effective in facilitating removal of the offending antigen than are the generalized antibodies released during the primary response.

Immunity may be either natural or acquired. *Natural immunity,* also called *innate immunity,* is genetically predetermined. It is present at birth and has no relation to previous exposure to a particular antigen. All humans are born with some innate immunity. *Acquired immunity* develops over time and results from exposure to an antigen. The immune system's production of antibodies specific for the antigens to which the body is exposed protects the organism, as subsequent exposure to the same antigen will result in a vigorous immune response. Naturally acquired immunity normally begins to develop after birth and is continually enhanced by exposure to new pathogens and antigens throughout life. For example, a child contracts chicken pox (varicella) at age 18 months. Following the infection, the child's immune system creates antibodies specific for the varicella virus. Repeated exposure to the varicella virus usually will not result in another infection. In fact, it is not unusual for a patient exposed to varicella to develop lifelong immunity to the infection.

Induced active immunity, also called *artificially acquired immunity,* is designed to provide protection from exposure to an antigen at some time in the future (Table 14–1). This is achieved through vaccination and provides relative protection against serious infectious agents. In vaccination, an antigen is injected into the body in order to generate an immune response. This results in the development of antibodies specific for the antigen and provides protection against future infection. Most vaccines contain antigenic proteins from a particular virus

TABLE 14-1   *Common Immunizations*

IMMUNIZATION TARGET	TYPE OF IMMUNITY PROVIDED	VACCINE TYPE	REMARKS
VIRUSES			
Poliovirus	Active	Live, attenuated	Oral
	Active	Killed	Boosters every 2–3 years
Rubella	Active	Live, attenuated	
	Passive	Human antibodies (pooled)	
Mumps	Active	Live, attenuated	
Measles (rubeola)	Active	Live, attenuated	May need second booster
Varicella	Active	Live, attenuated	
Hepatitis A	Passive	Human antibodies (pooled)	
Hepatitis B	Active	Killed	May need periodic boosters
	Passive	Human antibodies (pooled)	
Smallpox	Active	Live, related virus	Boosters every 3–5 years (no longer required as disease appears to have been eliminated)
Yellow fever	Active	Live, attenuated	Boosters every 10 years
Herpes zoster	Passive	Human antibodies (pooled)	
Rabies	Passive	Human antibodies (pooled)	
	Passive	Horse antibodies	
	Active	Killed	Boosters required
BACTERIA			
Typhoid	Active	Killed	Boosters every 2, 3, or 5 years, depending on vaccine type
Tuberculosis	Active	Live, attenuated	
Plague	Active	Killed	Boosters every 1–2 years
Tetanus	Active	Toxins only	Boosters every 5–10 years
	Passive	Human antibodies (pooled)	
Diptheria	Active	Toxins only	Boosters every 10 years
	Passive	Horse antibodies	
Streptococcal pneumonia	Active	Bacteria and cell wall components	
Botulism	Passive	Horse antibodies	
Rickettsia: Typhus	Active	Killed	Boosters yearly
Hemophilus influenza H (HIB)	Active	Killed	May need periodic boosters
OTHER TOXINS			
Snake bite	Passive	Horse antibodies	
Spider bite	Passive	Horse antibodies	
Venomous fish spine	Passive	Horse antibodies	

or bacterium. Later, when the individual is actually exposed to the pathogen, the immune response will be vigorous and will often be enough to prevent the infection from developing.

An example of a commonly used vaccine is the DPT (diphtheria/pertussis/tetanus) vaccine. This vaccine contains antigenic proteins from the bacteria that cause diphtheria, whooping cough, and tetanus. It is administered at several intervals during the first five years of life and provides protection against infection from these bacteria. Some vaccinations will impart lifelong immunity, while others must be periodically followed with a "booster dose" to ensure continued protection.

Acquired immunity can be either active or passive. Active immunity occurs following exposure to an antigen and results in the production of antibodies specific for the antigen. Most vaccinations result in the development of active immunity. However, it takes some time for a patient to develop specific antibodies. In certain cases, it is necessary to administer antibodies to provide protection until the active immunity can take effect. Passive immunity results from the introduction of antibodies. There are two types of passive immunity. Natural passive immunity occurs when antibodies cross the placental barrier from the mother to the infant to provide protection against embryonic or fetal infections. Induced passive immunity is the administration of antibodies to an individual to help fight infection or prevent diseases.

An example of the clinical use of both active and passive immunity is the regimen used for the prevention of tetanus. Most people from developed countries have received some form of tetanus vaccination during their life. These people typically have some antibodies to tetanus and often need nothing more than a tetanus booster. However, some people have never received any sort of tetanus vaccination. When they seek treatment for a tetanus-prone wound, they must receive prophylaxis for tetanus in addition to care for their wound. This is best achieved by providing both passive and active immunity. To provide immediate protection, the patient is administered antibodies specific for tetanus (tetanus immune globulin [TIG], Hypertet). Then, the patient is also administered a tetanus vaccination (Td or Dt). The tetanus immune globulin (TIG) provides passive immunity until the body's immune system can respond to the vaccination and develop antibodies specific for tetanus. This should be followed by periodic tetanus boosters until the patient's immunization program is complete.

Immunization

Immunization is the manipulation of the immune system by administering antigens under controlled conditions or by providing antibodies that can combat an existing infection. Active immunization is the process of inducing the immune system to produce specific antibodies through the administration of a *vaccine*. A vaccine is a preparation of antigens derived from a specific pathogen (Table 14–1). The vaccine can be given orally or by injection. Most vac-

cines contain the pathogenic organism. They may contain either the entire organism or just the antigenic proteins found on the organism. Some vaccines may contain the living organism, while others achieve satisfactory results with a dead organism. If a living organism is used, it is usually weakened, or *attenuated,* to lessen the chances of inducing a real infection. Despite attenuation, some vaccines may contain enough of the pathogenic organism to cause mild signs or symptoms. Regardless, the risks of developing a serious illness are quite low compared to those of contracting the disease because of lack of vaccination.

Inactivated, or "killed," vaccines consist of bacterial cell walls or viral protein coats. These vaccines cannot cause the development of even mild disease symptoms. However, they do not induce as strong an immune response as live or attenuated vaccines. Some attenuated vaccines do not induce long-lasting immunity but require periodic booster injections to keep antibody titers at a protective level.

The administration of a vaccine induces the immune system to produce antibodies specific for the antigen or antigens contained within the vaccine. This is a form of active immunity, and protective antibody levels may take several weeks to develop. Immediate protection can, in some cases, be provided by the administration of antibodies. This process, called *passive immunization,* is effective only for certain infections. Antibody preparations may be general or may be specific for a particular disease.

The development of specific antibodies is limited primarily to life-threatening diseases such as tetanus, rabies, diphtheria, and hepatitis that cannot be treated by other methods. Usually, specific passive immunizations are very expensive. General passive immunization involves the injection of various antibodies obtained from the blood donor pool. Referred to as gamma globulin, these agents can provide limited protection until the immune system responds to an administered vaccine.

More and more vaccines are being developed. It is important that emergency personnel be vaccinated against diseases for which they are deemed to be at increased risk. The most important of these is hepatitis B. The hepatitis B vaccine is effective and should be started as soon as possible. It is usually administered in three doses over six months. Hepatitis B vaccine is now a part of most infant immunization programs.

Influenza vaccines are important in the prevention of that disease. However, they must be administered on a yearly basis. Because the influenza viruses often have different antigenic proteins on their surfaces, epidemiologists must try to predict which influenza viruses are going to be a problem in the coming year. They then prepare an influenza vaccine that contains antigenic portions of the predicted influenza viruses. The epidemiologists' predictions are usually accurate, but in some years, inaccurate predictions have resulted in outbreaks of particular strains of influenza. ■

An Overview of the Immune Response

The purpose of the **immune response** is to inactivate or destroy pathogens, abnormal cells, and foreign molecules (such as toxins). It is based on the activation of lymphocytes by specific antigens through the process of *antigen recognition*. Figure 14–13● presents an overview of the immune response. When an antigen triggers an immune response, it usually activates T cells first and then B cells. T cells are typically activated by phagocytes that have engulfed the antigen. (Sometimes T cells are activated by abnormal body cells.) Once activated, the T cells both attack the antigen and stimulate the activation of B cells. The activated B cells mature into cells that produce antibodies. The circulating antibodies bind to and attack the antigen.

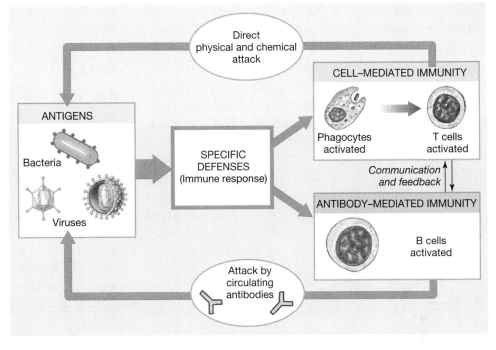

● **FIGURE 14–13** An Overview of the Immune Response.

T Cells and Cell-Mediated Immunity

Before an immune response can begin, T cells must be activated by exposure to an antigen. However, this activation seldom results from direct lymphocyte-antigen interaction, and the mere presence of foreign compounds or pathogens in a tissue rarely stimulates an immediate response.

T cells recognize antigens when those antigens are bound to membrane receptors of other cells. The structure of these antigen-binding membrane receptors is genetically determined and differs among individuals. The membrane receptors are called *major histocompatibility complex (MHC) proteins* and are grouped into two classes. An antigen bound to a Class I MHC protein acts like a red flag that in effect tells the immune system "Hey, I'm an abnormal cell—kill me!" An antigen bound to a Class II MHC protein tells the immune system "Hey, this antigen is dangerous—get rid of it!"

Class I MHC proteins are found on the surfaces of all of our nucleated cells. These MHC proteins bind small peptide molecules (chains of amino acids) that are normally produced in the cell and carried to the cell membrane. If the cell is healthy and the peptides are normal, T cells ignore them. If the cell contains abnormal (nonself), viral, or bacterial peptides, their appearance in the cell membrane will activate T cells, leading to destruction of the abnormal cell. The recognition of non-self peptides in transplanted tissue is the primary reason why donated organs are commonly rejected by the recipient. In the case of viruses or bacteria, T cells can be activated by contact with viral or bacterial antigens bound to Class I MHC proteins on the surface of infected cells. The activation of T cells by these antigens results in the destruction of the infected cells.

Class II MHC proteins are found in the membranes of lymphocytes and phagocytes called *antigen-presenting cells (APCs)*. Examples of antigen-presenting cells include the monocyte-macrophage group (such as the free and fixed macrophages in connective tissues) and macrophages in the liver (Kupffer cells) and central nervous system (microglia). APCs are specialized for activating T cells to attack foreign cells (including bacteria) and foreign proteins. After APCs engulf and break down pathogens or foreign antigens, fragments of the foreign antigens are displayed on their cell surfaces, bound to Class II MHC proteins. T cells that contact this APC's membrane become activated, initiating an immune response.

Inactive T cells have receptors that recognize either Class I or Class II MHC proteins. The receptors also have binding sites that detect specific bound antigens. T cell activation occurs when the MHC protein contains the specific antigen that the T cell is programmed to detect. On activation, T cells divide and differentiate into cells with specific functions in the immune response. The major cell types are *cytotoxic T cells, helper T cells, memory T cells,* and *suppressor T cells*.

Cytotoxic T Cells

Cytotoxic T cells, or *killer T cells,* are responsible for cell-mediated immunity. They are activated by exposure to antigens bound to Class I MHC proteins (Figure 14–14●). The activated cells undergo cell divisions that produce more active cytotoxic cells and memory cells. The activated cytotoxic T cells track down and attack the bacteria, fungi,

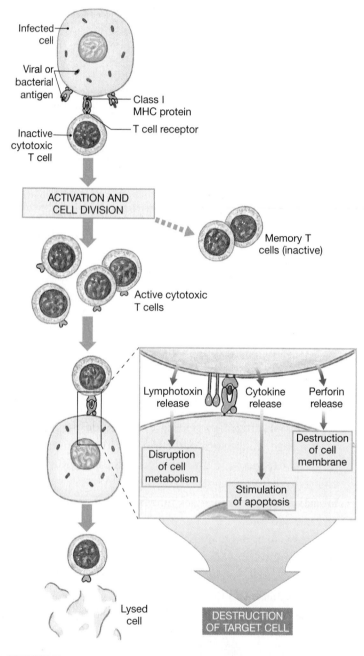

protozoa, or foreign transplanted tissues that contain the target antigen.

A cytotoxic T cell may destroy its target in several ways (see Figure 14–14):

- By secreting a *lymphotoxin* (lim-fō-TOK-sin), which kills the target cell by disrupting its metabolism.
- By secreting *cytokines* that activate target cell genes that tell the cell to die. This process of genetically programmed cell death is called *apoptosis.* ∞ p. 86
- By releasing *perforin,* which ruptures the target cell's membrane. ∞ p. 520

Helper T Cells

Activated **helper T cells** perform two functions, achieved by releasing various cytokines: they coordinate specific and nonspecific defenses, and they stimulate both cell-mediated immunity and antibody-mediated immunity. Helper T cells are activated by exposure to antigens bound to Class II MHC proteins on antigen-presenting cells. Upon activation, they divide to produce memory cells and more active helper T cells. We will learn more about the functions of activated helper T cells in the upcoming section on B cell activation.

Memory T Cells

As previously noted, some of the cells produced following the activation of cytotoxic T cells and helper T cells develop into **memory T cells.** Memory T cells remain "in reserve." If the same antigen appears a second time, these cells will immediately differentiate into cytotoxic T cells and helper T cells, enhancing the speed and effectiveness of the immune response.

Suppressor T Cells

Activated **suppressor T cells** dampen the responses of other T cells and of B cells by secreting cytokines called *suppression factors.* Suppression does not occur immediately, because suppressor T cells are activated more slowly than other types of T cells. As a result, suppressor T cells act *after* the initial immune response. In effect, these cells "put on the brakes," and limit the degree of the immune response to a single stimulus.

● **FIGURE 14–14 The Activation of Cytotoxic T cells.** For activation to occur, an inactive cytotoxic T cell must encounter an appropriate antigen bound to MHC proteins. Once activated, the T cell undergoes divisions that produce memory T cells and active cytotoxic T cells. When one of these active cytotoxic T cells encounters a cell membrane that displays the target antigen, the T cell will use one of several methods to destroy the cell.

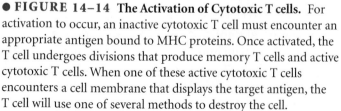

Key Note

Cell-mediated immunity involves close physical contact between activated cytotoxic T cells and foreign, abnormal, or infected cells. T cell activation usually involves antigen presentation by a phagocytic cell. Cytotoxic T cells may destroy target cells by the local release of cytokines, lymphotoxins, or perforin.

B Cells and Antibody-Mediated Immunity

B cells launch a chemical attack on antigens by participating in a series of events that results in the production of appropriate antibodies. The first event in antibody-mediated immunity is the sensitization of B cells (Figure 14–15●).

B Cell Sensitization and Activation

The body has millions of populations of B cells. Each B cell carries its own particular antibody molecules in its cell membrane. If corresponding antigens appear in the interstitial fluid, they will be bound by those antibodies. The antigens enter the B cell by endocytosis and become displayed on Class II MHC proteins on its surface. The B cell is now *sensitized*. Once a helper T cell has become activated to the same antigen and has attached to the MHC protein-antigen complex of the sensitized B cell, the helper T cell secretes cytokines. The cytokines have several effects, including promoting B cell activation, stimulating B cell division, and accelerating B cell development into plasma cells.

As Figure 14–15 shows, the activated B cell divides repeatedly, producing daughter cells that differentiate into **plasma cells** and **memory B cells.** Plasma cells begin synthesizing and secreting large quantities of antibodies that have the same target as the an-

tibodies on the surface of the sensitized B cell. Memory B cells, like memory T cells, remain in reserve to respond to subsequent exposure to the same antigen. At that time, they respond by differentiating into antibody-secreting plasma cells.

Antibody Structure

An antibody molecule consists of two parallel pairs of polypeptide chains: one pair of long *heavy chains* and one pair of shorter *light chains* (Figure 14–16a●). Each chain contains *constant* and *variable segments.* The constant segments of the heavy chains form the base of the antibody molecule. B cells produce only five types of constant segments. (These form the basis of the antibody classification scheme described in the following section.)

The specificity of an antibody molecule depends on the structure of the variable segments of the light and heavy chains. The free tips of the two variable segments form the **antigen binding sites** of the antibody molecule. Small differences in the amino acid sequence of the variable segments affect the precise shape of the antigen binding sites. The different shapes of these sites account for the differences in specificity among the antibodies produced by different B cells. It has been estimated that the approximately 10 trillion B cells of a normal adult can produce 100 million different antibodies.

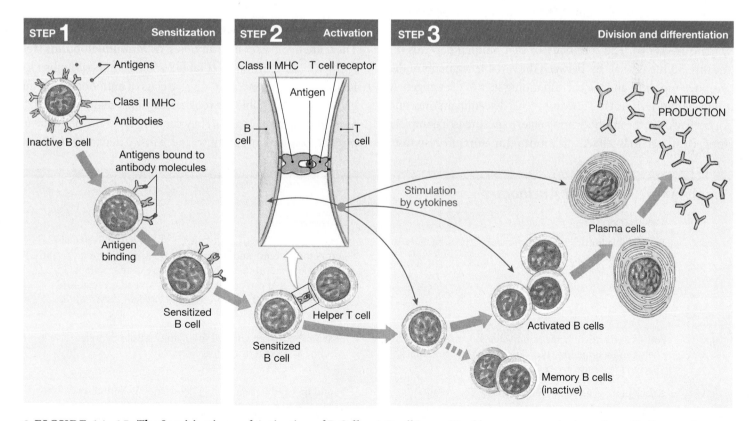

● **FIGURE 14–15 The Sensitization and Activation of B Cells.** A B cell is sensitized by exposure to antigens (step 1). Once antigens are bound to antibodies in the B cell membrane, the B cell displays those antigens in its cell membrane. Activated helper T cells that encounter the antigens on the sensitized B cell membrane then release cytokines that trigger the activation of the B cell (step 2). The activated B cell then divides, producing memory B cells and plasma cells that secrete antibodies (step 3).

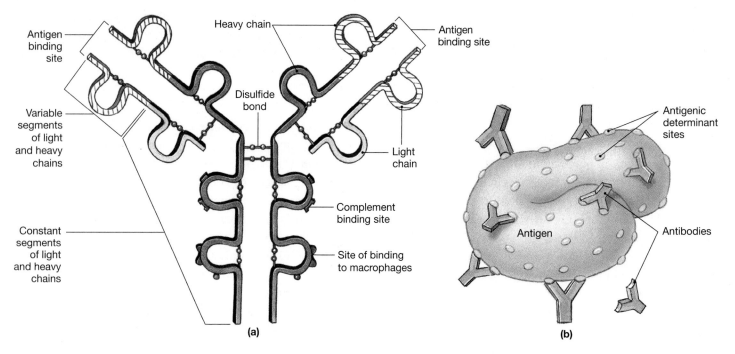

● **FIGURE 14–16 Antibody Structure.** (**a**) This diagram depicts the structural features of an antibody molecule. (**b**) Antigen-antibody binding occurs between the antigen binding sites on the antibody and antigenic determinant sites on the antigen.

When an antibody molecule binds to its specific antigen, an **antigen-antibody complex** is formed. Antibodies do not bind to the entire antigen as a whole. They bind to certain portions of its exposed surface, regions called *antigenic determinant sites* (Figure 14–16b●). The specificity of that binding depends on the three-dimensional "fit" between the variable segments of the antibody molecule and the corresponding sites of the antigen. A *complete antigen* has at least two antigenic determinant sites, one for each arm of the antibody molecule. Exposure to a complete antigen can lead to B cell sensitization and an immune response.

Most environmental antigens have multiple antigenic determinant sites; entire microorganisms may have thousands.

Classes of Antibodies

There are five classes of antibodies, or **immunoglobulins (Igs):** *IgG, IgM, IgA, IgE,* and *IgD* (Table 14–2). Immunoglobulin G, or IgG, is the largest and most diverse class of antibodies. IgG antibodies are responsible for resistance against many viruses, bacteria, and bacterial toxins. They can also cross the placenta and provide passive immunity to the fetus. Circulating *IgM* anti-

TABLE 14–2 *Classes of Antibodies*

CLASS	FUNCTION	REMARKS
IgG	Responsible for defense against many viruses, bacteria, and bacterial toxins	Largest class (80%) of antibodies, with several subtypes; also cross the placenta and provide passive immunity to fetus; anti-Rh antibodies produced by Rh-negative mothers are IgG antibodies that can cross the placenta and attack fetal Rh-positive red blood cells, producing *hemolytic disease of the newborn.* ∞ p. 419
IgM	Anti-A and anti-B forms responsible for cross-reactions between incompatible blood types; other forms attack bacteria insensitive to IgG	First antibody type secreted following arrival of antigen; levels decline as IgG production accelerates
IgA	Attack pathogens before they enter the body tissues	Found in glandular secretions (tears, mucus, and saliva)
IgE	Accelerate inflammation on exposure to antigen	Bound to surfaces of mast cells and basophils and stimulates release of histamine and other inflammatory chemicals; also important in allergic response
IgD	Bind antigens in the extracellular fluid to B cells	Binding can play a role in activation of B cells

Clinical Note

AIDS

Acquired immune deficiency syndrome (AIDS), or late-stage HIV disease, is caused by the *human immunodeficiency virus (HIV).* HIV is a *retrovirus,* which is a virus that carries its genetic information in RNA rather than DNA. The virus enters human leukocytes through receptor-mediated endocytosis. ∞ p. 73 In the human body, the virus binds to CD4, which is a membrane protein characteristic of helper T cells. Several types of antigen-presenting cells, including those of the monocyte-macrophage line, are also infected by HIV, but it is the infection of helper T cells that leads to clinical problems.

Cells infected with HIV ultimately die from the infection. The gradual destruction of helper T cells impairs the immune response because these cells play a central role in coordinating cell-mediated and antibody-mediated responses to antigens. To make matters worse, suppressor T cells are relatively unaffected by the virus, and over time the excess of suppressing factors "turns off" the normal immune response. Circulating antibody levels decline, cell-mediated immunity is reduced, and the body is left without defenses against a wide variety of microbial invaders. With immune function so reduced, ordinarily harmless microorganisms can initiate lethal *opportunistic infections.* Because immunological surveillance is also depressed, the risk of cancer increases.

HIV infection occurs through intimate contact with the body fluids of infected individuals. Although all body fluids may contain the virus, the major routes of transmission involve contact with blood, semen, or vaginal secretions. Most AIDS patients become infected through sexual contact with an HIV-infected person (who may not necessarily be showing the clinical signs of AIDS). The next largest group of patients consists of intravenous drug users who shared contaminated needles. Relatively few individuals have become infected with HIV after receiving a transfusion of contaminated blood or blood products. Finally, an increasing number of infants are born with the disease, having acquired it from their infected mothers.

AIDS is a public health problem of massive proportions. Nearly half a million people have already died of AIDS in the U.S. alone. Estimates of the total number of individuals infected with HIV in the U.S. range up to 1 million. The numbers worldwide are even more frightening. The World Health Organization (WHO) estimates that 40 million individuals were already infected as of 2003. *Every six seconds, another person becomes infected with HIV.* The total death toll for 2003 was estimated to be 3 million people, including 500,000 children under age 15.

The best defense against AIDS is to abstain from sexual contact or the sharing of needles. All forms of sexual intercourse carry the risk of viral transmission. The use of natural latex (rubber) condoms and synthetic (polyurethane) condoms provide protection from HIV (and other viral STDs). Although natural lambskin condoms are effective in preventing pregnancy, they do not block the passage of viruses.

Clinical signs of AIDS may not appear for 5–10 years or more following infection. When they do appear, they are often mild, consisting of lymphadenopathy and chronic but nonfatal infections. So far as is known, however, AIDS is almost always fatal, and most people infected with the virus will eventually die of the disease. (A handful of infected individuals have been able to tolerate the virus without apparent illness for many years.)

Despite intensive efforts, a vaccine has yet to be developed that prevents HIV infection in an uninfected person exposed to the virus. While efforts continue to prevent the spread of HIV, the survival rate for AIDS patients has been steadily increasing because new drugs that slow the progression of the disease are available, and improved antibiotic therapies help combat secondary infections. This combination is extending the life span of patients while the search for more effective treatment continues. ∎

---

bodies attack bacteria and are responsible for the cross-reactions between incompatible blood types, described in Chapter 11. ∞ p. 418 IgA occurs in exocrine secretions, such as mucus, tears, and saliva, and attacks pathogens before they enter the body. IgE that has bound to antigens stimulates basophils and mast cells to release chemicals that stimulate inflammation. IgD is attached to B cells and can be involved in their activation.

Antibody Function

The function of antibodies is to eliminate antigens. The formation of an antigen-antibody complex may cause their elimination in several ways:

1. *Neutralization.* Antibodies can bind to viruses or bacterial toxins, which makes them incapable of attaching to a cell. This mechanism is called **neutralization.**

2. *Agglutination and precipitation.* When a large number of antigens are close together, one antibody molecule can bind to antigenic sites on two different antigens. In this way, antibodies can tie antigens together and create large complexes. When the antigen is a soluble molecule (such as a bacterial toxin), the complex may then be too large to stay in solution. The resulting insoluble complex settles out of body fluids in a process called **precipitation.** The formation of large complexes is called **agglutination.** The clumping of red blood cells that occurs when incompatible blood types are mixed is an agglutination reaction. ∞ p. 418

3. *Activation of complement.* Upon binding to an antigen, portions of the antibody molecule change shape, and expose areas of the constant segments that bind complement proteins (see Figure 14–16a). The bound complement molecules then activate the complement system, which destroys the antigen.

4. *Attraction of phagocytes.* Antigens covered with antibodies attract eosinophils, neutrophils, and macrophages—cells that phagocytize pathogens and destroy cells with foreign or abnormal cell membranes.

5. *Enhancement of phagocytosis.* A coating of antibodies and complement proteins makes some pathogens easier to phagocytize. The antibodies involved are called *opsonins,* and the effect is known as *opsonization.*

6. *Stimulation of inflammation.* Antibodies may promote inflammation by stimulating basophils and mast cells. This action can help mobilize nonspecific defenses and slow the spread of the infection to other tissues.

🔒 Key Note

Antibody-mediated immunity involves the production of specific antibodies by plasma cells derived from activated B cells. B cell activation usually involves (1) antigen recognition, through binding to surface antibodies, and (2) stimulation by a helper T cell activated by the same antigen. The antibodies produced by active plasma cells bind to the target antigen and either inhibit its activity, destroy it, remove it from solution, or promote its phagocytosis by other defense cells.

Primary and Secondary Responses to Antigen Exposure

The body's initial response to antigen exposure is called the **primary response.** When an antigen appears a second time, it triggers a more extensive **secondary response.** The secondary response reflects the presence of large numbers of memory cells that are already "primed" for the arrival of the antigen.

Because the antigen must activate the appropriate B cells, and the B cells must then respond by differentiating into plasma cells, the primary response takes time to develop (Figure 14–17●). As plasma cells begin secreting antibodies, the concentration of circulating antibodies undergoes a gradual, sustained rise, and the antibody levels in the blood do not peak until one to two weeks after the initial exposure. IgM molecules are the first to appear in the bloodstream, followed by a slow rise in IgG. Thereafter, antibody concentrations decline (assuming that the individual is no longer being exposed to the antigen).

Memory B cells do not differentiate into plasma cells unless they are exposed to the same antigen a second time. If and when that exposure occurs, memory B cells respond immediately by dividing and differentiating into plasma cells that secrete antibodies in massive quantities. This antibody secretion is the secondary response to antigen exposure.

The secondary response produces an immediate rise in IgG concentrations to levels many times higher than those of the

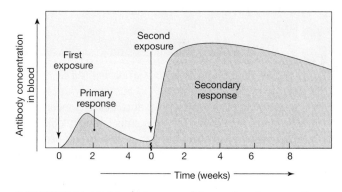

● **FIGURE 14–17 The Primary and Secondary Immune Responses.** The primary response takes about two weeks to develop peak antibody levels, and antibody concentrations do not remain elevated. The secondary response is characterized by a very rapid increase in antibody concentrations to levels much higher than those of the primary response. Antibody levels remain elevated for an extended period.

primary response. This response is much faster and stronger than the primary response because the numerous memory cells are activated by relatively low levels of antigen, and they give rise to plasma cells that secrete massive quantities of antibodies. The secondary response appears even if the second exposure occurs years after the first because memory cells may survive for 20 years or more.

The relatively slow primary response is much less effective at preventing disease than the more rapid and intense secondary response. Immunization is effective because it stimulates the production of memory B cells under controlled conditions; it is the secondary response that prevents disease.

🔒 Key Note

Immunization produces a primary response to a specific antigen under controlled conditions. If the same antigen is encountered at a later date, it triggers a powerful secondary response that is usually sufficient to prevent infection and disease.

Hormones of the Immune System

The body's specific and nonspecific defenses are coordinated by physical interaction and by the release of chemical messengers. An example of physical interaction is the displaying of antigens by antigen-presenting macrophages. An example of chemical messenger release is the secretion of cytokines by activated helper T cells and other cells involved in the immune response. Cytokines of the immune response are often classified according to their sources: *lymphokines,* which are secreted by lymphocytes, and *monokines,* which are released by active macrophages and other antigen-presenting cells. (These terms do not necessarily reflect

TABLE 14–3 *Examples of Cytokines of the Immune Response*

COMPOUND	FUNCTIONS
INTERLEUKINS	
IL-1	Stimulates T cells to produce IL-2, promotes inflammation, causes fever; stimulates the secreting cell in a positive feedback loop that recruits more immune cells
IL-2, -12	Stimulates T cells and NK cells; stimulates the secreting cell in a positive feedback loop that recruits more immune cells
IL-3	Stimulates production of mast cells and other blood cells
IL-4, -5, -6, -7, -10, -11	Promote differentiation and growth of B cells, and stimulate plasma cell formation and antibody production
IL-8	Stimulates blood vessel formation
IL-13	Suppresses production of several cytokines (IL-1, IL-8, TNF)
INTERFERONS	Activate other cells to prevent viral entry and replication, stimulate NK cells and macrophages
TUMOR NECROSIS FACTORS (TNFs)	Kill tumor cells, slow tumor growth; stimulate activities of T cells and eosinophils, inhibit parasites and viruses
PHAGOCYTIC REGULATORS	
Monocyte-chemotactic factor (MCF)	Attracts monocytes, transforms them into macrophages
Migration-inhibitory factor (MIF)	Prevents macrophage migration from the area
COLONY-STIMULATING FACTORS (CSFs)	Stimulates RBC and WBC production
M-CSF	Stimulates production of monocytes
GM-CSF	Stimulates production of both granulocytes (neutrophils, eosinophils, and basophils) and monocytes
Multi-CSF	Stimulates production of granulocytes, monocytes, and RBCs

different types of cytokines because lymphocytes and macrophages may secrete the same chemical messenger, as may cells involved with nonspecific defenses and tissue repair.) Table 14–3 contains examples of some of the cytokines identified to date.

Interleukins (IL) may be the most diverse and important chemical messengers in the immune system. Interleukins have widespread effects that include increasing T cell sensitivity to antigens presented on macrophage membranes, stimulating B cell activity and antibody production, and enhancing nonspecific defenses, such as inflammation or fever. Some interleukins help suppress immune function and shorten the duration of an immune response.

Interferons released by the synthesizing cell make its neighbors resistant to viral infection, thereby slowing the spread of the virus. In addition to their antiviral activity, interferons attract and stimulate NK cells and macrophages. Interferons are used to treat some cancers.

Tumor necrosis factors (TNFs) slow tumor growth and kill sensitive tumor cells. In addition to these effects, TNFs stimulate the production of neutrophils, eosinophils, and basophils; promote eosinophil activity; cause fever; and increase T cell sensitivity to interleukins.

Phagocytic regulators include several cytokines that coordinate the specific and nonspecific defenses by adjusting the activities of phagocytic cells. These cytokines include factors that attract free macrophages and microphages to the area and prevent their premature departure.

Colony-stimulating factors (CSFs) are produced by a wide variety of cells, including active T cells, cells of the monocyte-macrophage group, endothelial cells, and fibroblasts. CSFs stimulate the production of blood cells in the bone marrow and of lymphocytes in lymphoid tissues and organs.

→ CONCEPT CHECK QUESTIONS

1. How can the presence of an abnormal peptide within a cell initiate an immune response?
2. A decrease in the number of cytotoxic T cells would affect what type of immunity?
3. How would a lack of helper T cells affect the antibody-mediated immune response?
4. A sample of lymph contains an elevated number of plasma cells. Would you expect the amount of antibodies in the blood to be increasing or decreasing? Why?

Answers begin on p. 792.

Key Note

Viruses replicate inside of cells, whereas bacteria may live independently. Antibodies (and administered antibiotics) work outside of cells, so they are primarily effective against bacteria rather than viruses. (That is why antibiotics cannot fight the common cold or flu.) T cells, NK cells, and interferons are the primary defenses against viral infection.

■ Patterns of Immune Response

We have discussed the patterns of the body's normal responses to the appearance of a foreign antigen. A summary of the events in the immune response, and its relationship to nonspecific defenses, is presented in Figure 14–18●. Next we discuss two topics that involve variations on the normal pattern of the immune response: immune disorders and age-related changes in immune function.

Immune Disorders

Because the immune response is so complex, there are many opportunities for things to go wrong. A variety of clinical conditions result from disorders of immune function. General classes of such disorders include *autoimmune disorders, immunodeficiency diseases,* and *allergies.* Immunodeficiency diseases and autoimmune disorders are relatively rare conditions—clear evidence of the effectiveness of the immune system's control mechanisms. Allergies constitute a far more common (and usually a far less dangerous) class of immune disorders.

Autoimmune Disorders

Autoimmune disorders develop when the immune response mistakenly targets normal body cells and tissues. The immune system usually recognizes and ignores the antigens normally

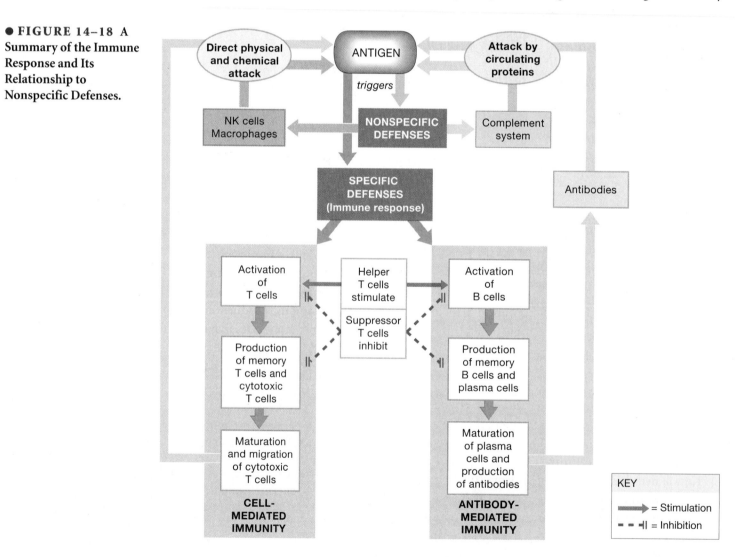

● **FIGURE 14–18 A Summary of the Immune Response and Its Relationship to Nonspecific Defenses.**

found in the body—self antigens. When the recognition system malfunctions, however, activated B cells begin to manufacture antibodies against normal body cells and tissues. These misguided antibodies are called *autoantibodies.* The resulting conditions depend on the identity of the antigen attacked. For example, *rheumatoid arthritis* occurs when autoantibodies attack connective tissues around the joints, and *insulin-dependent diabetes mellitus (IDDM)* is caused by autoantibodies that attack cells in the pancreatic islets.

Many autoimmune disorders appear to be cases of mistaken identity. In one example, proteins associated with several viruses (including the measles and influenza viruses) contain amino acid sequences that resemble those of the nervous system protein myelin. When antibodies that target these viruses mistakenly attack myelin sheaths, neurological complications sometimes associated with a vaccination or viral infection can result.

For unknown reasons, the risk of autoimmune problems increases for individuals with unusual types of MHC proteins. At least 50 clinical conditions have been linked to specific variations in MHC structure. Examples of such conditions include psoriasis, rheumatoid arthritis, myasthenia gravis, narcolepsy, Graves' disease, Addison's disease, pernicious anemia, systemic lupus erythematosus, and chronic hepatitis.

Immunodeficiency Diseases

In an **immunodeficiency disease,** either the immune system fails to develop normally or the immune response is blocked in some way. Individuals born with **severe combined immunodeficiency disease** (SCID) fail to develop either cellular or antibody-mediated immunity. Such infants cannot produce an immune response, so even a mild infection can prove fatal. Total isolation offers protection at great cost, with severe restrictions on lifestyle. Bone marrow transplants and gene-splicing techniques have been used to treat some types of SCID.

AIDS, an immunodeficiency disease (p. 531), is the result of a viral infection that targets primarily helper T cells. As the number of T cells falls, the normal immune response breaks down.

Allergies

Allergies are inappropriate or excessive immune responses to antigens. The sudden increase in leukocyte activity or antibody levels can have several unpleasant side effects. Neutrophils or cytotoxic T cells may destroy normal cells while attacking the antigen, or antigen-antibody complexes may trigger a massive inflammatory response. Antigens that trigger allergic reactions are often called **allergens.**

Four categories of allergies are recognized: *immediate hypersensitivity (Type I), cytotoxic reactions (Type II), immune complex disorders (Type III),* and *delayed hypersensitivity (Type IV).* Immediate hypersensitivity (discussed shortly) is probably the most common type; it includes "hay fever" and environmental allergies that may affect 15 percent of the U.S. population. The cross-reactions that occur following the transfusion of an incompatible blood type are an example of Type II (cytotoxic) hypersensitivity. ∞ p. 418 Type III allergies result if phagocytes are not able to rapidly remove circulating antigen-antibody complexes. The presence of these complexes leads to inflammation and tissue damage, especially within blood vessels and the kidneys. Type IV (delayed) hypersensitivity is an inflammatory response that occurs two to three days after exposure to an antigen, as in the itchy rash that may follow contact with poison ivy or poison oak.

Immediate hypersensitivity is a rapid and especially severe response to the presence of an antigen. Sensitization to an allergen during the initial exposure leads to the production of large quantities of IgE antibodies. Due to the lag time needed to activate B cells, produce plasma cells, and make antibodies, the first exposure to an allergen does not produce an allergic reaction. However, the IgE antibodies produced at this time become attached to the cell membranes of basophils and mast cells throughout the body. When exposed to the same allergen later, these cells are stimulated to release histamine, heparin, several cytokines, prostaglandins, and other chemicals into the surrounding tissues. The result is sudden inflammation of the affected tissues.

The severity of an immediate hypersensitivity allergic reaction depends on the person's sensitivity and the location involved. If allergen exposure occurs at the body surface, the response is usually restricted to that area. If the allergen enters the bloodstream, the response could be lethal.

In **anaphylaxis** (an-a-fi-LAK-sis; *ana-,* again + *phylaxis,* protection), another Type I allergy, a circulating allergen affects mast cells throughout the body. A wide range of signs and symptoms can develop within minutes. Changes in capillary permeability produces swelling and edema in the dermis, and raised welts, or *hives,* appear on the surface of the skin. Smooth muscles along the respiratory passageways contract; the narrowed passages make breathing extremely difficult. In severe cases, extensive peripheral vasodilation occurs, which produces a fall in blood pressure that can lead to a circulatory collapse. This response is **anaphylactic shock.** Many of the signs and symptoms of immediate hypersensitivity can be prevented by the prompt administration of **antihistamines** (an-tē-HIS-ta-mēnz)—drugs that block the action of histamine.

Clinical Note
ALLERGIES

An *allergic reaction* is an exaggerated response by the immune system to a foreign substance. Allergic reactions can range from mild skin rashes to severe, life-threatening reactions that involve virtually every body system. The most severe type of allergic reaction is called anaphylaxis. Anaphylaxis is a life-threatening emergency that requires prompt recognition and treatment.

The immune system is the principal body system involved in allergic reactions, although other body systems are also affected. These include the cardiovascular system, the respiratory system, the gastrointestinal system, and the nervous system, among others.

An individual's initial exposure, referred to as *sensitization,* results in an immune response. Subsequent exposure induces a much stronger secondary response. Some individuals can become hypersensitive (overly sensitive) to a particular antigen. *Hypersensitivity* is an unexpected and exaggerated reaction to a particular antigen. In many instances, hypersensitivity is used synonymously with the term *allergy.* There are two types of hypersensitivity reactions, delayed and immediate.

Delayed Hypersensitivity

Delayed hypersensitivity results from cellular immunity and therefore does not involve antibodies. It usually occurs in the hours and days following exposure and is the sort of allergy that occurs in normal people. Delayed hypersensitivity most commonly results in a skin rash and is often due to exposure to certain drugs and chemicals. The rash associated with poison ivy is an example of delayed hypersensitivity (Figure 14–19●).

Immediate Hypersensitivity

When people use the term *allergy* they usually are referring to immediate hypersensitivity reactions. Examples of immediate hypersensitivity reactions include hay fever, drug allergies, food allergies, eczema, and asthma. Some individuals have an allergic tendency. This allergic tendency is usually genetic, which means it is passed from parent to child and is characterized by the presence of large quantities of IgE antibodies. Any antigen that causes release of the IgE antibodies is referred to as an *allergen.* Common allergens include:

- Drugs
- Foods and food additives
- Animals
- Insects (Hymenoptera stings) and insect parts
- Fungi and molds
- Radiology contrast materials

Allergens can enter the body through various routes, including oral ingestion, inhalation, topical absorption, injection, and envenomation. The vast majority of anaphylactic reactions result from injection or envenomation.

(a)

(b)

(c)

● **FIGURE 14–19 Poisonous Plants.** Plants of the genus *Toxicodendron* contain the irritative resin urushiol, which causes a cell-mediated reaction and rash. They are: (**a**) poison ivy, (**b**) poison sumac, and (**c**) poison oak.

Parenteral penicillin injections are the most common cause of fatal anaphylactic reactions. Insect stings are the second most frequent cause of fatal anaphylactic reactions. The most frequent offending insects are those in the order *Hymenoptera.* There are three families in this order: fire ants (*Formicoidea*); wasps, yellow jackets, and hornets (*Vespidae*); and honey bees (*Apoidea*). Although these families' venoms all have similar components, each is unique.

Following the body's exposure to a particular allergen, large quantities of IgE antibodies are released. These antibodies attach to the membranes of basophils and mast cells—specialized cells of the immune system containing chemicals that assist in the immune response. When the allergen binds to IgE attached to the basophils and mast cells, those cells release histamine, heparin, and other substances into the surrounding tissues. Histamine and other substances are stored in granules found within the basophils and mast cells. (Because of this, basophils and mast cells are often called *granulocytes.*) The process of releasing these substances from the cells, termed *degranulation,* results in what people call an allergic reaction, which can range from very mild to very severe.

The principal chemical mediator of an allergic reaction is *histamine,* which is a potent substance that activates specialized histamine receptors present throughout the body. This causes bronchoconstriction, increased intestinal motility, vasodilation, and increased vascular permeability. Increased vascular permeability causes fluid from the circulatory system to leak into the surrounding tissues. Thus, a common manifestation of severe allergic reactions and anaphylaxis is angioneurotic edema, a marked edema of the skin that usually involves the head, neck, face, and upper airway (Figure 14–20●).

There are two classes of histamine receptors. H_1 receptors, when stimulated, cause bronchoconstriction and contraction of the intestines. H_2 receptors cause peripheral vasodilation and secretion of gastric acids. The goal of histamine release is to minimize the body's exposure to the antigen. Bronchoconstriction decreases the antigen's possibility of entering through the respiratory tract, while increased gastric acid production helps destroy an ingested antigen. Increased intestinal motility serves to move the antigen quickly through the gastrointestinal system with minimal absorption of the antigen into the body. Vasodilation and capillary permeability help remove the allergen from the circulation, where it has the potential to do the most harm. ■

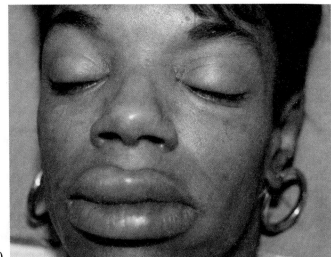

(a)

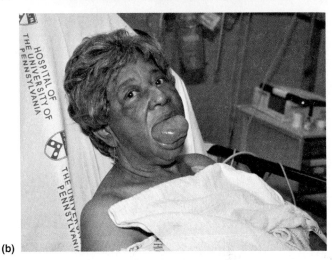

(b)

● **FIGURE 14–20 Examples of Angioneurotic Edema.**
(**a**) Swelling of the face and lips can be severe. (**b**) Marked swelling of the tongue can compromise or obstruct the airway.

Aging and the Immune Response

With advancing age, the immune system becomes less effective at combating disease. T cells become less responsive to antigens, so fewer cytotoxic T cells respond to an infection. This effect may, at least in part, be associated with the gradual shrinkage of the thymus and reduced levels of circulating thymic hormones. Because the number of helper T cells is also reduced, B cells are less responsive, and antibody levels rise more slowly after antigen exposure. The net result is an increased susceptibility to viral and bacterial infections. For this reason, vaccinations for acute viral diseases, such as the flu (influenza), are strongly recommended for elderly people. The increased incidence of cancer in the elderly reflects the fact that immunological surveillance declines, so that tumor cells are not eliminated as effectively.

■ Integration with Other Systems

Figure 14–21● summarizes the interactions between the lymphatic system and other organ systems. Two sets of particularly close relationships—between cells of the immune response and the endocrine system, and between those cells and the nervous system—are now the focus of intense research. In the first case, thymic hormones and cytokines stimulate TRH production by the hypothalamus, which leads to

The Lymphatic System in Perspective

For All Systems

Provides specific defenses against infection; immune surveillance eliminates cancer cells; returns tissue fluid to circulation

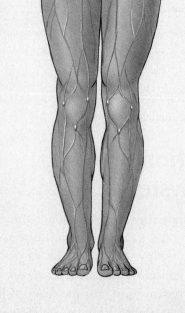

The Integumentary System

- Provides physical barriers to pathogen entry; macrophages in dermis resist infection and present antigens to trigger immune response; mast cells trigger inflammation, mobilize cells of lymphatic system
- Provides IgA antibodies for secretion onto integumentary surfaces

The Skeletal System

- Lymphocytes and other cells involved in the immune response are produced and stored in bone marrow
- Assists in repair of bone after injuries; macrophages fuse to become osteoclasts

The Muscular System

- Protects superficial lymph nodes and the lymphatic vessels in the abdominopelvic cavity; muscle contractions help propel lymph along lymphatic vessels
- Assists in repair after injuries

The Nervous System

- Microglia present antigens that stimulate specific defenses; glial cells secrete cytokines; innervation stimulates antigen-presenting cells
- Cytokines affect hypothalamic production of CRH and TRH

The Endocrine System

- Glucocorticoids have anti-inflammatory effects; thymosins stimulate development and maturation of lymphocytes; many hormones affect immune function
- Thymus secretes thymosins; cytokines affect cells throughout the body

The Cardiovascular System

- Distributes WBCs; carries antibodies that attack pathogens; clotting response helps restrict spread of pathogens; granulocytes and lymphocytes produced in bone marrow
- Fights infections of cardiovascular organs; returns tissue fluid to circulation

The Respiratory System

- Alveolar phagocytes present antigens and trigger specific defenses; provides oxygen required by lymphocytes and eliminates carbon dioxide generated during their metabolic activities
- Tonsils protect against infection at entrance to respiratory tract

The Digestive System

- Provides nutrients required by lymphatic tissues; digestive acids and enzymes provide nonspecific defense against pathogens
- Tonsils and lymphoid nodules of the intestine defend against infection and toxins absorbed from the digestive tract; lymphatics carry absorbed lipids to venous system

The Urinary System

- Eliminates metabolic wastes generated by cellular activity; acid pH of urine provides nonspecific defense against urinary tract infection
- Provides specific defenses against urinary tract infections

The Reproductive System

- Lysozymes and bactericidal chemicals in secretions provide nonspecific defense against reproductive tract infections
- Provides IgA antibodies for secretion by epithelial glands

● **FIGURE 14–21** Functional Relationships Between the Lymphatic System and Other Systems.

the release of TSH by the pituitary gland. As a result, circulating thyroid hormone levels increase and stimulate cell and tissue metabolism when an immune response is under way. But the nervous system can also adjust the level of the immune response. For example, some antigen-presenting cells are innervated, and neurotransmitter release exaggerates the local immune response. It is also known that the immune response can decline suddenly after a brief period of emotional distress, such as a heated argument.

→ **CONCEPT CHECK QUESTIONS**

1. Which would be more affected by a lack of memory B cells for a particular antigen: the primary response or the secondary response?
2. Which type of immunity protects a growing fetus, and how does that immunity develop?
3. How does the cardiovascular system aid the body's defense mechanisms?

Answers begin on p. 792.

Clinical Note
ANAPHYLAXIS

The most severe type of allergic reaction is anaphylaxis. Anaphylaxis usually occurs when a specific allergen is injected directly into the circulation. Thus, anaphylaxis is more common following injections of drugs and diagnostic agents and following bee stings (Table 14–4). When the allergen enters the circulation, it is rapidly distributed throughout the body. It interacts with both basophils and mast cells, resulting in the massive dumping of histamine and other substances associated with anaphylaxis. The principal body systems affected by anaphylaxis are the cardiovascular system, the respiratory system, the gastrointestinal system, and the skin. Histamine causes widespread peripheral vasodilation as well as increased permeability of the capillaries. In turn, increased capillary permeability results in marked loss of plasma from the circulation. People who sustain anaphylaxis can die from circulatory shock.

Also released from the basophils and mast cells is a substance called *slow-reacting substance of anaphylaxis (SRS-A)*. This substance causes spasm of the bronchial smooth muscle, resulting in an asthma-like attack and occasionally asphyxia. SRS-A potentiates the effects of histamine, especially on the respiratory system.

The signs and symptoms of anaphylaxis typically begin within 30–60 seconds following exposure to the offending allergen. In a small percentage of patients, their onset may be delayed more than an hour. The signs and symptoms of anaphylaxis can vary significantly. The severity of the reaction is often proportional to the speed of onset: reactions that develop very quickly tend to be much more severe. Patients who suffer an anaphylactic reaction often have a sense of impending doom. In addition to angioneurotic edema that involves the face and neck, laryngeal edema is a frequent complication and can threaten the airway. Initially, laryngeal edema will cause hoarseness, and as the edema worsens, the patient may develop stridor. Finally, this may culminate in complete airway obstruction from either massive laryngeal edema, laryngospasm, oropharyngeal edema, or any combination of these.

The respiratory system is significantly involved in anaphylactic reactions. Initially, the patient will become tachypneic. Later, as lower airway edema and bronchospasm develop, respirations will become labored as evidenced by retractions, accessory muscle usage, and prolonged expiration. Wheezing commonly results from bronchospasm and edema of the smaller airways and may be so pronounced that it can be heard without the aid of a stethoscope. Ultimately, anaphylaxis can result in markedly diminished lung sounds, which reflect decreased air movement and hypoventilation.

The skin is typically involved early in severe allergic reactions and anaphylaxis. Generally, a fine red rash will appear diffusely on the body. As histamine is released, fluid will diffuse from leaky capillaries, which results in urticaria. *Urticaria,* also called *hives,* is a wheal and flare reaction characterized by red raised bumps that may appear and disappear across the body (Figure 14–22●). As cardiovascular collapse

TABLE 14–4 *Agents That May Cause Anaphylaxis*
Antibiotics and other drugs
Foreign proteins (e.g., horse serum, streptokinase)
Foods (nuts, eggs, shrimp)
Allergen extracts (allergy shots)
Hymenoptera stings (bees, wasps)
Hormones (insulin)
Blood products
Aspirin
Nonsteroidal anti-inflammatory drugs (NSAIDs)
Preservatives (sulfiting agents)
X-ray contrast media
Dextran

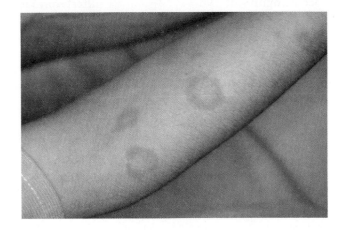

● **FIGURE 14–22 Urticaria (Hives) on the Arm.**

(continued)

Clinical Note—*continued*
ANAPHYLAXIS

and dyspnea progress, the patient will become diaphoretic. This may, if untreated, progress to cyanosis and pallor.

Histamine's affect on the gastrointestinal system is pronounced. Initially, the patient may note a rumbling sensation in the abdomen as gastrointestinal motility increases. Later, nausea, vomiting, cramping, and diarrhea develop as the body tries to rid itself of the offending allergen.

Emergency treatment of anaphylaxis includes monitoring the patient with all available devices, including the cardiac monitor, the pulse oximeter, and if the patient is intubated, an end-tidal carbon dioxide detector. As anaphylaxis progresses, the end-tidal carbon dioxide level may climb due to the development of both respiratory and metabolic acidosis, which results in increased carbon dioxide elimination. High concentrations of oxygen should be administered and intravenous fluids should be initiated. Emergency medications used in the treatment of anaphylaxis include oxygen, epinephrine, antihistamines, corticosteroids, and vasopressors. Occasionally, inhaled beta agonists, such as albuterol, may be required. Of these, epinephrine remains the primary treatment.

A severe allergic or anaphylactic reaction is a harrowing experience for the patient. Many of the patients you treat will be suffering from forms of allergic reaction less severe than anaphylaxis. An allergic reaction, as contrasted with an anaphylactic reaction, will have a more gradual onset with milder signs and symptoms, and the patient will have a normal mental status. Common manifestations of mild (nonanaphylactic) allergic reactions include itching, rash, and urticaria. Patients with simple itching and nonurticarial rashes may be treated with antihistamines alone. In addition to antihistamines, epinephrine is often necessary for the treatment of urticaria.

Lesser allergic reactions that are not accompanied by hypotension or airway problems can be adequately treated with epinephrine 1:1000 administered subcutaneously or intramuscularly. ■

Chapter Review

Access more review material online at *www.prenhall.com/bledsoe*. There you will find quiz questions, labeling activities, animations, essay questions, and web links.

Key Terms

acquired immunity 522
agglutination 531
allergen 535
antibody 514
antigen 514
antigen-antibody complex 530
antigen recognition 523
B cells 513

chemotaxis 519
cytokines 520
immunity 512
immunoglobulin 530
immunological surveillance 520
inflammation 521
interferon 520
lymphocyte 512

monocyte-macrophage system 519
NK cells 514
passive immunity 523
plasma cells 514
pyrogen 522
T cells 513
thymus 516

Related Clinical Terms

allergy An inappropriate or excessive immune response to antigens.
anaphylactic shock A drop in blood pressure that may lead to circulatory collapse, resulting from a severe case of anaphylaxis.
anaphylaxis (an-a-fi-LAK-sis) An allergic reaction in which a circulating allergen (antigen) affects mast cells throughout the body, producing numerous signs and symptoms very quickly.
appendicitis Inflammation of the lymphoid nodules in the appendix, often in response to an infection.
bone marrow transplantation The infusion of bone marrow from a compatible donor after the destruction of the host's marrow by radiation or chemotherapy; a treatment option for acute, late-stage lymphoma.
graft-versus-host disease (GVH) A condition that results when T cells in donor tissues, such as bone marrow, attack the tissues of the recipient.
immunodeficiency disease A disease in which either the immune system fails to

develop normally or the immune response is blocked.

immunology The study of the functions and disorders of the body's defense mechanisms.

immunosuppression A reduction in the sensitivity of the immune system.

lymphadenopathy (lim-fad-e-NOP-a-thē) An excessive enlargement of lymph nodes.

lymphedema A painless accumulation of lymph in a region in which lymphatic drainage has been blocked.

lymphomas Cancers that consist of abnormal lymphocytes or lymphoid stem cells; examples include *Hodgkin's disease* and *non-Hodgkin's lymphoma.*

mononucleosis A condition that results from chronic infection by the *Epstein-Barr virus (EBV)*; signs and symptoms include enlargement of the spleen, fever, sore throat, widespread swelling of lymph nodes, increased numbers of circulating lymphocytes, and the presence of circulating antibodies to the virus.

splenomegaly (splen-ō-MEG-a-lē) Enlargement of the spleen.

systemic lupus erythematosus (LOO-pus e-rith-ē-ma-TŌ-sis) **(SLE)** An autoimmune disorder that results from a breakdown in the antigen recognition mechanism, leading to the production of antibodies that destroy healthy cells and tissues.

tonsillectomy The removal of an inflamed tonsil.

tonsillitis Inflammation of one or more tonsils, typically in response to an infection; signs and symptoms include a sore throat, high fever, and leukocytosis (an elevated white blood cell count).

vaccine (vak-SĒN) A preparation of antigens derived from a specific pathogen; administered during *immunization,* or *vaccination.*

viruses Noncellular pathogens that replicate only within living tissue cells by directing the cells to synthesize virus-specific proteins and nucleic acids.

Summary Outline

1. The cells, tissues, and organs of the **lymphatic system** play a central role in the body's defenses against a variety of **pathogens,** or disease-causing organisms.

ORGANIZATION OF THE LYMPHATIC SYSTEM 512

1. The lymphatic system includes a network of **lymphatic vessels** called **lymphatics,** which carry **lymph** (a fluid similar to plasma but with a lower concentration of proteins). A series of **lymphoid organs** is connected to the lymphatic vessels. (*Figure 14–1*)

Functions of the Lymphatic System 512

2. The lymphatic system produces, maintains, and distributes lymphocytes (cells that attack invading organisms, abnormal cells, and foreign proteins). The system also helps maintain blood volume and eliminate local variations in the composition of the interstitial fluid.

Lymphatic Vessels 513

3. Lymph flows along a network of lymphatics that originates in the **lymphatic capillaries.** The lymphatic vessels empty into the **thoracic duct** and the **right lymphatic duct.** (*Figures 14–1 to Figure 14–3*)

Lymphocytes 513

4. The three classes of lymphocytes are **T cells** (*t*hymus-dependent), **B cells** (*b*one marrow-derived), and **natural killer (NK) cells.**

5. *Cytotoxic T cells* attack foreign cells or body cells infected by viruses; they provide cell-mediated immunity. **Regulatory T cells** (*helper T cells* and *suppressor T cells*) regulate and coordinate the immune response.

6. B cells can differentiate into **plasma cells,** which produce and secrete antibodies that react with specific chemical targets, or **antigens.** Antibodies in body fluids are also called **immunoglobulins.** B cells are responsible for *antibody-mediated immunity,* or *humoral immunity.*

7. NK cells attack foreign cells, normal cells infected with viruses, and cancer cells. They provide a monitoring service called *immunological surveillance.*

8. Lymphocytes continuously migrate in and out of the blood through the lymphoid tissues and organs. *Lymphopoiesis* (lymphocyte production and development) involves the bone marrow, thymus, and peripheral lymphoid tissues. (*Figure 14–4*)

Lymphoid Nodules 515

9. A **lymphoid nodule** consists of loose connective tissue that contains densely packed lymphocytes. **Tonsils** are lymphoid nodules in the pharynx wall. (*Figure 14–5*)

Lymphoid Organs 515

10. Important lymphoid organs include the *lymph nodes,* the *thymus,* and the *spleen.* Lymphoid tissues and organs are distributed in areas especially vulnerable to invasion by pathogens.

11. **Lymph nodes** are encapsulated masses of lymphoid tissue that contain lymphocytes. Lymph nodes monitor and filter the lymph before it drains into the venous system, removing antigens and initiating appropriate immune responses. (*Figure 14–6*)

12. The **thymus** lies behind the sternum. T cells become mature in the thymus. (*Figure 14–7*)

13. The adult **spleen** contains the largest mass of lymphoid tissue in the body. *Red pulp* contains large numbers of red blood cells, and areas of *white pulp* resemble lymphoid nodules. The spleen removes antigens and damaged blood cells from the circulation, initiates appropriate immune responses, and stores iron obtained from recycled red blood cells. (*Figures 14–8, 14–9*)

The Roles of the Lymphatic System in Body Defenses 518

14. The lymphatic system is a major component of the body's defenses, which fall into two categories: (1) **nonspecific defenses,** which do not discriminate between one threat and another; and (2) **specific defenses,** which protect against particular threats.

NONSPECIFIC DEFENSES 519

1. Nonspecific defenses prevent the approach, deny the entrance, or limit the spread of living or nonliving hazards. *(Figure 14–10)*

Physical Barriers 519

2. Physical barriers include the skin (including hair and secretions of dermal glands) and the epithelia and various secretions of the digestive system.

Phagocytes 519

3. **Phagocytes** include **microphages** (neutrophils and eosinophils) and **macrophages** (cells of the *monocyte-macrophage system*).

4. Phagocytes move between cells through *diapedesis,* and they show *chemotaxis* (sensitivity and orientation to chemical stimuli).

Immunological Surveillance 520

5. **Immunological surveillance** involves constant monitoring of normal tissues by NK cells sensitive to abnormal antigens on the surfaces of otherwise normal cells. NK cells kill both cancer cells that display tumor-specific surface antigens and virus-infected cells.

Interferons 520

6. **Interferons,** which are small proteins released by cells infected with viruses, trigger the production of antiviral proteins that interfere with viral replication inside other cells. Interferons are *cytokines,* which are chemical messengers released by tissue cells to coordinate local activities.

Complement 521

7. The 11 *complement proteins* of the **complement system** interact with each other in chain reactions to destroy target cell membranes, stimulate inflammation, attract phagocytes, and enhance phagocytosis.

Inflammation 521

8. **Inflammation** represents a coordinated nonspecific response to tissue injury. *(Figure 14–11)*

Fever 522

9. A **fever** (body temperature greater than 37.2°C or 99°F) can inhibit pathogens and accelerate metabolic processes.

SPECIFIC DEFENSES: THE IMMUNE RESPONSE 522

1. Specific defenses are provided by T cells and B cells. T cells provide cell-mediated immunity; B cells provide antibody-mediated immunity.

Types of Immunity 522

2. Specific immunity may involve **innate immunity** (genetically determined and present at birth) or **acquired immunity.** The two types of acquired immunity are **active immunity** (which appears following exposure to an antigen) and **passive immunity** (produced by the transfer of antibodies from another source). *(Figure 14–12; Table 14–1)*

Properties of Immunity 523

3. Lymphocytes provide specific immunity, which has four general characteristics: specificity, versatility, memory, and tolerance. **Specificity** occurs because receptors on T cell and B cell membranes can bind only to specific antigens. The immune system has **versatility** in that it can respond to any of the tens of thousands of antigens it encounters. **Memory cells** enable the immune system to "remember" previously encountered antigens. **Tolerance** refers to the ability of the immune system to ignore some antigens, such as those of normal body cells.

An Overview of the Immune Response 527

4. The purpose of the **immune response** is to inactivate or destroy pathogens, abnormal cells, and foreign molecules. It is based on the activation of lymphocytes by specific antigens through the process of **antigen recognition.** *(Figure 14–13)*

T Cells and Cell-Mediated Immunity 527

5. T cells recognize antigens bound to Class I and Class II MHC proteins. Foreign antigens must usually be processed by macrophages and incorporated into their cell membranes bound to *MHC proteins* before they can activate T cells. T cells can also be activated by viral antigens displayed on the surfaces of virus-infected cells.

6. Activated T cells may differentiate into *cytotoxic T cells, memory T cells, suppressor T cells,* or *helper T cells.*

7. Cell-mediated immunity results from the activation of **cytotoxic,** or *killer,* **T cells.** Activated **memory T cells** remain on reserve to respond to future exposures to the specific antigen involved. *(Figure 14–14)*

8. **Suppressor T cells** depress the responses of other T cells and B cells.

9. **Helper T cells** secrete cytokines that help coordinate specific and nonspecific defenses and regulate cellular and humoral immunity.

Key Note 528

B Cells and Antibody-Mediated Immunity 529

10. B cells, which are responsible for antibody-mediated immunity, undergo sensitization by a specific antigen before they become activated by helper T cells sensitive to the same antigen.

11. An activated B cell divides and produces plasma cells and **memory B cells.** Antibodies are produced by plasma cells. *(Figure 14–15)*

12. An antibody molecule consists of two parallel pairs of polypeptide chains that contain *fixed segments* and *variable segments.* *(Figure 14–16)*

13. The binding of an antibody molecule and an antigen forms an **antigen-antibody complex.** Antibodies bind to specific *antigenic determinant sites.*

14. Five classes of antibodies exist in body fluids: (1) **immunoglobulin G (IgG),** which is responsible for resistance against many viruses, bacteria, and bacterial toxins; (2) **IgM,** which is the first antibody type secreted in response to an antigen; (3) **IgA,** which is found in glandular secretions; (4) **IgE,** which stimulates the re-

lease of chemicals that accelerate local inflammation; and (5) **IgD,** which is found on the surfaces of B cells. *(Table 14–2)*

Key Note 532

15. Antibodies can eliminate antigens through **neutralization, precipitation, agglutination,** the activation of complement, the attraction of phagocytes, the enhancement of phagocytosis, and the stimulation of inflammation.

Key Note 532

16. The antibodies produced by plasma cells upon first exposure to an antigen are the agents of the **primary response.** Maximum antibody levels appear during the **secondary response,** which follows subsequent exposure to the same antigen. *(Figure 14–17)*

Key Note 532

Hormones of the Immune System 532

17. **Interleukins (IL)** increase T cell sensitivity to antigens; stimulate B cell activity, plasma cell formation, and antibody production; and enhance nonspecific defenses.

18. Interferons slow the spread of a virus by making the synthesizing cell's neighbors resistant to viral infections.

19. **Tumor necrosis factors (TNFs)** slow tumor growth and kill tumor cells.

20. Several **phagocytic regulators** adjust the activities of phagocytic cells to coordinate specific and nonspecific defenses. *(Table 14–3)*

Key Note 534

PATTERNS OF IMMUNE RESPONSE 534

1. The body's specific and nonspecific defenses cooperate in an integrated way to eliminate foreign antigens. *(Figures 14–18 through 14–20)*

Immune Disorders 534

2. **Autoimmune disorders** develop when the immune response mistakenly targets normal body cells and tissues.

3. In an **immunodeficiency disease,** either the immune system does not develop normally or the immune response is somehow blocked.

4. **Allergies** are inappropriate or excessive immune responses to **allergens** (antigens that trigger allergic reactions). The four types of allergies are *immediate hypersensitivity (Type I), cytotoxic reactions (Type II), immune complex disorders (Type III),* and *delayed hypersensitivity (Type IV). (Table 14–4)*

Aging and the Immune Response 537

5. As individuals age, the immune system becomes less effective at combating disease.

INTEGRATION WITH OTHER SYSTEMS 537

1. The lymphatic system has extensive interactions with the nervous and endocrine systems. *(Figures 14–21, 14–22)*

Review Questions

Level 1: Reviewing Facts and Terms

Match each item in column A with the most closely related item in column B. Place letters for answers in the spaces provided.

COLUMN A

____ 1. humoral immunity
____ 2. lymphoma
____ 3. complement
____ 4. microphages
____ 5. macrophages
____ 6. microglia
____ 7. interferon
____ 8. pyrogens
____ 9. innate immunity
____ 10. active immunity
____ 11. passive immunity
____ 12. apoptosis

COLUMN B

a. induce fever
b. system of circulating proteins
c. CNS macrophages
d. monocytes
e. genetically programmed cell death
f. transfers of antibodies
g. neutrophils, eosinophils
h. secretion of antibodies
i. present at birth
j. cytokine
k. lymphatic system cancer
l. exposure to antigen

13. Lymph collected from the lower abdomen, pelvis, and lower limbs is carried by the:
    (a) right lymphatic duct.
    (b) inguinal duct.
    (c) thoracic duct.
    (d) aorta.

14. Lymphocytes responsible for providing cell-mediated immunity are called:
    (a) macrophages.
    (b) B cells.
    (c) plasma cells.
    (d) cytotoxic T cells.

15. B cells are responsible for:
    (a) cellular immunity.
    (b) immunological surveillance.
    (c) antibody-mediated immunity.
    (d) a, b, and c are correct.

16. Lymphoid stem cells that can form all types of lymphocytes occur in the:
    (a) bloodstream.
    (b) thymus.
    (c) bone marrow.
    (d) spleen.

17. Lymphatic vessels are found in all portions of the body except the:
    (a) lower limbs.
    (b) central nervous system.
    (c) head and neck region.
    (d) hands and feet.

18. The largest collection of lymphoid tissue in the body is contained in the:
    (a) adult spleen.
    (b) adult thymus.
    (c) bone marrow.
    (d) tonsils.

19. Red blood cells that are damaged or defective are removed from the circulation by the:
    (a) thymus.
    (b) lymph nodes.
    (c) spleen.
    (d) tonsils.

20. Phagocytes move through capillary walls by squeezing between adjacent endothelial cells, which is a process known as:
    (a) diapedesis.
    (b) chemotaxis.
    (c) adhesion.
    (d) perforation.

21. Perforins are destructive proteins associated with the activity of:
    (a) T cells.
    (b) B cells.
    (c) macrophages.
    (d) plasma cells.

22. Complement activation:
    (a) stimulates inflammation.
    (b) attracts phagocytes.
    (c) enhances phagocytosis.
    (d) a, b, and c are correct.

23. Inflammation:
    (a) aids in temporary repair at an injury site.
    (b) slows the spread of pathogens.
    (c) facilitates permanent repair.
    (d) a, b, and c are correct.

24. Memory B cells:
    (a) respond to an initial threat.
    (b) secrete large numbers of antibodies into the interstitial fluid.
    (c) respond to subsequent infections that involve the same antigen.
    (d) contain binding sites that can activate the complement system.

25. Which two large collecting vessels are responsible for returning lymph to the veins of the circulatory system? What areas of the body does each serve?

26. Give a function for each of the following:
    (a) cytotoxic T cells
    (b) helper T cells
    (c) suppressor (regulatory) T cells
    (d) plasma cells
    (e) NK cells
    (f) interferons
    (g) T cells
    (h) B cells
    (i) interleukins

27. What seven defenses, present at birth, provide the body with the defensive capability known as nonspecific resistance?

Level 2: Reviewing Concepts

28. Compared with nonspecific defenses, specific defenses:
    (a) do not discriminate between one threat and another.
    (b) are always present at birth.
    (c) provide protection against threats on an individual basis.
    (d) deny entry of pathogens to the body.

29. T cells and B cells can be activated only by:
    (a) pathogenic microorganisms.
    (b) interleukins, interferons, and colony-stimulating factors.
    (c) cells infected with viruses, bacterial cells, or cancer cells.
    (d) exposure to a specific antigen at a specific site on a cell membrane.

30. List and explain the four general properties of immunity.

31. How does the formation of an antibody-antigen complex cause elimination of an antigen?

32. What effects follow activation of the complement system?

Level 3: Critical Thinking and Clinical Applications

33. Sylvia's grandfather is diagnosed with lung cancer. His physician orders biopsies of several lymph nodes from neighboring regions of the body. Sylvia wonders why, since his cancer is in his lungs. What would you tell her?

34. Ted finds out that he has been exposed to the measles. He is concerned that he might have contracted the disease. His physician takes a blood sample and sends it to a lab to measure antibody levels. The results show an elevated level of IgM antibodies to rubella (measles) virus but very few IgG antibodies to the virus. Did Ted contract the disease?

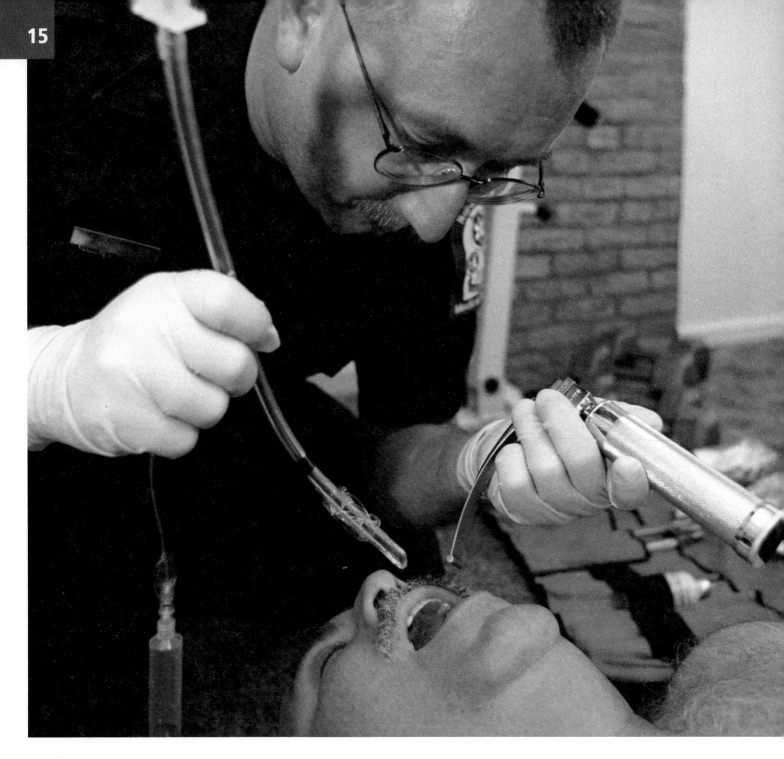

15 The Respiratory System

ONE OF THE PRIMARY ROLES of EMS is to evaluate and treat for respiratory problems. However, it is important to remember that the airway, ventilation, and circulation are all intertwined. Opening an airway is useless without ventilation, and so is ventilation without an open airway. Ventilation is use-less unless the circulatory system is functioning adequately in order to transport oxygen to the tissues. Likewise, circulation is useless without ventilation. Remember, all body systems are interdependent. When you assess and treat one, you are assessing and treating many of the others as well.

Chapter Outline

Chapter Objectives

1. Describe the primary functions of the respiratory system. (pp. 547–548)
2. Explain how the delicate respiratory exchange surfaces are protected from pathogens, debris, and other hazards. (p. 548)
3. Relate respiratory functions to the structural specializations of the tissues and organs in the respiratory system. (pp. 548–560)
4. Describe the physical principles that govern the movement of air into the lungs and the diffusion of gases into and out of the blood. (p. 560)
5. Describe the actions of respiratory muscles on respiratory movements. (pp. 560–565)
6. Describe how oxygen and carbon dioxide are transported in the blood. (pp. 565–572)
7. Describe the major factors that influence the rate of respiration. (p. 573)
8. Identify the reflexes that regulate respiration. (pp. 574–575, 577)
9. Describe the changes that occur in the respiratory system at birth and with aging. (p. 578)
10. Discuss the interrelationships between the respiratory system and other systems. (pp. 578–579)

Vocabulary Development

alveolus a hollow cavity; *alveolus, alveolar duct*
ateles imperfect; *atelectasis*
bronchus windpipe, airway; *bronchus*
cricoid ring-shaped; *cricoid cartilage*

ektasis expansion; *atelectasis*
-ia condition; *pneumonia*
kentesis puncture; *thoracentesis*
oris mouth; *oropharynx*
pneuma air; *pneumothorax*

pneumon lung; *pneumonia*
stoma mouth; *tracheostomy*
thorac- chest; *thoracentesis*
thyroid shield-shaped; *thyroid cartilage*

LIVING CELLS NEED ENERGY for maintenance, growth, defense, and reproduction. Our cells obtain that energy through an aerobic process that requires oxygen and produces carbon dioxide. ∞ p. 80 The respiratory system provides the body's cells with the means for obtaining oxygen and eliminating carbon dioxide. This exchange takes place within the lungs at air-filled pockets called **alveoli** (al-VĒ-ō-lī singular, *alveolus*). The gas-exchange surfaces of the alveoli are relatively delicate—they must be very thin to enable rapid diffusion between the air and the blood. The cardiovascular system provides the link between the interstitial fluids and the exchange surfaces of the lungs. Circulating blood carries oxygen from the lungs to peripheral tissues; it also accepts and then delivers to the lungs the carbon dioxide generated by those tissues.

Our discussion of the respiratory system begins by following air as it travels from outside the body to the alveoli of the lungs.

Next we consider the mechanics of breathing—the physical movement of air into and out of the lungs. Then we examine the physiology of respiration, which includes the process of breathing and the processes of gas exchange and gas transport between the air and blood and between the blood and tissues.

■ The Functions of the Respiratory System

The **respiratory system** has five basic functions:

1. Providing a large area for gas exchange between air and circulating blood.
2. Moving air to and from the gas-exchange surfaces of the lungs.

3. Protecting the respiratory surfaces from dehydration and temperature changes and defending against invading pathogens.
4. Producing sounds that permit speech, singing, and nonverbal auditory communication.
5. Providing olfactory sensations to the central nervous system for the sense of smell.

■ The Organization of the Respiratory System

The major anatomical structures of the respiratory system are the nose (including the nasal cavity and paranasal sinuses), pharynx (throat), larynx (voice box), trachea (windpipe), bronchi, and the lungs, which contain the bronchioles (conducting passageways) and the alveoli (exchange surfaces) (Figure 15–1●). Before we examine each structure in detail, we briefly consider the **respiratory tract,** which is comprised of the airways that carry air to and from the exchange surfaces of the lungs.

The Respiratory Tract

The respiratory tract can be divided into a *conducting portion* and a *respiratory portion.* The conducting portion begins at the entrance to the nasal cavity and continues through the pharynx, larynx, trachea, bronchi, and the larger bronchioles. The respiratory portion includes the smallest and most delicate bronchioles and the alveoli.

In addition to delivering air to the lungs, the conducting passageways filter, warm, and humidify the air, and thereby protect the alveoli from debris, pathogens, and environmental extremes. By the time inhaled air reaches the alveoli, most foreign particles and pathogens have been removed, and the humidity and temperature are within acceptable limits.

The Nose

Air normally enters the respiratory system through the paired **external nares** (NĀ-rēz), or nostrils, which open into the **nasal cavity** (Figure 15–2●). The **nasal vestibule** (VES-ti-būl)

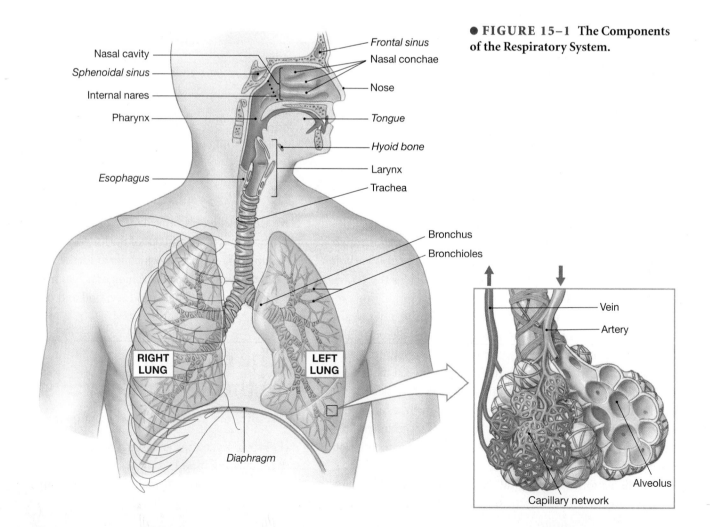

● **FIGURE 15–1 The Components of the Respiratory System.**

Nasal cavity
Sphenoidal sinus
Internal nares
Pharynx
Esophagus

Frontal sinus
Nasal conchae
Nose
Tongue
Hyoid bone
Larynx
Trachea

RIGHT LUNG
LEFT LUNG

Bronchus
Bronchioles

Diaphragm

Vein
Artery

Alveolus
Capillary network

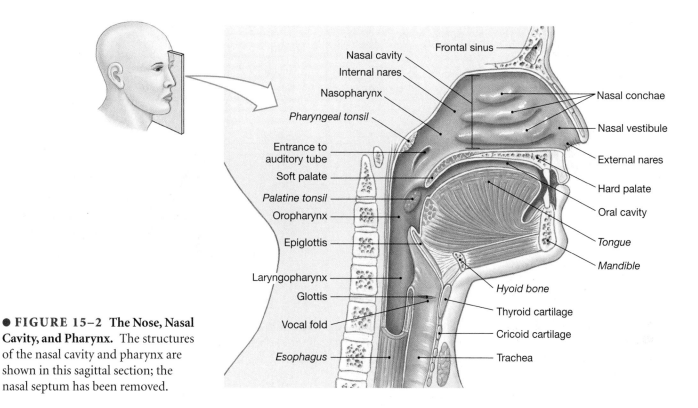

● **FIGURE 15–2 The Nose, Nasal Cavity, and Pharynx.** The structures of the nasal cavity and pharynx are shown in this sagittal section; the nasal septum has been removed.

is the space enclosed within the flexible tissues of the nose. Here coarse hairs extend across the nostrils and guard the nasal cavity from large airborne particles such as sand, dust, and even insects.

The maxillary, nasal, frontal, ethmoid, and sphenoid bones form the lateral and superior walls of the nasal cavity (see Figure 6–27b, c, p. 172). The *nasal septum* divides the nasal cavity into left and right sides. The anterior portion of the nasal septum is formed of hyaline cartilage. The bony posterior septum includes portions of the vomer and the ethmoid bone (see Figure 6–26a, p. 171). A bony **hard palate,** which is formed by the palatine and maxillary bones, forms the floor of the nasal cavity and separates the oral and nasal cavities (see Figure 15–2). A fleshy **soft palate** extends behind the hard palate and underlies the **nasopharynx** (nā-zō-FAR-inks). The nasal cavity opens into the nasopharynx at the **internal nares.**

The superior, middle, and inferior *nasal conchae* project toward the nasal septum from the lateral walls of the nasal cavity (see Figure 15–2). To pass from the nasal vestibule to the internal nares, air tends to flow in narrow grooves between adjacent conchae. As the air eddies and swirls like water flowing over rapids, small airborne particles come in contact with the mucus that coats the lining of the nasal cavity. In addition to promoting filtration, the turbulent flow allows extra time for warming and humidifying the incoming air.

The nasal cavity and much of the rest of the respiratory tract are lined by a protective, mucous membrane, or *respiratory mucosa.* ∞ p. 112 This membrane is made up of the *respiratory epithelium,* a ciliated columnar epithelium that contains many *goblet cells,* and an underlying loose connective tissue layer (the *lamina propria*) that contains mucous glands (Figure 15–3●). The goblet cells and mucous glands produce mucus that bathes the exposed surfaces of the nasal cavity and lower respiratory tract. Cilia sweep that mucus and any trapped debris or microorganisms toward the pharynx, where they can be swallowed and exposed to the acids and enzymes of the stomach. The respiratory surfaces of the nasal cavity are also flushed by mucus produced in the *paranasal sinuses* (the *frontal, sphenoid, ethmoid,* and *maxillary sinuses*) (see Figure 6–28, p. 174), and by tears flowing through the nasolacrimal duct (see Figure 9–9b, p. 336). Exposure to noxious vapors, large quantities of dust and debris, allergens, or pathogens usually causes a rapid increase in the rate of mucus production, and a "runny nose" develops.

The Pharynx

The **pharynx** (FAR-inks), or throat, is a chamber shared by the digestive and respiratory systems. It extends between the internal nares and the entrances to the larynx and esophagus and consists of three subdivisions: the nasopharynx, the oropharynx, and the laryngopharynx (see Figure 15–2).

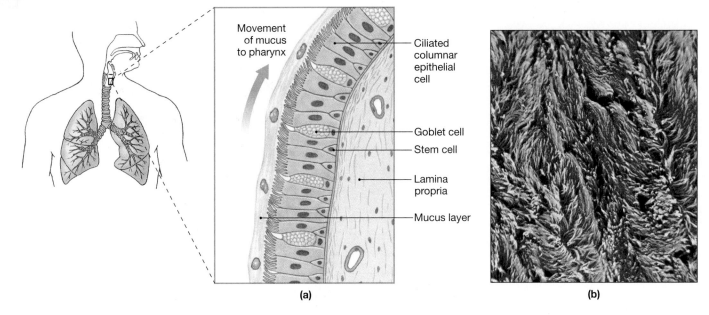

● FIGURE 15–3 The Respiratory Epithelium. **(a)** The appearance of the respiratory epithelium and its role in mucus transport are depicted in this sectional drawing. **(b)** The cilia of the epithelial cells form a dense layer that resembles a shag carpet. The movement of these cilia propels mucus across the epithelial surface. (SEM × 1614)

The **nasopharynx** is connected to the nasal cavity by the internal nares and extends to the posterior edge of the soft palate. The nasopharynx, which is lined by a typical respiratory epithelium, contains the *pharyngeal tonsil* on its posterior wall and entrances to the *auditory tubes*. The **oropharynx** extends between the soft palate and the base of the tongue at the level of the hyoid bone. The palatine tonsils lie in the lateral walls of the oropharynx. The narrow **laryngopharynx** (la-rin-gō;-FAR-inks) extends between the level of the hyoid bone and the entrance to the esophagus. Materials that enter the digestive tract pass through both the oropharynx and laryngopharynx. These regions are lined by a stratified squamous epithelium that can resist mechanical abrasion, chemical attack, and pathogenic invasion.

The Larynx

Inhaled air leaves the pharynx and enters the larynx through a narrow opening called the **glottis** (GLOT-is) (see Figure 15–2). The **larynx** (LAR-inks), or *voice box,* consists of nine cartilages stabilized by ligaments, skeletal muscles, or both. The three largest cartilages are the *epiglottis, thyroid cartilage,* and *cricoid cartilage* (Figure 15–4●).

The shoehorn-shaped **epiglottis** (ep-i-GLOT-is) projects above the glottis. During swallowing, the larynx is elevated and the elastic epiglottis folds back over the glottis, prevent-

ing the entry of liquids or solid food into the respiratory tract. The curving **thyroid** (*thyroid;* shield-shaped) **cartilage** forms much of the anterior and lateral surfaces of the larynx. A prominent ridge on the anterior surface of this cartilage forms the "Adam's apple." The thyroid cartilage sits superior to the **cricoid** (KRĪ-koyd; ring-shaped) **cartilage,** which provides posterior support to the larynx. The thyroid and cricoid car-

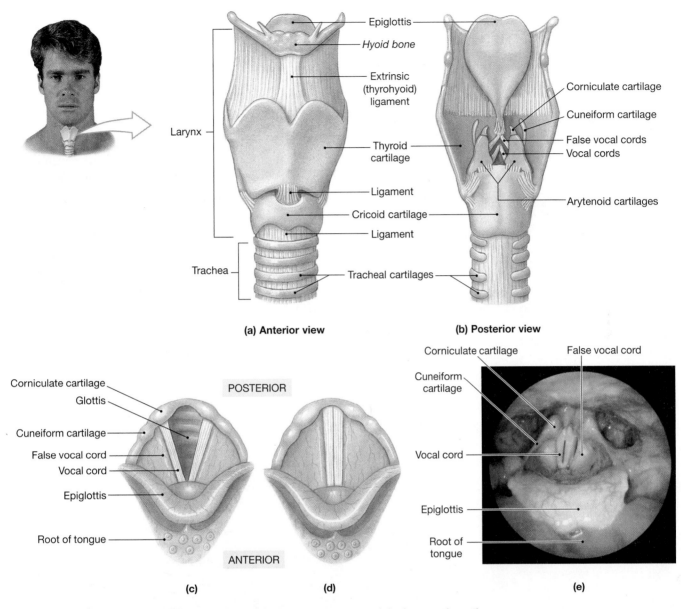

(a) Anterior view

(b) Posterior view

POSTERIOR

ANTERIOR

(c)   (d)   (e)

● **FIGURE 15–4 The Anatomy of the Larynx and Vocal Cords.** Many of the laryngeal cartilages are visible in (**a**) an anterior view of the larynx and (**b**) a posterior view of the larynx. Two superior views show the larynx with (**c**) the glottis open and (**d**) the glottis closed. (**e**) In this fiber-optic view of the larynx, the glottis is almost completely closed.

tilages protect the glottis and the entrance to the trachea, and their broad surfaces provide sites for the attachment of important laryngeal muscles and ligaments.

The larynx also contains three pairs of smaller cartilages—the *arytenoid, corniculate,* and *cuneiform cartilages*—that are supported by the cricoid cartilage. Two pairs of ligaments, enclosed by folds of epithelium, extend across the larynx between the thyroid cartilage and these smaller cartilages, which considerably reduces the size of the glottis. The ligaments of the upper pair, known as the **false vocal cords,** are relatively inelastic. They help prevent foreign objects from entering the

glottis, and they protect a more delicate pair of folds. These lower folds, the **true vocal cords,** contain elastic ligaments that extend between the thyroid cartilage and the arytenoid cartilages. Small muscles that insert on these cartilages change their position and alter the tension in these ligaments.

Food or liquids that touch the vocal cords trigger the *coughing reflex.* In a cough, the glottis is kept closed while the chest and abdominal muscles contract, compressing the lungs. When the glottis is opened suddenly, the resulting blast of air through the trachea ejects material that is blocking the entrance to the glottis.

The Vocal Cords and Sound Production

Air that passes through the glottis vibrates the vocal cords, which produces sound waves. The pitch of the sound produced, like the pitch of a vibrating harp string, depends on the diameter, length, and tension of the vibrating vocal cords. Short, thin strings vibrate rapidly, and produce a high-pitched sound; long, thick strings vibrate more slowly, and produce a low-pitched tone. The diameter and length of the vocal cords are directly related to larynx size. Children have small larynxes with slender, short vocal cords, so their voices tend to be high-pitched. At puberty, the larynx of males enlarges more than that of females; because the vocal cords of adult males are thicker and longer, they produce lower tones than those of adult females. The amount of tension in the vocal cords is controlled by skeletal muscles that change the position of the arytenoid cartilages. Increased tension in the vocal cords raises the pitch; decreased tension lowers the pitch.

The distinctive sound of your voice does not depend solely on the sounds produced by the larynx. Further amplification and resonance occur in the pharynx, the oral cavity, the nasal cavity, and the paranasal sinuses. The final production of distinct words further depends on voluntary movements of the tongue, lips, and cheeks.

The Trachea

The **trachea** (TRĀ-kē-uh), or *windpipe,* is a tough, flexible tube that is about 2.5 cm (1 in.) in diameter and approximately 11 cm (4.25 in.) long (Figure 15–5●). The trachea begins at the level of the sixth cervical vertebra, where it attaches to the cricoid cartilage of the larynx. It ends in the mediastinum, at the level of the fifth thoracic vertebra, where it branches to form the right and left primary bronchi.

The walls of the trachea are supported by 15–20 **tracheal cartilages.** These C-shaped cartilages protect the airway; by stiffening the tracheal walls, they prevent the trachea's collapse or overexpansion as pressures change in the respiratory system. The open portions of the C-shaped tracheal cartilages face pos-

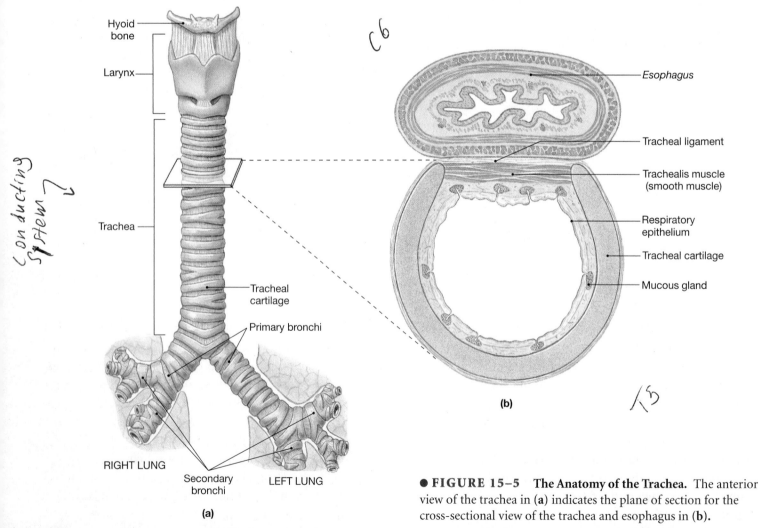

● **FIGURE 15–5** **The Anatomy of the Trachea.** The anterior view of the trachea in (**a**) indicates the plane of section for the cross-sectional view of the trachea and esophagus in (**b**).

teriorly, toward the esophagus, so the posterior tracheal wall can easily distort, and allow large masses of food to pass along the esophagus. The ends of each tracheal cartilage are connected by an elastic ligament and the *trachealis muscle,* which is a band of smooth muscle. The diameter of the trachea is adjusted by the contractions of these muscles, which are under autonomic control. Sympathetic stimulation increases the diameter of the trachea, making it easier to move large volumes of air along the respiratory passageways.

The Bronchi

Within the mediastinum the trachea branches into the **right** and **left primary bronchi** (BRONG-kī) (see Figure 15–5). The walls of the primary bronchi resemble that of the trachea, including a ciliated epithelium and C-shaped cartilaginous rings. The right primary bronchus supplies the right lung, and the left supplies the left lung. Because the right primary bronchus is larger in diameter and descends toward the lung at a steeper angle, most foreign objects that enter the trachea find their way into the right primary bronchus rather than the left.

In each lung, the primary bronchi branch into smaller and smaller airways that form the **bronchial tree** (Figure 15–6a●). As each primary bronchus enters the lung, it gives rise to **secondary bronchi,** which enter the lobes of that lung. The secondary bronchi divide to form 9–10 **tertiary bronchi** in each lung. Each tertiary bronchus branches repeatedly into smaller bronchi.

The cartilages of the secondary bronchi are relatively massive, but farther along the branches of the bronchial tree they become smaller and smaller. When the diameter of the passageway has narrowed to about 1 mm (0.04 in.), cartilages disappear completely. This narrow passage is a **bronchiole.**

The Bronchioles

Bronchioles are to the respiratory system what arterioles are to the circulatory system. Just as adjustments to the diameter of arterioles regulate blood flow into capillary beds, adjustments to the diameter of bronchioles control the amount of resistance to airflow and the distribution of air in the lungs. Such adjustments result from autonomic nervous system regulation. Sympathetic activation leads to a relaxation of smooth muscles in the walls of bronchioles, causing *bronchodilation,* which is the enlargement of airway diameter. Parasympathetic stimulation leads to contraction of these smooth muscles and *bronchoconstriction,* which is a reduction in the diameter of the airway. Extreme bronchoconstriction can almost completely block the passageways, making breath-

Clinical Note
BRONCHIAL ANGLES

The division of the trachea into the right and left primary bronchi occurs at the carina. The *right primary bronchus* angulates less acutely from the trachea than does the *left primary bronchus.* Because of this, foreign objects that inadvertently enter the respiratory tract usually enter the right primary bronchus and the right lung. The aspirated foreign body can act as a one-way valve, which allows air to enter the lung but not exit. This results in overexpansion of the lung and can eventually lead to a pneumothorax.

In an emergency, it is common for rescue personnel to place a breathing tube into the trachea (endotracheal intubation). Accidentally inserting the tube too far can cause the tip of the tube to enter the right primary bronchus. This results in only the right lung being effectively ventilated and can lead to hypoxia. Endobronchial intubation results in the absence of breath sounds over the left chest, decreased compliance, and possibly hypoxia. The situation can be remedied by simply withdrawing the tube until bilateral breath sounds return. It is important to frequently reassess tube placement to ensure that dislodgement or endobronchial intubation does not occur. ■

ing difficult or impossible. This can occur during an *asthma* (AZ-muh) attack or during allergic reactions in response to inflammation of the bronchioles.

Bronchioles branch further into *terminal bronchioles* with diameters of 0.3–0.5 mm. Each terminal bronchiole supplies air to a lobule of the lung. A **lobule** (LOB-ūl) is a segment of lung tissue that is bounded by connective tissue partitions and supplied by a single bronchiole, accompanied by branches of the pulmonary arteries and pulmonary veins (Figure 15–6b●). Within a lobule, a terminal bronchiole divides to form several *respiratory bronchioles.* These passages, the thinnest branches of the bronchial tree, deliver air to the gas-exchange surfaces of the lungs.

The Alveolar Ducts and Alveoli

Respiratory bronchioles open into passageways called **alveolar ducts** (Figure 15–7a●). The ducts end at **alveolar sacs,** common chambers connected to multiple individual alveoli—the exchange surfaces of the lungs. Each lung contains about 150 million alveoli, and their abundance gives the lung an open, spongy appearance (Figure 15–7b●).

To meet our metabolic requirements, the alveolar exchange surfaces of the lungs must be very large, equal to approximately 140 square meters—roughly one-half of a tennis court. The alveolar epithelium primarily consists of an unusually thin simple squamous epithelium (Figure 15–7c●).

● **FIGURE 15–6 The Bronchial Tree and Lobules of the Lung. (a)** This illustration shows a simplified version of the branching pattern of the airways in the left lung. **(b)** This drawing of a pulmonary lobule shows the lung tissue supplied by a single respiratory bronchiole. Note that a capillary bed surrounds each alveolus.

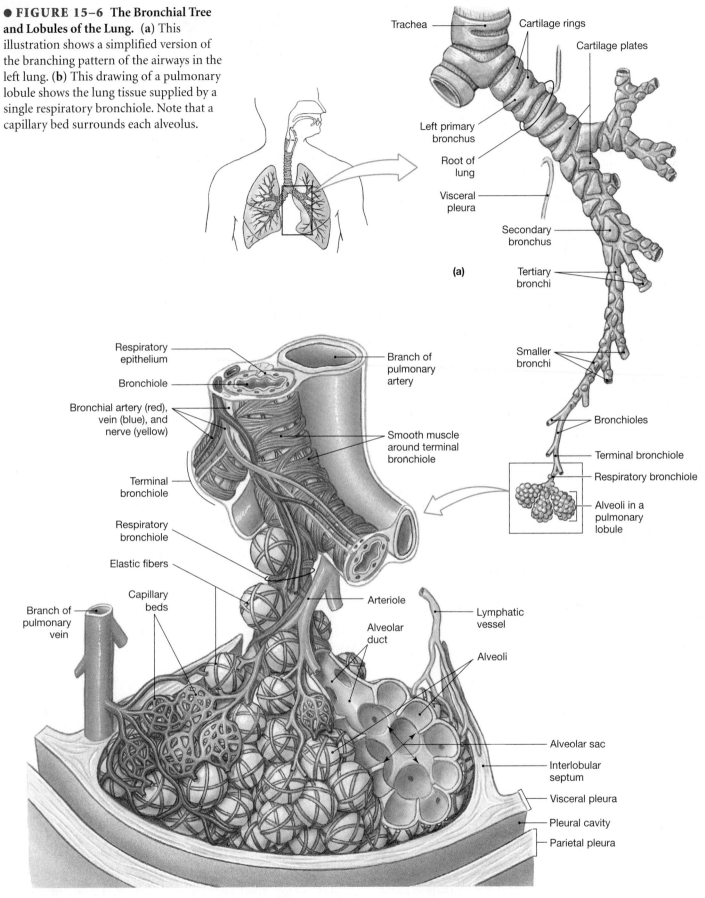

Trachea

Cartilage rings

Cartilage plates

Left primary bronchus

Root of lung

Visceral pleura

Secondary bronchus

Tertiary bronchi

Smaller bronchi

Bronchioles

Terminal bronchiole

Respiratory bronchiole

Alveoli in a pulmonary lobule

(a)

Respiratory epithelium

Bronchiole

Bronchial artery (red), vein (blue), and nerve (yellow)

Terminal bronchiole

Respiratory bronchiole

Elastic fibers

Capillary beds

Branch of pulmonary vein

Branch of pulmonary artery

Smooth muscle around terminal bronchiole

Arteriole

Lymphatic vessel

Alveolar duct

Alveoli

Alveolar sac

Interlobular septum

Visceral pleura

Pleural cavity

Parietal pleura

(b)

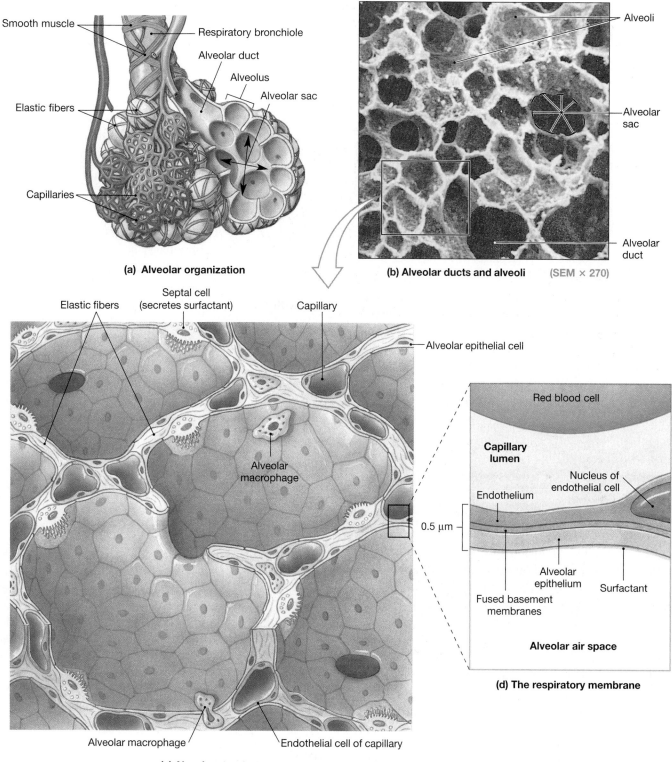

(a) Alveolar organization

(b) Alveolar ducts and alveoli   (SEM × 270)

(c) Alveolar structure

(d) The respiratory membrane

● **FIGURE 15–7 Alveolar Organization.** (**a**) Several alveoli open off of a single alveolar duct. (**b**) This SEM reveals the open, spongy texture of lung tissue. (**c**) This diagram provides some details of alveolar structure. (**d**) The respiratory membrane is made up of an alveolar epithelial cell, a capillary endothelial cell, and their fused basement membranes.

Clinical Note
BRONCHIOLITIS AND RESPIRATORY SYNCTIAL VIRUS INFECTIONS

Bronchiolitis, an infection of the smaller airways, is a common disease in children less than two years of age. Most cases of bronchiolitis are caused by the *respiratory syncytial virus (RSV).* It often occurs in epidemics during the midwinter and primarily affects children between the ages of two and six months. Typically, the child with an RSV infection will develop a runny nose, low-grade fever, and decreased appetite. Then, there will be an increase in the respiratory rate and depth and in the work of breathing. As the disease progresses, there is often a fall in the oxygen level, wheezing, and trapping of air within the lungs, which causes hyperexpansion of the chest. Treatment includes intensive monitoring, supplemental oxygen, bronchodilators, and occasionally, ventilatory support. Usually, children develop lifelong immunity following infection. ■

Roaming **alveolar macrophages** *(dust cells)* patrol the epithelium, phagocytizing dust or debris that has reached the alveolar surfaces.

Scattered among the squamous cells are larger **septal cells,** which secrete onto the alveolar surfaces an oily secretion called **surfactant** (sur-FAK-tant). Surfactant is important because by forming a superficial coating over the thin layer of water that coats the alveolar surface, it reduces surface tension, which results from the attraction between water molecules at an air-water boundary. Without surfactant, the surface tension would collapse the delicate alveolar walls. When surfactant levels are inadequate (as a result of injury or genetic abnormalities), each inhalation must be forceful enough to pop open the alveoli. An individual with this condition—called **respiratory distress syndrome**—is soon exhausted by the effort required to keep inflating the deflated lungs.

The Respiratory Membrane

Gas exchange occurs across the **respiratory membrane** of the alveoli. The respiratory membrane (Figure 15–7d●) consists of three components:

1. The squamous alveolar epithelium.
2. The endothelial cells that line an adjacent capillary.
3. The fused basement membranes that lie between the alveolar and endothelial cells.

At the respiratory membrane, the distance that separates alveolar air from blood can be as little as 0.1 μm. Diffusion across the respiratory membrane proceeds very rapidly because the distance is small and because both oxygen and carbon dioxide are lipid-soluble. The membranes of the epithelial and endothelial cells, thus, do not pose a barrier to the movement of oxygen and carbon dioxide between blood and the alveolar air spaces.

Circulation to the Respiratory Membrane

The respiratory exchange surfaces receive blood from arteries of the *pulmonary circuit.* ∞ p. 484 The pulmonary arteries enter the lungs and branch, then follow the bronchi and their branches to the lobules. Each lobule receives an arteriole, and a network of capillaries surrounds each alveolus directly beneath the alveolar epithelium. After passing through the capillaries and into pulmonary venules, blood enters the pulmonary veins, which deliver it to the left atrium.

Blood pressure in the pulmonary circuit is usually relatively low, with systemic pressures of 30 mmHg or less. With pressures that low, pulmonary arteries can easily become blocked by small blood clots, fat masses, or air bubbles. Because the lungs receive the entire cardiac output, any drifting masses in the blood are likely to cause problems almost at once. The blockage of a branch of a pulmonary artery will stop blood flow to a group of lobules or alveoli, which is a condition called **pulmonary embolism.**

Clinical Note
PULMONARY EMBOLISM

Pulmonary embolism (PE) is a common and potentially deadly disorder that can be extremely difficult to diagnose. It usually arises from the large veins of the leg or pelvis. A blood clot in the leg, referred to as *deep venous thrombosis, (DVT)* usually develops over a period of minutes to hours. Prolonged immobilization, such as occurs during hospitalization or a long plane or car ride, has been identified as a risk factor for the development of DVT. Ultimately, the clot breaks loose, travels through the circulatory system, and enters the pulmonary circulation. When it reaches the part of the blood vessel that is smaller than itself, the clot lodges there. Blood flow to the lung is restricted and oxygenation is impaired. Large clots can obstruct a significant portion of the lung and may be rapidly fatal. ■

The Lungs

Each of the two **lungs** has distinct **lobes** that are separated by deep fissures (Figure 15–8●). The right lung has three lobes (*superior, middle,* and *inferior*), and the left lung has two (*superior* and *inferior*). The bluntly rounded *apex* of each lung extends into the base of the neck above the first rib, and the concave base rests on the superior surface of the

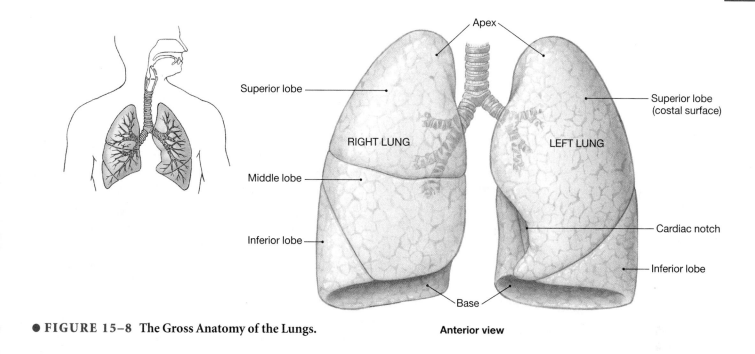

● **FIGURE 15–8** The Gross Anatomy of the Lungs.

Anterior view

Clinical Note
PNEUMONIA

Pneumonia is an infection of the lungs and a common medical problem, especially in the aged and those infected with the human immunodeficiency virus (HIV). In fact, pneumonia is one of the leading causes of death in both groups of patients and is the fifth leading overall cause of death in the U.S.

Patients with HIV infection and those on immune suppressive therapy (cancer patients) are at high risk of developing pneumonia. In addition, the very young and very old are at higher risk of acquiring pneumonia because of ineffective protective mechanisms. Other risk factors include a history of alcoholism, cigarette smoking, and exposure to cold temperatures.

Pneumonia is a collection of related respiratory diseases caused when a variety of infectious agents invade the lungs. It is crucial to remember the importance of adequate mucous production and the action of respiratory tract cilia in protecting the body against bacterial invasion. When considering which patients are at risk, the unifying concept is that there is a defect in mucous production, ciliary action, or both.

Bacterial and viral pneumonias occur the most frequently, although fungal and other forms of pneumonia do exist. More unusual forms of pneumonia are seen in those patients who are currently or recently have been hospitalized, where they are exposed to a more unusual variety of microorganisms. This is referred to as *hospital-acquired pneumonia.* (Cases that develop in the out-of-hospital setting are described as *community-acquired pneumonia.*)

The infection begins in one part of the lung and often spreads to nearby alveoli. The infection may ultimately involve the entire lung. As the disease progresses, fluid and inflammatory cells collect in the alveoli, and alveolar collapse may occur. Pneumonia is primarily a ventilation disorder. Occasionally, the infection will extend beyond the lungs into the bloodstream and to more distant sites in the body. This systemic spread may lead to septic shock.

A patient with pneumonia will generally appear ill. He may report a recent history of fever and chills, commonly described as "bed shaking." There is usually a generalized weakness and malaise. The patient will tend to complain of a deep, productive cough and may expel yellow to brown sputum, often streaked with blood. Many cases involve associated pleuritic chest pain; therefore, pneumonia should be considered in any patient who presents complaining of chest pain, especially if accompanied by fever or chills. In pneumonia that involves the lower lobes of the lungs, a patient may complain of nothing more than upper abdominal pain.

In the forms of pneumonia that involve viral, fungal, and rare bacterial causes, the typical symptoms described above are not seen. Instead, these patients may report a nonproductive cough with less prominent lung findings. Systemic symptoms such as headache, malaise, fatigue, muscle aches, sore throat, and abdominal complaints including nausea, vomiting, and diarrhea are more prominent. Fever and chills are not as impressive as in bacterial pneumonia. ■

diaphragm, which is the muscular sheet that separates the thoracic and abdominopelvic cavities. The curving *costal surface* follows the inner contours of the rib cage. The *mediastinal surfaces* of both lungs bear grooves that mark the passage of large blood vessels and indentations of the pericardium. In anterior view, the medial edge of the right lung forms a vertical line, whereas the medial margin of the left lung is indented at the *cardiac notch.*

The lungs have a light and spongy consistency because most of the actual volume of each lung consists of air-filled passageways and alveoli. An abundance of elastic fibers gives the lungs the ability to tolerate large changes in volume.

The Pleural Cavities

The thoracic cavity has the shape of a broad cone. Its walls are the rib cage, and its floor is the muscular diaphragm. Within the thoracic cavity, each lung occupies a single pleural cavity, which is lined by a serous membrane called the **pleura** (PLOOR-uh). ∞ p. 113 The *parietal pleura* covers the inner surface of the body wall and extends over the diaphragm and mediastinum. The *visceral pleura* covers the outer surfaces of the lungs, extending into the fissures between the lobes. The two pleural cavities are separated by the mediastinum (Figure 15–9●).

Each pleural cavity actually represents a potential space rather than an open chamber because the parietal and visceral layers are usually in close contact. During breathing, friction between the pleural surfaces is reduced through the lubricating action of *pleural fluid*, which is secreted by both pleural layers. Pleural fluid is sometimes obtained for diagnostic purposes using a long needle inserted between the ribs. This procedure is called *thoracentesis* (thōr-a-sen-TĒ-sis; *thorac-*, chest + *kentesis*, puncture). The fluid is examined for the presence of bacteria, blood cells, and other abnormal components.

An injury to the chest wall that penetrates the parietal pleura or damages the alveoli and the visceral pleura can allow air into the pleural cavity. This condition—**pneumothorax** (noo-mō-THŌ-raks; *pneuma*, air)—breaks the fluid bond between the pleurae and allows the elastic fibers to contract. The result is a collapsed lung, or *atelectasis* (at-e-LEK-ta-sis; *ateles*, imperfect + *ektasis*, expansion). Treatment involves removing as much of the air as possible before sealing the opening. This procedure restores the fluid bond and reinflates the lung. Lung volume can also be reduced by the accumulation of blood in the pleural cavity. This condition is called **hemothorax.**

→ CONCEPT CHECK QUESTIONS

1. When tension in the vocal cords increases, what happens to the pitch of the voice?
2. In the trachea, why are C-shaped cartilages functionally better than completely circular cartilages?
3. What would happen to the alveoli if surfactant were not produced?

Answers begin on p. 792.

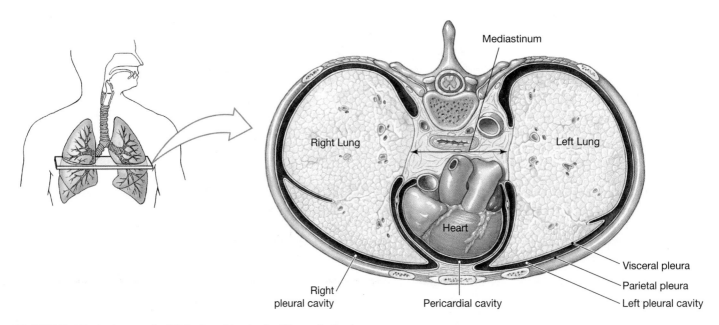

● **FIGURE 15–9** **Anatomical Relationships in the Thoracic Cavity.**

Clinical Note
TENSION PNEUMOTHORAX AND CHEST DECOMPRESSION

In certain thoracic emergencies, decompression of the chest by emergency personnel can be life-saving. The most frequently encountered condition that requires pleural decompression is *tension pneumothorax.* With tension pneumothorax, a leak or tear develops in one of the lungs. Air begins to leak into the pleural space, and because it has no way to exit the chest, it begins to accumulate. Eventually, it will start to compress the affected lung, which leads to decreased ventilation and decreased perfusion. If allowed to progress untreated, the expanding mass of air will start to displace the mediastinum away from the injured side. This can decrease ventricular filling, which in turn decreases cardiac output. To prevent continued deterioration, emergency personnel must equalize the pressure in the chest with that of the environment. This will allow removal of the pressurized air mass, re-expansion of the lung, and movement of the mediastinum back to its normal position.

The simplest way to decompress the chest is to place a needle through the chest wall into the pleural space. This procedure is safest at the fifth intercostal space in a midaxillary line or at the second intercostal space in a midclavicular line. Remember that the intercostal neurovascular bundle runs immediately beneath and slightly behind each rib. Accidental puncture of the intercostal artery or vein can cause significant bleeding, and puncture or laceration of the intercostal nerve can cause pain or a loss in sensation to the dermatome it supplies. Because of this, it is important to palpate the rib and insert the needle along the superior border of the rib. The same warning is especially important when placing a thoracostomy tube (chest tube). The incision and tube insertion should be guided across the superior surface of the rib (Figure 15–10●).

Remember, too, that the pleural pain fibers are located in the parietal pleura and not the visceral pleura. The patient will experience the most pain as the needle exits the chest wall. When placing larger tubes, such as chest tubes, it is important to inject an adequate amount of local anesthetic into the parietal pleura to ensure that the patient is as comfortable as possible. ■

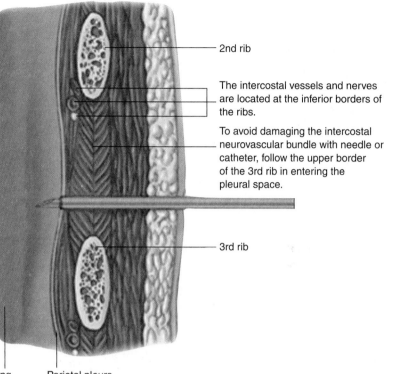

2nd rib

The intercostal vessels and nerves are located at the inferior borders of the ribs.

To avoid damaging the intercostal neurovascular bundle with needle or catheter, follow the upper border of the 3rd rib in entering the pleural space.

3rd rib

Lung   Parietal pleura

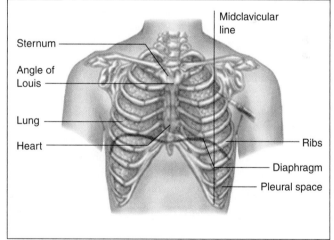

Midclavicular line

Sternum

Angle of Louis

Lung

Heart

Ribs

Diaphragm

Pleural space

● **FIGURE 15–10 Proper Needle Placement for Decompression of the Chest.** The needle should be inserted along the upper border of the rib to avoid puncturing the intercostal neurovascular bundle.

Clinical Note
TUBERCULOSIS

Tuberculosis (too-ber-kū-LŌ-sis), or *TB*, results from a bacterial infection of the lungs, although other organs may be invaded as well. The bacterium, *Mycobacterium tuberculosis,* may colonize respiratory passageways, interstitial spaces, or alveoli, or a combination of all three. Signs and symptoms are variable but generally include coughing and chest pain with fever, night sweats, fatigue, and weight loss.

Tuberculosis is a major worldwide health problem. With roughly three million deaths each year from TB, it is the leading cause of death among infectious diseases. An estimated *two billion* people are currently infected, and eight million new cases are diagnosed each year. Unlike other deadly infectious diseases, such as AIDS, TB is transmitted through casual contact. Coughing, sneezing, or speaking by an infected individual can spread the bacterium through the air in tiny droplets that can be inhaled by other people. ■

Clinical Note
TOBACCO ABUSE

Cigarette smoking and the abuse of tobacco products are significant risk factors for the development of respiratory and cardiovascular illnesses. Always question the patient about cigarette and tobacco usage. Patients will generally underreport it. A better picture of tobacco usage might be obtained from the patient's spouse or another family member.

A patient's cigarette smoking history is generally reported in pack/years. If possible, determine the number of cigarette packs (20 cigarettes/pack) smoked per day and the number of years the patient has smoked. Multiply the number of packs smoked per day by the number of years. For example, a man who has smoked two packs per day for 15 years would have a 30-pack/year smoking history. Medical problems related to smoking, such as emphysema, chronic bronchitis, and lung cancer, usually begin after a patient surpasses a 20-pack/year history, although this can vary significantly. This is an important part of the history and should be determined if the patient's condition allows it.

Patients should also be questioned about the use of smokeless tobacco. Although smokeless tobacco has less impact on the respiratory system than smoke products, it still increases the patient's risk for developing cancers of the mouth and throat. ■

■ Respiratory Physiology

The process of respiration involves three integrated steps:

Step 1: *Pulmonary ventilation,* or breathing, which involves the physical movement of air into and out of the lungs.

Step 2: *Gas exchange,* which involves gas diffusion at two sites: across the respiratory membrane between alveolar air spaces and alveolar capillaries, and across capillary cell membranes between blood and other tissues.

Step 3: *Gas transport,* which involves the transport of oxygen and carbon dioxide to and from the alveolar capillaries and the capillary beds in other tissues.

Abnormalities that affect any of these processes will ultimately affect the gas concentrations of the interstitial fluids and, thus, cellular activities as well. If oxygen concentrations decline, the affected tissues will become oxygen-starved. **Hypoxia** (hī-POKS-ē-uh), or *low tissue oxygen levels,* places severe limits on the metabolic activities of the affected area. If the supply of oxygen gets cut off completely, **anoxia** (an-ok-SĒ-uh) results, and cells die very quickly. Much of the damage caused by strokes and heart attacks is the result of localized anoxia.

Pulmonary Ventilation

Pulmonary ventilation is the physical movement of air into and out of the respiratory tract. A single breath, or **respiratory cycle,** consists of an inhalation (or *inspiration*) and an exhalation (or *expiration*). The **respiratory rate** is the number of breaths per minute. This rate in normal adults at rest ranges from 12 to 18 breaths per minute. Children breathe more rapidly, about 18 to 20 breaths per minute.

Breathing functions to maintain adequate **alveolar ventilation,** which is the movement of air into and out of the alveoli. Alveolar ventilation prevents the buildup of carbon dioxide in the alveoli and ensures a continuous supply of oxygen that keeps pace with absorption by the bloodstream. Next we consider the factors involved in achieving airflow to the lungs.

Factors in Achieving Airflow to the Lungs

As we know from television weather reports, air will flow from an area of higher pressure to an area of lower pressure. This difference between the high and low pressures is called a *pressure gradient,* and it applies both to the movement of atmospheric winds and to the movement of air into and out of the lungs (pulmonary ventilation). In a closed, flexible container (such as a lung), the pressure on a gas (such as air) can be changed by increasing or decreasing the container's volume: as the volume of the container (the lungs) increases, the pressure of the gas (air) decreases; as volume decreases, pressure increases.

The volume of the lungs depends on the volume of the pleural cavities. As we have seen, the parietal and pleural membranes are separated by only a thin film of pleural fluid, and although the two membranes can slide across each other,

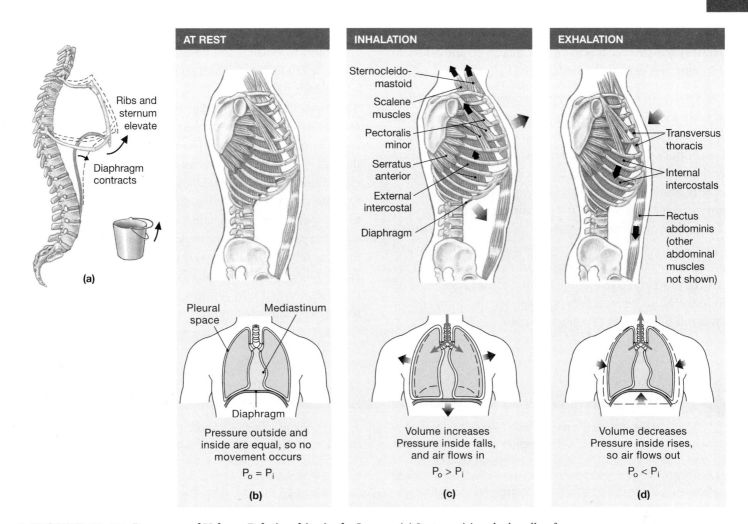

● FIGURE 15–11 Pressure and Volume Relationships in the Lungs. (a) Just as raising the handle of a bucket increases the amount of space between it and the bucket, the volume of the thoracic cavity increases when the ribs are elevated and when the diaphragm is depressed during contraction. (b) When the rib cage and diaphragm are at rest, the pressures inside and outside are equal, and no air movement occurs. (c) During inhalation, elevation of the rib cage and depression of the diaphragm increase the volume of the thoracic cavity. Pressure in the lungs decreases, and air flows into the lungs. (d) During exhalation, the rib cage returns to its original position or the diaphragm relaxes, which reduces the volume of the thoracic cavity. Pressure in the lungs rises, and air flows out of the lungs. During both inhalation and exhalation, contraction of accessory muscles may assist movements of the rib cage to increase the depth and rate of respiration.

they are held together by that fluid film. (You encounter the same principle when you set a wet glass on a smooth surface. You can slide the glass easily, but when you try to lift it, you encounter considerable resistance from this fluid bond.) Because a fluid bond exists between the parietal pleura and the visceral pleura that covers the lungs, the surface of each lung sticks to the inner wall of the chest and to the superior surface of the diaphragm. Thus, any expansion or contraction of the thoracic cavity directly affects the volume of the lungs.

Changes in the volume of the thoracic cavity result from movements of the diaphragm and rib cage, as shown in Figure 15–11a●:

- *Diaphragm.* The diaphragm forms the floor of the thoracic cavity. When relaxed, the diaphragm is dome-shaped and projects upward into the thoracic cavity, compressing the lungs. When the diaphragm contracts, it flattens, which increases the volume of the thoracic cavity and expands the lungs. When the diaphragm relaxes, it returns to its original position, which decreases the volume of the thoracic cavity.

- *Rib cage.* Because of the way the ribs and the vertebrae articulate, elevation of the rib cage increases the volume of the thoracic cavity, whereas lowering of the rib cage decreases the volume of the thoracic cavity. The external intercostal muscles and accessory muscles (such as the sternocleidomastoid) elevate the rib cage. The internal intercostal muscles and accessory muscles (such as the rectus abdominis and other abdominal muscles) lower the rib cage.

At the start of a breath, pressures inside and outside the lungs are identical, and there is no movement of air (Figure 15–11b●). When the diaphragm contracts and the movement of respiratory muscles enlarges the thoracic cavity, the pleural cavities and lungs expand to fill the additional space, so the pressure inside the lungs decreases. Air now enters the respiratory passageways because the pressure inside the lungs (P_i) is lower than atmospheric pressure (pressure outside, or P_o) (Figure 15–11c●). Downward movement of the rib cage and upward movement of the diaphragm during exhalation reverse the process and reduce the volume of the lungs. Pressure inside the lungs now exceeds atmospheric pressure, and air moves out of the lungs (Figure 15–11d●).

The **compliance** of the lungs is an indication of their resilience and ability to expand. The lower the compliance, the greater is the force required to fill and empty the lungs; the greater the compliance, the easier it is to fill and empty the lungs. Various disorders affect compliance. For example, the loss of supporting tissues due to alveolar damage, as occurs in *emphysema*, increases compliance (see p. 566). Compliance is reduced if surfactant production is insufficient to prevent the alveoli from collapsing on exhalation, as occurs in *respiratory distress syndrome* (see p. 556). Arthritis or other skeletal disorders that affect the joints of the ribs or spinal column also reduce compliance.

When you are at rest, the muscular activity involved in pulmonary ventilation accounts for 3–5 percent of your resting energy demand. If compliance is reduced, the energy demand increases dramatically, and you can become exhausted simply trying to continue breathing.

Clinical Note
RESPIRATORY SYSTEM INTERVENTIONS

Several emergency interventions are used in the care of respiratory system emergencies. Oftentimes it is necessary to actually take over a patient's breathing until the underlying problem is corrected. This is done with mechanical ventilation, which uses a device to generate a volume of air that can be administered to a patient. Initially, this is done with a bag-valve-mask unit, which is a common emergency device that can provide adequate respirations. However, if it is necessary to provide respirations over a prolonged period, a mechanical ventilator is used.

With volume-cycled ventilators, inspiration is terminated when a pre-set tidal volume is reached. The gas is usually delivered from compressible bellows. Most volume-cycled respirators are powered by an external electrical source.

With time-cycled ventilators, inspiration is terminated and expiration begins after a pre-set time has expired. Time-cycled ventilators are like volume-cycled ventilators in that they deliver a fairly constant tidal volume despite changes in the patient's airway compliance. They can also function as pressure-cycled ventilators when the secondary pressure limits are adjusted. Time-cycled ventilators are becoming increasingly popular.

Mechanical Ventilation

Patients who require prolonged ventilation are usually placed on a mechanical ventilator. The mechanical ventilator is a device that provides ventilatory support for patients in respiratory failure.

Mechanical ventilators can be classified as: *pressure-cycled, volume-cycled,* or *time-cycled.* In pressure-cycled ventilators, the inspiratory phase is terminated when a pre-set pressure limit is reached. This type of ventilator works well if the patient's airway compliance remains constant. The airway compliance is the respiratory system's resistance to airflow. The greater the airway resistance, the lower the compliance. Conversely, the less the airway resistance, the greater will be the airway compliance. In the emergency setting, a patient's airway compliance can change. An increase in airway resistance, or a decrease in airway compliance, can cause a decrease in tidal volume (V_T). In severe cases, this may lead to hypoventilation. Because of this, most pressure-cycled ventilators have been replaced with volume-cycled ventilators. Pressure-cycled ventilators nonetheless have several distinct advantages. First, they are more compact and can be powered by compressed gas without the need for electrical power. This makes them suitable for ambulance and helicopter usage.

VENTILATOR SETTINGS

Important ventilator parameters can be controlled on most mechanical ventilators. These include: respiratory rate, tidal volume, inspired oxygen concentration, positive end-expiratory pressure, and ventilation mode. The *respiratory rate* is the number of ventilatory cycles per minute. The *tidal volume* (V_T) is the amount of air delivered during each ventilatory cycle. The tidal volume usually is set initially at 10–15 milliliters per kilogram of body weight. The inspired oxygen concentration, or *FiO₂*, can also be set. This is usually expressed in percentages or in decimals (i.e., FiO_2 of 0.5 = 50% inspired oxygen concentration or FiO_2 of 1.0 = 100% inspired oxygen concentration). The *positive end-expiratory pressure (PEEP)* is the pressure within the airway at the end of expiration. PEEP is usually expressed in centimeters of water (cm/H_2O) and can be adjusted to meet the patient's needs. Normal PEEP ranges from 0 cm/H_2O to 2 cm/H_2O. Increasing the PEEP improves oxygenation by keeping alveoli open during expiration. It also helps to re-expand any collapsed alveoli, which in turn will help to decrease shunting and improve the PaO_2.

Finally, the ventilatory mode can be set on most ventilators. There are several ventilator modes including:

- *Controlled mechanical ventilation (CMV).* Usually used in situations where the patient is apneic. In CMV mode, the patient is ventilated at the rate set by the operator. The patient cannot breathe between machine breaths.
- *Assist control mode ventilation (ACMV).* With ACMV, the operator sets the minimum rate at which the patient is to be ventilated. If the patient makes no respiratory effort, then only the prescribed number of breaths will be delivered. If the patient tries to breathe, the machine will deliver an extra breath with the same tidal volume as has been set. The amount of negative inspiratory pressure necessary to trigger a ventilation can be adjusted by the operator.
- *Intermittent mandatory ventilation (IMV).* With IMV, as with ACMV, the patient may breathe at a rate faster than set on the ventilator. However, the machine offers no assistance to the patient-generated ventilation, and the patient receives only the tidal volume that is self-generated. This allows medical personnel to determine the rate and depth of the patient's native respiratory efforts.
- *Synchronized intermittent mandatory ventilation (SIMV).* A problem with IMV is that the ventilator sometimes delivers a ventilation just as the patient has completed a spontaneous inspiration. Because of this, SIMV was developed. With SIMV, the mode is the same as IMV except that the ventilator times the machine breaths to fall in a pause between the patient's spontaneous respiratory cycle or to coincide with the initiation of a spontaneous breath.
- *Pressure support ventilation (PSV).* PSV mode was the basic mode used by intermittent positive-pressure breathing machines. When the patient initiates a breath, the machine delivers a constant inspiratory pressure until inspiratory flow drops below 25 percent of the peak level. Thus, the patient determines the rate, and tidal volume is dependent on patient airway compliance. PSV is usually used when weaning a patient from the ventilator.

PREHOSPITAL MECHANICAL VENTILATION

Several mechanical ventilators have been developed for use in prehospital care. Most of these are pressure-cycled and powered by compressed oxygen. With most units, the respiratory rate and V_T can be adjusted. With some units, the inspiratory-to-expiratory ratio can be adjusted for use with pediatric patients. Prehospital mechanical ventilators, which are also called *automatic transport* ventilators, are common on EMS units that provide interhospital transport, particularly critical care interhospital transport (Figure 15–12●). It is important not to become overly reliant on mechanical ventilators. Bag-valve mask (BVM) units should be immediately available in case of respiratory failure.

COMPLICATIONS OF MECHANICAL VENTILATION

Mechanical ventilation is safe and effective; however, several complications can develop, especially with prolonged mechanical ventilation and use of positive end-expiratory pressure (PEEP). A relatively common complication of ventilator therapy is barotrauma. Pneumothorax is the most common form of barotrauma. Typically, the pneumothorax is simple, but failure to recognize and treat a ventilator-induced

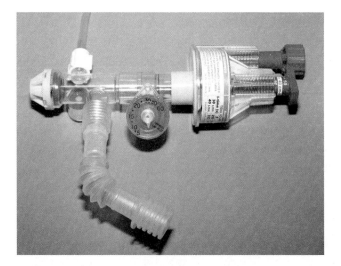

● **FIGURE 15–12 Prehospital Mechanical Ventilator.** New technologies have allowed the development of small, portable ventilators that can be used in out-of-hospital settings. These units allow the selection of several ventilatory parameters.

pneumothorax can potentially lead to a tension pneumothorax and cardiovascular collapse. A less common manifestation of barotrauma is pneumoperitoneum. This is often mistaken as a ruptured abdominal viscus and can result in unnecessary surgery. Other complications of mechanical ventilation include diminished cardiac output, pneumonia, and oxygen toxicity.

Positive End-Expiratory Pressure (PEEP)

Positive end-expiratory pressure (PEEP), or continuous positive airway pressure (CPAP), can be used to reclaim lost lung volumes and to increase oxygenation. PEEP or CPAP should be considered in cases where decreased pulmonary compliance prevents adequate tidal volumes or when hypoxemia exists despite delivery of 100% oxygen.

PEEP and CPAP are usually measured in centimeters of water (cm/H_2O). Initially, a PEEP of 2.5–5.0 cm/H_2O should be tried. This can be slowly increased to 10–15 cm/H_2O. A PEEP of greater than 12–15 cm/H_2O will usually affect cardiac output. PEEP pressures greater than 20 cm/H_2O affect ventricular filling to the point where the benefits of PEEP are not outweighed by the risks. High levels of PEEP result in increased intrathoracic pressure, which decreases venous return to the heart. This reduces ventricular filling and, ultimately, cardiac output. Thus, PEEP should be used with caution in any patient with a head injury, as increased intrathoracic pressure will impair venous return from the brain, and effectively increase intracranial pressure.

Devices are available that allow delivery of CPAP through a tightly fitted facemask. These were initially developed for treatment of obstructive sleep apnea. However, it has been found that use of CPAP can enhance oxygenation and, in many cases, prevent the need for endotracheal intubation. As CPAP devices have become more compact, they are used with increasing frequency in prehospital care. ■

Modes of Breathing

The respiratory muscles are used in various combinations, depending on the volume of air that must be moved into or out of the system. Respiratory movements are classified as quiet breathing or forced breathing.

In *quiet breathing,* inhalation involves muscular contractions, but exhalation is passive. Inhalation involves the contraction of the diaphragm and the external intercostal muscles. Diaphragm contraction normally accounts for around 75 percent of the air movement in normal quiet breathing, and the external intercostal muscles account for the remaining 25 percent. These percentages can change, however. For example, pregnant women increasingly rely on movements of the rib cage as expansion of the uterus forces abdominal organs against the diaphragm.

In *forced breathing,* both inhalation and exhalation are active. Forced breathing involves the accessory muscles during inhalation and the internal intercostal muscles and abdominal muscles during exhalation.

Lung Volumes and Capacities

As noted earlier, a respiratory cycle is a single cycle of inhalation and exhalation. The amount of air moved into or out of the lungs during a single respiratory cycle is the **tidal volume.** Only a small proportion of the air in the lungs is exchanged during a single quiet respiratory cycle; the tidal volume can be increased by inhaling more vigorously and exhaling more completely.

The amounts of air in the lungs can be expressed as various *volumes* and *capacities* (Figure 15–13●):

■ *Expiratory reserve volume.* During a normal, quiet respiratory cycle under resting conditions, the tidal volume (V_T) averages about 500 mL. The amount of air that could be voluntarily expelled at the end of such a respiratory cycle—about 1000 mL—is the **expiratory reserve volume (ERV).**

■ *Inspiratory reserve volume.* The **inspiratory reserve volume (IRV)** is the amount of air that can be taken in over and above the resting tidal volume. Because the lungs of

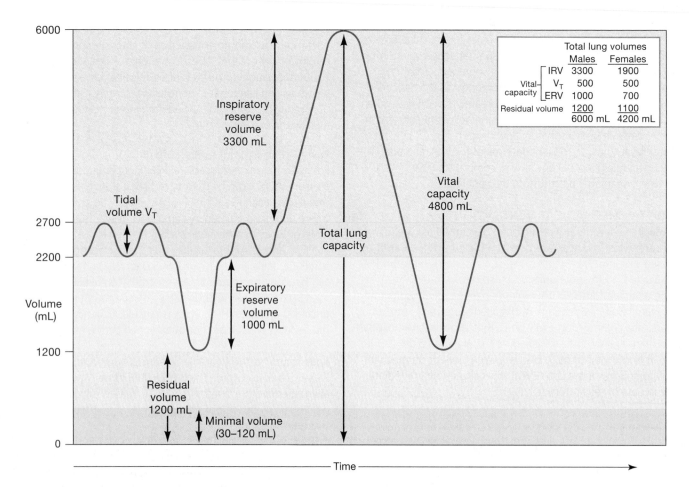

		Total lung volumes	
		Males	Females
Vital capacity	IRV	3300	1900
	V_T	500	500
	ERV	1000	700
Residual volume		1200	1100
		6000 mL	4200 mL

● **FIGURE 15–13 Respiratory Volumes and Capacities.** This graph diagrams the relationships between the respiratory volumes and capacities of an average male. The table compares the values for males and females. The red line indicates the volume of air within the lungs as respiratory movements are performed.

males are larger than those of females, the IRV of males averages 3300 mL versus 1900 mL in females.

- *Vital capacity.* The sum of the inspiratory reserve volume, the expiratory reserve volume, and the tidal volume is the **vital capacity**—the maximum amount of air that can be moved into and out of the respiratory system in a single respiratory cycle.

- *Residual volume.* The **residual volume** is the amount of air that remains in your lungs even after a maximal exhalation—typically about 1200 mL in males and 1100 mL in females. Most of this residual volume exists because the lungs are held against the thoracic wall, which prevents their elastic fibers from contracting further.

- *Minimal volume.* When the chest cavity has been penetrated, as in a pneumothorax, the lungs collapse, and the amount of air in the respiratory system is reduced to the **minimal volume.** Some air remains in the lungs, even at minimal volume, because the surfactant that coats the alveolar surfaces prevents their collapse.

Not all of the inspired air reaches the alveolar exchange surfaces within the lungs. A typical inhalation pulls around 500 mL of air into the respiratory system. The first 350 mL travels along the conducting passageways and enters the alveolar spaces, but the last 150 mL never gets farther than the conducting passageways and does not take part in gas exchange with the blood. The total volume of these passageways (150 mL) is known as the *anatomic dead space* of the lungs.

CONCEPT CHECK QUESTIONS

1. Mark breaks a rib and it punctures the chest wall on his left side. What will happen to his left lung?
2. In pneumonia, fluid accumulates in the alveoli of the lungs. How would vital capacity be affected?

Answers begin on p. 792.

Gas Exchange

During pulmonary ventilation, the alveoli are supplied with oxygen, and carbon dioxide is removed from the bloodstream. The actual process of gas exchange with the external environment occurs between the blood and alveolar air across the respiratory membrane. This process depends on (1) the partial pressures of the gases involved and (2) the diffusion of molecules between a gas and a liquid. ∞ p. 64

Mixed Gases and Partial Pressures

The air we breathe is not a single gas but a mixture of gases. Nitrogen molecules (N_2) are the most abundant and account for about 78.6 percent of the atmospheric gas molecules. Oxy-

---

gen molecules (O_2), the second most abundant, make up roughly 20.9 percent of the atmospheric content. Most of the remaining 0.5 percent consists of water vapor molecules; carbon dioxide (CO_2) contributes a mere 0.04 percent.

Atmospheric pressure at sea level is approximately 760 mmHg. Each of the gases in air contributes to the total atmospheric pressure in proportion to its relative abundance. The pressure contributed by a single gas is the **partial pressure** of that gas, abbreviated as P. All the partial pressures added together equal the total pressure exerted by the gas mixture. For the atmosphere, this relationship can be summarized as follows:

$$P_{N_2} + P_{O_2} + P_{H_2O} + P_{CO_2} = 760 \text{ mmHg}$$

Because each gas contributes to the total pressure in proportion to its relative abundance, the partial pressure of oxygen, P_{O_2}, is 20.9 percent of 760 mmHg, or approximately 159 mmHg. The partial pressures of other atmospheric gases are listed in Table 15–1. These values are important because the partial pressure of each gas determines its rate of diffusion between the alveolar air and the bloodstream. Note that whereas the partial pressure of oxygen determines how much oxygen enters solution, it has no effect on the diffusion rates of nitrogen or carbon dioxide.

Alveolar Air Versus Atmospheric Air

As soon as air enters the respiratory tract, its characteristics begin to change. For example, in passing through the nasal cavity, the air becomes warmer and the amount of water vapor increases. On reaching the alveoli, the incoming air mixes with air that remained in the alveoli after the previous respiratory cycle. The resulting alveolar gas mixture, thus, contains more carbon dioxide and less oxygen than does atmospheric air. As noted earlier, the last 150 mL of

TABLE 15–1 *Partial Pressures (mmHg) and Normal Gas Concentrations (%) in Air*

SOURCE OF SAMPLE	NITROGEN (N_2)	OXYGEN (O_2)	WATER VAPOR (H_2O)	CARBON DIOXIDE (CO_2)
Inhaled air (dry)	597 (78.6%)	159 (20.9%)	3.7 (0.5%)	0.3 (0.04%)
Alveolar air (saturated)	573 (75.4%)	100 (13.2%)	47 (6.2%)	40 (5.2%)
Exhaled air (saturated)	569 (74.8%)	116 (15.3%)	47 (6.2%)	28 (3.7%)

inhaled air (about 30 percent of the tidal volume) never gets farther than the conducting passageways, which are the anatomic dead space of the lungs. During expiration, the departing alveolar air mixes with air in the dead space to produce yet another mixture that differs from both atmospheric and alveolar samples. The differences in composition between atmospheric (inhaled) and alveolar air can be seen in Table 15–1.

Clinical Note
CHRONIC RESPIRATORY SYSTEM DISEASES

A multitude of problems can arise from the respiratory system. Many of these can develop over weeks or months, while others develop over hours. The following discussion details the more common types of respiratory system problems seen in the emergency setting.

Obstructive Lung Disease

Obstructive lung disease is widespread in our society. The most common obstructive lung diseases encountered in prehospital care are asthma, emphysema, and chronic bronchitis (the last two are often discussed together as chronic obstructive pulmonary disease, or COPD). Asthma afflicts 4–5 percent of the U.S. population, and COPD is found in 25 percent of all adults. Chronic bronchitis alone affects one in five adult males. Patients with COPD have a 50 percent mortality within 10 years of the diagnosis.

Although asthma may have a genetic predisposition, COPD is known to be caused directly by cigarette smoking and environmental toxins. Other factors have been shown to precipitate symptoms in patients who already have obstructive airway disease. Intrinsic factors include stress, upper respiratory infections, and exercise. Extrinsic factors include tobacco smoke, drugs, occupational hazards (chemical fumes, dust, etc.), and allergens such as foods, animal dander, dusts, and molds.

Abnormal ventilation is a common feature of all obstructive lung diseases. This abnormal ventilation is a result of obstruction that occurs primarily in the bronchioles, where several changes occur. One of these changes is bronchospasm (sustained smooth muscle contraction), which may be reversed by beta-adrenergic receptor stimulation. Agents such as terbutaline, albuterol, and epinephrine are used to accomplish this. Increased mucous production by goblet cells that line the respiratory tree also contribute to obstruction. This effect may be worsened by the fact that in many patients the cilia are destroyed, which results in poor clearance of excess mucus. Finally, inflammation of the bronchial passages results in the accumulation of fluid and inflammatory cells. Depending on the underlying cause, some elements of bronchial obstruction are reversible, whereas others are not.

During inspiration, the bronchioles will naturally dilate, allowing air to be drawn into the alveoli. As the patient begins to exhale, the bronchioles constrict. When this natural constriction occurs—in addition to the underlying bronchospasm, increased mucous production, and inflammation in patients with obstructive airway disease—the result is significant air trapping distal to the obstruction. This is one of the hallmarks of obstructive lung disease.

Emphysema

Emphysema results from destruction of the alveolar walls distal to the terminal bronchioles. It is more common in men than in women. The major factor that contributes to emphysema in our society is cigarette smoking. Significant exposure to environmental toxins is another contributing factor.

Continued exposure to noxious substances, such as cigarette smoke, results in the gradual destruction of the walls of the alveoli. This process decreases the alveolar membrane surface area, thus lessening the area available for gas exchange. The progressive loss of the respiratory membrane results in an increased ratio of air to lung tissue. The result is diffusion defects. Additionally, the number of pulmonary capillaries in the lung is decreased, which thus increases resistance to pulmonary blood flow. This condition ultimately causes pulmonary hypertension, which in turn may lead to right-heart failure, cor pulmonale, and death.

Emphysema also causes weakening of the walls of the small bronchioles. When the walls of the alveoli and small bronchioles are destroyed, the lungs lose their capacity to recoil and air becomes trapped in the lungs. Thus, residual volume increases while vital capacity remains relatively normal. The destroyed lung tissue (called *blebs*) results in alveolar collapse. To counteract this effect, patients tend to breathe through pursed lips. This creates continued positive pressure similar to PEEP (positive end-expiratory pressure) and prevents alveolar collapse.

As the disease progresses, the PaO_2 further decreases, which may lead to increased red blood cell production and polycythemia (an excess of red blood cells that results in an abnormally high hematocrit). The $PaCO_2$ also increases and becomes chronically elevated, which forces the body to depend upon hypoxic drive to control respirations. Finally, remember that emphysema is characterized by irreversible airway obstruction.

Patients with emphysema are more susceptible to acute respiratory infections, such as pneumonia, and to cardiac dysrhythmias. Chronic emphysema patients ultimately become dependent on bronchodilators, corticosteroids, and in the final stages, supplemental oxygen.

Chronic Bronchitis

Chronic bronchitis results from an increase in the number of the goblet (mucus-secreting) cells in the respiratory tree. It is characterized by the production of a large quantity of sputum. This often occurs after prolonged exposure to cigarette smoke.

Unlike emphysema, in chronic bronchitis the alveoli are not severely affected and diffusion remains normal. Gas exchange is decreased because alveolar ventilation is lowered, which ultimately results in hypoxia and hypercarbia. Hypoxia may increase red blood cell production, which in turn leads to polycythemia (as occurs in emphysema). Increased $PaCO_2$ levels may lead to irritability, somnolence, decreased intellectual abilities, headaches, and personality changes. Physiologically, an increased $PaCO_2$ causes pulmonary vasoconstriction, which results in pulmonary hypertension and, eventually, cor pulmonale. Unlike emphysema, the vital capacity is decreased, while the residual volume is normal or decreased.

Asthma

Asthma is a common respiratory illness that affects many persons. Although deaths from other respiratory diseases are steadily declining, deaths from asthma have significantly increased during the last decade. Most of the increased asthma deaths have occurred in patients who are 45 years of age or older. In addition, the death rate for black asthmatics has been twice as high as for their white counterparts. Approximately 50 percent of patients who die from asthma do so before reaching the hospital. Thus, EMS personnel are frequently called upon to treat patients suffering an asthma attack. Prompt recognition followed by appropriate treatment can significantly improve the patient's condition and enhance his chance of survival.

Asthma is a chronic inflammatory disorder of the airways. In susceptible individuals, this inflammation causes symptoms usually associated with widespread but variable airflow obstruction. In addition to airflow obstruction, the airway becomes hyperresponsive. The airflow obstruction and hyperresponsiveness are often reversible with treatment. These conditions may also reverse spontaneously.

Asthma may be induced by one of many different factors. These factors, commonly referred to as triggers or inducers, vary from one individual to the next. In allergic individuals, environmental allergens are a major cause of inflammation. These may occur both indoors and outdoors. In addition to allergens, asthma may be triggered by cold air, exercise, foods, irritants, stress, and certain medications. Often, no specific trigger can be identified. Extrinsic triggers tend predominantly to affect children, whereas intrinsic factors trigger asthma in adults.

Within minutes of exposure to the offending trigger, a two-phase reaction occurs. The first phase of the reaction is characterized by the release of chemical mediators such as histamine. These mediators cause contraction of the bronchial smooth muscle and leakage of fluid from peribronchial capillaries. This results in both bronchoconstriction and bronchial edema. These two factors can significantly decrease expiratory air flow and cause the typical asthma attack.

Often, the asthma attack will resolve spontaneously in one to two hours or may be aborted by the use of inhaled bronchodilator medications such as albuterol. However, within six to eight hours after exposure to the trigger, a second reaction occurs. This late phase is characterized by inflammation of the bronchioles as cells of the immune system (eosinophils, neutrophils, and lymphocytes) invade the mucosa of the respiratory tract. This leads to additional edema and swelling of the bronchioles and a further decrease in expiratory airflow.

The second phase reaction will not typically respond to inhaled beta-agonist drugs such as metaproterenol or albuterol. Instead, anti-inflammatory agents such as corticosteroids are often required. It is important to point out that the severe inflammatory changes seen in an acute asthma attack do not develop over a few hours or even a few days. The inflammation will often begin several days or several weeks before the onset of the actual asthma attack. Many asthmatic patients will wait before summoning EMS. The longer the time interval from the onset of the asthma attack until treatment, the less likely it will be that bronchodilator medications will work. Often, after a prolonged asthma attack, the patient may become fatigued. A fatigued patient can quickly develop respiratory failure and subsequently require intubation and mechanical ventilation. Always be prepared to provide airway and respiratory support for the asthmatic.

STATUS ASTHMATICUS

Status asthmaticus is a severe, prolonged asthma attack that cannot be broken by repeated doses of bronchodilators. It is a serious medical emergency that requires prompt recognition, treatment, and transport. The patient who suffers from status asthmaticus frequently will have a greatly distended chest from continued air trapping. Breath sounds, and often wheezing, may be absent. The patient is usually exhausted, severely acidotic, and dehydrated. The management of status asthmaticus is basically the same as for asthma. Recognize that respiratory arrest is imminent and be prepared for endotracheal intubation. The patient should be transported immediately with aggressive treatment continued en route. ■

Partial Pressures Within the Circulatory System

The processes of gas exchange may be divided into external respiration and internal respiration. *External respiration* is the diffusion of gases between the blood and alveolar air across the respiratory membrane. *Internal respiration* is the diffusion of gases between blood and interstitial fluid across the endothelial cells of capillary walls. Figure 15–14● indicates the partial pressures of oxygen and carbon dioxide in alveolar air and capillaries and in the arteries and veins of the pulmonary and systemic circuits.

The deoxygenated blood delivered by the pulmonary arteries has lower P_{O_2} and a higher P_{CO_2} than does alveolar air (Figure 15–14a). Diffusion between the alveolar air and the pulmonary capillaries, thus, elevates the P_{O_2} of the blood while lowering its P_{CO_2}. By the time it enters the pulmonary venules, the oxygenated blood has reached equilibrium with the alveolar air, so it departs the alveoli with a P_{O_2} of about 100 mmHg and a P_{CO_2} of roughly 40 mmHg.

Normal interstitial fluid has a P_{O_2} of 40 mmHg and a P_{CO_2} of 45 mmHg. As a result, oxygen diffuses out of the capillaries, and carbon dioxide diffuses in, until the capillary partial pressures are the same as those in the adjacent tissues (Figure 15–14b). When the blood returns to the alveolar capillaries, external respiration will replace the oxygen released into the tissues at the same time that the excess CO_2 is lost.

Gas Transport

Oxygen and carbon dioxide have limited solubilities in blood plasma. The limited extent to which these gases dissolve in plasma is a problem because peripheral tissues need more oxygen and generate more carbon dioxide than the plasma can ab-

● **FIGURE 15–14 An Overview of Respiration and Respiratory Processes.** **(a)** External respiration is the diffusion of gases between the blood and alveolar air across the respiratory membrane. Diffusion between the alveolar air and the pulmonary capillaries elevates the P_{O_2} of the blood and lowers its P_{CO_2}. **(b)** Internal respiration is the diffusion of gases between blood and interstitial fluid across capillary cell membranes. Diffusion between the systemic capillaries and interstitial fluid lowers the P_{O_2} of the blood and increases its P_{CO_2}.

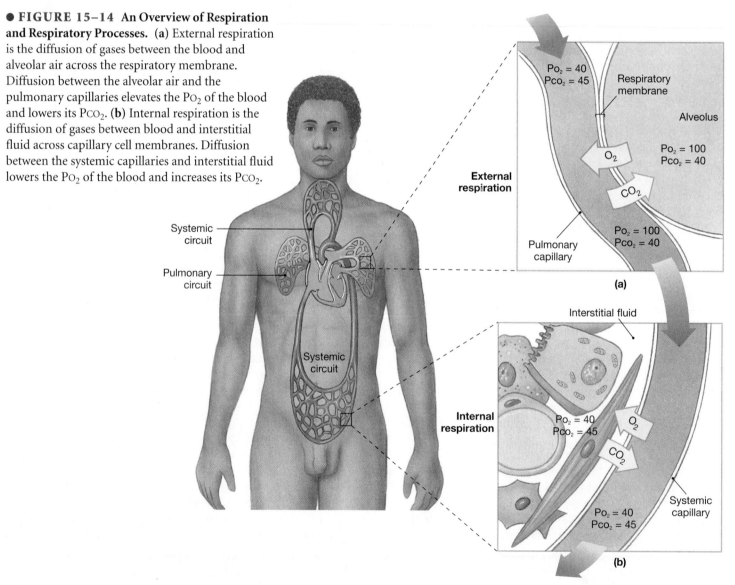

sorb and transport. The problem is solved by red blood cells (RBCs), which take up dissolved oxygen and carbon dioxide molecules from plasma and either bind them (in the case of oxygen) or use them to manufacture soluble compounds (in the case of carbon dioxide). Because these reactions remove dissolved gases from the blood plasma, gases continue to diffuse into the blood and never reach equilibrium. These reactions are also *temporary* and *completely reversible*. When plasma oxygen or carbon dioxide concentrations are high, the excess molecules are removed by RBCs; when the plasma concentrations are low, the RBCs release their stored reserves.

Clinical Note
UPPER RESPIRATORY INFECTION

Among the most common infections for which patients seek medical attention are those that involve the upper airway and respiratory tract. Although these conditions are rarely life threatening, *upper respiratory infections (URIs)* can make many existing pulmonary diseases worse or lead to direct pulmonary infection. The best defense against the spread of upper respiratory infection is to practice common hygiene such as good hand washing and covering the mouth during coughing and sneezing. Attention to such details is important when caring for patients with underlying pulmonary disease or those who are immunosuppressed (HIV infection, cancer) because URIs are more severe in these populations. Due to the prevalence of such infections, complete protection is impossible.

Remember that the upper airway begins at the nose and mouth, passes through the pharynx, and ends at the larynx. Other related structures are the paranasal sinuses and the eustachian tubes that connect the pharynx and the middle ear. In addition, several collections of lymphoid tissue found in the pharynx (palatine, pharyngeal, and lingual tonsils) produce antibodies and provide immune protection.

Viruses cause the vast majority of upper respiratory infections (URIs). A variety of bacteria may also produce infection of the upper respiratory tract. The most significant is *group A streptococcus*, which is the causative organism in "strep throat" and accounts for up to 30 percent of URIs. These bacteria are also implicated in sinusitis and middle-ear infections. Up to 50 percent of patients who have pharyngitis (inflammation of the pharynx) are not found to have a viral or bacterial cause. Fortunately, most URIs are self-limiting illnesses that resolve after several days of symptoms.

The major symptoms of URI are dependent upon the portion of the upper respiratory tract that is predominantly affected. Patients with URIs will often have accompanying symptoms such as fever, chills, myalgias (muscle pains), and fatigue. Most upper respiratory infections are treated symptomatically. Acetaminophen or ibuprofen is prescribed for fever, headache, and myalgias. Encourage patients to drink plenty of fluids. Saltwater gargles may be used for throat discomfort. Decongestants and antihistamines may be used to reduce mucous secretion. Encourage patients who are being treated with antibiotics for bacterial causes of URI to continue these agents. ■

Oxygen Transport

Only about 1.5 percent of the oxygen content of arterial blood consists of oxygen molecules in solution. The rest of the oxygen molecules are bound to hemoglobin (Hb) molecules—specifically, to the iron ions in the center of heme units. ∞ p. 412 This process occurs through a reversible reaction that can be summarized as follows:

$$Hb + O_2 \leftrightarrow HbO_2$$

The amount of oxygen bound (or released) by hemoglobin depends primarily on the P_{O_2} in its surroundings. The lower the oxygen content of a tissue, the more oxygen is released by hemoglobin molecules that pass through local capillaries. For example, inactive tissues have little demand for oxygen, and the local P_{O_2} is about 40 mmHg. Under these conditions, hemoglobin releases about 25 percent of its stored oxygen. In contrast, if the local P_{O_2} of active tissues declines to 15–20 mmHg (about one-half of the P_{O_2} of normal tissue), hemoglobin then releases up to 80 percent of its stored oxygen. In practical terms, this means that active tissues will receive roughly three times as much oxygen as will inactive tissues.

In addition to the effect of P_{O_2}, the amount of oxygen released by hemoglobin is influenced by pH and temperature. Active tissues generate acids that lower the pH of the interstitial fluids. When the pH declines, hemoglobin molecules release their bound oxygen molecules more readily. Hemoglobin also releases more oxygen when body temperature rises.

All three of these factors (P_{O_2}, pH, and temperature) are important during periods of maximal exertion. When a skeletal muscle works hard, its temperature rises and the local pH and P_{O_2} decline. The combination makes the hemoglobin that enters the area release much more oxygen that can be used by active muscle fibers. Without this automatic adjustment, tissue P_{O_2} would fall to very low levels almost immediately, and the exertion would come to a premature halt.

Key Note

Hemoglobin within RBCs carries most of the oxygen in the bloodstream, and it releases it in response to changes in the oxygen partial pressure in the surrounding plasma. If the P_{O_2} increases, hemoglobin binds oxygen; if the P_{O_2} decreases, hemoglobin releases oxygen. At a given P_{O_2} hemoglobin will release additional oxygen if the pH decreases or the temperature increases.

Clinical Note
PULSE OXIMETRY

The measurement of oxygen levels in the body through *pulse oximetry* has become commonplace in emergency medicine. In fact, the oxygen saturation level, as determined through pulse oximetry, is often referred to as the "fifth vital sign." A pulse oximeter measures the hemoglobin oxygen saturation in peripheral tissues. It is noninvasive, rapidly applied, and easy to operate. Pulse oximetry readings are accurate and continually reflect any changes in peripheral oxygen delivery. In fact, oximetry often detects problems with oxygenation faster than standard physical assessment techniques.

Approximately 98 percent of oxygen is transported to the peripheral tissues bound to hemoglobin. Only 2 percent of oxygen is transported dissolved in the plasma. Normally, there is a fixed relationship between the partial pressure of oxygen and hemoglobin saturation. However, in certain disease processes, this relationship can be impaired. Pulse oximetry measures only the oxygen bound to hemoglobin.

Peripheral oxygen saturation is measured by placing a probe on a peripheral capillary bed such as the fingertip, toe, or earlobe. In infants, the sensor can be placed on the heel of the foot and secured with tape. The sensor contains two light-emitting diodes (LEDs) and two sensors (photodetectors). One LED emits light at 660 nm (red) and the other emits light at 940 nm (infrared). Photodetectors placed on the opposite side of a capillary bed (such as a fingertip) detect the amount of light transmitted through the capillary bed. The two wavelengths were chosen because one is absorbed by oxyhemoglobin (hemoglobin with oxygen bound) and the other is absorbed by reduced hemoglobin (hemoglobin without oxygen bound). The amount of light absorbed by these substances is constant with time and does not vary during the cardiac cycle. A small increase in arterial blood flow occurs with each heartbeat, which results in increased light absorption. By comparing the ratio of pulsatile and baseline absorption of light at these two wavelengths, the ratio of oxyhemoglobin to reduced hemoglobin can be calculated (Figure 15–15●). This figure is the oxygen-saturation percentage or (SpO_2).

Pulse oximeters display the SpO_2 and the pulse rate as detected by the sensors. They show the SpO_2 either as a number or as a visual display that also shows the pulse's waveform. The relationship between the SpO_2 and the PaO_2 is very complex. However, the SpO_2 generally correlates with the PaO_2. The greater the PaO_2, the greater will be the oxygen saturation. Since hemoglobin carries 98 percent of oxygen in the blood while plasma carries only 2 percent, pulse oximetry accurately analyzes peripheral oxygen delivery (Figure 15–16●).

False readings with pulse oximetry are infrequent. When they do occur, the oximeter often generates an error signal or a blank screen. Causes of false readings include carbon monoxide poisoning, high-intensity lighting, and certain hemoglobin abnormalities. Nail polish,

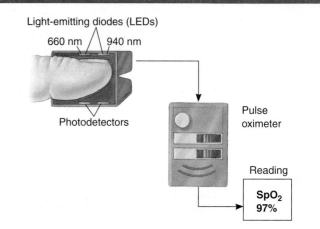

● **FIGURE 15–15 Pulse Oximetry.** Oxygen saturation can be determined by measuring the amounts of light absorbed by oxyhemoglobin and by reduced hemoglobin and then calculating the difference.

● **FIGURE 15–16 Pulse Oximeter.** Pulse oximeter technology now allows accurate measurements of SpO_2 levels with devices smaller than a matchbox.

in certain cases, can interfere with oximetry function. This is particularly problematic with blue nail polish, which absorbs light at 960 nm, close to the wavelengths monitored by the oximeter. The absence of a pulse in an extremity also will cause a false reading. In hypovolemia and in severely anemic patients, the pulse oximetry reading can be misleading. While the SpO_2 reading may be normal, the total amount of hemoglobin available to carry oxygen may be so markedly decreased that the patient will remain hypoxic at the cellular level. ■

Clinical Note

Clinical Note
CARBON MONOXIDE POISONING

Carbon monoxide (CO) is an odorless, tasteless gas that is often the by-product of incomplete combustion. Because of its chemical structure, CO has more than 200 times the affinity of oxygen to bind with hemoglobin in the red blood cells. The binding of CO to hemoglobin causes hypoxia as the oxygen-carrying capacity of the blood is markedly decreased.

Causes of carbon monoxide poisoning include improperly ventilated heating systems, enclosed structure fires, and automobile exhaust fumes. Signs and symptoms of CO poisoning depend on the severity. Initially, the signs are mild and nonspecific. They include headache, nausea, vomiting, altered mental status, and rapid breathing. With severe poisonings, coma and death can ensue.

Treatment includes maximizing oxygen delivery to ensure that all available hemoglobin is oxygenated. Some experts advocate placing the patient into a hyperbaric chamber. Increasing the environmental pressure to several atmospheres can drive oxygen to unbound hemoglobin. Often, however, the patient must wait on new red blood cell production for complete recovery. ■

Carbon Dioxide Transport

Carbon dioxide is generated by aerobic metabolism in peripheral tissues. After entering the bloodstream, a CO_2 molecule may (1) dissolve in the plasma, (2) bind to hemoglobin within red blood cells, or (3) be converted to a molecule of carbonic acid (H_2CO_3) (Figure 15–17●). All three processes are completely reversible.

TRANSPORT IN PLASMA. Plasma becomes saturated with carbon dioxide quite rapidly, and about 7 percent of the carbon dioxide absorbed by peripheral capillaries is transported as dissolved gas molecules. The rest diffuses into RBCs.

HEMOGLOBIN BINDING. Once in red blood cells, some of the carbon dioxide molecules are bound to the protein "globin" portions of hemoglobin molecules, which forms **carbaminohemoglobin** (kar-bām-i-nō-hē-mō-GLŌ-bin). Such binding does not interfere with the binding of oxygen to heme units, so hemoglobin can transport both oxygen and carbon dioxide simultaneously. Normally, about 23 percent of the carbon dioxide that enters the blood in peripheral tissues is transported as carbaminohemoglobin.

CARBONIC ACID FORMATION. About 70 percent of all carbon dioxide molecules in the body are ultimately transported in the plasma as bicarbonate ions. First, carbon dioxide is converted to carbonic acid by the enzyme carbonic anhydrase in RBCs. However, the carbonic acid molecules do not remain intact; almost immediately, each of these molecules dissoci-

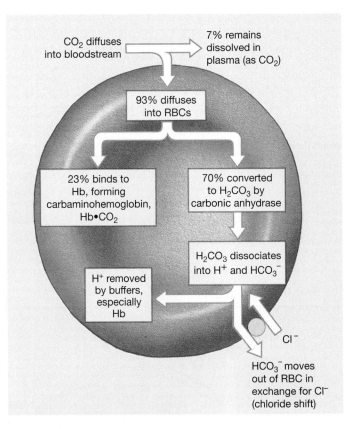

● **FIGURE 15–17 Carbon Dioxide Transport in the Blood.**

ates into a hydrogen ion and a bicarbonate ion. The reactions can be summarized as follows:

$$CO_2 + H_2O \xleftarrow{\text{carbonic anhydrase}} H_2CO_3 \leftrightarrow H^+ + HCO_3^-$$

The reactions occur very rapidly and are completely reversible. Because most of the carbonic acid formed immediately dissociates into bicarbonate and hydrogen ions, we can ignore the intermediary step and summarize the reaction as follows:

$$CO_2 + H_2O \xleftarrow{\text{carbonic anhydrase}} H^+ + HCO_3^-$$

In peripheral capillaries, this reaction rapidly ties up large numbers of carbon dioxide molecules. The reaction is driven from left to right because carbon dioxide continues to arrive, diffusing out of the interstitial fluids, and the hydrogen ions and bicarbonate ions are removed continuously. Most of the hydrogen ions bind to hemoglobin molecules, which prevents their release from the RBCs and a lowering of plasma pH. The bicarbonate ions diffuse into the surrounding plasma. The exit of the bicarbonate ions is matched by the entry of chloride ions from the plasma, thus trading one anion for another. This mass movement of chloride ions into RBCs is known as the *chloride shift*.

When venous blood reaches the alveoli, carbon dioxide diffuses out of the plasma, and the P_{CO_2} declines. Because all of the carbon dioxide transport mechanisms are reversible, when carbon dioxide diffuses out of the red blood cells, the processes shown in Figure 15–17 proceed in the opposite direction: hydrogen ions leave the hemoglobin molecules, and bicarbonate ions diffuse into the cytoplasm of the RBCs and are converted to water and CO_2.

Figure 15–18● summarizes the events by which oxygen and carbon dioxide are transported and exchanged between the respiratory and cardiovascular systems.

Key Note

Carbon dioxide primarily travels in the bloodstream as bicarbonate ions, which form through dissociation of the carbonic acid produced by carbonic anhydrase inside RBCs. Lesser amounts of CO_2 are bound to hemoglobin or dissolved in plasma.

Clinical Note
ARTERIAL BLOOD GAS MEASUREMENTS

The partial pressure of the respiratory gases can be determined with an *arterial blood gas (ABG)* measurement. For this, a sample of arterial blood is obtained from the radial, brachial, or femoral artery. The sample is immediately placed into a blood-gas machine and measured. Most blood-gas machines will provide readings of the partial pressure of oxygen (PO_2), the partial pressure of carbon dioxide (PCO_2), the pH, the bicarbonate level (HCO_3^-), and the hemoglobin (Hg). These parameters provide a great deal of information about the efficiency of ventilation and oxygenation and will readily detect acid-base abnormalities. ABGs are an essential tool in the treatment of most severe respiratory disease processes. ■

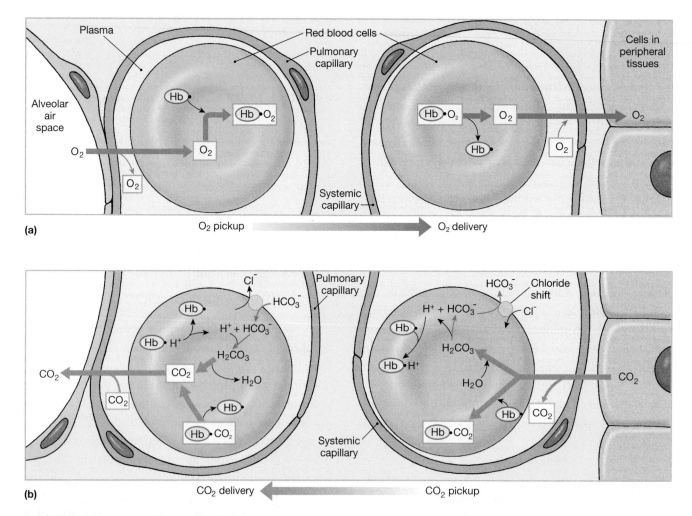

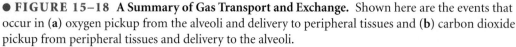

● **FIGURE 15–18 A Summary of Gas Transport and Exchange.** Shown here are the events that occur in (**a**) oxygen pickup from the alveoli and delivery to peripheral tissues and (**b**) carbon dioxide pickup from peripheral tissues and delivery to the alveoli.

1. Why does hemoglobin release more oxygen to active skeletal muscles during exercise than it does when the muscles are at rest?
2. How would blockage of the trachea affect blood pH?

Answers begin on p. 792.

■ The Control of Respiration

Cells continuously absorb oxygen from the interstitial fluids and generate carbon dioxide. Under normal conditions, cellular rates of absorption and generation are matched by the rates of delivery and removal at the capillaries. Moreover, those rates are identical to those of oxygen absorption and carbon dioxide excretion at the lungs. If these rates become unbalanced, the activities of the cardiovascular and respiratory systems must be adjusted. Equilibrium is restored through homeostatic mechanisms that involve (1) changes in blood flow and oxygen delivery under local control and (2) changes in the depth and rate of respiration under the control of the brain's respiratory centers.

The Local Control of Respiration

Both the rate of oxygen delivery at each tissue and the efficiency of oxygen pickup at the lungs are regulated at the local level. If a peripheral tissue becomes more active, the interstitial P_{O_2} falls and the P_{CO_2} rises. These changes increase the difference between partial pressures in the tissues and arriving blood, so more oxygen is delivered and more carbon dioxide is carried away. In addition, rising P_{CO_2} levels cause the relaxation of smooth muscles in the walls of arterioles in the area, which increases blood flow.

Local adjustments in blood flow, or of the flow of air into alveoli, also improve the efficiency of gas transport. For example, as blood flows to alveolar capillaries, it is directed to pulmonary lobules in which P_{O_2} is relatively high. This occurs because precapillary sphincters in alveolar capillary beds constrict when the local P_{O_2} is low. (This response is the opposite of that seen in peripheral tissues. ∞ p. 471) Smooth muscles in the walls of bronchioles are sensitive to the P_{CO_2} of the air they contain. When the P_{CO_2} increases, the bronchioles dilate; when the P_{CO_2} declines, the bronchioles constrict. Airflow is, therefore, directed to lobules in which the P_{CO_2} is high.

Control by the Respiratory Centers of the Brain

Respiratory control has both involuntary and voluntary components. The brain's involuntary respiratory centers (in the medulla oblongata and pons) regulate the respiratory muscles and control the frequency (respiratory rate) and the depth of breathing. They do so in response to sensory information that arrives from the lungs and other portions of the respiratory tract, as well as from a variety of other sites. The voluntary control of respiration reflects activity in the cerebral cortex that affects the output of the respiratory centers or of motor neurons that control respiratory muscles.

The **respiratory centers** are three pairs of nuclei in the reticular formation of the pons and medulla oblongata. The **respiratory rhythmicity centers** of the medulla oblongata set the pace for respiration. Each center can be subdivided into a *dorsal respiratory group (DRG)*, which contains an *inspiratory center*, and a *ventral respiratory group (VRG)*, which contains an *expiratory center*. Their output is adjusted by the two pairs of nuclei that comprise the respiratory centers of the pons. The centers in the pons adjust the respiratory rate and the depth of respiration in response to sensory stimuli, emotional states, or speech patterns.

The Activities of the Respiratory Rhythmicity Centers

The DRG's inspiratory center functions in every respiratory cycle. It contains neurons that control the external intercostal muscles and the diaphragm (inspiratory muscles). During quiet breathing, the neurons of the inspiratory center gradually increase stimulation of the inspiratory muscles for two seconds, and then the inspiratory center becomes silent for the next three seconds. During that period of inactivity, the inspiratory muscles relax and passive exhalation occurs. The inspiratory center will maintain this basic rhythm even in the absence of sensory or regulatory stimuli. The VRG functions only during forced breathing when it activates the accessory muscles involved in inhalation and exhalation. The relationships between the inspiratory and expiratory centers during quiet breathing and forced breathing are diagrammed in Figure 15–19●.

The performance of these respiratory centers can be affected by any factor that alters the metabolic or chemical activities of neural tissues. For example, elevated body temperatures or CNS stimulants (such as amphetamines or caffeine) increase the respiratory rate. Conversely, decreased body temperature or CNS depressants (such as barbiturates

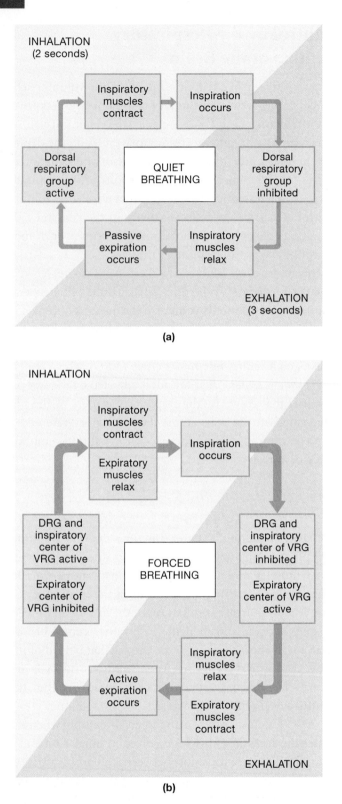

INHALATION
(2 seconds)

Inspiratory muscles contract → Inspiration occurs

Dorsal respiratory group active

QUIET BREATHING

Dorsal respiratory group inhibited

Passive expiration occurs ← Inspiratory muscles relax

EXHALATION
(3 seconds)

(a)

INHALATION

Inspiratory muscles contract / Expiratory muscles relax → Inspiration occurs

DRG and inspiratory center of VRG active

Expiratory center of VRG inhibited

FORCED BREATHING

DRG and inspiratory center of VRG inhibited

Expiratory center of VRG active

Active expiration occurs ← Inspiratory muscles relax / Expiratory muscles contract

EXHALATION

(b)

● **FIGURE 15–19 Basic Regulatory Patterns of Respiration.**
This figure depicts the events that occur during (**a**) quiet breathing and (**b**) forced breathing.

or opiates) reduce the respiratory rate. Respiratory activities are also strongly influenced by reflexes that are triggered by mechanical or chemical stimuli.

The Reflex Control of Respiration

Normal breathing occurs automatically without conscious control. The activities of the respiratory centers are modified by sensory information from mechanoreceptors (such as stretch and pressure receptors) and chemoreceptors. Information from these receptors alters the pattern of respiration. The induced changes are called *respiratory reflexes.*

Mechanoreceptor Reflexes

Mechanoreceptors respond to changes in the volume of the lungs or to changes in arterial blood pressure. Several populations of baroreceptors that are involved in respiratory function are described in Chapter 9. ∞ p. 332

The **inflation reflex** prevents the lungs from overexpanding during forced breathing. The mechanoreceptors involved are stretch receptors that are stimulated when the lungs expand. Sensory fibers leaving the lungs reach the respiratory rhythmicity centers through the vagus nerves. As the volume of the lungs increases, the DRG inspiratory center is gradually inhibited, and the VRG expiratory center is stimulated. Thus, inhalation stops as the lungs near maximum volume, and active exhalation then begins. In contrast, the **deflation reflex** inhibits the expiratory center and stimulates the inspiratory center when the lungs are collapsing. The smaller the volume of the lungs, the greater the inhibition of the expiratory center.

Although neither the inflation reflex nor the deflation reflex is involved in normal quiet breathing, both are important in regulating the forced inhalations and exhalations that accompany strenuous exercise. Together, the inflation and deflation reflexes are known as the *Hering-Breuer reflexes,* after the physiologists who described them in 1865.

The effects of the carotid and aortic baroreceptors on systemic blood pressure were described in Chapter 13. ∞ p. 479 The output from these baroreceptors also affects the respiratory centers. When blood pressure falls, the respiratory rate increases; when blood pressure rises, the respiratory rate declines. This adjustment results from the stimulation or inhibition of the respiratory centers by sensory fibers in the glossopharyngeal (IX) and vagus (X) nerves.

Chemoreceptor Reflexes

Chemoreceptors respond to chemical changes in the blood and cerebrospinal fluid. ∞ p. 332 Their stimulation leads to an increase in the depth and rate of respiration. Centers in the carotid bodies (adjacent to the carotid sinus) and the aortic bodies (near the aortic arch) are sensitive to the pH, P_{CO_2}, and P_{O_2} in arterial blood. Receptors in the medulla oblongata respond to the pH and P_{CO_2} in cerebrospinal fluid.

Clinical Note
HYPOXIC DRIVE

Respiratory drive is controlled primarily by the amount of carbon dioxide in the blood (P_{CO_2}). An increase in the P_{CO_2} stimulates respirations, while a fall in P_{CO_2} inhibits respirations. In chronic obstructive pulmonary diseases, such as emphysema or chronic bronchitis, the level of carbon dioxide in the blood gradually rises (hypercapnia). As the disease progresses, the chemoreceptors become accustomed to chronic hypercapnia.

When this occurs, the body begins to rely on oxygen levels (P_{O_2}) instead of P_{CO_2} levels to regulate respirations. This change, referred to as hypoxic drive, occurs only in advanced and severe disease. Administration of supplemental oxygen, which is a routine part of emergency care, can significantly increase P_{O_2} levels and can inhibit respirations. In severe cases, administration of high levels of supplemental oxygen can cause respiratory arrest. ■

Clinical Note
LUNG CANCER

Lung cancer (neoplasm) is the leading cause of cancer-related death in the U.S. in both men and women. Most patients with lung cancer are between the ages of 55 and 65 years. The mortality rate for patients with lung cancer is high after only one year with the disease.

There are currently four major types of lung cancer based on the predominant cell type. Twenty percent of cases involve only the lung tissue. Another 35 percent spread to the lymphatic system, and 45 percent have distant metastases (cancer cells that spread to other tissues). In cases that involve lung tissue invasion, the primary problem is disruption of diffusion. In some larger cancers, there may also be alterations in ventilation by obstruction of the conducting bronchioles.

Cigarette smoking has long been known to be a risk factor for development of lung cancer. Environmental exposure to asbestos, hydrocarbons, radiation, and fumes from metal production have also been identified as risk factors. Finally, home exposure to radon has been implicated in the development of lung cancer. Preventive strategies include educating teenagers about the dangers of cigarette smoking and encouraging current smokers to quit. Implementing environmental safety standards that reduce the risk of exposure to such substances as asbestos will also reduce the risk of lung cancer. Finally, cancer screening of populations at risk is encouraged.

Although cancers that start elsewhere in the body can spread to the lungs, the vast majority of lung cancers are caused by carcinogens (cancer-producing substances) from cigarette smoking. A small portion of lung cancers are caused by inhalation of occupational agents such as asbestos and arsenic. These substances irritate and adversely affect the various tissues of the lung, and ultimately lead to the development of abnormal (cancerous) cells.

There are four major types of lung cancers depending upon the type of lung tissue involved. The most common type, *adenocarcinoma,* arises from glandular-type (i.e., mucus-producing) cells found in the lungs and bronchioles. The next most frequently encountered type of lung cancer, *small cell carcinoma* (also called *oat cell carcinoma*), arises from bronchial tissues. The third type is *epidermoid carcinoma,* and the fourth is *large cell carcinoma.* Like small cell carcinoma, epidermoid and large cell carcinomas typically arise from the bronchial tissues. Lung cancers generally have a bad prognosis, and most patients die within a year of the diagnosis.

Patients with lung cancer will present with a variety of complaints, depending on whether they are related to direct lung involvement, invasion of local structures, or metastatic spread. Patients with localized disease will present with cough, dyspnea, hoarseness, vague chest pain, and hemoptysis (coughing up blood). Fever, chills, and pleuritic chest pain are seen in patients who develop pneumonia. Symptoms related to local invasion include pain on swallowing (dysphagia), weakness or numbness in the arm, and shoulder pain. Metastatic symptoms are related to the area of spread and include headache, seizures, bone pain, abdominal pain, nausea, and malaise.

Physical findings are nonspecific. Patients with advanced disease have profound weight loss and cachexia (general physical wasting and malnutrition). Crackles (rales), rhonchi, wheezes, and diminished breath sounds may be heard in the affected lung. If the superior vena cava is occluded, venous distention in the arms and neck (superior vena cava syndrome) may be present. The rapid progression of lung cancer can be striking. ■

Clinical Note
CAPNOGRAPHY

End-tidal carbon dioxide ($ETCO_2$) monitoring is a non-invasive method of measuring the levels of carbon dioxide (CO_2) in the exhaled breath. Recordings or displays of exhaled CO_2 measurements are called capnography.

Various terms have been applied to capnography, and a review of them may help you to understand the material in this section. These terms include:

- *Capnometry.* Capnometry is the measurement of expired CO_2. It typically provides a numeric display of the partial pressure of CO_2 (in Torr or mmHg) or the percentage of CO_2 present.
- *Capnography.* Capnography is a graphic recording or display of the capnometry reading over time.
- *Capnograph.* A capnograph is a device that measures expired CO_2 levels.
- *Capnogram.* A capnogram is the visual representation of the expired CO_2 waveform.
- *End-tidal CO_2 ($ETCO_2$).* End-tidal CO_2 is the measurement of the CO_2 concentration at the end of expiration (maximum CO_2).
- *$PETCO_2$.* $PETCO_2$ is the partial pressure of end-tidal CO_2 in a mixed gas solution.
- *$PaCO_2$.* The $PaCO_2$ represents the partial pressure of CO_2 in the arterial blood.

CO_2 is a normal end product of metabolism and is transported by the venous system to the right side of the heart. It is then pumped from the right ventricle to the pulmonary artery and eventually enters the pulmonary capillaries. There it diffuses into the alveoli and is removed from the body through exhalation. When circulation is normal, $ETCO_2$ levels change with ventilation and are a reliable estimate of the partial pressure of carbon dioxide in the arterial system ($PaCO_2$). Normal $ETCO_2$ is 1–2 mm less than the $PaCO_2$, or approximately 5 percent. A normal $PETCO_2$ is approximately 38 mmHg (0.05×760 mmHg = 38 mmHg). When perfusion decreases, as occurs in shock or cardiac arrest, $ETCO_2$ levels reflect pulmonary blood flow and cardiac output, not ventilation.

Decreased $ETCO_2$ levels can be found in shock, cardiac arrest, pulmonary embolism, bronchospasm, and with incomplete airway obstruction (such as mucus plugging). Increased $ETCO_2$ levels are found with hypoventilation, respiratory depression, and hyperthermia. Capnometry provides a non-invasive measure of $ETCO_2$ levels, and provides medical personnel with information about the status of systemic metabolism, circulation, and ventilation (Figure 15–20●). The use of capnography has become commonplace in the operating room, in the emergency department, and in the prehospital setting (Figure 15–21●).

When first introduced into prehospital care, $ETCO_2$ monitoring was used exclusively to verify proper endotracheal tube placement in the trachea. The presence of adequate CO_2 levels following intubation confirms that the tube is in the trachea through the presence of

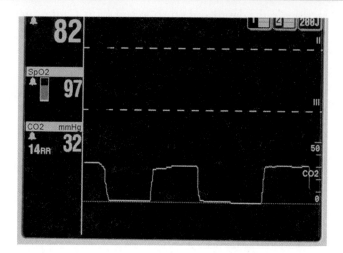

● **FIGURE 15–20** **Continuous Waveform Capnography.**

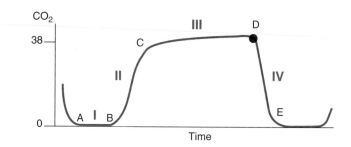

● **FIGURE 15–21** **Normal Capnogram.** AB = *Phase I:* late inspiration, early exhalation (no CO_2). BC = *Phase II:* appearance of CO_2 in exhaled gas. CD = *Phase III:* plateau (constant CO_2). D = highest point ($ETCO_2$). DE = *Phase IV:* rapid descent during inspiration. EA = respiratory pause.

exhaled CO_2. CO_2 is detected by using either a colorimetric or an infrared device.

Colorimentric Devices

The colorimetric device is a disposable $ETCO_2$ detector that contains pH-sensitive, chemically-impregnated paper encased within a plastic chamber. It is placed in the airway circuit between the patient and the ventilation device. When the paper is exposed to CO_2, hydrogen ions (H^+) are generated, which causes a color change in the paper. The color change is reversible and changes breath to breath. A color scale on the device estimates the $ETCO_2$ level. Colorimetric devices cannot detect hyper- or hypocarbia (increased or decreased CO_2 levels). If gastric contents or acidic drugs (e.g., endotracheal epinephrine) contact the paper in the device, subsequent readings may be unreliable. ■

Carbon dioxide levels have a much more powerful effect on respiratory activity than do oxygen levels. The reason is that a relatively small increase in arterial P_{CO_2} stimulates CO_2 receptors, but arterial P_{O_2} does not usually decline enough to activate oxygen receptors. Carbon dioxide levels are, therefore, responsible for regulating respiratory activity under normal conditions. However, when arterial P_{O_2} does fall, the two types of receptors cooperate. Carbon dioxide is generated during oxygen consumption, so when oxygen concentrations fall rapidly, carbon dioxide levels usually increase. As a result, you cannot hold your breath "until you turn blue." Once the P_{CO_2} rises to critical levels, you will be forced to take a breath.

The cooperation between the carbon dioxide and oxygen receptors breaks down only under unusual circumstances. For example, an individual can hold his or her breath longer than normal by taking deep, full breaths, but the practice is very dangerous. The danger lies in the fact that the increased ability is due not to extra oxygen but to the loss of carbon dioxide. If the P_{CO_2} is reduced enough, breath-holding ability may increase to the point that an individual becomes unconscious from oxygen starvation in the brain without ever feeling the urge to breathe.

The chemoreceptors that monitor CO_2 levels are also sensitive to pH, and any condition that affects the pH of blood or CSF will affect respiratory performance. For example, the rise in lactic acid levels after exercise causes a drop in pH that helps stimulate respiratory activity.

Clinical Note
SIDS

Sudden infant death syndrome (SIDS) is the sudden death of an infant under one year of age that remains unexplained after a thorough case investigation including a complete autopsy, an examination of the death scene, and a review of the clinical history. In the U.S. SIDS is the leading cause of death among infants between one month and one year of age. SIDS deaths occur quickly, with no signs of suffering, and are often associated with sleep. SIDS deaths occur more frequently in the fall and winter, and most take place in children between two and four months of age. Boys are affected more often than girls.

SIDS occurs in all types of families and is largely indifferent to race or socioeconomic status. The mother's health and behavior during her pregnancy and the baby's health before birth seem to influence the occurrence of SIDS. Maternal risk factors include cigarette smoking during pregnancy, age less than 20 years, poor prenatal care, low weight gain, anemia, and use of illegal drugs. SIDS is extremely tragic and extremely difficult for the family. ■

Control by Higher Centers

Higher centers influence respiration through their effects on the respiratory centers of the pons and by the direct control of respiratory muscles. For example, the contractions of respiratory muscles can be voluntarily suppressed or exaggerated; this control is necessary during talking or singing. The depth and rate of respiration also change following the activation of centers involved with rage, eating, or sexual arousal. These changes, directed by the limbic system, occur at an involuntary level. Figure 15–22● summarizes the factors involved in the regulation of respiration.

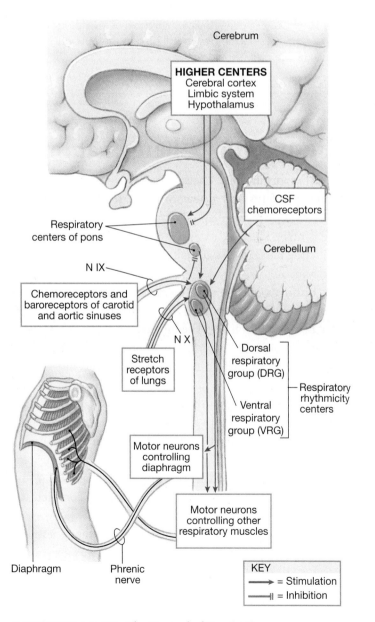

● **FIGURE 15–22 The Control of Respiration.**

A basic pace of respiration is established by the interplay between respiratory centers in the pons and medulla oblongata. That pace is modified in response to input from chemoreceptors, baroreceptors, and stretch receptors. In general, carbon dioxide levels, rather than oxygen levels, are the primary drivers for respiratory activity. Respiratory activity can also be interrupted by protective reflexes and adjusted by the conscious control of respiratory muscles.

→ CONCEPT CHECK QUESTIONS

1. Are peripheral chemoreceptors more sensitive to carbon dioxide levels or to oxygen levels?
2. Strenuous exercise stimulates which set of respiratory reflexes?
3. Johnny tells his mother he will hold his breath until he turns blue and dies. Should she worry?

Answers begin on p. 792.

■ Respiratory Changes at Birth

Several important differences exist between the respiratory systems of a fetus and a newborn. Before delivery, the pulmonary vessels are collapsed, so pulmonary arterial resistance is high. The rib cage is compressed, and the lungs and conducting passageways contain only small amounts of fluid and no air.

At birth, the newborn takes a truly heroic first breath through powerful contractions of the diaphragm and the external intercostal muscles. The inhaled air enters the passageways with enough force to push the contained fluids out of the way and to inflate the entire bronchial tree and most of the alveoli. The same drop in pressure that pulls air into the lungs pulls blood into the pulmonary circulation. The exhalation that follows fails to empty the lungs completely, because the rib cage does not return to its former, fully compressed state. Cartilages and connective tissues keep the conducting passageways open, and the surfactant that covers the alveolar surfaces prevents their collapse. Subsequent breaths complete the inflation of the alveoli.

Pathologists sometimes use these physical changes to determine whether a newborn died before delivery or shortly thereafter. Before the first breath, the lungs are completely filled with fluid, and they will sink if placed in water. After the infant's first breath, even the collapsed lungs contain enough air to keep them afloat.

■ Aging and the Respiratory System

Of the many factors that interact to reduce the efficiency of the respiratory system in elderly individuals, two are noteworthy:

1. Chest movements are restricted by arthritic changes in rib articulations, decreased flexibility at the costal cartilages, and age-related muscular weakness. (These skeletal and muscular restrictions counterbalance an increase in compliance of the lungs that occurs with the deterioration of elastic tissue with age.) Together, the stiffening and reduction in chest movement limit pulmonary ventilation and vital capacity, and contribute to the reduction in exercise performance and capabilities with increasing age.

2. Some degree of emphysema is normal in individuals over age 50. However, the extent varies widely with lifetime exposure to cigarette smoke and other respiratory irritants. Comparative studies of nonsmokers and those who have smoked for various lengths of time clearly show the negative effects of smoking on respiratory performance.

■ Integration with Other Systems

The respiratory system has extensive structural and functional connections to the cardiovascular system. Its many functional links to other organ systems are depicted in Figure 15–23●.

→ CONCEPT CHECK QUESTIONS

1. Describe two age-related changes that combine to reduce the efficiency of the respiratory system.
2. What homeostatic functions of the nervous system support the functional role of the respiratory system?

Answers begin on p. 792.

The Respiratory System in Perspective

For All Systems

Provides oxygen and eliminates carbon dioxide

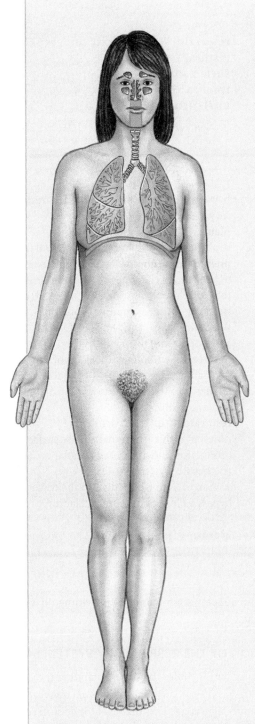

The Integumentary System

- Protects portions of upper respiratory tract; hairs guard entry to external nares

The Skeletal System

- Movements of ribs important in breathing; axial skeleton surrounds and protects lungs

The Muscular System

- Muscular activity generates carbon dioxide; respiratory muscles fill and empty lungs; other muscles control entrances to respiratory tract; intrinsic laryngeal muscles control airflow through larynx and produce sounds

The Nervous System

- Monitors respiratory volume and blood gas levels; controls pace and depth of respiration

The Endocrine System

- Epinephrine and norepinephrine stimulate respiratory activity and dilate respiratory passageways

The Cardiovascular System

- Red blood cells transport oxygen and carbon dioxide between lungs and peripheral tissues
- Bicarbonate ions contribute to buffering capability of blood

The Lymphatic System

- Tonsils protect against infection at entrance to respiratory tract; lymphatic vessels monitor lymph drainage from lungs and mobilize specific defenses when infection occurs
- Alveolar phagocytes present antigens to trigger specific defenses; mucous membrane lining the nasal cavity and upper pharynx traps pathogens, protects deeper tissues

The Digestive System

- Provides substrates, vitamins, water, and ions that are necessary to all cells of the respiratory system
- Increased thoracic and abdominal pressure through contraction of respiratory muscles can assist in defecation

The Urinary System

- Eliminates organic wastes generated by cells of the respiratory system; maintains normal fluid and ion balance in the blood
- Assists in the regulation of pH by eliminating carbon dioxide

The Reproductive System

- Changes in respiratory rate and depth occur during sexual arousal

● **FIGURE 15–23** Functional Relationships Between the Respiratory System and Other Systems.

Chapter Review

Access more review material online at *www.prenhall.com/bledsoe.* There you will find quiz questions, labeling activities, animations, essay questions, and web links.

Key Terms

alveolus/alveoli 547
bronchial tree 553
bronchiole 553
bronchus/bronchi 553
larynx 550

lungs 556
nasal cavity 548
partial pressure 565
pharynx 549
respiratory membrane 556

respiratory rhythmicity centers 573
respiratory system 547
surfactant 556
trachea 552
vital capacity 565

Related Clinical Terms

anoxia (an-ok-SĒ-uh) A lack of oxygen in a tissue.

apnea (AP-nē-uh) The cessation of breathing.

asthma (AZ-muh) An acute respiratory disorder characterized by unusually sensitive, irritated conducting airways.

atelectasis (at-e-LEK-ta-sis) A collapsed lung.

bronchitis (brong-KĪ-tis) Inflammation of the bronchial lining.

bronchoscope A fiber-optic bundle small enough to be inserted into the trachea and finer airways; the procedure is called *bronchoscopy.*

chronic obstructive pulmonary disease (COPD) A condition characterized by chronic bronchitis and chronic airway obstruction.

cystic fibrosis (CF) A relatively common lethal inherited disease; respiratory mucosa secretions become too thick to be transported easily, which leads to respiratory problems.

dyspnea (DISP-nē-uh) Difficult or labored breathing.

emphysema (em-fi-ZĒ-ma) A chronic, progressive condition characterized by shortness of breath and an inability to tolerate physical exertion.

epistaxis (ep-i-STAK-sis) A nosebleed.

hypercapnia (hī-per-KAP-nē-uh) An abnormally high level of carbon dioxide in the blood.

hypocapnia An abnormally low level of carbon dioxide in the blood.

hypoxia (hī-POKS-ē-uh) A low oxygen concentration in a tissue.

influenza A viral infection of the respiratory tract; "the flu."

pleurisy Inflammation of the pleurae and secretion of excess amounts of pleural fluid.

pneumonia (noo-MŌ-nē-uh) A respiratory disorder characterized by fluid leakage into the alveoli and/or swelling and constriction of the respiratory bronchioles.

pneumothorax (noo-mō-THŌ-raks) The entry of air into the pleural cavity.

pulmonary embolism Blockage of a branch of a pulmonary artery that produces an interruption of blood flow to a group of lobules and/or alveoli.

respiratory distress syndrome A condition that results from inadequate surfactant production and associated alveolar collapse.

tracheostomy (trā-kē-OS-to-mē) The insertion of a tube directly into the trachea to bypass a blocked or damaged larynx.

Summary Outline

1. To continue functioning, body cells must obtain oxygen and eliminate carbon dioxide.

THE FUNCTIONS OF THE RESPIRATORY SYSTEM 547

1. The functions of the **respiratory system** include (1) providing an area for gas exchange between air and circulating blood; (2) moving air to and from exchange surfaces; (3) protecting exchange surfaces from environmental variations and defending the respiratory system and other tissues from pathogens; (4) producing sound; and (5) providing olfactory sensations to the CNS.

THE ORGANIZATION OF THE RESPIRATORY SYSTEM 548

1. The respiratory system includes the nose (including the nasal cavity and paranasal sinuses), the pharynx, larynx, trachea, and the various conducting passageways that lead to the surfaces of the lungs. *(Figure 15–1)*

The Respiratory Tract 548

2. The **respiratory tract** consists of the conducting passageways that carry air to and from the alveoli.

The Nose 548

3. Air normally enters the respiratory system through the **external nares,** which open into the **nasal cavity.** The **nasal vestibule** (entrance) is guarded by hairs that screen out large particles. *(Figure 15–2)*

4. The **hard palate** separates the oral and nasal cavities. The **soft palate** separates the superior nasopharynx from the rest of the pharynx. The **internal nares** connect the nasal cavity and nasopharynx. *(Figure 15–2)*

5. Much of the respiratory epithelium is ciliated and produces mucus that traps incoming particles. *(Figure 15–3)*

The Pharynx 549

6. The **pharynx** (throat) is a chamber shared by the digestive and respiratory systems.

The Larynx 550

7. Inhaled air passes through the **glottis** en route to the lungs; the **larynx** surrounds and protects the glottis. The **epiglottis** projects into the pharynx. Exhaled air that passes through the glottis vibrates the **true vocal cords** and produces sound. *(Figure 15–4)*

The Trachea 552

8. The wall of the **trachea** ("windpipe") contains C-shaped tracheal cartilages, which protect the airway. The posterior tracheal wall can distort to permit large masses of food to pass. *(Figure 15–5)*

The Bronchi 553

9. The trachea branches within the mediastinum to form the **right** and **left primary bronchi.** *(Figure 15–5)*

10. The primary bronchi, **secondary bronchi,** and their branches form the *bronchial tree.* As the **tertiary bronchi** branch within the lung, the amount of cartilage in their walls decreases, and the amount of smooth muscle increases. *(Figure 15–6a)*

The Bronchioles 553

11. Each terminal **bronchiole** delivers air to a single pulmonary **lobule.** Within the lobule, the terminal bronchiole branches into *respiratory bronchioles. (Figure 15–6b)*

The Alveolar Ducts and Alveoli 553

12. The respiratory bronchioles open into **alveolar ducts,** which end at **alveolar sacs.** Many alveoli are interconnected at each alveolar sac. *(Figure 15–7a, b)*

The Respiratory Membrane 556

13. The **respiratory membrane** consists of (1) a simple squamous alveolar epithelium, (2) a capillary endothelium, and (3) their fused basement membranes. **Septal cells** produce **surfactant,** an oily secretion that keeps the alveoli from collapsing. **Alveolar macrophages** engulf foreign particles. *(Figure 15–7c,d)*

The Lungs 556

14. The **lungs** are made up of five **lobes:** three in the right lung and two in the left lung. *(Figure 15–8)*

The Pleural Cavities 558

15. Each lung occupies a single pleural cavity lined by a **pleura** (serous membrane). *(Figures 15–9, 15–10)*

RESPIRATORY PHYSIOLOGY 560

1. Respiratory physiology focuses on a series of integrated processes: *pulmonary ventilation,* or breathing (movement of air into and out of the lungs); *gas exchange,* or diffusion, between the alveoli and circulating blood, and between the blood and interstitial fluids; and *gas transport,* between the blood and interstitial fluids.

Pulmonary Ventilation 560

2. A single breath, or **respiratory cycle,** consists of an inhalation *(inspiration)* and an exhalation *(expiration).*

3. The relationship between the pressure inside the respiratory tract and atmospheric pressure determines the direction of airflow. *(Figures 15–11, 15–12)*

4. The diaphragm and the external intercostal muscles are involved in *quiet breathing,* in which exhalation is passive. Accessory muscles become active during the active inspiratory and expiratory movements of *forced breathing,* in which exhalation is active. *(Figure 15–11)*

5. The **vital capacity** includes the **tidal volume** plus the **expiratory reserve volume** and the **inspiratory reserve volume.** The air left in the lungs at the end of maximum expiration is the **residual volume.** *(Figure 15–13)*

Gas Exchange 565

6. Gas exchange involves *external respiration,* which is the diffusion of gases between the blood and alveolar air across the respiratory membrane, and *internal respiration,* which is the diffusion of gases between blood and interstitial fluid across the endothelial cells of capillary walls. *(Figures 15–14, 15–15, 15–16; Table 15–1)*

Gas Transport 568

7. Blood that enters peripheral capillaries delivers oxygen and takes up carbon dioxide. The transport of oxygen and carbon dioxide in the blood involves reactions that are completely reversible.

8. Over the range of oxygen pressures normally present in the body, a small change in plasma P_{O_2} will result in a large change in the amount of oxygen bound or released by hemoglobin.

Key Note 569

9. Aerobic metabolism in peripheral tissues generates carbon dioxide. Roughly 7 percent of the CO_2 transported in the blood is dissolved in the plasma; another 23 percent is bound as

carbaminohemoglobin in RBCs; 70 percent is converted to carbonic acid, which dissociates into a hydrogen ion and a bicarbonate ion. The bicarbonate ion exits the RBC into the plasma. *(Figures 15–17, 15–18)*

Key Note 572

THE CONTROL OF RESPIRATION 573

1. Large-scale changes in oxygen demand require the integration of cardiovascular and respiratory responses.

The Local Control of Respiration 573

2. Arterioles that lead to alveolar capillaries constrict when oxygen is low, and bronchioles dilate when carbon dioxide is high.

Control by the Respiratory Centers of the Brain 573

3. The **respiratory centers** include three pairs of nuclei in the reticular formation of the pons and medulla oblongata. These nuclei regulate the respiratory muscles and control the respiratory rate and the depth of breathing. The **respiratory rhythmicity centers** in the medulla oblongata set the basic pace for respiration. *(Figures 15–19, 15–20, 15–21)*

The Reflex Control of Respiration 574

4. The **inflation reflex** prevents overexpansion of the lungs during forced breathing; the **deflation reflex** stimulates inspiration when the lungs are collapsing. Chemoreceptor reflexes respond to changes in the pH, P_{O_2}, and P_{CO_2} of the blood and cerebrospinal fluid.

Control by Higher Centers 577

5. Conscious and unconscious thought processes can affect respiration by affecting the respiratory centers or the motor neurons that control respiratory muscles. *(Figure 15–22)*

Key Note 578

RESPIRATORY CHANGES AT BIRTH 578

1. Before delivery, the fetal lungs are fluid-filled and collapsed. After the first breath, the alveoli normally remain inflated for the life of the individual.

AGING AND THE RESPIRATORY SYSTEM 578

1. The respiratory system is generally less efficient in the elderly because (1) movements of the thoracic cage are restricted by arthritic changes, decreased flexibility of costal cartilages, and age-related muscle weakness interact to lower pulmonary ventilation and vital capacity of the lungs and (2) some degree of emphysema is normal in the elderly.

INTEGRATION WITH OTHER SYSTEMS 578

1. The respiratory system has extensive anatomical connections to the cardiovascular system. *(Figure 15–23)*

Review Questions

Level 1: Reviewing Facts and Terms

Match each item in column A with the most closely related item in column B. Place letters for answers in the spaces provided.

COLUMN A
- ____ 1. nasopharynx
- ____ 2. laryngopharynx
- ____ 3. thyroid cartilage
- ____ 4. septal cells
- ____ 5. dust cells
- ____ 6. parietal pleura
- ____ 7. visceral pleura
- ____ 8. hypoxia
- ____ 9. anoxia
- ____ 10. collapsed lung
- ____ 11. inhalation
- ____ 12. exhalation

COLUMN B
- a. no O_2 supply to tissues
- b. alveolar macrophages
- c. produce oily secretion
- d. covers inner surface of thoracic wall
- e. low O_2 content in tissue fluids
- f. inferior portion of pharynx
- g. inspiration
- h. superior portion of pharynx
- i. Adam's apple
- j. covers outer surface of lungs
- k. expiration
- l. atelectasis

13. The structure that prevents the entry of liquids or solid food into the respiratory passageways during swallowing is the:
    (a) glottis.
    (b) arytenoid cartilage.
    (c) epiglottis.
    (d) thyroid cartilage.

14. The amount of air moved into or out of the lungs during a single respiratory cycle is the:
    (a) respiratory rate.
    (b) tidal volume.
    (c) residual volume.
    (d) inspiratory capacity.

Level 2: Reviewing Concepts

15. When the diaphragm contracts, it tenses and moves inferiorly, which causes:
    (a) an increase in the volume of the thoracic cavity.
    (b) a decrease in the volume of the thoracic cavity.
    (c) decreased pressure on the contents of the abdominopelvic cavity.
    (d) increased pressure in the thoracic cavity.

16. Gas exchange at the respiratory membrane is efficient because:
    (a) the differences in partial pressure are substantial.
    (b) the gases are lipid soluble.
    (c) the total surface area is large.
    (d) a, b, and c are correct.

17. What is the functional significance of the decreased amount of cartilage and the increased amount of smooth muscle in the lower respiratory passageways?

18. Why is breathing through the nasal cavity more desirable than breathing through the mouth?

19. Justify the statement: "The bronchioles are to the respiratory system what the arterioles are to the cardiovascular system."

20. What path does air take in flowing from the glottis to the respiratory membrane?

Level 3: Critical Thinking and Clinical Applications

21. A decrease in blood pressure will trigger a baroreceptor reflex that leads to increased ventilation. What is the possible advantage of this reflex?

22. You spend the night at a friend's house during the winter. Your friend's home is quite old, and the hot-air furnace lacks a humidifier. When you wake up in the morning, you have a fair amount of nasal congestion and decide you might be coming down with a cold. After you take a steamy shower and drink some juice for breakfast, the nasal congestion disappears. Explain.

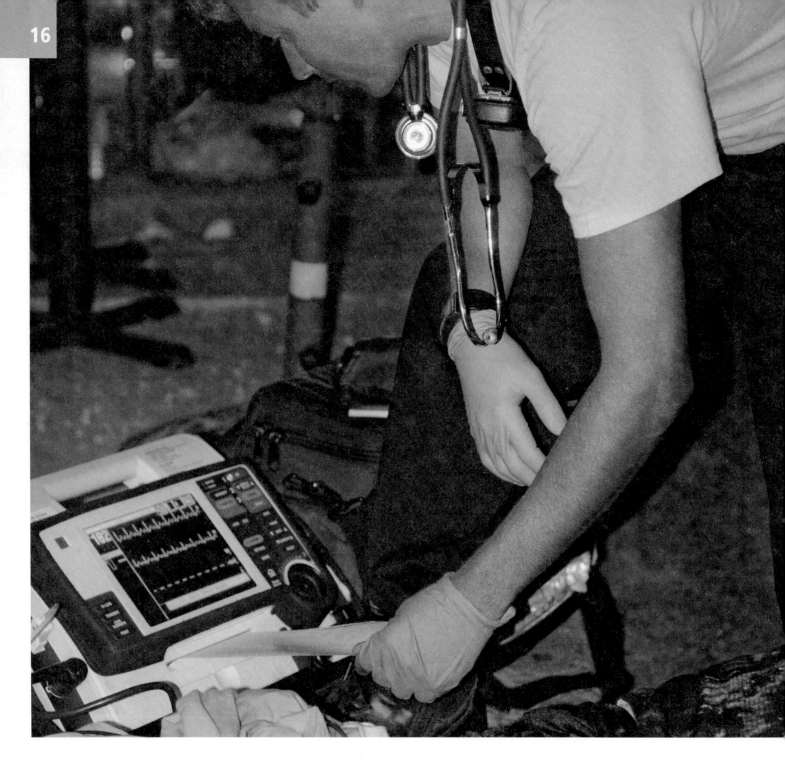

16 The Digestive System

THERE HAVE BEEN TREMENDOUS *advances in technology for the prehospital environment. Many diagnostic tests and procedures, once limited to the hospital setting, are now available in the EMS world. Now, 12-lead (diagnostic) electrocardiograms, capnography (exhaled carbon dioxide monitoring), and contin-* *uous oxygen saturation levels (pulse oximetry) can be monitored. In addition, technology is now available in the field to detect the presence of carbon monoxide in the body. However, with the emergence of these technologies, it is important to remember that they augment and not replace the physical examination.*

Chapter Outline

Chapter Objectives

1. Identify the organs of the digestive tract and the accessory organs of digestion. (p. 586)
2. List the functions of the digestive system. (p. 586)
3. Describe the histology of each digestive tract segment in relation to its function. (pp. 587–588)
4. Explain how ingested materials are propelled through the digestive tract. (pp. 588–589)
5. Describe how food is processed in the mouth and describe the key events of the swallowing process. (pp. 589–593)
6. Describe the anatomy of the stomach, its histological features, and its roles in digestion and absorption. (pp. 593–598)
7. Explain the functions of intestinal secretions and discuss the significance of digestion in the small intestine. (pp. 598–603)
8. Describe the structure and functions of the pancreas, liver, and gallbladder and explain how their activities are regulated. (pp. 603–612)
9. Describe the structure of the large intestine, its movements, and its absorptive functions. (pp. 613–617)
10. Describe the digestion and absorption of carbohydrates, lipids, and proteins. (pp. 617–621)
11. Describe the changes in the digestive system that occur with aging. (p. 621)
12. Discuss the interactions between the digestive system and other organ systems. (pp. 621–623)

Vocabulary Development

chymos juice; *chyme*
deciduus falling off; *deciduous*
enteron intestine; *myenteric plexus*
frenulum small bridle; *lingual frenulum*
gaster stomach; *gastric juice*
hepaticus liver; *hepatocyte*

hiatus gap or opening; *esophageal hiatus*
lacteus milky; *lacteal*
nutrients nourishing; *nutrient*
odonto- tooth; *periodontal ligament*
omentum fat skin; *greater omentum*
pyle gate; *pyloric sphincter*

rugae wrinkles; *rugae*
sigmoides Greek letter S; *sigmoid colon*
stalsis constriction; *peristalsis*
vermis worm; *vermiform appendix*
villus shaggy hair; *intestinal villus*

FEW PEOPLE GIVE any serious thought to the digestive system unless it malfunctions. Still, we spend hours of conscious effort filling and emptying it. References to this system are part of our everyday language. We "have a gut feeling," "want to chew on" something, or find someone's opinions "hard to swallow." When something does go wrong with the digestive system, even something minor, most people seek relief imme-

diately. For this reason, every hour of television programming contains advertisements that promote toothpaste and mouthwash, dietary supplements, antacids, and laxatives.

In our bodies, the respiratory system works with the cardiovascular system to supply the oxygen needed to "burn" metabolic fuels. The digestive system provides the fuel that keeps all the body's cells functioning, plus the building blocks needed for cell growth and repair. The digestive system consists of a muscular tube—the **digestive tract**—and **accessory organs,** including the salivary glands, gallbladder, liver, and pancreas.

Digestive functions involve six related processes:

1. **Ingestion** occurs when foods enter the digestive tract through the mouth.
2. **Mechanical processing** is the physical manipulation of solid foods, first by the tongue and the teeth and then by swirling and mixing motions of the digestive tract.
3. **Digestion** refers to the chemical breakdown of food into small organic fragments that can be absorbed by the digestive epithelium.
4. **Secretion** is the release of water, acids, enzymes, and buffers by the digestive tract and by the accessory organs.
5. **Absorption** is the movement of small organic molecules, electrolytes, vitamins, and water across the digestive epithelium and into the interstitial fluid of the digestive tract.

6. **Excretion** is the removal of waste products from body fluids. Within the digestive tract, these waste products are compacted and discharged as *feces* through the process of *defecation* (def-e-KĀ-shun).

The lining of the digestive tract also plays defensive roles: it protects surrounding tissues from the corrosive effects of digestive acids and enzymes, and it protects against bacteria that either are swallowed with food or reside in the digestive tract. The digestive epithelium and its secretions constitute a nonspecific defense against these bacteria; any bacteria that reach the underlying tissues are attacked by macrophages and by other cells of the immune system.

■ An Overview of the Digestive Tract

The major components of the digestive tract are shown in Figure 16–1●. The digestive tract begins with the oral cavity

● **FIGURE 16–1 The Components of the Digestive System and Their Functions.**

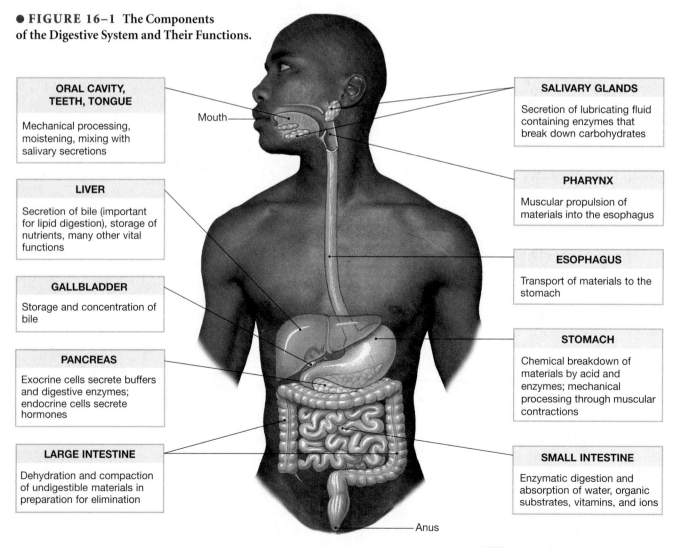

ORAL CAVITY, TEETH, TONGUE

Mechanical processing, moistening, mixing with salivary secretions

LIVER

Secretion of bile (important for lipid digestion), storage of nutrients, many other vital functions

GALLBLADDER

Storage and concentration of bile

PANCREAS

Exocrine cells secrete buffers and digestive enzymes; endocrine cells secrete hormones

LARGE INTESTINE

Dehydration and compaction of undigestible materials in preparation for elimination

SALIVARY GLANDS

Secretion of lubricating fluid containing enzymes that break down carbohydrates

PHARYNX

Muscular propulsion of materials into the esophagus

ESOPHAGUS

Transport of materials to the stomach

STOMACH

Chemical breakdown of materials by acid and enzymes; mechanical processing through muscular contractions

SMALL INTESTINE

Enzymatic digestion and absorption of water, organic substrates, vitamins, and ions

Mouth

Anus

and continues through the pharynx, esophagus, stomach, small intestine, and large intestine before ending at the rectum and anus. Although these subdivisions of the digestive tract have overlapping functions, each region has certain areas of specialization and shows distinctive histological specializations. After we examine the tissues of the digestive tract wall, we will follow the path of ingested materials from the mouth to the anus.

Histological Organization

The digestive tract has four major layers (Figure 16–2●): the *mucosa,* the *submucosa,* the *muscularis externa,* and the *serosa.*

The Mucosa

The **mucosa,** or inner lining of the digestive tract, is an example of a *mucous membrane.* ∞ p. 112 It consists of a mucosal epithelium (an epithelial surface moistened by glandular secretions) and an underlying layer of loose connective tissue, which is the *lamina propria.* Along most of the length of the digestive tract, the mucosa is thrown into folds that both increase the surface area available for absorption and permit expansion after a large meal. In the small intestine, the mucosa forms finger-like projections, called *villi* (*villus,* shaggy hair), that further increase the area for absorption.

The oral cavity, pharynx, esophagus, and anus (where mechanical stresses are most severe) are lined by a stratified squamous epithelium. The remainder of the digestive tract is lined by a simple columnar epithelium; this often contains various types of secretory cells. Ducts that open onto the epithelial surfaces carry the secretions of glands located in the lamina propria, in the surrounding submucosa, or within accessory glandular organs. In most regions of the digestive tract, the outer portion of the mucosa contains a narrow band of smooth muscle and elastic fibers. Contractions of this layer, the *muscularis* (mus-kū-LĀ-ris) *mucosae* (mū-KŌ-sē) (see Figure 16–2), move the mucosal folds and villi.

The Submucosa

The **submucosa** is a second layer of loose connective tissue that is immediately deep to the muscularis mucosae. It contains large blood vessels and lymphatic vessels, as well as a network of nerve fibers, sensory neurons, and parasympathetic motor neurons. This neural tissue—the *submucosal plexus*—is involved in controlling and coordinating contractions of the smooth muscle layers and in regulating the secretion of digestive glands.

The Muscularis Externa

The **muscularis externa** is a band of smooth muscle cells arranged in an inner circular layer and an outer longitudinal

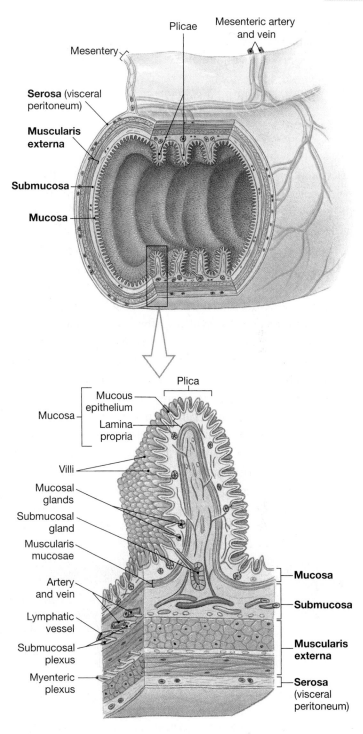

● **FIGURE 16–2 The Structure of the Digestive Tract.** llustrated here is a representative portion of the digestive tract—the small intestine. The wall of the digestive tract is made up of four layers: the mucosa, submucosa, muscularis externa, and serosa.

layer (see Figure 16–2). Contractions of these layers in various combinations both agitate materials and propel them along the digestive tract. Both actions are autonomic reflex movements controlled primarily by another network of nerves—

the *myenteric plexus* (*mys,* muscle + *enteron,* intestine)—sandwiched between the circular and longitudinal smooth muscle layers. Parasympathetic stimulation increases muscular tone and activity, whereas sympathetic stimulation promotes muscular inhibition and relaxation.

The Serosa

The **serosa,** which is a serous membrane, covers the muscularis externa along most portions of the digestive tract within the peritoneal cavity. This *visceral peritoneum* is continuous with the *parietal peritoneum,* which lines the inner surfaces of the body wall. ∞ p. 113 In some areas within the peritoneal cavity, portions of the digestive tract are suspended by **mesenteries** (MEZ-en-ter-ēz)—double sheets of serous membrane composed of the parietal peritoneum and visceral peritoneum. The loose connective tissue sandwiched between the epithelial surfaces of the mesenteries provides a pathway for the blood vessels, nerves, and lymphatic vessels that service the digestive tract. The mesenteries also stabilize the positions of the attached organs and prevent the intestines from becoming entangled during digestive movements or sudden changes in body position.

There is no serosa that covers the muscularis externa of the oral cavity, pharynx, esophagus, and rectum. Instead, the muscularis externa is surrounded by a dense network of collagen fibers that firmly attaches these regions of the digestive tract to adjacent structures. This fibrous wrapping is called an *adventitia* (ad-ven-TISH-ē-uh).

Clinical Note
ASCITES

The peritoneal lining continuously produces peritoneal fluid, which lubricates the opposing parietal and visceral surfaces. Even though about seven liters of fluid is secreted and reabsorbed each day, the volume within the cavity at any moment is very small. Several conditions, including liver disease, kidney disease, and heart failure, can cause an increase in the rate of fluid movement into the peritoneal cavity. The resulting accumulation of fluid creates a characteristic abdominal swelling called *ascites* (a-SĪ-tēz). The distortion of internal organs by the accumulated fluid can result in symptoms such as heartburn, indigestion, and low back pain. ∎

The Movement of Digestive Materials

As previously noted, the muscular layers of the digestive tract consist of smooth muscle tissue. ∞ p. 228 *Pacesetter cells* in the smooth muscle of the digestive tract trigger waves of contraction, resulting in rhythmic cycles of activity. The coordinated contractions in the walls of the digestive tract play a vital role in

two processes: *peristalsis* (*peri-,* around + *stalsis,* constriction), which is the movement of material along the tract, and *segmentation,* which is the mechanical mixing of the material.

Peristalsis and Segmentation

The muscularis externa propels materials from one part of the digestive tract to another by means of **peristalsis** (per-i-STAL-sis), waves of muscular contractions that move along the length of the digestive tract (Figure 16–3●). During a peristaltic movement, the circular muscles first contract behind the digestive contents. Then longitudinal muscles contract, shortening adjacent segments of the tract. A wave of contraction in the circular muscles then forces the materials in the desired direction.

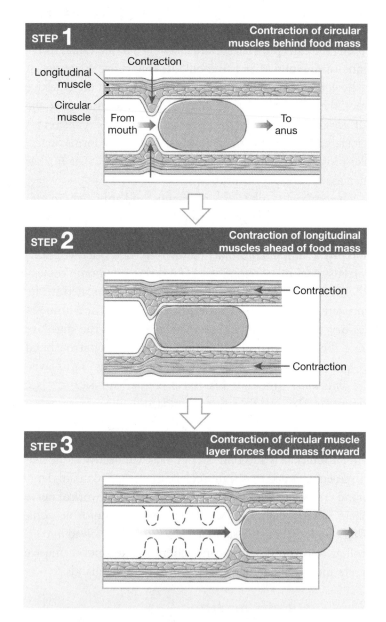

● **FIGURE 16–3 Peristalsis.** Peristalsis is the process that propels materials along the length of the digestive tract.

Regions of the small intestine also undergo **segmentation,** which is movements that churn and fragment digestive materials. Over time, this action results in a thorough mixing of the contents with intestinal secretions. Because they do not follow a set pattern, segmentation movements do not propel materials in any particular direction.

→ **CONCEPT CHECK QUESTIONS**

1. What is the importance of the mesenteries?
2. Which process is more efficient in propelling intestinal contents from one place to another—peristalsis or segmentation?
3. What effect would a drug that blocks parasympathetic stimulation of the digestive tract have on peristalsis?

Answers begin on p. 792.

■ The Oral Cavity

The mouth opens into the **oral cavity,** which is the part of the digestive tract that receives food. The oral cavity is lined by a mucous membrane that has a stratified squamous epithelium. The oral cavity (1) senses and analyzes material before swallowing; (2) mechanically processes material through the actions of the teeth, tongue, and surfaces of the palate; (3) lubricates material by mixing it with mucus and salivary secretions; and (4) begins the digestion of carbohydrates and lipids with salivary enzymes.

Figure 16–4● shows the boundaries of the oral cavity, also known as the **buccal** (BUK-al) **cavity.** The *cheeks* form the lateral walls of this chamber; anteriorly they are continuous with the lips, or **labia** (LĀ-bē-uh; singular, *labium*). The **vestibule** is the space between the cheeks or lips and the teeth. A pink ridge—the gums or **gingivae** (JIN-ji-vē; singular, *gingiva*)—surrounds the bases of the teeth. The gums cover the tooth-bearing surfaces of the upper and lower jaws.

The **hard palate** and **soft palate** form a roof for the oral cavity; the tongue dominates its floor. The free anterior portion of the tongue is connected to the underlying epithelium by a thin fold of mucous membrane, which is the **lingual frenulum** (FREN-ū-lum; *frenulum*, a small bridle). The imaginary dividing line between the oral cavity and the oropharynx extends between the base of the tongue and the dangling *uvula* (ō-vū-luh).

The Tongue

The muscular **tongue** manipulates materials inside the mouth and is occasionally used to bring foods (such as ice cream) into the oral cavity. The primary functions of the tongue are (1) mechanical processing by compression, abrasion, and distortion; (2) manipulation to assist in chewing and to prepare the material for swallowing; and (3) sensory analysis by touch, temperature, and taste receptors. Most of the tongue lies within the oral cavity, but the base of the tongue extends into the oropharynx. A pair of prominent lateral swellings at the base of the tongue marks the location of the *lingual tonsils,* which are lymphoid nodules that help resist infections. ∞ p. 515

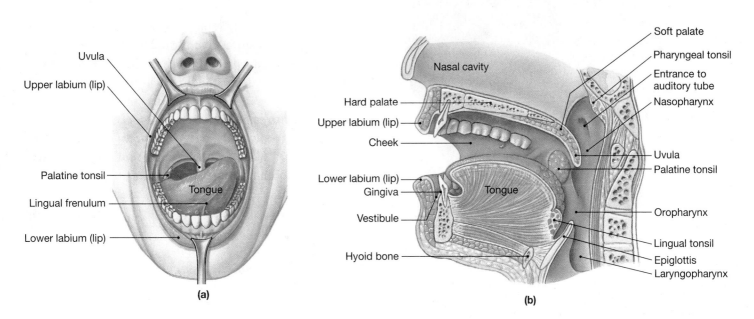

● **FIGURE 16–4 The Oral Cavity.** These drawings show the structures of the oral cavity in (**a**) an anterior view, as seen through the open mouth, and (**b**) a sagittal section.

Salivary Glands

Three pairs of salivary glands secrete into the oral cavity (Figure 16–5●). On each side, a large **parotid salivary gland** lies under the skin covering the lateral and posterior surface of the mandible. ∞ p. 174 The **parotid duct** empties into the vestibule at the level of the second upper molar. The **sublingual salivary glands** are located beneath the mucous membrane of the floor of the mouth, and numerous sublingual ducts open along either side of the lingual frenulum. The **submandibular salivary glands** are in the floor of the mouth along the inner surfaces of the mandible; their ducts open into the mouth behind the teeth on either side of the lingual frenulum.

These salivary glands produce 1.0–1.5 liters of saliva each day. Saliva is 99.4 percent water, plus *mucins* and an assortment of ions, buffers, waste products, metabolites, and enzymes. Mucins absorb water and form mucus. At mealtimes, large quantities of saliva lubricate the mouth and dissolve chemicals that stimulate the taste buds. Coating the food with slippery mucus reduces friction and makes swallowing possible. A continuous background level of secretion flushes and cleans the oral surfaces, and salivary antibodies (IgA) and lysozymes help control populations of oral bacteria. When salivary secretions are reduced, such as by radiation exposure, emotional distress, or other factors, the bacterial population in the oral cavity explodes. This bacterial increase soon leads to recurring infections and the progressive erosion of the teeth and gums.

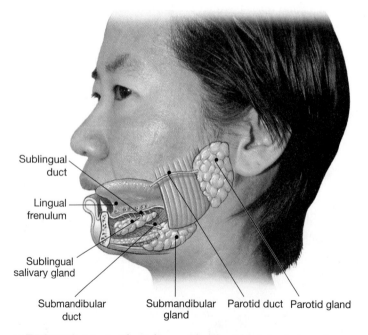

● **FIGURE 16–5 The Salivary Glands.** This lateral view shows the relative positions of the salivary glands and ducts on the left side of the head. The left half of the mandible has been removed to show deeper structures.

Labels:
Sublingual duct
Lingual frenulum
Sublingual salivary gland
Submandibular duct
Submandibular gland
Parotid duct
Parotid gland

Clinical Note
SALIVARY GLAND DISORDERS

Salivary gland disorders usually cause pain and swelling of the affected gland. Infections of the salivary glands *(siloadenitis)* are the most common disorders and can be either viral or bacterial. Mumps is the most common viral infection and primarily occurs in children. Bacterial infections are seen most frequently in dehydrated or debilitated patients and often result from slowing of the flow of saliva. Stones *(salivary calculi)* can develop in the salivary glands and can block the flow of saliva through the duct, which results in gland enlargement. Treatment includes the use of substances such as hard lemon-drop candy to increase salivary flow. ■

Salivary Secretions

Each of the salivary glands produces a slightly different kind of saliva. The parotid glands produce a secretion rich in **salivary amylase,** which is an enzyme that breaks down starches (complex carbohydrates) into smaller molecules that can be absorbed by the digestive tract. Saliva that originates in the submandibular and sublingual salivary glands contains fewer enzymes but more buffers and mucus. During eating, all three salivary glands increase their rates of secretion, and salivary production may reach 7 mL per minute, with about 70 percent of that volume provided by the submandibular glands. The pH of the saliva also rises, and shifts from slightly acidic (pH 6.7) to slightly basic (pH 7.5). Salivary secretions are normally controlled by the autonomic nervous system.

Teeth

Movements of the tongue are important in passing food across the opposing surfaces of the **teeth.** These surfaces perform chewing, or **mastication** (mas-ti-KĀ-shun), of food. Mastication breaks down tough connective tissues in meat and the plant fibers in vegetable matter, and it helps saturate the materials with salivary secretions.

Figure 16–6a● shows the parts of a tooth. The **neck** of the tooth marks the boundary between the **root** and the **crown.** The crown is covered by a layer of **enamel,** which contains a crystalline form of calcium phosphate, the hardest biologically manufactured substance. Adequate amounts of calcium, phosphates, and vitamin D_3 during childhood are essential if the enamel coating is to be complete and resistant to decay.

The bulk of each tooth consists of **dentin** (DEN-tin), which is a mineralized matrix similar to that of bone. Dentin differs from bone in that it does not contain cells. Instead, cytoplasmic processes extend into the dentin from cells within the central **pulp cavity.** The pulp cavity receives blood vessels and

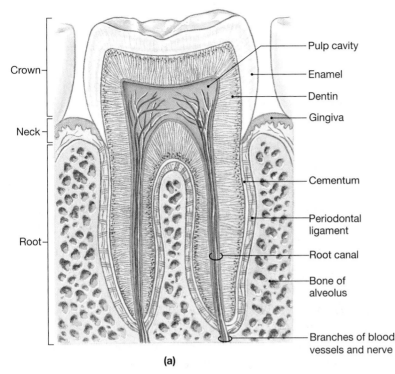

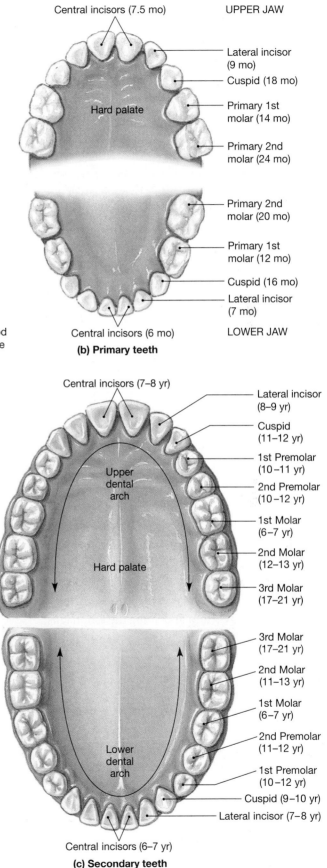

(a)

(b) Primary teeth

● **FIGURE 16–6** **Teeth: Structural Components and Dental Succession.** (**a**) The structures of a typical adult tooth are labeled in this diagrammatic section. Humans have two sets of teeth: (**b**) the primary (deciduous) teeth and (**c**) the adult teeth. The ages at eruption are indicated for the teeth in each set.

nerves through a narrow **root canal** at the root (base) of the tooth. The root sits within a bony socket, or *alveolus* (a hollow cavity). Collagen fibers of the **periodontal ligament** (*peri-*, around + *odonto-*, tooth) extend from the dentin of the root to the surrounding bone. A layer of **cementum** (se-MEN-tum) covers the dentin of the root; this provides protection and firmly anchors the periodontal ligament. Cementum is similar in structure to bone, but it is softer, and remodeling does not occur. Where the tooth penetrates the gum surface, epithelial cells form tight attachments to the tooth and prevent bacterial access to the easily eroded cementum of the root.

A set of adult teeth is shown in Figure 16–6c●. Each of the four types of teeth has a specific function. **Incisors** (in-SĪ-zerz), which are blade-shaped teeth found at the front of the mouth, are useful for clipping or cutting, as when you nip off the tip of a carrot stick. **Cuspids** (KUS-pidz), or *canines*, are conical, with a sharp ridge-line and a pointed tip. They are used for tearing or slashing. You might weaken a tough piece of celery by the clipping action of the incisors and then take advantage of the shearing action provided by the cuspids. **Bicuspids** (bi-KUS-pidz), or *premolars*, and **molars** have flattened crowns with prominent ridges. They are used for crushing, mashing, and grinding. You might shift a tough nut or piece of meat to the premolars and molars for crushing.

(c) Secondary teeth

Dental Succession

During development, two sets of teeth begin to form. The first to appear are the **deciduous teeth** (de-SID-ū-us; *deciduus,* falling off), also known as *primary teeth, milk teeth,* or *baby teeth.* Most children have 20 deciduous teeth (Figure 16–6b●). These teeth are later replaced by the **secondary dentition,** or *permanent dentition* (see Figure 16–6c), which permits the processing of a wider variety of foods. As replacement proceeds, the periodontal ligaments and roots of the deciduous teeth erode until these teeth either fall out or are pushed aside by the **eruption** (emergence) of the secondary teeth. Adult jaws are larger and can accommodate more than 20 teeth. As a person ages, three additional teeth appear on each side of the upper and lower jaws, which brings the permanent tooth count to 32. The last teeth to appear are the *third molars,* or *wisdom teeth.* Wisdom teeth (or any other teeth) that develop in locations that do not permit their eruption are called *impacted teeth.* Impacted teeth can be surgically removed to prevent the formation of abscesses.

■ The Pharynx

The **pharynx** serves as a common passageway for solid food, liquids, and air. The three major subdivisions of the pharynx were discussed in Chapter 15. ∞ p. 549 Food normally passes through the oropharynx and laryngopharynx on its way to the esophagus. Both of these pharyngeal regions have a stratified squamous epithelium similar to that of the oral cavity. The underlying lamina propria contains mucous glands plus the pharyngeal, palatal, and lingual tonsils. The pharyngeal muscles cooperate with muscles of the oral cavity and esophagus to initiate the process of swallowing (described shortly). The muscular contractions during swallowing force the food mass into and along the esophagus.

■ The Esophagus

The **esophagus** (see Figure 16–1) is a muscular tube, about 25 cm (10 in.) long and about 2 cm (0.75 in.) in diameter, that conveys solid food and liquids to the stomach. It begins at the pharynx, runs posterior to the trachea in the neck, passes through the mediastinum in the thoracic cavity, and enters the peritoneal cavity through an opening in the diaphragm—the *esophageal hiatus* (hī-Ā-tus; a gap or opening)—before emptying into the stomach.

The esophagus is lined with a stratified squamous epithelium that resists abrasion, hot or cold temperatures, and chemical attack. The secretions of mucous glands lubricate this epithelial surface and prevent materials from sticking to the sides of the esophagus during swallowing. The upper third of the muscularis externa contains skeletal muscle, the lower third contains smooth muscle, and a mixture of each comprises the middle third. Regions of circular muscle in the superior and inferior ends of the esophagus comprise the *upper esophageal sphincter* and the *lower esophageal sphincter.* The lower sphincter is normally in active contraction, which is a condition that prevents the backflow of materials from the stomach into the esophagus.

Swallowing

Swallowing, or **deglutition** (de-gloo-TISH-un), is a complex process that can be initiated voluntarily but proceeds automatically once it begins. Although you take conscious control over swallowing when you eat or drink, swallowing is also controlled at the subconscious level. Before food can be swallowed, it must have the proper texture and consistency. Once the material has been shredded or torn by the teeth, moistened with salivary secretions, and "approved" by the taste receptors, the tongue begins compacting the debris into a small mass, or **bolus.**

The process of swallowing occurs in a series of phases (Figure 16–7●). Swallowing begins in the **oral phase,** with the compression of the bolus against the hard palate. The tongue retracts, which forces the bolus into the pharynx and helps to elevate the soft palate, so that the bolus does not enter the nasopharynx. The oral phase is the only phase of swallowing that can be consciously controlled.

In the **pharyngeal phase,** the bolus comes in contact with sensory receptors around the pharynx and the posterior pharyngeal wall, which initiates the involuntary *swallowing reflex.* The larynx elevates, and the epiglottis folds to direct the bolus past the closed glottis. In less than a second, the contraction of pharyngeal muscles forces the bolus through the entrance to the esophagus, which is guarded by the upper esophageal sphincter.

The **esophageal phase** begins as the bolus enters the esophagus. During this phase, the bolus is pushed toward the stomach by a peristaltic contraction. The approach of the bolus triggers the opening of the lower esophageal sphincter, and the bolus enters the stomach.

For a typical bolus, the entire trip from the oral cavity to the esophagus takes about 9 seconds. Fluids may make the journey in a few seconds, and flow ahead of the peristaltic contractions; a relatively dry or bulky bolus travels much more slowly, and repeated peristaltic waves may be required to drive it into the stomach. A completely dry bolus cannot be swallowed at all, for friction with the walls of the esophagus will make peristalsis ineffective.

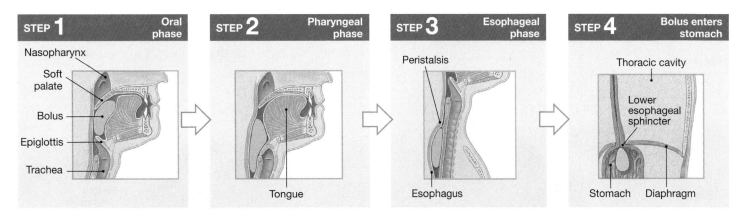

STEP **1** Oral phase	STEP **2** Pharyngeal phase	STEP **3** Esophageal phase	STEP **4** Bolus enters stomach

● **FIGURE 16–7 The Swallowing Process.** This sequence, based on a series of X-rays, shows the phases of swallowing and the movement of materials from the mouth to the stomach.

Clinical Note
SWALLOWED FOREIGN BODIES

Swallowed foreign bodies are a common reason people seek emergency care. Approximately 80 percent of ingested foreign bodies occur in children. Dentures in adults account for the second most frequent type of swallowed foreign body. The presence of dentures impairs the ability to determine how well food is chewed.

Most swallowed foreign bodies pass spontaneously. However, 10–20 percent require some medical intervention. Once an object has passed the pylorus and enters the stomach, it usually passes without further incident. Objects that lodge in the esophagus, however, usually require intervention.

A barium swallow and esophageal X-ray can aid in the diagnosis of an esophageal foreign body. Alternatively, direct examination with a fiber-optic endoscope will confirm the diagnosis.

The type of foreign body present dictates treatment. Food impactions often pass following administration of medications that relax smooth muscle and decrease lower esophageal sphincter pressures. These include glucagon, nifedipine, and nitroglycerin. Sharp objects may require surgical removal. Swallowed batteries are a true emergency, as the alkaline substance in the battery rapidly burns the esophageal mucosa, and thus may result in perforation. ■

Clinical Note
ESOPHAGITIS AND DIAPHRAGMATIC (HIATAL) HERNIAS

A weakened or permanently relaxed lower esophageal sphincter can cause inflammation of the esophagus, or *esophagitis* (ē-sof-a-JĪ-tis), as powerful gastric acids enter the lower esophagus. The esophageal epithelium has few defenses against acids and enzymes, and inflammation, epithelial erosion, and intense discomfort are the result. Occasional incidents of reflux, or backflow, from the stomach are responsible for the symptoms of "heartburn." This relatively common problem supports a multimillion-dollar industry devoted to producing and promoting antacids.

The esophagus and major blood vessels pass from the thoracic cavity to the abdominopelvic cavity through an opening in the diaphragm called the *esophageal hiatus.* In a *diaphragmatic hernia,* or *hiatal* (hī-Ā-tal) *hernia,* abdominal organs slide into the thoracic cavity through the esophageal hiatus. The severity of the condition depends on the location and size of the herniated organ or organs. Hiatal hernias are actually very common, and most go unnoticed. When clinical problems develop, they usually occur because the intruding abdominal organs are exerting pressure on structures or organs in the thoracic cavity. ■

→ CONCEPT CHECK QUESTIONS

1. Which type of epithelium lines the oral cavity?
2. The digestion of which nutrient would be affected by damage to the parotid salivary glands?
3. Which type of tooth is most useful for chopping off bits of relatively rigid foods?
4. What is occurring when the soft palate and larynx elevate and the glottis closes?

Answers begin on p. 792.

■ The Stomach

The **stomach,** which is located within the left upper quadrant of the abdominopelvic cavity (see Figure 1–7c, p. 17), receives food from the esophagus. The stomach has four primary functions: (1) the temporary storage of ingested food, (2) the mechanical breakdown of ingested food, (3) the breakdown of chemical bonds in food items through the action of acids and enzymes, and (4) the production of *intrinsic factor,* which is a compound necessary for the absorption of vitamin B_{12}.

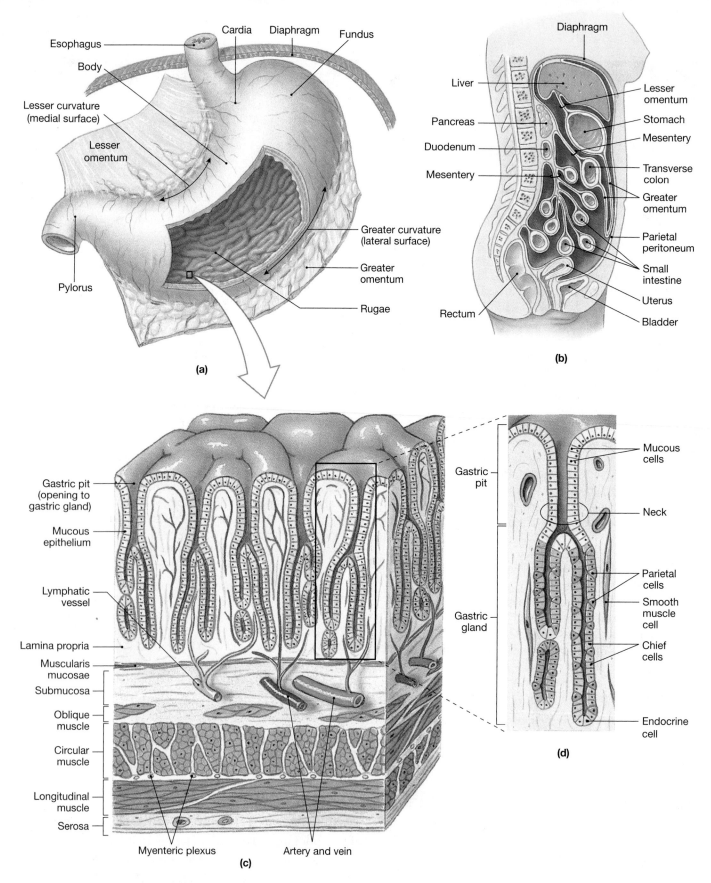

● FIGURE 16–8 The Anatomy of the Stomach. (**a**) This anterior view of the stomach shows important superficial landmarks. (**b**) The stomach's position in the peritoneal cavity is maintained by the greater and lesser omenta. (**c**) The organization of the stomach wall is shown in this diagrammatic section. (**d**) This diagram of a gastric gland reveals the sites of parietal cells and chief cells, which are present in all gastric glands. Note that endocrine cells are also present in glands that line the lower portion (pylorus) of the stomach.

Ingested materials mix with secretions of the glands of the stomach, and produce a viscous, highly acidic, soupy mixture of partially digested food called **chyme** (kīm).

The stomach is a muscular, J-shaped organ with four main regions (Figure 16–8●). The esophagus connects to the smallest part of the stomach, which is the **cardia** (KAR-dē-uh). The bulge of the stomach superior to the cardia is the **fundus** (FUN-dus) of the stomach, and the large area between the fundus and the curve of the J is the **body.** The distal part of the J, the **pylorus** (pī-LOR-us; *pyle*, gate + *ouros*, guard), connects the stomach with the small intestine. A muscular **pyloric sphincter** regulates the flow of chyme between the stomach and small intestine.

The stomach's volume increases when food enters it, then decreases as chyme enters the small intestine. When the stomach is relaxed, or empty, its mucosa contains a number of prominent ridges and folds, called **rugae** (ROO-gē; wrinkles). As the stomach expands, the rugae gradually flatten out until, at maximum distension, they almost disappear. When empty, the stomach resembles a muscular tube with a narrow and constricted lumen. When fully expanded, it can contain 1.0–1.5 liters of material.

Unlike the two-layered muscularis externa of other portions of the digestive tract, that of the stomach contains a longitudinal layer, a circular layer, and an inner oblique layer. This extra layer of smooth muscle adds strength and assists in the mixing and churning essential to forming chyme.

The visceral peritoneum that covers the outer surface of the stomach is continuous with a pair of mesenteries. The **greater omentum** (ō-MEN-tum; *omentum*, a fatty skin) extends below the *greater curvature* and forms an enormous pouch that hangs over and protects the abdominal viscera (Figure 16–8b). The much smaller **lesser omentum** extends from the *lesser curvature* to the liver.

The Gastric Wall

The stomach is lined by a simple columnar epithelium dominated by mucous cells. The alkaline mucus that this *mucous epithelium* secretes covers and protects epithelial cells from acids, enzymes, and abrasive materials. Shallow depressions, called **gastric pits,** open onto the gastric surface (Figure 16–8c). The mucous cells at the base, or *neck,* of each gastric pit actively divide and replace superficial cells of the mucous epithelium shed into the chyme.

In the fundus and body of the stomach, each gastric pit communicates with **gastric glands** that extend deep into the underlying lamina propria (Figure 16–8c). Each day the cells in these gastric glands secrete about 1500 mL of **gastric juice.** The cells that produce the components of gastric juice are *parietal cells* and *chief cells* (Figure 16–8d).

Gastric glands within the lower stomach (the *pylorus*) also contain endocrine cells that are involved in regulating gastric activity. Their role in the regulation of gastric activity will be discussed shortly.

Parietal Cells

Parietal cells secrete intrinsic factor and hydrochloric acid (HCl). **Intrinsic factor** facilitates the absorption of vitamin B_{12} across the intestinal lining. Hydrochloric acid lowers the pH of the gastric juice, keeping the stomach contents at a pH of 1.5–2.0. The acidity of gastric juice kills microorganisms, breaks down plant cell walls and connective tissues in meat, and activates the enzyme secretions of chief cells.

Chief Cells

Chief cells secrete a protein called **pepsinogen** (pep-SIN-ō-jen) into the stomach lumen. When it contacts the hydrochloric acid released by the parietal cells, pepsinogen is converted to **pepsin,** which is a *proteolytic* (protein-digesting) *enzyme.* In newborns (but not adults), the stomach produces *rennin* and *gastric lipase,* which are enzymes important for the digestion of milk. Rennin coagulates milk, which slows its passage through the stomach and allows more time for its digestion. Gastric lipase initiates the digestion of milk fats.

Clinical Note
PEPTIC ULCERS

Peptic ulcers are erosions caused by gastric acid (Figure 16–9●). They can occur anywhere in the gastrointestinal tract; terminology is based on the portion of the GI tract affected. Duodenal ulcers most frequently occur in the proximal portion of the duodenum; gastric ulcers occur exclusively in the stomach. Overall, peptic ulcers occur in males four times more frequently than in females, and duodenal ulcers occur from two to three times more frequently than do gastric ulcers. Current statistics place the number of peptic ulcers at 4–5 million, with approximately 500,000 new cases diagnosed yearly. Those patients who are more likely to have gastric ulcers are over 50 years old and work in jobs that require physical activity. Their pain usually increases after eating or with a full stomach, and they usually have no pain at night. Duodenal ulcers are more common in patients from 25 to 50 years old who are executives or leaders under high stress.

(continued next page)

Clinical Note—*continued*
PEPTIC ULCERS

There is also some familial tendency toward duodenal ulcer, which suggests genetic predisposition. Patients with duodenal ulcers commonly have pain at night or whenever their stomach is empty.

Nonsteroidal anti-inflammatory medications (aspirin, Motrin, Advil, Naprosyn), acid-stimulating products (alcohol, nicotine), or *Helicobacter pylori* bacteria are the most common causes of peptic ulcers. To help break down food boluses, the stomach secretes hydrochloric acid. One of the enzymes that control this secretion is pepsinogen. The hydrochloric acid helps to convert pepsinogen into its active form, pepsin. Between them, the pepsin and the hydrochloric acid can make the digestive enzymes very irritating to the GI tract's mucosal lining. Ordinarily, mucous gland secretions protect the stomach's mucosal barrier from these irritants. But when nonsteroidal anti-inflammatory medications, acid stimulators, or *H. pylori* damage the barrier, the mucosa is exposed to the highly acidic fluid, and peptic ulcers result. Prostaglandin, which is an important locally acting hormone, decreases the stimulation for blood flow through the gastric mucosa, and allows its further destruction.

The recent discovery that *Helicobacter pylori* bacteria appear in over 80 percent of gastric and duodenal ulcers has enabled physicians to treat the disease by eliminating its cause with antacids and antibiotics rather than merely treating its symptoms.

A blocked pancreatic duct can also contribute to duodenal ulcers. As chyme passes through the pyloric sphincter from the stomach into the duodenum, the pancreas secretes an alkalotic solution laden with bicarbonate ions that neutralize the acidic hydrogen ions in the chyme. If the pancreatic duct is blocked, however, the acidic chyme can cause ulcerations throughout the intestine. One other cause of duodenal ulcers is Zollinger-Ellison syndrome, in which an acid-secreting tumor provokes the ulcerations.

Acute, severe pain is probably due to a rupture of the ulcer into the peritoneal cavity that causes hemorrhage. Depending on the ulcer's location, the patient may have hematemesis or may have melena. Bouts of nausea and vomiting due to the irritation of the mucosa are common. If the ulcer has eroded through a highly vascular area, massive

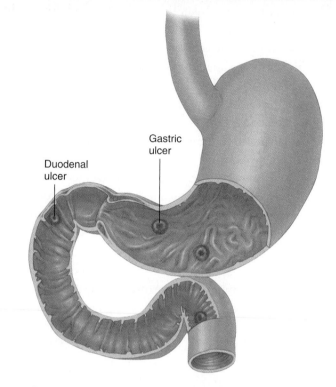

Duodenal ulcer

Gastric ulcer

● **FIGURE 16–9 Types of Peptic Ulcers.** Peptic ulcers are erosions in the mucosa of the stomach or duodenum. They can erode completely through the wall of the organ, which results in perforation and life-threatening hemorrhage.

hemorrhage can occur. Along with the signs of hemorrhage on visual inspection, these patients will appear very ill and have signs of hemodynamic instability, such as pale, cool, and clammy skin, tachycardia, decreased blood pressure, and possibly, altered mental status. Most patients will lie still to decrease the pain. They may have surgical scars from previous ulcer repair. Bowel sounds will usually be absent. ∎

The Regulation of Gastric Activity

The production of acid and enzymes by the stomach mucosa is controlled by the central nervous system and is regulated by reflexes within the walls of the digestive tract and by hormones of the digestive tract. Control of gastric secretion proceeds in three overlapping phases (Figure 16–10●), which are named according to the location of the control center:

1. *Cephalic phase.* The sight, smell, taste, or thought of food initiates the **cephalic phase** of gastric secretion. This stage, which is directed by the CNS, prepares the stomach to receive food. Under the control of the vagus nerves, parasympathetic fibers of the submucosal

plexus innervate mucous cells, parietal cells, chief cells, and endocrine cells of the stomach. In response to stimulation, the production of gastric juice accelerates, and reaches rates of about 500 mL per hour. This phase usually lasts only minutes before the gastric phase commences.

2. *Gastric phase.* The **gastric phase** begins with the arrival of food in the stomach. The stimulation of stretch receptors in the stomach wall and of chemoreceptors in the mucosa triggers local reflexes controlled by the submucosal and myenteric plexuses. The myenteric plexus stimulates mixing waves in the stomach wall. The submucosal plexus stimulates the parietal cells and

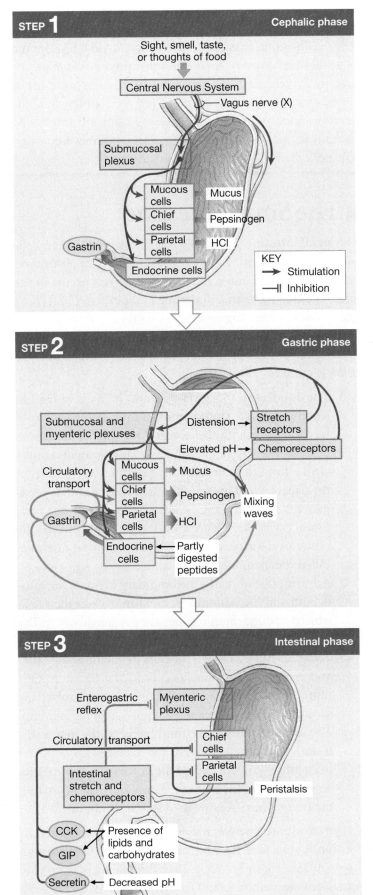

STEP 1 Cephalic phase

● **FIGURE 16–10** The Phases of Gastric Secretion.

chief cells, as well as stimulating endocrine cells in gastric glands within the pylorus to release the hormone **gastrin** into the circulatory system. Proteins, alcohol in small doses, and caffeine are potent stimulators of gastric secretion because they excite the chemoreceptors in the gastric lining. Both parietal and chief cells respond to the presence of gastrin by accelerating their secretory activities. The effect on the parietal cells is the most pronounced, and the pH of the gastric juice drops sharply. This phase may continue for several hours while the ingested materials are processed by the acids and enzymes.

During this period, gastrin stimulates stomach contractions, which begin to swirl and churn the gastric contents, mixing the ingested materials with the gastric secretions to form chyme. As mixing proceeds, the contractions begin sweeping down the length of the stomach, and each time the pylorus contracts, a small quantity of chyme squirts through the pyloric sphincter.

3. *Intestinal phase.* The **intestinal phase** begins when chyme starts to enter the small intestine. Most of the regulatory controls for this phase (whether neural or endocrine) are inhibitory; by controlling the rate of gastric emptying, they ensure that the secretory, digestive, and absorptive functions of the small intestine can proceed efficiently. For example, the movement of chyme temporarily reduces the stimulation of stretch receptors in the stomach wall and increases their stimulation in the wall of the small intestine. This produces the *enterogastric reflex,* which temporarily inhibits neural stimulation of gastrin production and gastric motility, and further movement of chyme. At the same time, the entry of chyme stimulates the release of the intestinal hormones *secretin, cholecystokinin (CCK),* and *gastric inhibitory peptide (GIP);* the resulting reduced gastric activity gives the small intestine time to adjust to the arriving acids.

Inhibitory reflexes that depress gastric activity are stimulated when the proximal portion of the small intestine becomes too full, too acidic, unduly irritated by chyme, or filled with partially digested proteins, carbohydrates, or fats.

In general, the rate of movement of chyme into the small intestine is highest when the stomach is greatly distended and the meal contains relatively little protein. A large meal that contains small amounts of protein, large amounts of carbohydrates (such as rice or pasta), wine (alcohol), and after-dinner coffee (caffeine) will leave your stomach extremely quickly because both alcohol and caffeine stimulate gastric secretion and motility.

Clinical Note
STOMACH CANCER

Stomach (or *gastric*) *cancer* is one of the most common lethal cancers, responsible for roughly 12,000 deaths in the U.S. each year. The incidence is higher in Japan and Korea, where the typical diet includes large quantities of pickled foods. Because the signs and symptoms can resemble those of gastric ulcers, the condition may not be reported in its early stages. Diagnosis usually involves X-rays of the stomach at various degrees of distension. The gastric mucosa can also be visually inspected using a flexible instrument called a *gastroscope.* Attachments permit the collection of tissue samples for histological analysis.

The treatment of stomach cancer involves a *gastrectomy* (gas-TREK-to-mē), the surgical removal of part or all of the stomach. People can survive even a total gastrectomy because the loss of such functions as food storage and acid production is not life threatening. Protein breakdown can still be performed by the small intestine, although at reduced efficiency, and the lack of intrinsic factor (produced by the stomach) can be overcome by supplementing the diet with high daily doses of vitamin B_{12}. ∎

Digestion in the Stomach

Within the stomach, pepsin performs the preliminary digestion of proteins, and for a variable time salivary amylase continues the digestion of carbohydrates. Salivary amylase remains active until the pH of the stomach contents falls below 4.5, usually within one to two hours after a meal.

As the stomach contents become more fluid and the pH approaches 2.0, pepsin activity increases and protein disassembly begins. Protein digestion is not completed in the stomach, but pepsin typically breaks down complex proteins into smaller peptide and polypeptide chains before chyme enters the small intestine.

Although digestion occurs in the stomach, nutrients are not absorbed there because (1) the epithelial cells are covered by a blanket of alkaline mucus and are not directly exposed to chyme, (2) the epithelial cells lack the specialized transport mechanisms found in cells that line the small intestine, (3) the gastric lining is impermeable to water, and (4) digestion has not proceeded to completion by the time chyme leaves the stomach. At this stage, most carbohydrates, lipids, and proteins are only partially broken down.

Key Note

The stomach is a storage site that provides time for the physical breakdown of food that must precede chemical digestion. Protein digestion begins in the acid environment of the stomach through the action of pepsin. Carbohydrate digestion, which began with the release of salivary amylase by the salivary glands prior to swallowing, continues for a variable period after food arrives in the stomach.

■ The Small Intestine

The **small intestine** plays a key role in the digestion and absorption of nutrients. Ninety percent of nutrient absorption occurs in the small intestine. (Most of the rest occurs in the large intestine.) The small intestine is about 6 m (20 ft) long and has a diameter ranging from 4 cm (1.6 in.) at the stomach to about 2.5 cm (1 in.) at the junction with the large intestine. It has three segments: the *duodenum,* the *jejunum,* and the *ileum* (Figure 16–11●):

1. The **duodenum** (doo-AH-de-num) is 25 cm (10 in.) in length and is the closest segment to the stomach. From its connection with the stomach, the duodenum curves in a C that encloses the pancreas. This segment receives chyme from the stomach and digestive secretions from the pancreas and liver. Unlike the stomach and its proximal 2.5 cm (1 in.), the duodenum lies outside the peritoneal cavity (see Figure 16–8b, p. 594). Organs that lie posterior to (instead of within) the peritoneal cavity are called *retroperitoneal* (*retro,* behind).

2. An abrupt bend marks the boundary between the duodenum and the **jejunum** (je-JOO-num) and the site at which the duodenum reenters the peritoneal cavity. The jejunum, which is supported by a sheet of **mesentery,** is about 2.5 m (8 ft) long. The bulk of chemical digestion and nutrient absorption occurs in the jejunum. One rather drastic approach to weight control involves the surgical removal of a significant portion of the jejunum.

3. The jejunum leads to the third segment, the **ileum** (IL-ē-um). It is also the longest, and averages 3.5 m (12 ft) in length. The ileum ends at the *ileocecal valve,* a sphincter that controls the flow of material from the ileum into the *cecum,* the first portion of the large intestine.

The small intestine fills much of the peritoneal cavity. Its position is stabilized by a mesentery attached to the dorsal body wall (see Figure 16–8b, p. 594). Blood vessels, lymphatic vessels, and nerves run to and from the segments of the small intestine within the connective tissue of the mesentery.

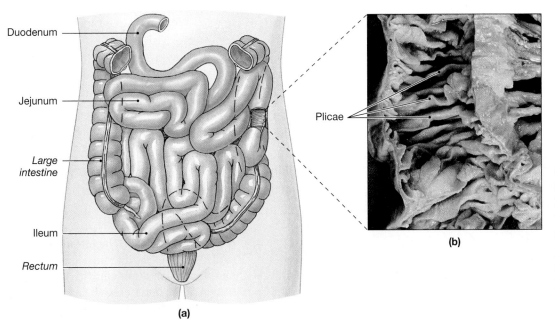

Duodenum

Jejunum

Large intestine

Ileum

Rectum

Plicae

(a)

(b)

● **FIGURE 16–11 The Segments of the Small Intestine. (a)** In this diagram, the three segments of the small intestine—the duodenum, jejunum, and ileum—are shown in different colors. **(b)** This photograph provides a representative view of the lumen of the jejunum.

Clinical Note
UPPER GI BLEEDING

Upper gastrointestinal bleeding can be defined as bleeding within the gastrointestinal tract proximal to the ligament of Treitz, which supports the duodenojejunal junction, which is the point where the first two sections of the small intestine (the duodenum and the jejunum) meet.

Upper gastrointestinal bleeds account for over 300,000 hospitalizations per year. The mortality rate has remained fairly steady at approximately 10 percent over the past years. Many factors contribute to this high mortality. First, the number of patients who treat their symptoms with home remedies and over-the-counter medications is increasing rapidly. Many of these patients come under medical care only when their disease has caused significant damage, such as large-scale hemorrhage from an ulcerated lesion. Second, the overall age of the population is increasing. The infirmities of age and its greater likelihood of coexisting illnesses, such as hypertension, atherosclerosis, diabetes, and substance abuse (including abuse of medications), make this older population more vulnerable to the effects of upper gastrointestinal bleeds. The mortality rate is highest in those over 60 years of age. One prevention strategy for the field is to check for such coexisting problems, especially in elderly patients, and to treat accordingly. In particular, look at the history and physical for evidence of tobacco or alcohol use, or both.

The six major identifiable causes of upper GI hemorrhage, in descending order of frequency, are peptic ulcer disease, gastritis, variceal rupture, Mallory-Weiss syndrome (esophageal laceration, usually secondary to vomiting), esophagitis, and duodenitis. Peptic ulcer disease accounts for approximately 50 percent of upper GI bleeds; gastritis accounts for an additional 25 percent. Overall, irritation or erosion of the gastric lining of the stomach causes more than 75 percent of upper GI bleeds. Most cases of upper GI bleeding are chronic irritations or inflammations that cause minimal discomfort and minor hemorrhage. Physicians can manage these conditions on an outpatient basis; however, if a peptic ulcer erodes through the gastric mucosa, if the esoph-

agus is lacerated in Mallory-Weiss syndrome, or if varices (often secondary to alcoholic liver damage) rupture, an acute-onset, life-threatening, and difficult-to-control hemorrhage can result.

Upper GI bleeds may be obvious, or they may present quite subtly. Most often patients will complain of some type of abdominal discomfort that ranges from a vague burning sensation to an upset stomach, gas pain, or tearing pain in the upper quadrants. Because blood severely irritates the GI system, most cases present with nausea and vomiting. If the bleeding is in the upper GI tract, the patient may experience hematemesis (bloody vomitus) or, if it passes through the lower GI tract, melena. The partially digested blood will turn the stool black and tarry. For melena to be recognizable, approximately 150 cc of blood must drain into the GI tract and remain there for from five to eight hours. Blood in emesis may be bright red (new, fresh blood) or look like coffee grounds (old, partially digested blood).

Upper GI bleeding may be light or it may be brisk and life threatening. Patients who suffer a rupture of an esophageal varix or a tear or disruption in the esophageal or gastric lining may vomit copious amounts of blood. These hemorrhages can cause the classic signs and symptoms of shock, including alteration in mental status, tachycardia, peripheral vasoconstriction, diaphoresis (sweating that produces pale, cool, clammy skin), and hemodynamic instability. Besides shock, the vomitus itself can compromise the airway, and result in impaired respirations, aspiration, and ultimately, respiratory arrest.

A frequently employed clinical indicator is the tilt test, which indicates if the patient has orthostatic hypotension (a 10-mmHg change in blood pressure or a 20-bpm change in heart rate when the patient rises from supine to standing). Hypotension suggests a decreased circulating volume. The human body can compensate for a circulating volume deficit of approximately 15 percent before clinical indicators such as the tilt test show positive results. Thus, those patients whose systolic blood pressure drops 10 mmHg or whose heart rate increases 20 bpm or more need aggressive fluid resuscitation. ■

The Intestinal Wall

The intestinal lining bears a series of transverse folds called **plicae,** or *plicae circulares* (PLĪ-sē sir-kū-lar-ēs) (Figure 16–12a●). The lining of the intestine is composed of a multitude of finger-like projections called **villi** (Figure 16–12b●). These structures are covered by a simple columnar epithelium carpeted with microvilli (Figure 16–12c●). Because the microvilli project from the epithelium like the bristles on a brush, these cells are said to have a *brush border.*

If the small intestine were a simple tube with smooth walls, it would have a total absorptive area of about 3300 cm^2 (13.6 ft^2). Instead, the epithelium contains several plicae; each plica supports a forest of villi, and each villus is covered by epithelial cells blanketed in microvilli. This arrangement increases the total area for absorption to approximately 2 million cm^2, or more than 2200 ft^2.

Each villus contains a network of capillaries (see Figure 16–12c) that transports respiratory gases and carries absorbed

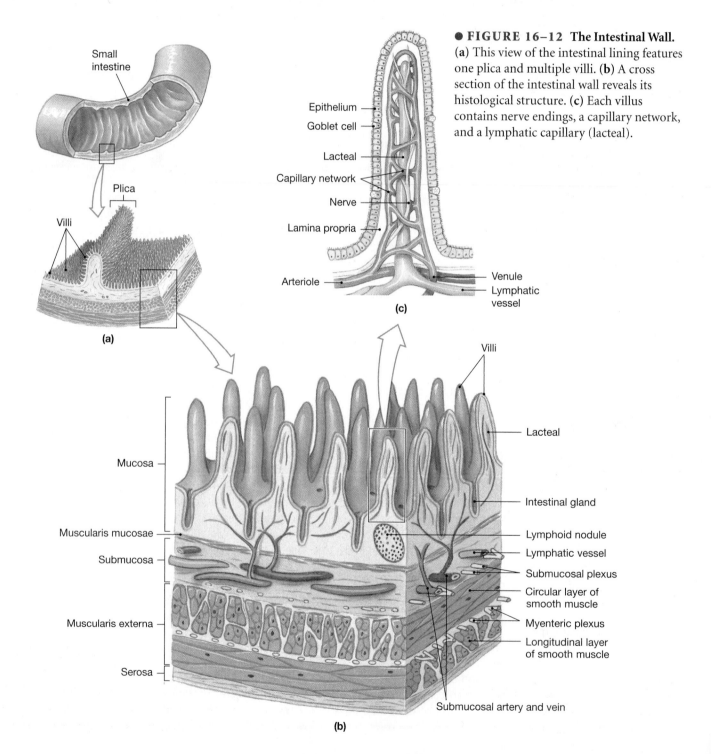

● **FIGURE 16–12 The Intestinal Wall.** (**a**) This view of the intestinal lining features one plica and multiple villi. (**b**) A cross section of the intestinal wall reveals its histological structure. (**c**) Each villus contains nerve endings, a capillary network, and a lymphatic capillary (lacteal).

nutrients to the hepatic portal circulation for delivery to the liver. ∞ p. 499 In addition to capillaries, each villus contains nerve endings and a lymphatic capillary called a **lacteal** (LAK-tē-ul; *lacteus*, milky). Lacteals transport materials that cannot enter blood capillaries. For example, absorbed fatty acids are assembled into protein-lipid packages that are too large to diffuse into the bloodstream. These packets, called *chylomicrons* (*chylos*, juice), reach the venous circulation by way of the lymphatic system. The name *lacteal* refers to the pale, milky appearance of lymph that contains large quantities of lipids.

At the bases of the villi are entrances to *intestinal glands* (see Figure 16–12b). Stem cells within the bases of the intestinal glands divide continuously to replenish the intestinal epithelium. Other cells secrete a watery *intestinal juice*. In addition to the intestinal glands, the duodenum contains large *duodenal glands,* or *submucosal glands,* which secrete an alkaline mucus that helps buffer the acids in chyme. Intestinal glands also contain endocrine cells responsible for the production of intestinal hormones discussed in a later section.

Intestinal Movements

Once the chyme has entered the small intestine, segmentation contractions mix it with mucous secretions and enzymes. As absorption subsequently occurs, weak peristaltic contractions slowly move the remaining materials along the length of the small intestine. These contractions are local reflexes not under CNS control, and the effects are limited to within a few centimeters of the site of the original stimulus.

More elaborate reflexes coordinate activities along the entire length of the small intestine. Distension of the stomach initiates the *gastroenteric* (gas-trō-en-TER-ik) *reflex,* which immediately accelerates glandular secretion and peristaltic activity in all intestinal segments. The increased peristalsis moves materials along the length of the small intestine and empties the duodenum. The *gastroileal* (gas-trō-IL-ē-al) *reflex* is a response to circulating levels of the hormone gastrin. The entry of food into the stomach triggers the release of gastrin, which relaxes the ileocecal valve at the entrance to the large intestine. Because the valve is relaxed, peristalsis pushes materials from the ileum into the large intestine. On average, it takes about five hours for ingested food to pass from the duodenum to the end of the ileum, so the first of the materials to enter the duodenum after breakfast may leave the small intestine during lunch.

Intestinal Secretions

Roughly 1.8 liters of watery **intestinal juice** enters the intestinal lumen each day. Much of this fluid arrives by osmosis (as water flows out of the mucosa), and the rest is secreted by in-

testinal glands stimulated by the activation of touch and stretch receptors in the intestinal walls. Intestinal juice moistens the intestinal contents, helps buffer acids, and keeps both the digestive enzymes and the products of digestion in solution.

Hormonal and CNS controls are important in regulating the secretions of the digestive tract and accessory organs. Because the duodenum is the first segment of the intestine to receive chyme, it is the focus of these regulatory mechanisms. Here the acid content of the chyme must be neutralized and the appropriate enzymes added. The duodenal glands protect the duodenal epithelium from gastric acids and enzymes. They increase their secretions in response to local reflexes and also to parasympathetic stimulation carried by the vagus nerves. As a result of parasympathetic stimulation, the duodenal glands begin secreting during the cephalic phase of gastric secretion, long before chyme reaches the pyloric sphincter. Sympathetic stimulation inhibits their activation, which leaves the duodenal lining relatively unprepared for the arrival of the acid chyme. This is probably why duodenal ulcers can be caused by chronic stress or by other factors that promote sympathetic activation.

Clinical Note
VOMITING

Vomiting, also called *emesis,* is a forceful emptying of the stomach and intestinal contents through the mouth. There are several things that can stimulate the vomiting reflex. Locally, distention of the stomach or duodenum can cause vomiting. On a systemic level, vomiting can result from activation of the *chemoreceptor trigger zone (CTZ)* in the medulla of the brain by the neurotransmitter serotonin. Serotonin appears to be released from cells within the intestinal wall and from neurons in the brain stem which then stimulate the vomiting center.

Nausea and retching usually precede vomiting. Nausea is a subjective sensation that is seen with many medical conditions. Increased salivation and an increase in heart rate are often seen. Retching involves contraction of the abdominal muscles and movement of stomach contents into the esophagus, but not into the mouth.

Vomiting usually follows retching. Peristalsis in the duodenum, stomach, and esophagus is reversed. Accompanied by contraction of the abdominal muscles, this results in forceful emptying of the stomach. Medications are available that help alleviate vomiting. Most appear to block serotonin receptors in the chemoreceptor trigger zone. ■

Intestinal Hormones

Duodenal endocrine cells produce various peptide hormones that coordinate the secretory activities of the stomach, duodenum, pancreas, and liver. These hormones were introduced in the discussion on the regulation of gastric activity, and Figure 16–10 indicates the factors that stimulate their secretion.

Gastrin is secreted by duodenal cells in response to large quantities of incompletely digested proteins. Gastrin promotes increased stomach motility and stimulates the production of acids and enzymes. (As previously noted, gastrin is also secreted by endocrine cells in the distal portion of the stomach.)

Secretin (sē-KRĒ-tin) is released when the pH falls in the duodenum as acidic chyme arrives from the stomach. The primary effect of secretin is to increase the secretion of bile and buffers by the liver and pancreas.

Cholecystokinin (kō-lē-sis-tō-KĪ-nin), or **CCK,** is secreted when chyme arrives in the duodenum, especially when the chyme contains lipids and partially digested proteins. CCK also targets the pancreas and gallbladder. In the pancreas, CCK accelerates the production and secretion of all types of digestive enzymes. At the gallbladder, it causes the ejection of *bile* into the duodenum. The presence of either secretin or CCK in high concentrations also reduces gastric motility and secretory rates.

Gastric inhibitory peptide (GIP), is released when fats and carbohydrates (especially glucose) enter the small intestine. GIP inhibits gastric activity and causes the release of insulin from the pancreatic islets. The functions of the major gastrointestinal hormones are summarized in Table 16–1, and their interactions are diagrammed in Figure 16–13●.

Digestion in the Small Intestine

In the stomach, food becomes saturated with gastric juices and exposed to the digestive effects of a strong acid (HCl) and a proteolytic enzyme (pepsin). Most of the important digestive processes are completed in the small intestine, where the final products of digestion—simple sugars, fatty acids, and amino acids—are absorbed, along with most of the water content. However, the small intestine produces only a few of the enzymes needed to break down the complex materials found in the diet. Most of the enzymes and buffers are contributed by the liver and pancreas, which are discussed in the next section.

Key Note

The small intestine receives and raises the pH of materials from the stomach. It then absorbs water, ions, vitamins, and the chemical products released by the action of digestive enzymes from intestinal glands and exocrine glands of the pancreas.

TABLE 16–1 *Important Gastrointestinal Hormones and Their Primary Effects*

HORMONE	STIMULUS	ORIGIN	TARGET	EFFECTS
Gastrin	Vagus nerve stimulation or arrival of food in the stomach	Stomach	Stomach	Stimulates production of acids and enzymes, increases motility
	Arrival of chyme that contains large quantities of undigested proteins	Duodenum	Stomach	As above
Secretin	Arrival of chyme in the duodenum	Duodenum	Pancreas	Stimulates production of alkaline buffers
			Stomach	Inhibits gastric secretion and motility
			Liver	Increases rate of bile secretion
Cholecystokinin (CCK)	Arrival of chyme that contains lipids and partially digested proteins	Duodenum	Pancreas	Stimulates production of pancreatic enzymes
			Gallbladder	Stimulates contraction of gallbladder
			Duodenum	Causes relaxation of sphincter at base of bile duct
			Stomach	Inhibits gastric secretion and motility
			CNS	May reduce hunger
Gastric Inhibitory Peptide (GIP)	Arrival of chyme that contains large quantities of fats and glucose	Duodenum	Pancreas	Stimulates release of insulin by pancreatic islets
		Stomach		Inhibits gastric secretion and motility

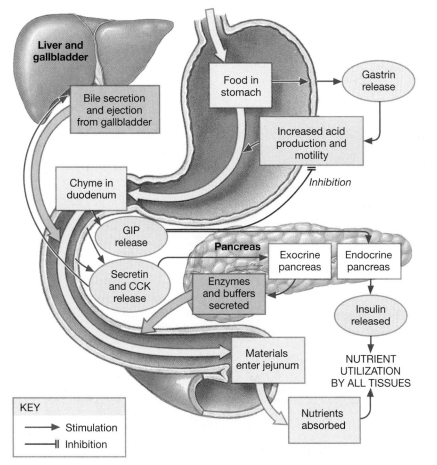

● FIGURE 16–13 The Activities of Major Digestive Tract Hormones. Depicted are the primary actions of gastrin, GIP, secretin, and CCK.

> ### CONCEPT CHECK QUESTIONS
>
> 1. Which ring of muscle regulates the flow of chyme from the stomach to the small intestine?
> 2. Why does treatment of chronic stomach ulcers sometimes involve cutting the branches of the vagus nerve that innervate the stomach?
> 3. How is the small intestine adapted for the absorption of nutrients?
> 4. How would a meal that is high in fat affect the level of cholecystokinin in the blood?
>
> *Answers begin on p. 792.*

■ The Pancreas

The **pancreas,** shown in Figure 16–14a●, lies behind the stomach, and extends laterally from the duodenum toward the spleen. It is an elongate, pinkish-gray organ with a length of about 15 cm (6 in.) and a weight of around 80 g (3 oz). The surface of the pancreas has a lumpy texture, and its tissue is soft and easily torn. Like the duodenum, the pancreas is retroperitoneal (lies outside the peritoneal cavity); only its anterior surface is covered by peritoneum (see Figure 16–8b).

Histological Organization

In Chapter 10 we learned that collections of endocrine cells called **pancreatic islets** secrete the hormones insulin and glucagon, but these cells account for only about 1 percent of the cellular population of the pancreas. ∞ p. 389 Exocrine cells and their associated ducts account for the rest, as the pancreas is primarily an exocrine organ that produces **pancreatic juice,** a mixture of digestive enzymes and buffers. The numerous ducts that branch throughout the pancreas end at sac-like pouches called **pancreatic acini** (AS-i-nī; singular *acinus,* grape) (Figure 16–14b●). Enzymes and buffers are secreted by the *acinar cells* of these pouches and by the epithelial cells that line the ducts. The smaller ducts converge to form larger ducts that ultimately fuse to form the **pancreatic duct,** which carries these secretions to the duodenum. The pancreatic duct penetrates the duodenal wall with the *common bile duct* from the liver and gallbladder.

Pancreatic enzymes do most of the digestive work in the small intestine. Pancreatic enzymes are broadly classified according to their intended targets. **Carbohydrases** (kar-bō-HĪ-drā-sez) digest sugars and starches, **lipases** (LĪ-pā-sez) break down lipids, **nucleases** break down nucleic acids, and **proteases** (prō-tē-ā-sez) (proteolytic enzymes) break proteins apart.

The Control of Pancreatic Secretion

Each day the pancreas secretes about 1000 mL (1 qt) of pancreatic juice. The secretions are controlled primarily by hormones from the duodenum.

When acidic chyme arrives in the duodenum, secretin is released, which triggers the pancreas to secrete a watery, alkaline fluid with a pH between 7.5 and 8.8. Among its other components, this secretion contains buffers, primarily *sodium bicarbonate,* that increase the pH of the chyme.

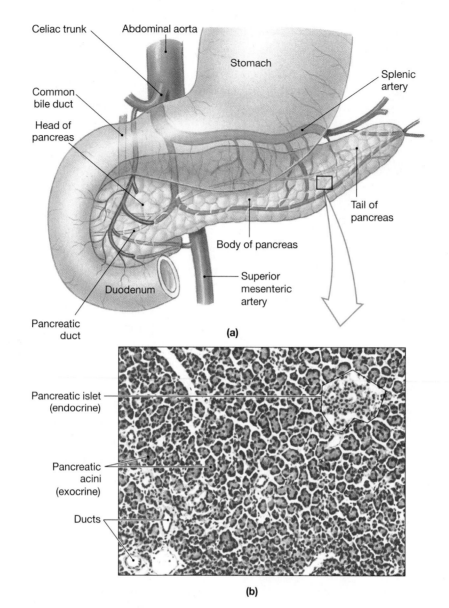

(a)

(b)

● **FIGURE 16–14 The Pancreas.** (**a**) The head of the pancreas is tucked into a curve of the duodenum that begins at the pylorus of the stomach. (**b**) Pancreatic acini and their ducts constitute the vast majority of all pancreatic cells. (LM × 168)

Clinical Note
PANCREATITIS

Pancreatitis is an inflammation of the pancreas. There are four main categories of pancreatitis: metabolic, mechanical, vascular, and infectious. Metabolic causes, especially alcoholism, account for 80 percent of all cases. Pancreatitis is common in the U.S. because of the high incidence of alcoholism. Mechanical obstruction caused by gallstones or elevated serum lipids account for another 9 percent of cases. Overall, mortality in acute pancreatitis is high; it approaches 30 to 40 percent, mainly due to accompanying sepsis and shock.

As discussed in the text, the pancreas produces digestive enzymes that empty into the duodenum at the ampulla of Vater, near the junc-

tion with the stomach. The other function of the pancreas is endocrine. The islets of Langerhans secrete glucagon, insulin, and somatostatin directly into the circulatory system. Occasionally, gallstones leaving the common bile duct become lodged at the ampulla of Vater and obstruct the pancreatic duct. These obstructions back up pancreatic digestive enzymes into the pancreatic duct and the pancreas itself. The digestive enzymes then inflame the pancreas and cause edema. This reduces blood flow, similar to the pathogenesis of acute appendicitis. In turn, the decreased blood flow causes ischemia and, finally, acinar destruction. This process is called acute pancreatitis because of the rapidity of onset.

Chronic pancreatitis results from acinar tissue destruction. This results from chronic alcohol intake, drug toxicity, ischemia, and infectious disease. Alcohol ingestion results in the formation and deposit of platelet plugs in the acinar tissue. The plugs disrupt the enzyme flow from the pancreas. When digestive juices back up into the pancreas from the ampulla of Vater, the digestive enzymes become activated and begin to digest the pancreas itself. This autodigestion causes lesions and fatty tissue changes to appear in the pancreas. Chronic pancreatitis can result in destruction of a significant portion of the pancreas, affecting both endocrine and exocrine tissues. In these cases, patients may be required to take digestive enzyme supplements. If a significant portion of the endocrine tissue is destroyed, diabetes mellitus can occur, which requires insulin replacement.

As tissue digestion continues, the lesion can erode and begin to hemorrhage (hemorrhagic pancreatitis). This causes severe abdominal pain, usually located in the epigastrium or the left upper quadrant. The pain often radiates straight through to the back and often requires high doses of narcotics to control. Morphine is contraindicated because it causes spasm of the sphincter of Oddi. The patient will appear acutely ill with diaphoresis, tachycardia, and possible hypotension if massive hemorrhage occurs. Intractable vomiting may be present and will require placement of a nasogastric tube. Cessation of alcohol intake must occur for recovery from pancreatitis. ∎

Key Note

The exocrine pancreas produces a mixture of buffers and enzymes essential for normal digestion. Pancreatic secretion occurs in response to the release of regulatory hormones (secretin and CCK) by the duodenum.

∎ The Liver

The firm, reddish-brown **liver** is the largest visceral organ; it weighs about 1.5 kg (3.3 lb) and accounts for roughly 2.5 percent of total body weight. Most of the liver lies in the right hypochondriac and epigastric abdominopelvic regions. ∞ p. 17

Anatomy of the Liver

The liver is wrapped in a tough fibrous capsule and covered by a layer of visceral peritoneum. The liver is divided into four unequal lobes: the large **left** and **right lobes** and the smaller **caudate** and **quadrate lobes** (Figure 16–15●). A tough connective tissue fold, the *falciform ligament,* marks the division between the left and right lobes. The thickened posterior margin of the falciform ligament is the *round ligament,* a fibrous remnant of the fetal umbilical vein.

Lodged within a recess under the right lobe of the liver is the *gallbladder,* which is a muscular sac that stores and concentrates

● **FIGURE 16–15 The Surface Anatomy of the Liver.**

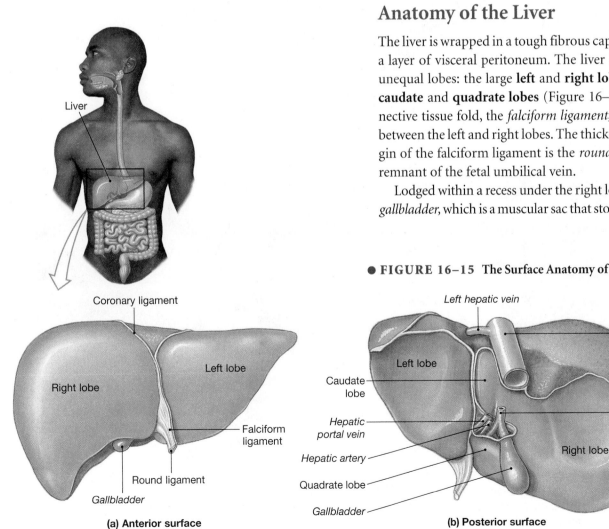

(a) Anterior surface

(b) Posterior surface

bile before it is excreted into the small intestine. The gallbladder and associated structures will be described in a later section.

Histological Organization of the Liver

The lobes of the liver are divided by connective tissue into about 100,000 **liver lobules,** which are the basic functional units of the liver. The histological organization and structure of a typical liver lobule is shown in Figure 16–16●.

Within a lobule, liver cells—called **hepatocytes** (hep-a-to-sīts)—are arranged into a series of irregular plates like the spokes of a wheel. The plates are only one cell thick and, where they are exposed, covered with microvilli. *Sinusoids,* which are specialized and highly permeable capillaries, form passageways between the adjacent plates that empty into the *central vein.* The sinusoidal lining includes a large number of phagocytic *Kupffer* (KOOP-fer) *cells.* These cells, part of the

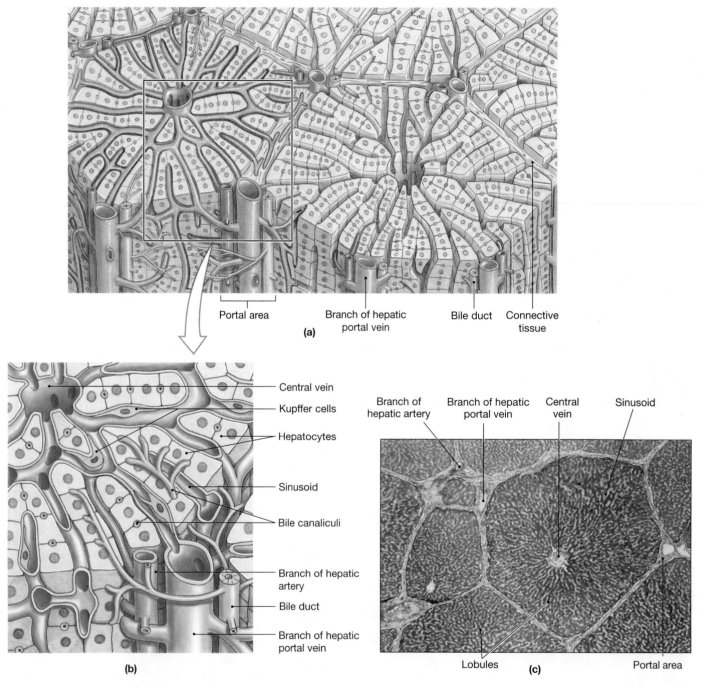

(a)

Portal area — Branch of hepatic portal vein — Bile duct — Connective tissue

(b)

Central vein
Kupffer cells
Hepatocytes
Sinusoid
Bile canaliculi
Branch of hepatic artery
Bile duct
Branch of hepatic portal vein

(c)

Branch of hepatic artery — Branch of hepatic portal vein — Central vein — Sinusoid

Lobules — Portal area

●**FIGURE 16–16 Liver Histology. (a)** A diagrammatic view of liver structure reveals the relationships among lobules. **(b)** This enlargement highlights the major structures of a liver lobule. **(c)** This photomicrograph shows a section of a pig liver because the boundaries of human liver lobules are very difficult to see at comparable magnification. (LM × 38)

monocyte-macrophage system, engulf pathogens, cell debris, and damaged blood cells.

The circulation of blood to the liver was discussed in Chapter 13 (p. 499) and illustrated in Figure 13–28 (p. 499). Blood enters the sinusoids from branches of the hepatic portal vein and hepatic artery. These two branches, plus a small branch of the bile duct, form a *portal area*, or *hepatic triad*, at each of the six corners of a lobule (see Figure 16–16a). As blood flows through the sinusoids, the hepatocytes absorb solutes from the plasma and secrete materials such as plasma proteins. Blood then leaves the sinusoids and enters the **central vein** of the lobule. The central veins of all of the lobules ultimately merge to form the *hepatic veins,* which empty into the inferior vena cava. Liver diseases (including the various forms of *hepatitis*) and conditions such as alcoholism can lead to degenerative changes in liver tissue and reduction of the blood supply.

Clinical Note
PORTAL HYPERTENSION

The portal system is a specialized component of the circulatory system. Veins from the spleen, stomach, pancreas, gallbladder, and intestines do not drain directly into the inferior vena cava, as do the veins from other abdominal organs. Instead, they drain into the portal vein that delivers the blood to the liver. In the liver, blood from the portal circulation mixes with the arterial blood in the hepatic capillaries and is eventually drained from the liver by the hepatic veins. The hepatic veins drain into the inferior vena cava.

Blood in the hepatic portal system contains substances absorbed by the digestive tract. Blood that enters the liver via the portal vein contains greater concentrations of glucose, amino acids, and fats than does blood that leaves the liver via the hepatic vein. The liver regulates the concentration of nutrients, such as glucose or amino acids, in the circulating blood.

The pressure within the portal system is normally 3 mmHg. An increase in portal pressure to at least 10 mmHg is referred to as *portal hypertension*. Portal hypertension is caused by disorders that impede or obstruct blood flow through any part of the portal system or the vena cava. The obstruction can occur in the liver or in the hepatic veins that drain the liver. The most common cause of portal hypertension is obstruction caused by cirrhosis of the liver. *Cirrhosis* is an irreversible inflammatory disease that disrupts the structure and function of the liver. The most common cause of cirrhosis in the U.S. is chronic alcohol abuse (Figure 16–17●). Chronic infectious hepatitis also can cause cirrhosis that results in portal hypertension.

Increased pressure within the portal system causes collateral blood vessels between the portal veins and the systemic veins to open. Blood pressure in the systemic veins is considerably lower than that in the portal system, which enables blood to bypass the obstructed portal vessels. The collateral veins develop in the esophagus, anterior abdominal wall, and rectum. High pressure and increased blood flow are transmitted through these veins from the portal to the systemic venous circulation. Blood that is shunted into the systemic circulation bypasses the liver where toxic metabolic waste products are usually removed. This results in an accumulation of these waste products in the systemic circulation.

Long-term portal hypertension results in numerous problems that are quite difficult to treat. These include:

- *Varices.* Varices are tortuous, distended collateral veins that develop secondary to long-standing increased portal pressure. They are found in the lower esophagus, upper stomach, and

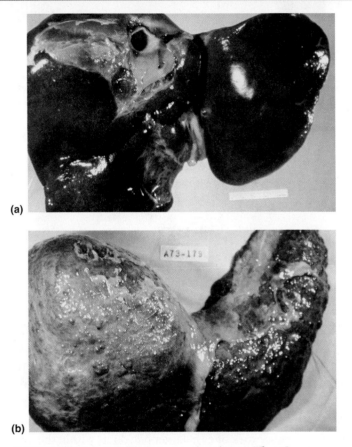

(a)

(b)

● **FIGURE 16–17 Postmortem Specimens That Compare (a) Normal Liver and (b) Cirrhotic Liver.** Note the cirrhotic liver is irregular and smaller than the normal specimen.

rectum. They are prone to bleeding that can be difficult to control (Figure 16–18●).

- *Ascites.* Ascites is the accumulation of fluid in the space between the parietal and visceral peritoneum. It is caused by increased pressure in the mesenteric tributaries of the portal vein. Hydrostatic pressure within the veins forces water out of these vessels and into the peritoneal cavity.

- *Splenomegaly.* Splenomegaly, an enlargement of the spleen, results from increased pressure in the splenic vein, which is a branch of the portal vein.

(continued next page)

Clinical Note—continued
PORTAL HYPERTENSION

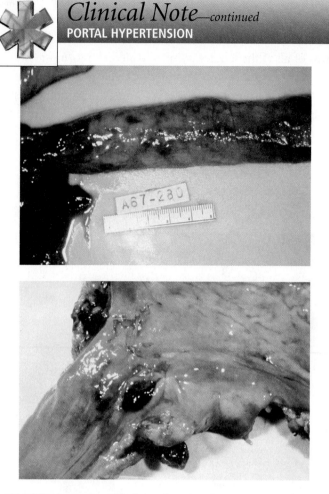

● **FIGURE 16–18 Esophageal Varices in a Patient with Portal Hypertension Secondary to Alcoholic Cirrhosis.**

■ *Hepatic encephalopathy.* Hepatic encephalopathy is characterized by central nervous system disturbances such as confusion, somnolence, and unconsciousness. Hepatic encephalopathy results from increased amounts of metabolic waste products in the blood, especially ammonia.

Portal hypertension develops over years. The most common clinical manifestation is vomiting of blood from bleeding esophageal varices. Slow, chronic variceal bleeding can cause anemia or melena. Rupture of esophageal varices is painless and can cause massive hemorrhage that is notoriously difficult to control and often fatal. Variceal bleeding can sometimes be controlled by injecting a sclerosing agent into the bleeding varix or banding of the bleeding varix with a fiber-optic esophagogastroduodenoscope (EGD). Placing a Sengstaken-Blakemore tube can sometimes control massive hemorrhages. This is a cylindrical balloon with a bulb at the end that is inserted into the distal esophagus and inflated. Inflation of the balloon and the bulb compresses the bleeding varices, slowing or stopping bleeding. Unfortunately, there are numerous, potentially lethal, complications associated with the Sengstaken-Blakemore tube use. Because of this, it is rarely used except as a last resort.

The viable treatment options for portal hypertension are exceedingly limited. Surgical construction of a portacaval shunt (connection of the portal vein to the inferior vena cava) can reduce portal pressure. However, it can cause liver failure or encephalopathy due to reduced hepatic blood flow. In selected cases, liver transplantation can be curative if the disease is not too advanced. Overall, there is no effective, definitive treatment for portal hypertension. ■

The hepatocytes also secrete a fluid called **bile.** Bile is released into a network of narrow channels, called **bile canaliculi,** between adjacent liver cells (see Figure 16–16b). These canaliculi extend outward from the central vein, and carry bile toward a network of ever-larger bile ducts within the liver until it eventually leaves the liver through the **common hepatic duct** (see Figure 16–16a). Bile in the common hepatic duct may either flow into the **common bile duct,** which empties into the duodenum, or enter the **cystic duct,** which leads to the gallbladder.

Clinical Note
HEPATIC ABSCESSES

The liver is the organ most commonly subject to the development of abscesses. Liver abscesses are uncommon, especially in industrialized countries. When they do occur, they most frequently result from parasitic infection. The amoeba *Entamoeba histolytica* is the most common etiological agent, especially in developing countries. Approximately half of all patients with amoebic liver abscess will be asymptomatic. The other half will have low-grade fever, nausea, vomiting, diarrhea, and abdominal pain. The abscesses appear like "anchovy paste" on gross examination. The diagnosis can usually be established by identifying the parasite through stool testing.

In the U.S., hepatic abscesses are usually the result of a surgical procedure, especially one that involves the biliary tract and gallbladder. The principle treatment is drainage of the abscess. This may be accomplished through open surgical drainage or placement of a catheter into the abscess. ■

Liver Functions

The liver is responsible for three general functional roles: (1) *metabolic regulation*, (2) *hematological regulation*, and (3) *bile production*. Because the liver has over 200 known functions, only a general overview is provided here.

Metabolic Regulation

The liver is the primary organ involved in regulating the composition of circulating blood. All blood that leaves the absorptive areas of the digestive tract flows through the liver before reaching the general circulation. Thus, hepatocytes can (1) extract absorbed nutrients or toxins from the blood before they reach the rest of the body and (2) monitor and adjust the circulating levels of organic nutrients. Excesses are removed and stored, and deficiencies are corrected by mobilizing stored reserves or synthesizing the necessary compounds. When blood glucose levels rise, for example, the liver removes glucose and synthesizes the storage compound glycogen. When blood glucose levels fall, the liver breaks down stored glycogen and releases glucose into the circulation. Circulating toxins and metabolic wastes are also removed for later inactivation or excretion. Additionally, fat-soluble vitamins (A, D, K, and E) are absorbed and stored.

Hematological Regulation

The liver, the largest blood reservoir in the body, receives about 25 percent of cardiac output. As blood passes through the liver, phagocytic Kupffer cells remove aged or damaged red blood cells, debris, and pathogens from the circulation. Kupffer cells are antigen-presenting cells that can stimulate an immune response. ∞ p. 527 Equally important, hepatocytes synthesize the plasma proteins that determine the osmotic concentration of the blood, transport nutrients, and make up the clotting and complement systems. ∞ p. 410

The Production and Role of Bile

As previously noted, bile is synthesized in the liver and excreted into the lumen of the duodenum. Bile consists mostly of water, ions, *bilirubin* (a pigment derived from hemoglobin), cholesterol, and an assortment of lipids collectively known as **bile salts.** The water and ions in bile help dilute and buffer acids in chyme as it enters the small intestine. Bile salts are synthesized from cholesterol in the liver and are required for the normal digestion and absorption of fats.

Most dietary lipids are not water soluble. Mechanical processing in the stomach creates large droplets that contain various lipids. Pancreatic lipase is not lipid soluble and can interact with lipids only at the surface of the droplet. The larger the droplet, the more lipids are inside it, isolated and protected from these digestive enzymes. Bile salts break the droplets apart through a process called **emulsification** (ē-mul-si-fi-KĀ-shun), which creates tiny droplets with a superficial coating of bile salts. The formation of tiny droplets increases the surface area available for enzymatic attack. In addition, the layer of bile salts facilitates interaction between the lipids and lipid-digesting enzymes from the pancreas. (We will return to the mechanism of lipid digestion later.)

Table 16–2 provides a summary of the liver's major functions.

Clinical Note
LIVER TRANSPLANT

Liver transplantation is the replacement of a native, diseased liver with a liver from a brain-dead donor *(allograft)*. This operation allows a patient who otherwise would have died from liver failure to live a relatively full and normal life. The donor liver contains several tissue antigens that can induce an immune response in the recipient. Because of this, the donor and recipient must be checked for tissue antigen compatibility. A good match will decrease the likelihood of *organ rejection*. Patients who receive a liver transplant will be placed on *immunosuppressive drugs* and will remain on them for the rest of their lives.

Patients with *fulminate liver failure*, regardless of the cause, will die within hours or days if a suitable organ donor cannot be located. In an extreme situation, a liver from a lower animal *(xenograft)*, most commonly a pig, can be used temporarily until a human donor becomes available.

On rare occasions, liver tissue may be harvested from a suitable living donor. Although the procedure is technically more complicated and places a second patient at risk, it is being used with increasing frequency when a suitable donor cannot be found. A lobe of the liver is taken from the donor and placed in the recipient. The liver is unique in that it will regenerate. Thus, in a living donor operation, the livers will grow to normal size in both the donor and the recipient within six to eight weeks. ∎

TABLE 16–2 *Major Functions of the Liver*

Digestive and Metabolic Functions

Synthesis and secretion of bile

Storage of glycogen and lipid reserves

Maintenance of normal blood levels of glucose, amino acids, and fatty acids

Synthesis and interconversion of nutrient types (e.g., transamination of amino acids or conversion of carbohydrates to lipids)

Synthesis and release of cholesterol bound to transport proteins

Inactivation of toxins

Storage of iron reserves

Storage of fat-soluble vitamins

Other Major Functions

Synthesis of plasma proteins

Synthesis of clotting factors

Synthesis of the inactive hormone angiotensinogen

Phagocytosis of damaged red blood cells (by Kupffer cells)

Blood storage (major contributor to venous reserve)

Absorption and breakdown of circulating hormones (insulin, epinephrine) and immunoglobulins

Absorption and inactivation of lipid-soluble drugs

Clinical Note
HEPATITIS

Hepatitis is an inflammation of the liver and can result from both infectious and noninfectious causes. Medications, toxins, chemicals, and autoimmune disorders may cause *noninfectious hepatitis. Infectious hepatitis* can result from infection with viruses, bacteria, fungi, and parasites. The vast majority of infectious hepatitis cases are viral (Figure 16–19●).

Several viruses have been identified as causative agents for hepatitis. These include: hepatitis A virus (HAV), hepatitis B virus (HBV), hepatitis C virus (HCV), hepatitis D virus (HDV), hepatitis E virus (HEV), hepatitis F virus (HFV), and hepatitis G virus (HGV). HAV, HBV, and HCV cause more than 90 percent of cases of acute viral hepatitis in the U.S. Emergency personnel are at particular risk for exposure to HBV and HCV because of the increased risk of exposure to body fluids. HAV, HBV, HCV, and HDV are the only hepatitis viruses endemic to the U.S.

The hepatitis viruses impair liver function by attacking and destroying liver cells. Hepatitis may be either acute or chronic. All hepatitis viruses cause an acute infection. The severity of the infection can vary from asymptomatic to fulminant liver failure. Patients who contract hepatitis A and hepatitis E usually do not develop a chronic infection. However, patients who are infected with hepatitis B, hepatitis C, hepatitis D, and hepatitis G are at risk of developing chronic active hepatitis. Chronic active hepatitis is progressive; it leads to deterioration in liver function and eventually cirrhosis. Many patients will develop hepatocellular cancer. Complications associated with chronic active hepatitis include portal hypertension, ascites, and eventually, hepatic encephalopathy (CNS dysfunction). As liver failure progresses, toxic metabolic waste products are not effectively cleared by the liver and start to accumulate in the blood. Among the more important of these waste products is ammonia. An increase in serum ammonia can cause hepatic encephalopathy which results in disorientation, confusion, somnolence, and eventually unconsciousness. As hepatic encephalopathy worsens, the patient will develop asterixis, which is an involuntary "flapping" of the hands when the patient holds his hands up, flexed at the wrist (such as the motion for stopping traffic).

The signs and symptoms of infectious hepatitis depend on the type of hepatitis involved. Typically, patients will develop a low-grade fever, loss of appetite, and malaise. The skin, sclera, and mucous membranes may become jaundiced (icteric) due to the accumulation of bilirubin in the body. In more severe cases, the patient may develop significant nausea and vomiting that can lead to dehydration as evidenced by tachycardia, dry mucous membranes, and decreased skin turgor. The liver may be diffusely enlarged and tender to palpation.

There are several different types of infectious viral hepatitis depending on the virus involved. These include:

■ *Hepatitis A.* Hepatitis A, often called *infectious hepatitis,* is the primary cause of viral hepatitis in the U.S. It is transmitted via the fecal-oral route, often through food, water, milk, and shellfish contaminated by fecal wastes. The incubation period is typically two to six weeks. Hepatitis A often occurs in epidemics that can be attributed to a community source such as a restaurant. Hepatitis A is usually a mild, self-limited disease. Infection with HAV confers life-long immunity, and chronic infection with HAV does not occur. A vaccine is available for those deemed to be at increased risk.

■ *Hepatitis B.* Hepatitis B, also called *serum hepatitis,* is a major cause of hepatitis worldwide. Hepatitis B is transmitted via blood or other body products, often through mucous membranes. Saliva, serum, and semen have all been demonstrated to be infectious. In addition, HBV can be transmitted perinatally from an infected mother to her unborn child. The incubation period for hepatitis B is considerably longer than for hepatitis A, and ranges from one to six months. Hepatitis B is a serious infection and can develop into a chronic infection. Chronic hepatitis B can lead to cirrhosis of the liver and, eventually, hepatocellular cancer. An effective vaccine against hepatitis B is available and should be administered to emergency personnel.

■ *Hepatitis C.* Hepatitis C, formerly referred to as non-A, non-B hepatitis, is a serious infection and the most common cause of chronic viral hepatitis in the U.S. Emergency personnel are at increased risk of infection by hepatitis C due to exposure to infected blood or body fluids. The incubation period can range from one to six months. The signs and symptoms of hepatitis C are similar to those of hepatitis B. Approximately 80 percent of those infected with HCV will go on to develop chronic HCV infection. Chronic hepatitis C infection is the leading reason for liver transplantation in the U.S. There is no effective vaccine against hepatitis C or post-exposure prophylaxis.

■ *Hepatitis D.* Hepatitis D, also called delta hepatitis, is a unique type of hepatitis. It requires the presence of hepatitis B virus in order to replicate. Thus, hepatitis D is often considered to be a coinfection or superinfection of hepatitis B. The transmission of hepatitis D is similar to that for hepatitis B. Patients with both hepatitis B and hepatitis D tend to have a more severe course than those with hepatitis B alone. Hepatitis D tends to lead to chronic infection. A vaccine for hepatitis D is not yet available.

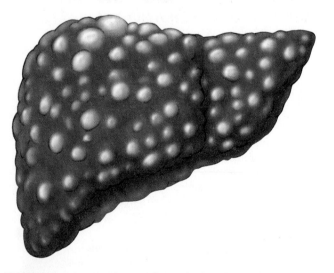

● **FIGURE 16–19 Liver with Marked Inflammation from Hepatitis.** Hepatitis can lead to liver failure, and some types can lead to chronic infection that causes cirrhosis and, in some cases, liver cancer.

■ *Hepatitis E.* Hepatitis E is similar to hepatitis A and is transmitted via the fecal-oral route. The incubation period for hepatitis E ranges from two to nine weeks. Like hepatitis A, the disease is usually self-limited and chronic infection does not occur. Hepatitis E is the most common cause of hepatitis worldwide, but seldom seen in the U.S. A vaccine against hepatitis E is not yet available.

■ *Hepatitis F.* Hepatitis F is proposed as another hepatitis virus transmitted by the fecal-oral route. A small number of cases have been reported in France. Little else is known about the infection.

■ *Hepatitis G.* Hepatitis G virus was identified in 1996 and is associated with both acute and chronic liver disease. The incidence is approximately 0.3 percent of all cases of acute viral hepatitis. Transmission of HGV is blood borne. Chronic infection is common, and occurs in 90–100 percent of infected persons. Much remains to be learned about hepatitis G. A vaccine is not yet available. ■

■ The Gallbladder

The **gallbladder** is a hollow, muscular, pear-shaped organ that stores and concentrates bile prior to its excretion into the small intestine (Figure 16–20a●). The cystic duct extends from the gallbladder to the point where its union with the common hepatic duct forms the common bile duct. The common bile duct and the pancreatic duct join and share a passageway that enters the duodenum at the *duodenal papilla.* The muscular **hepatopancreatic sphincter** surrounds their shared passageway (Figure 16–20b●).

A major function of the gallbladder is *bile storage.* Bile is secreted continuously—roughly 1 liter each day—but it is re-

leased into the duodenum only under the stimulation of the intestinal hormone cholecystokinin (CCK). In the absence of CCK, the hepatopancreatic sphincter remains closed, so bile that leaves the liver in the common hepatic duct cannot flow through the common bile duct and enter the duodenum. Instead, it enters the cystic duct and is stored within the expandable gallbladder. Whenever chyme enters the duodenum, CCK is released, which relaxes the hepatopancreatic sphincter and stimulates contractions within the walls of the gallbladder that push bile into the small intestine. The amount of CCK secreted increases if the chyme contains large amounts of fat.

Another function of the gallbladder is *bile modification.* When filled to capacity, the gallbladder contains 40–70 mL

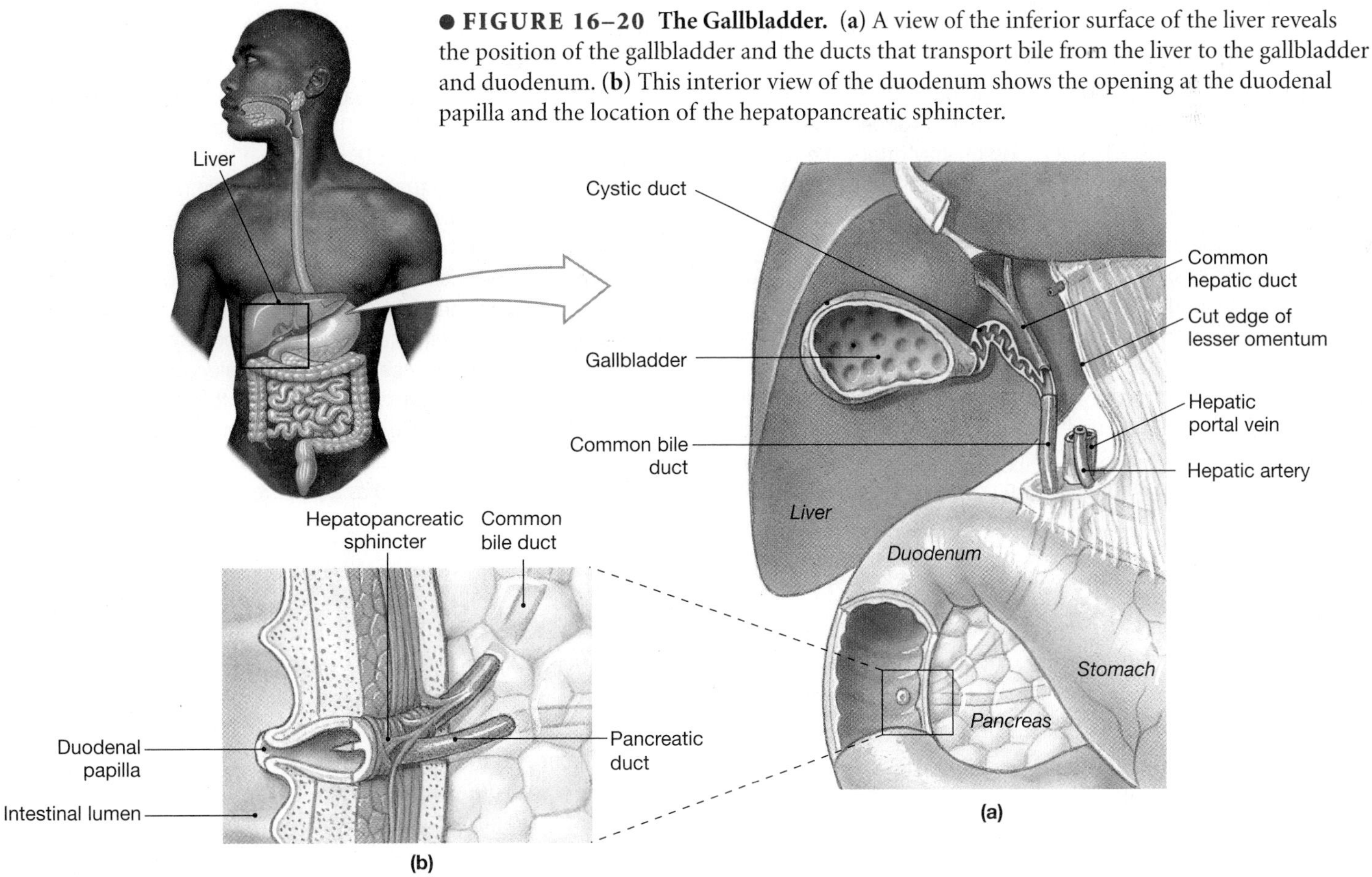

● **FIGURE 16–20 The Gallbladder. (a)** A view of the inferior surface of the liver reveals the position of the gallbladder and the ducts that transport bile from the liver to the gallbladder and duodenum. **(b)** This interior view of the duodenum shows the opening at the duodenal papilla and the location of the hepatopancreatic sphincter.

of bile. The composition of bile gradually changes as it remains in the gallbladder. Water is absorbed, and the bile salts and other components of bile become increasingly concentrated. If the bile salts become too concentrated, they may precipitate, and form *gallstones* that can cause a variety of clinical problems.

Clinical Note
BILIARY COLIC AND CHOLECYSTITIS

Cholecystitis is an inflammation of the gallbladder. Cholelithiasis (the formation of gallstones), which causes 90 percent of cholecystitis cases, occurs in approximately 15 percent of the adult population in the U.S., with over one million new cases diagnosed annually. There are two types of gallstones, cholesterol-based and bilirubin-based. Cholesterol-based stones are far more common and are associated with a specific risk profile: obese, middle-aged women with more than one biological child.

Definitive treatment of acute cholecystitis includes antibiotic therapy, laparoscopic surgery, lithotripsy (ultrasound treatment to break up the stones), and surgery if the other, less invasive, therapies fail. With the advent of laparoscopic surgery, mortality has fallen to less then 1 percent, with an overall morbidity of approximately 6 percent.

Cholecystitis caused by gallstones can be chronic or acute (Figure 16–21●). The liver produces bile, the primary vehicle for removing cholesterol from the body. The bile travels down the common bile duct to empty into the small intestine at the sphincter of Oddi. The sphincter of Oddi opens when chyme exits the stomach through the cardiac sphincter. When the sphincter of Oddi closes, the flow of bile backs up into the gallbladder via the cystic duct. The bile remains in the gallbladder until the sphincter of Oddi opens again.

The bile can become supersaturated and calculi—stone-like masses based on bilirubin, cholesterol, or both—form. These calculi travel down the cystic duct, and frequently lodge in the common bile duct. When they obstruct the flow of bile, gallbladder inflammation and irritation result. The bile salts subsequently attack the mucosal membrane that lines the gallbladder, which leaves the underlying epithelial tissue without protection. Prostaglandins are also released, which further irritates the epithelial wall. As irritation continues, the inflammation grows; this increases intraluminal pressure and ultimately reduces blood flow to the epithelium.

Other causes of cholecystitis include acalculus cholecystitis (cholecystitis without associated stones) and chronic inflammation caused by bacterial infection. Acalculus cholecystitis usually results from burns, sepsis, diabetes, and multiple organ failure. Chronic cholecystitis that results from a bacterial infection (*Escherichia coli* and enterococci) presents with an inflammatory process similar to cholelithiasis.

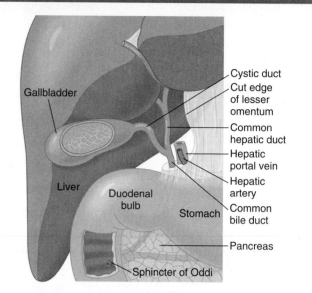

● **FIGURE 16–21 Cholecystitis.** The gallbladder is located immediately under the liver in the right upper abdominal quadrant. Gallstones can enter the cyctic duct or the common bile duct. The latter can result in partial or complete obstruction of the duct.

An inflamed gallbladder usually causes an acute attack of upper right quadrant abdominal pain. The inflammation can cause an irritation of the diaphragm with referred pain in the right shoulder. If the gallstones are lodged in the cystic duct, the pain may be colicky, due to expansion and contraction of the duct. Often the pain occurs after a meal that is high in fat content because of the secondary release of bile from the gallbladder. The right subcostal region may be tender because of abdominal muscle spasms. Patients may experience extreme pain as the epithelium in the gallbladder erodes away. Sympathetic stimulation because of the pain may cause pale, cool, clammy skin. If peritonitis occurs, the skin may be warm due to increased blood flow to the inflamed peritoneum. Nausea and vomiting are common, due to cystic duct spasm. Many patients will have tenderness under the right costal margin referred to as a positive Murphy's sign. ■

Key Note

The liver is the center for metabolic regulation in the body. It also produces bile that is stored in the gallbladder and ejected into the duodenum under stimulation of CCK. Bile is essential for the efficient digestion of lipids; it breaks down large lipid droplets so that the individual lipid molecules can be attacked by digestive enzymes.

CONCEPT CHECK QUESTIONS

1. A narrowing of the ileocecal valve would hamper movement of materials between what two organs?
2. The digestion of which nutrient would be most impaired by damage to the exocrine pancreas?
3. How would a decrease in the amount of bile salts in bile affect the digestion and absorption of fat?

Answers begin on p. 792.

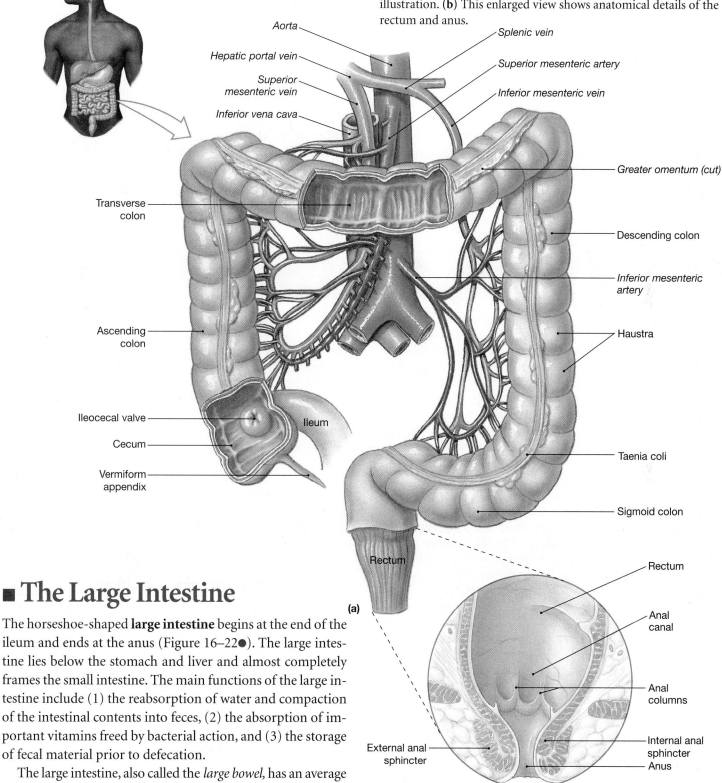

● **FIGURE 16–22 The Large Intestine.** (**a**) The gross anatomy and circulatory supply of the large intestine are presented in this illustration. (**b**) This enlarged view shows anatomical details of the rectum and anus.

Aorta

Hepatic portal vein

Superior mesenteric vein

Inferior vena cava

Splenic vein

Superior mesenteric artery

Inferior mesenteric vein

Greater omentum (cut)

Transverse colon

Descending colon

Inferior mesenteric artery

Ascending colon

Haustra

Ileum

Ileocecal valve

Cecum

Taenia coli

Vermiform appendix

Sigmoid colon

Rectum

(a)

Rectum

Anal canal

Anal columns

External anal sphincter

Internal anal sphincter

Anus

(b)

■ The Large Intestine

The horseshoe-shaped **large intestine** begins at the end of the ileum and ends at the anus (Figure 16–22●). The large intestine lies below the stomach and liver and almost completely frames the small intestine. The main functions of the large intestine include (1) the reabsorption of water and compaction of the intestinal contents into feces, (2) the absorption of important vitamins freed by bacterial action, and (3) the storage of fecal material prior to defecation.

The large intestine, also called the *large bowel,* has an average length of approximately 1.5 m (5 ft) and a width of 7.5 cm (3 in.). It can be divided into three parts: (1) the pouch-like *cecum,* which is the first portion; (2) the *colon,* which is the largest portion; and (3) the *rectum,* which is the last 15 cm (6 in.) of the large intestine and the end of the digestive tract.

The Cecum

Material that arrives from the ileum first enters an expanded pouch, the **cecum** (SĒ-kum), where compaction begins. A muscular sphincter, the **ileocecal** (il-ē-ō-SĒ-kal) **valve,** guards the connection between the ileum and the cecum. The slender, hollow **appendix,** or *vermiform* (*vermis,* worm) *appendix*, attaches to the cecum along its posteromedial surface. The appendix is generally about 9 cm (3.5 in.) long, but its size and shape are quite variable. The walls of the appendix are dominated by lymphoid nodules, and it functions primarily as an organ of the lymphatic system. Inflammation of the appendix is known as *appendicitis.*

The Colon

The **colon** has a larger diameter and a thinner wall than the small intestine. The most striking external feature of the colon is the presence of pouches, or **haustra** (HAWS-truh; singular, *haustrum*), that permit considerable distension and elongation (Figure 16–22a). Three longitudinal bands of smooth muscle—the **taeniae coli** (TĒ-nē-ē KŌ-lē)—run along the outer surface of the colon just beneath the serosa. Muscle tone within these bands creates the haustra.

The colon can be divided into four segments. The **ascending colon** begins at the ileocecal valve. It ascends along the right side of the peritoneal cavity until it reaches the infe-

---

Clinical Note
APPENDICITIS

Appendicitis is an inflammation of the vermiform appendix, which is located at the junction of the small intestine and the large intestine (ileocecal junction). Appendicitis occurs in approximately 10 to 20 percent of the U.S. population. The maximum incidence occurs in the second and third decades of life and is relatively rare at the extremes of age.

The most common cause of acute appendicitis is obstruction of the lumen of the appendix, usually by fecal material (fecalith). The shape and location of the appendix makes it particularly vulnerable to obstruction by feces or other material, such as food particles or tumor. This inflames the lymphoid tissue and often leads to bacterial infection that subsequently ulcerates the mucosa. The inflammation causes the appendix's internal diameter to expand, which can block blood flow through the appendicular artery and cause thrombosis. With its blood supply cut off, the appendix becomes ischemic, and infarction, tissue necrosis, and gangrene follow. At this point, the walls of the appendix weaken to the point of rupture, spilling the appendiceal contents into the peritoneal cavity. Rupture of the appendix can lead to peritonitis and systemic infection.

The signs and symptoms of appendicitis can vary. Older patients and diabetics tend to have less classical symptoms. Initially, the patient with appendicitis will develop diffuse, colicky abdominal pain, usually associated with nausea and vomiting and low-grade fever. Often the pain is located in the periumbilical region. As the illness progresses, the pain localizes to the right lower quadrant. A common location of appendicitis pain is McBurney's point. McBurney's point is 1.5 to 2 inches above the anterior iliac crest along a direct line from the anterior iliac crest to the umbilicus (Figure 16–23●). Once the appendix ruptures, the pain becomes diffuse due to the development of peritonitis. Occasionally, the appendix can be affixed to the posterior aspect of the large intestine (cecum). This condition, referred to as retrocecal appendicitis, often causes low-back pain or flank pain instead of pain over McBurney's point. Retrocecal appendicitis can be more difficult to diagnose.

The diagnosis of appendicitis can often be made based on the history and physical examination. Usually, the patient will have an elevated white blood cell count. In uncertain cases, ultrasound examination of the abdomen can aid in diagnosis. Normally, the appendix cannot be seen on ultrasound examination. If seen, then it is highly suggestive of acute appendicitis. Computed tomography (CT) of the abdomen is very helpful in confirming the diagnosis of acute appendicitis.

Treatment of appendicitis is surgical removal of the appendix (appendectomy). Today, most appendectomies can be performed using a laparoscope that markedly decreases pain and recovery time. Surgery for appendicitis should be prompt to prevent rupture of the appendix. If appendiceal rupture occurs, the patient often develops bacterial peritonitis. This usually requires a prolonged hospital stay and intravenous antibiotics. ■

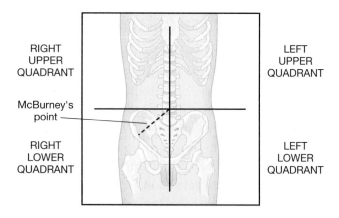

● **FIGURE 16–23 McBurney's Point.** McBurney's point is located along a line approximately $1\frac{1}{2}$ to 2 inches above the right anterior iliac crest along an imaginary line drawn between the right anterior iliac crest and the umbilicus.

rior margin of the liver. It then turns horizontally, and becomes the **transverse colon.** The transverse colon continues toward the left side, passing below the stomach and following the curve of the body wall. Near the spleen, it turns inferiorly to form the **descending colon.** The descending colon continues along the left side until it curves and forms the S-shaped **sigmoid** (SIG-moyd; *sigmoides,* the Greek letter *S*) **colon.** The sigmoid colon empties into the rectum.

The Rectum

The **rectum** (REK-tum) forms the end of the digestive tract and is an expandable organ for the temporary storage of feces (Figure 16–22b). The last portion of the rectum, the **anal canal,** contains small longitudinal folds called *anal columns.* The distal margins of these columns are joined by transverse folds that mark the boundary between the columnar epithelium of the rectum and a stratified squamous epithelium like that found in the oral cavity. Very close to the **anus,** which is the exit of the anal canal, the epidermis becomes keratinized and identical to that on the skin surface.

The circular muscle layer of the muscularis externa in this region forms the **internal anal sphincter,** the smooth muscle cells of which are not under voluntary control. The **external anal sphincter,** which encircles the anus, consists of skeletal muscle fibers and is under voluntary control.

Clinical Note
HEMORRHOIDS

The *rectum* is drained by the *internal* and *external hemorrhoidal veins.* Swelling of these veins is referred to as *hemorrhoids. Internal hemorrhoids* may be asymptomatic or may cause rectal pain, itching, or bright red bleeding (usually with defecation). *External hemorrhoids* tend to thrombose, especially after lifting, and cause severe rectal pain and pressure. Internal hemorrhoids usually respond to rectal creams and analgesics. External hemorrhoids require drainage of the thrombosed hemorrhoid, which usually provides immediate relief. ■

The Functions of the Large Intestine

The major functions of the large intestine are absorption and preparation of the fecal material for elimination.

Absorption in the Large Intestine

The reabsorption of water is an important function of the large intestine. Although roughly 1500 mL of watery material enters the colon each day, only about 200 mL of feces is

Clinical Note
COLON AND RECTUM CANCERS

Colon and *rectum cancers* are relatively common in both men and women. Approximately 105,000 new colon cancer cases and 40,000 new rectal cancer cases are expected in the U.S. in 2005. It is estimated that about 56,000 deaths from colorectal cancer will also occur the same year. The mortality rate for these cancers remains high, and the best defense appears to be early detection and prompt treatment. The standard screening test—checking the feces for blood—is a simple procedure that can easily be performed on a stool (fecal) sample as part of a routine physical examination. ■

ejected. The remarkable efficiency of digestion can best be appreciated by considering the average composition of feces: 75 percent water, 5 percent bacteria, and the rest a mixture of indigestible materials, small quantities of inorganic matter, and the remains of intestinal epithelial cells.

In addition to reabsorbing water, the large intestine absorbs a variety of other substances. Examples include useful compounds (including bile salts and vitamins), organic waste products (such as bilirubin, derived from the breakdown of hemoglobin), and various toxins generated by bacterial action.

BILE SALTS. Most of the bile salts remaining in the material that reaches the cecum are reabsorbed and transported to the liver for secretion in bile.

VITAMINS. **Vitamins** are organic molecules related to lipids and carbohydrates that are essential to many metabolic reactions. Many enzymes require the binding of an additional ion or molecule, called a *cofactor,* before substrates can also bind. *Coenzymes* are nonprotein molecules that function as cofactors, and many vitamins are essential coenzymes.

Bacteria that reside within the colon generate three vitamins that supplement our dietary supply:

- *Vitamin K,* a fat-soluble vitamin needed by the liver to synthesize four clotting factors, including *prothrombin.*
- *Biotin,* a water-soluble vitamin important in glucose metabolism.
- *Vitamin B$_5$* (pantothenic acid), a water-soluble vitamin required in the manufacture of steroid hormones and some neurotransmitters.

Vitamin K deficiencies lead to impaired blood clotting. Intestinal bacteria produce roughly half of our daily vitamin K requirements. Deficiencies of biotin or vitamin B$_5$ are extremely rare after infancy because the intestinal bacteria produce enough to make up for any shortage in the diet.

ORGANIC WASTES. The breakdown of heme and its release as bilirubin in the bile was discussed in Chapter 11. ∞ p. 413 Within the large intestine, bacteria convert bilirubin into other products, some of which are absorbed into the bloodstream and excreted in the urine, which produces its yellow color. Others remain in the colon and, upon exposure to oxygen, are further modified into the pigments that give feces a brown color.

TOXINS. Bacterial action breaks down peptides that remain in the feces and generates (1) ammonia, (2) nitrogen-containing compounds that are responsible for the odor of feces, and (3) hydrogen sulfide (H_2S), a gas that produces a "rotten egg" odor. Much of the ammonia and other toxins are absorbed into the hepatic portal circulation and are removed by the liver. The liver processes them into relatively nontoxic compounds that are excreted at the kidneys.

Indigestible carbohydrates are not altered by intestinal enzymes and arrive in the colon intact. These molecules provide a nutrient source for resident bacteria, whose metabolic activities are responsible for intestinal gas, or *flatus*, in the large intestine. Meals that contain large amounts of indigestible carbohydrates (such as beans) stimulate the production of gas by bacteria.

Movements of the Large Intestine

The gastroileal and gastroenteric reflexes move material into the cecum while you eat. Movement from the cecum to the transverse colon is very slow, and allows hours for the reabsorption of water. Movement from the transverse colon through the rest of the large intestine results from powerful peristaltic contractions called *mass movements*, which occur a few times a day. The normal stimulus for mass movements is distension of the stomach and duodenum. In response to commands relayed over the intestinal nerve plexuses, contractions force fecal material into the rectum, and cause the urge to defecate.

Clinical Note
DIVERTICULOSIS

In *diverticulosis* (dī-ver-tik-ū-LŌ-sis), pockets (*diverticula*) form in the mucosa, usually in the sigmoid colon. These get forced outward, probably by the pressures generated during defecation. If the pockets push through weak points in the muscularis externa, they form semi-isolated chambers that are subject to recurrent infection and inflammation. The infections cause pain and occasional bleeding, a condition known as *diverticulitis* (dī-ver-tik-ū-LĪ-tis). Inflammation of other portions of the colon is called *colitis* (ko-LĪ-tis). ∎

Defecation

The rectum is usually empty until a powerful peristaltic contraction forces fecal material out of the sigmoid colon. Distension of the rectal wall then triggers the **defecation reflex,** which involves two positive feedback loops:

1. In the shorter loop, stretch receptors in the rectal walls stimulate a series of increased local peristaltic contractions in the sigmoid colon and rectum. The contractions move feces toward the anus and increase distension of the rectum.
2. The stretch receptors in the rectal walls also stimulate parasympathetic motor neurons in the sacral spinal cord. These neurons stimulate increased peristalsis (mass movements) in the descending colon and sigmoid colon that push feces toward the rectum, which further increases distension there.

The passage of feces through the anal canal requires relaxation of the internal anal sphincter, but when it relaxes, the external sphincter automatically closes. Thus, the actual release

Clinical Note
DIARRHEA AND CONSTIPATION

Diarrhea (dī-a-RĒ-uh) exists when an individual has frequent, watery bowel movements. Diarrhea results when the mucosa of the colon becomes unable to maintain normal levels of absorption, or when the rate of fluid entry into the colon exceeds the colon's maximum reabsorptive capacity. Bacterial, viral, or protozoan infection of the colon or small intestine can cause acute bouts of diarrhea that last several days. Severe diarrhea is life-threatening due to cumulative fluid and ion losses. In *cholera* (KOL-e-ruh), bacteria bound to the intestinal lining release a toxin that stimulates a massive secretion of fluid across the intestinal epithelium. Without treatment, a person with cholera can die of acute dehydration in a matter of hours.

Constipation is infrequent defecation, and generally involves dry, hard feces. Constipation occurs when fecal material is moving through the colon so slowly that excessive water reabsorption occurs. The feces then become extremely compact, difficult to move, and highly abrasive. Inadequate dietary fiber and fluids, coupled with a lack of exercise, are common causes. Constipation can usually be treated by oral administration of stool softeners such as Colace™, laxatives, or *cathartics* (ka-THAR-tiks), which promote defecation. These compounds either promote water movement into the feces, increase fecal mass, or irritate the lining of the colon to stimulate peristalsis. The promotion of peristalsis is one benefit of "high fiber" cereals. Indigestible fiber adds bulk to the feces, which aids moisture retention and stimulates stretch receptors that promote peristalsis. Active movement during exercise also assists in the movement of fecal material through the colon. ∎

Clinical Note
DIVERTICULITIS

Diverticulitis is a relatively common complication of diverticulosis. Diverticulosis is a condition characterized by the presence in the intestine of diverticula, which are small outpouchings of mucosal and submucosal tissue that push through the outermost layer of the intestine, the muscularis. Colonic diverticula are far more common in developed countries such as the U.S. and increase markedly in prevalence with increased age. They are present in more than half of patients over 60 years of age. Diverticulitis is an inflammation of diverticula secondary to infection. Unlike diverticulosis, it is symptomatic; patients will complain of lower left-sided pain (because most diverticula are in the sigmoid colon); exam and testing will show fever and an increased white blood cell count.

The pathogenesis of an acquired diverticulum is twofold. First, stool passes sluggishly through the colon, a condition associated with the relatively low fiber diets common in developed countries. The colon responds with muscle spasms that increase bulk movement by raising the pressure on the contents inside the colon and pushing the fecal material forward. Second, the outermost layer of colon tissue is made up of fibrous bands of muscle wrapped around one another. Among them are muscles called the *taenia coli*. Nerves and blood vessels enter the colon through small openings within the taenia coli. These openings become weakened with age, and the increased pressure of muscle spasms can cause the inner layers of tissue, the mucosa and submucosa, to herniate through the openings, forming diverticula (Figure 16–24●).

These diverticula commonly trap small amounts of fecal material, including sunflower seeds, popcorn fragments, okra seeds, sesame seeds, and others. The entrapped feces may allow bacteria other then the normal flora to grow and cause an infection. The problem is compounded when the diverticula become inflamed, and cause diverticulitis. Complications secondary to diverticulitis include possible hemorrhage or larger perforations of the colon wall through which the infected fecal contents can spill into the peritoneal cavity and cause peritonitis.

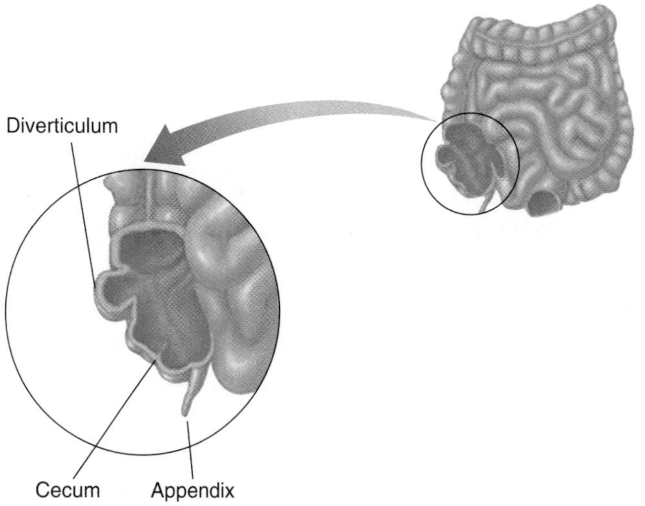

● **FIGURE 16–24 Diverticulum.** Diverticula, which are outpouchings of the wall of the colon, can become infected (diverticulitis) or bleed (diverticulosis).

The most common presentation of diverticulitis is colicky pain associated with a low-grade fever, nausea and vomiting, and tenderness upon palpation. The pain is usually localized to the lower left side because the sigmoid colon is involved in 95 percent of reported cases. Thus diverticulitis is often called *left-sided appendicitis*. If the diverticula begin to bleed significantly, the usual signs and symptoms associated with severe lower GI bleeding may be present: cool, clammy skin, tachycardia, and diaphoresis. Bleeding diverticula can also result in bright red and bloody feces (hematochezia) because of their close proximity to the rectum. Patients may additionally complain of the perception that they cannot empty their rectums, even after defecation. ■

of feces requires conscious effort to open the external sphincter voluntarily. If the conscious commands do not arrive, peristaltic contractions cease until additional rectal expansion triggers the defecation reflex a second time.

In addition to opening the external sphincter, consciously directed activities, such as tensing the abdominal muscles or exhaling while closing the glottis (called the *Valsalva maneuver*), elevate intra-abdominal pressures and help to force fecal material out of the rectum. Such pressures also force blood into the network of veins in the lamina propria and submucosa of the anal canal, which causes them to stretch. Repeated incidents of straining to force defecation can cause the veins to be permanently distended, and produce *hemorrhoids*.

Key Note
The large intestine stores digestive wastes and reduces the volume of the waste by reabsorbing water. Bacterial organisms living in the large intestine are an important source of vitamins, especially vitamin K, biotin, and vitamin B5.

■ Digestion and Absorption

A typical meal contains a mixture of carbohydrates, proteins, lipids, water, electrolytes, and vitamins. The digestive system handles each of these components differently. Large organic molecules must be broken down through digestion before absorption can occur. Water, electrolytes, and vitamins can be absorbed without preliminary processing, but special transport mechanisms may be involved.

The Processing and Absorption of Nutrients

Food contains large organic molecules, many of them insoluble. The digestive system first breaks down the physical structure of the ingested material and then disassembles the component molecules into smaller fragments. This disassembly produces small organic molecules that can be released into the bloodstream. Once absorbed by cells, they are used to generate ATP or to synthesize complex carbohydrates, proteins, and lipids. This section will focus on the mechanics of digestion and absorption; the fates of the compounds inside cells will be considered in Chapter 17.

Foods are usually complex chains of simpler molecules. In a typical dietary carbohydrate, the basic molecules are simple sugars. In a protein, the building blocks are amino acids, and in lipids they are usually fatty acids. Digestive enzymes break the bonds between the component molecules in a process called *hydrolysis.* The hydrolysis of carbohydrates, lipids, and proteins was detailed in Chapter 2. ∞ pp. 42, 43, 45

Digestive enzymes differ in their specific targets. Carbohydrases break the bonds between sugars; lipases separate fatty acids from glycerides; and proteases split the linkages between amino acids. Specific enzymes in each class may be even more selective, and break bonds between specific molecular participants. For example, a given carbohydrase

might ignore all bonds except those that connect two glucose molecules. Figure 16–25● summarizes the chemical events in the digestion of carbohydrates, lipids, and proteins. Table 16–3 reviews the major digestive enzymes and their functions.

Carbohydrate Digestion and Absorption

Carbohydrate digestion begins in the mouth through the action of salivary amylase (Figure 16–25a). Amylase breaks down complex carbohydrates into smaller fragments, which produces a mixture primarily composed of disaccharides (two simple sugars) and trisaccharides (three simple sugars). Salivary amylase continues to digest the starches and glycogen in the meal for an hour or two before stomach acids render it inactive. In the duodenum, the remaining complex carbohydrates are broken down through the action of pancreatic amylase.

Before disaccharides and trisaccharides are absorbed, they are fragmented into simple sugars (monosaccharides) by brush border enzymes found on the surfaces of the intestinal microvilli. ∞ p. 600 The intestinal epithelium then absorbs the resulting simple sugars through carrier-mediated transport mechanisms, such as facilitated diffusion or cotransport. ∞ p. 69 Glucose uptake, for example, occurs through cotransport with sodium ions. (The sodium ions are then ejected by the sodium-potassium exchange pump.)

Simple sugars that enter an intestinal cell diffuse through the cytoplasm and cross the basement membrane by facilitated diffusion to enter the interstitial fluid. They then enter intestinal capillaries for delivery to the hepatic portal vein and liver.

TABLE 16–3 *Digestive Enzymes and Their Functions*

ENZYME	SOURCE	OPTIMAL pH	TARGET	PRODUCTS
CARBOHYDRASES				
Amylase	Salivary glands, pancreas	6.7–7.5	Complex carbohydrates	Disaccharides and trisaccharides
Maltase, sucrase, lactase	Small intestine	7–8	Maltose, sucrose, lactose	Monosaccharides
LIPASES				
Pancreatic lipase	Pancreas	7–8	Triglycerides	Fatty acids and monoglycerides
PROTEASES				
Pepsin	Stomach	1.5–2.0	Proteins, polypeptides	Short polypeptides
Trypsin, chymotrypsin, carboxypeptidase	Pancreas	7–8	Proteins, polypeptides	Short peptide chains
Peptidases	Small intestine	7–8	Dipeptides, tripeptides	Amino acids
NUCLEASES	Pancreas	7–8	Nucleic acids	Nitrogenous bases and simple sugars

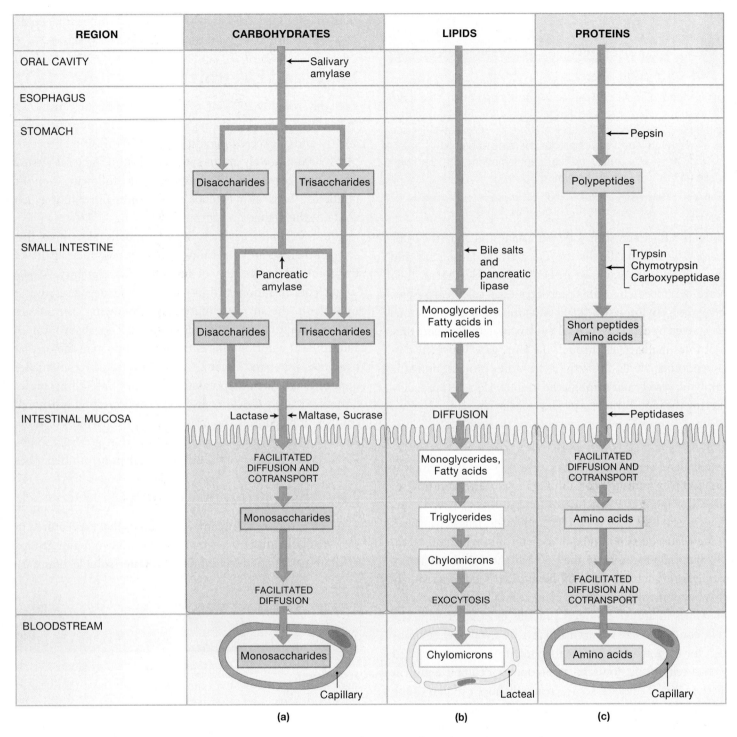

(a) (b) (c)

● **FIGURE 16–25** The Digestion and Absorption of Carbohydrates, Lipids, and Proteins.

Lipid Digestion and Absorption

The structure of fats, or triglycerides, which are the most abundant dietary lipids, was introduced in Chapter 2. ∞ p. 43 A triglyceride molecule consists of three fatty acids attached to a single molecule of glycerol. Triglycerides and other dietary fats are relatively unaffected by conditions in the stomach and enter the duodenum in the form of large lipid droplets.

As noted previously, in the duodenum bile salts emulsify these drops into tiny droplets that can be attacked by pancreatic lipase (Figure 16–25b). This enzyme breaks the triglycerides apart, and the resulting mixture of fatty acids and monoglycerides interacts with bile salts to form small lipid-bile salt complexes called **micelles** (mī-SELZ). When a micelle contacts the intestinal epithelium, the enclosed triglyceride

Clinical Note
LACTOSE INTOLERANCE

Have you ever wondered why there is no cheese in Chinese food? People of Asian and African descent develop a deficiency in the enzyme *lactase* during puberty. Because of this, they inadequately break down *lactose* (mild sugar). People with *lactose intolerance* who ingest a milk product develop abdominal cramps, bloating, distension, and diarrhea. Supplementing the diet with lactase tablets helps minimize symptoms. ■

products diffuse across the cell membrane and enter the cytoplasm. The intestinal cells use the arriving fatty acids and monoglycerides to manufacture new triglycerides that are then coated with proteins. This step creates a soluble complex known as a **chylomicron** (kī-lō-MĪ-kron). The chylomicrons are secreted by exocytosis into the interstitial fluids, where they enter intestinal lacteals through the large gaps between adjacent endothelial cells. From the lacteals they proceed along the lymphatic vessels and through the thoracic duct before finally entering the bloodstream at the left subclavian vein.

Protein Digestion and Absorption

Proteins have very complex structures, so protein digestion is both complex and time consuming. Protein digestion first requires disrupting the structure of food so that proteolytic enzymes can attack individual protein molecules. This step involves mechanical processing in the oral cavity, through mastication, and chemical processing in the stomach, through the action of hydrochloric acid. Exposure of the ingested food to a strongly acid environment breaks down plant cell walls and the connective tissues in animal products and kills most pathogens. The acidic contents of the stomach also provide the proper environment for the activity of pepsin, the proteolytic enzyme secreted by chief cells of the stomach. Pepsin does not complete protein digestion, but it does reduce the relatively huge proteins of the chyme into smaller polypeptide fragments (Figure 16–25c).

When chyme enters the duodenum and the pH has risen, pancreatic proteolytic enzymes can now begin working. Trypsin, chymotrypsin, and carboxypeptidase each break peptide bonds between different amino acids and complete the disassembly of the polypeptide fragments into a mixture of short peptide chains and individual amino acids. *Peptidases,* enzymes on the surfaces of the intestinal microvilli, complete the process by breaking the peptide chains into individual amino acids. The amino acids are absorbed into the intestinal epithelial cells through both facilitated diffusion and cotransport. Carrier proteins at the inner surface of the cells release the absorbed amino

acids into the interstitial fluid. Once within the interstitial fluids, most of the amino acids diffuse into intestinal capillaries.

Water and Electrolyte Absorption

Each day, roughly 2000 mL of water enters the digestive tract in the form of food or drink. Salivary, gastric, intestinal, pancreatic, and bile secretions add about 7000 mL. Out of that total of 9000 mL, only about 150 mL is lost in the fecal wastes. This water conservation occurs passively, following osmotic gradients; water always tends to flow into the solution that contains the higher concentration of solutes.

Intestinal epithelial cells are continually absorbing dissolved nutrients and ions, and these activities gradually lower the solute concentration of the intestinal contents. As the solute concentration within the intestine decreases, water moves into the surrounding tissues, "following" the solutes and maintaining osmotic equilibrium. The absorption of sodium and chloride ions is the most important factor that promotes water movement. Other ions absorbed in smaller quantities are calcium, potassium, magnesium, iodine, bicarbonate, and iron. Calcium absorption occurs under hormonal control, and requires the presence of parathyroid hormone and calcitriol. Regulatory mechanisms that govern the absorption or excretion of the other ions are poorly understood.

The Absorption of Vitamins

Vitamins are essential organic compounds that are required in very small quantities. ∞ p. 615 There are two major groups of vitamins: fat-soluble vitamins and water-soluble vitamins.

Clinical Note
EMERGENCY VITAMINS?

Vitamins are not generally thought of as emergency medications. However, two vitamins play an important role in emergency and critical care: vitamin B₁ and K.

Vitamin B₁, which is commonly referred to as *thiamine,* is important in many of the body's biochemical systems. It is a coenzyme for the first step of Kreb's cycle and plays an important role in several other metabolic processes. Thiamine deficiency is usually seen in chronic alcoholics and can result in altered mental status and other problems. Because of this, administration of thiamine is often a part of the emergency treatment of patients with altered mental status.

Vitamin K is necessary for blood coagulation. Deficiency can occur in chronic liver disease and causes bleeding. Vitamin K administration stimulates the coagulation system, causing blood clotting. In the U.S., all hospital-born babies are prophylactically treated with intramuscular vitamin K. ■

The four **fat-soluble vitamins**—vitamins A, D, E, and K—enter the duodenum in fat droplets, mixed with dietary lipids. The vitamins remain in association with those lipids when micelles form. The fat-soluble vitamins are then absorbed from the micelles along with the products of lipid digestion. Vitamin K is also produced by the action of resident bacteria in the colon. ∞ p. 616

The nine **water-soluble vitamins** include the B vitamins, which are common in milk and meats, and vitamin C, which is found in citrus fruits. All but one, vitamin B_{12}, are easily absorbed by the digestive epithelium. Vitamin B_{12} cannot be absorbed by the intestinal mucosa unless it has been bound to intrinsic factor, a protein secreted by the parietal cells of the stomach. ∞ p. 595 The bacteria that reside in the intestinal tract are an important source of several water-soluble vitamins. In Chapter 17 we will consider the functions of vitamins and associated nutritional problems (see Tables 17–2 and 17–3; pp. 644–645).

Clinical Note
MALABSORPTION SYNDROMES

Malabsorption is a disorder characterized by abnormal nutrient absorption. The disorder may affect the absorption of only one nutrient or many. A genetic inability to manufacture specific enzymes will result in discrete patterns of malabsorption—*lactose intolerance* is a good example.

Difficulties in the absorption of all classes of compounds will result from damage to the accessory glands or the intestinal mucosa. If the accessory organs are functioning normally but their secretions cannot reach the duodenum, the condition is either *biliary obstruction* (bile duct blockage) or *pancreatic obstruction* (pancreatic duct blockage). Alternatively, the ducts may remain open but the glandular cells are damaged and unable to continue normal secretory activities. Two examples, *pancreatitis* and *cirrhosis,* were noted earlier in the chapter.

Even when fully functional enzymes are in the lumen, absorption will not occur if the mucosa cannot function properly. Mucosal damage due to ischemia (an interruption of the blood supply), radiation exposure (such as from radiation therapy or contaminated food), toxic compounds, or infection will affect absorption and will deplete nutrient and fluid reserves as a result. ■

→ CONCEPT CHECK QUESTIONS

1. An increase in which component of a meal would increase the number of chylomicrons in the lacteals?
2. The absorption of which vitamin would be impaired by the removal of the stomach?
3. Why is diarrhea potentially life threatening but constipation is not?

Answers begin on p. 792.

■ Aging and the Digestive System

Essentially normal digestion and absorption occur in elderly individuals. However, many changes in the digestive system parallel age-related changes already described for other systems:

■ *The division rate of epithelial stem cells declines.* Because the digestive epithelium becomes more susceptible to damage by abrasion, acids, or enzymes, peptic ulcers become more likely. In the mouth, esophagus, and anus, the stratified epithelium becomes thinner and more fragile.

■ *Smooth muscle tone decreases.* General motility decreases, and peristaltic contractions are weaker. This change slows the rate of intestinal movement and promotes constipation. Sagging and inflammation of the pouches (haustra) in the walls of the colon can occur. Straining to eliminate compacted fecal material can stress the less resilient walls of blood vessels, which produces hemorrhoids. Problems are not restricted to the lower digestive tract. For example, weakening of muscular sphincters can lead to esophageal reflux and frequent bouts of "heartburn."

■ *The effects of cumulative damage become apparent.* One example is the gradual loss of teeth due to *dental caries* ("cavities") or *gingivitis* (inflammation of the gums). Cumulative damage can involve internal organs as well. Toxins such as alcohol and other injurious chemicals absorbed by the digestive tract are transported to the liver for processing. Liver cells are not immune to these compounds. Chronic exposure can lead to cirrhosis or other types of liver disease.

■ *Cancer rates increase.* Cancers are most common in organs in which stem cells divide to maintain epithelial cell populations. Rates of colon cancer and stomach cancer rise in the elderly; oral and pharyngeal cancers are particularly common in elderly smokers.

■ *Changes in other systems have direct or indirect effects on the digestive system.* For example, a reduction in bone mass and calcium content in the skeleton is associated with erosion of the tooth sockets and eventual tooth loss. The decline in smell and taste sensitivity with age can lead to dietary changes that affect the entire body.

■ Integration with Other Systems

The digestive system is functionally linked to all other systems, and it has extensive anatomical connections to the nervous, cardiovascular, endocrine, and lymphatic systems. Figure 16–26●

The Digestive System in Perspective

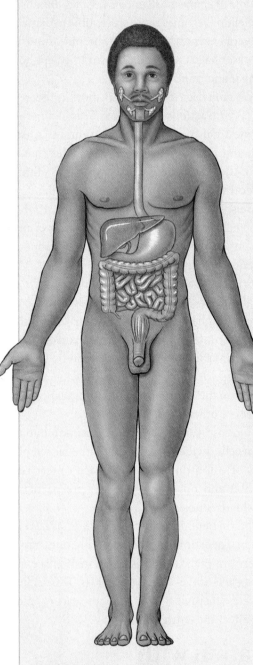

For All Systems

Absorbs organic substrates, vitamins, ions, and water required by all living cells

The Integumentary System

- Provides vitamin D_3 needed for the absorption of calcium and phosphorus
- Provides lipids for storage by adipocytes in subcutaneous layer

The Skeletal System

- Skull, ribs, vertebrae, and pelvic girdle support and protect parts of digestive tract; teeth important in mechanical processing of food
- Absorbs calcium and phosphate ions for incorporation into bone matrix; provides lipids for storage in yellow marrow

The Muscular System

- Protects and supports digestive organs in abdominal cavity; controls entrances and exits of digestive tract
- Liver regulates blood glucose and fatty acid levels, metabolizes lactic acid from active muscles

The Nervous System

- ANS regulates movement and secretion; reflexes coordinate passage of materials along tract; control over skeletal muscles regulates ingestion and defecation; hypothalamic centers control hunger, satiation, and feeding behaviors
- Provides substrates essential for neurotransmitter synthesis

The Endocrine System

- Epinephrine and norepinephrine stimulate constriction of sphincters and depress digestive activity; hormones coordinate activity along tract
- Provides nutrients and substrates to endocrine cells; endocrine cells of pancreas secrete insulin and glucagon; liver produces angiotensinogen

The Cardiovascular System

- Distributes hormones of the digestive tract; carries nutrients, water, and ions from sites of absorption; delivers nutrients and toxins to liver
- Absorbs fluid to maintain normal blood volume; absorbs vitamin K; liver excretes heme (as bilirubin), synthesizes coagulation proteins

The Lymphatic System

- Tonsils and other lymphoid nodules along digestive tract defend against infection and toxins absorbed from the tract; lymphatic vessels carry absorbed lipids to venous system
- Secretions of digestive tract (acids and enzymes) provide nonspecific defense against pathogens

The Respiratory System

- Increased thoracic and abdominal pressure through contraction of respiratory muscles can assist in defecation
- Pressure of digestive organs against the diaphragm can assist in exhalation and limit inhalation

The Urinary System

- Excretes toxins absorbed by the digestive epithelium; excretes some bilirubin produced by liver
- Absorbs water needed to excrete waste products at the kidneys; absorbs ions needed to maintain normal body fluid concentrations

The Reproductive System

- Provides additional nutrients required to support gamete production and (in pregnant women) embryonic and fetal development

●**FIGURE 16–26** Functional Relationships Between the Digestive System and Other Systems.

summarizes the functional relationships between the digestive system and other organ systems.

CONCEPT CHECK QUESTIONS

1. What factor is primarily responsible for the increased susceptibility of the lining of the digestive tract to damage by abrasion, acids, or enzymes as an individual ages?
2. How is the digestive system functionally related to the cardiovascular system?

Answers begin on p. 792.

Clinical Note
ABDOMINAL PAIN

Abdominal pain is one of the most frequent reasons people seek emergency care. The cause of abdominal pain can be difficult to determine. Inflammation, distension, or an interruption in blood supply to an organ causes a pain signal to be transmitted to the spinal cord and brain. Hollow organs, such as the stomach, gallbladder, small intestine, and large intestine tend to cause diffuse, poorly localized pain. However, pain from solid organs, such as the liver, pancreas, and kidneys, tends to be more localized. Finally, problems in other parts of the body, such as pneumonia, can cause abdominal pain. This is called *referred pain* and can further complicate the diagnostic workup. The patient with abdominal pain can be a challenge for emergency personnel. A systematic approach and a good knowledge of the pathophysiology of abdominal disorders will help identify the cause. ■

Chapter Review

Access more review material online at *www.prenhall.com/bledsoe*. There you will find quiz questions, labeling activities, animations, essay questions, and web links.

Key Terms

bile 608
chylomicrons 620
chyme 586
defecation reflex 616
digestion 586
duodenum 598
esophagus 592

gallbladder 611
gastric glands 595
lacteal 601
liver 605
mesentery 598
mucosa 587
pancreas 603

pancreatic juice 603
peristalsis 588
stomach 593
teeth 590
villus/villi 600

Related Clinical Terms

ascites (a-SĪ-tēz) The accumulation of fluid in the peritoneal cavity following its leakage across the serous membranes of the liver and viscera.

achalasia (ak-a-LĀ-zē-uh) A condition that results when a bolus cannot reach the stomach due to constriction of the lower esophageal sphincter.

cholecystitis (kō-lē-sis-TĪ-tis) Inflammation of the gallbladder due to a blockage of the cystic or common bile duct by gallstones.

cholelithiasis (ko-lē-li-THĪ-a-sis) The presence of gallstones in the gallbladder.

cirrhosis (sir-Ō-sis) A disease characterized by the widespread destruction of hepatocytes that results from exposure to drugs (especially alcohol), viral infection, ischemia, or blockage of the hepatic ducts.

colectomy (ko-LEK-to-mē) The removal of all or a portion of the colon.

colonoscope (ko-LON-o-skōp) A fiberoptic device for examining the interior of the colon.

colostomy (ko-LOS-to-mē) The attachment of the cut end of the colon to an opening in the body wall after a colectomy.

esophagitis (ē-sof-a-JĪ-tis) Inflammation of the esophagus.

gallstones Deposits of minerals, bile salts, and cholesterol that form if bile becomes too concentrated.

gastrectomy (gas-TREK-to-mē) The surgical removal of the stomach, generally to treat advanced stomach cancer.

gastroenteritis (gas-trō-en-ter-Ī-tis) A condition characterized by vomiting and diarrhea; results from bacterial toxins, viral infections, or various poisons.

gastroenterology (gas-trō-en-ter-OL-o-jē) The study of the digestive system and its diseases and disorders.

hepatitis (hep-a-TĪ-tis) A virus-induced disease of the liver; the most common forms include *hepatitis A, B,* and *C.*

inflammatory bowel disease (*ulcerative colitis*) A chronic inflammation of the digestive tract; most commonly affects the colon.

irritable bowel syndrome A disorder characterized by diarrhea, constipation, or both alternately. When constipation is the primary problem, this condition may be called a *spastic colon* or *spastic colitis.*

laparoscopy (lap-a-ROS-ko-pē) The use of a flexible fiber-optic instrument introduced through the abdominal wall to permit direct visualization of the viscera, tissue sampling, and limited surgical procedures.

liver biopsy A sample of liver tissue, generally taken by inserting a long needle through the anterior abdominal wall.

perforated ulcer A particularly dangerous ulcer in which gastric acids erode through the wall of the digestive tract, which allows its contents to enter the peritoneal cavity.

periodontal disease A loosening of the teeth within the bony sockets (alveolar sockets)

caused by erosion of the periodontal ligaments by acids produced through bacterial action.

peritonitis (per-i-tō-NĪ-tis) Inflammation of the peritoneal membrane.

polyps (POL-ips) Small mucosal tumors that grow from the intestinal wall.

Summary Outline

1. The digestive system consists of the muscular **digestive tract** and various **accessory organs.**

2. Digestive functions include **ingestion, mechanical processing, digestion, secretion, absorption,** and **excretion.**

AN OVERVIEW OF THE DIGESTIVE TRACT 586

1. The digestive tract includes the oral cavity, pharynx, esophagus, stomach, small intestine, large intestine, rectum, and anus. *(Figure 16–1)*

Histological Organization 587

2. The epithelium and underlying connective tissue, the *lamina propria,* form the **mucosa** (mucous membrane) of the digestive tract. Next, moving outward, are the **submucosa,** the **muscularis externa,** and the *adventitia,* which is a layer of loose connective tissue. In the peritoneal cavity, the muscularis externa is covered by the **serosa,** which is a serous membrane. *(Figure 16–2)*

3. Double sheets of peritoneal membrane called **mesenteries** suspend portions of the digestive tract.

The Movement of Digestive Materials 588

4. The neurons that innervate the smooth muscle of the muscularis externa are not under voluntary control.

5. The muscularis externa propels materials through the digestive tract by means of the contractions of **peristalsis. Segmentation** movements in areas of the small intestine churn digestive materials. *(Figure 16–3)*

THE ORAL CAVITY 589

1. The functions of the **oral cavity** are: (1) sensory analysis of potential foods; (2) mechanical processing using the teeth, tongue, and palatal surfaces; (3) lubrication of food by mixing with mucus and salivary secretions; and (4) digestion by salivary enzymes.

2. The oral cavity, or **buccal cavity,** is lined by oral mucosa. The **hard palate** and **soft palate** form its roof, and the tongue forms its floor. *(Figure 16–4)*

The Tongue 589

3. The primary functions of the **tongue** include (1) mechanical processing, (2) manipulation to assist in chewing and swallowing, and (3) sensory analysis.

Salivary Glands 590

4. The **parotid, sublingual,** and **submandibular salivary glands** discharge their secretions into the oral cavity. Saliva lubricates the mouth, dissolves chemicals, flushes the oral surfaces, and

helps control bacteria. Salivation is usually controlled by the ANS. *(Figure 16–5)*

Teeth 590

5. **Mastication** (chewing) occurs through the contact of the opposing surfaces of the **teeth.** The **periodontal ligament** anchors each tooth in a bony socket. **Dentin** forms the basic structure of a tooth. The **crown** is coated with **enamel,** and the **root** is covered with **cementum.** *(Figure 16–6a)*

6. The 20 primary teeth, or **deciduous teeth,** are replaced by the 32 teeth of the **secondary dentition** during development. *(Figure 16–6b,c)*

THE PHARYNX 592

1. The **pharynx** serves as a common passageway for solid food, liquids, and air. Pharyngeal muscle contractions during swallowing propel the food mass along the esophagus and into the stomach.

THE ESOPHAGUS 592

1. The **esophagus** carries solids and liquids from the pharynx to the stomach through an opening in the diaphragm, which is the *esophageal hiatus.*

Swallowing 592

2. **Deglutition** (swallowing) can be divided into **oral, pharyngeal,** and **esophageal phases.** Swallowing begins with the compaction of a **bolus** and its movement into the pharynx, followed by the elevation of the larynx, reflection of the epiglottis, and closure of the glottis. After opening of the *upper esophageal sphincter,* peristalsis moves the bolus down the esophagus to the *lower esophageal sphincter. (Figure 16–7)*

THE STOMACH 593

1. The **stomach** has four major functions: (1) temporary storage of ingested food, (2) the mechanical breakdown of food, (3) the breakage of chemical bonds by acids and enzymes, and (4) the production of intrinsic factor. **Chyme** forms in the stomach as gastric and salivary secretions are mixed with food.

2. The four regions of the stomach are the **cardia, fundus, body,** and **pylorus.** The **pyloric sphincter** guards the exit from the stomach. In a relaxed state the stomach lining contains numerous **rugae** (ridges and folds). *(Figures 16–8, 16–9)*

The Gastric Wall 595

3. Within the **gastric glands, parietal cells** secrete **intrinsic factor** and hydrochloric acid. **Chief cells** secrete **pepsinogen,** which

acids in the gastric lumen convert to the enzyme **pepsin.** Gastric gland endocrine cells secrete the hormone **gastrin.**

The Regulation of Gastric Activity 596

4. Gastric secretion includes (1) the **cephalic phase,** which prepares the stomach to receive ingested materials; (2) the **gastric phase,** which begins with the arrival of food in the stomach; and (3) the **intestinal phase,** which controls the rate of gastric emptying. *(Figure 16–10)*

Key Note 598

THE SMALL INTESTINE 598

1. The **small intestine** includes the **duodenum,** the **jejunum,** and the **ileum.** The *ileocecal valve,* which is a sphincter, marks the junction between the small and large intestines. *(Figure 16–11)*

The Intestinal Wall 600

2. The intestinal mucosa bears transverse folds called **plicae,** or *plicae circulares,* and small projections called intestinal **villi.** Both structures increase the surface area for absorption. Each villus contains a lymphatic capillary called a **lacteal.** *(Figure 16–12)*

3. Some of the smooth muscle cells in the musularis externa of the small intestine contract periodically, without stimulation, to produce brief localized peristaltic contractions that slowly move materials along the tract. More extensive peristaltic activities are coordinated by the *gastroenteric* and the *gastroileal reflexes.*

Intestinal Secretions 601

4. Intestinal glands secrete **intestinal juice,** mucus, and hormones. Intestinal juice moistens the chyme, helps buffer acids, and dissolves digestive enzymes and the products of digestion.

5. Intestinal hormones include **gastrin, secretin, cholecystokinin (CCK),** and **gastric inhibitory peptide (GIP).** *(Figure 16–13; Table 16–1)*

Digestion in the Small Intestine 602

6. Most of the important digestive and absorptive functions occur in the small intestine. Digestive enzymes and buffers are produced by the pancreas, liver, and gallbladder.

Key Note 602

THE PANCREAS 603

1. The **pancreatic duct** penetrates the wall of the duodenum, where it delivers the secretions of the **pancreas.** *(Figure 16–14a)*

Histological Organization 603

2. Exocrine gland ducts branch repeatedly before ending in the **pancreatic acini** (blind pockets). *(Figure 16–14b)*

3. The pancreas has both an endocrine function (secretes insulin and glucagon into the blood) and an exocrine function (secretes water, ions, and digestive enzymes into the small intestine). The enzymes include **carbohydrases, lipases, nucleases,** and **proteases.**

The Control of Pancreatic Secretion 603

4. Pancreatic exocrine cells produce a watery **pancreatic juice** in response to hormonal instructions from the duodenum. When chyme arrives in the small intestine, secretin and CCK are released.

5. The release of secretin triggers the pancreatic production of a fluid that contains buffers (primarily sodium bicarbonate) that help bring the pH of the chyme under control. CCK stimulates the pancreas to produce and secrete **pancreatic amylase, pancreatic lipase,** nucleases, and several proteolytic enzymes—notably, **trypsin, chymotrypsin,** and **carboxypeptidase.**

Key Note 605

THE LIVER 605

1. The **liver,** which is the largest visceral organ in the body, performs over 200 known functions.

Anatomy of the Liver 605

2. The liver is made up of four unequally sized lobes: the **left, right, caudate,** and **quadrate lobes.** *(Figure 16–15)*

3. The **liver lobule** is the organ's basic functional unit. Blood is supplied to the lobules by branches of the hepatic artery and hepatic portal vein. Within the lobules, blood flows past *hepatocytes* through *sinusoids* to the **central vein. Bile canaliculi** carry bile away from the central vein and toward bile ducts. *(Figure 16–16)*

4. The bile ducts from each lobule unite to form the **common hepatic duct,** which meets the **cystic duct** to form the **common bile duct,** which empties into the duodenum. *(Figure 16–16a)*

Liver Functions 609

5. The liver performs several major functions, including metabolic regulation, hematological regulation, and the production of **bile.** *(Figures 16–17 through 16–19; Table 16–2)*

THE GALLBLADDER 611

1. The **gallbladder** stores and concentrates bile for release into the duodenum. Relaxation of the **hepatopancreatic sphincter** by cholecystokinin (CCK) permits bile to enter the duodenum. *(Figures 16–20, 16–21)*

Key Note 612

THE LARGE INTESTINE 613

1. The main functions of the **large intestine** are to (1) reabsorb water and compact the feces, (2) absorb vitamins made by bacteria, and (3) store fecal material prior to defecation. The large intestine has three parts: the cecum, the colon, and the rectum. *(Figure 16–22a)*

The Cecum 614

2. The **cecum** collects and stores material from the ileum and begins the process of compaction. The **vermiform appendix** is attached to the cecum. *(Figure 16–23)*

The Colon 614

3. The **colon** has a larger diameter and a thinner wall than the small intestine. It bears **haustra** (pouches) and **taeniae coli** (longitudinal bands of muscle).

The Rectum 615

4. The **rectum** terminates in the **anal canal,** which leads to the **anus.** *(Figure 16–22b)*

The Functions of the Large Intestine 615

5. The large intestine reabsorbs water and other substances, such as *vitamins, organic wastes, bile salts,* and *toxins.* Bacteria that reside in the large intestine are responsible for intestinal gas, or *flatus.*

6. Distension of the stomach and duodenum stimulates peristalsis, or *mass movements,* of feces from the colon into the rectum. Muscular sphincters control the passage of fecal material to the anus. Distension of the rectal wall triggers the *defecation reflex.* Under normal circumstances, the release of feces cannot occur unless the **external anal sphincter** is voluntarily relaxed. *(Figure 16–24)*

Key Note 617

DIGESTION AND ABSORPTION 617

The Processing and Absorption of Nutrients 618

1. The digestive system first breaks down the physical structure of the ingested material, and then digestive enzymes break the component molecules into smaller fragments through a process called *hydrolysis.* *(Table 16–3)*

2. Amylases break down complex carbohydrates into *disaccharides* and *trisaccharides.* Enzymes at the epithelial surface break these molecules into *monosaccharides* that are absorbed by the intestinal epithelium through facilitated diffusion or cotransport. *(Figure 16–25a)*

3. *Triglycerides* are emulsified into large lipid droplets. The resulting fatty acids and other lipids interact with bile salts to form **micelles** from which they diffuse across the intestinal epithe-lium. The intestinal cells absorb fatty acids and synthesize new triglycerides. These are packaged in **chylomicrons,** which are released into the interstitial fluid and transported to the venous system by lymphatics. *(Figure 16–25b)*

4. Protein digestion involves low pH and the enzyme pepsin in the stomach and various pancreatic proteases in the small intestine. Peptidases liberate amino acids that are absorbed by the intestinal epithelium and released into the interstitial fluids. *(Figure 16–25c)*

Water and Electrolyte Absorption 620

5. About 2 liters of water are ingested each day, and digestive secretions provide another 7 liters. All but about 150 mL is reabsorbed through osmosis.

6. Various processes are responsible for the movement of ions (such as sodium, calcium, chloride, and bicarbonate).

The Absorption of Vitamins 620

7. The **fat-soluble vitamins** are enclosed within fat droplets and are absorbed with the products of lipid digestion. The **water-soluble vitamins** (except B_{12}) diffuse easily across the digestive epithelium.

AGING AND THE DIGESTIVE SYSTEM 621

1. Age-related digestive system changes include a thinner and more fragile epithelium due to a reduction in epithelial stem cell division and weaker peristaltic contractions as smooth muscle tone decreases.

INTEGRATION WITH OTHER SYSTEMS 621

1. The digestive system has extensive anatomical connections to the nervous, cardiovascular, endocrine, and lymphatic systems. *(Figure 16–26)*

Review Questions

Level 1: Reviewing Facts and Terms

Match each item in column A with the most closely related item in column B. Place letters for answers in the spaces provided.

COLUMN A
____ 1. pyloric sphincter
____ 2. liver cells
____ 3. mucosa
____ 4. mesentery
____ 5. chief cells
____ 6. palate
____ 7. parietal cells
____ 8. parasympathetic stimulation
____ 9. sympathetic stimulation
____ 10. peristalsis
____ 11. bile salts
____ 12. salivary amylase

COLUMN B
a. serous membrane sheet
b. moves materials along digestive tract
c. regulates flow of chyme
d. increases muscular activity of digestive tract
e. starch digestion
f. inhibits muscular activity of digestive tract
g. inner lining of digestive tract
h. roof of oral cavity
i. pepsinogen
j. hydrochloric acid
k. hepatocytes
l. emulsification of fats

13. The enzymatic breakdown of large molecules into their basic building blocks is called:
    (a) absorption.
    (b) secretion.
    (c) mechanical digestion.
    (d) chemical digestion.

14. The activities of the digestive system are regulated by:
    (a) hormonal mechanisms.
    (b) local mechanisms.
    (c) neural mechanisms.
    (d) a, b, and c are correct.

15. The layer of the peritoneum that lines the inner surfaces of the body wall is the:
    (a) visceral peritoneum.
    (b) parietal peritoneum.
    (c) greater omentum.
    (d) lesser omentum.

16. Protein digestion in the stomach results primarily from secretions released by:
    (a) hepatocytes.
    (b) parietal cells.
    (c) chief cells.
    (d) goblet cells.

17. The part of the gastrointestinal tract that plays the primary role in the digestion and absorption of nutrients is the:
    (a) large intestine.
    (b) small intestine.
    (c) stomach.
    (d) cecum and colon.

18. The duodenal hormone that stimulates the production and secretion of pancreatic enzymes is:
    (a) pepsinogen.
    (b) gastrin.
    (c) secretin.
    (d) cholecystokinin.

19. The essential physiological service(s) provided by the liver is (are):
    (a) metabolic regulation.
    (b) hematological regulation.
    (c) bile production.
    (d) a, b, and c are correct.

20. Bile release from the gallbladder into the duodenum occurs only under the stimulation of:
    (a) cholecystokinin.
    (b) secretin.
    (c) gastrin.
    (d) pepsinogen.

21. The major function(s) of the large intestine is (are):
    (a) reabsorption of water and compaction of feces.
    (b) absorption of vitamins produced by bacterial action.
    (c) storage of fecal material prior to defecation.
    (d) a, b, and c are correct.

22. The part of the colon that empties into the rectum is the:
    (a) ascending colon.
    (b) descending colon.
    (c) transverse colon.
    (d) sigmoid colon.

23. What are the primary digestive functions?

24. What is the purpose of the transverse or longitudinal folds in the mucosa of the digestive tract?

25. Name and describe the layers of the digestive tract, proceeding from the innermost layer to the outermost layer.

26. What are the four primary functions of the oral (buccal) cavity?

27. What specific function does each of the four types of teeth perform in the oral cavity?

28. What three segments of the small intestine are involved in the digestion and absorption of food?

29. What are the primary functions of the pancreas, liver, and gallbladder in the digestive process?

30. What are the three major functions of the large intestine?

31. What five age-related changes occur in the digestive system?

Level 2: Reviewing Concepts

32. If the lingual frenulum is too restrictive, an individual:
    (a) has difficulty tasting food.
    (b) cannot swallow properly.
    (c) cannot control movements of the tongue.
    (d) cannot eat or speak normally.

33. The gastric phase of secretion is initiated by:
    (a) distension of the stomach.
    (b) an increase in the pH of the gastric contents.
    (c) the presence of undigested materials in the stomach.
    (d) a, b, and c are correct.

34. A drop in pH to 4.0 in the duodenum stimulates the secretion of:
    (a) secretin.
    (b) cholecystokinin.
    (c) gastrin.
    (d) a, b, and c are correct.

35. Describe how the action and outcome of peristalsis differ from those of segmentation.

36. How does the stomach promote and assist in the digestive process?

37. What changes in gastric function occur during the three phases of gastric secretion?

Level 3: Critical Thinking and Clinical Applications

38. Some patients with gallstones develop pancreatitis. How could this occur?

39. Barb suffers from Crohn's disease, which is a regional inflammation of the intestine thought to have some genetic basis, although the actual cause remains unknown. When the disease flares up, she experiences abdominal pain, weight loss, and anemia. Which part(s) of the intestine is (are) probably involved, and what is the cause of her signs and symptoms?

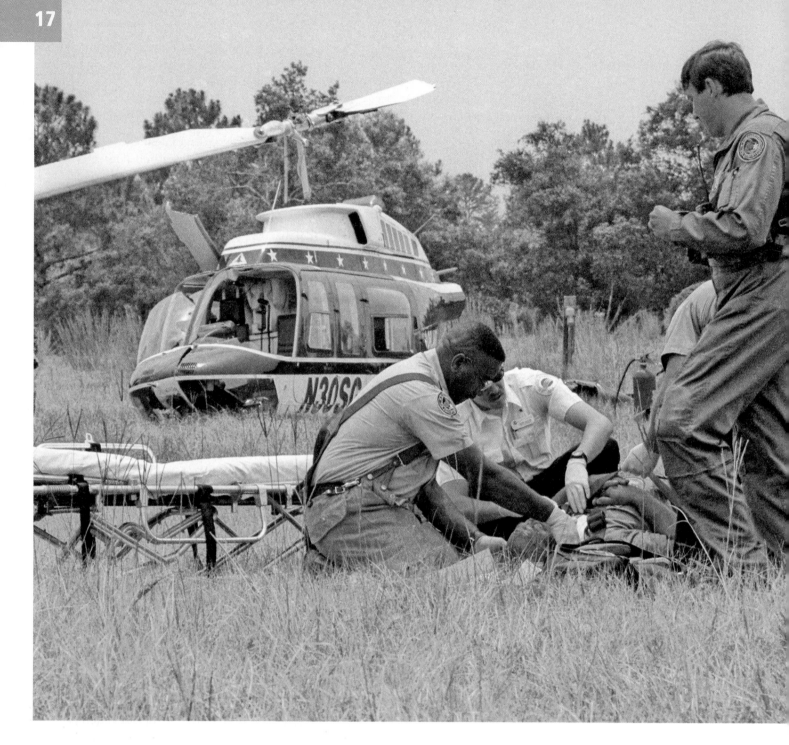

17 Nutrition and Metabolism

ACCESS TO THE EMERGENCY scene is a common problem for rural EMS providers. Likewise, removal of the patient from the scene to a hospital can be equally problematic. The use of medical helicopters can provide rapid access and removal of critically ill or critically injured patients. However, it is important to always weigh the potential benefits against any possible risks when considering rescue and transportation options. Helicopters are complex machines and the demands of EMS sometimes can result in a crash as shown here.

Chapter Outline

Chapter Objectives

1. Define metabolism and explain why cells need to synthesize new organic structures. (pp. 630–631)
2. Describe the basic steps in glycolysis, the TCA cycle, and the electron transport system. (pp. 631–635)
3. Describe the pathways involved in lipid metabolism. (pp. 636–638)
4. Discuss protein metabolism and the use of proteins as an energy source. (pp. 638, 640)
5. Discuss nucleic acid metabolism. (pp. 640–641)
6. Explain what constitutes a balanced diet and why it is important. (pp. 642–644)
7. Discuss the functions of vitamins, minerals, and other important nutrients. (pp. 644–648)
8. Describe the significance of the caloric value of foods. (p. 648)
9. Define metabolic rate and discuss the factors involved in determining an individual's metabolic rate. (pp. 648–649)
10. Discuss the homeostatic mechanisms that maintain a constant body temperature. (pp. 649–651)
11. Describe the age-related changes in nutritional requirements. (pp. 651–652)

Vocabulary Development

anabole a building up; *anabolism*
genesis an origin; *thermogenesis*
glykus sweet; *glycolysis*

katabole a throwing down; *catabolism*
lipos fat; *lipogenesis*
lysis breakdown; *glycolysis*

neo- new; *gluconeogenesis*
therme heat; *thermogenesis*
vita life; *vitamin*

CELLS ARE CHEMICAL FACTORIES that break down organic molecules to obtain energy, usually in the form of ATP. Chemical reactions within mitochondria provide most of the energy needed by a typical cell for its varied activities. ∞ p. 80 To carry out their energy-generating processes, cells in the human body must also obtain oxygen and nutrients. Whereas oxygen is absorbed at the lungs, **nutrients**—essential substances such as water, vitamins, ions, carbohydrates, lipids, and proteins—are normally obtained from the diet by absorption at the digestive tract. The cardiovascular system distributes oxygen and nutrients to cells throughout the body.

The ATP energy produced in a cell supports growth, cell division, contraction, secretion, and all the other special functions that vary from cell to cell and tissue to tissue. Because each tissue type contains different populations of cells, the energy and nutrient requirements of any two tissues (such as loose connective tissue and cardiac muscle) are typically quite different. When cells, tissues, and organs change their patterns or levels of activity, the body's metabolic needs change. Thus, our nutrient requirements can vary from moment to moment (resting versus active), hour to hour (asleep versus awake), and year to year (child versus adult).

When organic nutrients such as carbohydrates or lipids are abundant, the body's energy reserves are built up. Different tissues and organs are specialized to store excess nutrients; the storage of lipids in adipose tissue is one familiar example. These reserves can then be called on when the diet cannot provide a sufficient quantity or quality of nutrients. The endocrine `system, with the assistance of the nervous system, adjusts and coordinates the metabolic activities of the body's tissues and controls the storage and release of nutrient reserves.

The absorption of nutrients from food is called *nutrition*. The mechanisms involved in absorption of nutrients through the lining of the digestive tract were detailed in Chapter 16. ∞ p. 620 This chapter considers what happens to nutrients once they are inside the body.

■ Cellular Metabolism

The term **metabolism** refers to all the chemical reactions that occur in the body. ∞ p. 34 *Cellular metabolism*—chemical reactions within cells—provides the energy needed to maintain homeostasis and to perform essential functions. Figure 17–1● provides an overview of the processes involved in cellular metabolism. Amino acids, lipids, and simple sugars cross the cell membrane and join the other nutrients already in the cytoplasm. All of the cell's metabolic operations rely on this *nutrient pool.*

The breakdown of organic molecules is called *catabolism,* which is a process that releases energy for synthesizing ATP or other high-energy compounds. ∞ p. 36 Catabolism proceeds in a series of steps. In general, the initial steps occur in the cytosol, where enzymes break down large organic molecules into smaller fragments. Carbohydrates are broken down into short carbon chains, triglycerides are split into fatty acids and glycerol, and proteins are broken down into individual amino acids.

Relatively little ATP is formed during these initial steps. However, the simple molecules produced can be absorbed and processed by mitochondria, and the mitochondrial steps release significant amounts of energy. As mitochondrial enzymes break the covalent bonds that hold these molecules together, they cap-

ture roughly 40 percent of the energy released. The captured energy is used to convert ADP to ATP, and the rest escapes as heat that warms the interior of the cell and the surrounding tissues.

Anabolism, which is the synthesis of new organic molecules, involves the formation of new chemical bonds. ∞ p. 36 The ATP produced by mitochondrial processes provides energy to support anabolism and other cell functions. Those additional functions, including ciliary or cell movement, contraction, active transport, and cell division, vary from one cell to another. For example, muscle fibers need ATP to provide energy for contraction, whereas gland cells need ATP to synthesize and transport their secretions.

Cells synthesize new organic components for four basic reasons:

1. *To perform structural maintenance and repairs.* All cells must expend energy for ongoing maintenance and repairs because most structures in the cell are temporary, not permanent. The continuous removal and replacement of these structures are part of the process of **metabolic turnover.**
2. *To support growth.* Cells preparing to divide enlarge and synthesize extra proteins and organelles.
3. *To produce secretions.* Secretory cells must synthesize their products and deliver them to the interstitial fluid.

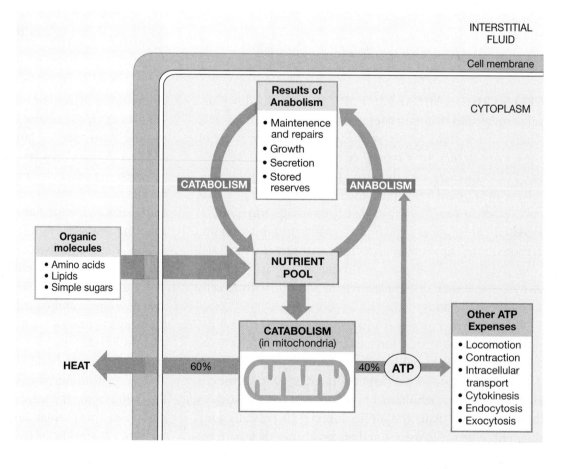

● **FIGURE 17–1 Cellular Metabolism.** Cells obtain organic molecules from the extracellular fluid and break them down to obtain ATP. Only about 40 percent of the energy released through catabolism is captured in ATP; the rest is radiated as heat. The ATP generated by catabolism provides energy for all vital cellular activities, including anabolism.

4. *To build nutrient reserves.* Most cells "prepare for a rainy day"—some emergency, an interval of extreme activity, or a time when the nutrient supply in the bloodstream is inadequate—by storing nutrients in a form that can be mobilized as needed. For example, muscle cells store glucose in the form of glycogen, adipocytes store triglycerides, and liver cells store both molecules.

The nutrient pool is the source of organic molecules for both catabolism and anabolism (see Figure 17–1). As you might expect, cells tend to conserve materials needed to build new compounds and tend to break down the rest. Cells continuously replace membranes, organelles, enzymes, and structural proteins. These anabolic activities require more amino acids than lipids and few carbohydrates. Catabolic activities, however, tend to process these organic molecules in the reverse order. In general, when a cell with excess carbohydrates, lipids, and amino acids needs energy, it will break down carbohydrates first. Lipids are the second choice as an energy source, and amino acids are seldom broken down if other energy sources are available.

Activities within mitochondria provide the energy that supports cellular operations. In effect, the cell feeds its mitochondria from its nutrient pool, and in return the cell gets the ATP it needs. However, only specific organic molecules are suitable to the mitochondria for processing and energy production. Thus, chemical reactions in the cytoplasm take organic nutrients in the pool and break them into smaller carbon chains that the mitochondria can use (Figure 17–2●). Chemical reactions within the mitochondria then break down the fragments further, and generate carbon dioxide, water, and ATP. This mitochondrial activity involves two pathways: the TCA cycle and the electron transport system. In the next section we examine the important catabolic and anabolic reactions that occur in our cells.

Key Note

There is an energy cost to staying alive, even at rest. All cells must expend ATP to perform routine maintenance, to remove and replace intracellular and extracellular structures and components. In addition, cells must spend additional energy doing other vital functions, such as growth, secretion, and contraction.

Carbohydrate Metabolism

Carbohydrates, most familiar to us as sugars and starches, are important sources of energy. Most cells generate ATP and other high-energy com-

pounds by breaking down carbohydrates, especially glucose. The complete reaction sequence can be summarized as:

$$C_6H_{12}O_6 + 6\,O_2 \rightarrow 6\,CO_2 + 6\,H_2O$$

glucose   oxygen   carbon dioxide   water

The breakdown occurs in a series of small steps, several of which release enough energy to support the conversion of ADP to ATP. During the complete catabolism of glucose, a typical cell gains 36 ATP molecules.

Although most ATP production occurs inside mitochondria, the first steps take place in the cytosol as a sequence of reactions called *glycolysis*. The steps were outlined in Chapter 7; because those steps do not require oxygen, they are said to be *anaerobic*. ∞ p. 223 The subsequent reactions, which occur within mitochondria, consume oxygen and are thus *aerobic*. The mitochondrial activity responsible for ATP production is called **aerobic metabolism,** or *cellular respiration.*

Glycolysis

Glycolysis (glī-KOL-i-sis; *glykus,* sweet + *lysis,* breakdown) is the breakdown of glucose to *pyruvic acid.* In this process, a se-

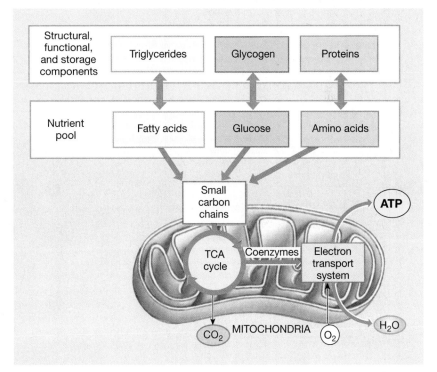

● **FIGURE 17–2 Nutrient Use in Cellular Metabolism.** Cells use molecules in the nutrient pool to build up reserves and to manufacture cellular structures. Catabolism within mitochondria provides the ATP needed to sustain cell functions. Mitochondria absorb small carbon chains produced by the breakdown of fatty acids, glucose, and amino acids from the nutrient pool. The small carbon chains are broken down further by means of the tricarboxylic acid (TCA) cycle and the electron transport system.

ries of enzymatic steps breaks the six-carbon glucose molecule ($C_6H_{12}O_6$) into two three-carbon molecules of pyruvic acid (CH_3—CO—COOH). Glycolysis requires (1) glucose molecules, (2) appropriate cytoplasmic enzymes, (3) ATP and ADP, and (4) **NAD** (**n**icotinamide **a**denine **d**inucleotide), a coenzyme that removes hydrogen atoms. *Coenzymes* are organic molecules, usually derived from vitamins, that must be present for an enzymatic reaction to occur. If the cell lacks any of these four participants, glycolysis cannot occur.

The basic steps of glycolysis are summarized in Figure 17–3●. This reaction sequence yields a net gain of two ATP molecules for each glucose molecule converted to two pyruvic acid molecules. A few highly specialized cells, such as red blood cells, lack mitochondria and derive all of their ATP by glycolysis. Skeletal muscle fibers rely on glycolysis for energy production during periods of active contraction, and most cells can survive brief periods of hypoxia (low oxygen levels) by using the ATP provided by glycolysis alone. When oxygen is readily available, however, mitochondrial activity provides most of the ATP required by body cells.

Energy Production Within Mitochondria

Even though glycolysis yields an immediate net gain of two ATP molecules for the cell, a great deal of additional energy is still stored in the chemical bonds of pyruvic acid. The cell's ability to capture that energy depends on the availability of oxygen. If oxygen supplies are adequate, mitochondria will absorb the pyruvic acid molecules and break them down completely. The hydrogen atoms of pyruvic acid are removed by coenzymes and are ultimately the source of most of the

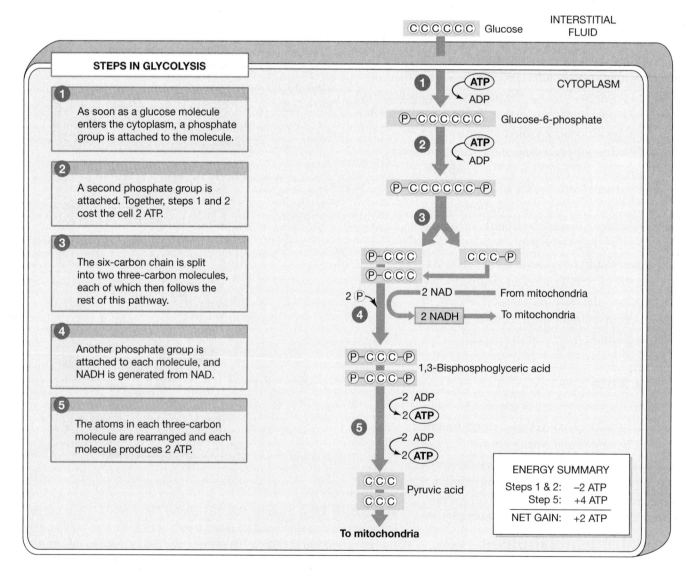

● **FIGURE 17–3 Glycolysis.** Within a cell's cytoplasm, glycolysis breaks down a six-carbon glucose molecule into two three-carbon pyruvic acid molecules. This process involves a series of enzymatic steps. A net gain of two ATPs results for each glucose molecule converted to pyruvic acid.

cell's energy gain. The carbon and oxygen atoms are removed and released as carbon dioxide.

Two membranes surround each mitochondrion. ∞ p. 80 The outer membrane is permeable to pyruvic acid, and a carrier protein in the inner membrane transports the pyruvic acid into the mitochondrial matrix. Once inside the mitochondrion, each pyruvic acid molecule participates in a reaction that leads to a sequence of enzymatic reactions called the **tricarboxylic** (trī-kar-bok-SIL-ik) **acid (TCA) cycle** (Figure 17–4●). The function of the TCA cycle is to remove hydrogen atoms from organic molecules and transfer them to coenzymes in the electron transport system.

THE TCA CYCLE. Within a mitochondrion, a complex reaction that involves pyruvic acid, NAD, and another coenzyme called *coenzyme A* (or *CoA*) yields one molecule of carbon dioxide, one molecule of NADH, and one molecule of **acetyl-CoA** (as-Ē-til-KŌ-ā). Acetyl-CoA consists of a two-carbon *acetyl group* (CH_3CO) bound to coenzyme A. When the acetyl group is transferred from CoA to a four-carbon molecule, a six-carbon molecule called *citric acid* is produced. The formation of citric acid is the basis for another name for these reactions: the *citric acid cycle*. (This sequence is also known as the *Krebs cycle* in honor of Hans Krebs, the biochemist who described these reactions in 1937.)

As citric acid is produced, CoA is released intact to bind with another acetyl group. A complete revolution of the TCA cycle removes the two added carbon atoms, and regenerates the four-carbon chain. (This is why this reaction sequence is called a *cycle.*) The two removed carbon atoms generate two molecules of carbon dioxide (CO_2), which is a metabolic waste product. The hydrogen atoms of the acetyl group are removed by coenzymes.

The only immediate energy benefit of one revolution of the TCA cycle is the formation of a single molecule of *GTP (guanosine triphosphate),* which is a high-energy compound readily converted into ATP. The real value of the TCA cycle can be seen by following the fate of the hydrogen atoms that are removed by the coenzymes NAD and **FAD** (**f**lavine **a**denine **d**inucleotide). The two coenzymes form NADH and $FADH_2$, and transfer the hydrogen atoms to the *electron transport system* (see Figure 17–4).

THE ELECTRON TRANSPORT SYSTEM. The **electron transport system (ETS)** is embedded in the inner mitochondrial membrane (Figure 17–5●). The ETS consists of an electron transport chain comprised of a series of protein-pigment complexes called *cytochromes.* The ETS does not produce ATP directly. Instead, it creates the conditions necessary for ATP production.

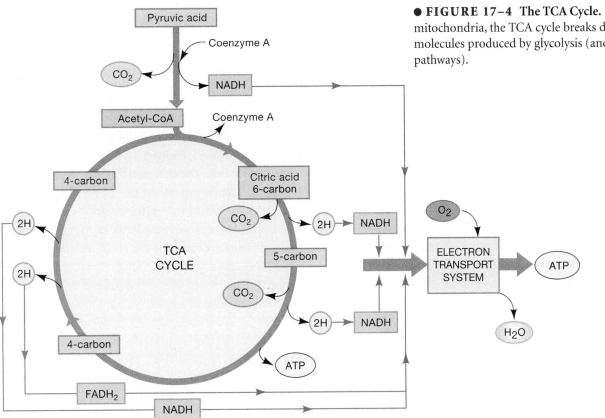

●**FIGURE 17–4 The TCA Cycle.** Within mitochondria, the TCA cycle breaks down pyruvic acid molecules produced by glycolysis (and other catabolic pathways).

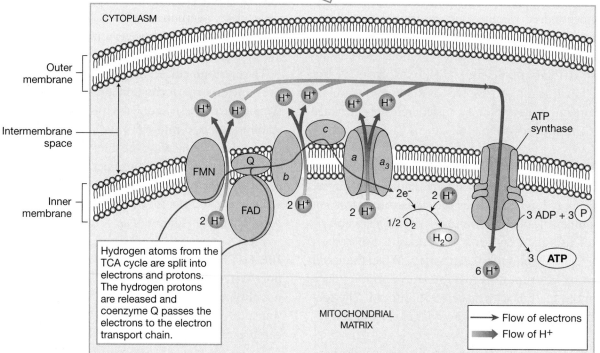

Hydrogen atoms from the TCA cycle are split into electrons and protons. The hydrogen protons are released and coenzyme Q passes the electrons to the electron transport chain.

→ Flow of electrons

➡ Flow of H$^+$

● **FIGURE 17–5   The Electron Transport System (ETS) and ATP Formation.** This diagrammatic view shows the locations of the coenzymes and the electron transport system in the inner mitochondrial membrane. The electrons of hydrogen atoms from the TCA cycle are transferred by coenzyme Q to the ETS (a series of cytochrome molecules), and the hydrogen ions (H$^+$) remain in the matrix. The energy-carrying electrons are passed from one cytochrome to another. Energy released by the passed electrons is used to pump H$^+$ from the matrix into the intermembrane space. This creates a difference in the concentration of H$^+$ across the inner membrane. The hydrogen ions then diffuse through ATP synthase in the inner membrane, and their kinetic energy is used to generate ATP.

The hydrogen atoms from the TCA cycle do not enter the ETS intact. Only the electrons (which carry the energy) enter the ETS; the protons that accompany them are released into the mitochondrial matrix. As indicated by the red line in Figure 17–5, the path of the electrons from NADH involves the *coenzyme FMN,* whereas the path from FADH$_2$ moves directly to *coenzyme Q* (see Figure 17–5). The electrons from both paths are passed from coenzyme Q to the first cytochrome and then from cytochrome to cytochrome, losing energy in a series of small steps. At several steps along the way, this energy is used to drive hydrogen ion pumps that move hydrogen ions from the mitochondrial matrix into the intermembrane space between the two mitochondrial membranes. This creates a large concentration gradient of hydrogen ions across the inner membrane, so the hydrogen ions then diffuse back into the matrix through a membrane enzyme called *ATP synthase.* The kinetic energy of the passing hydrogen

ions is used to attach a phosphate group to ADP, which forms ATP. This process is called *chemiosmosis* (kem-ē-oz-MŌ-sis), which is a term that links the chemical formation of ATP with transport across a membrane. At the end of the electron transport system, an oxygen atom accepts the electrons and combines with two hydrogen ions to form a molecule of water.

The electron transport system is the most important mechanism for the generation of ATP; in fact, it provides roughly 95 percent of the ATP needed to keep our cells alive. Halting or significantly slowing the rate of mitochondrial activity will usually kill a cell. If many cells are affected, the individual may die. If, for example, the cell's supply of oxygen is cut off, mitochondrial ATP production will cease because the ETS will be unable to pass along its electrons. With the last reaction in the chain stopped, the entire ETS comes to a halt, like a line of cars at a washed-out bridge. The affected cell quickly dies of energy deprivation.

Energy Yield of Glycolysis and Cellular Respiration

For most cells, the series of chemical reactions that begins with glucose and ends with carbon dioxide and water is the main method of generating ATP. A cell gains ATP at several steps along the way:

- During glycolysis in the cytoplasm, the cell gains two molecules of ATP for each glucose molecule broken down to pyruvic acid.
- Inside the mitochondria, the two pyruvic acid molecules derived from each glucose molecule are fully broken down in the TCA cycle. Two revolutions of the TCA cycle, each of which yield a molecule of ATP, provide a net gain of two additional molecules of ATP.
- For each molecule of glucose broken down, activity at the electron transport chain in the inner mitochondrial membrane provides 32 molecules of ATP.

In summary, for each glucose molecule processed, a typical cell gains 36 molecules of ATP. *All but two of them are produced within mitochondria.*

Clinical Note
CELLULAR HYPOXIA

Oxygen is essential for normal glucose metabolism. Glucose breakdown and energy production begins in the cytosol with *glycolysis.* Glycolysis, which does not require oxygen, produces two molecules of ATP as it converts the glucose molecule to *pyruvic acid.* If oxygen supplies are adequate, pyruvic acid enters the *mitochondria* and enters the *TCA cycle.* The TCA cycle breaks down the pyruvic acid to *carbon dioxide,* and generates two additional molecules of ATP. The hydrogen ions removed through the TCA cycle then enter the *electron transport system (ETS).* There, the hydrogen ions eventually bind with oxygen, and form *water.* However, through the ETS, 32 molecules of ATP are produced. When completely processed, one glucose molecule yields 36 molecules of ATP. A lack of oxygen, referred to as *hypoxia,* inhibits or stops the TCA cycle and electron transport, which results in the accumulation of pyruvic acid. Glycolysis continues unheeded. As pyruvic acid accumulates, it is converted to *lactic acid* and released into the extracellular fluid where it can cause dangerous shifts in body pH. If oxygen is restored, lactic acid is converted back to pyruvic acid and the normal processes resume. ■

Alternate Catabolic Pathways

Aerobic metabolism is relatively efficient and capable of generating large amounts of ATP. It is the cornerstone of normal cellular metabolism, but it has one obvious limitation—cells must have adequate supplies of both oxygen and glucose.

Cells can survive only for brief periods without oxygen. Low glucose concentrations have a much smaller effect on most cells, because cells can break down other nutrients to provide organic molecules for the TCA cycle, as shown in Figure 17–6●. Many cells can switch from one nutrient source to another as the need arises. For example, many cells can shift from glucose-based ATP production to lipid-based ATP production when necessary. When actively contracting, skeletal muscles catabolize glucose, but at rest they rely on fatty acids.

Cells break down proteins for energy only when lipids or carbohydrates are unavailable; this makes sense because the enzymes and organelles that cells need to survive are composed of proteins. Nucleic acids are present only in small amounts, and they are seldom catabolized for energy, even when the cell is dying of acute starvation. This constraint makes sense, too, because the DNA in the nucleus determines all of the structural and functional characteristics of the cell. We will consider the catabolism of other compounds in later sections as we discuss the metabolism of lipids, proteins, and nucleic acids.

Carbohydrate Synthesis

Because some of the steps in glycolysis are not reversible, cells cannot generate glucose by performing glycolysis in reverse, using the same enzymes. Therefore, glycolysis and the production of glucose require different sets of regulatory enzymes, and the two processes are independently regulated. Pyruvic acid or other three-carbon molecules can be used to

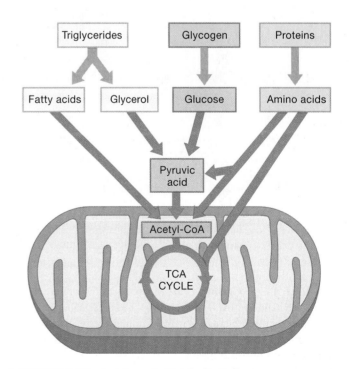

● **FIGURE 17–6 Alternate Catabolic Pathways.**

synthesize glucose; as a result, a cell can synthesize glucose from noncarbohydrate precursors, such as lactic acid, glycerol, or amino acids (Figure 17–7●). However, acetyl-CoA cannot be used to make glucose because the reaction that removes the carbon dioxide molecule (a *decarboxylation*) between pyruvic acid and acetyl-CoA cannot be reversed. The synthesis of glucose from noncarbohydrate (protein or lipid) precursor molecules is called **gluconeogenesis** (gloo-kō-nē-ō-JEN-e-sis; *glykus,* sweet + *neo-,* new + *genesis,* an origin). Fatty acids and many amino acids cannot be used for gluconeogenesis because their breakdown produces acetyl-CoA.

Glucose molecules synthesized during gluconeogenesis can be used to manufacture other simple sugars, complex carbohydrates, or nucleic acids. In the liver and in skeletal muscle, glucose molecules are stored as **glycogen.** Glycogen is an important energy reserve that can be broken down when the cell cannot obtain enough glucose from the inter-

stitial fluid. Although glycogen molecules are large, glycogen reserves take up very little space because they form compact, insoluble granules.

→ **CONCEPT CHECK QUESTIONS**

1. What is the primary role of the TCA cycle in the production of ATP?
2. Hydrogen cyanide gas is a poison that produces its lethal effect by binding to the last cytochrome molecule in the electron transport system. What effect would this have at the cellular level?

Answers begin on p. 792.

Lipid Metabolism

Like carbohydrates, lipid molecules contain carbon, hydrogen, and oxygen but in different proportions. Because triglycerides are the most abundant lipid in the body, our discussion will focus on pathways of triglyceride breakdown and synthesis. ∞ p. 45

Lipid Catabolism

During lipid catabolism, or **lipolysis** (lip-OL-i-sis), lipids are broken down into pieces that can be converted to pyruvic acid or channeled directly into the TCA cycle (see Figure 17–6). A triglyceride is first split into its component parts by hydrolysis, which yields one molecule of glycerol and three fatty acid molecules. Glycerol enters the TCA cycle after cytoplasmic enzymes convert it to pyruvic acid. The catabolism of fatty acids, known as *beta-oxidation,* involves a different set of enzymes that breaks the fatty acids down into two-carbon fragments. The fragments enter the TCA cycle or combine to form *ketone bodies,* which are short carbon chains discussed in a later section (p. 638). Beta-oxidation occurs inside mitochondria, so the two-carbon fragments can enter the TCA cycle immediately. A cell generates 144 ATP molecules from the breakdown of one 18-carbon fatty acid molecule—almost 1.5 times the energy obtained from the breakdown of three six-carbon glucose molecules.

Lipids and Energy Production

Lipids are important energy reserves because their breakdown provides large amounts of ATP. Because they are insoluble in water, lipids are stored in compact droplets in the cytosol. However, if the droplets are large, it is difficult for water-soluble enzymes to get at them. This makes lipid reserves more difficult to

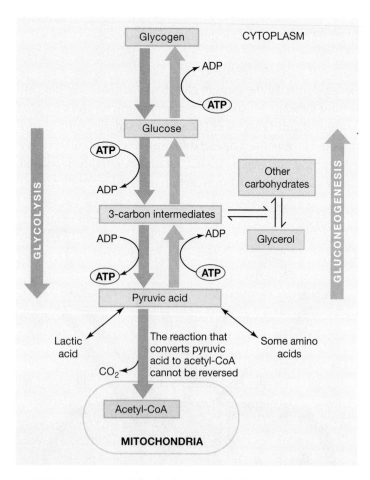

● **FIGURE 17–7 Carbohydrate Metabolism.** This flow chart presents the major pathways of glycolysis and gluconeogenesis. Some amino acids, other carbohydrates, and glycerol can be converted to glucose. The reaction that converts pyruvic acid to acetyl-CoA cannot be reversed.

access than carbohydrate reserves. In addition, most lipids are processed inside mitochondria, and mitochondrial activity is limited by the availability of oxygen. The net result is that lipids cannot provide large amounts of ATP quickly. However, cells with modest energy demands can shift to lipid-based energy production when glucose supplies are limited. Skeletal muscle fibers normally cycle between lipid metabolism and carbohydrate metabolism. At rest (when energy demands are low), these cells break down fatty acids. During activity (when energy demands are high and immediate), skeletal muscle fibers shift to glucose metabolism.

Lipid Synthesis

The synthesis of lipids is known as **lipogenesis** (lip-O-JEN-e-sis; *lipos*, fat). Glycerol is synthesized from an intermediate three-carbon product of glycolysis. The synthesis of most other types of lipids, including steroids and almost all fatty acids, begins with acetyl-CoA. Lipogenesis can use almost any organic molecule because lipids, amino acids, and carbohydrates can be converted to acetyl-CoA (Figure 17–8●). Body cells cannot *build* every fatty acid they can break down. For example, neither *linoleic acid* nor *linolenic acid,* both 18-carbon unsaturated fatty acids, can be synthesized. Because they must be included in your diet, they are called **essential fatty acids.** (They are synthesized by plants.) These fatty acids are also needed to synthesize prostaglandins and phospholipids for cell membranes.

Lipid Transport and Distribution

All cells need lipids to maintain their cell membranes, and steroid hormones must reach their target cells in many different tissues. Because most lipids are not soluble in water, special mechanisms are required to transport them around the body. Most lipids circulate in the bloodstream as **lipoproteins**—lipid-protein complexes that contain triglycerides and cholesterol within an outer coating of phospholipids and proteins. The proteins and phospholipids make the entire complex soluble. The exposed proteins, which bind to specific membrane receptors, determine which cells absorb the associated lipids.

Lipoproteins are classified by size and by their relative proportions of lipid and protein. One group, the *chylomicrons,* forms in the intestinal tract. ∞ p. 619 Chylomicrons are the largest lipoproteins, and 95 percent of their weight consists of triglycerides. Chylomicrons transport triglycerides absorbed from the intestinal tract to the bloodstream, from which they are absorbed by skeletal muscle, cardiac muscle, adipose tissue, and the liver.

Two other major groups of lipoproteins are the **low-density lipoproteins (LDLs)** and **high-density lipoproteins (HDLs).** These lipoproteins are formed in the liver and contain few triglycerides. Their main roles are to shuttle cholesterol between the liver and other tissues. LDLs deliver cholesterol to peripheral tissues. Because LDL cholesterol may end up in arterial plaques, it is often called "bad cholesterol." HDL cholesterol transports excess cholesterol from

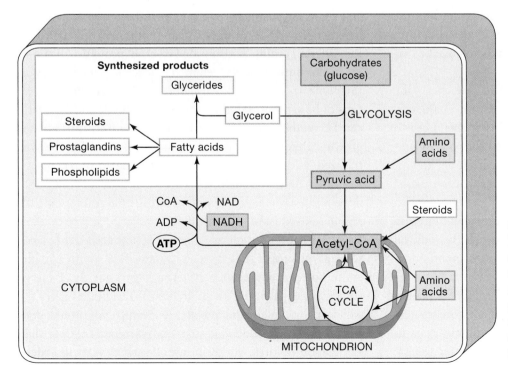

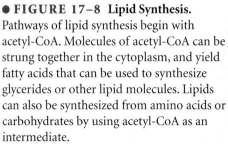

● **FIGURE 17–8 Lipid Synthesis.** Pathways of lipid synthesis begin with acetyl-CoA. Molecules of acetyl-CoA can be strung together in the cytoplasm, and yield fatty acids that can be used to synthesize glycerides or other lipid molecules. Lipids can also be synthesized from amino acids or carbohydrates by using acetyl-CoA as an intermediate.

peripheral tissues to the liver for storage or excretion in the bile. Because HDL cholesterol does not cause circulatory problems, it is called "good cholesterol."

Free fatty acids (FFA) are lipids that can diffuse easily across cell membranes. A major source of these fatty acids is the breakdown of fat stored in adipose tissue. When released into the blood, the fatty acids bind to albumin, which is the most abundant plasma protein. Liver cells, cardiac muscle cells, skeletal muscle fibers, and many other body cells can metabolize free fatty acids. They are an important energy source during periods of starvation, when glucose supplies are limited.

Protein Metabolism

Even though the body can synthesize at least 400,000 different proteins, each with varied forms, functions, and structures, all proteins are composed of some combination of only 20 amino acids. Under normal conditions, a continuous recycling of cellular proteins occurs in the cytoplasm. Peptide bonds are broken, and the free amino acids are used to manufacture new proteins. If other energy sources are inadequate, mitochondria can break down amino acids in the TCA cycle to generate ATP. Not all amino acids enter the TCA cycle at the same point, so the ATP benefits vary. However, the average ATP yield is comparable to that of carbohydrate catabolism.

Amino Acid Catabolism

The first step in amino acid catabolism is the removal of the amino group, which requires a coenzyme derived from **vitamin B$_6$** *(pyridoxine).* The amino group is removed by either transamination or deamination.

Transamination (trans-am-i-NĀ-shun) attaches the amino group of an amino acid to another carbon chain, creating a "new" amino acid. Transaminations enable a cell to synthesize many of the amino acids needed for protein synthesis. Cells of the liver, skeletal muscles, heart, lung, kidney, and brain, which are particularly active in protein synthesis, perform many transaminations.

Deamination (dē-am-i-NĀ-shun) prepares an amino acid for breakdown in the TCA cycle. Deamination is the removal of an amino group in a reaction that generates an ammonia molecule (NH$_3$). Ammonia is highly toxic, even in low concentrations. The liver, the primary site of deamination, has the enzymes needed to deal with the problem of ammonia generation. Liver cells combine carbon dioxide with ammonia to produce **urea,** which is a relatively harmless, water-soluble compound that is excreted in the urine.

The fate of the carbon chain that remains after deamination in the liver depends on its structure. The carbon chains of some amino acids can be converted to pyruvic acid and then used in gluconeogenesis. Other carbon chains are converted to acetyl-CoA and broken down in the TCA cycle. Still others are converted to **ketone bodies,** which are organic acids that are also produced during lipid catabolism. One example of a ketone body generated in the body is *acetone,* which is a small molecule that can diffuse into the alveoli of the lungs, and give the breath a distinctive odor.

Ketone bodies diffuse into the general circulation, but they are not metabolized by the liver. Instead, other body cells reconvert them into acetyl-CoA for breakdown in the TCA cycle and the production of ATP. The increased production of ketone bodies that occurs during protein and lipid catabolism by the liver results in high ketone body concentrations in body fluids, a condition called **ketosis** (kē-TŌ-sis).

Several factors make protein catabolism an impractical source of quick energy:

- Proteins are more difficult to break apart than are complex carbohydrates or lipids.
- One of the by-products, ammonia, is toxic to cells.
- Because proteins form the most important structural and functional components of any cell, extensive protein catabolism threatens homeostasis at the cellular and systems levels.

Amino Acids and Protein Synthesis

The basic mechanism of protein synthesis was detailed in Chapter 3. p. 83 Your body can synthesize roughly half of the different amino acids needed to build proteins. Of the 10 **essential amino acids,** eight (*isoleucine, leucine, lysine, threonine, tryptophan, phenylalanine, valine,* and *methionine*) cannot be synthesized; the other two (*arginine* and *histidine*) can be synthesized but in amounts that are insufficient for growing children. The other amino acids, which can be synthesized on demand, are called **nonessential amino acids.**

Protein deficiency diseases develop when an individual does not consume adequate amounts of all essential amino acids. All amino acids must be available if protein synthesis is to occur. If every transfer RNA molecule does not appear at the active ribosome at the proper time bearing its individual amino acid, the entire process comes to a halt. Regardless of the diet's energy content, if it is deficient in essential amino acids, the individual will be malnourished to some degree. Examples of protein deficiency diseases include *marasmus* and *kwashiorkor.* More than 100 million children worldwide are affected by these disorders, although neither condition is common in the U.S. today.

Clinical Note
COMPLICATIONS OF DIABETES MELLITUS

Diabetes mellitus is a disorder of glucose metabolism. The diabetic under treatment can develop complications that result from an inadequate amount of available glucose *(hypoglycemia)* or an excess amount of available glucose *(hyperglycemia)*. Patients with Type I diabetes require insulin. An excess dose of insulin or inadequate food intake after a standard dose of insulin can cause life-threatening hypoglycemia. Likewise, a lack of insulin or increased food intake with a standard dose of insulin can cause hyperglycemia. Type II diabetics usually do not require insulin. Instead, they are able to manage their blood-sugar levels through diet, exercise, or use of an *oral hypoglycemic agent*. Although much less common, a Type II diabetic can develop hypoglycemia if an excess of oral hypoglycemic medication is taken or if food intake decreases significantly. Like the Type I diabetic, the Type II diabetic can develop hyperglycemia if food intake is increased or if the oral hypoglycemic agent being used does not adequately lower blood-glucose levels.

The extremes of blood sugar can affect many body systems. Prolonged poor control of blood-sugar levels markedly increases the patient's chances of developing many of the long-term complications associated with diabetes. The extremes of blood sugar require prompt treatment by emergency personnel.

■ *Hypoglycemia.* The blood-glucose level at which hypoglycemia occurs varies from individual to individual. However, a blood-glucose level less than 50 mg/dL results in hypoglycemia in most diabetics. Hypoglycemia is a true medical emergency and the most life-threatening complication of diabetes mellitus. The central nervous system (CNS) relies almost exclusively on glucose as its sole source of energy. Thus, an inadequate blood-glucose level can cause CNS injury. The symptoms of hypoglycemia are due to CNS dysfunction. Prompt recognition of hypoglycemia and administration of glucose is essential if CNS injury is to be prevented.

■ *Hyperglycemia.* An abnormal elevation in blood glucose is termed *hyperglycemia*. Unlike hypoglycemia, which can develop in minutes, hyperglycemia can take hours or even days to develop. In most patients, there is an inadequate level of the insulin necessary for glucose entry into the cells. When glucose entry is impaired, cellular starvation occurs. In addition to its ef-

fect on glucose metabolism, insulin is also responsible for the manufacture and storage of *lipids* (fats) by the body. Inadequate insulin levels cause the breakdown of lipids into glucose, which further increases blood-glucose levels. As a by-product of lipid breakdown, free-fatty acids are converted to *ketone bodies*. As ketone bodies accumulate, systemic acidosis occurs. The sequence of events results in *diabetic ketoacidosis (DKA)*. In severe DKA, the pH can fall to 7.0 or lower.

— *Diabetic ketoacidosis.* Most Type I diabetics develop DKA if their blood-glucose levels are allowed to rise unchecked. The signs and symptoms of DKA are related to the various biochemical derangements described above. They include dehydration, hypotension, and reflex tachycardia. As the disease progresses, the patient will develop nausea, vomiting, and abdominal pain. Because of the acidosis, hyperventilation occurs as a compensatory mechanism. The hyperventilation is quite exaggerated and is referred to as *Kussmaul's respiration*. The sweet smell of ketones can sometimes be detected in the patient's breath. Eventually, the patient will develop altered mental status and, in time, unconsciousness. Treatment includes massive intravenous fluid replacement and the administration of insulin.

— *Non-ketotic hyperosmolar coma.* A certain subset of patients, most of whom have Type II diabetes, will not develop ketones as a complication of hyperglycemia. In these patients, there appears to be enough insulin present to prevent ketone formation. As their blood glucose levels rise, they can develop *non-ketotic hyperosmolar coma (NKHC)*. In NKHC, the blood glucose level can rise to 1,000 mg/dL or more. As the glucose is spilled into the urine, the resultant *osmotic diuresis* causes severe dehydration. The *osmolarity* of the blood, which is a measure of concentration of molecules in the blood, climbs significantly. NKHC most commonly occurs in middle-aged or elderly diabetics and is often associated with another disease process such as infection. It typically takes days for NKHC to occur, and the signs and symptoms are similar to those seen in DKA. Kussmaul's respirations are not seen. Treatment is similar to DKA. However, the mortality rate for NHKC is higher than for DKA. ■

Clinical Note
DIETARY FATS AND CHOLESTEROL

Elevated cholesterol levels are associated with the development of *atherosclerosis* and *coronary artery disease* (CAD). ∞ pp. 470, 445 Nutritionists currently recommend that you limit cholesterol intake to under 300 mg per day. This amount represents a 40 percent reduction for the average American adult. As a result of rising concerns about cholesterol, such phrases as "low in cholesterol," "contains no cholesterol," and "cholesterol free" are now widely used in the advertising and packaging of foods. Cholesterol content alone,

however, does not tell the entire story. Consider the following basic information about cholesterol and about lipid metabolism in general:

■ *Cholesterol has many vital functions in the human body.* It serves as a waterproofing agent in the epidermis, and it is a lipid component of all cell membranes, a key constituent of bile, and the precursor of several steroid hormones and vitamin D_3. Because cholesterol is so important, the goal of dietary restrictions

(continued next page)

Clinical Note—*continued*
DIETARY FATS AND CHOLESTEROL

is *not* to eliminate cholesterol from the diet or from the circulating blood but instead to keep cholesterol levels within acceptable limits.

■ *The cholesterol content of the diet is not the only source for circulating cholesterol.* The human body can manufacture cholesterol from acetyl-CoA produced during glycolysis or by the breakdown (beta-oxidation) of other lipids. If the diet contains an abundance of saturated fats, blood cholesterol levels will rise because excess lipids are broken down to acetyl-CoA, which can be used to make cholesterol. This means that a person trying to lower serum cholesterol by dietary control must also restrict other lipids—especially saturated fats.

■ *Genetic factors affect each individual's cholesterol level.* If you reduce your dietary intake of cholesterol, your body will synthesize more to maintain "acceptable" concentrations in the blood. What is an "acceptable" level depends on your genetic makeup. Because individuals have different genes, their cholesterol levels can vary, even on similar diets. In virtually all instances, however, dietary restrictions can lower blood cholesterol significantly.

■ *Cholesterol levels vary with age and physical condition.* At age 19, three out of four males have cholesterol levels below 170 mg/dL. Cholesterol levels in females of this age are slightly higher, typically at or below 175 mg/dL. As age increases, cholesterol values gradually climb; over age 70, typical values are 230 mg/dL (males) and 250 mg/dL (females). Cholesterol levels are considered unhealthy if they are higher than those of 90 percent of the population in a given age group. For males, this value ranges from 185 mg/dL at age 19 to 250 mg/dL at age 70. For females, the comparable values are 190 mg/dL and 275 mg/dL.

To determine whether or not you need to reduce your cholesterol level, remember three simple rules:

1. Individuals of any age with total cholesterol values below 200 mg/dL probably need not change their lifestyle unless they have a family history of coronary artery disease and atherosclerosis.
2. Those with cholesterol levels between 200 and 239 mg/dL should modify their diet, lose weight (if overweight), and have annual checkups.
3. Cholesterol levels over 240 mg/dL warrant drastic changes in dietary lipid consumption, perhaps coupled with drug treatment. Drug therapies are always recommended when serum cholesterol levels exceed 350 mg/dL. Examples of drugs used to lower cholesterol levels are *cholestyramine, colestipol,* and *lovastatin.*

Most physicians, when ordering a blood test for cholesterol, also request information on circulating triglycerides. When cholesterol levels are high, or when an individual has a family history of atherosclerosis or CAD, the HDL level may be measured, and the LDL may be calculated from an equation that relates the levels of cholesterol, HDL, and triglycerides. A high total cholesterol value linked to a high LDL spells trouble. In effect, an unusually large amount of cholesterol is being exported to peripheral tissues. Problems can also exist if the individual has high total cholesterol—or even normal total cholesterol—and HDL levels below 35 mg/dL. In this case, excess cholesterol delivered to the tissues cannot easily be returned to the liver for excretion. In either event, the amount of cholesterol in peripheral tissues, and especially in arterial walls, is likely to increase.

For years, LDL/HDL ratios were used to predict the risk of developing atherosclerosis. Risk-factor analysis and LDL levels are now thought to be more accurate indicators. Many clinicians recommend dietary restrictions and drug therapy for males with more than one risk factor and LDL levels that exceed 130 mg/dL, regardless of total cholesterol or HDL levels. ■

Several inherited metabolic disorders result from an inability to produce specific enzymes involved in amino acid metabolism. Individuals with **phenylketonuria** (fen-il-kē-to-NOO-rē-uh), or **PKU,** cannot convert the amino acid phenylalanine to the amino acid tyrosine, because of a defect in the enzyme phenylalanine hydroxylase. This reaction is an essential step in the synthesis of norepinephrine, epinephrine, and melanin. If PKU is not detected in infancy, central nervous system development is inhibited, and severe brain damage results.

Nucleic Acid Metabolism

Living cells contain both DNA and RNA. The genetic information contained in the DNA of the nucleus is absolutely essential to the long-term survival of a cell. As a result, nuclear DNA is never catabolized for energy, even if the cell is dying of starvation. By contrast, the RNA molecules involved in protein synthesis are broken down and replaced regularly.

RNA Catabolism

In the breakdown of RNA, the molecule is disassembled into individual nucleotides. Although most nucleotides are recycled into new nucleic acids, they can be broken down to simple sugars and nitrogen bases. When nucleotides are broken down, only the sugars, cytosine, and uracil can enter the TCA cycle and be used to generate ATP. Adenine and guanine cannot be catabolized; instead, they undergo deamination and are excreted as **uric acid.** Like urea, uric acid is a relatively nontoxic waste product, but it is far less soluble than urea. Urea and uric acid are called *nitrogenous wastes,* because they contain nitrogen atoms.

An elevated level of uric acid in the blood is called *hyperuricemia* (hī-per-ū-ri-SĒ-mē-uh). Uric acid saturates body fluids and, although symptoms may not appear immediately, uric acid crystals may begin to form. The condition that then develops is called *gout*. Most cases of hyperuricemia and gout are linked to problems with the excretion of uric acid by the kidneys.

Nucleic Acid Synthesis

Most cells synthesize RNA, but DNA synthesis occurs only in cells preparing for mitosis (cell division) or meiosis (gamete production). The process of DNA replication was described in Chapter 3. ∞ p. 87 Messenger RNA (mRNA), transfer RNA (tRNA), and ribosomal RNA (rRNA) are transcribed by different forms of the enzyme RNA polymerase. Messenger RNA is manufactured as needed, when specific genes are activated. A strand of mRNA has a life span measured in minutes or hours. Ribosomal RNA and tRNA are more durable than mRNA. For example, the average strand of rRNA lasts just over five days. However, because a typical cell contains roughly 100,000 ribosomes and many times that number of tRNA molecules, their replacement involves a considerable amount of synthetic activity.

→ CONCEPT CHECK QUESTIONS

1. How would a diet deficient in vitamin B_6 affect protein metabolism?
2. Elevated levels of uric acid in the blood could indicate that the individual has an increased metabolic rate for which macromolecule?
3. Why are high-density lipoproteins (HDLs) considered beneficial?

Answers begin on p. 792.

A Summary of Cellular Metabolism

Figure 17–9● summarizes the major pathways of cellular metabolism. Although this diagram presents the reactions in a "typical" cell, no one cell can perform all of the anabolic and catabolic operations required by the body as a whole. As differentiation proceeds, each cell type develops its own complement of enzymes that determines its metabolic capabilities. In the presence of such cellular diversity, homeostasis can be preserved only when the metabolic activities of tissues, organs, and organ systems are coordinated.

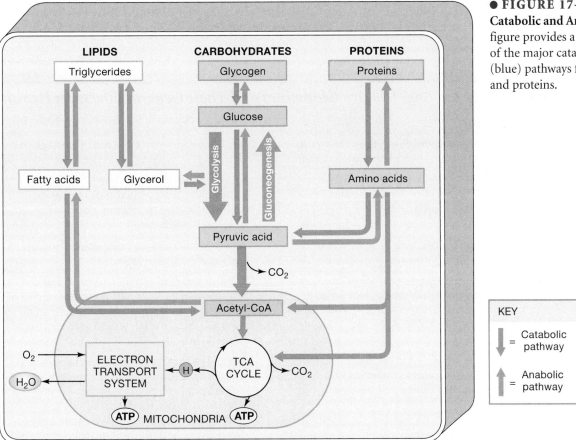

● **FIGURE 17–9 A Summary of Catabolic and Anabolic Pathways.** This figure provides a diagrammatic overview of the major catabolic (red) and anabolic (blue) pathways for lipids, carbohydrates, and proteins.

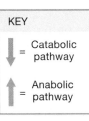

■ Diet and Nutrition

Homeostasis can be maintained indefinitely only if the digestive tract absorbs fluids, organic substrates, minerals, and vitamins at a rate that keeps pace with cellular demands. The absorption of essential nutrients from food is called **nutrition.**

The body's requirement for each nutrient varies from day to day and from person to person. *Nutritionists* attempt to analyze a diet in terms of its ability to prevent and treat illnesses for specific individuals and population groups. A **balanced diet** contains all the nutrients needed to maintain homeostasis, including adequate substrates for generating energy, essential amino acids and fatty acids, minerals, and vitamins. In addition, the diet must include enough water to replace losses in urine, feces, and evaporation. A balanced diet prevents **malnutrition,** which is an unhealthy state that results from the inadequate or excessive intake of one or more nutrients.

Food Groups and the Food Pyramid

One way of maintaining good health and preventing malnutrition and chronic diseases is to include in the diet members of different food groups. In 2005, new dietary guidelines were released as part of a revised Federal nutrition policy. The dietary guide consists of five **basic food groups:** (1) the *grains* group; (2) the *vegetables* group; (3) the *fruits* group;

(4) the *milk* group; and (5) the *meat and beans* group. Each group differs from the others in terms of protein, carbohydrate, and lipid content, as well as in the amounts and types of vitamins and minerals. Table 17–1 summarizes what each of these nutrient groups provide in the diet, as well as recommended choices from the dietary guidelines and their general effects on health.

The five groups, including a fats, sugars, and salt category, are arranged within a *food pyramid* as six bands of differing color and width (Figure 17–10●). The widths of the bands are designed to be a general guide to how much of each group should be consumed each day. (Specific recommendations based on an individual's sex, age, and activity level are available at *http://www.mypyramid.gov.*) The new food pyramid also includes a set of stairs on its side to stress that daily physical activity and weight control are also important aspects of maintaining good health.

Such groupings can help in planning a balanced and healthful diet. It is important to obtain nutrients not only in sufficient *quantity* (to meet energy needs) but also in adequate *quality* (including essential amino acids, fatty acids, vitamins, and minerals). There is nothing magical about the number five—since 1940, the U.S. government has at various times advocated 11, 7, 4, and, most recently, 6 food groups. The key is to make intelligent choices about what you eat. Poor choices can lead to malnutrition even when all five groups are represented.

TABLE 17–1 *Basic Food Groups of the 2005 Dietary Guidelines and Their General Effects on Health*

NUTRIENT GROUP	PROVIDES	HEALTH EFFECTS
Grains (recommended: at least half of the total eaten as whole grains)	Carbohydrates; vitamins E, thiamine, niacin, folacin; calcium; phosphorus; iron; sodium; dietary fiber	Whole grains prevent rapid rise in blood glucose levels, and consequent rapid rise in insulin levels
Vegetables (recommended: especially dark-green and orange vegetables)	Carbohydrates; vitamins A, C, E, folacin; dietary fiber; potassium	Reduce risk of cardiovascular disease; protect against colon cancer (folacin) and prostate cancer (lycopene in tomatoes)
Fruits (recommended: a variety of fruit each day)	Carbohydrates; vitamins A, C, E, folacin; dietary fiber; potassium	Reduce risk of cardiovascular disease; protect against colon cancer (folacin)
Milk (recommended: low-fat or fat-free milk, yogurt, and cheese)	Complete proteins; fats; carbohydrates; calcium; potassium; magnesium; sodium; phosphorus; vitamins A, B_{12}, pantothenic acid, thiamine, riboflavin	Excellent source of calcium, minerals, and carbohydrates. Whole milk is excellent source of calories for those who need it. Reduced fat milk (2% and skim) has less fat and decreases risk of heart disease
Meat and Beans (recommended: ☐an meats, fish, poultry, eggs, ☐ beans, nuts, legumes)	Complete proteins; fats; calcium; potassium; phosphorus; iron; zinc; vitamins E, thiamine, B_6	Fish and poultry lower risk of heart disease and colon cancer (compared to red meat); consumption of up to one egg per day does not appear to increase incidence of heart disease. Nuts and legumes improve blood cholesterol ratios, lower risk of heart disease and diabetes

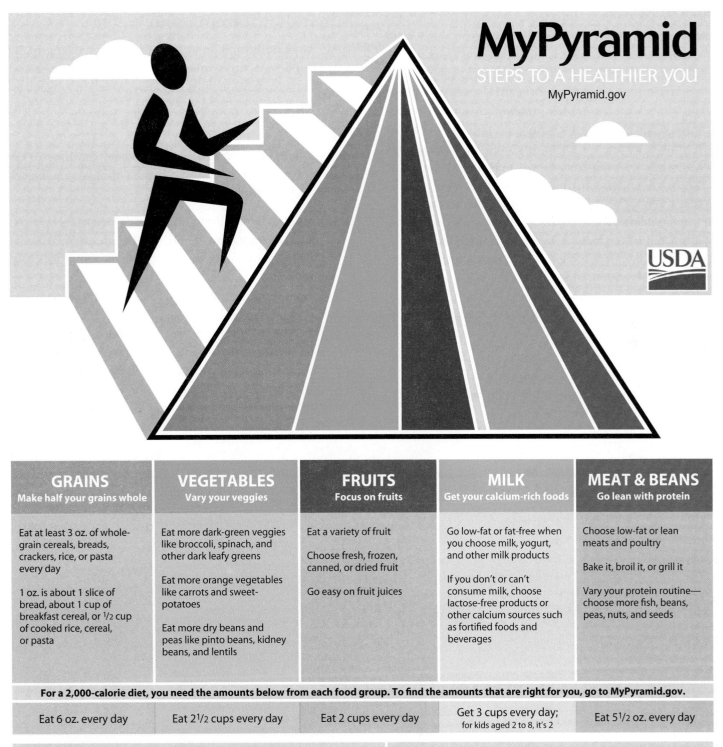

MyPyramid
STEPS TO A HEALTHIER YOU
MyPyramid.gov

USDA

GRAINS	VEGETABLES	FRUITS	MILK	MEAT & BEANS
Make half your grains whole	Vary your veggies	Focus on fruits	Get your calcium-rich foods	Go lean with protein
Eat at least 3 oz. of whole-grain cereals, breads, crackers, rice, or pasta every day 1 oz. is about 1 slice of bread, about 1 cup of breakfast cereal, or ½ cup of cooked rice, cereal, or pasta	Eat more dark-green veggies like broccoli, spinach, and other dark leafy greens Eat more orange vegetables like carrots and sweet-potatoes Eat more dry beans and peas like pinto beans, kidney beans, and lentils	Eat a variety of fruit Choose fresh, frozen, canned, or dried fruit Go easy on fruit juices	Go low-fat or fat-free when you choose milk, yogurt, and other milk products If you don't or can't consume milk, choose lactose-free products or other calcium sources such as fortified foods and beverages	Choose low-fat or lean meats and poultry Bake it, broil it, or grill it Vary your protein routine—choose more fish, beans, peas, nuts, and seeds

For a 2,000-calorie diet, you need the amounts below from each food group. To find the amounts that are right for you, go to MyPyramid.gov.

Eat 6 oz. every day	Eat 2½ cups every day	Eat 2 cups every day	Get 3 cups every day; for kids aged 2 to 8, it's 2	Eat 5½ oz. every day

Find your balance between food and physical activity
- Be sure to stay within your daily calorie needs.
- Be physically active for at least 30 minutes most days of the week.
- About 60 minutes a day of physical activity may be needed to prevent weight gain.
- For sustaining weight loss, at least 60 to 90 minutes a day of physical activity may be required.
- Children and teenagers should be physically active for 60 minutes every day, or most days.

Know the limits on fats, sugars, and salt (sodium)
- Make most of your fat sources from fish, nuts, and vegetable oils.
- Limit solid fats like butter, margarine, shortening, and lard, as well as foods that contain these.
- Check the Nutrition Facts label to keep saturated fats, trans fats, and sodium low.
- Choose food and beverages low in added sugars. Added sugars contribute calories with few, if any, nutrients.

● **FIGURE 17–10 The Food Pyramid.** (*Source: USDA, http://www.mypyramid.gov*)

Consider, for example, the essential amino acids. The liver cannot synthesize any of these amino acids, so you must obtain them from your diet. In general, animal proteins are **complete proteins** because they contain all of the essential amino acids in sufficient quantities. Many plants contain adequate *amounts* of protein, but these are **incomplete proteins** because they are deficient in one or more essential amino acids. Vegetarians who largely restrict themselves to the fruit and vegetable groups (with or without whole grain foods) must become adept at varying their food choices to include a combination of ingredients that meets all of their amino acid requirements. Even if their diets contain a proper balance of amino acids, vegetarians who avoid animal products altogether face a significant problem because vitamin B_{12} is obtained only from animal products or from fortified cereals or tofu.

Key Note

A balanced diet contains all the ingredients needed to maintain homeostasis, including adequate substrates for energy generation, essential amino acids and fatty acids, minerals, vitamins, and water.

Minerals, Vitamins, and Water

Minerals, vitamins, and water are essential components of the diet. The body cannot synthesize minerals, and our cells can generate only a small quantity of water and very few vitamins.

Minerals

Minerals are inorganic ions released through the dissociation of electrolytes, such as sodium chloride. Minerals are important for the following reasons:

1. Ions such as sodium and chloride determine the osmotic concentration of body fluids. Potassium is important in maintaining the osmotic concentration inside body cells.
2. Ions in various combinations play major roles in important physiological processes. As we have seen, these processes include the maintenance of membrane potentials, the construction and maintenance of the skeleton, muscle contraction, action potential generation, neurotransmitter release, blood clotting, the transport of respiratory gases, buffer systems, fluid absorption, and waste removal.

TABLE 17-2 *Minerals and Mineral Reserves*

MINERAL	SIGNIFICANCE	TOTAL BODY CONTENT	PRIMARY ROUTE OF EXCRETION	RECOMMENDED DAILY INTAKE
BULK MINERALS				
Sodium	Major cation in body fluids; essential for normal membrane function	110 g, primarily in body fluids	Urine, sweat, feces	0.5–1.0 g
Potassium	Major cation in cytoplasm; essential for normal membrane function	140 g, primarily in cytoplasm	Urine	1.9–5.6 g
Chloride	Major anion in body fluids	89 g, primarily in body fluids	Urine, sweat	0.7–1.4 g
Calcium	Essential for normal muscle and neuron function, and normal bone structure	1.36 kg, primarily in skeleton	Urine, feces	0.8–1.2 g
Phosphorus	As phosphate in high-energy compounds, nucleic acids, and bone matrix	744 g, primarily in skeleton	Urine, feces	0.8–1.2 g
Magnesium	Cofactor of enzymes, required for normal membrane functions	29 g (skeleton, 17 g; cytoplasm and body fluids, 12 g)	Urine	0.3–0.4 g
TRACE MINERALS				
Iron	Component of hemoglobin, myoglobin, cytochromes	3.9 g (1.6 g stored as ferritin or hemosiderin)	Urine (traces)	10–18 mg
Zinc	Cofactor of enzyme systems, notably carbonic anhydrase	2 g	Urine, hair (traces)	15 mg
Copper	Required as cofactor for hemoglobin synthesis	127 mg	Urine, feces (traces)	2–3 mg
Manganese	Cofactor for some enzymes	11 mg	Feces, urine (traces)	2.5–5.0 mg

3. Ions are essential cofactors in a variety of enzymatic reactions. For example, the enzyme that breaks down ATP in a contracting skeletal muscle requires the presence of calcium and magnesium ions, and an enzyme required for the conversion of glucose to pyruvic acid needs both potassium and magnesium ions.

The most important minerals and a summary of their functions are presented in Table 17–2. Your body contains significant reserves of several important minerals that help reduce the effects of variations in dietary supply. These reserves are often relatively small, however, and chronic dietary deficiencies can lead to various clinical problems. Alternatively, because storage capabilities are limited, a dietary excess of mineral ions can prove equally dangerous.

Vitamins

Vitamins (*vita*, life) are essential organic nutrients related to lipids and carbohydrates. They can be assigned to either of two groups: fat-soluble vitamins and water-soluble vitamins.

Clinical Note
VITAMINS

Vitamins are organic compounds needed in small quantities for normal body metabolism. However, they cannot be manufactured by the cells of the body and must be obtained from the diet. Physiological processes that require vitamins include metabolism, growth, development, and tissue repair. The body absorbs most vitamins through the gastrointestinal tract following dietary ingestion. Vitamins are stored in the liver and, to a lesser extent, in the cells. In developed countries, healthy adults usually receive adequate amounts of vitamins and do not need supplements. Vitamin supplements may, however, be indicated for special populations including pregnant and nursing women, patients with absorption disorders, the chronically ill, surgery patients, alcoholics, and the malnourished. Additionally, people on a strict vegetarian diet may need supplemental vitamins.

Vitamins are classified as either fat-soluble or water-soluble (Tables 17–3 through 17–6). The liver stores fat-soluble vitamins (A, D, E, and K), so the patient will become deficient only after long periods of inadequate vitamin intake. Vitamin D is unique in that skin produces it with exposure to sunlight. The water-soluble vitamins (C and those in the B complex) must be routinely ingested, as the body does not store them. After short periods of deprivation, patients may begin to experience vitamin deficiency. The B complex vitamins are grouped only because they occur together in foods; otherwise they share no significant characteristics. The individual B vitamins are named for the order in which they were discovered (B_1, B_2, B_6, B_{12}, and so forth). These vitamins also have specific names. For example, B_1 is also known as thiamine, which is a vitamin that plays a key role in carbohydrate metabolism. ∎

TABLE 17–3 *The Fat-Soluble Vitamins*

VITAMIN	SIGNIFICANCE	SOURCES	DAILY REQUIREMENT	EFFECTS OF DEFICIENCY	EFFECTS OF EXCESS
A	Maintains epithelia; required for synthesis of visual pigments	Leafy green and yellow vegetables	1 mg	Retarded growth, night blindness, deterioration of epithelial membranes	Liver damage, skin peeling, CNS effects (nausea, anorexia)
D (steroid-like compounds, including cholecalciferol or D_3)	Required for normal bone growth, calcium and phosphorus absorption at gut and retention at kidneys	Synthesized in skin exposed to sunlight	5–15 mcg*	Rickets, skeletal deterioration	Calcium deposits in many tissues, disrupting functions
E (tocopherols)	Prevents breakdown of vitamin A and fatty acids	Meat, milk, vegetables	30 IU	Anemia, other problems suspected	Nausea, stomach cramps, blurred vision, fatigue
K	Essential for liver synthesis of prothrombin and other clotting factors	Vegetables; production by intestinal bacteria	0.7–0.14 mg	Bleeding disorders	Liver dysfunction, jaundice

*Unless sunlight exposure is inadequate for extended periods and alternative sources (fortified milk products) are unavailable.

TABLE 17–4 *The Water-Soluble Vitamins*

VITAMIN	SIGNIFICANCE	SOURCES	DAILY REQUIREMENT	EFFECTS OF DEFICIENCY	EFFECTS OF EXCESS
B_1 (thiamine)	Coenzyme in decarboxylation reactions (removal of a carbon dioxide molecule)	Milk, meat, bread	1.9 mg	Muscle weakness, CNS and cardiovascular problems including heart disease; called *beriberi*	Hypotension
B_2 (riboflavin)	Part of FAD	Milk, meat	1.7 mg	Epithelial and mucosal deterioration	Itching, tingling
Niacin (nicotinic acid)	Part of NAD	Meat, bread, potatoes	14.6 mg	CNS, GI, epithelial, and mucosal deterioration; called *pellagra*	Itching, burning; vasodilation, death after large dose
B_5 (pantothenic acid)	Part of acetyl-CoA	Milk, meat	10 mg	Retarded growth, CNS disturbances	None reported
B_6 (pyridoxine)	Coenzyme in amino acid and lipid metabolism	Meat	1.42 mg	Retarded growth, anemia, convulsions, epithelial changes	CNS alterations, perhaps fatal
Folacin (folic acid)	Coenzyme in amino acid and nucleic acid metabolisms	Vegetables, cereal, bread	0.1 mg	Retarded growth, anemia, gastrointestinal disorders, developmental abnormalities	Few noted except at massive doses
B_{12} (cobalamin)	Coenzyme in nucleic acid metabolism	Milk, meat	6 mcg	Impaired RBC production causing *pernicious anemia*	Polycythemia (elevated hematocrit)
Biotin	Coenzyme in decarboxylation reactions	Eggs, meat, vegetables	0.1–0.2 mg	Fatigue, muscular pain, nausea, dermatitis	None reported
C (ascorbic acid)	Coenzyme; delivers hydrogen ions, antioxidant	Citrus fruits	60 mg	Epithelial and mucosal deterioration; called *scurvy*	Kidney stones

FAT-SOLUBLE VITAMINS. Vitamins A, D, E, and K are **fat-soluble vitamins,** because they are absorbed primarily from the digestive tract along with the lipid contents of micelles.

The term *vitamin D* refers to a group of steroid-like molecules, including vitamin D_3, or cholecalciferol. ∞ p. 45 Unlike the other fat-soluble vitamins, which must be obtained by absorption across the digestive tract, vitamin D_3 can usually be synthesized in adequate amounts by skin exposed to sunlight. Current information concerning the fat-soluble vitamins is summarized in Table 17–3.

Because fat-soluble vitamins dissolve in lipids, they normally diffuse into cell membranes, including the lipid inclusions in the liver and adipose tissue. Your body, therefore, contains a significant reserve of these vitamins, and normal metabolic operations can continue for several months after dietary sources have been cut off. For this reason, symptoms of **avitaminosis** (ā-vī-ta-min-Ō-sis), or *vitamin deficiency disease,* rarely result from dietary insufficiency of fat-soluble

TABLE 17–5 *Reference Daily Intake of Essential Vitamins*

ESSENTIAL VITAMINS	DAILY REQUIREMENTS
A	5000 IU
Thiamine	1.5 mg
Riboflavin	1.7 mg
Niacin	20 mg
Ascorbic acid	60 mg
D	400 IU
E	30 IU
K	70 mcg
Folic acid	0.4 mg
B_{12}	6 mcg
Pyridoxine	2 mg
Pantothenic acid	10 mg

vitamins. However, avitaminosis that involves either fat-soluble or water-soluble vitamins can result from factors other than dietary deficiencies, including an inability to absorb a vitamin from the digestive tract, inadequate storage, or excessive demand.

As Table 17–3 points out, *too much* of a vitamin can produce effects just as unpleasant as *too little*. **Hypervitaminosis** (hī-per-vī-ta-min-Ō-sis) occurs when dietary intake exceeds the ability to store, utilize, or excrete a particular vitamin. This condition most often involves one of the fat-soluble vitamins because the excess vitamins are retained and stored in body lipids.

WATER-SOLUBLE VITAMINS. Most of the water-soluble vitamins are components of coenzymes (Table 17–4). For example, NAD is derived from niacin, and coenzyme A from vitamin B_5 (pantothenic acid). Water-soluble vitamins are rapidly exchanged between the digestive tract and the circulating blood, and excessive amounts are readily excreted in the urine. For this reason, hypervitaminosis that involves water-soluble vitamins is relatively uncommon, except among individuals taking large doses of vitamin supplements.

The bacteria that reside in the intestines help prevent deficiency diseases by producing five of the nine water-soluble vitamins, in addition to fat-soluble vitamin K. The intestinal epithelium can easily absorb all of the water-soluble vitamins except B_{12}. The B_{12} molecule is large, and it must be bound to intrinsic factor, which is secreted by the gastric mucosa, before absorption can occur. ∞ p. 596

Clinical Note
BERIBERI

A deficiency of *thiamine (vitamin B_1)* causes *beriberi*. In developing countries, beriberi is due to consumption of milled (polished) rice. In developed nations, thiamine deficiency is due to inadequate thiamine intake and absorption in chronic alcoholics.

Thiamine deficiency primarily affects the cardiovascular system *(wet beriberi)* and the nervous system *(dry beriberi)*. Beriberi heart disease includes peripheral vasodilation, retention of sodium and water that leads to edema, and biventricular myocardial failure. *Acute fulminate cardiovascular beriberi* can end in cardiovascular collapse. Improvement occurs with thiamine replacement.

Two primary nervous system diseases are due to thiamine deficiency. *Wernicke's encephalopathy (WE)*, or cerebral beriberi, causes vomiting, dysfunction of the extraocular muscles, fever, ataxia, and mental deterioration. *Korsakoff's syndrome (KS)*, also called *Korsakoff's psychosis (KP)*, is a continuation of WE and includes retrograde amnesia and impaired ability to learn. Thiamine will completely reverse the effects of WE but will only partially reverse the symptoms of KS. ■

Water

Daily water requirements average 2500 mL (10 cups), or roughly 40 mL/kg body weight. The specific requirement varies with environmental conditions and metabolic activities. For example, exercise increases metabolic energy requirements and accelerates water losses due to evaporation and perspiration. The temperature rise that accompanies a fever has a similar effect; for each degree (°C) temperature rises above normal, daily water loss increases by 200 mL. Thus, the

TABLE 17-6 *Vitamin Sources and Common Vitamin Deficiencies*

VITAMIN	PROBLEMS THAT RESULT FROM DEFICIENCY	SOURCE
FAT-SOLUBLE		
A	Night blindness, skin lesions	Butter, yellow fruit, green leafy vegetables, milk
D	Bone and muscle pain, weakness, softening of bones	Fish, fortified milk, exposure to sunlight
E	Hyporeflexia, ataxia, anemia	Nuts, green leafy vegetables, wheat
K	Increased bleeding	Liver, green leafy vegetables
WATER-SOLUBLE		
B_1 (thiamine)	Peripheral neuritis, depression, anorexia, poor memory	Whole grains, beef, pork, peas, beans
B_2 (riboflavin)	Sore throat, stomatitis, painful or swollen tongue, anemia	Milk, eggs, cheese, green leafy vegetables
B_3 (niacin)	Skin eruptions, diarrhea, enteritis, headache, dizziness, insomnia	Meat, eggs, milk
B_6 (pyridoxine)	Skin lesions, seizures, peripheral neuritis	Liver, meats, eggs, vegetables
B_9 (folic acid)	Megaloblastic anemia	Liver, fresh green vegetables, yeast
B_{12} (cyanocobalamin)	Irreversible nervous system damage, pernicious anemia	Fish, egg yolk, milk
C	Scurvy	Citrus fruits, tomatoes, strawberries

advice "Drink plenty of fluids" when you are sick has a solid physiological basis.

Most of your daily water ration is obtained by eating or drinking. The food you consume provides roughly 48 percent, and another 40 percent is obtained by drinking fluids. But a small amount of water—called *metabolic water*—is produced in mitochondria during the operation of the electron transport system. ∞ p. 633 The actual amount produced per day varies with the composition of the diet. A typical mixed diet in the U.S. contains 46 percent carbohydrates, 40 percent lipids, and 14 percent protein. This diet would produce roughly 300 mL of water per day (slightly more than one cup), about 12 percent of the average daily water requirement.

Diet and Disease

Diet has a profound influence on general health. We have already considered the effects of too many and too few nutrients, above-normal or below-normal concentrations of minerals, and hypervitaminosis and avitaminosis. More subtle long-term problems can occur when the diet includes the wrong proportions or combinations of nutrients. The average diet in the U.S. contains too many calories, and lipids provide too great a proportion of those calories. Such a diet increases the incidence of obesity, heart disease, atherosclerosis, hypertension, and diabetes in the U.S. population.

→ **CONCEPT CHECK QUESTIONS**

1. Which one of the five basic food groups should be consumed in the smallest quantities each day?
2. What is the difference between foods described as containing complete proteins and those described as containing incomplete proteins?
3. How would a decrease in the amount of bile salts in the bile affect the amount of vitamin A in the body?

Answers begin on p. 792.

■ Bioenergetics

Scientists in the field of **bioenergetics** study how organisms acquire and use energy. When chemical bonds are broken, energy is released. Inside cells, some of that energy may be captured as ATP, but much of it is lost to the environment as heat. The unit of energy measurement is the **calorie (cal)** (KAL-o-rē), which is the amount of energy required to raise the temperature of 1 g of water 1° centigrade. One gram of water is not a very practical measure when you are interested in the meta-

bolic operations that keep a 70-kg human alive, however, so the **kilocalorie (kcal)** (KIL-ō-kal-o-rē), or **Calorie (Cal),** is used instead. One Calorie is the amount of energy needed to raise the temperature of 1 *kilo*gram of water 1° centigrade. Calorie-counting guides that give the caloric value of various foods list Calories (kilocalories), not calories.

The Energy Content of Food

In cells, organic molecules combine with oxygen and are broken down to carbon dioxide and water. Oxygen is also consumed when something burns, and this process of combustion can be experimentally observed and measured. A known amount of food is placed in a chamber, called a **calorimeter** (kal-o-RIM-e-ter), which is filled with oxygen and surrounded by a known volume of water. Once the food is inside, the chamber is sealed and the contents are electrically ignited. When the food is completely burned and only ash remains in the chamber, the number of Calories released can be determined by comparing the water temperatures before and after the test. Such measurements show that the burning, or catabolism, of lipids releases a considerable amount of energy—roughly 9.46 Calories per gram (Cal/g). In contrast, the catabolism of carbohydrates releases 4.18 Cal/g, and the catabolism of protein releases 4.32 Cal/g. Most foods are mixtures of fats, proteins, and carbohydrates, so the values in a "Calorie counter" vary as a result.

Metabolic Rate

Clinicians can assess your metabolic state to learn how many Calories your body is utilizing. The result can be expressed as Calories per hour, Calories per day, or Calories per unit of body weight per day. What is actually measured is the sum of all the various anabolic and catabolic processes occurring in your body—its **metabolic rate** at that time. Metabolic rate varies with the activity under way; for instance, measurements of someone sprinting and someone sleeping are quite different. To reduce such variations, the testing conditions are standardized so as to determine the **basal metabolic rate (BMR).** Ideally, the BMR reflects the minimum, resting energy expenditures of an awake, alert person. An average individual has a BMR of 70 Cal per hour, or about 1680 Cal per day. Although the test conditions are standardized, other uncontrollable factors influence the BMR, including age, sex, physical condition, body weight, and genetic differences.

The daily energy expenditure for each individual varies with the activities undertaken. For example, a person who leads a sedentary life may have near-basal energy demands, but one hour of swimming can increase the daily caloric requirements by 500 Cal or more. If daily energy intake exceeds the body's total energy demands, the excess energy will be stored, primarily as triglycerides in adipose tissue. If daily caloric expenditures exceed dietary intake, a net reduction in the body's energy reserves will occur, with a corresponding loss in weight. This relationship explains the importance of both Calorie counting and daily exercise in a weight-control program.

Thermoregulation

The BMR (basal metabolic rate) provides only an estimate of the rate of energy use. Body cells capture only a part of that energy as ATP, and the rest is "lost" as heat. This heat loss serves an important homeostatic purpose. Even though humans are subjected to vast changes in environmental temperatures, our complex biochemical systems have a major limitation: enzymes operate over only a relatively narrow range of temperatures. Fortunately, certain anatomical and physiological mechanisms keep body temperatures within acceptable limits, regardless of environmental conditions. This homeostatic process is called **thermoregulation** (*therme*, heat). ∞ p. 14 Failure to control body temperature can result in serious physiological effects (Figure 17–11●). For example, a body temperature below 36°C (97°F) or above 40°C (104°F) can cause disorientation, and a temperature above 42°C (108°F) can cause convulsions and permanent cell damage.

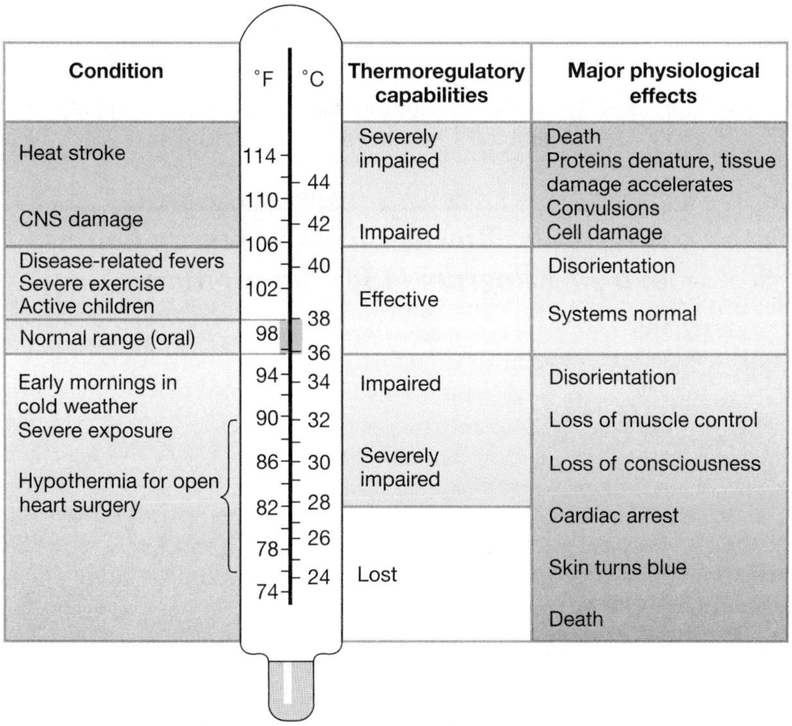

Condition	°F	°C	Thermoregulatory capabilities	Major physiological effects
Heat stroke	114		Severely impaired	Death Proteins denature, tissue damage accelerates
	110	44		Convulsions
CNS damage	106	42	Impaired	Cell damage
Disease-related fevers Severe exercise Active children	102	40		Disorientation
		38	Effective	Systems normal
Normal range (oral)	98	36		
Early mornings in cold weather Severe exposure	94	34	Impaired	Disorientation
	90	32		Loss of muscle control
	86	30	Severely impaired	Loss of consciousness
Hypothermia for open heart surgery	82	28		Cardiac arrest
	78	26	Lost	Skin turns blue
	74	24		Death

● **FIGURE 17–11 Normal and Abnormal Variations in Body Temperature.**

Mechanisms of Heat Transfer

Heat exchange with the environment involves four basic processes—*radiation, conduction, convection,* and *evaporation.*

1. *Radiation.* Warm objects lose heat energy as infrared radiation. When we feel the sun's heat, we are experiencing radiant heat. Your body loses heat the same way. More than half of the heat you lose occurs by radiation.

2. *Conduction.* Conduction is the direct transfer of energy through physical contact. When you sit on a cold plastic chair in an air-conditioned room, you are immediately aware of this process. Conduction is generally not an effective mechanism of gaining or losing heat.

3. *Convection.* Convection is the result of conductive heat loss to the air that overlies the surface of an object. Warm air rises because it is lighter than cool air. As your body conducts heat to the air next to your skin, that air warms and rises, which moves the air away from your skin surface. Cooler air replaces it, and as this air in turn warms, the pattern repeats.

4. *Evaporation.* When water evaporates, it changes from a liquid to a vapor. This process absorbs energy—roughly 580 calories (0.58 Cal) per gram of water evaporated—and, thus, cools any surface on which it occurs. The rate of evaporation and heat loss that occurs at your skin is highly variable. Each hour, 20–25 mL of water crosses epithelia and evaporates from the alveolar surfaces of the lungs and the surface of the skin. This *insensible perspiration* remains relatively constant; it accounts for roughly one-fifth of the average heat loss from a body at rest. The sweat glands responsible for *sensible perspiration* have a tremendous scope of activity, and range from virtual inactivity to secretory rates of 2–4 liters (or 2–4 kg) per hour. This is equivalent to an entire day's resting water loss in under an hour.

To maintain a constant body temperature, an individual must lose heat as fast as it is generated by metabolic operations. Altering the rates of heat loss and heat gain requires the coordinated activity of many different systems. That activity is coordinated by the **heat-loss center** and **heat-gain center** of the hypothalamus. The heat-loss center adjusts activity through the parasympathetic division of the autonomic nervous system, whereas the heat-gain center directs its responses through the sympathetic division. The overall effect is to control temperature fluctuations by influencing two events: the rate of heat production and the rate of heat loss to the environment. These events may be further supported by behavioral changes or modifications, such as moving into the shade or sunlight, or the addition or removal of clothing.

PROMOTING HEAT LOSS. When the temperature at the heat-loss center exceeds its set point, three responses occur:

1. Peripheral blood vessels dilate, which sends warm blood to the surface of the body. The skin takes on a reddish color and rises in temperature; heat loss through radiation and convection increases.
2. Sweat glands are stimulated, and as perspiration flows across the skin, heat loss through evaporation accelerates.
3. The respiratory centers are stimulated, and the depth of respiration increases. The individual often begins respiring through the mouth, which enhances heat loss through increased evaporation from the lungs.

The efficiency of heat loss by evaporation varies with environmental conditions, especially the "relative humidity" of the air. At 100 percent humidity, the air is saturated; it is holding as much water vapor as it can at that temperature. Under these conditions, evaporation is ineffective as a cooling mechanism. This is why humid, tropical conditions can be so uncomfortable—people perspire continuously but remain warm and wet.

PROMOTING HEAT GAIN: HEAT CONSERVATION AND GENERATION. The function of the heat-gain center of the brain is to prevent **hypothermia** (hī-pō-THER-mē-uh), or below-normal body temperature. When body temperature falls below acceptable levels, the heat-loss center is inhibited and the heat-gain center is activated. Its activation results in responses that conserve body heat and promote heat generation.

Heat is conserved by decreasing blood flow to the skin, which thereby reduces losses by radiation, convection, and conduction. The skin cools, and with blood flow restricted, it may take on a bluish or pale coloration. In addition, blood that returns from the limbs is shunted into a network of deep veins that lies beneath an insulating layer of subcutaneous fat. ∞ p. 497 (Under warm conditions, blood flows through a superficial venous network, through which heat can be lost.)

In addition to conserving heat, the heat-gain center stimulates two mechanisms that generate heat. In *shivering thermogenesis* (ther-mō-JEN-e-sis), muscle tone is gradually increased until stretch receptors stimulate brief, oscillatory contractions of antagonistic skeletal muscles. The resulting shivering stimulates energy consumption by skeletal muscles, and the generated heat warms the deep vessels to which the blood has been diverted. Shivering can increase the rate of heat generation by as much as 400 percent.

Clinical Note
FIRE-GROUND REHABILITATION

Firefighting is one of the most dangerous and physically demanding occupations today. It is common for EMS personnel to establish and staff rehabilitation units on the fire ground. For years, firefighter rehabilitation at the fire scene consisted of coffee and doughnuts provided by volunteer organizations. However, because it has been recognized that stress- and heat-related emergencies are the primary causes of on-duty firefighter deaths, the concept of *Emergency Incident Rehabilitation (EIR)* was developed. Large fires, such as wildfires, can involve hundreds of firefighters from multiple departments. In these situations, multiple rehab/EIR sectors must be established as a functional part of the Incident Command System (ICS).

It is important for all fire-ground personnel to report to the Rehab Sector immediately after any of the following activities:

- Strenuous activity such as forcible entry, advancing hose lines, closed space search and rescue, and/or ventilation
- The use and depletion of two self-contained breathing apparatus (SCBA) bottles
- Thirty (30) minutes of operation within a hazardous/dangerous environment
- Failure of an SCBA unit

Incoming personnel should be carefully evaluated including vital signs, breath sounds, examination of skin color and condition, and body core temperature. It is important to try and to determine the firefighters' hydration status. Remember, over 60 percent of body weight is water (Table 17–7). Water is the best rehydration agent. Alternatively, water can be mixed with a commercial sport hydration beverage in a 50/50 mixture and administered at about 40°F (4.5°C).

Oral hydration with water is indicated in almost every fire-ground operation, regardless of the season. Large fires, or those that occur during summer months, especially in the South and the Southwest, can cause significant fluid loss. In many cases, firefighting personnel may need several liters of intravenous fluids to restore lost water.

TABLE 17–7 *Biochemical Content in Grams of a 70-Kilogram (154-Pound) Man*

CONTENT	GRAMS
Water	41,400
Fat	12,600
Protein	12,600
Carbohydrate	300
Na	63
K	150
Ca	1,160
Mg	21
Cl	85
P	670
S	112
Fe	3
L	0.014

● **FIGURE 17–12 Fire-ground Operations.** Large fires require multiple rehab sectors. Firefighting personnel should not be released to return to firefighting until cleared by rehab sector personnel.

There should be established protocols for intravenous hydration in the field. Rehab sector personnel should decide whether a firefighter can return to fire fighting or remain for additional rehabilitation based on protocols (Figure 17–12●). In addition to hydration, personnel should be provided some form of nutrition while in the rehab sector. Cut fresh fruit, such as apples or oranges, are best. Salty products and salt tablets should not be provided. ■

In *nonshivering thermogenesis,* hormones are released that increase the metabolic activity of cells in all tissues. Epinephrine from the adrenal gland immediately increases the breakdown of glycogen and glycolysis in the liver and in skeletal muscles and increases the metabolic rate in most tissues. The heat-gain center also stimulates the release of thyroxine by the thyroid gland, which accelerates carbohydrate use and the breakdown of all other nutrients. These effects develop gradually over a period of days to weeks.

→ **CONCEPT CHECK QUESTIONS**

1. How would a pregnant woman's BMR (basal metabolic rate) compare with her BMR when she is not pregnant?
2. Under what conditions would evaporative cooling of the body be ineffective?
3. What effect would the vasoconstriction of peripheral blood vessels have on body temperature on a hot day?

Answers begin on p. 792.

■ Aging and Nutritional Requirements

Nutritional requirements do not change drastically with age. However, changes in lifestyle, eating habits, and income that often accompany aging can directly affect nutrition and health.

The recommended proportions of calories provided by different foods do not change with advancing age; current guidelines indicate that for individuals of all ages, proteins should provide 11 percent to 12 percent of daily caloric intake; carbohydrates 55 percent to 60 percent; and fats less than 30 percent. Caloric *requirements,* however, do change with aging. For each decade after age 50, caloric requirements decrease by 10 percent. These decreases are associated with reductions in metabolic rates, body mass, activity levels, and exercise tolerance.

Clinical Note
NUTRITION AND EMERGENCY MEDICAL SERVICES

The job demands and scheduling of emergency medical services work make it difficult to eat a balanced diet and obtain adequate exercise. For the most part, EMS work is fairly sedentary. However, at certain times, it can require significant physical exertion and stamina. Because of this, it is essential that emergency personnel eat a balanced diet, obtain moderate aerobic exercise on a regular basis, and maintain normal body weight. Unfortunately, more and more EMS workers can be classified as obese. The definition of obesity varies, but it is generally defined as weighing at least 20 percent or more in excess of your ideal body weight. For example, patients with an ideal body weight of 178 pounds would be considered obese when their weight exceeded 214 pounds. A better measurement of obesity is through determination of the *body mass index (BMI).* The BMI takes the patient's height into consideration and is calculated by dividing the body weight (in kilograms) by the patient's height (in meters). Many of the health problems associated with obesity begin to cause problems when the person's body weight exceeds 120 percent or more of ideal weight. Low back strain, one of the most common on-the-job injuries in EMS, occurs more frequently in personnel who are overweight.

Unfortunately, a few emergency workers can be classified as being morbidly obese. To be labeled as *morbidly obese,* the individuals must weigh more than two times their ideal body weight. For example, a patient with an ideal body weight of 178 pounds would be considered morbidly obese when their weight exceeds exceeded 356 pounds. People who are morbidly obese have serious risk factors for developing multiple life-threatening problems. Regardless of their overall general health, a morbidly obese person generally cannot perform all of the tasks expected of an EMT or paramedic during the course of routine EMS work.

Many EMS systems have health maintenance programs that identify personnel at risk for illness or injury, including obesity. When employees are identified as being obese, they may be asked to participate in developing a plan that will help them to lose weight and restore them to health. Such plans usually include an exercise regimen, dietary counseling, and wellness education. Physical fitness is more ingrained in the fire service, where most fire-based EMS operations have ongoing fitness programs. Although EMS units roll considerably more often than engine and truck companies, the EMS crew can usually get needed exercise over the course of the day. In busy systems, it may be prudent simply to block out time for exercise with the understanding that an MCI or high system usage would require interruption of exercise regimens. ■

With age, several factors combine to result in an increased need for calcium. Some degree of osteoporosis is a normal consequence of aging; a sedentary lifestyle contributes to the problem. The rate of bone loss decreases if calcium levels are kept elevated. The elderly are also likely to require supplemental vitamin D_3 if they are to absorb the calcium they need. Many elderly people spend most of their time indoors and avoid the sun when outdoors. Although this behavior slows sun damage to their skin, which is thinner than that of younger people, it also eliminates vitamin D_3 production by the skin. ∞ p. 128 This vitamin is converted to the hormone calcitriol, which stimulates calcium absorption by the small intestine.

Maintaining a healthy diet becomes more difficult with age as a result of changes in the senses of smell and taste and in the structure of the digestive system. With age, the number and sensitivity of olfactory and gustatory receptors decrease. ∞ p. 324 As a result, food becomes less appetizing, so less food is eaten. Making matters worse, the mucosal lining of the digestive tract becomes thinner with age, so nutrient absorption becomes less efficient. Thus, what food the elderly do eat is not utilized very efficiently. Elderly people on fixed budgets may reduce their consumption of animal protein, which is the primary source of dietary iron. The combination of small quantities plus inefficient absorption makes them prone to iron deficiency, which causes anemia.

Clinical Note
NUTRITION

Many believe that it is impossible to maintain an adequate diet while working in EMS. However, with a little planning and awareness of your options, you can eat sensibly and avoid the fast-food habit. Unfortunately, high performance EMS System Status Management systems often requires EMS crews to remain in the ambulance for the entire shift due to "System Status Management." This makes it very difficult to plan meals and eat sensibly. However, with a little ingenuity and planning, this too can be mastered.

The most difficult part of improving nutrition is altering established bad habits. Changing your behavior requires some commitment and self-discipline, an understanding of the change process, and patience with what will become long-term self-improvement. You must set realistic goals, and understand that backsliding occasionally happens. Whatever your goals may be, such as reducing excess weight, gaining weight, or regularly eating more wholesome foods, it is helpful to be able to analyze your progress by using charts or daily food diaries.

Good nutrition is fundamental to your well-being because your food is your body's fuel. In addition to eating balanced meals, you must also eat in moderation, limit fat consumption, and make time for exercise. If you eat and treat yourself poorly, both your short- and long-term well-being will be jeopardized.

The topic of nutrition can seem, at times, overwhelming, but if you rely on sound sources of information and build your knowledge gradually, you will benefit in many ways. One key to eating well is to learn the major food groups and eat a variety of foods from them daily (Figure 17–13●). Those food groups and the recommended number of daily servings are:

- *Grains/breads.* 6–11 servings per day, for complex carbohydrates, B vitamins, and fiber.
- *Vegetables.* 3–5 servings per day, for fiber, iron, vitamins A and C, and folate.

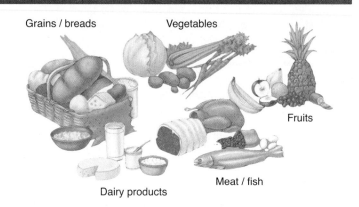
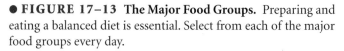

● **FIGURE 17–13 The Major Food Groups.** Preparing and eating a balanced diet is essential. Select from each of the major food groups every day.

- *Fruits.* 2–4 servings per day, for vitamins A and C, potassium, and fiber.
- *Dairy products.* 2–3 servings per day, for calcium, protein, and vitamins A and D.
- *Meat/fish.* 2–3 servings per day, for protein, zinc, iron, and B vitamins.

Avoid or minimize the intake of fat, salt, sugar, cholesterol, and caffeine. For example, you can avoid a dose of fat by eating lean meat instead of marbled meat. An apple is far more nutritious than a slice of apple pie, which has a filling that is high in sugar and a crust that is saturated with fat. Food labels contain abundant information about nutritional content. Learn to read them. Standardization of food labels has reduced much of the confusion (Figure 17–14●). Be sure to check the serving size to avoid misinterpreting the food's overall nutritional value. In general, aim for a diet that is approximately 40 percent carbohydrates, 40 percent protein, and 20 percent fat.

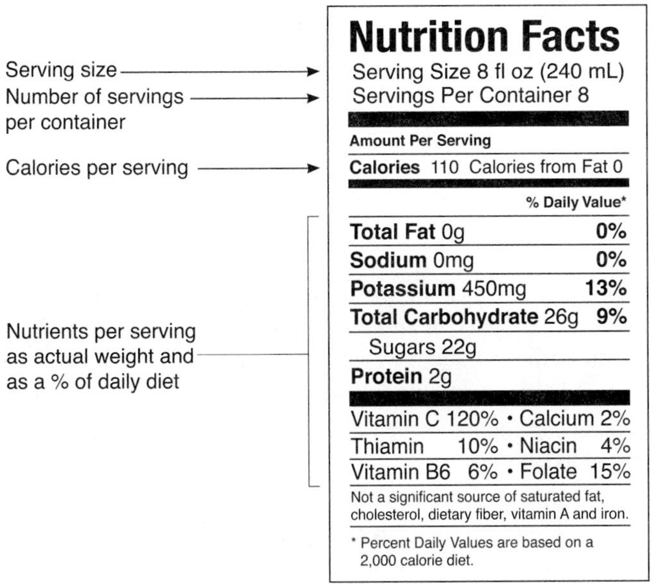

Food portions also have a significant impact on body weight. Even a well-planned, healthy diet can result in weight gain if the portions are too large. Note that snacking is a weight-gain trap. Plan to eat low-calorie snacks, and buy them before you get hungry.

Eating on the run, as EMS providers must often do, can be less detrimental if you plan ahead and carry a small cooler filled with whole-grain sandwiches, cut vegetables, fruit, and other wholesome foods. If you must, stop at a local market instead of the fast-food place next door. Buy fresh fruit, yogurt, and sensible deli selections. They are more nutritious and less expensive than "fast foods."

Finally, monitor your fluid intake. Your body needs plenty of fluids to properly maintain the internal environment. Pay attention to what you are drinking. Fill a "go-cup" with fresh ice water when you stop by the emergency department instead of spending your money on soft drinks. Water is more thirst-quenching, cheaper, and much better for you. ■

Nutrition Facts

Serving Size 8 fl oz (240 mL)
Servings Per Container 8

Amount Per Serving

Calories 110   Calories from Fat 0

% Daily Value*

Total Fat 0g	**0%**
Sodium 0mg	**0%**
Potassium 450mg	**13%**
Total Carbohydrate 26g	**9%**
Sugars 22g	
Protein 2g	

Vitamin C 120% • Calcium 2%
Thiamin   10% • Niacin   4%
Vitamin B6   6% • Folate   15%

Not a significant source of saturated fat, cholesterol, dietary fiber, vitamin A and iron.

* Percent Daily Values are based on a 2,000 calorie diet.

Serving size →
Number of servings per container →
Calories per serving →
Nutrients per serving as actual weight and as a % of daily diet →

● **FIGURE 17–14 Example of Standardized Food Label.**

Chapter Review

Access more review material online at *www.prenhall.com/bledsoe*. There you will find quiz questions, labeling activities, animations, essay questions, and web links.

Key Terms

aerobic metabolism 631
basal metabolic rate (BMR) 648
Calorie 648
electron transport system (ETS) 633
glycogen 636

glycolysis 631
metabolic turnover 630
metabolism 630
nutrient 629
nutrition 642

thermoregulation 649
tricarboxylic (TCA, citric acid, or Krebs) cycle 633
vitamin 645

Related Clinical Terms

antipyretic drugs Drugs administered to control or reduce fever.

avitaminosis (a-vī-ta-min-Ō-sis) A vitamin deficiency disease.

eating disorders Psychological problems that result in inadequate or excessive food consumption. Examples include anorexia nervosa and bulimia.

gout A metabolic disorder characterized by the precipitation of uric acid crystals within joint cavities.

heat exhaustion A malfunction of the thermoregulatory system caused by excessive fluid loss in perspiration.

heat stroke A condition in which the thermoregulatory center stops functioning and body temperature rises uncontrollably.

hyperuricemia (hī-per-ū-ri-SĒ-mē-uh) Plasma levels of uric acid above 7.4 mg/dL; can result in the condition called *gout*.

hypervitaminosis (hī-per-vī-ta-mi-NŌ-sis) A disorder caused by the ingestion of excessive quantities of one or more vitamins.

hypothermia (hī-pō-THER-mē-uh) Below-normal body temperature.

ketoacidosis (kē-tō-as-i-DŌ-sis) Reduced blood pH due to the presence of ketone bodies.

ketonemia (kē-tō-NĒ-mē-uh) Elevated levels of ketone bodies in blood.

ketonuria (kē-tō-NOO-rē-uh) The presence of ketone bodies in urine.

ketosis (kē-TŌ-sis) Abnormally high concentration of ketone bodies in body fluids.

liposuction The removal of adipose tissue by suction through an inserted tube.

obesity Body weight more than 20 percent above the ideal weight for a given individual.

phenylketonuria (fen-il-kē-to-NOO-rē-uh) An inherited metabolic disorder characterized by an inability to convert phenylalanine to tyrosine.

protein deficiency diseases Nutritional disorders that result from a lack of essential amino acids.

pyrexia (pī-REK-sē-uh) Elevated body temperature; a fever is a body temperature maintained at greater than 99°F (37.2°C).

Summary Outline

1. Body cells are chemical factories that break down organic molecules and their building blocks to obtain energy for growth, cell division, and other tissue specific activities.

CELLULAR METABOLISM 630

1. In general, cells break down excess carbohydrates first, then lipids, while conserving amino acids. Only about 40 percent of the energy released through catabolism is captured in ATP; the rest is released as heat. *(Figure 17–1)*

2. Cells synthesize new compounds (1) to perform structural maintenance and repair, (2) to support growth, (3) to produce secretions, and (4) to build nutrient reserves.

3. Cells feed small organic molecules to their mitochondria to obtain ATP to perform cellular functions. *(Figure 17–2)*

Key Note 631

Carbohydrate Metabolism 631

4. Most cells generate ATP and other high-energy compounds through the breakdown of carbohydrates.

5. **Glycolysis** and **aerobic metabolism** provide most of the ATP used by typical cells. In glycolysis, each molecule of glucose yields two molecules of pyruvic acid and two molecules of ATP. *(Figure 17–3)*

6. In the presence of oxygen, the pyruvic acid molecules enter the mitochondria, where they are broken down completely in the **tricarboxylic acid (TCA) cycle.** The carbon and oxygen atoms are lost as carbon dioxide, and the hydrogen atoms are passed by *coenzymes* to the *electron transport system. (Figure 17–4)*

7. *Cytochromes* of the **electron transport system (ETS)** pass along electrons to oxygen to form water and generate ATP. *(Figure 17–5)*

8. For each glucose molecule completely broken down by aerobic pathways, a typical cell gains 36 ATP molecules.

Alternate Catabolic Pathways 635

9. When supplies of glucose are limited, cells can break down other nutrients to provide molecules for the TCA cycle. *(Figure 17–6)*

10. **Gluconeogenesis,** the synthesis of glucose, enables a cell to manufacture glucose molecules from other carbohydrates, glycerol, or some amino acids. *Glycogen* is an important energy reserve when extracellular glucose is low. *(Figure 17–7)*

Lipid Metabolism 636

11. During **lipolysis** (lipid catabolism), lipids are broken down into pieces that can be converted into pyruvic acid or channeled into the TCA cycle.

12. Triglycerides, which are the most abundant lipids in the body, are split into glycerol and fatty acids. Glycerol enters the glycolytic pathways, and fatty acids enter the mitochondria.

13. *Beta-oxidation* is the breakdown of fatty acid molecules into two-carbon fragments that can enter the TCA cycle or be converted to ketone bodies.

14. Lipids cannot provide large amounts of ATP in a short amount of time. However, cells can shift to lipid-based energy production when glucose reserves are limited.

15. In **lipogenesis,** the synthesis of lipids, almost any organic molecule can be used to form glycerol. **Essential fatty acids** cannot be synthesized and must be included in the diet. *(Figure 17–8)*

16. Lipids circulate as **lipoproteins** (lipid-protein complexes that contain triglycerides and cholesterol) or as **free fatty acids (FFA)** (lipids associated with albumin that can diffuse easily across cell membranes).

Protein Metabolism 638

17. If other energy sources are inadequate, mitochondria can break down amino acids. In the mitochondria, the amino group may be removed by *transamination* or *deamination*. The resulting carbon skeleton may enter the TCA cycle to generate ATP or be converted to ketone bodies.

18. Protein catabolism is impractical as a source of quick energy.

19. The body can synthesize roughly half of the amino acids needed to build proteins. The 10 *essential amino acids* must be acquired through the diet.

Nucleic Acid Metabolism 640

20. DNA in the nucleus is never catabolized for energy. RNA molecules are broken down and replaced regularly; usually they are recycled as new nucleic acids.

A Summary of Cellular Metabolism 641

21. No one cell can perform all of the anabolic and catabolic operations necessary to support life. Homeostasis can be preserved only when the metabolic activities of different tissues are coordinated. *(Figure 17–9)*

DIET AND NUTRITION 642

1. **Nutrition** is the absorption of essential nutrients from food. A *balanced diet* contains all of the ingredients needed to maintain homeostasis; it prevents **malnutrition.**

Food Groups and the Food Pyramid 642

2. The *food pyramid* includes five **basic food groups:** grains; vegetables; fruits; milk; meat and beans; as well as a fats, sugars and salt category. *(Figure 17–10, Table 17–1)*

Key Note 644

Minerals, Vitamins, and Water 644

3. **Minerals** act as cofactors in various enzymatic reactions. They also contribute to the osmotic concentration of body fluids, and they play a role in transmembrane potentials, action potentials, neurotransmitter release, muscle contraction, skeletal construction and maintenance, gas transport, buffer systems, fluid absorption, and waste removal. *(Tables 17–2, 17–3, 17–4)*

4. **Vitamins** are needed in very small amounts. Vitamins A, D, E, and K are **fat-soluble vitamins;** taken in excess, they can lead to **hypervitaminosis. Water-soluble vitamins** are not stored in the body; a lack of adequate dietary supplies can lead to **avitaminosis** *(deficiency disease). (Tables 17–4, 17–5, 17–6)*

5. Daily water requirements average about 40 mL/kg body weight. Water is obtained from food, drink, and metabolic generation.

Diet and Disease 648

6. A balanced diet can improve general health. Most Americans consume too many calories, mostly in the form of lipids.

BIOENERGETICS 648

1. The energy content of food is usually expressed as **Calories** per gram (Cal/g). Less than half of the energy content of glucose or any other organic nutrient can be captured by body cells.

The Energy Content of Food 648

2. The catabolism of each gram of lipid releases 9.46 C, about twice the Calories released by the breakdown of the same amount of carbohydrate or protein.

Metabolic Rate 648

3. The total of all the body's anabolic and catabolic processes over a given period of time is an individual's **metabolic rate.** The **basal metabolic rate (BMR)** is the rate of energy utilization at rest.

Thermoregulation 649

4. The homeostatic regulation of body temperature is **thermoregulation.** Heat exchange with the environment involves four processes: **radiation, conduction, convection,** and **evaporation.** *(Figure 17–11)*

5. The hypothalamus acts as the body's thermostat; it contains the **heat-loss center** and the **heat-gain center.**

6. Mechanisms for increasing heat loss include both physiological mechanisms (superficial blood vessel dilation, increased perspiration, and accelerated respiration) and behavioral adaptations.

7. Body heat may be conserved by reducing blood flow to the skin. Heat can be generated by *shivering thermogenesis* and *nonshivering thermogenesis. (Figure 17–12; Table 17–7)*

AGING AND NUTRITIONAL REQUIREMENTS 651

1. Caloric requirements drop by 10 percent each decade after age 50. Changes in the senses of smell and taste dull appetite, and changes to the digestive system decrease the efficiency of nutrient absorption from the digestive tract. *(Figures 17–13, 17–14)*

Review Questions

Level 1: Reviewing Facts and Terms

Match each item in column A with the most closely related item in column B. Place letters for answers in the spaces provided.

COLUMN A
- ____ 1. glucose formation
- ____ 2. lipid catabolism
- ____ 3. synthesis of lipids
- ____ 4. linoleic acid
- ____ 5. deamination
- ____ 6. phenylalanine
- ____ 7. ketoacidosis
- ____ 8. A, D, E, K
- ____ 9. B complex and vitamin C
- ____ 10. calorie
- ____ 11. uric acid
- ____ 12. hypothermia

COLUMN B
- a. gluconeogenesis
- b. essential amino acid
- c. below-normal body temperature
- d. unit of energy
- e. fat-soluble vitamins
- f. water-soluble vitamins
- g. lipolysis
- h. nitrogenous waste
- i. essential fatty acid
- j. removal of an amino group
- k. decrease in pH
- l. lipogenesis

13. Cells synthesize new organic components to:
    (a) perform structural maintenance and repairs.
    (b) support growth.
    (c) produce secretions.
    (d) a, b, and c are correct.

14. During the complete catabolism of one molecule of glucose, a typical cell gains _____ ATP.
    (a) 4
    (b) 18
    (c) 36
    (d) 144

15. The breakdown of glucose to pyruvic acid is:
    (a) glycolysis.
    (b) gluconeogenesis.
    (c) cellular respiration.
    (d) beta-oxidation.

16. Glycolysis yields an immediate net gain of _____ ATP molecules for the cell.
    (a) 1
    (b) 2
    (c) 4
    (d) 36

17. The electron transport chain yields a total of _____ molecules of ATP in the complete catabolism of one glucose molecule.
    (a) 2
    (b) 4
    (c) 32
    (d) 36

18. The synthesis of glucose from simpler molecules is called:
    (a) glycolysis.
    (b) lipolysis.
    (c) gluconeogenesis.
    (d) beta-oxidation.

19. The lipoproteins that transport excess cholesterol from peripheral tissues back to the liver for storage or excretion in the bile are the:
    (a) chylomicrons.
    (b) FFA.
    (c) LDLs.
    (d) HDLs.

20. The removal of an amino group in a reaction that generates an ammonia molecule is called:
    (a) ketoacidosis.
    (b) transamination.
    (c) deamination.
    (d) denaturation.

21. A complete protein contains:
    (a) the proper balance of amino acids.
    (b) all the essential amino acids in sufficient quantities.
    (c) a combination of nutrients selected from the food pyramid.
    (d) N compounds produced by the body.

22. All minerals and most vitamins:
    (a) are fat-soluble.
    (b) cannot be stored by the body.
    (c) cannot be synthesized by the body.
    (d) must be synthesized by the body because they are not present in adequate amounts in the diet.

23. The basal metabolic rate represents the:
    (a) maximum energy expenditure when exercising.
    (b) minimum, resting energy expenditure of an awake, alert person.
    (c) minimum amount of energy expenditure during light exercise.
    (d) muscular energy expenditure added to the resting energy expenditure.

24. Over half of the heat loss from our bodies is attributable to:
    (a) radiation.
    (b) conduction.
    (c) convection.
    (d) evaporation.

25. Define the terms *metabolism, anabolism,* and *catabolism.*

26. What is a lipoprotein? What are the major groups of lipoproteins, and how do they differ?

27. Why are vitamins and minerals essential components of the diet?

28. What energy yields (in Calories per gram) are associated with the catabolism of carbohydrates, lipids, and proteins?

29. What is the basal metabolic rate (BMR)?

30. What four mechanisms are involved in thermoregulation?

Level 2: Reviewing Concepts

31. The function of the TCA cycle is to:
    (a) produce energy during periods of active muscle contraction.
    (b) break six-carbon chains into three-carbon fragments.
    (c) prepare the glucose molecule for further reactions.
    (d) remove hydrogen atoms from organic molecules and transfer them to coenzymes.

32. During periods of fasting or starvation, the presence of ketone bodies in the circulation causes:
    (a) an increase in blood pH.
    (b) a decrease in blood pH.
    (c) a neutral blood pH.
    (d) diabetes insipidus.

33. What happens during the process of glycolysis? What conditions are necessary for this process to take place?

34. Why is the TCA cycle called a cycle? What substance(s) enter(s) the cycle, and what substance(s) leave(s) it?

35. How does beta-oxidation function in lipid catabolism?

36. How can the food pyramid be used as a tool for developing a healthy lifestyle?

37. How is the brain involved in the regulation of body temperature?

38. Articles in popular magazines sometimes refer to "good cholesterol" and "bad cholesterol." To what types and functions of cholesterol might these terms refer? Explain your answer.

Level 3: Critical Thinking and Clinical Applications

39. Why is an individual who is starving more susceptible to infectious disease than an individual who is well-nourished?

40. The drug Colestipol™ binds bile salts in the intestine, and forms complexes that cannot be absorbed. How would this drug affect cholesterol levels in the blood?

18 The Urinary System

THE BEAUTY OF MODERN emergency care is that the capabilities of the modern emergency department can be brought to the patient's side. Here, a team from Fire Department New York (FDNY) EMS works to stabilize a patient they have resuscitated from cardiac arrest. Generally speaking, cardiac arrest patients who are not resuscitated in the field are unlikely to be resuscitated when they arrive at the emergency department.

Chapter Outline

Chapter Objectives

1. Identify the components of the urinary system and list their functions. (pp. 659–660)
2. Describe the structural features of the kidneys. (pp. 660–661)
3. Trace the path of blood flow through a kidney. (pp. 662–663)
4. Describe the structure of the nephron and the processes involved in urine formation. (pp. 663–666)
5. List and describe the factors that influence filtration pressure and the rate of filtrate formation. (pp. 666–668)
6. Describe the changes that occur in the tubular fluid as it moves through the nephron and exits as urine. (pp. 668–671)
7. Describe the structures and functions of the ureters, urinary bladder, and urethra. (pp. 675–680)
8. Discuss the process of urination and how it is controlled. (pp. 672–674)
9. Explain how the urinary system interacts with other body systems to maintain homeostasis in body fluids. (pp. 689, 690)
10. Describe how water and electrolytes are distributed within the body. (p. 680)
11. Explain the basic concepts involved in the control of fluid and electrolyte regulation. (pp. 681–684)
12. Explain the buffering systems that balance the pH of the intracellular and extracellular fluids. (pp. 684–685)
13. Identify the most frequent threats to acid-base balance. (pp. 686–687)
14. Describe the effects of aging on the urinary system. (p. 688)

Vocabulary Development

calyx a cup of flowers; *minor calyx*
detrudere to push down; *detrusor muscle*
fenestra a window; *fenestrated capillaries*
glomus a ball; *glomerulus*
gonion angle; *trigone*

juxta near; *juxtaglomerular apparatus*
micturire to urinate; *micturition*
nephros kidney; *nephron*
papillae small, nipple-shaped projections; *renal papillae*

podon foot; *podocyte*
rectus straight; *vasa recta*
ren kidney; *renal artery*
retro- behind; *retroperitoneal*
vasa vessels; *vasa recta*

THE HUMAN BODY CONTAINS trillions of cells bathed in extracellular fluid. In previous chapters we compared these cells to factories that burn nutrients to obtain energy. Imagine what would happen if real factories were as close together as cells in the body. Each factory would generate significant quantities of solid wastes and noxious gases, and create a serious pollution problem.

Comparable problems do not develop within the body as long as the activities of the digestive, cardiovascular, respiratory, and urinary systems are coordinated. The digestive tract absorbs nutrients from food and excretes solid wastes, and the liver adjusts the nutrient concentration of the circulating blood. The cardiovascular system delivers these nutrients, plus oxygen from the respiratory system, to peripheral tissues. As blood leaves these tissues, it carries the waste gas carbon dioxide and organic waste products to sites of excretion. The carbon dioxide is eliminated at the lungs, as described in Chapter 15. Most of the organic waste products in the blood are removed by the urinary system.

The **urinary system** performs the vital function of removing the organic waste products generated by cells throughout the body. It also has other essential functions that are often overlooked. A more complete list of its activities includes the following:

■ *Regulate blood volume and blood pressure.* Blood volume and blood pressure are regulated by (1) adjusting the volume of water lost in the urine, (2) releasing erythropoietin, and (3) releasing renin. ∞ p. 394

- *Regulate plasma concentrations of ions.* The plasma concentrations of sodium, potassium, chloride, and other ions are regulated by controlling the quantities lost in the urine. The plasma concentration of calcium ions is regulated by the synthesis of calcitriol. ∞ p. 394
- *Help to stabilize blood pH.* Blood pH is stabilized by controlling the loss of hydrogen ions (H^+) and bicarbonate ions HCO_3^- in the urine.
- *Conserve valuable nutrients.* Nutrients such as glucose and amino acids are conserved by preventing their excretion in the urine while organic waste products (especially the nitrogenous wastes *urea* and *uric acid*) are eliminated.

These activities are carefully regulated to keep the composition of the blood within acceptable limits. A disruption of any of these functions has immediate and potentially fatal consequences. This chapter examines the organization of the urinary system and describes the major regulatory mechanisms that control urine production and concentration.

■ The Organization of the Urinary System

The components of the urinary system are illustrated in Figure 18–1●. The two *kidneys* produce *urine,* a liquid that contains water, ions, and small soluble compounds. Urine leaving the kidneys travels along the paired *ureters* to the *urinary bladder* for temporary storage. When *urination* occurs, contraction of the muscular bladder forces the urine through the urethra and out of the body.

■ The Kidneys

The **kidneys** are located on either side of the vertebral column between the last thoracic and third lumbar vertebrae. The right kidney often sits slightly lower than the left (Figure 18–2a●), and

both lie between the muscles of the dorsal body wall and the peritoneal lining (Figure 18–2b●). This position is called *retroperitoneal* (re-trō-per-i-tō-NĒ-al; *retro-,* behind) because the organs are behind the peritoneum.

The position of the kidneys is maintained by (1) the overlying peritoneum, (2) contact with adjacent organs, and (3) supporting connective tissues. Each kidney is covered by a dense, fibrous

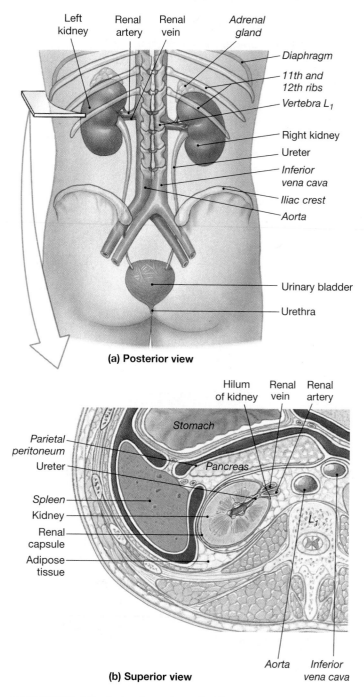

(a) Posterior view

(b) Superior view

● **FIGURE 18–2 An Overview of Kidney Anatomy. (a)** This posterior view of the trunk shows the positions of the kidneys and other components of the urinary system. **(b)** A superior view of a section at the level indicated in part (a) reveals the kidney's retroperitoneal position.

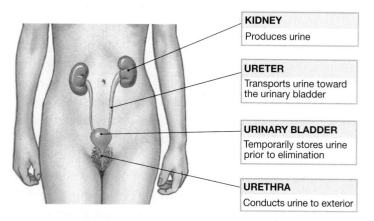

| **KIDNEY** |
| Produces urine |

| **URETER** |
| Transports urine toward the urinary bladder |

| **URINARY BLADDER** |
| Temporarily stores urine prior to elimination |

| **URETHRA** |
| Conducts urine to exterior |

● **FIGURE 18–1 The Components of the Urinary System.**

renal capsule and is packed in a soft cushion of adipose tissue. These connective tissues, along with suspensory collagen fibers, help prevent the jolts and shocks of day-to-day existence from disturbing normal kidney function. Damage to the suspensory fibers may cause the kidney to be displaced. This condition, called a *floating kidney,* is dangerous because the ureters or renal blood vessels may become twisted or kinked during movement.

Superficial and Sectional Anatomy

A typical kidney is reddish-brown and about 10 cm (4 in.) long, 5.5 cm (2.2 in.) wide, and 3 cm (1.2 in.) thick in adults. Each kidney weighs about 150 g (5.25 oz). An indentation called the **hilum** is the site of exit for the ureter (see Figure 18–2b), as well as the site at which the renal artery and renal nerve enter and the renal vein exits. (The adjective *renal* is derived from *ren,* which means "kidney" in Latin.) The **renal capsule** covers the surface of the kidney and lines the *renal sinus,* which is an internal cavity within the kidney.

The kidney is divided into an outer **renal cortex** and an inner **renal medulla** (Figure 18–3a,b●). The medulla contains 6 to 18 conical **renal pyramids.** The tip of each pyramid, known as the **renal papilla,** projects into the renal sinus. Bands of cortical tissue called *renal columns* extend toward the renal sinus between adjacent renal pyramids.

Urine production occurs in the renal pyramids and overlying areas of renal cortex. Ducts within each renal papilla discharge urine into a cup-shaped drain called a **minor calyx** (KĀ-liks; *calyx,* a cup of flowers; plural *calyces*). Four or five minor calyces (KAL-i-sēz) merge to form the **major calyces,** both of which combine to form a large, funnel-shaped chamber, which is the **renal pelvis.** The renal pelvis is connected to the ureter, through which urine drains out of the kidney.

Urine production begins in the renal cortex, in microscopic tubular structures called **nephrons** (NEF-ronz) (Figure 18–3c●). Each kidney has roughly 1.25 million nephrons, with a combined length of about 145 kilometers (85 miles).

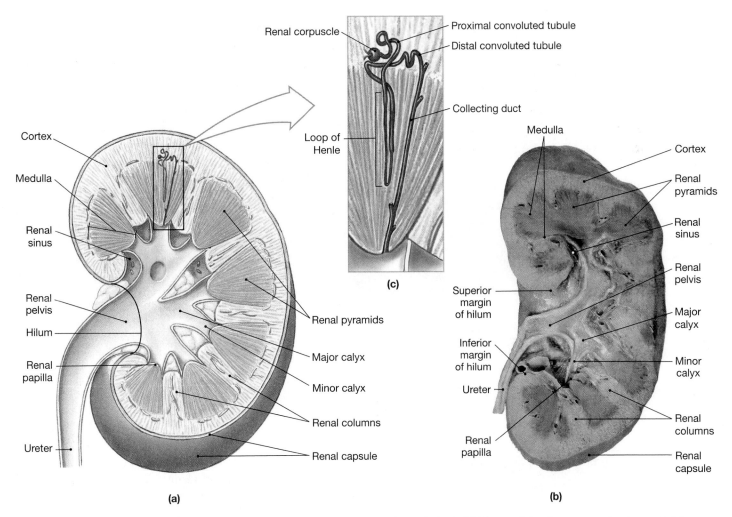

● **FIGURE 18–3 The Structure of the Kidney.** Two frontal sections through the left kidney—(**a**) a diagrammatic view and (**b**) a photograph—reveal the internal anatomy of the kidney. (**c**) This enlarged view shows the location and general structure of a nephron.

The Blood Supply to the Kidneys

Because the kidneys function to filter out wastes in the blood and excrete them in the urine, it's not surprising that the kidneys are well-supplied with blood. In healthy individuals, about 1200 mL of blood flows through the kidneys each minute, or approximately 20–25 percent of the cardiac output—a phenomenal amount of blood for organs with a combined weight of less than 300 g (10.5 oz)!

Figure 18–4a● diagrams the path of blood flow to, within, and out of the kidney. Each kidney receives blood from a

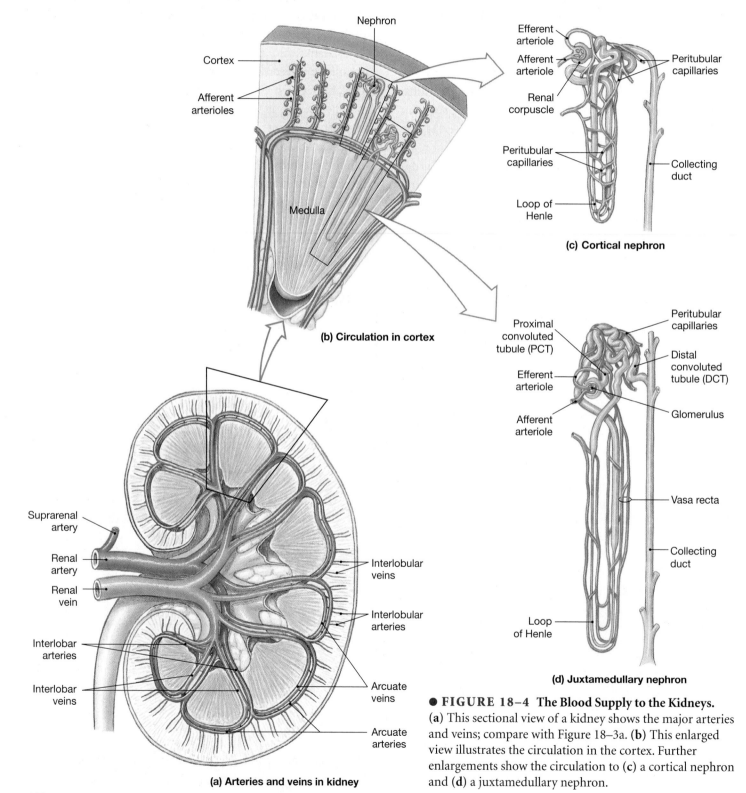

(b) Circulation in cortex

(c) Cortical nephron

(d) Juxtamedullary nephron

(a) Arteries and veins in kidney

● **FIGURE 18–4 The Blood Supply to the Kidneys.**
(**a**) This sectional view of a kidney shows the major arteries and veins; compare with Figure 18–3a. (**b**) This enlarged view illustrates the circulation in the cortex. Further enlargements show the circulation to (**c**) a cortical nephron and (**d**) a juxtamedullary nephron.

renal artery that originates from the abdominal aorta. As the renal artery enters the renal sinus, it divides into branches that supply a series of **interlobar arteries** that radiate outward between the renal pyramids. They then turn, and arch along the boundary lines between the cortex and medulla as the **arcuate** (AR-kū-āt) **arteries.** Each arcuate artery gives rise to a number of **interlobular arteries** that supply the cortex. **Afferent arterioles** that branch from each interlobular artery deliver blood to the capillaries that supply individual nephrons (Figure 18–4b●).

Blood reaches each nephron through an afferent arteriole and leaves in an **efferent arteriole** (Figure 18–4c●). It then travels to the **peritubular capillaries** that surround the proximal and distal convoluted tubules. The peritubular capillaries provide a route for the pickup or delivery of substances that are reabsorbed or secreted by these portions of the nephron.

The path of blood from the peritubular capillaries differs in **cortical nephrons,** which are located mostly within the cortex (Figure 18–4c), and in **juxtamedullary** (juks-ta-MED-ū-lar-ē) **nephrons** (*juxta,* near), which are located near the renal medulla. In juxtamedullary nephrons, the peritubular capillaries are connected to the **vasa recta** (*rectus,* straight)— long, straight capillaries that parallel the loop of Henle deep into the medulla (Figure 18–4d●). As we will see later, it is the juxtamedullary nephrons that enable the kidneys to produce concentrated urine.

Blood from the peritubular capillaries and vasa recta enters a network of venules and small veins that converge on the **interlobular veins.** In a mirror image of the arterial distribution, blood continues to converge and empty into the **arcuate, interlobar,** and **renal veins** (see Figure 18–4a).

The Nephron

The nephron is the basic functional unit in the kidney. Each nephron consists of two main parts: (1) a *renal corpuscle,* and (2) a 50-mm-long (2-inch-long) **renal tubule** composed of two *convoluted* (coiled or twisted) segments separated by a simple U-shaped tube (see Figure 18–3c). The convoluted segments are in the cortex, and the U-shaped tube extends partially or completely into the medulla.

An Overview of the Nephron

A schematic diagram of a representative nephron is shown in Figure 18–5●. The nephron begins at the **renal corpuscle** (KOR-pus-ul), which is a round structure that consists of a cup-shaped chamber (called *Bowman's capsule*) that contains a capillary network, or *glomerulus* (glo-MER-ū-lus; *glomus,* a ball). As previously described, blood arrives at the glomerulus by way of an

afferent arteriole and departs in an *efferent arteriole.* In the renal corpuscle, blood pressure forces fluid and dissolved solutes out of the glomerular capillaries and into the surrounding *capsular space.* This process is called *filtration.* ∞ p. 62 Filtration produces a protein-free solution known as a **filtrate.**

From the renal corpuscle, the filtrate enters the renal tubule. The major segments of the renal tubule are the *proximal convoluted tubule (PCT),* the *loop of Henle* (HEN-lē), and the *distal convoluted tubule (DCT).* As the filtrate travels along the tubule, its composition gradually changes, and it is then called **tubular fluid.** The changes that occur and the urine that results depend on the specialized activities under way in each segment of the nephron.

Each nephron empties into a *collecting duct,* which is the start of the **collecting system.** The collecting duct leaves the cortex and descends into the medulla, and carries tubular fluid from many nephrons toward a *papillary duct* that delivers the fluid, now called *urine,* into the calyces and on to the renal pelvis.

Key Note

The kidneys remove waste products from the blood; they also assist in the regulation of blood volume and blood pressure, ion levels, and blood pH. Nephrons are the primary functional units of the kidneys.

Functions of the Nephron

Urine has a very different composition from the filtrate produced at the renal corpuscle. The role of each segment of the nephron in converting filtrate to urine is indicated in Figure 18–5. The renal corpuscle is the site of filtration. The functional advantage of filtration is that it is passive; it does not require an expenditure of energy. The disadvantage of filtration is that any filter with pores large enough to permit the passage of organic waste products cannot *prevent* the passage of water, ions, and nutrients such as glucose, fatty acids, and amino acids. These substances, along with most of the water, must be reclaimed or they are lost in the urine.

Filtrate leaves the renal corpuscle and enters the renal tubule. The renal tubule is responsible for:

- Reabsorbing all of the useful organic molecules from the filtrate.
- Reabsorbing over 90 percent of the water in the filtrate.
- Secreting into the tubular fluid any waste products that were missed by the filtration process.

Additional water and salts will be removed in the collecting system before the urine is released into the renal sinus.

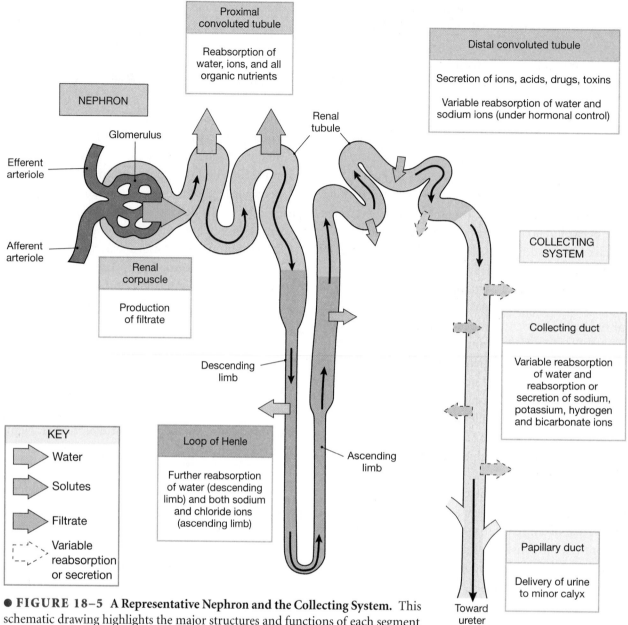

● **FIGURE 18–5 A Representative Nephron and the Collecting System.** This schematic drawing highlights the major structures and functions of each segment of the nephron (purple) and collecting system (tan).

Labels in figure:

Proximal convoluted tubule — Reabsorption of water, ions, and all organic nutrients

Distal convoluted tubule — Secretion of ions, acids, drugs, toxins / Variable reabsorption of water and sodium ions (under hormonal control)

NEPHRON

Glomerulus

Efferent arteriole

Renal tubule

Afferent arteriole

Renal corpuscle — Production of filtrate

COLLECTING SYSTEM

Collecting duct — Variable reabsorption of water and reabsorption or secretion of sodium, potassium, hydrogen and bicarbonate ions

Descending limb

Ascending limb

KEY
Water
Solutes
Filtrate
Variable reabsorption or secretion

Loop of Henle — Further reabsorption of water (descending limb) and both sodium and chloride ions (ascending limb)

Papillary duct — Delivery of urine to minor calyx

Toward ureter

The Renal Corpuscle

The **renal corpuscle** (Figure 18–6●) consists of (1) the capillary network of the **glomerulus** and (2) a structure known as **Bowman's capsule.** Bowman's capsule forms the outer wall of the renal corpuscle and encapsulates the glomerular capillaries. The glomerulus projects into Bowman's capsule much as the heart projects into the pericardial cavity (Figure 18–6a●). A *capsular epithelium* makes up the wall of the capsule and is continuous with a specialized epithelium that covers the glomerular capillaries. The two epithelia are separated by the **capsular space,** which receives the filtrate and empties into the renal tubule.

The epithelium that covers the capillaries consists of cells called **podocytes** (PŌ-dō-sīts, *podon,* foot) (Figure 18–6b●).

Podocytes have long cellular processes called *pedicels* that wrap around individual capillaries. A thick basement membrane separates the endothelial cells of the capillaries from the podocytes. The glomerular capillaries are said to be *fenestrated* (FEN-e-strā-ted; *fenestra,* a window) because their endothelial cells contain pores (see Figure 18–6b). To enter the capsular space, a solute must be small enough to pass through (1) the pores of the endothelial cells, (2) the fibers of the basement membrane, and (3) the *filtration slits* between the slender processes of the podocytes. The fenestrated capillary, basement membrane, and filtration slits form a **filtration membrane** that prevents the passage of blood cells and most plasma proteins but permits the movement of water, meta-

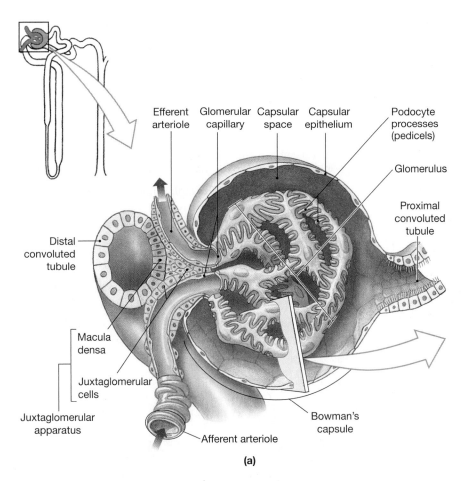

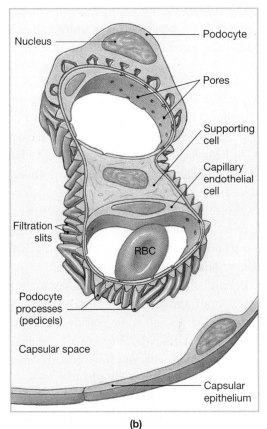

(a)

(b)

bolic wastes, ions, glucose, fatty acids, amino acids, vitamins, and other solutes into the capsular space. Most of the valuable solutes will be reabsorbed by the proximal convoluted tubule.

The Proximal Convoluted Tubule

The filtrate next moves into the first segment of the renal tubule, which is the **proximal convoluted tubule (PCT)** (see Figure 18–5). The cells that line the PCT absorb organic nutrients, plasma proteins, and ions from the tubular fluid and release them into the interstitial fluid that surrounds the renal tubule. As a result of this transport, the solute concentration of the interstitial fluid increases while that of the tubular fluid decreases. Water then moves out of the tubular fluid by osmosis, which reduces the volume of tubular fluid.

The Loop of Henle

The last portion of the PCT bends sharply toward the renal medulla and connects to the **loop of Henle** (see Figure 18–5). This loop is composed of a *descending limb* that travels toward the renal pelvis and an *ascending limb* that returns to the cortex. The ascending limb, which is not permeable to water and solutes, actively transports sodium and chloride ions out of the tubular fluid. As a result, the interstitial fluid of the

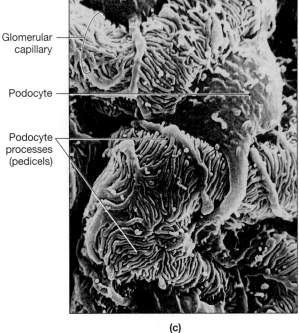

(c)

● **FIGURE 18–6 The Renal Corpuscle. (a)** This sectional view of a renal corpuscle illustrates its important structural features. **(b)** A cross section through a segment of the glomerulus reveals the components of the filtration membrane. **(c)** This colorized photomicrograph shows the glomerular surface, including individual podocytes and their processes. (SEM × 27,248)

medulla contains an unusually high solute concentration. The descending limb is permeable to water, and as it descends into the medulla, water moves out of the tubular fluid by osmosis.

The Distal Convoluted Tubule

The ascending limb of the loop of Henle ends where it bends and comes in close contact with the glomerulus and its vessels (see Figure 18–3c). At this point, the **distal convoluted tubule** (**DCT**) begins, and it passes immediately adjacent to the afferent and efferent arterioles (see Figure 18–6a).

The distal convoluted tubule is an important site for (1) the active secretion of ions, acids, drugs, and toxins and (2) the selective reabsorption of sodium ions from the tubular fluid. In the final portions of the DCT, an osmotic flow of water may assist in concentrating the tubular fluid.

The epithelial cells of the DCT closest to the glomerulus are unusually tall, and their nuclei are clustered together. This region of the DCT is called the *macula densa* (MAK-ū-la DEN-sa) (see Figure 18–6a). The cells of the macula densa are closely associated with unusual smooth muscle fibers—the *juxtaglomerular* (*juxta,* near) *cells*—in the wall of the afferent arteriole. Together, the macula densa and juxtaglomerular cells form the **juxtaglomerular apparatus,** which is an endocrine structure that secretes the hormone erythropoietin and the enzyme renin, introduced in Chapter 10. ∞ p. 394

The Collecting System

The distal convoluted tubule, which is the last segment of the nephron, opens into the collecting system. The collecting system consists of collecting ducts and papillary ducts (see Figure 18–5). Each **collecting duct** receives tubular fluid from many nephrons, and several collecting ducts merge to form a **papillary duct,** which delivers urine to a minor calyx. In addition to transporting tubular fluid from the nephron to the renal pelvis, the collecting system can make final adjustments to the composition of the urine by reabsorbing water and by reabsorbing or secreting sodium, potassium, hydrogen, and bicarbonate ions.

Table 18–1 summarizes the functions of the different regions of the nephron and collecting system.

CONCEPT CHECK QUESTIONS

1. How is the position of the kidneys different from most other organs in the abdominal region?
2. Why don't plasma proteins pass into the capsular space under normal circumstances?
3. Damage to which part of the nephron would interfere with the control of blood pressure?

Answers begin on p. 792.

TABLE 18–1 *The Functions of the Nephron and Collecting System in the Kidney*

REGION	PRIMARY FUNCTION
Renal corpuscle	Filtration of plasma to initiate urine formation
Proximal convoluted tubule (PCT)	Reabsorption of ions, organic molecules, vitamins, water
Loop of Henle	Descending limb: reabsorption of water from tubular fluid Ascending limb: reabsorption of ions; creates the concentration gradient in the medulla, which enables the kidney to produce concentrated urine
Distal convoluted tubule (DCT)	Reabsorption of sodium ions; secretion of acids, ammonia, drugs
Collecting duct	Reabsorption of water and of sodium and bicarbonate ions
Papillary duct	Conduction of urine to minor calyx

■ Basic Principles of Urine Production

The primary purpose of **urine** production is to maintain homeostasis by regulating the volume and composition of the blood. This process involves the excretion of dissolved solutes, especially the following three metabolic waste products:

1. *Urea.* Urea is the most abundant organic waste. You generate about 21 grams of urea each day, most of it during the breakdown of amino acids.
2. *Creatinine.* Creatinine is generated in skeletal muscle tissue through the breakdown of creatine phosphate, which is a high-energy compound that plays an important role in muscle contraction. Your body generates roughly 1.8 g of creatinine each day.
3. *Uric acid.* Uric acid is produced during the breakdown and recycling of RNA. You generate about 480 mg of uric acid each day.

These waste products must be excreted in solution, and their elimination is accompanied by an unavoidable water loss. The kidneys can minimize this water loss by producing urine that is four to five times more concentrated than normal body fluids. If the kidneys could not concentrate the filtrate produced by glomerular filtration, water losses would lead to fatal dehydration within hours. At the same time, the kidneys ensure that the excreted urine does not contain potentially useful organic substrates present in blood plasma, such as sugars or amino acids.

To accomplish these goals, the kidneys perform three distinct processes:

1. *Filtration.* In filtration, blood pressure forces water across the filtration membrane in the renal corpuscle. Solute molecules small enough to pass through the membrane are carried into the filtrate by the surrounding water molecules.

2. *Reabsorption.* Reabsorption is the removal of water and solute molecules from the filtrate and their reentry into the circulation at the peritubular capillaries. Reabsorption occurs after the filtrate enters the renal tubule. Whereas filtration occurs solely on the basis of size, reabsorption of solutes is a selective process that involves simple diffusion or the activity of carrier proteins in the tubular epithelium. Water reabsorption occurs passively through osmosis.

3. *Secretion.* Secretion is the transport of solutes out of the peritubular capillaries, across the tubular epithelium, and into the filtrate. This process is necessary because filtration does not force all of the dissolved materials out of the blood. Secretion can further lower the plasma concentration of undesirable materials, including many drugs.

Together, these three processes produce a fluid that is very different from other body fluids. Table 18–2 indicates the efficiency of the renal system by comparing the solute compositions of urine and plasma. The kidneys can continue to work efficiently only so long as filtration, reabsorption, and secretion proceed in proper balance. Any disruption in this balance has immediate and potentially disastrous effects on the com-

TABLE 18-2 *Significant Differences in Solute Concentrations Between Urine and Plasma*

COMPONENT	URINE	PLASMA
Ions (mEq/L)		
Sodium (Na$^+$)	147.5	138.4
Potassium (K$^+$)	47.5	4.4
Chloride (Cl$^-$)	153.3	106
Bicarbonate (HCO$_3^-$)	1.9	27
METABOLITES AND NUTRIENTS (mg/dL)		
Glucose	0.009	90
Lipids	0.002	600
Amino acids	0.188	4.2
Proteins	0.000	7.5 g/dL
NITROGENOUS WASTES (mg/dL)		
Urea	1800	10–20
Creatinine	150	1–1.5
Ammonia	60	<0.1
Uric acid	40	3

position of the circulating blood. If both kidneys fail to perform their assigned roles, death will occur within a few days unless medical assistance is provided.

As we have seen, all segments of the nephron and collecting system participate in the process of urine formation. Most regions perform a combination of reabsorption and secretion, but the balance between these two processes varies from one region to another. As indicated in Table 18–1:

■ Filtration occurs exclusively in the renal corpuscle, across the capillary walls of the glomerulus.

■ Reabsorption of nutrients occurs primarily at the proximal convoluted tubule.

■ Active secretion occurs primarily at the distal convoluted tubule.

■ Regulation of the amount of water, sodium ions, and potassium ions lost in the urine results from interactions between the loop of Henle and the collecting system.

Next we will take a closer look at events in each of the segments of the nephron and collecting system.

Filtration at the Glomerulus

Filtration Pressure

The forces that act across capillary walls were introduced in Chapter 13. (You may find it helpful to review Figure 13–7 before you proceed.) ∞ p. 475 Blood pressure at the glomerulus tends to force water and solutes out of the bloodstream and into the capsular space. For filtration to occur, this outward force must exceed any opposing pressures, such as the osmotic pressure of the blood. The net force that promotes filtration is called the **filtration pressure.** Filtration pressure at the glomerulus is higher than capillary blood pressure elsewhere in the body because of the slight difference in the diameters of afferent and efferent arterioles (see Figure 18–6a). The diameter of the efferent arteriole is slightly smaller, so it offers more resistance to blood flow than does the afferent arteriole. As a result, blood "backs up" in the afferent arteriole, which increases the blood pressure in the glomerular capillaries.

Filtration pressure is very low (around 10 mmHg), and kidney filtration will stop if glomerular blood pressure falls significantly. Reflexive changes in the diameters of the afferent arterioles, the efferent arterioles, and/or the glomerular capillaries can compensate for minor variations in blood pressure. These changes can occur automatically or in response to sympathetic stimulation. More serious declines in systemic blood pressure can reduce or even stop glomerular filtration. As a result, hemorrhaging, shock, or dehydration can cause a dangerous or even fatal reduction in kidney function. Because the kidneys are more sensitive to blood pressure than are other organs, it is not surprising to find that they control many of the homeostatic mechanisms responsible for regulating blood pressure and blood volume. One example of such a mechanism—the renin-angiotensin system—is considered later in this chapter.

The Glomerular Filtration Rate

The process of filtrate production at the glomerulus is called *glomerular filtration.* The **glomerular filtration rate (GFR)** is the amount of filtrate produced in the kidneys each minute. Because each kidney contains about 6 square meters of filtration surface, the GFR averages an astounding *125 mL per minute.* This means that almost 20 percent of the fluid delivered to the kidneys by the renal arteries leaves the bloodstream and enters the capsular spaces. In the course of a single day, the glomeruli generate about 180 liters (50 gal) of filtrate, roughly 70 times the total plasma volume. But as the filtrate passes through the renal tubules, over 99 percent of it is reabsorbed.

Tubular reabsorption is obviously an extremely important process. An inability to reclaim the water that enters the filtrate, as in *diabetes insipidus,* can quickly cause death by dehydration. (This condition, caused by inadequate ADH secretion, was discussed in Chapter 10.) ∞ p. 391

Glomerular filtration is the vital first step essential to all kidney functions. If filtration does not occur, waste products are not excreted, pH control is jeopardized, and an important mechanism for blood volume regulation is eliminated. Filtration depends on the maintenance of adequate blood flow to the glomerulus and of normal filtration pressures. The regulatory factors involved in maintaining a stable GFR are discussed later in the section on the control of kidney function.

🔒 Key Note

Roughly 180 L of filtrate is produced at the glomeruli each day, and that represents 70 times the total plasma volume. Almost all of that fluid volume must be reabsorbed to avoid fatal dehydration.

Reabsorption and Secretion Along the Renal Tubule

Reabsorption and secretion in the kidney involve a combination of diffusion, osmosis, and carrier-mediated transport. Recall that in carrier-mediated transport, a specific substrate binds to a carrier protein that facilitates its movement across the cellular membrane. ∞ p. 69 This movement may or may not require energy from ATP molecules.

Events at the Proximal Convoluted Tubule

The cells of the PCT actively reabsorb organic nutrients, plasma proteins, and ions from the filtrate and then transport them into the interstitial fluid that surrounds the renal tubule. Osmotic forces pull water across the wall of the PCT and into the surrounding interstitial fluid. The reabsorbed materials and water diffuse into peritubular capillaries.

The PCT usually reclaims 60–70 percent of the volume of filtrate produced at the glomerulus, along with virtually all of the glucose, amino acids, and other organic nutrients. The PCT also actively reabsorbs ions, including sodium, potassium, calcium, magnesium, bicarbonate, phosphate, and sulfate ions. The ion pumps involved are individually regulated and may be influenced by circulating ion or hormone levels. For example, the presence of parathyroid hormone stimulates calcium ion reabsorption. ∞ p. 384

Although reabsorption represents the primary function of the PCT, a few substances (such as hydrogen ions) can be actively secreted into the tubular fluid. Such active secretion can play an important role in the regulation of blood pH, a topic considered in a later section. A few compounds in the tubular fluid, including urea and uric acid, are ignored by the PCT and by other segments of the renal tubule. As water and other nutrients are removed, the concentrations of these waste products gradually rise in the tubular fluid.

Events at the Loop of Henle

Roughly 60–70 percent of the volume of the filtrate produced at the glomerulus has been reabsorbed before the tubular fluid reaches the loop of Henle. In the process, useful organic molecules and many mineral ions have been reclaimed. The loop of Henle reabsorbs more than half of the remaining water, as well as two-thirds of the sodium and chloride ions that remain in the tubular fluid.

The descending and ascending limbs of the loop of Henle have different permeability characteristics. Because the descending limb is permeable to water but not to solutes, water can flow in or out by osmosis, but solutes cannot cross the tubular epithelium. The ascending limb is impermeable both to water and to solutes. However, the ascending limb actively pumps sodium and chloride ions out of the tubular fluid and into the interstitial fluid of the renal medulla (Figure 18–7●). Over time, a *concentration gradient* is created in the medulla; the highest concentration of solutes (roughly four times that of plasma) occurs near the bend in the loop of Henle. Because the descending limb is freely permeable to water, as tubular fluid flows along that limb, water continually flows out of the tubular fluid and into the interstitial fluid by osmosis.

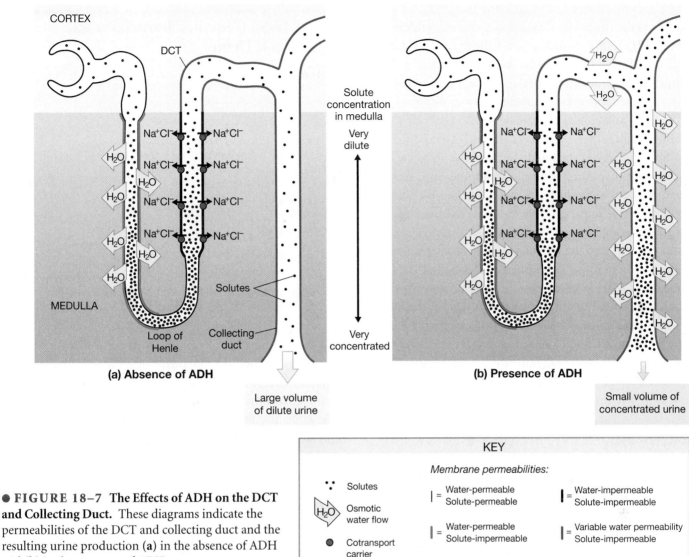

(a) Absence of ADH

Large volume of dilute urine

(b) Presence of ADH

Small volume of concentrated urine

Solute concentration in medulla

Very dilute

Very concentrated

KEY

Membrane permeabilities:

| = Water-permeable Solute-permeable

| = Water-impermeable Solute-impermeable

| = Water-permeable Solute-impermeable

| = Variable water permeability Solute-impermeable

⋰ Solutes

H₂O Osmotic water flow

● Cotransport carrier

● **FIGURE 18–7   The Effects of ADH on the DCT and Collecting Duct.** These diagrams indicate the permeabilities of the DCT and collecting duct and the resulting urine production (**a**) in the absence of ADH and (**b**) in the presence of ADH.

Roughly half the volume of filtrate that enters the loop of Henle is reabsorbed in the descending limb, and most of the sodium and chloride ions are removed in the ascending limb. With the loss of the sodium and chloride ions, the solute concentration of the filtrate declines to around one-third that of plasma. However, waste products such as urea now make up a significant percentage of the remaining solutes. In essence, most of the water and solutes have been removed, which leaves relatively highly concentrated waste products behind.

Events at the Distal Convoluted Tubule and the Collecting System

By the time the filtrate reaches the distal convoluted tubule, roughly 80 percent of the water and 85 percent of the solutes have already been reabsorbed. The DCT is connected to a collecting duct that drains into the renal pelvis.

As filtrate passes through the DCT and collecting duct, final adjustments are made in its composition and concentration. Its composition depends on the types of solutes present; its concentration depends on the volume of water in which the solutes are dissolved. Because the DCT and collecting duct are impermeable to solutes, changes in filtrate composition can occur only through active reabsorption or secretion. The primary function of the DCT is active secretion.

Throughout most of the DCT, the tubular cells actively transport sodium ions out of the tubular fluid in exchange for potassium ions or hydrogen ions. The DCT and collecting ducts contain ion pumps that respond to the hormone **aldosterone,** which is produced by the adrenal cortex. Aldosterone secretion occurs in response to lowered sodium ion concentrations or elevated potassium ion concentrations in the blood. The higher the aldosterone levels, the more sodium ions are reclaimed, and the more potassium ions are lost.

The amount of water reabsorbed along the DCT and collecting duct is controlled by circulating levels of *antidiuretic hormone (ADH)*. In the absence of ADH, the distal convoluted tubule and collecting duct are impermeable to water. The higher the level of circulating ADH, the greater the water permeability and the more concentrated the urine. Water moves out of the DCT and collecting duct because in each case the tubular fluid contains fewer solutes than the surrounding interstitial fluid. As previously noted, the fluid that arrives at the DCT has a solute concentration only about one-third that of the interstitial fluid in the surrounding cortex because the ascending limb of the loop of Henle has removed most of the sodium and chloride ions. The fluid that passes along the collecting duct travels into the medulla, where it passes through the concentration gradient established by the loop of Henle.

If circulating ADH levels are low, little water reabsorption will occur, and virtually all of the water that reaches the DCT will be lost in the urine (Figure 18–7a). If circulating ADH levels are high, the DCT and collecting duct will be very permeable to water (Figure 18–7b). In this case, the individual will produce a small quantity of urine with a solute concentration four to five times that of extracellular fluids.

Key Note

Reabsorption involves a combination of diffusion, osmosis, and active transport. Many of these processes are independently regulated by local or hormonal mechanisms. The primary mechanism that governs water reabsorption is "water follows salt." Secretion is a selective, carrier-mediated process.

The Properties of Normal Urine

The general characteristics of normal urine are listed in Table 18–3. Note, however, that the composition of the 1200 mL of urine excreted each day depends on the metabolic and hormonal events under way. Because the *composition* and *concentration* of the urine vary independently, an individual can produce either a small quantity of concentrated urine or a large quantity of dilute urine and still excrete the same amount of dissolved materials. For this reason, physicians often analyze the urine produced over a 24-hour period rather than testing a single sample. This enables them to assess both quantity and composition accurately.

A Summary of Kidney Function and Urine Formation

Figure 18–8● provides a functional overview of the major steps involved in the reabsorption of water and the production of concentrated urine.

TABLE 18-3 *General Characteristics of Normal Urine*

CHARACTERISTIC	NORMAL RANGE
pH	4.5–8 (average: 6.0)
Specific gravity (density of urine/density of pure water)	1.003–1.030
Osmotic concentration (Osmolarity) (number of solute particles per Liter; for comparison, fresh water ≈ 5 mOsm/L, body fluids ≈ 300 mOsm/L, and sea water ≈ 1000 mOsm/L)	855–1335 mOsm/L
Water content	93–97%
Volume	700–2000 mL/day
Color	Clear yellow
Odor	Varies with composition
Bacterial content	None (sterile)

Step 1: Glomerular filtration produces a filtrate that resembles blood plasma but contains few plasma proteins. This filtrate has the same osmotic, or solute, concentration as plasma or interstitial fluid.

Step 2: In the proximal convoluted tubule (PCT), 60–70 percent of the water and almost all of the dissolved nutrients are reabsorbed. The osmotic concentration of the tubular fluid remains unchanged.

Step 3: In the PCT and descending limb of the loop of Henle, water moves into the surrounding interstitial fluid, and leaves a small fluid volume (roughly 20 percent of the original filtrate) of highly concentrated tubular fluid.

Step 4: The ascending limb is impermeable to water and solutes. The tubular cells actively pump sodium and chloride ions out of the tubular fluid. Because only sodium and chloride ions are removed, urea now accounts for a greater proportion of the solutes in the tubular fluid.

Step 5: The final composition and concentration of the tubular fluid are determined by events under way in the DCT and the collecting ducts. These segments are impermeable to solutes, but ions may be actively transported into or out of the filtrate under the control of hormones such as aldosterone.

Step 6: The concentration of urine is controlled by variations in the water permeability of the DCT and the collecting ducts. These segments are impermeable to water unless exposed to antidiuretic hormone (ADH). In the absence of ADH, no water reabsorption occurs, and the individual produces a large volume of dilute urine. At high concentrations of ADH, the collecting ducts become freely permeable to water, and the individual produces a small volume of highly concentrated urine.

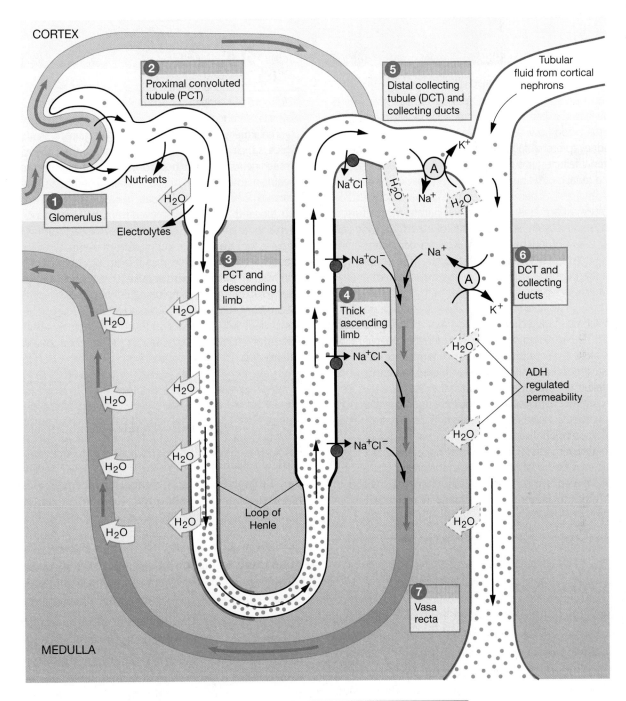

● **FIGURE 18–8** A Summary of Kidney Function and Urine Formation.

Acute renal failure (ARF) is a deterioration in renal function over hours or days that causes the accumulation of toxic wastes and causes the loss of internal homeostasis. ARF usually results from a decline in renal blood flow. ARF can result from problems in blood flow to the kidney (prerenal failure), a problem within the kidney itself (intrinsic renal failure), or a problem in urine flow (postrenal failure). Depressed renal blood flow and substances that are toxic to the kidney cause renal cell ischemia and death. Damage to the renal tubules can cause a back leak of glomerular filtrate, which further decreases perfusion pressure. Finally, the sloughing-off of dying cells can obstruct the renal tubules. Recovery depends upon restoration of renal blood flow. In prerenal failure, this involves the replacement of circulating blood volume. Rapid relief of the urinary obstruction in postrenal failure cases will restore glomerular flow as postrenal pressures are reduced. Restoration of renal blood flow and clearance of tubular toxins will aid in correcting intrinsic disease. Once blood supply is restored, the remaining nephrons will increase their filtration rate and eventually hypertrophy. If a critical number of nephrons have been lost, the patient will ultimately deteriorate and develop chronic renal failure that requires hemodialysis.

An all-too-common complication of multiple trauma is the development of acute tubular necrosis. This results from renal ischemia due to prolonged hypotension. If enough nephrons remain functioning, the patient can recover with relatively normal kidney function. If too many nephrons have been lost, then chronic renal failure will eventually develop. Prolonged hypotension due to multiple trauma should be avoided in order to prevent the development of acute tubular necrosis. ■

The Control of Kidney Function

Normal kidney function depends on an adequate glomerular filtration rate (GFR). GFR is regulated in three ways: (1) by local automatic adjustments in glomerular pressures; (2) through activities of the sympathetic division of the autonomic nervous system (ANS); and (3) through the effects of various hormones. The hormonal mechanisms result in long-term adjustments in blood pressure and blood volume that stabilize the GFR.

The Local Regulation of Kidney Function

Local, automatic changes in the diameters of the afferent arterioles, the efferent arterioles, and the glomerular capillaries can compensate for minor variations in blood pressure. For example, a reduction in blood flow and a decline in glomerular filtration pressure trigger dilation of the afferent arteriole and glomerular capillaries and constriction of the efferent arteriole. This combination keeps glomerular blood pressure and blood flow within normal limits. As a result, glomerular

Renal failure (RF) can be either *acute (ARF)* or *chronic (CRF)*. ARF is a sudden drop in urine output to less than half a liter per day. It is not uncommon in the very ill, including the victim of multiple trauma. *Emergency hemodialysis* may be initiated. With hemodialysis, which is the most common type of dialysis, the patient is attached to a machine where his blood comes in contact with a large semipermeable membrane. On the opposite side of the semipermeable membrane is a specialized dialysate solution that is hypo-osmolar for substances that must be removed from the blood. Blood contacts this membrane, and targeted substances diffuse into the dialysate. The net effect of dialysis is the correction of electrolyte abnormalities and blood volume and the removal of toxic substances such as urea or creatinine. For patients who require chronic hemodialysis, a vascular shunt, capable of handling blood flow of 300–400 mL per minute, is placed between a suitable vein and artery for dialysis access. Typically, the patient will dialyze for 2–3 hours, two or three times per week. Prior to the 1960s, dialysis was unavailable and patients with ARF or CRF died. ■

filtration rates remain relatively constant. If blood pressure rises, the afferent arteriole walls become stretched, and smooth muscle cells respond by contracting. The resulting reduction in the diameter of the afferent arterioles decreases glomerular blood flow and keeps the GFR within normal limits.

Sympathetic Activation and Kidney Function

Autonomic regulation of kidney function occurs primarily through the sympathetic division of the ANS. Sympathetic activity primarily serves to shift blood flow away from the kidneys, which lowers the glomerular filtration rate. Sympathetic activation has both direct and indirect effects on kidney function. The direct effect of sympathetic activation is a powerful constriction of the afferent arterioles, which decreases the GFR and slows production of filtrate. Triggered by a sudden crisis, such as an acute reduction in blood pressure or a heart attack, sympathetic activation can override the local regulatory mechanisms that act to stabilize the GFR. As the crisis passes and sympathetic activity decreases, the GFR returns to normal.

When the sympathetic division changes the regional pattern of blood circulation, blood flow to the kidneys is often affected. This change can further reduce the GFR. For example, the dilation of superficial vessels in warm weather shunts blood away from the kidneys, and glomerular filtration declines temporarily. The effect becomes especially pronounced during periods of strenuous exercise. As blood flow increases to the skin and skeletal muscles, it decreases to the kidneys. At maximal levels of exertion, renal blood flow may be less than one-quarter of normal resting levels.

Such a reduction can create problems for distance swimmers and marathon runners, whose glomerular cells may be damaged by low oxygen levels and the buildup of metabolic wastes over the course of a long event. After such events, protein is commonly lost in the urine, and in some cases, blood appears in the urine. Such problems generally disappear within 48 hours, although a small number of runners experience kidney failure (renal failure) and permanent impairment of kidney function.

Clinical Note
THE TREATMENT OF KIDNEY FAILURE

Kidney failure, or *renal failure,* occurs when the kidneys become unable to perform the excretory functions needed to maintain homeostasis. Many different conditions can result in renal failure. *Acute renal failure* occurs when exposure to toxic drugs, renal ischemia, urinary obstruction, or trauma causes a sudden reduction or stoppage of filtration. In *chronic renal failure,* kidney function deteriorates gradually, and the associated problems accumulate over time. The condition generally cannot be reversed, only prolonged, and symptoms of acute renal failure eventually develop.

Management of chronic kidney failure typically involves restricting dietary protein, water, and salt intake and reducing caloric intake to a minimum. This combination reduces strain on the urinary system by (1) minimizing the volume of urine produced and (2) preventing the generation of large quantities of nitrogenous wastes. Acidosis—blood plasma pH below 7.35—is a common problem in patients with kidney failure; it can be countered with infusions of bicarbonate ions. If drugs, infusions, and dietary controls cannot stabilize the composition of the blood, more drastic measures are taken, such as dialysis or kidney transplantation.

In *dialysis,* the functions of damaged kidneys are performed by a machine that facilitates diffusion between the patient's blood and a *dialysis fluid* whose composition is carefully regulated. The process, which takes several hours, must be repeated two or three times each week.

In a *kidney transplant,* the kidney of a healthy compatible donor is surgically inserted into the patient's body and connected to the circulatory system. If the surgical procedure is successful, the transplanted kidney(s) can take over all of the normal kidney functions. ∎

The Hormonal Control of Kidney Function

The major hormones involved in regulating kidney function are angiotensin II, ADH, aldosterone, and ANP. These hormones have been discussed in earlier chapters, so only a brief overview will be provided here. ∞ pp. 482, 483 The secretion of angiotensin II, aldosterone, and ADH is integrated by the *renin-angiotensin system*.

THE RENIN-ANGIOTENSIN SYSTEM. If glomerular pressures remain low because of a decrease in blood volume, a fall in systemic pressures, or a blockage in the renal artery or its tributaries, the juxtaglomerular apparatus releases the enzyme renin into the circulation. **Renin** converts inactive *angiotensinogen* to *angiotensin I,* which a *converting enzyme* activates to **angiotensin II.** Figure 18–9● diagrams the primary effects of this potent hormone in regulating GFR.

Angiotensin II has the following effects:

- *In peripheral capillary beds,* it causes a brief but powerful vasoconstriction, which elevates blood pressure in the renal arteries.
- *At the nephron,* it triggers constriction of the efferent arterioles, which elevates glomerular pressures and filtration rates.
- *In the CNS,* it triggers the release of ADH, which in turn stimulates the reabsorption of water and sodium ions and induces the sensation of thirst.
- *At the adrenal gland,* it stimulates the secretion of aldosterone by the adrenal cortex and of epinephrine and norepinephrine (NE) by the adrenal medullae. The result is a sudden, dramatic increase in systemic blood pressure. At the kidneys, aldosterone stimulates sodium reabsorption along the DCT and collecting system.

ADH. Antidiuretic hormone (1) increases the water permeability of the DCT and collecting duct, which stimulates the reabsorption of water from the tubular fluid; and (2) induces the sensation of thirst, which leads to the consumption of additional water. ADH release occurs under angiotensin II stimulation; it also occurs independently, when hypothalamic neurons are stimulated by a reduction in blood pressure or an increase in the solute concentration of the circulating blood. The nature of the receptors involved has not been determined, but these specialized hypothalamic neurons are called *osmoreceptors.*

ALDOSTERONE. Aldosterone secretion stimulates the reabsorption of sodium ions and the secretion of potassium ions along the DCT and collecting duct. Aldosterone secretion primarily occurs (1) under angiotensin II stimulation and (2) in response to a rise in the potassium ion concentration of the blood. ∞ p. 385

ATRIAL NATRIURETIC PEPTIDE. The actions of atrial natriuretic peptide (ANP) oppose those of the renin-angiotensin system (see Figure 13–12b, p. 481). This hormone is released by atrial cardiac muscle cells when blood volume and blood pressure are too high. The actions of ANP that affect the kidneys include (1) a decrease in the rate of sodium ion

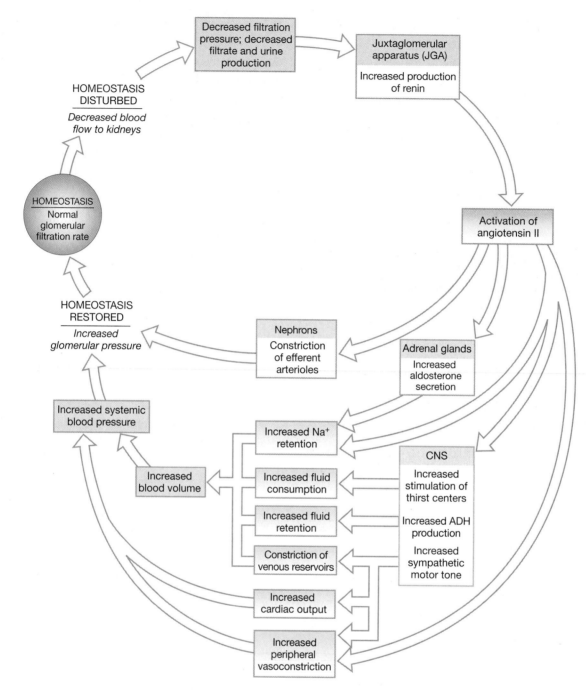

● **FIGURE 18–9** The Roles of the Renin-Angiotensin System in the Regulation of GFR.

reabsorption in the DCT, which leads to increased sodium ion loss in the urine; (2) dilation of the glomerular capillaries, which results in increased glomerular filtration and urinary water loss; and (3) the inactivation of the renin-angiotensin system through the inhibition of renin, aldosterone, and ADH secretion. The net result is an accelerated loss of sodium ions and an increase in the volume of urine produced. This combination lowers blood volume and blood pressure.

Urine Transport, Storage, and Elimination

Filtrate modification and urine production end when the fluid enters the renal pelvis. The structures of the urinary tract—the ureters, urinary bladder, and urethra—are responsible for the transport, storage, and elimination of the urine.

The Ureters and Urinary Bladder

The **ureters** (ū-RĒ-terz) are a pair of muscular tubes that conduct urine from the kidneys to the urinary bladder, a distance of about 30 cm (12 in.) (see Figure 18–1). Each ureter begins at the funnel-shaped renal pelvis and ends at the posterior wall of the bladder; the ureters never enter the peritoneal cavity.

Their **ureteral openings** within the urinary bladder are slit-like, which is a shape that prevents the backflow of urine into the ureters or kidneys when the urinary bladder contracts.

The wall of each ureter contains an inner expandable transitional epithelium, a middle layer of longitudinal and circular bands of smooth muscle, and an outer connective tissue layer continuous with the renal capsule. About every 30 seconds, a peristaltic contraction begins at the renal pelvis and sweeps along the ureter, which forces urine toward the urinary bladder.

Occasionally, solids composed of calcium deposits, magnesium salts, or crystals of uric acid form within the kidney, ureters, or urinary bladder. These solids are called *calculi* (KAL-kū-lī), or *kidney stones,* and their presence results in a painful condition known as *nephrolithiasis* (nef-rō-li-THĪ-a-sis). Kidney stones not only obstruct the flow of urine but may also reduce or prevent filtration in the affected kidney.

Clinical Note
KIDNEY STONES

Kidney stones, or renal calculi (singular, calculus), represent crystal aggregation in the kidney's collecting system (Figure 18–10●). This condition is also called *nephrolithiasis* (from Greek *lithos,* stone). Kidney stones affect about 500,000 people each year. Brief hospitalization is common due to the severity of pain as the stone travels from the renal pelvis, through the ureter to the bladder, and is eventually eliminated in urine. If necessary, additional inpatient treatment may include shock-wave lithotripsy, which is a procedure that uses sound waves to break large stones into smaller ones, and other treatment modalities. Overall morbidity and mortality are low, however, unless a complication such as hemorrhage or urinary tract obstruction results.

Kidney stones occur more frequently in the southern and southeastern U.S. which is a region referred to as the Kidney Stone Belt (Figure 18–11●). North Carolina has more kidney stones per capita than any other state. Several factors appear to be involved in this phenomenon. First, the typical southern diet is high in green vegetables and brewed tea—both of which are high in oxalates, which is a key ingredient for kidney stones. Another factor is the warm climate, which

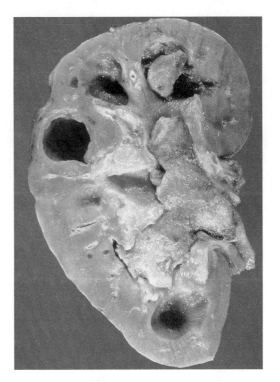

● **FIGURE 18–10 Kidney Stones.** Cross section through a kidney that shows multiple stones, including one "staghorn" stone in the renal pelvis.

FIGURE 18–11 The Kidney Stone Belt. The incidence of kidney stones is markedly higher in the southern U.S.

(continued next page)

Clinical Note—*continued*
KIDNEY STONES

causes increased sweating and increased loss of body fluids. Finally, modern lifestyles often reduce the level of physical activity. Together, these risk factors increase the likelihood of kidney stone development.

Stones form more commonly in men than women, although the ratio varies for different types of stones with different compositions. Certain stones also occur in familial patterns, which suggests hereditary factors. Another risk factor for calculus formation is immobilization due to surgery or injury, with the latter including immobilization secondary to paraplegia or other paralysis syndromes that involve the absence of motor impulses, sensation, or both. Last, the use of certain medications, including anesthetics, opiates, and psychotropic drugs, increases the risk for stones. Historically, about 60 percent of individuals who have experienced one kidney stone will develop another within seven years. Fortunately, with modern therapy, recurrence can be prevented in more than 95 percent of sufferers.

Stones may form in several metabolic disorders including gout or primary hyperparathyroidism, which produce excessive amounts of uric acid and calcium, respectively. More often, they occur when the general balance between water conservation and dissolution of relatively insoluble substances such as mineral ions and uric acid is lost and excessive amounts of the insolubles aggregate into stones. The problem boils down to "too much insoluble stuff" and "too-concentrated urine," which is a situation that may more likely arise with a change in diet, climate, or physical activity.

Stones that consist of calcium salts (namely, calcium oxalate and calcium phosphate) are by far the most common. These compounds are found in 75 to 85 percent of all stones. Calcium stones are from two to three times more common in men than in women, and the average age at onset is between 20 and 30 years. Their formation frequently runs in families, and anyone who has had a calcium stone is at fairly high risk to form another within two to three years.

Struvite stones (chemically denoted $MgNH_4PO_4$) are also common, and represent about 10 to 15 percent of all stones. The pathophysiology of struvite stones differs from that of calcium stones. Their formation is associated with chronic urinary tract infection (UTI) or frequent bladder catheterization. The association with bacterial UTI makes struvite stones much more common in women than in men. These stones can grow to fill the renal pelvis, which produces a characteristic "staghorn" appearance on X-rays.

Far less common are stones made of uric acid or cystine. Uric acid stones form more often in men than in women and tend to occur in families; about half of all patients with uric acid stones have gout. Cystine stones are the least common. They are associated with excess levels of the amino acid cystine in filtrate and are probably due at least in part to hereditary factors, as they often run in families.

The largest kidney stone ever reported weighed a fraction under 3 pounds (1.36 kilograms). The smallest stones are the size of grains of sand. Kidney stones come in virtually every color although most are yellow to brown. It is important to try to collect a passed kidney stone for laboratory analysis. Once the chemical composition of the stone has been determined, then prevention strategies can be initiated.

The pain from kidney stones is generally conceded to be among the most painful of human medical conditions. Typically, the patient first notes discomfort as a vague, visceral pain in one flank. Within 30 to 60 minutes it progresses to an extremely sharp pain that may remain in the flank or migrate downward and anteriorly toward the groin. Migrating pain indicates that the stone has passed into the lowest third of the ureter. Kidney stone pain can be colicky in nature. As a rule, kidney-stone patients cannot lie still. They continuously move or pace due to the unrelenting nature of the pain. Stones that lodge in the lowest part of the ureter, within the bladder wall, often cause characteristic bladder symptoms such as frequency during the day or during the night (nocturia), urgency, and painful urination. Because these latter three symptoms far more frequently suggest bladder infection, making the tentative diagnosis may be difficult, particularly in women. Visible blood in the urine (hematuria) is not uncommon in urine specimens taken during passage of a stone. However, hematuria may be microscopic, which means that it is not visible to the naked eye but can be detected with reagent strips dipped into a urine specimen. Fever, however, is not common unless concomitant infection is present. Whenever kidney stones are suspected, be sure to obtain the patient's personal medical history and family history, because both will often provide useful information.

As mentioned earlier, the patient may be agitated or physically restless; walking sometimes reduces the pain. Vital signs will vary with the level of discomfort experienced by the patient, with highest blood pressure and heart rate associated with the greatest pain. The skin will typically be pale, cool, and clammy. Most patients are nauseated and many vomit because of the severe, unrelenting pain.

Emergency treatment of kidney stones should include hydration and analgesia. Often, the pain is so severe that patients will require large doses of intravenous narcotics for pain control (morphine, meperidine, fentanyl). Virtually all patients with a symptomatic kidney stone will be nauseated or vomiting. Administration of medicine to treat nausea and vomiting will also help to increase the effectiveness of the analgesic medications.

In the hospital, kidney stone patients will usually receive intravenous fluids and parenteral analgesics. The kidney stone can be diagnosed with spiral computerized tomography (CT) scanning. Alternatively, physicians can order an intravenous pyelogram (IVP), which involves the intravenous administration of a radiopaque contrast dye. The dye collects in the kidneys, which allows the kidneys, ureters, and bladder to be visualized. On occasion, ultrasound can detect a kidney stone.

Many patients will spontaneously pass smaller stones with adequate hydration and analgesia. If the stones do not pass, the urologist can place a stent in the affected ureter and remove the stone with a basket-like device inserted through the cystoscope. Stones that cannot be removed with basket extraction may be treated with other therapies. These include lithotripsy, which uses shock waves to break up stones so that they can pass with little problem. Another technology is lasertripsy, where a laser is threaded through a cystoscope and, once properly positioned, pulverizes the stone. Another available technology is electrohydraulic lithotripsy (EHL), where a special probe breaks up small stones with shock waves generated by electricity. The gravel that remains following lithotripsy or lasertripsy easily passes through the ureters into the bladder. ■

The **urinary bladder** is a hollow, muscular organ that stores urine prior to urination. Its dimensions vary depending on the state of distension, but a full urinary bladder can contain up to a liter of urine. The urinary bladder lies in the pelvic cavity, and only its superior surface is covered by a layer of peritoneum. It is held in position by peritoneal folds *(umbilical ligaments)* that extend to the umbilicus (navel) and by bands of connective tissue attached to the pelvic and pubic bones. In males, the base of the urinary bladder lies between the rectum and the pubic symphysis (Figure 18–12a●). In females, the urinary bladder sits inferior to the uterus and anterior to the vagina (Figure 18–12b●).

The triangular area within the urinary bladder that is bounded by the ureteral openings and the entrance to the urethra forms the *trigone* (TRĪ-gōn) of the bladder (Figure 18–12c●). The urethral entrance lies at the apex of this triangle at the lowest point in the bladder. The area that surrounds the urethral entrance, called the *neck* of the urinary bladder, contains a muscular **internal urethral sphincter.** The smooth muscle of this sphincter provides involuntary control over the discharge of urine from the bladder.

A *transitional epithelium* that is continuous with the renal pelvis and the ureters also lines the urinary bladder. This stratified epithelium can tolerate a considerable amount of stretching, as indicated in Figure 4–5b. ∞ p. 102 The middle layer of the bladder wall consists of inner and outer layers of longitudinal smooth muscle with a circular layer in between. These three layers of smooth muscle form the powerful *detrusor* (de-TROO-sor) *muscle* of the bladder. Contraction of this muscle compresses the urinary bladder and expels its contents into the urethra.

The Urethra

The urethra extends from the neck of the urinary bladder to the exterior of the body. In females, the **urethra** is very short, and extends 2.5–3.0 cm (about 1 in.) from the bladder to its opening in the vestibule anterior to the vagina. In males, the urethra extends from the neck of the urinary bladder to the tip of the penis, about 18–20 cm (7–8 in.) in length. In both sexes, as the urethra passes through the muscular floor of the pelvic cavity, a circular band of skeletal muscle forms the **external urethral sphincter** (see Figure 18–12). This sphincter consists of skeletal muscle fibers, and its contractions are under voluntary control.

● **FIGURE 18–12 Organs for the Conduction and Storage of Urine.** The organs of urinary transport and storage are shown in lateral views of **(a)** a male and **(b)** a female. **(c)** The urinary bladder of a male is shown in a frontal view.

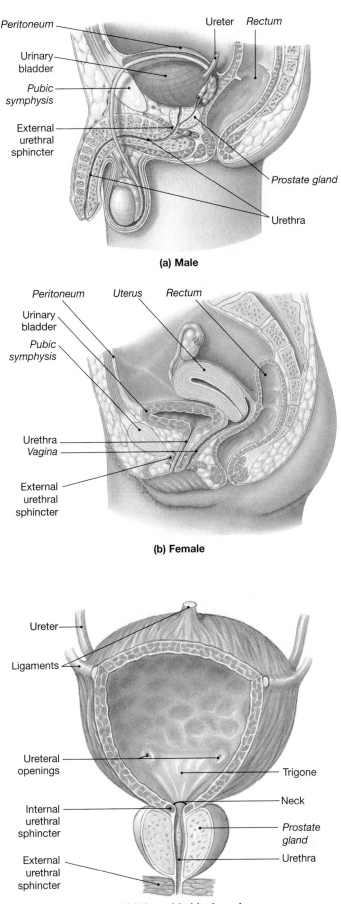

(a) Male

(b) Female

(c) Urinary bladder in male

Clinical Note
URINARY TRACT INFECTIONS

Urinary tract infection, or UTI, affects the urethra, bladder, ureter, or kidney, as well as the prostate gland in men. UTIs are extremely common, and account for over 6 million medical office visits yearly (Figure 18–13●). Almost all UTIs start with pathogenic colonization of the bladder by bacteria that enter through the urethra. Thus, females in general are at higher risk because of their relatively short urethra. Other groups at risk for UTI are paraplegic patients or patients with nerve disruption to the bladder, which includes some diabetic persons. Any condition that promotes urinary stasis (incomplete urination with urine that remains in the bladder) places a person at increased risk. Pregnant women often have urinary stasis due to pressure from the gravid uterus. People with neurological impairment (some patients with spina bifida or with diabetic neuropathy, for example) also tend to have urinary stasis, which predisposes them to infection. The use of instrumentation in patients who require bladder catheterization places them at even higher risk of developing a UTI.

Associated conditions such as scarring, abscesses, or eventual development of chronic renal failure are most likely in persons with anatomic abnormalities of the urinary system or chronic calculi (the latter acts as a focus for continuing infection and inflammation), those who are immunocompromised, or those who have renal disease due to diabetes mellitus or another condition.

UTIs are generally divided into those of the lower urinary tract, namely, urethritis (urethra), cystitis (bladder), and prostatitis (prostate gland), and those of the upper urinary tract, pyelonephritis (kidney). Lower UTIs are far more common than upper UTIs, for two reasons. First, seeding of infection via the bloodstream is rare. Second, asymptomatic bacterial colonization of the urethra, especially in females, is very common, and can predispose a person to infection by other, pathogenic bacteria. In females, infection may begin when gram-negative bacteria normally found in the bowel (that is, the enteric floras) colonize the urethra and bladder. Symptomatic urethritis, which is inflammation secondary to urethral infection, is very uncommon. More often you will see joint symptomatic infection of the urethra and bladder (urethritis and cystitis, respectively). Sexually active females are at higher risk, which may be attributed to use of contraceptive devices or agents, to the introduction of enteric floras (bacteria from the GI tract) during intercourse, or both. Recently, homosexually active men who engage in anal sex have also been found to be at higher risk for bacterial cystitis, possibly due to introduction of enteric bacterial floras during rectal intercourse. In any case, sexually active persons

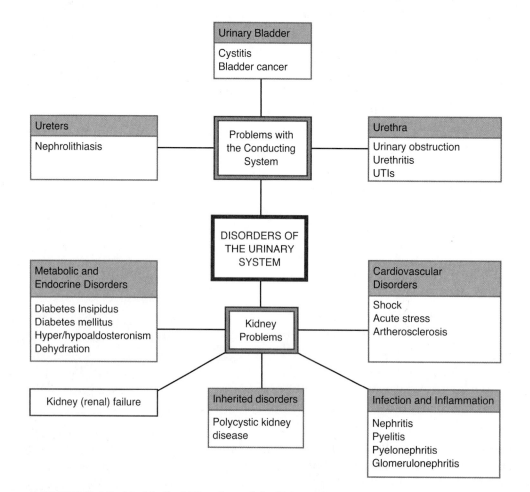

● **FIGURE 18–13** **Medical Disorders of the Urinary System.**

who suffer from urinary stasis are at even higher risk for infection. Persons who urinate after intercourse might lower their risk because voiding eliminates some bacteria. The pathophysiology for persons using bladder catheterization probably differs only in that pathogenic bacteria are introduced directly into the bladder via the catheter. In general, the likelihood that active cystitis will develop, that antibiotic treatment will clear such infections, and that reinfection will occur is determined by the interplay of the pathogen's virulence, the size of its colony, its sensitivity to antibiotic treatment, and the strength of the host's local and systemic immune functions.

Prostatitis, which is inflammation of the prostate gland, is secondary to bacterial infection, as well as any general inflammatory condition. Men with acute bacterial prostatitis, which is the closest parallel to acute cystitis in women, also tend to show evidence of associated urethritis, and the same bowel flora tends to be involved. The major difference between acute bacterial prostatitis and acute cystitis is the much lower incidence of prostatitis among men who do not require bladder catheterization.

Upper UTIs usually evolve from infection that spreads upward into the kidney. Pyelonephritis is an infectious inflammation of the renal parenchyma: nephrons, interstitial tissue, or both. Acute pyelonephritis is 10 times more common in women than in men. Its incidence is highest in pregnancy and during periods of sexual activity, which reflects the epidemiology of lower UTIs. If the infection of pyelonephritis persists, intrarenal or perinephric abscesses may occur, but these complications are uncommon. Intrarenal abscesses form within the renal parenchyma. If they rupture and spill their contents into the adjacent fatty tissue, perinephric abscesses may result. From 20 to 60 percent of patients who develop perinephric abscesses have a clear predisposing factor such as renal calculi, anatomic abnormalities of the kidney, history of urologic surgery or injury, or diabetic renal disease.

Urinary tract infections may be community-acquired or nosocomial. Among community-acquired infections, gram-negative enteric bacteria are the most common. In fact, *E. coli* accounts for roughly 80 percent of infections in persons without the complicating factors of bladder catheterization, renal calculi, or anatomic abnormalities. In nosocomial infections, which are cases acquired in an inpatient setting or related to catheterization, *Proteus, Klebsiella,* and *Pseudomonas* are commonly identified. Less common, but still important, are sexually transmitted pathogens (among women and men) such as *Chlamydia* and *N. gonorrhoeae.* Fungi such as Candida are rarely seen except in catheterized or immunocompromised patients or patients with diabetes mellitus.

The symptoms of lower UTI typically include: painful urination, frequent urge to urinate, and difficulty in beginning and continuing to void. Pain often begins as visceral discomfort that progresses to severe, burning pain, particularly during and just after urination. The evolution of pain corresponds roughly to the degree of epithelial damage caused by the pathogen. In both men and women, pain is often localized to the pelvis and perceived as in the bladder (in women) or in the bladder and prostate (in men). The patient may complain of a strong or foul odor in the urine. Many women will give a history of similar episodes, which may or may not have been diagnosed or treated. Patients with pyelonephritis are more likely to feel generally ill or feverish. They typically complain of constant, moderately severe or severe pain in a flank or lower back (just under the rib cage). The pain may be referred to the shoulder or neck. The triad of urgency, pain, and difficulty may or may not be present or included in the past history.

On physical exam, patients with UTI appear restless and uncomfortable. Typically, patients with pyelonephritis appear more ill and are far more likely to have a fever. Skin will often be pale, cool, and moist (in lower UTI) or warm and dry (in febrile upper UTI). Vital signs will vary with the degree of illness and pain, but in an otherwise healthy individual they should not be far from normal. Inspect and auscultate the abdomen to document findings, but neither procedure is likely to be very useful, because visible appearance and bowel sounds are usually within normal limits. Percussion and palpation will probably reveal painful tenderness over the pubis in lower UTI and at the flank in upper UTI. Lloyd's sign, which is tenderness to percussion of the lower back at the costovertebral angle (CVA), suggests pyelonephritis.

The best prevention technique is hydration to increase blood flow through the kidneys and to produce a more dilute, but voluminous, urine. In many cases, this is better accomplished by IV fluid administration, which eliminates the risk of vomiting and satisfies the guidelines for possible surgical cases. ■

The Micturition Reflex and Urination

Peristaltic contractions of the ureters move the urine into the urinary bladder. The process of **urination,** or **micturition** (mik-tu-RI-shun), is coordinated by the **micturition reflex** (Figure 18–14●). Stretch receptors in the wall of the urinary bladder are stimulated as the bladder fills with urine. Afferent sensory fibers in the pelvic nerves carry the resulting impulses to the sacral spinal cord. The fibers' increased level of activity (1) brings parasympathetic motor neurons in the sacral spinal cord close to threshold and (2) stimulates interneurons that relay sensations to the cerebral cortex. As a result, we become consciously aware of the fluid pressure within the urinary bladder.

The urge to urinate usually occurs when the bladder contains about 200 mL of urine. The micturition reflex begins to function when the stretch receptors have provided adequate stimulation to the parasympathetic motor neurons. At this time, the motor neurons stimulate the detrusor muscle in the bladder wall. These commands travel over the pelvic nerves and produce a sustained contraction of the urinary bladder.

This contraction elevates fluid pressures inside the bladder, but urine ejection cannot occur unless both the internal and external sphincters are relaxed. The relaxation of the external sphincter occurs under voluntary control. Once the external sphincter relaxes, so does the internal sphincter. If the external sphincter does not relax, the internal sphincter remains

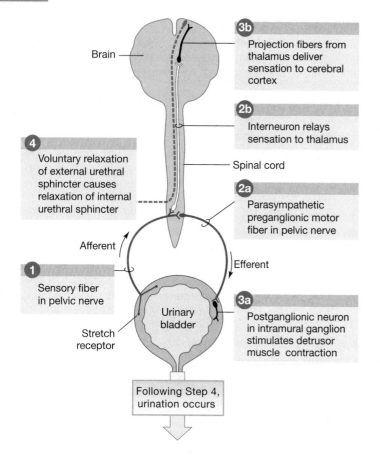

3b Projection fibers from thalamus deliver sensation to cerebral cortex

Brain

2b Interneuron relays sensation to thalamus

4 Voluntary relaxation of external urethral sphincter causes relaxation of internal urethral sphincter

Spinal cord

2a Parasympathetic preganglionic motor fiber in pelvic nerve

Afferent

Efferent

1 Sensory fiber in pelvic nerve

Urinary bladder

3a Postganglionic neuron in intramural ganglion stimulates detrusor muscle contraction

Stretch receptor

Following Step 4, urination occurs

● **FIGURE 18–14 The Micturition Reflex.** This diagram illustrates the components of the reflex arc that stimulates smooth muscle contractions in the urinary bladder. Micturition occurs after voluntary relaxation of the external urethral sphincter.

closed, and the bladder gradually relaxes. A further increase in bladder volume begins the cycle again, usually within an hour. Each increase in urinary volume leads to an increase in stretch receptor stimulation that makes the sensation more acute. Once the volume of the urinary bladder exceeds 500 mL, the micturition reflex may generate enough pressure to force open the internal sphincter. A reflexive relaxation of the external sphincter follows, and urination occurs despite voluntary opposition or potential inconvenience. At the end of normal micturition, less than 10 mL of urine remains in the bladder.

CONCEPT CHECK QUESTIONS

1. What process is responsible for the movement of urine from the kidneys to the urinary bladder?
2. An obstruction of a ureter by a kidney stone would interfere with the flow of urine between what two structures?
3. The ability to control the micturition reflex depends on the ability to control which muscle?

Answers begin on p. 792.

Clinical Note
INCONTINENCE

Incontinence (in-KON-ti-nens) is the inability to control urination voluntarily. Infants lack voluntary control over urination because the necessary corticospinal connections have yet to be established. Accordingly, "toilet training" before age 2 often involves training the parent to anticipate the timing of the micturition reflex rather than training the child to exert conscious control.

Trauma to the internal or external urethral sphincter can contribute to incontinence in otherwise normal adults. Some new mothers, for example, develop *stress incontinence* after childbirth has stretched and damaged the sphincter muscles. In this condition, elevated intra-abdominal pressures, caused simply by a cough or sneeze, can overwhelm the sphincter muscles, and cause urine to leak out. Incontinence may also develop in older individuals because of a general loss of muscle tone.

Damage to the CNS, the spinal cord, or the nerve supply to the bladder or external sphincter may also produce incontinence. For example, incontinence often accompanies Alzheimer's disease or spinal cord injury. In most cases, the affected individual develops an *automatic bladder*. The micturition reflex remains intact, but voluntary control of the external sphincter is lost, and the person cannot prevent the reflexive emptying of the bladder. Damage to the pelvic nerves can eliminate the micturition reflex entirely because these nerves carry both afferent and efferent fibers that control this reflex. The urinary bladder becomes greatly distended with urine and remains filled to capacity while excess urine flows into the urethra in an uncontrolled stream. The insertion of a catheter is commonly required to facilitate the discharge of urine. ■

■ Fluid, Electrolyte, and Acid-Base Balance

Most of your body weight is water. Water accounts for up to 99 percent of the volume of the fluid outside cells, and it is an essential ingredient of cytoplasm. All of a cell's operations rely on water as a diffusion medium for the distribution of gases, nutrients, and waste products. If the water content of the body declines too far, cellular activities are jeopardized: proteins denature, enzymes cease functioning, and cells ultimately die. To survive, the body must maintain a normal volume and composition in both the **extracellular fluid** or **ECF** (the interstitial fluid, plasma, and other body fluids) and the **intracellular fluid** or **ICF** (the cytosol).

The concentrations of various ions and the pH of the body's water are as important as its absolute quantity. If concentrations of calcium or potassium ions in the ECF become too high, cardiac arrhythmias develop, and place the individual's life in jeopardy. A pH outside the normal range can lead to a variety of dangerous conditions. Low pH is especially dangerous because hydrogen ions break chemical bonds, change the shapes of complex molecules, disrupt cell membranes, and impair tissue functions.

In this section we consider the dynamics of exchange between the various body fluids, such as blood plasma and interstitial fluid, and between the body and the external environment. Homeostasis of fluid volumes, solute concentrations, and pH involves three interrelated factors:

■ *Fluid balance.* You are in **fluid balance** when the amount of water you gain each day is equal to the amount you lose to the environment. Maintaining normal fluid balance involves regulating the content and distribution of water in the ECF and ICF. Because your cells and tissues cannot transport water, fluid balance primarily reflects the creation of ion concentration gradients that are then eliminated by osmosis.

■ *Electrolyte balance.* **Electrolytes** are ions released through the dissociation of inorganic compounds; they are so named because when in solution they can conduct an electrical current. ⚯ p. 41 Each day, your body fluids gain electrolytes from the food and drink you consume, and they lose electrolytes in urine, sweat, and feces. **Electrolyte balance** exists when there is neither a net gain nor a net loss of any ion in body fluids. Electrolyte balance primarily involves balancing the rates of absorption across the digestive tract with rates of loss at the kidneys.

■ *Acid-base balance.* You are in **acid-base balance** when the production of hydrogen ions is equal to their loss. When acid-base balance exists, the pH of body fluids remains within normal limits. Preventing a reduction in pH is a primary problem because normal metabolic operations generate a variety of acids. The kidneys and lungs play key roles in maintaining the acid-base balance of body fluids.

This section provides an overview that integrates earlier discussions that involve fluid, electrolyte, and acid-base balance in the body. Few other topics have such wide-ranging clinical importance: *treatment of any serious illness that affects the nervous, cardiovascular, respiratory, urinary, or digestive system must always include steps to restore normal fluid, electrolyte, and acid-base balance.*

Fluid and Electrolyte Balance

Water accounts for roughly 60 percent of the total body weight of an adult male and 50 percent of that of an adult female. This sex-related difference primarily reflects the relatively larger mass of adipose tissue in adult females and the greater average muscle mass in adult males. (Adipose tissue is 10 percent water, whereas skeletal muscle is 75 percent water.) In both sexes, intracellular fluid contains more of the total body water than does extracellular fluid.

Figure 18–15● illustrates the distribution of water in the body. Nearly two-thirds of the total body water content is found inside living cells, as the fluid medium of the intracellular fluid (ICF), introduced in Chapter 3. ⚯ p. 60 The extracellular fluid (ECF) contains the rest of the body's water. The largest subdivisions of the ECF are the interstitial fluid in peripheral tissues, the plasma of the circulating blood, and water in the ground substance of bone and dense connective tissue. Minor components of the ECF include lymph, cerebrospinal fluid (CSF), synovial fluid, serous fluids (pleural, pericardial, and peritoneal fluids), aqueous humor, and the fluids of the inner ear (perilymph and endolymph). Exchange between the ICF and ECF occurs across cell membranes by osmosis, diffusion, and carrier-mediated processes (see Table 3–4, p. 75).

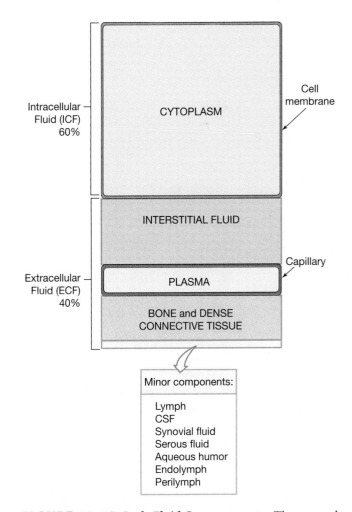

● **FIGURE 18–15 Body Fluid Compartments.** The rectangles in this diagram represent the major body fluid compartments (and their relative proportions) in a healthy adult.

Exchange among the subdivisions of the ECF occurs primarily across the endothelial lining of capillaries. Fluid may also travel from the interstitial spaces to the plasma through lymphatic vessels that drain into the venous system. The kinds and amounts of dissolved electrolytes, proteins, nutrients, and waste products within each ECF subdivision or component are not the same. However, these variations are relatively minor compared with the major differences *between* the ECF and the ICF.

The ICF and ECF are called **fluid compartments** because they commonly behave as distinct entities. As noted in earlier chapters, the principal ions in the ECF are sodium, chloride, and bicarbonate. The ICF contains an abundance of potassium, magnesium, and phosphate ions, plus large numbers of negatively charged proteins. Figure 18–16● compares the ions in the ICF with those in plasma and interstitial fluid, which are the two main subdivisions of the ECF.

Despite these differences in the concentrations of specific substances, the osmotic concentrations of the ICF and ECF are identical. Osmosis eliminates minor differences in concentration almost at once because most cell membranes are freely permeable to water. Because changes in solute concentrations lead

to immediate changes in water distribution, the regulation of water balance and electrolyte balance is tightly intertwined.

Fluid Balance

Water circulates freely within the ECF compartment. At capillary beds throughout the body, capillary blood pressure forces water out of the plasma and into the interstitial spaces. Some of that water is reabsorbed along the distal portion of the capillary bed, and the rest circulates into lymphatic vessels for transport to the venous circulation (see Figure 13–7, p. 475).

Water also moves back and forth across the epithelial surfaces that line the peritoneal, pleural, and pericardial cavities and through the synovial membranes that line joint capsules. The flow rate is significant; for example, roughly 7 liters of peritoneal fluid is produced and reabsorbed each day. Small volumes of water move between the blood and the cerebrospinal fluid, the aqueous and vitreous humors of the eye, and the perilymph and endolymph of the inner ear.

Roughly 2500 mL of water is lost each day in urine, feces, and perspiration (Table 18–4). The water lost in perspiration varies with the level of activity; in individuals who exercise vigorously, the additional losses can be considerable, and

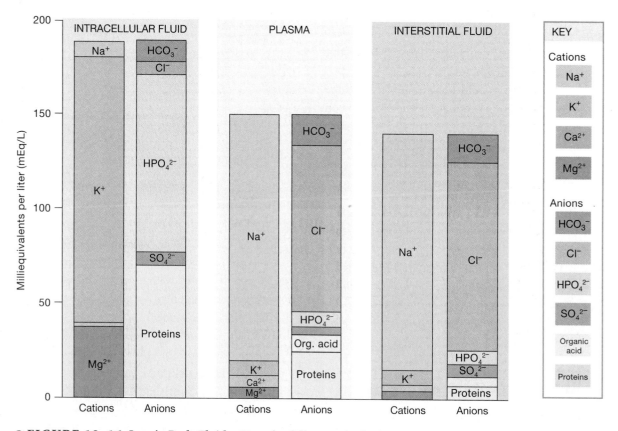

● **FIGURE 18–16 Ions in Body Fluids.** Note the differences in the ion concentrations between the plasma and interstitial fluid of the ECF and those in the ICF. For information concerning the chemical composition of other body fluids, see Appendix IV.

TABLE 18–4 *Water Balance*

SOURCE	DAILY INPUT (mL)
Water content of food	1000
Water consumed as liquid	1200
Metabolic water produced during catabolism	300
Total	2500

METHOD OF ELIMINATION	DAILY OUTPUT (mL)
Urination	1200
Evaporation at skin	750
Evaporation at lungs	400
Loss in feces	150
Total	2500

reach well over 4 liters an hour. ∞ p. 137 Water losses are normally balanced by the gain of fluids through eating (40 percent), drinking (48 percent), and metabolic generation (12 percent). *Metabolic generation* of water occurs primarily as a result of mitochondrial ATP production. ∞ p. 648

Fluid Shifts

Water movement between the ECF and ICF is called a **fluid shift.** Fluid shifts occur relatively rapidly, and reach equilibrium within a period of minutes to hours. These shifts occur in response to changes in the osmotic concentration, or *osmolarity,* of the extracellular fluid.

■ If the ECF becomes more concentrated (hypertonic) with respect to the ICF, water will move from the cells into the ECF until equilibrium is restored.
■ If the ECF becomes more dilute (hypotonic) with respect to the ICF, water will move from the ECF into the cells, and the volume of the ICF will increase accordingly.

In summary, if the osmolarity of the ECF changes, a fluid shift between the ICF and ECF will tend to oppose the change. Because the volume of the ICF is much greater than that of the ECF, the ICF acts as a "water reserve." In effect, instead of a large change in the osmotic concentration of the ECF, smaller changes occur in both the ECF and the ICF.

Electrolyte Balance

You are in electrolyte balance when the rates of gain and loss are equal for each electrolyte in your body. Electrolyte balance is important because:

■ A gain or loss of electrolytes can cause a gain or loss in water.
■ The concentrations of individual electrolytes affect a variety of cell functions. Many examples of the effects of ions

on cell function were described in earlier chapters. For example, the effects of high or low calcium and potassium ion concentrations on cardiac muscle tissue were noted in Chapter 12. ∞ p. 461

Two cations, Na^+ and K^+, deserve attention because (1) they are major contributors to the osmotic concentrations of the ECF and ICF, and (2) they directly affect the normal functioning of all cells. Sodium is the dominant cation within the extracellular fluid. More than 90 percent of the osmotic concentration of the ECF results from the presence of sodium salts—principally, sodium chloride (NaCl) and sodium bicarbonate ($NaHCO_3$)—so changes in the osmotic concentration of extracellular fluids usually reflect changes in the concentration of sodium ions. Potassium is the dominant cation in the intracellular fluid; extracellular potassium concentrations are normally low. In general:

■ The most common problems involving electrolyte balance are caused by an imbalance between sodium gains and losses.
■ Problems with potassium balance are less common but significantly more dangerous than those related to sodium balance.

SODIUM BALANCE. The amount of sodium in the ECF represents a balance between sodium ion absorption at the digestive tract and sodium ion excretion at the kidneys and other sites. The rate of uptake varies directly with the amount included in the diet. Sodium losses occur primarily by excretion in urine and through perspiration. The kidneys are the most important sites for regulating sodium ion losses. In response to circulating aldosterone, the kidneys reabsorb sodium ions (which decreases sodium loss), and in response to atrial natriuretic peptide, the kidneys increase the loss of sodium ions. ∞ p. 673

Whenever the rate of sodium intake or output changes, a corresponding gain or loss of water tends to keep the Na^+ concentration constant. For example, eating a heavily salted meal will not raise the sodium ion concentration of body fluids because as sodium chloride crosses the digestive epithelium, osmosis brings additional water into the ECF. This is why individuals with high blood pressure are told to restrict their salt intake; dietary salt is absorbed, and because "water follows salt," blood volume—and, thus, blood pressure—increases.

POTASSIUM BALANCE. Roughly 98 percent of the potassium content of the body lies within the ICF. Cells expend energy to recover potassium ions as the ions diffuse out of the cytoplasm and into the ECF. The K^+ concentration of the ECF is relatively low and represents a balance between (1) the rate of gain across the digestive epithelium and (2) the rate of loss in urine. The rate of gain is proportional to the amount of potassium in the diet. The rate of loss is strongly affected by aldosterone. Urinary potassium losses are controlled through adjustments in the rate of active secretion along the distal convoluted tubules of the kidneys. The ion pumps sensitive to aldosterone reabsorb sodium ions from the filtrate in exchange for potassium ions from the interstitial fluid. High plasma concentrations of potassium ions also stimulate aldosterone secretion directly. When potassium levels rise in the ECF, aldosterone levels climb, and additional potassium ions are lost in the urine. When potassium levels fall in the ECF, aldosterone levels decline, and potassium ions are conserved.

Key Note

Fluid balance and electrolyte balance are interrelated. Small water gains or losses affect electrolyte concentrations only temporarily. The impacts are reduced by fluid shifts between the ECF and ICF, and by hormonal responses that adjust the rates of water intake and excretion. Similarly, electrolyte gains or losses produce only temporary changes in solute concentration. These changes are opposed by fluid shifts, adjustments in the rates of ion absorption and secretion, and adjustments to the rates of water gain and loss.

CONCEPT CHECK QUESTIONS

1. How would eating a meal high in salt content affect the amount of fluid in the intracellular fluid compartment?
2. What effect would being lost in the desert for a day without water have on your blood osmotic concentration?

Answers begin on p. 792.

Acid-Base Balance

The pH of your body fluids represents a balance among the acids, bases, and salts in solution. The pH of the ECF normally remains within relatively narrow limits, usually 7.35–7.45. Any deviation from the normal range is extremely dangerous because changes in hydrogen ion concentrations disrupt the stability of cell membranes, alter protein structure, and change the activities of important enzymes. You could not survive for long with a pH below 6.8 or above 7.7.

When the pH of blood falls below 7.35, the physiological state called **acidosis** exists. **Alkalosis** exists if the pH exceeds 7.45. These conditions affect virtually all systems, but the nervous and cardiovascular systems are particularly sensitive to

pH fluctuations. Severe acidosis (pH below 7.0) can be deadly because (1) CNS function deteriorates, and the individual becomes comatose; (2) cardiac contractions grow weak and irregular, and signs of heart failure develop; and (3) peripheral vasodilation produces a dramatic drop in blood pressure, and circulatory collapse can occur.

Acidosis and alkalosis are both dangerous, but in practice, problems with acidosis are much more common because several acids (including carbonic acid) are generated by normal cellular activities.

Acids in the Body

Carbonic acid (H_2CO_3) is an important acid in body fluids. At the lungs, carbonic acid breaks down into carbon dioxide and water; the carbon dioxide diffuses into the alveoli. In peripheral tissues, carbon dioxide in solution interacts with water to form carbonic acid. As noted in Chapter 15, the carbonic acid molecules then dissociate to produce hydrogen ions and bicarbonate ions. ∞ p. 571 The complete reaction sequence is:

$$\underset{\substack{\text{carbon} \\ \text{dioxide}}}{CO_2} + \underset{\text{water}}{H_2O} \leftrightarrow \underset{\substack{\text{carbonic} \\ \text{acid}}}{H_2CO_3} \leftrightarrow \underset{\substack{\text{hydrogen} \\ \text{ion}}}{H^+} + \underset{\substack{\text{bicarbonate} \\ \text{ion}}}{HCO_3^-}$$

This reaction occurs spontaneously in body fluids, but it occurs very rapidly in the presence of *carbonic anhydrase*, which is an enzyme found in many cell types, including red blood cells, liver and kidney cells, and parietal cells of the stomach.

Because most of the carbon dioxide in solution is converted to carbonic acid, and most of the carbonic acid dissociates, the partial pressure of carbon dioxide P_{CO_2} and the pH are inversely related (Figure 18–17●). When carbon dioxide concentrations rise, additional hydrogen ions and bicarbonate ions are released, and the pH goes down. (Recall that the greater the concentration of hydrogen ions, the lower the pH value.) The P_{CO_2} is the most important factor that affects the pH in body tissues.

At the alveoli, carbon dioxide diffuses into the atmosphere, the number of hydrogen ions and bicarbonate ions drops, and the pH rises. This process, which effectively removes hydrogen ions from solution, will be considered in more detail in the next section.

Organic acids, or metabolic acids, are generated during normal metabolism. Some are generated during the catabolism of amino acids, carbohydrates, or lipids. Examples are lactic acid produced during anaerobic metabolism of pyruvic acid, and ketone bodies produced in the breakdown of fatty acids. Under normal conditions, metabolic acids are recycled or excreted rapidly, and significant accumulations do not occur.

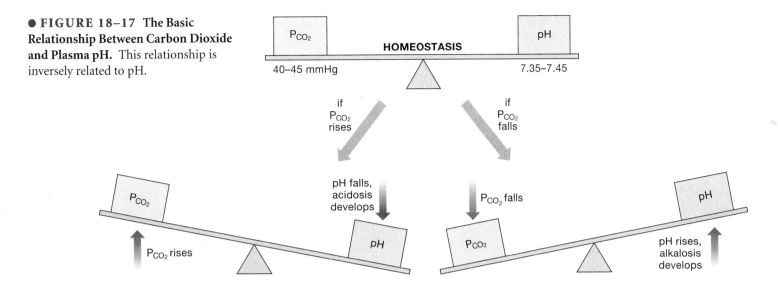

● **FIGURE 18–17** The Basic Relationship Between Carbon Dioxide and Plasma pH. This relationship is inversely related to pH.

Buffers and Buffer Systems

The acids produced in the course of normal metabolic operations are temporarily neutralized by buffers and buffer systems in body fluids. *Buffers,* which are introduced in Chapter 2, are dissolved compounds that can provide or remove hydrogen ions (H^+), thereby stabilizing the pH of a solution. ⚬ p. 39 Buffers include *weak acids* that can donate H^+ and *weak bases* that can absorb H^+. A **buffer system** consists of a combination of a weak acid and its dissociation products: a hydrogen ion and an anion. The body has three major buffer systems, each with slightly different characteristics and distributions: the *protein buffer system,* the *carbonic acid-bicarbonate* buffer system, and the *phosphate buffer system.*

Protein buffer systems contribute to the regulation of pH in the ECF and ICF. Protein buffer systems depend on the ability of amino acids to respond to changes in pH by accepting or releasing hydrogen ions. If pH climbs, the carboxyl group (—COOH) of the amino acid can dissociate, which releases a hydrogen ion. If pH drops, the amino group (—NH_2) can accept an additional hydrogen ion, which forms an amino ion (—NH_3^+).

The plasma proteins and hemoglobin in red blood cells contribute to the buffering capabilities of the blood. Interstitial fluids contain extracellular protein fibers and dissolved amino acids that also help regulate pH. In the ICF of active cells, structural and other proteins provide an extensive buffering capability that prevents destructive pH changes when organic acids, such as lactic acid, are produced by cellular metabolism.

The **carbonic acid-bicarbonate buffer system** is an important buffer system in the ECF. With the exception of red blood cells, your cells generate carbon dioxide 24 hours a day. As noted earlier, most of the carbon dioxide is converted to carbonic acid, which then dissociates into a hydrogen ion and a bicarbonate ion. The carbonic acid and its dissociation products constitute the carbonic acid-bicarbonate buffer system; the carbonic acid acts as a weak acid, and the bicarbonate ion acts as a weak base. The net effect of this buffer system is that $CO_2 + H_2O \leftrightarrow H^+ + HCO_3^-$. If hydrogen ions are removed, they will be replaced through the combining of water with carbon dioxide; if hydrogen ions are added, most will be removed through the formation of carbon dioxide and water.

The primary role of the carbonic acid-bicarbonate buffer system is to prevent pH changes caused by organic (metabolic) acids. The hydrogen ions released through the dissociation of these acids combine with bicarbonate ions, producing water and carbon dioxide. The carbon dioxide can then be excreted at the lungs. This buffering system can cope with large amounts of acid because body fluids contain an abundance of bicarbonate ions, which is known as the *bicarbonate reserve.* When hydrogen ions enter the ECF, the bicarbonate ions that combine with them are replaced by the bicarbonate reserve.

The **phosphate buffer system** consists of an anion, *dihydrogen phosphate* ($H_2PO_4^-$), which is a weak acid. The dihydrogen phosphate ion and its dissociation products constitute the phosphate buffer system:

$$H_2PO_4^- \leftrightarrow H^+ + HPO_4^{2-}$$

dihydrogen   hydrogen   monohydrogen
phosphate   ion   phosphate

In solution, dihydrogen phosphate ($H_2PO_4^-$) reversibly dissociates into a hydrogen ion and monohydrogen phosphate (HPO_4^{2-}). In the ECF, the phosphate buffer system plays only a supporting role in the regulation of pH, primarily because the concentration of bicarbonate ions far exceeds that of phosphate ions. However, the phosphate buffer system is quite important in buffering the pH of the ICF, where the concentration of phosphate ions is relatively high.

Maintaining Acid-Base Balance

Buffer systems can tie up excess hydrogen ions (H^+), but because the H^+ ions have not been eliminated, buffer systems provide only a temporary solution. For homeostasis to be preserved, the captured H^+ must ultimately be removed from body fluids. The problem is that the supply of buffer molecules is limited; once a buffer binds a H^+, it cannot bind any more H^+. With the buffer molecules tied up, the capacity of the ECF to absorb more H^+ is reduced, and pH control is impossible.

The maintenance of acid-base balance involves balancing hydrogen ion losses and gains. In this "balancing act," the respiratory and renal mechanisms support the buffer systems by (1) secreting or absorbing hydrogen ions; (2) controlling the excretion of acids and bases; and (3) generating additional buffers. It is the *combination* of buffer systems and these respiratory and renal mechanisms that maintain body pH within narrow limits.

RESPIRATORY CONTRIBUTIONS TO pH REGULATION. **Respiratory compensation** is a change in the respiratory rate that helps stabilize the pH of the ECF. Respiratory compensation occurs whenever pH exceeds normal limits. Respiratory activity has a direct effect on the carbonic acid-bicarbonate buffer system. Increasing or decreasing the rate of respiration alters pH by lowering or raising the partial pressure of CO_2 (P_{CO_2}). Changes in P_{CO_2} have a direct effect on the concentration of hydrogen ions in the plasma. When the P_{CO_2} rises, the pH declines, and when the P_{CO_2} decreases, the pH increases (see Figure 18–17).

Mechanisms responsible for controlling the respiratory rate were discussed in Chapter 15, p. 573 and only a brief summary will be presented here. A rise in P_{CO_2} stimulates chemoreceptors in the carotid and aortic bodies and within the CNS; a fall in the P_{CO_2} inhibits them. Stimulation of the chemoreceptors leads to an increase in the respiratory rate. As the rate of respiration increases, more CO_2 is lost at the lungs, so the P_{CO_2} returns to normal levels. When the P_{CO_2} of the blood or CSF declines, respiratory activity becomes depressed, the breathing rate falls, and the P_{CO_2} in the extracellular fluids rises.

RENAL CONTRIBUTIONS TO pH REGULATION. **Renal compensation** is a change in the rates of hydrogen ion and bicarbonate ion secretion or absorption by the kidneys in response to changes in plasma pH. Under normal conditions, the body generates H^+ through the production of metabolic acids. The H^+ these acids release must be excreted in the urine to maintain acid-base balance. Glomerular filtration puts hydrogen ions, carbon dioxide, and the other components of the carbonic acid-bicarbonate and phosphate buffer systems into the filtrate. The kidney tubules then modify the pH of the filtrate by secreting hydrogen ions or reabsorbing bicarbonate ions.

ACID-BASE DISORDERS. Together, buffer systems, respiratory compensation, and renal compensation maintain normal acid-base balance. These mechanisms are usually able to control pH very precisely, so that the pH of the extracellular fluids seldom varies more than 0.1 pH units, from 7.35 to 7.45. When buffering mechanisms are severely stressed, however, pH wanders outside these limits, which produces symptoms of alkalosis or acidosis.

Respiratory acid-base disorders result when abnormal respiratory function causes an extreme rise or fall in CO_2 levels in the ECF. Metabolic acid-base disorders result from the generation of organic acids or by conditions that affect the concentration of bicarbonate ions in the ECF. Table 18–5 summarizes the general causes and treatments of acid-base disorders.

TABLE 18–5 *Acid-Base Disorders*

DISORDER	pH (NORMAL = 7.35–7.45)	GENERAL CAUSES	TREATMENT
Respiratory acidosis	Decreased (below 7.35)	Generally caused by hypoventilation and CO_2 buildup in tissues and blood	Improve ventilation; in some cases, with bronchodilation and mechanical assistance
Metabolic acidosis	Decreased (below 7.35)	Caused by buildup of metabolic acid, impaired H^+ excretion at kidneys, or bicarbonate loss in urine or feces	Administration of bicarbonate (gradual) with other steps as needed to correct primary cause
Respiratory alkalosis	Increased (above 7.45)	Generally caused by hyperventilation and reduction in plasma CO_2 levels	Reduce respiratory rate, allow rise in P_{CO_2}
Metabolic alkalosis	Increased (above 7.45)	Usually caused by prolonged vomiting and associated acid loss	For pH below 7.55, no treatment; pH above 7.55 may require administration of ammonium chloride

Clinical Note
DISTURBANCES OF ACID-BASE BALANCE

The maintenance of normal acid-base balance is one of the body's most important homeostatic functions. A problem with acid-base balance can constitute a serious medical emergency that requires immediate treatment.

The body's pH is tightly regulated. Changes in pH seldom exceed 0.1 pH units. However, in disease, significant pH shifts can occur. Normal pH is considered 7.35 to 7.45. A pH less than 7.35 is considered *acidosis,* while a pH greater than 7.45 is considered *alkalosis.* The body controls pH principally through three systems: the *buffer system,* the *respiratory system,* and the *renal system.*

The buffer systems remove hydrogen ions, which prevents a significant change in pH. The three major buffer systems are the *protein buffer system,* the *carbonic acid-bicarbonate buffer system,* and the *phosphate buffer system.* The buffer systems react immediately to changes in pH.

The respiratory system compensates for changes in pH by increasing or decreasing respirations. It is primarily controlled by changes in the carbon dioxide (P_{CO_2}) level. An increase in the P_{CO_2} stimulates respirations, which causes the removal of CO_2 molecules. Likewise, a decrease in P_{CO_2} inhibits respirations, which causes CO_2 molecules to accumulate.

The renal system reacts to changes in pH by producing more acidic or more alkaline urine. When the body's pH falls, the distal tubule of the kidney begins excreting hydrogen ions into the urine and reabsorbing bicarbonate. Likewise, when the body's pH rises, the distal tubule slows hydrogen ion secretion and reabsorbs less bicarbonate.

Acid-base disorders are classified based upon the source of the problem. Disorders due to an abnormality in the level of CO_2 in the body are called *respiratory disorders. Metabolic disorders* are caused by the generation of organic or fixed acids. They can also be caused by conditions that affect the concentration of bicarbonate ions in the body.

■ *Respiratory acidosis.* Respiratory acidosis is due to a decrease in alveolar ventilation that results in CO_2 retention *(hypercapnia).* The excess CO_2 is converted to carbonic acid, which decreases the pH.

Causes of respiratory acidosis include sedative drugs, brain stem trauma, respiratory muscle paralysis, and disorders of the chest wall. Disorders of the lung parenchyma, such as pneumonia, asthma, emphysema, pulmonary edema, and bronchitis, can also cause respiratory acidosis. Respiratory acidosis can be either acute or chronic. Chronic respiratory acidosis is seen in chronic obstructive pulmonary disease (COPD) and deformities of the chest wall. In chronic respiratory acidosis, the renal system ultimately compensates by increasing hydrogen ion excretion. Treatment of respiratory acidosis involves increasing ventilation and correcting the underlying cause.

■ *Respiratory alkalosis.* Respiratory alkalosis occurs when alveolar ventilation is increased, which results in excess elimination of CO_2 *(hypocapnia)* and an increase in pH. Often, stimulation of ventilation is due to *hypoxemia* (low oxygen levels) from problems such as pulmonary disease, congestive heart failure, high altitudes, fever, and early salicylate intoxication. A common cause of hyperventilation and subsequent respiratory alkalosis is *hysteria* (hyperventilation). Treatment of respiratory alkalosis involves correcting the underlying cause. In selected cases, hyperventilation from hysteria can be remedied by having a patient rebreathe into a paper bag, which thus increases CO_2 levels.

■ *Metabolic acidosis.* In metabolic acidosis, there is an accumulation of noncarbonic acids or a loss of bicarbonate. Causes include lactic acidosis (from poor perfusion), renal failure, or diabetic ketoacidosis. The buffer systems attempt to compensate. Respirations are increased as the pH falls, which results in eliminations of CO_2. The renal system secretes excess hydrogen ions. Treatment is directed at correcting the underlying causes. In severe cases, administration of sodium bicarbonate ($NaHCO_2$) may be required to aid in buffering the excess acids.

■ *Metabolic alkalosis.* Metabolic alkalosis is due to an increase in bicarbonate or to excessive loss of metabolic acids. Causes include prolonged vomiting, gastrointestinal suctioning, excessive bicarbonate intake, or diuretic therapy. Vomiting and gastrointestinal suctioning can cause loss of gastric acids and ultimately lead to metabolic alkalosis. Treatment is directed at correcting the underlying cause. ■

Key Note
The most common and acute acid-base disorder is respiratory acidosis, which develops when respiratory activity cannot keep pace with the rate of carbon dioxide generation in peripheral tissues.

→ CONCEPT CHECK QUESTIONS

1. What effect would a decrease in the pH of body fluids have on the respiratory rate?
2. How would a prolonged fast affect the body's pH?
3. How does prolonged vomiting produce alkalosis?

Answers begin on p. 792.

■ Aging and the Urinary System

In general, aging is associated with an increased incidence of kidney problems. Age-related changes in the urinary system and in aspects of fluid, electrolyte, and acid-base balance include the following:

1. *A decline in the number of functional nephrons.* The total number of nephrons in the kidneys drops by 30–40 percent between ages 25 and 85.

2. *A reduction in the GFR.* This results from fewer glomeruli, cumulative damage to the filtration apparatus in the remaining glomeruli, and reductions in renal blood flow. Fewer nephrons and a reduced GFR also reduce the body's ability to regulate pH through renal compensation.

3. *Reduced sensitivity to ADH and aldosterone.* With age, the distal portions of the nephron and collecting system become less responsive to ADH and aldosterone. Reabsorption of water and sodium ions is reduced, and more potassium ions are lost in the urine.

4. *Problems with the micturition reflex.* Several factors are involved in such problems:
   - The sphincter muscles lose tone and become less effective at voluntarily retaining urine. Loss of muscle tone leads to incontinence, which often involves a slow leakage of urine.
   - The ability to control micturition is often lost after a stroke, Alzheimer's disease, or other CNS problems that affect the cerebral cortex or hypothalamus.
   - In males, **urinary retention** may develop secondary to enlargement of the prostate gland. In this condition, swelling and distortion of surrounding prostatic tissues compress the urethra, and restrict or prevent the flow of urine.

5. *A gradual decrease of total body water content with age.* Between ages 40 and 60, total body water content averages 60 percent for males and 50 percent for females. After age 60, the values decline to roughly 50 percent for males and 45 percent for females. Among other effects, such decreases result in less dilution of waste products, toxins, and administered drugs.

6. *A net loss in body mineral content in many people over age 60 as muscle mass and skeletal mass.* This loss can be prevented, at least in part, by a combination of exercise and increased dietary mineral intake.

7. *Increased incidence of disorders that affect major systems with increasing age.* Most of these disorders have some impact on fluid, electrolyte, and/or acid-base balance.

Clinical Note
RENAL TRAUMA

The renal system is fairly well-protected from trauma. Injuries to the genitourinary system occur in only 5 percent or less of all trauma victims. Approximately 10 percent of children with blunt abdominal trauma have renal system injury. Most patients with a kidney injury have other concurrent injuries. The kidneys and ureters are located in the retroperitoneal space and are thus well-protected. The bladder is well-protected by the bony structures of the pelvis. In males, the urethra is more vulnerable to trauma due to the anatomy of the penis. Fortunately, renal system injuries are rarely life threatening and usually do not require immediate intervention.

Trauma to the kidneys and ureters usually results from penetrating trauma such as gunshots or stabbings (Figure 18–18●). Rapid deceleration forces may cause injury to the renal pedicle or to the vascular structures. The signs and symptoms of urinary system trauma can be subtle. The most common indicator of renal system trauma is blood in the urine (hematuria). This may be gross (readily visible) or microscopic (seen only with a microscope or detected by urine test strips). Trauma to the prostate and urethra can result in blood at the urethral meatus (or on the underwear). Any of these findings warrant further investigation. In the emergency department, a rectal examination should be carried out. A high-riding or floating prostate indicates the need for further studies. After life-threatening problems have been addressed, the urinary system can be studied with contrast X-rays and CT scanning. ■

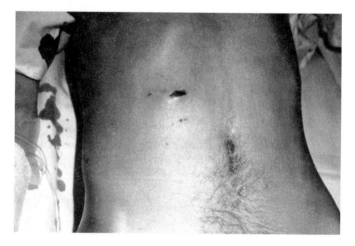

● **FIGURE 18–18 Renal Trauma.** Most cases of renal trauma are due to penetrating injuries. Although this wound is on the anterior abdomen, the right kidney and ureter are lacerated.

CONCEPT CHECK QUESTION

1. What effect does aging have on the GFR?

Answers begin on p. 792.

The Urinary System in Perspective

For All Systems
Excretes waste products; maintains normal body fluid pH and ion composition

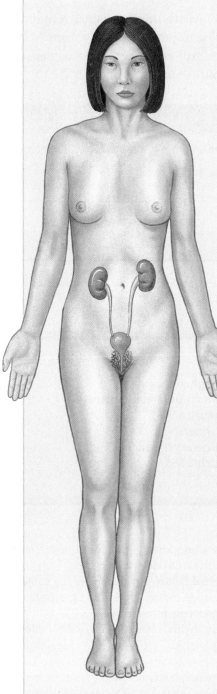

● **FIGURE 18–19 Functional Relationships Between the Urinary System and Other Systems.**

The Integumentary System
- Sweat glands assist in elimination of water and solutes, especially sodium and chloride ions; keratinized epidermis prevents excessive fluid loss through skin surface; epidermis produces vitamin D_3, important for the renal production of calcitriol

The Skeletal System
- Axial skeleton provides some protection for kidneys and ureters; pelvis protects urinary bladder and proximal portion of urethra
- Conserves calcium and phosphate needed for bone growth

The Muscular System
- Sphincter controls urination by closing urethral opening; muscle layers of trunk provide some protection for urinary organs
- Removes waste products of protein metabolism; assists in regulation of calcium and phosphate concentrations

The Nervous System
- Adjusts renal blood pressure; monitors distension of urinary bladder and controls urination

The Endocrine System
- Aldosterone and ADH adjust rates of fluid and electrolyte reabsorption in kidneys
- Kidney cells release renin when local blood pressure declines and erythropoietin (EPO) when renal oxygen levels decline

The Cardiovascular System
- Delivers blood to capillaries, where filtration occurs; accepts fluids and solutes reabsorbed during urine production
- Releases renin to elevate blood pressure and erythropoietin to accelerate red blood cell production

The Lymphatic System
- Provides specific defenses against urinary tract infections
- Eliminates toxins and wastes generated by cellular activities; acid pH of urine provides nonspecific defense against urinary tract infections

The Respiratory System
- Assists in the regulation of pH by eliminating carbon dioxide
- Assists in the elimination of carbon dioxide; provides bicarbonate buffers that assist in pH regulation

The Digestive System
- Absorbs water needed to excrete wastes at kidneys; absorbs ions needed to maintain normal body fluid concentrations; liver removes bilirubin
- Excretes toxins absorbed by the digestive epithelium; excretes bilirubin and nitrogenous wastes produced by the liver; calcitrol production by kidneys aids calcium and phosphate absorption along digestive tract

The Reproductive System

- Accessory organ secretions may have antibacterial action that helps prevent urethral infections in males
- Urethra in males carries semen to the exterior

■ Integration with Other Systems

The urinary system is not the only organ system involved in excretion. Along with the urinary system, the integumentary, respiratory, and digestive systems are considered to form an anatomically diverse *excretory system*. Examples of their excretory activities include:

1. *Integumentary system.* Water and electrolyte losses in perspiration affect plasma volume and composition. The effects are most apparent when losses are extreme, as occurs during maximum sweat production. Small amounts of metabolic wastes, including urea, are also excreted in perspiration.

2. *Respiratory system.* The lungs remove the carbon dioxide generated by cells. Small amounts of other compounds, such as acetone and water, evaporate into the alveoli and are eliminated during exhalation.

3. *Digestive system.* The liver excretes metabolic waste products in bile, and a variable amount of water is lost in feces.

All these nonurinary excretory activities affect the composition of body fluids. However, the excretory functions of these systems are not as closely regulated as are those of the kidneys, and the effects of integumentary and digestive excretory activities normally are minor compared with those of the urinary system.

Figure 18–19● summarizes the functional relationships between the urinary system and other organ systems. Many of these relationships were explored earlier in this chapter, which addressed fluid, pH, and electrolyte balance.

> **CONCEPT CHECK QUESTION**
>
> 1. What organ systems contribute to an overall excretory system of the body?
>
> *Answers begin on p. 792.*

Chapter Review

Access more review material online at *www.prenhall.com/bledsoe*. There you will find quiz questions, labeling activities, animations, essay questions, and web links.

Key Terms

acidosis 684
aldosterone 669
alkalosis 684
angiotensin II 673
buffer system 685
filtrate 663

glomerular filtration rate 668
glomerulus 664
juxtaglomerular apparatus 666
kidney 660
loop of Henle 665
micturition 679

nephron 661
peritubular capillaries 663
renin 673
ureters 675
urinary bladder 677
urine 666

Related Clinical Terms

calculi (KAL-kū-lī) Insoluble deposits within the urinary tract formed from calcium salts, magnesium salts, or uric acid.

cystitis Inflammation of the lining of the urinary bladder.

diuretics (dī-ū-RET-iks) Substances that promote fluid loss in urine.

dysuria (dis-ō-rē-uh) Painful urination.

glomerulonephritis (glo-mer-ū-lō-nef-RĪ-tis) Inflammation of the glomeruli (the filtration units of the nephron).

glycosuria (glī-kō-SOO-rē-uh) The presence of glucose in urine.

hematuria The presence of blood in urine.

hypernatremia A condition characterized by excess sodium in the blood.

hyponatremia A condition characterized by lower-than-normal sodium in the blood.

incontinence (in-KON-ti-nens) An inability to control urination voluntarily.

nephritis Inflammation of the kidneys.

nephrology (ne-FROL-o-jē) The medical specialty concerned with the kidney and its disorders.

proteinuria The presence of protein in urine.

pyelogram (PĪ-el-ō-gram) An X-ray image of the kidneys taken after a radiopaque compound has been administered.

pyelonephritis (pī-e-lō-nef-RĪ-tis) Inflammation of kidney tissues.

renal failure An inability of the kidneys to excrete wastes in sufficient quantities to maintain homeostasis.

urethritis Inflammation of the urethra.

urinalysis A physical and chemical assessment of urine.

urinary tract infection (UTI) An infection of the urinary tract by bacteria or fungi (yeast).

urology (u-ROL-o-jē) Branch of medicine concerned with the urinary system and its disorders and with the male reproductive tract and its disorders.

Summary Outline

1. The functions of the **urinary system** include (1) eliminating organic waste products, (2) regulating plasma concentrations of ions, (3) regulating blood volume and blood pressure by adjusting the volume of water lost and releasing hormones, (4) stabilizing blood pH, and (5) conserving nutrients.

THE ORGANIZATION OF THE URINARY SYSTEM 660

1. The urinary system includes the kidneys, the ureters, the urinary bladder, and the urethra. The kidneys produce urine (a fluid that contains water, ions, and soluble compounds); during urination, urine is forced out of the body. (*Figure 18–1*)

THE KIDNEYS 660

1. The left **kidney** extends superiorly slightly more than the right kidney. Both kidneys lie in a *retroperitoneal* position. (*Figure 18–2*)

Superficial and Sectional Anatomy 661

2. A fibrous renal capsule surrounds each kidney. The **hilum** provides entry for the *renal artery* and *renal nerve* and exit for the *renal vein* and *ureter*.

3. The ureter is continuous with the **renal pelvis.** This chamber branches into two **major calyces,** each connected to four or five **minor calyces,** which enclose the **renal papillae.** Urine production begins in **nephrons.** (*Figure 18–3*)

The Blood Supply to the Kidneys 662

4. The blood vessels of the kidneys include the **interlobar, arcuate,** and **interlobular arteries** and the **interlobar, arcuate,** and **interlobular veins.** Blood travels from the **afferent** and **efferent arterioles** to the **peritubular capillaries** and the **vasa recta.** Diffusion occurs between the capillaries of the vasa recta and the tubular cells through the interstitial fluid that surrounds the nephron. (*Figure 18–4*)

The Nephron 663

5. The **nephron** (the basic functional unit in the kidney) includes the *renal corpuscle* and a **renal tubule,** which empties into the **collecting system** through a *collecting duct*. From the renal corpuscle, filtrate travels through the *proximal convoluted tubule*, the *loop of Henle*, and the *distal convoluted tubule*. (*Figures 18–3c, 18–5*)

Key Note 663

6. Nephrons are responsible for (1) the production of **filtrate,** (2) the reabsorption of nutrients, and (3) the reabsorption of water and ions.

7. The **renal corpuscle** consists of a knot of intertwined capillaries called the **glomerulus** surrounded by the **Bowman's capsule.** Blood arrives from the *afferent arteriole* and departs in the *efferent arteriole*. (*Figure 18–6*)

8. At the glomerulus, **podocytes** cover the basement membrane of the capillaries that project into the **capsular space.** The processes of the podocytes are separated by narrow slits. (*Figure 18–6*)

9. The **proximal convoluted tubule (PCT)** actively reabsorbs nutrients, plasma proteins, and electrolytes from the filtrate. These substances are then released into the surrounding interstitial fluid. (*Figure 18–5*)

10. The **loop of Henle** includes a *descending limb* and an *ascending limb*. The ascending limb is impermeable to water and solutes; the descending limb is permeable to water. (*Figure 18–5*)

11. The ascending limb delivers fluid to the **distal convoluted tubule (DCT),** which actively secretes ions and reabsorbs sodium ions from the urine. The **juxtaglomerular apparatus,** which releases renin and erythropoietin, is located at the start of the DCT. (*Figure 18–6a*)

12. The **collecting ducts** receive urine from nephrons and merge into a **papillary duct,** which delivers urine to a minor calyx. The collecting system makes final adjustments to the urine by reabsorbing water or reabsorbing or secreting various ions.

BASIC PRINCIPLES OF URINE PRODUCTION 666

1. The primary purpose in **urine** production is the excretion and elimination of dissolved solutes, principally metabolic waste products such as **urea, creatinine,** and **uric acid.**

2. Urine formation involves **filtration, reabsorption,** and **secretion.** (*Tables 18–1, 18–2*)

Filtration at the Glomerulus 667

3. Glomerular filtration occurs as fluids move across the wall of the glomerular capillaries into the capsular space in response to blood pressure in those capillaries. The **glomerular filtration rate (GFR)** is the amount of filtrate produced in the kidneys each minute. Any factor that alters the **filtration** (blood) **pressure** will change the GFR and affect kidney function.

4. Declining filtration pressures stimulate the juxtaglomerular apparatus to release *renin*. Renin release results in increases in blood volume and blood pressure.

Key Note 668

Reabsorption and Secretion Along the Renal Tubule 668

5. The cells of the PCT normally reabsorb 60–70 percent of the volume of the filtrate produced in the renal corpuscle. The PCT reabsorbs nutrients, sodium and other ions, and water from the filtrate and transports them into the interstitial fluid. It also secretes various substances into the tubular fluid.

6. Water and ions are reclaimed from the filtrate by the loop of Henle. The ascending limb pumps out sodium and chloride ions, and the descending limb reabsorbs water. A concentration gradient in the medulla encourages the osmotic flow of water out of the tubular fluid and into the interstitial fluid. As water is lost by osmosis and the volume of tubular fluid decreases, the urea concentration rises.

7. The DCT performs final adjustments by actively secreting or absorbing materials. Sodium ions are actively absorbed in exchange for potassium and hydrogen ions discharged into the tubular fluid. **Aldosterone** secretion increases the rate of sodium reabsorption and potassium loss.

8. The amount of water in the urine of the collecting ducts is regulated by the secretion of *antidiuretic hormone (ADH)*. In the absence of ADH, the DCT, collecting tubule, and collecting duct are impermeable to water. The higher the ADH level in circulation, the more water is absorbed and the more concentrated the urine. *(Figure 18–7)*

9. More than 99 percent of the filtrate produced each day is reabsorbed before reaching the renal pelvis. Still, the water content of normal urine is 93–97 percent. *(Table 18–3)*

10. Each segment of the nephron and collecting system contributes to the production of hypertonic urine. *(Figure 18–8)*

Key Note 670

The Control of Kidney Function 672

11. Renal function may be regulated by automatic adjustments in glomerular pressures through changes in the diameters of the afferent and efferent arterioles.

12. Sympathetic activation produces powerful vasoconstriction of the afferent arterioles, which decreases the GFR and slows the production of filtrate. Sympathetic activation also alters the GFR by changing the regional pattern of blood circulation.

13. Hormones that regulate kidney function include angiotensin II, aldosterone, ADH, and atrial natriuretic peptide (ANP). *(Figure 18–9)*

URINE TRANSPORT, STORAGE, AND ELIMINATION 675

1. Filtrate modification and urine production end when the fluid enters the renal pelvis. The rest of the urinary system is responsible for transporting, storing, and eliminating the urine. *(Figures 18–10, 18–11)*

The Ureters and Urinary Bladder 675

2. The **ureters** extend from the renal pelvis to the urinary bladder. Peristaltic contractions by smooth muscles in the walls of the ureters move the urine. *(Figures 18–1, 18–12)*

3. Internal features of the **urinary bladder,** which is a distensible sac for urine storage, include the *trigone,* which is the neck, and the **internal urethral sphincter.** Contraction of the *detrusor muscle* compresses the bladder and expels the urine into the urethra. *(Figure 18–12)*

The Urethra 677

4. In both sexes, as the urethra passes through the muscular pelvic floor, a circular band of skeletal muscles forms the **external urethral sphincter,** which is under voluntary control. *(Figure 18–12)*

The Micturition Reflex and Urination 679

5. The process of **urination** is coordinated by the **micturition reflex,** which is initiated by stretch receptors in the bladder wall. Voluntary urination involves coupling this reflex with

the voluntary relaxation of the external urethral sphincter, which allows the opening of the internal urethral sphincter. *(Figures 18–13, 18–14)*

FLUID, ELECTROLYTE, AND ACID-BASE BLANCE 680

1. The maintenance of normal volume and composition in the extracellular and intracellular fluids is vital to life. Three types of homeostasis are involved: *fluid balance, electrolyte balance,* and *acid-base balance.*

Fluid and Electrolyte Balance 681

2. The **intracellular fluid (ICF)** contains about 60 percent of the total body water; the **extracellular fluid (ECF)** contains the rest. Exchange occurs between the ICF and ECF, but the two **fluid compartments** retain their distinctive characteristics. *(Figures 18–15, 18–16)*

3. Water circulates freely within the ECF compartment.

4. Water losses are normally balanced by gains through eating, drinking, and metabolic generation. *(Table 18–4)*

5. Water movement between the ECF and ICF is called a *fluid shift.* If the ECF becomes hypertonic relative to the ICF, water will move from the ICF into the ECF until osmotic equilibrium has been restored. If the ECF becomes hypotonic relative to the ICF, water will move from the ECF into the cells, and the volume of the ICF will increase accordingly.

6. Electrolyte balance is important because total electrolyte concentrations affect water balance, and because the levels of individual electrolytes can affect a variety of cell functions. Problems with electrolyte balance generally result from an imbalance between sodium gains and losses. Problems with potassium balance are less common but more dangerous.

7. The rate of sodium uptake across the digestive epithelium is directly related to the amount of sodium in the diet. Sodium losses occur mainly in the urine and through perspiration. The rate of sodium reabsorption along the DCT is regulated by aldosterone levels; aldosterone stimulates sodium ion reabsorption.

8. Potassium ion concentrations in the ECF are very low. Potassium excretion increases (1) when sodium ion concentrations decline and (2) as ECF potassium concentrations rise. The rate of potassium excretion is regulated by aldosterone; aldosterone stimulates potassium ion excretion.

Key Note 684

Acid-Base Balance 684

9. The pH of normal body fluids ranges from 7.35 to 7.45; variations outside this range produce **acidosis** or **alkalosis.**

10. Carbonic acid is the most important factor that affects the pH of the ECF. In solution, CO_2 reacts with water to form carbonic acid; the dissociation of carbonic acid releases hydrogen ions. An inverse relationship exists between the concentration of CO_2 and pH. *(Figure 18–17)*

11. Organic acids (metabolic acids) include products of metabolism such as lactic acid and ketone bodies.

12. A buffer system consists of a weak acid and its anion dissociation product, which acts as a weak base. The three major buffer systems are (1) *protein buffer systems* in the ECF and ICF; (2) the *carbonic acid-bicarbonate buffer system*, which is most important in the ECF; and (3) the *phosphate buffer system* in the intracellular fluids.

13. In **protein buffer systems,** the component amino acids respond to changes in H^+ concentrations. Blood plasma proteins and hemoglobin in red blood cells help prevent drastic changes in pH.

14. The **carbonic acid-bicarbonate buffer system** prevents pH changes due to organic acids in the ECF.

15. The **phosphate buffer system** is important in preventing pH changes in the intracellular fluid.

16. In **respiratory compensation,** the lungs help regulate pH by affecting the carbonic acid-bicarbonate buffer system; changing the respiratory rate can raise or lower the pH of body fluids, and affect the buffering capacity.

17. In **renal compensation,** the kidneys vary their rates of hydrogen ion secretion and bicarbonate ion reabsorption depending on the pH of extracellular fluids.

18. Respiratory acid-base disorders result when abnormal respiratory function causes an extreme rise or a fall in CO_2 levels.

Metabolic acid-base disorders are caused by the formation of organic acids or conditions that affect the levels of bicarbonate ions. *(Table 18–5)*

Key Note 687

AGING AND THE URINARY SYSTEM 688

1. Aging is usually associated with increased kidney problems. Age-related changes in the urinary system include (1) loss of functional nephrons, (2) reduced GFR, (3) reduced sensitivity to ADH, (4) problems with the micturition reflex (urinary retention may develop in men whose prostate gland is inflamed), (5) declining body water content, (6) a loss of mineral content, and (7) disorders that affect either fluid, electrolyte, or acid-base balance. *(Figure 18–18)*

INTEGRATION WITH OTHER SYSTEMS 690

1. The urinary, integumentary, respiratory, and digestive systems are sometimes considered an anatomically diverse *excretory system.* The systems' components work together to perform all of the excretory functions that affect the composition of body fluids. *(Figure 18–19)*

Review Questions

Level 1: Reviewing Facts and Terms

Match each item in column A with the most closely related item in column B. Place letters for answers in the spaces provided.

COLUMN A

____ 1. urination
____ 2. renal capsule
____ 3. hilum
____ 4. medulla
____ 5. nephrons
____ 6. renal corpuscle
____ 7. external urethral sphincter
____ 8. internal urethral sphincter
____ 9. aldosterone
____ 10. podocytes
____ 11. efferent arteriole
____ 12. afferent arteriole
____ 13. vasa recta
____ 14. ADH
____ 15. ECF
____ 16. sodium
____ 17. potassium

COLUMN B

a. site of urine production
b. capillaries around loop of Henle
c. causes sensation of thirst
d. accelerated sodium reabsorption
e. voluntary control
f. filtration slits
g. fibrous covering
h. blood leaves glomerulus
i. dominant cation in ICF
j. renal pyramids
k. blood to glomerulus
l. interstitial fluid
m. contains glomerulus
n. exit for ureter
o. dominant cation in ECF
p. involuntary control
q. micturition

18. The filtrate that leaves Bowman's capsule empties into the:
    (a) distal convoluted tubule.
    (b) loop of Henle.
    (c) proximal convoluted tubule.
    (d) collecting duct.

19. The distal convoluted tubule is an important site for:
    (a) active secretion of ions.
    (b) active secretion of acids and other materials.
    (c) selective reabsorption of sodium ions from the tubular fluid.
    (d) a, b, and c are correct.

20. The endocrine structure that secretes renin and erythropoietin is the:
    (a) juxtaglomerular apparatus.
    (b) vasa recta.
    (c) Bowman's capsule.
    (d) adrenal gland.

21. The primary purpose of the collecting system is to:
    (a) transport urine from the bladder to the urethra.
    (b) selectively reabsorb sodium ions from tubular fluid.
    (c) transport urine from the renal pelvis to the ureters.
    (d) make final adjustments to the osmotic concentration and volume of urine.

22. A person is in fluid balance when:
    (a) the ECF and ICF are isotonic.
    (b) no fluid movement occurs between compartments.
    (c) the amount of water gained each day is equal to the amount lost to the environment.
    (d) a, b, and c are correct.

Level 2: Reviewing Concepts

30. The urinary system regulates blood volume and pressure by:
    (a) adjusting the volume of water lost in the urine.
    (b) releasing erythropoietin.
    (c) releasing renin.
    (d) a, b, and c are correct.

31. The balance of solute and water reabsorption in the renal medulla is maintained by the:
    (a) segmental arterioles and veins.
    (b) interlobar arteries and veins.
    (c) vasa recta.
    (d) arcuate arteries.

23. The primary components of the extracellular fluid are:
    (a) lymph and cerebrospinal fluid.
    (b) blood plasma and serous fluids.
    (c) interstitial fluid and plasma.
    (d) a, b, and c are correct.

24. All the homeostatic mechanisms that monitor and adjust the composition of body fluids respond to changes:
    (a) in the ICF.
    (b) in the ECF.
    (c) inside the cell.
    (d) a, b, and c are correct.

25. The most common problems with electrolyte balance are caused by an imbalance between gains and losses of _____ ions.
    (a) calcium
    (b) chloride
    (c) potassium
    (d) sodium

26. What is the primary function of the urinary system?

27. What are the structural components of the urinary system?

28. What are fluid shifts? What is their function, and what factors can cause them?

29. What three major hormones mediate major physiological adjustments that affect fluid and electrolyte balance? What are the primary effects of each hormone?

32. The higher the plasma concentration of aldosterone, the more efficiently the kidney will:
    (a) conserve sodium ions.
    (b) retain potassium ions.
    (c) stimulate urinary water loss.
    (d) secrete greater amounts of ADH.

33. When pure water is consumed:
    (a) the ECF becomes hypertonic with respect to the ICF.
    (b) the ECF becomes hypotonic with respect to the ICF.
    (c) the ICF becomes hypotonic with respect to the plasma.
    (d) water moves from the ICF into the ECF.

34. Increasing or decreasing the rate of respiration alters pH by:
    (a) lowering or raising the partial pressure of carbon dioxide.
    (b) lowering or raising the partial pressure of oxygen.
    (c) lowering or raising the partial pressure of nitrogen.
    (d) a, b, and c are correct.

35. What interacting controls operate to stabilize the glomerular filtration rate (GFR)?

36. Describe the micturition reflex.

37. Differentiate among fluid balance, electrolyte balance, and acid-base balance, and explain why each is important to homeostasis.

38. Why should a person with a fever drink plenty of fluids?

39. Exercise physiologists recommend that adequate amounts of fluid be ingested before, during, and after exercise. Why is adequate fluid replacement during extensive sweating important?

Level 3: Critical Thinking and Clinical Applications

40. Long-haul truck drivers are on the road for long periods of time between restroom stops. Why might that lead to kidney problems?

41. For the past week, Susan has felt a burning sensation in the area of her urethra when she urinates. She checks her temperature and finds that she has a low-grade fever. What unusual substances are likely to be present in her urine?

42. *Mannitol* is a sugar that is filtered but not reabsorbed by the kidneys. What effect would drinking a solution of mannitol have on the volume of urine produced?

19 The Reproductive System

IT IS IMPORTANT TO remember that what appears to be a trauma case may have started as a medical problem. Always consider the possibility of a medical problem in a driver. There have been several instances where collisions, such as shown here, were due to hypoglycemia, drug use, bee stings, or other medical conditions and not trauma. Always consider all possible causes and avoid tunnel vision when confronted with a prehospital emergency.

Chapter Outline

Chapter Objectives

Vocabulary Development

andro- male; *androgen*
crypto hidden; *cryptorchidism*
diplo double; *diploid*
follis a leather bag; *follicle*
genesis generation; *oogenesis*
gyne woman; *gynecologist*
haplo single; *haploid*

labium lip; *labium minus*
lutea yellow; *corpus luteum*
meioun to make smaller; *meiosis*
men month; *menopause*
metra uterus; *endometrium*
myo- muscle; *myometrium*

oon an egg; *oocyte*
orchis testis; *cryptorchidism*
pausis cessation; *menopause*
pellucidus translucent; *zona pellucida*
rete a net; *rete testis*
tetras four; *tetrad*

AN INDIVIDUAL LIFE SPAN can be measured in decades, but the human species has perpetuated itself for hundreds of thousands of years through the activities of the reproductive system. The entire process of reproduction seems almost magical; many aboriginal societies even failed to discover the basic link between sexual activity and childbirth and assumed that supernatural forces were responsible for producing new individuals. Our society has a much clearer understanding of the reproductive process, but the events of procreation—that the fusion of two reproductive cells, one produced by a man, the other by a woman, starts a chain of events that leads to the birth of an infant—still produces a sense of wonder.

This chapter and the next will consider the mechanisms involved in this remarkable process. We will begin by examining the anatomy and physiology of the reproductive system in this chapter. **Gonads** (GŌ-nadz) are reproductive organs that produce hormones and reproductive cells, or **gametes** (GAM-ēts)—

sperm in males and *ova* (Ō-va; singular, *ovum*) in females. Other components of the reproductive system store, nourish, and transport the gametes. Chapter 20 begins with **fertilization,** also known as *conception,* in which a male gamete and female gamete unite. All the cells in the body are the mitotic descendants of a **zygote** (ZĪ-gōt), which is the single cell created at fertilization. The gradual transformation of that single cell into a functional adult occurs through the process of *development,* which is the topic of Chapter 20. From fertilization to birth, development occurs within specialized organs of the female reproductive system.

■ The Reproductive System of the Male

The principal structures of the male reproductive system are shown in Figure 19–1●. Each male gonad, or *testis* (TES-tiss;

697

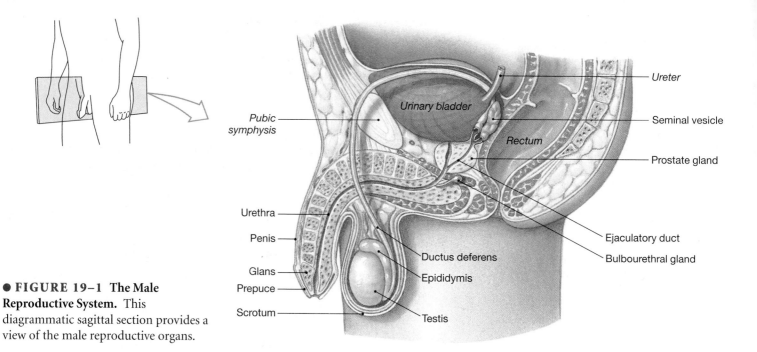

Pubic symphysis

Urinary bladder

Ureter

Seminal vesicle

Rectum

Prostate gland

Urethra

Penis

Glans

Prepuce

Scrotum

Ductus deferens

Epididymis

Testis

Ejaculatory duct

Bulbourethral gland

● **FIGURE 19–1 The Male Reproductive System.** This diagrammatic sagittal section provides a view of the male reproductive organs.

plural, *testes*), produces reproductive cells called sperm, or **spermatozoa** (sper-ma-tō-ZŌ-uh; singular, *spermatozoon*). The spermatozoa that leave each testis travel within structures that compose the *male reproductive tract*—the *epididymis* (ep-i-DID-i-mus), the *ductus deferens* (DUK-tus DEF-e-renz), the *ejaculatory* (ē-JAK-ū-la-tō-rē) *duct*, and the *urethra*—before leaving the body. Accessory organs, which include the *seminal* (SEM-i-nal) *vesicles*, the *prostate* (PROS-tāt) *gland*, and the *bulbourethral* (bul-bō-ū-RĒ-thral) *glands*, secrete their products into the ejaculatory ducts and urethra. The externally visible structures of the reproductive system constitute the *external genitalia* (jen-i-TĀ-lē-uh). The external genitalia of the male include the *scrotum* (SKRŌ-tum), which encloses the testes, and the *penis* (PĒ-nis), an erectile organ through which the distal portion of the urethra passes.

The Testes

The *primary sex organs* of the male system are the **testes** (TES-tēz). The testes hang within the **scrotum,** which is a fleshy pouch suspended from the perineum posterior to the base of the penis. The scrotum is subdivided into two chambers, or *scrotal cavities*, each of which contain a testis. Each testis has the shape of a flattened egg roughly 5 cm (2 in.) long, 3 cm (1.2 in.) wide, and 2.5 cm (1 in.) thick and weighs 10–15 g (0.35–0.53 oz). A serous membrane lines the scrotal cavity, which reduces friction between the inner surface of the scrotum and the outer surface of the testis.

The scrotum consists of a thin layer of skin that contains smooth muscle (Figure 19–2a●). Sustained contractions of the

smooth muscle layer, the *dartos* (DAR-tōs), cause the characteristic wrinkling of the scrotal surface. Beneath the dermis is a layer of skeletal muscle, the **cremaster** (krē-MAS-ter) **muscle,** which can contract to pull the testes closer to the body. Normal sperm development in the testes requires temperatures about 1.1°C (2°F) lower than those elsewhere in the body. When air or body temperatures rise, the cremaster relaxes, and the testes move away from the body. Sudden cooling of the scrotum, as occurs during entry into a cold swimming pool, results in contractions of the cremaster muscle that pull the testes closer to the body and keep testicular temperatures from falling.

Clinical Note
CRYPTORCHIDISM

During normal development of a male fetus, the testes descend from inside the body cavity and pass through the inguinal canal of the abdominal wall into the scrotum. In *cryptorchidism* (krip-TOR-ki-dizm; *crypto,* hidden + *orchis,* testis), one or both of the testes have not completed this process *(descent of the testes)* by the time of birth. This condition occurs in about 3 percent of full-term deliveries and in roughly 30 percent of premature births. In most instances, normal descent occurs a few weeks later, but the condition can be surgically corrected if it persists. Corrective measures should be taken before *puberty* (sexual maturation) because cryptorchid (abdominal) testes will not produce sperm, and the individual will be *sterile (infertile)* and unable to father children. If the testes cannot be moved into the scrotum, they are usually removed, because about 10 percent of men with uncorrected cryptorchid testes eventually develop testicular cancer. This surgical procedure is called a *bilateral orchiectomy* (or-kē-EK-to-mē; *ectomy,* excision). ■

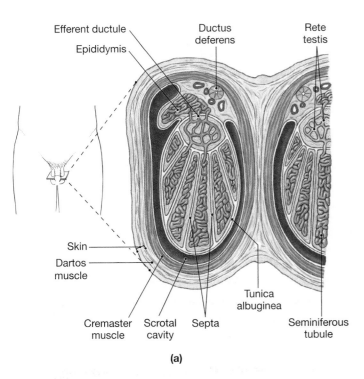

Efferent ductule
Epididymis
Ductus deferens
Rete testis
Skin
Dartos muscle
Cremaster muscle
Scrotal cavity
Septa
Tunica albuginea
Seminiferous tubule

(a)

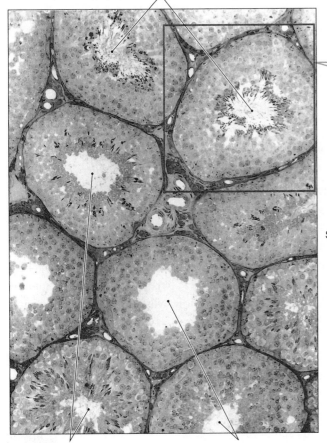

Seminiferous tubules containing nearly mature spermatozoa about to be released into the lumen

Seminiferous tubules containing late spermatids

Seminiferous tubules containing early spermatids

(b)

Each testis is wrapped in a dense fibrous capsule, the **tunica albuginea** (TŪ-ni-ka al-bū-JIN-ē-uh). Collagen fibers from this wrapping extend into the testis, and form partitions, or *septa*, that subdivide the testis into roughly 250 *lobules*. Distributed among the lobules are approximately 800 slender, tightly coiled **seminiferous** (sem-i-NIF-er-us) **tubules** (Figure 19–2b●). Each tubule averages around 80 cm (32 in.) in length, and a typical testis contains nearly half a mile of seminiferous tubules. Sperm produced within the seminiferous tubules (by a process discussed shortly) leave the tubules and pass through a maze of passageways—the **rete** (RĒ-tē *rete*, a net) **testis** and *efferent ductules*—before entering the epididymis, which is the beginning of the male reproductive tract (see Figure 19–2a).

The spaces between the tubules are filled with loose connective tissue, numerous blood vessels, and large **interstitial cells** (Figure 19–2c●) that produce male sex hormones, or *androgens*. The steroid **testosterone** is the most important androgen. ∞ p. 395

Each seminiferous tubule contains numerous cells. Among them are large **sustentacular** (sus-ten-TAK-ū-lar) **cells** (or *Sertoli cells*) that extend from the perimeter of the

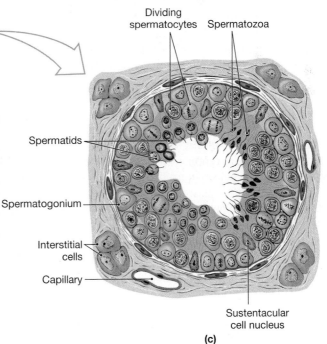

Dividing spermatocytes
Spermatozoa
Spermatids
Spermatogonium
Interstitial cells
Capillary
Sustentacular cell nucleus

(c)

●**FIGURE 19–2 The Scrotum, Testes, and Seminiferous Tubules.** (**a**) This horizontal section reveals the anatomical relationships of the testes within the scrotum. (**b**) This photomicrograph shows a section through a coiled seminiferous tubule. (**c**) This diagrammatic view details the cellular organization of a seminiferous tubule.

tubule to the lumen (see Figure 19–2c). Sustentacular cells nourish the developing sperm cells. Between and adjacent to the sustentacular cells are the various cells involved in **spermatogenesis** (sper-ma-tō-JEN-e-sis), a series of cell divisions that ultimately produces sperm cells (spermatozoa). With each successive division, the daughter cells move closer to the lumen.

Clinical Note
TESTICULAR CANCER

Although cancer of the testicle is relatively rare in men overall, it is the most common cancer in men from 15 to 35 years old. Testicular cancer is highly treatable and usually curable. It is generally divided into seminoma and nonseminoma types for treatment planning because seminomas are more sensitive to radiation therapy. The cure rate for patients with seminoma-type cancers exceeds 90 percent. The cure rate for nonseminoma type cancers approaches 100 percent. Men who have an undescended testicle (cryptorchidism) are at a higher risk of developing cancer of the testicle than those where the testicle has descended properly. This is true even if surgery was performed to place the testicle at the appropriate place in the scrotum.

Treatment of testicular cancer includes surgery, chemotherapy, radiation therapy, and/or bone marrow transplantation. Typically, both testicles are surgically removed. The lymph nodes in the abdomen are often sampled during surgery. Radiation therapy is used to kill cancer cells and shrink tumors. Chemotherapy kills cancer cells that are located outside of the testicle. Autologous bone marrow transplantation is a newer type of treatment. Bone marrow is taken from the patient and treated with drugs to kill any cancer cells. The marrow is then frozen while the patient is given high-dose chemotherapy, either with or without radiation therapy, to destroy all of the remaining marrow. Following this, the marrow that was previously removed and treated is thawed and given back to the patient, which replaces the marrow that was destroyed. The marrow grows and begins to function by producing essential blood cells.

The best treatment for testicular cancer is prevention. The earlier testicular cancer is detected, the better are the chances of obtaining a cure. All men should perform testicular self-examination at least once a month. This is especially important in the 15- to 40-year-old age group. The American Cancer Society and the National Cancer Institute have several tapes and brochures that detail this safe and easy examination. ■

Spermatogenesis

Spermatogenesis involves three processes:

1. *Mitosis.* Spermatogenesis begins with the mitotic divisions of stem cells called **spermatogonia** (sper-ma-tō-GŌ-nē-uh; singular, *spermatogonium*), which are located in the outermost layer of cells in the seminiferous tubules (see Figure 19–2c). (See Chapter 3 for a review of mitosis and cell division.) ∞ p. 87 Spermatogonia undergo mitosis throughout the individual's adult life. One daughter

Clinical Note
TESTICULAR TORSION

Testicular torsion is a major cause of scrotal pain. It occurs when a testicle twists on the spermatic cord. The testicle almost always torses lateral to medial. This usually compromises blood supply to the testicle, which results in severe scrotal and abdominal pain. If emergency surgery is not performed within six hours, the testicle may be lost. Testicular torsion most often occurs at puberty, but can occur at any age. Often there is a history of an athletic event or strenuous physical exercise prior to the onset. The diagnosis is made by physical exam and by scrotal ultrasound. The scrotal ultrasound will detect any interruptions in blood supply to the testicle. Treatment requires surgical exploration and detorsion of the testicle. The urologist will then secure *(pex)* both testicles so that additional torsions cannot occur. ■

cell from each mitotic division is pushed toward the lumen of the seminiferous tubule. These daughter cells differentiate into *spermatocytes* (sper-MA-tō-sīts), which prepare to begin the second process in spermatogenesis—meiosis.

2. *Meiosis.* Meiosis (mī-Ō-sis; *meioun*, to make smaller) is a special form of cell division that produces gametes, which are cells that contain half the number of chromosomes found in other cells. In the seminiferous tubules, the meiotic divisions of spermatocytes produce immature gametes called *spermatids* (see Figure 19–2c).

3. *Spermiogenesis.* In *spermiogenesis*, the small, relatively unspecialized spermatids differentiate into physically mature spermatozoa, which enter the fluid within the lumen of the seminiferous tubule.

Next we consider in more detail the roles of mitosis and meiosis in spermatogenesis.

MITOSIS AND MEIOSIS. In both males and females, mitosis and meiosis differ significantly in terms of the events that take place in the nucleus. **Mitosis** is part of the process of *somatic* (nonreproductive) cell division, which involves one division that produces two daughter cells, each of which contain the same number of chromosomes as the original cell. In humans, each somatic cell contains 23 pairs of chromosomes. Each pair consists of one chromosome provided by the father and another by the mother at the time of fertilization. Because each cell contains both members of each chromosome pair, these cells, and each of their daughter cells, are described as **diploid** (DIP-loyd; *diplo,* double).

By contrast, **meiosis** involves two cycles of cell division (*meiosis I* and *meiosis II*) and produces four cells, or gametes, each of which contain 23 individual chromosomes. Because these gametes contain only one member of each chromosome pair, gametes are described as **haploid** (HAP-loyd; *haplo,* single). Thus, the fusion of a haploid sperm and a haploid egg yields a single cell with the normal number of chromosomes.

Figure 19–3● illustrates the role of meiosis in spermatogenesis. (To make the chromosomal events easier to follow, only 3 of the 23 pairs of chromosomes present in each spermatogonium are shown.) Each mitotic division of spermatogonia produces two primary spermatocytes. Like spermatogonia, primary spermatocytes are diploid cells, but they divide by meiosis rather than mitosis. As a primary spermatocyte prepares to begin meiosis, DNA replication occurs within the nucleus, just as it does in a cell preparing to undergo mitosis. As prophase of the first meiotic division (meiosis I) occurs, the chromosomes condense and become visible. As in mitosis, each chromosome consists of two duplicate *chromatids* (KRŌ-ma-tidz).

At this point, the close similarities between meiosis and mitosis end. As prophase of meiosis I unfolds, the corresponding maternal and paternal chromosomes come together. This event, known as **synapsis** (si-NAP-sis), produces 23 pairs of chromosomes; each member of the pair consists of two identical chromatids. A matched set of four chromatids is called a **tetrad** (TET-rad; *tetras,* four). An exchange of genetic material can occur between the chromatids of a chromosome pair at this time. This exchange, called *crossing-over,* increases genetic variation among offspring. Prophase ends with the disappearance of the nuclear envelope.

During metaphase of meiosis I, the tetrads line up along the metaphase plate. As anaphase begins, the tetrads break up, and the maternal and paternal chromosomes separate. This is a major difference between mitosis and meiosis: in mitosis, each daughter cell receives one of the two copies of *every* chromosome, maternal and paternal; whereas in meiosis, each daughter cell receives both copies of *either* the maternal chromosome or the paternal chromosome from each tetrad.

As anaphase proceeds, the maternal and paternal components are randomly distributed. For example, most of the maternal chromosomes may go to one daughter cell and most of the paternal chromosomes to the other. As a result, telophase I ends with the formation of two daughter cells that contain unique combinations of maternal and paternal chromosomes. In the testes, the daughter cells produced by the first meiotic division (meiosis I) are called **secondary spermatocytes.**

Every secondary spermatocyte contains 23 chromosomes, each of which consists of two duplicate chromatids. The duplicate chromatids separate during **meiosis II.** The interphase that separates meiosis I and meiosis II is very brief, and the secondary spermatocyte soon enters meiosis II. The completion of meiosis II produces four immature gametes, or **spermatids**

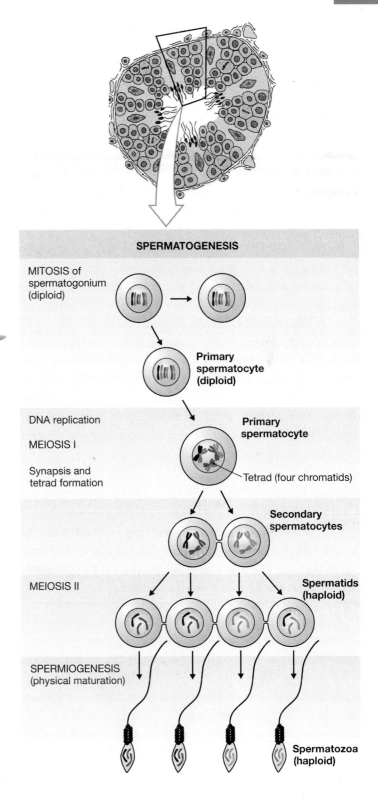

● FIGURE 19–3 Spermatogenesis. In this process, which occurs within the seminiferous tubules, each diploid primary spermatocyte that undergoes meiosis produces four haploid spermatids. Each spermatid then develops into a spermatozoon.

(SPER-ma-tidz), which are identical in size and each with 23 chromosomes. In summary, for every diploid primary spermatocyte that enters meiosis, four haploid spermatids are produced (see Figure 19–3). (Meiosis in females produces one huge ovum and three tiny, nonfunctional cells that disintegrate. We will consider ovum production in a later section.)

SPERMIOGENESIS. In **spermiogenesis,** each spermatid matures into a single **spermatozoon** (sper-ma-tō-ZŌ-on), or **sperm cell.** The entire process of spermatogenesis, from spermatogonial division to the release of a physically mature spermatozoon, takes approximately 9 weeks.

As previously noted, sustentacular cells play a key role in spermatogenesis and spermiogenesis. Developing spermatocytes undergoing meiosis and spermatids are not free in the seminiferous tubules. Instead, they are surrounded by the cytoplasm of sustentacular cells. Because there are no blood vessels inside the seminiferous tubules, all nutrients must enter by diffusion from the surrounding interstitial fluids. The large sustentacular cells control the chemical environment inside the seminiferous tubules and provide nutrients and chemical stimuli that promote the production and differentiation of spermatozoa. They also help regulate spermatogenesis by producing *inhibin,* a hormone introduced in Chapter 10. p. 395

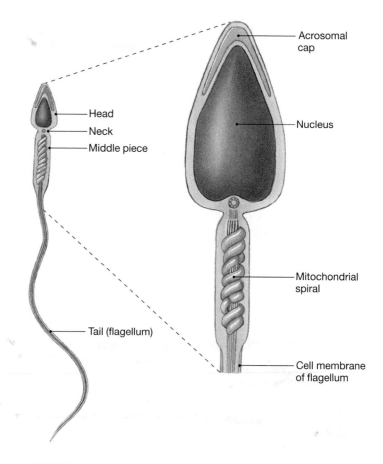

● **FIGURE 19–4 Spermatozoon Structure.** A spermatozoon measures only 60 μm (0.06 mm) in total length.

> ## Key Note
> Meiosis produces gametes that contain half the number of chromosomes found in somatic cells. For each cell that enters meiosis, the testes produce four spermatozoa; whereas the ovaries produce a single ovum.

> ## Key Note
> Spermatogenesis begins at puberty (sexual maturation) and continues until relatively late in life (above age 70). It is a continuous process, and all stages of meiosis can be observed within the seminiferous tubules.

Anatomy of a Spermatozoon

A sperm cell has three distinct regions: the head, the middle piece, and the tail (Figure 19–4●). The **head** contains a nucleus filled with densely packed chromosomes. At the tip of the head is the **acrosomal** (ak-rō-SŌ-mal) **cap,** which contains enzymes essential for fertilization. A short **neck** attaches the head to the **middle piece,** which is dominated by a spiral arrangement of mitochondria that provides the ATP energy for moving the tail. The **tail** is the only example of a *flagellum* in the human body. It moves the sperm cell from one place to another. p. 77

Unlike most other cells, a mature spermatozoon lacks an endoplasmic reticulum, a Golgi apparatus, lysosomes, peroxisomes, and many other intracellular structures. Because the cell does not contain glycogen or other energy reserves, it must absorb nutrients (primarily fructose) from the surrounding fluid.

The Male Reproductive Tract

The testes produce physically mature spermatozoa that are not yet capable of fertilizing an ovum. The other portions of the male reproductive system are responsible for the functional maturation, nourishment, storage, and transport of spermatozoa.

The Epididymis

Late in their development, spermatozoa detach from the sustentacular cells and lie within the lumen of the seminiferous tubule. Although they have most of the physical characteristics of mature sperm cells, they are still functionally immature and incapable of coordinated locomotion or fertilization. At this point, cilia-driven fluid currents within the efferent ducts transport them into the **epididymis** (see Figure 19–1). This elongate tubule, which is almost 7 meters (23 ft) long, is so twisted and coiled that it takes up very little space.

The functions of the epididymis include adjusting the composition of the fluid from the seminiferous tubules, acting as a recycling center for damaged spermatozoa, and storing the maturing spermatozoa. Cells that line the epididymis absorb cellular debris from damaged or abnormal spermatozoa and release it for pickup by surrounding blood vessels; the cells also absorb organic nutrients. Spermatozoa complete their physical maturation during the two weeks it takes for them to travel through the epididymis and arrive at the ductus deferens.

Although the spermatozoa that leave the epididymis are physically mature, they remain immobile. To become motile (actively swimming) and fully functional, they must undergo **capacitation.** Capacitation occurs after the spermatozoa (1) mix with secretions of the seminal vesicles and (2) are exposed to conditions inside the female reproductive tract. The epididymis secretes a substance that prevents premature capacitation.

The Ductus Deferens

Each **ductus deferens,** or *vas deferens,* is 40–45 cm (16–18 in.) long (see Figure 19–1). It ascends into the abdominal cavity

within the *spermatic cord,* which is a sheath of connective tissue and muscle that also encloses the blood vessels, nerves, and lymphatics that serve the testis. (The passageway through the abdominal musculature is called the *inguinal canal.*)

Once within the abdominal cavity, each ductus deferens passes lateral to the urinary bladder (see Figure 19–1); then each curves downward past the ureter on its way toward the prostate gland (Figure 19–5●). Peristaltic contractions in the muscular walls of the ductus deferens propel spermatozoa and fluid along the length of the duct. The ductus deferens can also store spermatozoa for up to several months. During this period the spermatozoa are inactive and have low metabolic rates.

The junction of the ductus deferens with the duct of the seminal vesicle creates the *ejaculatory duct,* which is a relatively short (2 cm, or less than 1 in.) passageway (see Figure 19–5a). This duct penetrates the muscular wall of the prostate gland and empties into the urethra near the opening of the ejaculatory duct from the other side.

The Urethra

In males, the urethra extends 18–20 cm (7–8 in.) from the urinary bladder to the tip of the penis (see Figure 19–1). The male urethra is a passageway that functions in both the urinary and reproductive systems.

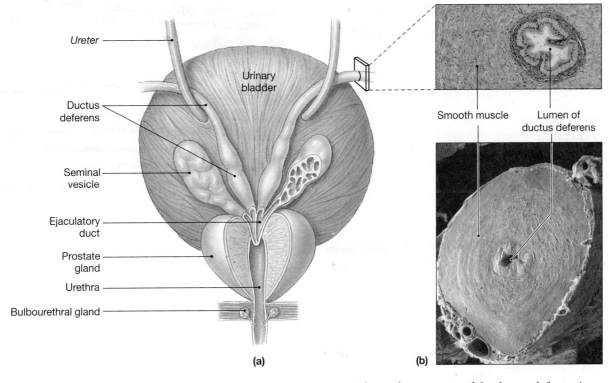

(a)   (b)

● **FIGURE 19–5 The Ductus Deferens.** **(a)** This posterior view shows the segments of the ductus deferens in relation to nearby structures. **(b)** These micrographs show extensive layering with smooth muscle around the lumen of the ductus deferens. (Reproduced from R. G. Kessel and R. H. Kardon, *Tissues and Organs: A Text-Atlas of Scanning Electron Microscopy,* W. H. Freeman & Co., 1979. All Rights Reserved.) (LM × 34; SEM × 42)

The Accessory Glands

The fluids contributed by the seminiferous tubules and the epididymis account for only about 5 percent of the volume of *semen*, which is the fluid that transports and nourishes sperm. The majority of seminal fluid is composed of secretions from the *seminal vesicles*, the *prostate gland*, and the *bulbourethral glands* (see Figure 19–5a). Primary functions of these glands include (1) activating the spermatozoa, (2) providing the nutrients spermatozoa need for motility, (3) generating peristaltic contractions that propel spermatozoa and fluids along the reproductive tract, and (4) producing buffers that counteract the acidity of the urethral and vaginal environments.

The Seminal Vesicles

Each **seminal vesicle** is a tubular gland that is about 15 cm (6 in.) long (see Figures 19–1 and 19–5a). The body of the gland is coiled and folded into a compact, tapered mass roughly 5 cm by 2.5 cm (2 in. by 1 in.).

The seminal vesicles are extremely active secretory glands that contribute about 60 percent of the volume of semen. Their secretions contain (1) fructose, which is a six-carbon sugar easily metabolized by spermatozoa; (2) prostaglandins, which can stimulate smooth muscle contractions along the male and female reproductive tracts; and (3) fibrinogen, which after ejaculation forms a temporary clot within the vagina. The slight alkalinity of the secretions helps neutralize acids in the secretions of the prostate gland and within the vagina. When mixed with the secretions of the seminal vesicles, mature but previously inactive spermatozoa begin beating their flagella and become highly motile.

The Prostate Gland

The **prostate gland** is a small, muscular, rounded organ, about 4 cm (1.6 in.) in diameter, that surrounds the urethra as it leaves the urinary bladder (see Figure 19–5a). The *prostatic fluid* produced by the prostate gland is a slightly acidic secretion that contributes 20–30 percent of the volume of semen. In addition to several other compounds of uncertain significance, prostatic fluid contains **seminalplasmin** (sem-i-nal-PLAZ-min), which is an antibiotic that may help prevent urinary tract infections in males. These secretions are ejected into the urethra by peristaltic contractions of the muscular prostate wall.

The Bulbourethral Glands

The paired bulbourethral glands, or Cowper's glands, are spherical structures almost 10 mm (less than 0.5 in.) in diameter (see Figure 19–5a). These glands secrete a thick, sticky, alkaline mucus that helps neutralize urinary acids in the urethra and has lubricating properties.

Clinical Note
PROSTATITIS

Prostatic inflammation, or *prostatitis* (pros-ta-TĪ-tis), can occur in males of any age but most often afflicts older men. Prostatitis can result from bacterial infections, but it also occurs in the apparent absence of pathogens. Signs and symptoms can resemble those of prostate cancer. Individuals with prostatitis may complain of pain in the lower back, perineum, or rectum, sometimes accompanied by painful urination and the discharge of mucous secretions from the urethral opening. Antibiotic therapy is usually effective in treating most cases that result from bacterial infection. ∎

Semen

Semen (SĒ-men) is the fluid that contains sperm and the secretions of the accessory glands of the male reproductive tract. In a typical *ejaculation* (ē-jak-ū-LĀ-shun), 2–5 mL of semen is expelled from the body. This volume of fluid, called an **ejaculate,** contains three major components:

- *Spermatozoa.* A normal **sperm count** ranges from 20 million to 100 million spermatozoa per milliliter of semen.
- *Seminal fluid.* **Seminal fluid,** which is the fluid component of semen, is a mixture of glandular secretions with a distinct ionic and nutrient composition. Of the total volume of seminal fluid, the seminal vesicles contribute about 60 percent; the prostate, 30 percent; the sustentacular cells and epididymis, 5 percent; and the bulbourethral glands, less than 5 percent.
- *Enzymes.* Several important enzymes are present in the seminal fluid, including (1) a protease that helps dissolve mucous secretions in the vagina; (2) *seminalplasmin,* which is a prostatic antibiotic enzyme that kills a variety of bacteria; (3) a prostatic enzyme that causes the semen to clot in the vagina; and (4) an enzyme that subsequently liquefies the clotted semen.

The Penis

The **penis** is a tubular organ that both introduces semen into the female's vagina during sexual intercourse and conducts urine to the exterior (through the urethra; see Figure 19–1). As shown in Figure 19–6a●, the penis is composed of three regions: (1) the **root,** which is the fixed portion that attaches the penis to the body wall; (2) the **body (shaft),** which is the tubular portion that contains masses of erectile tissue; and (3) the **glans,** which is the expanded distal portion that surrounds the external urethral opening, or *external urethral meatus.*

● **FIGURE 19–6** **The Penis.** (**a**) This anterolateral view of a penis shows the positions of the erectile tissues. (**b**) A frontal section through the penis and associated organs provides another view of the erectile tissues. (**c**) This sectional view through the penis reveals the texture of the erectile tissues.

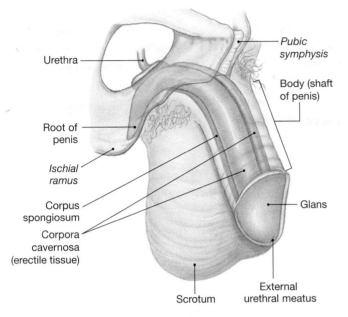

(a) Anterior and lateral view of penis

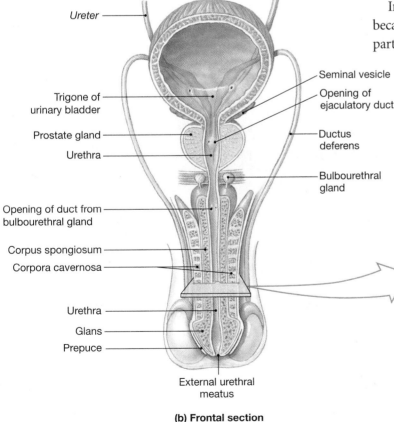

(b) Frontal section

The skin that overlies the penis resembles that of the scrotum. A fold of skin, the **prepuce** (PRĒ-pūs), or *foreskin,* surrounds the tip of the penis (Figure 19–6b●). The prepuce attaches to the relatively narrow neck of the penis and continues over the glans. *Preputial glands* in the skin of the neck and inner surface of the prepuce secrete a waxy material called *smegma* (SMEG-ma). Unfortunately, smegma can be an excellent nutrient source for bacteria. Mild inflammation and infections in this region are common, especially if the area is not washed frequently. One way to avoid these conditions is *circumcision* (ser-kum-SIZH-un), the surgical removal of the prepuce. In Western societies (especially the U.S.), this procedure is generally performed shortly after birth.

Most of the body, or shaft, of the penis consists of three columns of **erectile tissue** (Figure 19–6c●). Erectile tissue consists of a maze of vascular channels incompletely separated by partitions of elastic connective tissue and smooth muscle. The anterior surface of the penis covers two cylindrical **corpora cavernosa** (KOR-por-a ka-ver-NŌ-suh). Their bases are bound to the pubis and ischium of the pelvis (see Figure 19–6a). The corpora cavernosa extend along as far as the glans of the penis. The relatively slender **corpus spongiosum** (spon-jē-Ō-sum) surrounds the urethra and extends all the way to the tip of the penis, where it forms the glans.

In the resting state, little blood flows into the erectile tissue because the arterial branches are constricted and the muscular partitions are tense. Parasympathetic innervation of the penile

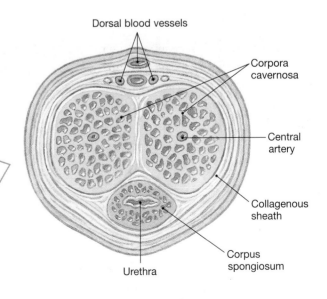

(c) Section through shaft of penis

arteries involves neurons that release nitric oxide (NO) at their synaptic knobs. In response to the NO, the smooth muscle in the arterial walls relaxes and the vessels dilate, which results in increased blood flow and **erection** of the penis.

Clinical Note
PRIAPISM

Priapism is a prolonged, usually painful, penile erection that is not associated with sexual arousal. In many cases, the cause of priapism is unclear *(idiopathic)*. However, it is associated with sickle cell disease, spinal cord injury, spinal anesthesia, leukemia, and drugs. Most cases of priapism result from intracavernous injection of medications to treat impotence. In high spinal cord injury, priapism can occur if there is unopposed parasympathetic stimulation.

The erection seen in priapism is due to engorgement of the *corpora cavernosa*. The *corpora spongiosum* and the glans are not engorged as with a normal erection. Priapism is a urological emergency. Treatment within hours is necessary to prevent permanent injury. ∎

Hormones and Male Reproductive Function

The major reproductive hormones were introduced in Chapter 10. ⚬ p. 395 The interactions among the reproductive hormones in males are diagrammed in Figure 19–7●. In the presence of **gonadotropin-releasing hormone (GnRH)**—a peptide synthesized in the hypothalamus and carried to the anterior pituitary by the hypophyseal portal system—the anterior pituitary gland releases two hormones: **follicle-stimulating hormone (FSH)** and **luteinizing hormone (LH).**

The Role of FSH in Spermatogenesis

In males, FSH targets primarily the sustentacular cells of the seminiferous tubules. Under FSH stimulation, and in the presence of testosterone from the interstitial cells, sustentacular cells promote spermatogenesis and spermiogenesis.

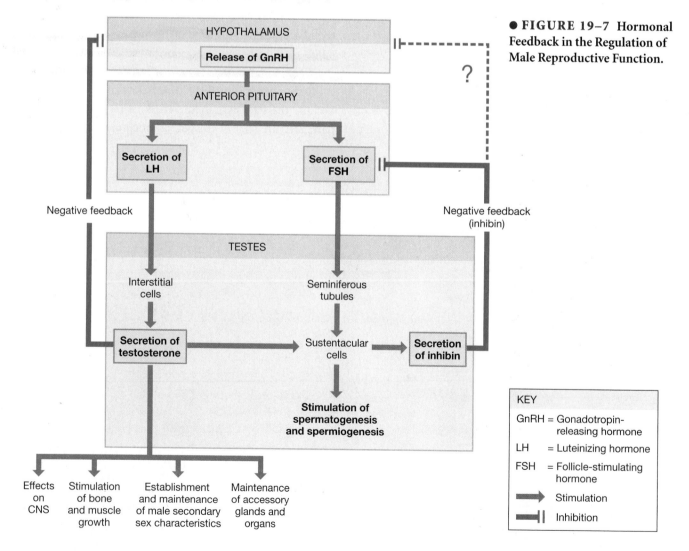

● **FIGURE 19–7** Hormonal Feedback in the Regulation of Male Reproductive Function.

The rate of spermatogenesis is regulated by a negative-feedback mechanism that involves GnRH, FSH, testosterone, and inhibin. Under GnRH stimulation, FSH promotes spermatogenesis along the seminiferous tubules. As spermatogenesis accelerates, however, so does the rate of inhibin secretion by the sustentacular cells of the testes. Inhibin inhibits FSH production in the anterior pituitary and may also inhibit secretion of GnRH at the hypothalamus. Rising levels of testosterone also reduce GnRH secretion and pituitary gland responsiveness to GnRH, which thereby inhibits FSH (and LH) secretion.

The net effect is that when FSH levels become elevated, inhibin production increases until FSH levels return to normal. If FSH levels decline, inhibin production falls, so the rate of FSH production accelerates.

The Role of LH in Androgen Production

LH in males—which was once called *interstitial cell-stimulating hormone (ICSH)* before it was found to be identical to the hormone in females—causes the secretion of testosterone and other androgens by the interstitial cells of the testes. Testosterone, which is the most important androgen, has numerous functions, including (1) stimulating spermatogenesis and promoting the functional maturation of spermatozoa; (2) affecting central nervous system (CNS) function, including the influence of sexual drive (libido) and related behaviors; (3) stimulating metabolism throughout the body, especially pathways concerned with protein synthesis and muscle and bone growth; (4) determining and maintaining secondary sex characteristics, such as facial hair, increased muscle mass and body size, and the quantity and location of adipose tissue; (5) maintaining the accessory glands and organs of the male reproductive tract; and (6) regulating LH and FSH secretion at the hypothalamus.

Testosterone production begins around the seventh week of development and reaches a peak after roughly 6 months of development. The early surge in testosterone levels stimulates the differentiation of the male duct system and accessory organs and affects CNS development. Testosterone secretion accelerates markedly at puberty, which initiates sexual maturation and the appearance of secondary sex characteristics. In adult males, negative-feedback mechanisms control the level of testosterone production. Above-normal testosterone levels inhibit the release of GnRH by the hypothalamus. This inhibition causes a reduction in LH (and FSH) secretion and lowers testosterone levels.

Clinical Note
MALE GENITAL TRAUMA

The male genitalia are, for the most part, located outside of the body. Despite this, they are rarely injured. The testicles are highly mobile within the scrotum, and the external capsule of the testicle (tunica albuginea) is very tough. These features protect the testes from injury. The most common cause of testicular injury is a direct blow. Most testicular injuries are contusions, although ruptures can occur.

Injuries to the penis can result from several causes. Zippers can trap the penile skin, and cause pain and bleeding. Mineral oil and ice can aid in removing the entrapped skin. In severe cases, a local anesthetic must be injected so that a wire cutter can be used to open the zipper.

Self-inflicted injuries of the penis can range from the common to the bizarre. Vacuum cleaners can cause extensive injury to the glans penis. Blade injuries can cause lacerations or even amputations. Amputation of the penis is best managed by replantation, if possible.

Traumatic rupture of the corpus cavernosum of the penis, which is also called a fracture of the penis, occurs when the erect penis impacts forcibly on a hard object, receives a direct blow, or is subjected to abnormal bending. A cracking sound may be heard. This is usually followed by penile pain, loss of the erection, rapid swelling, discoloration, and deformity. Penile ruptures must be carefully managed surgically to avoid permanent loss of function. ∎

CONCEPT CHECK QUESTIONS

1. On a warm day, would the cremaster muscle be contracted or relaxed? Why?
2. What will occur if the arteries within the penis dilate?
3. What effect would low levels of FSH have on sperm production?

Answers begin on p. 792.

∎ The Reproductive System of the Female

A woman's reproductive system must not only produce sex hormones and gametes but also be able to protect and support a developing embryo and nourish an infant. The principal organs of the female reproductive system are the *ovaries*, the *uterine tubes*, the *uterus* (womb), the *vagina*, and the components of the external genitalia (Figure 19–8a●). The female reproductive system also includes the mammary glands and a variety of accessory glands that secrete their products into the reproductive tract.

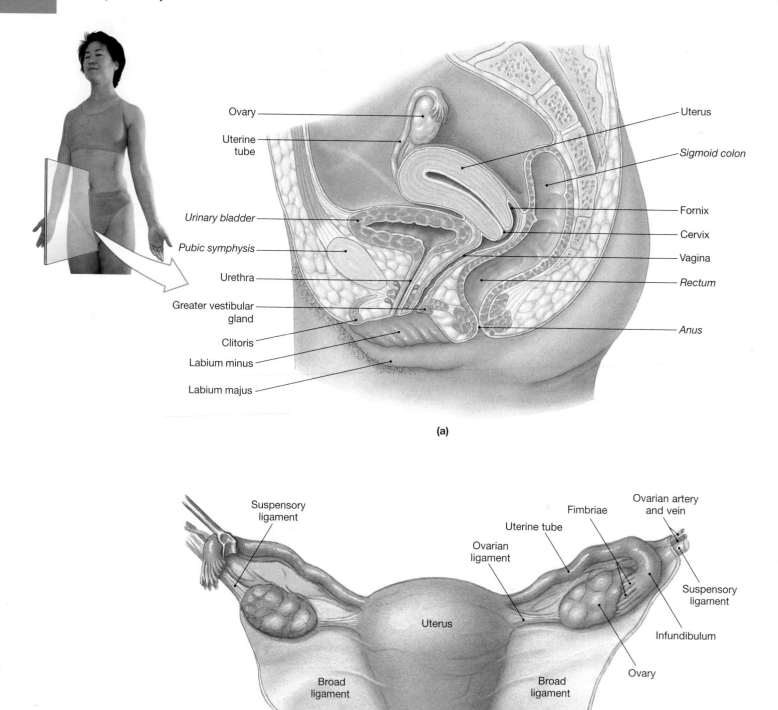

(a)

(b)

● **FIGURE 19–8 The Female Reproductive System. (a)** This sagittal section shows the female reproductive organs. **(b)** The locations of the uterus, uterine tubes, and ovaries are apparent in this posterior view.

The Ovaries

The paired ovaries are small, lumpy, almond-shaped organs near the lateral walls of the pelvic cavity (Figure 19–8b●). The ovaries are responsible for (1) the production of female gametes, or **ova** (singular *ovum*); (2) the secretion of female sex hormones, including *estrogens* and *progestins*; and (3) the secretion of inhibin, which is involved in the negative-feedback control of pituitary FSH production.

A typical **ovary** is a flattened oval that measures approximately 5 cm long, 2.5 cm wide, and 8 mm thick (2 in. × 1 in. × 0.33 in.). It has a pale white or yellowish coloration and a consistency that resembles cottage cheese or lumpy oatmeal.

The position of each ovary is stabilized by a mesentery known as the *broad ligament* and by a pair of supporting ligaments. The mesentery also encloses the uterine tubes and uterus. The ligaments attached to each ovary extend to the uterus and pelvic wall. The latter contain the major blood vessels of each ovary, which are the *ovarian artery* and *ovarian vein.*

Oogenesis

Ovum production, or **oogenesis** (ō-ō-JEN-e-sis; *oon*, egg), is a process in females that begins before birth, accelerates at puberty, and ends at *menopause* (*men*, month + *pausis*, cessation). Between puberty and menopause, oogenesis occurs in the ovaries each month, as part of the *ovarian cycle* (discussed shortly). The events in oogenesis are summarized in Figure 19–9●; as was the case for spermatogenesis, only 3 of the 23 pairs of chromosomes are illustrated.

In the ovaries, stem cells, or **oogonia** (ō-ō-GŌ-nē-uh), complete their mitotic divisions before birth (see Figure 19–9). Between the third and seventh months of fetal development, the daughter cells, or **primary oocytes** (Ō-ō-sīts), begin to undergo meiosis. They proceed as far as prophase of meiosis I, but then the process stops. The primary oocytes remain in a state of suspended development for years—until the individual reaches puberty—and await the hormonal signal to complete meiosis. Not all of the primary oocytes survive; of the roughly 2 million in the ovaries at birth, about 400,000 remain at the start of puberty. The rest of the primary oocytes degenerate in a process called *atresia* (a-TRĒ-zē-uh).

Although the nuclear events under way during meiosis in the ovary are the same as those in the testis, the process differs in two important ways:

1. The cytoplasm of the original oocyte is not evenly distributed during the meiotic divisions. Oogenesis produces one functional ovum, which contains most of that

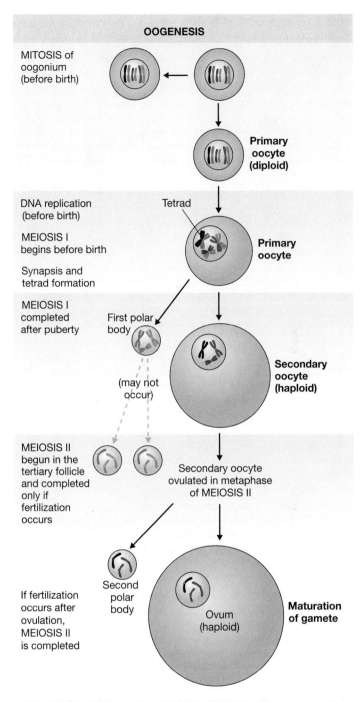

● **FIGURE 19–9 Oogenesis.** In oogenesis, a primary oocyte produces an ovum plus two or three nonfunctional polar bodies. Most of the events depicted here occur in the ovary as part of the ovarian cycle, shown in Figure 19–10.

cytoplasm, and up to three nonfunctional **polar bodies** that later disintegrate.

2. The ovary does not release a mature gamete; instead of a mature ovum, a *secondary oocyte* is released. Moreover, the second meiotic division is not completed until *after* fertilization.

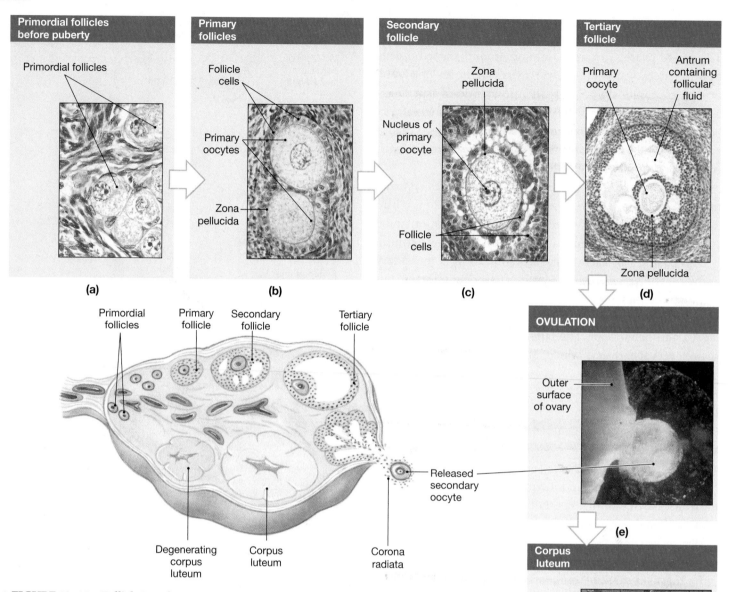

Primordial follicles before puberty	Primary follicles	Secondary follicle	Tertiary follicle

Primordial follicles

Follicle cells

Primary oocytes

Zona pellucida

(a)

(b)

Zona pellucida

Nucleus of primary oocyte

Follicle cells

(c)

Primary oocyte

Antrum containing follicular fluid

Zona pellucida

(d)

Primordial follicles

Primary follicle

Secondary follicle

Tertiary follicle

Released secondary oocyte

Degenerating corpus luteum

Corpus luteum

Corona radiata

OVULATION

Outer surface of ovary

(e)

Corpus luteum

(f)

● **FIGURE 19–10   Follicle Development and the Ovarian Cycle.** The photomicrographs show the changes in a follicle as it develops and enters the ovarian cycle. The drawing of the ovary illustrates the sequence and relative sizes of the various stages in the development, ovulation, and degeneration of an ovarian follicle; follicles do not physically move around the periphery of the ovary.

Follicle Development

Ovarian follicles (ō-VAR-ē-an FOL-i-klz) are specialized structures in the ovaries where both oocyte growth and meiosis I of oogenesis occur. In the outer portion of each ovary are clusters of primary oocytes, each surrounded by a single layer of *follicle cells.* The combination is known as a **primordial** (prī-MOR-dē-al) **follicle** (Figure 19–10a●). Beginning at puberty, primordial follicles are continuously activated to join other follicles already in development. The activating mechanism is unknown, although local hormones or growth factors within the ovary may be involved. The activated primordial follicle will either mature and be released as a secondary oocyte or degenerate (atresia).

The preliminary steps in follicle development (Figure 19–10a–c) are of variable length, but may take almost a year to complete. Follicle development begins with the activation of primordial follicles into **primary follicles** (Figure 19–10b●). The follicular cells enlarge, divide, and form several layers of cells around the growing primary oocyte. Microvilli from the surrounding follicular cells intermingle with microvilli that originate at the surface of the oocyte. This region is called the zona pellucida (ZŌ-na pe-LOO-sid-uh; *pellucidus,* translucent). The microvilli increase the surface area available for the transfer of materials from the follicular cells to the growing oocyte.

Although many primordial follicles develop into primary follicles, only a few of the primary follicles mature further. This process is apparently under the control of a growth factor produced by the oocyte. The transformation begins as the wall of the follicle thickens and the deeper follicular cells begin secreting small amounts of fluid. This *follicular fluid* accumulates in small pockets that gradually expand and separate the inner and outer layers of the follicle. At this stage, the complex is known as a **secondary follicle** (Figure 19–10c●). Although the primary oocyte continues to grow slowly, the follicle as a whole enlarges rapidly because follicular fluid accumulates.

Clinical Note
MITTELSCHMERZ

Occasionally, ovulation is accompanied by midcycle abdominal pain known as *mittelschmerz.* It is thought that this pain is related to peritoneal irritation due to follicle rupture or bleeding at the time of ovulation. The unilateral lower quadrant pain is usually self-limited and may be accompanied by midcycle spotting. While some woman may report a low-grade fever, it should be noted that body temperature normally increases at the time of ovulation and remains elevated until the day prior to the onset of the menstrual period. Treatment is symptomatic. ■

The Ovarian Cycle

The follicles are then ready to complete their maturation as part of the **ovarian cycle,** which is a 28-day cycle that includes follicle maturation (the *follicular phase*), ovulation, and the subsequent release of hormones by the remaining follicle cells (the *luteal phase*). Each of these phases lasts approximately 14 days, and the changes in follicle structure that occur as part of the ovarian cycle are shown in Figure 19–10d–f●.

THE FOLLICULAR PHASE. At the start of each ovarian cycle, an ovary contains only a few secondary follicles destined for further development; by day 5 of the cycle, there is usually only one. Stimulated by FSH, that follicle forms a **tertiary follicle,** or mature *Graafian* (GRAF-ē-an) *follicle,* which is roughly 15–20 mm in diameter (Figure 19–10d). The tertiary follicle is formed by days 10–14 of the ovarian cycle. Its large size creates a prominent bulge in the surface of the ovary. The oocyte and its covering of follicular cells projects into the expanded central chamber of the follicle, which is the **antrum** (AN-trum).

Until this time, the primary oocyte has been suspended in prophase of meiosis I. As the development of the tertiary follicle ends, rising LH levels prompt the primary oocyte to complete meiosis I. The completion of the first meiotic division produces a secondary oocyte (and also a nonfunctional polar body; see Figure 19–9). The secondary oocyte begins meiosis II but stops short of dividing. Meiosis II will not be completed unless fertilization occurs.

Generally, on day 14 of a 28-day ovarian cycle, the secondary oocyte and its surrounding follicular cells lose their connections with the follicular wall and float within the antrum. The follicular cells that surround the oocyte are now known as the *corona radiata* (ko-RŌ-nuh rā-dē-A-tuh).

OVULATION. At **ovulation** (ōv-ū-LĀ-shun), the tertiary follicle releases the secondary oocyte (an immature gamete) (Figure 19–10e). The distended follicular wall then ruptures, and releases the follicular contents, including the secondary oocyte, into the pelvic cavity. The sticky follicular fluid keeps the corona radiata attached to the surface of the ovary near the ruptured wall of the follicle. Contact with projections of the uterine tube or with fluid currents established by its ciliated epithelium then sweeps the secondary oocyte into the uterine tube.

THE LUTEAL PHASE. The 14-day luteal phase of the ovarian cycle begins at ovulation. The empty follicle collapses, and the remaining follicular cells invade the resulting cavity and multiply to create an endocrine structure known as the **corpus luteum** (LOO-tē-um; *lutea,* yellow) (Figure 19–10f). Unless fertilization occurs, the corpus luteum begins to degenerate roughly 12 days after ovulation. The disintegration of the corpus luteum marks the end of an ovarian cycle. A new ovarian cycle begins with the selection of another secondary follicle and its formation into a tertiary follicle under the stimulation of FSH.

Clinical Note
RUPTURED OVARIAN CYSTS

Cysts are fluid-filled pockets. When they develop in the ovary, they can rupture and be a source of abdominal pain. When an egg is released from the ovary, a cyst, which is known as a *corpus luteum cyst,* is often left in its place. Occasionally, cysts develop independent of ovulation. When the cysts rupture, a small amount of blood spills into the abdomen. Because blood irritates the peritoneum, it can cause abdominal pain and rebound tenderness. Ovarian cysts may be found during a routine pelvic examination. In the field setting, however, your patient is likely to complain of moderate to severe unilateral abdominal pain, which may radiate to her back. She may also report a history of dyspareunia, irregular bleeding, or a delayed menstrual period. It is not uncommon for patients to rupture ovarian cysts during intercourse or physical activity. This often results in immediate, severe abdominal pain that causes the patient to immediately stop intercourse or other physical activity. Ruptured ovarian cysts may be associated with vaginal bleeding. ■

Oogenesis begins during embryonic development, and primary oocyte production is completed before birth. After puberty, each month the ovarian cycle produces one or more secondary oocytes from the pre-existing population of primary oocytes. The number of viable and responsive primary oocytes declines markedly over time, until ovarian cycles end at age 45–55.

The Uterine Tubes

Each **uterine tube** (*Fallopian tube,* or *oviduct*) measures roughly 13 cm (5 in.) in length. The end closest to the ovary forms an expanded funnel, or **infundibulum** (in-fun-DIB-ū-lum; *infundibulum,* a funnel), with numerous finger-like projections that extend into the pelvic cavity (see Figure 19–8b). Both the projections, called **fimbriae** (FIM-brē-ē), and the inner surfaces of the infundibulum are carpeted with cilia that beat toward the broad entrance to the uterine tube.

Once inside the uterine tube, the secondary oocyte is transported by ciliary movement and peristaltic contractions by smooth muscles in the walls of the uterine tubes. It normally takes 3–4 days for the oocyte to travel from the infundibulum to the uterine cavity. *If fertilization is to occur, the secondary oocyte must encounter spermatozoa during the first 12–24 hours of its passage.* Unfertilized oocytes will degenerate in the uterine tubes or uterus without completing meiosis.

In addition to ciliated cells, the epithelium that lines the uterine tubes contains *Peg cells* and scattered mucin-secreting cells. The Peg cells project into the lumen of the uterine tube and secrete a fluid that completes the capacitation of spermatozoa, and supplies nutrients to spermatozoa and the developing *pre-embryo* (the cluster of cells produced by the initial mitotic divisions following fertilization).

The Uterus

The **uterus** (Ū-ter-us) is a muscular chamber that provides mechanical protection and nutritional support for the developing *embryo* (weeks 1–8) and *fetus* (from week 9 to delivery). In addition, contractions in the muscular wall of the uterus are important in ejecting the fetus at the time of birth.

A typical uterus is a small, pear-shaped organ that is about 7.5 cm (3 in.) long and 5 cm (2 in.) in diameter (see Figure 19–8b). It weighs 30–40 g (1–1.4 oz) and is stabilized by various ligaments. In its normal position, the uterus bends anteriorly near its base (see Figure 19–8a).

The uterus consists of two regions: the body and the cervix (Figure 19–11●). The **body** is the largest division of the uterus.

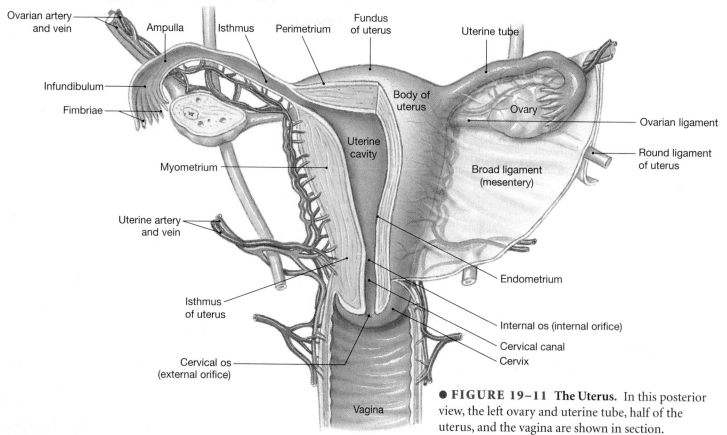

● **FIGURE 19–11 The Uterus.** In this posterior view, the left ovary and uterine tube, half of the uterus, and the vagina are shown in section.

The *fundus* is the rounded portion of the body superior to the attachment of the uterine tubes. Laterally, the body ends at a constriction known as the **isthmus.** The tubular **cervix** (SER-viks), which is the inferior portion of the uterus, projects a short distance into the vagina, where its surface surrounds the **external orifice,** or *cervical os.* The cervical canal opens into the **uterine cavity** at the **internal orifice,** or *internal os.*

The uterine wall is made up of an inner **endometrium** (en-dō-MĒ-trē-um) and a muscular **myometrium** (mī-ō-MĒ-trē-um; *myo-*, muscle + *metra*, uterus), covered by the **perimetrium,** which is a layer of visceral peritoneum (see Figure 19–11). The endometrium includes the epithelium that lines the uterine cavity and the underlying connective tissues. Uterine glands that open onto the endometrial surface extend deep into the connective tissue layer almost to the myometrium. The myometrium consists of a thick mass of interwoven smooth muscle cells. In adult women of reproductive age who have not given birth, the uterine wall is about 1.5 cm (0.5 in.) thick.

The endometrium consists of a superficial *functional zone* and a deeper *basilar zone* that is adjacent to the myometrium. The structure of the basilar layer remains relatively constant over time, but that of the functional zone undergoes cyclical changes in response to sex hormone levels. These alterations produce the characteristic features of the monthly uterine cycle.

Clinical Note
ENDOMETRITIS

An infection of the uterine lining called *endometritis* is an occasional complication of miscarriage, childbirth, or gynecological procedures such as dilatation and curettage (D and C). Commonly reported signs and symptoms include mild to severe lower abdominal pain; a bloody, foul-smelling discharge; and fever (101° to 104°F). The onset of symptoms is usually 48 to 72 hours after the gynecological procedure or miscarriage. These infections often mimic the presentation of PID and can be quite serious if not quickly treated with the appropriate antibiotics. Complications of endometritis may include sterility, sepsis, or even death. ■

The Uterine Cycle

The **uterine cycle,** or *menstrual* (MEN-stroo-al) *cycle,* is a repeating series of changes in the structure of the endometrium. The first uterine cycle occurs with the **menarche** (me-NAR-kē), or first menstrual period at puberty, typically age 11–12. The cycles continue until age 45–55, when **menopause** (MEN-ō-pawz), which is the last menstrual cycle, occurs. In the interim, the regular appearance of menstrual cycles is interrupted only by circumstances such as illness, stress, starvation, or pregnancy.

The uterine cycle averages 28 days in length, but it can range from 21 to 35 days in normal individuals. It consists of three stages: *menses,* the *proliferative phase,* and the *secretory phase.*

Clinical Note
ENDOMETRIOSIS

Endometriosis is a condition in which endometrial tissue is found outside of the uterus. Most commonly it is found in the abdomen and pelvis, although it has been found virtually everywhere in the body, including the central nervous system and lungs. Regardless of its site, the tissue responds to the hormonal changes associated with the menstrual cycle and thus bleeds cyclically. This bleeding causes inflammation, scarring of adjacent tissues, and the subsequent development of adhesions, particularly in the pelvic cavity.

Endometriosis is usually seen in women between the ages of 30 and 40 and is rarely seen in postmenopausal women. The exact cause is unknown. The most common symptom is dull, cramping pelvic pain that is usually related to menstruation. Dyspareunia and abnormal uterine bleeding are also commonly reported. Painful bowel movements have also been reported when the endometrial tissue has invaded the gastrointestinal tract. Endometriosis is commonly diagnosed when the patient is being evaluated for infertility. Definitive treatment may include medical management with hormones, analgesics, and anti-inflammatory drugs, and/or surgery to remove the excessive endometrial tissue or adhesions from other organs. ■

MENSES. The menstrual cycle begins with the onset of **menses** (MEN-sēz), which is a period marked by the degeneration of the superficial *functional zone* of the endometrium. The process is triggered by a decline in progesterone and estrogen levels as the corpus luteum disintegrates. The endometrial arteries constrict, which reduces blood flow to this region, and the secretory glands, epithelial cells, and other tissues of the functional zone die of oxygen and nutrient deprivation. Eventually the weakened arterial walls rupture, and blood pours into the connective tissues of the functional zone. Blood cells and degenerating tissues break away and enter the uterine cavity to be lost by passage into the vagina. This sloughing of tissue, which continues until the entire functional zone has been lost, is called **menstruation** (men-stroo-Ā-shun). Menstruation usually lasts 1 to 7 days, and roughly 35–50 mL of blood is lost. Painful menstruation, or *dysmenorrhea,* can result from uterine inflammation, myometrial contractions ("cramps"), or from conditions that involve adjacent pelvic structures.

THE PROLIFERATIVE PHASE. The **proliferative phase** begins in the days that follow the completion of menses as the surviving epithelial cells multiply and spread across the surface of the endometrium. This repair process is stimulated by rising

Clinical Note
PELVIC INFLAMMATORY DISEASE (PID)

Probably the most common cause of nontraumatic gynecological pain is pelvic inflammatory disease (PID). PID is an infection of the female reproductive tract that can be caused by bacteria, viruses, or fungi. The organs most commonly involved are the uterus, fallopian tubes, and ovaries. Occasionally the adjoining structures, such as the peritoneum and intestines, also become involved. PID is the most common cause of abdominal pain in women in the childbearing years, and occurs in one percent of that population. The highest rate of infection occurs in sexually active women ages 15 to 24. The most common causes of PID are gonorrhea (*Neisseria gonorrhoeae*) or chlamydia (*Chlamydia trachomatis*), although rarely streptococcus or staphylococcus bacteria may cause it. Commonly, gonorrhea or chlamydia progresses undetected in a female until frank PID develops (Figure 19–12●).

Predisposing factors include multiple sexual partners, prior history of PID, recent gynecological procedure, or an IUD. Postinfection damage to the fallopian tubes is a common cause for infertility. PID may be either acute or chronic. If it is allowed to progress untreated, sepsis may develop. Additionally, PID may cause adhesions, in which the pelvic organs "stick together." These adhesions are a common cause of chronic pelvic pain and increase the frequency of infertility and ectopic pregnancies.

While it is possible for a patient with pelvic inflammatory disease to be asymptomatic, most patients with PID complain of abdominal pain. It is often diffuse and located in the lower abdomen. It may be moderate to severe, which occasionally makes distinguishing it from appendicitis difficult. Pain may intensify either before or after the menstrual period. It may also worsen during sexual intercourse (dyspareunia), as movement of the cervix tends to cause increased discomfort. Patients with PID tend to walk with a shuffling gait, since walking often intensifies their pain. In severe cases, fever, chills, nausea, vomiting, or even sepsis may accompany PID. Occasionally, patients have a foul-smelling, often yellow, vaginal discharge, as well as irregular menses. Midcycle bleeding is also common.

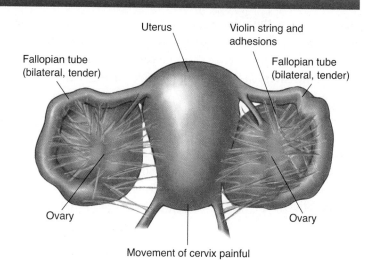

● **FIGURE 19–12 Pelvic Inflammatory Disease.** PID is an infection of the female reproductive organs. It typically causes adhesions of all infected organs.

The patient with PID may appear acutely ill or toxic. The blood pressure is normal, although the pulse rate may be slightly increased. Fever may or may not be present. Palpation of the lower abdomen generally elicits moderate to severe pain. In the emergency department, the patient will undergo a pelvic examination. In PID, movement of the cervix causes severe pain, which is referred to as a "chandelier sign" as the patient reaches for the ceiling when the cervix is moved. It is often difficult to distinguish PID from appendicitis in young females.

The primary treatment for PID is antibiotics. Toxic patients may require antibiotics administered intravenously. Once the causative organism is determined, the sexual partner may also require treatment. ■

---

estrogen levels that accompany the growth of another set of ovarian follicles. By the time ovulation occurs, the functional zone is several millimeters thick, and its new set of uterine glands secretes a mucus that is rich in glycogen. In addition, the entire functional zone is filled with small arteries that branch from larger trunks in the myometrium.

THE SECRETORY PHASE.   During the **secretory phase** of the cycle, the uterine glands enlarge, and increase their rates of secretion as the endometrium prepares for the arrival of a developing embryo. This activity is stimulated by progestins and estrogens from the corpus luteum. This phase begins at the time of ovulation and persists as long as the corpus luteum remains intact. Se-

cretory activities peak about 12 days after ovulation. Over the next day or two glandular activity declines, and the uterine cycle comes to a close. A new cycle then begins with the onset of menses and the disintegration of the functional zone.

The Vagina

The **vagina** (va-JĪ-nuh) is an elastic, muscular tube that extends between the uterus and the vestibule, which is a space bounded by the external genitalia (see Figure 19–8a). The vagina is typically 7.5–9 cm (3–3.5 in.) long, but its diameter varies because it is highly distensible. The cervix of the uterus projects into the vagina (see Figure 19–11). The shal-

Clinical Note
MENSTRUAL IRREGULARITIES

Menstrual irregularities are common. Normal menstrual flow usually lasts less than a week. On the average, about 35 mL (a little more than an ounce) of blood is lost. A disturbance in the menstruation that results in abnormal uterine bleeding is called *dysfunctional uterine bleeding (DUB).* It is most commonly associated with a cycle where ovulation has not occurred. The absence of a period is termed amenorrhea. Pregnancy is the most common cause of amenorrhea, although hormonal factors are also frequently implicated. Pelvic pain associated with the onset of menses is called *dysmenorrhea.* Usually, the discomfort begins shortly before the onset of menstruation and ends by the second day. Dysmenorrhea improves in most women with use of oral contraceptives. ■

low recess that surrounds the cervical protrusion is known as the **fornix** (FOR-niks) (see Figure 19–8a). The vagina lies parallel to the rectum, and the two are in close contact posteriorly. The urethra extends along the superior wall of the vagina from the urinary bladder to the urethral opening, or external urethral meatus.

Clinical Note
AMENORRHEA

If menarche does not occur by age 16, or if a woman's normal menstrual cycle becomes interrupted for 6 months or more, the condition of *amenorrhea* (ā-men-ō-RĒ-uh) exists. *Primary amenorrhea* is the failure to begin menses. This condition may indicate developmental abnormalities, such as nonfunctional ovaries, the absence of a uterus, or some endocrine or genetic disorder. It can also result from malnutrition; puberty is delayed if leptin levels are too low. ∞ p. 628 Transient *secondary amenorrhea* may be caused by severe physical or emotional stresses. In effect, the reproductive system gets "switched off." Factors that cause either type of amenorrhea include drastic weight reduction programs, anorexia nervosa, and severe depression or grief. Amenorrhea has also been observed in marathon runners and other women engaged in training programs that require sustained high levels of exertion and severely reduce body lipid reserves. ■

The vaginal walls contain a network of blood vessels and layers of smooth muscle. The lining is moistened by the mucous secretions of the cervical glands and by the movement of water across the permeable epithelium. The vagina and vestibule are separated by an elastic epithelial fold,

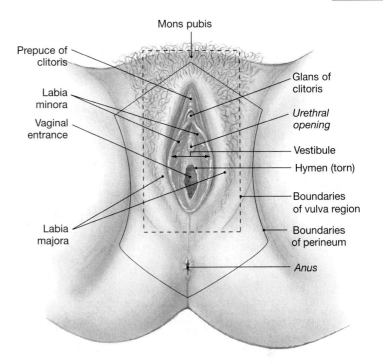

● **FIGURE 19–13** The Female External Genitalia.

which is the **hymen** (HĪ-men) (see Figure 19–13●), which partially or completely blocks the entrance to the vagina before the first occasion of sexual intercourse or tampon usage. The two bulbospongiosus muscles lie on either side of the vaginal entrance, which is constricted by their contractions. ∞ p. 238

The vagina has three major functions: it (1) serves as a passageway for the elimination of menstrual fluids; (2) receives the penis during sexual intercourse and holds spermatozoa prior to their passage into the uterus; and (3) forms the lower portion of the birth canal, through which the fetus passes during delivery.

The vagina normally contains resident bacteria supported by nutrients in the cervical mucus. The metabolic activity of these bacteria creates an acidic environment, which restricts the growth of many pathogens. Inflammation of the vaginal canal, known as *vaginitis* (vaj-i-NĪ-tis), is typically caused by fungi, bacteria, or parasites. In addition to any discomfort that may result, the condition may lower the survival of sperm and thereby reduce fertility.

The External Genitalia

The **perineum,** which is the muscular floor of the pelvic cavity, includes the anus and structures associated with the reproductive system called the **external genitalia.** ∞ p. 238 The perineal region that encloses the female external genitalia is the **vulva** (VUL-vuh), or *pudendum* (Figure 19–13). The vagina opens into the **vestibule,** which is a central space bounded by the **labia minora** (LĀ-bē-uh mi-NOR-uh; *labia*, lips; singular *labium minus*). The labia minora are covered with a smooth, hairless skin. The urethral opening lies in the vestibule just anterior to the vaginal entrance. Anterior to the urethral opening, the **clitoris** (KLIT-o-ris) projects into the vestibule. The clitoris is derived from the same embryonic structures as the penis in males. Internally it contains erectile tissue comparable to the corpora cavernosa of the penis. The clitoris engorges with blood during sexual arousal. A small erectile *glans* sits atop the organ, and extensions of the labia minora encircle the body of the clitoris, and form the *prepuce.*

A variable number of small **lesser vestibular glands** discharge secretions onto the exposed surface of the vestibule, which keep it moist. During sexual arousal, a pair of ducts discharges the secretions of the **greater vestibular glands** into the vestibule near the vaginal entrance (see Figure 19–8). These mucous glands resemble the bulbourethral glands of males.

The outer limits of the vulva are formed by the mons pubis and labia majora. The prominent bulge of the **mons pubis** is created by adipose tissue beneath the skin anterior to the pubic symphysis. Adipose tissue also accumulates in the fleshy **labia majora** (singular *labium majus*), which encircle and partially conceal the labia minora and vestibular structures.

The Mammary Glands

A newborn cannot fend for itself, and several of its key systems have not yet developed fully. During its adjustment to an independent existence, the infant gains nourishment from the milk secreted by the **mammary glands** of the breasts (Figure 19–14●). Milk production, or **lactation** (lak-TĀ-shun), occurs in these glands. In females, the mammary glands are specialized organs of the integumentary system that are controlled by hormones of the reproductive system and the *placenta*, which is a temporary structure that provides the embryo and fetus with nutrients.

Each breast contains a mammary gland within the subcutaneous tissue of the *pectoral fat pad* beneath the skin. The ducts of the underlying mammary gland open onto the body surface at a small conical projection—the **nipple.** The reddish-brown skin that surrounds each nipple is the **areola** (a-RĒ-ō-luh). Large sebaceous glands beneath the areolar surface give it a grainy texture.

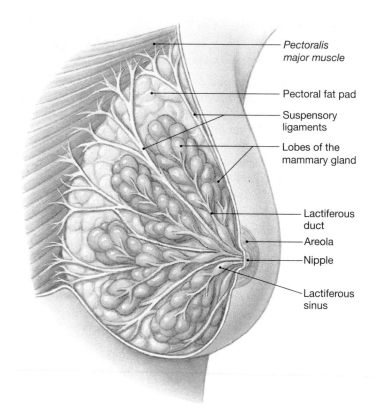

Pectoralis major muscle

Pectoral fat pad

Suspensory ligaments

Lobes of the mammary gland

Lactiferous duct

Areola

Nipple

Lactiferous sinus

● **FIGURE 19–14 The Mammary Gland of the Left Breast.** Lobes of glandular tissue within each mammary gland are responsible for the production of breast milk.

Clinical Note
ENDOCRINE EFFECTS ON THE BREAST

At puberty, increasing levels of *estrogen* stimulate growth of the breast's *mammary gland.* In pregnancy, the placenta secretes large quantities of estrogens that stimulate growth of the ductal system. Once the ductal system has developed, *progesterone* causes final development of the breasts into milk-secreting glands. While estrogens and progesterone stimulate development of the breasts, they actually inhibit the secretion of milk.

Prolactin causes the production of milk. The prolactin level begins to rise around the fifth week of pregnancy and continues until delivery. When the placenta is delivered, the levels of estrogen and progesterone drop markedly, and the breasts begin to secrete milk. Finally, milk is "let down," or ejected, by the nipples due to stimulation by oxytocin. Suckling of the breast by the baby sends sensory impulses to the hypothalamus, which results in oxytocin secretion. ■

The glandular tissue of a mammary gland consists of a number of separate lobes, each made up of several lobules that contain milk glands. Within each lobe, the ducts that leave the lobules converge, and give rise to a single **lactiferous** (lak-TIF-er-us) **duct.** Near the nipple, that lactiferous duct expands, forming an expanded chamber called a **lactiferous sinus.** Some 15–20 lactiferous sinuses open onto the surface of each nipple. Dense connective tissue surrounds the duct system and forms partitions that extend between the lobes and lobules. These bands of connective tissue, the *suspensory ligaments of the breast,* originate in the dermis of the overlying skin. A layer of loose connective tissue separates the mammary gland complex from the underlying muscles, and the two can move relatively independently.

Clinical Note
BREAST CANCER

Breast cancer is a malignant, metastasizing tumor of the mammary gland. Almost 90 percent of breast cancers begin in the ducts and lobes of the mammary glands and are called *ductal carcinomas* and *lobular carcinomas,* respectively. If the tumor cells have spread outside of a duct or lobule and into the surrounding tissue, the cancer is called an *invasive ductal* or *lobular carcinoma.* Cancers that have not spread are called *in situ,* which means "in place," and are known as *ductal carcinoma in situ (DCIS)* and *lobular carcinoma in situ (LCIS).*

Breast cancer is the leading cause of death in women between the ages of 35 and 45, but it is most common in women over age 50. For 2005, it is estimated that there will be 40,410 female deaths and approximately 211,240 new cases of breast cancer in the U.S. An estimated 12 percent of U.S. women will develop breast cancer at some point in their lifetime, and the rate is steadily rising. The incidence is highest among Caucasian Americans, somewhat lower in African Americans, and lowest in Asian Americans and American Indians. Notable risk factors include (1) a family history of breast cancer, (2) a first pregnancy after age 30, and (3) early menarche (first menstrual period) or late menopause (last menstrual period). Breast cancers in males are very rare, but about 400 men die from the disease each year in the U.S. ∎

→ CONCEPT CHECK QUESTIONS

1. As the result of infections such as gonorrhea, scar tissue can block both uterine tubes. How would this blockage affect a woman's ability to conceive?
2. What advantage does the acidic pH of the vagina confer?
3. Which layer of the uterus sloughs off during menstruation?
4. Would the blockage of a single lactiferous sinus interfere with delivery of milk to the nipple? Explain.

Answers begin on p. 792.

Hormones and the Female Reproductive Cycle

As is the case for males, the activity of the female reproductive system is under hormonal control by both pituitary and gonadal secretions. However, the regulatory pattern in females is much more complicated than in males because circulating hormones must coordinate the ovarian and uterine cycles to ensure that the **female reproductive cycle** results in the proper functioning of all reproductive activities. If the ovarian and uterine cycles are not coordinated normally, infertility results. A woman who fails to ovulate will be unable to conceive, even if her uterus is perfectly normal. A woman who ovulates normally but whose uterus is not ready to support an embryo will be just as infertile.

Figure 19–15● presents the events of a single female reproductive cycle. Figure 19–15a–c plots events in the ovarian cycle, which is divided into the follicular phase and the luteal phase. Figure 19–15d depicts the endometrial changes of the uterine cycle, which include menses, the proliferative phase, and the secretory phase. Figure 19–15e plots body temperature during the cycle.

Clinical Note
DERMOID CYSTS

An interesting class of tumors of the female reproductive system is the teratomas. Teratomas are divided into three categories: (1) mature (benign), (2) immature (malignant), and (3) monodermal, or highly specialized. The vast majority of benign teratomas are cystic and are more commonly referred to as *dermoid cysts.* These cysts are lined by skin and are usually filled with a sebaceous cheesy substance. Often, matted hair and teeth can be found within the cyst. The cyst wall arises from stratified squamous epithelium. In more complex cysts, layers of germ cells can give rise to cartilage, bone, thyroid tissue, or organoid formation.

The origination of dermoid tumors has been a matter of fascination for centuries. Some common beliefs blamed witches, nightmares, or adultery with the devil as possible causes. Interestingly, the karyotype of all benign ovarian teratomas is 46,XX (normal human female). Dermoid tumors are usually easily removed, sparing the ovary and uterine tube. ∎

Hormones and the Follicular Phase

The **follicular phase** (or *preovulatory phase*) of the ovarian cycle begins each month when FSH from the anterior pituitary stimulates some of the secondary follicles to begin their development into tertiary (Graafian) follicles (Figure 19–16●). As the follicular cells enlarge and multiply (see Figure 19–15b), they release steroid hormones collectively known as **estrogens** (see Figure 19–15c). The most important estrogen is **estradiol** (es-tra-DĪ-ol). Estrogens have multiple functions, including (1) affecting central nervous system (CNS) activity (especially in the hypothalamus, where estrogens increase sexual drive); (2) stimulating bone and muscle growth; (3) establishing and maintaining female secondary sex characteristics, such as body hair distribution and the location of adipose tissue deposits;

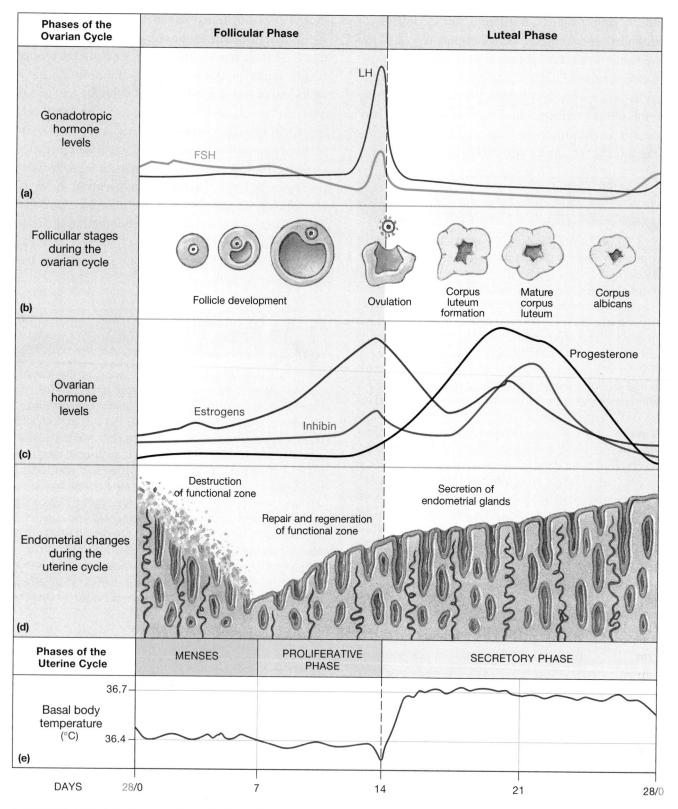

Phases of the Ovarian Cycle	Follicular Phase	Luteal Phase

(a) Gonadotropic hormone levels

LH

FSH

(b) Follicular stages during the ovarian cycle

Follicle development · Ovulation · Corpus luteum formation · Mature corpus luteum · Corpus albicans

(c) Ovarian hormone levels

Progesterone · Estrogens · Inhibin

(d) Endometrial changes during the uterine cycle

Destruction of functional zone · Repair and regeneration of functional zone · Secretion of endometrial glands

Phases of the Uterine Cycle	MENSES	PROLIFERATIVE PHASE	SECRETORY PHASE

(e) Basal body temperature (°C)

36.7
36.4

DAYS 28/0   7   14   21   28/0

● **FIGURE 19–15** Hormonal Regulation of the Female Reproductive Cycle.

(4) maintaining functional accessory reproductive glands and organs; (5) initiating the repair and growth of the endometrium; and (6) regulating GnRH and gonadotropin secretion through feedback to the hypothalamus and pituitary,

respectively. (Because the feedback mechanisms are complex, they are not included in Figure 19–16.)

Early in the follicular phase, estrogen and inhibin levels are low (see Figure 19–15c). The estrogens and inhibin have

● **FIGURE 19–16** Hormonal
Regulation of Ovarian Activity.

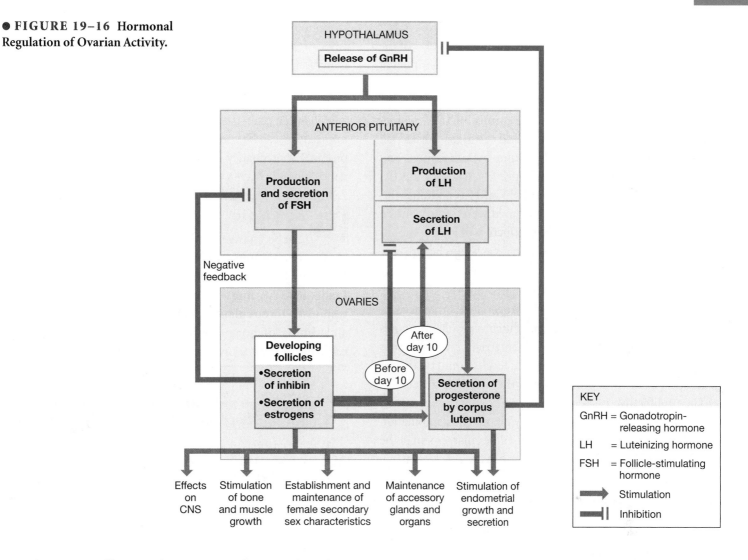

complementary effects on the secretion of FSH and LH (see Figure 19–16): low levels of estrogen inhibit LH and FSH secretion. As follicular development proceeds, the concentration of circulating estrogens and inhibin rises (see Figure 19–15c) because the follicular cells are increasing in number and secretory activity. As secondary follicles develop, FSH levels decline due to the negative-feedback effects of inhibin, and estrogen levels continue to rise. Despite the slow decline in FSH concentrations, the combination of estrogens, FSH, and LH continues to support follicular development and maturation.

Estrogen concentrations take a sharp upturn in the second week of the ovarian cycle, with the development of a tertiary follicle as it enlarges in preparation for ovulation (see Figure 19–15). The rapid increase in estrogen leads to the secretion of FSH and LH by acting on the hypothalamus and stimulating the production of GnRH (see Figure 19–16). The pituitary gland also becomes more sensitive to GnRH and contributes to the rise in FSH and LH. At roughly day 10 of the cycle, the effect of estrogen on LH secretion also changes from inhibi-

tion to stimulation. At about day 14, estrogen levels peak, and accompany the maturation of the tertiary follicle. The high estrogen concentration then triggers a massive outpouring, or surge, of LH from the anterior pituitary, which triggers the rupture of the follicular wall and ovulation.

Hormones and the Luteal Phase

The **luteal phase,** or *postovulatory phase,* of the ovarian cycle begins as the high LH levels that triggered ovulation also stimulate the remaining follicular cells to form a corpus luteum (see Figure 19–15a–c). The yellow color of the corpus luteum results from its lipid reserves, which are used to manufacture steroid hormones known as **progestins** (prō-JES-tinz), predominantly the steroid **progesterone** (prō-JES-ter-ōn). Progesterone, which is the principal hormone of the luteal phase, prepares the uterus for pregnancy by stimulating the growth and development of the blood supply and secretory glands of the endometrium (see Figure 19–15d). Progesterone also stimulates metabolic activity and elevates basal body temperature (see Figure 19–15e).

Luteinizing hormone (LH) levels remain elevated for only 2 days, but that is long enough to stimulate the formation of the functional corpus luteum. Progesterone secretion continues at relatively high levels for the next week. Unless pregnancy occurs, however, the corpus luteum then begins to degenerate. Roughly 12 days after ovulation, the corpus luteum becomes nonfunctional, and progesterone and estrogen levels fall markedly. This decline stimulates the hypothalamus, and GnRH production increases (see Figure 19–16). With this decline the hypothalamus production of GnRH is no longer inhibited (see Figure 19–16). The increased secretion of GnRH, in turn, leads to an increase in the production of FSH in the anterior pituitary gland, and the ovarian cycle begins again.

Hormones and the Uterine Cycle

The declines in progesterone and estrogen levels that accompany the breakdown of the corpus luteum result in menses (see Figure 19–15d). The loss of endometrial tissue continues for several days, until rising estrogen levels stimulate the regeneration of the functional zone of the endometrium.

The proliferative phase continues until rising progesterone levels mark the arrival of the secretory phase. The combination of estrogen and progesterone then causes the enlargement of the endometrial glands as well as an increase in their secretions.

Hormones and Body Temperature

The monthly hormonal fluctuations also cause physiological changes that affect core body temperature. During the follicular phase, when estrogen is the dominant hormone, the resting, or basal, body temperature measured on awakening in the morning

Clinical Note
INFERTILITY

Infertility (sterility) is usually defined as an inability to achieve pregnancy after 1 year of appropriately timed sexual intercourse. Problems with infertility are relatively common. An estimated 10–15 percent of married couples in the U.S. are infertile, and another 10 percent are unable to have as many children as they desire. It is, thus, not surprising that reproductive physiology has become a popular field of medicine and that the treatment of infertility has become a major medical industry. Recent advances in our understanding of reproductive physiology are providing new solutions to fertility problems as varied as low sperm count, abnormal spermatozoa, inadequate maternal hormone levels, problems with oocyte production or oocyte transport from ovary to uterine tube, blocked uterine tubes, abnormal oocytes, and an abnormal uterine environment. Procedures meant to resolve these reproductive difficulties are known as *assisted reproductive technologies (ART).* ■

is about 0.3°C (or 0.5°F) lower than it is during the luteal phase, when progesterone dominates (see Figure 19–15e). At the time of ovulation, basal temperature declines sharply, making the temperature rise over the following day even more noticeable. By keeping records of body temperature over a few menstrual cycles, a woman can often determine the precise day of ovulation. This information can be very important for those wishing to avoid or promote a pregnancy because fertilization that leads to pregnancy typically occurs within a day of ovulation.

Key Note

Cyclic changes in FSH and LH levels are responsible for the maintenance of the ovarian cycle; the hormones produced by the ovaries in turn regulate the uterine cycle. Inadequate hormone levels, inappropriate or inadequate responses to circulating hormones, or poor coordination and timing of hormone production or secondary oocyte release will reduce or eliminate the chances of pregnancy.

CONCEPT CHECK QUESTIONS

1. What changes would you expect to observe in the ovarian cycle if the LH surge did not occur?
2. What effect would blockage of progesterone receptors in the uterus have on the endometrium?
3. What event occurs in the uterine cycle when estrogen and progesterone levels decline?

Answers begin on p. 792.

■ The Physiology of Sexual Intercourse

Sexual intercourse, also known as **coitus** (KŌ-i-tus) or *copulation*, introduces semen into the female reproductive tract. The following sections consider the physiological bases for the sexual responses of males and females.

Male Sexual Function

Male sexual function is coordinated by reflex pathways that involve both divisions of the ANS. During sexual **arousal**, erotic thoughts or the stimulation of sensory nerves in the genital region increase the parasympathetic outflow over the pelvic nerves. This outflow leads to erection of the penis (discussed on p. 704). The skin of the glans of the penis contains numerous sensory receptors, and erection tenses the skin and increases their sensitivity. Subsequent stimulation may initiate the secretion of the bulbourethral glands, lubricating the urethra and the surface of the glans.

During intercourse, the sensory receptors in the penis are rhythmically stimulated, eventually resulting in emission and ejaculation. **Emission** occurs under sympathetic stimulation. The process begins with peristaltic contractions of the ductus deferens, which push fluid and spermatozoa through the ejaculatory ducts and into the urethra. The seminal vesicles then contract, followed by waves of contraction in the prostate gland. While these contractions are proceeding, sympathetic commands close the sphincter at the entrance to the urinary bladder, and prevent the passage of semen into the bladder.

Ejaculation occurs as powerful, rhythmic contractions appear in the *ischiocavernosus* and *bulbospongiosus* muscles, two superficial skeletal muscles of the pelvic floor. (The positions of these muscles can be seen in Figure 7–20b, p. 239.) Ejaculation is associated with intensely pleasurable sensations, which is an experience known as male **orgasm** (OR-gazm). Several other physiological changes also occur at this time, including temporary increases in heart rate and blood pressure. After ejaculation, blood begins to leave the erectile tissue, and the erection begins to subside. This subsidence, called *detumescence* (de-tū-MES-ens), is mediated by the sympathetic nervous system. An inability to achieve or maintain an erection is called **impotence.**

Clinical Note
DATE RAPE

Date rape, also called *acquaintance rape,* is defined as forced, unwanted intercourse with a person known by the victim. Sometimes, physical force or threats are used to force the victim to cooperate. In many instances, the victim is given a sedative drug that allows the assailant to carry out the assault.

One of the most frequently used drugs in date rape is the benzodiazepine *flunitrazepam.* Also known as *Rohypnol,* or "roofies," flunitrazepam is used in the short-term management of insomnia and as a sedative-hypnotic and preanesthetic medication. It has pharmacological effects similar to those of diazepam (Valium), although Rohypnol is approximately 10 times more potent. The effects begin within 30 minutes and peak within 2 hours. It has significant amnestic properties, similar to those seen with the sedative midazolam (Versed). Rohypnol is neither manufactured nor sold legally in the U.S. It is produced and sold by prescription in Europe and Latin America. It is supplied in 1- and 2-milligram tablets. Rohypnol enters the U.S. through several routes. It is transported across the Mexican border to Texas and then delivered to other parts of the country. It also enters South Florida from Columbia via international mail services or commercial airlines.

The effects of Rohypnol can be reversed with flumazenil (Romazicon) if needed. Because of Rohypnol's amnestic properties, victims may not even remember the assault. They may wake up in a strange place with their clothing in disarray. Many are too embarrassed to seek help after the assault. ■

Female Sexual Function

The phases of female sexual function are comparable to those of male sexual function. During sexual arousal, parasympathetic activation leads to an engorgement of the erectile tissues of the clitoris and increased secretion of cervical mucous glands and the greater vestibular glands. Clitoral erection increases the receptors' sensitivity to stimulation, and the cervical and vestibular glands provide lubrication for the vaginal walls. A network of blood vessels in the vaginal walls becomes filled with blood at this time, and the vaginal surfaces are also moistened by fluid from underlying connective tissues. Parasympathetic stimulation also causes engorgement of blood vessels at the nipples, which makes them more sensitive to touch and pressure.

During intercourse, rhythmic contact of the penis with the clitoris and vaginal walls, reinforced by touch sensations from the breasts and other stimuli (visual, olfactory, and auditory), provides stimulation that can lead to orgasm. Female orgasm is accompanied by peristaltic contractions of the uterine and vaginal walls and by rhythmic contractions of the bulbospongiosus and ischiocavernosus muscles. The latter contractions give rise to the sensations of orgasm.

Clinical Note
SEXUALLY TRANSMITTED DISEASES

Sexually transmitted diseases (STDs) are infections contracted by intimate, as well as sexual, contact. The causative organism may be bacterial, viral, protozoan, parasitic, or fungal. The incidence of STDs is high. In the U.S. the most common STDs are *chlamydia, gonorrhea, syphilis, genital warts,* and *herpes simplex virus, type 2.* These infections primarily affect the genitourinary system. However, STDs such as syphilis and gonorrhea can have systemic effects. Some of the most feared infectious diseases can be sexually transmitted, including *HIV infection (AIDS)* and *hepatitis B (HBV).* The incidence of STDs is highest in adolescents and declines exponentially with age. ■

■ Aging and the Reproductive System

Aging affects all body systems, including the reproductive systems of men and women alike. After puberty, when these systems become fully functional in both sexes, the most striking age-related changes in the female reproductive system occur at menopause. Comparative changes in the male reproductive system occur more gradually and over a longer period of time.

Menopause

Menopause is usually defined as the time that ovulation and menstruation cease. It typically occurs at age 45–55, but in the years preceding it, the ovarian and menstrual cycles become irregular. The transition period from normal menstrual cycles to none at all is called *perimenopause,* and it normally begins at age 40. A shortage of follicles is the underlying cause of cycle irregularities. Of an estimated 2 million potential oocytes present at birth, only a few hundred thousand remain at puberty. By age 50, there are often no secondary follicles left to respond to FSH. In *premature menopause,* this depletion occurs before age 40.

Menopause is accompanied by a decline in circulating concentrations of estrogens and progesterone and a sharp and sustained rise in the production of GnRH, FSH, and LH. The decline in estrogen levels leads to reductions in the sizes of the uterus and breasts, accompanied by a thinning of the urethral and vaginal walls. The reduced estrogen concentrations have also been linked to the development of osteoporosis, presumably because bone deposition proceeds at a slower rate. A variety of neural effects are also reported, including "hot flashes," anxiety, and depression. Hot flashes typically begin while estrogen levels are declining and cease when estrogen levels reach minimal levels. These intervals of elevated body temperature are associated with surges in LH production. The hormonal mechanisms involved in other CNS effects of menopause are not well understood. In addition, the risks of atherosclerosis and other forms of cardiovascular disease increase after menopause.

The majority of women experience only mild symptoms, but some individuals experience unpleasant symptoms in perimenopause or during or after menopause. For most of those individuals, hormone replacement therapies that involve a combination of estrogens and progestins can prevent osteoporosis and the neural and vascular changes associated with menopause.

The Male Climacteric

Changes in the male reproductive system occur more gradually than do those in the female reproductive system. The period of declining reproductive function is known as the **male climacteric.** Levels of circulating testosterone begin to decline between ages 50 and 60, and levels of FSH and LH increase. Although sperm production continues (men well into their eighties can father children), older men experience a gradual reduction in sexual activity. This decrease may be linked to declining testosterone levels, and some clinicians suggest the use of testosterone replacement therapy to enhance libido (sexual drive) in elderly men.

🔒 Key Note

Sex hormones have widespread effects on the body. They affect brain development and behavioral drives, muscle mass, bone mass and density, body proportions, and the patterns of hair and body fat distribution. As aging occurs, reduction in sexual hormone levels affect appearance, strength, and a variety of physiological functions.

→ CONCEPT CHECK QUESTIONS

1. An inability to contract the ischiocavernosus and bulbospongiosus muscles would interfere with which part of the sexual response in males?
2. What changes occur in females during sexual arousal as the result of increased parasympathetic stimulation?
3. Why does the level of FSH rise and remain high during menopause?

Answers begin on p. 792.

■ Integration with Other Systems

Normal human reproduction is a complex process that requires the participation of several organ systems. The hormones discussed in this chapter (Table 19–1) play a major role in coordinating reproductive events. Reproduction depends on various physical, physiological, and psychological factors. For example, the male's sperm count must be adequate, the semen must have the correct pH and nutrients, and erection and ejaculation must occur in the proper sequence; the woman's ovarian and uterine cycles must be properly coordinated, ovulation and oocyte transport must occur normally, and her reproductive tract must be suitable for sperm survival, movement, and fertilization. Many of these events require the participation and normal functioning of the digestive, endocrine, nervous, cardiovascular, and urinary systems. Figure 19–17● summarizes the interactions between these organ systems and the reproductive system.

The Reproductive System in Perspective

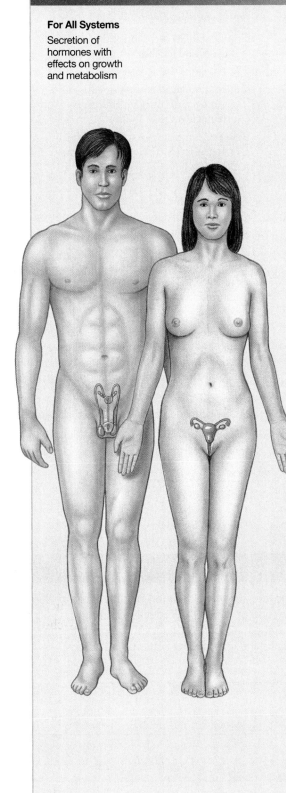

For All Systems
Secretion of hormones with effects on growth and metabolism

The Integumentary System
- ⟲ • Covers external genitalia; provides sensations that stimulate sexual behaviors; mammary gland secretions provide nourishment for newborn
- ⟳ • Reproductive hormones affect distribution of body hair and subcutaneous fat deposits

The Skeletal System
- ⟲ • Pelvis protects reproductive organs of females, portion of ductus deferens and accessory glands in males
- ⟳ • Sex hormones stimulate growth and maintenance of bones; sex hormones at puberty accelerate growth and closure of epiphyseal cartilages

The Muscular System
- ⟲ • Contractions of skeletal muscles eject semen from male reproductive tract; muscle contractions during sexual act produce pleasurable sensations in both sexes
- ⟳ • Reproductive hormones, especially testosterone, accelerate skeletal muscle growth

The Nervous System
- ⟲ • Controls sexual behaviors and sexual function
- ⟳ • Sex hormones affect CNS development and sexual behaviors

The Endocrine System
- ⟲ • Hypothalamic regulatory hormones and pituitary hormones regulate sexual development and function; oxytocin stimulates smooth muscle contractions in uterus and mammary glands
- ⟳ • Steroid sex hormones and inhibin inhibit secretory activities of hypothalamus and pituitary gland

The Cardiovascular System
- ⟲ • Distributes reproductive hormones; provides nutrients, oxygen, and waste removal for fetus; local blood pressure changes responsible for physical changes during sexual arousal
- ⟳ • Estrogens may help maintain healthy vessels and slow development of atherosclerosis

The Lymphatic System
- ⟲ • Provides IgA for secretions by epithelial glands; assists in repairs and defense against infection
- ⟳ • Lysosomes and bactericidal chemicals in secretions provide nonspecific defense against reproductive tract infections

The Respiratory System
- ⟲ • Provides oxygen and removes carbon dioxide generated by tissues of reproductive system
- ⟳ • Changes in respiratory rate and depth occur during sexual arousal, under control of the nervous system

The Digestive System
- ⟲ • Provides additional nutrients required to support gamete production and (in pregnant women) embryonic and fetal development

The Urinary System
- ⟲ • Urethra in males carries semen to exterior; kidneys remove wastes generated by reproductive tissues and (in pregnant women) by a growing embryo and fetus
- ⟳ • Accessory organ secretions may have antibacterial action that helps prevent urethral infections in males

● **FIGURE 19–17 Functional Relationships Between the Reproductive System and Other Systems.**

TABLE 19-1 *Hormones of the Reproductive System*

HORMONE	SOURCE	REGULATION OF SECRETION	PRIMARY EFFECTS
Gonadotropin-releasing hormone (GnRH)	Hypothalamus	*Males:* inhibited by testosterone *Females:* inhibited by estrogens and/or progestins	Stimulates FSH secretion and LH synthesis in males and females
Follicle-stimulating hormone (FSH)	Anterior pituitary gland	*Males:* stimulated by GnRH, inhibited by inhibin and testosterone *Females:* stimulated by GnRH, inhibited by estrogens and/or progestins	*Males:* stimulates spermatogenesis and spermiogenesis through effects on sustentacular cells *Females:* stimulates follicle development, estrogen production, and oocyte maturation
Luteinizing hormone (LH)	Anterior pituitary gland	*Males:* stimulated by GnRH *Females:* production stimulated by GnRH and secretion by estrogens	*Males:* stimulates interstitial cells to secrete testosterone *Females:* stimulates ovulation, formation of corpus luteum, and progestin secretion
Androgens (primarily testosterone)	Interstitial cells of testes	Stimulated by LH	Establishes and maintains secondary sex characteristics and sexual behavior, promotes maturation of spermatozoa, inhibits GnRH secretion
Estrogens (primarily estradiol)	Follicular cells of ovaries; corpus luteum	Stimulated by FSH	Stimulates LH secretion (at high levels), establishes and maintains secondary sex characteristics and behavior, stimulates repair and growth of endometrium, inhibits secretion of GnRH
Progestins (primarily progesterone)	Corpus luteum	Stimulated by LH	Stimulates endometrial growth and glandular secretion, inhibits GnRH secretion
Inhibin	Sustentacular cells of testes and follicle cells of ovaries	Stimulated by FSH and factors released by developing sperm (*male*) or developing follicles (*female*)	Inhibits secretion of FSH (and possibly GnRH)

Even when all is normal, and fertilization occurs at the proper time and place, a healthy infant will still not be produced unless the zygote—a single cell the size of a pinhead—manages to develop into a full-term fetus that typically weighs about 3 kg. Chapter 20 considers the process of development and the mechanisms that determine both body structure and the distinctive characteristics of each individual.

→ CONCEPT CHECK QUESTIONS

1. Describe the functional relationships between the integumentary system and the reproductive system.
2. Describe the functional relationships between the endocrine system and the reproductive system.

Answers begin on p. 792.

Clinical Note
BIRTH CONTROL STRATEGIES

For physiological, logistical, financial, or emotional reasons, most adults practice some form of conception control during their reproductive years. Two out of three U.S. women ages 15–44 are practicing some method of contraception. When the simplest and most obvious method—sexual abstinence—is unsat-isfactory for some reason, another method of contraception must be used to avoid unwanted pregnancies. Many methods are available, so the selection process can be quite involved. Because each has specific strengths and weaknesses, the potential risks and benefits must be carefully analyzed on an individual basis.

Hormonal contraceptives manipulate the female hormonal cycle so that ovulation does not occur. The contraceptive pills produced in the 1950s combined quantities of estrogens and progestins sufficient to suppress pituitary production of GnRH, so FSH was not released and ovulation did not occur. Most of the oral contraceptives developed subsequently contain much smaller amounts of estrogens, or only progesterone. Current *combination* hormone contraceptives are administered in a cyclic fashion, using medication for 3 weeks. Then, during the fourth week, the woman takes no medication. For user convenience, monthly injections, weekly skin patches, and insertable vaginal rings that contain combined estrogen/progesterone products are available.

At least 20 brands of combination oral contraceptives are now available, and over 200 million women use them worldwide. In the U.S. 33 percent of women under age 45 use the combination pill to prevent conception. The failure rate for the combination oral contraceptives, when used as prescribed, is 0.24 percent over a 2-year period. (*Failure* for a birth control method is defined as a pregnancy.) Birth control pills are not risk-free: combination pills can worsen problems associated with severe hypertension, diabetes mellitus, epilepsy, gallbladder disease, heart trouble, and acne. Women taking oral contraceptives are also at increased risk for venous thrombosis, strokes, pulmonary embolism, and (for women over 35) heart disease. However, pregnancy has similar or higher risks.

Hormonal postcoital contraception, or the emergency "morning after" pill, involves taking either combination estrogen/progesterone birth control pills or progesterone-only pills in two large doses 12 hours apart within 72 hours of unprotected sexual intercourse. Particularly useful when barrier methods malfunction or coerced intercourse occurs, it reduces expected pregnancy rates by up to 89 percent. The progesterone-only version is considered safe for nonprescription use and may become available for purchase over the counter in the future.

Progesterone-only forms of birth control—Depo-provera®, the Norplant® system, and the progesterone-only pill—are now available. Depo-provera is injected every 3 months. The Silastic® (silicon rubber) tubes of the Norplant system are saturated with progesterone and inserted under the skin. This method provides birth control for a period of approximately 5 years, but to date the relatively high cost has limited the use of this contraceptive method. Both Depo-provera and the Norplant system can cause irregular menstruation and temporary amenorrhea, but they are extremely convenient to use. The progesterone-only pill is taken daily and may cause irregular uterine cycles. Skipping just one pill may result in pregnancy.

The **condom,** also called a *prophylactic* or "rubber," covers the glans and body of the penis during intercourse and keeps sperm from reaching the female reproductive tract. Latex condoms also reduce the spread of STDs, such as syphilis, gonorrhea, and HIV. The reported condom failure rate varies from 6 percent to 17 percent.

Vaginal barriers such as the *diaphragm* and *cervical cap* rely on similar principles. A diaphragm, the most popular form of vaginal barrier in use today, consists of a dome of latex rubber with a small metal hoop that supports the rim. Because vaginas vary in size, women who choose this method must be individually fitted. Before intercourse, the diaphragm is inserted so that it covers the cervical os, and it is usually coated with a small amount of spermicidal (sperm-killing) jelly or cream, which adds to the effectiveness of the barrier. The failure rate of a properly fitted diaphragm is estimated at 5–6 percent. A cervical cap is smaller and lacks the metal rim. It, too, must be fitted carefully, but unlike the diaphragm, it can be left in place for several days. Its failure rate (8 percent) is higher than that for diaphragm use.

An **intrauterine device (IUD)** consists of a small plastic loop or T that can be inserted into the uterine cavity. The mechanism of action remains uncertain, but IUDs are known to stimulate prostaglandin production in the uterus. The resulting change in the chemical composition of uterine secretions lowers the chances of fertilization and subsequent *implantation* of the zygote in the uterine lining. (Implantation is discussed in Chapter 20.) In the U.S. IUDs are in limited use today, but they remain popular in many other countries. The failure rate is estimated at 5–6 percent.

The **rhythm method** involves abstaining from sexual intercourse on the days ovulation might occur. The timing is estimated on the basis of previous patterns of menstruation and, in some cases, by monitoring changes in basal body temperature. The failure rate for the rhythm method is very high—almost 25 percent.

Sterilization is a surgical procedure that makes an individual unable to provide functional gametes for fertilization. Members of either sex may be sterilized. In a **vasectomy** (vaz-EK-to-mē) a segment of the ductus deferens is removed, which makes it impossible for sperm to pass from the epididymis to the distal portions of the reproductive tract. The surgery can be performed in a physician's office in a matter of minutes. The spermatic cords are located as they ascend from the scrotum on either side, and after each cord is opened, the ductus deferens is severed. After a 1 cm section is removed, the cut ends are usually tied shut (Figure 19–18a●). The cut ends cannot reconnect; in time, scar tissue forms a permanent seal. Alternatively, the cut ends of the ductus deferens are blocked with silicone plugs that can later be removed. This more recent vasectomy procedure may make it possible to restore fertility at a later date. After a vasectomy, the man experiences normal sexual function, for epididymal and testicular secretions account for only about 5 percent of the volume of the semen. Sperm continue to develop, but they remain within the epididymis until they degenerate. The failure rate for this procedure is 0.08 percent.

The uterine tubes can be blocked through a surgical procedure known as a **tubal ligation** (Figure 19–18b●). The failure rate for this procedure is estimated at 0.45 percent. Because the surgery involves entering the abdominopelvic cavity, complications are more likely than with a vasectomy. As in a vasectomy, attempts may be made to restore fertility after a tubal ligation.

Oral contraceptives, condoms, vaginal barriers, and sterilization are the primary contraception methods for individuals of all age groups. But the proportion of the population using a particular method varies by age group. Sterilization is most popular among older women, who may already have had children. Relative availability also plays a role. For example, a sexually active female under age 18 can buy a condom more easily than she can obtain a prescription for an oral contraceptive. But many of the differences are attributable to the relationship between risks and benefits for each age group.

When considering the use and selection of contraceptives, many people simply examine the list of potential complications and make the "safest" choice. For example, media coverage of the potential risks associated with oral contraceptives made many

(continued next page)

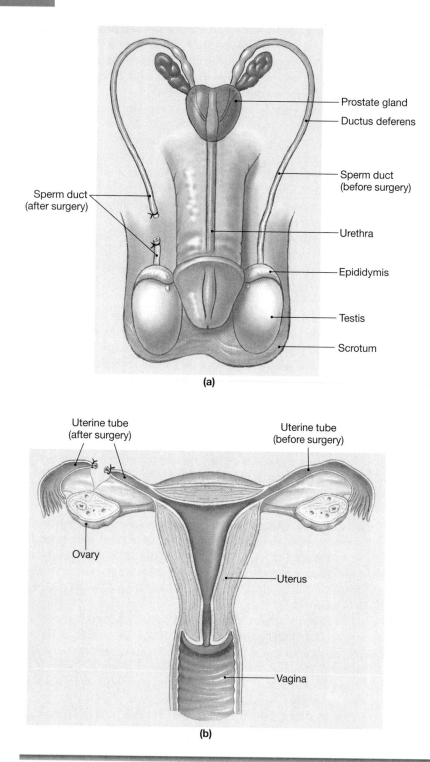

Sperm duct
(after surgery)

Prostate gland

Ductus deferens

Sperm duct
(before surgery)

Urethra

Epididymis

Testis

Scrotum

(a)

Uterine tube
(after surgery)

Uterine tube
(before surgery)

Ovary

Uterus

Vagina

(b)

● **FIGURE 19–18 Surgical Sterilization.** (a) In a vasectomy, the removal of a 1-cm section of the ductus deferens prevents the passage of sperm cells. (b) In a tubal ligation, the removal of a section of the uterine tube prevents the passage of sperm and the movement of an ovum or embryo into the uterus.

women reconsider their use. But complex decisions should not be made on such a simplistic basis, and the risks associated with contraceptive use must be considered in light of their relative efficiencies.

Although pregnancy is a natural phenomenon, it has risks, and the mortality rate for pregnant women in the U.S. averages about 8 deaths per 100,000 pregnancies. That average incorporates a broad range; the rate is 7 per 100,000 among women under 20 but 40 per 100,000 for women over 40. Although these risks are small, for pregnant women over age 35, the chances of dying from pregnancy-related complications are almost twice as great as the chances of being killed in an automobile accident, and they are many times greater than the risks associated with the use of oral contraceptives. For women in developing nations, the comparison is even more striking: the mortality rate for pregnant women in parts of Africa is approximately 1 per 150 pregnancies. In addition to preventing pregnancy, combination birth control pills have also been shown to reduce the risks of ovarian and endometrial cancers and of fibrocystic breast disease.

Before age 35, the risks associated with contraceptive use are lower than the risks associated with pregnancy. The notable exception involves individuals who take oral contraceptives but also smoke cigarettes. Younger women are more fertile, so despite a lower mortality rate for each pregnancy, they are likely to have more pregnancies. As a result, birth control failures imply a higher risk in the younger age groups.

After age 35, the risks of complications associated with oral contraceptive use increase, but the risks of using other methods remains relatively stable. Women over age 35 (smokers) or 40 (nonsmokers) are, therefore, typically advised to seek other forms of contraception. Because each contraceptive method has its own advantages and disadvantages, research on contraception control continues. ■

Chapter Review

Access more review material online at *www.prenhall.com/bledsoe*. There you will find quiz questions, labeling activities, animations, essay questions, and web links.

Key Terms

ductus deferens 703	**ovarian follicles** 710	**spermatogenesis** 700
endometrium 713	**ovary** 709	**spermatozoa** 698
estrogens 717	**ovulation** 711	**testes** 698
lactation 716	**perineum** 715	**testosterone** 699
meiosis 700	**prepuce** 705	**vulva** 716
menses 713	**progesterone** 719	
oogenesis 709	**seminiferous tubules** 699	

Related Clinical Terms

amenorrhea The failure of menarche to appear before age 16, or a cessation of menstruation for 6 months or more in an adult female of reproductive age.

cervical cancer A malignant, metastazing tumor of the cervix; the most common reproductive cancer in women.

cryptorchidism (krip-TOR-ki-dizm) The failure of one or both testes to descend into the scrotum by the time of birth.

dysmenorrhea Painful menstruation.

endometriosis (en-dō-mē-trē-Ō-sis) The growth of endometrial tissue outside the uterus.

gynecology (gī-ne-KOL-o-jē) The study of the female reproductive tract and its disorders.

mammography The use of X-rays to examine breast tissue.

mastectomy The surgical removal of part or all of a cancerous mammary gland.

oophoritis (ō-of-ō-RĪ-tis) Inflammation of the ovaries.

orchiectomy (or-kē-EK-to-mē) The surgical removal of a testis.

orchitis (or-KĪ-tis) Inflammation of the testis.

ovarian cancer A malignancy of the ovaries; the most dangerous reproductive cancer in women.

pelvic inflammatory disease (PID) An infection of the uterine tubes.

prostate cancer A malignant, metastazing tumor of the prostate gland; the second most common cause of cancer deaths in males.

prostatectomy (pros-ta-TEK-to-mē) The surgical removal of the prostate gland.

testicular torsion A condition in which the twisting of the spermatic cord obstructs the blood supply to a testis.

vaginitis (va-jin-ī-tis) Inflammation of the vaginal canal.

vasectomy (vaz-EK-to-mē) The surgical removal of a segment of the ductus deferens, which makes it impossible for spermatozoa to reach the distal portions of the male reproductive tract.

Summary Outline

1. The human reproductive system, including the **gonads** (reproductive organs), produces, stores, nourishes, and transports functional **gametes** (reproductive cells). **Fertilization** is the fusion of male and female gametes to create a **zygote** (fertilized egg).

THE REPRODUCTIVE SYSTEM OF THE MALE 697

1. In males, the *testes* produce **sperm** cells. The **spermatozoa** travel along the *epididymis*, the *ductus deferens*, the *ejaculatory duct*, and the *urethra* before leaving the body. Accessory organs (notably, the *seminal vesicles, prostate gland*, and *bulbourethral glands*) secrete products into the ejaculatory ducts and urethra. The scrotum encloses the testes, and the penis is an erectile organ. *(Figure 19–1)*

The Testes 698

2. The **testes,** which are the primary sex organ of males, hang within the **scrotum.** The *dartos* muscle layer gives the scrotum a wrinkled appearance. The **cremaster muscle** pulls the testes closer to the body. The **tunica albuginea** surrounds each testis. Septa subdivide each testis into a series of lobules. **Seminiferous tubules** within each lobule are the sites of sperm production. Between the seminiferous tubules are **interstitial cells,** which secrete sex hormones. *(Figure 19–2)*

3. Seminiferous tubules contain **spermatogonia,** which are stem cells involved in **spermatogenesis,** and **sustentacular cells,** which sustain and promote the development of spermatozoa. *(Figure 19–3)*

Key Note 702

4. Each spermatozoon has a **head, middle piece,** and **tail.** (*Figure 19–4*)

Key Note 702

The Male Reproductive Tract 702

5. From the testis, the spermatozoa enter the **epididymis,** which is an elongate tubule that regulates the composition of the tubular fluid and serves as a recycling center for damaged spermatozoa. Spermatozoa that leave the epididymis are functionally mature, yet immobile.

6. The **ductus deferens,** or *vas deferens,* begins at the epididymis and passes through the inguinal canal within the **spermatic cord.** The junction of the base of the seminal vesicle and the ductus deferens creates the **ejaculatory duct,** which penetrates the prostate gland and empties into the urethra. (*Figure 19–5*)

7. The **urethra** extends from the urinary bladder to the tip of the penis and serves as a passageway that functions in both the urinary and reproductive systems.

The Accessory Glands 704

8. Each **seminal vesicle** is an active secretory gland that contributes about 60 percent of the volume of semen; its secretions contain fructose, which is easily metabolized by spermatozoa. The **prostate gland** secretes fluid that makes up about 30 percent of seminal fluid. Alkaline mucus secreted by the **bulbourethral glands** has lubricating properties. (*Figure 19–5*)

Semen 704

9. A typical **ejaculation** expels 2–5 mL of semen (an **ejaculate**), which contains 20–100 million sperm per milliliter.

The Penis 704

10. The skin that overlies the **penis** resembles that of the scrotum. Most of the body of the penis consists of three masses of **erectile tissue.** Beneath the superficial layers are two **corpora cavernosa** and a single **corpus spongiosum,** which surrounds the urethra. Dilation of the erectile tissue with blood produces an **erection.** (*Figure 19–6*)

Hormones and Male Reproductive Function 706

11. Important regulatory hormones of males include **follicle stimulating hormone (FSH), luteinizing hormone (LH),** and **gonadotropin-releasing hormone (GnRH).** Testosterone is the most important *androgen.* (*Figure 19–7*)

THE REPRODUCTIVE SYSTEM OF THE FEMALE 707

1. Principal organs of the female reproductive system include the *ovaries, uterine tubes, uterus, vagina,* and *external genitalia.* (*Figure 19–8*)

The Ovaries 709

2. The **ovaries** are the primary sex organs of females. Ovaries are the site of **ovum** production, or **oogenesis,** which occurs monthly in **ovarian follicles** as part of the **ovarian cycle.** As development proceeds, **primordial, primary, secondary,** and **tertiary follicles** form. At **ovulation,** an oocyte and the surrounding follicular walls of the **corona radiata** are released through the ruptured ovarian wall. (*Figures 19–9, 19–10*)

Key Note 712

The Uterine Tubes 712

3. Each **uterine tube** has an **infundibulum,** which is a funnel that opens into the pelvic cavity. For fertilization to occur, a secondary oocyte must encounter spermatozoa during the first 12–24 hours of its passage from the infundibulum to the uterine cavity. (*Figure 19–11*)

The Uterus 712

4. The **uterus** provides mechanical protection and nutritional support to the developing embryo. It is stabilized by various ligaments. Major anatomical landmarks of the uterus include the **body, cervix, external orifice,** and **uterine cavity.** The uterine wall consists of an inner **endometrium,** a muscular **myometrium,** and a superficial **perimetrium.** (*Figures 19–11, 19–12*)

5. A typical 28-day **uterine cycle,** or *menstrual cycle,* begins with the onset of **menses** and the destruction of the *functional zone* of the endometrium. This process of menstruation continues from 1 to 7 days.

6. After menses, the **proliferative phase** begins, and the functional zone undergoes repair and thickens. During the **secretory phase,** the endometrial glands are active and the uterus is prepared for the arrival of an embryo. Menstrual activity begins at **menarche** and continues until **menopause.**

The Vagina 714

7. The **vagina** is a muscular tube that extends between the uterus and the external genitalia. A thin epithelial fold, the **hymen,** partially blocks the entrance to the vagina.

The External Genitalia 715

8. The components of the **vulva,** or perineal region that encloses the female external genitalia, include the **vestibule,** the **labia minora,** the **clitoris,** the **labia majora,** the **mons pubis,** and the **lesser** and **greater vestibular glands.** (*Figure 19–13*)

The Mammary Glands 716

9. A newborn infant gains nourishment from milk secreted by maternal **mammary glands. Lactation** is the process of milk production. (*Figure 19–14*)

Hormones and the Female Reproductive Cycle 717

10. Regulation of the **female reproductive cycle** involves the coordination of the ovarian and uterine cycles by circulating hormones. (*Figure 19–15*)

11. **Estradiol,** one of the estrogens, is the dominant hormone of the **follicular phase** of the ovarian cycle. Ovulation occurs in response to a midcycle surge in LH secretion. *(Figure 19–15)*

12. The hypothalamic secretion of GnRH triggers the pituitary secretion of FSH and the synthesis of LH. FSH stimulates secondary follicle development, and activated follicles and ovarian interstitial cells produce estrogens. **Progesterone,** which is one of the steroid hormones called **progestins,** is the principal hormone of the **luteal phase** of the ovarian cycle. Hormonal changes regulate the uterine, or menstrual, cycle. *(Figures 19–15, 19–16)*

Key Note 720

THE PHYSIOLOGY OF SEXUAL INTERCOURSE 720

Male Sexual Function 720

1. During **arousal** in males, erotic thoughts, sensory stimulation, or both lead to parasympathetic activity that ultimately produces erection. Stimuli that accompany **coitus** (sexual intercourse) lead to **emission** and ejaculation. Strong muscle contractions are associated with **orgasm.**

Female Sexual Function 721

2. The phases of female sexual function resemble those of the male, with parasympathetic arousal and muscular contractions associated with orgasm.

AGING AND THE REPRODUCTIVE SYSTEM 721

Menopause 722

1. Menopause (the time that ovulation and menstruation cease in women) typically occurs around age 50. Production of GnRH, FSH, and LH rise while circulating concentrations of estrogen and progesterone decline.

The Male Climacteric 722

2. During the **male climacteric,** at ages 50–60, circulating testosterone levels decline while levels of FSH and LH rise.

Key Note 722

INTEGRATION WITH OTHER SYSTEMS 722

1. Hormones play a major role in coordinating reproduction. *(Table 19–1)*

2. In addition to the endocrine and reproductive systems, reproduction requires the normal functioning of the digestive, nervous, cardiovascular, and urinary systems. *(Figures 19–17, 19–18)*

Review Questions

Level 1: Reviewing Facts and Terms

Match each item in column A with the most closely related item in column B. Place letters for answers in the spaces provided.

COLUMN A
- ____ 1. gametes
- ____ 2. gonads
- ____ 3. interstitial cells
- ____ 4. seminal vesicles
- ____ 5. prostate gland
- ____ 6. bulbourethral glands
- ____ 7. prepuce
- ____ 8. corpus luteum
- ____ 9. endometrium
- ____ 10. myometrium
- ____ 11. dysmenorrhea
- ____ 12. menarche
- ____ 13. clitoris
- ____ 14. lactation
- ____ 15. coitus

COLUMN B
- a. production of androgens
- b. outer muscular uterine wall
- c. high concentration of fructose
- d. female erectile tissue
- e. secretes thick, sticky, alkaline mucus
- f. painful menstruation
- g. sexual intercourse
- h. uterine lining
- i. reproductive cells
- j. female puberty
- k. milk production
- l. secretes antibiotic
- m. reproductive organs
- n. foreskin of penis
- o. endocrine structure

16. Perineal structures associated with the reproductive system are collectively known as:
    (a) gonads.
    (b) sex gametes.
    (c) external genitalia.
    (d) accessory glands.

17. Meiosis in males produces four spermatids, each of which contains:
    (a) 23 chromosomes.
    (b) 23 pairs of chromosomes.
    (c) the diploid number of chromosomes.
    (d) 46 pairs of chromosomes.

18. Erection of the penis occurs when:
    (a) sympathetic activation of penile arteries occurs.
    (b) arterial branches are constricted and muscular partitions are tense.
    (c) the vascular channels become engorged with blood.
    (d) a, b, and c are correct.

19. In males, the primary target of FSH is the:
    (a) sustentacular cells of the seminiferous tubules.
    (b) interstitial cells of the seminiferous tubules.
    (c) prostate gland.
    (d) epididymis.

20. The ovaries are responsible for:
    (a) the production of female gametes.
    (b) the secretion of female sex hormones.
    (c) the secretion of inhibin.
    (d) a, b, and c are correct.

21. Ovum production, or oogenesis, begins:
    (a) before birth.
    (b) after birth.
    (c) at puberty.
    (d) after puberty.

22. In females, meiosis is not completed:
    (a) until birth.
    (b) until puberty.
    (c) unless and until fertilization occurs.
    (d) until ovulation occurs.

23. If fertilization is to occur, an ovum must encounter spermatozoa during the first _____ of its passage.
    (a) 1 to 5 hours
    (b) 6 to 11 hours
    (c) 12 to 24 hours
    (d) 25 to 36 hours

24. The part of the endometrium that undergoes cyclical changes in response to variations in sexual hormonal levels is the:
    (a) serosa.
    (b) basilar zone.
    (c) muscular myometrium.
    (d) functional zone.

25. A sudden surge in LH concentration causes:
    (a) the onset of menses.
    (b) the rupture of the follicular wall and ovulation.
    (c) the beginning of the proliferative phase.
    (d) the end of the uterine cycle.

26. At the time of ovulation, the basal body temperature:
    (a) is not affected.
    (b) increases noticeably.
    (c) declines sharply.
    (d) may increase or decrease a few degrees.

27. Menopause is accompanied by:
    (a) sustained rises in GnRH, FSH, and LH.
    (b) declines in circulating levels of estrogen and progesterone.
    (c) thinning of the urethral and vaginal walls.
    (d) a, b, and c are correct.

28. Which reproductive structures are common to both males and females?

29. Which accessory organs and glands contribute to the composition of semen? What are the functions of each?

30. What are the primary functions of the epididymis?

31. What are the primary functions of the ovaries?

32. What are the three major functions of the vagina?

Level 2: Reviewing Concepts

33. How does the human reproductive system differ functionally from all other systems in the body?

34. How is meiosis involved in the development of spermatozoa and ova?

35. Using an average uterine cycle of 28 days, describe each of the three phases of the menstrual cycle.

36. Describe the hormonal events associated with the uterine cycle.

37. How does aging affect the reproductive systems of men and women?

Level 3: Critical Thinking and Clinical Applications

38. Diane has peritonitis (inflammation of the peritoneum), which her physician says resulted from a urinary tract infection. Why could this situation occur in females but not in males?

39. Rod injures the sacral region of his spinal cord. Will he still be able to achieve an erection? Explain.

40. Female bodybuilders and women who suffer from eating disorders such as anorexia nervosa often cease having menstrual cycles, a condition known as amenorrhea. What does this relationship suggest about the role of body fat in menstruation? How might the body benefit from the discontinuance of menstruation under such circumstances?

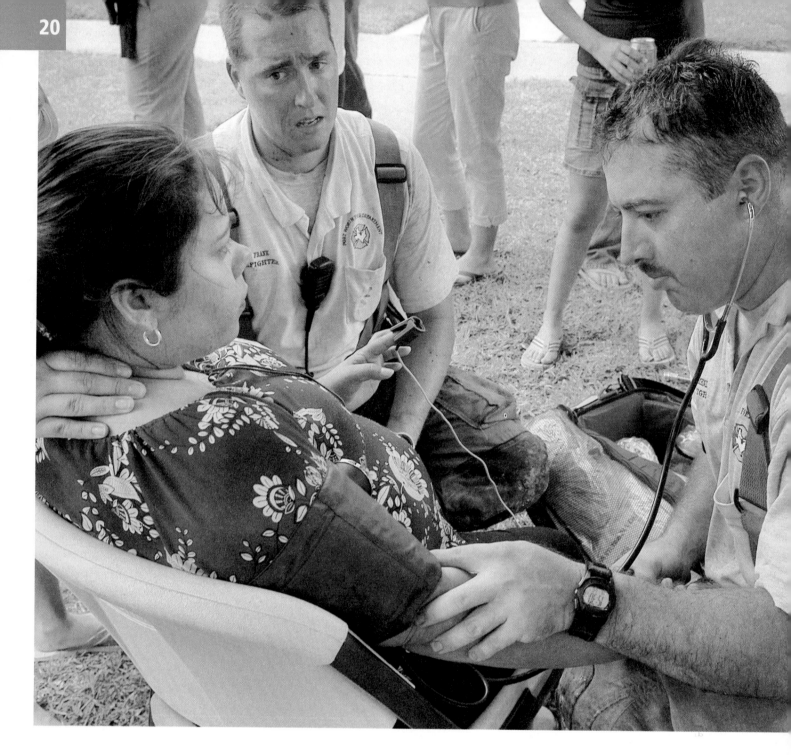

20 Development and Inheritance

EMS IS ALL ABOUT taking care of people. The purpose of EMS is to aid the ill and the infirm. Sometimes, the best treatment is simply listening. One of the hardest decisions you must make as an EMT or paramedic is not determining when to do a skill or procedure. Instead, it is determining when not to do a skill or procedure. The fundamental tenet of medicine is "First, do no harm." This admonition should always be in your mind as a prehospital provider.

Chapter Outline

Chapter Objectives

1. Describe the process of fertilization. (pp. 734–735)
2. List the three prenatal periods and describe the major events associated with each. (pp. 735–751)
3. Describe the origins of the three primary germ layers and their participation in the formation of extraembryonic membranes. (pp. 738–741)
4. Describe the interplay between maternal organ systems and the developing embryo. (pp. 746–750)
5. List and discuss the events that occur during labor and delivery. (pp. 751–752)
6. Contrast the development of fraternal and identical twins. (pp. 752–753)
7. Discuss the major stages of life after delivery. (pp. 753–756)
8. Relate the basic principles of genetics to the inheritance of human traits. (pp. 757–763)

Vocabulary Development

allanto- sausage; *allantois*
amphi two-sided; *amphimixis*
blast precursor; *trophoblast*
gestare to bear; *gestation*
heteros other; *heterozygous*
homos same; *homozygous*

karyon nucleus; *karyotyping*
koiloma hollow; *blastocoele*
meros part; *blastomere*
meso- middle; *mesoderm*
mixis mixing; *amphimixis*
morula mulberry; *morula*

phainein to display; *phenotype*
praegnans with child; *pregnant*
tropho- food; *trophoblast*
typos mark; *phenotype*
vitro glass; *in vitro*

TIME REFUSES TO STAND STILL; today's infant will be tomorrow's adult. The gradual modification of anatomical structures and physiological characteristics during the period from conception to maturity is called **development.** The changes are truly remarkable—what begins as a single cell slightly larger than the period at the end of this sentence becomes an individual whose body contains trillions of cells organized into tissues, organs, and organ systems. The creation of different cell types in this process is called **differentiation.** Differentiation occurs through selective changes in genetic activity. As development proceeds, some genes are turned off and others are turned on. The kinds of genes turned off or on vary from one cell type to another.

A basic understanding of human development provides important insights into anatomical structures. In addition, many of the mechanisms of development and growth are similar to those responsible for the repair of injuries. This chapter will focus on major aspects of development and will present highlights of the developmental process. We will also consider the regulatory mechanisms involved and how developmental patterns can be modified—for good or ill. Few topics in the biological sciences are so fascinating, and fewer still confront investigators with so daunting an array of scientific, technological, and ethical challenges.

■ An Overview of Topics in Development

Development involves (1) the division and differentiation of cells and (2) the changes that produce and modify anatomical structures. Development begins at **fertilization,** or **conception.** We can divide development into several periods. **Embryological development** involves events that occur in the first 2 months after fertilization. The study of these events in the developing organism, or **embryo,** is called **embryology** (em-brē-OL-ō-jē).

After 2 months, the developing embryo becomes a **fetus. Fetal development** begins at the start of the ninth week and continues until birth. Together, embryological development and fetal development are referred to as **prenatal development,** which is the primary focus of this chapter. **Postnatal development** begins at birth and continues to maturity, when the aging process begins.

Although all human beings go through the same developmental stages, differences in genetic makeup produce distinctive individual characteristics. **Inheritance** refers to the transfer of genetically determined characteristics from generation to generation. **Genetics** is the study of the mechanisms responsible for inheritance. This chapter considers basic genetics as it applies to inherited characteristics such as sex, hair color, and various diseases.

■ Fertilization

Fertilization involves the fusion of two haploid gametes, which produces a *zygote* that contains 46 chromosomes, the normal number for a *somatic* (nonreproductive) cell. The roles and contributions of the male and female gametes are very different. Whereas the spermatozoon simply delivers the paternal chromosomes to the site of fertilization, the female gamete—a secondary oocyte ∞ p. 709—must provide all of the nourishment and genetic programming to support development of the embryo for nearly a week after conception. At fertilization, the diameter of the secondary oocyte is more than twice the length of the spermatozoon (Figure 20–1a●). The ratio of their volumes is even more striking—roughly 2000 to 1.

The sperm deposited in the vagina are already motile, as a result of mixing with secretions of the seminal vesicles—the first step of *capacitation.* ∞ p. 703 But they cannot accomplish fertilization until they have been exposed to conditions in the female reproductive tract, specifically, a uterine tube. Capacitation is completed when sperm are exposed to a fluid secreted by *Peg cells,* which are nonciliated cells within the epithelial lining of a uterine tube.

Fertilization typically occurs in the upper one-third of the uterine tube within a day after ovulation. It takes sperm between 30 minutes and 2 hours to pass from the vagina to the upper portion of a uterine tube. Of the 200 million spermatozoa introduced into the vagina in a typical ejaculation, only about 10,000 enter the uterine tube, and fewer than 100 actually reach the secondary oocyte. In general, a male with a sperm count below 20 million per milliliter is sterile because too few sperm survive to reach the oocyte. Dozens of sperm cells are required for successful fertilization, because, as we will see, a single sperm cannot penetrate the *corona radiata,* which is the layer of follicle cells that surrounds the oocyte. ∞ p. 710

Activation of the Oocyte

Ovulation occurs before the oocyte is completely mature. The secondary oocyte that leaves the follicle is in metaphase of the second meiotic division (meiosis II). The cell's metabolic operations have been discontinued, and the oocyte drifts in a sort of suspended animation, and awaits the stimulus for further development. If fertilization does not occur, the oocyte disintegrates without completing meiosis.

Fertilization and the events that follow are diagrammed in Figure 20–1b●. The corona radiata protects the oocyte as it passes through the ruptured follicular wall and into the infundibulum of the uterine tube. Although the physical process of fertilization requires that only a single sperm contact the oocyte membrane, that spermatozoon must first penetrate the corona radiata (the layer of follicular cells that surround the oocyte). The corona radiata protects the oocyte as it passes through the ruptured follicular wall and into the infundibulum of the uterine tube. The acrosomal cap of the sperm contains several enzymes, including *hyaluronidase* (hi-a-loo-RON-a-dās), which breaks down the bonds between adjacent follicle cells. Dozens of sperm cells must release hyaluronidase before an opening forms between the follicular cells.

No matter how many sperm slip through the resulting gap in the corona radiata, only a single spermatozoon will accomplish fertilization and activate the oocyte (Step 1). That sperm cell first binds to sperm receptors in the zona pellucida. ∞ p. 711 Next, its acrosomal cap ruptures, and releases hyaluronidase and another proteolytic enzyme, which digests a path through the zona pellucida to the oocyte membrane. Upon contact, the membranes of the sperm cell and oocyte begin to fuse, triggering **oocyte activation** (see Figure 20–1b). As the membranes fuse, the entire sperm enters the cytoplasm of the oocyte.

Activation of the oocyte involves a series of sudden changes in its metabolism. For example, vesicles located just interior to the oocyte membrane undergo exocytosis, and release enzymes that prevent an abnormal process called *polyspermy* (fertilization by more than one sperm). Polyspermy produces a zygote that is incapable of normal development. Other important changes include the completion of meiosis II and a rapid increase in the oocyte's metabolic rate.

After oocyte activation has occurred and meiosis has been completed, the nuclear material that remains within the ovum reorganizes into a *female pronucleus* (Step 2, Figure 20–1b). At the same time, the nucleus of the spermatozoon swells, and becomes the *male pronucleus* (Step 3). The male pronucleus and female pronucleus then fuse in a process called **amphimixis** (am-fi-MIK-sis; see Step 4). Fer-

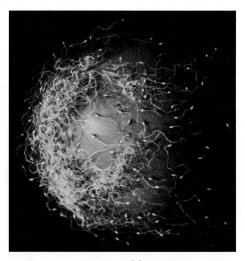

(a)

tilization is now complete, with the formation of a zygote that has the normal complement of 46 chromosomes. The first cleavage division yields two daughter cells (Step 5).

An Overview of Prenatal Development

During prenatal development, a single cell ultimately forms a 3–4 kg (6.6–8.8 lb) infant. The time spent in prenatal development is known as the period of **gestation** (jes-TĀ-shun), or

● **FIGURE 20–1 Fertilization.** (**a**) An oocyte and numerous sperm cells are shown at the time of fertilization. Notice the difference in size between the gametes. (**b**) Fertilization and the preparations for cleavage occur in the sequence of events diagrammed here.

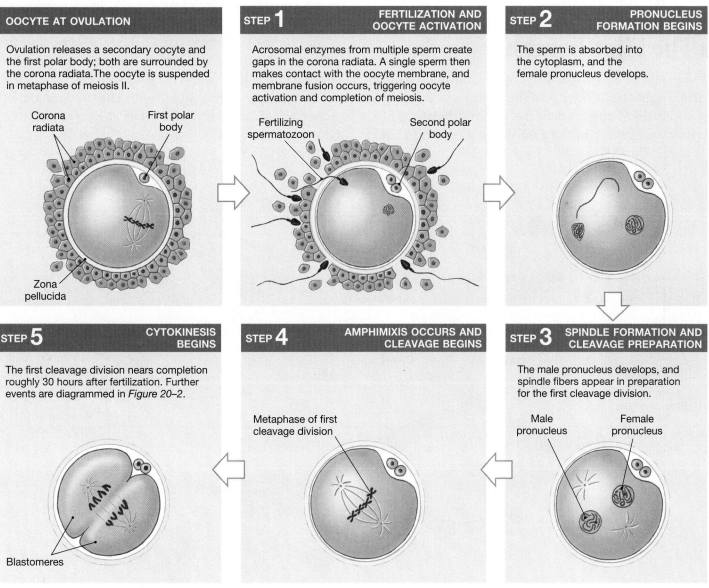

OOCYTE AT OVULATION

Ovulation releases a secondary oocyte and the first polar body; both are surrounded by the corona radiata. The oocyte is suspended in metaphase of meiosis II.

Corona radiata
First polar body
Zona pellucida

STEP 1 FERTILIZATION AND OOCYTE ACTIVATION

Acrosomal enzymes from multiple sperm create gaps in the corona radiata. A single sperm then makes contact with the oocyte membrane, and membrane fusion occurs, triggering oocyte activation and completion of meiosis.

Fertilizing spermatozoon
Second polar body

STEP 2 PRONUCLEUS FORMATION BEGINS

The sperm is absorbed into the cytoplasm, and the female pronucleus develops.

STEP 5 CYTOKINESIS BEGINS

The first cleavage division nears completion roughly 30 hours after fertilization. Further events are diagrammed in *Figure 20–2*.

Blastomeres

STEP 4 AMPHIMIXIS OCCURS AND CLEAVAGE BEGINS

Metaphase of first cleavage division

STEP 3 SPINDLE FORMATION AND CLEAVAGE PREPARATION

The male pronucleus develops, and spindle fibers appear in preparation for the first cleavage division.

Male pronucleus
Female pronucleus

(b)

pregnancy. Gestation occurs within the uterus over a period of 9 months. For convenience, prenatal development is usually considered to consist of three **trimesters,** each 3 months in duration:

1. The **first trimester** is the period of embryological and early fetal development. During this period, the basic components of each of the major organ systems appear.
2. The **second trimester** is dominated by the development of organs and organ systems to near completion. The fetus's body proportions change, and by the end of the second trimester it looks distinctively human.
3. The **third trimester** is characterized by rapid fetal growth. Early in the third trimester most of the major organ systems become fully functional. An infant born 1 month or even 2 months prematurely has a reasonable chance of survival.

■ The First Trimester

At the moment of conception, the fertilized ovum is a single cell that is about 0.135 mm (0.005 in.) in diameter and weighs approximately 150 mg. By the end of the first trimester, the fetus is almost 75 mm (3 in.) long and weighs about 14 g (0.5 oz).

Many complex and vital developmental events accompany this increase in size and weight. Perhaps because the events are so complex, *the first trimester is the most dangerous period in prenatal life*—only about 40 percent of conceptions produce embryos that survive this period. For that reason, pregnant women are advised to take great care to avoid drugs and other disruptive stresses during the first trimester.

In the sections that follow we will focus on four general processes that occur during the first trimester: *cleavage and blastocyst formation, implantation, placentation,* and *embryogenesis.*

Cleavage and Blastocyst Formation

Cleavage (KLĒV-ij) is a series of cell divisions that begins immediately after fertilization. This process produces an ever-increasing number of smaller and smaller, genetically identical daughter cells called **blastomeres** (BLAS-tō-mērz; *blast,* precursor + *meros,* part) (Figure 20–2●). The first cleavage division produces two daughter cells, each of which is one-half the size of the original zygote. The first division is completed roughly 30 hours after fertilization, and subsequent cleavage divisions occur at intervals of 10–12 hours.

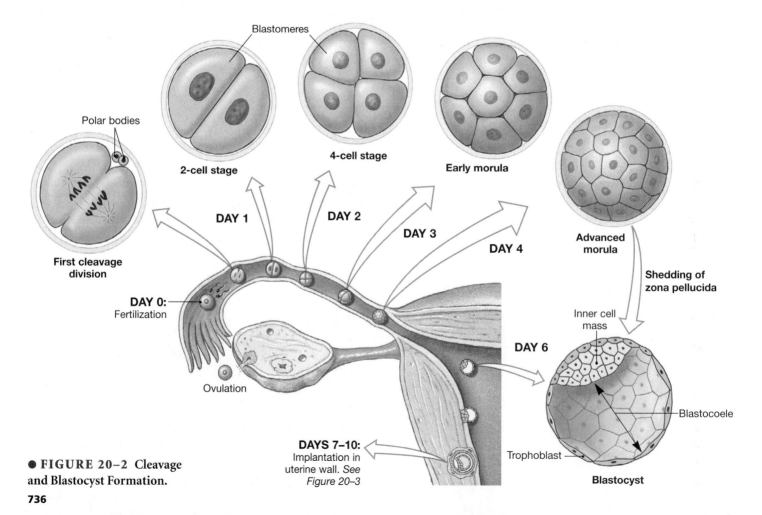

● **FIGURE 20–2** Cleavage and Blastocyst Formation.

After 3 days of cleavage, the embryo—now well on its way along the uterine tube—is a solid ball of cells that resemble a mulberry (see Figure 20–2). This stage is called the **morula** (MOR-ū-la; *morula,* mulberry). Over the next 2 days, as the embryo enters the uterine cavity, the blastomeres form a **blastocyst,** which is a hollow ball with an inner cavity known as the *blastocoele* (BLAS-tō-sēl; *koiloma,* cavity). At this stage, differences among the blastomeres of the blastocyst become apparent. The outer layer of cells that separate the outside world from the blastocoele is called the **trophoblast** (TRŌ-fō-blast). As implied by its name—*tropho,* food + *blast,* precursor—this layer of cells provides food to the developing embryo. A second group of cells, which is the **inner cell mass,** lies clustered in one portion of the blastocyst. In time, the inner cell mass will form the embryo.

Clinical Note
ECTOPIC PREGNANCY

The fertilized egg normally is implanted in the endometrial lining of the uterine wall. The term *ectopic pregnancy* refers to the abnormal implantation of the fertilized egg outside of the uterus. Approximately 95 percent of ectopic pregnancies are implanted in the fallopian tube. Occasionally (less than 1 percent), the egg is implanted in the abdominal cavity. Current research indicates that the incidence of ectopic pregnancy is one for every 44 live births. Improved diagnostic technology is credited with an increased incidence, as most are detected between the 2nd and 12th weeks. Ectopic pregnancy accounts for approximately 10 percent of maternal mortality.

Predisposing factors in the development of ectopic pregnancy include scarring of the fallopian tubes due to pelvic inflammatory disease (PID), a previous ectopic pregnancy, or previous pelvic or tubal surgery, such as a tubal ligation. Other factors include endometriosis or use of an intrauterine device (IUD) for birth control.

Ectopic pregnancy most often presents as abdominal pain, which starts as diffuse tenderness and then localizes as a sharp pain in the lower abdominal quadrant on the affected side. This pain is due to rupture of the fallopian tube when the fetus outgrows the available space. The woman often reports that she missed a period or that her LMP occurred 4 to 6 weeks ago, but with decreased menstrual flow that was brownish in color and of shorter duration than usual. As the intra-abdominal bleeding continues, the abdomen becomes rigid and the pain intensifies and is often referred to the shoulder on the affected side. The pain is often accompanied by syncope, vaginal bleeding, and shock.

Assume that any female of childbearing age with lower abdominal pain is experiencing an ectopic pregnancy. Ectopic pregnancy poses a significant life threat to the mother. Surgery is required to resolve the situation. ■

During this time, the zona pellucida is shed. The blastocyst is now freely exposed to the fluid contents of the uterine cavity. This glycogen-rich fluid, which is secreted by the endometrial glands of the uterus, provides nutrients to the blastocyst. When fully formed, the blastocyst contacts the endometrium; at that time, cleavage ends and implantation begins.

Implantation

Implantation (Figure 20–3●) begins as the surface of the blastocyst closest to the inner cell mass touches and adheres to the uterine lining (see day 7, Figure 20–3). At the area of contact, the superficial cells undergo rapid divisions, which make the trophoblast several layers thick. The cells closest to the interior of the blastocyst remain intact and form a layer called the *cellular trophoblast* (see day 8, Figure 20–3). Near the uterine wall, the cell membranes that separate the trophoblast cells disappear, which creates a layer of cytoplasm that contains multiple nuclei. This outer layer, called the *syncytial* (sin-SISH-al) *trophoblast,* erodes a path through the uterine epithelium. At first, this erosion creates a gap in the uterine lining, but the division and migration of epithelial cells soon repair the surface. By day 10, the repairs are complete, and the blastocyst no longer contacts the uterine cavity. Further development occurs entirely within the functional zone of the endometrium.

In most cases, implantation occurs in the fundus or elsewhere in the body of the uterus. In an **ectopic pregnancy,** implantation occurs somewhere other than within the uterus, such as in one of the uterine tubes. Approximately 0.6 percent of pregnancies are ectopic pregnancies, which do not produce a viable embryo and can be life threatening.

As implantation proceeds, the syncytial trophoblast continues to enlarge into the surrounding endometrium (see day 9, Figure 20–3). The erosion of uterine gland cells releases nutrients that are absorbed by the trophoblast and spread by diffusion to the inner cell mass. These nutrients provide the energy needed to support the early stages of embryo formation. Extensions of the trophoblast grow around endometrial capillaries. As the capillary walls are destroyed, maternal blood begins to flow through trophoblastic channels known as *lacunae* (singular *lacuna*). Finger-like *villi* extend away from the trophoblast into the surrounding endometrium; these extensions gradually increase in size and complexity as development proceeds.

Formation of the Amniotic Cavity

By the time of implantation, the inner cell mass has separated from the trophoblast. The separation gradually enlarges, which creates a fluid-filled chamber called the **amniotic** (am-nē-OT-ik) **cavity** (see day 9, Figure 20–3). (The amniotic cavity during

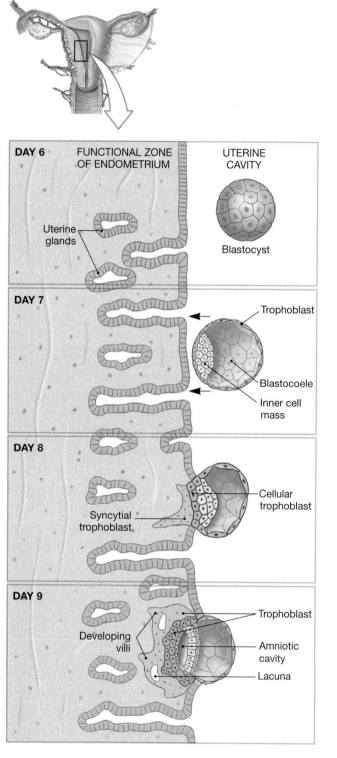

● **FIGURE 20–3** Events in Implantation.

days 10–12 is shown in Figure 20–4●.) When the amniotic cavity first appears, the cells of the inner cell mass are organized into an oval sheet that is two cell layers thick—a superficial layer that faces the amniotic cavity and a deeper layer that is exposed to the fluid contents of the blastocoele.

Gastrulation and Germ Layer Formation

By day 12, a third layer of cells begins forming through the process of **gastrulation** (gas-troo-LĀ-shun) (see day 12, Figure 20–4). During gastrulation, some of the cells of the inner cell mass exposed to the amniotic cavity move toward a center line, called the *primitive streak,* and migrate down and in between the two existing layers. This movement creates three distinct embryonic layers of cells, or **primary germ layers**—each with markedly different fates. The superficial layer in contact with the amniotic cavity is called the **ectoderm,** the layer facing the blastocoele is known as the **endoderm,** and the poorly organized layer of migrating cells is the **mesoderm** (*meso-,* middle). Table 20–1 lists the contributions each germ layer makes to the body systems described in previous chapters. Thus, gastrulation produces an oval sheet that contains the three primary germ layers, which is a structure called the *embryonic disc.* The embryonic disc will form the body of the embryo, whereas the rest of the blastocyst will be involved in forming the extraembryonic membranes.

The Formation of Extraembryonic Membranes

The germ layers also participate in the formation of four membranes that support embryonic and fetal development. Figure 20–5● illustrates the stages in the development of these **extraembryonic membranes**—the *yolk sac,* the *amnion,* the *allantois,* and the *chorion.* Few traces of their existence remain in adult systems.

YOLK SAC. The first extraembryonic membrane to appear is the **yolk sac.** The yolk sac, which is already present 10 days after fertilization, forms a pouch within the blastocoele (see Figure 20–4). As gastrulation proceeds, mesodermal cells migrate around this pouch and complete the formation of the yolk sac (see week 2, Figure 20–5). Blood vessels soon appear within the mesoderm, and the yolk sac becomes an important site of blood cell formation.

AMNION. The **amnion** (AM-nē-on) is composed of both ectoderm and mesoderm. Ectodermal cells first spread over the inner surface of the amniotic cavity and, soon after, mesodermal cells follow and create a second, outer layer. As the embryo and later the fetus enlarges, the amnion continues to expand, which increases the size of the amniotic cavity (see week 3, Figure 20–5). The amnion encloses *amniotic fluid* that surrounds and cushions the developing embryo and fetus (see week 10, Figure 20–5).

THE ALLANTOIS. The **allantois** (a-LAN-tō-is) is a sac of endoderm and mesoderm that extends away from the embryo (see week 3, Figure 20–5). The base of the allantois later gives

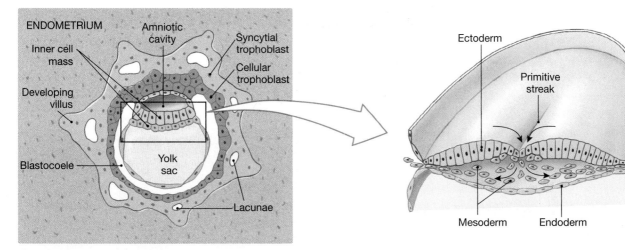

Day 10: The inner cell mass begins as two layers: a superficial layer, facing the amniotic cavity, and a deep layer, exposed to the blastocoele. Migration of cells around the amniotic cavity is the first step in the formation of the amnion. Migration of cells around the edges of the blastocoele is the first step in yolk sac formation.

Day 12: Migration of superficial cells creates a third layer. From the time this process (gastrulation) begins, the surface facing the amniotic cavity is called *ectoderm*, the layer facing the yolk sac is called *endoderm*, and the migrating cells form the *mesoderm*.

● **FIGURE 20–4** The Inner Cell Mass and Gastrulation.

TABLE 20–1 *The Fates of the Primary Germ Layers*

GERM LAYER	DEVELOPMENTAL CONTRIBUTIONS TO THE BODY
Ectoderm	*Integumentary system:* epidermis, hair follicles and hairs, nails, and glands that communicate with the skin (sweat glands, mammary glands, and sebaceous glands) *Skeletal system:* pharyngeal cartilages of the embryo develop into portions of sphenoid and hyoid bones, auditory ossicles, and styloid processes of temporal bones *Nervous system:* all neural tissue, including brain and spinal cord *Endocrine system:* pituitary gland and the adrenal medullae *Respiratory system:* mucous epithelium of nasal passageways *Digestive system:* mucous epithelium of mouth and anus, salivary glands
Mesoderm	*Integumentary system:* dermis (and hypodermis) *Skeletal system:* all components except some pharyngeal cartilage derivatives *Muscular system:* all components *Endocrine system:* adrenal cortex, endocrine tissues of heart, kidneys, and gonads *Cardiovascular system:* all components *Lymphatic system:* all components *Urinary system:* the kidneys, including the nephrons and the initial portions of the collecting system *Reproductive system:* the gonads and the adjacent portions of the duct systems *Miscellaneous:* the lining of the body cavities (pleural, pericardial, and peritoneal) and the connective tissues that support all organ systems
Endoderm	*Endocrine system:* thymus, thyroid gland, and pancreas *Respiratory system:* respiratory epithelium (except nasal passageways) and associated mucous glands *Digestive system:* mucous epithelium (except mouth and anus), exocrine glands (except salivary glands), liver, and pancreas *Urinary system:* urinary bladder and distal portions of the duct system *Reproductive system:* distal portions of the duct system, stem cells that produce gametes

WEEK 2

Migration of mesoderm around the inner surface of the trophoblast creates the chorion. Mesodermal migration around the outside of the amniotic cavity, between the ectodermal cells and the trophoblast, forms the amnion. Mesodermal migration around the endodermal pouch creates the yolk sac.

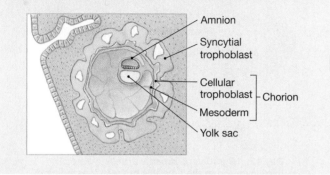

- Amnion
- Syncytial trophoblast
- Cellular trophoblast
- Mesoderm
— Chorion
- Yolk sac

WEEK 3

The embryonic disc bulges into the amniotic cavity at the head fold. The allantois, an endodermal extension surrounded by mesoderm, extends toward the trophoblast.

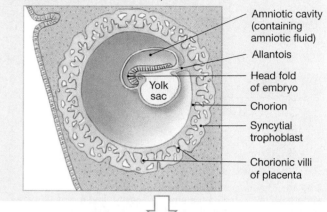

- Amniotic cavity (containing amniotic fluid)
- Allantois
- Head fold of embryo
- Chorion
- Syncytial trophoblast
- Chorionic villi of placenta
- Yolk sac

WEEK 5

The developing embryo and extraembryonic membranes bulge into the uterine cavity. The trophoblast pushing out into the uterine cavity remains covered by endometrium but no longer participates in nutrient absorption and embryo support. The embryo moves away from the placenta, and the body stalk and yolk stalk fuse to form an umbilical stalk.

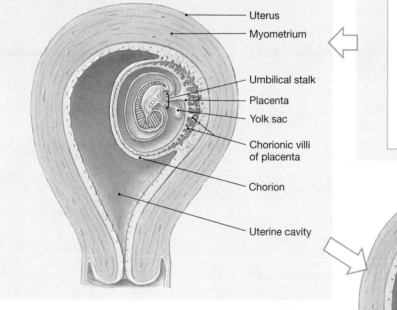

- Uterus
- Myometrium
- Umbilical stalk
- Placenta
- Yolk sac
- Chorionic villi of placenta
- Chorion
- Uterine cavity

WEEK 4

The embryo now has a head fold and a tail fold. Constriction of the connections between the embryo and the surrounding trophoblast narrows the yolk stalk and body stalk.

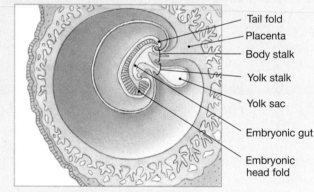

- Tail fold
- Placenta
- Body stalk
- Yolk stalk
- Yolk sac
- Embryonic gut
- Embryonic head fold

WEEK 10

The amnion has expanded greatly, filling the uterine cavity. The fetus is connected to the placenta by an elongated umbilical cord that contains a portion of the allantois, blood vessels, and the remnants of the yolk stalk.

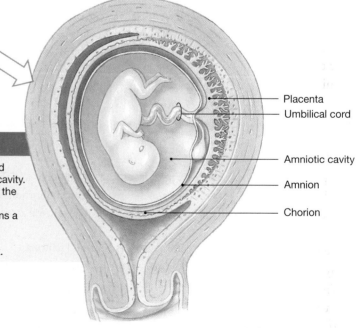

- Placenta
- Umbilical cord
- Amniotic cavity
- Amnion
- Chorion

● **FIGURE 20–5** Extraembryonic Membranes and Placenta Formation.

rise to the urinary bladder. The allantois accumulates some of the small amount of urine produced by the kidneys during embryological development.

THE CHORION. The **chorion** (kō-rē-on) is created as migrating mesodermal cells form a layer beneath the trophoblast, which separates it from the blastocoele (see weeks 2 and 3, Figure 20–5). When implantation first occurs, the nutrients absorbed by the trophoblast can easily reach the inner cell mass by diffusion. But as the embryo and trophoblast enlarge, the distance between them increases, and diffusion alone cannot keep pace with the demands of the embryo. Blood vessels now begin to develop within the mesoderm of the chorion, which creates a "rapid-transit system" that links the embryo with the trophoblast.

Placentation

The **placenta** is a temporary structure in the uterine wall that provides a site for diffusion between the fetal and maternal circulatory systems. **Placentation** (pla-sen-TĀ-shun), or *placenta formation*, takes place when blood vessels form in the chorion around the periphery of the blastocyst (see Figure 20–5). By week 3 of development, the mesoderm extends along each of the trophoblastic villi, and forms *chorionic villi* in contact with maternal tissues. Embryonic blood vessels develop in each villus, and circulation through these chorionic vessels begins early in week 3, when the heart starts beating. These villi continue to enlarge and branch, and form an intricate network within the endometrium. Blood vessels continue to be eroded, and maternal blood flows slowly through the lacunae. Chorionic blood vessels pass close by, and gases and nutrients diffuse between the embryonic and maternal circulations across the trophoblast layers.

Initially, the entire blastocyst is surrounded by chorionic villi. The chorion continues to enlarge, expanding like a balloon within the endometrium, and by week 4 the embryo, amnion, and yolk sac are suspended within an expansive, fluid-filled chamber. The *body stalk,* which is the connection between the embryo and the chorion, contains the distal portions of the allantois and blood vessels that carry blood to and from the placenta. The narrow connection between the endoderm of the embryo and the yolk sac is called the *yolk stalk.* As the end of the first trimester approaches, the fetus moves farther away from the placenta. The yolk stalk and body stalk begin to fuse, which forms an *umbilical stalk* (see week 5, Figure 20–5). By week 10, the fetus floats free within the amniotic cavity. The fetus remains connected to the placenta by the elongate **umbilical cord,** which contains the allantois, placental blood vessels, and the yolk stalk.

Placental Circulation

Figure 20–6● diagrams circulation at the placenta near the end of the first trimester. Deoxygenated blood flows from the developing embryo or fetus to the placenta through the paired **umbilical arteries,** and oxygenated blood returns to the developing embryo or fetus in a single **umbilical vein.** ∞ p. 500 The chorionic villi provide the surface area for the active and passive exchanges of gases, nutrients, and wastes between the fetal and maternal bloodstreams.

Placental Hormones

In addition to its role in the nutrition of the fetus, the placenta functions as an endocrine organ. Several placental hormones—including *human chorionic gonadotropin, progesterone, estrogens, human placental lactogen, placental prolactin,* and *relaxin*—are synthesized by the syncytial trophoblast and released into the maternal circulation.

Human chorionic (kō-rē-ON-ik) **gonadotropin (hCG)** appears in the maternal bloodstream soon after implantation has occurred. The presence of hCG in blood or urine samples is a reliable indication of pregnancy. Kits sold for the early detection of a pregnancy are sensitive to the presence of this hormone. Like luteinizing hormone, hCG maintains the corpus luteum and promotes the continued secretion of progesterone. As a result, the endometrial lining remains perfectly functional, and menses does not occur. In the absence of hCG, the pregnancy ends, because another uterine cycle begins and the endometrial lining disintegrates.

In the presence of hCG, the corpus luteum persists for 3–4 months before gradually shrinking. The decline of the corpus luteum does not trigger the return of menstrual periods, because by the end of the first trimester, the placenta is actively secreting both **progesterone** and **estrogens.**

After the first trimester, the placenta produces sufficient amounts of progesterone to maintain the endometrial lining and continue the pregnancy. As the end of the third trimester approaches, estrogen production accelerates. The rising estrogen levels play a role in stimulating labor and delivery.

Human placental lactogen (hPL) and **placental prolactin** help prepare the mammary glands for milk production. The conversion of the mammary glands from resting to active status requires the presence of both placental hormones (hPL, placental prolactin, estrogen, and progesterone) and maternal hormones (growth hormone, prolactin, and thyroid hormones).

Relaxin is a hormone secreted by both the placenta and the corpus luteum. Relaxin (1) increases the flexibility of the pubic symphysis, which permits the pelvis to expand during delivery; (2) causes the dilation of the cervix, which makes it

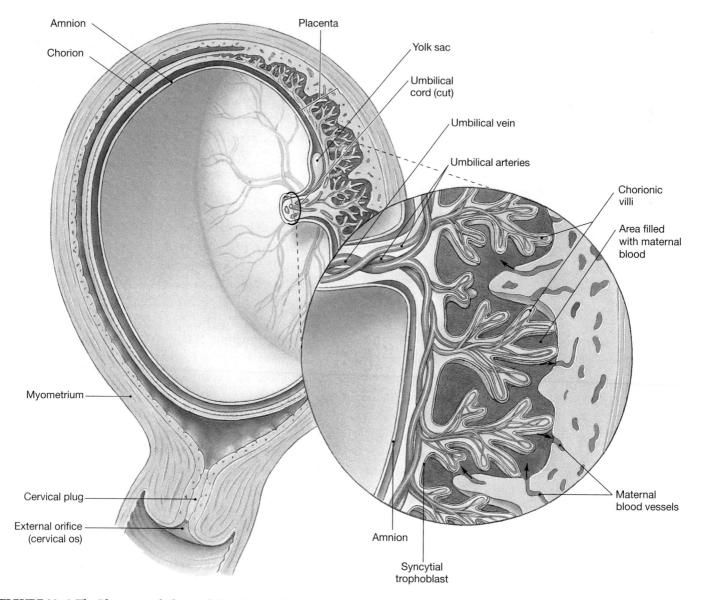

● FIGURE 20–6 The Placenta and Placental Circulation. For clarity, the uterus is shown after the embryo has been removed and the umbilical cord cut. Oxygenated blood flows from the mother into the placenta through ruptured maternal arteries. It then flows around chorionic villi that contain fetal blood vessels. Deoxygenated fetal blood enters the chorionic villi in paired umbilical arteries, and oxygenated blood then leaves in a single umbilical vein. Deoxygenated maternal blood reenters the mother's venous system through the broken walls of small uterine veins. Note that no mixing of maternal and fetal blood occurs.

easier for the fetus to enter the vaginal canal; and (3) delays the onset of labor contractions by suppressing the release of oxytocin by the hypothalamus.

Embryogenesis

Shortly after gastrulation begins, the body of the embryo begins to separate itself from the rest of the embryonic disc, and the embryo's internal organs start to form. The process of embryo formation is called **embryogenesis** (em-brē-ō-JEN-e-sis). Embryogenesis begins as folding and differential growth of the embryonic disc produce a bulge—the *head fold*—that

projects into the amniotic cavity (see week 3, Figure 20–5). Similar movements lead to the formation of a *tail fold* (see week 4, Figure 20–5). By this time, the embryo can be seen to have dorsal and ventral surfaces and left and right sides. The changes in proportions and appearance that occur between week 2 of development and the end of the first trimester are presented in Figure 20–7●.

The first trimester is a critical period for development, because events during the first 12 weeks establish the basis for **organogenesis,** which is the process of organ formation. Table 20–2 includes important developmental milestones, including those during the first trimester.

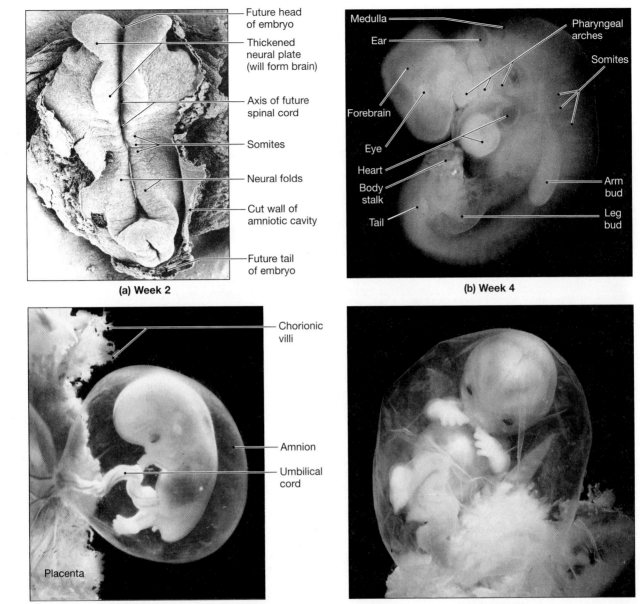

Future head of embryo

Thickened neural plate (will form brain)

Axis of future spinal cord

Somites

Neural folds

Cut wall of amniotic cavity

Future tail of embryo

(a) Week 2

Medulla

Ear

Forebrain

Eye

Heart

Body stalk

Tail

Pharyngeal arches

Somites

Arm bud

Leg bud

(b) Week 4

Chorionic villi

Amnion

Umbilical cord

Placenta

(c) Week 8

(d) Week 12

● **FIGURE 20–7 Development During the First Trimester.** (**a**) This SEM shows the superior surface of a monkey embryo after 2 weeks of development. A human embryo at this stage would look essentially the same. (**b–d**) Fiber-optic technology provided these views of human embryos (b, c) and a human fetus (d). For actual sizes, see Figure 20–9 (p. 747).

Clinical Note
TERATOGENICITY

The developing fetus is most vulnerable to birth defects during the first 60 days of pregnancy. The first stage of gestation, *embryogenesis,* begins at conception and continues to day 18. Exposure to a *teratogen* during this time can threaten the entire embryo. (A teratogen is a chemical or disease that causes malformation of a fetus.) The second stage of gestation, *organogenesis,* extends from day 18 to day 60. This is the stage when the fetus is most susceptible to teratogens and when most gross anatomical malformations develop. The third stage, which is the *fetal period,* extends from day 60 to birth. Exposure during this stage can cause abnormal cell growth and differentiation. ■

CONCEPT CHECK QUESTIONS

1. What two important roles do acrosomal enzymes of spermatozoa play in fertilization?
2. What is the developmental fate of the inner cell mass of the blastocyst?
3. Sue's pregnancy test indicates elevated levels of the hormone hCG (human chorionic gonadotropin). Is she pregnant?
4. What are two important functions of the placenta?

Answers begin on p. 792.

TABLE 20–2 *An Overview of Prenatal and Early Postnatal Development*

GESTATIONAL AGE (MONTHS)	SIZE AND WEIGHT	INTEGUMENTARY SYSTEM	SKELETAL SYSTEM	MUSCULAR SYSTEM	NERVOUS SYSTEM	SPECIAL SENSE ORGANS
1	5 mm 0.02 g		(b) Somite formation	(b) Somite formation	(b) Neural tube	(b) Eye and ear formation
2	28 mm 2.7 g	(b) Nail beds, hair follicles, sweat glands	(b) Axial and appendicular cartilage formation	(c) Rudiments of axial musculature	(b) CNS, PNS organization, growth of cerebrum	(b) Taste buds, olfactory epithelium
3	78 mm 26 g	(b) Epidermal layers appear	(b) Ossification centers spreading	(c) Rudiments of appendicular musculature	(c) Basic spinal cord and brain structure	
4	133 mm 0.15 kg	(b) Hair, sebaceous glands (c) Sweat glands	(b) Articulations (c) Facial and palatal organization	Movements of fetus can be felt by the mother	(b) Rapid expansion of cerebrum	(c) Basic eye and ear structure (b) Peripheral receptor formation
5	185 mm 0.46 kg	(b) Keratin production, nail production			(b) Myelination of spinal cord	
6	230 mm 0.64 kg			(c) Perineal muscles	(b) CNS tract formation (c) Layering of cortex	
7	270 mm 1.492 kg	(b) Keratinization, nail formation, hair formation				(c) Eyelids open, retinae sensitive to light
8	310 mm 2.274 kg		(b) Epiphyseal cartilage formation			(c) Taste receptors functional
9	346 mm 3.2 kg					
Postnatal development		Hair changes in consistency and distribution	Formation and growth of epiphyseal cartilages continue	Muscle mass and control increase	Myelination, layering, CNS tract formation continue	

*(b) = beginning to form; (c) = completed

ENDOCRINE SYSTEM	CARDIOVASCULAR AND LYMPHATIC SYSTEMS	RESPIRATORY SYSTEM	DIGESTIVE SYSTEM	URINARY SYSTEM	REPRODUCTIVE SYSTEM
	(b) Heartbeat	(b) Trachea and lung formation	(b) Intestinal tract, liver, pancreas (c) Yolk sac	(c) Allantois	
(b) Thymus, thyroid, pituitary, adrenal glands	(c) Basic heart structure, major blood vessels, lymph nodes and ducts (b) Blood formation in liver	(b) Extensive bronchial branching into mediastinum (c) Diaphragm	(b) Intestinal subdivisions, villi, salivary glands	(b) Kidney formation (adult form)	(b) Mammary glands
(c) Thymus, thyroid gland	(b) Tonsils, blood formation in bone marrow		(c) Gallbladder, pancreas		(b) Gonads, ducts, and genitalia; oogonia in female
	(b) Migration of lymphocytes to lymphoid organs, blood formation in spleen			(b) Degeneration of embryonic kidneys	
	(c) Tonsils	(c) Nostrils open	(c) Intestinal subdivisions		
(c) Adrenal glands	(c) Spleen, liver, bone marrow	(b) Formation of alveoli	(c) Epithelial organization, glands		
(c) Pituitary gland			(c) Intestinal plicae		(b) Testes descend
		Complete pulmonary branching and alveolar formation		(c) Nephron formation	Descent of testes complete at or near time of delivery
	Cardiovascular changes at birth; immune response gradually becomes operative				

■ The Second and Third Trimesters

By the end of the first trimester (week 12), the basic elements of all the major organ systems have formed. Over the next 3 months, the fetus will grow to a weight of about 0.64 kg (1.4 lb). During this second trimester, the fetus, which is encircled by the amnion, grows faster than the surrounding placenta. When the outer surface of the amnion contacts the inner surface of the chorion, these layers fuse. Figure 20–8a● shows a 4-month fetus; Figure 20–8b● shows a 6-month fetus. The changes in body form that occur during the first trimester, and during part of the second trimester, are shown in the upper portion of Figure 20–9●.

During the third trimester, the basic components of all the organ systems appear, and most become ready to perform their normal functions. The rate of growth starts to decrease, but in absolute terms, the largest weight gain occurs in this trimester. In the last 3 months of gestation, the fetus gains about 2.6 kg (5.7 lb), and reaches a full-term weight of about 3.2 kg (7 lb). Important events in organ system development during the second and third trimesters are summarized in Table 20–2.

 Key Note

The basic body plan, the foundations of all of the organ systems, and the four extraembryonic membranes appear during the first trimester. These are complex and delicate processes; not every zygote starts cleavage, and fewer than half of the zygotes that do begin cleavage survive until the end of the first trimester. The second trimester is a period of rapid growth, accompanied by the development of fetal organs that will then become fully functional by the end of the third trimester.

The Effects of Pregnancy on Maternal Systems

The developing fetus is totally dependent on maternal organ systems for nourishment, respiration, and waste removal. The mother must absorb enough oxygen, nutrients, and vitamins for herself and her fetus, and she must eliminate all of the wastes generated by both of them. Although this is not a burden over the initial weeks of gestation, the demands placed on the mother become significant as the fetus grows larger. For the mother to survive under these conditions, maternal sys-

(a)

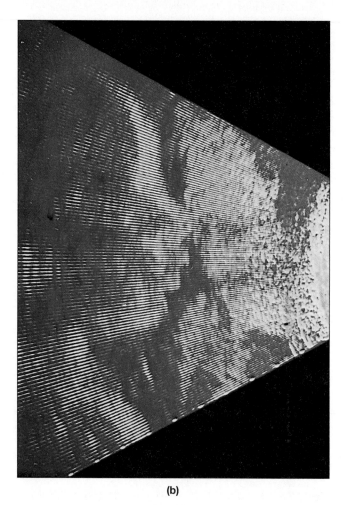

(b)

● **FIGURE 20–8 The Fetus During the Second and Third Trimesters.** (a) This photograph of a 4-month fetus was taken through a fiber-optic endoscope. (b) This image of a 6-month fetus was obtained during an ultrasound procedure.

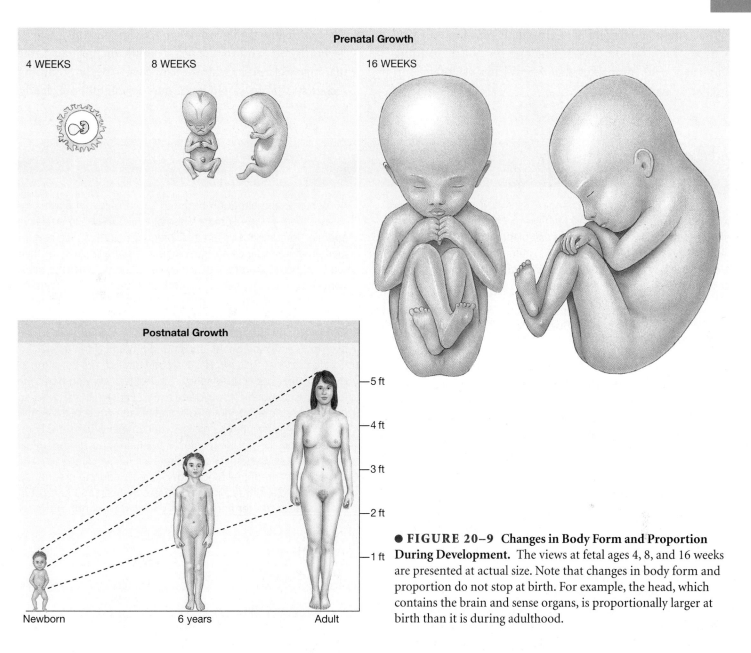

● FIGURE 20–9 **Changes in Body Form and Proportion During Development.** The views at fetal ages 4, 8, and 16 weeks are presented at actual size. Note that changes in body form and proportion do not stop at birth. For example, the head, which contains the brain and sense organs, is proportionally larger at birth than it is during adulthood.

tems must make major adjustments. In practical terms, the mother must breathe, eat, and excrete for two.

The major changes that take place in maternal systems include the following:

■ *Maternal respiratory rate goes up and tidal volume increases.* As a result, the mother's lungs deliver the extra oxygen the fetus requires and remove the excess carbon dioxide the fetus generates.

■ *Maternal blood volume increases.* This increase occurs because (1) blood flowing into the placenta reduces the volume in the rest of the systemic circuit, and (2) fetal activity lowers blood P_{O_2} and elevates blood P_{CO_2}. The combination stimulates the production of renin and erythropoietin (EPO) by

the kidneys, which leads to an increase in maternal blood volume (see Figure 13–12a, p. 481). By the end of gestation, maternal blood volume has increased by almost 50 percent.

■ *Maternal requirements for nutrients and vitamins climb 10–30 percent.* Because pregnant women must nourish both themselves and their fetus, they tend to have increased sensations of hunger.

■ *Maternal glomerular filtration rate increases by roughly 50 percent.* This increase, which corresponds to the increase in blood volume, accelerates the excretion of metabolic wastes generated by the fetus. Because the volume of urine produced increases and the weight of the uterus presses down on the urinary bladder, pregnant women need to urinate frequently.

■ *The uterus undergoes a tremendous increase in size.* Structural and functional changes in the expanding uterus are so important that we will discuss them in a separate section.

■ *The mammary glands increase in size, and secretory activity begins.* By the end of the sixth month of pregnancy, the mammary glands are fully developed and begin producing secretions that are stored in the duct system of those glands.

Clinical Note
PHYSIOLOGIC CHANGES OF PREGNANCY

The physiologic changes associated with pregnancy are due to an altered hormonal state, the mechanical effects of the enlarging uterus and its significant vascularity, and the increasing metabolic demands on the maternal system. Paramedics must understand the physiologic changes associated with pregnancy in order to better assess pregnant patients.

■ *Reproductive System.* Understandably, the most significant pregnancy-related changes occur in the uterus. In its nonpregnant state, the uterus is a small pear-shaped organ that weighs about 60 g (2 ounces) with a capacity of approximately 10 cc. By the end of pregnancy, its weight has increased to 1000 g (slightly more than 2 pounds), while its capacity is now approximately 5,000 mL (Figure 20–10●). Another notable change is that during pregnancy the vascular system of the uterus contains about one-sixth (16 percent) of the mother's total blood volume. Other changes that occur in the reproductive system include the formation of a mucous plug in the cervix that protects the developing fetus and helps to prevent infection. This plug will be expelled when cervical dilation begins prior to delivery. Estrogen causes the vaginal mucosa to thicken, vaginal secretions to increase, and the connective tissue to loosen to allow for delivery. The breasts enlarge and become more nodular as the mammary glands increase in number and size in preparation for lactation.

■ *Respiratory System.* During pregnancy, maternal oxygen demands increase. To meet this need, progesterone causes a decrease in airway resistance. This results in a 20 percent increase in oxygen consumption and a 40 percent increase in tidal volume. There is only a slight increase in respiratory rate. The enlarging uterus pushes up the diaphragm, which results in flaring of the rib margins to maintain intrathoracic volume.

■ *Cardiovascular System.* Many changes take place in the cardiovascular system during pregnancy (Figure 20–11●). Cardiac output increases throughout pregnancy, and peaks at 6–7 liters/minute by the time the fetus is fully developed. The maternal blood volume increases by

45 percent, and although both red blood cells and plasma increase, there is slightly more plasma, which results in a relative anemia. To combat this anemia, pregnant women receive supplemental iron to increase the oxygen-carrying capacity of their red blood cells. Due to the increase in blood volume, the pregnant female may suffer an acute blood loss of 30–35 percent without a significant change in vital signs. The maternal heart rate increases by 10–15 beats/minute. Blood pressure decreases slightly during the first two trimesters of pregnancy and then rises to near nonpregnant levels during the third trimester.

■ *Gastrointestinal System.* Nausea and vomiting are common in the first trimester of pregnancy as a result of varying hormone levels and changed carbohydrate needs. Peristalsis is slowed, so delayed gastric emptying is likely and bloating or constipation is common. As the uterus enlarges, abdominal organs are compressed, and the resulting compartmentalization of abdominal organs makes assessment difficult.

■ *Urinary System.* Renal blood flow increases during pregnancy. The glomerular filtration rate increases by nearly 50 percent in the second trimester and remains elevated throughout the re-

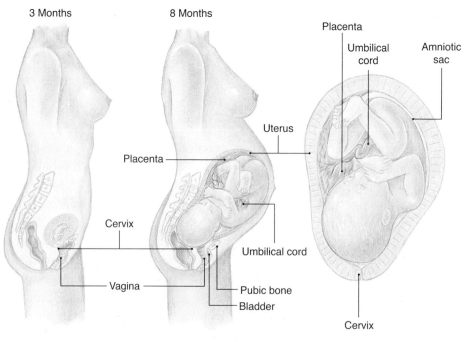

3 Months

8 Months

Placenta
Umbilical cord
Amniotic sac
Uterus
Placenta
Placenta
Cervix
Umbilical cord
Vagina
Pubic bone
Bladder
Cervix

● **FIGURE 20–10 Uterus During Pregnancy.** Significant uterine changes are associated with pregnancy. By term, the uterus is the largest organ in the abdomen.

Blood volume usually increases by about 45%. Dilution resulting from the disproportionate increase of plasma volume over the red cell mass is responsible for the so-called "anemia of pregnancy."

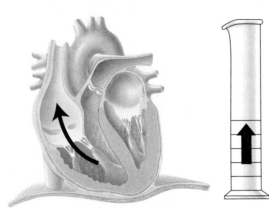

Cardiac output increases by 1.0 to 1.5 L/min during the 1st trimester, reaches 6 to 7 L/min by the late 2nd trimester, and is maintained essentially at this level until delivery.

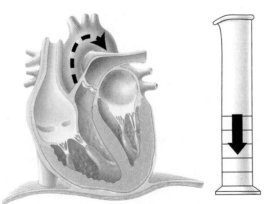

The stroke volume progressively declines to term following a rise early in pregnancy. Heart rate, however, increases by an average of 10 to 15 beats/min.

● **FIGURE 20–11 The Hemodynamic Changes Associated with Pregnancy.** Pregnancy causes the heart rate to be faster and the blood pressure to be lower. These findings can easily be interpreted as early signs of shock if unaware of the physiological changes of pregnancy.

mainder of the pregnancy. As a result, the renal tubular absorption also increases. Occasionally glucosuria (large amounts of sugar in the urine) may result from the kidney's inability to reabsorb all of the glucose being filtered. Glucosuria may be normal or may indicate the development of gestational diabetes. The urinary bladder gets displaced anteriorly and superiorly, which increases the potential for rupture. Urinary

frequency is common, particularly in the first and third trimesters, due to uterine compression of the bladder.

■ *Musculoskeletal System.* Loosened pelvic joints caused by hormonal influences account for the waddling gait that is often associated with pregnancy. As the uterus enlarges and the mother's center of gravity shifts, posture changes to compensate for anterior growth, which often causes low back pain. ■

Structural and Functional Changes in the Uterus

At the end of gestation, a typical uterus will have grown from 7.5 cm (3 in.) long to 30 cm (12 in.) long, and from 60 g (2 oz) to 1100 g (2.4 lb). It may then contain almost 5 liters of fluid, so the organ and its contents weigh roughly 10 kg (22 lb). This remarkable expansion occurs through the enlargement of existing cells (especially smooth muscle cells) rather than by an increase in the total number of cells.

The tremendous stretching of the uterus is accompanied by a gradual increase in the rates of spontaneous smooth muscle contractions in the myometrium. In the early stages of pregnancy, the contractions are weak, painless, and brief. Evidence indicates that progesterone released by the placenta has an inhibitory effect on the uterine smooth muscle, which prevents more extensive and powerful contractions.

Three major factors oppose the calming action of progesterone:

1. *Rising estrogen levels.* Estrogens, also produced by the placenta, increase the sensitivity of the uterine smooth muscles and make contractions more likely. Throughout pregnancy, progesterone exerts the dominant effect, but as the time of delivery approaches, estrogen production accelerates and the myometrium becomes more sensitive to stimulation. Estrogens also increase the sensitivity of smooth muscle fibers to oxytocin.

2. *Rising oxytocin levels.* Rising oxytocin levels stimulate increases in the force and frequency of uterine contractions. Oxytocin release is stimulated by high estrogen levels and by distortion of the cervix.

3. *Prostaglandin production.* Estrogens and oxytocin stimulate the production of prostaglandins in the endometrium. These prostaglandins further stimulate smooth muscle contractions.

After 9 months of gestation, several factors interact to produce **labor contractions** in the myometrium of the uterine wall. Once begun, positive feedback ensures that the contractions continue until delivery has been completed.

Figure 20–12● diagrams important factors that stimulate and sustain labor. The actual trigger for the onset of labor may be events in the fetus rather than in the mother. When labor begins, the fetal pituitary gland secretes oxytocin that is released into the maternal bloodstream at the placenta. The resulting increase in myometrial contractions and prostaglandin production, on top of the priming effects of estrogens and maternal oxytocin, may be the extra stimulus that finally initiates labor and delivery.

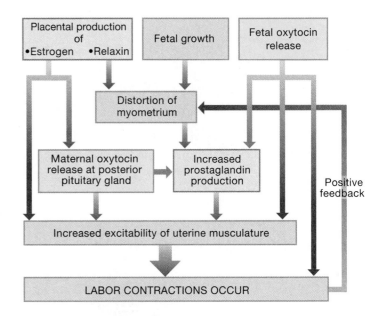

● **FIGURE 20–12** Factors Involved in Initiating and Sustaining Labor and Delivery.

Clinical Note
FETAL DEVELOPMENT

Fetal development begins immediately after fertilization and is quite complex. The time at which fertilization occurs is called *conception.* Since conception occurs approximately 14 days after the first day of the last menstrual period (LMP), it is possible to calculate, with fair accuracy, the approximate date the baby should be born. This estimate is usually made during the mother's first prenatal visit. The normal duration of pregnancy is 40 weeks from the first day of the mother's last menstrual period. This is equal to 280 days, which is 10 lunar months, or roughly 9 calendar months. This estimated birth date is commonly called the *due date.* Medically, it is known as the *estimated date of confinement (EDC).* Generally, pregnancy is divided into trimesters. Each trimester is approximately 13 weeks, or 3 calendar months, long.

Several different terms are used to describe the stages of fetal development during the course of pregnancy. The pre-embryonic stage covers the first 14 days following conception; the embryonic stage begins at day 15 and ends at approximately 8 weeks; and the fetal stage lasts from 8 weeks of age until delivery. Emergency personnel should be familiar with some of the significant developmental milestones that occur during these three periods (Table 20–3). During normal fetal development, the sex of the infant can usually be determined by 16 weeks gestation. By the 20th week, fetal heart tones (FHTs) can be detected by stethoscope. The mother also has generally felt fetal movement. By 24 weeks, the baby may be able to survive if born prematurely. Fetuses born after 28 weeks have an excellent chance of survival. By the 38th week the baby is considered term, or fully developed. Most of the fetus's organ systems develop during the first trimester. Therefore, this is when the fetus is most vulnerable to the development of birth defects. ■

TABLE 20-3 *Significant Fetal Development Milestones*

PRE-EMBRYONIC STAGE

| 2 weeks | Rapid cellular multiplication and differentiation |

EMBRYONIC STAGE

| 4 weeks | Fetal heart begins to beat |
| 8 weeks | All body system and external structures are formed
Size: approximately 3 centimeters (1.2 inches) |

FETAL STAGE

8–12 weeks	Fetal heart tones audible with Doppler Kidneys begin to produce urine Fetus most vulnerable to toxins Size: 8 centimeters (3.2 inches) Weight: about 1.6 ounces
16 weeks	Sex can be determined visually Swallowing amniotic fluid and producing meconium Looks like a baby, although thin
20 weeks	Fetal heart tones audible with stethoscope Mother able to feel fetal movement Baby develops schedule of sucking, kicking, and sleeping Hair, eyebrows, and eyelashes present Size: 19 centimeters (8 inches) Weight: approximately 16 ounces
24 weeks	Increased activity Begins respiratory movement Size: 28 centimeters (11.2 inches) Weight: 1 pound 10 ounces
28 weeks	Surfactant, which is necessary for lung function, is formed Eyes begin to open and close Weight: 2 to 3 pounds
32 weeks	Bones are fully developed but soft and flexible Subcutaneous fat being deposited Fingernails and toenails present
38–40 weeks	Considered to be full term Baby fills uterine cavity Baby receives maternal antibodies

■ Labor and Delivery

The goal of labor is **parturition** (par-tūr-ISH-un), which is the forcible expulsion of the fetus from the uterus. During labor, each uterine contraction begins near the top of the uterus and sweeps in a wave toward the cervix. These contractions are strong and occur at regular intervals. As parturition approaches, increases in the force and frequency of contractions change the position of the fetus, and move it toward the cervical canal.

The Stages of Labor

Labor consists of three stages: the *dilation stage*, the *expulsion stage*, and the *placental stage* (Figure 20–13●).

1. The **dilation stage** begins with the onset of labor, as the cervix dilates and the fetus begins to slide down the cervical canal (Figure 20–13a). This stage is highly variable in length but typically lasts 8 or more hours. At the start of this stage, labor contractions occur once every 10–30 minutes; their frequency increases steadily. Late in this stage, the amnion usually ruptures, which is an event sometimes referred to as "having one's water break."

2. The **expulsion stage** begins as the cervix, which is pushed open by the approaching fetus, dilates completely (Figure 20–13b). Expulsion continues until the fetus has emerged from the vagina, which is a period that usually takes less than 2 hours. The arrival of the newborn into the outside world is **delivery,** or birth.

   If the vaginal canal is too small to permit the passage of the fetus, and poses acute danger of perineal tearing, a physician may temporarily enlarge the passageway by performing an **episiotomy** (e-pēz-ē-OT-o-mē), which is an incision through the perineal musculature. After delivery, this surgical cut can be repaired with sutures, which is a much simpler procedure than suturing the jagged edges associated with an extensive perineal tear. If complications arise during the dilation or expulsion stage, the infant can be surgically removed by **cesarean section,** or "C-section." In such cases, an incision is made through the abdominal wall and the uterus is opened just enough to allow passage of the infant's head. This procedure is performed during 15–25 percent of the deliveries in the U.S.—more often than necessary, according to some studies. Over the past decade, efforts have been made to reduce the frequency of both episiotomies and cesarean sections.

3. During the **placental stage** of labor, muscle tension builds in the walls of the uterus (which still contains the placenta), and the organ gradually decreases in size (Figure 20–13c). This contraction of the uterus tears the connections between the endometrium and the placenta. Usually within an hour after delivery, the placental stage ends with the ejection of the placenta, or "afterbirth." The disruption of the placenta is accompanied by a loss of blood. Because the maternal blood volume has increased greatly during pregnancy, this loss can be tolerated without difficulty.

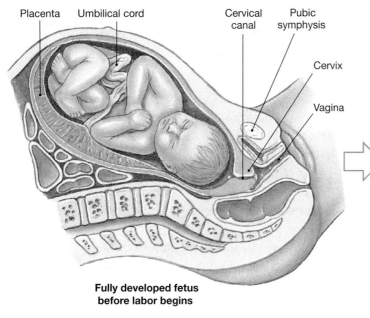

Placenta Umbilical cord Cervical canal Pubic symphysis Cervix Vagina

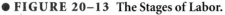

Fully developed fetus before labor begins

● **FIGURE 20–13** The Stages of Labor.

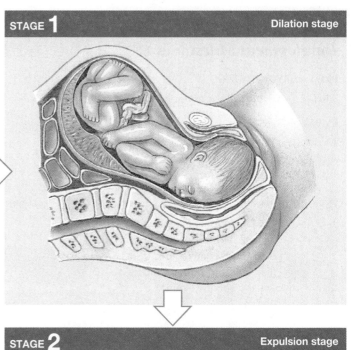

STAGE **1** Dilation stage

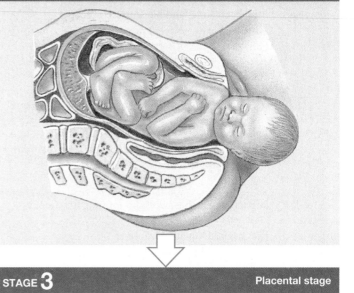

STAGE **2** Expulsion stage

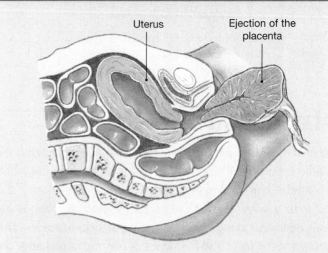

STAGE **3** Placental stage

Uterus Ejection of the placenta

Premature Labor

Premature labor occurs when labor contractions begin before the fetus has completed normal development. The chances of a newborn's survival are directly related to its body weight at delivery. Even with massive supportive efforts, newborns who weigh less than 400 g (14 oz) at birth will not survive, primarily because their respiratory, cardiovascular, and urinary systems are unable to support life. As a result, the dividing line between spontaneous abortion and *immature delivery* is usually set at a body weight of 500 g (17.6 oz), which is the normal weight near the end of the second trimester.

Most fetuses born at 25–27 weeks of gestation (birth weight under 600 g) die despite intensive neonatal care; survivors have a high risk of developmental abnormalities. A **premature delivery** usually refers to the birth of infants at 28–36 weeks (birth weight over 1 kg). Given the proper medical care, these infants have a good chance of surviving and developing normally.

Multiple Births

Of all *multiple births* (twins, triplets, quadruplets, and so forth), twins are most common. In the U.S. population, one of every 89 births contains twins, whereas triplets occur in one of every 89^2 (or 7921) births. About 70 percent of all twins are "fraternal," or **dizygotic** (dī-zī-GOT-ik). Fraternal

Clinical Note
OBSTETRICAL TERMINOLOGY

The field of obstetrics has its own unique terminology. You should be familiar with this terminology, since patient documentation and communications with other health care workers and physicians often require it (Figure 20–14●). ■

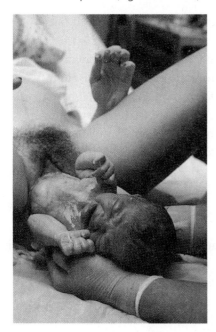

● **FIGURE 20–14 Childbirth.** The field of obstetrics has its own unique terminology. Emergency personnel should be familiar with this system, because they must interact with other members of the health care team.

antepartum	time interval prior to delivery of the fetus
postpartum	time interval after delivery of the fetus
prenatal	time interval prior to birth, synonymous with antepartum
natal	relating to birth or the date of birth
gravidity*	number of times a woman has been pregnant
parity*	number of pregnancies carried to full term
primigravida	woman who is pregnant for the first time
primipara	woman who has given birth to her first child
multigravida	woman who has been pregnant more than once
nulligravida	woman who has not been pregnant
multipara	woman who has delivered more than one baby
nullipara	woman who has yet to deliver her first child
grand multiparity	woman who has delivered at least seven babies
gestation	period of time for intrauterine fetal development

*Gravidity and parity are expressed in the following shorthand: G_4P_2. The letter G refers to gravidity, and the letter P refers to parity. The woman referred to by "G_4P_2" would have had four pregnancies and two births.

twins develop when two eggs are fertilized at the same time, and form two separate zygotes. Fraternal twins can be of the same or different sexes.

"Identical," or **monozygotic,** twins result from the separation of blastomeres early in cleavage or from the splitting of the inner cell mass before gastrulation. In either event, the genetic makeup and sex of the pair are identical because both twins are formed from the same pair of gametes. Identical twins occur in about 30 percent of all twin births. Triplets and larger multiple births can result from the same processes that produce twins.

If the splitting of the blastomeres or of the embryonic disc is not complete, **conjoined** (Siamese) **twins** may develop. These genetically identical twins typically share some skin, a portion of the liver, and perhaps other internal organs as well. When the fusion is minor, the infants can be surgically separated with some success. Most conjoined twins with more extensive fusions fail to survive delivery.

CONCEPT CHECK QUESTIONS

1. Why does a mother's blood volume increase during pregnancy?
2. What effect would a decrease in progesterone have on the uterus during late pregnancy?
3. By what means does the uterus greatly increase in size and weight during pregnancy?

Answers begin on p. 792.

■ Postnatal Development

Developmental processes do not cease at delivery. A newborn has few of the anatomical, functional, or physiological characteristics of mature adults. In postnatal development, each individual passes through a number of **life stages**—*the neonatal period, infancy, childhood, adolescence,* and *maturity*—each with a distinctive combination of characteristics and abilities.

The Neonatal Period, Infancy, and Childhood

The **neonatal period** extends from the moment of birth to 1 month thereafter. **Infancy** then continues to 2 years of age, and **childhood** lasts until **adolescence,** which is the period of sexual and physical maturation.

During these developmental stages, the major nonreproductive organ systems become fully operational and gradually acquire the functional characteristics of adult structures. Additionally, the individual grows rapidly, and body proportions change significantly.

Pediatrics is the medical specialty that focuses on postnatal development from infancy through adolescence. Infants and young children often cannot clearly describe the problems they are experiencing, so pediatricians and parents must be skilled observers. Standardized tests are used to compare developmental progress with average values.

The Neonatal Period

Physiological and anatomical changes occur as the fetus completes the transition to the status of a newborn, or **neonate.** Before delivery, dissolved gases, nutrients, waste products, hormones, and antibodies were transferred across the placenta. At birth, the newborn must come to rely on its own specialized organs and organ systems to perform respiration, digestion, and excretion. The neonate must make numerous adjustments to life outside the womb, including the following:

- At birth, the lungs are collapsed and filled with fluid. Filling them with air requires a powerful inhalation. ∞ p. 578
- When the lungs expand, the pattern of cardiovascular circulation changes due to alterations in blood pressure and flow rates. The circulation changes result in the separation of the pulmonary and systemic circuits. ∞ p. 500
- Typical neonatal heart rates (120–140 beats per minute) are lower than fetal heart rates (average 150 beats per minute). Both neonatal heart rates and respiratory rates (30 breaths per minute) are considerably higher than those of adults.
- Before birth, the digestive system remains relatively inactive, although it does accumulate a mixture of bile secretions, mucus, and epithelial cells. The collected debris is excreted in the first few days of life. Over that period the newborn begins to nurse.
- As waste products build up in the arterial blood, they are excreted at the kidneys. Glomerular filtration is normal, but the neonate cannot concentrate urine to any significant degree. As a result, urinary water losses are high, so neonatal fluid requirements are high (greater than those of adults).
- The neonate has little ability to control body temperature, particularly in the first few days after delivery. As the infant grows larger and its insulating subcutaneous fat "blanket" gets thicker, its metabolic rate also rises. Daily and even hourly shifts in body temperature continue throughout childhood.

Clinical Note
POSTNATAL DEVELOPMENT

Developmental processes do not end at delivery. The newborn has few of the anatomical or physiological characteristics of mature adults. In postnatal development, the individual passes through several life stages. These include: *neonatal, infancy, childhood, adolescence,* and *maturity.* The neonatal period extends from birth to 1 month of age. Infancy is the period from 1 month of age to 2 years of age. Childhood lasts from two years of age until the onset of puberty.

Neonates typically lose up to 10 percent of their birth weight as they adjust to extrauterine life. However, this weight is usually recovered within 10 days. Infants should have doubled their birth weight by 5 or 6 months of age. Development occurs in a cephalo-caudal direction: muscle control begins at the head and later spreads toward the lower extremities. The personality begins to develop in early infancy and definite behaviors and characteristics are present by 5 to 6 months of age (Figure 20–15●).

It is important for emergency personnel to be familiar with many of the anatomical and physiological differences of children, because they can affect patient care. Table 20–4 illustrates important anatomical and physiological characteristics of infants and children. ∎

● **FIGURE 20–15 Early Infancy.** The period from 1 month to 2 years of age is a time of rapid development. The birth weight usually doubles, the infant exhibits improved muscle control, and the personality begins to develop.

TABLE 20-4 *Anatomical and Physiological Characteristics of Infants and Children*

DIFFERENCES IN INFANTS AND CHILDREN AS COMPARED TO ADULTS	POTENTIAL EFFECTS THAT MAY IMPACT ASSESSMENT AND CARE
Tongue proportionately larger	More likely to block airway
Smaller airway structures	More easily blocked
Abundant secretions	Can block the airway
Deciduous (baby) teeth	Easily dislodged; can block the airway
Flat nose and face	Difficult to obtain good face mask seal
Head heavier relative to body and less developed neck structures and muscles	Head may be propelled more forcefully than body, which produces a higher incidence of head injury
Fontanelle and open sutures (soft spots) palpable on top of young infant's head	Bulging fontanelle can be a sign of increased intracranial pressure (but may be normal if infant is crying); shrunken fontanelle may indicate dehydration
Thinner, softer brain tissue	Susceptible to serious brain injury
Head larger in proportion to body	Tips forward when supine; possible flexion of neck, which makes neutral alignment of airway difficult
Shorter, narrower, more elastic (flexible) trachea	Can close off trachea with hyperextension of neck
Short neck	Difficult to stabilize or immobilize
Abdominal breathers	Difficult to evaluate breathing
Faster respiratory rate	Muscles easily fatigued, which causes respiratory distress
Newborns breathe primarily through the nose (obligate nose breathers)	May not automatically open mouth to breathe if nose is blocked; airway more easily blocked
Larger body surface relative to body mass	Prone to hypothermia
Softer bones	More flexible, less easily fractured; traumatic forces may be transmitted to internal organs, and cause injury without fracturing the ribs; lungs easily damaged with trauma
Spleen and liver more exposed	Organ injury likely with significant force to abdomen

LACTATION AND THE MAMMARY GLANDS. By the end of the sixth month of pregnancy, the expectant mother's mammary glands are fully developed, and the gland cells begin producing a secretion known as **colostrum** (ko-LOS-trum). Ingested by the newborn during the first 2 or 3 days of life, colostrum contains more proteins and far less fat than breast milk. Many of the proteins are antibodies that help the infant ward off infections until its own immune system becomes fully functional. As colostrum production declines, the mammary glands convert to milk production. Breast milk consists of water, proteins, amino acids, lipids, sugars, and salts. It also contains large quantities of *lysozymes,* which are enzymes with antibiotic properties.

Milk becomes available to infants through the **milk let-down reflex** (Figure 20–16●). Mammary gland secretion is triggered when the infant begins to suck on the nipple. The stimulation of tactile receptors there leads to the release of oxytocin at the posterior pituitary gland. The arrival of cir-culating oxytocin at the mammary gland stimulates the contraction of contractile cells in the walls of the lactiferous ducts and sinuses, and results in the ejection of milk. The milk let-down reflex continues to function until *weaning,* which is typically 1–2 years after birth. Milk production ceases soon after, and the mammary glands gradually return to a resting state.

Infancy and Childhood

The rate of growth is greatest during prenatal development and declines after delivery. Postnatal growth during infancy and childhood occurs under the direction of circulating hormones, notably pituitary growth hormone, adrenal steroids, and thyroid hormones. These hormones affect each tissue and organ in specific ways, depending on the sensitivities of the individual cells. As a result, growth does not occur uniformly, and body proportions gradually change (see Figure 20–9, p. 747).

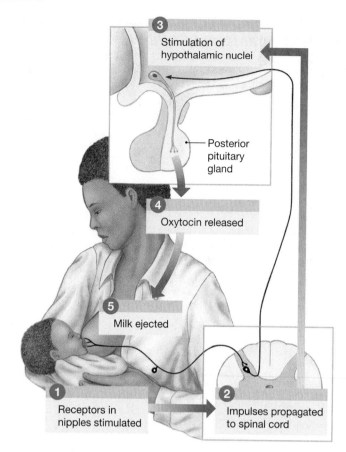

● **FIGURE 20–16** The Milk Let-Down Reflex.

Adolescence and Maturity

Adolescence begins at **puberty,** which is the period of sexual maturation, and ends when growth is completed. Three hormonal events interact at the onset of puberty:

1. The hypothalamus increases its production of gonadotropin-releasing hormone (GnRH).
2. The anterior pituitary becomes more sensitive to the presence of GnRH, and the circulating levels of FSH and LH rise rapidly.
3. Ovarian or testicular cells become more sensitive to FSH and LH. These changes initiate (1) gamete formation,

(2) the production of sex hormones that stimulate the appearance of secondary sex characteristics and behaviors, and (3) a sudden acceleration in the growth rate, which ends in the closure of the epiphyseal cartilages.

The age at which puberty begins varies. In the U.S. today, puberty generally occurs at about age 14 in boys and at 13 in girls, but the normal ranges are broad (9–14 in boys, 8–13 in girls). The combination of sex hormones, growth hormone, adrenal steroids, and thyroxine leads to a sudden acceleration in the growth rate. The timing of the growth spurt varies between the sexes, and corresponds to different ages at the onset of puberty. In girls, the growth rate is maximal between ages 10 and 13; boys grow most rapidly between ages 12 and 15. Growth continues at a slower pace until ages 18 to 21. By that time, most of the epiphyseal cartilages have closed. ∞ p. 158

The boundary between adolescence and maturity is very hazy, for it encompasses physical, emotional, behavioral, and legal implications. Adolescence is often said to be over when growth ends, typically in the late teens or early twenties. The individual is then considered physically mature. Although physical growth ends at maturity, physiological changes continue. The sex-specific differences produced at puberty are retained, but further changes occur when sex hormone levels decline at menopause or the male climacteric. ∞ p. 722 All these changes are part of the aging process, or **senescence.** As we have seen, aging reduces an individual's functional capabilities. Even in the absence of such factors as disease or injury, aging-related changes at the molecular level ultimately lead to death.

→ **CONCEPT CHECK QUESTION**

1. An increase in the levels of GnRH, FSH, LH, and sex hormones in the circulation mark the onset of which stage of development?

Answers begin on p. 792.

Clinical Note
ABORTION

Abortion, which is the expulsion of the fetus prior to 20 weeks gestation, is the most common cause of bleeding in the first and second trimesters of pregnancy. The terms *abortion* and *miscarriage* can be used interchangeably. Generally, the lay public thinks of *abortion* as termination of pregnancy at maternal request and of miscarriage as an accident of nature. Medically, the term abortion applies to both kinds of fetal loss.

Spontaneous abortion, which is the naturally occurring termination of pregnancy that is often called *miscarriage*, is most commonly seen between the 12th and 14th weeks of gestation. It is estimated that 10 to 20 percent of all pregnancies end in spontaneous abortion. If the pregnancy has not yet been confirmed, the mother often assumes she is merely having a period with unusually heavy flow.

About half of all abortions are due to fetal chromosomal anomalies. Other causes include maternal reproductive system abnormalities, maternal use of drugs, placental defects, or maternal infections. Although many people believe that trauma and psychological stress can cause abortion, research does not support that belief.

Since you will be interacting with other healthcare professionals, you should be familiar with the variety of terms used to describe the classifications of abortion.

- *Complete Abortion.* Abortion in which all of the uterine contents including the fetus and placenta have been expelled.
- *Incomplete Abortion.* Abortion in which some, but not all, fetal tissue has been passed. Incomplete abortions are associated with a high incidence of infection.
- *Threatened Abortion.* Potential abortion characterized by unexplained vaginal bleeding during the first half of pregnancy in which the cervix is slightly open and the fetus remains in the uterus and is still alive. In some cases of threatened abortion, the fetus still can be saved.
- *Inevitable Abortion.* Potential abortion characterized by vaginal bleeding accompanied by severe abdominal cramping and cervical dilatation, in which the fetus has not yet passed from the uterus, but the fetus cannot be saved.
- *Spontaneous Abortion.* Naturally occurring expulsion of the fetus prior to viability, generally as a result of chromosomal abnormalities. Most spontaneous abortions occur before the twelfth week of pregnancy. Many occur within two weeks after conception and are mistaken for menstrual periods. Commonly called a miscarriage.

- *Elective Abortion.* Abortion in which the termination of pregnancy is desired and requested by the mother. Elective abortions during the first and second trimesters of pregnancy have been legal in the U.S. since 1973. Most elective abortions are performed during the first trimester. Some clinics perform second-trimester abortions. Second-trimester abortions have a higher complication rate than first-trimester abortions. Third-trimester elective abortions are generally illegal in this country.
- *Criminal Abortion.* Intentional termination of a pregnancy, under any condition, that is not allowed by law. It is usually the attempt to destroy a fetus by a person who is not licensed or permitted to do so. Criminal abortions are often attempted by amateurs and they are rarely performed in aseptic surroundings.
- *Therapeutic Abortion.* Termination of a pregnancy deemed necessary by a physician, usually to protect maternal health and well being.
- *Missed Abortion.* Abortion in which fetal death occurs but the fetus is not expelled. This poses a potential threat to the life of the mother if the fetus is retained beyond six weeks.
- *Habitual Abortion.* Spontaneous abortions that occur in three or more consecutive pregnancies.

The patient experiencing an abortion is likely to report crampy abdominal pain and a backache. She is also likely to report vaginal bleeding, which is often accompanied by the passage of clots and tissue. If the abortion was not recent, then frank signs and symptoms of infection may be present. Any tissue or large clots should be retained and given to emergency department personnel. If the abortion occurs during or after the late first trimester, a fetus may be passed. ■

■ Genetics, Development, and Inheritance

Chromosomes contain DNA, and genes are segments of DNA. Each gene carries the information needed to direct the synthesis of a specific polypeptide. Chromosome structure and the functions of genes were introduced in Chapter 3. p. 81 Every nucleated somatic cell in your body carries copies of the original 46 chromosomes present when you were a zygote. Those chromosomes and their component genes represent your **genotype** (JĒN-ō-tīp).

Through development and differentiation, the instructions contained within the genotype are expressed in many ways. No single living cell or tissue makes use of all the information contained within the genotype. For example, in muscle fibers the genes important for excitable membrane formation and contractile proteins are active; whereas, a different set of genes is operating in cells of the pancreatic islets. Collectively, however, the instructions contained within the genotype determine the anatomical and physiological characteristics that make you a unique individual. Collectively those characteristics make up your **phenotype** (FĒ-nō-tīp; *phainein*, to display + *typos*, mark). Specific elements in your phenotype, such as your hair and eye color, skin tone, and foot size, are called phenotypic *characters*, or *traits*.

Your genotype is derived from those of your parents, but not in a simple way. You are not an exact copy of either parent, nor are you an easily identifiable mixture of their characteristics. Our discussion of genetics will begin with an overview of the basic patterns of inheritance and their implications. We will then examine the mechanisms responsible for regulating the activities of the genotype during prenatal development.

Genes and Chromosomes

In humans, every somatic cell contains 46 chromosomes, arranged in 23 pairs. One member of each pair was contributed by the sperm, the other by the ovum. The two members of each pair are known as **homologous** (hō-MOL-o-gus) **chromosomes.** Twenty-two of those pairs are known as **autosomal** (aw-tō-SŌ-mal) **chromosomes.** The twenty-third

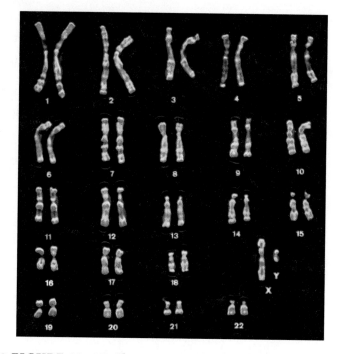

● **FIGURE 20–17** Chromosomes of a Normal Male.

pair is called the *sex chromosomes* because this pair differs between the two sexes. Figure 20–17● shows the **karyotype**, or entire set of chromosomes, of a normal male.

Autosomal Chromosomes

The two chromosomes in a homologous autosomal pair have the same structure and carry genes that affect the same traits. Suppose one member of the pair contains three genes in a row, with number 1 determining hair color, number 2 eye color, and number 3 skin pigmentation. The other member of the pair contains genes that affect the same traits, and the genes are in the same positions and sequence along the chromosomes.

The two chromosomes in a pair may not carry the same *form* of each gene. The various forms of any one gene are called **alleles** (a-LĒLZ; *allelon*, of one another). If both chromosomes of one of your homologous pairs carry the same allele of a particular gene, you are **homozygous** (hō-mō-ZĪ-gus; *homos*, same) for the trait affected by that gene. That allele's trait will be expressed in your phenotype. For example, if you receive a gene for curly hair from your father and one for curly hair from your mother, you will be homozygous for curly hair and will have curly hair. Because the chromosomes of a homologous pair have different origins, one paternal and the other maternal, they need not carry identical alleles. When you have two different alleles of the same gene, you are **heterozygous** (het-er-ō-ZĪ-gus; *heteros*, other) for the trait determined by that gene. In that case, your phenotype will be determined by the interactions between the corresponding alleles:

- An allele that is **dominant** will be expressed in the phenotype regardless of any conflicting instructions carried by the other allele.
- An allele that is **recessive** will be expressed in the phenotype only if it is present on both chromosomes of a homologous pair. For example, albinism is characterized by an inability to synthesize the yellow-brown pigment *melanin*. ∞ p. 127 The presence of a single dominant allele results in normal skin coloration; two recessive alleles must be present to produce albinism.

PREDICTING INHERITANCE. Not every allele can be neatly characterized as dominant or recessive. For the traits listed as dominant in Table 20–5, you can predict the characteristics of individuals on the basis of the parents' alleles.

TABLE 20–5 *The Inheritance of Selected Phenotypic Characteristics*

DOMINANT TRAITS

One allele determines phenotype; the other is suppressed:
- normal skin pigmentation
- lack of freckles
- brachydactyly (short fingers)
- ability to taste phenylthiocarbamate (PTC)
- free earlobes
- curly hair
- color vision
- presence of Rh factor on red blood cell membranes

Both dominant alleles are expressed (codominance):
- presence of A or B antigens on red blood cell membranes
- structure of serum proteins (albumins, transferrins)
- structure of hemoglobin molecule

RECESSIVE TRAITS
- albinism
- freckles
- normal digits
- attached earlobes
- straight hair
- blond hair
- red hair (expressed only if individual is also homozygous for blond hair)
- lack of A, B surface antigens (Type O blood)
- inability to roll the tongue into a U-shape

SEX-LINKED TRAITS
- color blindness
- hemophilia

POLYGENIC TRAITS
- eye color
- hair colors other than pure blond or red

Dominant alleles are traditionally indicated by capitalized abbreviations, and recessives are abbreviated in lowercase letters. For a gene designated *A*, the possible genotypes are indicated by *AA* (homozygous dominant), *Aa* (heterozygous), or *aa* (homozygous recessive). Each gamete involved in fertilization contributes a single allele for a given trait. That allele must be one of the two contained in all somatic cells in the parent's body. Consider, for example, the offspring of an albino mother and a father with normal skin pigmentation. Because albinism is a recessive trait, the maternal alleles are abbreviated *aa*. No matter which of her oocytes gets fertilized, it will carry the recessive *a* allele. The father has normal pigmentation, a dominant trait. He is, therefore, either homozygous dominant or heterozygous for this trait, because both *AA* and *Aa* will produce the same phenotype—normal skin pigmentation.

A simple box diagram known as a *Punnett square* enables us to predict the probabilities that a given child will have particular characteristics. In the Punnett squares in Figure 20–18●, the two maternal alleles are listed along the horizontal axis and the two paternal alleles along the vertical axis. The possible combinations of alleles a child can inherit are indicated in the small boxes. Figure 20–18a shows the possible offspring of an *aa* (albino) mother and an *AA* father. Every child must have the genotype *Aa*, so every child will have normal skin pigmentation. Compare these results with those of Figure 20–18b, which involves a heterozygous father *(Aa)*. The heterozygous male produces two types of gametes, *A* and *a*, and either one may fertilize the oocyte. As a result, the probability is 50 percent that a child of such a father will inherit the genotype *Aa* and so have normal skin pigmentation. The probability of inheriting the genotype *aa*, and thus having the albino phenotype, is also 50 percent.

A Punnett square can also be used to draw conclusions about the identity and genotype of a given child's parent. For example, a man with the genotype *AA* cannot be the father of an albino child *(aa)*.

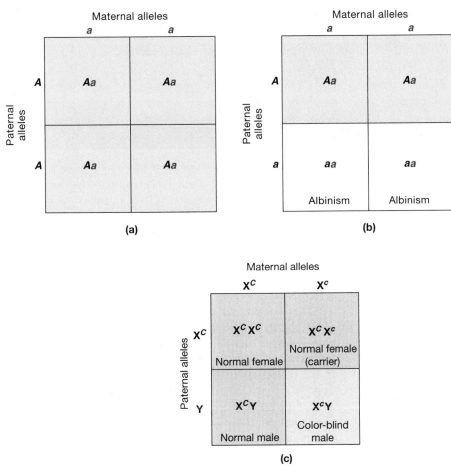

● FIGURE 20–18 Predicting Genotypes and Phenotypes with Punnett Squares. The Punnett squares in parts (a) and (b) concern the inheritance of albinism. **(a)** All the offspring of a homozygous dominant (normally pigmented) father and a homozygous recessive (albino) mother will be heterozygous for that trait, and all will have normal pigmentation. **(b)** The offspring of a heterozygous father and a homozygous recessive mother will either be heterozygous or homozygous for the recessive trait. In this example, half of the offspring will have normal pigmentation and half will be albinos. **(c)** This Punnett square shows the inheritance of color blindness, which is a sex-linked trait. A father with normal vision (X^{C}Y) and a heterozygous mother (X^{C}X^c) will produce daughters with normal vision, but half of the sons will be color-blind.

SIMPLE INHERITANCE. In **simple inheritance,** phenotypes are determined by interactions between a single pair of alleles. The frequency of appearance of an inherited disorder that results from simple inheritance can be predicted using a Punnett square. Although they are rare disorders in terms of overall numbers, more than 1200 inherited conditions are known to reflect the presence of one or two abnormal alleles for a single gene. A partial listing of inherited disorders is given in Table 20–6.

POLYGENIC INHERITANCE. Many phenotypic characters are determined by interactions among several genes. Such interactions are called **polygenic inheritance.** Because multiple alleles are involved, the presence or absence of phenotypic traits cannot easily be predicted using a simple Punnett square. The risks of developing several important adult

TABLE 20-6 *Fairly Common Inherited Disorders*

DISORDER	PAGE IN TEXT
AUTOSOMAL DOMINANTS	
Marfan's syndrome	pp. 107, 760
Huntington's disease	pp. 319, 760
AUTOSOMAL RECESSIVES	
Deafness	p. 359
Albinism	p. 127
Sickle-cell disease	pp. 412, 760
Cystic fibrosis	pp. 550, 760
Phenylketonuria	p. 640
X-LINKED	
Duchenne's muscular dystrophy	pp. 211, 760
Hemophilia (one form)	pp. 429, 761
Color blindness	p. 347

disorders, including hypertension and coronary artery disease, fall within this category.

Many of the developmental disorders responsible for fetal deaths and congenital malformations result from polygenic inheritance. In these cases, the particular genetic composition of the individual does not by itself determine the onset of the disease. Instead, the conditions regulated by these genes establish a susceptibility to particular environmental influences. This means that not every individual with the genetic tendency for a certain condition will actually develop that condition. It is, therefore, difficult to track polygenic conditions through successive generations. However, because many inherited polygenic conditions are likely (but not guaranteed) to occur, steps can be taken to prevent a crisis. For example, you can reduce the likelihood of developing hypertension by controlling your diet and fluid volume, and you can prevent coronary artery disease by lowering your serum cholesterol levels.

Clinical Note
INHERITED DISORDERS: EMERGENCY IMPLICATIONS

Inherited diseases are those diseases that are passed from parents to child through the genetic material (DNA). Many diseases are thought to have an inheritable component. It is clear that some disease processes, such as heart disease, diabetes, asthma, and Alzheimer's disease, tend to cluster in families. However, some diseases are inherited directly. While they are somewhat uncommon, several of them develop conditions that require emergency treatment.

- *Marfan's syndrome.* Marfan's syndrome (MS) is a connective tissue disease that is transmitted through autosomal dominant inheritance. MS is characterized by long, thin extremities that are frequently associated with other skeletal changes, reduced vision due to dislocation of the lenses, and aortic aneurysms. Most persons with MS are tall and thin (some historians have suggested that Abraham Lincoln had MS). Dilation and rupture of an aortic aneurysm is the cause of death in many patients with MS and often occurs in the third or fourth decade of life.

- *Huntington's disease.* Huntington's disease (HD), also called *Huntington's chorea,* is a debilitating disease that is transmitted by autosomal dominant patterns. HD develops at about 35–40 years of age and is characterized by progressive dementia and involuntary movements. The disease advances slowly; death occurs, on average, 15–20 years after the onset of symptoms. HD accounts for many of the nursing home admissions of patients in their fourth or fifth decade of life.

- *Sickle cell disease.* Sickle cell disease (SCD) is a group of disorders characterized by the presence of an abnormal form of hemoglobin *(hemoglobin S).* It is inherited in an autosomal recessive pattern. Patients who receive the gene for SCD from only one parent tend to develop *sickle cell trait (SCT).* SCT occurs in 7–13 percent of black Americans. In East Africa, the incidence of SCT may be as

high as 45 percent. It appears that SCT may provide protection against lethal forms of malaria. Persons who receive the gene for SCD from both parents develop SCD. SCD causes the red blood cells to *sickle* (assume a crescent shape) due to the abnormal hemoglobin. Sickling causes the clinical signs and symptoms of SCD. Patients with SCD are often frequent visitors to hospital emergency departments as they periodically develop *sickle cell crisis,* where blood flow to small blood vessels in organs such as the bones and spleen become occluded. Sickle cell crisis can be extremely painful, requiring high doses of narcotics to achieve pain control.

- *Cystic fibrosis.* Cystic fibrosis (CF) is an autosomal recessive disease of the exocrine glands that causes production of excess, thick mucus that obstructs the gastrointestinal system and lungs. One of the classic signs of CF is elevated concentrations of sodium chloride in sweat and other secretions. The excess mucus coagulates in the ducts of organs or in the airways, which causes dilation and obstruction. Bronchial obstruction predisposes the patient to the development of lung infection, primarily with *Staphylococcal aureus* and *Pseudomonas aeruginosa.* Often the CF patient becomes permanently colonized with one of these bacteria. Progressive obstruction of the respiratory system eventually leads to respiratory failure. CF also impacts the pancreas, and causes obstruction and destruction of the exocrine glands. The mean life expectancy of the CF patient is about 30 years. The need for medical intervention increases significantly as the patient ages.

- *Muscular dystrophy.* Muscular dystrophy (MD) is a group of disorders that cause degeneration of skeletal muscle fibers. The major type of MD is *pseudohypertrophic MD,* often called *Duchenne's MD,* which causes an abnormality in the intracellular metabolism of the muscle fibers. Duchenne's MD is usually detected in children around 3 years of age. Muscle weakness begins at the pelvic girdle and spreads. Eventually, the pulmonary and cardiovascular systems are affected. Duchenne's MD follows an X-linked recessive pattern and is thought to be caused by a single-gene defect.

■ *Hemophilia.* Hemophilia is a disease characterized by abnormal bleeding due to a genetic deficiency in one of the coagulation factors. The most common form of hemophilia is *hemophilia A (classic hemophilia).* It causes a deficiency in factor VII and is inherited as an X-linked recessive disorder that affects males and is transmitted by females. *Hemophilia B,* also called *Christmas disease,* is due to a deficiency in factor IX. It too is an X-linked recessive trait and is clinically indistinguishable from hemophilia A. *Hemophilia C* is an autosomal recessive disease that causes a factor XI deficiency. It occurs equally in males and females. *Von Willebrand disease* is an autosomal dominant trait that causes a defect in factor VIII that differs from the defect caused by hemophilia A. The hemophilias cause bleeding. Spontaneous bleeding into a joint *(hemarthrosis)* is not uncommon and is a frequent reason for hemophiliacs to seek emergency care. ■

Sex Chromosomes

The **sex chromosomes** determine an individual's biological sex. Unlike the other 22 chromosomal pairs, the sex chromosomes are not identical in appearance and gene content. There are two different sex chromosomes: an **X chromosome** and a **Y chromosome.** X chromosomes are considerably larger and have more genes than Y chromosomes. The Y chromosome includes dominant alleles that specify that an individual with that chromosome will be male. The normal pair of sex chromosomes in males is XY. Females do not have a Y chromosome; their sex chromosome pair is XX.

All ova carry an X chromosome, because the only sex chromosomes females have are X chromosomes. But each sperm carries either an X chromosome or a Y chromosome. Thus, using a Punnett square you can show that the ratio of males to females in offspring should be 1:1.

The X chromosome also carries genes that affect somatic structures. These characteristics are called **X-linked** (or *sex-linked*) because in most cases there are no corresponding alleles on the Y chromosome. The best-known X-linked traits are associated with noticeable diseases or functional deficits.

The inheritance of color blindness demonstrates the differences between sex-linked and autosomal inheritance. Normal color vision is determined by the presence of a dominant allele, C, on the X chromosome (designated X^C), whereas red-green color blindness results from the presence of a recessive allele c, on the X chromosome (X^c). A woman, with her two X chromosomes, can be either homozygous dominant, X^CX^C, or heterozygous, X^CX^c, and still have normal color vision. She will be unable to distinguish reds from greens only if she carries two recessive alleles, X^cX^c. But a male has only one X chromosome, so whichever allele that chromosome carries will determine whether he has normal color vision or is red-green color-blind. The Punnett square in Figure 20–18c reveals that each son produced by a father with normal vision and a heterozygous mother will have a 50 percent chance of being red-green color-blind, whereas all daughters will have normal color vision.

A number of other clinical disorders involve X-linked traits, including certain forms of hemophilia, diabetes in- sipidus, and muscular dystrophy. In several instances, advances in molecular genetics techniques have enabled geneticists to locate specific genes on the X chromosome. These techniques provide a relatively direct method of screening for the presence of a particular condition before signs and symptoms appear and even before birth.

The Human Genome Project

Until recently, few of the genes responsible for inherited disorders had been identified or even localized to a specific chromosome. That situation is changing rapidly, however, due to the **Human Genome Project (HGP).** Funded by the National Institutes of Health and the Department of Energy, the project's goal was to describe the entire human **genome**—that is, the full set of DNA, or genetic material—chromosome by chromosome, gene by gene, and nucleotide by nucleotide. The project began in October 1990 and was expected to take 15 years. Progress was more rapid than expected. In terms of size, the human genome was found to consist of some 3.2 billion base pairs, or 3200 Mb (1Mb = 1 megabase, or 1 million base pairs). A working draft of the nucleotide sequence was published in 2001, and 99 percent of the entire genome was listed as finished, "high-quality sequence" as of May 2003. A high-quality sequence is defined as a complete sequence of nucleotides, with no gaps or ambiguities and an error rate of no more than one base per 10,000.

The first step in understanding the human genome was to prepare a map of the individual chromosomes. **Karyotyping** (KAR-ē-ō-tīp-ing; *karyon,* nucleus + *typos,* mark) is the isolation and identification of an individual's complete set of chromosomes, as shown in Figure 20–17. Each chromosome has characteristic banding patterns when stained with special dyes. The banding patterns are useful as reference points when more detailed genetic maps are prepared. The banding patterns themselves can be useful, because abnormal banding patterns are characteristic of some genetic disorders and several cancers.

As of 2005, a progress report included the following:

- Sixteen chromosomes—chromosomes 2, 4, 5, 6, 7, 9, 10, 13, 14, 16, 19, 20, 21, 22, and the X and Y sex chromosomes—have been mapped completely, and several have been completely sequenced. Preliminary maps have been made of all the other chromosomes.

- Roughly 25,000 genes have been tentatively identified, and over 15,000 have been mapped. The first number may represent most of the genes in the human genome.

- The genes responsible for roughly 1500 disorders, including those that cause 60 inherited disorders, have been identified. Examples are included in Figure 20–19●. Genetic screening is now performed for many of these disorders.

It was originally thought that the relationship between genes and proteins was 1:1, and as a result, investigators were anticipating the discovery of as many as 140,000 genes in the human genome. Instead, the total number of protein-coding genes appears to be fewer than 30,000, which represents just 2 percent of the total genome. This stands in stark contrast with a minimal estimate of 400.000 different proteins in the human body. The realization that one gene can carry instructions for more than one protein has revolutionized thinking about genetic diseases and potential therapies. A whole new set of questions has arisen as a result. For example, what factors and enzymes control mRNA processing? How are these factors regulated? Although we may be close to unraveling the human genome, we are still many years from the answers to these questions—and they must be an-

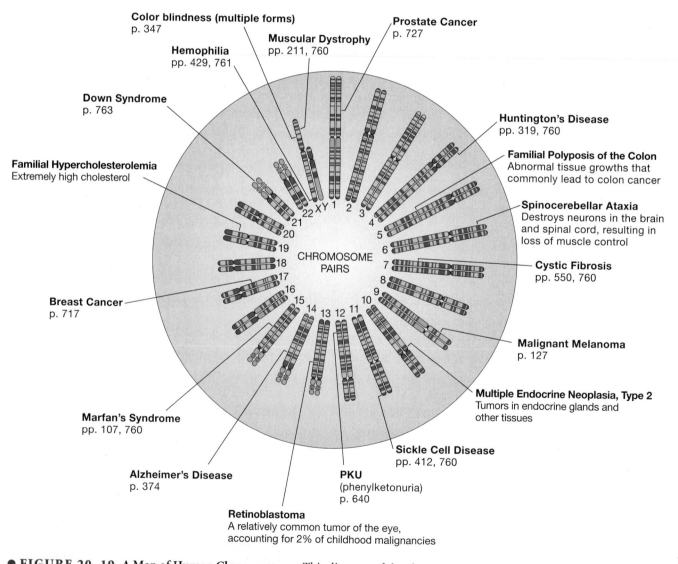

●**FIGURE 20–19** **A Map of Human Chromosomes.** This diagram of the chromosomes of a normal male individual shows typical banding patterns and the locations of the genes responsible for specific inherited disorders. The chromosomes are not drawn to scale.

swered if we are to manipulate genes effectively to treat many congenital (present at birth) diseases.

Of course, controversy remains over the advisability of tinkering with our genetic makeup. The Human Genome Project is attempting to determine the normal genetic composition of a "typical" human. Yet we all are variations on a basic theme. How do we decide what set of genes to accept as "normal?" Moreover, as we improve our abilities to manipulate our own genetic foundations, we will face many additional troubling ethical and legal decisions. For example, few people object to the insertion of a "correct" gene into somatic cells to cure a specific disease. But what if we could insert that modified gene into a gamete and change not only that individual but all of his or her descendants as well? And what if the goal of manipulating the gene was not to correct or prevent any disorder, but instead to "improve" the individual by increasing his or her

intelligence, height, or vision, or by altering some other phenotypic characteristic? Such difficult questions will not go away. In the years to come, we will have to find answers that are acceptable to us all.

→ **CONCEPT CHECK QUESTIONS**

1. Curly hair is an autosomal dominant trait. What would be the phenotype of a person who is heterozygous for this trait?
2. Joe has three daughters and complains that it's his wife's "fault" that he has no sons. What would you tell him?
3. The human genome consists of approximately 3200 Mb. What is a genome and how many nucleotide base pairs does 3200 Mb represent?

Answers begin on p. 792.

Clinical Note
CHROMOSOMAL ABNORMALITIES AND GENETIC ANALYSIS

Embryos that have abnormal autosomal chromosomes rarely survive. However, *translocation defects* and *trisomy* are two types of autosomal chromosome abnormalities that do not invariably kill the individual before birth.

In a *translocation defect,* crossing-over occurs between different chromosome pairs such that, for example, a piece of chromosome 8 may become attached to chromosome 14. The genes moved to their new position may function abnormally, and become inactive or overactive. A translocation between chromosomes 8 and 14 is responsible for *Burkitt's lymphoma,* which is a type of lymphatic system cancer.

In *trisomy,* something goes wrong in meiosis. One of the gametes involved in fertilization carries an extra copy of one chromosome, so the zygote then has three copies of this chromosome rather than two. (The nature of the trisomy is indicated by the number of the chromosome involved. Thus, individuals with trisomy 13 have three copies of chromosome 13.) Zygotes with extra copies of chromosomes seldom survive. Individuals with trisomy 13 and trisomy 18 may survive until delivery but rarely live longer than a year. The notable exception is trisomy 21.

Trisomy 21, or *Down syndrome,* is the most common viable chromosomal abnormality. Estimates of the frequency of appearance range from 1.5 to 1.9 per 1000 births for the U.S. population. Affected individuals exhibit mental retardation and characteristic physical malformations, including a facial appearance that gave rise to the term *mongolism,* which was once used to describe this condition. The degree of mental retardation ranges from moderate to severe. Few individuals with this condition lead independent lives. Anatomical problems that affect the cardiovascular system often prove fatal during childhood or early adulthood. Although some individuals survive to moderate old age, many develop Alzheimer's disease while still relatively young (before age 40).

For unknown reasons, there is a direct correlation between maternal age and the risk of having a child with trisomy 21. For a maternal age below 25, the incidence of Down syndrome approaches 1 in 2000 births, or 0.05 percent. For maternal ages 30–34, the odds increase to 1 in 900, and over the next decade they go from 1 in 290 to 1 in 46, or more than 2 percent. These statistics are becoming increasingly significant because many women are delaying childbearing until their mid-thirties or later.

Abnormal numbers of sex chromosomes do not produce effects as severe as those induced by extra or missing autosomal chromosomes. In *Klinefelter syndrome,* the individual carries the sex chromosome pattern XXY. The phenotype is male, but the extra X chromosome causes reduced androgen production. As a result, the testes fail to mature so the individuals are sterile, and the breasts are slightly enlarged. The incidence of this condition among newborn males averages 1 in 750 births.

Individuals with *Turner syndrome* have only a single, female sex chromosome; their sex chromosome complement is abbreviated XO. This kind of chromosomal deletion is known as *monosomy.* The incidence of this condition at delivery has been estimated as 1 in 10,000 live births. At birth, the condition may not be recognized, because the phenotype is normal female. But maturational changes do not appear at puberty. The ovaries are nonfunctional, and estrogen production occurs at negligible levels.

Fragile-X syndrome causes mental retardation, abnormal facial development, and enlarged testes in affected males. The cause is an abnormal X chromosome that contains a *genetic stutter,* which is an abnormal repetition of a single nucleotide triplet. The presence of the stutter in some way disrupts the normal functioning of adjacent genes and so produces the signs and symptoms of the disorder.

(continued next page)

Clinical Note—*continued*
CHROMOSOMAL ABNORMALITIES AND GENETIC ANALYSIS

Many of these conditions can be detected before birth through the analysis of fetal cells. In *amniocentesis,* a sample of amniotic fluid is removed and the fetal cells it contains are analyzed. This procedure permits the identification of more than 20 congenital conditions, including Down syndrome. The needle inserted to obtain a fluid sample is guided into position during an ultrasound procedure. ∞ p. 23 Unfortunately, amniocentesis has two major drawbacks:

1. Because the sampling procedure represents a potential threat to the health of fetus and mother alike, amniocentesis is performed only when known risk factors are present. Examples of risk factors are a family history of specific conditions, or in the case of Down syndrome, maternal age over 35.

2. Sampling cannot safely be performed until the volume of amniotic fluid is large enough that the fetus will not be injured during the process. The usual time for amniocentesis is at 14–15 weeks of gestation. It may take several weeks to obtain results once samples have been collected, and by the time the results are received, an induced or therapeutic abortion may no longer be a viable option.

An alternative procedure known as *chorionic villus sampling* analyzes cells collected from the chorionic villi during the first trimester. Although it can be performed at an earlier gestational age, this technique has largely been abandoned because it is associated with an increased risk of spontaneous abortion (miscarriage). ■

Clinical Note
TRAUMA DURING PREGNANCY

Trauma is the number-one killer of pregnant females. Penetrating abdominal trauma alone accounts for as much as 36 percent of overall maternal mortality. Gunshot wounds to the abdomen of the pregnant female also account for fetal mortality rates of between 40 and 70 percent. In blunt trauma, auto collisions are the leading cause of maternal and fetal mortality and morbidity. Proper seat belt placement can significantly reduce injury to the pregnant mother and fetus, while improper placement increases the incidence of both uterine rupture and separation of the placenta from the wall of the uterus. Unrestrained mothers in serious auto collisions are four times more likely to suffer fetal mortality than those who are restrained.

The physiologic changes associated with pregnancy protect both the mother and her abdominal organs. With the increasing size of the uterus, most of the abdominal organs are displaced higher in the abdomen (Figure 20–20●). This generally protects them unless blunt or penetrating trauma impacts the upper abdomen. If that happens, then the injury may involve numerous organs with increased morbidity and mortality. Direct penetrating injury to the central and lower abdomen of the late-pregnancy mother often spares her from serious injury; the resulting injury, however, often damages the uterus and endangers the fetus.

The female in the later part of her pregnancy is at additional risk of vomiting and possible aspiration. Increasing uterine size increases intra-abdominal pressure, while the hormones of pregnancy relax the cardiac sphincter (the valve that prevents reflux of the stomach contents). The bladder is displaced superiorly early in pregnancy and then becomes more prone to injury and, when injured, bleeds more heavily.

The increasing size and weight of the uterus and its contents have several effects on the mother, especially when trauma strikes. The uterus of a supine patient in late pregnancy may compress the inferior vena cava and reduce the venous return to the heart. This may induce hypotension in the uninjured patient and have severe consequences in the hemorrhaging trauma patient (Figure 20–21●). The increased intra-abdominal pressure, along with the compression of the inferior vena cava by the gravid uterus, can raise venous pressure in the pelvic region and the lower extremities. This pressure engorges the vessels and increases the rate of venous hemorrhage from pelvic fractures or lower extremity wounds.

The increased maternal vascular volume (up by 45 percent) helps protect the mother from hypovolemia. However, this protection does not extend to the fetus because fetal blood flow is affected well before there are changes in the maternal blood pressure or pulse rate. In fact, changes in maternal blood pressure or heart rate may not become evident until maternal blood loss reaches between 30 and 35 percent. Therefore, it becomes very important to ensure early and aggressive resuscitation of the potentially hypotensive pregnant mother.

In the pregnant female, the thick and muscular uterus contains both the developing fetus and amniotic fluid. This container is strong, and distributes the forces of trauma uniformly to the fetus, which and thereby reduces chances for injury. Significant blunt trauma may cause the uterus to rupture, or penetrating trauma may perforate or tear it. The dangers of severe maternal hemorrhage and disruption of the blood supply to the fetus threaten the lives of the mother and the fetus. The potential release of amniotic fluid into the abdomen is also of great concern. The risk of uterine and fetal injury increases with the length of gestation and is greatest during the third trimester of pregnancy.

If an open wound to the uterus does occur, there may be added risk to the mother, in addition to hemorrhage, if she is Rh negative and the fetus is Rh positive. Penetrating or severe blunt trauma may permit some fetal/maternal blood mixing and lead to compatibility problems. Frank uterine rupture is a rare complication of trauma, but it does occur with severe blunt impact, pelvic fracture, and—very infrequently—with stab or shotgun wounds.

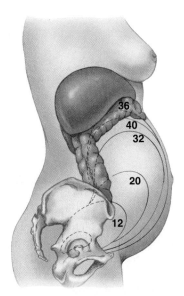

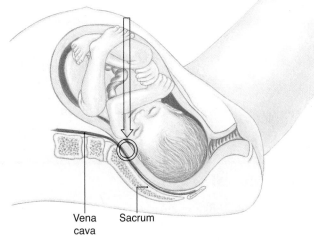

Vena   Sacrum
cava

● **FIGURE 20–21 Supine Hypotensive Syndrome.** When the mother lies flat on her back as the pregnancy nears term, the weight of the gravid uterus compresses the inferior vena cava. This markedly reducing cardiac preload, which leads to hypotension. This situation can be corrected by having the patient turn on her side or tilt the backboard to the left.

ABRUPTIO PLACENTAE

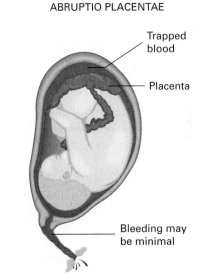

Trapped
blood

Placenta

Bleeding may
be minimal

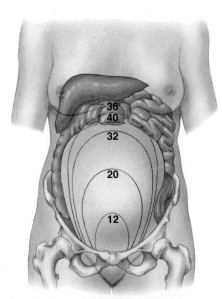

● **FIGURE 20–20 Expansion of the Gravid Uterus by Weeks.** Note that the other abdominal organs are displaced until the uterus becomes the largest organ in the abdomen.

Blunt trauma to the uterus may cause the rather inelastic placenta to detach from the very flexible uterine wall. This condition, called *abruptio placentae*, presents a life-threatening risk to both mother and fetus because the separation causes both maternal and fetal hemorrhage (Figure 20–22●). More frequently than not, this hemorrhage is contained within the uterus and does not extend to the

● **FIGURE 20–22 Abruptio Placentae.** Premature separation of the placenta from the wall of the uterus can result from trauma and threatens the life of both the mother and the fetus.

vaginal outlet. Blunt trauma may also cause the premature rupture of the amniotic sac (breaking of the "membranes" or "bag of waters") and may induce an early labor. ■

Chapter Review

Access more review material online at *www.prenhall.com/bledsoe*. There you will find quiz questions, labeling activities, animations, essay questions, and web links.

Key Terms

amnion 738	gestation 735	parturition 751
blastocyst 737	heterozygous 758	phenotype 757
embryo 733	homozygous 758	placenta 741
fetus 734	implantation 737	trimester 736
genotype 757	neonate 751	trophoblast 737

Related Clinical Terms

abruptio placentae (ab-RUP-shē-ō pla-SEN-tē) A tearing away of the placenta from the uterine wall after the fifth gestational month.

amniocentesis A procedure for the collection and genetic analysis of fetal cells taken from a sample of amniotic fluid.

Apgar rating A method of evaluating newborns for developmental problems and neurological damage.

breech birth A delivery during which the legs or buttocks of the fetus enter the vaginal canal first.

chorionic villus sampling A procedure for the genetic analysis of cells collected from the chorionic villi during the first trimester.

congenital malformation A severe structural abnormality, present at birth, that affects major systems.

ectopic pregnancy A pregnancy in which implantation occurs somewhere other than the uterus.

fetal alcohol syndrome (FAS) A neonatal condition that results from maternal alcohol consumption; characterized by developmental defects that typically involve the skeletal, nervous, and/or cardiovascular systems.

geriatrics A medical specialty concerned with aging and its medical problems.

infertility The inability to achieve pregnancy after 1 year of appropriately timed sexual intercourse.

in vitro fertilization Fertilization outside the body, generally in a petri dish.

pediatrics A medical specialty that focuses on postnatal development from infancy through adolescence.

placenta previa A condition that results from implantation in or near the cervix, in which the placenta covers the cervix and prevents normal birth.

teratogens (TER-a-tō-jenz) Agents or factors that disrupt normal development by damaging cells, altering chromosome structure, or altering the chemical environment of the embryo.

Summary Outline

1. **Development** is the gradual modification of physical and physiological characteristics from **conception** to physical maturity. The creation of different cell types during development is **differentiation.**

AN OVERVIEW OF TOPICS IN DEVELOPMENT 733

1. **Prenatal development** occurs before birth; **postnatal development** begins at birth and continues to maturity, when senescence (aging) begins. **Inheritance** is the transfer of genetically determined characteristics from generation to generation. **Genetics** is the study of the mechanisms of inheritance.

FERTILIZATION 734

1. **Fertilization** normally occurs in the uterine tube within a day after ovulation. Sperm cannot fertilize an egg until they have undergone **capacitation.**

Activation of the Oocyte 734

2. The acrosomal caps of spermatozoa release *hyaluronidase*, which is an enzyme that separates cells of the *corona radiata*. Another acrosomal enzyme digests the zona pellucida and ex-

poses the oocyte membrane. When a single spermatozoon contacts that membrane, fertilization occurs and **oocyte activation** follows.

3. During activation, the secondary oocyte completes meiosis, and the penetration of additional sperm is prevented.

4. After activation, the *female pronucleus* and *male pronucleus* fuse in a process called **amphimixis.** *(Figure 20–1)*

An Overview of Prenatal Development 735

5. The 9-month period of **gestation,** or *pregnancy,* can be divided into three **trimesters.**

THE FIRST TRIMESTER 736

1. The **first trimester** is the most dangerous period of prenatal development. The processes of *cleavage and blastocyst formation, implantation, placentation,* and *embryogenesis* take place during this critical period.

Cleavage and Blastocyst Formation 736

2. **Cleavage** is a series of cell divisions that subdivides the cytoplasm of the zygote. The zygote becomes a hollow ball of

blastomeres called a **blastocyst.** The blastocyst consists of an outer **trophoblast** and an **inner cell mass.** *(Figure 20–2)*

Implantation 737

3. During **implantation,** the blastocyst burrows into the uterine endometrium. Implantation occurs about 7 days after fertilization. *(Figure 20–3)*

4. As the trophoblast enlarges and spreads, maternal blood flows through open *lacunae.* After **gastrulation,** there is an embryonic disc composed of **endoderm, ectoderm,** and **mesoderm.** It is from these three **germ layers** that the body systems differentiate. *(Figure 20–4; Table 20–1)*

5. Germ layers help form four **extraembryonic membranes:** the *yolk sac, amnion, allantois,* and *chorion. (Figure 20–5)*

6. The **yolk sac** is an important site of blood cell formation. The **amnion** encloses fluid that surrounds and cushions the developing embryo. The base of the **allantois** later gives rise to the urinary bladder. Circulation within the vessels of the **chorion** provides a "rapid-transport system" that links the embryo with the trophoblast.

Placentation 741

7. **Placentation** occurs as blood vessels form around the blastocyst and the **placenta** develops. *Chorionic villi* extend outward into the maternal tissues, and form a branching network through which maternal blood flows. As development proceeds, the **umbilical cord** connects the fetus to the placenta. *(Figure 20–6)*

8. The trophoblast synthesizes **human chorionic gonadotropin (hCG),** estrogens, progestins, **human placental lactogen (hPL), placental prolactin,** and **relaxin.**

Embryogenesis 742

9. The first trimester is critical because events in the first 12 weeks establish the basis for **organogenesis** (organ formation). *(Figure 20–7; Table 20–2)*

THE SECOND AND THIRD TRIMESTERS 746

1. In the **second trimester,** the organ systems increase in complexity. During the **third trimester,** these organ systems become functional. *(Figures 20–8, 20–9; Table 20–2)*

Key Note 746

The Effects of Pregnancy on Maternal Systems 746

2. The developing fetus is totally dependent on maternal organs for nourishment, respiration, and waste removal. Maternal adaptations include increases in blood volume, respiratory rate, tidal volume, nutrient intake, and glomerular filtration.

Structural and Functional Changes in the Uterus 750

3. Progesterone produced by the placenta has an inhibitory effect on uterine muscles; its calming action is opposed by estrogens, oxytocin, and prostaglandins. At some point, multiple factors interact to produce **labor contractions** in the uterine wall. *(Figures 20–10, 20–11, 20–12)*

LABOR AND DELIVERY 751

1. The goal of labor is **parturition,** which is the forcible expulsion of the fetus from the uterus.

The Stages of Labor 751

2. Labor can be divided into three stages: the **dilation stage, expulsion stage,** and **placental stage.** *(Figure 20–13)*

Premature Labor 752

3. **Premature labor** results in the delivery of a newborn that has not completed normal development.

Multiple Births 752

4. Twins are either **dizygotic** (fraternal) or **monozygotic** (identical).

POSTNATAL DEVELOPMENT 753

1. Postnatal development involves a series of five **life stages:** the *neonatal period, infancy, childhood, adolescence,* and *maturity. Senescence* begins at maturity and ends in the death of the individual. *(Tables 20–3, 20–4)*

The Neonatal Period, Infancy, and Childhood 754

2. The **neonatal period** extends from birth to 1 month of age. **Infancy** then continues to 2 years of age, and **childhood** lasts until puberty commences. During these stages, major nonreproductive organ systems become fully operational and gradually acquire adult characteristics, and the individual grows rapidly.

3. In the transition from fetus to **neonate,** the respiratory, circulatory, digestive, and urinary systems begin functioning independently. The newborn must also begin to perform thermoregulation.

4. Mammary glands produce protein-rich **colostrum** during the neonate's first few days and then convert to milk production. These secretions are released as a result of the *milk let-down reflex. (Figures 20–14, 20–15, 20–16)*

Adolescence and Maturity 756

5. **Adolescence** begins at **puberty:** (1) the hypothalamus increases its production of GnRH, (2) circulating levels of FSH and LH rise rapidly, and (3) ovarian or testicular cells become more sensitive to FSH and LH. These changes initiate gametogenesis, which is the production of sex hormones, and a sudden acceleration in growth rate. Adolescence continues until growth is completed.

6. **Maturity,** which is the end of growth and adolescence, occurs by the early twenties. Postmaturational changes in physiological processes are part of aging, or **senescence.**

GENETICS, DEVELOPMENT, AND INHERITANCE 757

1. Every somatic cell carries copies of the zygote's original 46 chromosomes; these chromosomes and their genes comprise the individual's genotype. The physical expression of the individual's **genotype** is the **phenotype.**

Genes and Chromosomes 757

2. Every somatic human cell contains 23 pairs of chromosomes; each pair consists of **homologous chromosomes.** Twenty-two pairs are **autosomal chromosomes.** The chromosomes of the twenty-third pair are the sex chromosomes, which differ between the sexes. *(Figure 20–17)*

3. Chromosomes contain DNA, and genes are functional segments of DNA. The various forms of a gene are called **alleles.** If both homologous chromosomes carry the same allele of a particular gene, the individual is **homozygous;** if they carry different alleles, the individual is **heterozygous.**

4. Alleles are either **dominant** or **recessive** depending on how their traits are expressed. *(Table 20–5)*

5. Combining maternal and paternal alleles in a *Punnett square* helps us to predict the characteristics of offspring. *(Figure 20–18)*

6. In **simple inheritance,** phenotypic traits are determined by interactions between a single pair of alleles. **Polygenic inheritance** involves interactions among alleles on several chromosomes. *(Table 20–6)*

7. The two types of **sex chromosomes** are an **X chromosome** and a **Y chromosome.** The normal sex chromosome complement of males is XY; that of females is XX. The X chromosome carries **X-linked** (sex-linked) **genes,** which affect somatic structures but have no corresponding alleles on the Y chromosome.

The Human Genome Project 761

8. The **Human Genome Project** has mapped more than 15,000 genes, including some of those responsible for inherited disorders. *(Figures 20–19 through 20–22; Table 20–6)*

Review Questions
Level 1: Reviewing Facts and Terms

Match each item in column A with the most closely related item in column B. Place letters for answers in the spaces provided.

COLUMN A
- ____ 1. gestation
- ____ 2. cleavage
- ____ 3. gastrulation
- ____ 4. chorion
- ____ 5. human chorionic gonadotropin
- ____ 6. birth
- ____ 7. episiotomy
- ____ 8. afterbirth
- ____ 9. senescence
- ____ 10. neonate
- ____ 11. phenotype
- ____ 12. homozygous recessive
- ____ 13. heterozygous
- ____ 14. male genotype
- ____ 15. female genotype
- ____ 16. trisomy 21

COLUMN B
- a. blastocyst formation
- b. ejection of placenta
- c. germ-layer formation
- d. indication of pregnancy
- e. embryo-maternal circulatory exchange
- f. visible characteristics
- g. time of prenatal development
- h. *aa*
- i. Down syndrome
- j. newborn infant
- k. *Aa*
- l. XY
- m. XX
- n. parturition
- o. process of aging
- p. perineal musculature incision

17. The gradual modification of anatomical structures during the period from conception to maturity is:
    (a) development.   (c) embryogenesis.
    (b) differentiation.   (d) capacitation.

18. Human fertilization involves the fusion of two haploid gametes, which produce a zygote that contains:
    (a) 23 chromosomes.
    (b) 46 chromosomes.
    (c) the normal haploid number of chromosomes.
    (d) 46 pairs of chromosomes.

19. The secondary oocyte that leaves the follicle is in:
    (a) interphase.
    (b) metaphase of the first meiotic division.
    (c) telophase of the second meiotic division.
    (d) metaphase of the second meiotic division.

20. The process that establishes the foundation of all major organ systems is:
    (a) cleavage.   (c) placentation.
    (b) implantation.   (d) embryogenesis.

21. The zygote arrives in the uterine cavity as a:
    (a) morula.
    (b) trophoblast.
    (c) lacuna.
    (d) blastomere.

22. The surface that enables active and passive exchange between the fetal and maternal bloodstreams is the:
    (a) yolk stalk.
    (b) chorionic villi.
    (c) umbilical veins.
    (d) umbilical arteries.

23. Milk let-down is associated with:
    (a) events that occur in the uterus.
    (b) placental hormonal influences.
    (c) circadian rhythms.
    (d) a reflex action triggered by suckling.

24. If an allele must be present on both the maternal and paternal chromosomes to affect the phenotype, the allele is said to be:
    (a) dominant.
    (b) recessive.
    (c) complementary.
    (d) heterozygous.

25. Summarize the developmental changes that occur during the first, second, and third trimesters.

26. Identify the three stages of labor, and describe the events that characterize each stage.

27. Identify the three life stages that occur between birth and approximately age 10. Describe the characteristics of each stage and when it occurs.

Level 2: Reviewing Concepts

28. Relaxin is a peptide hormone that:
    (a) increases the flexibility of the symphysis pubis.
    (b) causes dilation of the cervix.
    (c) suppresses the release of oxytocin by the hypothalamus.
    (d) a, b, and c are correct.

29. During adolescence, the events that interact to promote increased hormone production and sexual maturation result from activity of the:
    (a) hypothalamus.
    (b) anterior pituitary.
    (c) ovaries and testicular cells.
    (d) a, b, and c are correct.

30. In addition to its role in the nutrition of the fetus, what are the primary endocrine functions of the placenta?

31. Discuss the changes that occur in maternal systems during pregnancy. Why are these changes functionally significant?

32. During labor, what physiological mechanisms ensure that uterine contractions continue until delivery has been completed?

33. During the process of labor, to what does the phrase "having one's water break" refer?

34. What would you conclude about a trait in each of the following situations?
    (a) Children who exhibit this trait have at least one parent who exhibits the same trait.
    (b) Children exhibit this trait even though neither of the parents exhibits it.
    (c) The trait is expressed more frequently in sons than in daughters.
    (d) The trait is expressed equally in both daughters and sons.

35. Explain why more men than women are color-blind. Which type of inheritance is involved?

36. Explain the goals and possible benefits of the Human Genome Project.

Level 3: Critical Thinking and Clinical Applications

37. Hemophilia A, a condition in which the blood does not clot properly, is a recessive trait located on the X chromosome (X^h). A woman heterozygous for the trait marries a normal male. What is the probability that this couple will have hemophiliac daughters? What is the probability that this couple will have hemophiliac sons?

38. Explain why the normal heart and respiratory rates of neonates are so much higher than those of adults, even though adults are so much larger.

39. Sally gives birth to a baby with a congenital deformity of the stomach. She swears that it is the result of a viral infection that she suffered during the third trimester of pregnancy. Do you think this is a possibility? Explain.

Roles and Responsibilities of Emergency Medical Personnel

A SIGNIFICANT PORTION OF THE EDUCATION of emergency medical personnel occurs in various hospital settings. Because of this, it is important for emergency care providers to understand the roles and responsibilities of other medical personnel as well as their own. The following discussion outlines the health care system and the roles and responsibilities of various health care providers.

◼ The House of Medicine

Medicine is the science and art of diagnosing, treating, and preventing disease and injury. Disease has been one of humanity's greatest problems. Only during the last 100 years has medicine developed weapons to effectively fight disease. Physicians and other health care professionals use clues to identify, or diagnose, a specific disease or injury. This is usually accomplished by taking a medical history, performing a physical examination, and by assessing the results of laboratory tests and imaging exams. After making a diagnosis, physicians choose the best treatment for the patient. Some treatments cure a disease, others only relieve symptoms and do not cure the underlying disease.

Physicians

Physicians are practitioners who diagnose diseases and injuries, administer treatment, and provide advice on ways to stay healthy. The two kinds of physicians in the U.S. are the doctor of medicine (MD) and the doctor of osteopathy (DO). Both use medicines, surgery, and other standard methods of treating disease. DOs place special emphasis on problems that involve the musculoskeletal system.

Patients may receive care from primary care physicians, specialists, or both. Primary care physicians include general practitioners, family practitioners, general internists, and general pediatricians. Obstetricians/gynecologists are sometimes considered primary care physicians because many women use them as their primary health care providers. Patients usually consult a primary care doctor when they first become ill or injured. Primary care physicians can treat most common disorders, and provide comprehensive care for their patients.

Medical knowledge and technology have advanced so far that no one physician can master the entire field of medicine. Because of this, primary care physicians may refer patients with unusually complicated problems to specialists who have advanced training in a particular disease process or field of medicine. Specialists may even concentrate on one particular area and become subspecialists.

Medical Education

The education of a physician is long and demanding. In the U.S., physicians usually complete four years of college before applying to medical school. The actual courses of study that students pursue in college vary, but students must complete a number of science and mathematics classes before applying. Of the 144 medical schools in the U.S., 125 award the MD degree and 28 award the DO degree. Entrance to medical or osteopathic school is very competitive. Only students with excellent grades and high scores on the Medical College Admissions Test (MCAT) are accepted. In a recent year, close to 47,000 people applied for admission to medical school, but only 17,000 were accepted. Medical school consists of four years of intense study. Typically, the first two years emphasize basic science education including anatomy, physiology, pharmacology, pathology, microbiology, and much more. First- and second-year medical students are slowly introduced to the practice of medicine through classes in physical examination and through observation of practicing physicians. The last two years of medical school are devoted to learning the clinical sciences and are largely spent in a teaching hospital or clinic. Upon completion of medical school, the physician will be awarded the MD or DO degree. Medical graduates must successfully complete a medical licensure examination. In most states, MDs and DOs take the same exam. However, the physician cannot be licensed until he or she has successfully completed a year of postgraduate training, which was formerly called an *internship*.

Upon graduation, physicians can choose to become primary care physicians or specialist physicians. They learn their chosen area of medicine by completing a residency in that field. Residencies may last from three to six years, depending on the field of study chosen. Residencies in the surgical fields are usually the longest. Some physicians still refer to the first year of residency training as an internship and refer to physicians in their first year of postgraduate training as interns. For the most part, however, physicians in postgraduate residencies are referred to as *residents*. They may also be classified by their year of postgraduate training (i.e., PGY-1, PGY-2). At the completion of residency, physicians take an examination in their chosen field and obtain board certification. Some physicians may choose to further specialize following residency training. In this case, they will enter a fellowship that may last from one to five years. Physicians in fellowship programs are referred to as *fellows*. Cardiothoracic surgeons, for example, must complete four years of college, four years of medical school, six years of surgical residency, and four years of cardiothoracic surgical fellowship. This totals 18 years of education after high school before they practice independently. Upon completion of their fellowship, they will take a board certification examination in their specialty.

In England, Australia, India, and many other foreign countries, students who have been accepted to medical school enter medical school immediately after high school. The medical school program in those countries is six years long. The first two years are similar to regular college courses, although the students begin to have some exposure to clinical medicine. The last four years are similar to U.S. medical schools. At the completion of medical school, students are awarded a bachelor of medicine, bachelor of surgery (MBBS) degree. Following this, they can enter general practice or enter into postgraduate training. The MBBS degree is not recognized in the U.S. Thus, when physicians who have the MBBS enter the U.S., they are labeled and licensed as MDs and use the MD degree after their name instead of MBBS.

Upon completion of residency and/or fellowship training, physicians enter the independent practice of medicine. Physicians who have completed their postgraduate training are referred to as *attending physicians*. In teaching hospitals, attending physicians supervise and are responsible for all patient care. Medical students report to residents, residents report to fellows, and fellows report to attendings. The level of training can sometimes be determined by the length of the lab coat worn by the physician. By tradition, medical students wear short lab coats (similar in look to a dinner jacket). Residents wear lab coats that extend to halfway down the thigh. Attending physicians wear full lab coats that extend nearly to the knee.

Medical Specialties

There are many different medical specialties. The American Board of Medical Specialties (ABMS), which is the principal accrediting board of MD training programs, recognizes 17 specialties. The American Osteopathic Association (AOA), which is the principal accrediting board of DO training programs, recognizes 18 major medical specialties and numerous subspecialties. A DO may be certified by an ABMS board if he or she completed an MD residency program or by an AOA board if he or she completed a DO residency. Some DOs are

certified by both boards. The major medical specialties as recognized by the ABMS are:

- *Allergy and immunology.* Physicians who specialize in allergy and immunology study the diagnosis and treatment of allergic and immunological disorders. The physician must complete a three-year residency in either internal medicine or pediatrics and become board certified in that field. Then, he or she must complete a two-year fellowship in allergy and immunology before becoming board certified.
- *Anesthesiology.* Anesthesiologists are physicians who specialize in providing anesthesia for surgical patients. Many anesthesiologists also specialize in the management of acute and chronic pain. Anesthesiology residencies typically last four years. In addition, anesthesiologists can obtain subspecialty certification in critical care medicine and pain management.
- *Colon and rectal surgery.* Colon and rectal surgeons specialize in surgery of the large intestine and rectum. They must first complete a six-year general surgery residency and become board certified in general surgery. Then, they must complete a one-year fellowship in colon and rectal surgery.
- *Dermatology.* A dermatologist is a physician who specializes in diseases of the skin. Dermatologists must complete a four-year residency. In addition, they can obtain subspecialty certification in dermatopathology and dermatological immunology.
- *Emergency medicine.* Physicians who specialize in emergency medicine train to treat and stabilize emergency injuries and illnesses. Emergency physicians must complete a three- or four-year residency. They can obtain certificates of added qualification in medical toxicology, undersea and hyperbaric medicine, pediatric emergency medicine, and sports medicine through completion of a fellowship.
- *Family practice.* A family practitioner is a primary care physician trained in all aspects of medicine. He or she can manage the majority of problems encountered in general medical practice. Some family practitioners practice uncomplicated obstetrics. The family practice residency program is three years long. In addition, family practitioners can obtain certificates of added qualification in adolescent medicine, geriatric medicine, and sports medicine.
- *Internal medicine.* An internist is a physician who specializes in the diagnosis and treatment of diseases of the adult. Residency training for internal medicine is usually three years. There are 10 subspecialties of internal medicine:
  - *Allergy and immunology.* Allergists must complete a two-year fellowship that emphasizies treatment of allergic and immunological diseases.

■ *Cardiology.* Cardiologists must complete a three-year fellowship that emphasizes the treatment of heart and related disorders. Cardiology is subdivided into clinical cardiac electrophysiology and interventional cardiology.

■ *Endocrinology, diabetes, and metabolism.* Endocrinologists must complete a two-year fellowship that concentrates on diseases of the endocrine system, especially diabetes mellitus.

■ *Gastroenterology.* Gastroenterologists undergo a three-year fellowship that concentrates on diseases of the gastrointestinal system and related organs. Significant emphasis is placed upon obtaining competence at fiberoptic endoscopy of the colon and stomach.

■ *Hematology.* Hematologists complete a two-year fellowship that studies diseases of the blood and related disorders.

■ *Infectious disease.* Physicians interested in infectious disease must complete a two-year fellowship that concentrates on disease spread by infectious agents such as bacteria or viruses.

■ *Medical oncology.* Medical oncologists must complete a two-year fellowship that concentrates on treatment, including chemotherapy, of malignant and similar diseases.

■ *Nephrology.* Nephrologists are physicians who study diseases of the kidneys and genitourinary system. The fellowship program lasts two years.

■ *Pulmonary disease.* Pulmonologists are physicians who complete a two-year fellowship that concentrates on the treatment of diseases of the lungs and respiratory tract.

■ *Rheumatology.* Rheumatologists are physicians who treat diseases of the joints and diseases that are autoimmune in nature. The fellowship program is two years.

In addition to those subspecialties, internists can seek certificates of added qualification in adolescent medicine, critical care medicine, geriatric medicine, and sports medicine.

■ *Medical genetics.* Medical geneticists are physicians and scientists who specialize in treating inherited diseases and advising patients about disease possibilities. The residency program is four to five years long.

■ *Neurological surgery.* Neurosurgeons specialize in the surgical treatment of injuries and illnesses that involve the brain, spinal cord, or peripheral nerves. Most neurosurgical residencies are five years long.

■ *Nuclear medicine.* Nuclear medicine physicians use radioactive and stable substances in the diagnosis and treatment of disease. They must complete a two-year residency.

■ *Obstetrics and gynecology.* Physicians who specialize in the treatment of diseases of women are called obstetricians and gynecologists. The residency program is usually four years.

Subspecialty certification is available in gynecological oncology, maternal/fetal medicine, and reproductive endocrinology and fertility.

■ *Ophthalmology.* Ophthalmologists are physicians who specialize in the medical and surgical treatment of eye disorders. They must complete a four-year residency program before sitting for the board certification exam.

■ *Orthopedic surgery.* Surgeons who specialize in the treatment of bone and joint disorders are called orthopedic surgeons. They must complete an accredited orthopedic surgical residency, which usually lasts five to six years. A certificate of added qualification is available in hand surgery (Figure A1–1●).

■ *Otolaryngology.* Physicians who specialize in the medical and surgical treatment of diseases and injuries of the ears, nose, and throat are otolaryngologists. The residency program is five years long.

■ *Pathology.* Physicians who specialize in the laboratory study of disease are pathologists. Pathology residencies are at least four years long. Subspecialty certification is available for: blood banking/transfusion medicine, chemical pathology, cytopathology, dermatopathology, forensic pathology, hematology, medical microbiology, molecular/genetic pathology, neuropathology, and pediatric pathology.

■ *Pediatrics.* Pediatrics is the area of medicine devoted to the care of infants and children. The primary pediatric residency

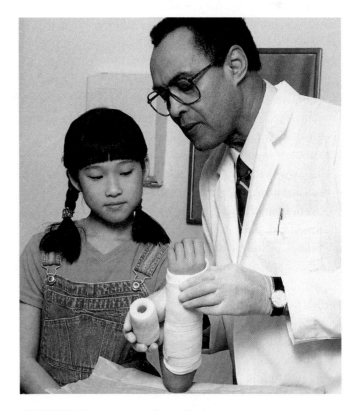

● **FIGURE A1–1 An Orthopedist.**

is three years. Subspecialty certification is available in: adolescent medicine, pediatric cardiology, pediatric critical care medicine, development/behavioral pediatrics, pediatric emergency medicine, pediatric endocrinology, pediatric gastroenterology, pediatric hematology/oncology, pediatric infectious diseases, neonatal/perinatal medicine, pediatric nephrology, neurodevelopmental disorders, pediatric pulmonology, and pediatric rheumatology.

- *Physical medicine and rehabilitation.* Physiatrists are physicians who have completed a four-year residency in physical medicine and rehabilitation. They concentrate on the assessment and rehabilitation of patients with physical disabilities. Subspecialty certification is available in spinal cord injury medicine, pain management, and pediatric rehabilitation medicine.
- *Plastic surgery.* Plastic surgeons are physicians who perform restorative or cosmetic surgery on the skin and associated structures. Residency training is five to six years. Subspecialty certification is available in hand surgery.
- *Preventative medicine.* Preventative medicine physicians typically will concentrate on aerospace medicine, occupational medicine, or general/preventive medicine. Residency programs are typically four years long. Subspecialty certification is available in undersea and hyperbaric medicine and medical toxicology.
- *Psychiatry and neurology.* Psychiatrists are physicians who specialize in the treatment of behavioral and mental disease. Neurologists treat medical conditions of the nervous system. Psychiatry residencies and neurology residencies are typically four years. Subspecialty certification for psychiatrists is available in addiction medicine, child and adolescent psychiatry, clinical neurophysiology, forensic psychiatry, geriatric psychiatry, and pain management. Neurologists can obtain subspecialty certification in clinical neurophysiology and pain management.
- *Radiology.* Radiology is the specialty of medicine that deals with diagnostic imaging through X-rays, ultrasound, magnetic resonance, and similar technologies. Residency programs are usually four years long. The subspecialties of radiology include: diagnostic radiology with special competence in nuclear radiology; certificates of added qualification in vascular and interventional radiology, neuroradiology, and pediatric radiology.
- *Surgery.* Physicians who provide surgical treatment of injuries and illness are called surgeons. Surgical residency programs are usually six years. The subspecialties of surgery include: pediatric surgery, vascular surgery, surgical critical care, and surgery of the hand.

- *Thoracic surgery.* Surgery of the thorax and its contents, including the heart, is the domain of a thoracic surgeon. Thoracic surgery residencies are typically six years long.
- *Urology.* Urology is the surgical treatment of diseases and illnesses of the genitourinary tract. Urologists must complete a six-year residency program.

Osteopathic Medicine

As previously discussed, MDs and DOs are the only complete physicians in the U.S. Medicine as practiced by MDs is often referred to as *allopathic medicine*, while that practiced by DOs is called *osteopathic medicine*. Today, the difference between the two disciplines is minimal. DOs obtain additional undergraduate education in the musculoskeletal system and learn to perform osteopathic manipulation. In addition, DOs are trained to treat the "whole patient" instead of some disease entity.

The system of bones and muscles comprises about two-thirds of the body's mass. DOs know that the body's structure plays a critical role in its ability to function. They can use their eyes and hands to identify structural problems and to support the body's natural tendency toward health and self-healing.

Andrew Taylor Still, MD, developed osteopathic medicine in 1874. Dr. Still became dissatisfied with the effectiveness of nineteenth-century medicine after several of his children died of meningitis. He believed that many of the medications of his day were useless or even harmful. In response, he founded a philosophy of medicine based on ideas that dated back to Hippocrates. The philosophy focused on the unity of all body parts. He identified the musculoskeletal system as a key element of health. Recognizing the body's ability to heal itself, he stressed preventive medicine, eating well, and staying fit. He introduced the concept of osteopathic manipulative therapy where problems with the musculoskeletal system were corrected by applying various manual treatment techniques.

Osteopathic medicine was regarded as a cult until the middle of the twentieth century. Because DOs and MDs did not practice together, DOs established a parallel system of osteopathic hospitals where DOs could practice and schools where DOs could train. Following the Vietnam war, during which the U.S. military commissioned DOs as medical officers, the barriers between MDs and DOs were slowly broken down. Now, DOs have served as surgeon general of the army and as personal physicians for several presidents. Today, DOs and MDs practice together, and in many states, they take the same licensure examination. Unfortunately, because of the demands of medicine today, much of the uniqueness of osteopathic medicine has been lost. Many DOs complete MD residency programs and practice identically to MDs. Osteopathic medical schools now emphasize primary care, and because of this, approximately 64 percent of all DOs practice in primary care, many in rural areas.

Other Doctoral Health Care Providers

Although DOs and MDs are the only complete physicians in U.S. health care, there are several other doctoral level health care providers. These include dentists, podiatrists, optometrists, psychologists, chiropractors, and naturopaths.

Dentists

Dentists treat problems related to the teeth, gums, and related structures. They look beyond the mouth and treat people as individuals. Many systemic disease processes can be identified by examination of the mouth and teeth. Dentists detect and diagnose disease, provide for the aesthetic appearance of the teeth, provide surgical restoration of the teeth, and provide public information about prevention. Dentists perform surgery related to the mouth and can administer and prescribe medications as needed.

Dental school is four years long. Like medical and osteopathic school, students usually enter dental school after completion of a bachelor's degree, although some enter after two years of college. Dental school admission is highly selective. Students must have good college grades, perform well on the Dental Admissions Test (DAT), and show aptitude toward dentistry. Upon completion of dental school, students are awarded the doctor of dental surgery (DDS) or doctor of dental medicine (DMD) degree. Dentists who desire to practice general dentistry can enter practice after dental school. Those who desire to specialize will enter postgraduate training programs, which usually last two years. The recognized specialties of dentistry include:

- *Dental public health.* The science and art of preventing and controlling dental diseases.
- *Endodontics.* The branch of dentistry that is concerned with the morphology, physiology, and pathology of the human dental pulp and periadicular tissues.
- *Oral and maxillofacial pathology.* The branch of dentistry that deals with the nature, identification, and management of oral and maxillofacial diseases.
- *Oral and maxillofacial surgery.* The specialty of dentistry that includes the diagnosis, surgical and adjunctive treatment of diseases, injuries, and defects that involve the oral and maxillofacial region (also called oral surgeons).

- *Orthodontics and dentofacial orthopedics.* The area of dentistry that corrects the movement and location of teeth and related structures.
- *Pediatric dentistry.* The age-defined specialty of dentistry that is limited to infants and children.
- *Periodontics.* The branch of dentistry that treats the gums and other structures that surround the teeth.
- *Prosthodontics.* The branch of dentistry that pertains to restoration or maintenance of oral functions, comfort, and appearance.
- *Oral and maxillofacial radiology.* Specialty of dentistry that involves imaging of the teeth and associated structures.

General dentists and most specialists are office-based. Some of the surgical specialties of dentistry, such as oral and maxillofacial surgery, are hospital-based. A physician will usually work with the oral surgeon to manage nondental-related problems in hospitalized patients.

Podiatrists

Podiatrists are doctoral health care providers who concentrate on medical and surgical care of the feet. Most students who enter podiatric medical school hold a bachelor's degree. Most podiatric medical schools require that students also write the Medical College Admissions Test (MCAT). Podiatric medical school is four years and similar to medical school, except that the emphasis is on the feet. Upon graduation from podiatric medical school, podiatrists are awarded the doctor of podiatric medicine (DPM) degree. Some will enter general practice and others will complete a one- to two-year residency, often in a teaching hospital. Podiatrists perform surgery on the feet and can prescribe medications for foot and related problems. Podiatrists can obtain specialty certification in podiatric orthopedics, podiatric surgery, or primary podiatric medicine.

Optometrists

Doctors of optometry (ODs) are health care providers who examine, diagnose, treat, and manage diseases and disorders of the visual system, the eye, and associated structures. Optometrists enter optometry school after two to four years of college. Optometry school is four years, and students earn the doctor of optometry (OD) degree upon graduation. Optometrists primarily perform refractive examinations of the eye. In some states, optometrists can prescribe a limited number of medications related to the eye. In other states they cannot prescribe. Optometrists are often confused with ophthalmologists. Ophthalmologists are MDs or DOs who have completed a four-year residency in the medical and surgical care of the eyes and related structures. Optometrists do not perform surgery.

Psychologists

Psychologists have completed a clinical psychology program and have been awarded the doctor of philosophy (PhD) degree. They provide individual and family counseling for emotional and mental diseases. They also provide diagnostic testing and assessment. Psychologists cannot prescribe medications. The difference between psychiatrists and psychologists is often misunderstood. Psychiatrists are physicians (MD or DO) who have completed a four-year psychiatry residency. They treat emotional and mental illness with multiple modalities including medications. Often, psychiatrists and psychologists work together, with the psychologist providing much of the counseling while the psychiatrist manages medications and other problems.

Chiropractors

Chiropractic is a branch of the healing arts concerned with human health and disease processes. Central to chiropractic is the "vertebral subluxation," a condition in which a vertebra becomes slightly misaligned with an adjacent segment in such a way as to disturb nerve function. The objective of chiropractic is to locate, analyze, and correct vertebral subluxations, usually through spinal manipulation. One criticism of chiropractic is that the vertebral subluxation has never been scientifically proven to exist.

Some chiropractors limit their practice to spinal manipulation for back pain and related problems. Others believe that chiropractic manipulation can treat other conditions by removing impediments to nerve conduction. This is a major source of discord within the chiropractic profession. "Straight" chiropractors limit their practice to treatment of musculoskeletal problems and back pain. They refer patients to medical or osteopathic doctors for problems that are not of a musculoskeletal origin. "Mixer" chiropractors use spinal manipulation, also called chiropractic adjustments, to treat problems outside of the musculoskeletal system such as ear infections, ulcer disease, and gallbladder disease.

Chiropractic was founded in 1878 when Daniel David Palmer, a magnetic healer, applied force to a bump that he detected on the back of a janitor who had recently lost his hearing after working in a cramped, stooped position. As he applied the force, Palmer felt a "pop," and the bump disappeared. Several days later, the janitor's hearing returned and chiropractic was born.

Chiropractic colleges are usually three to four years long. Students usually enter chiropractic school after two years of college. Upon completion of the chiropractic curriculum,

they are granted a doctor of chiropractic (DC) degree. They then take a state licensure exam and can begin independent practice. Chiropractors cannot perform surgery or prescribe medication. They can use X-rays for evaluation of the spine and can use various physical therapy modalities. In some states in the U.S., chiropractors can refer to themselves as chiropractic physicians.

Spinal manipulation has been found effective for the treatment of mechanical musculoskeletal back pain and similar conditions. Its effectiveness in treating other medical conditions has never been scientifically documented.

Naturopathic Physicians

A relatively new doctoral level health care provider is the naturopathic physician. Naturopathic schools have a four-year program that contains the same basic science courses as traditional medical schools. However, they also study clinical nutrition, acupuncture, homeopathic medicine, botanical medicine, psychology, and counseling. At the completion of their education, students receive a doctor of naturopathic medicine (ND) degree. Naturopathic physicians use holistic and nontoxic approaches to therapy with an emphasis on disease prevention and optimizing wellness. They cannot prescribe medications or practice surgery.

Midlevel Practitioners

With the advent of managed care, a need has arisen to use nonphysicians to assist in primary care roles. These health care providers, often called midlevel practitioners, work with licensed physicians. Physician assistants (PAs), nurse practitioners (NPs), and nurse anesthetists (CRNAs) are the most common midlevel practitioners.

PAs attend physician assistant programs that are usually located in medical schools. They enter after they have completed two years of college. PA school is usually two years long, and some PA programs now offer master's degrees. PAs practice under the direction of a licensed physician. However, the physician does not have to be physically present in the clinic. PAs can specialize in areas of medicine such as family practice, emergency medicine, and pediatrics.

Nurse practitioners are advanced nurses who have completed an approved nurse practitioner program. Nurse practitioners must have a bachelor's degree in nursing and experience as a registered nurse prior to entering NP school. NP school is usually two years long and trains the nurse for practice, usually in a clinic setting. NP programs can specialize in family practice, geriatrics, internal medicine, or pediatrics. NPs work under the direction of a licensed physician. However, the physician does not have to be physically present. NPs often staff rural and public health clinics and can consult with physicians by telephone if needed.

Certified registered nurse anesthetists (CRNAs) are nurses who have completed a two-year program in anesthesia. To enter the program, the nurse must have a bachelor's degree in nursing and several years of clinical nursing experience. Most CRNA programs award a master's degree upon completion. CRNAs work under the supervision of an MD or DO anesthesiologist (although this is being challenged). People often confuse anesthesiologists and anesthetists. Anesthesiologists are MD or DO physicians who have completed a residency program in anesthesiology. Anesthetists are nurses who have completed a nurse anesthesia program.

Nursing

Nurses provide the vast majority of actual patient care in this country. There are two major categories of nurses: registered nurses (RNs) and licensed practical nurses (LPNs). In some states, LPNs are referred to as licensed vocational nurses (LVNs) (Figure A1–2●).

LPNs undergo an intense one-year program that prepares them to provide bedside nursing care. The first part of the program is classroom-based and includes the basic sciences and nursing science. The latter parts of the class are conducted in the hospital, where the students learn and practice their skills. Upon completion of their training, they receive a certificate and are able to sit for the state licensure examination.

Registered nurses can complete one of three different program types. The shortest is a two-year associate's degree program. In this case, the student completes required prenursing classes, then studies nursing science in the classroom, and finally finishes by learning in the hospital setting. Upon completion of the program, students earn an associate degree in nursing (ADN). Although not common today, some programs still offer a diploma program. This is a three-year program that concentrates on prenursing, classroom, and clinical nursing instruction. Finally, many nursing programs offer a bachelor's degree. The first two years are prenursing courses, and the last two years are classroom and clinical nursing classes. Upon completion, the nurse receives a bachelor of science degree in nursing (BSN).

Graduate courses that lead to a master's of science in nursing (MSN) and a PhD in nursing are available. Many community colleges have begun to offer a paramedic-to-nurse bridge program for paramedics who want to become a nurse. Several programs provide nursing education by correspondence and over the Internet. These programs are popular with EMS personnel because of the difficult work schedules inherent in EMS.

Allied Health Personnel

For one person to be well-versed in all aspects of health care is impossible. Because of this, numerous allied health personnel have evolved to assist physicians and nurses in providing patient care. Examples of allied health personnel include:

- *Anesthesiologist assistant.* Assists anesthesiologist by preparing equipment and supplies and by monitoring patients.
- *Art therapist.* Provides art therapy as a rehabilitation tool.
- *Athletic trainer.* Provides preventive and field care for various sporting teams and events.
- *Audiologist.* Provides hearing testing and fits hearing aids and similar devices.
- *Blood bank technologist.* Ensures blood is gathered, stored, tested, and administered in a safe and proper manner.
- *Cardiovascular technologist.* Performs cardiovascular diagnostic testing such as electrocardiograms, echo-cardiography, stress testing, and cardiac monitoring.
- *Clinical laboratory science/medical technology.* Staffs the medical laboratory and performs essential testing of body products.
- *Counseling-related occupations.* Provide counseling for mental health, substance abuse, rehabilitation, and similar patients.
- *Cytotechnologist.* Studies and prepares cells and tissues such as Pap smears.
- *Dental assistant.* Chair-side assistant to a dentist in day-to-day practice.
- *Dental hygienist.* Evaluates and cleans teeth in dental office and instructs the patient in preventive dental care.
- *Dental laboratory technician.* Works in dental lab making dentures, crowns, implants, and bridges.
- *Diagnostic medical sonographer.* Performs diagnostic ultrasound, usually in a hospital radiology department.

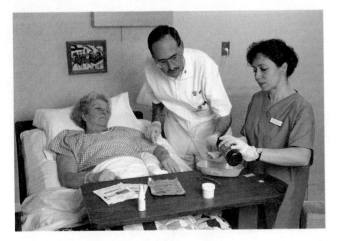

● **FIGURE A1–2** **The Nursing Assistant Helps the Nurse Care for the Patient.**

- *Dietetic technician, dietician.* Instructs patients in dietary health and supervises hospital nutrition programs.
- *Emergency medical technician, paramedic.* Provides advanced emergency care to ill or injured patients.
- *Genetic counselor.* Assists patients and families in regard to diagnosed or possible genetic or hereditary problems.
- *Health information management.* Organizes and controls all patient medical records in hospital or office setting.
- *Histologic technician/histotechnologist.* Prepares tissues for examination by pathologist.
- *Kinesiotherapist.* Assists patients in movement following injury or illness, often working with physical therapists.
- *Medical assistant.* Medical office assistant who provides medical services for a practicing physician or midlevel provider.
- *Music therapist.* Works with other rehabilitative specialists by providing music therapy to rehabilitation and mental health patients.
- *Nuclear medical technologist.* Prepares, administers, and monitors nuclear radioisotopes in a hospital nuclear medicine department.
- *Occupational therapy.* Provides instruction and rehabilitation to patients following injury or illness so they can return to their chosen occupation (Figure A1–3●).
- *Ophthalmic dispensing optician.* Prepares and dispenses eyewear and contact lenses based upon prescription by an optometrist or ophthalmologist.
- *Ophthalmic laboratory technician/technologist.* Technician who prepares prescription lenses in an ophthalmic laboratory.
- *Orthotist.* Prepares braces and similar devices for patients who have had injuries or illnesses.

- *Orthotist and prosthetic technician.* Prepares artificial limbs and other body parts for patients who have undergone amputation or another body changing procedure.
- *Pathologist's assistant.* Assists pathologist in the hospital laboratory and in the autopsy theater.
- *Perfusionist.* Operates cardiac bypass pump for patients undergoing open-heart surgery.
- *Physical therapist/physical therapy assistant.* A physical therapist is a person with a bachelor's or master's degree in physical therapy who prepares and supervises physical rehabilitation for a patient. A physical therapy assistant is a person with an associate's degree who assists a physical therapist in patient care including use of modalities.
- *Radiation therapist/radiographer.* Radiation therapist provides therapeutic radiation therapy to cancer patients. Radiographer or radiological technician performs medical imaging techniques such as X-rays, computed tomography (CT) scans, and magnetic resonance imaging (MRI) scans (Figure A1–4●).
- *Rehabilitation counselor.* Oversees emotional and mental component of injury or illness rehabilitation (Figure A1–5●).
- *Respiratory therapist.* Responsible for assessing and providing ordered respiratory procedures for patients. Is also responsible for the preparation and operation of ventilators and all respiratory devices found in a hospital.

● **FIGURE A1–3** Occupational Therapist.

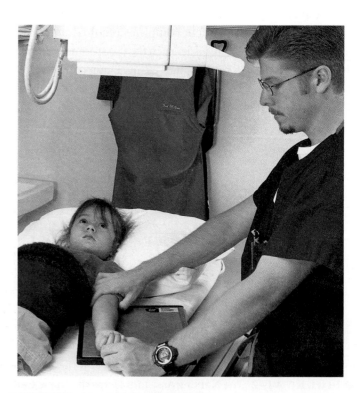

● **FIGURE A1–4** X-Ray Technician.

● **FIGURE A1–5 Assisting a Patient at Home.**

■ *Speech-language pathologist.* Assists patients, especially those who have had a stroke or head injury, in learning to speak and communicate.

■ *Surgical technologist.* Technologist who works in hospital setting assisting surgeons in performing surgical procedures.

■ *Therapeutic recreation specialist.* Instructs and oversees patients' recreation needs, especially those patients who are hospitalized for a prolonged period.

The house of medicine is indeed complex, and this discussion has not covered every aspect of health care provision. However, emergency medical personnel should be familiar with the various health care personnel they will encounter during the course of their hospital rotations, and later as they interact with hospital personnel as members of the emergency medical services system. Remember, emergency medical technicians and paramedics are essential parts of the health care system.

A Periodic Chart of the Elements

The **periodic table** presents the known elements in order of their atomic weights. Each horizontal row represents a single electron shell. The number of elements in that row is determined by the maximum number of electrons that can be stored at that energy level. The element at the left end of each row contains a single electron in its outermost electron shell; the element at the right end of the row has a filled outer electron shell. Organizing the elements in this fashion highlights similarities that reflect the composition of the outer electron shell. These similarities are evident when you examine the vertical columns. All the gases of the rightmost column—helium, neon, argon, krypton, xenon, and radon—have full electron shells; each is a gas at normal atmospheric temperature and pressure, and none reacts readily with other elements. These elements, highlighted in blue, are known as the *noble*, or *inert*, *gases*. In contrast, the elements of the left-most column—lithium, sodium, potassium, and so forth—are silvery, soft metals that are so highly reactive that pure forms cannot be found in nature. The fourth and fifth electron levels can hold 18 electrons. Table inserts are used for the *lanthanide* and *actinide series* to save space, as higher levels can store up to 32 electrons. Elements of particular importance to our discussion of human anatomy and physiology are highlighted in pink.

Atomic number—1
Atomic weight—1.01
H—Chemical symbol
Hydrogen—Element name

1 H																		2 He
Hydrogen 1.01																		Helium 4.00
3 Li	4 Be											5 B	6 C	7 N	8 O	9 F	10 Ne	
Lithium 6.94	Beryllium 9.01											Boron 10.81	Carbon 12.01	Nitrogen 14.01	Oxygen 16.00	Fluorine 19.00	Neon 20.18	
11 Na	12 Mg											13 Al	14 Si	15 P	16 S	17 Cl	18 Ar	
Sodium 22.99	Magnesium 24.31											Aluminum 26.98	Silicon 28.09	Phosphorus 30.97	Sulfur 32.07	Chlorine 35.45	Argon 39.95	
19 K	20 Ca	21 Sc	22 Ti	23 V	24 Cr	25 Mn	26 Fe	27 Co	28 Ni	29 Cu	30 Zn	31 Ga	32 Ge	33 As	34 Se	35 Br	36 Kr	
Potassium 39.10	Calcium 40.08	Scandium 44.96	Titanium 47.88	Vanadium 50.94	Chromium 52.00	Manganese 54.94	Iron 55.85	Cobalt 58.93	Nickel 58.69	Copper 63.55	Zinc 65.39	Gallium 69.72	Germanium 72.61	Arsenic 74.92	Selenium 78.96	Bromine 79.90	Krypton 83.80	
37 Rb	38 Sr	39 Y	40 Zr	41 Nb	42 Mo	43 Tc	44 Ru	45 Rh	46 Pd	47 Ag	48 Cd	49 In	50 Sn	51 Sb	52 Te	53 I	54 Xe	
Rubidium 85.47	Strontium 87.62	Yttrium 88.91	Zirconium 91.22	Niobium 92.91	Molybdenum 95.94	Technetium (98)	Ruthenium 101.07	Rhodium 102.91	Palladium 106.42	Silver 107.87	Cadmium 112.41	Indium 114.82	Tin 118.71	Antimony 121.76	Tellurium 127.60	Iodine 126.90	Xenon 131.29	
55 Cs	56 Ba	57 La *	72 Hf	73 Ta	74 W	75 Re	76 Os	77 Ir	78 Pt	79 Au	80 Hg	81 Tl	82 Pb	83 Bi	84 Po	85 At	86 Rn	
Cesium 132.91	Barium 137.33	Lanthanum 138.91	Hafnium 178.49	Tantalum 180.95	Tungsten 183.85	Rhenium 186.21	Osmium 190.2	Iridium 192.22	Platinum 195.08	Gold 196.97	Mercury 200.59	Thallium 204.38	Lead 207.2	Bismuth 208.98	Polonium (209)	Astatine (210)	Radon (222)	
87 Fr	88 Ra	89 Ac †	104 Db	105 Jl	106 Rf	107 Bh	108 Hn	109 Mt	110	111	112							
Francium (223)	Radium 226.03	Actinium 227.03	Dubnium (261)	Joliotium (262)	Ruther-fordium (263)	Bohrium (262)	Hahnium (265)	Meitnerium (266)	Unnamed (269)	Unnamed (272)	Unnamed (272)							

Lanthanide series

58 Ce	59 Pr	60 Nd	61 Pm	62 Sm	63 Eu	64 Gd	65 Tb	66 Dy	67 Ho	68 Er	69 Tm	70 Yb	71 Lu
Cerium 140.12	Praseo-dymium 140.91	Neodymium 144.24	Promethium (145)	Samarium 150.36	Europium 151.96	Gadolinium 157.25	Terbium 158.93	Dysprosium 162.50	Holmium 164.93	Erbium 167.26	Thulium 168.93	Ytterbium 173.04	Lutetium 174.97

Actinide series

90 Th	91 Pa	92 U	93 Np	94 Pu	95 Am	96 Cm	97 Bk	98 Cf	99 Es	100 Fm	101 Md	102 No	103 Lr
Thorium 232.04	Protactinium 231.04	Uranium 238.03	Neptunium 237.05	Plutonium (244)	Americium (243)	Curium (247)	Berkelium (247)	Californium (251)	Einsteinium (252)	Fermium (257)	Mendelevium (258)	Nobelium (259)	Lawrencium (260)

Weights and Measures

Accurate descriptions of physical objects would be impossible without a precise method of reporting the pertinent data. Dimensions such as length and width are reported in standardized units of measurement, such as inches or centimeters. These values can be used to calculate the **volume** of an object, which is a measurement of the amount of space the object fills. **Mass** is another important physical property. The mass of an object is determined by the amount of matter the object contains; on Earth the mass of an object determines the object's weight.

In the U.S., length and width are typically described in inches, feet, or yards; volumes in pints, quarts, or gallons; and weights in ounces, pounds, or tons. These are units of the **U.S. system** of measurement. Table A3–1 summarizes terms used in the U.S. system. For reference purposes, this table also includes a definition of the "household units," which are popular in recipes. The U.S. system can be very difficult to work with because there is no logical relationship between the various units. For example, there are 12 inches in a foot, 3 feet in a yard, and 1760 yards in a mile. Without a clear pattern of organization, converting feet to inches or miles to feet can be confusing and time-consuming. The relationships among ounces, pints, quarts, and gallons are no more logical than those among ounces, pounds, and tons.

In contrast, the **metric system** has a logical organization based on powers of 10, as indicated in Table A3–2. For example, a **meter** (**m**) is the basic unit for the measurement of size. For measuring larger objects, data can be reported in **dekameters** (*deka*, ten), **hectometers** (*hekaton*, hundred), or **kilometers** (**km**; *chilioi*, thousand); for smaller objects, data can be reported in **decimeters** (0.1 m; *decem*, ten), **centimeters** (**cm** = 0.01 m; *centum*, hundred), **millimeters** (**mm** = 0.001 m; *mille*, thousand), and so forth. In the metric system, the same prefixes are used to report weights, based on the **gram** (**g**), and volumes, based on the **liter** (**L**).

This text reports data in metric units, usually with U.S. equivalents. Use this opportunity to become familiar with the metric system because most technical sources report data only in metric units, and most of the world outside the U.S. uses the metric system exclusively. Conversion factors are included in Table A3–2.

The U.S. and metric systems also differ in their methods of reporting temperatures: in the U.S., temperatures are usually reported in degrees Fahrenheit (°F), whereas scientific literature and individuals in most other countries report temperature in degrees centigrade or degrees Celsius (°C). The relationship between temperatures in degrees Fahrenheit and those in degrees centigrade is indicated in Table A3–2.

TABLE A3–1   *The U.S. System of Measurement*

PHYSICAL PROPERTY	UNIT	RELATIONSHIP TO OTHER U.S. UNITS	RELATIONSHIP TO HOUSEHOLD UNITS
Length	inch (in.)	1 in. = 0.083 ft	
	foot (ft)	1 ft = 12 in.	
		= 0.33 yd	
	yard (yd)	1 yd = 36 in.	
		= 3 ft	
	mile (mi)	1 mi = 5280 ft	
		= 1760 yd	
Volume	fluidram (fl dr)	1 fl dr = 0.125 fl oz	
	fluid ounce (fl oz)	1 fl oz = 8 fl dr	= 6 teaspoons (tsp)
		= 0.0625 pt	= 2 tablespoons (tbsp)
	pint (pt)	1 pt = 128 fl dr	= 32 tbsp
		= 16 fl oz	= 2 cups (c)
		= 0.5 qt	
	quart (qt)	1 qt = 256 fl dr	= 4c
		= 32 fl oz	
		= 2 pt	
		= 0.25 gal	
	gallon (gal)	1 gal = 128 fl oz	
		=8 pt	
		= 4 qt	
Mass	grain (gr)	1 gr = 0.002 oz	
	dram (dr)	1 dr = 27.3 gr	
		= 0.063 oz	
	ounce (oz)	1 oz = 437.5 gr	
		= 16 dr	
	pound (lb)	1 lb = 7000 gr	
		= 256 dr	
		= 16 oz	
	ton (t)	1 t = 2000 lb	

TABLE A3-2 *The Metric System of Measurement*

PHYSICAL PROPERTY	UNIT	RELATIONSHIP TO STANDARD METRIC UNITS	CONVERSION TO U.S. UNITS	
Length	nanometer (nm)	1 nm = 0.000000001m (10^{-9})	= 4×10^{-8} in.	25,400,000 nm = 1 in.
	micrometer (μm)	1 μm = 0.000001 m (10^{-6})	= 4×10^{-5} in.	25,400 μm = 1 in.
	millimeter (mm)	1 mm = 0.001 m (10^{-3})	= 0.0394 in.	25.4 mm = 1 in.
	centimeter (cm)	1 cm = 0.01 m (10^{-2})	= 0.394 in.	2.54 cm = 1 in.
	decimeter (dm)	1 dm = 0.1 m (10^{-1})	= 3.94 in.	0.254 dm = 1 in.
	meter (m)	Standard unit of length	= 39.4 in.	0.0254 m = 1 in.
			= 3.28 ft	0.3048 m = 1 ft
			= 1.09 yd	0.914 m = 1 yd
	dekameter (dam)	1 dam = 10 m		
	hectometer (hm)	1 hm = 100 m		
	kilometer (km)	1 km = 1000 m	= 3280 ft	
			= 1093 yd	
			= 0.62 mi	1.609 km = 1 mi
Volume	microliter (μL)	1 μL = 0.000001 L (10^{-6})		
		= 1 cubic millimeter (mm^3)		
	milliliter (mL)	1 mL = 0.001 L (10^{-3})	= 0.03 fl oz	5 mL = 1 tsp
		= 1 cubic centimeter (cm^3 or cc)		15 mL = 1 tbsp
				30 mL = 1 fl oz
	centiliter (cL)	1 cL = 0.01 L (10^{-2})	= 0.34 fl oz	3 cL = 1 fl oz
	deciliter (dL)	1 dL = 0.1 L (10^{-1})	= 3.38 fl oz	0.29 dL = 1 fl oz
	liter (L)	Standard unit of volume	= 33.8 fl oz	0.0295 L = 1 fl oz
			= 2.11 pt	0.473 L = 1 pt
			= 1.06 qt	0.946 L = 1 qt
Mass	picogram (pg)	1 pg = 0.000000000001 g (10^{-12})		
	nanogram (ng)	1 ng = 0.000000001 g (10^{-9})	= 0.000000015 gr	66,666,666 ng = 1 gr
	microgram (μg)	1 μg = 0.000001 g (10^{-6})	= 0.000015 gr	66,666 μg = 1 gr
	milligram (mg)	1 mg = 0.001 g (10^{-3})	= 0.015 gr	66.7 mg = 1 gr
	centigram (cg)	1 cg = 0.01 g (10^{-2})	= 0.15 gr	6.67 cg = 1 gr
	decigram (dg)	1 dg = 0.1 g (10^{-1})	= 1.5 gr	0.667 dg = 1 gr
	gram (g)	Standard unit of mass	= 0.035 oz	28.4 g = 1 oz
			= 0.0022 lb	454 g = 1 lb
	dekagram (dag)	1 dag = 10 g		
	hectogram (hg)	1 hg = 100 g		
	kilogram (kg)	1 kg = 1000 g	= 2.2 lb	0.454 kg = 1 lb
	metric ton (mt)	1 mt = 1000 kg	= 1.1 t	
			= 2205 lb	0.907 mt = 1 t

TEMPERATURE	CENTIGRADE	FAHRENHEIT
Freezing point of pure water	0°	32°
Normal body temperature	36.8°	98.6°
Boiling point of pure water	100°	212°
Conversion	°C → °F: °F = (1.8 × °C) + 32	°F → °C: °C = (°F − 32) × 0.56

The following illustration spans the entire range of measurements that we consider in this book. Gross anatomy traditionally deals with structural organization as seen with the naked eye or with a simple hand lens. A microscope can provide higher levels of magnification and reveal finer details. Before the 1950s, most information was provided by *light microscopy*. A photograph taken through a light microscope is called a **light micrograph** (**LM**). Light microscopy can magnify cellular structures about 1000 times and show details as fine as 0.25 μm. The symbol **μm** stands for **micrometer**; 1 μm = 0.001 mm, or 0.00004 inches. With a light microscope, we can identify cell types, such as muscle cells or neurons, and see large structures within a cell. Because individual cells are relatively transparent, thin sections cut through a cell are treated with dyes that stain specific structures, which makes them easier to see.

Although special staining techniques can show the general distribution of proteins, lipids, carbohydrates, and nucleic acids in a cell, many fine details of intracellular structure remained a mystery until investigators began using *electron microscopy*. This technique uses a focused beam of electrons, rather than a beam of light, to examine cell structure. In *transmission electron microscopy*, electrons pass through an ultrathin section to strike a photographic plate. The result is a **transmission electron micrograph** (**TEM**). Transmission electron microscopy shows the fine structure of cell membranes and intracellular structures. In *scanning electron microscopy*, electrons bouncing off exposed surfaces create a **scanning electron micrograph** (**SEM**). Although it cannot achieve as much magnification as transmission microscopy, scanning microscopy provides a three-dimensional perspective of cell structure.

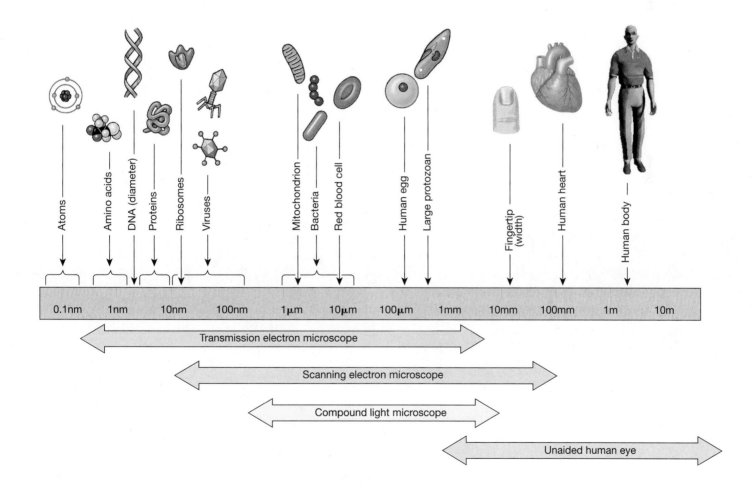

Normal Physiological Values

Tables A4–1 and –2 present normal averages or ranges for the chemical composition of body fluids. These values are approximations rather than absolute values, because test results vary from laboratory to laboratory due to differences in procedures, equipment, normal solutions, and so forth. Sources used in the preparation of these tables are indicated on p. 785. Blanks in the tabular data appear where data are not available.

TABLE A4–1 *The Chemistry of Blood, Cerebrospinal Fluid, and Urine*

TEST	NORMAL RANGES		
	BLOOD[a]	CSF	URINE
pH	S: 7.38–7.44	7.31–7.34	4.5–8.0
OSMOLARITY (mOsm/L)	S: 280–295	292–297	855–1335
ELECTROLYTES	(mEq/L unless noted)		(urinary loss per 24-hour period[b])
Bicarbonate	P: 21–28	20–24	
Calcium	S: 4.5–5.3	2.1–3.0	6.5–16.5 mEq
Chloride	S: 100–108	116–122	120–240 mEq
Iron	S: 50–150 μg/L	23–52 μg/L	40–150 μg
Magnesium	S: 1.5–2.5	2–2.5	4.9–16.5 mEq
Phosphorus	S: 1.8–2.6	1.2–2.0	0.8–2 g
Potassium	P: 3.8–5.0	2.7–3.9	35–80 mEq
Sodium	P: 136–142	137–145	120–220 mEq
Sulfate	S: 0.2–1.3		1.07–1.3 g
METABOLITES	(mg/dL unless noted)		(urinary loss per 24-hour period[c])
Amino acids	P/S: 2.3–5.0	10.0–14.7	41–133 mg
Ammonia	P: 20–150 μg/dL	25–80 μg/dL	340–1200 mg
Bilirubin	S: 0.5–1.0	<0.2	0.02–1.9 mg
Creatinine	P/S: 0.6–1.2	0.5–1.9	1.01–2.5
Glucose	P/S: 70–110	40–70	16–132 mg
Ketone bodies	S: 0.3–2.0	1.3–1.6	10–100 mg
Lactic acid	WB: 5–20[d]	10–20	100–600 mg
Lipids (total)	S: 400–1000	0.8–1.7	0–31.8 mg
Cholesterol (total)	S: 150–300	0.2–0.8	1.2–3.8 mg
Triglycerides	S: 40–150	0–0.9	
Urea	P/S: 23–43	13.8–36.4	12.6–28.6
Uric acid	S: 2.0–7.0	0.2–0.3	80–976 mg
PROTEINS	(g/dL)	(mg/dL)	(urinary loss per 24-hour period[c])
Total	S: 6.0–7.8	2.0–4.5	47–76.2 mg
Albumin	S: 3.2–4.5	10.6–32.4	10–100 mg
Globulins (total)	S: 2.3–3.5	2.8–15.5	7.3 mg (average)
Immunoglobulins	S: 1.0–2.2	1.1–1.7	3.1 mg (average)
Fibrinogen	P: 0.2–0.4	0.65 (average)	

[a]S = serum, P = plasma, WB = whole blood

[b]Because urinary output averages just over 1 liter per day, these electrolyte values are comparable to mEq/L.

[c]Because urinary metabolite and protein data approximate mg/L or g/L, they must be divided by 10 for comparison with CSF or blood concentrations.

[d]Venous blood sample

TABLE A4-2 *The Composition of Minor Body Fluids*

TEST	PERILYMPH	ENDOLYMPH	NORMAL AVERAGES OR RANGES			
			SYNOVIAL FLUID	SWEAT	SALIVA	SEMEN
pH			7.4	4–6.8	6.4[a]	7.19
SPECIFIC GRAVITY			1.008–1.015	1.001–1.008	1.007	1.028
ELECTROLYTES (mEq/L)						
Potassium	5.5–6.3	140–160	4.0	4.3–14.2	21	31.3
Sodium	143–150	12–16	136.1	0–104	14[a]	117
Calcium	1.3–1.6	0.05	2.3–4.7	0.2–6	3	12.4
Magnesium	1.7	0.02		0.03–4	0.6	11.5
Bicarbonate	17.8–18.6	20.4–21.4	19.3–30.6		6[a]	24
Chloride	121.5	107.1	107.1	34.3	17	42.8
PROTEINS (mg/dL)						
Total	200	150	1.72 g/dL	7.7	386[b]	4.5 g/dL
METABOLITES (mg/dL)						
Amino acids				47.6	40	1.26 g/dL
Glucose	104		70–110	3.0	11	224 (fructose)
Urea				26–122	20	72
Lipids (total)	12		20.9	[c]	25–500[d]	188

[a]Increases under salivary stimulation

[b]Primarily alpha-amylase, with some lysozymes

[c]Not present in eccrine secretions

[d]Cholesterol

Sources

Braunwauld, Eugene, Kurt J. Isselbacher, Dennis L. Kasper, Jean D. Wilson, Joseph B. Martin, and Anthony S. Faber, eds. 1998. *Harrison's Principles of Internal Medicine*, 14th ed. New York: McGraw-Hill.

Halsted, James A. 1976. *The Laboratory in Clinical Medicine Interpretation and Application*. Philadelphia: W.B. Saunders Company.

Lentner, Cornelius, ed. 1981. *Geigy Scientific Tables*, 8th ed. Basel, Switzerland: Ciba-Geigy Limited.

Wintrobe, Maxwell, G. Richard Lee, Dane R. Boggs, Thomas C. Bitnell, John Foerster, John W. Athens, and John N. Lukens. 1981. *Clinical Hematology*, Philadelphia: Lea and Febiger.

Appendix V

Key Notes

■ Chapter 1

- All physiological functions are performed by anatomical structures. These functions follow the same physical and mechanical principles that can be seen in the world at large.
- The body can be divided into 11 organ systems, but all work together and the boundaries between them are not absolute.
- Physiological systems work together to maintain a stable internal environment, which is the foundation of homeostasis. In doing so they monitor and adjust the volume and composition of body fluids and keep body temperature within normal limits. If they cannot do so, internal conditions become increasingly abnormal and survival becomes uncertain.
- Anatomical descriptions refer to an individual in the anatomical position: standing, with the hands at the sides, palms facing forward, and feet together.

■ Chapter 2

- All matter is composed of atoms in various combinations. The chemical rules that govern the interactions among atoms alone and in combination establish the foundations of physiology at the cellular level.
- When energy is exchanged, heat is produced. Heat raises local temperatures, but cells cannot capture it or use it to perform work.
- In biological reactions, things tend to even out, unless something prevents this from happening. Most reversible reactions quickly reach equilibrium, where opposing reaction rates are balanced. If reactants are added or removed, reaction rates change until new equilibrium is established.
- Most of the chemical reactions that sustain life cannot occur under homeostatic conditions unless appropriate enzymes are present.
- Water accounts for most of your body weight; proteins, which are the key structural and functional components of cells, and nucleic acids, which control cell structure and function, work only in solution.
- Proteins are the most abundant organic components of the body, and they are the key to both anatomical structure and physiological function. Proteins determine cell shape and tissue properties, and almost all cell functions are performed by proteins and by interactions between proteins and their immediate environment.

■ Chapter 3

- Things tend to even out, unless something—like a cell membrane—prevents this from happening. In the absence of a cell membrane, or across a freely permeable membrane, diffusion will quickly eliminate concentration gradients. Osmosis will attempt to eliminate concentration gradients across membranes that are permeable to water but not to the solutes involved.
- Cells respond directly to their environment and help maintain homeostasis at the cellular level. They can also change their internal structure and physiological functions over time.
- Mitochondria provide most of the energy needed to keep your cells (and you) alive. They require oxygen and organic substrates and they generate carbon dioxide and ATP.
- The nucleus contains the genetic instructions needed to synthesize the proteins that determine cell structure and function. This information is stored in chromosomes, which consist of DNA and various proteins involved in controlling and accessing the genetic information.
- Genes are the functional units of DNA that contain the instructions for making one or more proteins. The creation of specific proteins involves multiple enzymes and three types of RNA.
- Mitosis is the separation of duplicated chromosomes into two identical sets and nuclei in the process of somatic cell division.
- Cancer results from mutations that disrupt the control mechanism that regulates cell growth and division. Cancers most often begin where cells divide rapidly, because the more chromosomes are copied, the greater the chances of error.

■ Chapter 4

- Tissues are collections of cells and extracellular material that perform a specific but limited range of functions. There are four tissue types in varying combinations that form all of the structures of the human body: epithelial, connective, muscle, and neural tissue.

■ Chapter 5

- The epidermis is a multi-layered, flexible, self-repairing barrier that prevents fluid loss, provides protection from UV radiation, produces vitamin D_3, and resists damage from abrasion, chemicals, and pathogens.

- The dermis provides mechanical strength, flexibility, and protection for underlying tissue. It is highly vascular and contains a variety of sensory receptors that provide information about the external environment.
- The skin plays a major role in controlling body temperature. It acts as a radiator: the heat is delivered by the dermal circulation and removed primarily by the evaporation of sweat, or perspiration.

■ Chapter 6

- What you don't use, you lose. The stresses applied to bones during exercise are essential to maintaining bone strength and bone mass.
- A joint cannot be both highly mobile and very strong. The greater the mobility, the weaker the joint, because mobile joints rely on support from muscles and ligaments rather than solid bone-to-bone connections.

■ Chapter 7

- Skeletal muscle fibers shorten as thin filaments interact with thick filaments and sliding occurs. The trigger for contraction is the appearance of free calcium ions in the sarcoplasm; the calcium ions are released by the sarcoplasmic reticulum when the muscle fiber is stimulated by the associated motor neuron. Contraction is an active process; relaxation and the return to resting length are entirely passive.
- All voluntary (intentional) movements involve the sustained contractions of skeletal muscle fibers in incomplete tetanus. The force exerted can be increased by increasing the frequency of action potentials or the number of stimulated motor units (recruitment).
- Skeletal muscles at rest metabolize fatty acids and store glycogen. During light activity, muscles can generate ATP through the aerobic breakdown of carbohydrates, lipids, or amino acids. At peak levels of activity, most of the energy is provided by anaerobic reactions that generate lactic acid as a byproduct.
- What you don't use, you lose. Muscle tone is an indication of the background level of activity in the motor units in skeletal muscles. When inactive for days or weeks, muscles become flaccid, and the muscle fibers break down their contractile proteins and grow smaller and weaker. If inactive for long periods, muscle fibers may be replaced by fibrous tissue.

■ Chapter 8

- Neurons perform all of the communication, information processing, and control functions of the nervous system. Neuroglia outnumber neurons and have functions that are essential to preserving the physical and biochemical structure of neural tissue and the survival of neurons.
- A transmembrane potential exists across the cell membrane. It is there because (1) the cytosol differs from extracellular fluid in its chemical and ionic composition and (2) the cell membrane is selectively permeable. The transmembrane potential can change from moment to moment, as the cell membrane changes its permeability in response to chemical or physical stimuli.
- "Information" travels within the nervous system primarily in the form of propagated electrical signals known as action potentials. The most important information, including vision and balance sensations and the motor commands to skeletal muscles, is carried by myelinated axons.
- At a chemical synapse a synaptic terminal releases a neurotransmitter that binds to the postsynaptic cell membrane. The result is a temporary, localized change in the permeability or function of the postsynaptic cell. This change may have broader effects on the cell, depending on the nature and number of the stimulated receptors. Many drugs affect the nervous system by stimulating receptors that otherwise respond only to neurotransmitters. These drugs can have complex effects on perception, motor control, and emotional states.
- The spinal cord has a narrow central canal surrounded by gray matter that contains sensory and motor nuclei. Sensory nuclei are dorsal; motor nuclei are ventral. The gray matter is covered by a thick layer of white matter that consists of ascending and descending axons. These axons are organized in columns that contain axon bundles with specific functions. Because the spinal cord is so highly organized, it is often possible to predict the results of injuries to localized areas.
- The brain is a large, delicate mass of neural tissue that contains internal passageways and chambers filled with cerebrospinal fluid. Each of the six major regions of the brain has specific functions. As you ascend from the medulla oblongata, which connects to the spinal cord to the cerebrum, those functions become more complex and variable. Conscious thought and intelligence are provided by the neural cortex of the cerebral hemispheres.
- There are 12 pairs of cranial nerves. They are responsible for the special senses of smell, sight, and hearing/balance, and for control over the muscles of the eye, jaw, face, and tongue, and superficial muscles of the neck, back, and shoulders. The cranial nerves also provide sensory information from the

face, neck, and upper chest and autonomic innervation to organs in the thoracic and abdominopelvic cavities.

- Reflexes are rapid, automatic responses to stimuli that "buy time" for the planning and execution of more complex responses that are often consciously directed.

- The autonomic nervous system operates largely outside of our conscious awareness. It has two divisions: a sympathetic division concerned with increasing alertness, metabolic rate, and muscular abilities; and a parasympathetic division concerned with reducing metabolic rate and promoting visceral activities such as digestion.

■ Chapter 9

- Stimulation of a receptor produces action potentials along the axon of a sensory neuron. The frequency or pattern of action potentials contains information about the stimulus. Your perception of the nature of that stimulus depends on the path it takes inside the CNS and the region of the cerebral cortex it stimulates.

- Olfactory information is routed directly to the cerebrum, and olfactory stimuli have powerful effects on mood and behavior. Gustatory sensations are strongest and clearest when integrated with olfactory sensations.

- Light passes through the conjunctiva and cornea, crosses the anterior cavity to reach the lens, transits the lens, crosses the posterior chamber, and then penetrates the neural tissue of the retina before reaching and stimulating the photoreceptors. Cones are most abundant at the fovea and macula lutea, and they provide high-resolution color vision in brightly lit environments. Rods dominate the peripheral areas of the retina, and they provide relatively low-resolution black and white vision in dimly lit environments.

- Balance and hearing rely on the same basic types of sensory receptors (hair cells). The nature of the stimulus that stimulates a particular group of hair cells depends on the structure of the associated sense organ. In the semicircular ducts, the stimulus is fluid movement caused by head rotation in the horizontal, sagittal, or frontal planes. In the utricle and saccule, the stimulus is gravity-induced shifts in the position of attached otoliths. In the cochlea, the stimulus is movement of the basilar membrane by pressure waves.

■ Chapter 10

- Hormones coordinate cell, tissue, and organ activities on a sustained basis. They circulate in the extracellular fluid and bind to specific receptors on or in target cells. They then modify cellular activities by altering membrane permeability, activating or inactivating key enzymes, or changing genetic activity.

- The hypothalamus produces regulatory factors that adjust the activities of the anterior pituitary gland, which produces seven hormones. Most of these hormones control other endocrine organs, including the thyroid gland, adrenal gland, and gonads. It also produces growth hormone, which stimulates cell growth and protein synthesis. The posterior pituitary gland releases two hormones produced in the hypothalamus. ADH restricts water loss and promotes thirst, and oxytocin stimulates smooth muscle contractions in the mammary glands and uterus (in females) and the prostate gland (in males).

- The thyroid gland produces (1) hormones that adjust tissue metabolic rates and (2) a hormone that usually plays a minor role in calcium ion homeostasis by opposing the action of parathyroid hormone.

- The adrenal glands produce hormones that adjust metabolic activities at specific sites, which affects either the pattern of nutrient utilization, mineral ion balance, or the rate of energy consumption by active tissues.

- The pancreatic islets release insulin and glucagon. Insulin is released when blood glucose levels rise, and it stimulates glucose transport into, and utilization by, peripheral tissues. Glucagon is released when blood glucose levels decline, and it stimulates glycogen breakdown, glucose synthesis, and fatty acid release.

■ Chapter 11

- Approximately half of the volume of whole blood consists of cells and cell products. Plasma resembles interstitial fluid but it contains a unique mixture of proteins not found in other extracellular fluids.

- Red blood cells (RBCs) are the most numerous cells in the body. They remain in circulation for approximately four months before being recycled; several million are produced each second. The hemoglobin inside RBCs transports oxygen from the lungs to peripheral tissues; it also carries carbon dioxide from those tissues to the lungs.

- White blood cells (WBCs) are usually outnumbered by RBCs by a ratio of 1000:1. WBCs are responsible for defending the body against infection, foreign cells, or toxins, and for assisting in the cleanup and repair of damaged tissues. The most numerous are neutrophils, which engulf bacteria, and lymphocytes, which are responsible for the specific defenses of the immune response.

■ Platelets are involved in the coordination of hemostasis (blood clotting). When platelets are activated by abnormal changes in their local environment, they release clotting factors and other chemicals. Hemostasis is a complex cascade that establishes a fibrous patch that can subsequently be remodeled and then removed as the damaged area is repaired.

■ Chapter 12

■ The heart has four chambers, two associated with the pulmonary circuit (right atrium and right ventricle) and two with the systemic circuit (left atrium and left ventricle). The left ventricle has a greater workload and is much more massive than the right ventricle, but the two chambers pump equal amounts of blood. AV valves prevent backflow from the ventricles into the atria, and semilunar valves prevent backflow from the aortic and pulmonary trunks into the ventricles.

■ The heart rate is normally established by the cells of the SA node, but that rate can be modified by autonomic activity, hormones, and other factors. From the SA node the stimulus is conducted to the AV node, the AV bundle, the bundle branches, and Purkinje fibers before reaching the ventricular muscle cells. The electrical events associated with the heartbeat can be monitored in an electrocardiogram (ECG).

■ Cardiac output is the amount of blood pumped by the left ventricle each minute. It is adjusted on a moment-to-moment basis by the ANS, and in response to circulating hormones, changes in blood volume, and alterations in venous return. Most healthy people can increase cardiac output by 300–500 percent.

■ Chapter 13

■ Blood flow is the goal, and total peripheral blood flow is equal to cardiac output. Blood pressure is needed to overcome friction and elastic forces and sustain blood flow. If blood pressure is too low, vessels collapse, blood flow stops, and tissues die; if blood pressure is too high, vessel walls stiffen and capillary beds may rupture.

■ Cardiac output cannot increase indefinitely, and blood flow to active versus inactive tissues must be differentially controlled. This is accomplished by a combination of autoregulation, neural regulation, and hormone release.

■ Chapter 14

■ Cell-mediated immunity involves close physical contact between activated cytotoxic T cells and foreign, abnormal, or infected cells. T cell activation usually involves antigen presentation by a phagocytic cell. Cytotoxic T cells may destroy target cells by the local release of cytokines, lymphotoxins, or perforin.

■ Antibody-mediated immunity involves the production of specific antibodies by plasma cells derived from activated B cells. B cell activation usually involves (1) antigen recognition, through binding to surface antibodies, and (2) stimulation by a helper T cell activated by the same antigen. The antibodies produced by active plasma cells bind to the target antigen and either inhibit its activity, destroy it, remove it from solution, or promote its phagocytosis by other defense cells.

■ Immunization produces a primary response to a specific antigen under controlled conditions. If the same antigen is encountered at a later date, it triggers a powerful secondary response that is usually sufficient to prevent infection and disease.

■ Viruses replicate inside of cells, whereas bacteria may live independently. Antibodies (and administered antibiotics) work outside of cells, so they are primarily effective against bacteria rather than viruses. (That is why antibiotics cannot fight the common cold or flu.) T cells, NK cells, and interferons are the primary defenses against viral infection.

■ Chapter 15

■ Hemoglobin within RBCs carries most of the oxygen in the bloodstream, and it releases it in response to changes in the oxygen partial pressure in the surrounding plasma. If the P_{O_2} increases, hemoglobin binds oxygen; if the P_{O_2} decreases, hemoglobin releases oxygen. At a given P_{O_2} hemoglobin will release additional oxygen if the pH decreases or the temperature increases.

■ Carbon dioxide primarily travels in the bloodstream as bicarbonate ions, which form through dissociation of the carbonic acid produced by carbonic anhydrase inside RBCs. Lesser amounts of CO_2 are bound to hemoglobin or dissolved in plasma.

■ A basic pace of respiration is established by the interplay between respiratory centers in the pons and medulla oblongata. That pace is modified in response to input from chemoreceptors, baroreceptors, and stretch receptors. In general, carbon dioxide levels, rather than oxygen levels, are the primary drivers for respiratory activity. Respiratory activity can also be interrupted by protective reflexes and adjusted by the conscious control of respiratory muscles.

Chapter 16

- The stomach is a storage site that provides time for the physical breakdown of food that must precede chemical digestion. Protein digestion begins in the acid environment of the stomach through the action of pepsin. Carbohydrate digestion, which began with the release of salivary amylase by the salivary glands prior to swallowing, continues for a variable period after food arrives in the stomach.

- The small intestine receives and raises the pH of materials from the stomach. It then absorbs water, ions, vitamins, and the chemical products released by the action of digestive enzymes from intestinal glands and exocrine glands of the pancreas.

- The exocrine pancreas produces a mixture of buffers and enzymes essential for normal digestion. Pancreatic secretion occurs in response to the release of regulatory hormones (secretin and CCK) by the duodenum.

- The liver is the center for metabolic regulation in the body. It also produces bile that is stored in the gallbladder and ejected into the duodenum under stimulation of CCK. Bile is essential for the efficient digestion of lipids; it breaks down large lipid droplets so that the individual lipid molecules can be attacked by digestive enzymes.

- The large intestine stores digestive wastes and reduces the volume of the waste by reabsorbing water. Bacterial organisms living in the large intestine are an important source of vitamins, especially vitamin K, biotin, and vitamin B_5.

Chapter 17

- There is an energy cost to staying alive, even at rest. All cells must expend ATP to perform routine maintenance, to remove and replace intracellular and extracellular structures and components. In addition, cells must spend additional energy doing other vital functions, such as growth, secretion, and contraction.

- A balanced diet contains all the ingredients needed to maintain homeostasis, including adequate substrates for energy generation, essential amino acids and fatty acids, minerals, vitamins, and water.

Chapter 18

- The kidneys remove waste products from the blood; they also assist in the regulation of blood volume and blood pressure, ion levels, and blood pH. Nephrons are the primary functional units of the kidneys.

- Roughly 180 L of filtrate is produced at the glomeruli each day, and that represents 70 times the total plasma volume. Almost all of that fluid volume must be reabsorbed to avoid fatal dehydration.

- Reabsorption involves a combination of diffusion, osmosis, and active transport. Many of these processes are independently regulated by local or hormonal mechanisms. The primary mechanism that governs water reabsorption is "water follows salt." Secretion is a selective, carrier-mediated process.

- Fluid balance and electrolyte balance are interrelated. Small water gains or losses affect electrolyte concentrations only temporarily. The impacts are reduced by fluid shifts between the ECF and ICF, and by hormonal responses that adjust the rates of water intake and excretion. Similarly, electrolyte gains or losses produce only temporary changes in solute concentration. These changes are opposed by fluid shifts, adjustments in the rates of ion absorption and secretion, and adjustments to the rates of water gain and loss.

- The most common and acute acid-base disorder is respiratory acidosis, which develops when respiratory activity cannot keep pace with the rate of carbon dioxide generation in peripheral tissues.

Chapter 19

- Meiosis produces gametes that contain half the number of chromosomes found in somatic cells. For each cell that enters meiosis, the testes produce four spermatozoa; whereas the ovaries produce a single ovum.

- Spermatogenesis begins at puberty (sexual maturation) and continues until relatively late in life (above age 70). It is a continuous process, and all stages of meiosis can be observed within the seminiferous tubules.

- Oogenesis begins during embryonic development, and primary oocyte production is completed before birth. After puberty, each month the ovarian cycle produces one or more secondary oocytes from the pre-existing population of primary oocytes. The number of viable and responsive primary oocytes declines markedly over time, until ovarian cycles end at age 45–55.

- Cyclic changes in FSH and LH levels are responsible for the maintenance of the ovarian cycle; the hormones produced by the ovaries in turn regulate the uterine cycle. Inadequate hormone levels, inappropriate or inadequate responses to circulating hormones, or poor coordination and timing of hormone production or secondary oocyte release will reduce or eliminate the chances of pregnancy.
- Sex hormones have widespread effects on the body. They affect brain development and behavioral drives, muscle mass, bone mass and density, body proportions, and the patterns of hair and body fat distribution. As aging occurs, reduction in sexual hormone levels affects appearance, strength, and a variety of physiological functions.

■ Chapter 20

- The basic body plan, the foundations of all of the organ systems, and the four extraembryonic membranes appear during the first trimester. These are complex and delicate processes; not every zygote starts cleavage, and fewer than half of the zygotes that do begin cleavage survive until the end of the first trimester. The second trimester is a period of rapid growth, and is accompanied by the development of fetal organs that will then become fully functional by the end of the third trimester.

Answers to Concept Check and Chapter Review Questions

■ Chapter 1

Concept Check Answers, page 5

1. *Metabolism* refers to all the chemical operations under way in the body. Organisms rely on complex chemical reactions to provide the energy required for responsiveness, growth, reproduction, and movement.

2. *Histologists* specialize in *histology*, which is the study of the structure and properties of tissues and the cells that compose tissues. Because histologists must use microscopes to observe cells, they are specialists in microscopic anatomy.

Concept Check Answers, page 15

1. Physiological systems can function normally only under carefully controlled conditions. Homeostatic regulation prevents potentially disruptive changes in the body's internal environment.

2. Positive feedback is useful in processes that must move quickly to completion once they have begun, such as blood clotting. It is harmful in situations in which a stable condition must be maintained, because it increases any departure from the desired condition. For example, positive feedback in the regulation of body temperature would cause a slight fever to spiral out of control, with fatal results. For this reason, most physiological systems exhibit negative feedback, which tends to oppose any departure from the norm.

3. When homeostasis fails, organ systems function less efficiently or begin to malfunction. The result is the state we call *disease*. If the situation is not corrected, death can result.

Concept Check Answers, page 22

1. The two eyes would be separated by a *midsagittal section*.

2. The body cavity inferior to the diaphragm is the *abdominopelvic* (or *peritoneal*) cavity.

Review Question Answers

Level 1: Reviewing Facts and Terms

1. g 2. d 3. a 4. j 5. b 6. l 7. n 8. f 9. h 10. e 11. c
12. o 13. k 14. i 15. m 16. c 17. a 18. d 19. b
20. c 21. b

Level 2: Reviewing Concepts

22. responsiveness, adaptability, growth, reproduction, movement, metabolism, absorption, respiration, excretion

23. molecule-cell-tissue-organ-organ system-organism

24. Homeostatic regulation refers to adjustments in physiological systems that are responsible for the preservation of homeostasis.

25. In negative feedback, a variation outside normal ranges triggers an automatic response that corrects the situation. In positive feedback, the initial stimulus produces a response that exaggerates the stimulus.

26. The body is erect and the hands are at the sides with the palms facing forward.

27. Stomach. (You would cut the pericardium to access the heart.)

28. (a) ventral cavity (b) thoracic cavity (c) abdominopelvic cavity

Level 3: Critical Thinking and Clinical Applications

29. Since calcitonin is controlled by negative feedback, it should bring about a decrease in blood calcium level, and thus decrease the stimulus for its release.

30. To see a complete view of the medial surface of each half of the brain, midsagittal sections are needed.

■ Chapter 2

Concept Check Answers, page 34

1. Atoms combine with each other such that their outer electron shells have a complete set of electrons. Oxygen atoms do not have a full outer electron shell and so will readily react with many other elements to attain this stable arrangement. Neon already has a full outer shell and, thus, has little tendency to combine with other elements.

2. Hydrogen can exist as three different isotopes: hydrogen-1, with a mass of 1; hydrogen-2 with a mass of 2; and hydrogen-3, with a mass of 3. The heavier sample must contain a higher proportion of one or both of the heavier isotopes.

3. A water molecule is formed by polar covalent bonds. Water molecules are attracted to one another by hydrogen bonds.

Concept Check Answers, page 37

1. Since this reaction involves a large molecule being broken down into two smaller ones, it is a decomposition reaction. Because energy is released in the process, the reaction can also be classified as exergonic.

2. Removing the product of a reversible reaction would keep its concentration low compared to the concentration of the reactants. Thus, the formation of product molecules would continue, but the reverse reaction would slow down, and result in a shift in the equilibrium toward the product.

3. Enzymes in our cells promote chemical reactions by lowering the activation energy requirements. Enzymes make it possible for chemical reactions to proceed under conditions compatible with life.

Concept Check Answers, page 41

1. Water has an unusually high heat capacity because water molecules in the liquid state are attracted to one another through hydrogen bonding. Heat is an increase in the random motion of molecules. Hydrogen bonds retard such motion of water molecules and must be broken to increase the temperature of liquid water. Temperature must be quite high before individual molecules have enough energy to break free of all hydrogen bonds and become water vapor.

2. An acid is a solute that releases H^+ in a solution; a base is a solute that removes H^+ from a solution.

3. The normal pH range of body fluids is 7.35 to 7.45. Fluctuations in pH outside this range can break chemical bonds, alter the shape of molecules, and affect the functioning of cells, which thereby harms cells and tissues.

4. Stomach discomfort is often the result of excess stomach acidity ("acid indigestion"). Antacids contain a weak base that neutralizes the excessive acid.

Concept Check Answers, page 45

1. AC:H:O ratio of 1:2:1 indicates that the molecule is a carbohydrate. The body uses carbohydrates chiefly as an energy source.

2. When two monosaccharides undergo a dehydration synthesis reaction, they form a disaccharide.

3. The most abundant lipid in a sample taken from beneath the skin would be a triglyceride.

4. An analysis of the lipid content of cell membranes would indicate the presence of mostly phospholipids and small amounts of cholesterol.

Concept Check Answers, page 48

1. Proteins are chains of small organic molecules called amino acids.

2. The heat of boiling will break bonds that maintain the three-dimensional shape and/or binding with other proteins or polypeptides. The resulting change in shape will affect the ability of the protein molecule to perform its normal biological functions. These alterations are known as denaturation.

Concept Check Answers, page 54

1. Because this nucleic acid contains the sugar ribose, it is RNA (ribonucleic acid).

2. Both DNA and RNA are nucleic acids composed of sequences of nucleotides. Each nucleotide consists of a five-carbon sugar, a phosphate group (PO_4^{3-}) and a nitrogenous base.

3. Hydrolysis is a decomposition reaction that involves water. The hydrolysis of ATP (adenosine triphosphate) yields ADP (adenosine diphosphate) and P (a phosphate group). It would also release energy for cellular activities.

Review Question Answers

Level 1: Reviewing Facts and Terms

1. f 2. c 3. i 4. g 5. a 6. l 7. d 8. k 9. b 10. e
11. h 12. j 13. a 14. d 15. a

16. Enzymes are specialized protein catalysts that lower the activation energy of chemical reactions. Enzymes speed up chemical reactions but are not used up or changed in the process.

17. carbon, hydrogen, oxygen, nitrogen, calcium, phosphorus
18. carbohydrates, lipids, proteins, nucleic acids
19. support: structural proteins; movement: contractile proteins; transport: transport proteins; buffering; metabolic regulation; coordination and control; defense

Level 2: Reviewing Concepts

20. b **21.** a **22.** a
23. (1) covalent bond: equal sharing of electrons (2) polar covalent bond: unequal sharing of electrons (3) ionic bond: loss and/or gain of electrons
24. A solution such as pure water with a pH of 7 is neutral because it contains equal numbers of hydrogen and hydroxyl ions.
25. A nucleic acid. Carbohydrates and lipids do not contain the element nitrogen. Although both proteins and nucleic acids contain nitrogen, only nucleic acids contain phosphorus.

Level 3: Critical Thinking and Clinical Applications

26. The number of neutrons in an atom is equal to the atomic mass minus the atomic number. In the case of sulfur, this would be $32 - 16 = 16$ neutrons. Since the atomic number of sulfur is 16, the neutral sulfur atom contains 16 protons and 16 electrons. The electrons would be distributed as follows: 2 in the first level, 8 in the second level, and 6 in the third level. To achieve a full 8 electrons in the third level, the sulfur atom could accept 2 electrons in an ionic bond or share 2 electrons in a covalent bond. Since hydrogen atoms can share 1 electron in a covalent bond, the sulfur atom would form 2 covalent bonds, 1 with each of 2 hydrogen atoms.
27. If a person exhales large amounts of CO_2 the equilibrium will shift to the left and the level of H^+ in the blood will decrease. A decrease in the amount of H^+ will cause the pH to rise.

■ Chapter 3

Concept Check Answers, page 64

1. The phospholipid bilayer of the cell membrane forms a physical barrier between the internal environment of the cell and the external environment.
2. Channel proteins are transmembrane proteins that allow water and small ions to pass through the cell membrane.

Concept Check Answers, page 76

1. Active transport processes require the expenditure of cellular energy in the form of the high-energy bonds of ATP molecules. Passive transport processes (*diffusion, osmosis, filtration,* and *facilitated diffusion*) move ions and molecules across the cell membrane without any energy expenditure by the cell.
2. Energy must be expended to transport H^+ ions against their concentration gradient—that is, from a region where they are less concentrated (the cells that line the stomach) to a region where they are more concentrated (the interior of the stomach). An active transport process must be involved.
3. This process is an example of phagocytosis.

Concept Check Answers, page 78

1. The finger-like projections on the surface of intestinal cells are *microvilli*. They increase the cells' surface area so they can absorb nutrients more efficiently.
2. Cells that lack centrioles are unable to divide.

Concept Check Answers, page 81

1. The SER functions in the synthesis of lipids, including steroids. Cells of the ovaries and testes would be expected to have a great deal of SER because these organs produce large amounts of steroid hormones.
2. The function of mitochondria is to produce energy for the cell in the form of ATP molecules. A large number of mitochondria in a cell would indicate a high demand for energy.

Concept Check Answers, page 86

1. The nucleus of a cell contains DNA that codes for the production of all of the cell's proteins. Some of these proteins are structural proteins responsible for the shape and other physical characteristics of the cell. Other proteins are enzymes that govern cellular metabolism, direct the production of cell proteins, and control all of the cell's activities.
2. If a cell lacked the enzyme RNA polymerase, it would not be able to transcribe RNA from DNA.
3. The deletion of a base from a coding sequence of DNA during transcription would alter the entire mRNA base sequence after the deletion point. This would result in different codons on the messenger RNA that was

transcribed from the affected region, and this, in turn, would result in the incorporation of a different series of amino acids into the protein. It is almost certain that the protein product would not be functional.

Concept Check Answers, page 88

1. Cells preparing to undergo mitosis manufacture additional organelles and duplicate sets of their DNA.
2. The four stages of mitosis are prophase, metaphase, anaphase, and telophase.
3. If spindle fibers failed to form during mitosis, the cell would not be able to separate the chromosomes into two sets. If cytokinesis occurred, the result would be one cell with two sets of chromosomes and one cell with none.

Review Question Answers

Level 1: Reviewing Facts and Terms

1. e 2. d 3. h 4. a 5. f 6. b 7. g 8. c 9. m 10. k
11. q 12. i 13. o 14. j 15. l 16. n 17. p 18. b
19. d 20. c 21. d 22. c 23. a 24. c

25. physical isolation; regulation of exchange with the environment; sensitivity; structural support
26. diffusion; filtration; carrier-mediated transport; vesicular transport
27. synthesis of proteins, carbohydrates, and lipids; storage of absorbed molecules; transport of materials; detoxification of drugs or toxins
28. prophase; metaphase; anaphase; telophase

Level 2: Reviewing Concepts

29. b 30. b 31. c 32. c

33. similarities: both processes utilize carrier proteins; differences:

Facilitated diffusion	Active transport
passive	active
no ATP expended	ATP expended
concentration gradient	no concentration gradient

34. Cytosol has a high concentration of K^+; interstitial fluid has a high concentration of Na^+. Cytosol also contains a high concentration of suspended proteins, small quantities of carbohydrates, and large reserves of amino acids and lipids. Cytosol may also contain insoluble materials known as inclusions.
35. In transcription, RNA polymerase uses genetic information to assemble a strand of mRNA. In translation, ribosomes use information carried by the mRNA strand to assemble functional proteins.
36. Prophase: chromatin condenses and chromosomes become visible; centrioles migrate to opposite poles of the cell and spindle fibers develop; nuclear membrane disintegrates. Metaphase: chromatids attach to spindle fibers and line up along the metaphase plate. Anaphase: chromatids separate and migrate toward opposite poles of the cell. telophase: The nuclear membrane re-forms; chromosomes disappear as chromatin relaxes; nucleoli reappear.
37. Cytokinesis is the cytoplasmic movement that separates two daughter cells, which thereby completes mitosis.

Level 3: Critical Thinking and Clinical Applications

38. Facilitated transport, which requires a carrier molecule but not cellular energy. The energy for the process is provided by the diffusion gradient for the substance being transported. When all of the carriers are actively involved in transport, the rate of transport plateaus and cannot be increased further.
39. Solution A must have initially had more solutes than solution B. As a result, water moved by osmosis across the semipermeable membrane from side B to side A, which increases the fluid level on side A.

■ Chapter 4

Concept Check Answers, page 104

1. No. A simple squamous epithelium is not found at the skin surface because it cannot provide enough protection against infection, abrasion, and dehydration.
2. The process described is holocrine secretion.
3. The presence of microvilli on the free surface of epithelial cells greatly increases the surface area for absorption. Cilia function to move materials over the surface of epithelial cells.

Concept Check Answers, page 112

1. The two connective tissues that contain a fluid matrix are blood and lymph.
2. The tissue is adipose (fat) tissue.
3. Collagen fibers add strength to connective tissue. Thus, a vitamin C deficiency might result in connective tissue that is weak and prone to damage.

4. Cartilage lacks a direct blood supply, which is necessary for rapid healing to occur. Materials that are needed to repair damaged cartilage must diffuse from the blood to the chondrocytes. Such diffusion takes a long time and retards the healing process.

Concept Check Answers, page 113

1. Cell membranes are composed of lipid bilayers. Tissue membranes consist of a layer of epithelial tissue and a layer of connective tissue.
2. Serous fluid minimizes the friction between the serous membranes that cover the surfaces of organs and the surrounding body cavity.
3. The lining of the nasal cavity is a mucous membrane.

Concept Check Answers, page 116

1. Given that both cardiac and skeletal muscles are striated (banded), this must be *smooth muscle tissue*.
2. Only skeletal muscle tissue is voluntary.
3. Both skeletal muscle cells and neurons are called fibers because they are relatively long and slender.

Concept Check Answers, page 116

1. Redness, warmth, swelling, and pain are familiar indications of inflammation.
2. Fibrosis is the permanent replacement of normal tissues by fibrous tissue.

Review Question Answers

Level 1: Reviewing Facts and Terms

1. g 2. d 3. j 4. i 5. a 6. f 7. c 8. e 9. b 10. h 11. l
12. k 13. a 14. c 15. d 16. b 17. d 18. c 19. a
20. b 21. b 22. a
23. provide physical protection; control permeability; provide sensations; produce specialized secretions
24. simple, stratified, transitional
25. specialized cells, extracellular protein fibers, fluid ground substance
26. fluid connective tissues: blood and lymph; supporting connective tissues: bone and cartilage
27. mucous, serous, cutaneous, synovial
28. Neurons and neuroglia. The neurons transmit electrical impulses. The neuroglia comprise several kinds of supporting cells and play a role in providing nutrients to neurons.

Level 2: Reviewing Concepts

29. a
30. Holocrine secretion destroys the gland cell. During holocrine secretion, the entire cell becomes packed with secretory products and then bursts, which releases the secretion but kills the cell. The gland cells must be replaced by the division of stem cells.
31. Exocrine secretions are secreted onto a surface or outward through a duct. Endocrine secretions are secreted by ductless glands into surrounding tissues. The secretions are called hormones, which usually diffuse into the blood for distribution to other parts of the body.
32. Tight junctions block the passage of water or solutes between cells. In the digestive tract, these junctions keep enzymes, acids, and wastes from damaging delicate underlying tissues.
33. The extensive connections between cells formed by tight junctions, intercellular cement, and physical interlocking hold skin cells together and can deny access to chemicals or pathogens that may cover their free surfaces. If the skin is damaged and the connections are broken, infection can easily occur.
34. Cutaneous membranes are thick, relatively waterproof, and usually dry.

Level 3: Critical Thinking and Clinical Applications

35. Since animal intestines are specialized for absorption, you would look for a slide that shows a single layer of epithelium that lines the cavity. The cells would be cuboidal or columnar and would probably have microvilli on the surface to increase surface area. With the right type of microscope, you could also see tight junctions between the cells. Since the esophagus receives undigested food, it would have a stratified epithelium that consists of squamous cells to protect it against damage.
36. Step 1: Check for striations. (If striations are present, the choices are skeletal muscle or cardiac muscle. If striations are absent, the tissue is smooth muscle.) Step 2: Check for the presence of intercalated discs. (If the discs are present, the tissue is cardiac muscle. If they are absent, the tissue is skeletal muscle.)

■ Chapter 5

Concept Check Answers, page 128

1. Cells are constantly shed from the *stratum corneum.*
2. Sanding the tips of the fingers will not permanently remove fingerprints. Because the ridges of the fingerprints are formed in layers of the skin that are constantly regenerated, the ridges will eventually reappear.
3. When exposed to the ultraviolet radiation in sunlight or tanning lamps, melanocytes in the epidermis (and dermis) synthesize the pigment melanin, which darkens the skin.

Concept Check Answers, page 136

1. The subcutaneous layer stabilizes the position of the skin relative to underlying tissues and organs, stores fat, and provides a useful site for the injection of drugs (because its deeper region contains few capillaries and no vital organs).
2. Contraction of the arrector pili muscles pulls the hair follicles erect, which depresses the area at the base of the hair and makes the surrounding skin appear higher. The result is known as "goose bumps" or "goose pimples."
3. Hair is a derivative of the epidermis, but the follicles are in the dermis. Where the epidermis and deep dermis are destroyed, no new hair will grow.

Concept Check Answers, page 138

1. Sebaceous gland secretions, called *sebum,* lubricate both the hair and skin and inhibit the growth of bacteria.
2. Deodorants are used to mask the odor of apocrine sweat gland secretions, which contain several kinds of organic compounds. These compounds either have an odor or produce an odor when metabolized by skin bacteria.

Concept Check Answers, page 145

1. Skin can regenerate effectively because stem cells are present in both the epithelial and connective tissues of the skin. After an injury, cells of the stratum germinativum replace epithelial cells, and connective tissue stem cells replace cells lost from the dermis.
2. It is more difficult for the elderly to cool themselves in hot weather because, as a person ages, the blood supply to the dermis decreases and merocrine sweat glands become less active.

Concept Check Answers, page 145

1. Vitamin D_3 production in the skin is interrelated with the functions of the endocrine, digestive, skeletal, and muscular systems. Vitamin D_3 production is the starting point in the final synthesis of the hormone *calcitriol* by the endocrine system. Calcitriol, in turn, is essential for the absorption of calcium and phosphorus by the digestive system. The skeletal system is affected because bone maintenance and growth depend on the availability of calcium and phosphorus. The muscular system is affected because calcium is essential for muscle contraction.

Review Question Answers

Level 1: Reviewing Facts and Terms

1. f 2. g 3. i 4. j 5. a 6. e 7. b 8. d 9. c 10. h
11. a 12. d 13. b 14. c 15. a 16. a
17. carotene and melanin
18. Papillary layer: consists of loose connective tissue and contains capillaries and sensory neurons; reticular layer: consists of dense irregular connective tissue and bundles of collagen fibers. The reticular and papillary layers of the dermis contain blood vessels, lymphatic vessels, and nerve fibers.
19. apocrine sweat glands and merocrine sweat glands

Level 2: Reviewing Concepts

20. Substances that are lipid-soluble pass through the permeability barrier easily, because the barrier is composed primarily of lipids that surround the epidermal cells. Water-soluble drugs are hydrophobic to the permeability barrier.
21. A tan is a result of the synthesis of melanin in the skin. Melanin helps prevent skin damage by absorbing ultraviolet radiation before it reaches the deep layers of the epidermis and dermis. Within the epidermal cells, melanin concentrates around the outer wall of the nucleus, so it absorbs the UV light before it can damage the nuclear DNA.
22. The hormone calcitriol is essential for the absorption of calcium and phosphorus, which are elements needed to form strong bones and teeth. The production of this hormone involves the exposure of the deepest layers of the epidermis to UV light where a cholesterol-related steroid is converted into vitamin D_3. Vitamin D_3 is then modified by the liver and

converted by the kidneys into calcitriol. When the body surface is covered, UV light cannot penetrate deep enough into the skin to begin vitamin D_3 production, which results in fragile bones.

23. The subcutaneous layer is not highly vascular and does not contain major organs; thus, this method reduces the potential for tissue damage.

24. The dermis becomes thinner and the elastic fiber network decreases in size, which weakens the integument and causes loss of resilience.

Level 3: Critical Thinking and Clinical Applications

25. The child probably has a fondness for vegetables high in carotene, such as sweet potatoes, squash, and carrots. It is not uncommon for parents to feed a baby foods that he or she prefers to eat. If the child consumes large amounts of carotene, the yellow-orange pigment will be stored in the skin, which produces a yellow-orange skin color.

26. Like most elderly people, Vanessa's grandmother has poor circulation to the skin. As a result, the temperature receptors in the skin do not sense as much warmth as when there is a rich blood supply. The sensory information is relayed to the brain. The brain interprets this as being cool or cold, and thus causes Vanessa's grandmother to feel cold.

■ Chapter 6

Concept Check Answers, page 156

1. If the ratio of collagen to calcium in a bone increased, the bone would be more flexible and less strong.

2. Concentric layers of bone around a central canal are indicative of osteons, which make up compact bone. Given that the ends (epiphyses) of long bones are primarily cancellous (spongy) bone, this sample most likely came from the shaft (diaphysis) of a long bone.

3. Osteoclasts function in breaking down or demineralizing bone, so the bone would have a lower mineral content and, thus, would be weaker.

Concept Check Answers, page 159

1. During intramembranous ossification, fibrous connective tissue is replaced by bone.

2. Long bones of the body (including the femur) have a plate of cartilage, called the epiphyseal cartilage, that separates the epiphysis from the diaphysis so long as the bone is still growing longer. An X-ray would indicate whether the epiphyseal cartilage is still present. If it is, then growth is still occurring; if not, the bone has reached its full length.

3. Castration (removal of the testes) removes the source of the male sex hormone testosterone, which during puberty contributes to a spurt of bone growth and the closure of the epiphyseal cartilages. Thus, we would expect these castrated boys to have a longer (though slower) growth period and be taller than they would have been if they had not been castrated.

4. Pregnant women need large amounts of calcium to support bone growth in the developing fetus. If an expectant mother does not take in enough calcium in her diet, her body will mobilize the calcium reserves of her own skeleton to provide for the needs of the fetus, which results in weakened bones and an increased risk of fracture.

Concept Check Answers, page 166

1. The larger arm muscles of the weight lifter apply more mechanical stress to the bones of the arms. In response to the stress, the bones grow thicker.

2. A simple, or closed, fracture is completely internal; there is no break in the skin. A compound, or open, fracture projects through the skin. Compound fractures are more dangerous because of the possibility of infection or uncontrolled bleeding.

3. The sex hormones known as estrogens play an important role in moving calcium into bones. After menopause, the levels of these hormones decrease dramatically. As a result, it is difficult to replace the calcium that is being lost from bones due to normal aging. Males do not show a decrease in sex hormone (androgen) levels until much later in life.

Concept Check Answers, page 173

1. The mastoid and styloid processes are projections on the temporal bones of the skull.

2. The sella turcica is located in the sphenoid bone; it contains the pituitary gland.

3. The occipital bone of the cranium (specifically, the occipital condyles) articulates with the vertebral column.

Concept Check Answers, page 175

1. The three bones fractured by the ball are the frontal bone (which forms the superior portion of the orbit) and the maxillary and zygomatic bones (which form the inferior portion of the orbit).
2. The paranasal sinuses make some of the heavier skull bones lighter and contain a mucous epithelium that releases mucous secretions into the nasal cavities. The mucus warms, moistens, and filters particles out of the incoming air.
3. A fracture of the coronoid process would make it difficult to close the mouth because such a fracture would hinder the functioning of muscles that attach to the mandible (lower jaw) at the coronoid process.
4. Because many muscles that move the tongue and the larynx are attached to the hyoid bone, you would expect a person with a fractured hyoid bone to have difficulty moving the tongue, breathing, and swallowing.

Concept Check Answers, page 179

1. The dens is part of the second cervical vertebra, or axis, which is in the neck.
2. In adults, the five sacral vertebrae fuse to form a single sacrum.
3. The lumbar vertebrae must support a great deal more weight than do more superior vertebrae. The large vertebral bodies allow the weight to be distributed over a larger area.
4. True ribs are each attached directly to the sternum by their own costal cartilage. False ribs either do not attach to the sternum (the floating ribs) or attach by means of a shared, common costal cartilage.
5. Improper compression of the chest during CPR could (and frequently does) result in a fracture of the sternum or ribs.

Concept Check Answers, page 183

1. The clavicle attaches the scapula to the sternum and, thus, restricts the scapula's range of movement. If the clavicle is broken, then the scapula will have a greater range of movement and will be less stable.
2. The two rounded prominences on either side of the elbow are the lateral and medial epicondyles of the humerus.
3. The radius is lateral when the forearm is in the anatomical position.

Concept Check Answers, page 187

1. The three bones that make up the hip are the ilium, the ischium, and the pubis.
2. Although the fibula is not part of the knee joint and does not bear weight, it is an important site of attachment for many leg muscles. When the fibula is fractured, these muscles cannot function properly, so moving the leg and walking are difficult and painful. The fibula also helps stabilize the ankle joint.
3. Cesar has most likely fractured his calcaneus (heel bone).

Concept Check Answers, page 194

1. In a newborn, this is an amphiarthrotic (slightly movable) joint. Structurally, it is a type of syndesmosis. When the skull bones interlock, they form a synarthrotic (immovable) joint. This type of fibrous joint is called a suture.
2. (a) Moving the humerus away from the body's longitudinal axis is abduction. (b) Turning the palms forward is supination. (c) Bending the elbow is flexion.
3. Flexion and extension are associated with hinge joints.

Concept Check Answers, page 196

1. The subscapular bursa is located in the shoulder joint, so the tennis player is more likely to have inflammation of this structure (bursitis) than the jogger.
2. Daphne has most likely fractured her ulna.

Concept Check Answers, page 197

1. A complete dislocation of the knee is rare because the joint is stabilized by ligaments on its anterior, posterior, medial, and lateral surfaces, as well as by a pair of ligaments within the joint capsule.
2. Damage to the menisci in the knee joint would decrease the joint's stability. The individual would have a more difficult time locking the knee in place while standing and would have to use muscle contractions to stabilize the joint. When the individual stands for long periods, the muscles would fatigue and the knee would "give out." The individual is also likely to experience pain.

Concept Check Answers, page 199

1. The skeletal system provides calcium for the muscular system (normal muscle contraction), the nervous system (normal neural function), and the cardiovascular system (normal contraction of cardiac muscle of the heart).

Review Question Answers

Level 1: Reviewing Facts and Terms

1. i 2. h 3. m 4. l 5. j 6. k 7. f 8. c 9. o 10. b
11. g 12. n 13. d 14. a 15. e 16. b 17. a 18. c
19. b 20. d 21. a 22. a 23. a 24. a 25. b 26. c
27. d 28. b 29. c

30. support, storage of minerals and lipids, blood cell production, protection, leverage

31. In intramembranous ossification, bone develops from fibrous connective tissue. In endochondral ossification, bone develops from a cartilage model.

32. The hyoid is the only bone in the body that does not articulate with another bone.

33. (1) It protects the heart, lungs, thymus, and other structures in the thoracic cavity. (2) It serves as an attachment point for muscles involved with respiration, the position of the vertebral column, and movements of the pectoral girdle and upper extremities.

34. the acromion and coracoid processes

Level 2: Reviewing Concepts

35. The osteons are parallel to the long axis of the shaft, which does not bend when forces are applied to either end. An impact to the side of the shaft can lead to a fracture.

36. The chondrocytes of the epiphyseal cartilage enlarge and divide, which increases the thickness of the cartilage. On the shaft side, the chondrocytes become ossified, and "chase" the expanding epiphyseal cartilage away from the shaft.

37. The lumbar vertebrae have massive bodies and carry a large amount of weight—both causative factors related to rupturing a disc. The cervical vertebrae are more delicate and have small bodies, which increases the possibility of dislocations and fractures in this region compared with other regions of the vertebral column.

38. The clavicles are small and fragile, so fractures are quite common. Their position also makes them vulnerable to injury and damage.

39. The pelvic girdle consists of the ossa coxae (two coxal bones). The pelvis is a composite structure that includes the ossa coxae of the appendicular skeleton and the sacrum and coccyx of the axial skeleton.

40. Articular cartilages do not have perichondrium, and their matrix contains more water than do other cartilages.

41. The slight movability of the pubic symphysis joint facilitates childbirth by spreading the pelvis to ease movement of the baby through the birth canal.

Level 3: Critical Thinking and Clinical Applications

42. The fracture might have damaged the epiphyseal cartilage in Yasmin's right leg. Even though the bone healed properly, the damaged leg did not produce as much cartilage as did the undamaged leg. The result would be a shorter bone on the side of the injury.

43. The virus could have been inhaled through the nose and passed by way of the cribriform plate of the ethmoid bone into the cranium.

44. The large bones of a child's cranium are not yet fused; they are connected by areas of connective tissue called fontanels. By examining the bones, the archaeologist could readily see if sutures had formed yet. By knowing approximately how long it takes for the various fontanels to close and by determining the sizes of the fontanels, she could make a good estimation of the child's age.

45. Frank may have suffered a shoulder dislocation, which is quite a common injury due to the weak nature of the scapulohumeral joint.

46. Ed has a sprained ankle. This condition occurs when ligaments are stretched to the point at which some of the collagen fibers are torn. Stretched ligaments in joints can cause the release of synovial fluid, which results in swelling and pain in the affected area.

■ Chapter 7

Concept Check Answers, page 212

1. Tendons attach muscles to bones, so severing the tendon would disconnect the muscle from the bone. When the muscle contracted, neither the bone nor the body part would be moved.

2. Skeletal muscle appears striated when viewed under the microscope because its myofibrils are composed of

the myofilaments actin and myosin, which have an arrangement that produces a banded appearance.

3. You would expect to find the greatest concentration of calcium ions in the terminal cisternae of the sarcoplasmic reticulum of the muscle fiber.

Concept Check Answers, page 216

1. The ability of a muscle to contract depends on the formation of cross-bridges between the myosin and actin myofilaments, so a drug that interferes with cross-bridge formation would prevent the muscle from contracting.

2. Because the amount of cross-bridge formation is proportional to the amount of available calcium ions, increased permeability of the sarcolemma to calcium ions would lead to an increased intracellular concentration of calcium and a greater degree of contraction. In addition, because relaxation depends on decreasing the amount of calcium in the sarcoplasm, an increase in the permeability of the sarcolemma to calcium could prevent the muscle from being able to relax completely.

3. Without acetylcholinesterase, the motor end plate would be continuously stimulated by the acetylcholine, and the muscle would be locked into contraction.

Concept Check Answers, page 221

1. The amount of tension produced in a skeletal muscle depends on (1) the frequency of stimulation and (2) the number of activated muscle fibers.

2. A motor unit with 1500 fibers is most likely from a large muscle involved in powerful, gross body movements. Muscles that control fine or precise movements (such as movement of the eye or the fingers) have only a few fibers per motor unit, whereas muscles involved in more powerful contractions (such as those that move the leg) have hundreds of fibers per motor unit.

3. Yes. A skeletal muscle undergoing an isometric contraction does not shorten, even though tension in the muscle increases. By contrast, in an isotonic contraction, tension remains constant and the muscle shortens.

Concept Check Answers, page 225

1. The sprinter requires large amounts of energy for a relatively short burst of activity. To supply this demand for energy, the muscles switch to anaerobic

metabolism. Anaerobic metabolism is not as efficient as aerobic metabolism in producing energy, and the process also produces acidic waste products. The lower energy and the waste products contribute to fatigue. Marathon runners, conversely, derive most of their energy from aerobic metabolism, which is more efficient and does not produce the level of waste products that anaerobic respiration does.

2. We would expect activities that require short periods of strenuous activity to produce a greater oxygen debt because this type of activity relies heavily on energy production by anaerobic respiration. Since lifting weights is more strenuous over the short term, we would expect this type of exercise to produce a greater oxygen debt than would swimming laps, which is an aerobic activity.

3. Individuals who are naturally better at endurance activities, such as cycling and marathon running, have a higher percentage of slow muscle fibers, which are physiologically better adapted to this type of activity than are fast fibers, which are less vascular and fatigue faster.

Concept Check Answers, page 227

1. The cell membranes of cardiac muscle cells are extensively interwoven and are bound tightly to each other at intercalated discs, which allows these muscles to "pull together" efficiently. The intercalated discs also contain gap junctions, which allow ions and small molecules to flow directly between cells. This flow results in the rapid passage of action potentials from cell to cell, so their contraction is simultaneous.

2. Cardiac muscle and smooth muscle require extracellular calcium ions for contraction. In skeletal muscle, the calcium ions come from the sarcoplasmic reticulum.

3. Smooth muscle cells can contract over a relatively large range of resting lengths because their actin and myosin filaments are not as rigidly organized as in skeletal muscle.

Concept Check Answers, page 231

1. The opening between the stomach and the small intestine is guarded by a circular muscle known as a sphincter muscle. The concentric circles of muscle fibers in sphincter muscles are ideally suited for opening and closing passageways and for acting as valves in the body.

2. The *triceps brachii* extends the forearm and is an antagonist of the biceps brachii.

3. The name *flexor carpi radialis longus* tells you that this muscle is a long muscle that lies next to the radius and produces flexion at the wrist (carpus).

Concept Check Answers, page 234

1. Contraction of the masseter muscle raises the mandible, whereas relaxation of this muscle depresses the mandible. If you were contracting and relaxing the masseter, you would likely be chewing.

2. You would expect the buccinator muscle, which shapes the mouth for blowing, to be well-developed in a trumpet player.

Concept Check Answers, page 238

1. Damage to the external intercostal muscles would interfere with the process of breathing.

2. A blow to the rectus abdominis would cause that muscle to contract forcefully, and result in flexion of the torso. In other words, you would "double up."

Concept Check Answers, page 246

1. When you shrug your shoulders, you are contracting your levator scapulae muscles.

2. The rotator cuff muscles include the supraspinatus, infraspinatus, subscapularis, and teres minor.

3. Injury to the flexor carpi ulnaris would impair the ability to flex and adduct the wrist.

Concept Check Answers, page 249

1. A "pulled hamstring" is a muscle strain that affects one of the three muscles that comprise the hamstrings— the biceps femoris, the semimembranosus, and the semitendinosus, which collectively function in flexion at the knee and extension at the hip.

2. The calcaneal (Achilles) tendon attaches the soleus and gastrocnemius muscles to the calcaneus (heel bone). When these muscles contract, they produce extension (plantar flexion) at the ankle. A torn Achilles tendon would make ankle extension difficult, and the opposite action (flexion) would be more pronounced as a result of reduced antagonism from the soleus and gastrocnemius.

Concept Check Answers, page 253

1. The number of myofibrils in a muscle fiber decreases with age. The overall effects of such a change are a reduction in muscle strength and endurance and a tendency to fatigue rapidly.

Concept Check Answers, page 253

1. The recovery of skeletal muscle requires the removal of lactic acid from both muscle tissue and body fluids and restoration of energy reserves in the muscle tissue. The cardiovascular system absorbs lactic acid from actively contracting muscle and carries it to the liver, an organ of the digestive system. Within the liver, lactic acid is converted to pyruvic acid. The liver utilizes some of the pyruvic acid to produce ATP, which is then used to combine other pyruvic acid molecules to produce glucose. The cardiovascular system carries the glucose molecules back to the muscles, where they may be combined and stored as glycogen.

Review Question Answers

Level 1: Reviewing Facts and Terms

1. f 2. i 3. d 4. p 5. l 6. c 7. h 8. n 9. b 10. k
11. o 12. a 13. m 14. j 15. g 16. e 17. d 18. d

19. produce skeletal movement, maintain body posture and body position, support soft tissue, guard entrances and exits, maintain body temperature

20. Step 1: active site exposure. Step 2: cross-bridge attachment. Step 3: pivoting of myosin head (power stroke). Step 4: cross-bridge detachment. Step 5: myosin head activation (cocking).

21. ATP, creatine phosphate, glycogen

22. aerobic metabolism and glycolysis

23. The axial musculature positions the head and spinal column and moves the rib cage, which assists in the movements that make breathing possible. The appendicular musculature stabilizes or moves components of the appendicular skeleton.

Level 2: Reviewing Concepts

24. b

25. Acetylcholine released by the motor neuron at the neuromuscular junction changes the permeability of the cell membrane at the motor end plate. The permeability change allows the influx of positive charges, which in

turn trigger an electrical event called an action potential. The action potential spreads across the entire surface of the muscle fiber and into the interior via the T tubules. The cytoplasmic concentration of calcium ions (released from sarcoplasmic reticulum) increases, which triggers the start of a contraction. The contraction ends when the ACh has been removed from the synaptic cleft and motor end plate by AChE.

26. Metabolic turnover is rapid in skeletal muscle cells. The genes contained in multiple nuclei direct the production of enzymes and structural proteins required for normal contraction, and the presence of multiple gene copies speeds up the process.

27. The spinal column does not need a massive series of flexors because many of the large trunk muscles flex the spine when they contract. In addition, most of the body weight lies anterior to the spinal column and gravity tends to flex the spine.

28. The urethral and anal sphincter muscles are usually in a constricted state, which prevents the passage of urine and feces. The muscles are innervated by nerves that are under conscious control, so the sphincters normally relax to allow the passage of wastes only when the individual decides so.

29. flexion of the leg and extension of the hip

Level 3: Critical Thinking and Clinical Applications

30. Since organophosphates block the action of the enzyme acetylcholinesterase, acetylcholine released into the neuromuscular cleft would not be inactivated. This would allow the acetylcholine to continue to stimulate the muscles, and cause a state of persistent contraction (spastic paralysis). If this were to affect the muscles of respiration (which is likely), Terry would die of suffocation. Prior to death, the most obvious symptoms would be uncontrolled tetanic contractions of the skeletal muscles.

31. In rigor mortis, the muscles lock in the contracted position, which makes the body extremely stiff. The membranes of the dead muscle cells are no longer selectively permeable and calcium leaks in, and triggers contraction. Contraction persists because the dead cells can no longer make ATP, which is necessary for cross-bridge detachment from the active sites. Rigor mortis begins a few hours after death and lasts until 15 to 25 hours later, when the lysosomal enzymes released by autolysis break down the myofilaments.

32. Makani should do squatting exercises. If he places a weight on his shoulders as he does these, he will notice better results, since the quadriceps muscles would be working against a greater resistance.

■ Chapter 8

Concept Check Answers, page 266

1. The afferent division of the PNS is composed of nerves that carry sensory information to the brain and spinal cord, so damage to it would interfere with a person's ability to experience a variety of sensory stimuli.

2. Unipolar neurons are most likely sensory neurons of the peripheral nervous system.

3. Infected (or damaged) areas of the CNS typically contain increased numbers of the small phagocytic glial cells called microglia.

Concept Check Answers, page 270

1. Depolarization of the neuron membrane involves the opening of the sodium channels and the rapid influx of sodium ions into the cell. If a chemical blocked the sodium channels, a neuron would not be able to depolarize and propagate an action potential.

2. Action potentials are propagated along myelinated axons by saltatory propagation at speeds much higher than those by continuous propagation along unmyelinated axons. An axon with a propagation speed greater than about 10 m/sec must be myelinated.

Concept Check Answers, page 274

1. A neurotransmitter that opens the potassium channels but not the sodium channels would cause a hyperpolarization at the postsynaptic membrane. The transmembrane potential would increase, and it would be more difficult to bring the membrane to threshold.

2. When an action potential reaches the presynaptic terminal of a cholinergic synapse, calcium channels are opened, and the influx of calcium triggers the release of acetylcholine into the synapse to stimulate the postsynaptic neuron. If the calcium channels were blocked, the acetylcholine would not be released, and transmission across the synapse would cease.

3. Convergence permits both conscious and subconscious control of the same motor neurons.

Concept Check Answers, page 282

1. Damage to the ventral root of spinal nerves, which is composed of visceral and somatic motor fibers, would interfere with motor function.
2. You would find the polio virus in the anterior gray horns of the spinal cord because that is where the cell bodies of somatic motor neurons are located.
3. All spinal nerves are classified as mixed nerves because they contain both sensory and motor fibers.

Concept Check Answers, page 284

1. The six regions in the adult brain and their major functions are (1) the *cerebrum*: conscious thought processes; (2) the *diencephalon*: the thalamic portion contains relay and processing centers for sensory information, and the hypothalamic portion contains centers involved with emotions, autonomic function, and hormone production; (3) the *midbrain*: processes visual and auditory information and generates involuntary motor responses; (4) the *pons*: contains tracts and relay centers that connect the brain stem to the cerebellum; (5) the *medulla oblongata*: contains major centers concerned with the regulation of autonomic function, such as heart rate, blood pressure, respiration, and digestive activities; and (6) the *cerebellum*: adjusts voluntary and involuntary motor activities.
2. The pituitary gland is attached to the hypothalamus, or the floor of the diencephalon.

Concept Check Answers, page 292

1. Cerebrospinal fluid re-enters the bloodstream by diffusion across the arachnoid granulations. If this diffusion were reduced, the volume of cerebrospinal fluid in the ventricles would increase.
2. The primary motor cortex is located in the precentral gyrus of the frontal lobe of the cerebrum.
3. Damage to the temporal lobes of the cerebrum would interfere with the processing of olfactory (smell) and auditory (sound) sensations.

Concept Check Answers, page 294

1. All ascending sensory information other than olfactory information passes through the thalamus before reaching our conscious awareness.

2. Changes in body temperature stimulate the hypothalamus, which is a division of the diencephalon.
3. Damage to the medulla oblongata can have fatal results because, despite its small size, it contains many vital reflex centers, including those that control breathing and regulate the heart and blood pressure.

Concept Check Answers, page 297

1. The abducens nerve (N VI) controls lateral movements of the eyes through the lateral rectus muscles, so an individual with damage to this nerve would be unable to move his or her eyes laterally (to the side).
2. Control of the voluntary muscles of the tongue occurs through the hypoglossal nerve (N XII).

Concept Check Answers, page 303

1. Physicians use the sensitivity of stretch reflexes (such as the knee jerk reflex or patellar reflex) to test the general condition of the spinal cord, peripheral nerves, and muscles.
2. In a monosynaptic reflex, a sensory neuron synapses directly on a motor neuron and produces a rapid, stereotyped movement. More involved responses occur with polysynaptic reflexes because the interneurons between the sensory and motor neurons may control several muscle groups simultaneously. In addition, some interneurons may stimulate a muscle group or groups, whereas others may inhibit other muscle groups.
3. A positive Babinski reflex is abnormal for an adult and indicates possible damage of descending tracts in the spinal cord.

Concept Check Answers, page 305

1. A tract within the posterior column of the spinal cord that carries information about touch and pressure from the lower part of the body to the brain is being compressed.
2. The primary motor cortex of the right cerebral hemisphere controls motor function on the left side of the body.
3. An injury to the superior portion of the motor cortex would affect the ability to control the muscles in the hand, arm, and upper portion of the leg.

Concept Check Answers, page 313

1. The sympathetic division of the autonomic nervous system is responsible for the physiological changes that occur in response to stress and increased activity.

2. The parasympathetic division is sometimes referred to as "the anabolic system" because parasympathetic stimulation leads to a general increase in the nutrient content of the blood. Cells throughout the body respond to the increase by absorbing the nutrients and using them to support growth and other anabolic activities.

3. A decrease in sympathetic stimulation would result in a decrease in the diameter of respiratory airways to the lungs and reduced airflow because parasympathetic effects would dominate.

4. Anxiety or stress causes an increase in sympathetic stimulation, so a person who is anxious about an impending root canal might exhibit some or all of the following: a dry mouth, increased heart rate, increased blood pressure, increased breathing rate, cold sweats, an urge to urinate or defecate, change in motility of the digestive tract ("butterflies" in the stomach), and dilated pupils.

Concept Check Answers, page 314

1. Age-related reduction in brain size and weight results from a decrease in the volume of the cerebral cortex due to the loss of cortical neurons.

Concept Check Answers, page 319

1. The nervous system controls contraction of the arrector pili muscles and secretion of the sweat glands within the integumentary system. The integumentary system provides the nervous system with the sensations of touch, pressure, pain, vibration, and temperature through sensory receptors. The integumentary system also protects peripheral nerves and hair provides some protection and insulation for the skull and brain.

Review Question Answers

Level 1: Reviewing Facts and Terms

1. f 2. g 3. p 4. s 5. m 6. a 7. b 8. j 9. h 10. c
11. d 12. o 13. t 14. q 15. l 16. i 17. n 18. e
19. r 20. k 21. c 22. a 23. a 24. d 25. b 26. b
27. a 28. b 29. c 30. a 31. d

32. The properties of the action potential are independent of the relative strength of the depolarization stimulus.

33. I: olfactory; II: optic; III: oculomotor; IV: trochlear; V: trigeminal; VI: abducens; VII: facial; VIII: vestibulocochlear; IX: glossopharyngeal; X: vagus; XI: accessory; XII: hypoglossal

34. The sympathetic preganglionic fibers emerge from the thoracolumbar area (T_1 through L_2) of the spinal cord. The parasympathetic fibers emerge from the brain stem and the sacral region of the spinal cord (craniosacral).

Level 2: Reviewing Concepts

35. d 36. c

37. Since the ventral roots contain axons of motor neurons, those muscles controlled by the neurons of the damaged root would be paralyzed.

38. Centers in the medulla oblongata are involved in respiratory and cardiac activity.

39. Transmission across a chemical synapse always involves a synaptic delay, but with only one synapse (monosynaptic), the delay between stimulus and response is minimized. In a polysynaptic reflex, the length of delay is proportional to the number of synapses involved.

40.

	Sympathetic	Parasympathetic
mental alertness	increased	decreased
metabolic rate	increased	decreased
digestive/urinary function	inhibited	stimulated
use of energy reserves	stimulated	inhibited
respiratory rate	increased	decreased
heart rate/blood pressure	increased	decreased
sweat glands	stimulated	inhibited

Level 3: Critical Thinking and Clinical Applications

41. Brain tumors result from uncontrolled division of neuroglial cells. Unlike neurons, neuroglial cells are capable of cell division. In addition, cells of the meningeal membranes can give rise to tumors.

42. The officer is testing the function of Bill's cerebellum. Many drugs, including alcohol, have pronounced effects on the function of the cerebellum. A person who is under the influence of alcohol is not able to properly anticipate the range and speed of limb movement because of slow processing and correction by the cerebellum. As a result, Bill would have a difficult time performing simple tasks such as walking a straight line or touching his finger to his nose.

43. Stress-induced stomach ulcers are due to excessive sympathetic stimulation. The sympathetic division causes the vasoconstriction of vessels that supply the digestive organs, which leads to an almost total shutdown of blood supply to the stomach. Lack of blood leads to the death of cells and tissues (necrosis), which causes the ulcers.

44. The radial nerve is involved.

45. Ramon damaged the femoral nerve. Since this nerve also supplies the sensory innervation of the skin on the anteromedial surface of the thigh and medial surfaces of the leg and foot, Ramon may also experience numbness in these regions.

■ Chapter 9

Concept Check Answers, page 333

1. Receptor A has the smaller receptive field and, thus, provides more precise sensory information.

2. Nociceptors are pain receptors, so when they are stimulated, you perceive a painful sensation.

3. Proprioceptors relay information about limb position and movement to the central nervous system, especially the cerebellum. Blockage of this information would result in uncoordinated movements, and the individual probably would be unable to walk.

Concept Check Answers, page 336

1. Repeated sniffing increases the perception of faint odors by increasing the flow of air and the number of odorant molecules that pass over the olfactory epithelium.

2. Drying the surface of the tongue removes moisture needed to dissolve the sugar molecules or salt ions. Because taste receptors (taste buds) are sensitive only to molecules and ions that are in solution, the taste buds will not be stimulated.

Concept Check Answers, page 349

1. The first layer of the eye to be affected by inadequate tear production is the conjunctiva. Drying of this layer produces an irritated, scratchy feeling.

2. When the lens is round, you are looking at something close to you.

3. Individuals with a congenital lack of cone cells would be able to see so long as they had functioning rod cells. Because cones function in color vision, such individuals would see in black and white only.

4. A deficiency or lack of vitamin A in the diet would affect the quantity of retinal the body could produce and, thus, would interfere with night vision.

Concept Check Answers, page 360

1. If the round window could not move, vibration of the stapes at the oval window would not move the perilymph, and there would be little or no perception of sound.

2. Loss of stereocilia (as a result of constant exposure to loud noises, for instance) would reduce hearing sensitivity and could eventually result in deafness.

Concept Check Answers, page 361

1. Because the number and sensitivity of taste buds declines dramatically after age 50, older individuals experience a reduced ability to taste food.

2. With age, the lens loses its elasticity and stiffens, which results in a flatter lens that cannot round up to focus the image of a close object on the retina. This condition, called presbyopia, may be corrected with a converging lens.

Review Question Answers

Level 1: Reviewing Facts and Terms

1. k 2. c 3. a 4. e 5. f 6. j 7. i 8. b 9. h 10. m
11. g 12. l 13. n 14. d 15. b 16. d 17. b 18. d
19. b 20. b 21. d 22. a 23. c 24. b 25. b 26. a 27. d

28. tactile receptors, baroreceptors, and proprioceptors

29. (1) free nerve endings: sensitive to touch and pressure; (2) root hair plexuses: monitor distortions and movements across the body surface; (3) tactile discs: detect fine touch and pressure; (4) tactile corpuscles: detect fine touch and pressure; (5) lamellated corpuscles: sensitive to pulsing or vibrating stimuli (deep pressure); and (6) Ruffini corpuscles: sensitive to pressure and distortion of the skin

30. (a) the sclera and the cornea (b) provides mechanical support and some degree of physical protection; serves as an attachment site for the extrinsic muscles; contains structures that assist in focusing

31. iris, ciliary body, and choroid

32. Step 1: Sound waves arrive at the tympanum. Step 2: Movement of the tympanum causes displacement of the auditory ossicles. Step 3: Movement of the stapes at the oval window establishes pressure waves in the perilymph of the vestibular duct. Step 4: The pressure waves distort the basilar membrane on their way to the round window of the tympanic duct. Step 5: Vibration of the basilar membrane causes hair cells to vibrate against the tectorial membrane. Step 6: Information concerning the region and intensity of stimulation is relayed to the CNS over the cochlear branch of N VIII.

Level 2: Reviewing Concepts

33. c **34.** c

35. The general senses include somatic and visceral sensation. The special senses are those whose receptors are confined to the head.

36. Regardless of the type of stimulus, the CNS receives the sensory information in the form of action potentials.

37. The olfactory system has extensive limbic system connections, which accounts for its effect on memories and emotions.

38. A visual acuity of 20/15 means that Jane can discriminate images at a distance of 20 feet, whereas someone with "normal" vision must be 5 feet closer (15 ft) to see the same details. Jane's visual acuity is better than the acuity of someone with 20/20 vision.

Level 3: Critical Thinking and Clinical Applications

39. As you turn to look at your friend, your medial rectus muscles will contract, which directs your gaze more medially. In addition, your pupils will constrict and the lenses will become more spherical.

40. The loud noises from the fireworks have transferred so much energy to the endolymph in the cochlea that the fluid continues to move for a long period of time. As long as the endolymph is moving, it will vibrate the tectorial membrane and stimulate the hair cells. This stimulation produces the "ringing" sensation that Millie perceives. She finds it difficult to hear normal conversation because the vibrations associated with it are not strong enough to overcome the currents already moving through the endolymph, so the pattern of vibrations is difficult to discern against the background "noise."

41. The rapid descent in the elevator causes the maculae in the saccule of the vestibule to slide upward, which produces the sensation of downward vertical motion.

When the elevator abruptly stops, the maculae do not. It takes a few seconds for them to come to rest in the normal position. As long as the maculae are displaced, the perception of movement will remain.

■ Chapter 10

Concept Check Answers, page 373

1. A cell's sensitivity to any hormone is determined by the presence or absence of the specific receptor molecule for that hormone.

2. Adenylate cyclase is the enzyme that converts ATP to cAMP, so a molecule that blocks adenylate cyclase would block the action of any hormone that required cAMP as a second messenger.

3. Cyclic-AMP is considered a second messenger because it is a second molecule required for converting the binding of epinephrine, norepinephrine, and peptide hormones (first messengers that cannot enter their target cells) into some effect on the metabolic activity of the target cell.

4. The three types of stimuli that control endocrine activity are (1) humoral (changes in the composition of the extracellular fluid), (2) hormonal (changes in the levels of circulating hormones), and (3) neural (the arrival of neurotransmitter at a neuroglandular junction).

Concept Check Answers, page 379

1. Dehydration increases the electrolyte concentration in the blood, which leads to an increase in osmotic pressure that stimulates the posterior pituitary gland to release more ADH.

2. Elevated levels of somatomedins, which are the mediators of growth hormone action, would lead us to expect an elevated level of growth hormone as well.

3. Because increased levels of cortisol inhibit the cells that control ACTH release from the pituitary gland, the level of ACTH would decrease. This is an example of a negative feedback mechanism.

Concept Check Answers, page 385

1. Because an individual who lacks iodine would not be able to form the hormone thyroxine, we would expect to see signs associated with thyroxine deficiency, such as decreased metabolic rate, lower body temperature, poor response to physiological stress, and an enlarged thyroid gland (goiter).

2. Most of the thyroid hormone in the blood is bound to transport proteins, which constitute a large reservoir of thyroxine that would not be depleted until days after the thyroid gland had been removed.

3. Removal of the parathyroid glands would result in a decrease in the blood levels of calcium ions. This decrease could be counteracted by increasing the dietary intake of vitamin D_3 and calcium.

Concept Check Answers, page 389

1. Because cortisol decreases the cellular use of glucose while increasing the available glucose (by promoting the breakdown of glycogen and the conversion of amino acids to carbohydrates), an elevation in cortisol levels would elevate blood glucose levels.

2. Increased amounts of light inhibit the production and release of melatonin from the pineal gland.

Concept Check Answers, page 396

1. Insulin increases the rate of conversion of glucose to glycogen in skeletal muscle and liver cells.

2. Glucagon stimulates the conversion of glycogen to glucose in the liver, so increased amounts of glucagon lead to decreased amounts of liver glycogen.

3. Calcitriol targets cells that line the digestive tract, and stimulates their absorption of calcium and phosphate. Erythropoietin (EPO) targets bone marrow cells that produce red blood cells. An increase in the number of red blood cells improves the delivery of oxygen to body tissues.

Concept Check Answers, page 401

1. The hormonal interaction exemplified by insulin and glucagon is antagonistic because the two hormones have opposite effects on their target tissues.

2. The lack of growth hormone, thyroid hormone, parathyroid hormone, or the reproductive hormones would inhibit the normal formation and development of the skeletal system.

Review Question Answers

Level 1: Reviewing Facts and Terms

1. h 2. e 3. g 4. l 5. b 6. k 7. c 8. a 9. j 10. i 11. f
12. d 13. d 14. a 15. c 16. d 17. b 18. d 19. c 20. b

21. (1) thyroid-stimulating hormone (TSH); (2) adrenocorticotropic hormone (ACTH); (3) follicle-stimulating hormone (FSH); (4) luteinizing hormone (LH); (5) prolactin (PRL); (6) growth hormone (GH); and (7) melanocyte-stimulating hormone (MSH)

22. The overall effect of calcitonin is to decrease the concentration of calcium ions in body fluids. The overall effect of parathyroid hormone is to cause an increase in the concentration of calcium ions in body fluids.

23. (a) The GAS phase includes alarm, resistance, and exhaustion phases. (b) Epinephrine is the dominant hormone of the alarm phase. Glucocorticoids are the dominant hormones of the resistance phase.

Level 2: Reviewing Concepts

24. In communication by the nervous system, the source and destination are quite specific and the effects are short-lived. In endocrine communication, the effects are slow to appear and often persist for days. A single hormone can alter the metabolic activities of multiple tissues and organs simultaneously.

25. Hormones direct the synthesis of an enzyme (or other protein) that is not already present in the cytoplasm. They also turn an existing enzyme "on" or "off" and increase the rate of synthesis of a particular enzyme or other protein.

26. The two hormones may have opposing, or antagonistic, effects; the hormones may have additive or synergistic effects; one hormone may have a permissive effect on another (the first hormone is needed for the second to produce its effect); the hormones may produce different but complementary effects in specific tissues and organs.

27. Phosphodiesterase is the enzyme that converts cAMP to AMP, which thus inactivates it. If this enzyme were blocked, the effect of the hormone would be prolonged.

Level 3: Critical Thinking and Clinical Applications

28. Extreme thirst and frequent urination are characteristics of both diabetes insipidus and diabetes mellitus. To distinguish between the two, glucose levels in the blood and urine could be measured. A high glucose concentration would indicate diabetes mellitus.

29. Julie should exhibit elevated levels of parathyroid hormone in her blood. Her poor diet does not supply enough calcium for her developing fetus. The fetus removes large amounts of calcium from the maternal

blood, which lowers the mother's blood calcium levels. This would lead to an increase in the blood level of parathyroid hormone and increased mobilization of stored calcium from the maternal skeletal reserves.

■ Chapter 11

Concept Check Answers, page 410

1. Venipuncture is a common sampling technique because superficial veins are easy to locate, the walls of veins are thinner than those of arteries, and blood pressure in veins is relatively low, so the puncture wound seals quickly.

2. A decrease in the amount of plasma proteins could cause (1) decreased plasma osmotic pressure, (2) a decreased ability to fight infections, and (3) decreases in the transport and binding of some ions, hormones, and other molecules.

Concept Check Answers, page 417

1. The hematocrit is the percentage of total blood volume occupied by the formed elements (mostly red blood cells). The loss of water during dehydration decreases plasma volume, resulting in an increased hematocrit.

2. Diseases that damage the liver impair its ability to excrete bilirubin, so bilirubin accumulates in the blood. (The result is jaundice.)

3. A decreased oxygen supply to the kidneys triggers the release of erythropoietin. (The elevated erythropoietin would lead to an increase in erythropoiesis, or red blood cell formation.)

Concept Check Answers, page 419

1. A person with Type AB blood can accept Type A, Type B, Type AB, or Type O blood.

2. If a person with Type A blood received a transfusion of Type B blood, the anti-B antibodies in the plasma of the Type A person would attack the B antigens on the surfaces of the Type B RBCs. Thus, the transfused RBCs would clump or agglutinate, and potentially block blood flow to various organs and tissues.

Concept Check Answers, page 425

1. Because neutrophils are usually the first WBCs to arrive at the site of an injury and are specialized to

phagocytize invading bacteria, you would expect to find a large number of neutrophils in an infected cut.

2. The white blood cells that produce circulating antibodies are lymphocytes.

3. During inflammation, basophils release chemicals, including histamine and heparin, that enhance inflammation and attract other types of white blood cells.

Concept Check Answers, page 428

1. Megakaryocytes are the precursors of platelets. A decreased number of megakaryocytes would impair the clotting process.

2. The faster extrinsic pathway is initiated by the release of a lipoprotein called tissue factor by damaged endothelial cells and damaged tissues. The slower intrinsic pathway is initiated by the activation of clotting protein proenzymes exposed to damaged collagen fibers and by the release of platelet factor by aggregating platelets.

3. A vitamin K deficiency would reduce the production of clotting factors necessary for normal blood coagulation.

Review Question Answers

Level 1: Reviewing Facts and Terms

1. f 2. k 3. a 4. l 5. c 6. j 7. e 8. h 9. g 10. b 11. d
12. i 13. c 14. c 15. a 16. c 17. d 18. a 19. d
20. d 21. b

22. (1) the transportation of dissolved gases, nutrients, hormones, and metabolic wastes; (2) the regulation of pH and electrolyte composition of interstitial fluids throughout the body; (3) the restriction of fluid losses through damaged vessels or at other injury sites; (4) defense against toxins and pathogens; and (5) the stabilization of body temperature

23. Albumins maintain the osmotic pressure of plasma and are important in the transport of fatty acids. Transport globulins bind small ions, hormones, or compounds that might otherwise be filtered out of the blood at the kidneys or have very low solubility in water, and immunoglobulins attack foreign proteins and pathogens. Fibrinogen functions in blood clotting.

24. (a) anti-B antibodies (b) anti-A antibodies (c) neither anti-A nor anti-B antibodies (d) anti-A antibodies and anti-B antibodies

25. (1) amoeboid movement, the extension of a cellular process; (2) emigration (diapedesis), squeezing between adjacent endothelial cells in the capillary wall;

and (3) positive chemotaxis, attraction to specific chemical stimuli

26. The common pathway begins when Factor X activator from either the extrinsic or intrinsic pathway appears in the plasma.

27. An embolus is a drifting blood clot. A thrombus is a blood clot that sticks to the wall of an intact blood vessel.

Level 2: Reviewing Concepts

28. a **29.** d **30.** c **31.** d

32. Red blood cells are biconcave discs that lack mitochondria, ribosomes, and nuclei and contain a large amount of the protein hemoglobin.

Level 3: Critical Thinking and Clinical Applications

33. After donating a pint of blood (or after any other loss of blood), you would expect to see a substantial increase in the number of reticulocytes. Since there is not enough time for large numbers of erythrocytes to mature, the bone marrow releases large numbers of reticulocytes (immature cells) in an effort to maintain a constant number of formed elements.

34. In many cases of kidney disease, the cells responsible for producing erythropoietin are either damaged or destroyed. The reduction in erythropoietin levels leads to reduced erythropoiesis and fewer red blood cells, which results in anemia.

■ Chapter 12

Concept Check Answers, page 450

1. Damage to the semilunar valves on the right side of the heart would interfere with the blood flow to the pulmonary artery.

2. When the ventricles begin to contract, they force the AV valves to close, which tenses the chordae tendineae. The chordae tendineae are attached to the papillary muscles. The papillary muscles respond by contracting, which counteracts the force that is pushing the valves upward.

3. The wall of the left ventricle is more muscular than that of the right ventricle because the left ventricle must generate enough force to propel blood throughout the body (except the lungs). The right ventricle must generate only enough force to propel the blood the few centimeters to the lungs.

Concept Check Answers, page 455

1. The longer refractory period in cardiac muscle cells prevents tetanus and results in a relatively long relaxation period, during which the heart's chambers can refill with blood. A heart in tetany could not fill with blood.

2. If the cells of the SA node were not functioning, the cells of the AV node would become the pacemaker cells. The heart would still continue to beat but at a slower rate.

3. If the impulses from the atria were not delayed at the AV node, they would be conducted through the ventricles so quickly by the bundle branches and Purkinje fibers that the ventricles would begin contracting immediately, before the atria had finished contracting. As a result, the ventricles would not be as full of blood as they could be, and the pumping of the heart would be less efficient, especially during activity.

Concept Check Answers, page 458

1. No. During the initial phase of contraction, pressure in the left ventricle is rising as tension in the contracting heart muscle increases. But no blood is leaving the heart because both the AV valves and the semilunar valves are closed. (When the pressure in the ventricle exceeds the pressure in the aorta, the aortic semilunar valves are forced open, and the blood is rapidly ejected from the ventricle.)

2. The first sound is produced by the simultaneous closing of the AV valves and the opening of the semilunar valves. The second sound is produced when the semilunar valves close.

Concept Check Answers, page 462

1. Stimulating the acetylcholine receptors of the heart lowers heart rate. Given that cardiac output equals stroke volume times heart rate, then (assuming no change in the stroke volume) cardiac output declines.

2. Venous return fills the heart with blood, which stretches the heart muscle. According to the Frank-Starling principle, the more the heart muscle is stretched, the more forcefully (to a point) it will contract, and the more blood it will eject (stroke volume) with each beat. Therefore, if all other factors are constant, increased venous return increases stroke volume.

3. Increased sympathetic stimulation of the heart results in increased heart rate and increased force of contraction.

Review Question Answers

Level 1: Reviewing Facts and Terms

1. i 2. e 3. g 4. h 5. a 6. b 7. k 8. l 9. d 10. f 11. c 12. j 13. c 14. b 15. b 16. a 17. a 18. b 19. a

20. During ventricular contraction, tension in the papillary muscles pulls against the chordae tendineae, which keep the cusps of the AV valve from swinging into the atrium. This action prevents the backflow, or regurgitation, of blood into the atrium as the ventricle contracts.

21. The atrioventricular (AV) valves prevent backflow of blood from the ventricles into the atria. The right AV valve is the tricuspid valve, and the left AV valve is the bicuspid (mitral) valve. The pulmonary and aortic semilunar valves prevent the backflow of blood from the pulmonary trunk and aorta into the right and left ventricles.

22. SA node → AV node → AV bundle (bundle of His) → right and left bundle branches → Purkinje fibers (into the mass of ventricular muscle tissue)

23. (a) The cardiac cycle is a complete heartbeat, including a contraction/relaxation period for both atria and ventricles. (b) The cycle begins with atrial systole as the atria contract and push blood into the relaxed ventricles. As the atria relax (atrial diastole), the ventricles contract (ventricular systole), which forces blood through the semilunar valves into the pulmonary trunk and aorta. The ventricles then relax (ventricular diastole). For the rest of the cardiac cycle, both the atria and ventricles are in diastole; passive filling occurs.

Level 2: Reviewing Concepts

24. c 25. d 26. a

27. The right atrium receives blood from the systemic circuit and passes it to the right ventricle. The right ventricle discharges blood into the pulmonary circuit. The left atrium collects blood. Contraction of the left ventricle ejects blood into the systemic circuit.

28. Listening to the heart sounds (auscultation) is a simple, effective method of cardiac diagnosis. The first and second heart sounds accompany the action of the heart valves. The first sound ("lubb") marks the start of ventricular contraction and is produced as the AV valves close and the semilunar valves open. The second sound ("dupp") occurs when the semilunar valves close, which marks the start of ventricular diastole.

29. (a) Sympathetic activation causes the release of norepinephrine by postganglionic fibers and the secretion of norepinephrine and epinephrine by the adrenal medullae. These hormones stimulate the metabolism of cardiac muscle cells and increase the force and degree of contraction. (b) Parasympathetic stimulation causes the release of ACh at membrane surfaces, where it produces hyperpolarization and inhibition. The result is a decrease in the heart rate and in the force of cardiac contractions.

Level 3: Critical Thinking and Clinical Applications

30. This patient has second-degree heart block, a condition in which not every signal from the SA node reaches the ventricular muscle. In this case, for every two action potentials generated by the SA node, only one is reaching the ventricles. This accounts for the two P waves but only one QRS complex. This condition frequently results from damage to the internodal pathways of the atria or problems with AV node.

31. Blocking the calcium channels in myocardial cells would lead to a decrease in the force of cardiac contraction. Since the force of cardiac contraction is directly proportional to the stroke volume, you would expect a reduced stroke volume.

■ Chapter 13

Concept Check Answers, page 472

1. The blood vessels are veins. Arteries and arterioles have a relatively large amount of smooth muscle tissue in a thick, well-developed tunica media.

2. The relaxation of precapillary sphincters increases the blood flow to a tissue.

3. Because the blood pressure in veins is very low, the venous circulation relies on the contractions of skeletal muscles to keep the blood moving against the pull of gravity. Valves prevent blood from flowing backward whenever venous pressure is too low.

Concept Check Answers, page 476

1. In a normal individual, the pressure is greatest in the aorta and least in the venae cavae. Blood, like other fluids,

moves along a pressure gradient from high pressure to low pressure. If the pressure were higher in the inferior vena cava, blood would flow in the reverse direction.

2. When a person stands for a long time, blood tends to pool in the lower extremities. Venous return to the heart decreases, and in turn cardiac output decreases, sending less blood to the brain, causing light-headedness and fainting. A hot day adds to the effect, because the loss of body water through sweating reduces blood volume and, thus, venous return.

Concept Check Answers, page 483

1. Compression of the carotid artery would increase heart rate. The compression decreases blood pressure at the carotid sinus, where the carotid baroreceptors are located. The resulting decreased baroreceptor activity stimulates the cardioacceleratory center, which produces sympathetic stimulation that elevates the heart rate. (Additionally, stimulation of the vasomotor centers causes vasoconstriction of arterioles and increase in blood pressure.)

2. Blood pressure increases during exercise because (1) blood flow to muscles increases, (2) cardiac output increases, and (3) resistance in visceral tissues increases.

3. Vasoconstriction of the renal artery would lead to increased blood pressure and blood volume. In response to vasoconstriction-induced declines in blood flow and blood pressure, the kidney releases more renin, which leads to elevated angiotensin II levels. Angiotensin II then produces increases in blood pressure and blood volume.

Concept Check Answers, page 500

1. Blockage of the left subclavian artery would interfere with blood flow to the left arm.

2. Compression of one of the common carotid arteries would lead to loss of consciousness (or even death) by cutting off blood flow to the brain.

3. Rupture of the celiac trunk would most directly affect the stomach, spleen, liver, and pancreas.

4. Whereas arteries in the neck and limbs are located deep beneath the skin, protected by bones and surrounding soft tissues, veins in these sites follow two courses: one superficial and one deep. This dual venous drainage helps control body temperature; blood flow through superficial veins promotes heat loss, and blood flow through deep veins conserves body heat.

Concept Check Answers, page 503

1. This blood must have come from the umbilical vein, which carries oxygenated, nutrient-rich blood from the placenta to the fetus.

2. The cardiovascular system is most closely intertwined with the lymphatic system. The two systems are physically connected (lymphatic vessels drain into veins), and cell populations of the lymphatic system are carried from one part of the body to another within the cardiovascular system.

Review Question Answers

Level 1: Reviewing Facts and Terms

1. d 2. l 3. a 4. j 5. i 6. h 7. k 8. f 9. e 10. c 11. g
12. b 13. d 14. b 15. b 16. a 17. d 18. b 19. c 20. b
21. c 22. b 23. d 24. c 25. b 26. c 27. c

28. (a) Fluid leaves the capillary at the arteriole end primarily in response to hydrostatic pressure. (b) Fluid returns to the capillary at the venous end primarily in response to osmotic pressure.

29. The cardiac output decreases due to parasympathetic stimulation and inhibition of sympathetic activity, and peripheral vasodilation is widespread due to the inhibition of excitatory neurons in the vasomotor center.

30. These chemoreceptors are sensitive to changes in the carbon dioxide, oxygen, or pH levels in the blood and cerebrospinal fluid.

31. When an infant takes its first breath, the lungs expand, and so do the pulmonary vessels. The smooth muscles in the ductus arteriosus contract, which isolates the pulmonary and aortic trunks, and blood begins flowing through the pulmonary circuit. As pressure rises in the left atrium, the valvular flap closes the foramen ovale, which completes the circulatory remodeling.

32. blood: decreased hematocrit, formation of thrombin, and valvular malfunction; heart: reduction in maximum cardiac output and changes in the activities of the nodal and conducting fibers, reduction in the elasticity of the fibrous skeleton, progressive atherosclerosis, and replacement of damaged cardiac muscle fibers by scar tissue; blood vessels: progressive inelasticity in arterial walls, deposition of calcium salts on weakened vascular walls, and formation of thrombi at atherosclerotic plaques

Level 2: Reviewing Concepts

33. d **34.** c **35.** b

36. Artery walls are generally thicker. They contain more smooth muscle and elastic fibers, which enables them to resist and adjust to the pressure generated by the heart. Venous walls are thinner and the pressure in the veins is less than that in the arteries. Arteries constrict more than veins do when not expanded by blood pressure, due to a greater degree of elastic tissue.

37. Capillaries are only one cell thick. Small gaps between adjacent endothelial cells permit the diffusion of water and small solutes into the surrounding interstitial fluid but prevent the loss of blood cells and plasma proteins. Some capillaries also contain pores that permit very rapid exchange of fluids and solutes between the interstitial fluid and the plasma. Conversely, the walls of arteries and veins are several cell layers thick and are not specialized for diffusion.

38. The brain receives arterial blood from four arteries. Because these arteries form anastomoses inside the cranium, interruption of any one vessel will not compromise the circulatory supply to the brain.

39. The accident victim is suffering from shock and acute circulatory crisis characterized by hypotension and inadequate peripheral blood flow. The hypotension results from the loss of blood volume and decreased cardiac output. Her skin is pale and cool due to peripheral vasoconstriction; the moisture results from the sympathetic activation of sweat glands. Falling blood pressure to the brain causes confusion and disorientation. If you took her pulse, you would find it to be rapid and weak, which reflects the heart's response to reduced blood flow and volume.

Level 3: Critical Thinking and Clinical Applications

40. Three factors contribute to Bob's elevated blood pressure. (1) The loss of water through sweating increases blood viscosity. The number of red blood cells remains about the same, but because there is less plasma volume, the concentration of red cells is increased, which increases the blood viscosity. Increased viscosity increases peripheral resistance and contributes to increased blood pressure. (2) To cool Bob's body, blood flow to the skin is increased. This in turn increases venous return, which increases stroke volume and cardiac output (Frank-Starling's law of the heart). The increased cardiac output can also

contribute to increased blood pressure. (3) The heat stress that Bob is experiencing leads to increased sympathetic stimulation (the reason for the sweating). Increased sympathetic stimulation of the heart will increase heart rate and stroke volume, which increases his cardiac output and blood pressure.

41. Antihistamines and decongestants are drugs that have the same effects on the body as stimulating the sympathetic nervous system. In addition to the desired effects of counteracting the symptoms of the allergy, these medications can produce an increased heart rate, stroke volume, and peripheral resistance, all of which will contribute to elevating blood pressure. If a person has hypertension (high blood pressure), these drugs will aggravate this condition with potentially hazardous consequences.

42. When Gina rapidly moved from a lying position to a standing position, gravity caused her blood volume to move to the lower parts of her body away from the heart, which decreased venous return. The decreased venous return resulted in a decreased volume of blood at the end of diastole, which led to a decreased stroke volume and cardiac output. These resulted in decreased blood flow to the brain, where the diminished oxygen supply caused her to be light-headed and feel faint. This usually does not occur, because as soon as the pressure drops due to blood moving inferiorly, the baroreceptor reflex should be triggered. Normally, a rapid change in blood pressure is sensed by baroreceptors in the aortic arch and carotid sinus. Action potentials from these areas are carried to the medulla oblongata, where appropriate responses are integrated. In this case, we would expect an increase in peripheral resistance to compensate for the decreased blood pressure. If this does not compensate enough for the drop, then an increase in heart rate and force of contraction would occur. Normally, these responses occur so quickly that no changes in pressure are noticed after body position is changed.

■ Chapter 14

Concept Check Answers, page 518

1. Blockage of the thoracic duct would impair circulation of lymph through all of the body inferior to the diaphragm and the left side of the head and thorax. The accumulation of fluid in the involved extremities (lymphedema) could also result.

2. A lack of thymic hormones would result in an absence of T lymphocytes.

3. During an infection, lymphocytes and phagocytes in lymph nodes in the affected region multiply in response to the infectious agent. This increase in the number of cells in the nodes causes the nodes to become enlarged or swollen.

Concept Check Answers, page 522

1. A decrease in the number of monocyte-forming cells in the bone marrow would reduce the abundance of macrophages, including microglia of the CNS, Kupffer's cells of the liver, and alveolar macrophages in the lungs.

2. A rise in interferon levels indicates a viral infection. (Interferon is released from cells that are infected with viruses. It does not help the infected cell but "interferes" with the virus's ability to infect other cells.)

3. Pyrogens stimulate the temperature control area within the hypothalamus to raise body temperature, or to produce a fever.

4. Fever increases the metabolic rate and speeds cell movement (which enhances phagocytosis) and speeds enzymatic reactions.

Concept Check Answers, page 533

1. Once abnormal peptides within a cell become attached to MHC proteins and displayed on the cell's surface, the peptides can be recognized by T cells, which triggers an immune response.

2. A decrease in the number of cytotoxic T cells would interfere with cell-mediated immunity, which is the body's response against foreign cells and virus-infected cells.

3. A lack of helper T cells would retard the antibody-mediated immune response because they promote B cell division, the maturation of plasma cells, and the production of antibodies by plasma cells.

4. Plasma cells produce and secrete antibodies, so we would expect to see increased levels of antibodies in the blood if the number of plasma cells were elevated.

Concept Check Answers, page 539

1. The secondary response would be more affected by the lack of memory B cells because it depends on the presence of memory cells produced during the primary response to an antigen.

2. A developing fetus is protected primarily by passive immunity that results from IgG antibodies that cross the placenta from the mother's circulation.

3. The cardiovascular system aids the body's nonspecific and specific defenses by distributing white blood cells, transporting antibodies, and restricting the spread of pathogens (through the clotting response).

Review Question Answers

Level 1: Reviewing Facts and Terms

1. h 2. k 3. b 4. g 5. d 6. c 7. j 8. a 9. i 10. l 11. f 12. e 13. c 14. d 15. c 16. c 17. b 18. a 19. c 20. a 21. a 22. d 23. d 24. c

25. Left thoracic duct: collects lymph from the body below the diaphragm and from the left side of the body above the diaphragm; right thoracic duct: collects lymph from the right side of the body above the diaphragm.

26. (a) responsible for cell-mediated immunity, which defends against abnormal cells and pathogens inside living cells; (b) stimulate the activation and function of T cells and B cells; (c) inhibit the activation and function of both T cells and B cells; (d) produce and secrete antibodies; (e) recognize and destroy abnormal cells; (f) interfere with viral replication inside the cell and stimulate the activities of macrophages and NK cells; (g) provide cell-mediated immunity; (h) provide antibody-mediated immunity, which defends against antigens and pathogenic organisms in the body; (i) enhance nonspecific defenses and increase T cell sensitivity and stimulate B cell activity

27. (1) physical barriers; (2) phagocytic cells; (3) immunological surveillance; (4) interferons; (5) complement; (6) inflammation; and (7) fever

Level 2: Reviewing Concepts

28. c 29. d

30. Specificity: The immune response is triggered by a specific antigen and defends against only that antigen. Versatility: The immune system can differentiate from among tens of thousands of antigens it may encounter during a normal lifetime. Memory: The immune response following second exposure to a particular antigen is stronger and lasts longer. Tolerance: Some antigens, such as those on an individual's own cells, do not elicit an immune response.

31. The antigen may be destroyed by neutralization, agglutination and precipitation, activation of complement, attraction of phagocytes, stimulation of inflammation, or prevention of bacterial and viral adhesion.

32. destruction of target cell membranes, stimulation of inflammation, attraction of phagocytes, and enhancement of phagocytosis

Level 3: Critical Thinking and Clinical Applications

33. A key characteristic of cancer cells is their ability to break free of a tumor, migrate to other tissues of the body, and form new tumors. This process is called metastasis. The primary route for the spread of cancer cells is the lymphatic system, and cancer cells may remain in a lymph node for a period of time before moving on to other tissues. Examination of regional lymph nodes for the presence of cancer cells can help the physician determine if the cancer was caught in an early stage or whether it has started to spread to other tissues. It can also give the physician an idea of what other tissues may be affected by the cancer, which would help him to decide on the proper treatment.

34. It appears that Ted has contracted the disease. On initial contact with a virus, the first type of antibody to be produced is IgM. The response is fairly rapid but short-lived. About the time IgM peaks, IgG levels are beginning to rise. IgG plays the more important role in eventually controlling the disease. The fact that Ted's blood sample has an elevated level of IgM antibodies would indicate that he is in the early stages of a primary response to the measles virus.

■ Chapter 15
Concept Check Answers, page 558

1. Increased tension in the vocal cords raises the pitch of the voice.

2. The C-shaped tracheal cartilages allow room for the esophagus to expand when food or drink is swallowed.

3. Without surfactant, surface tension in the thin layer of water that moistens alveolar surfaces would cause the alveoli to collapse.

Concept Check Answers, page 565

1. When the rib penetrates the chest wall, atmospheric air enters the thoracic cavity, which produces pneumothorax. The ensuing breakage of the fluid bond between the pleurae allows contraction of the elastic fibers in the lung. The result is atelectasis, or a collapsed lung.

2. The fluid in the alveoli takes up space that would normally be occupied by air, so the vital capacity decreases.

Concept Check Answers, page 573

1. Hemoglobin releases more oxygen during exercise because active skeletal muscles produce changes—elevated body temperature and reduced pH (due to acidic waste products)—that cause the release of oxygen.

2. Obstruction of the trachea would lower blood pH by interfering with the body's ability to eliminate carbon dioxide. The resulting excess carbon dioxide in the blood is converted to carbonic acid, which then dissociates into bicarbonate ions and hydrogen ions, and lowers blood pH.

Concept Check Answers, page 578

1. Chemoreceptors are more sensitive to carbon dioxide levels than to oxygen levels. When carbon dioxide goes into solution, it produces hydrogen ions, which lower pH and alter cell or tissue activity.

2. Strenuous exercise stimulates the inflation and deflation reflexes, also known as the Hering-Breuer reflexes. In the inflation reflex, the stimulation of stretch receptors in the lungs results in an inhibition of the inspiratory center and a stimulation of the expiratory center. In contrast, reducing the volume of the lungs initiates the deflation reflex. This reflex results in an inhibition of the expiratory center and a stimulation of the inspiratory center.

3. Johnny's mother should not worry. When Johnny holds his breath, the increasing level of carbon dioxide in his blood will lead to increased stimulation of the inspiratory center, and force Johnny to breathe again.

Concept Check Answers, page 578

1. Two age-related changes that reduce the efficiency of the respiratory system are (1) reduced chest movements that result from arthritic changes in rib

joints and stiffening of the costal cartilages and (2) some degree of emphysema that stems from the gradual destruction of alveolar surfaces.

2. The nervous system controls the pace and depth of respiration and monitors the respiratory volume of the lungs and levels of blood gases.

Review Question Answers

Level 1: Reviewing Facts and Terms

1. h 2. f 3. i 4. c 5. b 6. d 7. j 8. e 9. a 10. l
11. g 12. k 13. c 14. b

Level 2: Reviewing Concepts

15. a 16. d

17. With less cartilaginous support in the lower respiratory passageways, the amount of tension in the smooth muscles has a greater effect on bronchial diameter and on the resistance to air flow.

18. The nasal cavity is designed to cleanse, moisten, and warm inspired air; whereas the mouth is not. Air that enters through the mouth is drier and as a result can irritate the trachea, and cause soreness of the throat.

19. The walls of bronchioles, like the walls of arterioles, are dominated by smooth muscle tissue. Varying the diameter (bronchodilation or bronchoconstriction) of the bronchioles provides control over the amount of resistance to air flow and over the distribution of air in the lungs, just as vasodilation and vasoconstriction of the arterioles regulate blood flow/distribution.

20. Air that passes through the glottis flows into the larynx and through the trachea. From there, the air flows into a primary bronchus, which supplies the lungs. In the lungs, the air passes to bronchi, bronchioles, a terminal bronchiole, a respiratory bronchiole, an alveolar duct, an alveolar sac, and lastly to the respiratory membrane.

Level 3: Critical Thinking and Clinical Applications

21. An increase in ventilation will increase the movement of venous blood back to the heart. (Recall the respiratory pump.) Increasing venous return would in turn help increase the blood pressure (according to the Frank-Starling law of the heart).

22. While you were sleeping, the air that you were breathing was so dry it absorbed more than the normal amount of moisture as it passed through the nasal cavity. The loss of moisture made the mucous secretions quite viscous and harder for the cilia to move. Your nasal epithelia continued to secrete mucus, but very little of it moved. This ultimately produced the nasal congestion. After the shower and juice, more moisture was transferred to the mucus, which loosened it and made it easier to move, and cleared up the problem.

■ Chapter 16

Concept Check Answers, page 589

1. The mesenteries support and stabilize the positions of organs in the abdominopelvic cavity and provide a route for the blood vessels, nerves, and lymphatic vessels associated with the digestive tract.

2. Peristalsis is more efficient in propelling intestinal contents down the tract. Segmentation is essentially a churning action that mixes intestinal contents with digestive fluids.

3. Given that parasympathetic stimulation increases muscle tone and motility in the digestive tract, a drug that blocks this activity would decrease the rate of peristalsis.

Concept Check Answers, page 593

1. The oral cavity is lined by stratified squamous epithelium, which is typically located in sites that receive a great deal of friction or abrasion.

2. Because the parotid salivary glands secrete the enzyme salivary amylase, which digests complex carbohydrates (starches), damage to those glands would interfere with the digestion of carbohydrates.

3. Incisors are the type of tooth best suited for chopping (or cutting or shearing) pieces of relatively rigid food, such as raw vegetables.

4. The process described is swallowing.

Concept Check Answers, page 603

1. The pyloric sphincter regulates the flow of chyme into the small intestine.

2. Cutting the branches of the vagus nerve that supply the stomach severs parasympathetic motor fibers that can stimulate gastric secretions even when the stomach is empty (the cephalic phase of gastric secretion). The prevention of parasympathetic secretion reduces the likelihood of new ulcer formation.

3. Several adaptations increase the small intestine's surface area, which thereby increases its absorptive capacity. The walls of the small intestine contain folds called plicae, each of which is covered by finger-like projections called villi. The exposed surface of each villus is in turn covered by smaller projections called microvilli. In addition, the small intestine has extensive blood and lymphatic supplies to transport absorbed nutrients.

4. A high-fat meal would increase CCK blood levels.

Concept Check Answers, page 612

1. A narrowing of the ileocecal valve would interfere with the flow of chyme from the small intestine to the large intestine.

2. Damage to the exocrine pancreas would most impair the digestion of fats (lipids) because that organ is the primary source of lipases.

3. A decrease in the amount of bile salts would reduce the effectiveness of fat digestion and absorption.

Concept Check Answers, page 621

1. An increase in the fat content of a meal would increase the number of chylomicrons in the lacteals.

2. The removal of the stomach would interfere with the absorption of vitamin B_{12}, which requires intrinsic factor, a molecule produced by parietal cells in the stomach.

3. Diarrhea can be life threatening because the loss of fluid and electrolytes faster than they can be replaced may result in potentially fatal dehydration. Constipation, although uncomfortable, does not interfere with any major body process. The few toxic waste products that are normally eliminated by the digestive system can move into the blood and be eliminated by the kidneys.

Concept Check Answers, page 623

1. The digestive epithelium becomes more susceptible to damage in older individuals because the rate of epithelial stem cell division declines with age.

2. Digestive system functions related to the cardiovascular system include the absorption of water to maintain blood volume; the absorption of vitamin K (produced by intestinal bacteria, and vital to blood clotting); the excretion of bilirubin (a product of the breakdown of the heme portion of hemoglobin) by the liver; and synthesis of blood clotting factors by the liver.

Review Question Answers

Level 1: Reviewing Facts and Terms

1. c 2. k 3. g 4. a 5. i 6. h 7. j 8. d 9. f 10. b 11. l 12. e 13. d 14. d 15. b 16. c 17. b 18. d 19. d 20. a 21. d 22. d

23. ingestion, mechanical processing, secretion, digestion, absorption, and excretion

24. The folds increase the surface area available for absorption and may permit expansion of the lumen after a large meal.

25. Mucosa: This innermost layer is a mucous membrane that consists of epithelia and loose connective tissue (the lamina propria). Submucosa: This layer, which surrounds the mucosa, contains blood vessels, lymphatic vessels, and neural tissue (the submucosal nerve plexus). Muscularis externa: This layer is made up of two layers of smooth muscle tissue—longitudinal and circular—whose contractions agitate and propel materials along the digestive tract and neural tissue (the myenteric nerve plexus). Serosa: This outermost layer is a serous membrane that protects and supports the digestive tract inside the peritoneal cavity.

26. (1) analysis of material before swallowing; (2) mechanical processing through the actions of the teeth, tongue, and palatal surfaces; (3) lubrication by mixing with mucus and salivary secretions; and (4) limited digestion of carbohydrates and lipids

27. incisors: clipping or cutting; cuspids: tearing or slashing; bicuspids: crushing, mashing, grinding; molars: crushing and grinding

28. duodenum, jejunum, and ileum

29. The pancreas provides digestive enzymes as well as buffers that assist in the neutralization of acid chyme. The liver and gallbladder provide bile, which is a solution that contains additional buffers and bile salts that facilitate the digestion and absorption of lipids. The liver is responsible for metabolic regulation, hematological regulation, and bile production. It is the primary organ involved in regulating the composition of the circulating blood.

30. (1) reabsorption of water and compaction of chyme into feces; (2) absorption of important vitamins generated by bacterial action; (3) storage of fecal material prior to defecation

31. (1) the rate of epithelial stem cell division declines; (2) smooth muscle tone decreases; (3) the effects of cumulative damage become apparent; (4) cancer rates increase; and (5) changes in other systems have direct or indirect effects on the digestive system.

Level 2: Reviewing Concepts

32. d **33.** d **34.** a

35. Peristalsis consists of waves of muscular contractions that move along the length of the digestive tract. During a peristaltic movement, the circular muscles contract behind the digestive contents. Longitudinal muscles contract next, and shorten adjacent segments. A wave of contraction in the circular muscles then forces the materials in the desired direction. Segmentation movements churn and fragment the digestive materials, which mixes the contents with intestinal secretions. Because they do not follow a set pattern, segmentation movements do not produce directional movement of materials along the tract.

36. The stomach performs four major functions: the bulk storage of ingested food, the mechanical breakdown of ingested food, the disruption of chemical bonds through the actions of acids and enzymes, and the production of intrinsic factor.

37. The cephalic phase begins with the sight or thought of food. Directed by the CNS, this phase prepares the stomach to receive food. The gastric phase begins with the arrival of food in the stomach. The gastric phase is initiated by distension of the stomach, an increase in the pH of the gastric contents, and the presence of undigested materials in the stomach. The intestinal phase begins when chyme starts to enter the small intestine. This phase controls the rate of gastric emptying and ensures that the secretory, digestive, and absorptive functions of the small intestine can proceed at reasonable efficiency.

Level 3: Critical Thinking and Clinical Applications

38. If the gallstone is small enough, it can pass through the common bile duct and block the pancreatic duct. Enzymes from the pancreas will not be able to reach the small intestine; as the enzymes accumulate, they will irritate the duct and ultimately the exocrine pancreas, which produces pancreatitis.

39. The small intestine, especially the jejunum and ileum, are probably involved. Regional inflammation is the cause of Barb's pain. The inflamed tissue will not absorb nutrients; this accounts for her weight loss. Among the nutrients that are not absorbed are iron and vitamin B_{12}, which are necessary for formation of hemoglobin and red blood cells; this accounts for her anemia.

■ Chapter 17

Concept Check Answers, page 636

1. The primary role of the TCA cycle is to transfer electrons from organic substrates to coenzymes for use in the electron transport system, where the electrons provide an energy source for the production of ATP.

2. The binding of hydrogen cyanide molecules to the final cytochrome of the electron transport system would prevent the transfer of electrons to oxygen. As a result, cells would be unable to produce ATP in the mitochondria and would die from energy starvation.

Concept Check Answers, page 641

1. A vitamin B_6 (pyridoxine) deficiency would interfere with protein metabolism because this vitamin is an important coenzyme in the processes of deamination and transamination, which are the first steps in the processing of amino acids.

2. Elevated blood uric acid levels could indicate increased breakdown of nucleic acids. Uric acid is the product of the degradation of the nucleotides adenine and guanine, which are components of nucleic acids.

3. HDLs are considered beneficial because they reduce the amount of cholesterol in the bloodstream by transporting it to the liver for storage or for excretion in the bile.

Concept Check Answers, page 648

1. Foods from the meat and beans group should constitute the fewest number of daily servings of any of the food groups.

2. Foods that contain complete proteins contain all of the essential amino acids in nutritionally required amounts; foods that contain incomplete proteins are deficient in one or more of the essential amino acids.

3. Bile salts are necessary for the digestion and absorption of fats and fat-soluble vitamins. A decrease in the amount of bile salts in the bile would cause vitamin A deficiency that results from a reduced ability to absorb the fat-soluble vitamin A from food.

Concept Check Answers, page 651

1. A pregnant woman's BMR would be higher than her BMR when she is not pregnant because her metabolic activity increases to support the fetus and because of the added effects of fetal metabolism.

2. Evaporation is ineffective as a cooling mechanism under conditions of high relative humidity, when the air is holding large amounts of water vapor.

3. Vasoconstriction of peripheral vessels would cause body temperature to rise because the decreased blood flow to the skin reduces the amount of heat the body can lose to the environment.

Review Question Answers

Level 1: Reviewing Facts and Terms

1. a 2. g 3. l 4. i 5. j 6. b 7. k 8. e 9. f 10. d 11. h 12. c 13. d 14. c 15. a 16. b 17. c 18. c 19. d 20. c 21. b 22. c 23. b 24. a

25. Metabolism is the sum of all of the chemical reactions that occur in the cells of the body. Anabolism refers to chemical reactions that result in the synthesis of complex molecules from simpler reactants. Products of anabolism are used for maintenance/repair, growth, and secretion. Catabolism is the breakdown of complex molecules into their building block molecules, which results in the release of energy for the synthesis of ATP and related molecules.

26. Lipoproteins are lipid-protein complexes that contain large; insoluble glycerides and cholesterol, with a superficial coating of phospholipids and proteins. The major groups are chylomicrons (the largest lipoproteins, which are 95 percent triglyceride and carry absorbed lipids from the intestinal tract to the circulation), low-density lipoproteins (LDLs, which are mostly cholesterol and deliver cholesterol to peripheral tissues; sometimes this cholesterol gets deposited in arteries, hence the designation of LDLs as "bad cholesterol"), and high-density lipoproteins (HDLs, which are known as "good cholesterol," contain equal parts protein and lipid—cholesterol and phospholipids—and transport excess cholesterol back to the liver for storage or excretion in the bile).

27. Most vitamins and all minerals must be provided in the diet because the body cannot synthesize these nutrients.

28. carbohydrates: 4.18 Cal/g; lipids: 9.46 Cal/g; proteins: 4.32 Cal/g

29. The BMR is the minimum, resting energy expenditures of an awake, alert person.

30. (1) radiation: heat loss as infrared waves; (2) conduction: heat loss to surfaces in physical contact; (3) convection: heat loss to the air; and (4) evaporation: heat loss with water that becomes gas

Level 2: Reviewing Concepts

31. d 32. b

33. Glycolysis results in the breakdown of glucose to pyruvic acid through a series of enzymatic steps. Two other components, 4 ATP and 2 NADH, are also produced. Glycolysis requires glucose, specific cytoplasmic enzymes, ATP and ADP, inorganic phosphates, and NAD (nicotinamide adenine dinucleotide), which is a coenzyme.

34. The TCA reaction sequence is a cycle because the four-carbon starting compound (oxaloacetic acid) is regenerated at the end. Acetyl-CoA and oxaloacetic acid enter the cycle and CO_2, NADH, ATP, $FADH_2$, and oxaloacetic acid leave the cycle.

35. A triglyceride is hydrolyzed, which yields glycerol and fatty acids. Glycerol is converted to pyruvic acid and enters the TCA cycle. Fatty acids are broken into two-carbon fragments by beta-oxidation, which is a process that occurs inside mitochondria. The two-carbon compounds then enter the TCA cycle.

36. The food pyramid indicates the relative amounts of each of the five basic food groups an individual should choose to consume each day to ensure adequate intake of nutrients and calories. It also is a reminder that daily physical activity is an important aspect of weight control.

37. The brain contains the "thermostat" of the body, which is a region known as the hypothalamus. The hypothalamus regulates the ANS control of such mechanisms as sweating and shivering thermogenesis through negative-feedback homeostatic mechanisms.

38. These terms refer to the HDL and LDL, which are lipoproteins in the blood that transport cholesterol. HDL ("good cholesterol") transports excess cholesterol to the liver for storage or breakdown; whereas LDL ("bad cholesterol") transports cholesterol to peripheral tissues, which unfortunately may include the arteries. The buildup of cholesterol in the arteries is linked to cardiovascular disease.

Level 3: Critical Thinking and Clinical Applications

39. During starvation, the body must use fat and protein reserves to supply the energy necessary to sustain life. Some of the protein that is metabolized for energy is the gamma globulin fraction of the blood, which is mostly composed of antibodies. This loss of antibodies coupled with a lack of amino acids to synthesize new ones, as well as protective molecules such as interferon and complement proteins, renders an individual more susceptible to contracting a disease and less likely to recover from it.

40. The drug *colestipol* would lead to a decrease in the plasma levels of cholesterol. Bile salts are necessary for the absorption of fats. If the bile salts cannot be absorbed, the amount of fat absorption, namely, cholesterol and triglycerides, will decrease. This would in turn lead to a decrease in cholesterol from a dietary source as well as a decrease in fatty acids that could be used to synthesize new cholesterol. In addition, the body will have to replace the bile salts that are being lost with the feces. Since bile salts are formed from cholesterol, this will also contribute to a decline in cholesterol levels.

■ Chapter 18

Concept Check Answers, page 666

1. Unlike most other organs in the abdominal region, the kidneys lie behind the peritoneal lining (that is, they are retroperitoneal).
2. Plasma proteins do not normally pass into the capsular space because they are too large to pass through the filtration slits between the processes of the podocytes.
3. Damage to the juxtaglomerular apparatus would interfere with the normal control of blood pressure.

Concept Check Answers, page 674

1. Lowered blood pressure would reduce the blood hydrostatic pressure within the glomerulus, which would decrease the GFR.
2. If nephrons lacked a loop of Henle, the production of concentrated urine would not be possible.

Concept Check Answers, page 680

1. Peristaltic contractions move urine from the kidneys to the urinary bladder.

2. Obstruction of the ureters would interfere with the passage of urine from the renal pelvis to the urinary bladder.
3. Controlling the micturition reflex requires the ability to control the external urinary sphincter, which is a ring of skeletal muscle that acts as a valve.

Concept Check Answers, page 684

1. Consuming a meal high in salt would cause a reduction of fluid in the ICF; the ingested salt would temporarily increase the osmolarity of the ECF, so water would shift from the ICF to the ECF.
2. Fluid loss through perspiration, urine formation, and respiration would increase the osmolarity of blood and other body fluids.

Concept Check Answers, page 687

1. A decrease in the pH of body fluids stimulates the respiratory center in the medulla oblongata, and results in an increased breathing rate. As a result, more CO_2 is eliminated, and pH rises.
2. In a prolonged fast, fatty acids are catabolized, producing ketone bodies, which are acids that lower the body's pH. The eventual result is called ketoacidosis.
3. In vomiting, large amounts of stomach acid (HCl) produced by parietal cells are lost from the body. To replace the lost H^+, hydrogen ions and bicarbonate ions are formed from CO_2 and H_2O within the parietal cells. The release of bicarbonate ions into the blood (in exchange for chloride ions) raises the body's pH, which causes metabolic alkalosis.

Concept Check Answers, page 688

1. GFR declines with age due to a loss of nephrons, cumulative damage to the filtration mechanism within the remaining glomeruli, and reduced blood flow to the kidneys.

Review Question Answers

Level 1: Reviewing Facts and Terms

1. q 2. g 3. n 4. j 5. a 6. m 7. e 8. p 9. d 10. f
11. h 12. k 13. b 14. c 15. l 16. o 17. i 18. c 19. d
20. a 21. d 22. c 23. c 24. b 25. d

26. The urinary system performs vital excretory functions and eliminates the organic waste products generated by cells throughout the body.

27. The urinary system includes the kidneys, ureters, urinary bladder, and urethra.

28. A fluid shift is a water movement between the ECF and ICF; this movement helps prevent drastic variations in the volume of the ECF. Changes in the osmolarity of the ECF can cause fluid shifts. If the ECF becomes more concentrated (hypertonic) with respect to the ICF, water will move from the cells into the ECF until equilibrium is restored. If the ECF becomes more dilute (hypotonic), water will move from the ECF into the cells. Changes in osmolarity can result from water loss (such as excessive perspiration, dehydration, vomiting, or diarrhea), water gain (drinking pure water, administering hypotonic solutions through an IV), or changes in electrolyte concentrations (such as sodium).

29. (1) antidiuretic hormone: stimulates the thirst center and water conservation at the kidneys; (2) aldosterone: determines the rate of sodium absorption along the DCT and collecting system of the kidneys; and (3) atrial natriuretic peptide: reduces thirst and blocks the release of ADH and aldosterone

Level 2: Reviewing Concepts

30. d 31. c 32. a 33. b 34. a

35. autoregulation at the local level; hormonal regulation initiated by the kidneys; autonomic regulation (sympathetic division of the ANS)

36. The urge to urinate usually appears when the bladder contains about 200 ml of urine. The micturition reflex begins to function when the stretch receptors have provided adequate stimulation to the parasympathetic motor neurons. The activity in the motor neurons generates action potentials that reach the smooth muscle in the bladder wall. These efferent impulses travel over the pelvic nerves, producing a sustained contraction of the urinary bladder.

37. Fluid balance is a state in which the amount of water gained each day is equal to the amount lost to the environment. The water content of the body must remain stable because water is an essential ingredient of cytoplasm and accounts for about 99 percent of the volume of extracellular fluid. Electrolyte balance exists when there is neither a net gain nor a net loss of any ion in body fluids. The ionic concentrations in body water must remain within normal limits; if levels of calcium or potassium become too high, for instance, cardiac arrhythmias can develop. Acid-base balance exists when the production of hydrogen ions precisely offsets their loss. The pH of body fluids must remain within a relatively narrow range; variations outside this range can be life threatening.

38. "Drink plenty of fluids" is physiologically sound advice because the temperature rise that accompanies a fever can also increase water losses. For each degree the temperature rises above normal, the daily water loss increases by 200 ml.

39. Since sweat is usually hypotonic, the loss of a large volume of sweat causes hypertonicity in body fluids. The loss of fluid volume is primarily from the interstitial space, which leads to a reduction in plasma volume and an increase in the hematocrit. Severe dehydration can cause the blood viscosity to increase substantially, which results in an increased workload on the heart, and ultimately increases the probability of heart failure.

Level 3: Critical Thinking and Clinical Applications

40. Truck drivers may not urinate as frequently as they should. Resisting the urge to urinate can result in urine backing up into the kidneys. This puts pressure on the kidney tissues, which can lead to tissue death and ultimately to kidney failure.

41. Susan may have a urinary tract infection; her urine may contain blood cells and bacteria. She is more likely to have this problem than her husband because the urethral orifice in females is closer to the anus, and the urethral canal is short and opens near the vagina. Both regions normally harbor bacteria, which can easily reach the urethral entrance (often during sexual intercourse).

42. Because mannitol is filtered but not reabsorbed, drinking a mannitol solution would lead to an increase in the osmolarity of the filtrate. Less water would be absorbed from the filtrate and an increased volume of urine would be produced.

■ Chapter 19

Concept Check Answers, page 707

1. On a warm day, the cremaster muscle (and the dartos muscle as well) would be relaxed, which allows the scrotal sac to descend away from the body and cools the testes.

2. Dilation of the arteries within the penis increases blood flow within the penis; the filling of the vascular channels with blood results in an erection.

3. Low levels of FSH would lower sperm production by inhibiting the actions of sustentacular cells in providing nutrients and chemical stimuli to developing sperm.

Concept Check Answers, page 717

1. Blockage of both uterine tubes would cause sterility.
2. The acidic pH of the vagina helps prevent local bacterial, fungal, and parasitic infections.
3. The functional layer of the endometrium sloughs off during menstruation.
4. The blockage of a single lactiferous sinus would have little effect on the delivery of milk to the nipple because each breast usually has 15–20 lactiferous sinuses.

Concept Check Answers, page 720

1. If the LH surge did not occur during an ovarian cycle, ovulation and the subsequent formation of the corpus luteum would not occur.
2. Blocking progesterone receptors in the uterus would prevent the functional maturation and secretory activities of the endometrium, causing the uterus to be unprepared for pregnancy.
3. Declines in estrogen and progesterone levels signal the end of the uterine cycle and the beginning of menses.

Concept Check Answers, page 722

1. An inability to contract the ischiocavernosus and bulbospongiosus muscles would interefere with a male's ability to ejaculate and to experience orgasm.
2. As the result of parasympathetic stimulation in females during sexual arousal, the erectile tissues of the clitoris engorge with blood, the secretion of cervical and vaginal glands increases, blood flow to the walls of the vagina increases, and the blood vessels in the nipples engorge.
3. Declines in circulating estrogen levels at menopause reduce the inhibitory effect of estrogens on FSH (and on GnRH as well), causing FSH levels to rise and remain elevated.

Concept Check Answers, page 724

1. The integumentary system covers the external genitalia; sensory receptors in the skin provide sensations that stimulate sexual behaviors; and the mammary glands provide nourishment for the infant. The reproductive system produces hormones (primarily, testosterone in males and estrogens in females) that affect secondary sex characteristics such as the distribution of hair and subcutaneous fat deposits.

2. The endocrine system regulates sexual development and function by means of hypothalamic regulatory hormones and pituitary hormones; oxytocin in females stimulates smooth muscle contractions in the uterus and mammary glands. The reproductive system produces steroid sex hormones and inhibin, which inhibit secretory activities of the hypothalamus and pituitary gland.

Review Question Answers

Level 1: Reviewing Facts and Terms

1. i 2. m 3. a 4. c 5. l 6. e 7. n 8. o 9. h 10. b
11. f 12. j 13. d 14. k 15. g 16. c 17. a 18. c 19. a
20. d 21. a 22. c 23. c 24. d 25. b 26. c 27. d

28. The reproductive systems of both males and females include reproductive organs (gonads) that produce gametes and hormones, ducts that receive and transport gametes, accessory glands and organs that secrete fluids into these or other excretory ducts, and perineal structures collectively known as external genitalia.

29. The accessory organs and glands include the seminal vesicles, prostate gland, and the bulbourethral glands. The major functions of these glands are activating the spermatozoa, providing the nutrients sperm need for motility, propelling sperm and fluids along the reproductive tract, and producing buffers that counteract the acidity of the urethral and vaginal contents.

30. monitors and adjusts the composition of the tubular fluid, acts as a recycling center for damaged spermatozoa, and stores spermatozoa and facilitates their functional maturation

31. produces female gametes (ova); secretes female sex hormones, including estrogens and progestins; and secretes inhibin, which is involved in the feedback control of pituitary FSH production

32. serves as passageway for the elimination of menstrual fluids; receives the penis during sexual intercourse and

holds spermatozoa prior to their passage into the uterus; and, in childbirth, forms the lower portion of the birth canal through which the fetus passes during delivery

Level 2: Reviewing Concepts

33. The reproductive system is the only physiological system that is not required for the survival of the individual.

34. Meiosis is the two-step nuclear division that results in the formation of 4 haploid cells from 1 diploid cell. In males, 4 sperm are produced from each diploid cell, whereas in females only 1 ovum (plus 3 polar bodies) is produced from each diploid cell.

35. (1) Menses: This phase is marked by the degeneration and loss of the functional zone of the endometrium; usually lasts 1 to 7 days, and approximately 35–50 mL of blood is lost. (2) Proliferative phase: Growth and vascularization result in the complete restoration of the functional zone; lasts from the end of the menses until the beginning of ovulation around day 14. (3) Secretory phase: The endometrial glands enlarge, and accelerate their rates of secretion; the arteries elongate and spiral through the tissues of the functional zone, under the combined stimulatory effects of progestins and estrogens from the corpus luteum; begins at ovulation and persists as long as the corpus luteum remains intact.

36. The corpus luteum degenerates and a decline in progesterone and estrogen levels results in endometrial breakdown of menses. Next, rising levels of FSH, LH, and estrogen stimulate the repair and regeneration of the functional zone of endometrium. During the postovulatory phase, the combination of estrogen and progesterone cause the enlargement of the endometrial glands and an increase in their secretory activity.

37. In women, menopause is defined as the time that ovulation and menstruation cease. Menopause is accompanied by a sharp and sustained rise in the production of GnRH, FSH, and LH, while concentrations of circulating estrogen and progesterone decline. The decline in estrogen levels leads to reductions in the size of the uterus and breasts, accompanied by a thinning of the urethral and vaginal walls. In addition to a variety of neural and cardiovascular effects, reduced estrogen concentrations have been linked to the development of osteoporosis, presumably because bone deposition proceeds at a slower rate. During the male climacteric, circulating testosterone levels begin to decline between ages 50 and 60, coupled with increases in circulating levels of FSH and LH. Although sperm production continues, there is a gradual reduction in sexual activity in older men.

Level 3: Critical Thinking and Clinical Applications

38. There is no direct entry into the abdominopelvic cavity in males as there is in females. In females, the urethral opening is in close proximity to the vaginal orifice; therefore, infectious organisms can exit from the urethral meatus and enter the vagina. They can then proceed through the vagina to the uterus, then into the uterine tubes and finally into the peritoneal cavity.

39. The sacral region of the spinal cord contains the parasympathetic centers that control the genitals. Damage to this area of the spinal cord would interfere with the ability to achieve an erection by way of parasympathetic stimuli. However, erection can also occur by way of sympathetic centers in the lower thoracic region of the spinal cord. Visual, auditory, or cerebral stimuli can result in decreased tone in the arteries that serve the penis. This results in increased blood flow and erection. Tactile stimulation of the penis, however, would not generate an erection.

40. It appears that a certain amount of body fat is necessary for menstrual cycles to occur. The ratio of body fat to muscle tissue is somehow monitored by the nervous system; when the ratio falls below a certain set point, menstruation ceases. The actual mechanism appears to be a change in the levels of hypothalamic releasing hormone and pituitary gonadotropins. Possibly without proper fat reserves, a woman could not have a successful pregnancy. To avoid damage to the female body and death of a fetus, the body prevents pregnancy from occurring by shutting down the ovarian cycle and thus the menstrual cycle. When appropriate energy reserves are available, the cycles begin again.

■ Chapter 20

Concept Check Answers, page 743

1. Hyaluronidase released from dozens of spermatozoa breaks down the connections between the follicular cells of the corona radiata; another acrosomal enzyme, which is released after the binding of a single spermatozoon to the zona pellucida, digests a path through the zona pellucida to the oocyte membrane.

2. The inner cell mass of the blastocyst eventually develops into the embryo.

3. Sue is pregnant. The hCG detected by the test was produced by the developing trophoblasts and/or the placenta.

4. The placenta (1) supplies the developing fetus with a route for gas exchange, nutrient transfer, and waste product elimination and (2) produces hormones that affect maternal systems.

Concept Check Answers, page 753

1. During pregnancy, blood flow through the placenta reduces blood volume in the mother's systemic circuit, and the fetus adds carbon dioxide to the maternal circulation. The result is the release of renin and EPO, which stimulate an increase in maternal blood volume.

2. A decrease in progesterone levels in late pregnancy would remove its inhibition of uterine contractions, which could stimulate labor.

3. The uterus increases in size and weight during gestation through enlargement of the uterine cells, primarily smooth muscle cells of the myometrium.

Concept Check Answers, page 756

1. An increase in the circulating levels of GnRH, FSH, LH, and sex hormones marks the onset of puberty.

Concept Check Answers, page 763

1. A person who is heterozygous for curly hair would have one dominant gene and one recessive gene. The person's phenotype would be "curly hair."

2. Because females are XX and males are XY, only a male can provide a gamete that contains a Y chromosome. Thus, the sex of each of Joe's children depends on the genetic makeup of the sperm cell that fertilizes his wife's oocyte.

3. A genome is the full complement of an organism's genetic material (DNA). One Mb is equal to 1 million base pairs, so the 3200 Mb of the human genome represents 3,200,000,000 base pairs.

Review Question Answers

Level 1: Reviewing Facts and Terms

1. g 2. a 3. c 4. e 5. d 6. n 7. p 8. b 9. o 10. j 11. f
12. h 13. k 14. l 15. m 16. i 17. a 18. b 19. d 20. d
21. a 22. b 23. d 24. b

25. First trimester: The rudiments of all major organ systems appear. Second trimester: The organs and organ systems complete most of their development, and the body proportions change to become more human. Third trimester: The fetus grows rapidly, and most of the major organ systems become fully functional.

26. (1) Dilation stage: It begins with the onset of true labor, as the cervix dilates and the fetus begins to slide down the cervical canal; late in this stage, the amnion usually ruptures. (2) Expulsion stage: It begins as the cervix dilates completely and continues until the fetus has completely emerged from the vagina (delivery). (3) Placental stage: The uterus gradually contracts, which tears the connections between the endometrium and the placenta and ejects the placenta.

27. (1) Neonatal period (birth to 1 month): The newborn becomes relatively self-sufficient and begins performing respiration, digestion, and excretion for itself. Heart rates and fluid requirements are higher than those of adults. Neonates have little ability to thermoregulate. (2) Infancy (1 month to 2 years): Major organ systems (other than those related to reproduction) become fully operational and start to take on the functional characteristics of adult structures. Daily, even hourly, variations in body temperature continue throughout childhood. (3) Childhood (2 years to puberty): The child continues to grow, and significant changes in body proportions occur.

Level 2: Reviewing Concepts

28. d 29. d

30. The placenta produces human chorionic gonadotropin (HCG), which maintains the integrity of the corpus luteum and promotes the continued secretion of progesterone (keeping the endometrial lining functional); human placental lactogen (HPL) and placental prolactin which help prepare the mammary glands for milk production; and relaxin which increases the flexibility of the symphysis pubis, causes the dilation of the cervix, and suppresses the release of oxytocin by the hypothalamus, delaying the onset of labor contractions.

31. The respiratory rate and tidal volume increase, which allows the lungs to obtain the extra oxygen and to remove the excess carbon dioxide generated by the fetus. Maternal blood volume increases, which compensates for blood that will be lost during delivery.

Requirements for nutrients and vitamins climb 10–30 percent, which reflects the fact that part of the mother's nutrients must nourish the fetus. The glomerular filtration rate increases by about 50 percent, which corresponds to the increased blood volume, and accelerates the excretion of metabolic wastes generated by the fetus.

32. Positive-feedback mechanisms ensure that labor contractions continue until delivery is complete.

33. The amnion generally ruptures late in the dilation stage of labor.

34. (a) dominant (b) recessive (c) X-linked (d) autosomal

35. The trait of color blindness is carried on the X chromosome. Men are, thus, more likely to inherit it, because they have only one X chromosome. So whatever that chromosome carries will determine whether he is color blind or has normal vision. Since women have two X chromosomes, they will be color blind only if they are homozygous recessive. This is an X-linked inheritance.

36. The goal of the Human Genome Project is to determine the normal genetic composition of a "typical" human being. The project will help identify the genes responsible for inherited disorders and their locations on specific chromosomes. Other benefits include gaining an understanding of the genetic basis of diseases such as cancer and why the effectiveness of drugs varies between individuals.

Level 3: Critical Thinking and Clinical Applications

37. None of the couple's daughters will be hemophiliacs, since each will receive a normal allele from her father. There is a 50 percent chance that a son will be hemophiliac since there is a 50 percent chance of receiving either the mother's normal allele or her recessive allele.

38. The fact that the adults are larger is precisely why their rates are lower. Heat is lost across the skin. In infants, the surface-area-to-volume ratio is high, so they lose heat very quickly. To maintain a constant body temperature in the face of the heat loss, cellular metabolism must be high. Cellular metabolism requires oxygen; thus, increased metabolism demands an increased respiratory rate. There must also be an increase in cardiac output to move the blood from the lungs to the tissues. Since the range of contraction in the neonate heart is limited, the greatest increase in cardiac output is achieved by increased heart rate.

39. The baby's condition is almost certainly not the result of a virus contracted during the third trimester. The development of organ systems occurs during the first trimester. By the end of the second trimester, almost all of the organ systems are fully formed. During the third trimester, the fetus undergoes tremendous growth but very little new organ formation.

A

A band, 210, 211
Abdominal aorta, 490
Abdominal aortic aneurysm, 488
Abdominal cavity, 22
Abdominal pain, 623
Abdominal reflex, 302
Abdominopelvic cavity: Portion of the ventral body cavity that contains abdominal and pelvic subdivision, 22
Abdominopelvic quadrants, 17
Abdominopelvic regions, 17
Abducens nerve, 295–97
Abduction: Movement away from the midline, 190, 191
Abnormal hemostasis, 429
Abortion, 756–57
Abrasion, 141
Abruptio placentae, 765
Abscess, 522
Absence seizures, 291
Absorption: The active or passive uptake of gases, fluids, or solutes, 617–21
Accessory nerve, 295, 297
Accommodation: Alteration in the curvature of the lens to focus an image on the retina, 345
Acetabulum: Fossa on lateral aspect of pelvis that accommodates the head of the femur, 184
Acetylcholine (ACh): Chemical neurotransmitter in the brain and PNS; dominant neurotransmitter in the PNS, released at neuromuscular junctions and synapses of the parasympathetic division, 212, 216, 271, 272, 312
Acetylcholinesterase (AChE): Enzyme found in the synaptic cleft, bound to the postsynaptic membrane, and in tissue fluids; breaks down and inactivates ACh molecules, 212, 272, 312
Acid: A compound whose dissociation in solution releases a hydrogen ion and an anion; an acid solution has a pH below 7.0 and contains an excess of hydrogen ions, 38
Acid-base balance, 40, 684–87
Acid-base disorders, 686, 687
Acidic, 39
Acidosis: An abnormal physiological state characterized by a plasma pH below 7.35, 684
ACMV, 563
Acquaintance rape, 721
Acquired immunity, 522, 524
Acromegaly, 399
Acromion, 182
ACTH. *See* Adrenocorticotropic hormone (ACTH)
Actin: Protein component of microfilaments; form thin filaments in skeletal muscles and produce contractions of all muscles through interaction with thick (myosin) filaments, 76
Action potential: A conducted change in the membrane potential of excitable cells, initiated by a change in the membrane permeability to sodium ions, 268–70
Activation, 37
Active immunity, 522, 526
Active processes, 64
Active transport: The ATP-dependent absorption or excretion of solutes across a cell membrane, 70
Acute arterial occlusion, 495
Acute coronary syndrome, 226, 430, 447
Acute glaucoma, 351
Acute ischemic stroke, 430
Acute myocardial infarction (AMI), 447
Acute renal failure (ARF), 672
Acute tubular necrosis, 672
Adaptation, 328
Addison's disease, 388, 399
Adduction: Movement toward the axis or midline of the body as viewed in the anatomical position, 190, 191
Adductor brevis, 246
Adductor longus, 246
Adductor magnus, 246
Adenine: One of the nitrogenous bases in the nucleic acids RNA and DNA, 49
Adenocarcinoma, 575

Adenosine diphosphate (ADP): Adenosine with two phosphate groups attached, 51
Adenosine monophosphate (AMP): A nucleotide consisting of adenosine plus a phosphate group also known as adenosine phosphate, 50
Adenosine triphosphate (ATP): A high-energy compound consisting of adenosine with three phosphate groups attached; the third is attached by a high-energy bond, 50, 221
Adenylate cyclase: An enzyme bound to the inner surfaces of cell membranes that can convert ATP to cyclic-AMP; also called adenyl cyclase or adenylyl cyclase, 370
ADH. *See* Antidiuretic hormone (ADH)
Adipocytes, 106
Adipose tissue: Loose connective tissue dominated by adipocytes, 107, 108, 395–96
Adolescence, 756
ADP: Adenosine with two phosphate groups attached, 51
Adrenal cortex: Superficial portion of adrenal gland that produces steroid hormones, 385–86
Adrenal glands: Small endocrine glands secreting hormones, located superior to each kidney, 385–88
Adrenal insufficiency, 388
Adrenal medulla: Core of the adrenal gland; a modified sympathetic ganglion that secretes hormones into the blood following sympathetic activation, 308, 309, 386–87
Adrenergic: A synaptic terminal that releases norepinephrine when stimulated, 273
Adrenergic receptors, 310
Adrenocorticotropic hormone (ACTH): Hormone that stimulates the production and secretion of glucocorticoids by the adrenal cortex; released by the anterior pituitary, 375
Adult-onset diabetes, 391
Adventitia: Superficial layer of connective tissue surrounding an internal organ; fibers are continuous with those of surrounding tissues, providing support and stabilization, 588
AEDs, 454
Aerobic endurance, 225
Aerobic metabolism: The complete breakdown of organic substrates into carbon dioxide and water; a process that yields large amounts of ATP but requires mitochondria and oxygen, 80, 221
Afferent arteriole: An arteriole that carries blood to a glomerulus of the kidney, 663
Afferent division, 260
Agglutination: Aggregation of red blood cells due to interactions between surface agglutinogens and plasma agglutinins, 418, 531
Aging
    cancer, 117
    cardiovascular system, 506
    digestive system, 621
    hormones, 398–99
    immune response, 537
    muscular system, 251, 253
    nervous system, 314
    nutritional requirements, 651–52
    reproductive system, 721–22
    respiratory system, 578
    senses, 360–61
    skeletal system, 166
    tissues, 116–17
    urinary system, 688
Agonist: A muscle responsible for a specific movement, 230
Agranulocytes, 419
AIDS, 531
Albinism: Absence of pigment in hair and skin caused by inability of melanocytes to produce melanin, 127
Albumin, 68, 410
Aldosterone: A mineralocorticoid produced by the adrenal cortex; stimulates sodium and water conservation at the kidneys; secreted in response to the presence of angiotensin II, 385, 669, 673
Alkaline, 39
Alkalosis: Condition characterized by a plasma pH of greater than 7.45 and associated with relative deficiency of hydrogen ions or an excess of bicarbonate ions, 684
All-or-none principle, 268
Allantois: One of the four extraembryonic membranes; it provides blood vessels to the chorion and is, therefore, essential to placenta

Endocrine system, (cont.)
    endocrine regulation, 372–73
    gonads, 395
    growth, 396–97
    heart, 394
    hormonal action, 369–72
    intestines, 393
    kidneys, 394
    other systems, and, 400
    overview, 368, 369
    pancreas, 389–90
    parathyroid glands, 384
    pineal gland, 388–89
    pituitary gland, 373–78
    stress, 397–98
    thymus, 394
    thyroid gland, 379–84
Endocrinology diabetes and metabolism, 773
Endocytosis: The movement of relatively large volumes of extracellular material into the cytoplasm via the formation of a membranous vesicle at the cell surface; includes pinocytosis and phagocytosis, 73
Endoderm: One of the three primary germ layers; the layer on the undersurface of the embryonic disc that gives rise to the epithelia and glands of the digestive system, the respiratory system, and portions of the urinary system, 738, 739
Endolymph: Fluid contents of the membranous labyrinth (the saccule, utricle, semicircular canals, and cochlear duct) of the inner ear, 353
Endometriosis, 713
Endometrium: The mucous membrane lining the uterus, 713
Endomysium: A delicate network of connective tissue fibers that surrounds individual muscle cells, 208
Endoplasmic reticulum (ER): A network of membranous channels in the cytoplasm of a cell that function in intracellular transport, synthesis, storage, packaging, and secretion, 78
Endosteum, 153
Energy, 35
Energy concepts, 34–35
Enzyme: A protein that catalyzes a specific biochemical reaction, 37, 47–48, 64
Eosinophil: A granulocyte (WBC) with a lobed nucleus and red-staining granules; participates in the immune response and is especially important during allergic reactions, 421–22
Ependyma: Layer of cells lining the ventricles and central canal of the CNS, 264
Ependymal cells, 264
Epicardium: Serous membrane covering the outer surface of the heart; also called visceral pericardium, 438, 439, 441
Epicranial aponeurosis, 233
Epicranium, 233
Epidermal ridges, 125
Epidermis: The epithelium covering the surface of the skin, 124–28
Epidermoid carcinoma, 575
Epididymis: Coiled duct that connects the rete testis to the ductus deferens; site of functional maturation of spermatozoa, 702–3
Epidural block, 275
Epidural hematoma, 275, 316, 317
Epidural hemorrhage, 276
Epidural space: Space between the spinal dura mater and the walls of the vertebral foramen; contains blood vessels and adipose tissue; a frequent site of injection for regional anesthesia, 275
Epiglottis: Blade-shaped flap of tissue, reinforced by cartilage, that is attached to the dorsal and superior surface of the thyroid cartilage; it folds over the entrance to the larynx during swallowing, 550
Epilepsy, 291
Epimysium: A dense layer of collagen fibers that surrounds a skeletal muscle and is continuous with the tendons/aponeuroses of the muscle and with the perimysium, 208
Epinephrine, 387
Epiphyseal closure, 158
Epiphyseal line, 158
Epiphyses: The head of a long bone, 153
Episiotomy, 751
Epithalamus, 292
Epithelia: One of the four primary tissue types; a layer of cells that forms a superficial covering or an internal lining of a body cavity or vessel, 96

Epithelial tissue, 96–104
    basement membrane, 99
    classifying epithelia, 100–102
    epithelial renewal and repair, 99
    epithelial surface, 98
    functions, 96
    glandular epithelia, 103–4
    intercellular connections, 97–98
EPO. *See* Erythropoietin (EPO)
Eponychium, 138
EPS, 305
Epsilon receptors, 330
Equilibrium, 36, 353–56
ER. *See* Endoplasmic reticulum (ER)
Erector spinae, 235
Erosion, 130, 131
ERT, 395
ERV, 564
Erythroblastosis fetalis, 419
Erythroblasts, 414
Erythropoiesis: The formation of red blood cells, 413–16
Erythropoietin (EPO): Hormone released by kidney tissues exposed to low oxygen concentrations; stimulates erythropoiesis in bone marrow, 394, 416, 482
Esophagitis, 593
Esophagus: A muscular tube that connects the pharynx to the stomach, 592
Essential amino acids: Amino acids that cannot be synthesized in the body in adequate amounts and must be obtained from the diet, 638
Essential fatty acids: Fatty acids that cannot be synthesized in the body and must be obtained from the diet, 637
Estradiol, 717
Estrogen replacement therapy (ERT), 395
Estrogens: A class of steroid sex hormones that includes estradiol, 395, 717–19, 724, 741
Ether, 53
Ethmoid bone, 173
Ethmoidal sinuses, 173
ETS. *See* Electron transport system (ETS)
Evaporation: Movement of molecules from the liquid to the gaseous state, 649
Eversion: A turning outward, 191
Exanthems, 133
Excessive coagulation, 429
Exchange pump, 70
Exchange reaction, 36
Excoriation, 130, 131
Exercise, 482–83
Exergonic, 37
Exfoliative cytology, 104
Exocrine glands: A gland that secretes onto the body surface or into a passageway connected to the exterior, 104
Exocrine secretions, 96
Exocytosis: The ejection of cytoplasmic materials by fusion of a membranous vesicle with the cell membrane, 74
Expiratory reserve volume (ERV), 564
Expulsion stage, 751
Extension: An increase in the angle between two articulating bones; the opposite of flexion, 190
Extensor carpi radialis, 245
Extensor carpi ulnaris, 245
Extensor digitorum, 245
Extensor digitorum longus, 251
Extensor hallucis longus, 251
External acoustic canal: Passageway in the temporal bone that leads to the tympanum, 170
External anal sphincter, 240, 615
External callus, 160
External carotid artery, 490
External ear: The pinna, external acoustic canal, and tympanum, 351–52
External hemorrhoids, 615
External iliac vein, 499
External intercostals, 237
External jugular veins, 497
External nares: The nostrils; the external openings into the nasal cavity, 548

Hormones: Compound secreted by one cell that travel through the circulatory system to affect the activities of cells in another portion of the body, 45, 46, 56
    cardiovascular regulation, 481–82
    defined, 368
    endocrine system. *See* Endocrine system
    female reproductive cycle, 717–20, 724
    immune system, 532–33
    intestinal, 601–2
    kidney function, 673–74
    male reproductive system, 706–7, 724
    placental, 741–42
Hospital-acquired pneumonia, 557
hPL. *See* Human placental lactogen (hPL)
Human chorionic gonadotropin (hCG): Placental hormone that maintains the corpus luteum for the first three months of pregnancy, 741
Human Genome Project, 761–63
Human growth hormone (hGH), 376
Human physiology, 5
Human placental lactogen (hPL): Placental hormone that stimulates the functional development of the mammary glands, 741
Humerus, 182
Huntington's disease: An inherited disease marked by a progressive deterioration of mental abilities and by motor disturbances, 760
Hyaline cartilage, 109, 110
Hydrocephalus, 286
Hydrogen, 30
Hydrogen bond: Weak interaction between the hydrogen atom on one molecule and a negatively charged portion of another molecule, 33–34
Hydrogen ions, 39
Hydrolysis: The breakage of a chemical bond through the addition of a water molecule; the reverse of dehydration synthesis, 36, 618
Hydrophilic, 63
Hydrophobic, 63
Hydrostatic pressure, 68
Hyoid bone, 174
Hyperadrenalism, 387–88
Hyperextension, 191
Hyperglycemia, 639
Hyperopia, 346
Hyperpolarization, 267
Hypertension, 474
Hypertonic solutions, 66, 67, 71
Hypertrophy: Increase in the size of tissue without cell division, 224
Hyphema, 343, 350
Hypoglossal nerve, 295, 297
Hypoglycemia, 392, 639
Hypogonadism, 376
Hypophyseal portal system: Network of vessels that carry blood from capillaries in the hypothalamus to capillaries in the anterior pituitary (hypophysis), 374–75
Hypophysis: The anterior pituitary, 373
Hypothalamus: The floor of the diencephalon; region of the brain containing centers involved with the unconscious regulation of visceral functions, emotions, drives, and the coordination of neural and endocrine functions, 292–93, 373
Hypothyroidism, 383–84
Hypotonic solutions, 66, 67, 71
Hypovolemic shock, 483
Hypoxia: Low tissue oxygen concentrations, 416, 560
Hypoxic drive, 575

I

I band, 210, 211
IACD, 454
ICF, 680
ICH, 317
IgA, 530
IgD, 530
IgE, 530
IGFs, 376
IgG, 530
IgM, 530

IH, 373
IL. *See* Interleukins (IL)
Ileum: The last 2.5 m of the small intestine, 598
Iliocostalis, 235
Ilium: The largest of the three bones whose fusion creates a coxa, 184
IM administration, 243
Immediate hypersensitivity, 535, 536
Immune disorders, 534–37
Immunity, 522–26. *See also* Lymphatic system and immunity
Immunization: Developing immunity by the deliberate exposure to antigens under conditions that prevent the development of illness but stimulate the production of memory B cells, 525, 526
Immunodeficiency disease, 535
Immunoglobulin: A circulating antibody, 410, 530
Immunological escape, 520
Immunological surveillance, 520
Impacted fracture, 163
Implantation: The erosion of a blastocyst into the uterine wall, 737–41
Implanted automatic cardioverter-defibrillator (IACD), 454
IMV, 563
Inadequate coagulation, 429
Incisional hernia, 238
Incisors, 591
Incomplete abortion, 757
Incomplete proteins, 644
Incomplete tetanus, 218
Incontinence, 680
Incus: The central auditory ossicle, situated between the malleus and the stapes in the middle ear cavity, 352
Induced active immunity, 523, 524
Induced passive immunity, 526
Inevitable abortion, 757
Infectious disease, 773
Inferior, 19
Inferior mesenteric artery, 490
Inferior mesenteric vein, 499
Inferior nasal conchae, 174
Inferior vena cava: The vein that carries blood from the parts of the body below the heart to the right auricle, 442, 498–99
Infertility, 720
Inflammation: A nonspecific defense mechanism that operates at the tissue level, characterized by swelling, redness, warmth, pain, and some loss of function, 116, 521
Inflation reflex: A reflex mediated by the vagus nerve that prevents overexpansion of the lungs, 574
Infraorbital foramen, 170
Infraspinatus, 243
Infraspinous fossa, 182
Infundibulum: A tapering, funnel-shaped structure; in the nervous system, the connection between the pituitary gland and the hypothalamus; in the uterine tube, the entrance bounded by fimbriae that receives the oocytes at ovulation, 712
Ingestion: The introduction of materials into the digestive tract via the mouth, 586
Inguinal hernia: The area near the junction of the trunk and the thighs that contains the external genitalia, 238
Inheritance, 758–60
Inherited disorders, 760–61
Inhibin: A hormone produced by the sustentacular cells that inhibits the pituitary secretion of FSH, 395, 718, 724
Inhibiting hormone (IH), 373
Innate immunity, 522, 524
Inner cell mass: Cells of the blastocyst that will form the body of the embryo, 737, 739
Inner ear, 353
Inorganic compounds, 37–41
Inspiratory reserve volume (IRV): The maximum amount of air that can be drawn into the lungs over and above the normal tidal volume, 564
Insula, 286
Insulin: Hormone secreted by the beta cells of the pancreatic islets; causes a reduction in plasma glucose concentrations, 388, 390
Insulin-like growth factors (IGFs), 376
Insulin shock, 392
Integumentary system, 8, 122–49
    accessory structures, 134–38
    aging, 145

Lymphocyte: A cell of the lymphatic system that participates in the immune response, 422, 513–514
Lymphopoiesis: The production of lymphocytes, 422, 514

M

Macrophage: A phagocytic cell of the monocyte-macrophage system, 106, 519
Malleus: The first auditory ossicle, bound to the tympanum and the incus, 352
Mammary glands: Milk-producing glands of the female breast, 716–17, 755

Mast cells: A connective tissue cell that when stimulated releases histamine, serotonin, and heparin, initiating the inflammatory response, 106
Maxillary sinus: One of the paranasal sinuses; an air-filled chamber lined by a respiratory epithelium that is located in a maxillary bone and opens into the nasal cavity, 173
Medial: Toward the midline of the body, 19
Mediastinum: Central tissue mass that divides the thoracic cavity into two pleural cavities; includes the aorta and other great vessels, the esophagus, trachea, thymus, the pericardial cavity and heart, and a host of nerves, small vessels, and lymphatics, 21
Medulla: Inner layer or core of an organ, 135
Medulla oblongata: The most caudal of the brain regions, 294
Megakaryocytes: Bone marrow cells responsible for the formation of platelets, 424
Meiosis: Cell division that produces gametes with half the normal somatic chromosome complement, 700–701
Melanin: Yellow-brown pigment produced by the melanocytes of the skin, 126, 127
Melanocyte: Specialized cell in the deeper layers of the stratified squamous epithelium of the skin, responsible for the production of melanin, 126, 127
Membrane potential: The potential difference, in millivolts, measured across the cell membrane; a potential difference that results from the uneven distribution of positive and negative ions across a cell membrane, 266–68

Muscularis externa: Concentric layers of smooth muscle responsible for peristalsis, 587
Myelin: Insulating sheath around an axon consisting of multiple layers of glial cell membrane; significantly increases conduction rate along the axon, 264
Myocardium: The cardiac muscle tissue of the heart, 439, 441
Myofibril: Organized collections of myofilaments in skeletal and cardiac muscle cells, 209
Myoglobin: An oxygen-binding pigment especially common in slow skeletal and cardiac muscle fibers, 224, 226
Myometrium: The thick layer of smooth muscle in the wall of the uterus, 713
Myosin: Protein component of the thick myofilaments, 76

N

Nasal cavity: A chamber in the skull bounded by the internal and external nares, 548
Nasopharynx: Region posterior to the internal nares, superior to the soft palate, and ending at the oropharynx, 550
Negative feedback: Corrective mechanism that opposes or reverses a variation from normal limits and restores homeostasis, 7, 14–15

Nephron: Basic functional unit of the kidney, 663–66
Nerve impulse: An action potential in a neuron cell membrane, 270
Neuroglia: Cells of the CNS and PNS that support and protect the neurons, 115, 263–64
Neuromuscular junction: A specific type of neuroeffector junction, 212–14
Neuron: A cell in neural tissue specialized for intercellular communication by (1) changes in membrane potential and (2) synaptic connections, 115, 261–66
Neurotransmitter: Chemical compound released by one neuron to affect the membrane potential of another, 271–73
Neutrophil: A phagocytic microphage that is very numerous and usually the first of the mobile phagocytic cells to arrive at an area of injury or infection, 421
Nipple: An elevated epithelial projection on the surface of the breast, containing the openings of the lactiferous sinuses, 716
Nissl bodies: The ribosomes, Golgi, rough endoplasmic reticulum, and mitochondria of the cytoplasm of a typical nerve cell, 262

Supination: Rotation of the forearm such that the palm faces anteriorly, 183, 191

Supinator, 245

Supine: Lying face up, with palms facing anteriorly, 17

Supine hypotensive syndrome, 765

Supporting connective tissues, 109–11

Suppressor T cells: Lymphocytes that inhibit B cell activation and plasma cell secretion of antibodies, 528

Supracondylar fracture, 164

Supraorbital foramen, 169

Suprarenal arteries, 490

Supraspinatus, 243

Supraspinous fossa, 182

Surface anatomy, 4

Surfactant: Lipid secretion that coats alveolar surfaces and prevents their collapse, 34, 556

Surgery, 774

Suspensory ligaments, 340

Sustentacular cells: Supporting cells of the seminiferous tubules of the testis, responsible for the differentiation of spermatids and the secretion of inhibin, 699

Suture: Fibrous joint between flat bones of the skull, 188

SV, 458

Swallowed foreign bodies, 593

Swallowing, 592, 593

Sweat glands, 136–38

Swollen glands, 516

Sympathetic chain, 309

Sympathetic division: Division of the autonomic nervous system responsible for "fight or flight" reactions; primarily concerned with the elevation of metabolic rate and increased alertness, 308–10

Symphysis: A fibrous amphiarthrosis, such as that between adjacent vertebrae or between the pubic bones of the coxae, 188

Synapse, 271

Synapsis, 701

Synaptic cleft, 212, 271

Synaptic knob, 271

Synaptic terminal, 212, 261

Synarthrosis, 188

Synchondrosis, 188

Synchronized intermittent mandatory ventilation (SIMV), 563

Syndesmosis, 188

Synergist: A muscle that assists a prime mover in performing its primary action, 230

Synovial fluid: Substance secreted by synovial membranes that lubricates joints, 188

Synovial joints, 188, 189

Synovial membranes: An incomplete layer of fibroblasts lining the synovial cavity, plus the underlying loose connective tissue, 113

Synthesis: Manufacture; anabolism, 36

Systemic anatomy, 4

Systemic arteries, 487–93

Systemic circulation, 487–500

Systemic physiology, 5

Systemic veins, 495–500

Systems of measurement, 781–83

Systole: The period of cardiac contraction, 457

Systolic pressure: Peak arterial pressures measured during ventricular systole, 474

T

T_3, 379

T_4, 379

T cells: Lymphocytes responsible for cellular immunity and for the coordination and regulation of the immune response; includes regulatory T cells (helpers and suppressors) and cytotoxic (killer) T cells, 513, 527–28

T tubules: Transverse, tubular extensions of the sarcolemma that extend deep into the sarcoplasm to contact terminal cisternae of the sarcoplasmic reticulum, 209

T wave: Deflection of the ECG corresponding to ventricular repolarization, 455

Tachycardia, 453

Tactile corpuscles, 331

Tactile discs, 331

Tactile receptors, 329–32

Talus, 186

Tapeworms, 277

Tarsal bones, 186

Taste, 334–36

Taste buds, 335

TB, 560

TBI, 316

TCA cycle, 221, 633

Tectorial membrane: Gelatinous membrane suspended over the hair cells of the organ of Corti, 357

Teeth, 590–92

Telangiectasia, 130, 132

Telophase: The final stage of mitosis, characterized by the disappearance of the spindle apparatus, the reappearance of the nuclear membrane and the disappearance of the chromosomes, and the completion of cytokinesis, 87

TEM, 60

Temperature sensations, 329

Temporal bones, 170

Temporal lobe, 286

Temporalis, 233, 234

TEN, 134

Tendon: A collagenous band that connects a skeletal muscle to an element of the skeleton, 107

Tension, 218

Tension headache, 306

Tension pneumothorax, 559

Tensor fasciae latae, 246

Teratogenicity, 743

Teres major, 243

Teres minor, 243

Tertiary follicle: A mature ovarian follicle, containing a large, fluid-filled chamber, 711

Testes: The male gonads, sites of gamete production and hormone secretion, 395, 698–702

Testicular cancer, 700

Testicular torsion, 700

Testosterone: The principal androgen produced by the interstitial cells of the testes, 395, 699

Tetanus: A tetanic contraction; also used to refer to a disease state resulting from the stimulation of muscle cells by bacterial toxins, 219

Tetrad: Paired, duplicated chromosomes visible at the start of meiosis, 701

Thalamus: The walls of the diencephalon, 292

Therapeutic abortion, 757

Thermoreceptors, 329

Thermoregulation: Homeostatic maintenance of body temperature, 14, 649–51

Thermostat, 7

Theta waves, 290

Thiamine, 646, 647

Thick filament: A myosin filament in a skeletal or cardiac muscle cell, 76, 210, 211

Thick skin, 124

Thin filament: An actin filament in a skeletal or cardiac muscle cell, 210, 211

Thin skin, 124

Third-degree burns, 140, 144

Third trimester, 744–51

Thoracentesis, 558

Thoracic aorta, 490

Thoracic cavity, 21

Thoracic duct, 513, 514

Thoracic girdle, 179–82

Thoracic region, 175, 176

Thoracic surgery, 774

Thoracic vertebrae, 176, 178

Threatened abortion, 757

Threshold: The membrane potential at which an action potential begins, 268

Throat, 549

Thrombin: Enzyme that converts fibrinogen to fibrin, 427

Thrombocytes, 424

Thrombocytopenia, 424

Thrombocytosis, 424

Chapter 1 Chapter Opener © Jeff Forster **1-8** Edward T Dickinson, MD **01-07a,b,c** Custom Medical Stock Photo, INC. **01-12a,b left** Science Source/Photo Researchers, Inc. **01-11b right** Custom Medical Stock Photo, Inc. **01-13b** CNRI/Photo Researcher Inc. **01-22c** Photo Researchers Inc. **01-13d** Ben Edwards/Getty Images, Inc.

Chapter 2 Chapter Opener © Ken Kerr

Chapter 3 Chapter Opener Courtesy of Glen E. Ellman **03-07a,b,c** David M. Phillips/Visuals Unlimited **03-17b** L.A. Hufnagel, Ultrastructral Aspects of Chemoreception in Ciliated Protists (Ciliophora), Journal of Electron Microscopy Technique, 1919/Photomicrograph by Jurgen Bohmer and Linda Hufnagel, University of Rhode Island. **03-18** CNRI/Science Source/Photo Researcherss INC. **03-19** Don W, Fawcett, M.D., Harvard Medical School

Chapter 4 Chapter Opener © Joshua Menzies **4-11** Frederic H. Martini **4-14b** Frederic H. Martini **04-03** Custom Medical Stock Photo Inc. **04-04a** Ward's Natural Science Establishment, INC. **04-04b** Pearson Education/PH College **04-04c** Frederic H. Martini **04-05a,b right, left, c** Frederic H. Martini **04-06** Z.Legacy. Corporate Digital Archive

Chapter 5 Chapter Opener © Jeff Forster **05-22a, b** Kenneth Phillips, DO **05-22c** David Zohr, MD **05-24** Scott & White Hospital & Clinic **5-25** Scott & White Hospital & Clinic **05-26** Scott & White Hospital & Clinic **05-04** John D. Cunningham/Visuals Unlimited **05-06** Pearson Education/PH College **05-06a** Courtesy Elizabeth A. Abe., M.D., from the Leonard C. Winograd Memorial Slide Collection, Stanford University School of Medicine **05-06b** Courtesy Elizabeth A. Abe., M.D., from the Leonard C. Winograd Memorial Slide Collection, Standford University School of Medicine **05-011a** Manfred Kage/Peter Arnold, Inc. **05-12** Frederic H. Martini

Chapter 6 Chapter Opener Courtesy of Mark C. Ide **06-14** Charles Stewart MD and Asssociates **06-03** Visuals Unlimited/© R.G. Kessel and R.H. Kardon, "Tissues and Organs: A Text-Atlas of Scanning Electron Microscopy," W.H. Freeman & CO., 1979. All Rights Reserved. **06-09** Ralph T. Hutchings **06-32a** Ralph T. Hutchings **06-32b** Ralph T. Hutchings **06-29** Anita Impagliazzo **06-32a** Ralph T. Hutchings **06-32b** Ralph T. Hutchings **06-32c** Ralph T. Hutchings **06-34a** Ralph T. Hutchings **06-34b** Ralph T. Hutchings **06-35a** Ralph T. Hutchings **06-36** Ralph T. Hutchings **06-37** Ralph T. Hutchings **06-38a,b** Ralph T. Hutchings **06-39a** Ralph T. Hutchings **06-40** Ralph T. Hutchings **06-41c** Ralph T. Hutchings **06-43a,b** Ralph T. Hutchings **06-44** Ralph T. Hutchings **06-45a** Ralph T. Hutchings **06-47a,b** Anita Impagliazzo **06-47c,d** Ralph T. Hutchings **06-48** Anita Impagliazzo **06-49** Ralph T. Hutchings

Chapter 7 Chapter Opener © Courtesy of Mark C. Ide **07-01** Anita Impagliazzo **07-03** Don W. Fawcett/Photo Researchers, Inc. **07-04a** Don W. Fawcett/Science Source/Photo Researchers, Inc. **07-10a** G.W. Willis/Biological Photo Service **07-10b** Frederic H. Martini **07-21** Anita Impagliazzo

Chapter 8 Chapter Opener Courtesy of Glen E. Ellman **8-23** M.A. Ansary/Custom Medical Stock Photo, Inc. **08-02** Ward's Natural Science Establishment **08-05a** Biophoto Associates/Photo Researchers, Inc. **08-05b** Photo Researchers, Inc. **08-10** David Scott/phototake NYC **08-17** Michael J. Timmons **08-20** Ralph T. Hutchings **08-26** Larry Mulvehill/Photo Reasearchers, Inc. **08-30b** Ralph T. Hutchings

Chapter 9 Chapter Opener Chris Barry/Phototake NYC **09-05** Anita Impagliazzo **09-06** Anita Impagliazzo **09-08** Pearson Education/PH College **09-09** Ralph T. Hutchings **09-11 left** Diane Hirsch/Fundamental Photographs **09-12 right** Diane Schiumo/Fundamental Photographs **09-13a** ED Reschke/Peter Arnold, Inc. **09-13c** Custom Medical Stock Photo, Inc. **09-20** Richmond Products, Inc. **09-30e** Anita Impagliazzo **09-31** Ward's Natural Science Establishment, Inc.

Chapter 10 Chapter Opener © Ken Kerr **10-20** © Mark C. Ide **10-17** Photo Researchers, Inc. **10-01** Anita Impagliazzo **10-06** Manfred Kage/Peter Arnold, Inc. **10-10** Frederic H. Martini **10-15c** Frederic H. Martini **10-16a** Anita Impagliazzo **10-16c** Ward's Natural Science Establishment, Inc. **10-13a** Anita Impagliazzo **10-13b** Ward's Natural Science Establishment, Inc. **10-22a** Project Masters, Inc. /The Bergman Collection **10-22b** John Paul Kay/Peter Arnold, Inc. **10-22c** Project Masters, Inc. /The Bergman Collection **10-22d** Custom Medical Stock Photo, Inc. **10-22e** Biophoto Associates/Science Source/Photo Researchers, Inc.

Chapter 11 Chapter Opener Courtesy of Glen E. Ellman **11-9** Prentice Hall file photo **11-01** Martin M. Rotker **11-02a** David Scharf/Peter Arnold, Inc. **11-02b** Frederic H. Martini **11-03a** Stanley Fledger **11-03b** Visuals Unlimited **11-08a-e** Ed Reschke/Peter Arnold, Inc. **11-10** Custom Medical Stock Photo, Inc.

Chapter 12 Chapter Opener © Ken Kerr **12-11** Science Photo Library/Photo Researchers, Inc. **12-18** Michal Heron/Pearson Education/PH College **12-03a** Ralph T. Hutchings **12-04c** Ed Reschke/Peter Arnold, Inc. **12-20** Larry Mulvehill/Photo Researchers, Inc.

Chapter 13 Chapter Opener Jon Freilich **13-7** Jack Star/Photo Disc, Inc. **13-01** Michael J. Timmons **13-03a** B&B Photos/Custom Medical Stock Photo, Inc. **13-03b** William Ober/Visuals Unlimited **13-04** Biophoto Associates/Photo Researchers, Inc. **13-08b** Jack Star/Getty Images, INC. **13-19a** Anita Impagliazzo **13-26** Anita Impagliazzo

Chapter 14 Chapter Opener J.P. Moczulski/Canadian Press/Phototake NYC **14-9** © Edward T Dickinson, MD **14-02b** Frederic H. Martini **14-07c** Frederic H. Martini **14-08b** Anita Impagliazzo **14-08c** Frederic H. Martini

Chapter 15 Chapter Opener Craig Jackson/In the Dark Photography **15-12** Photo courtesy of Hartwell Medical, Carlsbad, CA **15-20** Scott Metcalfe Photography **15-03b** Photo Researchers, Inc. **15-04e** Phototake NYC **15-07b** Micrograph by P. Gehr from Bloom & Fawcett, "Textbook of Histology," W.B. Saunders Co.

Chapter 16 Chapter Opener Craig Jackson/In the Dark Photography **16-01** Anita Impagliazzo **16-05** Anita Impagliazzo **16-11b** Ralph T. Hutchings **16-14b** Frederic H. Martini **16-15a** Anita Impagliazzo **16-16c** © Michael J. Timmons **16-20a** Anita Impagliazzo

Chapter 17 Chapter Opener Craig Jackson/In the Dark Photography **17-10** Anita Hylton

Chapter 18 Chapter Opener Craig Jackson/In the Dark Photography **18-10** Photo Reseachers, Inc. **18-18** Shout Pictures, UK **18-03a** Ralph T. Hutchings **18-06c** David M. Phillips/Visuals Unlimited

Chapter 19 Chapter Opener Craig Jackson/In the Dark Photography **19-02b** Don W. Fawcett, M.D., Harvard Medical School **19-05a** Ward's Natural Science Establishment, Inc. **19-05b** Visuals Unlimited/© R.G. Kessel and R.H. Kardon, "Tissues and Organs: A Text Atlas of Scanning Electron Microscopy," W. H. Freeman & Co., 1979. All Rights Reserved. **19-10a-d** Frederic H. Martini **19-10f** G.W. Willis, M.D./Biological Photo Service **19-10e** C. Edelmann/La Villete/Photo Researchers, Inc.

Chapter 20 Chapter Opener Courtesy of Glen E. Ellman **20-01** Francis Leroy, Biocosmos/Science Photo Library/Custom Medical Stock, Inc. **20-07a** Dr. Arnold Tamarin/Arnold Tamarin **20-07b right**, Photo Lennart Nilsson/Albert Lennart Nilsson/Albert Bonniers Forlag **20-07c** Photo Lennart Nilsson/Albert Bonniers Forlag **20-07d** Photo Lennart Nilsson/Albert Bonniers Forlag **20-08a** Photo Lennart Nilsson/Albert Bonniers Forlag **20-08b** Photo Researchers, Inc. **20-17** CNRI/Science Photo Library/Photo Researchers, Inc.

CLINICAL NOTES IN ANATOMY & PHYSIOLOGY FOR EMERGENCY CARE
The following topics appear throughout the text as Clinical Notes

FOREIGN WORD ROOTS, PREFIXES, SUFFIXES, AND COMBINING FORMS

Each entry starts with the commonly encountered form or forms of the prefix, suffix, or combining form followed by the word root (shown in italics) with its English translation. One example is given to demonstrate the use of each entry.

a-, *a-*, without: avascular
ab-, *ab*, from: abduct
-ac, *-akos*, pertaining to: cardiac
acr-, *akron*, extremity: acromegaly
ad-, *ad*, to, toward: adduct
aden-, adeno-, *adenos*, gland: adenoid
adip-, *adipos*, fat: adipocytes
aer-, *aeros*, air: aerobic respiration
af-, *ad*, toward: afferent
-al, -alis, pertaining to: brachial
alb-, *albicans*, white: albino
-algia, *algos*, pain: neuralgia
allo-, *allos*, other: allograft
ana-, *ana*, up, back: anaphase
andro-, *andros*, male: androgen
angio-, *angeion*, vessel: angiogram
ante-, *ante*, before: antebrachial
anti-, ant-, *anti*, against: antibiotic
apo-, *apo*, from: apocrine
arachn-, *arachne*, spider: arachnoid
arter-, *arteria*, artery: arterial
arthro-, *arthros*, joint: arthroscopy
-asis, -asia, *asis*, state, condition: homeostasis
astro-, *aster*, star: astrocyte
atel-, *ateles*, imperfect: atelectasis
aur-, *auris*, ear: auricle
auto-, *auto*, self: autonomic
baro-, *baros*, pressure: baroreceptor
bi-, *bi-*, two: bifurcate
bili-, *bilis*, bile: bilirubin
bio-, *bios*, life: biology
blast-, -blast, *blastos*, precursor: blastocyst
brachi-, *brachium*, arm: brachiocephalic
brachy-, *brachys*, short: brachydactyly
brady-, *bradys*, slow: bradycardia
bronch-, *bronchus*, windpipe, airway: bronchial
carcin-, *karkinos*, cancer: carcinoma
cardi-, cardio-, -cardia, *kardia*, heart: cardiac
-cele, *kele*, tumor, hernia, or swelling: blastocoele
-centesis, *kentesis*, puncture: thoracocentesis
cephal-, *cephalos*, head: brachiocephalic
cerebr-, *cerebros*, brain: cerebral hemispheres
cerebro-, *cerebrum*, brain: cerebrospinal fluid
cervic-, *cervicis*, neck: cervical vertebrae
chole-, *chole*, bile: cholecystitis
chondro-, *chondros*, cartilage: chondrocyte
chrom-, chromo-, *chroma*, color: chromatin
circum-, *circum*, around: circumduction
-clast, *klastos*, broken: osteoclast
colo-, *kolon*, colon: colonoscopy
contra-, *contra*, against: contralateral
corp-, *corpus*, body: corpuscle
cortic-, *cortex*, rind or bark: corticospinal
cost-, *costa*, rib: costal
cranio-, *cranium*, skull: craniosacral
cribr-, *cribrum*, sieve: cribriform
-crine, *krinein*, to secrete: endocrine

cut-, *cutis*, skin: cutaneous
cyan-, *kyanos*, blue: cyanosis
cyst-, -cyst, *kystis*, sac: blastocyst
cyt-, cyto-, *kyton*, a hollow cell: cytology
de-, *de*, from, away: deactivation
dendr-, *dendron*, tree: dendrite
dent-, *dentes*, teeth: dentition
derm-, *derma*, skin: dermis
desmo-, *desmos*, band: desmosome
di-, *dis*, twice: disaccharide
dia-, *dia*, through: diameter
digit-, *digit*, a finger or toe: digital
dipl-, *diploos*, double: diploid
dis-, *des*, apart, away from: disability
diure-, *diourein*, to urinate: diuresis
dys-, *dys-*, painful: dysmenorrhea
-ectasis, *ektasis*, expansion: atelectasis
ecto-, *ektos*, outside: ectoderm
-ectomy, *ektome*, excision: appendectomy
ef-, *ex*, away from: efferent
emmetro-, *emmetros*, in proper measure: emmetropia
encephalo-, *enkephalos*, brain: encephalitis
end-, endo-, *endos*, inside: endometrium
entero-, *enteron*, intestine: enteric
epi-, *epi*, on: epimysium
erythema-, *erythema*, flushed (skin): erythematosis
erythro-, *erythros*, red: erythrocyte
ex-, *ex*, out, away from: exocytosis
extra-, outside, beyond, in addition: extracellular
ferr-, *ferrum*, iron: transferrin
fil-, *filum*, thread: filament
-form, *-formis*, shape: fusiform
gastr-, *gaster*, stomach: gastrointestinal
-gen, -genic, *gennan*, to produce: mutagen
genicula-, *geniculum*, kneelike structure: geniculate nuclei
gest-, *gesto*, to bear: gestation
glosso-, -glossus, *glossus*, tongue: hypoglossal
glyco-, *glykys*, sugar: glycogen
-gram, *gramma*, record: myogram
gran-, *granulum*, grain: granulocyte
-graph, -graphia, *graphein*, to write, record: electroencephalograph
gyne-, gyno-, *gynaikos*, woman: gynecologist
hem-, hemato-, *haima*, blood: hemopoiesis
hemi-, *hemi-*, half: hemisphere
hepato-, *hepaticus*, liver: hepatocyte
hetero-, *heteros*, other: heterozygous
histo-, *histos*, tissue: histology
holo-, *holos*, entire: holocrine
homeo-, homo-, *homos*, same: homeostasis
hyal-, hyalo-, *hyalos*, glass: hyaline
hydro-, *hydros*, water: hydrolysis
hyo-, *hyoeides*, U-shaped: hyoid
hyper-, *hyper*, above: hyperpolarization
hypo-, *hypo*, under: hypothyroid
hyster-, *hystera*, uterus: hysterectomy
-ia, state or condition: insomnia
idi-, *idios*, one's own: idiopathic